Noninfectious Diseases and Pathology of Reptiles

Noninfectious Diseases and Pathology of Reptiles

Color Atlas and Text

Volume 2

Edited by
Michael M. Garner
Elliott R. Jacobson

CRC Press
Taylor & Francis Group
Boca Raton London New York

CRC Press is an imprint of the
Taylor & Francis Group, an **informa** business

First edition published 2021
by CRC Press
6000 Broken Sound Parkway NW, Suite 300, Boca Raton, FL 33487-2742

and by CRC Press
2 Park Square, Milton Park, Abingdon, Oxon, OX14 4RN

ISBN: 978-1-4987-7158-0 (hbk)
ISBN: 978-0-429-15431-7 (ebk)

Typeset in ITC Garamond Std
by Nova Techset Private Limited, Bengaluru & Chennai, India

Courtesy of Karina Mathes.

As both a friend and colleague, we wish to recognize Professor Dr. Peernel Zwart for his numerous contributions to the reptile disease and pathology literature and presentations at national and international meetings that span his academic career at Utrecht University, where he worked for 35 years in the Division of Veterinary Pathology. He is an inspiration to all of us working on health issues of reptiles.

Michael M. Garner
Elliott R. Jacobson

CONTENTS

2020 PREFACE, VOLUME 2

It is a great honor for me to be a coeditor for this two-volume set, *Diseases and Pathology of Reptiles*. Collaborating with my coeditor, Elliott Jacobson and the folks at CRC Press/Taylor & Francis, especially Alice Oven, has been seamless. I think these two volumes hold the most information ever accrued into one publication on the diseases and pathology of this class of animals. The first edition, *Infectious Diseases and Pathology of Reptiles: Color Atlas and Text* (Jacobson, 2007) set the tone for this second edition and the addition of Volume 2, *Noninfectious Diseases and Pathology of Reptiles*. I have chosen authors for Volume 2 who have extensive experience in their areas of expertise, such that the depth and color of each chapter makes for a worthwhile and accessible experience to the reader. Subjects include Reptile Neoplasia (Chapter 1), Nutritional and Metabolic Diseases (Chapter 2), Depositional Diseases (Chapter 3), Normal and Abnormal Reptile Development (Chapter 4), Degenerative Diseases (Chapter 5), Trauma and Physical Diseases (Chapter 6), Environmental and Miscellaneous Toxicoses in Reptiles (Chapter 7), Physiology and Diseases of the Reproductive System (Chapter 8), Reptile Cardiovascular System (Chapter 9), and the unique contribution of Bone Alteration by Disease (Chapter 10). Each chapter is supported by numerous figures, many of which are unique and cannot be found in the published literature. Readers will note that some of the chapters are based on the organ system, a trend that will continue into the next edition to encompass all of the basic organ systems. Although timelines for various stages of chapter development and finalization were established within the first few months of the start of this project, throughout the process we mused that getting the chapter authors to finish their drafts on time was like herding cats; in fact, working with these gifted and patient veterinarians has been truly wonderful, and I look forward to working with them again on future editions.

Finally, I wish to thank my family, my staff at Northwest ZooPath, and my partner for tolerating my early morning and late night candle burning. I will make it up to you, I promise.

Michael M. Garner, DVM, DACVP

ACKNOWLEDGMENTS

First and foremost, I want to acknowledge all of the contributors to Northwest ZooPath who have allowed me to use their case materials and images. Without the collaboration of these dedicated clinicians, curators, keepers, and technicians, this volume would not have been possible. I would be remiss if I didn't recognize Leroy and Nancy Brown of Histologic Consulting Service for their 27 years of faithful service, patience, and expertise in preparing excellent histologic slides, and on species as diverse as coral and elephants. I must give a shout out to Drs. Elise LaDouceur and Nancy Stedman, pathologists extraordinaire in every way, who have covered for me the many times I was at yet another meeting, editing some book [sic], or on those rare vacations when I just needed to get away from it all. A heartfelt thanks to my trusted and long-term staff, Heather Cook, Christie Buie, and Cathy Minoque. These are the behind the scenes folks who manage the books, maintain the archives, and see that my daily reports get transcribed, reported, and filed. Without them, there would be no Northwest ZooPath. Finally, a tip of the hat to my coeditor, Elliott Jacobson, whom I have known over my entire career, and who inspired me to pay attention to reptiles. Our perspectives and vision for this two-volume set were seamless.

Michael M. Garner
Zanesville, Ohio

EDITORS

Michael M. Garner, DVM, DACVP, is a renowned veterinary pathologist who specializes in zoological and wildlife diseases. In 1994, Dr. Garner founded the consulting firm, Northwest ZooPath (Monroe, Washington). Since that time he has worked with numerous zoos, wildlife centers, universities, and veterinary clinics in the United States and abroad. He has authored or coauthored more than 200 papers and book chapters. Dr. Garner is also an enthusiastic teacher and is committed to training veterinary pathologists and residents. He established an outreach training program for foreign pathologists in 1999, and currently has housed and trained 22 externs and residents. A sought-after speaker, Dr. Garner has given hundreds of presentations and seminars at national and international meetings and veterinary colleges, including Washington State University where he has held an adjunct faculty appointment since 1998.

Elliott R. Jacobson, DVM, PhD, DACZM, was born in Brooklyn, New York. He attended Brooklyn College of the City University of New York, and in 1967 he earned his BS degree in Biology. Next, he earned an MS at New Mexico State University in 1969, working on physiological ecology of snakes. He enrolled in graduate school and veterinary school at the University of Missouri and earned his DVM and PhD in Zoology in 1975. He was a wildlife veterinarian for the State of Maryland from 1975 to 1977. He arrived at the College of Veterinary Medicine, University of Florida, Gainesville, in 1977, where he was a resident in Wildlife and Laboratory Animal Medicine. In 1979, he was appointed Assistant Professor and helped develop the Zoological Medicine Service. In 1986, he became a Diplomate of the American College of Zoological Medicine. He moved up through the ranks to full professor in 1989. During his career, he advised 31 residents and advised or served on the committee of 18 graduate students. Over the last 41 years, he has worked on health problems of a wide variety of amphibians, reptiles, birds, and mammals. His former laboratory specialized in infectious diseases of reptiles, including the development of serologic assays and molecular diagnostic assays used to determine exposure to and infection with certain pathogens. He has authored or coauthored 283 refereed scientific papers, 46 chapters in texts, edited and coedited four books, and has been either the principal or coprincipal investigator on 83 funded projects since 1978. He retired from the University of Florida in 2012, and is Professor Emeritus of Zoological Medicine. Since June/July 2013 he has cotaught a course titled "Conservation Medicine and Diseases of Herpetofauna" at the American Museum of Natural History Southwest Research Station, Portal, Arizona. This has resulted in a full-circle experience for him since first visiting this field station in 1967 when he was a graduate student at New Mexico State University. He is currently consulting on a project in Belize involving captive breeding of the Hicatee (*Dermatemys mawii*).

CONTRIBUTORS

James E. Bogan Jr.
Central Florida Zoo & Botanical Gardens
Sanford, Florida

Thomas H. Boyer
Pet Hospital of Penasquitos
San Diego, California

Rachel E. Burns
Disease Investigations
San Diego Zoo Global
San Diego, California

María Camacho
Toxicology Unit
Research Institute of Biomedical and Health Sciences (IUIBS)
Universidad de Las Palmas de Gran Canaria
Las Palmas de Gran Canaria, Spain

Marcelo P.N. de Carvalho
Clínica de Animais Silvestres e Exóticos
Departamento de Clínica e Cirurgia Veterinárias
Escola de Veterinária
Universidade Federal de Minas Gerais
Minas Gerais, Brasil

Erin A. Graham
Department of Pathology
University of Georgia College of Veterinary Medicine
Tifton Veterinary Diagnostic and Investigational Laboratory
Tifton, Georgia

Carles Juan-Sallés
Noah's Path
Elche (Alicante), Spain

Elise E.B. LaDouceur
Joint Pathology Center
Silver Spring, Mary Land

Gregory A. Lewbart
College of Veterinary Medicine
North Carolina State University
Raleigh, North Carolina

Octavio P. Luzardo
Toxicology Unit
Research Institute of Biomedical and Health Sciences (IUIBS)
Universidad de Las Palmas de Gran Canaria
Study Group on Wild Animal Conservation Medicine (GEMAS)
Spanish Biomedical Research Center in Physiopathology of Obesity and Nutrition (CIBERObn)
Las Palmas de Gran Canaria, Spain

Christine L. Miller
Suffolk Animal Hospital
Suffolk, Virginia

Jorge Orós
Department of Morphology (Veterinary Faculty)
University of Las Palmas de Gran Canaria
Las Palmas de Gran Canaria, Spain

Joaquín Ortega
Departamento PASAPTA
(Grupo Patología y Sanidad Animal)
Universidad Cardenal Herrera-CEU
CEU Universities
Moncada, Spain

Robert J. Ossiboff
Department of Comparative, Diagnostic, and Population Medicine
University of Florida
Gainesville, Florida

John F. Roberts
Department of Comparative, Diagnostic, and Population Medicine
University of Florida
Gainesville, Florida

Bruce Rothschild
Vertebrate Paleontology
Carnegie Museum of Natural History
Pittsburgh, Pennsylvania

and

Department of Medicine
IU Health
Muncie, Indiana

Nancy L. Stedman
Busch Gardens Tampa
Tampa, Florida

James R. Stewart
Department of Biological Sciences
East Tennessee State University
Johnson City, Tennessee

Daniel A. Warner
Department of Biological Sciences
Auburn University
Auburn, Alabama

Jeanette Wyneken
Departments of Biological Sciences
Florida Atlantic University
Boca Raton, Florida

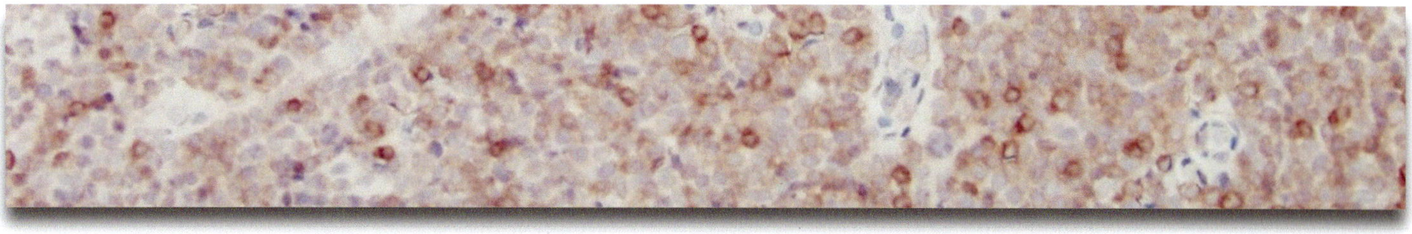

1

REPTILE NEOPLASIA

ELISE E.B. LADOUCEUR

Contents

1.1 Introduction

Neoplasia means "new growth." It is the growth of cells that have undergone heritable genetic changes, allowing them to become relatively unresponsive to normal growth control and to expand beyond their ordinary anatomic boundaries. Benign neoplasms do not invade into surrounding tissue and do not spread to new anatomic locations within the body. Malignant neoplasms have the ability to invade locally, and/or metastasize to other locations throughout the body (Kusewitt, 2012). This chapter reviews current knowledge of neoplasia in reptiles. The review is organized by body system (e.g., urinary, digestive, etc.), and subdivided by Order (e.g., Testudines (chelonians; turtles and tortoises), Crocodylia (crocodiles, alligators, caiman, and gharial), and Squamata (lizards and Snakes), where appropriate. The chapter closes with a discussion of ancillary diagnostic techniques, such as immunohistochemistry (IHC), that can aid in the diagnosis of neoplasia in reptiles.

Neoplasia is relatively common in most reptilian Orders, with reviews of reptile neoplasia reporting the low prevalence of neoplasia in crocodilians, followed by chelonians, and higher prevalence in lizards and snakes (Jacobson, 1981; Garner et al. 2004; Sykes and Trupkiewicz, 2006). There is only one report of a neoplasm in Rhynchocephalia, the Tuatara (*Sphenodon punctatus*) from New Zealand (Roe et al., 2002). Comparative studies suggest that the rate of neoplasia in reptiles is similar to that seen in mammals (Effron et al., 1977). Evidence of neoplasia in this group dates back to the dinosaur fossil record from the Cretaceous (Rothschild et al., 2003). The estimated overall prevalence of neoplasia in reptiles varies widely by study, with estimates ranging from 2.1% to 23.1% (Effron et al., 1977; Ramsay et al., 1996; Hernandez-Divers and Garner, 2003; Garner et al. 2004; Sykes and Trupkiewicz, 2006). Most cases of neoplasia in reptiles are believed to occur spontaneously, and multiple neoplasms can occur in a single animal, particularly when the animal is geriatric (Frye, 1994). There are few reports, however, of viral oncogenesis, genetic predisposition to neoplasia, and hormonal imbalances that may contribute to development of neoplasia in reptiles (Christman et al., 2017).

Diagnostic techniques that aid in identifying reptilian neoplasia include imaging, cytology, histology, special stains, immunohistochemical (IHC) stains, and electron microscopy. The fundamentals of these procedures are provided in other chapters of this text, and excellent reviews specific to reptiles are available (Hernandez-Divers and Garner, 2003; Christman et al., 2017).

1.2 Review by Systems and Organs

1.2.1 Integumentary System

Cutaneous tumors are common in reptiles (Garner et al. 2004; Sykes and Trupkiewicz, 2006). There are many cell types in reptilian skin that are prone to neoplastic transformation. Squamous epithelium, which forms the outermost layer of the skin, is the most commonly reported type of epithelium to undergo neoplastic transformation in reptiles. Neoplasia arising from the squamous epithelium includes squamous cell carcinoma, papilloma, and fibropapilloma (Garner et al. 2004). Several cell types in the dermis may give rise to neoplasia, including cells from the complex pigment system (e.g., melanophores, erythrophores, xanthophores, and iridophores), and various mesenchymal cells (e.g., fibroblasts, adipocytes, endothelial cells, etc.). A number of round cell neoplasms may involve skin, and these are included under the Hematopoietic and Lymphoid section.

1.2.1.1 Squamous Cell Carcinoma Squamous cell carcinoma (SCC) is a malignant tumor of squamous epithelial cells. It is relatively common in many animal species. Certain predisposing factors, including exposure to ultraviolet light, lack of epidermal pigmentation, and lack of hair or sparse hair coat, have been identified in association with SCC in mammals. Papillomaviruses are also a reported cause of SCC in mammals (Mauldin and Peters-Kennedy, 2016). Generally, SCC are common in reptiles, and have been reported mainly in snakes, occasionally in lizards, rarely in chelonians, and once each in a Tuatara and a Saltwater Crocodile (*Crocodylus porosus*) (Roe et al., 2002; Garner et al. 2004; Sykes and Trupkiewicz, 2006; Hannon et al., 2011; Hill et al., 2016). A predilection for mucocutaneous sites, primarily eyelids or periocular tissues, has been identified in Inland Bearded Dragons (*Pogona vitticeps*) (**Figure 1.1**). Other predilection sites in reptiles include oral cavity, cloaca (particularly in snakes), scent glands (a squamous epithelial/keratinaceous gland (**Figure 1.2**), and skin of the face (particularly in monitor lizards) (Ramsay et al., 1996; Abou-madi and Kern, 2002; Garner et al. 2004; Orós et al., 2004; Anderson et al., 2010; Steeil et al., 2013). Metastasis of SCC is rarely reported in reptiles, with metastatic cases reported in an Eastern Diamondback Rattlesnake (*Crotalus adamanteus*), Loggerhead Sea Turtles (*Caretta caretta*), Saltwater Crocodile, and European Pond Turtle (*Emys orbicularis*) (Orós et al., 2004; Anderson et al., 2010; Hill et al., 2016). Grossly, SCC are sometimes mistaken for chronic, non-healing wounds or abscesses, and can be secondarily infected (**Figure 1.3**) (Abou-madi and Kern, 2002; Anderson et al., 2010; Hill et al., 2016). Histologically, SCC in reptiles is usually well differentiated and is composed of packets of neoplastic epidermal cells with squamous differentiation. Frequently, packets have peripheral basaloid differentiation with central keratinization, occasionally forming compressed, central cores of keratin, also known as "keratin pearls" (**Figures 1.4–1.5**). Tumors may invoke a marked scirrhous response. The unique nature of innate shell repair in chelonians, with infolding fronds of hyperplastic and sometimes dysplastic epithelium, can resemble invasive SCC, and the diagnosis of neoplasia should be made conservatively in this regard (Garner et al., 1997).

1.2.1.2 Papilloma and Fibropapilloma Cutaneous papillomas are benign, proliferative, epithelial neoplasms. Many papillomas are associated with or caused by papillomaviruses of the family Papillomaviridae or herpesviruses from Herpesviridae. Papillomas that are not associated with viruses are considered idiopathic squamous papillomas. Papillomaviruses are discussed in detail in Volume 1, Chapter 10 of this two-volume book, but in brief are double-stranded DNA viruses that infect the stratum basale of the epidermis. Papillomas typically are one or multiple exophytic masses comprised of marked epidermal hyperplasia, particularly of the stratum spinosum, with abundant superficial hyperkeratosis (**Figure 1.6**). Other features of viral-induced papillomas include ballooning degeneration, koilocytes, and intranuclear viral inclusions; these changes are particularly pronounced in the stratum granulosum (**Figure 1.7**). Fibropapillomas (FP) are similar to papillomas in that they form nodules with hyperkeratosis and hyperplasia, and may be viral induced, but there is also a dermal proliferative fibroblast component (Mauldin and Peters-Kennedy, 2016).

Cutaneous papillomas have been described in lizards, crocodilians, and chelonians, and in one study were most prevalent in turtles (Garner et al. 2004). Multiple viruses have been associated with papillomas and fibropapillomas in reptiles. In European Green Lizards (*Lacerta viridis*), papillomas having intranuclear inclusions within epidermal cells were evaluated by transmission electron microscopy (TEM) and found to have viral particles with morphology suggestive of Herpesviridae, Reoviridae, and Papovaviridae (Raynaud and Adrian, 1976; Cooper, 1982). Herpesvirus was also detected using molecular methods in papillomas of European Green Lizards, though inclusions were not identified ultrastructurally in those cases (Literak et al., 2010). Viral particles resembling papillomaviruses have also been seen using electron microscopy in Bolivian Side-Neck Turtles (*Platemys platycephala*) (Jacobson et al., 1982). Herpesvirus-induced papillomas are likely to recur after surgical resection (Hellebuyck et al., 2012).

Fibropapillomas (FP) are reported in every species of marine turtle, though Green Turtles (*Chelonia mydas*) are most severely affected, with panzootics occurring in this species (Smith and Coates, 1983; Harshbarger, 1991; Limpus et al., 1993; Barragan and Sarti, 1994; Aguirre et al., 1999; D'Amato and Moraes-Neto, 2000; Huerta et al., 2000). The typical gross presentation of FP includes single or multiple masses on the skin, oral cavity, shell, eyes, and internal organs (**Figures 1.8–1.9**). Gross lesions of FP in sea turtles are considered pathognomonic by some (Jones et al., 2016). In sea turtles, FP often obscure vision, feeding, or locomotion. Green Turtles with FP have also been reported to have internal fibromas, fibrosarcomas, and myxofibromas (Norton et al., 1990; Garner et al. 2004; Work et al., 2004). In sea turtles with FP, chelonid herpesvirus 5 was identified and demonstrated to have lytic replication in neoplastic epidermal cells. Viral shedding from the skin is a postulated route of transmission (Work et al., 2014). A causative association between chelonid herpesvirus 5 and FP is difficult to definitively assert as the virus is resistant to culture, preventing the fulfillment of Koch's postulates (Work et al., 2009); however, the disease is transmissible in filtered extracts from macerated papillomas (Herbst et al., 1995), and herpesvirus inclusions can be found in early stages of cutaneous (Jacobson et al., 1991) and esophageal fibropapillomas.

1.2.1.3 Chromatophoroma Chromatophoromas are tumors of pigment-producing cells that largely occur in the skin. Three types of pigment cells (aka chromatophores) are present in the integument of reptiles, each of which gives rise to a type of chromatophoroma: (1) melanin-producing cells give rise to melanophoromas or melanomas, (2) carotenoid or pteridine-producing cells give rise to xanthophoromas, and (3) crystalline purine-producing cells give rise to iridophoromas. Mixed-type chromatopohromas are also possible in which multiple types of pigment producing cells are within a single neoplasm (Heckers et al., 2012). Mosaic chromatophoromas may also occur in which there is more than one pigment type in each cell (Jacobson et al., 1989a,). Chromatophores are most abundant in the skin, but also present in lesser numbers in other organs. As such, chromatophoromas rarely occur as primary neoplasms in locations outside of the integument in reptiles, but have been reported to cause a primary intestinal chromatophoroma in a Green Tree Python (*Chondropython viridis*) and primary pulmonary melanophoroma in a Beaded Lizard (*Heloderma horridum exasperatum*) (Catão-Dias and Nichols, 1999; Rivera et al., 2015). Chromatophoromas are common in reptiles, and one report found them to be the most common type of neoplasm in a retrospective study of 179 reptilian neoplasms (Heckers et al., 2012). Most reports of chromatophoromas are in snakes, followed by lizards, and rarely chelonians, with one case report of a malignant melanophoroma of the carapace in a tortoise (Garner et al. 2004; Heckers et al., 2011, 2012). Chromatophoromas are commonly invasive and difficult to completely excise. Metastasis to the viscera is also reported with some frequency, and has been documented most commonly for melanocytic neoplasms (including in a rosy boa [*Lichanura* sp.], Dumeril's Ground Boa [*Acrantophis dumerili*], Mojave Desert Sidewinder [*Crotalus cerastes*], Sierra Mountain Kingsnake [*Lampropeltis zonata*], Temple Viper [*Tropidolaemus wagleri*], monitor [*Varanus* sp.], Leopard Gecko [*Eublepharis macularius*], Yellow Anaconda [*Eunectes notaeus*], Hermann's Tortoise [*Testudo hermanni*], Veiled Chameleon [*Chamaeleo calyptratus*], multiple bearded dragons [*Pogona* spp.], and multiple species of rattlesnakes [*Crotalus* spp.]), with fewer reports of metastasis for iridophoromas (including in a Northern Pine Snake, Green Tree Python, and Tricolor Hognose Snake [*Xenodon pulcher*]), and mixed chromatophoromas (including in a Corn Snake [*Elaphe guttata*] and Veiled Chameleon) (Jacobson, Ferris et al., 1989a,b; Garner et al. 2004; Reavill et al., 2004; Heckers et al., 2012; Muñoz-Gutiérrez et al., 2016). In one report, multiple chromatophoromas were diagnosed in a captive Western Terrestrial Gartersnake (*Thamnophis elegans*). This animal presented for cutaneous ulcerations and vesicles, and histologically the masses were ulcerated and had a range of pigments (Frye et al., 1975). Immunohistochemical staining has been applied to chromatophoromas with some success. Melan A, PNL2, and S100 are reportedly positive in melanophoromas and iridophoromas (Heckers et al., 2012; Muñoz-Gutiérrez et al., 2016). Melan A has some reported specificity for chromatophoromas, while control tumors (i.e., fibrosarcoma, myxoma, myxosarcoma, and squamous cell carcinoma) were uniformly negative for melan A. S100 is reportedly expressed in other tumors, and its specificity for chromatophoromas is likely low (Heckers et al., 2012).

Iridophoromas are grossly pale, dermal masses. Both iridophoromas and melanophoromas have predominantly spindle-shaped cells histologically, though epithelioid variants are possible. Melanophoromas and iridophoromas can be distinguished by the nature of their cytoplasmic granules. Histologically, the diagnosis of iridophoromas is based on the detection of coarse, golden-brown to olive-green pigment granules in neoplastic cells that are birefringent under polarized light; pigment granules are slightly larger in iridophoromas than melanophoromas (**Figure 1.10**) (Heckers et al., 2012).

The terminology of neoplasms arising from melanin-producing cells is convoluted in the literature. Generally, the terms "melanophoroma" and "melanin-producing neoplasm" are used to indicate a benign neoplasm of melanophores, while "melanoma" and "malignant melanoma" indicate a malignant neoplasm arising from melanophores (Garner et al. 2004; Heckers et al., 2012; Muñoz-Gutiérrez et al., 2016). Melanophoromas do not have evidence of cellular atypia, mitotic figures, or local invasion; melanomas generally have one or more of these features.

Melanomas and melanophoromas are commonly reported in some reptiles, particularly snakes, and are reported in fewer lizards and rarely chelonians (Garner et al. 2004; Bielli et al., 2015). Grossly, melanophoromas and melanomas tend to be at least partially pigmented, and firm (**Figure 1.11**). Ulceration may also occur. Less common, mucinous variants are gelatinous on cut section. These neoplasms are composed of predominantly spindle-shaped or epithelioid cells with variable amounts of fine, granular, dark brown pigment. Amelanotic variants of melanoma are considered less common (**Figure 1.12**), but as immunohistochemical markers are not highly sensitive or routinely applied, the actual incidence of these neoplasms may be higher. One study reported that neoplastic cells in melanophoroma reportedly have less pleomorphism, fewer nucleoli, and fewer mitoses than in melanoma (Heckers et al., 2012). Another study found that neoplastic cells of melanoma had substantial nuclear atypia compared to melanophoromas, and that this feature may be useful as a predictor of metastasis (Muñoz-Gutiérrez et al., 2016). Local invasion is common, and melanomas are frequently difficult to completely excise surgically. Reported metastatic sites of melanoma in reptiles include liver and kidney (Schlumberger and Lucke, 1948; Sykes and Trupkiewicz, 2006), but also may be seen in coelomic surfaces or lung. Multiple cutaneous melanomas occurring in a single animal are also reported (Schlumberger and Lucke, 1948). Almost all reported cases of melanomas and melanophoromas in reptiles are in the skin. There is a single report of two melanophoromas in the lungs of a Beaded Lizard (*Heloderma horridum exasperatum*) in which a primary cutaneous neoplasm was not identified (Rivera et al., 2015). These masses were presumed to arise from pigmented cells of the lungs and coelomic membranes. The melanophoromas were surgically removed and had no evidence of recurrence at the time of death, 3 years later.

Xanthophoromas are rarely reported in reptiles. They are masses of neoplastic chromatophores containing red to brown pigment. This neoplasm was reported in a bush viper (*Atheris* sp.) (Muñoz-Gutiérrez et al., 2016).

1.2.1.4 Lipoma Lipomas are uncommon in reptiles, but have been reported in snakes, lizards, and crocodilians. Corn Snakes may be overrepresented in this group as there are several reports of subcutaneous lipomas in this species (Frye, 1994; Reavill and Schmidt, 2003; Garner et al. 2004; Dietz et al., 2016). Interestingly, in 7 out of 13 Corn Snakes reported with lipoma, the lipoma appeared to infiltrate into adjacent skeletal muscle and fascia, a feature that is consistent with infiltrative lipoma (**Figure 1.13**) (Reavill and Schmidt, 2003; Garner et al. 2004). Infiltrative lipomas are characteristically difficult to completely excise and are locally invasive, despite their benign histomorphology (Mauldin and Peters-Kennedy, 2016). Histologically, lipomas are composed of sheets of well-differentiated adipocytes, and can be difficult to differentiate from normal adipose. Lipomas in some snakes have robust proliferations of well-differentiated blood vessels and have been called angiolipomas (Dietz et al., 2016).

1.2.1.5 Miscellaneous Cutaneous Neoplasia There are several types of benign neoplasms that are rarely reported in the integument and body wall of reptiles. These include fibromas, myxomas, osteomas, and hemangiomas, among others. Only fibromas and myxomas are discussed here, as other types of benign mesenchymal neoplasms are included elsewhere in this chapter. Fibromas are aggregates of mature collagen with small numbers of well-differentiated spindle cells (i.e., fibroblasts). They are rarely reported in reptiles, including in the intercostal region in African Rock Python (*Python sebae*), in multiple Green Iguanas (*Iguana iguana*) with some ossification (i.e., ossifying fibroma), and in multiple chelonians, including a Common Snapping Turtle (*Chelydra serpentine*) and an Alligator Snapping Turtle (*Macrochelys temminckii*); vimentin was used in one case to confirm mesenchymal origin of the spindle cells (Effron et al., 1977; Frye, 1994; Garner et al. 2004; Gonzales-Viera et al., 2012). Large fibromas may have ulceration. Myxomas are similar to fibromas and are sometimes called myxofibromas. Microscopically, they are masses of low cellularity with well-differentiated mesenchymal cells (fibroblasts or multipotent mesenchymal cells) embedded in abundant myxomatous matrix with relatively few, fine, collagen fibers. Myxomas have been diagnosed in multiple Green Iguanas (Frye, 1994). Myxosarcoma is occasionally seen, particularly in Bearded Dragons, where it can be difficult to distinguish from amelanotic melanoma and edematous nerve sheath tumor (see neoplasms of the nervous system and respiratory system, this chapter, Sections 1.2.11.1 and 1.2.11.2) (**Figure 1.14**).

Reptilian skin has relatively few glands, and as such, glandular neoplasms arising from the skin are few. Neoplasia is reported to arise from the scent glands, tear glands, and venom glands in snakes and lizards, and from cutaneous apocrine glands in chelonians. Histologically, these are composed of tubules and packets of neoplastic polygonal cells in a variable amount of stroma (**Figure 1.15**). Squamous cell carcinoma is the most common tumor of scent glands, although scent gland adenoma was described in a California Kingsnake (*Lampropeltis getulus californiae*), in which cytokeratin confirmed epithelial origin (Gal and Mandoki, 2012). The authors have also seen a scent gland adenoma in a Crested Gecko (*Correlophus ciliatus*) (**Figure 1.16**). A carcinoma from the venom gland was reported in a Central African Forest Cobra (*Naja melanoleuca*); additional details were not provided (Machotka, 1984). Apocrine adenocarcinomas in the forelimb of an Agassiz's Desert Tortoise (*Gopherus agassizii*) and the hindlimb of an Asian Leaf Turtle (*Cyclemys dentata*) were described (Gál et al., 2010; Abu-Seida and Saeid, 2013). In the Agassiz's Desert Tortoise, the neoplasm was located on the anterior and axial portion of the limb. It was surgically removed but recurred 1 year after surgery (Abu-Seida and Saeid, 2013). Follow-up information was not provided in the Asian Leaf Turtle case. Details of neoplasms arising from the tear gland are provided in Section 1.2.14 (Special Senses) of this chapter (Garner et al. 2004).

1.2.2 Urinary System

Neoplasia arising from the urinary system is common in reptiles, particularly snakes. Almost all reports of neoplasia of the urinary system in reptiles arise from cells in the kidney, with epithelial cells of the nephron being the most common cell to undergo neoplastic transformation. One of the most common tumors in reptiles is renal adenocarcinoma; renal adenoma is also common. Other renal neoplasia in reptiles includes nephroblastoma, fibroma in marine turtles, and various metastatic tumors. Ureteral or urinary bladder tumors are very rarely reported in reptiles, partly because snakes, some lizards, and crocodilians lack urinary bladders. A single case of ureteral carcinoma with metastasis to the pancreas was diagnosed in a Mottled Rock Rattlesnake (*Crotalus lepidus lepidus*) (Sykes and Trupkiewicz, 2006). Renal tumors in snakes frequently present as a swelling in the caudal one-third of the body cavity. In lizards, renal tumors present as firm swelling in the caudal coelom that are occasionally palpable caudal to the rim of the pelvis (Reavill and Schmidt, 2007).

1.2.2.1 Renal Adenocarcinoma Neoplasia arising from the renal tubules is common in reptiles and widely reported in many species, particularly snakes (Jacobson et al., 1986), with fewer reports in lizards and rare reports in chelonians. The typical gross appearance for this neoplasm is a large, solitary, unilateral mass associated with the kidney (**Figure 1.17**). With large neoplasms, there is sometimes externally visible distention of the caudal aspect of the coelom (**Figures 1.18–1.19**). In snakes, multicentric and bilateral tumors are somewhat common, especially in the colubrids. Histologically, renal adenocarcinoma comprises irregular tubular structures in a moderate amount of fibrovascular stroma. Neoplastic cells are 1 or 2 layers thick, and cuboidal to columnar with a moderate amount of variably vacuolated eosinophilic cytoplasm. Nuclei are oval with stippled chromatin and usually one prominent nucleolus. Renal adenocarcinoma is frequently associated with development of substantial urate tophi in both renal tubules and adjacent scirrhous stroma where neoplastic tubules have ruptured (**Figure 1.20**). Criteria for malignancy are not well established, but are generally based on local invasion, including through the renal capsule (Garner et al. 2004). Because these tumors are well differentiated, produce urates, and are slow growing, there is a tendency to underestimate their biological behavior. These tumors may be slow to metastasize, and death is frequently attributed to renal failure. Reported sites of metastasis include liver, lung, body wall, contralateral kidney, and coelomic adipose tissue (Barten et al., 1994; Sykes and Trupkiewicz, 2006; Reavill and Schmidt, 2007; Keck et al., 2011; Belasco-Zeitz et al., 2013). These neoplasms can cause constipation or intestinal perforation due to compression and obstruction of the colon, respectively. Diagnosis is often straightforward, and pancytokeratin immunohistochemical stain has been used to confirm epithelial origin. In one case using IHC, pancytokeratin also labeled some mesenchymal cells in the surrounding scirrhous stroma, raising questions as to the specificity of this antibody (Kao et al., 2016). There are rare reports of primary renal sarcoma, which are variably associated with metastasis (Mauldin and Done, 2006).

Using TEM, C-type retroviruses were identified in renal epithelial cells of Lance-headed Vipers (*Bothrops moojeni*) having renal adenocarcinoma (Volume 1, **Figures 7.102–7.105**) (Hoge et al., 1995). Mature virus was also found in the lumen of tubules and most likely is eliminated with the urine.

1.2.2.2 Nephroblastoma Nephroblastomas are embryonal tumors arising from primitive blastema and foci of renal dysplasia. This neoplasm is rare and has only been reported in a few species of reptiles, including a Green Anaconda (*Eunectes murinus*), Leopard Gecko and Frilled Lizard (*Chlamydosaurus kingii*); a nephroblastoma was also diagnosed by the authors in a Ball Python (*Python regius*). In the Common Anaconda, the attachment to the kidney could not be identified, and it was speculated that this mass arose from ectopic renal tissue (Jacobson et al., 1986). The Leopard Gecko tumor was considered malignant based on effacement of renal architecture; this tumor did not have evidence of metastasis (Garner et al. 2004). Metastasis was identified in the Frilled Lizard, though the location of the metastatic focus was not specified (Ladds, 2009). Grossly, these masses are discrete, pale tan, and expansile (**Figure 1.21**). Histologically, nephroblastomas are comprised of primitive tubules/acini, primitive glomeruli, undifferentiated blastemal cells, and a mesenchymal component that may have differentiation toward a variety of tissues, including collagen, skeletal muscle, bone, cartilage, or adipose tissue (**Figure 1.22**). In dogs and pigs, the diagnosis can be confirmed in some cases by applying IHC to Wilms tumor protein gene product C-19. Immunoreactivity is frequently negative to glial fibrillary acidic protein and variably positive to cytokeratin and vimentin in dogs. Since so few cases have been reported, it is not surprising that IHC has not been applied to nephroblastomas in reptiles, to our knowledge. In domestic species, glomerular and tubular differentiation is associated with a more favorable prognosis, while sarcomatous differentiation is associated with metastasis and a less favorable prognosis (Cianciolo and Mohr, 2016).

1.2.2.3 Urothelial Carcinoma Urothelial carcinoma is also called transitional cell carcinoma (TCC), as the urothelium is composed of transitional epithelium. This is rarely reported in snakes, including a cloacal TCC in a Reticulated Python (*Malayopython [Python] reticulatus*), and a TCC arising from the renal pelvis in a Burmese Python (*Python morulus bivittatus*) (Machotka, 1984; Chandra et al., 2001). In the Burmese Python, type C retroviral particles were identified within the neoplasm, but are of uncertain importance in the oncogenesis of the tumor (Chandra et al., 2001).

1.2.3 Alimentary System

Neoplasia of the alimentary system is relatively common in reptiles. The oral and cloacal mucosae are lined by stratified squamous epithelium, which can undergo neoplastic transformation to squamous cell carcinoma, as reported in the skin.

The submucosa throughout the alimentary tract has numerous glands and gut-associated lymphoid tissue (GALT), both of which can give rise to neoplasia (see Hematopoietic and Lymphoid Systems). Most commonly, glandular neoplasms of the alimentary system are malignant, although benign variants and adenomatous polyps have also been reported in reptiles.

Neoplasia arising from the salivary glands is rare in reptiles, with a single report of parotid carcinoma in a Gold Tegu (*Tupinambis taguixin*), and an adenocarcinoma of the palatine gland in a Burmese Python (*Python bivittatus*) (Machotka, 1984; Mauldin and Done, 2006). Three additional cases of a salivary gland neoplasia were identified in the Northwest ZooPath (Monroe, WA) archives, including adenocarcinoma in a ball python, adenocarcinoma in a Green Iguana, and adenoma in a bearded dragon (*Pogona* sp.). Neoplasia arising from mesenchymal cells in the submucosa and muscularis of the alimentary tract is rarely reported in reptiles. Reports of odontogenic tumors are also rare in reptiles (Michael M. Garner et al. 2004; Heckers et al., 2012).

Fibromatous, inflammatory and adenomatous polyps and granulation tissue are relatively common, non-neoplastic lesions affecting the alimentary tract, especially the oral mucosa, of reptiles. Clinically, these lesions may resemble neoplastic processes, and thus are frequently biopsied.

1.2.3.1 Odontogenic Neoplasia Odontogenic neoplasms are rarely described in reptiles, with only five reports of ameloblastomas, including one report in a dinosaur, *Telmatasaurys transsylvanicus*, from the Cretaceous Period (Machotka, 1984; Comolli et al., 2015; Schmidt and Reavill, 2015; Dumbravă et al., 2016; Goe et al., 2016). In addition, the authors diagnosed a fibromatous epulis of periodontal ligament origin (FEPLO) in an Inland Bearded Dragon from the Northwest ZooPath archives. The histogenesis of ameloblastomas and FEPLO is linked to tooth development, and as such these lesions are not found outside the oral cavity. However, teratomas in any location can develop tooth-like components that resemble ameloblastoma or odontoma.

Ameloblastomas are neoplasms derived from odontogenic epithelium in the developing tooth germ (i.e., from ameloblasts). These tumors are locally invasive and can be associated with substantial facial deformity and bone/tooth destruction. Microscopically, most of the neoplasm is composed of islands of odontogenic epithelium with lesser amounts of fibrous stroma with or without cemento-osseous matrix. The cardinal features of odontogenic epithelium are (1) columnar cells, (2) peripheral palisading, (3) antibasilar nuclei, and (4) basilar clearing. Cells within the center of islands of odontogenic epithelium may become loosely arranged, resembling stellate reticulum (Jundt, 2005). Neoplasms consistent with ameloblastoma are described in a Banded Gila Monster (*Heloderma suspectum cinctum*), python (unknown species), gecko (unknown species), and wild Black (Eastern) Rat Snake (*Pantherophis alleghaniensis*). In the Black Rat Snake and Gila Monster, masses were evident grossly in the mandible, one of which was associated with demonstrable osteolysis, which was evident on radiographs. Microscopically, the masses were composed mostly of islands of odontogenic epithelium. In the Black Rat Snake, IHC revealed that odontogenic epithelium was reactive to pancytokeratin, and using TEM, morphology of odontogenic epithelium was identified. In the Gila Monster, a novel herpesvirus was identified in the neoplasm via polymerase chain reaction (PCR). However, the importance of this finding in the oncogenesis of the tumor is unknown. Further details on the python and gecko cases were not available (Machotka, 1984; Comolli et al., 2015; Schmidt and Reavill, 2015; Goe et al., 2016).

Fibromatous epulis of periodontal ligament origin are benign lesions with limited growth potential. The terminology of FEPLO is convoluted in the literature, with some pathologists using peripheral odontogenic fibroma synonymously with FEPLO. This lesion is composed predominantly of periodontal ligament-like stroma, which is dense, finely fibrillar, and has a vascularized stroma with stellate fibroblasts. Small islands of odontogenic epithelium and/or cemento-osseous matrix may also be present. The distinction between ameloblastoma and FEPLO is important as the former may be aggressive locally, and the latter lacks local invasion. A single case of FEPLO was diagnosed via biopsy in the rostral mandible of an Inland Bearded Dragon. Microscopically, the neoplasm was composed largely of periodontal ligament–like stroma with small foci of odontogenic epithelium and cemento-osseous matrix (**Figure 1.23**). Follow-up information was not available.

1.2.3.2 Squamous Cell Carcinoma In reptiles, SCC has been reported in all areas of the alimentary tract that are lined by squamous epithelium, including the oral cavity, pharynx, and cloaca. Histologically, SCC of the alimentary tract resembles that described in the skin. Probably the most common location for SCC within the alimentary tract is the oral cavity, and oral SCC was reported in lizards and snakes (Schlumberger and Lucke, 1948; Hill, 1977; Ramsay et al., 1996; Garner et al. 2004). This diagnosis is usually straightforward as this neoplasm is usually well differentiated. The primary differential diagnosis is squamous metaplasia, due to hypovitaminosis A or chronic inflammation/infection. Although unlikely to metastasize, oral SCC is locally invasive and can be difficult to completely excise surgically. A pharyngeal mass that was identified on physical examination of the oral cavity of a 15-year-old Madagascar Ground Boa (*Acrantophis madagascariensis*), was microscopically diagnosed as SCC. Radiation was attempted, which failed to reduce the size of the tumor, and the animal was euthanized (Steeil et al., 2013).

1.2.3.3 Gastric and Esophageal Neoplasia The most commonly reported gastric neoplasm in chelonians and snakes is gastric adenocarcinoma. This is a locally invasive neoplasm arising from the gastric mucosa that is frequently associated with a marked scirrhous response, and less commonly, mucinous matrix. Grossly, these neoplasms cause poorly demarcated, firm thickening of the gastric wall (**Figure 1.24**). The neoplasm is composed of abundant tubules and packets of frequently atypical epithelial cells embedded in scirrhous stroma (**Figure 1.25**). Metastasis is reported in an East African Black Mud Turtle (*Pelusios subniger*) and Red-Eared Slider Turtle (*Trachemys scripta elegans*), in which the neoplasms metastasized to the kidney and intestines, respectively (Cowan, 1968; Mauldin and Done, 2006). Complications from gastric adenocarcinoma include obstruction, perforation, and coelomitis. Gastric adenocarcinoma has been reported in an Aldabra Tortoise (*Aldabrachelys gigantea* [*Testudo elephantina*]), Eastern Indigo Snake (*Drymarchon couperi* [*corais couperi*]), East African Black Mud Turtle, Eastern Kingsnake (*Lampropeltis getula* [*getulus*]), Diamond Python (*Morelia spilota spilota*), rattlesnake (*Crotalus* sp.), Boa Constrictor (*Boa constrictor*), and an additional unidentified python (Schlumberger and Lucke, 1948; D. F. Cowan, 1968; Frye, 1994; Martin et al., 1994; Garner et al. 2004; Sykes and Trupkiewicz, 2006; Baron et al., 2014).

Gastric neuroendocrine carcinoma is reported in Inland Bearded Dragons, and this species appears to have a high prevalence of this aggressive malignancy. This tumor has been reported in 11 Inland Bearded Dragons (Ritter et al., 2009; Lyons et al., 2010). In six cases, tumors were immunoreactive to somatostatin, suggesting these tumors are likely somatostatinomas. These tumors also are sometimes associated with marked hyperglycemia, similar to the tumor counterpart in humans. In all cases, the tumors were located primarily within the submucosa of the pylorus with variable extension to the mucosa and muscularis. The tumors were well differentiated but had considerable cellular anaplasia, primarily anisokaryosis, with bizarre mitoses. Neoplastic cells were arranged in packets separated by fine fibrovascular stroma, typical of neuroendocrine cells (**Figure 1.26**). Most cases had intravascular tumor emboli. In contrast to gastric adenocarcinomas in reptiles, gastric neuroendocrine carcinomas frequently metastasize. Metastasis to the liver was identified in all animals; metastasis to other organs was reported in fewer cases (**Figure 1.27**). Electron microscopy was performed in two cases and revealed neurosecretory granules.

Esophageal adenocarcinomas are rarely reported in reptiles, with only two cases in the literature: one case in an unidentified rattlesnake and the other in a Mata Mata Turtle (*Chelus fimbriatus*) (Garner et al., 2004; Lombardini et al., 2013). Histology is similar to that described in gastric adenocarcinoma. Metastases were not reported. IHC with cytokeratins confirmed epithelial origin in the Mata Mata Turtle.

Adenomas and various polyps have been reported in the gastrointestinal tract of reptiles but are more frequently encountered in the oral cavity and cloaca. A cystic adenoma was described in the stomach of an African Rock Python (*Python sebae*), and a gastric polyp was described in an Emerald Tree Boa (*Corallus caninus*) (Machotka, 1984; Frye, 1994). Grossly, these frequently appear as discrete, exophytic neoplasms arising from the mucosa. Histologically, they are composed of either proliferations of well-differentiated glandular tissue without evidence of invasion (adenoma or adenomatous polyp), or fibroepithelial proliferation of a fibrous stalk overlain by well-differentiated epithelium (polyp; **Figure 1.28**).

Sarcomas in the esophagus and stomach are rarely described in reptiles. Gastric sarcoma was described in a Water Moccasin (Cottonmouth) (*Agkistrodon piscivorus*); metastases were not identified (Cowan, 1968). A hemangioma was identified in the esophageal wall of a Red-Eared Slider (Gál et al., 2009b); further detail on this case is provided in the Cardiovascular System section of this chapter.

1.2.3.4 Intestinal Neoplasia Intestinal and colonic adenocarcinomas are locally invasive and occasionally metastatic neoplasms (**Figure 1.29**). They cause gross thickening of the intestinal and colonic wall, and can be associated with impaction, perforation, and coelomitis. Additional complications arise from metastatic disease. Microscopically, some variants resemble gastric adenocarcinomas. They are relatively common in snakes, and have been described in colubrids, crotalids, and less commonly boas and pythons (Machotka, 1984; Catão-Dias and Nichols, 1999; Garner et al. 2004; Sykes and Trupkiewicz, 2006). In lizards, there is a single case of intestinal adenocarcinoma described in an unidentified iguana, and multiple cases of colonic adenocarcinoma described in a Mexican Beaded Lizard (*Heloderma horridum*), Komodo Dragon (*Varanus komodoensis*), and Leopard Gecko (Garner et al., 2004; Patterson-Kane and Redrobe, 2005; Sykes and Trupkiewicz, 2006). In the Mexican Beaded Lizard, the neoplasm metastasized to the liver and spleen, and in the Leopard Gecko to the oviduct. In one case of metastatic intestinal adenocarcinoma in an Emerald Tree Boa, neoplastic cells contained cytoplasmic viral particles resembling type A retrovirus; the relationship between the virus and neoplasia could not be definitively established (Oros et al., 2004).

Adenoma and adenomatous polyp are reported uncommonly in the intestines and colon of reptiles. Most epithelial proliferations in the intestines and colon are malignant. Benign intestinal and colonic epithelial proliferations should not have evidence of invasion. An intestinal adenoma was reported in a Timber Rattlesnake (*Crotalus horridus*), and multiple intestinal polyps and papillomas have been described in Bullsnakes (*Pituophis catenifer* [*melanoleucus*] *sayi*), Pacific Gopher Snake (*Pituophis catenifer* [*melanoleucus*] *catenifer*), and a Spotted Tree Monitor (*Varanus timorensis similis*) (Machotka, 1984; Frye, 1994; Sykes and Trupkiewicz, 2006).

Leiomyosarcoma or leiomyoma can arise from the smooth muscle of the intestines (**Figure 1.30**). This is relatively uncommon and has been reported in snakes and rarely lizards, including a Texas Indigo snake (*Drymarchon melanurus* [*corais*] *erebennus*), Corn Snake (*Pantherophis guttatus* [*Elaphe guttata*]), unidentified gecko, and Palm Viper (*Bothriechis marchi*) (Barten and Frye, 1981; Catão-Dias and Nichols, 1999; Garner et al., 2004). The latter case metastasized to other viscera. A leiomyoma was diagnosed in the intestines of a Green Turtle that developed volvulus and stricture secondary to the neoplasm (Mauldin and Done, 2006).

Intestinal leiomyoma was diagnosed in a monitor lizard, and intestinal leiomyosarcoma was diagnosed in a Green Turtle submitted to Northwest ZooPath (NWZP).

1.2.3.5 Rectal and Cloacal Neoplasia The most commonly reported neoplasm in the cloaca or rectum (terminal colon) of reptiles is cloacal adenocarcinoma. This neoplasm is uncommon, but has been diagnosed in multiple Corn Snakes, Everglades Rat Snake (*Pantherophis alleghaniensis* [*Elaphe obsolete*] *rossalleni*), Gopher Snake, unidentified kingsnake, unidentified viper, and unidentified chuckwalla (Machotka, 1984; Catão-Dias and Nichols, 1999; Garner et al., 2004; Mauldin and Done, 2006; Sykes and Trupkiewicz, 2006). Metastasis to the spleen was reported in the Everglades Rat Snake (Sykes and Trupkiewicz, 2006).

Leiomyosarcoma and undifferentiated sarcoma are reported rarely in the cloaca and rectum. Cloacal leiomyosarcoma was diagnosed in an unidentified indigo snake, cloacal sarcoma in a Western Hognose Snake (*Heterodon nasicus*), and rectal leiomyosarcoma with metastasis to the liver in an Indian Cobra (*Naja naja*) (Wadsworth, 1954; Machotka, 1984; Ramsay et al., 1996). Hemangioma of the cloaca was diagnosed in a Prairie Rattlesnake (*Crotalus viridis viridis*) (Machotka, 1984), as well as in an Angolan Python (*Python anchietae*) submitted to NWZP (**Figure 1.31**). Adenomatous polyps are seen somewhat frequently in the cloaca of lizards, and rarely reported chelonians, with one case in an Eastern Box Turtle (*Terrepene carolina*) (Frye, 1994; Garner et al., 2004). Grossly, these may resemble neoplasia, but histologically they lack invasion or cellular atypia. They are composed of noninvasive proliferations of well-differentiated adenomatous tissue that frequently form tubulopapillary structures. The most substantial complication from these lesions is due to ulceration with secondary bacterial invasion.

1.2.4 Pancreas

Among reptiles, pancreatic neoplasia is most commonly reported in snakes. Cells in the pancreas that undergo malignant transformation include both acinar and ductular cells of the exocrine pancreas, which are described here, and islet cells of the endocrine pancreas, which are described in the Endocrine System section of this chapter.

1.2.4.1 Pancreatic Adenocarcinoma Pancreatic adenocarcinoma can arise from pancreatic acinar cells or the ductular epithelium. There is some suggestion that pancreatic carcinomas with tubular patterns likely represent ductular origin, whereas tumors with brightly eosinophilic zymogen granules likely represent acinar origin. Even in domestic animals, the distinction between ductular and acinar origin is frequently impossible to discern histologically (Jubb and Stent, 2016). Here we will consider malignant acinar and ductular pancreatic neoplasia together as "pancreatic adenocarcinoma."

Pancreatic adenocarcinoma is frequently described in snakes from various groups. Metastases to spleen, liver, heart, muscle, lung, large intestines, mesovarium, and kidney are described uncommonly, with reports of metastatic disease in a Black Rat Snake, Western Hognose Snake (*Heterodon nasicus*), Northern Water Snake (*Nerodia sipedon*), unidentified viper, and unidentified rattlesnake (Ramsay et al., 1996; Garner et al., 2004; Stern et al., 2010; Matthews et al., 2014). Reports of pancreatic adenocarcinoma in reptiles outside of snakes are few, and include a chameleon (*Chamaeleo* sp.), Solomon Island Skink (*Corucia zebrata*), unidentified lizard, and unidentified chelonian (Frye, 1994; Garner et al., 2004; Sykes and Trupkiewicz, 2006). Grossly, these neoplasms can be difficult to differentiate from pancreatitis depending on the size of the neoplasm and if the associated inflammation or scirrhous response are severe. Histologically, these neoplasms are composed of packets of neoplastic cells embedded in a variable, frequently large amount of dense fibrous stroma, sometimes with granulocytic inflammation and occasional abscess formation. Neoplastic cells sometimes maintain cytoplasmic polarity with basilar basophilia and apical eosinophilia, occasionally with identifiable zymogen granules (**Figure 1.32**). Cystic formation of neoplastic cells is also sometimes observed (i.e., cystic adenocarcinoma). In a Northern Water Snake the pancreatic tumor and ducts labeled for cytokeratins, but no labeling was observed in acinar cells. Based on the pattern of staining and histomorphology (tubules lined by multiple layers of neoplastic cells without discernable zymogen granules), this was diagnosed as a pancreatic adenocarcinoma of ductular origin (i.e., ductular adenocarcinoma) (Matthews et al., 2014). Further studies would be required to determine whether differential cytokeratin expression can be reliably used to differentiate cell of origin in this neoplasm in reptiles. Pancreatic adenocarcinoma suggestive of ductular origin was also diagnosed in an Inland Bearded Dragon submitted to NWZP.

1.2.4.2 Pancreatic Adenoma Pancreatic acinar cell adenomas are small, well demarcated neoplasms that are frequently at least partially encapsulated and form single or multiple masses in the pancreas. Neoplastic cells resemble well differentiated acinar cells (**Figure 1.33**). Neither local invasion nor metastasis occur, although compression of surrounding pancreatic or splenic tissues may be seen. Pancreatic acinar adenoma was diagnosed in a Gila Monster, Banded Rock Rattlesnake (*Crotalus lepidus klauberi*), Brazilian Rainbow Boa (*Epicrates cenchria cenchria*), and four additional snakes from various groups (Frye, 1994; Catão-Dias and Nichols, 1999; Garner et al., 2004; Sykes and Trupkiewicz, 2006).

Pancreatic ductal adenomas form proliferative tubular structures lined by epithelium with little difference from normal pancreatic duct epithelium that lacks zymogen granules. Only a single case has been reported in a reptile, a Western Diamondback Rattlesnake (*Crotalus atrox*) (Frye, 1994). An additional pancreatic adenoma suggestive of ductular origin was diagnosed in an Copperhead (*Agkistrodon contortrix*) submitted to NWZP.

1.2.5 Hepatobiliary System

Hepatobiliary neoplasia is common in reptiles (Christman et al., 2017). Both hepatocytes and biliary epithelial cells undergo neoplastic transformation relatively frequently in lizards and snakes; neoplasia of these organs is rarely reported in chelonians (Garner et al., 2004; Wilkinson et al., 2004). In lizards and snakes, approximately half of all tumors arising from hepatocytes are benign (i.e., hepatoma) and the other half are malignant (i.e., hepatocellular carcinoma). Tumors arising from biliary epithelial cells are more frequently malignant (i.e., biliary adenocarcinoma) than benign (i.e., biliary cystadenoma) (Garner et al., 2004).

1.2.5.1 Hepatocellular Carcinoma Hepatocellular carcinoma is characterized grossly as single or multiple large masses within the liver. Microscopically, hepatocellular carcinoma is composed of cords frequently more than two cells thick (trabecular pattern) with little to no scirrhous response. Neoplastic cells have a hepatoid appearance with a variable amount of eosinophilic cytoplasm. Cellular atypia, primarily anisokaryosis, anisocytosis, and nucleolar enlargement, typically is mild to moderate (**Figure 1.34**). The mitotic index usually is low in well-differentiated tumors, but can be high, with bizarre mitoses in anaplastic tumors. This neoplasm is common in snakes, particularly aged colubrids, and is occasional in lizards (Machotka, 1984; Frye, 1994; Garner et al., 2004; Sykes and Trupkiewicz, 2006). Metastatic cases have not been reported in reptiles.

1.2.5.2 Hepatoma Hepatomas (hepatocellular adenomas) are benign neoplastic proliferations of hepatocytes that differ little from mature hepatocytes and lack features of malignancy (**Figure 1.35**). They are compressive, but not invasive. Grossly, these may be solitary or multiple and can become large. The primary differentials for this neoplasm include hepatocellular carcinoma and nodular hyperplasia, the latter of which is partially marginated by biliary ducts and is frequently seen in cirrhotic livers (Garner et al., 2004). As with hepatocellular carcinoma, hepatomas are common in snakes (especially colubrids) and occasionally seen in lizards (Cowan, 1968; Machotka, 1984; Frye, 1994; Ramsay et al., 1996; Garner et al., 2004; Sykes and Trupkiewicz, 2006). It is common to encounter a full spectrum of hepatocellular proliferative lesions multifocally in the liver of old snakes, including hyperplasia, adenoma, and carcinoma, although metastasis is uncommon.

1.2.5.3 Biliary Adenocarcinoma Biliary adenocarcinoma (also known as cholangiocarcinoma or cholangiocellular carcinoma) with or without metastasis is somewhat common in lizards and snakes, particularly colubrids and crotalids (Schlumberger and Lucke, 1948; Effron et al., 1977; Machotka, 1984; Frye, 1994; Ramsay et al., 1996; Garner et al., 2004; Sykes and Trupkiewicz, 2006). Biliary and hepatocellular carcinomas can cause external distention of the mid-coelomic cavity (**Figure 1.36**). Grossly, biliary adenocarcinoma may form one or multiple nodular foci in the liver or gallbladder and is usually firm and pale (**Figure 1.37**), as opposed to the dark parenchymal color of hepatocellular tumors. These neoplasms are composed of small tubules, ducts, and packets of polygonal cells that frequently have a moderate to large amount of associated scirrhous stroma, the latter accounting for the pale gross appearance of the tumor. Cellular atypia and mitotic rate are usually moderate (**Figure 1.38**). Variable features include occlusion or hyperplasia of the bile ducts, papillary growth, and cystic formation (Schlumberger and Lucke, 1948; Machotka, 1984; Sykes and Trupkiewicz, 2006). A single case was associated with a trematode infection (Schlumberger and Lucke, 1948).

1.2.5.4 Biliary Adenoma Biliary adenomas (also known as cholangiomas or cholangiocellular adenomas) are less commonly reported in reptiles than their malignant counterpart. They have been reported in many species of snakes, and only a few lizards—a Green Iguana, Ricord's Iguana (*Cyclura ricordii*), and a Panther Chameleon (*Furcifer pardalis*) (Cowan, 1968; Effron et al., 1977; Frye, 1994; Ramsay et al., 1996; Garner et al., 2004; Mauldin and Done, 2006; Sykes and Trupkiewicz, 2006). These neoplasms are well-demarcated proliferations of bile duct epithelium (**Figure 1.39**) that closely resembles normal bile duct epithelium (Volume 1, **Figures 1.171–1.174**). Biliary adenomas described in the above iguanas may in fact be hyperplastic lesions associated with cholelithiasis or other obstructive biliary processes, and these lesions are covered in Chapter 5 (Wilson et al., 2004).

1.2.5.5 Miscellaneous Hepatic Neoplasms Primary hepatic mesenchymal neoplasia is uncommon in reptiles as in domestic animals. There is a single case of a sarcoma with biphasic differentiation originating in the liver of a Madagascar Tree Boa (*Sanzinia madagascariensis*). In the referenced case, the neoplastic cells differentiated toward multiple lineages (Bera et al., 2008). In domestic animals, embryonal neoplasms of the liver are frequently termed hepatoblastomas, and can have mesenchymal differentiation, including fibroblastic differentiation (Cullen and Stalker, 2016). Hemangioma and hemangiosarcoma have been diagnosed in the liver of reptiles and are discussed in Section 1.2.14 on the Cardiovascular System in this chapter. The liver is also a common site for metastatic sarcomas in reptiles.

1.2.6 Reproductive System

Reproductive neoplasia is more common in lizards than snakes, with the exception of granulosa cell tumors (Garner et al., 2004). Reproductive tract tumors are rarely reported in chelonians and crocodilians, with a single report of a seminoma in an American Alligator, multiple cases of dysgerminomas in chelonians, a seminoma in a Spur-Thighed Tortoise

(*Testudo graeca*), and an ovarian teratoma in a Hermann's Tortoise (Wadsworth, 1954; Frye, 1994; Martorell et al., 2009; Pees et al., 2015). For additional comments on the reproductive neoplasia, see Chapter 8.

1.2.6.1 Ovarian Neoplasia Primary tumors of the ovary can be broadly subdivided based on their embryological origin: sex cord–stromal tumors, tumors of germ cells, tumors of the surface epithelium, tumors of nongonadal tissues, and vascular hamartoma. Ovarian neoplasia has not been reported in crocodilians (Stacy et al., 2004).

1.2.6.1.1 Sex Cord–Stromal Tumors These tumors often have mixtures of phenotypes with granulosa cells and theca cells. Grossly, tumors frequently have a smooth surface; on cut surface, they are often cystic (**Figure 1.40**). Histologically, these tumors contain areas with characteristic gland-like or rosette patterns, some of which may contain a secretory globule resembling an ovum called a Call-Exner body (**Figure 1.41**). Cyst formation and hemorrhage are common. Tumors that differentiate toward one cell type are either termed granulosa cell tumor (i.e., predominantly granulosa cells), or thecoma (i.e., predominantly theca cells; also termed a luteoma), while tumors that are a mixture of both cell types are termed sex cord–stromal tumor (Schlafer and Foster, 2016).

Granulosa cell tumors (GCT) are commonly reported in snakes, and less commonly in lizards (Garner et al., 2004). Although metastasis is rarely reported, a 6-year-old Common Garter Snake (*Thamnophis sirtalis*) had metastases to the kidneys, a 6-year-old Savanna Monitor (*Varanus exanthematicus*) had metastasis to unspecified viscera, and a 9-year-old Green Iguana in which neoplastic cells had metastasized to the fat body at the time of initial resection, had metastasized widely to viscera 11 months later (Onderka and Zwart, 1978; Cruz Cardona et al., 2011; Elliott et al., 2013). Interestingly, in the latter case, the GCT was associated with abundant mast cell proliferation, a feature that is unknown to occur in association with GCT. In addition, it was speculated that this GCT may have been induced by hormonal imbalances following incomplete ovariohysterectomy (Cruz Cardona et al., 2011). A mixed granulosa–theca tumor (sex cord–stromal tumor) was diagnosed in a Green Tree Viper (*Trimeresurius albolabui*) (Effron et al., 1977). Luteoma has been diagnosed a Green Iguana (*Iguana iguana*). The referenced case had an adenocarcinoma in the same ovary as the luteoma. Luteoma cells formed sheets of vacuolated cells with euchromatic nuclei (Garner et al., 2004).

1.2.6.1.2 Dysgerminoma Dysgerminoma is composed of cells that show morphologic similarity to germ cells resembling those in their male homolog, the seminoma. Grossly, these neoplasms are soft, pale tan, and expansile. This neoplasm is composed of diffuse sheets and tubular patterns of uniform round cells with large, central nuclei; stroma is usually scant. As in seminoma, mitotic figures and multinucleated cells are frequent (Schlafer and Foster, 2016). This tumor is rarely reported in reptiles, with two cases in Red-Eared Sliders (*Trachyms scripta elegans*), one in a Green Iguana (*Iguana iguana*), and one in a Snapping Turtle (*Chelydra serpentina*) (Machotka et al., 1992; Frye, 1994; Stacy et al., 2004).

1.2.6.1.3 Ovarian Epithelial Neoplasia Ovarian carcinoma arises from the outer surface epithelium of the ovary. Histologically, this neoplasm resembles adenocarcinoma of other sites, forming packets and small tubules of polygonal, occasionally ciliated cells, and occasionally forming papillary structures (**Figure 1.42**). Frequently, the diagnosis of ovarian carcinoma is prefaced with "cystic-," "papillary-," "adeno-," or combinations thereof to denote predominant morphologic patterns which have no known prognostic significance in reptiles. Ovarian carcinoma is reported in both lizards and snakes, including a Shingleback Skink (*Tiliqua rugosa*), several species of boas (*Boa constrictor ortoni, B. c. imperator, B. c. constrictor, Candoia bibroni*), an African Rock Python, a Reticulated Python, multiple Green Iguanas and additional unidentified lizards and snakes (Bland-Sutton, 1885; Schlumberger and Lucke, 1948; Michaels and Sanecki, 1988; Gibbons and Schiller, 2000; Garner et al., 2004; Stacy et al., 2004; Sykes and Trupkiewicz, 2006). Metastasis of ovarian carcinoma has been reported to occur both hematogenously and via implantation (i.e., carcinomatosis) in reptiles (Michaels and Sanecki, 1988; Stacy et al., 2004; Sykes and Trupkiewicz, 2006). Immunoreactivity to pancytokeratin confirmed the diagnosis of ovarian carcinoma in a Green Iguana (*Iguana iguana*) (Stacy et al., 2004).

1.2.6.1.4 Teratoma Teratomas are comprised of totipotential germ cells that are differentiated toward more than one germ cell layer. Teratomas can be comprised of cells representing either two or three germ cell lines (i.e., endoderm, mesoderm, ectoderm). Broadly, endodermal cell lineages are those that make up epithelial components of the digestive, urinary, and reproductive systems; this lineage can frequently be identified by epithelial tubules. Mesodermal origins include musculoskeletal, adipose, and vascular systems and are recognized by stromal cells in the neoplasm. Ectodermal tissues include neural elements, epidermis, glands, hair follicles, and tooth elements; the histomorphology of this lineage varies widely and can range from stratified squamous epithelium to differentiated teeth (**Figure 1.43**) (Schlafer and Foster, 2016). Ovarian teratomas have been reported in a Fiji Island Banded Iguana (*Brachylophus fasciatus*), several Green Iguanas, multiple Red-Eared Sliders, Desert Grassland Whiptails (*Aspidoscelis* [*Cnemidophorus*] *uniparens*], and a Hermann's Tortoise. In the Hermann's Tortoise, Red-Eared Slider, Green Iguana, and the Fiji Island Banded Iguana, teratomas histologically had components from all three germ layers (i.e., ectoderm, endoderm, mesoderm); other cases had only derivatives from the endoderm and mesoderm (Anderson et al., 1996; Tocidlowski et al., 2001a,; Stacy et al., 2004; Hidalgo-Vila et al., 2006; Martorell et al., 2009; Wenger et al., 2010). Two teratomas, one in a Green Iguana and one in a Red-Eared Slider were diagnosed as malignant based on cellular atypia, necrosis, and mitotic figures. The referenced cases had no evidence of metastasis or local invasion (Newman et al., 2003; Bel et al., 2016). In multiple Green Iguanas, ovarian teratomas were resected and the animals were clinically healthy at 10 months after surgery. All tumors were cystic on imaging and/or gross examination (**Figure 1.44**).

1.2.6.1.5 Vascular Hamartomas Vascular hamartomas of the ovary have been observed in many domestic species, but never reported in reptiles. These are not true neoplasms, but rather tumor-like malformations that are present at birth. The distinction between vascular hamartomas and hemangiomas is difficult and there are possible misdiagnoses of hamartomas as hemangiomas (Schlafer and Foster, 2016). A single case of an ovarian hemangioma was reported in an Eastern Massasauga (*Sistrurus catenatus*), without including a differential of hamartoma (Effron et al., 1977).

1.2.6.1.6 Miscellaneous Ovarian Neoplasia Neoplasia arising from mesenchymal cells in the ovary is rarely reported in veterinary species. Benign and malignant neoplasia can arise from any supporting mesenchymal structures in the ovary, including blood vessels, nerves, and mesovarium. Several primary ovarian mesenchymal neoplasms have been reported in reptiles, including hemangiomas in an Eastern Massasauga and Timber Rattlesnake (*Crotalus horridus*), fibroma in an Elephant Trunk Snake (*Acrochordus javanicus*), sarcoma in a Turks Island Boa (*Chilabothrus* [*Epicrates*] *chrysogaster*), and fibroma in a Water Mocassin (Effron et al., 1977). In addition, a stromal cell neoplasm was identified in the ovary of a Savannah Monitor (**Figure 1.45**) (Elliott Jacobson, personal communiation), and a hemangiosarcoma was diagnosed in the ovary of a Black Tailed Horned Viper (*Mixcoatlus melanurus*) submitted to NWZP.

1.2.6.2 Oviductal Neoplasia The most commonly reported tumors of the oviduct in reptiles are of epithelial origin (i.e., adenoma or adenocarcinoma). Less commonly, sarcomas of smooth muscle are reported.

1.2.6.2.1 Oviductal Epithelial Tumors Oviductal adenocarcinoma has been uncommonly reported in snakes and lizards. Metastases were reported in some of the snake cases. When present, metastases were identified histologically in all cases and grossly in several cases. These tumors appear grossly as red to tan, exophytic, firm masses within the oviduct, and are sometimes only identified after microscopic examination. The microscopic appearance is typical of adenocarcinoma, specifically acini or nests of neoplastic cuboidal to elongate, occasionally ciliated cells with variable nuclear polarity, mild to moderate anisokaryosis, and occasional mitotic figures (**Figure 1.46**). These tumors are frequently associated with an abundant scirrhous reponse (Garner et al., 2004; Pereira and Viner, 2008). Metastic foci of poorly differentiated ovarian and oviduct tumors can be difficult to discern from other forms of metastatic adenocarcinoma.

Benign epithelial oviduct neoplasia is less commonly reported than malignant neoplasia in reptiles. Benign oviduct tumors resemble their malignant counterpart but lack features of malignancy, such as cellular atypia, invasion, and metastasis. Reported tumors in this group include oviductal papillary adenoma and oviductal cystadenoma, both of which were reported in kingsnakes (*Lampropeltis* sp.) (Frye, 1994).

1.2.6.2.2 Oviductal Smooth Muscle Neoplasia As in the ovary, any of the mesenchymal structures in the oviduct can give rise to neoplasia, include those associated with vessels, nerves, and musculature. Benign and malignant mesenchymal neoplasia have been reported in the oviduct, including a leiomyoma in an Agassiz's Desert Tortoise and leiomyosarcomas in Puerto Rican Boas (*Chilabothrus inornatus*) (Frye, 1994; Ramsay et al., 1996). No metastases were reported in these cases. Oviductal leiomyosarcoma with metastatis was diagnosed in a Mexican Milksnake (*Lampropeltis annulata*) and without metastasis in an Eastern Indigo Snake submitted to NWZP (**Figure 1.47**).

1.2.6.3 Testicular Neoplasia In reptiles, neoplasia of the male reproductive tract is less common than neoplasia of the female reproductive tract. All primary cell types in the testicle are reported to give rise to neoplasia in reptiles, including interstitial cells, Sertoli cells, and germ cells. In addition, mesenchymal cells of the interstitium can give rise to various mesenchymal tumors, as in the ovary. Although hormonal production is frequently associated with some testicular neoplasms in mammals, this has not yet been reported in reptiles.

1.2.6.3.1 Interstitial Cell Tumor Interstitial cell tumors, also called Leydig cell tumors or interstitial cell adenomas, have been reported in snakes, lizards, and chelonians. Histologically, there is expansion of the testicular interstitium by polygonal or occasionally elongate cells with eosinophilic cytoplasm that have a "ground glass" appearance, typical of interstitial, or "Leydig," cells; stroma is scant. Multinucleation of neoplastic cells and a variable mitotic rate are reported (**Figure 1.48**) (Frye et al., 1988). This neoplasm has been reported in a Burmese Python, a Western Fence Lizard (*Sceloporus occidentalis*), *Epicrates* sp., Palestine Viper (*Vipera palaestinae*), multiple Komodo Dragons, a Cuban Rock Iguana (*Cyclura nubile*), and multiple Agassizi's Desert Tortoises (Jacobson, 1981; Frye et al., 1988; Frye, 1994; Ramsay et al., 1996; Garner et al., 2004). A cystic variant of interstitial cell tumor has been described in multiple Komodo Dragons (Machotka, 1984; Hernandez-Divers and Garner, 2003).

1.2.6.3.2 Seminoma Seminomas are neoplasms of testicular germ cells and are uncommon in reptiles. Grossly, these are soft, white, and bulging masses within the testicular parenchyma. Microscopically, they are composed of intratubular or diffuse sheets of neoplastic round cells with scant stroma. Neoplastic cell morphology is typical of germ cells—large, round cells with a large round vesicular nucleus, large nucleolus, and a small amount of eosinophilic cytoplasm (**Figure 1.49**). Multinucleated cells and small lymphocytes are commonly scattered throughout the neoplasm. Three cases of seminoma have been described in an American Alligator (*Alligator mississippiensis*), Indian Cobra, and Spur-Thighed Tortoise (*Testudo graeca*), respectively (Wadsworth and Hill, 1956; Frye, 1994; Pees et al., 2015). The Indian Cobra with seminoma was presented for urocystolithiasis, which was an unusual presentation, as snakes lack urinary bladders. In this case, however, the seminoma exerted sufficient pressure on the ureter to cause the formation of a pseudobladder, in which urinary calculi formed (Frye, 1994). The authors have two cases of seminoma in Leopard Geckos, and single cases in an Inland Bearded Dragon, Haitian Boa (*Chilabothrus striatus*), Agassiz's (California) Desert Tortoise, Spiny-Tailed Lizard (*Uromastyx* sp.), Corn Snake, and Common Blue-Tongue Skink (*Tiliqua scincoides*) in the archives of NWZP. Metastatic seminoma has not been described for reptiles.

1.2.6.3.3 Sertoli Cell Tumoe Sertoli cell tumors, like seminomas, may have a tubular pattern, especially early in the course of disease. They are differentiated from other primary testicular neoplasms by the histomorphology of their cells, as well as the abundance of scirrhous stroma that is induced by this neoplasm. In well-differentiated variants, neoplastic cells resemble mature Sertoli cells that palisade along stroma, form small tubules with occasional cystic distention, and are elongate with foamy, pale eosinophilic cytoplasm, and small basal nuclei. In less-differentiated Sertoli cell tumors, nuclei are pleomorphic, and cells have less tendency to palisade. Grossly, these neoplasms are firm due to the large amount of scirrhous stroma. Very little data are available regarding Sertoli cell tumors in reptiles, and all reports in the literature are in snakes, although NWZP archives have one of these in an unidentified Bearded Dragon (**Figure 1.50**). This neoplasm has been reported in a Corn Snake, Honduran Milk Snake (*Lampropeltis triangulum hondurensis*), Common Garter Snake, and Egyptian Cobra (*Naja haje*) (Machotka, 1984; Frye, 1994; Sykes and Trupkiewicz, 2006). The Sertoli cell tumor metastasized to the lung of the Corn Snake, warranting a diagnosis of malignant Sertoli cell tumor (Sykes and Trupkiewicz, 2006).

1.2.6.3.4 Mesenchymal Cell Tumors of the Testes Any of the mesenchymal components of the testes can give rise to neoplasia, though this is rare. A single case of leiomyosarcoma was described in an Emerald Tree Boa (Catão-Dias and Nichols, 1999). The authors have a testicular soft tissue sarcoma in a California Kingsnake (*Lampropeltis californiae* [*getulus californiae*]) in the archives of NWZP. The authors could not identify any cases of testicular teratoma in a reptile.

1.2.7 Endocrine System

Endocrine organs of reptiles include the pituitary gland, thyroid gland, parathyroid gland, adrenal gland, ultimobranchial body, pineal gland, and islets of Langerhans. Neoplasia of endocrine organs is uncommon in reptiles. In addition, neuroendocrine cells are distributed diffusely in many organs as small clusters or individual cells (Banks, 1993), and may give rise to neoplasia. Gastric neuroendocrine carcinoma in Inland Bearded Dragons is an example and is discussed in Section 1.2.2 on the Alimentary System in this chapter.

IHC is the gold standard for diagnosing and/or differentiating neuroendocrine neoplasia, and electron microscopy showing neurosecretory granules is also supportive of this diagnosis. Antigens that have been targeted by IHC for neuroendocrine tumors in reptiles include protein gene product 9.5, neuron-specific enolase, endorphin, chromagranins A and B, synaptophysin, somatostatin, insulin, glucagon, gastrin, pancreatic polypeptide, and vasoactive intestinal peptide (Lyons et al., 2010). Positive immunoreactivity to porcine calcitonin has been reported in normal ultimobranchial glands of the Red-Eared Slider, Japanese Grass Lizard (*Takydromus tachydromoides*), and Japanese Striped Snake (*Elaphe quadrivirgata*), and it is possible that this antigen would be useful in identifying future cases of neoplasia of ultimobranchial glands (Boudbid et al., 1987; Yamada et al., 2001).

1.2.7.1 Thyroid Follicular Neoplasia Benign and malignant neoplasms of the thyroid have been reported in reptiles, most commonly in chelonians and lizards, with very few reports in snakes. Diagnostic criteria for differentiating benign from malignant endocrine neoplasms are not well established in reptiles. In mammals, necrosis, capsular or vascular invasion, hemorrhage, and large size are more common in thyroid carcinomas than in adenomas. In the author's experience, necrosis and capsular invasion are useful criteria in determining malignancy in thyroid follicular neoplasia, as both variants tend to have well differentiated epithelium with little anaplasia. Thyroid carcinoma was reported in a Red-Eared Slider, Rough Knob-Tail Gecko (*Nephrurus amyae*), Smooth Knob-Tail Gecko (*Nephrurus levis*), Painted Turtle (*Chrysemys picta*), Indian Black Turtle (*Melanochelys trijuga*), unidentified terrapin, and Crocodile Lizard (*Shinisaurus crocodilurus*) (Machotka, 1984; Whiteside and Garner, 2001; Ritter et al., 2009; Lyons et al., 2010; Hadfield et al., 2012). In all cases, neoplastic cells occasionally formed colloid-filled follicles and had minimal cellular atypia and rare mitoses (**Figure 1.51**). Metastasis to the liver and lungs was reported in the Rough Knob-Tail Gecko case, and to the mediastinum in the Indian Black Turtle (Hadfield et al., 2012).

Thyroid adenoma is most commonly reported in chelonians with fewer cases in lizards and a snake. Reported cases include a Geoffroy's Side Necked Turtle (*Platemys* sp.), Geoffroy's Toadhead Turtle (*Phrynops geoffroanus*), Karoo Girdled Lizard (*Cordylus polyzonus*), multiple Komodo dragons, Agassiz's Desert Tortoise, Gopher Snake, and Green Iguana (Schlumberger and Lucke, 1948; Machotka, 1984; Frye, 1994; Ramsay et al., 1996; Hernandez-Divers et al., 2001). Microscopically, thyroid follicular adenomas are composed of multiple follicles of varying size, sometimes with cystic formation, containing variable amounts of colloid and supported by a small amount of fibrovascular stroma. In the Green Iguana, the animal presented with symptomatic hyperthyroidism, which resolved after surgical removal of the thyroid follicular adenoma (Hernandez-Divers et al., 2001). Thyroid hormone derangements associated with thyroid neoplasms were not reported in the other cases.

1.2.7.2 Thyroid C Cell/Ultimobranchial Neoplasia Neoplasia of the ultimobranchial gland (i.e., C cells) is very rare in reptiles. There are two reported cases in lizards, including a thyroid C cell adenoma in a Green Iguana, and a thyroid C cell carcinoma in a Crocodile Lizard The carcinoma metastasized to the liver and mesentery (Sykes and Trupkiewicz, 2006). Endocrine abnormalities associated with C cell neoplasia were not reported.

1.2.7.3 Parathyroid Neoplasia Parathyroid neoplasms reported in reptiles have been uniformly benign (i.e., parathyroid adenomas). They are most common in chelonians. Parathyroid adenomas may be partially or circumferentially encapsulated. They are composed of nests and clusters of polygonal cells supported by fine, fibrovascular stroma. The cytoplasm of neoplastic cells is eosinophilic, granular, and occasionally vacuolated. Cellular and nuclear atypia are minimal. Mitoses are not reported in documented cases of parathyroid adenomas in reptiles, which include multiple Red-Footed Tortoises (*Chelonoidis carbonarius* [*Geochelone carbonaria*]), multiple Green Iguanas, Spur-Thighed Tortoise, Agassiz's Desert Tortoise, Indian Star Tortoise (*Geochelone* [*Testudo*] *elegans*), and Common Garter Snake (Frye, 1975; Machotka, 1984; Frye, 1994; Sykes and Trupkiewicz, 2006). In one of the Red-Footed Tortoises, the parathyroid adenoma was associated with calcium and phosphorous imbalances. Instead of increased calcium and decreased phosphorus, as would be expected in primary hyperparathyroidism, this animal had decreased calcium and increased phosphorus (Frye and Carney, 1975). Parathyroids become hyperplastic in reptiles that have renal disease, chronic osteoarthritis, dietary imbalances in calcium and vitamin D, and exposure to inappropriate ultraviolet lighting, and these should be ruled out as contributing factors before diagnosing a well-differentiated parathyroid neoplasm.

1.2.7.4 Adrenal Gland Neoplasia Adrenal gland neoplasia is rare in reptiles, but pheochromocytoma and interrenal cell are reported. Similar to thyroid neoplasms, features of malignancy are not well established, but hemorrhage, necrosis, and capsular invasion are frequently seen. Interrenal (also known as adrenal cortical) cell neoplasia is reported in snakes and a single lizard. Interrenal adenomas are reported with approximately equal frequency as interrenal carcinomas in reptiles. These neoplasms frequently cause gross enlargement of the adrenal gland. Interrenal neoplasms are composed of packets of slightly atypical interrenal cells supported by delicate fibrovascular stroma. These cells form a mass with variable encapsulation that compresses the surrounding parenchyma (**Figure 1.52**). Interrenal adenomas are reported in a Green Iguana, Asiatic (Indian) Cobra, unidentified python, and Yellow Anaconda (Frye, 1994; Ramsay et al., 1996; Garner et al., 2004; Sykes and Trupkiewicz, 2006) Interrenal carcinoma is described in snakes with cases in a Jaraca Pintada Pit Viper (*Bothrops neuwiedi*), Levantine Viper (*Macrovipera lebetina*), Woma Python (*Aspidites ramsayi*), Corn Snake, Emerald Tree Boa, and unidentified rattlesnake. There was local invasion in the Woma Python, and metastasis to multiple visceral organs and the mesentery in the rattlesnake (Catão-Dias and Nichols, 1999; Garner et al., 2004; Sykes and Trupkiewicz, 2006; Kaye et al., 2016). Colubrids seem overrepresented for these neoplasms in the archives of NWZP.

Pheochromocytomas have similar histologic organization as adrenal cortical neoplasms, except that the cells are similar to adrenal medullary cells with light eosinophilic or granulated cytoplasm. Reports in reptiles are sparse with a case in a California Glossy Snake (*Arizona elegans occidentalis*), Desert Cobra (*Walterinnesia aegyptia*), and Komodo Dragon (Effron et al., 1977; Mauldin and Done, 2006).

1.2.7.5 Neoplasia of the Pancreatic Islets Islet cell tumors are reported rarely in reptiles. Grossly, these may be indistinguishable from pancreatic adenocarcinoma, although the latter tends to be associated with more inflammation, necrosis, and metastasis. Islet cell tumors in reptiles are typically well circumscribed masses of packets of islet cells with delicate fibrovascular stroma (**Figure 1.53**). Islet cell tumors have been reported in Komodo Dragons, Exuma Island Iguana (*Cyclura c. figginsi*), hog nose snake (*Heterodon* sp.), Rough-Necked Monitor (*Varanus rudicollis*), and a Green Iguana (Machotka, 1984; Frye, 1994; Ramsay et al., 1996; Garner et al., 2004; Eustace et al., 2017). In three Komodo Dragons, islet cell neoplasms were metastatic, strongly and diffusely immunoreactive to somatostatin and glucagon, and variably immunoreactive to pancreatic polypeptide; these neoplasms were diagnosed as multihormonal islet cell neoplasms (**Figure 1.54**) (Eustace et al., 2017). Unlike their mammalian counterparts, these neoplasms are rarely known to be functional. The Exuma Island Iguana case was reportedly functional and diagnosed as an alpha cell glucagonoma (Frye, 1994). Metastatic islet cell carcinomas were diagnosed in an Aruba Island Rattlesnake (*Crotalus unicolor*) and an unidentified boa submitted to NWZP.

1.2.7.6 Pituitary Neoplasia Proliferative pituitary lesions are occasionally seen in reptiles, including cysts, nodular hyperplasia, and adenomas. Malignant pituitary tumors have not been reported. Nodular hyperplasia of chromophobe cells in the pars distalis is probably an aging change, as in mammals. Cysts may be developmental or acquired and also may be more common in aged animals. Three cases of pituitary adenoma have been reported in snakes—an Everglades Rat Snake, Dumeril's Ground Boa, and Black-Headed Python (*Aspidites melanocephalus*) (Linn et al., 1996; Gyimesi and Garner, 2007; Dadone et al., 2010). Clinical signs have included loss of righting reflex and seizures. Gross, asymmetrical enlargement of the pituitary gland can be seen upon removal of the brain from the calvarium. There can be some compression/invasion of the surrounding nervous parenchyma. Microscopically, pituitary adenomas are composed of papillary structures and packets of well differentiated cuboidal to columnar chromophobe cells with few granules (**Figure 1.55**). Cystic formation was reported in the Everglades Rat Snake. Pituitary tumors likely are underreported in reptiles, due to the difficulty in collecting this tissue and the expense of processing intact skull for histologic exam. More detail on the anatomic location of the pituitary gland is present in Volume 1, Chapter 1.

1.2.7.7 Miscellaneous Endocrine Neoplasia Intraceolomic endocrine carcinomas of undetermined origin have been reported in multiple snakes and one lizard (Garner et al., 2004). In addition, the author diagnosed a widely metastatic endocrine carcinoma that was immunoreactive to pancytokeratin Lu5 and chromagranin A in a bearded dragon (*Pogona* sp.). The specific endocrine cell of origin could not be determined in any of the aforementioned cases. It is possible that these neoplasms arose from diffuse endocrine tissue rather than from one of the discrete endocrine organs. There is a single reported case of a pinealoma in a Green Iguana (Frye, 1994).

1.2.8 Musculoskeletal System

A low prevalence of musculoskeletal neoplasia has been reported in lizards and snakes, and rarely crocodilians. Muscular neoplasms include rhabdomyosarcoma, rhabdomyoma, leiomyosarcoma, and leiomyoma. Neoplasms of smooth muscle (i.e., leiomyosarcoma and leiomyoma) are described under their host organ (see Reproductive and Alimentary systems). Striated muscle neoplasms (i.e., rhabdomyosarcoma, rhabdomyoma) are described here. Skeletal neoplasms include those arising from cartilage (i.e., chondrosarcoma, chondroma) and bone (i.e., osteosarcoma, osteoma), as well as some with both cartilaginous and osseous differentiation. An important differential for osteomas, chondromas, and related neoplasms is metabolic bone disease or spinal osteopathy of snakes, as these conditions can cause cartilaginous and bony proliferations. A single case of a periosteal fibrosarcoma was diagnosed in a Siamese Crocodile (*Crocodylus siamensis*) in which the neoplasm was locally invasive with some cellular pleomorphism and a low mitotic rate (Janert, 1998).

1.2.8.1 Neoplasia of Skeletal Muscle Rhabdomyosarcomas are uncommonly reported in reptiles. These are malignant neoplasms of striated muscle that are frequently composed of anaplastic mesenchymal cells with occasional rowing of nuclei and/or cross-striations (i.e., "strap cells") (**Figure 1.56**), the latter of which may be highlighted with phosphotungstic acid-hematoxylin (PTAH) stain. In mammals, rhabdomyosarcoma has several distinct histologic phenotypes, including embryonal, alveolar, and pleomorphic. Histologic subtyping is not routinely applied in reptile cases, and whether this classification scheme is appropriate in reptiles is unknown (Caserto, 2013). In reptiles, these neoplasms are apparently locally invasive and less commonly metastatic. Reported cases of rhabdomyosarcoma in reptiles include one on the ventral body wall in a Corn Snake, unspecific location in a python (*Morelia* sp.), maxilla in a Black Ratsnake, body wall in a Charles Island Giant Tortoise (*Chelonoidis niger* [*Geochelone nigra*]), oral cavity in a snake of unidentified species, widely metastatic of undetermined primary location in a snake of unidentified species, and five additional lizards, four of which were geckos in which the tumors were intracoelomic and associated with concurrent xanthomatosis (Ball, 1945; Lunger et al., 1974; Catão-Dias and Nichols, 1999; Garner et al., 2004; Eyarefe et al., 2012). Additional cases of rhabdomyosarcoma have been reported in the heart of reptiles (see Section 1.2.13 on the Cardiovascular System for additional details). In the aforementioned Corn Snake with rhabdomyosarcoma, the neoplasm did not have evident cross-striations histologically, and the diagnosis of rhabdomyosarcoma was made via electron microscopy and identification of actin-like filaments. In addition, C-type virus particles were observed in neoplastic cells on electron microscopy, the significance of which is unknown (Lunger et al., 1974). Benign variants of striated muscle tumors (i.e., rhabdomyoma) are rarely reported in reptiles and include a rhabdomyoma on the hard palate of a Northern Pine Snake (*Pituophis melanoleucus*) (Ball, 1945).

1.2.8.2 Neoplasia of Bone The terminology for bone neoplasms in reptiles is convoluted in the literature. While there are many straightforward diagnoses, such as osteosarcoma and chondrosarcoma, older literature also includes diagnoses such as osteochondrosarcoma, attributed to the presence of multiple types of matrix. Subtypes of osteosarcoma reflect the predominant type of matrix present, such as chondroblastic osteosarcoma, which produces cartilaginous matrix (Craig et al., 2016). For reported reptile cases of osseous neoplasia, the diagnosis will include the previous/historical term (e.g., osteochondrosarcoma), as well as current terminology (e.g., osteosarcoma) when possible. Osteosarcoma is a malignant neoplasm that frequently causes osteolytic lesions radiographically, and irregular, regionally extensive, thickening of bones grossly (**Figure 1.57**). Osteosarcoma is composed of streams or sheets of round, spindle, or stellate cells with variable amounts of neoplasm-associated osteoid production (**Figure 1.58**). Other types of matrix, such as chondroid or fibrous matrix, may also be present in small to large amounts. Osteosarcoma is reported occasionally in snakes and lizards, as in the maxilla of a Central African Forest Cobra (original diagnosis: osteochondrosarcoma), ribs of a Woma Python, pelvis of two related Spiny-Tailed Monitor Lizards (*Varanus acanthurus*), vertebrae and proximal ribs in a California (Sierra) Mountain Kingsnake (*Lampropeltis zonata*), mandible in a Desert Monitor (*Varanus griseus*) (original diagnosis: osteochondrosarcoma), mandible in a Burmese Python (original diagnosis: osteogenic sarcoma), unspecified location in a Bengal Monitor (*Varanus bengalensis*) (original diagnosis: osteochondrosarcoma), vertebrae in two unidentified snakes, and mesentery in a cobra (suspect extraskeletal osteosarcoma) (Wadsworth, 1954; Schonbauer et al., 1982; Machotka, 1984; Frye, 1994; Latimer et al., 2000; Garner et al., 2004; Cowan et al., 2011; Needle et al., 2013). In the referenced cases, production of chondroid matrix was common, and metastases were occasional. There are two anecdotal reports of "ossifying fibrosarcoma" in the oral cavity of a Ball Python and a Green Tree Python (*Morelia viridis*) (Mauldin and Done, 2006).

Osteoma is a benign neoplasm composed of well differentiated bone ranging from trabecular to lamellar, depending on maturity, typically delineated by fibrous periosteum and with cell-poor matrix (**Figure 1.59**). This is rarely reported in reptiles with multiple osteomas reportedly arising from the caudal vertebrae of a European Green Lizard, an osteoma arising de novo within the skin in a European Pond Turtle (*Emys orbicularis*), an osteoma arising in the cervical subcutis of a Charles Island Giant Tortoise, and an osteoma in the fossil records of the extinct aquatic reptile, the mosasaur (Stolk, 1958; Machotka, 1984; Frye et al., 2009; Christman et al., 2017). In addition, a "chondro-osteo-fibroma" was diagnosed in a Rhinocerous Iguana (*Cyclura cornuta*) (Machotka, 1984). The NWZP archives have osteomas in an Inland Bearded Dragon, Dwarf Caiman (*Paleosuchus palpebrosus*), and Amazon Tree Boa (*Corallus hortulanus*).

1.2.8.3 Neoplasia of Cartilage Chondrosarcomas are reportedly common in colubrids and rattlesnakes. Corn Snakes may have a particular predisposition to developing chondrosarcoma, as there are at least four separate case reports in this species, including a primary mandibular mass with metastasis to the heart, lung, kidney, pancreas, and eye, and multiple cases of vertebral chondrosarcoma with no evidence of metastasis (Machotka, 1984; Garner et al., 2004; Schmidt and Reavill, 2012). Other cases of chondrosarcoma are described in the subcutis of a Black Ratsnake with hepatic metastasis, in the proximal humerus of a plated lizard (*Gerrhosaurus* sp.) without metastasis, in an Indian Monitor (*Varanus dracoena*), and in the vertebral articulations of eight additional snakes of unidentified species in which local invasion occasionally caused vertebral fracture/paralysis and without evidence of metastasis (Schlumberger and Lucke, 1948; Honour et al., 1993; Garner et al., 2004). Typically, neoplastic cells have a moderate degree of atypia and moderate mitotic rate, but in one report of chondrosarcoma in a Corn Snake, neoplastic cells were fairly well differentiated without mitotic figures, yet metastasized widely (Schmidt and Reavill, 2012). Neoplastic cells are round to spindle shaped, sometimes binucleate within lacuna-like structures and associated with variable amounts of chondroid matrix with no identifiable neoplasm-associated osteoid (**Figure 1.60**).

Chondromas are benign neoplasms of cartilage composed of irregular lobules of hyaline cartilage with chondrocytic cells that have minimal atypia. These are rarely reported in reptiles, with several cases in lizards including one on the right shoulder in a Sudan Mastigure (*Uromastyx dispar maliensis*), and multiple cases of intratracheal chrondromas in Ball Pythons causing tracheal stenosis (**Figure 1.61**) (Diethelm et al., 1996).

1.2.9 Soft Tissue Sarcoma

There are many cell types that can give rise to soft tissue sarcomas (STS), including fibroblasts, smooth muscle cells, skeletal myocytes, cardiomyocytes, vascular mesenchymal cells, and adipocytes, among others. These tumors are common in reptiles and some reports indicate they are the most common type of neoplasm in snakes, particularly colubrids, crotalids, agamids, and geckos (Garner et al., 2004; Dietz et al., 2016). Neoplasms from these cell types are most commonly diagnosed in the skin or connective tissues of the body wall, but some soft tissue sarcomas occurring in other anatomic locations are also included in this section. Identification of cell lineage in reptilian STS relies heavily on histology, including the tissue of origin, organizational pattern of the cells, and cellular morphology. This contrasts with sarcomas in mammals (especially humans), whereby immunohistochemical profiles are well established for sarcomas, and often define cell lineage (Milovancev et al., 2015). IHC may be used to differentiate some reptilian sarcomas. Reported reptilian STS include fibrosarcoma, liposarcoma, myxosarcoma, peripheral nerve sheath tumor, neurofibrosarcoma, rhabdomyosarcoma, hemangiopericytoma, and hemangioendothelioma (Garner et al., 2004; Dietz et al., 2016). Cell lineage in most cases is presumptive based on histology. When cell lineage cannot be determined by either histology or ancillary diagnostic procedures, spindle cell neoplasms in the skin and subcutis are termed soft tissue sarcomas (STS), or spindle cell sarcomas. Soft tissue sarcomas are generally invasive, and have potential for metastasis, probably based on degree of anaplasia and cell type, as with mammals. Recently, a grading system has been suggested for cutaneous/subcutaneous STS of snakes. The referenced work proposed that grade may be associated with increased risk for metastasis. Grading was based on a system used for canine STS, and includes tumor differentiation, mitotic rate, and amount of tumor necrosis. Tumors that are poorly differentiated with increased mitoses and necrosis received a higher grade (Dietz et al., 2016).

Fibrosarcoma was historically considered to be the most common neoplasm of snakes. Although more recent retrospective reviews of reptile neoplasia have found other neoplasms to be more prevalent, fibrosarcoma is relatively common and has been reported in the oral cavity, skin, subcutis, body wall, musculoskeletal system, and endocardium of snakes (Garner et al., 2004; Gumber et al., 2010). Particular prevalence is reported in colubrids and boids, with oral tumors being more common in boids and cutaneous tumors more common in colubrids (Garner et al., 2004). Fibrosarcoma is a neoplasm comprised of interwoven bundles and streams of spindle cells with interweaving strands of collagenous stroma. Anisocytosis and anisokaryosis can be marked. Mitoses are present in variable amounts. Fibrosarcomas can be morphologically similar to other spindle cell tumors, and diagnosis is usually based on a significant amount of collagenous stroma identified histologically (**Figure 1.62**). In one case of endocardial fibrosarcoma in a Reticulated Python, the diagnosis was supported using electron microscopy that showed abundant collagen surrounding neoplastic cells (Gumber et al., 2010). Fibrosarcomas generally are locally invasive. Visceral metastasis has been reported in several cases of fibrosarcoma in which the primary neoplasm could not be identified. Cellular anaplasia is a reported feature of metastatic fibrosarcomas (Garner et al., 2004).

Liposarcomas are rarely reported in reptiles. Single cases have been reported in a Shingleback Skink, a Red-Tailed Boa, and a chameleon (unidentified genus) (Garner et al., 1994; Reavill et al., 2002). An additional case is in an Indigo Snake (species not specified) in the NWZP archives. Histologically, liposarcomas in reptiles are composed of neoplastic spindle cells with single to multiple clear cytoplasmic vacuoles (**Figure 1.63**). Mitotic figures and necrosis are frequently reported. Lipid within cytoplasmic vacuoles can be identified in frozen sections using oil red O staining technique.

Other types of STS are included under the system of their cell of origin (e.g., rhabdomyosarcoma under the Musculoskeletal System section of this chapter).

1.2.10 Hematopoietic and Lymphoid Systems

The hematopoietic system is likely the most commonly affected system for reptilian neoplasia (Christman et al., 2017). Lymphoma with or without leukemia is the most commonly reported hematopoietic neoplasm in reptiles (Catão-Dias and Nichols, 1999; Garner et al., 2004). Other types of hematoipoietic neoplasia (e.g., myeloid leukemia, myeloproliferative disease) are reported less commonly. There are two reports of thymoma in reptiles, including an Agassiz's Desert Tortoise and Green Tree Python; additional information on these tumors was not provided in the references (Frye, 1994; Catão-Dias and Nichols, 1999). A Columbian Rainbow Boa (*Epicrates maurus*) with a thymoma is in the NWZP archives. There is a single report of erythroid leukemia in a Green Iguana; aside from the diagnosis and host species, further information was not provided in the reference (Frye, 1994).

Immunophenotyping can be challenging due to the limited number of standardized round cell markers or markers that work on formalin-fixed, paraffin-embedded reptilian tissues. The CD3 immunomarker for T lymphocytes has been used in reptiles with good success. B cell markers are more challenging with few reported cases of successfully labeling B cells in tissue section, and most reported cases using immunocytochemistry. One report suggests that BLA36 may be a useful immunohistochemical B cell marker in reptiles (Silverstone et al., 2007). The literature on the utility of CD79a on B cells using immunocytochemistry in reptiles is conflicting (Tocidlowski et al., 2001a,b; Silverstone et al., 2007). The diagnosis of B cell/plasma cell lineage may be based on histomorphology and ultrastructural features resembling plasma cells, such as amphophilic cytoplasm with Golgi zone or apparent Russell bodies and "cartwheel nucleus" (Gyimesi et al., 2005). Further discussion on leukemia is provided in Volume 1, Chapter 3. More work is needed to further characterize the lineage of hematopoietic and "histiocytoid" neoplasia in reptiles.

1.2.10.1 Lymphoma Some reviews indicate that lymphoma is the most common type of neoplasia in reptiles, particularly in snakes, though lymphoma has also been reported in lizards, chelonians, and crocodilians (Effron et al., 1977; Machotka, 1984; Garner et al., 2004). Due to the large number of cases of lymphoma that have been reported in reptiles, this review will focus more on distribution and diagnosis in reptiles rather than specific host species. The most commonly affected organs include the lungs, liver, spleen, pancreas, gastrointestinal tract, and kidneys; less commonly the heart, skeletal muscle, and skin are affected (Effron et al., 1977). Lymphoma in reptiles is often multicentric with multiple masses or diffuse organomegaly with sheets of round cells (**Figures 1.64–1.65**). Lymphoma is occasionally reported to be associated with leukemia (Effron et al., 1977; Frye, 1994) (**Figure 1.66**). Large, blast-like cells are observed more commonly than small cell lymphoma (Garner et al., 2004). IHC for T cells using CD3 has been an effective marker in the author's experience (**Figures 1.67–1.68**). Electron microscopy has been performed in some cases to screen for viral particles (Effron et al., 1977; Gyimesi et al., 2005), and although C-type retroviral particles have been identified in some cases, their role in oncogenesis has not been determined.

Plasmacytoid lymphoma is characterized by cells with an eccentric nucleus and a perinuclear region of pallor/pale eosinophilia in the cytoplasm (i.e., Golgi body) (**Figure 1.69**). These may represent variations of B cell neoplasms, or true plasma cell tumors. In lizards with lymphoma, ultrastructural examination supported plasma cell morphology based on clumped nuclear chromatin in a "cartwheel" pattern (Schmidt, 1977; Gyimesi et al., 2005); IHC with CD3 and BLA36 was noncontributory in the Gyimesi et al. study. In these cases, plasma cell neoplasia was multicentric and affected many viscera. Plasma cell neoplasia was diagnosed based on histomorphology in a Nile Monitor (*Varanus niloticus*); further information on this case was not available (Machotka, 1984).

1.2.10.2 Lymphoid Leukemia Leukemia is a neoplasm that primarily arises in the bloodstream or bone marrow without a distinct tissue phase. Lymphoid leukemia is the most common type of leukemia in reptiles, with fewer cases of myelogenous leukemia. There are several names for large cell lymphocytic leukemia of reptiles in the literature, including lymphoid leukemia, acute lymphocytic leukemia, and lymphoblastic leukemia. In many reported cases, particularly in the older literature, it is difficult to discern if these are separate entities or all represent large cell lymphoid leukemia. Small cell lymphocytic leukemia is also referred to as chronic lymphocytic leukemia. Both acute (large cell) and chronic (small cell) lymphoid leukemia are reported in many reptilian species, and are most commonly reported in snakes with fewer cases in lizards and chelonians (Frye and Carney, 1973; Effron et al., 1977; Machotka, 1984; Frye, 1994; Garner et al., 2004; Christman et al., 2017). Generally, large cell lymphocytic leukemia is expected to behave more aggressively than small cell variants. Grossly, leukemia frequently causes a diffuse enlargement of the involved organs, such as hepatomegaly and splenomegaly. Histologically, neoplastic lymphocytes expand the vasculature, including red pulp of the spleen and hepatic sinusoids (**Figure 1.70**). Solid tumor foci may also contribute to organomegaly, effacement of tissue architecture and clinical morbidity.

1.2.10.3 Myelogenous Leukemia and Myeloproliferative Disease Myelogenous leukemia is a neoplasm arising in the bloodstream or bone marrow that is composed predominantly of myeloid precursors. Myeloproliferative disease is composed of the same neoplastic cell type as myelogenous leukemia but presents as one or multiple tissue masses rather than leukemia, although variants with both patterns occur, especially in snakes. Histologically, neoplastic cells either cluster in blood vessels (i.e., myelogenous leukemia) or form sheets (i.e., myeloproliferative disease) of round cells with eosinophilic cytoplasmic granules and round, reniform, or segmented nuclei. Cellular atypia is variable (**Figure 1.71**). Myelogenous leukemia is less common in reptiles than lymphoid leukemia. Among reptiles, myelogenous leukemia is reported mostly in snakes with fewer reports in chelonians and lizards, including a Red-Eared Slider, multiple African Helmeted Turtles (*Pelomedusa subrufa*), Rhinoceros Viper (*Bitis nasicornis*), Russell's Viper (*Vipera russelli*), Honduran Milk Snake, Rainbow Boa (*Epicrates cenchria*), Green Tree Python, Corn Snake, Copperhead, Eastern Hognose Snake (*Heterodon platirhinos*), bearded dragon (*Pogona* sp), and Solomon Island Ground Skink (*Eugongylus albofasciolatus*) (Machotka, 1984; Frye, 1994; Catão-Dias and Nichols, 1999; Tocidlowski et al., 2001a,b; Sykes and Trupkiewicz, 2006; Rousselet et al., 2017). Myeloproliferative disease is reported in an Aruba Island Rattlesnake and a Jungle Carpet Python (*Morelia spilota cheynei*) (Garner et al., 2004).

1.2.10.4 Histiocytic Neoplasia Only a few cases of histiocytic neoplasia are reported in reptiles. Some lymphoid neoplasms have occasional histiocytoid morphology, and without reliable IHC it is difficult to know if some lymphoid neoplasms are misdiagnosed as histiocytic, and vice-versa, or if some of the anaplastic soft tissue sarcomas commonly encountered in the body wall of snakes are histiocytic. Histiocytoid morphology includes large round cells with a moderate to large amount of pale eosinophilic cytoplasm, sometimes with some degree of phagocytosis of erythrocytes or cellular debris. Nuclei are mostly round to ovoid, and binucleated or multinucleated cells with anisokaryosis and bizarre mitotic figures may be seen (**Figure 1.72**). Most reports of histiocytic neoplasia in reptiles are reported as histiocytic sarcoma, with a single report of a histiocytoma. The histiocytoma was in an ulcerated cutaneous mass in a Rainbow Boa and had reportedly metastasized to the skeletal muscle (Frye, 1994). Histiocytic sarcoma has been reported in a Common Garter Snake, Boa Constrictor, Green (Carolina) Anole (*Anolis carolinensis*), and Bull Snake (Frye, 1994; Sykes and Trupkiewicz, 2006); minimal additional information is available regarding the histiocytic sarcomas in these cases.

1.2.10.5 Mast Cell Neoplasia Mast cell neoplasms are rarely reported in reptiles, with approximately equal numbers of reports in chelonians, snakes, and lizards. Cutaneous mast cell tumors may present as exophytic, sessile, or, rarely, pedunculated masses. Histology of neoplastic cells is characterized by sheets of round cells with abundant pale eosinophilic cytoplasm with variable numbers of granules, which are the diagnostic feature of these cells. Nuclei are round with coarse chromatin, and occasionally bilobed (**Figure 1.73**). Confirmation of mast cell origin can be accomplished with Giemsa or Toluidine blue stains, or ultrastructural features using TEM, in reptiles (Santoro et al., 2008; Irizarry et al., 2014). Malignant mast cell tumors have been identified in a Boa Constrictor, Eastern Kingsnake, and African Fat-Tail Gecko (*Hemitheconyx caudicinctus*) (Frye, 1994; Schumacher et al., 1998; Irizarry et al., 2014). In the African Fat-Tail Gecko, neoplastic masses were identified in multiple organs, but there were also neoplastic mast cells filling blood vessels, suggestive of mastocytic (basophilic) leukemia. In several other cases, malignancy was not indicated; these included a mast cell tumor in a Agassiz's Desert Tortoise, an Isabelle Island Giant Tortoise (*Chelonoidis vicina* [*Geochelona nigra vicina*]), and two Green Iguanas, one of which had an oral mast cell tumor and the other had multiple cutaneous mast cell tumors (Frye, 1994; Reavill et al., 2000; Santoro et al., 2008; Christman et al., 2017).

1.2.11 Nervous System

Neoplasia arising in the nervous system is rare in reptiles. Most reports are in snakes, with rare reports in a chelonian, crocodilian, and lizard. There are case reports of various types of tumors arising in the central nervous system, most of which are gliomas or presumed gliomas. Most of the reported neoplasms in the peripheral nervous system are probably peripheral nerve sheath tumors.

1.2.11.1 Neoplasia of the Central Nervous System Except for glioma, reported neoplasms in the central nervous system of reptiles are rare, and most consist of a single reported case. Grossly, gliomas tend to be very poorly demarcated and cause asymmetrical expansion of nervous tissue. Gliomas with predominant oligodendroglial morphology have a gelatinous appearance grossly, similar to mammals. Histologically, gliomas are composed of sheets of neoplastic cells with scant stroma. Neoplastic cells can vary widely in morphology from oligodendroglial-like to astrocyte-like. A spinal cord glioma was reported in a Ridge-Nosed Rattlesnake (*Crotalus willardi*). In this case, the majority of the neoplasm had oligodendrioglial morphology (i.e., round to polygonal cells with distinct cell borders, round nuclei, and clear cytoplasm), and was grossly gelatinous. Glial cell origin was confirmed with glial fibrillary acidic protein (GFAP) IHC (Craig et al., 2005). There are two additional reports of neoplasms that may have arisen from the glia in reptiles. The first case was a neoplasm replacing the midbrain, thalamus and pineal gland of a Red-Tailed Boa (*B. c. constrictor*). This neoplasm was identified as either ependymal or glial origin based on histomorphology and immunoreactivity to GFAP (Keller et al., 2016). The second case was invading the cerebellum of a Saltwater (Porose) Crocodile. This neoplasm had glial morphology with some round and spindle cells and was diagnosed as a potential glioma (Scott and Beattie, 1927). Additional reports of neoplasms in the central nervous system of reptiles are sparse. There is a case of a meningeal sarcoma with poorly differentiated cells in an unidentified snake; further information on either case was not provided in the references (Schmidt and Reavill, 2015). The NWZP archives have an ependymoma in a Green Iguana that was arising from the third ventricle and causing hydrocephalus.

1.2.11.2 Neoplasia of the Peripheral Nervous System Peripheral nerve sheath tumors (PNST) are the most common neoplasm arising from the peripheral nervous system in reptiles. These tumors arise from Schwann cells, fibroblasts, or perineural cells (Mauldin and Peters-Kennedy, 2016). Most cases reported in reptiles were considered malignant (i.e., malignant peripheral nerve sheath tumor [MPNST]) based on invasive behavior. PNST has been diagnosed in two colubrids of unidentified species, a Hermann's Tortoise, a Korean (Haly's) Viper (*Gloydius* [*Agkistrodon*] *halys*), multiple Corn Snakes, an Emerald Tree Boa, an African House Snake (*Lamprophis* sp.), a California Kingsnake, a Grass Snake (*Natrix natrix*), a Water Moccasin, a Boa Constrictor, and multiple Inland Bearded Dragons, in which these tumors may be particularly common (Effron et al., 1977; Cooper et al., 1983; Ramis et al., 1998; Mikaelian et al., 2001; Garner et al., 2004; Lemberger et al., 2005; Dietz et al., 2016). Some of the MPNST were graded according to tumor differentiation, mitotic rate, and tumor necrosis; MPNST represented all grades (i.e., 1–3), with most being grade 2 or 3 (Dietz et al., 2016). The diagnosis of PNST was typically based on features of Schwannoma, including uniform, fusiform cells arranged with nuclear palisading, herringbone pattern, or bands (Antoni A pattern), intermixed with regions of loosely packed cells in a loose, myxoid to amphophilic stroma (Antoni B pattern) (**Figure 1.74**). Most of the PNST were reported in the skin or subcutis, although the Korean Viper case (Effron et al., 1977) was adjacent to the spine. Two of the Inland Bearded Dragons were related animals, and the diagnosis was supported by positive immunolabeling for S100 and neuron-specific enolase (NSE,) a positive laminin stain, and characteristic ultrastructural features (i.e., continuous basal lamina and intertwining cell processes) (Lemberger et al., 2005).

Peripheral neuroblastic tumors, which arise from ganglia, are rarely reported in reptiles. There is a single report of a ganglioneuroblastoma reported in a Veiled (Yemen) Chameleon (Vasil'ev and Solov'ev, 2006).

1.2.12 Respiratory System

Neoplasia arising from the respiratory system is rarely reported in reptiles, and limited to a few snakes, chelonians, and lizards. With the exception of pulmonary fibromatosis as a manifestation of papillomatosis in green turtles, no primary respiratory neoplasm is common in any reptile. Pulmonary fibroadenoma of the lung was described in multiple chelonians, and pulmonary adenocarcinoma was described in multiple snakes (Schlumberger and Lucke, 1948; Effron et al., 1977; Machotka, 1984; Catão-Dias and Nichols, 1999). Due to the low number of reports, species trends were not identified.

1.2.12.1 Neoplasia of the Upper Respiratory Tract Neoplasia of the trachea has been described in snakes. Less is known about nasal cavity neoplasia, and there is a single report of a non-neoplastic nasal polyp in an anole (*Anolis* sp.) (Frye, 1994). Neoplasia of the trachea in snakes can arise from epithelial or mesenchymal tissues. An adenosquamous carcinoma was reported in the trachea of a California Kingsnake; further details on the case are not available (Machotka, 1984). In addition, multiple chondromas have been reported in the trachea of Ball pythons; these cases are discussed further in the Musculoskeletal System section of this chapter (Diethelm et al., 1996; Drew et al., 1999; Gál et al., 2007).

1.2.12.2 Neoplasia of the Lower Respiratory Tract Although metastatic neoplasia is common, primary neoplasia of the lower respiratory tract is more commonly reported than that of the upper respiratory tract in reptiles. There are two reports of pulmonary fibroadenomas in chelonians, one in a Horsefield's Tortoise (*Testudo horsfieldii*), and one in a European Pond Turtle (Schlumberger and Lucke, 1948; Machotka, 1984). Pulmonary fibroadenoma is rare in all species, and histologically consists of tubules and acini of well-differentiated epithelium embedded in a moderate to large amount of dense collagenous stroma, within which are few, well-differentiated spindle cells. Visceral involvement of fibropapillomatosis commonly occurs in Green Sea Turtles (*Chelona mYdas*) (**Figure 1.75**) and manifests as pulmonary fibroma or fibrosarcoma. Pulmonary adenocarcinoma is more common across species but is rarely reported in reptiles, with three reports in snakes, including a Cape Cobra (*Naja nivea*), California Kingsnake, and Corn Snake (Effron et al., 1977; Catão-Dias and Nichols, 1999; Mauldin and Done, 2006). These are composed of neoplastic epithelial cells forming alveolar-like or bronchiolar-like patterns. These neoplasms have historically been referred to as "bronchoalveolar carcinoma," a term that has been abandoned in favor of "pulmonary adenocarcinoma" with subclassification based on predominant histologic pattern (Caswell and Williams, 2016). In the Cape Cobra, the neoplasm was described as "bronchogenic" (Effron et al., 1977).

A case of multiple primary pulmonary melanophoromas was described in a in Beaded Lizard (Rivera et al., 2015). Reptiles differ from mammals in that they have pigment cells (i.e., chromatophores) in coelomic membranes and pulmonary tissue. These pulmonary chromatophores likely gave rise to the pulmonary melanophoromas in the Beaded Lizard. Histologically, the pulmonary neoplasms in the Beaded Lizard resembled melanophoromas as described in the skin. Primary, melanin-producing neoplasms are extremely rare in the lungs of people but are believed to originate from embryonic remnants of melanoblasts or represent metastasis of occult cutaneous melanomas (Ost et al., 1999).

Mesothelioma is a neoplasm arising from mesothelial cells, which line the pleural surfaces of the thoracic cavity. This neoplasm has been described in two unidentified colubrids as well as a Corn Snake and a Boa Constrictor; additional suspect cases were in a Timber Rattlesnake and Pine Snake (Machotka, 1984; Frye, 1994; Garner et al., 2004). Three cases are in the NWZP archives in a Dunn's Hognose Viper (*Porthidium dunni*), Corn Snake (red ratsnake), and Green Iguana (*Iguana iguana*), respectively. Grossly, mesothelioma forms nodular thickenings on serosal surfaces. Histologically, mesothelial surfaces are expanded by proliferative, neoplastic, polygonal cells resembling reactive mesothelial cells (**Figure 1.76**).

1.2.13 Cardiovascular System

Primary neoplasms of vessels are generally divided into those of blood vessels (hemangioma and hemangiosarcoma) and those of lymphatics (lymphangioma and lymphangiosarcoma). Lymphatic neoplasia is rarely reported. There is a single case of carcinoma in situ of the endolymphatic sac in a Tokay Gecko (*Gekko gecko*) (Sander et al., 2015). Neoplasia of the blood vessels and primary neoplasms of the heart are occasionally reported in reptiles.

1.2.13.1 Neoplasia of Blood Vessels Hemangioma and hemangiosarcoma are occasionally reported in reptiles. The term "hemangioendothelioma" has been historically used in reptiles to denote tumors of blood vessels that are of presumed intermediate biological behavior. For this chapter the preferred term for malignant vascular tumor is hemangiosarcoma. Neoplasms of blood vessels have been reported in many species of snakes, most of which were colubrids, and include several reports in Corn Snakes, several unidentified lizards, and in a Red-Eared Slider. Primary locations of hemangiomas and hemangiosarcomas reported in reptiles include liver, skin, heart, ovary, cloaca, spleen, lung, skeletal muscle, vena cava, and esophagus (**Figures 1.77–1.78**) (Effron et al., 1977; Machotka, 1984; Catão-Dias and Nichols, 1999; Garner et al., 2004; Sykes and Trupkiewicz, 2006; Tuttle et al., 2006; Gál et al., 2009a,b; Stumpel et al., 2012; Shoemaker et al., 2016). These tumors are comprised of irregular, blood-filled sinuses that are either lined by single or small streams of plump, malignant endothelial cells (i.e., hemangiosarcoma; **Figure 1.79**) or a single layer of flattened, neoplastic endothelial cells (i.e., hemangioma; **Figure 1.80**). Solid tumor foci with considerable cellular anaplasia may be seen with the former. Immunoreactivity using claudin 5- and factor VIII-supported endothelial origin in a Red-Eared Slider and Madagascar Giant Hognose Snake (*Leioheterodon madagascariensis*), respectively (Gál et al., 2009a,b; Shoemaker et al., 2016). Factor VIII was negative in a hemangiosarcoma in a Corn Snake, but in the referenced report it is not stated if controls were positive or negative (Tuttle et al., 2006)

1.2.13.2 Neoplasia of the Heart Multiple sarcomas have been reported in the heart of reptiles. Cardiac rhabdomyosarcoma has been reported in several animals, including a Mottled Rock Rattlesnake, Black Ratsnake and Boa Constrictor (**Figure 1.81**) (Elkan and Cooper, 1976; Catão-Dias and Nichols, 1999; Sykes and Trupkiewicz, 2006). Metastasis were not reported in any case, despite a high mitotic rate in at least one case (Boa Constrictor). Striated muscle origin was confirmed in the Boa Constrictor case with PTAH staining highlighting cross-striations (**Figure 1.82**) (Elkan and Cooper, 1976). Other sarcomas that are reported as primary neoplasms in the heart include fibrosarcomas in a Reticulated Python and Gaboon Viper (*Bitis gabonica*), and a sarcoma without further differentiation in a European Glass Lizard (*Pseudopus apodus*) (Mauldin and Done, 2006; Sykes and Trupkiewicz, 2006; Gumber et al., 2010). Round cell neoplasms are also occasionally reported in the heart and are discussed in Section 1.2.10 on the Hematopoietic and Lymphoid Systems of this chapter. "Precardial" neoplasms are rarely reported, including an edematous myxofibroma in an Asian pit viper (*Gloydius* sp.), a myofibroma in a Russell's Viper (*Daboia russelli*), and a pleomorphic sarcoma in a Green Anaconda

(*Eunectes murinus*) (Lunger et al., 1974; Ramsay and Fowler, 1992; Ramsay et al., 1996). The heart seems to be a common site of metastasis for various forms of neoplasia in reptiles, especially sarcomas, and interpretation of "primary cardiac sarcoma" should be conservative.

1.2.14 Special Senses

Special senses of reptiles include eyes, ears, lacrimal glands, vomeronasal organs, and infrared detection organs. Of the organs included in special senses, periocular glands most commonly give rise to neoplasia in reptiles, especially iguanas and chameleons (**Figure 1.83**). In marine turtles, FP can develop on the cornea, limbus, conjunctiva, or mucocutaneous junction (Jacobson et al., 1989a,b; Brooks et al., 1994). FP are discussed in detail in the Integumentary System section of this chapter. A single case of aural neoplasia was described in a Leopard Gecko that had SCC in the ear canal. SCC has also been described in the tissue around the eye and is particularly common in bearded dragons (Abou-madi and Kern, 2002).

In reptiles, various neoplasms from the glands of the eye have been described as arising from the "lacrimal gland," "tear gland," and "Harderian gland" in reptiles. The exact glandular origin for these neoplasms (i.e., Harderian vs. lacrimal gland) can be difficult or impossible to determine histologically, depending on the stage of tumor development. As such, some authors describe neoplasms as arising from "tear gland" unless lacrimal or Harderian origin can be definitively identified. An important differential for neoplasia arising from tear glands is benign proliferative change and/or squamous metaplasia secondary to obstruction, chronic inflammation, or hypovitaminosis A. In the latter case, changes are frequently bilateral and may be associated with squamous metaplasia in other regions, although bilateral tear gland tumors also occur. Neoplasms that are reported to arise from periocular glands in reptiles include tear gland adenomas in multiple species of iguanids, lacrimal gland papillary cystadenoma in a Green Iguana, Harderian gland adenocarcinoma in a Florida Red-Bellied Turtle (*Pseudemys nelsoni*), lacrimal cystadenoma in a Chinese Box Turtle (*Cuora flavomarginata*), and periorbital adenocarcinoma in an Inland Bearded Dragon (Garner et al., 2004; Kottwitz et al., 2008; Gál et al., 2009a,b; Darrow et al., 2013). These neoplasms have typical features of glandular neoplasia and form tubules and packets of polygonal cells (**Figure 1.84**). In both the reported Harderian adenocarcinoma and periorbital adenocarcinoma, malignancy was based on mitotic rate, cellular pleomorphism, and local expansion or invasion; there was no evidence of metastasis in either case, although necropsy and complete organ histology were only performed in the former case (Gál et al., 2009a,b; Darrow et al., 2013). The authors have multiple cases of tear gland tumors from chameleons, as well.

Primary ocular neoplasia is extremely rare in reptiles, with no reports of ocular neoplasia in multiple large retrospective studies on reptile neoplasia, or in a large retrospective study on ocular lesions in snakes (Garner et al., 2004; Sykes and Trupkiewicz, 2006; Hausmann et al., 2013). There is a single report of a keratoacanthoma of the spectacle in a Boa Constrictor. In the referenced case, there were occasional mitotic figures in the basal layer, but no evidence of local invasion or metastasis. Retained spectacle was speculated as a possible predisposing factor for the development of spectacular keratoacanthoma (Hardon et al., 2007).

1.2.15 Miscellaneous Intracoelomic Neoplasia

There are occasional reports of intracoelomic neoplasia for which a primary neoplasm could not be identified. Mesenchymal and epithelial neoplasms occurring within the coelomic cavity have been reported in reptiles. Metastatic intraceolomic adenocarcinoma has been reported in snakes and lizards. The cell of origin is frequently difficult to discern due to cellular anaplasia and widespread metastasis preventing localization of a primary tumor (Elkan and Cooper, 1976; Ramsay et al., 1996; Garner et al., 2004). In some cases, particularly in biopsies, insufficient tissue may have been submitted, hindering identification of parent tissue or cell lineage. Reports of intracoelomic sarcoma in reptiles are limited to snakes, and include spindle cell sarcomas in a Baird's Rat Snake (Pantherophis [*Elaphe*] *bairdi*), Cape Cobra, Saw-Scaled Viper (*Echis carinatus*), and Red-Tailed Green Rat Snake (*Gonyosoma oxycephala*), a fibrosarcoma in a Black Pine Snake (*Pituophis melanoleucus lodingi*), and a leiomyosarcoma in a Yellow Rat Snake (*Elaphe obsoleta quadrivittata*) (Ramsay and Fowler, 1992; Sykes and Trupkiewicz, 2006).

1.3 Ancillary Diagnostic Procedures

Detailed descriptions on ancillary diagnostic testing in reptiles are provided in other chapters of this text. This section is intended to supplement other chapters with neoplasia-specific reports of ancillary procedures. These lists are not comprehensive but may serve as points of reference for ancillary stains, antigens, and ultrastructural features that have been useful in diagnosing neoplasia in reptiles.

1.3.1 Immunohistochemistry

IHC has been used in many types of reptilian neoplasia. **Tables 1.1** and **1.2** provide lists of some of the reported antigens that have been expressed (**Table 1.1**) or not expressed (**Table 1.2**) in reptilian neoplasms.

1.3.2 Special Stains

Table 1.3 lists some cases in which special stains were useful in diagnosing reptilian neoplasms.

Table 1.1 Antigens That Have Been Expressed in Neoplasms of Reptiles Using Immunohistochemistry

Antigen	Neoplasm	Staining Pattern[a]	Species	Internal Control	Reference
α Smooth muscle actin	Leiomyosarcoma	Cytoplasmic	Boa Constrictor	Vascular smooth muscle	Gumber et al. (2012)
CD3	T cell lymphoma	Membranous	Many	Lymphocytes	
Chromogranin AB	Gastric neuroendocrine carcinoma (somatostatinoma)	Positive in 1/6 cases	Bearded Dragon	Pancreas	Lyons et al. (2010); Ritter et al. (2009)
Claudin 5	Hemangioma	Membranous	Red-Eared Slider	Esophageal mucosa epithelium	Gál et al. (2009a)
Cytokeratins	1. Ovarian cystadenocarcinoma 2. Renal adenocarcinoma 3. Scent gland adenoma 4. Ameloblastoma 5. Esophageal adenocarcinoma 6. Pancreatic ductular adenocarcinoma 7. Periocular adenocarcinoma (cytokeratins 5/6)	Positive, cytoplasmic	1. Green Iguana 2. Corn Snake 3. California Kingsnake 4. Black Rat Snake 5. Mata Mata 6. Northern Water Snake 7. Unidentified Bearded Dragon	1. Ovarian surface epithelium 2. NR 3. Scent gland 4. NR 5. NR 6. Pancreatic ductular epithelium 7. Glandular epithelium	1. Stacy et al. (2004) 2. Kao et al. (2016) 3. Gal and Mandoki, 2012 4. Comolli et al. (2015) 5. Lombardini et al. (2013) 6. and 7. Darrow et al. (2013)
Factor VIII	Hemangiosarcoma	Membranous	Madagascar Giant Hognose Snake	Endothelial cells	Shoemaker et al. (2016)
Gastrin	Gastric neuroendocrine carcinoma (somatostatinoma)	Positive in 3/5 cases	Inland Bearded Dragon	Stomach	Lyons et al. (2010)
Glial fibrillary acidic protein	1. Glioma 2. Suspect glioma or ependymoma	Cytoplasmic	Ridge-Nosed Rattlesnake	Astrocytes	1. Craig et al. (2005) 2. Keller et al. (2016)
Glucagon	1. Gastric neuroendocrine carcinoma (somatostatinoma) 2. Multihormonal islet cell tumor	1. Positive in 1/5 cases 2. Positive in 3/3 cases	1. Bearded Dragon 2. Komodo Dragon	Pancreas	1. Lyons et al. (2010) 2. Eustace et al. (2017)
HMB45	Mixed chromatophoroma	Positive	Eastern Kingsnake	NR	Muñoz-Gutiérrez et al. (2016)
Melan A	1. Melanophoroma 2. Iridophoroma 3. Mixed chromatophoroma	Cytoplasmic	Squamata and chelonian species	NR	1. Heckers et al. (2011, 2012) 2. Heckers et al. (2012) 3. Muñoz-Gutiérrez et al. (2016)
Neuron-specific enolase	1. Gastric neuroendocrine carcinoma (somatostatinoma) 2. Peripheral nerve sheath tumor	Positive	Bearded Dragons	Pancreatic islet cells	1. Lyons et al. (2010) 2. Lemberger, Manharth, and Pessier (2005)
Pancreatic polypeptide	Gastric neuroendocrine carcinoma (somatostatinoma)	Positive in 2/5 cases	Bearded Dragon	Pancreas	Lyons et al. (2010)
PNL2	Melanocytic neoplasms, mixed chromatophoromas, iridophoromas	Cytoplasmic	Multiple snake species	NR	Muñoz-Gutiérrez et al. (2016)
S100	1. Melanophoroma 2. Iridophoroma 3. Fibrosarcoma 4. Myxoma 5. Myxosarcoma 6. Peripheral nerve sheath tumor	Cytoplasmic	Squamata and chelonian species	NR	1. Heckers et al. (2011, 2012) 2. Heckers et al. (2012) 3. Heckers et al. (2012) 4. Heckers et al. (2012); 5. Heckers et al. (2012) 6. Lemberger et al. (2005)
Somatostatin	1. Gastric neuroendocrine carcinoma (somatostatinoma) 2. Multihormonal islet cell tumor	1. Cytoplasmic 2. Positive in 3/3 cases	1. Bearded Dragon 2. Komodo Dragon	1. Gastric pits 2. Pancreas	1. Lyons et al. (2010); Ritter et al. (2009); 2. Eustace et al. (2017)
Synaptophysin	Gastric neuroendocrine carcinoma (somatostatinoma)	Positive in 2/5 cases	Bearded Dragon	Stomach	Lyons et al. (2010)
Vimentin	Fibroma	Cytoplasmic	Common Snapping Turtle	NR	Gonzales-Viera et al. (2012)

Abbreviations: NR, not reported.

[a] Location of immunoreactivity in the neoplastic cell (i.e., membranous, cytoplasmic, or nuclear) was not always reported in the referenced literature. In cases where location was not reported, immunoreactivity is reported as "positive."

Table 1.2 Antigens That Were Not Expressed by IHC in Reptilian Neoplasms[a]

Antigen	Neoplasm	Animal Species	Internal Control	Reference
BLA36	Lymphoid malignancy	Egyptian Spiny-Tailed Lizards	Negative (spleen and gastrointestinal tract)	Gyimesi et al. (2005)
CD3	Lymphoid malignancy	Egyptian Spiny-Tailed Lizards	Negative (spleen and gastrointestinal tract)	Gyimesi et al. (2005)
CD31	Hemangioma	Red-Eared Slider	Negative	Gál et al. (2009a)
Chromogranin	Esophageal adenocarcinoma	Mata Mata Turtle	NR	Lombardini et al. (2013)
C-kit	Mastocytosis	African Fat-Tail Gecko	Negative (host tissues)	Irizarry et al. (2014)
Cytokeratins	1. SCC 2. Ovarian cystadenocarcinoma 3. Periocular adenocarcinoma (cytokeratins 8 and 18)	1. Loggerhead Turtle 2. Green Iguana 3. Bearded Dragon	Negative (host epithelium)	1. Orós et al. (2004) 2. Stacy et al. (2004) 3. Darrow et al. (2013)
Desmin	1. SCC 2. Peripheral nerve sheath tumor	1. Loggerhead Turtle 2. Bearded Dragon	1. Positive (host striated muscle) 2. Positive (mammalian skeletal muscle)	
Endorphin	Gastric neuroendocrine carcinoma (somatostatinoma)	Unidentified Bearded Dragon	Positive (host adrenal gland)	Lyons et al. (2010)
Factor VIII	Hemangiosarcoma	Corn Snake	NR	Tuttle et al. (2006)
Gastrin	Islet cell tumor	Komodo Dragon	Positive (gastric mucosa)	Eustace et al. (2017)
Iba-1	Suspect glioma	Red-Tailed Boa	Positive (host microglia and macrophages)	Keller et al. (2016)
Insulin	1. Gastric neuroendocrine carcinoma (somatostatinoma) 2. Islet cell tumor	1. Unspecified Bearded Dragon 2. Komodo Dragon	1. Positive (host pancreas) 2. Negative (pancreas)	1. Lyons et al. (2010) 2. Eustace et al. (2017)
Laminin	Suspect glioma	Red-Tailed Boa	Positive (host peripheral nerves and ganglia)	Keller et al. (2016)
Melan A	Fibrosarcoma, myxoma, myxosarcoma, SCC	Squamata and chelonian species	NR	Heckers et al. (2012)
Neuron-specific enolase	Gastric neuroendocrine carcinoma (somatostatinoma)	Unspecified Bearded Dragon	Positive (host pancreas)	Lyons et al. (2010)
Periaxin	Suspect glioma	Red-Tailed Boa	Positive (host peripheral nerves and ganglia)	Keller et al. (2016)
Protein gene product 9.5	Gastric neuroendocrine carcinoma (somatostatinoma)	Unspecified Bearded Dragon	Positive (host pancreas)	Lyons et al. (2010)
S100	SCC	Squamata and chelonian species	NR	Heckers et al. (2012)
Smoth muscle actin	Peripheral nerve sheath tumor	Unspecified Bearded Dragon	Positive (mammalian intestine)	Lemberger et al. (2005)
Synaptophysin	1. Esophageal adenocarcinoma 2. Suspect glioma	1. Mata Mata Turtle 2. Red-Tailed Boa	1. NR 2. negative	1. Lombardini et al. (2013) 2. Keller et al. (2016)
Thyroglobulin	1. Esophageal adenocarcinoma	Mata Mata Turtle	NR	Lombardini et al. (2013)
Vasoactive intestinal peptide	1. Gastric neuroendocrine carcinoma (somatostatinoma) 2. Islet cell tumor	1. Unspecified Bearded Dragon 2. Komodo Dragon	1. Positive (host pancreas) 2. Positive (pancreatic islets)	1. Lyons et al. (2010); 2. Eustace et al. (2017)
Vimentin	1. SCC 2. Ovarian cystadenocarcinoma 3. Esophageal adenocarcinoma	1. Loggerhead Turtle 2. Green Iguana 3. Mata Mata Turtle	1. Positive (host fibroblasts); 2. Negative 3. NR	1. Orós et al. (2004) 2. Stacy et al. (2004) 3. Lombardini et al. (2013)

Abbreviations: SCC, squamous cell carcinoma; NR, not reported.

[a] Including those in which the internal control was successful (listed as positive) and those in which the internal control did not work (listed as negative), the latter suggesting the IHC antigen or protocol was not appropriate for the host species.

Table 1.3 Special Stains That Have Been Applied to Neoplasms in Reptiles and the Staining Result

Stain	Neoplasm	Animal Species	Result	Internal Control	Reference
Churukian-Schenk	Gastric neuroendocrine carcinoma	Bearded Dragon	Negative	Negative	Lyons et al. (2010)
Giemsa	Mast cell tumor	Galapagos Tortoise	Positive for metachromatic granules	NR	Santoro et al. (2008)
Laminin	Peripheral nerve sheath tumor	Bearded Dragon	Continuous basal lamina	NR	Lemberger et al. (2005)
Oil red O	Liposarcoma	Shingleback Skink	Positive	NR	Garner et al. (1994)
Periodic acid-Schiff	1. Liposarcoma 2. Peripheral nerve sheath tumor	1. Shingleback Skink 2. Bearded Dragon	1. Negative 2. Faint cytoplasmic granules	NR	1. Garner et al. (1994) 2. Lemberger et al. (2005)
PTAH	Cardiac rhabdomyosarcoma	Boa constrictor	Positive	NR	Elkan and Cooper (1976)
Toluidine blue	1. Mastocytosis 2. Mast cell tumor	1. African Fat-Tail Gecko; Galapagos Tortoise	Positive for metachromatic granules	NR	1. Irizarry et al. (2014) Santoro et al. (2008)

Abbreviations: PTAH, phosphotungstic acid-hematoxylin; NR, none reported.

1.3.3 Transmission Electron Microscopy

TEM serves two main functions when evaluating reptilian neoplasia. First, this modality is used to search for pathogens that may be associated with oncogenesis. Examples in reptiles include the identification of viral particles in marine turtle fibropapilloma, snake intestinal adenocarcinoma with type A retrovirus particles, C-like retroviral particles in neoplasms from multiple Burmese Pythons, and C-type virus particles in a rhabdomyosarcoma from a Corn Snake (Lunger et al., 1974; Jacobson et al., 1991; Chandra et al., 2001; Oros et al., 2004). Second, ultrastructural features of neoplastic cells can help to determine histogenesis. This has been reported in unidentified bearded dragons with gastric neuroendocrine carcinomas that contained neurosecretory granules, bearded dragons with peripheral nerve sheath tumors and continuous basal lamina, Spiny-Tailed Lizards and an East Indian Water Dragon (*Physignathus cocincinus*) with plasma cell tumors and classic nuclear chromatin patterns, multiple species with mast cell neoplasms, a Red-Tailed Boa and chameleon with liposarcoma and lipid vacuoles, and a Black Rat Snake with an odontogenic neoplasm (Schmidt, 1977; Schumacher et al., 1998; Reavill et al., 2002; Gyimesi et al., 2005; Lemberger et al., 2005; Ritter et al., 2009; Lyons et al., 2010; Irizarry et al., 2014; Comolli et al., 2015).

References

Abou-madi N and Kern TJ. 2002. Squamous cell carcinoma associated with a periorbital mass in a veiled chameleon (*Chamaeleo calyptratus*). *Vet Ophth* 5(3):217–220.

Abu-Seida A and Saeid S. 2013. Cutaneous adenocarcinoma in a desert tortoise (*Gopherus agassizii*). *Int J Vet Sci Med* 1(1):48–50.

Aguirre AA, Spraker TR, Chaves A, Du Toit L, Eure W, and Balazs G. 1999. Pathology of fibropapillomatosis in Olive Ridley Turtles *Lepidochelys Olivacea* nesting in Costa Rica. *J Aquat Anim Health* 11:283–289.Anderson ET, Kennedy-Stoskopf S, Sandy JR, Dorn B, Boyette T, and Harms C. 2010. Squamous cell carcinoma with vascular invasion in a Diamondback Rattlesnake (*Crotalus adamanteus*). *J Zoo Wildl Med* 41(4):745–748.

Anderson NJ, Williams J, Sagartz JE, and Barnewell R. 1996. Ovarian teratoma in a Green Iguana (*Iguana iguana*). *J Zoo Wildl Med* 27(1):90–95.

Ball HA. 1945. Melanosarcoma and rhabdomyoma in two pine snakes. *Cancer Res* 6(3):134–138.

Banks WJ. 1993. *Endocrine system, in Applied Veterinary Histology*, 3rd ed. Reinhardt RW (Ed.), Mosby, St. Louis, Missouri, pp. 408–428.

Baron HR, Allavena R, Melville L, and Doneley RJ. 2014. Gastric adenocarcinoma in a diamond python (*Morelia spilota spilota*). *Aust Vet J* 92(10):405–409.

Barragan AR and Sarti LM. 1994. A possible case of fibropapilloma in Kemp's Ridley turtle (*Lepidochelys kempii*). *Marine Turtle Newsletter* 67:28.

Barten SA, Dvais K, Harris RK, and Jacobson ER. 1994. Renal cell carcinoma with metastases in a corn snake (*Elaphe guttata*). *J Zoo Wildl Med* 25(1):123–127.

Barten SD and Frye FL. 1981. Leiomyosarcoma and myxoma in a Texas Indigo snake. *J Am Vet Med Assoc* 179(11):1292–1295.

Bel L, Tecilla M, Borza G et al. 2016. Diagnosis and surgical management of malignant ovarian teratoma in a Green Iguana (*Iguana iguana*). *BMC Vet Res* 12(1):144.

Belasco-Zeitz M, Pye GW, Burns RE, and Pessier AP. 2013. Clinical challenge. *J Zoo Wild Med* 44(3):807–810. http://www.ncbi.nlm.nih.gov/pubmed/24063120.

Bera, MM, Veeramachaneni DNR, Pandher K. 2008. Characterization of a biphasic neoplasm in a Madagascar Tree Boa (*Sanzinia madagascariensis*). *Vet Pathol* 45(2):259–263.

Bielli M, Forlani A, Nardini G, and Avallone G. 2015. Mucinous melanophoroma in a northern red-bellied cooter (*Pseudemys rubriventris*). *J Exot Pet Med* 24(1). Elsevier: 71–75. doi:10.1053/j.jepm.2014.11.007.

Bland-Sutton J. 1885. Tumours in animals. *J Anat Physiol* 19: 415–475.

Boudbid H, Leger AE, Pidoux E et al. 1987. Localization of a Salmon Calcitonin-like ultimobranchial cells in the freshwater molecule in one type of turtle *Pseudemys scripta*. *Gen Comp Endocr* 422:415–422.

Brooks DE, Ginn PE, Miller TR, Bramson L, and Jacobson ER. 1994. Ocular fibropapillomas of Green Turtles (*Chelonia mydas*). *Vet Pathol* 31(3):335–339.

Caserto BG. 2013. A comparative review of canine and human rhabdomyosarcoma with emphasis on classification and pathogenesis. *Vet Pathol* 50(5):806–826.

Caswell JL and Williams KJ. 2016. Respiratory system, in *Jubb, Kennedy, and Palmer's Pathology of Domestic Animals*, 6th ed. Maxie GM (Ed.), Elsevier, St. Louis, Missouri, pp. 465–591.

Catão-Dias JL and Nichols D. 1999. Neoplasia in snakes at the National Zoological Park, Washington, DC (1978–1997). *J Comp Pathol* 120(1):89–95.

Chandra AMS, Jacobson ER, and Munn RJ. 2001. Retroviral particles in neoplasms of Burmese pythons (*Python molurus bivittatus*). *Vet Pathol* 38:561–564.

Christman J, Devau M, Wilson-Robles H, Hoppes S, Rech R, Russell K, and Heatley JJ. 2017. Oncology of reptiles. *Vet Clin North Am Exot Anim Pract* 20(1):87–110.

Cianciolo RE and Mohr FC. 2016. Urinary system, in *Jubb, Kennedy, and Palmer's Pathology of Domestic Animals*, 6th ed. Grant Maxie M (Ed.), Elsevier, St. Louis, Missouri, pp. 376–466.

Comolli JR, Olsen HMH, Seguel M et al. 2015. Ameloblastoma in a wild black rat snake (*Pantherophis alleghaniensis*). *J Vet Diagn Invest*, no. 27: 1–4. doi:10.1177/1040638715590652.

Cooper JE. 1982. Viral particles in a papilloma from a Green Lizard (*Lacerta viridis*). *Lab Anim* 16:12–13.

Cooper JE, Jackson OF, and Harshbarger JC. 1983. A neurilemmal sarcoma in a tortoise (*Testudo hemanni*). *J Comp Pathol* 93:541–545.

Cowan DF. 1968. Diseases of captive reptiles. *J Am Vet Med Assoc* 153(7):848–859.

Cowan ML, Monks DJ, and Raidal SF. 2011. Osteosarcoma in a woma python (*Aspidites ramsayi*). *Aust Vet J* 89(12):520–523.

Craig LE, Dittmer KE, and Thompson KG. 2016. Bones and joints, in *Jubb, Kennedy, and Palmer's Pathology of Domestic Animals*, 6th ed. Maxie GM (Ed.), Elsevier, St. Louis, Missouri, pp. 16–163.

Craig LE, Wolf JC, and Ramsay EC. 2005. Spinal cord glioma in a ridge-nosed rattlesnake (*Crotalus willardi*). *J Zoo Wildl Med* 36(2):313–315.

Cruz Cardona JA, Conley KJ, Wellehan JF et al. 2011. Incomplete ovariosalpingectomy and subsequent malignant granulosa cell tumor in a female Green Iguana (*Iguana iguana*). *J Am Vet Med Assoc* 239(2):237–242.

Cullen JM and Stalker MJ. 2016. Liver and biliary system, in *Jubb, Kennedy, and Palmer's Pathology of Domestic Animals*, 6th ed. Maxie GM (Ed.), Elsevier, St. Louis, Missouri, pp. 258–352.

D'Amato AF, and Moraes-Neto M. 2000. First documentation of fibropapillomas verified by histopathology in Eretmochelys Imbricata. *Marine Turtle Newsletter* 89:12–13.

Dadone LI, Klaphake E, Garner MM et al. 2010. Pituitary cystadenoma, enterolipidosis, and cutaneous mycosis in an everglades ratsnake (*Elaphe obsoleta rossalleni*). *J Zoo Wildl Med* 41(3):538–541.

Darrow BG, Johnstone McLean NS, Russman SE, and Schiller CA. 2013. Periorbital adenocarcinoma in a Bearded Dragon (*Pogona vitticeps*). *Vet Ophthalmol* 16(Suppl 1) (July):177–182.

Diethelm G, Stauber E, Tillson M, and Ridgley S. 1996. Tracheal resection and anastomosis for an intratracheal chondroma in ball python. *J Am Vet Med Assoc* 209:786–788.

Dietz J, Heckers KO, Aupperle H, and Pees M. 2016. Cutaneous and subcutaneous soft tissue tumours in snakes: A retrospective study of 33 cases. *J Comp Pathol* 155:76–87.

Drew ML, Phalen DN, Berridge BR et al. 1999. Partial tracheal obstruction due to chondromas in ball pythons (*Python regius*). *J Zoo Wildl Med* 30:151–157.

Dumbravă MD, Rothschild BM, Weishampel DB et al. 2016. A dinosaurian facial deformity and the first occurrence of ameloblastoma in the fossil record. *Sci Rep* 6(July):29271.

Effron M, Griner L, and Benirschke K. 1977. Nature and rate of neoplasia found in captive wild mammals, birds, and reptiles at necropsy. *J Natl Cancer Inst* 59(1):185–198.

Elkan E, and Cooper JE. 1976. Tumours and pseudotumours in some reptiles. *J Comp Pathol* 86:337–348.

Eustace R, Garner MM, Cook K, Miller C, and Kiupel M. 2017. Multihormonal Islet cell carcinomas in three Komodo dragons (*Varanus komodoensis*). *J Zoo Wildl Med* 48(1):241–244.

Eyarefe OD, Antia RE, Oguntoye CO, Abiola OO, Alaka OO, and Ogunsola JO. 2012. Rhabdomyosarcoma in a terrestrial tortoise (*Geochelone nigra*) in Nigeria: A case report. *J S Afr Vet Assoc* 83(1): Art. #300. 24 ref. doi:10.4102/jsava.v83i1.300.

Frye FL. 1975. Parathyroid adenoma in a tortoise. *Veterinary Med, Small Animal Clinician* 70(5):582–584.

Frye FL. 1994. Diagnosis and surgical treatment of reptilian neoplasms with a compilation of cases 1966–1993. *In Vivo* 8:885–892.

Frye FL, and Carney JD. 1973. Acute lymphatic leukemia in a Boa Constrictor. *J Am Vet Med Assoc* 163(6):653–654.

Frye FL, Carney JD, Harshbarger JC, and Zeigel RF. 1975. Malignant chromatophoroma in a Western Terrestrial Garter Snake. *J Am Vet Med Assoc* 167(7):557–558.

Frye FL, Dybdal NO, and Harshbarger JC. 1988. Testicular interstitial cell tumor in a Desert Tortoise (*Gopherus agassizi*). *J Zoo Wildl Med* 19:55–58.

Frye FL, Modry D, and Siroky P. 2009. Pathology in practice. *J Am Vet Med Assoc* 235(5):511–512.

Gál J, Demeter Z, Palade EA, Rusvai M, and Géczy C. 2009a. Harderian gland adenocarcinoma in a Florida Red-Bellied Turtle (*Pseudemys nelsoni*)—case report. *Acta Vet Hung* 57(2):275–281.

Gál J, Jakab C, Szabó Z et al. 2009b. Haemangioma in the oesophagus of a red-eared slider (*Trachemys scripta elegans*). *Acta Vet Hung* 57(4). Akadémiai Kiadó: 477–484. doi:10.1556/AVet.57.2009.4.2.

Gál J, Jakab C, Balogh B, Tóth T, and Farkas B. 2007. First occurrence of periosteal chondroma (Juxtacortical Chondroma) in Uromastyx Maliensis (Reptilia: Sauria: Agamidae). Acta Vet Hung, July. Akadémiai Kiadó.

Gal J and Mandoki M. 2012. Adenoma of the cloacal scent gland in a California Kingsnake (*Lampropeltis getulus californiae*). *Acta Vet Hung* 60(4):459–463. 7 ref.

Gál J, Mándoki M, Sátorhelyi T, and Jakab C. 2010. In situ complex adenocarcinoma on the femoral part of the hindlimb in an Asian Leaf Turtle (*Cyclemys dentata*). *Acta Vet Hung* 58(4):431–440.

Garner MM, Hernandez-Divers SM, and Raymond JT. 2004. Reptile neoplasia: A retrospective study of case submissions to a specialty diagnostic service. *Vet Clin North Am Exot Anim Pract* 7(3):653–671.

Garner MM, Herrington R, Howerth EW et al. 1997. Shell disease in river cooters (*Pseudemys concinna*) and Yellow-Bellied Turtles (*Trachemys scripta*) in a Georgia (USA) Lake. *J Wildl Dis* 33(1):78–86.

Garner MM, Johnson C, and Funk R. 1994. Liposarcoma in a Shingleback Skink (*Trachydosaurus rugosus*). *J Zoo Wildl Med* 25(1):150–153.

Gibbons PM and Schiller CA. 2000. What's your diagnosis? Ovarian adenocarcinoma in a Green Iguana (*Iguana iguana*). *J Herp Med Surg* 10:34–38.

Goe A, Darryl M, Heard J et al. 2016. Surgical management of an odontogenic Tumor in a Banded Gila Monster (*Heloderma suspectum cinctum*) with a novel herpesvirus. *Vet Quart* 36(2). Taylor & Francis:109–114. doi:10.1080/01652176.2016.1153169.

Gonzales-Viera O, Bauer G, Bauer A, Aguiar LS, Brito LT, and Catão-Dias JL. 2012. Cutaneous fibroma in a captive Common Snapping Turtle (*Chelydra serpentina*). *J Comp Pathol* 147(4):574–576. doi:10.1016/j.jcpa.2012.03.005.

Gumber S, Nevarez JG, and Cho DY. 2010. Endocardial fibrosarcoma in a reticulated python (*Python reticularis*). *J Vet Diagn Invest* 22(6):1013–1016.

Gumber S, Schnellbacher R, Orandle MS, and Wakamatsu N. 2012. Pathology in practice. *J Am Vet Med Assoc* 241(3):327–329.

Gyimesi SS and Garner MM. 2007. Pituitary adenoma in a Dumeril's Ground Boa, *Arcantophis dumerili*. *J Herp Med Surg* 17:16–18.

Gyimesi ZS, Garner MM, Burns RB et al. 2005. High incidence of lymphoid neoplasia in a colony of Egyptian Spiny-Tailed Lizards (*Uromastyx aegyptius*). *J Zoo Wildl Med* 36(1):103–110.

Hadfield CA, Clayton LA, Clancy MM et al. 2012. Proliferative Thyroid lesions in three Diplodactylid Geckos: Nephrurus Amyae, Nephrurus Levis, and Oedura Marmorata. *J Zoo Wildl Med* 43(1):131–140.

Hannon DE, Garner MM, and Reavill DR. 2011. Squamous cell carcinomas in Inland Bearded Dragons (*Pogona vitticeps*). *J Herp Med Surg* 21(4):101–106.

Hardon T, Fledelius B, and Heegaard S. 2007. Keratoacanthoma of the spectacle in a Boa Constrictor. *Vet Ophthalmol* 10(5):320–322.

Harshbarger JC. 1991. Sea turtle Fibropapilloma cases in the registry of tumors in lower animals, in *Research Plan for Marine Turtle Fibropapilloma: Results of a December 1990 Workshop. NOAA Technical Memorandum*, USA, pp. 63–70.

Hausmann JC, Hollingsworth SR, Hawkins MG, Kass PH, and Maggs DJ. 2013. Distribution and outcome of ocular lesions in snakes examined at a veterinary teaching hospital: 67 Cases (1985–2010). *J Am Vet Med Assoc* 243(2):252–260.

Heckers KO, Aupperle H, Schmidt V, and Pees M. 2012. Melanophoromas and Iridophoromas in Reptiles. *J Comp Pathol* 146(2–3). Elsevier Ltd: 258–268. doi:10.1016/j.jcpa.2011.07.003.

Heckers KO, Schmidt V, Krastel D, Hildebrandt G, and Kiefer I. 2011. Malignes melanophorom Bei Einer Griechischen Landschildkröte (*Testudo hermanni*). *Tierarztliche Praxis* 39 (K):45–50.

Hellebuyck T, Pasmans F, Haesebrouck F, and Martel A. 2012. Dermatological diseases in lizards. *Vet J* 193(1). *J Comp Pathol* 146(2–3). Elsevier Ltd: 258–268. doi:10.1016/j.jcpa.2011.07.003.

Herbst LH, Jacobson ER, Moretti R, Brown T, Sundberg JP and Klein PA. 1995. Experimental transmission of green turtle fibropapillomatosis using cell-free extracts. *Dis Aq Org* 22:1–12.

Hernandez-Divers SM, and Garner MM. 2003. Neoplasia of reptiles with an emphasis on lizards. *Vet Clin North Am Exot Anim Pract* 6(1):251–273.

Hernandez-Divers SJ, Knott CD, and Macdonald J. 2001. Diagnosis and surgical treatment of thyroid adenoma-induced hyperthyroidism in a Green Iguana (*Iguana iguana*). *J Zoo Wildl Med* 32(4):465–475.

Hidalgo-Vila J, Martinez-Silvestre A, and Diaz-Paniagua C. 2006. Benign ovarian teratoma in a Red-Eared Slider Turtle (*Trachemys scripta elegans*). *Vet Rec* 159:122–123.

Hill AG, Denis MM, and Pyne M. 2016. Squamous cell carcinoma with hepatic metastasis in a saltwater crocodile (*Crocodylus porosus*). *Aust Vet J* 94(3):83–86.

Hill JR. 1977. Oral squamous cell caricnoma in a California King snake. *J Am Vet Med Assoc* 171(9):981–982.

Hoge AYA, Tucker S, Williams DS, Ogata AS, Guerra JK, and Jacobson ER. 1995. Spontaneous renal tumors in *Bothrops moojeni*, in *Proceedings of the Fifth International Colloquium on Pathology of Reptiles and Amphibians*, Zwart P and Matz G (Eds.), Alphen a/d Rijn, The Netherlands, pp. 283–285.

Honour SM, Ayroud M, and Wheler C. 1993. Metastatic chondrosarcoma and subcutaneous granulomas in a grey rat snake (*Elaphe obsoleta obsoleta*). *Can Vet J* 34(April):238–240.

Huerta P, Pineda H, Aguirre A, Spraker T, Sarti L, and Barragan A. 2000. First confirmed case of fibropapilloma in a Leatherback Turtle (Dermochelys coriacea), in *Proceedings of the Twentieth Annual Symposium on Sea Turtle Biology and Conservation*, p. 193.

Irizarry Rovira AR, Holzer TR, and Credille KM. 2014. Systemic mastocytosis in an African fat-tail gecko (*Hemitheconyx caudicinctus*). *J Comp Pathol* 151(1). Elsevier Ltd: 130–134. doi:10.1016/j.jcpa.2014.03.006.

Jacobson ER. 1981. Neoplastic diseases, in *Diseases of the Reptilia*, Vol 2, John EC and Oliphant FJ (Eds), San Diego, Academic Press, pp. 429–468.

Jacobson ER, Buergelt C, Williams B, and Harris RK. 1991. Herpesvirus in cutaneous fibropapillomas of the green turtle *Chelonia mydas*. *Dis Aquat Org* 12(1):1–6.

Jacobson ER, Ferris R, Bagnara JR, and Iverson WO. 1989a. Chromatophoromas in a pine snake. *Pigm Cell Res* 2:26–33.

Jacobson ER, Reichmann ME, Mansell JL et al. 1989b. Cutaneous fibropapillomas of Green Turtles (*Chelonia mydas*). *J Comp Pathol* 101: 39–51.

Jacobson ER, Gaskin JM, Clubb S, and Calderwood MB. 1982. Papilloma-like virus infection in Bolivian Side-Neck Turtles. *J Am Vet Med Assoc* 181(11):1325–1328.

Jacobson ER, Long PH, Miller RE, Kramer LW, Beehler B, Harshbarger JC, and Buergelt CD. 1986. Renal Neoplasia of Snakes. *J Am Vet Med Assoc* 189(9):1134–1136.

Jacobson ER, Reese DJ, Berry CR et al. 2013. Vet med today: What is your diagnosis? *J Am Vet Med Assoc* 243(11):1533–1535.

Janert B. 1998. A fibrosarcoma in a Siamese Crocodile (*Crocodylus siamensis*). *J Zoo Wildl Med* 29(1):72–77.

Jones K, Ariel E, Burgess G, and Read M. 2016. A review of fibropapillomatosis in Green Turtles (*Chelonia mydas*). *The Vet J* 212. *J Comp Pathol* 146(2–3). Elsevier Ltd: 258–268. doi:10.1016/j.jcpa.2011.07.003.

Jubb K and Stent AW. 2016. Pancreas, in *Jubb, Kennedy, and Palmer's Pathology of Domestic Animals*, 6th ed. Maxie GM, (Ed.), Elsevier, St. Louis, Missouri, pp. 353–375.

Jundt G. 2005. Odontogenic tumors, in *World Health Organization Classification of Tumors: Pathology and Genetics of Tumors of Head and Neck Tumors*. Barnes L, Eveson JW, Reichart P, and Sidransky D, (Eds.), IARC Press, Lyon, France, pp. 283–318.

Kao CF, Chen JL, Tsao WT, Lee AH, Liu CH, and Wang FI. 2016. A renal adenocarcinoma in a corn snake (*Pantherophis guttatus*) resembling human collecting duct carcinoma. *J Vet Diagn Invest* 28(5):599–603.

Kaye SW, Daverio H, Eddy R, Ossiboff RJ, Peters-Kennedy J, and Morrisey JK. 2016. Surgical resection of an interrenal cell adenocarcinoma in a Woma Python (*Aspidites ramsayi*) with 18 month follow-up. *J Herp Med Surg* 26: 26–31.

Keck M, Zimmerman DM, Ramsay EC, Douglass M, and Reavill DR. 2011. Renal adenocarcinoma in Cape Coral Snakes (*Aspidelaps lubricus lubricus*). *J Herp Med Surg* 21(1):5–9.

Keller KA, Sanchez-migallon Guzman D, Sanders C et al. 2016. Clinical and pathological findings in a red-tailed Boa Constrictor (*Boa constrictor constrictor*) with a primary neural neoplasm within the diencephalon and mesencephalon. *J Herp Med Surg* 26(3–4):85–89.

Kottwitz J, Zehnder AM, Wyre N, and Aquino S. 2008. Lacrimal cystadenoma in a Chinese Box Turtle (*Cuora flavomarginata*). *J Zoo Wildl Med* 39(1):103–106.

Kusewitt DF. 2012. Neoplasia and tumor biology, in *Pathologic Basis of Veterinary Disease*, 5th ed. Zachary JF and Donald McGavin M (Eds.), Elsevier, pp. 289–321.

Ladds P. 2009. Neoplasia and related proliferations in reptiles, in *Pathology of Australian Native Wildlife*, 1st ed. Ladds P (Eds.). CSIRO Publishing, Collingwood, Australia, pp. 477–486.

Latimer KS, Rich GA, and Gregory CR. 2000. Osteosarcoma in a California Mountain Kingsnake (*Lampropeltis zonata*), in *International Virtual Conference in Veterinary Medicine*. University of Georgia College Of Veterinary Medicine, Athens, GA.

Lemberger KY, Manharth A, and Pessier AP. 2005. Multicentric benign peripheral nerve sheath tumors in two related Bearded Dragons, *Pogona vitticeps*. *Vet Pathol* 42(4):507–510.

Limpus CL, Couper PJ, and Couper KID. 1993. Crab island revisited: Reassessment of the world's largest flatback turtle rookery after twelve years. *Mem Queensl Mus* 33:277–289.

Linn MJ, McNamara T, and Steinerg J. 1996. Pituitary adenoma in a black-headed python (*Aspidites melanocephalus*), in *Proceedings of the American Association of Zoo Veterinarians*, Puerto Vallarta, Mexico, p. 449.

Literak I, Robesova B, Majlathova V et al. 2010. Herpesvirus-associated Papillomatosis in a Green Lizard. *J Wildl Dis* 46(1):257–261.

Lombardini ED, Desoutter AV, Montali RJ, and Piero FD. 2013. Esophageal Adenocarcinoma in a 53-year-old Mata Mata Turtle (*Chelus fimbriatus*). *J Zoo Wildl Med* 44(3):773–776.

Lunger PD, Hardy WG, and Clark H. 1974. C-type particles in a reptilian tumor. *J Natl Cancer Inst* 52(4):1231–1235.

Lyons JA, Newman SJ, Greenacre CB, and Dunlap J. 2010. A gastric neuroendocrine carcinoma expressing somatostatin in a Bearded Dragon (*Pogona vitticeps*). *J Vet Diagn Invest* 22(2):316–320.

Machotka SV. 1984. Neoplasia in reptiles, in *Diseases of Amphibians and Reptiles*. Hoff GL, Frye FL, and Jacobson ER (Eds.), Plenum Press, New York, New York, pp. 519–580.

Machotka SV, Wisser J, Ippen R, and Nawab E. 1992. Report of Dysgerminoma in the ovaries of a snapping turtle (*Chelydra serpentina*) with discussion of ovarian neoplasms reported in reptilians and women. *In Vivo* 6(4):349–354.

Martin JC, Schelling SH, and Pokras M. 1994. Gastric adenocarcinoma in a Florida Indigo Snake (*Drymarchon corais couperi*). *J Zoo Wildl Med* 25(1):133–137.

Martorell J, Soto S, Barrera S, and Ramis A. 2009. Case report: Ovarian teratoma in a mediterranean tortoise. *Compendium* (Yardley, PA) 31(4):193–196. http://europepmc.org/abstract/med/19517411.

Matthews ME, Eshar D, Delk KW, Pohlman L, and Andrews G. 2014. Splenopancreatic ductal adenocarcinoma with multiorgan metastasis in a Northern Water Snake (*Nerodia sipedon*). *J Zoo Wildl Med* 45(2):437–440.

Mauldin EA and Peters-Kennedy J. 2016. Integumentary system, in *Jubb, Kennedy, and Palmer's Pathology of Domestic Animals*, 6th ed. Maxie MG (Ed.), Elsevier, St. Louis, Missouri, pp. 509–736.

Mauldin GN and Done LB. 2006. Oncology, in *Reptile Medicine and Surgery*. Mader DR (Ed.), Saunders-Elsvier, St. Louis, Missouri, pp. 299–322. doi:10.1016/B0-7216-9327-X/50023-7.

Michaels SJ and Sanecki R. 1988. Undifferentiated carcinoma in the ovary of a Boa Constrictor (*Boa constrictor ortoni*). *J Zoo Wildl Med* 19(4):237–240.

Mikaelian I, Levine BS, Smith SG, Harshbarger JC, and Wong VJ. 2001. Malignant peripheral nerve sheath tumor in a Bearded Dragon, *Pogona Vitticeps*. *J Herp Med Surg* 11(1):9–12.

Milovancev M, Hauck M, Keller C, Stranahan LW, Mansoor A, and Malarkey DE. 2015. Comparative pathology of canine soft tissue sarcomas: Possible models of human non-rhabdomyosarcoma soft tissue sarcomas. *J Comp Pathol* 152(1). Elsevier Ltd: 22–27. doi:10.1016/j.jcpa.2014.09.005.

Muñoz-Gutiérrez JF, Garner MM, and Kiupel M. 2016. Cutaneous chromatophoromas in captive snakes. *Vet Pathol* 53(6):1213–1219.

Needle D, McKnight CA, and Kiupel M. 2013. Chondroblastic osteosarcoma in two related spiny-tailed monitor lizards (*Varanus acanthurus*). *J Exot Pet Med* 22(3). Elsevier: 265–269. doi:10.1053/j.jepm.2013.08.013.

Newman SJ, Brown CJ, and Patnaik AK. 2003. Malignant ovarian teratoma in a red-eared slider (*Trachemys scripta elegans*). *J Vet Diagn Invest* 15:77–81.

Norton TM, Jacobson ER, and Sundberg JP. 1990. Cutaneous fibropapillomas and renal myxofibroma in a Green Turtle, *Chelonia mydas*. *J Wildl Dis* 26(2):265–270.

Onderka DK and Zwart P. 1978. Granulosa cell tumor in a Garter Snake (*Thamnophis sirtalis*). *J Wildl Dis* 14(2):218–221.

Oros J, Lorenzo HL, Andrara M, and Recuero R. 2004. Type A – like retroviral particles in a metastatic intestinal adenocarcinoma in an Emerald Tree Boa (*Corallus caninus*). *Vet Pathol* 518:515–518.

Orós J, Tucker S, Fernández L, and Jacobson ER. 2004. Metastatic squamous cell carcinoma in two Loggerhead sea turtles Caretta Caretta. *Dis Aquat Org* 58:245–250.

Ost D, Joseph C, Sogoloff H, and Menezes G. 1999. Primary pulmonary melanoma: Case report and literature review. *Mayo Clin Proc* 74(1). Mayo Foundation for Medical Education and Research: 62–66. doi:S0025-6196(11)64561-8 [pii].

Patterson-Kane JC and Redrobe SP. 2005. Colonic adenocarcinoma in a Leopard Gecko (*Eublepharis macularius*). *The Vet Rec* 157(10):294–295.

Pees M, Ludewig E, Plenz B, and Schmidt V. 2015. Imaging diagnosis—Seminoma causing liver compression in a spur-thighed tortoise (*Testudo graeca*). *Vet Radiol Ultrasound* 56(2):E21–E24.

Pereira ME and Viner TC. 2008. Oviduct adenocarcinoma in some species of captive snakes. *Vet Pathol* 45(5):693–697.

Ramis A, Pumarola M, Fernandez-Moran J, Anor S, Majo N, and Zidan A. 1998. Malignant peripheral nerve sheath tumor in a water moccasin (*Agkistrodon piscivorus*). *J Vet Diagn Invest* 10:205–208.

Ramsay EC and Fowler ME. 1992. Reptile neoplasms at the Sacramento Zoo, 1981–1991, in *Proceedings of the Joint Conference of the American Association of Zoo Veterinarians and the American Association of Wildlife Veterinarians*, Oakland, pp. 153–155.

Ramsay EC, Munson L, Lowenstine L, and Fowler ME. 1996. A retrospective study of neoplasia in a collection of captive snakes *J Zoo Wildl Med* 27(1):28–34.

Raynaud A and Adrian M. 1976. Cutaneous lesions with papillomatous structure associated with viruses in the Green Lizard (*Lacerta viridis* Laur.). *C R Acad Sci Hebd Seances Acad Sci D* 283(7):845–847.

Reavill D, Dahlausen B, and Zaffarano B. 2002. Multiple cutaneous liposarcomas in a red-tailed Boa, Boa Constrictor, and Chameleon, in *Proceedings of the Association of Reptilian and Amphibian Veterinarians*, October: 5–6.

Reavill DR, Fassler SA, and Schmidt RE. 2000. Mast cell tumor in a common iguana (*Iguana iguana*), in *Proceedings of the Association of Reptilian and Amphibian Veterinarians*.

Reavill DR and Schmidt RE. 2003. Lipomas in corn snakes (*Elaphe guttata guttata*): A series of four cases, in *Proceedings of the Association of Reptilian and Amphibian Veterinarians*, Minneapolis, p. 60.

Reavill DR and Schmidt RE. 2007. Urogenital tract diseases of reptiles and amphibians, in *Proceedings Association of Reptile and Amphibian Veterinarians*, pp. 87–109.

Reavill DR, Schmidt RE, and Stevenson RE. 2004. Malignant chromatophoromas in three veiled chameleons (Chamaeleo Calyptratus), in *Proceedings Association of Reptile and Amphibian Veterinarians*, pp. 131–132.

Ritter JM, Garner MM, Chilton J, Jacobson ER, and Kiupel M. 2009. Gastric neuroendocrine carcinomas in Bearded Dragons (*Pogona vitticeps*). *Vet Pathol* 46(6):1109–1116.

Rivera S, Crane MM, Mcmanamon R, and Gregory CR. 2015. Surgical treatment of pulmonary melanophoroma in a Beaded Lizard (*Heloderma horridum exasperatum*). *J Zoo Wildl Med* 46(2):397–399.

Roe WD, Cooper SM, and Hazley L. 2002. Squamous cell carcinoma in a Tuatara (*Sphenodon punctatus*) case history. *New Zealand Vet J* 50(5):207–210.

Rothschild BM, Tanke DH, Helbling M, and Martin LD. 2003. Epidemiologic study of tumors in dinosaurs. *Naturwissenschaften* 90(11):495–500.

Rousselet E, Norton TM, Wellman M, Kohart N, and Stacy NI. 2017. Presumptive granulocytic leukemia in a Hognose Snake (*Heterodon platirhinos*). *Vet Quart* 37(1). Taylor & Francis: 43–47. doi:10.1080/01652176.2017.1281464.

Sander SJ, Ossiboff RJ, Stokol J, Steeil JC, and Neiffer DL. 2015. Endolymphatic sac carcinoma *in situ* in a Tokay Gecko (*Gekko gecko*). *J Herp Med Surg* 25(3–4):82–86.

Santoro M, Stacy B, Morales J, Gastezzi-Arias P, Landazuli S, and Jacobson ER. 2008. Mast cell tumour in a giant Galapagos Tortoise (*Geochelone nigra vicina*). *J Comp Pathol* 138(2–3):156–159.

Schlafer DH and RA. Foster. 2016. Female genital system, in *Jubb, Kennedy, and Palmer's Pathology of Domestic Animals*, 6th ed. Maxie GM, (Ed.), Elsevier, St. Louis, Missouri, pp. 358–464.

Schlumberger HG and Lucke B. 1948. Tumors of fishes, amphibians, and reptiles. *Cancer Res* 8(12):657–753.

Schmidt RE. 1977. Plasma cell tumor in an Est Indian Water Lizard (*Hydrosaurus amboinensis*). *J Wildl Dis* 13(1):47–48.

Schmidt RE and Reavill DR. 2012. Metastatic chondrosarcoma in a corn snake (*Pentherophis guttatus*). *J Herp Med Surg* 22(3–4):67–69.

Schmidt RE and Reavill D. 2015. Diseases of the head of amphibians and reptiles, in *Proceedings of Exotics Con*, pp. 571–582.

Schonbauer M, Loupal G, and Schonbauer-Langle A. 1982. Osteochondrosarcoma in a desert monitor (Varanus). *Berl Munch Tierarztl Wochenschr* 95(10):193–194.

Schumacher J, Bennett R, Fox LE, Deem SL, Neuwirth L, and Fox JH. 1998. Mast cell tumor in an Eastern Kingsnake (*Lampropeltis getulus getulus*). *J Vet Diagn Invest* 10(1):101–104.

Scott HH and Beattie J. 1927. Neoplasm on a porose crocodile. *J Pathol Bacteriol.* 30(1):61–66.

Shoemaker M, Barrie M, Holman H, Wolk KE, Stromberg P, and Aeffner F. 2016. Pathology in practice. *J Am Vet Med Assoc* 248(2):153–155.

Silverstone AM, Garner MM, Wojcieszyn JW, Guillermo Couto C, and Raskin RE. 2007. Acute lymphoblastic leukemia in a Diamondback Terrapin, *Malaclemys Terrapin*. *J Herp Med Surg* 17(3):92–99.

Smith GM and Coates CW. 1983. Fibro-epithelial growths of the skin in large marine turtles, *Chelonia mydas (Linnaeus)*. *Zoologica* 23:93–98.

Stacy BA, Vidal JD, Osofsky A, Terio K, Koski M, and De Cock HE. 2004. Ovarian papillary cystadenocarcinomas in a Green Iguana (*Iguana iguana*). *J Comp Pathol* 130(2–3):223–228.

Steeil JC, Schumacher J, Hecht S et al. 2013. Diagnosis and treatment of a pharyngeal squamous cell carcinoma in a Madagascar Ground Boa (*Boa madagascariensis*). *J Zoo Wildl Med* 44(1):144–151.

Stern AW, Velguth KE, and D'Agostino J. 2010. Metastatuc ductal adenocarcinoma in a Western Hognose Snake (*Heterodon nasicus*). *J Zoo Wildl Med* 41(2):320–324.

Stolk A. 1958. Tumors of reptiles 4, mutliple osteomas in the lizard lacerta viridis. *Beaufortia* 7(79):1–9.

Stumpel J, Gérard B, Del-Pozo J, French A, and Eatwell K. 2012. Cardiac hemangioma in a Corn Snake (*Pantherophis guttatus*). *J Zoo Wildl Med* 43(2):360–366.

Sykes JM and Trupkiewicz JG. 2006. Reptile neoplasia at the Philadelphia zoological garden, 1901–2002. *J Zoo Wildl Med* 37(1):11–19.

Tocidlowski ME, Merrill CL, Loomis MR, and Wright JF. 2001a. Teratoma in desert Grassland Whiptail Lizards (*Cnemidophorus uniparens*). *J Zoo Wildl Med* 32(2):257–259.

Tocidlowski MW, McNamara PL, and Wojcieszyn JW. 2001b. Myelogenous leukemia in a Bearded Dragon (*Acanthodraco vitticeps*). *J Zoo Wildl Med* 32(1):90–95. http://www.ncbi.nlm.nih.gov/pubmed/12790401.

Tuttle AD, Harms CA, Van Wettere A, Grafinger MS, and Lewbart GA. 2006. Splenic hemangiosarcoma in a corn snake, *Elaphe Guttata*. *J Herp Med Surg* 16(4):140–143.

Vasil'ev DB and Solov'ev L. 2006. Ganglioneuroblastoma in Yeminite Chameleon (*Chamaeleo calyptratus*): The first recorded case of a tumor of neuroectodermal histogenesis in reptiles. *Arkh Patol* 68(4):45–47.

Wadsworth JR. 1954. Some neoplasms of captive wild animals. *J Am Vet Med Assoc* 125(929):121–123.

Wadsworth JR and Hill WC. 1956. Selected tumors from the London Zoo Menagerie. *University of Pennsylvania Veterinary Extension Quarterly* 141:70–73.

Wenger S, Simova-Curd S, Grest P, Steinmetz H, and Hatt J. 2010. Ovarian teratoma in a Fiji Island Banded Iguana (*Brachylophus fasciatus*) and a Green Iguana (*Iguana iguana*). *J Herp Med Surg* 20(1):20–23.

Whiteside DP and Garner MM. 2001. Thyroid adenocarcinoma in a crocodile lizard, *Shinisaurus Crocodilurus*. *J Herp Med Surg* 11(1):13–16.

Wilkinson R, Hernandez-Divers S, Lafortune M, Calvert I, Gumpenberger M, McArthur S. 2004. Diagnostic imaging techniques, in *Medicine and Surgery of Tortoises and Turtles*. McArthur S, Wilkinson R, and Meyer J (Eds.), Blackwell Publishing, Ames, Iowa, pp. 187–328.

Wilson GH, Fontenot DK, Brown CA, Kling MA, Stedman N, and Greenacre CB. 2004. Pseudocarcinomatous biliary hyperplasia in two Green Iguanas (*Iguana iguana*). *J Herp Med Surg* 14(4):12–18.

Work TM, Balazs GH, Rameyer R, and Morris R. 2004. Retrospective pathology survey of green turtles *Chelonia mydas* with Fibropapillomatosis in the Hawaiian Islands, 1993–2003. *Dis Aquat Org* 62(1–2):163–176.

Work TM, Dagenais J, Balazs GH, Schettle N, and Ackermann M. 2014. Dynamics of virus shedding and *in situ* confirmation of chelonid herpesvirus 5 in Hawaiian green turtles with fibropapillomatosis. *Vet Pathol* 52(6):1195–1201.

Work TM, Dagenais J, Balazs GH et al. 2009. In vitro biology of fibropapilloma-associated turtle herpesvirus and host cells in Hawaiian Green Turtles (*Chelonia mydas*). *J General Virol* 90(8):1943–1950.

Yamada K, Sakai K, and Yamada S. 2001. A morphological and immunohistochemical study of the ultimobranchial body in the Japanese lizard and the snake. *Okajimas Folia Anat Jpn* 78(5):161–168.

Figure 1.1 Inland Bearded Dragon, *Pogona vitticeps*. Agamidae. Squamous cell carcinoma. Squamous cell carcinoma arising from the periocular tissues partially obstructs the globe. (Courtesy of L. Wright.)

Figure 1.2 Anaconda, *Eunectes* sp. Boidae. Squamous cell carcinoma. The region of the scent gland is expanded by a mass like effect, which was histologically confirmed as squamous cell carcinoma of the scent gland. (Courtesy of D. Collins.)

Figure 1.3 Red-Eared Slider, *Trachemys scripta elegans*. Emydidae. Squamous cell carcinoma. Ulcerated squamous cell carcinoma on the neck is visible upon extraction of the head. (Courtesy of R. Monaco.)

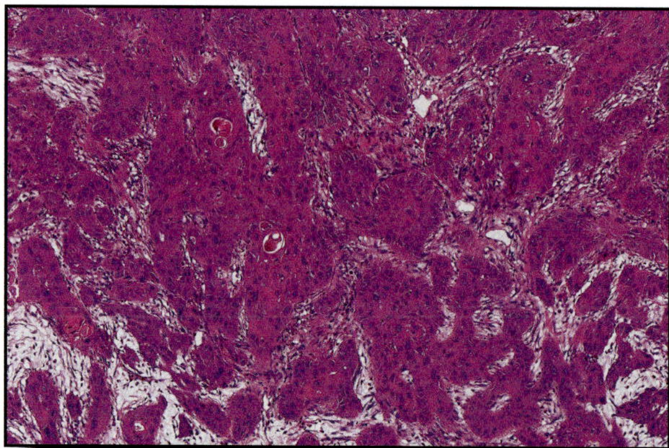

Figure 1.4 Panther Chameleon, *Furcifer pardalis*. Chamaeleonidae. Squamous cell carcinoma. Cutaneous squamous cell carcinoma forms abundant, invasive packets in the dermis with occasional central keratinization (keratin pearls). Inflammation and scirrhous responses are variable in the surrounding dermis. H&E stain.

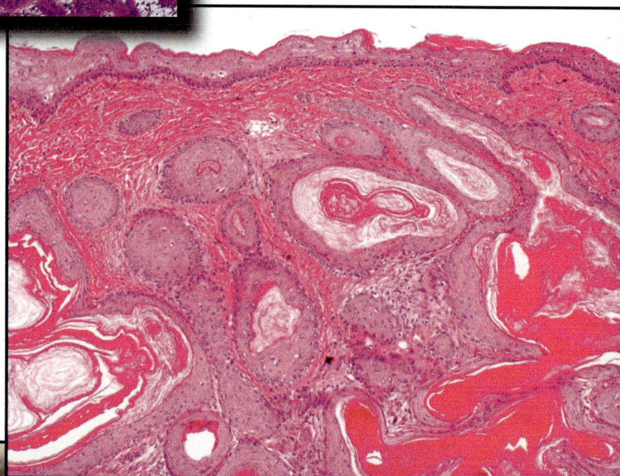

Figure 1.5 Inland Bearded Dragon, *Pogona vitticeps*. Agamidae. Squamous cell carcinoma. Squamous cell carcinoma arising from the periocular tissues forms numerous keratin pearls. H&E stain.

Figure 1.6 American Snapping Turtle, *Chelydra serpentine*. Chelydridae. Papilloma. Cutaneous papilloma is a papillary, exophytic, pale tan mass. (Courtesy of S. Barten.)

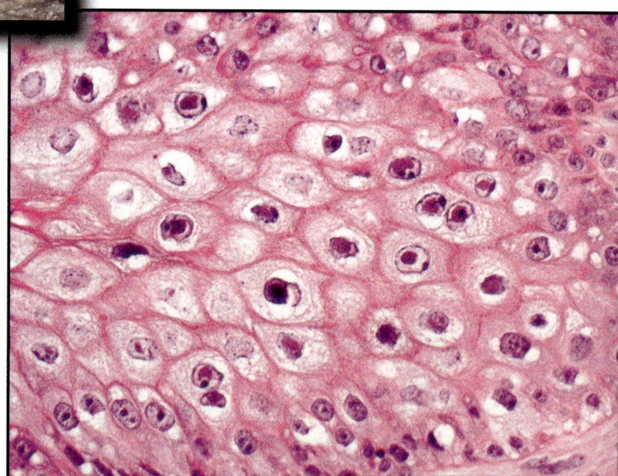

Figure 1.7 Green Turtle, *Chelonia mydas*. Cheloniidae. Papilloma. Cutaneous papilloma with proliferative keratinocytes. Note intranuclear amphophilic inclusions causing margination of chromatin. H&E stain.

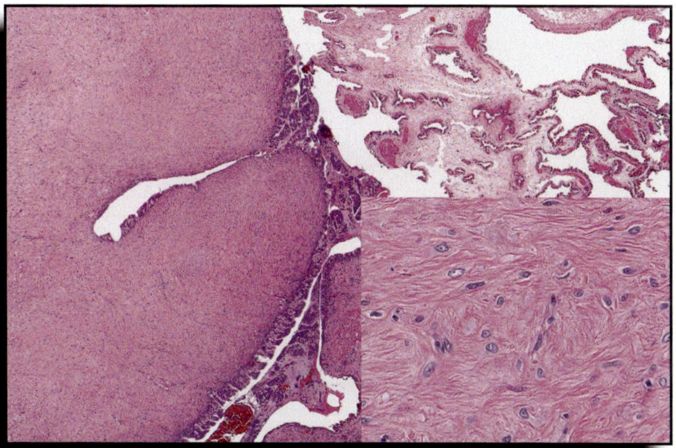

Figure 1.8 Green Turtle, *Chelonia mydas.* Cheloniidae. Fibromatosis. Pulmonary manifestation of fibromatosis in a green turtle with cutaneous fibropapillomatosis. Inset: higher magnification of fibropapilloma. H&E stain.

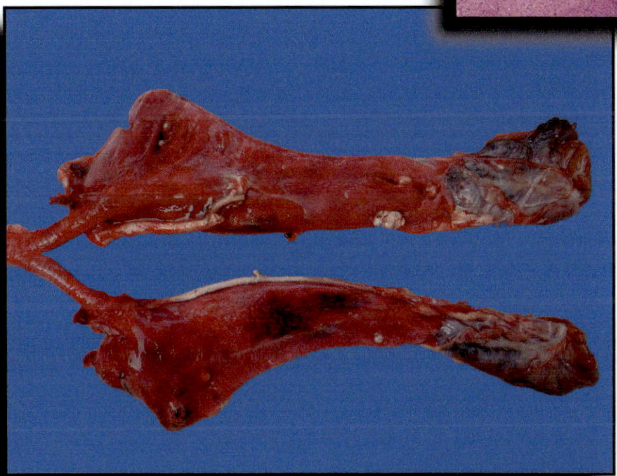

Figure 1.9 Green Turtle, *Chelonia mydas.* Cheloniidae. Papilloma. Pulmonary fibromatosis is represented by multiple, multinodular, pale tan masses. (Courtesy of D. Mader.)

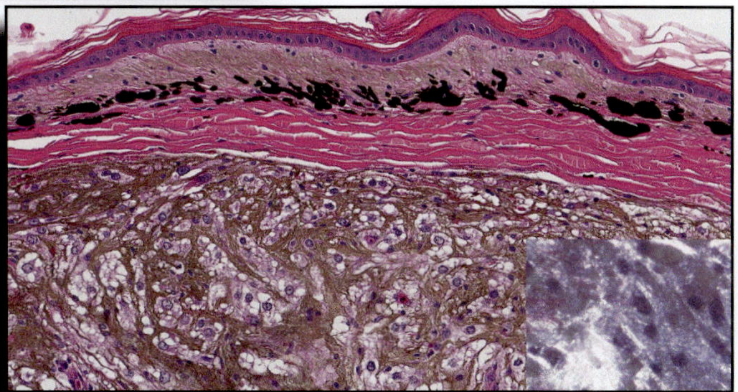

Figure 1.10 Veiled Chameleon, *Chamaeleo calyptratus.* Chamaeleonidae I. Iridophoroma. An iridophoroma arising from dermal iridophores markedly expands the dermis. Neoplastic cells are polygonal with moderate anisokaryosis, and contain brown, birefringent pigment (inset). H&E stain.

Figure 1.11 Inland Bearded Dragon, *Pogona vitticeps.* Agamidae. Melanoma. Multiple foci of dark grey to black discoloration on the lateral body wall represent melanoma. (Courtesy of A. Meyer.)

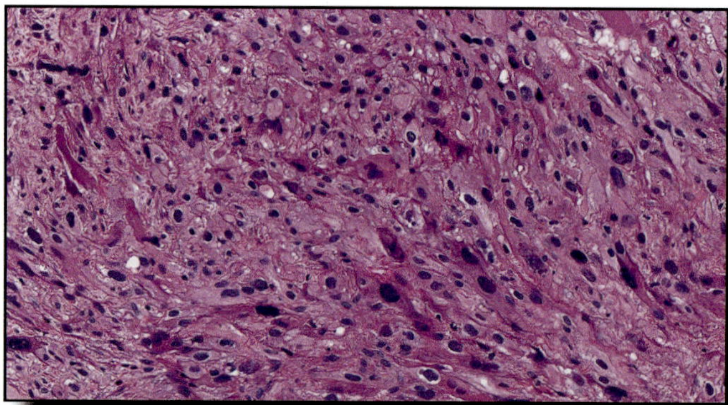

Figure 1.12 Veiled Chameleon, *Chamaeleo calyptratus*. Chamaeleonidae. Melanoma. Metastatic melanoma in the omentum. Neoplastic cells are poorly pigmented ("amelanotic melanoma"), pleomorphic, and have prominent nucleoli. H&E stain.

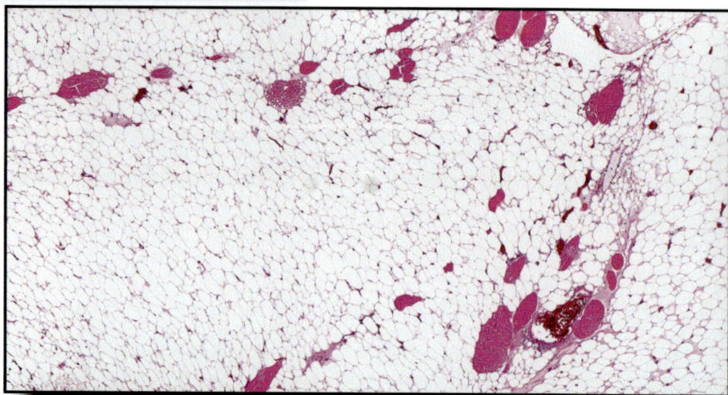

Figure 1.13 Spotted Python, *Antaresia maculosa*. Pythonidae. Lipoma. Neoplastic cells resemble mature adipocytes. There is some invasion of skeletal muscle with isolation of small clusters of skeletal myocytes, resembling infiltrative lipoma as reported in mammals. H&E stain.

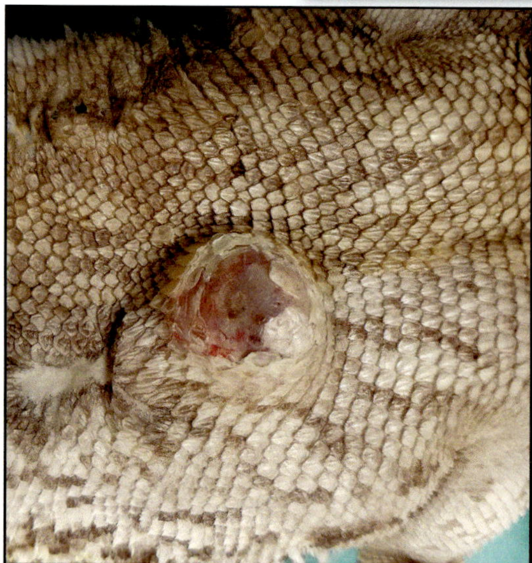

Figure 1.14 Inland Bearded Dragon, *Pogona vitticeps.* Agamidae. Myxosarcoma. The body wall is focally expanded by an ulcerated neoplasm, confirmed histologically as myxosarcoma. (Courtesy of K. Ramsell.)

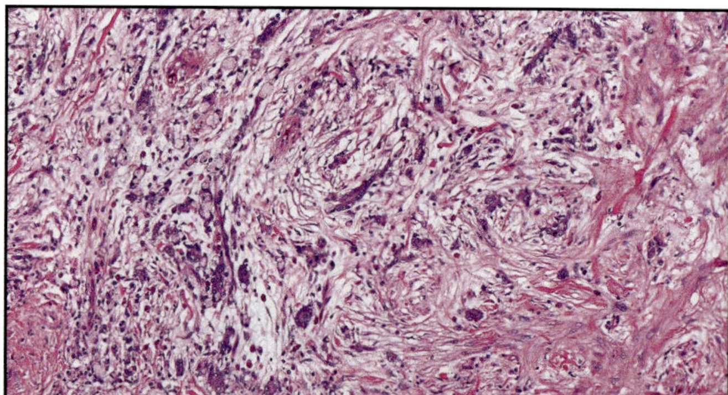

Figure 1.15 Indian Python, *Python morulus.* Pythonidae. Scent gland carcinoma. Small clusters of neoplastic epithelial packets and pseudoacini are embedded in a scirrhous stroma in the wall of the cloaca. These neoplastic cells extended from a scent gland carcinoma. H&E stain.

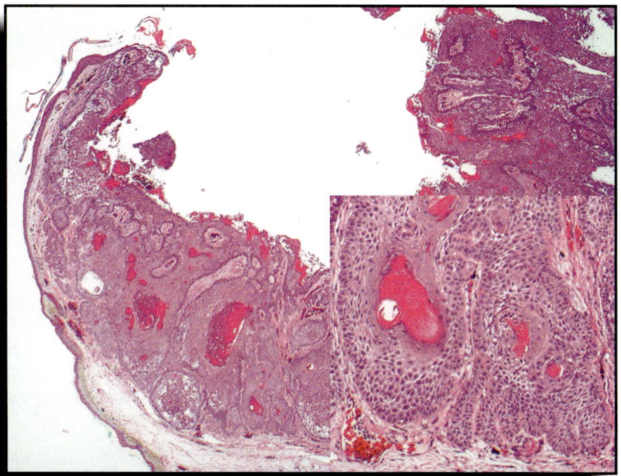

Figure 1.16 Crested Gecko, *Correlophus ciliatus*. Diplodactylidaae. Scent gland adenoma. The scent gland epithelium is neoplastic and proliferative without invasion of the basement membrane or substantial cellular atypia. H&E stain.

Figure 1.17 Inland Bearded Dragon, *Pogona vitticeps.* Agamidae. Renal tubular adenocarcinoma. One pole of the kidney is markedly expanded by a renal adenocarcinoma. (Courtesy of S. Rosenbloom.)

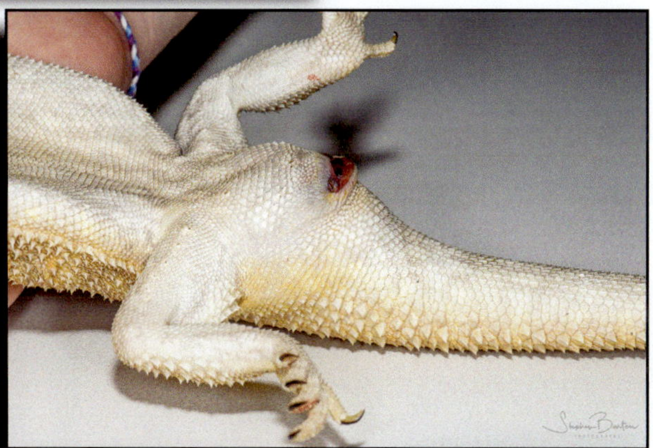

Figure 1.18 Inland Bearded Dragon, *Pogona vitticeps*. Agamidae. Renal tubular adenocarcinoma. The pericloacal region is markedly expanded. (Courtesy of S. Barten.)

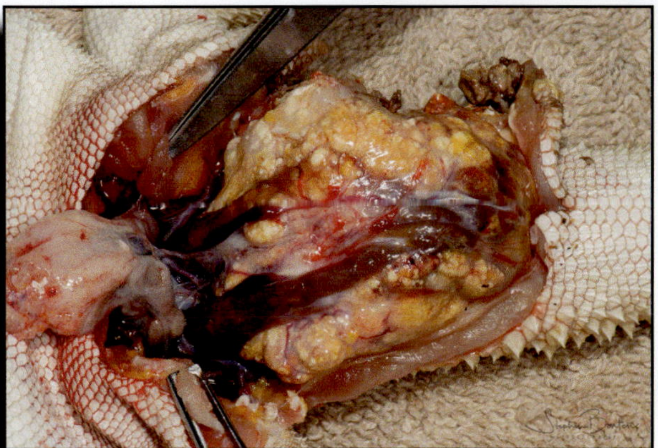

Figure 1.19 Inland Bearded Dragon, *Pogona vitticeps*. Agamidae. Renal tubular adenocarcinoma. Upon incision of the coelomic cavity, the externally appreciable distention of the coelom and cloaca is due to bilateral renal tubular adenocarcinomas that extend into surrounding connective tissues of the body wall. (Courtesy of S. Barten.)

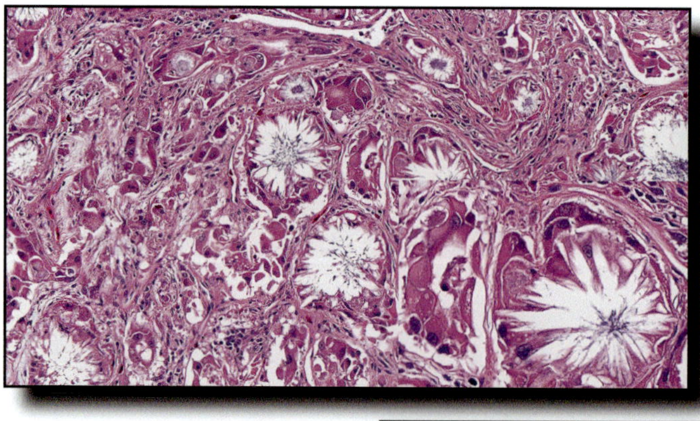

Figure 1.20 California Kingsnake, *Lampropeltis californiae*. Colubridae. Renal tubular adenocarcinoma. Renal tubular adenocarcinoma effaces a portion of the kidney. Neoplastic cells form tubules and packets of anaplastic epithelial cells resembling renal tubules (inset). Few tubules are ruptured due to formation of urate tophi. H&E stain.

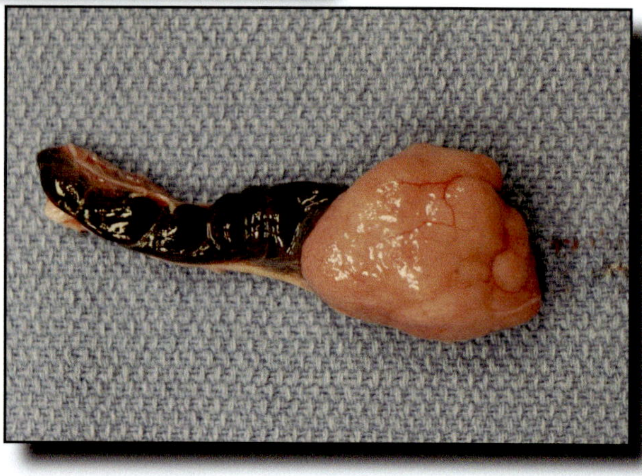

Figure 1.21 Ball Python, *Python regius*. Pythonidae. Nephroblastoma. One pole of the kidney is expanded by a mass confirmed histologically as a nephroblastoma. (Courtesy of T. Kruse and C. Buie.)

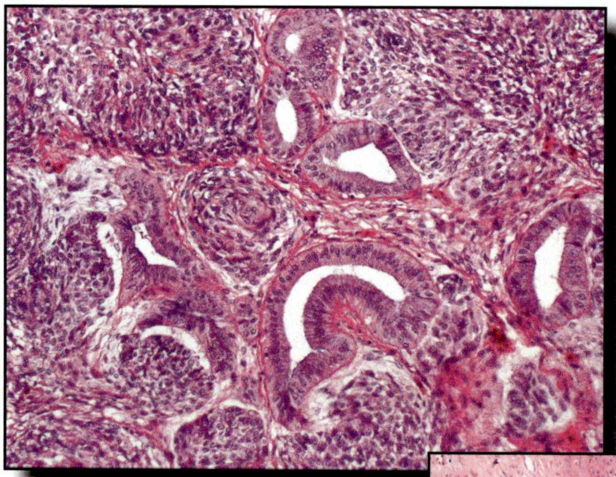

Figure 1.22 Ball Python, *Python regius.* Pythonidae. Nephroblastoma. Renal mass composed of branching and infolding tubules separated by densely packed blastemal cells with a high nuclear to cytoplasmic ratio, characteristic of nephroblastoma. H&E stain.

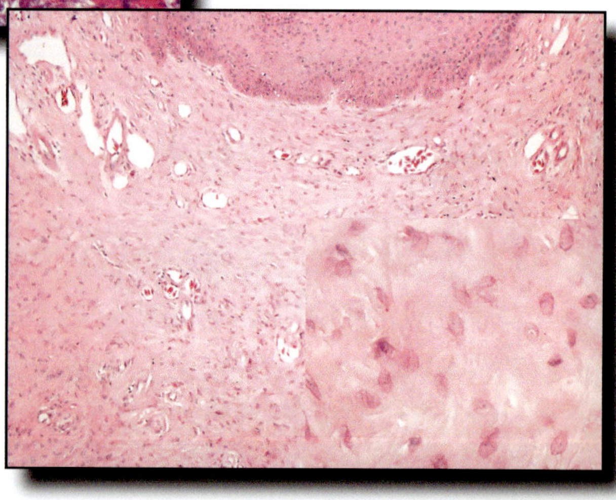

Figure 1.23 Inland Bearded Dragon, *Pogona vitticeps.* Agamidae. Fibromatous epulis of periodontal ligament origin. The gingiva is expanded by a mass composed mainly of regularly spaced spindle cells with a plump, oval to angular nucleus. Cytoplasm of these cells blends imperceptibly with the surrounding, dense, collagenous stroma (inset). H&E stain.

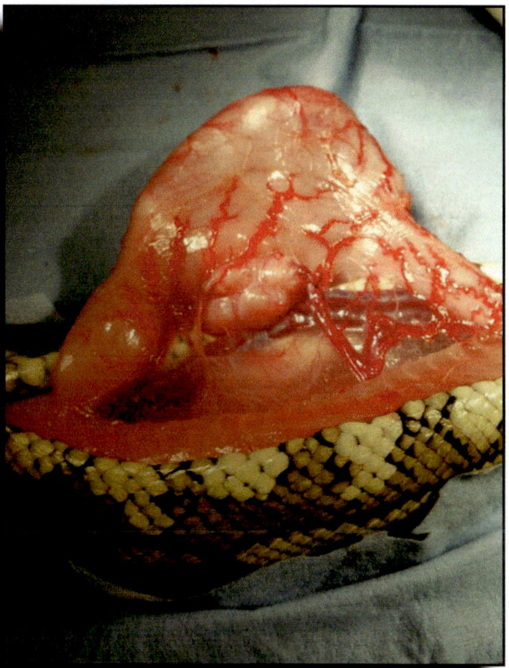

Figure 1.24 Unidentified Python. Pythonidae. Gastric adenocarcinoma. The lesser curvature of the stomach is expanded by a neoplasm confirmed histologically to be gastric adenocarcinoma. (Courtesy of Z. Steffe.)

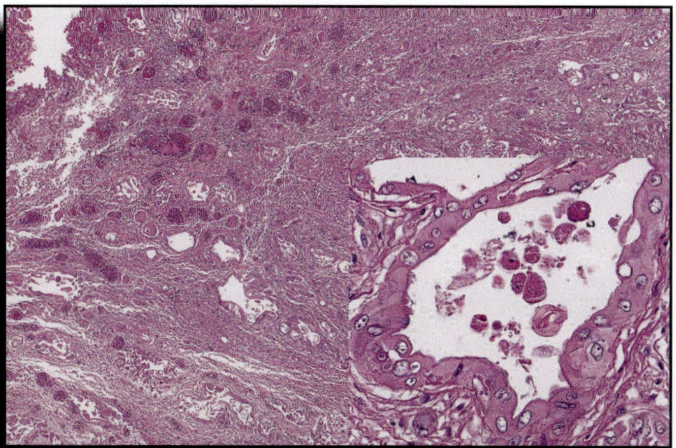

Figure 1.25 Desert Rosy Boa, *Lichanura trivirgata*. Boidae. Gastric adenocarcinoma. The gastric mucosa is markedly ulcerated and subtended by neoplastic packets and tubules (inset) originating from the gastric mucosa that markedly invade the subjacent submucosa and muscularis and induce a marked scirrhous response. H&E stain.

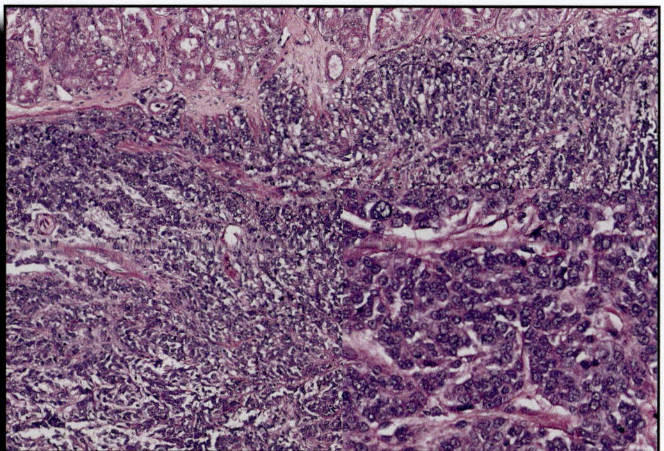

Figure 1.26 Inland Bearded Dragon, *Pogona vitticeps*. Agamidae. Gastric neuroendocrine carcinoma. Expanding the submucosa and infiltrating all layers of the stomach are abundant, closely associated packets of neoplastic cells (inset) associated with fine fibrovascular stroma. H&E stain.

Figure 1.27 Inland Bearded Dragon, *Pogona vitticeps*. Agamidae. Gastric neuroendocrine carcinoma. Arising from the wall of the stomach and infiltrating into/metastasizing to the liver is a nodular neoplasm confirmed histologically to be a gastric neuroendocrine carcinoma.

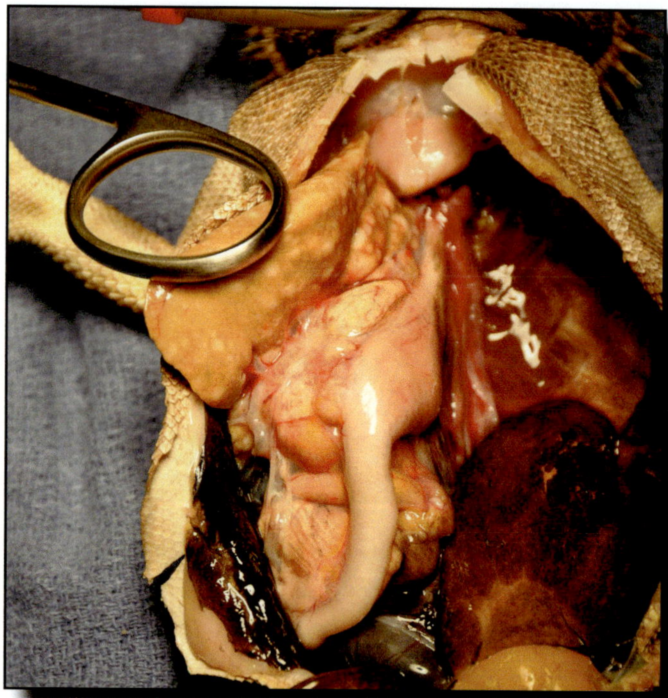

Figure 1.28 Inland Bearded Dragon, *Pogona vitticeps*. Agamidae. Oral adenomatous polyp. The mass is composed of numerous papillary projections lined by epithelium with little cellular atypia. H&E stain.

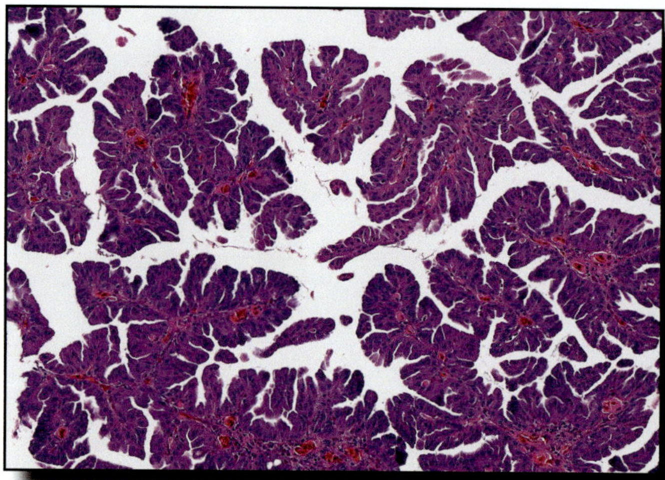

Figure 1.29 Corn Snake, *Pantherophis guttatus*. Colubridae. Intestinal adenocarcinoma. The wall of the intestine is markedly expanded by scirrhous tissue centered on well-defined neoplastic tubules of columnar and goblet cells. Inset: higher magnification of a neoplastic tubule. H&E stain.

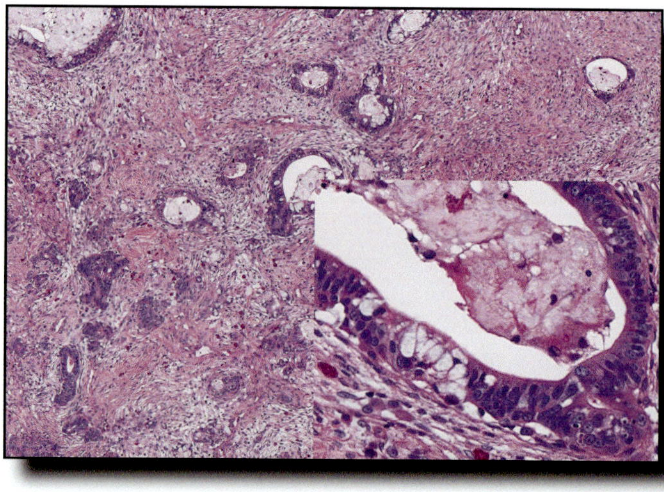

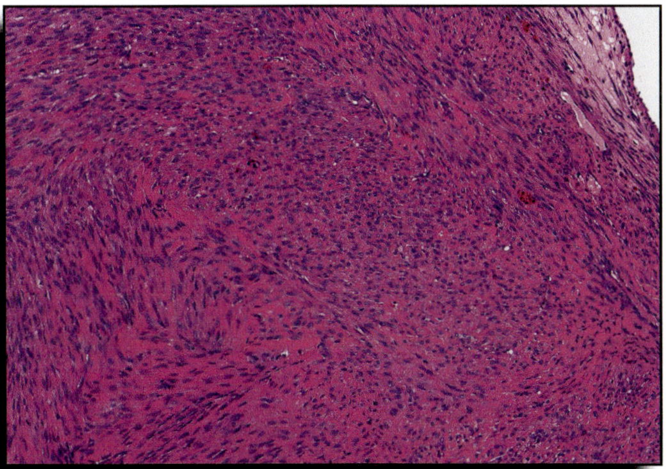

Figure 1.30 Green Turtle, *Chelonia mydas.* Cheloniidae. Intestinal leiomyosarcoma. Normal smooth muscle layering of the wall of the intestines is markedly distorted and expanded by intersecting streams of spindle cells resembling poorly defined smooth muscle. H&E stain.

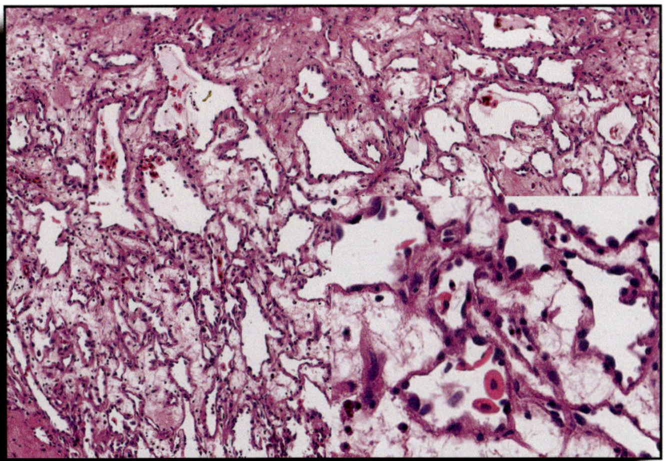

Figure 1.31 Angolan Python, *Python anchietae.* Pythonidae. Cloacal hemangioma. Expanding the submucosa of the cloaca, there is an edematous and mildly inflamed neoplasm composed of abundant, well-defined vascular structures that are lined by a single layer of flattened to slightly plump epithelial cells with minimal cellular atypia (inset). Vascular structures contain variable amounts of blood. H&E stain.

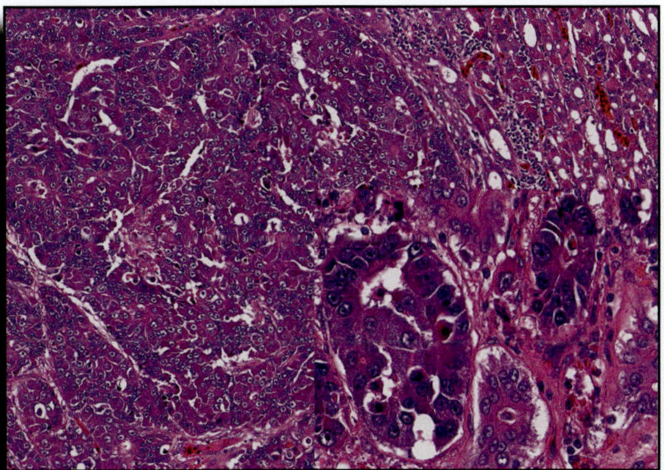

Figure 1.32 California Kingsnake, *Lampropeltis californiae.* Colubridae. Pancreatic acinar adenocarcinoma. Pancreas is compressed and effaced by packets and acini of neoplastic cells (inset) with occasional eosinophilic cytoplasmic granules resembling zymogen granules. These cells have moderate anisocytosis and anisokaryosis with prominent nucleoli, and variable nuclear polarity. H&E stain.

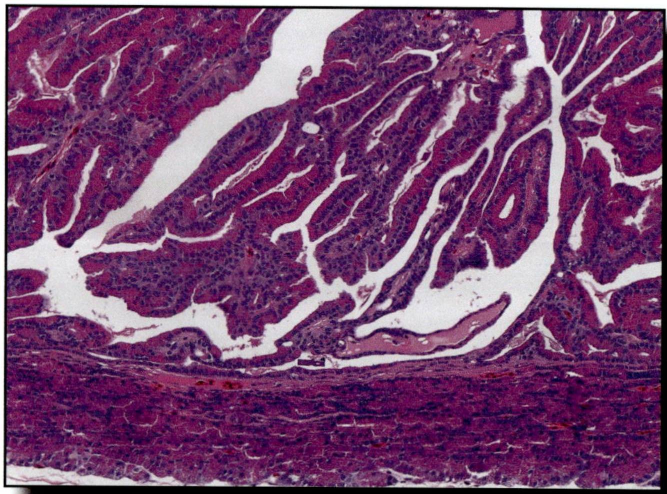

Figure 1.33 Inland Bearded Dragon, *Pogona vitticeps.* Agamidae. Pancreatic acinar adenoma. Pancreatic tissue is compressed by a partially encapsulated neoplasm composed of tubulopapillary projections lined by a single layer of polygonal cells that are well differentiated with minimal cellular atypia. These cells maintain polarity with basal nuclei and variable amounts of apical, eosinophilic, cytoplasmic granules (zymogen granules). H&E stain.

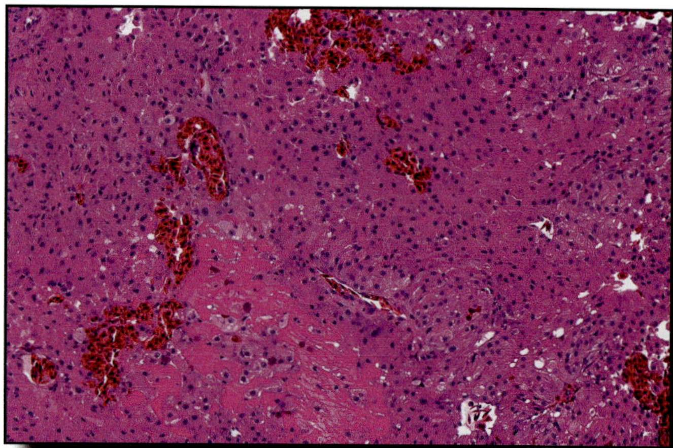

Figure 1.34 Corn Snake, *Pantherophis guttatus.* Colubridae. Hepatocellular carcinoma. This solidly cellular neoplasm is composed of polygonal cells with abundant, eosinophilic cytoplasm and moderate cellular atypia. Hemorrhage and fibrin are frequent throughout this hepatocellular carcinoma. H&E stain.

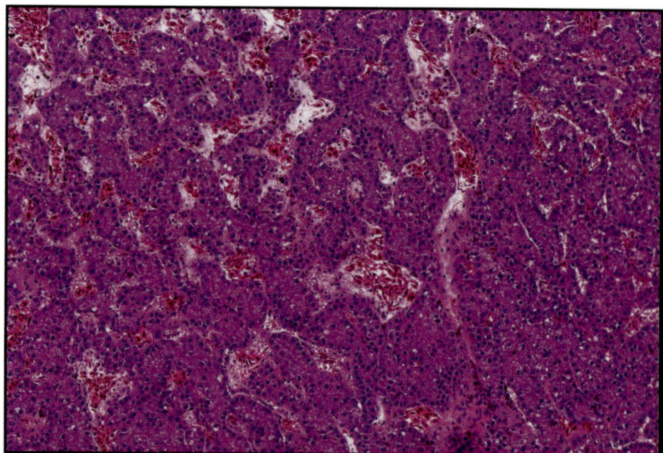

Figure 1.35 California Kingsnake, *Lampropeltis californiae.* Colubridae. Hepatoma. The liver is regionally expanded by a well-demarcated neoplasm composed of disorganized cords of well-differentiated polygonal cells resembling hepatocytes; portal triads are not present. H&E stain.

Figure 1.36 Desert Grassland Whiptail Lizard, *Aspidoscelis uniparens.* Teiidae. Biliary adenocarcinoma. Upon external examination, the mid-coelomic cavity is asymmetrically distended. Necropsy revealed a neoplasm originating from the liver that was confirmed histologically as a biliary adenocarcinoma. (Courtesy of D. Baumann and R. Kupronis.)

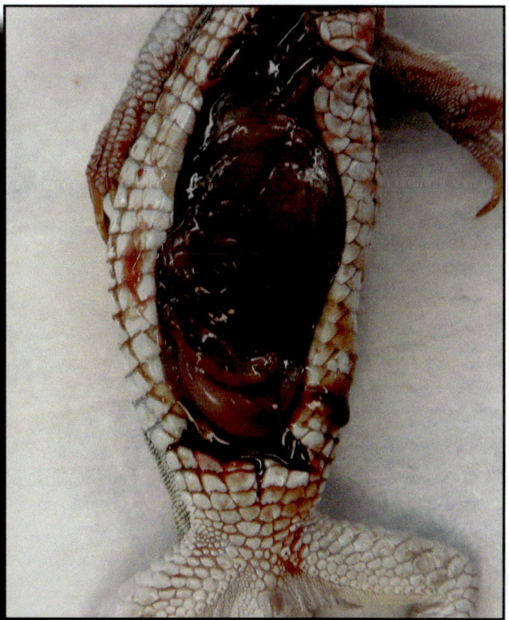

Figure 1.37 Desert Grassland Whiptail Lizard, *Aspidoscelis uniparens.* Teiidae. Biliary adenocarcinoma. The liver is markedly expanded by multiple, poorly demarcated masses that were confirmed histologically as biliary adenocarcinoma. (Courtesy of D. Baumann and R. Kupronis.)

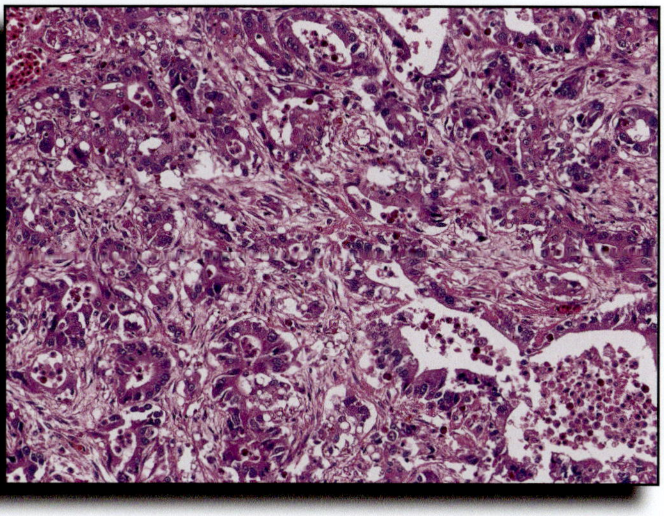

Figure 1.38 California Mountain Kingsnake, *Lampropeltis zonata.* Colubridae. Biliary adenocarcinoma. The liver has multiple invasive nodules of pleomorphic neoplastic cells that frequently form tubules and are associated with a moderate scirrhous response. Neoplastic cells show moderate to marked atypia with multinucleation and frequent mitoses. H&E stain.

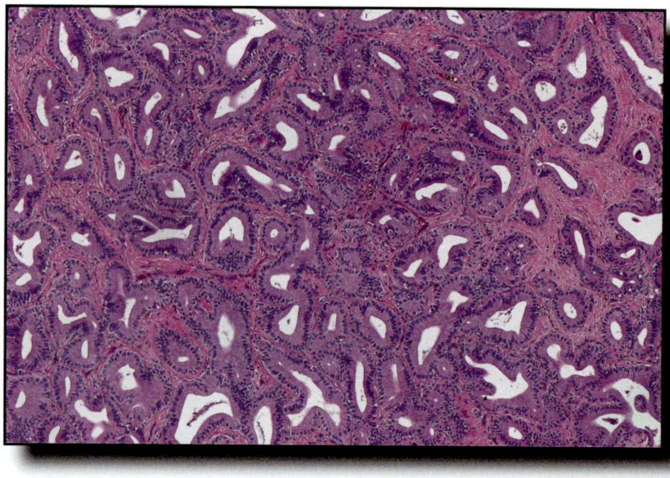

Figure 1.39 Chameleon (unidentified species). Chamaeleonidae. Biliary adenoma. Hepatic parenchyma is focally expanded by a neoplasm composed of tubules in a moderate amount of fibrous stroma. Tubules are lined by a single layer of well-differentiated columnar cells that have basilar nuclei. H&E stain.

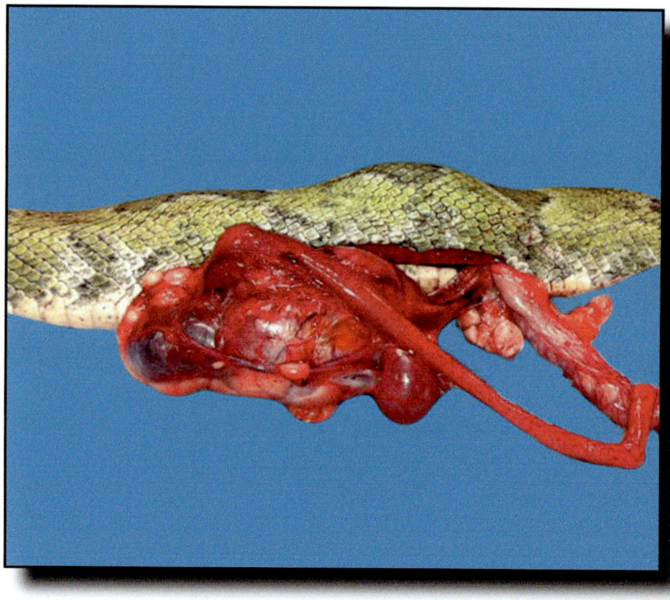

Figure 1.40 Madagascar Tree Boa, *Sanzinia madagascariensis.* Boidae. Granulosa cell tumor. The ovary is effaced and enlarged by a multicystic neoplasm confirmed histologically as a granulosa cell tumor. (Courtesy of K. Helmick.)

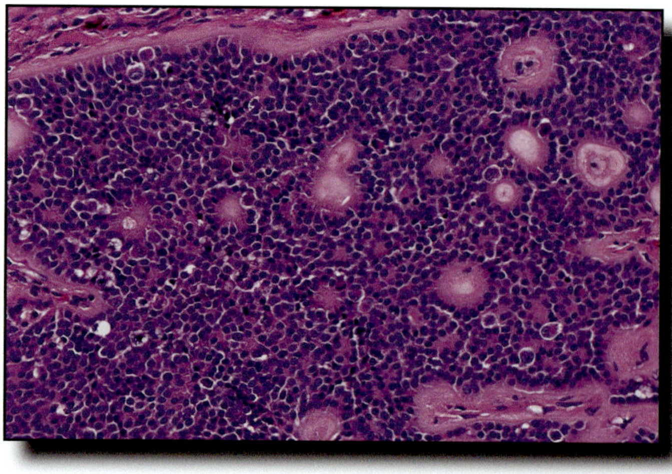

Figure 1.41 Corn Snake, *Pantherophis guttatus.* Colubridae. Granulosa cell tumor. Tightly packed cells in a granulosa cell tumor frequently form Call-Exner bodies in which a rosette arrangement of granulosa cells forms a microcavity with luminal, hyaline, eosinophilic material. H&E stain.

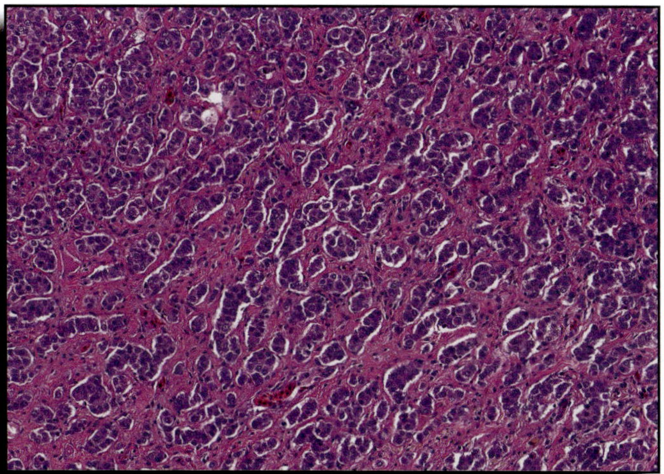

Figure 1.42 Shingleback Skink, *Tiliqua rugosa*. Scincidae. Ovarian carcinoma. Replacing much of the ovary are packets and occasional small tubules of neoplastic cells embedded in a dense, scirrhous stroma. H&E stain.

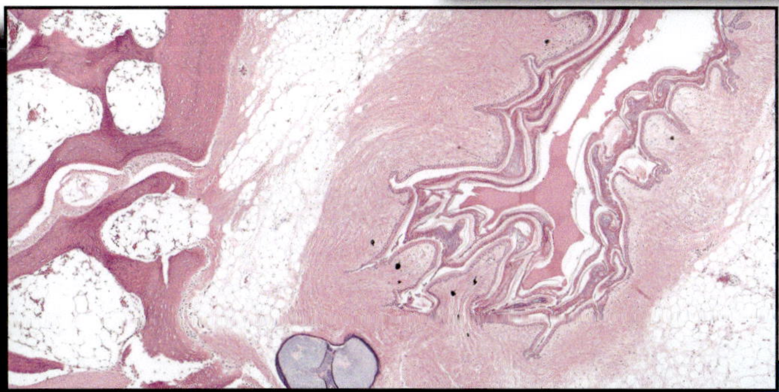

Figure 1.43 Inland Bearded Dragon, *Pogona vitticeps.* Agamidae. Ovarian teratoma. The neoplasm effacing the ovary is composed of tissues derived from all three germ layers, including ectoderm (e.g., squamous epithelium), endoderm (e.g., ciliated respiratory epithelium; not pictured), and mesoderm (e.g., bone). H&E stain. (Courtesy of J. Bernard.)

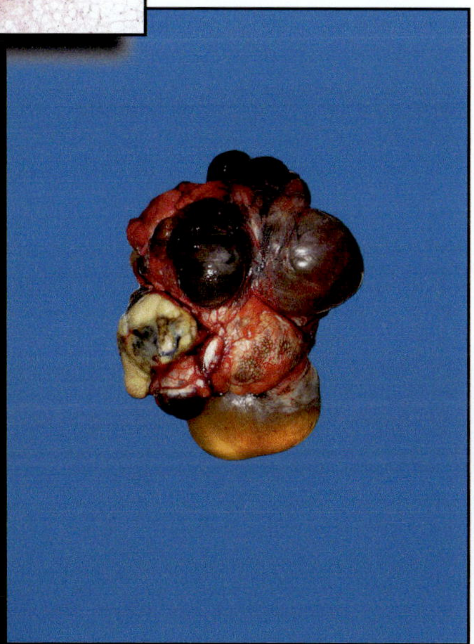

Figure 1.44 Inland Bearded Dragon, *Pogona vitticeps*. Agamidae. Ovarian teratoma. The ovary is effaced by a multicystic, multilobular neoplasm that was histologically confirmed to be a teratoma. (Courtesy of J. Bernard.)

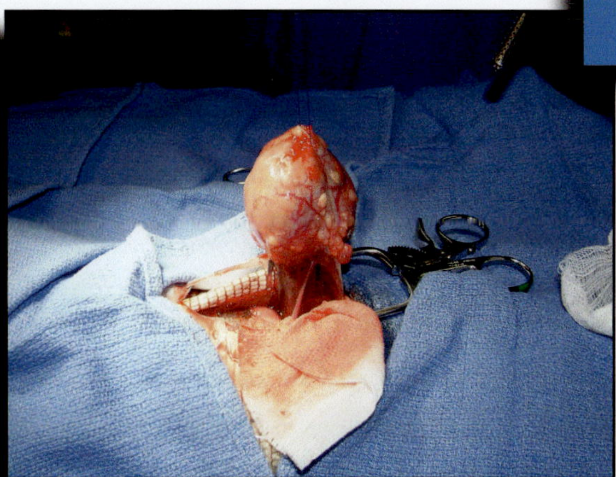

Figure 1.45 Savannah Monitor, *Varanus exanthematicus*. Varanidae. Ovarian stromal sarcoma. Intra-operative photograph of a retracted ovary that is expanded by a solid, pale tan mass confirmed histologically as an ovarian stromal sarcoma. (Courtesy of E. Jacobson.)

Figure 1.46 Boa. Boidae (genus/species not provided). Oviductal adenocarcinoma. The wall of the oviduct is markedly expanded by numerous packets and small tubules of neoplastic epithelial cells. The oviduct epithelium is ulcerated and there is frequent hemorrhage/fibrin throughout the neoplasm (not shown). H&E stain.

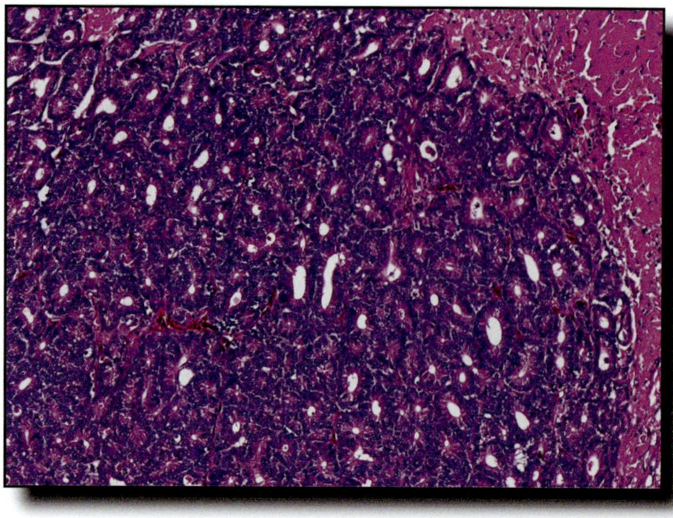

Figure 1.47 Eastern Indigo Snake, *Drymarchon couperi.* Colubridae. Oviductal leiomyosarcoma. The wall of the oviduct is markedly expanded by intersecting streams of neoplastic smooth muscle cells. H&E stain.

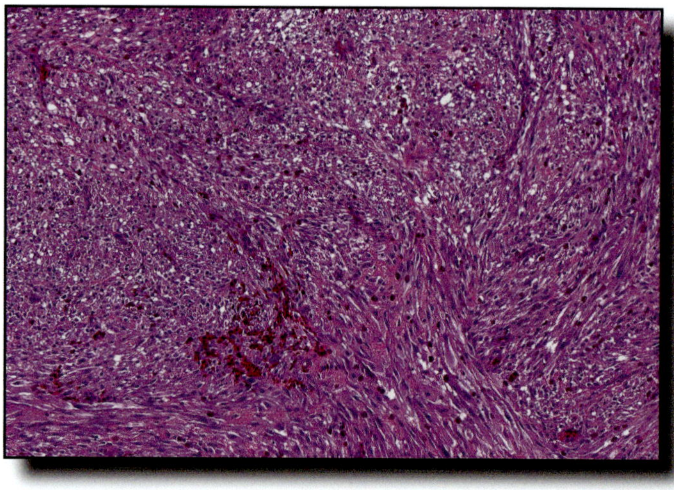

Figure 1.48 Tokay Gecko, *Gekko gecko.* Gekkonidae I. Interstitial cell tumor (Leydig cell tumor). The testicular interstitium is expanded by a neoplastic proliferation of polygonal cells with moderate to large amounts of slightly vacuolated, eosinophilic cytoplasm and oval nuclei with coarse chromatin. Neoplastic cells closely resemble interstitial cells (Leydig cells). Seminiferous tubules are at the margin of the neoplasm. H&E stain. (Courtesy of A. Cartoceti.)

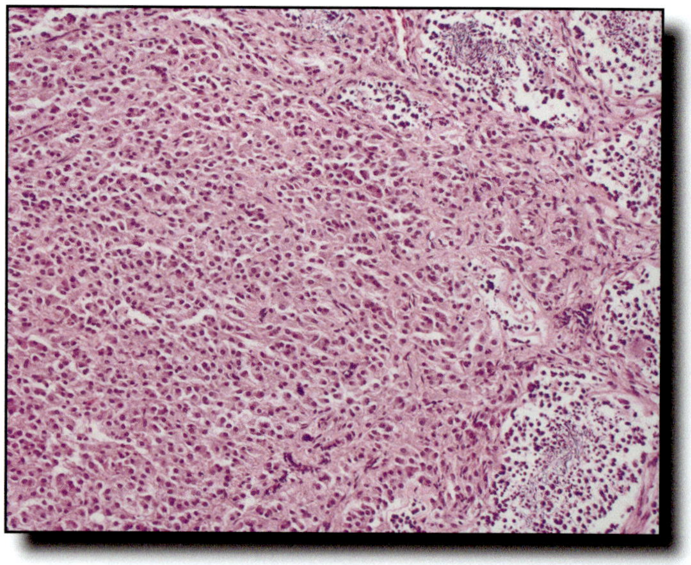

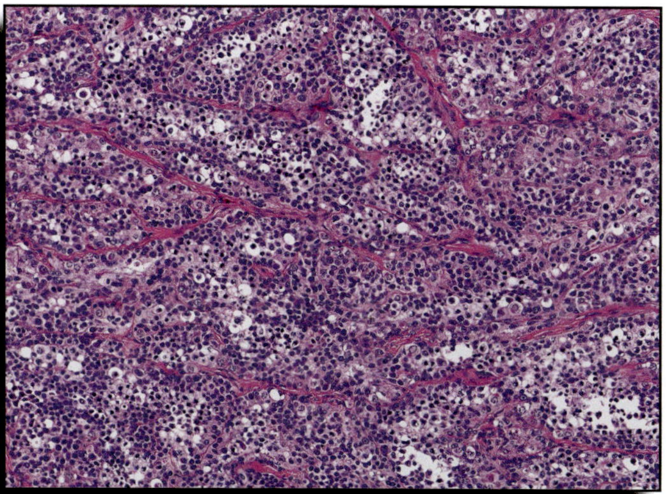

Figure 1.49 Black Speckled Palm Pit Viper, *Bothriechis nigroviridis*. Viperidae. Seminoma. The testis is completely replaced by tubules occluded with round cells that have minimal cytoplasm and rarely have spermatogenic maturation. H&E stain.

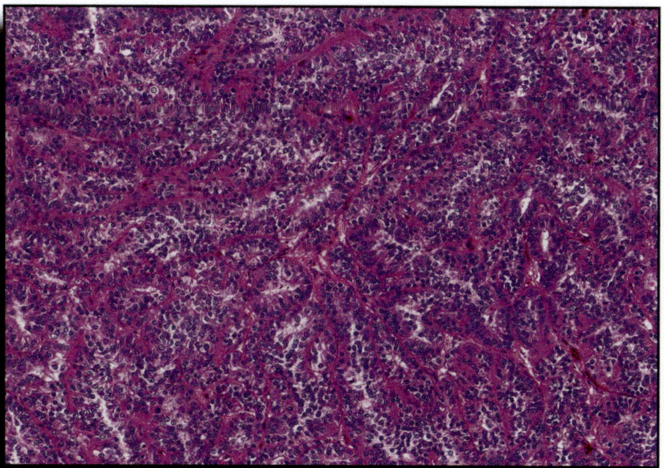

Figure 1.50 Inland Bearded Dragon, *Pogona vitticeps*. Agamidae. Sertoli cell tumor. A neoplasm replacing a large section of the testis is composed of well-formed tubules lined by a single layer of well-differentiated cells resembling Sertoli cells. H&E stain.

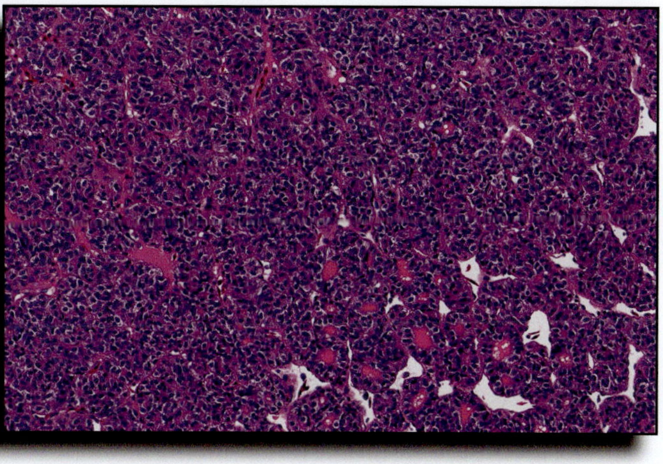

Figure 1.51 Mexican Beaded Lizard, *Heloderma horridum*. Helodermatidae. Thyroid adenocarcinoma. The thyroid is regionally expanded by an invasive neoplasm composed of packets of neoplastic epithelial cells resembling thyroid follicular cells. These cells occasionally form small follicles with central colloid. H&E stain.

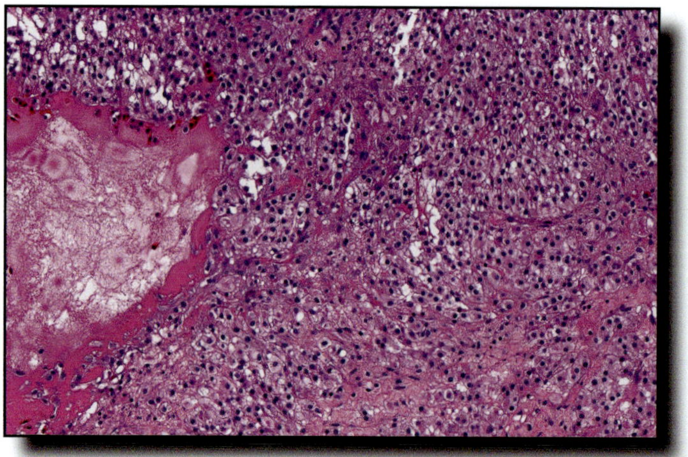

Figure 1.52 Iguana, *Iguana* spp. Iguanidae. Interrenal cell carcinoma. Neoplastic cells efface the adrenal gland, form packets, are highly vacuolated, and have mild atypia. There is abundant necrosis and hemorrhage in the neoplasm, as well as local invasion, consistent with malignancy (i.e., carcinoma). H&E stain.

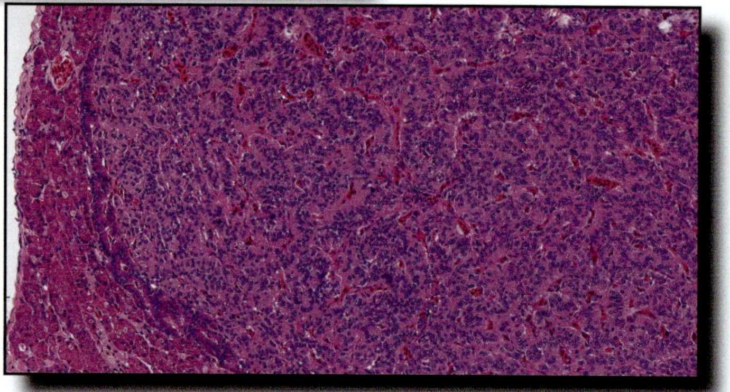

Figure 1.53 Taiwan Beauty Snake, *Elaphe taeniura*. Colubridae. Islet adenoma. Exocrine pancreas is regionally compressed by a proliferation of cells resembling well-differentiated islet cells. H&E stain.

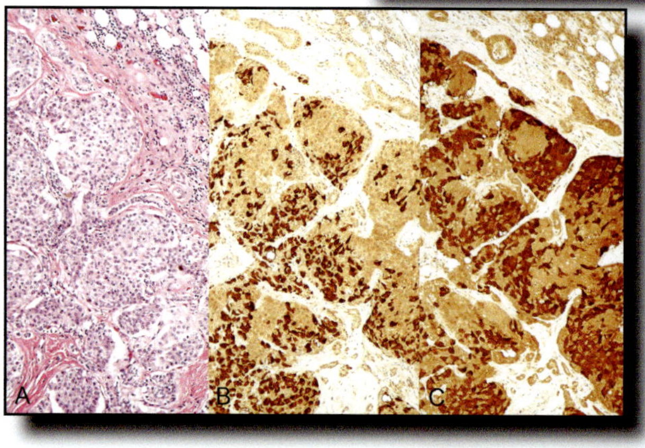

Figure 1.54 Komodo Dragon, *Varanus komodoensis*. Varanidae. Multihormonal islet cell tumor. A metastatic focus of neoplastic islet cells is in the spleen (A; H&E stain) and is immunoreactive to somatostatin (B) and glucagon (C). (Courtesy of M. Kuipel.)

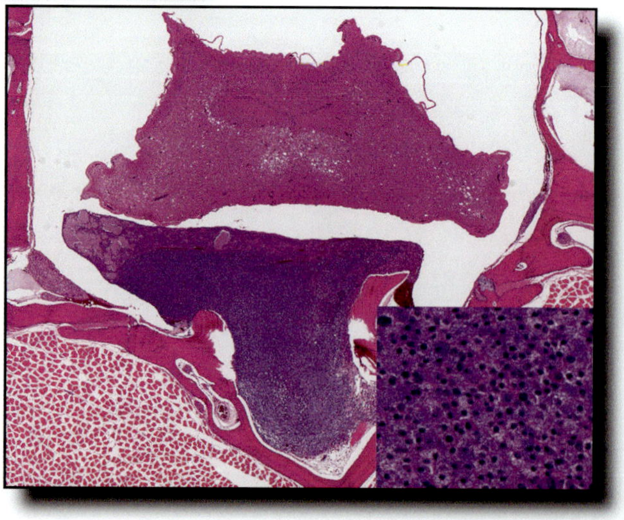

Figure 1.55 Prairie Rattlesnake, *Crotalus viridus*. Viperidae. Pituitary adenoma. The pituitary is asymmetrically and markedly expanded by a well-demarcated neoplasm that compresses the brain dorsally. Neoplastic cells form packets separated by delicate vascular stroma and have abundant, basophilic cytoplasmic granules (inset). H&E stain.

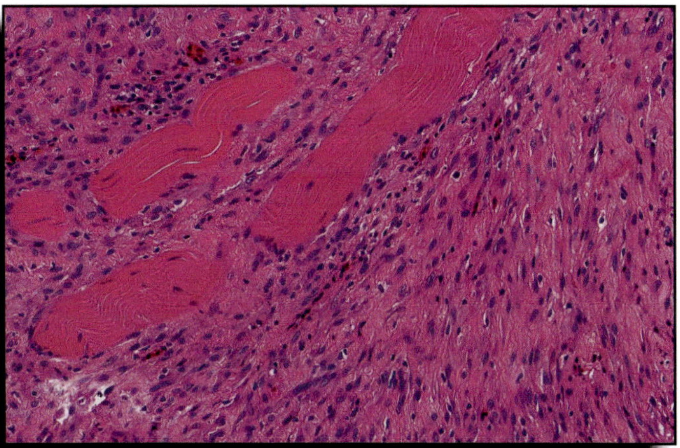

Figure 1.56 Hopi Rattlesnake, *Crotalus viridus nuntius.* Viperidae. Rhabdomyosarcoma. A mesenchymal neoplasm arises from and effaces skeletal muscle. Rowing nuclei, which are multifocally present here, are a characteristic feature of rhabdomyosarcoma, but are not always present. H&E stain.

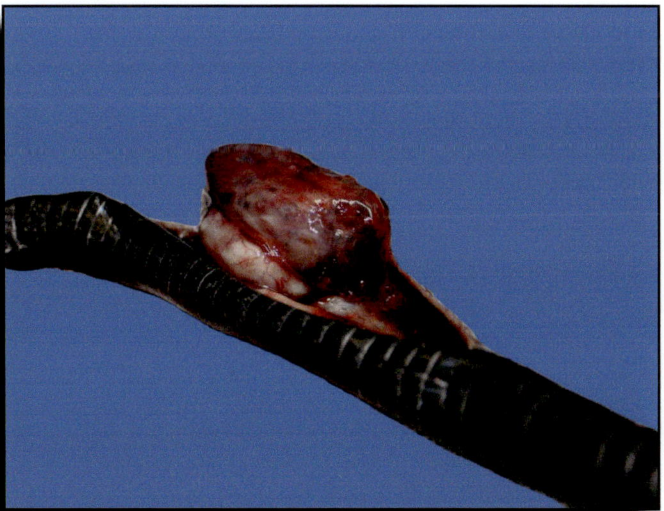

Figure 1.57 Garter Snake, *Thamnophis* spp. Colubridae. Osteosarcoma. Gross photograph showing osteosarcoma arising from the body wall/ribs. (Courtesy of B. Groves.)

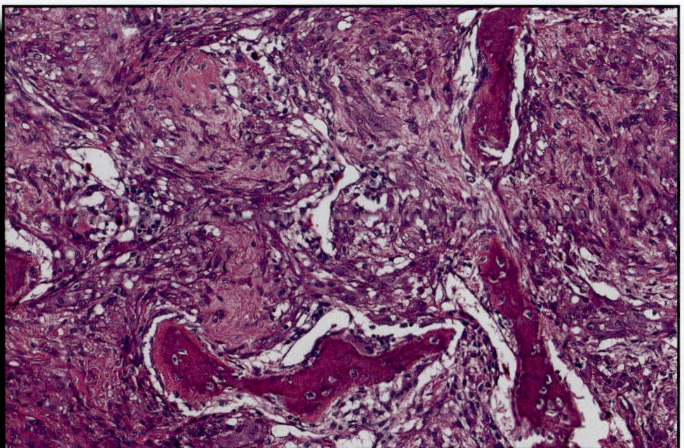

Figure 1.58 Garter Snake, *Thamnophis* spp. Colubridae. Osteosarcoma. Neoplastic spindle cells are associated with osteoid production. Neoplasm-associated osteoid frequently has osteocytes with lacunae. H&E stain.

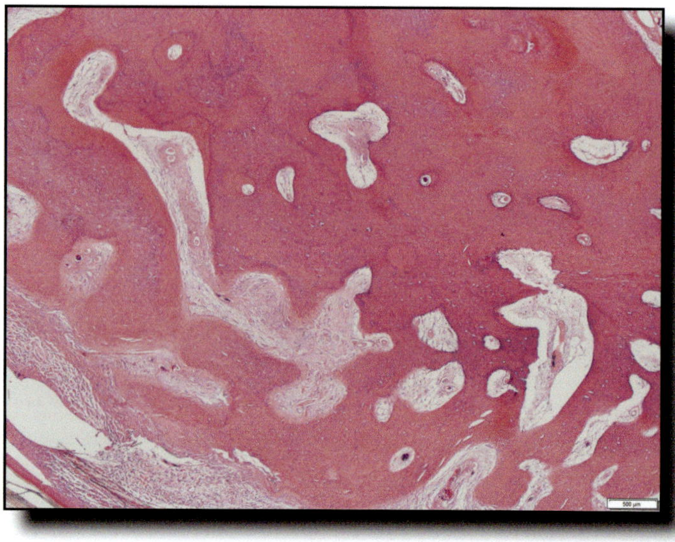

Figure 1.59 Dwarf Caiman, *Paleosuchus palpebrosus*. Alligatoridae. Osteoma. A hard mass arising from the forelimb is histologically composed mostly of woven bone with lacunae encircling osteocytes, cell-poor matrix, and peripheral periosteum. H&E stain.

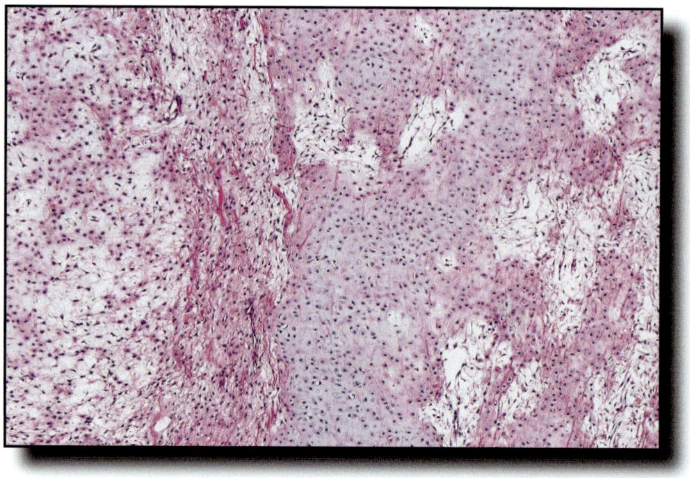

Figure 1.60 Spiny-Tailed Monitor, *Varanus acanthurus*. Varanidae. Vertebral chondrosarcoma. A neoplasm arising from the vertebral body is composed of streams of loosely aggregated neoplastic spindle cells interspersed with abundant chondroid matrix, within which are lacunae encircling chondrocytes. H&E stain.

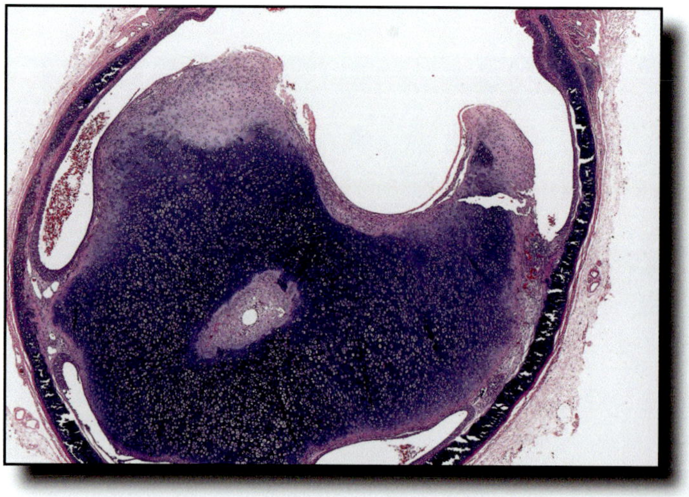

Figure 1.61 Ball Python, *Python regius*. Pythonidae. Tracheal chondroma. Arising from the tracheal cartilage, there is an exophytic mass composed largely of mature-appearing cartilage that subtotally occludes the tracheal lumen. H&E stain.

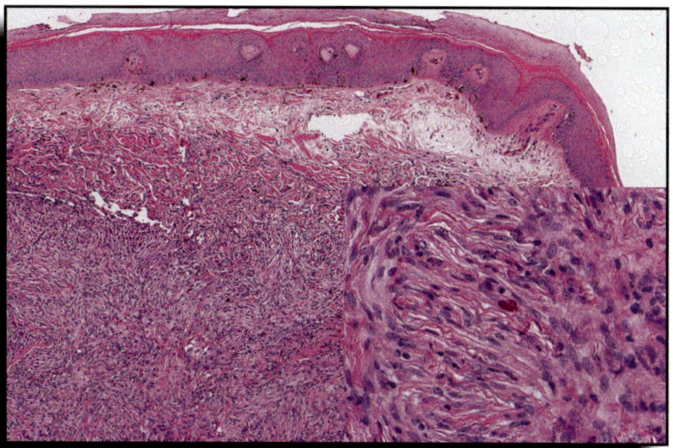

Figure 1.62 African Shield-Nosed Cobra, *Aspidelaps scutatus*. Elapidae. Fibrosarcoma. The dermis is expanded by an infiltrative neoplasm composed of intersecting bundles of anaplastic spindle cells (inset) interspersed with some collagenous matrix. H&E stain.

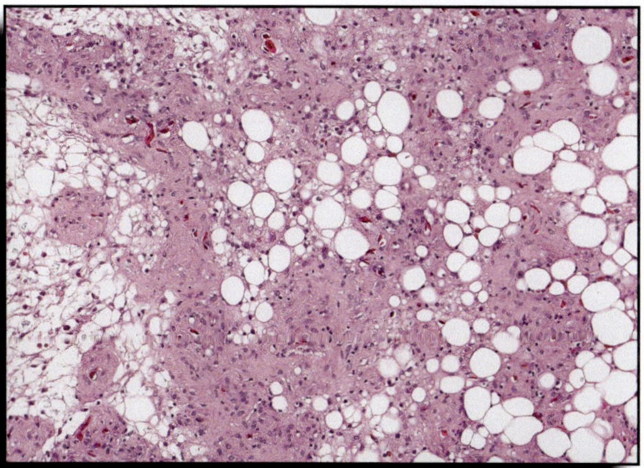

Figure 1.63 Eastern Indigo Snake, *Drymarchon couperi*. Colubridae. Liposarcoma. Arising from mature adipose tissue is an infiltrative neoplasm composed of mesenchymal cells that frequently have distinct, cytoplasmic vacuoles of varying sizes containing clear space (lipid vacuoles) and displaced but plump nuclei. H&E stain.

Figure 1.64 King Cobra, *Ophiophagus hannah*. Elapidae. Lymphoma. Multiple, poorly demarcated foci of scale thickening and dysplasia are present on the body wall due to cutaneous lymphoma. (Courtesy of J. D'Agostino.)

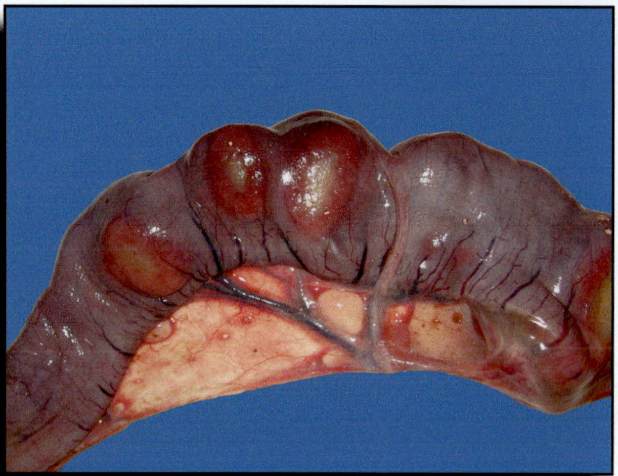

Figure 1.65 African Spurred Tortoise, *Centrochelys sulcata*. Testudinidae. Lymphoma. Multiple discrete, pale tan masses expand the intestinal wall. These masses were confirmed histologically as lymphoma. (Courtesy of E. Baitchman.)

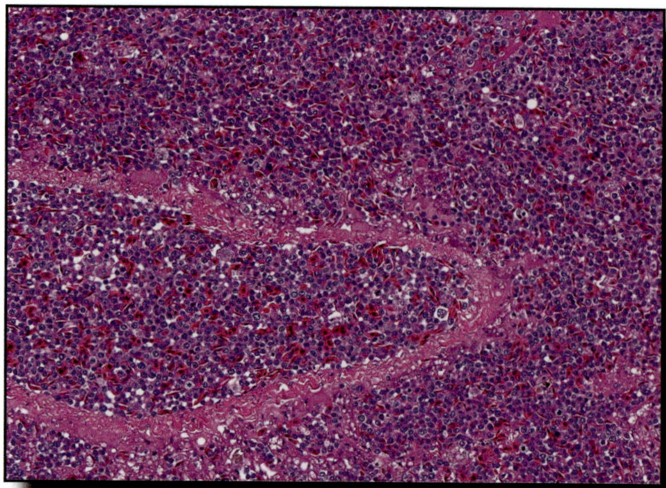

Figure 1.66 Boa Constrictor, *Boa constrictor.* Boidae. Lymphoma with leukemia. In a large mass of neoplastic lymphocytes, blood vessels also contain neoplastic lymphocytes, suggestive of lymphoma with leukemia. H&E stain.

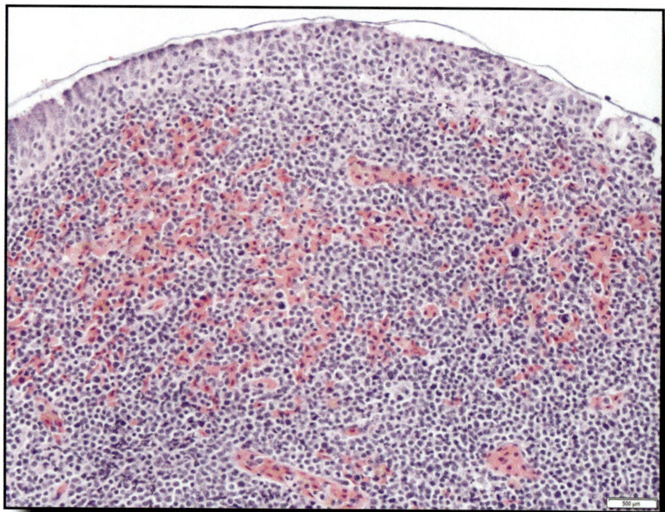

Figure 1.67 Boa Constrictor, *Boa constrictor.* Boidae. T cell lymphoma. The submucosa and mucosa of the intestine are infiltrated by sheets of neoplastic lymphocytes, which were identified as T cell origin with CD3 immunohistochemical stain (see **Figure 1.68**). H&E stain.

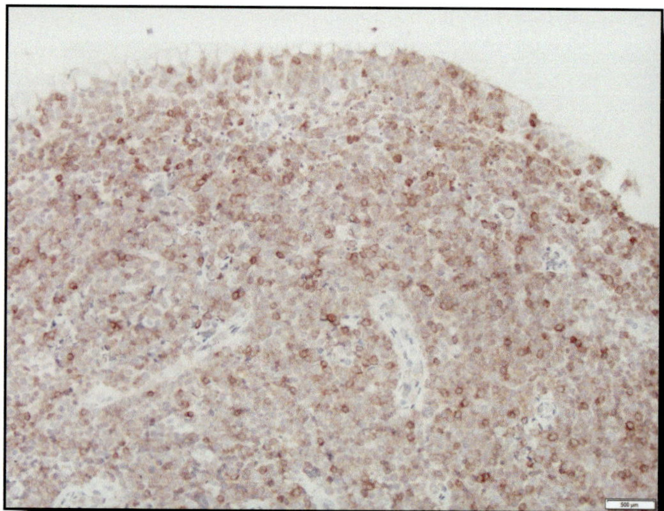

Figure 1.68 Boa Constrictor, *Boa constrictor.* Boidae. T cell lymphoma. CD3 immunohistochemical stain reveals strong, diffuse, membranous immunoreactivity of neoplastic lymphocytes, consistent with T cell lymphoma.

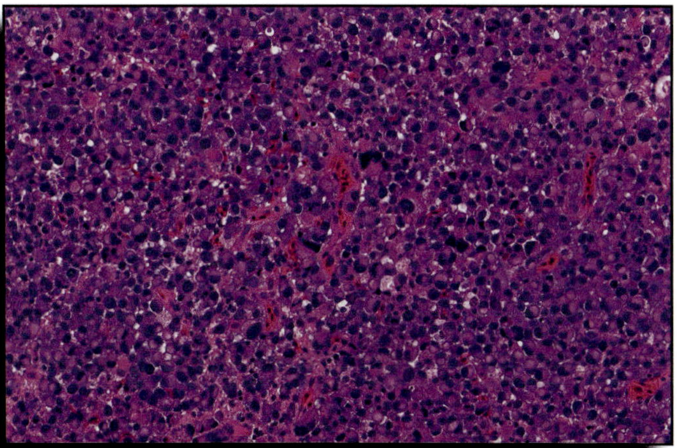

Figure 1.69 Corn Snake, *Pantherophis guttatus*. Colubridae. Plasmacytoid round cell neoplasia. A large round cell neoplasm is composed of sheets of neoplastic round cells that have plasmacytoid morphology, including an eccentric nucleus and perinuclear eosinophilia, reminiscent of a Golgi body. Anisokaryosis is marked and mitotic figures are common. H&E stain.

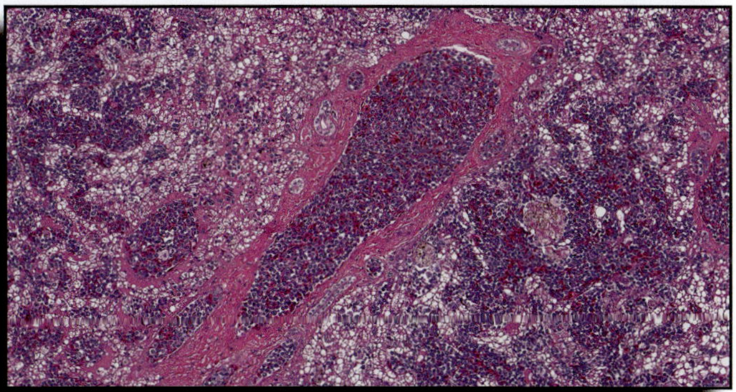

Figure 1.70 Carpet Python, *Morelia spilota*. Pythonidae. Lymphoid leukemia. Neoplastic round cells with lymphoid morphology expand vascular spaces, including hepatic sinusoids and blood vessels. H&E stain.

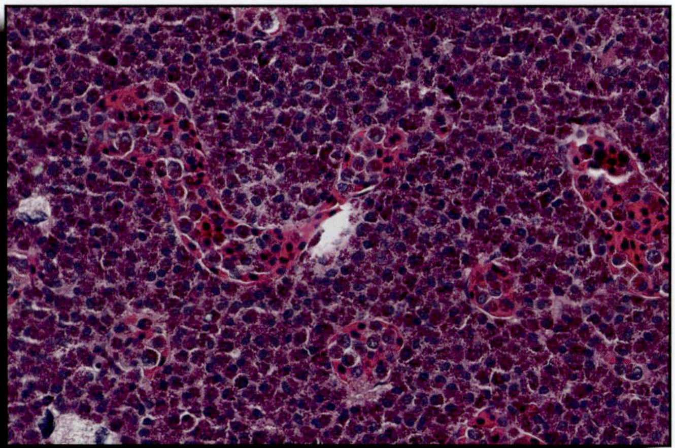

Figure 1.71 African Pancake Tortoise, *Malacochersus tornieri*. Testudinidae. Myelogenous leukemia. Splenic sinusoids and small-caliber vessels are markedly and diffusely distended by neoplastic round cells with myelogenous morphology, specifically eccentric nuclei and abundant, eosinophilic, cytoplasmic granules. H&E stain.

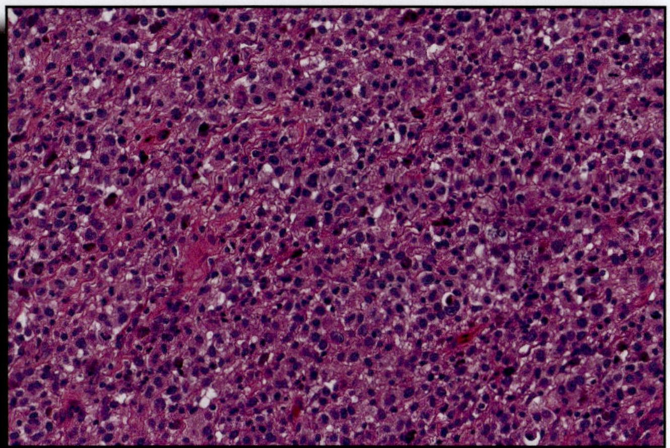

Figure 1.72 Desert Rosy Boa, *Lichanura trivirgata*. Boidae. Histiocytic sarcoma. A large mass is composed of sheets of neoplastic round cells that have a moderate to large amount of eosinophilic cytoplasm, and round to reniform nuclei with occasional binucleation. H&E stain.

Figure 1.73 Iguana, *Iguana* sp. Iguanidae. Mast cell tumor. The dermis is expanded by sheets of neoplastic mast cells that have a moderate amount of granular eosinophilic cytoplasm (inset). H&E stain.

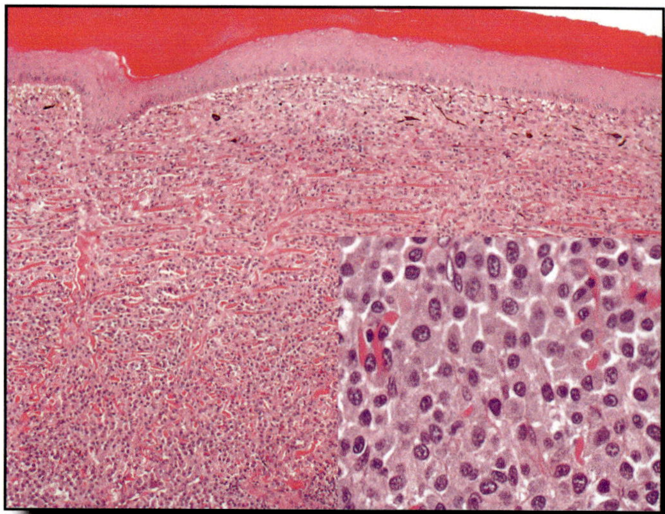

Figure 1.74 Inland Bearded Dragon, *Pogona vitticeps*. Agamidae. Peripheral nerve sheath tumor. (A) Antoni A pattern has closely packed spindle cells in a herringbone pattern with some nuclear bands. (B) Antoni B pattern has loosely packed cells in a myxoid stroma. H&E stain.

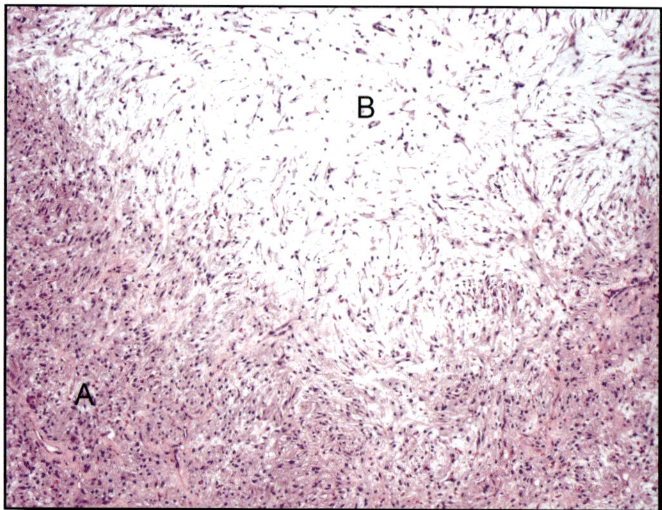

Figure 1.75 Green Turtle, *Chelonia mydas*. Cheloniidae. Pulmonary fibroma. The subepithelial interstitial tissue of the lung is markedly expanded by an orderly proliferation collagen and fibroblasts. H&E stain.

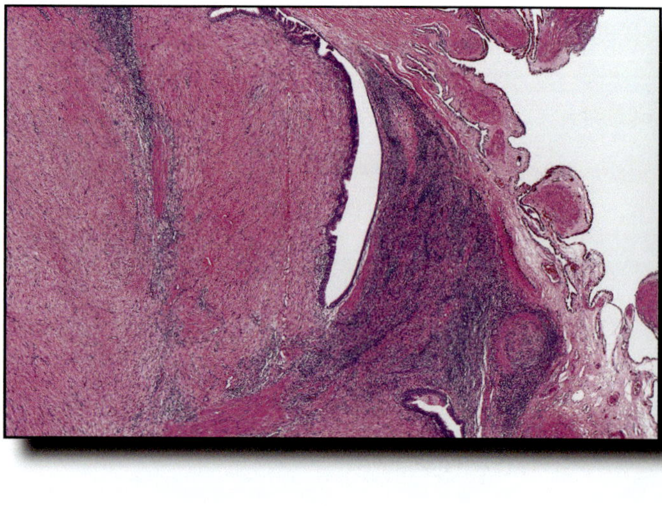

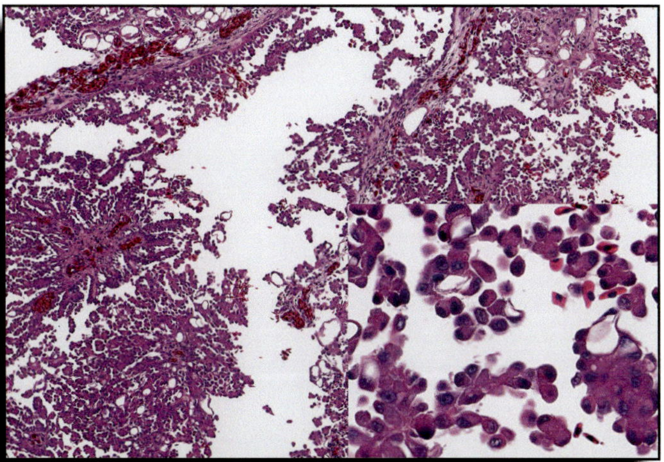

Figure 1.76 Dunn's Hognose Viper, *Porthidium dunni*. Viperidae. Mesothelioma. The mesothelial surface around the heart is expanded by branching papillary projections that have a fibrovascular core and are lined by neoplastic polygonal cells with mild cellular atypia, consistent with mesothelioma. Inset: high magnification of neoplastic cells. H&E stain.

Figure 1.77 Eastern Diamodback Rattlesnake, *Crotalus adamanteus*. Viperidae. Hemangiosarcoma. A large bulge is on the side of the face, posterior to the globe due to cutaneous hemangiosarcoma. (Courtesy of R. Junge.)

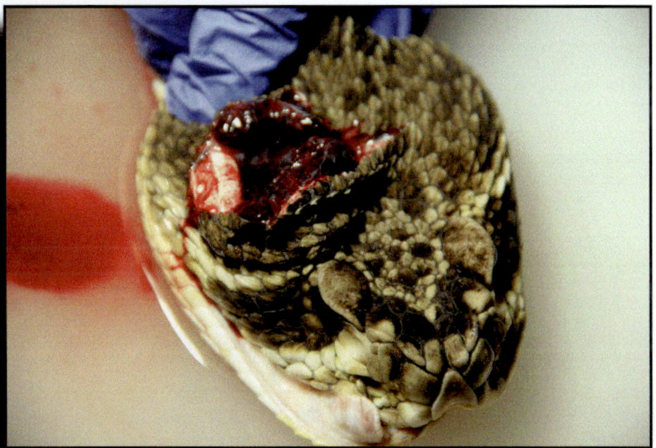

Figure 1.78 Eastern Diamondback Rattlesnake, *Crotalus adamanteus*. Viperidae. Hemangiosarcoma. Upon incising the mass from the snake in **Figure 1.77**, there is moderate hemorrhage centered on a fleshy, dark red neoplasm that was confirmed histologically as cutaneous hemangiosarcoma. (Courtesy of R. Junge.)

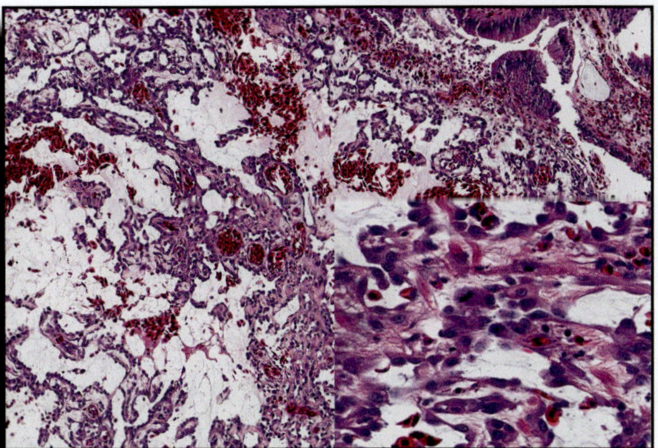

Figure 1.79 Green Iguana, *Iguana iguana*. Iguanidae. Hemangiosarcoma. The intestinal submucosa is expanded by an infiltrative neoplasm composed of abundant, irregular, vascular channels that contain blood and are lined by neoplastic spindle cells supported by a fibrovascular stroma. Neoplastic cells have moderate anisocytosis and anisokaryosis, and occasional mitotic figures. H&E stain.

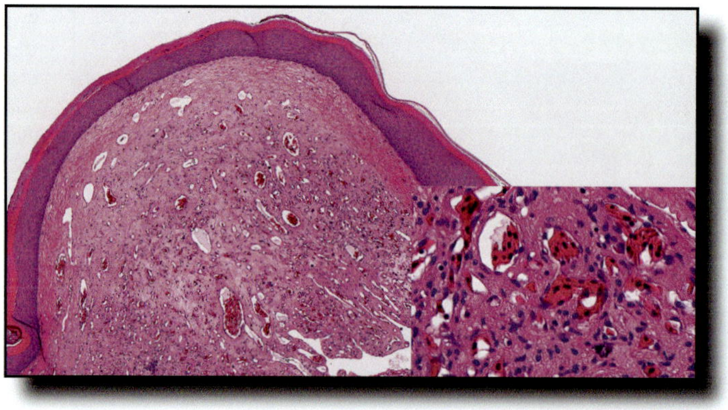

Figure 1.80 Boa Constrictor, *Boa constrictor.* Boidae. Hemangioma. The dermis is focally expanded by a proliferation of well-differentiated blood vessels embedded in a dense collagenous stroma. Blood vessels are lined by a single layer of flattened spindle cells, consistent with hemangioma. Inset: higher magnification of neoplastic cells. H&E stain.

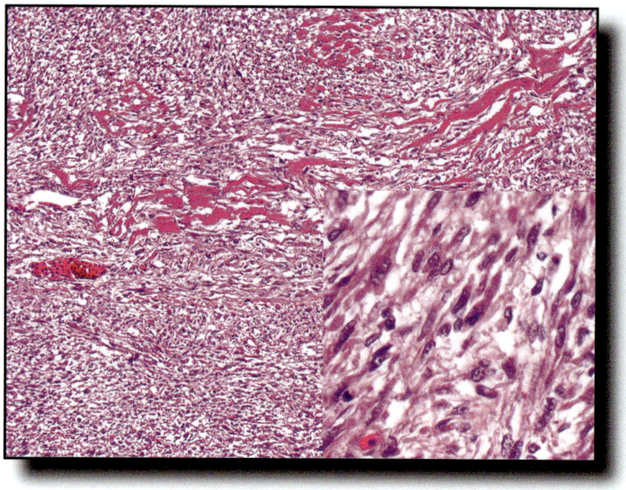

Figure 1.81 Kingsnake, *Lampropeltis* sp. Colubridae. Cardiac rhabdomyosarcoma. Interlacing streams of neoplastic spindle cells that rarely form strap cells efface the majority of this section of heart. Mitoses are present (inset). H&E stain.

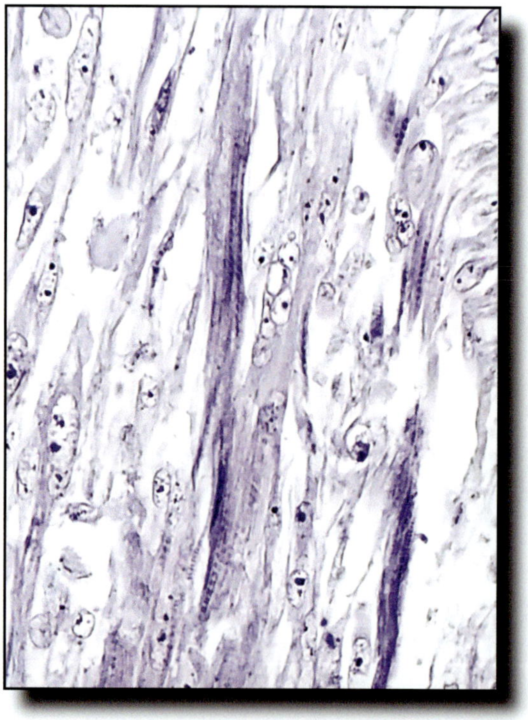

Figure 1.82 Kingsnake, *Lampropeltis* sp. Colubridae. Cardiac rhabdomyosarcoma. PTAH stain from the neoplasm pictured in **Figure 1.81** demonstrates cross striations.

Figure 1.83 Inland Bearded Dragon, *Pogona vitticeps*. Agamidae. Tear gland adenocarcinoma. The anterior aspect of the palpebrae is markedly swollen, causing distortion and anterior displacement of the palpebral fissure with conjunctival edema. Upon dissection (not shown), a mass was arising from the tear gland that was confirmed histologically as tear gland adenocarcinoma. (Courtesy of W. Kopit.)

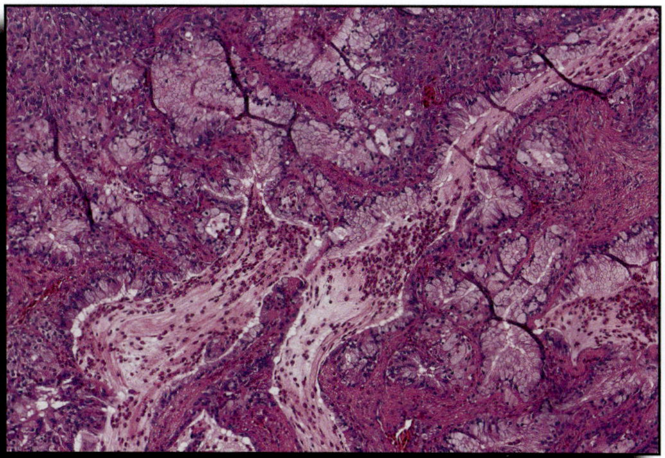

Figure 1.84 Bearded Dragon, *Pogona vitticeps.* Agamidae. Tear gland adenocarcinoma. This neoplasm arising from the periorbital tissue (as seen in **Figure 1.83**) is composed of abundant packets and tubules of neoplastic cells with frequent mucous production. Tubular lumina frequently contain secretory material admixed with granulocytes. H&E stain.

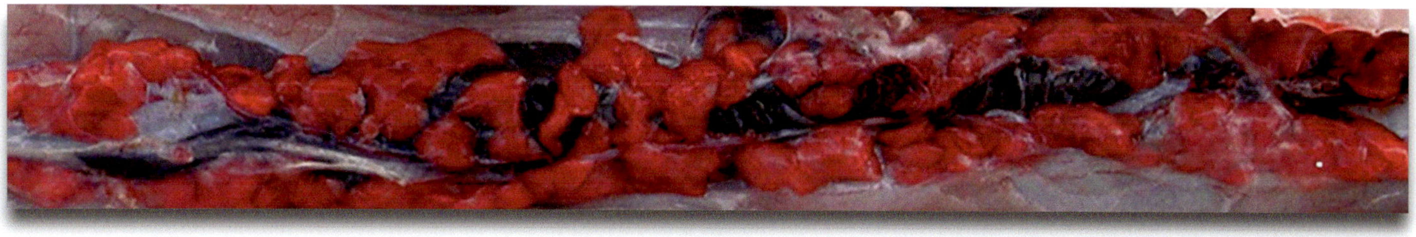

2

NUTRITIONAL AND METABOLIC DISEASES

CARLES JUAN-SALLÉS AND THOMAS H. BOYER

Contents

2.1 Introduction

Despite many advances over the last 20 years, nutritional disease remains a leading cause of morbidity and mortality in reptiles. This is because reptile nutrition is not as well studied as domestic animal nutrition, poor nutritional advice is widely disseminated, the nutritional requirements of relatively few species have been studied and identified, and their foods in the wild, for the most part, are commercially unavailable. Complete manufactured diets available for domestic species are just starting to be used for reptiles; some are good and some are not. Thus the reptile pathologist may confirm or diagnose many more nutritional diseases from deficiencies, derangements, and excesses in reptiles than they would in domestic and exotic mammals and birds. For a current review of reptile nutrition, see Boyer and Scott (2019a).

2.2 Obesity

Excessive caloric intake is a frequent problem in captive reptiles and may lead to obesity and lipid storage or deposition disorders, particularly in carnivorous species. Because the caloric requirements of reptiles in captivity are often not well known, regular monitoring of body weight may help in assessing the adequacy of caloric intake (Mans and Braun, 2014). A number of factors related to the diet, affected species, metabolic/physiologic status, and husbandry practices may contribute to the development of obesity. Some species, especially the large sedentary carnivorous reptiles, may be predisposed to obesity in captivity; examples include the large boids, Monitor Lizards (*Varanus* spp.), crocodilians, Tegu Lizards (*Tupinambis* spp.), and snapping turtles (*Chelydra* and *Macrochelys* spp.). Bearded Dragons (*Pogona vitticeps*) are frequently obese in captivity. Food intake should be closely monitored in their captive management (Donoghue, 2006). Excessive feeding frequency and caloric intake are major factors causing obesity in reptiles. A diet based on rats and mice that had been fed dog food can lead to obesity in reptiles (Rossi, 2006). Insectivorous reptiles can also become obese, particularly with certain husbandry practices and diet-related factors. For example, larval insects commonly fed in captivity such as Yellow Mealworms (*Tenebrio molitor*), Superworms (*Zophobas morio*) and Waxworms (Greater Wax Moth; *Galleria mellonella*) contain much more fat than the adult beetles (Oonincx and Dierenfeld, 2012) or House Crickets (*Acheta domesticus*) (Oonincx and Dierenfeld, 2012), and diets rich in such larval insects may predispose to obesity. Thus, for insectivores, feeding a wide variety of insects and not relying on a just few species is essential (Boyer and Scott, 2019a).

Enclosure design, such as larger enclosures, or more vertical enclosures for climbing species, and some husbandry practices such as the way in which food is offered (whether or not it promotes physical activity for food procurement) may reduce the caloric balance and help prevent obesity. An abnormally long photoperiod in species usually inactive/anorectic during the winter can cause obesity (Rossi, 2006). Species that normally hibernate, or estivate, should do so in captivity to decrease excess fat reserves. Consideration of metabolic/physiologic status is necessary for adequate control of the caloric balance of captive reptiles. Breeding activity, preparation for hibernation, and growth require a higher caloric intake.

Obesity may affect the general and reproductive fitness of captive reptiles as it does in other species. Lipid storage or deposition disorders, particularly hepatic or hepatorenal lipidosis, and more sporadically others such as atheromatosis, may occur as a result of obesity. Obese reptiles have hypertrophic adipose tissue, which can be noted clinically during examination based on external appearance (accumulation of markedly excessive subcutaneous fat or coelomic distension) (**Figures 2.1–2.2**) or at necropsy, particularly in the subcutis and intracoelomic fat bodies (**Figure 2.2**). Whitish foci or small nodular lesions corresponding to aggregates of well-differentiated adipocytes (lipomatosis or steatosis) may occur in the liver, kidney, heart, thyroid, and other organs of obese reptiles (**Figures 2.3–2.4**). Lipid storage (lipidosis) in obese reptiles is seen mainly within the cytoplasm of hepatocytes, but also in other tissues such as the renal tubular epithelium (see Lipidosis).

2.3 Lipidosis

The reversible accumulation of lipid occurs mainly in the liver (hepatocytes) of reptiles and can result from physiological processes, nutritional imbalances, impairment of the metabolic capacity of hepatocytes, or combinations thereof. Lipid accumulation in the liver can occur normally in reptiles preparing for brumation or folliculogenesis and therefore should be interpreted conservatively; some authors use the terminology "fatty change" instead of hepatic lipidosis for this physiological lipid storage (Hernandez-Divers and Cooper, 2006). However, lipid storage is dependent on a number of additional factors including ambient temperature, photoperiod, food availability, mating, or spermatogenesis (Xiang and Peichao, 1990). As a rule, females have the highest lipid contents in their livers just before brumation, while it is usually low at the time of maximum vitellogenic activity (Xiang and Peichao, 1990). Hepatic lipid may vary in males depending on the reproductive status as well (Lacy et al., 2002). Therefore, it is essential to know the physiological status, gender, and basic husbandry data for a proper evaluation and interpretation of hepatic lipidosis.

A number of disease processes can cause hepatic lipidosis in reptiles. Obesity is a major cause, particularly in crocodiles (**Figure 2.5**), large boids, lizards, and chelonians, or may exacerbate lipid storage due to other conditions. Lipid mobilization due to anorexia, inanition, or underlying disease is another common cause of hepatic lipidosis in reptiles, particularly in obese animals, as it can surpass the metabolic capacity of hepatocytes. Hepatotoxins, hypoxia resulting from respiratory or cardiac disease and severe anemia, hepatic infections, biliary cirrhosis, and any other process causing an impairment or failure of hepatocellular metabolism can lead to hepatic lipidosis. With severe lipid mobilization, lipidosis can also occur in other organs/cells, particularly the renal tubular epithelium, but also cardiomyocytes, enterocytes, and cells in the red pulp of the spleen. Severe hepatic or hepatorenal lipidosis may lead to metabolic derangements that can potentially cause clinical disease or death, especially if the patient is chronically anorexic.

Lipidosis can be diagnosed by signalment (hyporexia or anorexia in an animal that was eating well, lethargy), history (overfeeding [bearded dragons], fat-rich diet, overfeeding starved mealworms [leopard geckos], or poor diet, such as grocery store fruits and vegetables [tortoises]). Physical exam findings include obesity, lethargy, weakness (not lifting body, reluctance to move, head and legs easily extracted from shell) and edema as a late stage sign. Clinical pathology may reveal hypoalbuminemia, with or without anemia, hypercholesterolemia, hypertriglyceridemia, hyperglycemia, elevated fasting bile acids with normal levels of AST, ALT, and GGT (Dutra, 2014), enlarged hepatic silhouette in lateral radiographs (greater than half the dorsoventral space), normal bone density, and isoechoicity or hyperechoicity of liver compared to coelomic fat bodies. For accuracy, ultrasonic-guided hepatic aspirates are preferred, as fat bodies often envelope the caudolateral liver (particularly in obese reptiles). Hepatocytes represented in the smears will contain intracytoplasmic clear spherical vacuoles (**Figure 2.6**). Lipid-laden hepatocytes of reptiles seem particularly fragile and prone to lysis during aspiration or when smeared on a slide, and may take longer to fix to the slide, which should be considered in preparation and interpretation of the specimen. Extracellular lipid droplets may indicate lipid from lysed hepatocytes, or misguided aspiration of a fat body. The gold standard for diagnosis is hepatic endoscopic examination and biopsy; however, many owners are reluctant to pursue additional diagnostic confirmation in a patient with an already guarded to poor prognosis. Clinical signs noted in mammals, including vomiting and icterus, are usually not seen in reptiles with hepatic lipidosis.

Grossly, the liver is typically enlarged and has rounded edges. It may be pale, tan yellow to brown, but its color may vary depending on the amount of blood and melanin in the parenchyma at the time of assessment (**Figures 2.7–2.9**). The liver may be friable, ooze yellow brown fluid on cut surfaces, with or without clear vacuoles, and may float in formalin (normal liver sinks). The surface of the liver may be irregular in cases with fibrosis, biliary hyperplasia, or cirrhosis (**Figure 2.8**). The kidneys in reptiles with advanced hepatorenal lipidosis may also be pale. Microscopically, lipid accumulation within the cytoplasm of hepatocytes and other cells occurs as distinct clear vacuoles, which can peripherally displace and compress the nucleus (**Figures 2.10–2.11**); with severe swelling, vacuoles may not be distinct and it can be difficult to differentiate lipid accumulation from other forms of cytoplasmic vacuolation (e.g., edema associated with excessive glycogen accumulation, so-called "hydropic degeneration"). Differentiation can be accomplished by lipid stains (oil red O) on cryostat sections of formalin-fixed tissues or cytological preparations, as well as by electron microscopy. When severe and chronic, hepatic lipidosis may be complicated or associated with fibrosis and proliferation of well-differentiated bile ducts, leading to cirrhosis (**Figures 2.12–2.13**). Hepatic lipidosis can also be a secondary event associated with cirrhosis; for example, Green Iguanas (*Iguana iguana*) with cirrhosis associated with evidence of obstructive disease of the biliary tree may develop hepatic lipidosis.

2.4 Xanthomatosis

Xanthomatosis is a lipid deposition disorder characterized by the accumulation of cholesterol crystals in tissues, with varying degrees of granulomatous inflammation. Diagnosis can be obtained by cytologic evaluation, biopsy, or at necropsy. Although the etiopathogenesis is not fully elucidated, xanthomatosis can be caused by or associated with high cholesterol diets or derangements in cholesterol and triglyceride metabolism that may cause hypercholesterolemia or hypertriglyceridemia, such as hypothyroidism, hepatic disease, diabetes, pancreatitis, or obesity. Lesions associated with chronic hemorrhage, such as trauma, infection, ruptured ovarian follicle, or neoplasia can also result in localized xanthogranulomatous inflammation (Carpenter et al., 1986; Garner et al., 1999; Dötsch et al., 2001; Kummrow et al., 2010; Baila-Rueda et al., 2014; Solak et al., 2015). Experimentally, the feeding of high cholesterol diets can cause cholesterol deposition in arteries of rabbits (Yanni, 2004), xanthomatosis in Japanese Quail (*Coturnix japonica*) (Hoekstra et al., 1998), and lipid keratopathy (corneal lipidosis or corneal arcus) in Cuban Tree Frogs (*Osteopilus septentrionalis*) (Shilton et al., 2001).

In reptiles, xanthomatosis has been reported in geckos including Round Island Geckos (*Phelsuma guentheri*) (Cooper et al., 1998), Northern Green Geckos (*Naultinus grayi*), Leaf Tailed Geckos (*Uroplatus* spp.) (Garner et al., 1999; Raiti and Garner, 2000), and especially female Leopard Geckos (*Eublepharis macularius*) (T. Boyer and M. Garner, personal observations), but also other lizards, such as the Great Plated Lizard (*Gerrhosaurus major*) (Gyimesi et al., 2002), Cuban Anoles (*Anolis e. equestris*) (Schmidt and Hubbard, 1987), and Green Water Dragons (*Physignathus cocincinus*) (Kummrow et al., 2010). Rare cases have been documented in snakes as well (Ryan and Whitney, 1980; Schmidt and Hubbard, 1987). Rubber boas may be especially prone to developing cephalic xanthomatosis with associated signs of central nervous system disease (M. Garner, personal observation). Xanthomatosis in reptiles occurs most frequently in females (Cooper et al., 1998; Garner et al., 1999; Gyimesi et al., 2002; Kummrow et al., 2010). Diets rich in cholesterol and active folliculogenesis or follicular degeneration, and resorption may contribute to the occurrence of xanthomatosis in reptiles (Garner et al., 1999; Kummrow et al., 2010). The disease is apparently uncommon in chelonians, although can occur in marine turtles fed fish high in lipid content (M. Garner, personal observation).

Localized presentations of xanthomatosis in the brain and ovary predominate, but multicentric disease has also been reported, particularly involving the brain, fat bodies, ovary, and/or lungs (Schmidt and Hubbard, 1987; Cooper et al., 1998; Garner et al., 1999; Raiti and Garner, 2000; Gyimesi et al., 2002; Kummrow et al., 2010). In the brain, the lateral and third ventricles are most frequently affected, but lesions can involve the meninges and spinal cord as well (Schmidt and Hubbard, 1987), and advanced cases may extend into and replace neuropil. Cutaneous xanthomatosis has been documented in a Gopher Snake (*Pituophis catenifer sayi*) (Ryan and Whitney, 1980). The clinical presentation is most frequently neurological (Garner et al., 1999; Gyimesi et al., 2002; Kummrow et al., 2010). Head tilt is common in lizards and rolling, or knotting is seen in snakes. Affected specimens may be either emaciated, in good body condition, or obese (Cooper et al., 1998; Garner et al., 1999; Gyimesi et al., 2002; Kummrow et al., 2010).

A diagnosis of xanthomatosis can be obtained in some cases by cytological examination. Cytologically, cholesterol crystals are unstained rhomboid structures that may pile and be surrounded or associated with varying numbers of macrophages and multinucleate giant cells, which may contain clear intracytoplasmic vacuoles (**Figures 2.14–2.15**) (Campbell and Ellis, 2007).

Grossly, distinct masses may be observed in affected organs (**Figures 2.16–2.17**) but in some cases are not visible. Severe intracoelomic deposits may obscure organ identification. Histological evaluation of biopsy or necropsy tissues is generally diagnostic for xanthomatosis and may also reveal evidence of underlying diseases/lesions. Accumulations of stacked clear acicular cholesterol clefts multifocally surrounded by macrophages and multinucleate giant cells are noted (**Figures 2.18–2.21**). Clear lipid vacuoles may be present as well, particularly within the cytoplasm of macrophages, which may appear foamy. Fibrosis may accompany these lesions in chronic cases. In the brain, xanthogranulomas can compress or infiltrate the surrounding neuropil, in which malacia may be noted (Kummrow et al., 2010). At times, depending on the quality of the histologic section, differentiating visceral gout from xanthomatosis can be difficult. Urate tophi generally have a stellate formation of crystals, whereas xanthoma crystals have a stacked appearance. (For additional information on xanthomatosis see also Chapter 3.)

2.5 Atheromatosis

Although common in birds, atheromatosis appears to be rare in reptiles (Schilliger et al., 2010). Sporadic cases have been reported in a Central Bearded Dragon (*Pogona vitticeps*) (Schilliger et al., 2010) and a Green Water Dragon (*Physignathus cocincinus*) (Kummrow et al., 2010). Whether atheromatosis in reptiles results from dietary lipid imbalances or other forms of arterial damage (e.g., infection, hypertension, or aging) is not known. Isolated atheromatous plaques or multifocal to diffuse thickening of arteries, particularly the aorta, may be noted grossly.

Grossly, affected arteries have plaque-like, nodular, or diffuse yellowish thickenings (**Figures 2.22–2.23**). Histologically, atheromatous plaques are comprised of clear acicular clefts (cholesterol clefts), deposits of amorphous eosinophilic material, vacuolated cells, fibrous tissue, and on occasion mineral deposits, metaplastic cartilage, or bone (**Figures 2.24–2.27**). Concurrent thrombosis of affected arteries and xanthomatosis may be noted (**Figure 2.24**). (For additional information on vascular disease, see also Chapter 9.)

2.6 Emaciation

Causes of emaciation in captive and wild reptiles include prolonged inappetence or anorexia, underlying disease, or husbandry practices and other factors causing insufficient caloric intake during an extended period of time (Donoghue, 2006; Romero and Wikelski, 2010). Emaciation associated with helminths and pentastomes has been seen in wild marine turtles and snakes, respectively. Cryptosporidiosis is well documented for causing severe emaciation in squamates, as do hookworms in snakes, and flagellated protozoan infections in snakes and carnivorous lizards (T. Boyer, personal observation). Even though reptiles adapt remarkably to fasting, prolonged starvation or reduced caloric intake will cause exacerbated catabolism, weight loss, increased susceptibility to disease, and mortality. Anorexia and insufficient caloric intake in captive reptiles are most commonly due to improper husbandry, but a number of diseases can also cause anorexia (**Figure 2.28**). Numerous husbandry practices and failures can lead to anorexia, inanition, and insufficient caloric intake. Failures may involve inappropriate temperature and humidity gradients, lighting (particularly artificial light of poor quality), use of inappropriate terrarium elements, feeding time (e.g., feeding during daylight hours in nocturnal species), accessibility to food, diet design, frequency of feeding, form of presentation of diet (e.g., dead instead of live prey, excessively large food items, feeding cold or frozen food, or food thawed slowly at room temperature), food item variety, density of animals and feeding places, group composition that facilitates aggressive and dominant behaviors, and stress (e.g., excessive handling, reptiles on display, improper husbandry). Most of the failures in reptile husbandry will cause a stress response and decreased food consumption (Donoghue, 2006) (see also Chapter 6). Adaptations to stressful conditions vary with species and this may impact their tendency to thrive. In general, captive born reptiles adapt better to captivity than wild caught reptiles. Individual variation in the stress response is also important. Starvation during an el Niño phenomenon was associated with mortality in Galápagos Marine Iguanas (*Amblyrhynchus cristatus*), in which affected lizards had a suboptimal post-stress suppression of corticosterone release (Romero and Wikelski, 2010).

Proper documentation on the biology, natural behavior and diet, and captive husbandry recommendations for individual species kept in captivity is essential to prevent anorexia, inanition, and insufficient caloric intake in reptiles. Many reptile species undergo ontogenetic changes in their natural diet with age, such as Caicos Rock Iguanas (*Cyclura carinata*), Gray's Monitor Lizard (*Varanus olivaceous*), and aquatic turtles in the family Emydidae (*Chrysemys*, *Graptemys*, *Pseudemys*, *Trachemys*), all of which switch from a carnivorous juvenile to a herbivorous adult (Barten, 2006; Avery et al., 1993), or crocodiles that switch from a hatchling's insectivorous, molluscivorous, and piscivorous diet to an adult's vertebrate diet (Webb et al., 1991; Lane, 2006). These changes must be considered in dietary design.

Anorexia or decreased food intake can also occur during physiological processes, particularly in preparation for brumation in chelonians and reproductive activity (e.g., gestation in viviparous snakes, and oviparous or ovoviviparous lizards), but this is transitory, may not be associated with clinical morbidity, and does not lead to emaciation. Male Kenyan Sand Boas (*Eryx colubrinus*) may only feed a few months of the year. Mature Ball Pythons (*Python regius*) often do not eat from November through April or May. Most temperate colubrids will cease feeding in the winter. Most snakes will stop feeding as they enter a cycle of ecdysis; then will feed following the shed.

The effects of emaciation or suboptimal nutritional status can be assessed clinically and histologically. Grossly, the adipose tissue is atrophied. Diverse distribution and concentration of adipose tissue among reptilian taxa account for a diverse gross appearance of this lesion. For example, atrophy of fat in lizards will involve most notably the caudoventral coelomic fat bodies and, particularly in geckos, the adipose tissue stores in the tail (**Figure 2.29**). A striking aspect of this lesion occurs in cachectic snakes, in which severely atrophied intracoelomic adipose tissue presents as small isolated red plaques orderly placed along serosal surfaces (**Figures 2.28, 2.30**). In some cases, atrophied adipose tissue may be gelatinous as well (serous atrophy of adipose tissue). Clear, straw-colored fluid may occasionally accumulate within the coelomic cavity (**Figures 2.31–2.32**). Atrophy may also involve the skeletal muscles (**Figures 2.29, 2.33**), liver (**Figure 2.31**), kidneys, spleen, and other organs, which may be smaller and usually darker.

Microscopically, adipocytes are small and contain either very small clear vacuoles or no visible lipid, with indistinct cytoplasmic borders or a small amount of eosinophilic cytoplasm arranged around a prominent and dense capillary network, the latter accounting for the pink macroscopic appearance of the fat body (**Figures 2.34–2.35**). Atrophy of adipocytes in the tail of geckos results in a thin tail which can be used in estimating nutritional status of the lizard. Transverse sections through the tail identify the atrophied adipose tissue (**Figures 2.36–2.37**). Varying amounts of mucin-rich edema may be present in the interstitium in cases of serous atrophy of adipose tissue (**Figures 2.38–2.40**). In cachectic reptiles, a number of cells and tissues can be atrophied, and this is most obvious in cords of hepatocytes (**Figures 2.41–2.42**), pancreatic acinar cells (**Figures 2.43–2.44**), myocytes in the skeletal muscles, and renal tubular epithelium (**Figure 2.45**). A stress response may be noted in the adrenal glands, and this usually consists of hypertrophy and cytoplasmic vacuolation of interrenal cells (**Figure 2.46**; see also Chapter 6). Emaciation can also affect bones, but specific skeletal lesions—particularly the microscopic findings—in emaciated reptiles are rarely reported in the literature (this topic is addressed under Metabolic Bone Disease).

2.7 Pyramiding in Chelonians

Pyramiding refers to an abnormal growth of scutes and underlying bone on the carapace of tortoises and occasionally turtles, which results in a pyramidal shape for each scute (**Figures 2.47–2.48**), a disease process frequently observed in captive chelonians (Donoghue, 2006; Heinrich and Heinrich, 2016). Pyramidal shell growth is not seen in wild chelonians except for Indian Star Tortoises (*Geochelone elegans*) and the South African Geometric Tortoise (*Psammobates geometricus*), which might normally have some conical elevation of scutes, particularly the vertebral scutes (**Figure 2.49**) (Daniel, 2002; de Silva, 2003; Branch, 2008), but considerable variation may exist between individuals, age, geographic locations, and subspecies, and debate exists as to the extent to which this is truly normal.

Pyramiding is likely a multifactorial process (Donoghue, 2006) that is expected to occur more commonly with increased growth rates and low humidity (Wiesner and Iben, 2003; Ritz et al., 2010; Heinrich and Heinrich, 2016), although the precise etiopathogenesis has not been fully elucidated. In a study in captive Spur-Thighed Tortoises (*Testudo graeca ibera*) and Aldabra Giant Tortoises (*Aldabrachelys gigantea*) estimating the calcium and phosphorous contents of their diets, pyramiding was attributed to a low calcium:phosphorous ratio (due to diets rich in phosphorous, or calcium deprivation in excessively high protein diets); pyramiding was limited in diets rich in protein (Gerlach, 2004). Captive raised tortoises may have increased growth rates compared to wild conspecifics. An experiment with recently hatched African Spurred Tortoises (*Centrochelys sulcata*) demonstrated that humidity played a minor role in pyramiding (greater pyramiding was seen under drier conditions), while protein contents of the diet had little effect; the authors recommended providing hide areas with relatively high humidity (Wiesner and Iben, 2003). Higher protein diets did lead to more rapid growth rate and larger size, but without a substantial effect on pyramiding. The authors hypothesized that during humid seasons concurrent increase in quality and quantity of browse supports more rapid growth. During dry seasons, nutritional quality and quantity decreases, and growth slows. In captivity, continual access to large quantities of high-quality food, under dry conditions, may cause rapid and unnatural growth. Tortoise hatchlings with all-day access to food, opportunity to exercise, exposure to unfiltered sunlight, temperature gradients including the highest preferred temperatures for the species, adequate humidity, and continued access to water are considered to have a low risk for pyramiding (Donoghue, 2006). Recently, however, a study in African Spurred and Leopard Tortoises demonstrated that growth rate and carapacial scute pyramiding are directly related, and both decrease in tortoises with a normal drop in nighttime temperature; tortoises without a nighttime temperature drop grew faster and larger, which lead to statistically significant increases in pyramidal shell growth (Heinrich and Heinrich, 2016). The authors suggested that accelerated growth may lead to deposition of material between scutes faster than the underlying bone base can expand, thus keratin and bone are forced conically upward (convex upheaval). A consequence of pyramidal shell growth is separation of the underlying vertebrae from the carapace with extremely thin bridging of bone from the dorsal vertebral body to the carapace (Heinrich and Heinrich, 2016), which may lead to future spinal impairment from kyphosis, lordosis, or scoliosis. Shell growth is more normal in tortoises fed complete manufactured diets with a nutritional profile much closer to what the tortoise would consume in the wild (Boyer and Boyer, 2019).

2.8 Metabolic Bone Disease

2.8.1 Calcium and Phosphorus Homeostasis

Bone homeostasis is the result of a complex interaction of nutrients and compounds synthesized in the body through regulatory pathways that are subject to the influence of both intrinsic and extrinsic factors, particularly solar irradiation and genetics. The bulk of knowledge related to bone homeostasis comes from the human and experimental animal literature and, to a lesser extent, the domestic animal literature. A relative lack of equivalent information on the physiology of bone in reptiles affects our understanding of bone disease in these species, which are highly diverse biologically. However, a brief review of basic concepts in bone homeostasis is herein considered necessary.

Approximately 99% of calcium and 85% of phosphorous in the body are located in the bones. The remaining 1% calcium is either ionized (50%) or bound to protein (albumin) or other ions (40% and 10%, respectively), while the remaining 15% phosphorous is either ionized (55%), ion-bound (35%) or protein-bound (10%). Ionized calcium, the physiologically active form, is under the control of parathormone (PTH) from chief cells in the parathyroid gland, vitamin D and calcitonin in the intestine, kidney, and bone, with inverse relationships to ionized phosphorous, which is controlled by PTH, vitamin D, and the phosphatonins, a novel system of proteins (hormones; see below) playing an important role in the control of phosphate homeostasis (Masi, 2011; Craig et al., 2016). A decrease in ionized calcium is followed by an increase in PTH secretion, which in turn increases renal absorption of calcium while diminishing phosphorous absorption. PTH also potentiates 25-hydroxicholecalciferol hydroxylation in the kidney to the biologically active vitamin D. In bones, PTH increases osteoclastic resorption, which increases blood ionized calcium and phosphorous, and osteoblastic activity as well (Craig et al., 2016).

Vitamin D is a fat-soluble vitamin with a range of effects beyond those of bone homeostasis because of the expression of vitamin D receptors in a number of tissues. Evidence for a role in defense mechanisms, autoimmunity, neoplasia, and cardiovascular disease exists in humans (Craig et al., 2016). Vitamin D increases the intestinal and renal tubular absorption of calcium and phosphorous, inhibits PTH and hydroxylation of 25-hydroxycholecalciferol to 1,25-dihydroxycholecalciferol, and increases osteoclastic resorption in bones. Bones produce vitamin D locally, which contributes to bone synthesis and mineralization (Craig et al., 2016). Vitamin D levels depend on dietary sources, cutaneous synthesis of vitamin D precursor (previtamin D_3) from 7-dehydrocholesterol, or a combination of both depending on the species. Cutaneous production of vitamin D in the skin is subject to variations caused by latitude, season, amount of cutaneous pigmentation, time of day, species and behavior (diurnal versus nocturnal species), and type of artificial ultraviolet (UV) light or window glass used for animals maintained in captivity indoors (Mader, 2006; Mans and Braun, 2014; Craig et al., 2016).

Phosphatonins are a group of hormones involved in phosphorous homeostasis (Masi, 2011; Craig et al., 2016. They have been characterized in humans, and the main molecule in this group is fibroblast growth factor 23 (FGF23), which is produced by osteocytes. With elevations of ionized phosphorous and 1,25-dihydroxycholecalciferol, osteocytes increase the production of FGF23, which diminishes phosphorous absorption in the renal tubules, 1,25-dihydroxycholecalciferol production, and PTH secretion (Craig et al., 2016).

Calcitonin is synthesized by microscopic ultimobranchial glands, nestled among the thyroid and parathyroid glands (C-type cells). All adult reptiles possess one (on the left side in many squamates and all crocodilians), or two ultimobranchial glands (in some snakes and chelonians) (Jacobson, 2007). Calcitonin, in response to elevation of ionized calcium, inhibits osteoclast action and osteoclastic bone resorption in order to reduce the serum-ionized calcium (Srivastav et al., 1998; Craig et al., 2016).

In reptiles, the normal physiology of calcium homeostasis, vitamin D synthesis, and bone pathophysiology have not been studied in depth. Iguanid lizards have been shown to possess specific plasma binding of vitamin D metabolites with a capacity and binding characteristics more closely resembling those of birds than mammals; in addition, they store vitamin D in the egg yolk similarly to domestic hens and therefore vitamin D appears to be important for embryonic development in these species (Laing and Fraser 1999). Based on a study in Veiled Chameleons (*Chamaeleo calyptratus*) (Haxhiu et al., 2014), another similarity between reptiles and birds is the upregulation of the cytosolic calcium-binding protein calbindin D28k, which seems to play a role in calcium ion translocation within the duodenal enterocytes, in diets supplemented with calcium, vitamins D and A, as well as UVB exposure.

There is controversy as to the degree to which reptiles depend on dietary sources of vitamin D or cutaneous synthesis for normal calcium and bone homeostasis (Haxhiu et al., 2014; Mans and Braun, 2014). The great diversity of species, feeding strategies, behavior (e.g., diurnal vs. nocturnal species) and physiological adaptations, as well as the scarcity of studies of diet, vitamin D/calcium physiology, and sunlight versus artificial ultraviolet-B (UVB)-induced cutaneous synthesis of vitamin D in wild or captive specimens fuels this controversy. For example, chameleons are commonly kept in captivity and frequently affected with nutritional or husbandry-related metabolic bone disease (MBD), and some popular species such as the Veiled (*Chamaeleo calyptratus*) and Panther (*Furcifer pardalis*) Chameleons have bred well on diets adequate in cholecalciferol or with adequate artificial UVB (wavelengths of 280–315 nm) light exposure, respectively (Ferguson et al., 1996; Ferguson et al., 2003; Hoby et al., 2010). In addition, vitamin A status can also influence the development of MBD in chameleons (Ferguson et al., 1996; Stahl, 1997; Hoby et al., 2010). Green Iguanas and Rock Iguanas (*Cyclura* spp.) do not appear to be able to utilize dietary vitamin D without high doses likely to cause toxicity in other species, and may be dependent on UVB photochemical synthesis (Allen and Oftedal, 1989, 2003; Oftedal et al., 1997; Lemm and Alberts, 2011). The only reptiles that do not seem to require UVB irradiation, if dietary vitamin D levels are adequate, are snakes and nocturnal geckos, but even among these reptiles studies have demonstrated significantly increased plasma levels of calcidiol (25-hydroxyvitamin D3) after UVB exposure (Acierno et al., 2008; Gould et al., 2018). In general, broad-spectrum (including UVB) lighting is recommended for all reptiles and amphibians (Boyer and Scott, 2019a).

While commonly attributed to nutritional and husbandry-related disorders of calcium and vitamin D in reptiles (Donoghue, 2006; Mader, 2006; Mans and Braun, 2014), in most animals MBD encompasses a wider variety of disease processes that affect the structure and physiology of the skeleton with a more diverse range of potential etiologic factors and interactions worthy of review. Some of these processes may share clinical and pathologic features, and an overlap between different conditions may occur in the same patient. Osteoporosis, rickets, osteomalacia, and fibrous osteodystrophy/hyperparathyroidism are included in this category of skeletal diseases in animals. In reptiles, however, nutritional and husbandry-related MBD most commonly presents or is reported as fibrous osteodystrophy, with the exception of chelonians, in which fibrous osteodystrophy does not routinely occur. This discussion will focus on nutritional and husbandry-related MBD. Other forms of MBD, including renal secondary hyperparathyroidism, hypertrophic osteopathy, hypertrophic osteodystrophy, Paget's disease, and panosteitis will not be included here because they are not nutritional diseases.

2.8.2 Osteoporosis

Osteoporosis is due to an imbalance between bone production and resorption in favor of the latter, resulting in a reduction in the quantity of bone, which is otherwise structurally normal, and in the breaking strength (Craig et al., 2016). This differentiates it from osteomalacia, in which there is a decreased mineral-to-collagen ratio in the bones. Osteoporosis can occur in humans and other species due to aging, inanition and nutritional deficiencies, corticosteroid excess, gastrointestinal parasitism and inflammatory disease, lack of exercise, vitamin A toxicity, chronic metabolic acidosis, and hyperthyroidism (Craig et al., 2016). Nutritional diseases that can lead to osteoporosis in animals include deficiencies of calcium (with optimum phosphorous and vitamin D status), copper, and phosphorous (Craig et al., 2016).

Bones and areas of bone rich in cancellous bone such as vertebrae and flat bones are most severely affected. Grossly, cancellous bone is reduced and porous; transverse reinforcement trabeculae may be formed in the medullary cavity. In advanced cases, cortical bone may be thinner than normal and pathologic fractures may occur (Craig et al., 2016). Histologically, osteoporosis caused by increased resorption (e.g., secondary to uncomplicated dietary calcium deficiency) is characterized by decreased numbers of trabeculae, whereas osteoporosis resulting from decreased bone production (e.g., with starvation) is characterized by normal numbers of thinner trabeculae. Advanced osteoclastic resorption may involve the cortical bone along vascular channels. In growing animals, severe osteoporosis can cause narrowing or loss of the hypertrophic zone of the growth cartilage, and the physis may be sealed by a plate of bone. Chondrocytes in the growth plates may be smaller and disperse in a greater amount of cartilaginous matrix. With severe osteoporosis resulting from starvation or nutritional disease, growth arrest lines may occur in the metaphysis (Craig et al., 2016). Experimental nutritional secondary hyperparathyroidism (NSHP) in Green Iguanas presented as osteoporosis in individuals fed a low calcium diet with adequate phosphorus; osteoporosis was characterized by flattened inactive osteoblasts, marked resorption of endosteal bone by large multinucleated osteoclasts, and pronounced resorption of cortical bone by osteoblastic osteolysis (Anderson and Capen, 1976a; Anderson and Capen, 1976b). Osteoporosis secondary to starvation and chronic debilitating diseases will be accompanied by serous atrophy of the adipose tissue in the medullary cavity of bones. Gastrointestinal parasitism and gastroenteritis or other diseases of the digestive system causing malabsorption may be complicated by phosphorous deficiency, protein–calorie malnutrition, and other secondary nutritional imbalances leading to osteoporosis or nutritional secondary hyperparathyroidism (Craig et al., 2016). For example, juvenile and neonatal tortoises with gastroduodenal cryptosporidiosis and retarded growth may present with osteoporosis (**Figures 2.50–2.51**) as well as serous atrophy of adipose tissue, subcutaneous edema, and hydrocoelom. The carapace can be poorly mineralized or completely unmineralized. Osteoporosis has been documented in farmed Saltwater Crocodiles (*Crocodylus porosus*) that had a chronic stress response in the adrenal glands and elevated serum corticosterone (Shilton et al., 2014).

2.8.3 Rickets and Osteomalacia

Osteomalacia refers to defective bone mineralization that occurs in the newly formed bone during remodeling in adult animals, as opposed to rickets, which occurs in young individuals and therefore involves not only newly formed osteoid but also the physeal growth plates where the cartilaginous matrix undergoes endochondral ossification during growth (Craig et al., 2016). Both manifestations of MBD have the same pathogenesis: defective bone mineralization. The causes of rickets and osteomalacia are diverse, and include nutritional deficiencies (vitamin D, phosphorous), lack of exposure to sunlight/adequate UVB light, malabsorption, and hereditary disorders. There are species differences in the circumstances in which rickets and osteomalacia develop. In juvenile Veiled Chameleons with experimental MBD and subject to different manipulations involving calcium, vitamin D, vitamin A, and exposure to UVB light, rickets was not observed (Hoby et al., 2010).

In cases of osteomalacia, the cortical lesions are similar to those of rickets and consist of filling of Haversian systems with pale-staining, unmineralized osteoid, which is also deposited over bone trabeculae in the medullary cavity. However, osteoid seams are not always present, and, in addition, they may occur with other syndromes in areas of rapid bone formation (Craig et al., 2016). Green Iguanas experimentally fed a diet low in calcium and phosphorus developed osteoporosis and osteomalacia characterized by large active osteoblasts overlying wide osteoid seams, marked osteoclastic resorption of endosteal bone, and large osteocytic lacunae which had a prominent layer of osteoid (Anderson and Capen, 1976b).

2.8.4 Nutritional and Husbandry-Related Secondary Hyperparathyroidism

In reptiles, hyperparathyroidism is usually secondary to nutritional and husbandry-related factors; however, renal secondary hyperparathyroidism can also occur (Donoghue, 2006; Mader, 2006). Disorders of calcium and vitamin D are most commonly involved in nutritional or renal MBD in reptiles, but phosphorous or vitamin A may also play a role in some cases (Ferguson et al., 1996; Stahl, 1997; Donoghue, 2006; Mader, 2006; Hoby et al., 2010; Mans and Braun, 2014).

2.8.4.1 Dietary Factors The occurrence of nutritional and husbandry-related secondary hyperparathyroidism in reptiles depends on a variety of factors, including species-related differences (nutritional strategy, diurnal versus nocturnal habits, and any potential species-specific metabolic/physiological pathways involved in calcium/phosphorous/vitamin D metabolism), diet, and husbandry (management of exposure to sunlight or artificial light) (**Figure 2.52**). An experimental study in juvenile Veiled Chameleons (*Chamaeleo calyptratus*) demonstrated that bone homeostasis was best maintained with diets supplemented with calcium and vitamins D and A in specimens exposed to UVB light. However, calcium and vitamin A supplementation resulted in resistance to MBD regardless of UVB light exposure and/or vitamin D supplementation, unlike UVB light exposure without any dietary supplementation, which delayed but did not prevent MBD (Hoby et al., 2010).

Lizards and chelonians are most commonly affected with nutritional and husbandry-related secondary hyperparathyroidism, particularly the herbivorous and insectivorous species. Snakes and crocodilians in captivity are rarely affected if fed whole prey; if fed diets consisting of meat (without bone), viscera, pinkie mice, or day-old chicks, calcium deficiency can occur (Donoghue, 2006; Mans and Braun, 2014). Carnivores fed whole vertebrate diets are typically spared NSHP because bones are an excellent source of calcium. Rapidly growing juvenile Saltwater Crocodiles (*Crocodylus porosus*) select a diet with an overall Ca:P ratio of 6.7 to 1, primarily because of calcium-rich crustacean shells from the prawns and crabs they consume. Adult Saltwater Crocodiles consume more mammals and fish, but fewer crustaceans, and consequently the Ca:P ratio of their diet drops to 1.85 to 1 (Webb et al., 1991). Panther Chameleons (*Furcifer pardalis*) adapt their basking behavior to the vitamin D_3 contents of their diet; chameleons with a lower dietary intake obtain a higher UVB irradiation by increasing the time they spend basking (Ferguson et al., 2003). A study comparing two species of anoles, the Cuban Brown Anole (*Anolis sagrei*), which is heliophilic and thermophilic, and the Jamaican Grey Anole (*Anolis lineotopus merope*), which is shade-tolerant and thermal-conforming, pointed to a greater vitamin D_3 photobiosynthesis capacity in the latter that compensates for its reduced exposure to sunlight and lower dietary intake of vitamin D_3 (Ferguson et al., 2005). A greater vitamin D_3 photobiosynthesis capacity appears to allow the House Gecko (*Hemydactylus turcicus*) to obtain sufficient vitamin D_3 at dusk or dawn, despite its overall nocturnal behavior (Carman et al., 2000).

NSHP can be associated with diets low in calcium, decreased calcium absorption in the intestine, diets high in phosphorous, insufficient exposure to UVB (wavelengths of 280–315 nm), inadequate dietary vitamin D, or combinations thereof (Donoghue, 2006; Mader, 2006; Mans and Braun, 2014). Herbivores are fed diets too high in protein and carbohydrates and too low in calcium and fiber without an adequate calcium to phosphorus ratio. Wild Desert Tortoises (*Gopherus agassizii* and *G. morafkai*) select a diet with calcium:phosphorus ratio of 3.2 to 5.8:1 (Jarchow, 1984). Almost all insects are deficient in calcium, vitamins A and D_3, and have negative calcium:phosphorous ratios (Finke, 2002). Feeding insectivores to avoid NSHP involves a simple triad of feeding a wide diversity of insects, dusting insects with calcium carbonate, and making sure the insects are fed a diet with $\geq$8% calcium on dry matter basis (for more information see Boyer and Scott, 2019a). Note that feeding insects does little to change their nutritional profile, and is not the same as "gut loading"; the latter involves feeding a calcium and multivitamin–enriched diet to change the nutritional profile of the insect (Zwart and Rulkens, 1979; Strzelewicz et al., 1985; Allen and Oftedal, 1989; Latney LV et al., 2017).

Apart from the dietary calcium, phosphorous, and vitamin D, other diet-related factors need to be considered. Chameleons fed diets low in vitamin A or oversupplemented with vitamin A can develop metabolic bone disease (Ferguson et al., 1996; Stahl, 1997; Hoby et al., 2010). Foods rich in phytates (particularly soy, cereals [most notably the bran products but also whole wheat], and some beans), high fat or acid, or with a high oxalate:calcium ratio (e.g., spinach, rhubarb, leek, beet), interfere with calcium absorption in the intestine (Donoghue, 2006).

2.8.4.2 Ultraviolet Light UV light is critical to bone homeostasis in reptiles. A study in captive Tuataras (*Sphenodon punctatus*) in New Zealand revealed a wide variation in the UVB spectral irradiance, and fractional reduction in electromagnetic fluency with increasing distance from the UVB light source between participating institutions, with 29% of these having no measurable UVB light on the premises (Burgess et al., 2009). The quality of a UVB lamp can be evaluated by its influence on the plasma 25(OH)vitamin D_3 (25OHD$_3$) levels and effects on skin health (Gehrmann, 2006; Diehl et al., 2017). An increase in plasma levels of 25OHD$_3$ has been demonstrated in most crepuscular and nocturnal reptiles tested after exposure to artificial UVB light (Carman et al., 2000; Gillespie et al., 2000; Ferguson et al., 2003; Acierno et al., 2006; Acierno et al., 2008; Hoby et al., 2010; Oonincx et al., 2010; Selleri and di Girolamo, 2012). An experimental study in the nocturnal Ball Python (*Python regius*), however, revealed no significant changes in plasma 25OHD$_3$ levels after exposure to artificial UV light for 70 days (Hedley and Eatwell, 2013). Despite this capacity of reptile skin, UVB light bulbs cannot completely replace exposure to direct sunlight in some species (Laing et al., 2001; Mader, 2006; Diehl et al., 2017). A study in Bearded Dragons using five different commercially available UVB lamps showed that the plasma levels of 25OHD$_3$ were lower than those in wild lizards exposed to sunlight (Diehl et al., 2017). Similarly, captive Hermann's Tortoises (*Testudo hermanni*) exposed to two different artificial UVB lamps for 35 days had lower plasma 25OHD$_3$ levels than conspecifics exposed to sunlight (Selleri and di Girolamo, 2012).

A number of factors add difficulties to establishing the appropriate artificial UVB exposure regimes for captive reptiles (Diehl et al., 2017). These include the quality of lamps, the generalized unavailability of normal plasma 25OHD$_3$ levels for most wild reptile species, and the great diversity in reptilian species and their potential physiological and behavioral adaptations to the different habitats and vitamin D contents of their diets. The plasma 25OHD$_3$ levels have been determined in few wild reptilian species including Ricord's Iguanas (*Cyclura ricordii*), Rhinoceros Iguanas (*Cyclura cornuta cornuta*) (Ramer et al., 2005), Eastern Box Turtles (*Terrapene carolina carolina*) (Watson et al., 2017), Komodo Dragons (*Varanus komodoensis*) (Gillespie et al., 2000), Bearded Dragons (Diehl et al., 2017), Green Turtles (*Chelonia mydas*) (Stringer et al., 2010), Eastern Massasaugas (*Sistrurus catenatus*) (Slater et al., 2017), Alligator Snapping Turtles (*Macrochelys temminckii*) (Chaffin et al., 2008), and Common Chuckwallas (*Sauromalus ater* [*obesus*] (Aucone et al., 2003). In addition to this reduced availability of normal plasma vitamin D_3 levels in wild species, other factors that affect the interpretation of plasma levels obtained in captive reptiles for diagnostic purposes or to evaluate dietary adequacy and husbandry include the variations between different assays and laboratories for the same assay, as well as the lack of signs of bone disease in some reptiles with low vitamin D levels (Purgley et al., 2009; Ferguson et al., 2015; Diehl et al., 2017). In addition, it has been experimentally demonstrated in Bearded Dragons that plasma 25OHD$_3$ levels may not decrease significantly after discontinuing UVB exposure for periods up to 83 days; these dragons also maintained normal total serum calcium levels and there was only a tendency to decrease ionized calcium levels (Oonincx et al., 2013).

The spectral power distribution of artificial UVB lamps determines their potential for vitamin D_3 synthesis in the irradiated skin. This potential depends on the phosphor blend and transmission qualities of the outer glass tube of a UVB lamp (Diehl et al., 2017). With sunlight, the shorter the wavelength of spectrum, the greater the synthetizing potential of vitamin D_3 per microwatt, with the most effective wavelength at solar altitude 60° is 308 nm (Bernhard et al., 1997). Studies assessing the vitamin D_3 potential of different commercially available lamps by determining the effect on plasma 25OHD$_3$ levels are available for some reptilian species. However, most of these studies did not provide data regarding the UVB intensity, and spectral distribution (Diehl et al., 2017). Diehl et al. (2017) determined that the five different commercially available lamps had a relative spectral power distribution similar to that of sunlight, and postulated that UVB dose would be the main factor involved in the differences of plasma 25OHD$_3$ levels in the respective experimental Central Bearded Dragon groups and wild specimens exposed to sunlight (Diehl et al., 2017). A previous study with one of the lamp brands used by Diehl et al. (2017), with a larger proportion of short wavelength UVB, produced much higher plasma 25OHD$_3$ levels than other lamps at higher irradiances (Oonincx et al., 2010). Diehl et al. (2017) postulated that lamps emitting non-terrestrial short wavelength UVB (<290 nm) may affect the cutaneous self-regulation of vitamin D_3 synthesis determined by the percentage of conversion of vitamin D_3 precursor into the different photoproducts. Care must be taken with these lamps with a higher proportion of short wavelength UVB, as they can induce ulcerative photodermatitis and keratoconjunctivitis and even mortality in reptiles (Gardiner et al., 2009; Diehl et al., 2017). In addition to these factors, the potential for vitamin D_3 synthesis of a lamp may vary between lamps of the same brand and model and with the age of the lamp (Diehl et al., 2017). UVB lights should be replaced every 6 (compact fluorescent lights) to 12 (fluorescent tubes) months or when UVB output drops below 70% of the original output (Boyer and Scott, 2019a). UV meters are now available and are an important tool for testing bulb output and assessing efficacy of basking areas. UVB light intensity decreases with the inverse square law (1/distance^2), so a light that is twice as far away has one-fourth the UVB intensity. Consequently, lights, especially those of lower power like fluorescents, should be kept close to the animal, no more than 18–24 inches away (Boyer and Scott, 2019a). Not all reptile lights produce UVB; practitioners must check products individually. Many reptile lights state "daylight," "broad spectrum," or "UVA" (320–400 nm) but lack critical UVB (Boyer and Scott, 2019a). For a good review of UV light and reptiles see Baines and Cusack (2019).

2.8.4.3 Clinical and Pathological Findings Nutritional and husbandry–related secondary hyperparathyroidism most often affects juvenile animals and reproductively active females whenever diets, husbandry, or both do not meet their increased demand of calcium for a protracted period. As disease progresses, secondary (reactive) hyperparathyroidism with increased PTH production will cause bone resorption and overt bone disease in the axial and appendicular skeletons of lizards. Chelonians typically spare their appendicular skeleton and remove bone from shell, which results in a soft, pliable shell. In Hermann's Tortoises, decreased bone mineral density of the shell was apparent with dual-energy x-ray absorptiometry (DEXA) but not in the axial and appendicular skeleton (Gramanzini et al., 2013). A range of clinical signs can be noted in affected reptiles including sprawl (legs spread horizontally with an inability to lift the legs, shell, or body off the ground), lethargy, weakness, declining appetite to anorexia, reduced growth, lack of weight gain or weight loss, ataxia, hyperesthesia, abnormal postures (**Figure 2.53**), follicular stasis, egg retention, and colonic or cloacal prolapse. Lizards are prone to spinal damage from vertebral fractures, which may result in paraparesis or paraplegia of the rear legs, as well as horizontal flattening of the ribcage and scapuli (barrel chested). Both lizards and crocodilians may have bowed, thickened, or fractured long bones or mandibles, pliable mandibles and maxillae, retrognathism (overbite), kyphoscoliosis, lordosis, or rhoecosis (combination of the later three, which are secondary to fibrous osteodystrophy), and walk on the forelimbs rather than plantigrade. Chameleons may have a noticeably weakened grip, decreased prehensile ability of their tail, fall repeatedly, and have cloacal distention and scoliosis of the parietal crest or vertebrae (Boyer and Scott, 2019b).

Crocodilians with NSHP may present with teeth that protrude laterally, fall out, or are lucent around their edges or diaphanous (glassy). The upper jaw softens so much it can be bent upward. Crocodilians may move freely in water but are weak, sluggish, or ataxic on land and eventually are unable to come out of the water. Disturbance may result in tremors or seizures; seizures in water may lead to drowning (Huchzermeyer, 2003; Boyer and Scott, 2019b).

Juvenile chelonians may have soft shells (normally the shell should feel solid, like skull), splay out their legs, have prognathism (underbite), but do not typically have long bone fractures or fibrous osteodystrophy. Juvenile chelonians with pliable shells that are no longer eating typically die, despite treatment. In box turtles, the marginal scutes may curl dorsally, the body appears too small for the shell, and the phallus or cloaca may prolapse. Adult chelonians may have minimal clinical signs (other than sprawl) if they developed a normal shell prior to a calcium-deficient diet because shell is such a large calcium reservoir (Boyer, 1996), but eventually their body may appear too large for their shell. Chelonians typically present with demineralization and subsequent deformities of the carapace, especially scoliosis and lordosis of the caudal vertebral scutes (Anderson and Capen, 1976a; Donoghue, 2006; Mader, 2006; Hoby et al., 2010; Mans and Braun, 2014).

In lizards, radiographically and grossly, bones are demineralized, with increased flexibility due to decreased mineralization and reactive PTH-induced bone resorption (**Figure 2.54**). There may also be soft tissue enlargement associated with fibrous osteodystrophy. Only severe reductions in bone density result in decreased bone opacity, because 40%–50% of bone mineralization must be depleted before it is visible radiographically (Lauten et al., 2001). Thus radiographs are useful prognostically; prognosis worsens with decreasing bone opacity, and patients with poorly visible bone typically die (Boyer and Scott, 2019b). Bone deformities include thin cortices, bowing of long bones, scoliosis, kyphosis, lordosis, or rhoecosis, fibrous osteodystrophy and pathologic fractures of long bones and vertebrae, and bowing, swelling, and shortening of the mandible (**Figures 2.55–2.59**) (Boyer, 1996; Donoghue, 2006; Mader, 2006; Hoby et al., 2010; Mans and Braun, 2014). Nodular lesions and fractures of ribs can also occur, particularly in lizards (**Figure 2.60**). (See also Chapter 10.) In chelonians, decreased opacity is difficult to appreciate in the shell (occasionally one can appreciate wider lucent bone seams in areas of the carapace not overlapping the plastron) and limb bones, but may be present in the pectoral and pelvic girdles, which typically are in good contrast to the overlying shell (Boyer and Scott, 2019b).

The best ways to measure bone density are via quantitative computed tomography (Gramanzini et al., 2013) or DEXA (Zotti et al., 2004). DEXA has been described experimentally in Green Iguanas (Zotti et al., 2004), Leopard Tortoises (Raiti and Haramati, 1997) and Hermann's Tortoises (Gramanzini et al., 2013). Regression equations of bone density as a function of body weight are available for normal iguanas and iguanas with NSHP, with the skull, lumbar spine, and femurs being most useful (Zotti et al., 2004).

Usually, total calcium (Ca) is just below or within normal limits until skeletal reserves have been depleted, whereas ionized Ca levels are maintained until terminal (Boyer and Scott, 2019b). One study found total serum Ca levels of veiled chameleons with NSHP to be 2.0–2.2 mmol/L (8.0–8.8 mg/dL), slightly below the reference range of 2.3–3.5 mmol/L (9.2–14.00 mg/dL) (Hoby et al., 2010). Elevated nitrogenous byproducts (ammonia, urea, or uric acid) may indicate renal secondary hyperthyroidism or dehydration (Boyer and Scott, 2019b). Calcium-deficient females may have elevated Ca levels because of egg production. Conversely, a gravid oviparous female with normal Ca levels may be calcium deficient (Boyer and Scott, 2019b). With obvious clinical signs, further diagnostics, apart from radiography, are seldom required. The definitive diagnosis of hyperparathyroidism requires demonstration of elevated PTH, but commercially available mammalian assays do not appear to be accurate for reptiles and cannot be recommended. Plasma calcidiol values can be more easily measured to assess the potential deficiencies of UVB or dietary D_3 (Boyer and Scott, 2019b).

Microscopically, the hallmark of secondary hyperparathyroidism is fibrous metaplasia replacing poorly mineralized or lytic bone. In juvenile specimens, the metaphyseal trabecular bone may be decreased in density (**Figures 2.61–2.62**), with shorter trabeculae, replacement of the primary spongiosa with disorganized spicules of new woven bone without cartilage cores, or complete loss of the primary spongiosa. In these cases, the metaphyseal plate may be flattened and covered with new woven bone. Chondrocytes in the hypertrophic zone of the physis may be clumped rather than organized in vertical columns. Increased osteoclastic resorption of bone, as well as prominent osteoblastic hypertrophy, are accompanied by varying degrees of fibrovascular tissue proliferation in affected areas. Fibrous tissue replaces lytic bone, fills the medullary spaces, and can invade trabecular and cortical bone (**Figures 2.63–2.64**) (Hoby et al., 2010). In chameleons, areas of chondroid metaplasia may be noted as well (**Figure 2.64**). In chelonians, osteoporosis predominates, especially in the shells, with little or no fibrous reaction (**Figures 2.65–2.66**).

2.9 Soft Tissue Mineralization of Dietary and Metabolic Origin

Soft tissue mineralization (metastatic mineralization or metastatic calcification) is the hallmark of hypervitaminosis D, which can occur in mammals, birds, and reptiles as a result of overzealous oral supplementation with vitamin D, calcium, or both to prevent metabolic bone disease (e.g., dusting or gut-loading of prey items with excessive vitamin supplements), improper formulation of vitamin supplements, injections of toxic doses of vitamin D, or more rarely, consumption of cholecalciferol-based rodenticides (Fledelius et al., 2005; Swenson and Bradley, 2013; Lopez et al., 2016). Soft tissue mineralization is a frequent disease process in lizards and chelonians. Underlying renal disease, dehydration, and folliculostasis in females (**Figure 2.67**) are most commonly involved in its etiopathogenesis, and should be ruled out before considering nutritional, iatrogenic, and toxic causes. Suspected nutritional hypervitaminosis D has been reported in Green Iguanas with metastatic calcification (Pallaske, 1961; Wallach, 1966; Shively and Edwards, 1985; Frye, 1991); however, vitamin D metabolites were not measured. Metastatic mineralization has also been reported in Green Iguanas at the Smithsonian National Zoological Park (DC) with extremely low levels of the principal vitamin D metabolite, calcidiol (Richman et al., 1995a,b; Allen and Oftedal, 2003). It was thought iguanas did not have access to, or utilize, UV light in the exhibits. It is probably not appropriate to diagnose vitamin D toxicity when soft tissue calcification is present in reptiles without supporting evidence such as very high levels of dietary vitamin D and circulating calcidiol (Allen and Oftedal, 2003). Soft tissue mineralization has also been noted experimentally in Leopard Tortoises supplemented with dietary calcium at 3 and 9 times the recommended concentration (Fledelius et al., 2005). Soft tissue mineralization of metabolic origin is observed in reptiles with dehydration, renal disease, follicular stasis in the ovary, or combinations thereof due to resulting derangements in the calcium:phosphorous ratio. Green Iguanas presenting with hemoptysis often had radiographically visible metastatic calcification of pulmonary blood vessels, with rupture at necropsy (T. Boyer, personal observation).

If progressive, soft tissue mineralization is typically a fatal disease due to failure of organs severely affected by mineral deposits. Soft tissue mineralization can involve a number of organs in reptiles, most notably the elastic arteries and other blood vessels, kidneys, lung, gastrointestinal tract, trachea, heart, testis, cornea, and skin. Mineralization can be detected radiographically in advanced cases, particularly in the descending aorta. Grossly, mineralized tissues may be characterized by a whitish discoloration and granular or firm consistency. Affected arteries may be tortuous and thickened. The mucosal surface of the gastrointestinal tract or capsule of affected organs may be thickened and irregular (**Figure 2.68**). Distinct whitish poorly or well-circumscribed foci may be noted. In the skin, patchy areas or gray to dark red discoloration and necrosis can be observed in some patients. Vesicle formation or epithelial sloughing with ulceration may occur if the epidermal basement membrane is mineralized (**Figure 2.69**). In the lungs, deposits usually involve most notably the muscular trabeculae, and this can appear as a reticulated pattern grossly (**Figure 2.70**). Microscopically, deposition of a basophilic material is observed in affected tissues (**Figures 2.71–2.77**). The blood vessels and hepatic sinusoids, mucosae, and smooth muscle of the digestive and respiratory tract, basement membranes, renal tubules and glomeruli, corneas, and skin (dermis and epidermis) are frequently involved. Mineral deposits may be accompanied by fibrosis and infiltration of multinucleated giant cells at the margins of deposits, with partial loss of the normal tissue architecture. Necrosis and ulceration of the skin may be noted in reptiles with cutaneous involvement (**Figure 2.72**), and the corneas may be cloudy or scarred when the inner or outer basement membranes are mineralized. The parathyroid gland should be normal histologically in reptiles with nutritional, toxic, or iatrogenic hypervitaminosis D, but hyperplasia/hypertrophy with cytoplasmic vacuolation of chief cells will usually be observed in patients with soft tissue mineralization of renal origin.

Dystrophic mineralization refers to deposition of mineral in previously injured or diseased tissue, and generally is a localized event and not related to hypervitaminosis D. Mineralization of the aorta at the cardiac outflow tract is common in old reptiles, likely a manifestation of age-related arteriosclerosis, and should not be confused with metastatic mineralization. Various degrees of mineralization may be seen in localized enteric ulcers, cutaneous abscesses, old ischemic brain lesions, and in areas of necrosis associated with neoplastic processes. (For further information on soft tissue mineralization, see also Chapter 6.)

2.10 Vitamin A Deficiency

Vitamin A is a fat-soluble vitamin with physiologic functions that affect vision, cellular growth and differentiation, reproduction, bone and tooth development, antioxidant defense mechanisms, and immunity. The most obvious effects of a deficiency involve disruption of epithelia and their protective functions, particularly in the respiratory tract, conjunctiva, glands, digestive tract, skin, and genitourinary system. In addition, vitamin A deficiency or toxicosis can be teratogenic, and a number of malformations have been reported, particularly in mammals (Craig et al., 2016).

Vitamin A comprises several preformed biologically active compounds normally found in vertebrate tissues synthetized from dietary sources. The animal-derived retinyl esters and the provitamin precursors of vegetal origin (carotenoids) are the natural dietary sources of compounds with vitamin A activity. Beta-carotene is considered the main plant precursor (Donoghue, 2006), but this may not apply to all reptiles (Raila et al., 2002; Koelle et al., 2010). Differences exist between vertebrate species regarding their ability to absorb and convert beta-carotene into active vitamin A compounds. As a general rule, herbivores are considered more efficient in the process of absorption and conversion of beta-carotene than carnivores and omnivores; this may explain the rare occurrence of vitamin A deficiency in tortoises, iguanas, and Bearded Dragons, unless they are anorectic for months and subsequently utilize their hepatic stores of retinyl compounds (Boyer, 2006; Donoghue, 2006; Mans and Braun, 2014). Species and age-related variations may also exist. For example, an experimental study in Green Iguanas demonstrated that carotenoids other than beta-carotene including oxygenated carotenoids (xanthophylls), lutein, zeaxanthin, and canthaxanthin were absorbed and found in the plasma, while supplementation of beta-carotene caused no increase in plasma beta-carotene concentration. In addition, beta-carotene did not affect the concentrations of retinol in plasma. The authors concluded that Green Iguanas are selective accumulators of polar xanthophylls and not beta-carotene (Raila et al., 2002). However, this study only tested juvenile Green Iguanas; possible age-related differences associated with age-dependent changes in food preferences or dietary habits were not assessed.

In reptiles, vitamin A deficiency occurs most commonly in turtles in the family Emydidae, especially Box Turtles (*Terrapene* spp.) and Spotted Turtles (*Clemmys guttata*) (Ed. note: M. Garner, personal observation), but insectivorous lizards and crocodilians can also be affected (Ariel et al., 1997; Stahl, 1997; Boyer, 2006; Donoghue, 2006; Mans and Braun, 2014). Box Turtles are considered dependent on a diet continuously rich in vitamin A (Boyer and Boyer, 2006). In turtles, disease often involves young specimens fed diets of unsupplemented vegetables (lettuce), meat, and inadequately formulated commercial diets (Donoghue, 2006). However, the occurrence of vitamin A deficiency in chelonians has been reduced by the use of commercial diets with adequate vitamin A content. Among lizards, insectivorous species, such as anoles, chameleons, and geckos, are most commonly affected (Stahl, 1997; Boyer, 2006; Donoghue, 2006; Mans and Braun, 2014; Wiggans et al., 2018). Interestingly, although frequently referenced as a susceptible species, hypovitaminosis A seems to be uncommon or rare in Green Iguanas (T. Boyer and M. Garner, personal observations). Insects are deficient in vitamin A, or retinyl esters, and many reptile multivitamins lack preformed vitamin A, which leads to vitamin A deficiency in the insectivore (Stahl, 1997; Barker et al., 1998; Finke, 2002; Donoghue, 2006; Oonincx and Dierenfeld, 2012; Finke, 2015; Boyer and Scott, 2019b). In a recent retrospective study of ophthalmologic disease in captive Leopard Geckos (*Eublepharis macularius*), lack of vitamin A supplementation was significantly associated with the development of ocular disease in 52 affected geckos, as was conjunctival squamous metaplasia and cranial dysecdysis (Wiggans et al., 2018). Because of the potential interaction with vitamin D_3, vitamin A supplementation of such prey items should not be overzealous, as it can cause metabolic bone disease (Stahl, 1997). Hypervitaminosis A is also a potential effect of excessive supplementation.

Clinically, turtles with vitamin A deficiency present with anorexia, lethargy, retarded growth, weight loss, blepharedema (**Figure 2.78**), and sometimes nasal or ocular discharge (Boyer, 2006; Donoghue, 2006; Mans and Braun, 2014). Palpebral thickening may be more pronounced on one side and is due to squamous metaplasia of the Harderian gland and posterolateral lacrimal gland (Elkan and Zwart, 1967). As a result of palpebral closure, white to yellow material, or exudate, may accumulate in the conjunctival sac, with conjunctival eversion, particularly in chronic cases; secondary infection can occur in the conjunctiva and other affected mucosae/epithelia (Elkan and Zwart, 1967). This is due to keratin accumulation, loss of normal mucosal barrier, and vitamin A–dependent immunological defense mechanisms. In Box Turtles, an association of aural abscesses with hypovitaminosis A has been reported but the etiopathogenesis has not been fully elucidated and may involve other factors, including accumulation of organochlorine compounds in tissues (Brown et al., 2004; Sleeman et al., 2008). Lizards have blepharitis, blepharospasm, and anorexia. Leopard Geckos may present with mucoid-to-solid cellular debris under the eyelids, much like turtles. Chameleons may have mucoid buildup in the eye, thickened eyelids, difficulty capturing prey with the tongue (possibly from dysfunction of sticky mucus glands in the tongue tip), and dull coloration, which can be difficult to appreciate in a lone animal (Boyer, 2018). Chameleons frequently have a history of worsening eye disease unresponsive to antibiotic therapy, and eventually keep their eyes shut unless disturbed (Boyer and Scott, 2019b). Other findings in chameleons associated with experimental or suspected vitamin A deficiency included labial swelling with oozing of thick clear mucus upon firm digital pressure, poor reproductive performance both in males and females, necrosis of tail tip, vertebral kinking, dysecdysis, and impaction of the hemipenes with plugs of shed epithelium (Ferguson et al., 1996; Stahl, 1997; Boyer, 2006). However, squamous metaplasia of epithelia, the hallmark of vitamin A deficiency, was not mentioned among the microscopic lesions in chameleons dying during the experimental study reporting most of these clinical signs (Ferguson et al., 1996). Similar labial and palpebral swellings with cheilitis have been reported in a laboratory colony of Green Anoles (*Anolis carolinensis*) with vitamin A deficiency suspected on the sole basis of histopathologic findings (squamous metaplasia of mucosal epithelia) (Miller et al., 2001). Farmed-reared hatchlings of Freshwater (*Crocodylus johnstoni*) and Saltwater (*C. porosus*) Crocodiles fed a vitamin A–deficient diet developed pale brown nodular lesions on the surface of the tongue up to 5 mm in diameter (Ariel et al., 1997).

Microscopically, the hallmark of vitamin A deficiency is squamous metaplasia of mucosal epithelia (**Figures 2.79–2.82**) (Elkan and Zwart, 1967; Frye, 1984). The metaplastic squamous epithelium produces abundant keratin (hyperkeratosis), which accumulates within glands and over the surface of affected mucosae (**Figure 2.80**). Hyperkeratosis is noted in epithelia that are normally squamous (e.g., oral, cutaneous). This lesion is accompanied by infiltration of granulocytes, which transmigrate through the affected epithelia (granulocytic exocytosis) and accumulate within the exudate. Secondary bacterial and fungal infections can occur. Different epithelia can be affected. In chronically inflamed mucosae, reactive squamous metaplasia can occur, and this should be considered in the differential diagnosis. Squamous metaplasia can also be noted as a result of estrogen excess, but this has not been documented in reptiles. Involvement of the renal tubular epithelium in hypovitaminosis A can lead to obstructive renal disease and gout, as well as secondary ascending bacterial infections. Such renal disease has been described in farm-reared Freshwater (*Crocodylus johnstoni*) and Saltwater (*C. porosus*) Crocodile hatchlings fed a vitamin A–deficient diet; affected specimens also had squamous metaplasia of lingual glands (Ariel et al., 1997). Similar lesions have been noted in the labial salivary gland of a Jackson's Chameleon (*Trioceros [Chamaeleo] jacksonii*) that responded to a single vitamin A and D injection (Boyer, 2006).

The dietary history, compatible clinical findings, microscopic lesions (squamous metaplasia and hyperkeratosis of epithelia), and response to vitamin A treatment can support a presumptive diagnosis of vitamin A deficiency. Roughly 80%–90% of vitamin A is stored in the liver of vertebrates. Blood levels of vitamin A do not drop until the large hepatic reservoir of vitamin A is exhausted; thus blood levels may not reflect whole-body vitamin A status. Confirmation should include the determination of plasma retinol or hepatic vitamin A levels, but the availability of reference values, differences between assays, laboratories, degradation of vitamin A with light exposure, high temperatures, and patient autolysis are important limitations in this regard (Boyer, 2006; Mans and Braun, 2014; Boyer and Scott, 2019b). With vitamin A deficiency, blood concentrations of vitamin A drop below 20 micrograms/100 mL after hepatic reserves are emptied in domestic animals (National Research Council, 1987). Liver vitamin A levels of less than 5–10 micrograms are suspicious for vitamin A deficiency in amphibians (Pessier, 2014).

2.11 Hypervitaminosis A

Hypervitaminosis A is an iatrogenic disease most commonly observed in tortoises but can occur in other reptiles treated with excessive vitamin A (Boyer, 2006). Therefore, while excessive dietary intake, mostly feeding of excessive raw liver to semi-aquatic turtles, may contribute to the development of hypervitaminosis A (Mans and Braun, 2014), injection of formulas with excessive vitamin A is the primary cause. Hypervitaminosis A most commonly occurs in tortoises from the mistaken notion that vitamin A injections will help with upper respiratory tract disease (mycoplasmosis), boost the immune system, or improve appetite. Tortoises, being herbivores, are good at converting abundant carotenoids in their diet to vitamin A, and rarely, if ever, develop primary vitamin A deficiency (Boyer and Scott, 2019b). Emydid turtles, being carnivorous, do develop vitamin A deficiency.

Commercial formulations designed for large animals contain concentrated vitamin A in small volumes (100,000–500,000 IU/mL) that can cause hypervitaminosis A with small or repeated dosages. In addition, the empirically set dosages vary widely in the literature, and often were not related to body weight. Finally, it has been shown that aqueous vitamin A solutions can produce hypervitaminosis A experimentally at doses well below fat-soluble vitamin A preparations. For example, in an experimental study in Hermann's Tortoises, 100,000–400,000 IU/kg of water-soluble vitamin A palmitate induced epidermal sloughing in 2 weeks, but not with identical doses of fat-soluble vitamin A palmitate (Palmer et al., 1984). Other nutrients can interfere with vitamin A physiology, such as proteins and other fat-soluble vitamins (D, E, K). Protein deficiency, by causing a decrease in the plasma levels of retinol-binding protein, may provide protection against hypervitaminosis A. Hypervitaminosis A ensues when the storage and metabolic capacity of the liver is surpassed and retinol unbound to its plasma carrier protein reaches the tissues (Boyer, 2006).

The clinical presentation of tortoises with hypervitaminosis A is characterized by cutaneous disease. The earliest effect in the skin is erythema, followed by the appearance of vesicles, 2 weeks post injection. Subsequently, epidermal sloughing occurs in affected areas. Axillary, inguinal, cervical, and limb skin is most commonly affected, but other areas may be involved, particularly in the most severely diseased patients, in which most of the epidermis, including carapacial scutes, may be lost (**Figures 2.83–2.85**). As epidermal lesions progress, the loss of cutaneous integrity and anorexia can lead to secondary bacterial and fungal infections, dehydration, and death. Histologically, lesions initially consist of flattening of the stratum corneum and stratum germinativum, which is followed by acantholysis, spongiosis, and separation of both strata with formation of vesicles or bullae, as well as epidermal hyperplasia and acanthosis. Bullae eventually rupture, and epidermal ulceration and sloughing ensues (**Figure 2.86**) (Palmer et al., 1984; Frye, 1989). Essentially, hypervitaminosis A causes chelonians to slough their outer epidermis, much like a severe burn (Boyer and Scott, 2019b).

2.12 Pansteatitis

Steatonecrosis (necrosis of adipose tissue) and granulomatous-necrotizing pansteatitis (granulomatous inflammation around necrotic adipose tissue) can occur in a wide range of animal taxa as a result of diverse insults, including (1) ischemia (e.g., from trauma, tall fescue toxicosis, vasculitis, and toxic compounds in general causing vasoconstriction); (2) primary vitamin E and/or selenium deficiency; (3) feeding rancid/improperly frozen or thawed fish—particularly species rich in polyunsaturated fatty acids—in piscivorous animals; (4) pancreatitis; (5) exposure to toxins/pollutants with pro-oxidative effects; (6) maldigestion/malabsorption syndromes of gastrointestinal, hepatic, or pancreatic origin causing secondary deficiency of vitamin E and other antioxidants; and (7) infections (e.g., Coxsackie viruses) (Godman et al., 1952; Huchzermeyer, 2002; Donoghue, 2006; Lane et al., 2013; Orós et al., 2013). Interaction of multiple potential etiological factors may affect the balance between oxidation and antioxidant capacity or other aspects of the physiology of the adipose tissue. Potential interactions include, for example, the first known example of a virus undergoing mutations as a result of a nutritional deficiency. Vitamin E and selenium deficiencies each can induce the same six-point mutations in the genome of Coxsackie viruses isolated from humans with Keshan disease, and these mutations increase their pathogenicity in mouse experimental models (Levander and Beck, 1997). Coxsackie viruses are known to cause rhabdomyonecrosis and steatitis in mice (Godman et al., 1953), and have also been isolated from tamarins with rhabdomyonecrosis and steatitis (Baskin et al., 1983).

Steatonecrosis and pansteatitis have been described in reptiles, mostly captive crocodilians (Wallach and Hoessle, 1968; Larsen et al., 1983; Ladds et al., 1995; Huchzermeyer, 2002) but also wild Nile Crocodiles (*Crocodylus niloticus*) (Lane et al., 2013), captive Olive Ridley Sea Turtles (*Lepidochelys olivacea*) (Manawatthana and Kasorndorkbua, 2005), as well as a stranded Loggerhead (*Caretta caretta*) (Orós et al., 2013) and Kemp's Ridley (*Lepidochelys kempi*) (Mumford, 1999) Sea Turtles. In some of these reports, steatitis was associated with or due to feeding only-fish diets (Wallach and Hoessle, 1968; Larsen et al., 1983), such as freshwater fish scraps without vitamin supplements (Larsen et al., 1983), feeding dead and rancid fish (Huchzermeyer, 2002), and feeding frozen fish without vitamin E supplementation for years (Manawatthana and Kasorndorkbua, 2005). Vitamin E is abundant in fresh fish, but fish lipids have a lower freezing point than mammalian fats and much of the vitamin E may be destroyed by oxidation after only a few weeks of freezing (Allen and Oftedal, 1994; Bernard and Allen, 1997). Degradation and oxidation of fish lipids, resulting in rancidity, can be accelerated by inappropriately high standard freezer temperatures, and therefore fish should be frozen between −18° to −30°C (0° to −22°F). Frozen fish should be used within 4–6 months (Bernard and Allen, 1997). Suggested levels of supplementation are 100 IU of vitamin E per kg of wet fish or 1 IU/kg/day for reptiles (Bernard and Allen, 1997; Donoghue, 2006). If fresh fish is not available, frozen fish must be thawed rapidly in hot water, or under cool temperatures, not room temperature, and any fish with gross evidence of rancidity must be discarded (Donoghue, 2006).

In wild Nile Crocodiles with pansteatitis the etiopathogenesis was not determined but an interaction of environmental, nutritional, and toxic factors was considered (Lane et al., 2013). Interestingly, pansteatitis was also diagnosed in Sharptooth Catfish (*Clarias gariepinus*), caught in the same river and a component of the diet of these crocodiles. This latter study suggested that an alien-invasive Silver Carp (*Hypophthalmychthys molitrix*) with high levels of n-3 polyunsaturated fatty acids, which other fish and crocodiles feed upon, could be the culprit for steatitis (Huchzermeyer, 2012).

Exposure to toxic compounds and water pollutants increasing lipid peroxidation or affecting the physiology of adipose tissue by other mechanisms needs to be investigated in cases of this condition in reptiles. High levels of polychlorinated biphenyls were noted in a wild Loggerhead Sea Turtle with pansteatitis (Orós et al., 2013). (For additional information, see Chapter 7.)

Clinical findings of pansteatitis/steatonecrosis in reptiles include lethargy, inability or unwillingness to move, and buoyancy abnormalities (Lane et al., 2013; Orós et al., 2013). In crocodilians, large saponified intermuscular fat deposits in the tail may render the tail immobile, palpably firm, and the animal unable to swim. Saponification of intracoelomic fat may reduce intestinal motility (Huchzermeyer, 2003; Boyer and Scott, 2019b). Grossly, areas of yellow-orange to brownish discoloration are noted in the adipose tissue in any location (**Figure 2.87**). This lesion may be diffuse. Affected adipose tissue is typically hardened (Larsen et al., 1983; Lane et al., 2013; Orós et al., 2013). Histologic examination reveals foci of necrosis and saponification of adipocytes with infiltration of degenerate heterophils (**Figure 2.88**). These areas may be surrounded by vacuolated macrophages and multinucleate giant cells as well as fibrous tissue (**Figures 2.89–2.90**), and contain mineral deposits. A slightly refractile, bright yellow pigment may be found within the cytoplasm of adipocytes and phagocytic cells. This pigment (ceroid) is acid fast with modified Ziehl-Neelsen stain (Larsen et al., 1983; Lane et al., 2013; Orós et al., 2013).

2.13 Goiter

Goiter refers to the enlargement of the thyroid glands due to non-inflammatory and non-neoplastic disease processes. Thyroid function may be normal (nontoxic goiter) or abnormal (toxic goiter). In animals, goiter is most commonly due to dietary deficit or excess of iodine, exposure to goitrogenic substances in the diet or water, or combinations thereof. Hereditary goiter is rare in nonhuman species. Iodine-deficient soils are present in a number of geographic areas in the world but the incidence of goiter in these areas has been reduced by the use of iodine salts and kelp. Endemic goiter was reported by Zwart and Kok (1978), in reptiles in the Netherlands. Disease can occur with marginal iodine deficiency in animals exposed to goitrogenic compounds (Rosol and Gröne, 2016). Goitrogenic substances include numerous iodine-binding plants in the family Brassicaceae (e.g., broccoli, cauliflower, Brussels sprouts, mustard greens, kale, cabbage, bok choy, and turnips), walnuts, nitrites and nitrates (e.g., from soils rich in these substances or fertilized pastures), sulfonamides, or thiouracil, among others (Frye and Dutra, 1974; Donogue, 2006; Rosol and Gröne, 2016). Although Brassicas are wildly speculated to cause goiter in tortoises, this has never been validated. Feeding Brassicas to herbivorous reptiles is harmless in moderation, as part of a balanced diet, as long as Brassicas are not fed exclusively over a long period of time (Boyer and Scott, 2019b).

Goiter has been described in reptiles, including chelonians (Frye and Dutra, 1974), Kirtland's Snakes (*Clonophis kirtlandii*) (Gyimesi et al., 2008), and an Eastern Diamondback Rattlesnake (*Crotalus adamanteus*) (Topper et al., 1994). Toxic goiter was observed in a Green Iguana secondary to a thyroid adenoma (Hernandez-Divers et al., 2001). Two tortoises with intranuclear coccidiosis had nontoxic goiter, a Radiated Tortoise (*Astrochelys radiata*) (T. Boyer, personal observation) and a Bowsprit Tortoise (*Chersina angulata*) (Garner et al., 2006). In the Radiated Tortoise, the head, neck, and front legs (not the back legs) were edematous and T_4 was 11.5 nmol/L, 11 days prior to death (within normal limits for other chelonians).

Goiter is commonly misdiagnosed in Galapagos (*Chelonoidis nigra* complex) and Aldabra (*Aldabrachelys gigantea*) Tortoises with peripheral edema and a large ventral neck swelling that superficially resembles human goiter; many of these patients suffer from myocardial and hepatic fibrosis. Ventral neck swelling in giant tortoises often represents thymic hyperplasia (Fleming et al., 2004)—the chelonian thymus does not involute as in mammals (Boyer and Innis, 2019)—but other forms of swelling in the ventral neck or head should also be considered, such as edema due to heart, liver, or renal disease, abscess formation, neoplasia, or regional envenomation. The unpaired chelonian thyroid gland sits just cranial to the chelonian heart, near the bifurcation of the common carotid artery, well within the shell, caudal to the thymus, and not on the ventral neck (Boyer and Innis, 2019).

Buehler (2006) described an adult male Hermann's Tortoise making whistling respiratory sounds associated with a sonographically enlarged thyroid gland compressing both atria. Blood parameters and thyroid hormone concentrations were within established normal ranges. After 1 month of 1 mg/kg of potassium iodide orally daily, the whistling sound ceased, and the sonographic thyroid diameter had decreased from 34 mm to 23 mm and the atria were no longer compressed. Therapy was continued, and 3 months later the ultrasonic thyroid diameter was 16.8 mm. Over the next few months the thyroid diameter remained $17.0 \text{ mm} \pm 0.2 \text{ mm}$ and the potassium iodide was reduced to twice weekly doses, and the thyroid size remained unchanged sonographically for 6 additional years (Buehler, 2006; Boyer and Steffes, 2011).

In reptiles, a great diversity exists for anatomical and physiological aspects of thyroid glands, and a detailed review can be found elsewhere (Boyer and Steffes, 2011). (See also Volume 1, Chapter 1.) Plasma levels of thyroid hormones in reptiles are lower than those of mammals (roughly 20% and 25% for T_4 and T_3, respectively), which adds difficulty to thyroid hormone function testing and interpretation because assays are designed and standardized for thyroid hormone testing in mammals (Boyer and Steffes, 2011). Overall, there is a paucity of reference intervals for T_3/T_4 in reptiles (see Raiti, 2019, for normal values). A number of factors may affect the levels of thyroid hormones and glandular morphology, including species, gender, age, temperature/climate and geographic distribution, underlying disease, season, daily variations, shedding, stress, drugs such as propofol, reproductive activity (folliculogenesis and spermatogenesis), hibernation, or anorexia and inanition (Boyer and Steffes, 2011). For example, withholding food for 14 days significantly decreased T_4 levels in Mojave Desert Tortoises; once fed, T_4 levels rose within 36 hours (Kohel et al., 2001). These variations should be taken into account when assessing thyroidal diseases in reptiles. Low T_3 and T_4 levels may be noted in reptiles with goiter, as compared to unaffected reptiles, but this does not prove hypothyroidism has developed, as these values may actually be due to other diseases not involving the thyroid glands in euthyroid patients (euthyroid sick syndrome) (Donogue, 2006; Mans and Braun, 2014). Thyroid stimulation tests are necessary to properly document hypothyroidism and have been reported in Painted Turtles (*Chrysemys picta*) (Sawin et al., 1981). Imaging techniques may be useful in assessing alterations in thyroid gland morphology; in this regard, thyroid glands are easily measured in chelonians with ultrasound (Pajdak-Czaus et al., 2019).

Grossly, thyroid glands with goiter are enlarged and may have an irregular to nodular surface due to the presence of follicles that are markedly distended with colloid (**Figure 2.91**). In diffuse hyperplastic goiter or thyroid follicular hyperplasia, follicles are depleted of colloid and lined by tall cuboidal to cylindrical epithelium that may form papillary projections into the follicular lumen (**Figures 2.92–2.93**). An adenomatous appearance may be noted in advanced lesions. In colloid goiter, follicles are distended with excessive colloid and lined by a single layer of flattened follicular epithelium (**Figures 2.94–2.95**). In some cases of colloid goiter, even markedly distended follicles may retain papillary projections of cylindrical epithelium with prominent vacuolation of the adjacent colloid (**Figures 2.94–2.97**), which is consistent with an incomplete transition from diffuse hyperplastic to colloid goiter and ongoing TSH-mediated endocytosis of colloid.

2.14 Thiamine Deficiency

Thiamine (vitamin B_1) is a water-soluble, labile vitamin involved in oxidative decarboxylations and other metabolic reactions. Some enzymatic systems including α-ketoglutarate dehydrogenase, pyruvate dehydrogenase, or transketolase use thiamine as a cofactor. Thiamine may play a role in normal neurological functions such as axonal conduction and synaptic transmission (Cantile and Youssef, 2016). Unlike herbivores, which can obtain thiamine by bacterial synthesis in the gastrointestinal tract, carnivores are dependent on a nutritional source to meet thiamine requirements (Cantile and Youssef, 2016). A number of compounds may be involved in the metabolism of thiamine in herbivores and carnivores. Most notably, the thiamine-splitting enzymes (thiaminases) found normally in many fish species, as well as sulfur compounds in feeds and water, can cause thiamine deficiency and thiamine-responsive neurological disorders in fish-eating mammals, domestic carnivores, and ruminants (polioencephalomalacia) (Mans and Braun, 2014; Cantile and Youssef, 2016). Sulfite inactivates thiamine by cleaving it into pyrimidine and thiazole constituents. For example, sulfur compounds used to color meat can cause thiamine deficiency in dogs and cats, but several other sources of the sulfur-mediated disorder exist, including thiocyanate present in plants of the genus *Brassica* (Cantile and Youssef, 2016). Other causes of thiamine deficiency or alterations of thiamine metabolism include plants containing phytothiaminases (such as Bracken and Rock Ferns [*Pteridium aquilinum, Cheilanthes sieberi*], Nardoo [*Marsilea drummondi*], Horsetail [*Equisetum arvense*], and Kochia [*Kochia scoparia*]), thiaminase-containing bacteria particularly in an acidic environment, and the coccidiostat amprolium (Cantile and Youssef, 2016), none of which commonly apply to reptiles. Heat and acidic environments may inactivate thiamine (Cantile and Youssef, 2016). Feeding frozen fish after thawing for several hours at room temperature may cause thiamine deficiency because thiaminases are not inactivated/destroyed during freezing (Mans and Braun, 2014). An interaction of several of these etiological factors may be involved in some cases of thiamine deficiency.

In reptiles, thiamine deficiency or thiamine-responsive disorders have been documented mostly in fish-eating species or animals fed fish in captivity such as crocodilians, turtles, garter snakes (*Thamnophis* spp.), and Brown Water Snakes (*Nerodia taxisplilota*); however, herbivorous and insectivorous reptiles may also be rarely affected (Zwart and van Ham, 1980; Frye, 1991; Rossi and Rossi, 1992; Honeyfield et al., 2008; Feldman et al., 2011; Mans and Braun, 2014). Thiamine deficiency has been reported in farmed and wild crocodilians (Jubb, 1992; Schoeb et al., 2002; Honeyfield et al., 2008). Farmed American Alligators (alligators; *Alligator misssissippiensis*) fed shrimp bycatch had increased aggression and blindness and developed acute focal necrosis of the torus semicircularis in the midbrain (Schoeb et al., 2002). In Australia, farmed Saltwater Crocodiles were affected with a thiamine-responsive neurological disease characterized by open jaws, loss of righting reflex (often found lying on their sides or backs, or floating), or sluggishness (Jubb, 1992). Affected specimens were 2- to 3-month-old hatchlings, usually the largest in their groups; Freshwater Crocodiles bred under similar conditions on the same farm were not affected. The diet consisted of thawed raw donkey meat after frozen storage and a multivitamin supplement with a low content of thiamine (3.1 mg/kg). Affected hatchlings recovered after administration of thiamine hydrochloride (30 mg IM, SID for 2 days). No further cases occurred after the addition of dried brewer's yeast (0.3% of diet), with a concentration of thiamine (95.2 mg/kg) higher than the levels in the multivitamin supplement (3.1 mg/kg). Sulfur dioxide concentrations in the diet were low. The authors hypothesized the affected hatchlings possibly had higher thiamine requirements as a result of high growth rate as compared to smaller clutch mates and to unaffected Freshwater Crocodiles, characterized by an inherent lower growth rate.

In late 1990s and early 2000s, increased mortality in wild alligators from Lake Griffin (Florida, US) was reported. Affected specimens were adult or subadult and were found either dead or lethargic and neurologically impaired, with incoordination, swimming in circles, inability to submerge, unresponsiveness, slow reflexes, dragging the dorsal surfaces of the hind feet, and, rarely, anisocoria, head tilt, and reduced or absent hearing (Schoeb et al., 2002). Grossly, affected alligators were unremarkable except for a variety of parasites and, in a few specimens, trauma-related lesions, gastric ulcers, or the presence of water or mud within the lumen of lungs that suggested drowning as the ultimate cause of death (Schoeb et al., 2002; Honeyfield et al., 2008). Histologically, acute necrosis in the brain was the most consistent finding. The telencephalon, particularly the dorsal ventricular ridge, and torus semicircularis in the midbrain were the most frequently affected areas (Schoeb et al., 2002; Honeyfield et al., 2008), although others such as the thalamus, granular layer and white matter of the cerebellum, and medulla were also involved in some alligators (Honeyfield et al., 2008). Thiamine levels were lower in affected specimens than in unaffected alligators from other lakes and Lake Griffin as well (Honeyfield et al., 2008). Embryonic mortality was also associated with thiamine deficiency (Sepúlveda et al., 2004), and similar CNS lesions were noted in some embryos (Richey, 2001). Honeyfield et al. (2008) hypothesized that affected alligators developed thiamine deficiency as a result of consumption of Gizzard Shad

(*Dorosoma cepedianum*), which is rich in thiaminase and an important dietary component for alligators (Honeyfield et al., 2008). However, alligators in a different lake unaffected with mortality were shown to consume a higher proportion of gizzard shad than Lake Griffin alligators and therefore gizzard shad consumption was ruled out as the only cause of this thiamine deficiency–associated mortality (Rice et al., 2007).

For fish-eating species, long-term prevention includes feeding a better balanced, more diverse diet, and supplementation of 25–30 mg thiamine per kg of wet fish (Bernard and Allen, 1997). Thiaminases are not restricted to marine or freshwater fish; they can be found in either. Trout do not contain thiaminases and do not require thiamine supplementation (Boyer and Scott, 2019b).

A vitamin B complex–responsive neurological condition has also been reported in the insectivorous anoles (*Anolis spp.*) kept in a laboratory colony (Feldman et al., 2011). Reluctance to move, tremors of the distal appendages, and postural distortion of the axial skeleton, including hyperextension or twisting and serpentine distortion, were the reported clinical signs. One affected anole treated with vitamin B complex intraperitoneally recovered after the second injection. Based on these findings, and considering the half-life of B complex vitamins, the authors suspected thiamine deficiency and attributed it to the use of a multivitamin complex stored for over 2 years. However, supportive diagnostic data including tissue and diet thiamine levels, as well as compatible encephalic lesions, were not provided.

2.15 Hyperglycemia

Hyperglycemia is defined as a persistent elevation of blood glucose above 300 mg/dL (16.7 mmol/L) and is an uncommon clinical biochemical abnormality in reptiles (Stahl, 2003; Stahl, 2019). Hyperglycemia has not been established as a consistent or specific indicator of pancreatic disease or diabetes mellitus (DM) in reptiles (Stahl, 2003; Raiti, 2019; Stahl, 2019). Reports of spontaneous DM in reptiles are rare (Frye et al., 1976). Diabetes has been reproduced experimentally in reptiles (Edwards et al., 2018) and postoperatively after removing a neoplastic splenopancreas in a snake (Stern et al., 2010). The diagnosis of spontaneous cases of DM in reptiles was based mostly on the finding of severe or persistent hyperglycemia, glucosuria, and presence of lesions in the islets of Langerhans or the pancreas as a whole (e.g., hydropic degeneration or loss of cytoplasmic granulation of islet cells, islet amyloidosis, granulomatous pancreatitis, or fibrosis effacing the entire pancreas) or combinations thereof. However, environmental and physiologic/metabolic factors (e.g., brumation or reproductive activity) as well as stress or concurrent disease, especially neoplasia (e.g., hepatic lipidosis, hepatic carcinoma, renal or gastrointestinal adenocarcinoma, chronic glomerulonephritis, interstitial nephritis, granulomatous pancreatitis, pancreatic glucagonoma, gastric somatostatinoma), can cause hyperglycemia in reptiles (Stahl, 2003; Ritter JM et al., 2009; Raiti, 2019; Stahl, 2019). Furthermore, the significance of glucosuria in reptiles has not been consistently established (Stahl, 2003). In addition, the morphological and functional diversity of the endocrine pancreas among reptiles, with different localization and density of the islet tissue depending on the species, must be taken into account to avoid failures in the histopathologic evaluation of the pancreas, both for biopsy and postmortem samples. Finally, the clinical tests evaluating the glycemic control beyond the finding of severe or persistent hyperglycemia in mammals, most notably the determination of glycosylated hemoglobin, are not standardized in reptiles.

References

Acierno MJ, Mitchell MA, Roundtree MK, and Zachariah TT. 2006. Effects of ultraviolet radiation on 25-hydroxyvitamin D_3 synthesis in red-eared slider turtles (*Trachemys scripta elegans*). *Am J Vet Res* 67:2046–2049.

Acierno MJ, Mitchell MA, Zachariah TT, Roundtree MK, Kirchgessner MS, and Sanchez-Migallon GD. 2008. Effects of ultraviolet radiation on plasma 25-hydroxyvitamin D_3 concentrations in corn snakes (*Elaphe guttata*). *Am J Vet Res* 69:294–297.

Allen M, and Oftedal O. 1989. Dietary manipulation of the calcium content of feed crickets. *J Zoo Wildl Med* 20:27–33.

Allen M, and Oftedal O. 1994. The nutrition of carnivorous reptiles, in *Captive Management and Conservation of Amphibians and Reptiles*. Murphy JB, Adler K, and Collins JT (Eds.), Society for the Study of Amphibians and Reptiles, Washington, DC, pp. 71–82.

Allen M, and Oftedal O. 2003. Nutrition in captivity, in *Biology, Husbandry, and Medicine of the Green Iguana*. Jacobson E (ed.), Krieger Publishing Co, Malabar FL, 55.

Anderson MP, and Capen CC. 1976a. Nutritional osteodystrophy in captive green iguanas (*Iguana iguana*). *Virchows Arch B Cell Pathol* 21:229–247.

Anderson MP, and Capen CC. 1976b. Fine structural changes of bone cells in experimental nutritional osteodystrophy of green iguanas. *Virchows Arch B Cell Pathol* 20:169–184.

Ariel E, Ladds PW, and Buenviaje GN. 1997. Concurrent gout and suspected hypovitaminosis A in crocodile hatchlings. *Aust Vet J* 75:247–249.

Aucone BM, Gehrmann WH, Ferguson GW, Chen T, and Holick MF. 2003. Comparison of two artificial ultraviolet light sources used for Chuckwalla, *Sauromalus obesus*, husbandry. *J Herpetol Med Surg* 13:14–17.

Avery HW, Spotila JR, Congdon JD, Fischer RU, Standora EA, and Avery SB. 1993. Roles of diet protein and temperature in the growth and nutritional energetics of juvenile slider turtles, *Trachemys scripta*. *Phys Biochem Zool* 66:902–925.

Baila-Rueda L, Mateo-Gallego R, Lamiquiz-Moneo I, Cenarro A, and Civeira F. 2014. Severe hypercholesterolemia and phytosterolemia with extensive xanthomas in primary biliary cirrhosis: Role of biliary excretion on sterol homeostasis. *J Clin Lipidol* 8:520–524.

Baines F, and Cusack L. 2019. Environmental lighting, in *Mader's Reptile and Amphibian Medicine and Surgery*, third edition. Divers S, Stahl SJ (Eds.), Elsevier, St. Louis, MO, pp. 131–138

Barker D, Fitzpatrick MP, and Dierenfeld ES. 1998. Nutrient composition of selected whole invertebrates. *Zoo Biol* 17:123–134.

Barten SL. 2006. Lizards, in *Reptile Medicine and Surgery*, 2nd ed. Mader DR (Ed.), Saunders Elsevier, St. Louis, pp. 59–77.

Baskin GB, Wolf RH, Worth CL, Soike K, Gibson S, and Bieri JG. 1983. Anemia, steatitis and muscle necrosis in marmosets (*Saguinus labiatus*). *Lab Anim Sci* 33:74–80.

Bernard JB, and Allen ME. 1997. Feeding captive piscivorous animals: Nutritional aspects of fish as food, fact sheet 5, JB Bernard, Allen ME, and Ullrey DE (Eds.). *Nutrition Advisory Group Handbook*: 1–7.

Bernhard G, Mayer B, Seckmeyer G, and Moise A. 1997. Measurements of spectral solar UV irradiance in tropical-Australia. *J Geophys Res Atmos* 102:8719–8730.

Boyer TH. 1996. Metabolic bone disease, in *Reptile Medicine and Surgery*. Mader DR (Ed.), W. B. Saunders Company, Philadelphia, PA, 385–392.

Boyer TH. 2006. Hypovitaminosis A and hypervitaminosis A, in *Reptile Medicine and Surgery*, 2nd ed. Mader DR (Ed.), Saunders Elsevier, St. Louis, pp. 831–835.

Boyer TH. 2018. Vitamin A deficiency in insectivorous lizards. *Clinician's Brief*, 16(11):15–20.

Boyer TH, and Boyer DM. 2006. Turtles, tortoises and terrapins, in *Reptile Medicine and Surgery*, 2nd ed. Mader DR (Ed.), Saunders Elsevier, St. Louis, pp. 78–89.

Boyer TH, and Innis CJ. 2019. Chelonian taxonomy, anatomy and physiology. In *Mader's Reptile and Amphibian Medicine and Surgery*. Divers SJ, and Stahl SJ (Eds). Elsevier, St. Louis, pp. 31–49.

Boyer TH, and Steffes ZJ. 2011. Reptilian thyroid anatomy, physiology and disease. Master class. *Proc Assoc Reptilian Amphib Vet* 18–35.

Boyer T, and Boyer D. 2019. Tortoises, freshwater turtles and terrapins, in *Mader's Reptile and Amphibian Medicine and Surgery*, 3rd ed. Divers S and Stahl SJ (Eds.), Elsevier, St. Louis, MO, pp. 168–179.

Boyer T, and Scott P. 2019a. Nutrition, in *Mader's Reptile and Amphibian Medicine and Surgery*, 3rd ed. Divers S and Stahl SJ (Eds.), Elsevier, St. Louis, MO, pp. 201–223.

Boyer T, and Scott P. 2019b. Nutritional diseases, in *Mader's Reptile and Amphibian Medicine and Surgery*, 3rd ed. Divers S, and Stahl SJ (Eds.), Elsevier, St. Louis, MO, pp. 932–950.

Branch B 2008. *Tortoises, Terrapins and Turtles of Africa*. Struik Publishers, Cape Town.

Brown JD, Richards JM, Robertson J, Holladay S, and Sleeman JM. 2004. Pathology of aural abscesses in free-living Eastern box turtles (*Terrapene carolina carolina*). *J Wildl Dis* 40:704–712.

Burgess TL, Gartrell BD, and Blanchard B. 2009. A survey of the husbandry of captive tuatara (*Sphenodon* spp.) in relation to factors implicated in nutritional secondary hyperparathyroidism. *N Z Vet J* 57:378–382.

Campbell TW, and Ellis CK. 2007. *Avian and Exotic Animal Hematology and Cytology*, 3rd ed. Blackwell Publishing, Ames.

Cantile C, and Youssef S. 2016. Nervous system, in *Jubb, Kennedy, and Palmer's Pathology of Domestic Animals*, 6th ed. Maxie MG (Ed.), Elsevier, St Louis, pp. 250–406.

Carman EN, Ferguson GW, Gehrmann WH, Chen TC, Holick MF, and Douglas ME. 2000. Photobiosynthetic opportunity and ability for UV-B generated vitamin D synthesis in free-living house geckos (*Hemidactylus turcicus*) and Texas spiny lizards (*Sceloporus olivaceous*). *Copeia* 2000:245–250.

Carpenter JL, Bacharach A, Albert DM, Vainsi SJ, and Goldstein MA. 1986. Xanthomatous keratitis, disseminated xanthomatosis, and artherosclerosis in Cuban tree frogs. *Vet Pathol* 23:337–339.

Chaffin K, Norton TM, Gilardi K et al. 2008. Health assessment of free-ranging alligator snapping turtles (*Macrochelys temminckii*) in Georgia and Florida. *J Wildl Dis* 44:670–686.

Cooper JE, Bloxam QMC, and Tonge SJ. 1998. Pathology of Round Island geckos *Phelsuma guentheri*: Some unexpected findings. *Dodo* 34:153–158.

Craig LE, Dittmer KE, and Thompson KG. 2016. Bones and joints, in *Jubb, Kennedy, and Palmer's Pathology of Domestic Animals*, 6th ed. Maxie MG (Ed.), Elsevier, St Louis, pp. 16–163.

Daniel JC 2002. *The Book of Indian Reptiles and Amphibians*. Oxford University Press, Oxford.

De Silva A. 2003. *The Biology and Status of the Star Tortoise (Geochelone elegans Schoepff, 1795) in Sri Lanka*. AMP Print Shop, Gampola.

Diehl JJE, Baines FM, Heijboer AC, van Leeuwen JP, Kik M, Hendriks WH, and Oonincx DGAB. 2017. A comparison of UVb compact lamps in enabling cutaneous vitamin D synthesis in growing bearded dragons. *J Anim Physiol Anim Nutr* 102:308–316.

Donoghue S. 2006. Nutrition, in *Reptile Medicine and Surgery*, 2nd ed. Mader DR (Ed.), Saunders Elsevier, St. Louis, pp. 251–298

Dötsch J, Zepf K, Schellmoser S, Rascher W, and Dörr HG. 2001. Unmasking of childhood hypothyroidism by disseminated xanthomas. *Pediatrics* 108:E96.

Dutra G. 2014. Diagnostic value of hepatic enzymes, triglycerides and serum proteins for the detection of hepatic lipidosis in *Chelonoidis carbonaria* in captivity. *J Life Sci* 8:633–639.

Edwards TM, Hamlin HJ, Freymiller H, Green S, Thurman J, and Guillette LJ. 2018. Nitrate induces a type 1 diabetic profile in alligator hatchlings. *Ecotoxicol Environ Saf* 147:767–775.

Elkan E, and Zwart P. 1967. The ocular disease of young terrapins caused by vitamin A deficiency. *Pathol Vet* 4:201–222.

Feldman SH, Formica M, and Brodie ED III. 2011. Opisthotonus, torticollis and mortality in a breeding colony of *Anolis* sp. lizards. *Lab Anim (NY)* 40:107–108.

Ferguson GW, Gehrmann WH, Bradley KA, Lawrence B, Hartdegen R, Storms T, Chen TC, and Holick MF. 2015. Summer and winter seasonal changes in vitamin D status of captive rhinoceros iguanas (*Cyclura cornuta*). *J Herpetol Med Surg* 25:128–136.

Ferguson GW, Gehrmann WH, Karsten KB, Hammack SH, McRae M, Chen TC, Lung NP, and Holick MF. 2003. Do panther chameleons bask to regulate endogenous vitamin D_3 production?. *Physiol Biochem Zool* 76:52–59.

Ferguson GW, Gehrmann WH, Karsten KB, Landwer AJ, Carman EN, Chen TC, and Holick MF. 2005. Ultraviolet exposure and vitamin D synthesis in a sun-dwelling and a shade-dwelling species of *Anolis*: Are there adaptations for lower ultraviolet B and dietary vitamin D_3 availability in the shade?. *Physiol Biochem Zool* 78:193–200.

Ferguson GW, Jones JR, Gehrmann WH et al. 1996. Indoor husbandry of the panther chameleon *Chamaeleo* [*Furcifer*] *pardalis*: Effects of dietary vitamins A and D and ultraviolet irradiation on pathology and life-history traits. *Zoo Biol* 15:279–299.

Finke MD. 2002. Complete nutrient composition of commercially raised invertebrates used as food for insectivores. *Zoo Biol* 21:269–285.

Finke MD. 2015. Complete nutrient content of four species of commercially available feeder insects fed enhanced diets during growth. *Zoo Biol* 34:554–564.

Fledelius B, JØrgensen GW, Jensen HE, and Brimer L. 2005. Influence of the calcium content of the diet offered to leopard tortoises (*Geochelone pardalis*). *Vet Rec* 156:831–835.

Fleming GJ, Heard DJ, Uhl EW, and Johnson CM. 2004. Thymic hyperplasia in subadult Galapagos tortoises, *Geochelone nigra*. *J Herp Med Surg* 14:24–27.

Frye FL. 1984. Nutritional disorders in reptiles, in *Diseases of Amphibians and Reptiles*. Hoff GL, Frye FL, Jacobson, ER (Eds.), Plenum, New York, 654–656.

Frye FL. 1989. Vitamin A sources, hypovitaminosis A and iatrogenic hypervitaminosis A in captive chelonians, in *Current Veterinary Therapy X Small Animal Practice*. Kirk RW (Ed.), Saunders, Philadelphia, pp. 791–796.

Frye FL. 1991. *A Practical Guide for Feeding Captive Reptiles*. Krieger Publishing, Malabar, FL.

Frye FL, and Dutra FR. 1974. Hypothyroidism in turtles and tortoises. *Vet Med Small Anim Clin* 69:990–993.

Frye FL, Dutra FR, Carney JD, and Johnson B. 1976. Spontaneous diabetes mellitus in a turtle. *Vet Med Small Anim Clin* 71:935–939.

Gardiner DW, Baines FM, and Pandher K. 2009. Photodermatitis and photokeratoconjunctivitis in a ball python (*Python regius*) and a blue-tongue skink (*Tiliqua* spp.). *J Zoo Wildl Med* 40:757–766.

Garner MM, Gardiner CH, Wellehan JFX, Johnson AJ, McNamara T, Linn M, Terrell SP, Childress A, and Jacobson ER. 2006. Intranuclear coccidiosis in tortoises: Nine cases. *Vet Pathol* 43:311–320.

Garner MM, Lung NP, and Murray S. 1999. Xanthomatosis in geckos: Five cases. *J Zoo Wildl Med* 30:443–447.

Gehrmann WH. 2006. Artificial lighting, in *Reptile Medicine and Surgery*, 2nd ed. Mader DR (Ed.), Saunders Elsevier, St. Louis, pp. 1081–1084.

Gerlach J. 2004. Effects of diet on the systematic utility of the tortoise carapace. *Afr J Herpetol* 53:77–85.

Gillespie D, Frye FL, Stockham SL, and Fredeking T. 2000. Blood values in wild and captive Komodo dragons (*Varanus komodoensis*). *Zoo Biol* 19:495–509.

Godman GC, Bunting H, and Melnick JL. 1952. The histopathology of Coxsackie virus infection in mice. I. Morphologic observations with four different viral types. *Am J Pathol* 28:223–257.

Gould A, Molitor L, Rockwell K, Watson M, and Mitchell MA. 2018. Evaluating the physiologic effects of short duration ultraviolet B radiation exposure in leopard geckos (*Eublepharis macularius*). *J Herpetol Med Surg* 28:34–39.

Gramanzini M, Di Girolamo N, Gargiulo S, Greco A, Cocchia N, Delogu M, Rosapane I, Liuzzi R, Selleri P, and Brunetti A. 2013. Assessment of dual-energy x-ray absorptiometry for use in evaluating the effects of dietary and environmental management on Hermann's tortoises (*Testudo hermanni*). *Am J Vet Res* 74:918–924.

Gyimesi ZS, Garner MM, and Burns RB. 2008. Goiter and thyroid disease in captive kirtland's snakes, *Clonophis Kirtlandi*. *J Herpetol Med Surg* 18:75–80.

Gyimesi ZS, Stedman NL, and Crossett VR. 2002. Cholesterol granulomas in a great plated lizard, *Gerrhosaurus major*. *J Herpetol Med Surg* 12:36–39.

Haxhiu D, Hoby S, Wenker C, Boos A, Kowalewski MP, Lewis F, and Liesegang A. 2014. Influence of feeding and UVB exposition on the absorption mechanisms of calcium in the gastrointestinal tract of veiled chameleons (*Chamaeleo calyptratus*). *J Anim Physiol Anim Nutr (Berl)* 98:1021–1030.

Hedley J, and Eatwell K. 2013. The effects of UV light on calcium metabolism in ball pythons (*Python regius*). *Vet Rec* 173:345.

Heinrich ML, and Heinrich KK. 2016. Effect of supplemental heat in captive African leopard tortoises (*Stigmochelys pardalis*) and spurred tortoises (*Centrochelys sulcata*) on growth rate and carapacial scute pyramiding. *J Exot Pet Med* 25:18–25.

Hernandez-Divers SJ, and Cooper JE. 2006. Hepatic lipidosis, in *Reptile Medicine and Surgery*, 2nd ed. Mader DR (Ed.), Saunders Elsevier, St. Louis, pp. 806–813.

Hernandez-Divers SJ, Knott CD, and MacDonald J. 2001. Diagnosis and surgical treatment of thyroid adenoma-induced hyperthyroidism in a green iguana (*Iguana iguana*). *J Zoo Wildl Med* 32:465–475.

Hoby S, Wenker C, Robert N, Jermann T, Hartnack S, Segner H, Aebischer C-P, and Liesegang A. 2010. Nutritional metabolic bone disease in juvenile veiled chameleons *Chamaeleo calyptratus*) and its prevention. *J Nutr* 140:1923–1931.

Hoekstra KA, Nichols CR, Godin DV, and Cheng KM. 1998. Dietary cholesterol-induced xanthomatosis in atherosclerosis-susceptible Japanese quail (*Coturnix japonica*). *J Comp Pathol* 119:419–427.

Honeyfield DC, Ross JP, Carbonneau DA, Terrell SP, Woodward AR, Schoeb TR, Perceval HF, Hinterkopf JP. 2002. Pathology, physiologic parameters, tissue contaminants, and tissue thiamine in morbid and healthy central Florida adult American alligators (*Alligator mississippiensis*). *J Wildl Dis* 44:280–294.

Huchzermeyer FW. 2002. Diseases of farmed crocodiles and ostriches. *Rev Sci Tech* 21:265–276.

Huchzermeyer FW. 2003. *Crocodiles: biology, Husbandry and Diseases*. CABI Publishing, Oxon, UK.

Huchzermeyer KD. 2012. Prevalence of pansteatitis in African sharptooth catfish, *Clarias gariepinus* (Burchell), in the Kruger National Park, South Africa. *J S Afr Vet Assoc* 83:916.

Jacobson ER. 2007. Biology, anatomy, and histology, in *Infectious Diseases and Pathology of Reptiles: A Color Atlas and Text*. Jacobson ER (ed). CRC Press, Boca Raton, FL, pp. 1–130.

Jarchow J. 1984. Veterinary management of the desert tortoise, *Gopherus agassizii* at the Arizona-Sonora Desert Museum: A rational approach to diet. *Proc Desert Tortoise Council Symposium*: 83–84.

Jubb TF. 1992. A thiamine responsive nervous disease in saltwater crocodiles (*Crocodylus porosus*). *Vet Rec* 131:347–348.

Koelle P, Wiedemann A, and Kienze E. 2010. Vitamin A in plasma of reptiles. *Proc Am Assoc Rept Amphib Vet* 4–6.

Kohel KA, MacKenzie DS, Rostal DC, Grumbles JC, and Lance VA. 2001. Seasonality in plasma thyroxine in the desert tortoise, *Gopherus agassizii*. *Gen Comp Endocrinol* 121:214–222.

Kummrow MS, Berkvens CN, Paré JA, and Smith DA. 2010. Cerebral xanthomatosis in three green water dragons (*Physignathus cocincinus*). *J Zoo Wildl Med* 41:128–132.

Lacy EL, Sheridan MA, and Moore MC. 2002. Sex differences in lipid metabolism during reproduction in free-living tree lizards (*Urosaurus ornatus*). *Gen Comp Endocrinol* 128:180–192.

Ladds PW, Mangunwirjo H, Sebayang D, and Daniels PW. 1995. Diseases in young farmed crocodiles in Irian Jaya. *Vet Rec* 136:121–124.

Laing CJ, and Fraser DR. 1999. The vitamin D system in iguanid lizards. *Comp Biochem Physiol B* 123:373–379.

Laing CJ, Trube A, Shea GM, and Fraser DR. 2001. The requirement for natural sunlight to prevent vitamin D deficiency in iguanian lizards. *J Zoo Wildl Med* 32:342–348.

Lane T. 2006. Crocodilians, in *Reptile Medicine and Surgery*, 2nd ed. Mader DR (Ed.), Saunders Elsevier, St. Louis, pp. 100–117.

Lane EP, Huchzermeyer FW, Govender D, Bengis RG, Buss PE, Hofmeyr M, Myburgh JG, Steyl JCA, Pienaar DJ, and Kotze A. 2013. Pansteatitis of unknown etiology associated with large-scale Nile crocodile (*Crocodylus niloticus*) mortality in Kruger National Park, South Africa: Pathologic findings. *J Zoo Wildl Med* 44:899–910.

Larsen RE, Buergelt C, Cardeilhac PT, and Jacobson ER. 1983. Steatitis and fat necrosis in captive alligators. *J Am Vet Med Assoc* 183:1202–1204.

Latney LV, Toddes BD, Wyre NR, Brown DC, Michel KE, and Briscoe JA. 2017. Effects of various diets on the calcium and phosphorus composition of mealworms (*Tenebrio molitor* larvae) and superworms (*Zophobas morio* larvae). *Am J Vet Res* 78:178–185.

Lauten SD, Cox NR, Brawner WR, and Baker HJ. 2001. Use of dual energy x-ray absorptiometry for noninvasive body composition measurements in clinically normal dogs. *Am J Vet Res* 62:1295–1301.

Lemm J, and Alberts AC. 2011. Cyclura, in *Natural History, Husbandry, and Conservation of West Indian Rock Iguanas*. Elsevier Academic Press, New York.

Levander OA, and Beck MA. 1997. Interacting nutritional and infectious etiologies of Keshan disease. Insights from Coxsackie virus B-induced myocarditis in mice deficient in selenium or vitamin E. *Biol Trace Elem Res* 56:5–21.

Lopez I, Pineda C, Muñoz L, Raya A, Lopez G, and Aguilera-Tejero E. 2016. Chronic vitamin D intoxication in captive Iberian lynx (*Lynx pardinus*). *PLOS ONE* 11 e0156331. https://doi.org/10.1371/journal.pone.0156331

Mader DR. 2006. Metabolic bone diseases, in *Reptile Medicine and Surgery*, 2nd ed. Mader DR (Ed.), Saunders Elsevier, St. Louis, pp. 841–851.

Manawatthana S, and Kasorndorkbua C. 2005. Steatitis and vitamin E deficiency in captive olive ridley turtles (*Lepidochelys olivacea*). *Proc 6th Sea Star 2000 workshop* 85–87

Mans C, and Braun J. 2014. Update on common nutritional disorders of captive reptiles. *Vet Clin North Am Exot Anim Pract* 17:369–395.

Masi L. 2011. Phosphatonins: New hormones involved in numerous inherited bone disorders. *Clin Cases Miner Bone Metab* 8:9–13.

Miller EA, Green SL, Otto GM, and Bouley DM. 2001. Suspected hypovitaminosis A in a colony of captive green anoles (*Anolis carolinensis*). *Contemp Top Lab Anim Sci* 40:18–20.

Mumford SL, Johnson L, and Herbst L. 1999. Multifocal steatitis lesions in a Kemp's ridley sea turtle (*Lepidochelys kempi*). *Proc Intl Assoc for Aq Anim Med*, Boston, MA.

National Research Council. 1987. Vitamin A, in *Vitamin Tolerance of Animals. Subcommittee on Vitamin Tolerance, Committee on Animal Nutrition, Board of Agriculture*. National Academy Press, Washington, DC, pp. 3–10.

Oftedal OT, Chen TC, and Schulkin J. 1997. Preliminary observations on the relationship of calcium ingestion to vitamin D status in the green iguana *(Iguana iguana) Zoo Bio* 16:201–207.

Oonincx DGAB, and Dierenfeld ES. 2012. An investigation into the chemical composition of alternative invertebrate prey. *Zoo Biol* 31:40–54.

Oonincx DG, Stevens Y, van den Borne JJ, van Leeuwen JP, and Hendriks WH. 2010. Effects of vitamin D3 supplementation and UVb exposure on the growth and plasma concentration of vitamin D3 metabolites in juvenile bearded dragons (*Pogona vitticeps*). *Comp Biochem Physiol B Biochem Mol Biol* 156:122–128.

Oonincx DG, van de Wal MD, Bosch G et al. 2013. Blood vitamin D(3) metabolite concentrations of adult female bearded dragons (*Pogona vitticeps*) remain stable after ceasing UVb exposure. *Comp Biochem Physiol B Biochem Mol Biol* 165:196–200.

Orós J, Monagas P, Calabuig P, Luzardo OP, and Camacho M. 2013. Pansteatitis associated with high levels of polychlorinated biphenyls in a wild loggerhead sea turtle *Caretta caretta*. *Dis Aquat Organ* 102:237–242.

Pajdak-Czaus J, Terech-Majewska E, Będzłowicz D, Mączyński M, Krystkiewicz W, Łabuć S, Platt-Samoraj A, and Szweda W. 2019. Applicability of thyroxine measurements and ultrasound imaging in evaluations of thyroid function in turtles. *J Vet Res* 63:267–273.

Pallaske G. 1961. Changes of vessel caused by hypervitaminosis D in an iguana. *Berl Munch Tierarztl Wochenschr* 74:132–133.

Palmer DG, Rübel A, Mettler F, and Volker L. 1984. Experimentally induced skin changes in land tortoises by giving high doses of vitamin A parenterally. [Experimentell erzeugte hautveranderungen bei landschildkroten durch hohe parenterale gaben von vitamin A.] *Zebl Vet Med A* 31:625–633.

Pessier A. 2014. Short tongue syndrome and hypovitaminosis A, in *Current Therapy in Reptile Medicine and Surgery*. Mader D, Divers S (Eds.), Elsevier Saunders, St. Louis, pp. 271–276.

Purgley H, Jewell J, Deacon JE, Winokur RM, and Tripoli VM. 2009. Vitamin D3 in captive green sea turtles (*Chelonia mydas*). *Chelonian Conserv Biol* 8:161–167.

Raila J, Schuhmacher A, Gropp J, and Schweigert FJ. 2002. Selective absorption of carotenoids in the common green iguana (*Iguana iguana*). *Comp Biochem Physiol A Mol Integr Physiol* 132:513–518.

Raiti P. 2019. Endocrinology, in *Mader's Reptile and Amphibian Medicine and Surgery*, 3rd ed. Divers S, and Stahl SJ (Eds.), Elsevier, St. Louis, MO, pp. 837–838

Raiti P, and Garner MM. 2000. Uremic encephalitis in a leaf-tailed gecko, *Uroplatus Henkeli*. *Annu Meet Assoc Rept Amphib Vet* 29–30.

Raiti P, and Haramati N. 1997. Magnetic resonance imaging and computerized tomography of a gravid leopard tortoise (*Geochelone pardalis pardalis*) with metabolic bone disease. *J Zoo Wildl Med.* 28:189–197.

Ramer JC, Maria R, Reichard T, Tolson PJ, Chen TC, and Holick MF. 2005. Vitamin D status of wild Ricord's iguanas (*Cyclura ricordii*) and captive and wild rhinoceros iguanas (*Cyclura cornuta cornuta*) in the Dominican Republic. *J Zoo Wildl Med* 36:188–191.

Rice AN, Ross JP, Woodward AR, Carbonneau DA, and Franklin PH. 2007. Alligator diet in relation to alligator mortality on Lake Griffin, Florida. *Southeastern Naturalist* 6:97–110.

Richey LJ. 2001. Effects of endocrine-disrupting contaminants on the immune system of hatchling American alligators. *PhD diss.*, University of Florida.

Richman LK, Montali RJ, Allen ME, and Oftedal OT. 1995a. Paradoxical changes in vitamin D deficient green iguanas (*Iguana iguana*). *Proc Amer Assoc Zoo Vet* 231–232.

Richman LK, Montali RJ, Allen ME, and Oftedal OT. 1995b. Widespread metastatic soft tissue mineralization in vitamin D deficient green iguanas (*Iguana iguana*). *Vet Pathol* 32:560.

Ritter JM, Garner MM, Chilton JA, Jacobson ER, and Kiupel M. 2009. Gastric neuroendocrine carcinomas in bearded dragons (*Pogona vitticeps*). *Vet Pathol* 46:1109–1116.

Ritz J, Hammer C, and Clauss M. 2010. Body size development of captive and free-ranging leopard tortoises (*Geochelone pardalis*). *Zoo Biol* 29:517–525.

Romero LM, and Wikelski M. 2010. Stress physiology as a predictor of survival in Galapagos marine iguanas. *Proc Biol Sci* 277:3157–3162.

Rosol TJ, and Gröne A. 2016. Bones and joints, in *Jubb, Kennedy, and Palmer's Pathology of Domestic Animals*, 6th ed. Maxie MG (Ed.), Elsevier, St. Louis, pp. 269–357.

Rossi JV. 2006. General husbandry and management, in *Reptile Medicine and Surgery*, 2nd ed. Mader DR (Ed.), St. Louis, Saunders Elsevier, pp. 25–41

Rossi J, and Rossi R. 1992. *Snakes of the United States and Canada*, vol. 1. Krieger Publishing, Malabar, FL, p. 124.

Ryan MJ, and Whitney GD. 1980. Xanthoma in a gopher snake. *Vet Med Small Anim Clin* 75:503–507.

Sawin CT, Bacharach P, and Lance V. 1981. Thyrotropin-releasing hormone and thyrotropin in the control of thyroid function in the turtle, *Chrysemys picta*. *Gen Comp Endocrinol*, 45:7–11.

Schilliger L, Lemberger K, Chai N, Bourgeois A, and Charpentier M. 2010. Atherosclerosis associated with pericardial effusion in a central bearded dragon (*Pogona vitticeps*). *J Vet Diagn Invest* 22:789–792.

Schmidt RE, and Hubbard GB. 1987. *Atlas of Zoo Animal Pathology*, vol. 2. CRC Press, Boca Raton.

Schoeb TR, Heaton-Jones TG, Clemmons RM, Carbonneau DA, Woodward AR, Shelton D, and Poppenga RH. 2002. Clinical and necropsy findings associated with increased mortality among American alligators of Lake Griffin, Florida. *J Wildl Dis* 38:320–337.

Selleri P, and di Girolamo N. 2012. Plasma 25-hydroxyvitamin D_3 concentrations in Hermann's tortoises (*Testudo hermanni*) exposed to natural sunlight and two artificial ultraviolet radiation sources. *Am J Vet Res* 73:1781–1786.

Sepúlveda MS, Wiebe JJ, Honeyfield DC, Rauschenberger HR, Hinterkopf JP, Johnson WE, and Gross TS. 2004. Organochlorine pesticides and thiamine in eggs of largemouth bass and American alligators and their relationship with early life-stage mortality. *J Wildl Dis* 40:781–785.

Shilton C, Brown GP, Chambers L, Benedict S, Davis S, Aumann S, and Isberg SR. 2014. Pathology of runting in farmed saltwater crocodiles (*Crocodylus porosus*) in Australia. *Vet Pathol* 51:1022–1034.

Shilton CM, Smith DA, Crawshaw GJ, Valdes E, Keller CB, Maguire GF, Connelly PW, and Atkinson J. 2001. Corneal lipid deposition in Cuban tree frogs (*Osteopilus septentrionalis*) and its relationship to serum lipid: An experimental study. *J Zoo Wildl Med* 32:305–319.

Shively MJ, and Edwards K. 1985. Pathological calcification in a green iguana (*Iguana iguana*). *Avian/Exotic Practice* 2:32–33.

Slater OM, Faust LJ, Hileman ET, and Lavin SR. 2017. Plasma vitamin, mineral, and biochemistry concentrations in free-ranging, adult Eastern massasauga (*Sistrurus catenatus*) during spring egress. *J Herpetol Med Surg* 27:29–35.

Sleeman JM, Brown J, Steffen D, Jones D, Roberston J, and Holladay S. 2008. Relationships among aural abscesses, organochlorine compounds, and vitamin a in free-ranging eastern box turtles (*Terrapene carolina carolina*). *J Wildl Dis* 44:922–929.

Solak B, Kara RO, Acikgoz SB, and Kosem M. 2015. First and only symptom of undiagnosed diabetes mellitus: Eruptive xanthoma. *BMJ Case Rep* 2015. pii:bcr2015212160. doi: 10.1136/bcr-2015-212160.

Srivastav AK, Das VK, Srivastav SK, Sasayama Y, and Suzuki N. 1998. Morphological and functional aspects of reptilian ultimobranchial gland. *Anatom Histol Embryol* 27:359–364.

Stahl SJ. 1997. Captive management, breeding and common medical problems of the veiled chameleon (*Chamaeleo calyptratus*). *Proc Assoc Rept Amphib Vet* 29–40.

Stahl SJ. 2003. Diseases of the reptile pancreas. *Vet Clin Exot Anim* 6:191–212.

Stahl SJ. 2019. Hyperglycemia, in *Mader's Reptile and Amphibian Medicine and Surgery*, 3rd ed. Divers S and Stahl SJ (Eds.), Elsevier, St. Louis, MO, pp. 1314–1315.

Stern AW, Velguth KE, and D'Agostino J. 2010. Metastatic ductal adenocarcinoma in a Western Hognose snake (*Heterodon nasicus*). *J Zoo Wildl Med* 41:320–324.

Stringer EM, Harms CA, Beasley JF, and Anderson ET. 2010. Comparison of ionized calcium, parathyroid hormone, and 25-hydroxyvitamin D in rehabilitating and healthy wild green sea turtles (*Chelonia mydas*). *J Herpetol Med Surg* 20:122–127.

Strzelewicz MA, Ullrey DE, Schafer ST, and Bacon JP. 1985. Feeding insectivores: Increasing the calcium content of wax moth (*Galleria mellonella*) larvae. *J Zoo Anim Med* 16:25–27.

Swenson J, and Bradley GA. 2013. Suspected cholecalciferol rodenticide toxicosis in avian species at a zoological institution. *J Avian Med Surg* 27:136–147.

Topper MJ, Latimer KS, McManamon R, and Thorstad CL. 1994. Colloid goiter in an eastern diamondback rattlesnake (*Crotalus adamanteus*). *Vet Pathol* 31:380–382.

Wallach JD. 1966. Hypervitaminosis D in green iguanas. *J Am Vet Med Assoc* 149:912–914.

Wallach JD, and Hoessle C. 1968. Steatitis in captive crocodilians. *J Am Vet Med Assoc* 153:845–847.

Watson MK, Byrd J, Phillips CA, and Allender MC. 2017. Characterizing the 25-hydroxyvitamin D status of two populations of free-ranging eastern box turtles (*Terrapene carolina carolina*). *J Zoo Wildl Med* 48:742–747.

Webb GJW, Hollis GJ, and Manolis SC. 1991. Feeding, growth, and food conversion rates of wild juvenile saltwater crocodiles (*Crocodylus porosus*). *J Herp* 25:462–473.

Wiesner CS, and Iben C. 2003. Influence of environmental humidity and dietary protein on pyramidal growth of carapaces in African spurred tortoises (*Geochelone sulcata*). *J Anim Physiol Anim Nutr (Berl)* 87:66–74.

Wiggans KT, Sanchez-Migallon Guzman D, Reilly CM, Vergneau-Grosset C, Kass PH, and Hollingsworth SR. 2018. Diagnosis, treatment, and outcome of and risk factors for ophthalmic disease in leopard geckos (*Eublepharis macularius*) at a veterinary teaching hospital: 52 cases (1985-2013). *J Am Vet Med Assoc* 252:316–323.

Xiang J, and Peichao W. 1990. Annual cycles of lipid contents and caloric values of carcass and some organs of the gecko, *Gekko japonicus*. *Comp Biochem Physiol* 96A:267–271.

Yanni AE. 2004. The laboratory rabbit: An animal model of atherosclerosis research. *Lab Anim* 38:246–256.

Zotti P, Selleri P, Carnier P, Morgante M, and Bernardini D. 2004. Relationship between metabolic bone disease and bone mineral density measured by dual-energy x-ray absorptiometry in the green iguana (*Iguana iguana*). *Vet Radiol Ultrasound* 45:10–16.

Zwart P, and Kok AGG. 1978. Endemic goiter in reptiles in the Netherlands. *Berh Ber Erkrg Zoo Anim* 20:373–382.

Zwart P, and Rulkens RJ. 1979. Improving the calcium content of mealworms. *Int Zoo Yearb* 19:254–255.

Zwart P, and van Ham B. 1980. Keeping, breeding and raising garter snakes (*Thamnophis radix*), in *The Care and Breeding of Captive Reptiles*. Townson S, Millichamp NJ, Lucas DGD, and Millwood AJ (Eds.), British Herpetological Society, London, pp. 81–85.

Figure 2.1 Central Bearded Dragon, *Pogona vitticeps*. Agamidae. Obesity. Note prominent dilatation of the coelomic cavity (due to excessive adipose tissue with hypertrophy of fat bodies as shown in Figure 2.2). Bar = 1 cm.

Figure 2.2 Central Bearded Dragon, *Pogona vitticeps*. Agamidae. Obesity. Same animal as in Figure 2.1 showing marked hypertrophy of fat bodies with distention of the coelomic cavity. The left fat body has been displaced to expose the underlying liver, which is pale due to lipidosis. Bar = 1 cm.

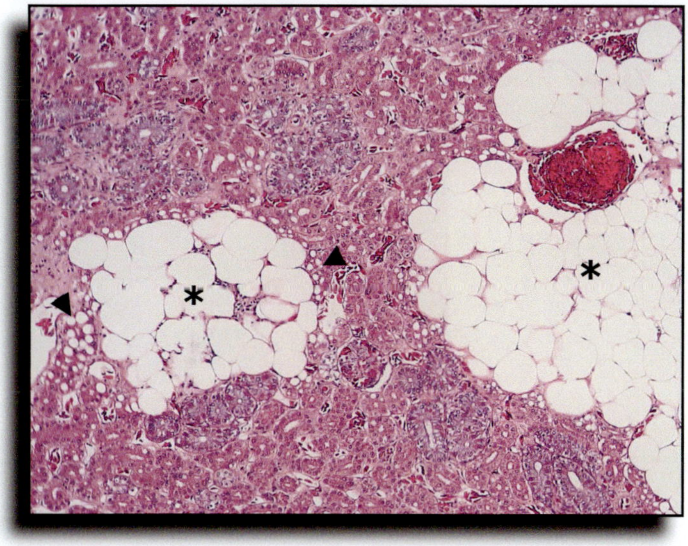

Figure 2.3 Central Bearded Dragon, *Pogona vitticeps*. Agamidae. Obesity. Large aggregates of well-differentiated adipocytes are present in the renal interstitium. Few tubular epithelial cells contain clear cytoplasmic vacuoles. H&E stain.

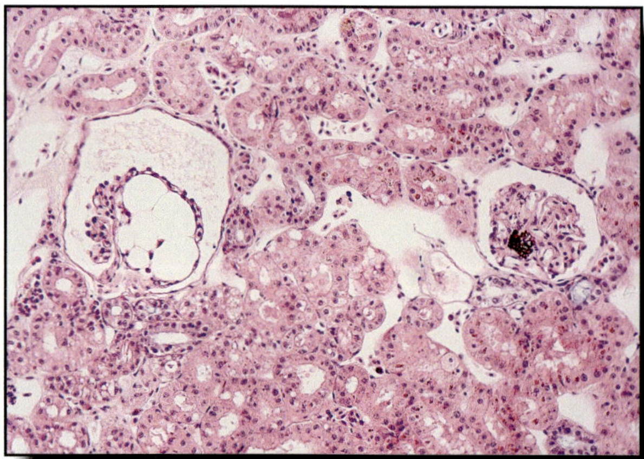

Figure 2.4 Western Bearded Anole, *Anolis* (*Xiphosurus*, *Anolis, Chamaeleolis*) *barbatus*. Dactyloidae. Obesity. A renal glomerulus (left) has a segmental aggregate of well-differentiated adipocytes. H&E stain.

Figure 2.5 Nile's Crocodile, *Crocodylus niloticus*. Crocodylidae. Obesity. Large carnivorous reptiles such as crocodiles are predisposed to obesity in captivity.

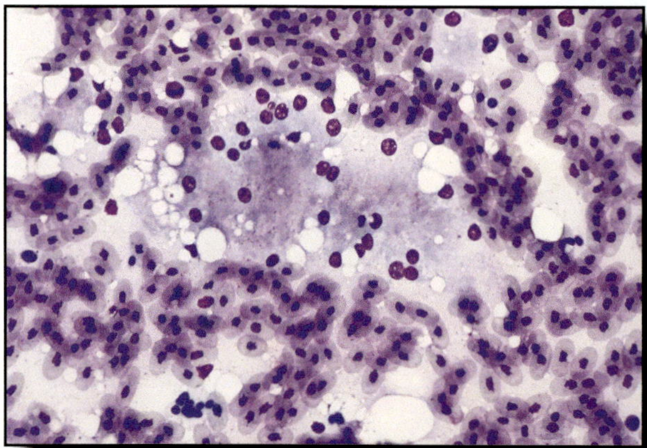

Figure 2.6 Green Iguana, *Iguana iguana*. Iguanidae. Hepatic lipidosis. Cytoplasmic clear vacuoles are noted in hepatocytes obtained from a fine-needle aspirate. Extracellular lipid droplets are also present, likely from hepatocytes that ruptured during aspiration. Modified Wright-Giemsa stain.

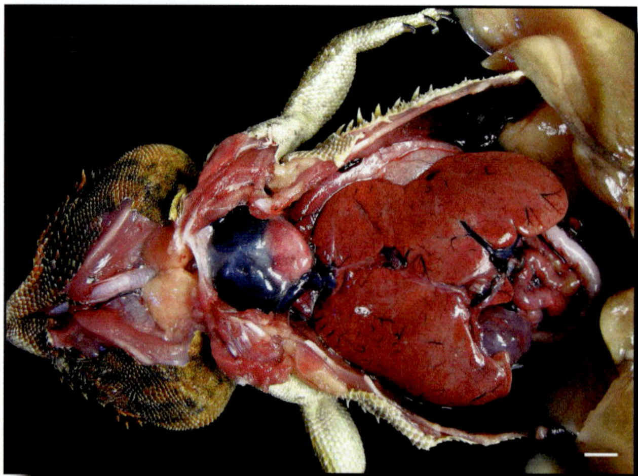

Figure 2.7 Central Bearded Dragon, *Pogona vitticeps*. Agamidae. Hepatic lipidosis. The liver is pale due to lipidosis. The margins are slightly rounded. Bar = 1 cm.

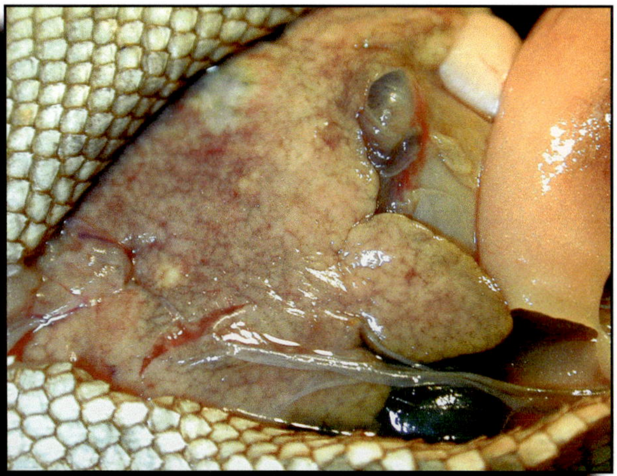

Figure 2.8 Central Bearded Dragon, *Pogona vitticeps*. Agamidae. Hepatic lipidosis. Note the pale liver, which has a slightly irregular surface due to fibrosis. The few nodular white foci represent steatosis. (Photo courtesy of Andrés Montesinos.)

Figure 2.9 Common Snapping Turtle, *Chelydra serpentine*. Chelydridae. Hepatic lipidosis. Transverse section of formalin-fixed liver showing pale parenchyma due to lipidosis. Bar = 1 cm.

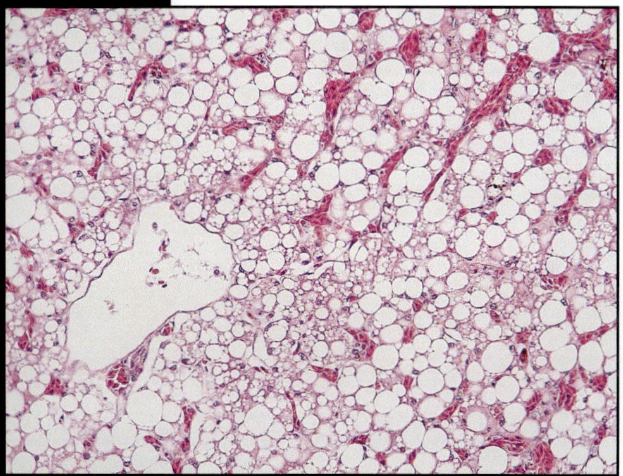

Figure 2.10 Central Bearded Dragon, *Pogona vitticeps*, Agamidae. Hepatic lipidosis. Diffusely, hepatocytes contain one or more intracytoplasmic clear, spherical vacuoles that displace and compress the nucleus. H&E stain.

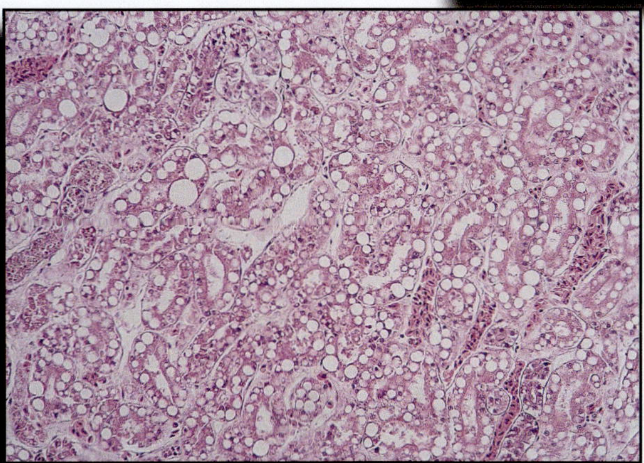

Figure 2.11 Central Bearded Dragon, *Pogona vitticeps*. Agamidae. Renal tubular lipidosis. Diffusely, medium or large-sized clear vacuoles are present within the cytoplasm of epithelial cells lining renal tubules. H&E stain.

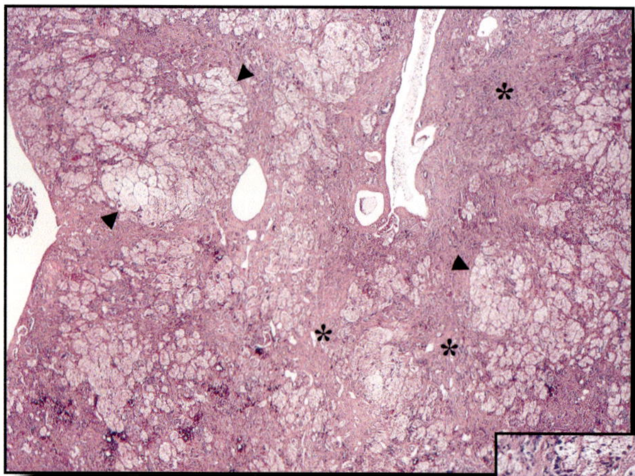

Figure 2.12 Veiled Chameleon, *Chamaeleo calyptratus*. Chamaeleonidae. Hepatic lipidosis and cirrhosis. Note hepatic lipidosis (arrowheads) associated with bridging fibrosis and biliary hyperplasia (asterisks). H&E stain.

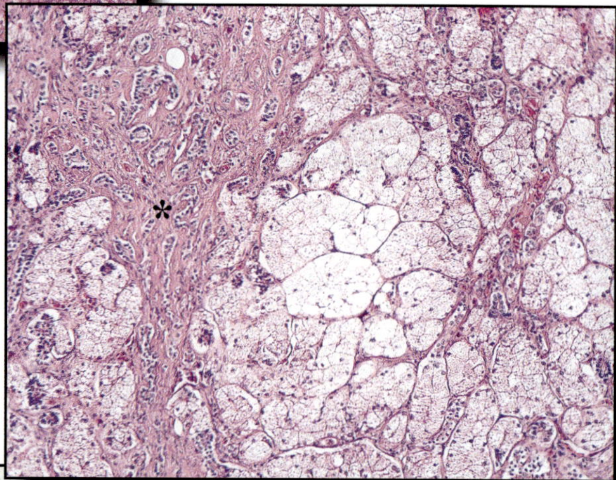

Figure 2.13 Veiled Chameleon, *Chamaeleo calyptratus*, Chamaeleonidae, Hepatic lipidosis and cirrhosis. A higher magnification of Figure 2.12 illustrates severe hepatocellular lipidosis with bridging fibrosis and proliferation of bile ducts (asterisk). H&E stain.

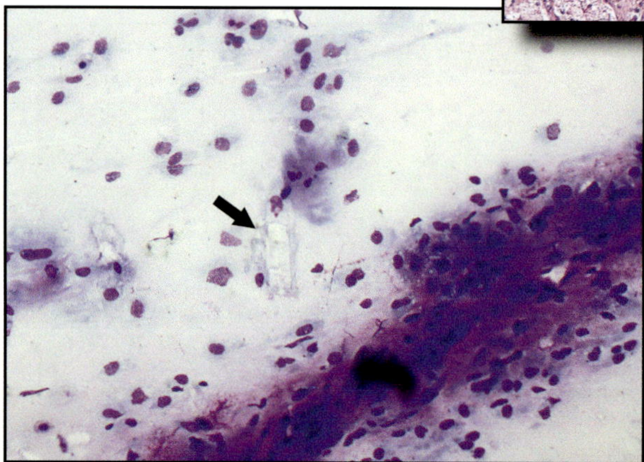

Figure 2.14 Leopard Gecko, *Eublepharis macularius*, Eublepharidae. Xanthomatosis. Note macrophages and fibrohistiocytoid cells associated with clear angular cholesterol crystals (arrow). Modified Wright-Giemsa stain. (Photo courtesy of Michael M. Garner.)

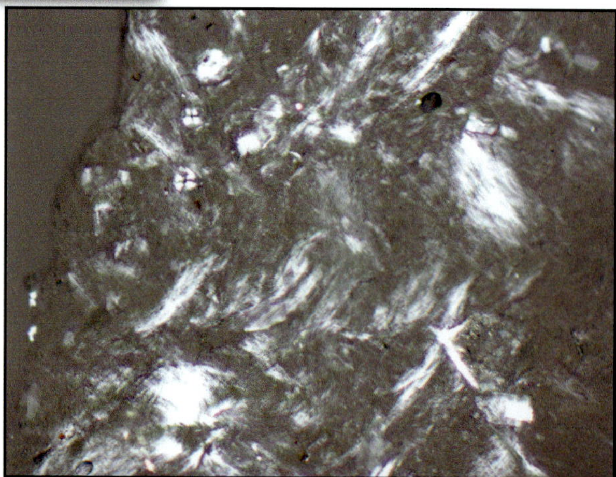

Figure 2.15 Leopard Gecko, *Eublepharis macularius*, Eublepharidae. Xanthomatosis. Cholesterol crystals are birefringent with polarized light. (Photo courtesy of Michael M. Garner.)

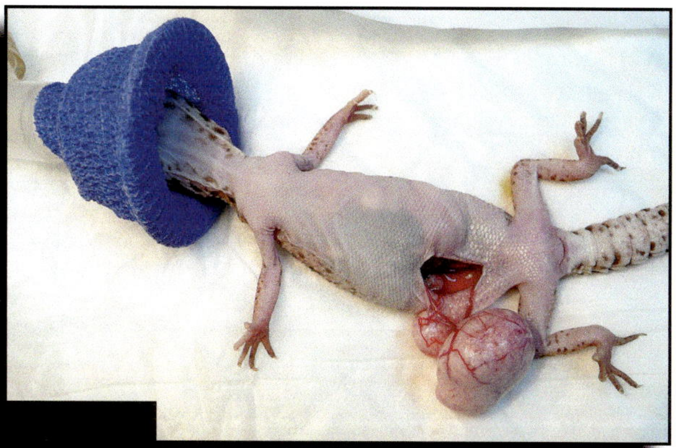

Figure 2.16 Female Leopard Gecko, *Eublepharis macularius*. Eublepharidae. Xanthomatosis. A large mass is present in the caudal coelomic cavity. (Photo courtesy of Xavier Valls, reprinted with permission of Editorial Servet. Valls X, Bueno JV. 2012. *Casos clínicos de animals exóticos*.)

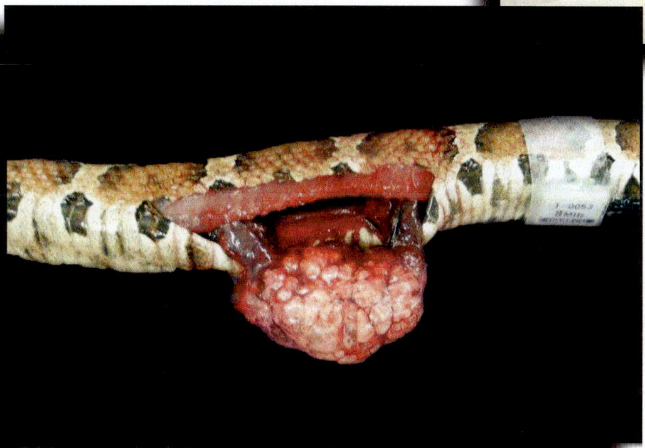

Figure 2.17 Copperhead, *Agkistrodon contortrix*. Viperidae. Ovarian xanthomatosis. Note a coelomic mass involving the ovary. (Photo courtesy of Michelle Bowman and Christie Buie.)

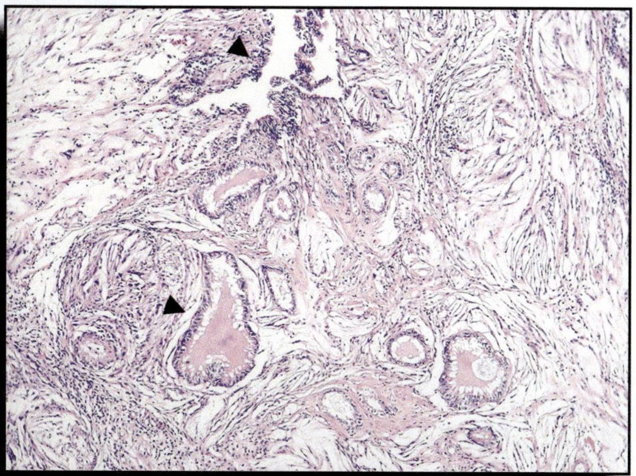

Figure 2.18 Leopard Gecko, *Eublepharis macularius*. Eublepharidae. Xanthomatosis. Histologically, the mass in Figure 2.16 corresponds to a xanthogranuloma. Note abundant closely stacked clear acicular (cholesterol) clefts multifocally surrounded by low numbers of multinucleate giant cells and fewer macrophages. This xanthogranuloma involved or extended into the oviduct (mucosa pointed out with arrowheads). H&E stain.

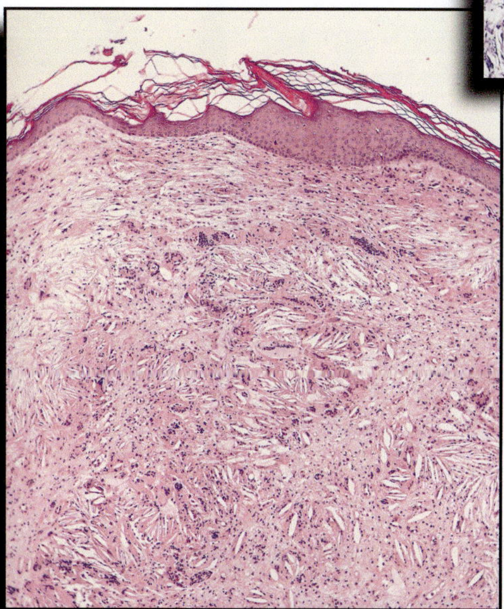

Figure 2.19 Sudan Plated Lizard, *Gerrhosaurus major*. Gerrhosauridae. Xanthomatosis. Cutaneous xanthogranulomatous inflammation presenting as multiple cutaneous nodules in the tail and forelimbs in a specimen fed mainly cat food. H&E stain.

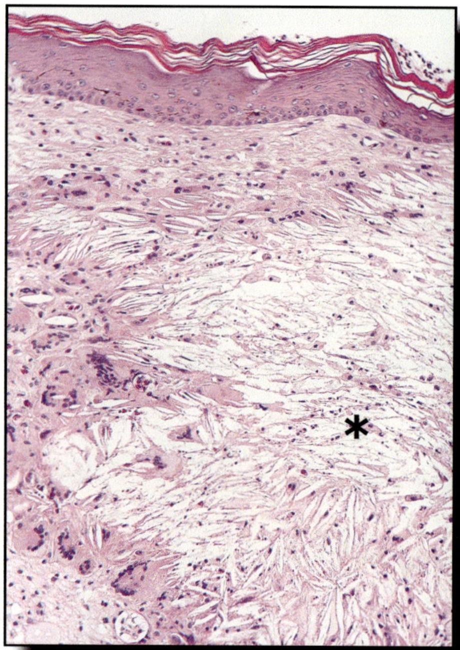

Figure 2.20 Sudan Plated Lizard, *Gerrhosaurus major.* Gerrhosauridae. Xanthomatosis. A higher magnification of Figure 2.19 demonstrates dermal stacked cholesterol clefts (asterisk) delineated by macrophages and multinucleate giant cells. H&E stain.

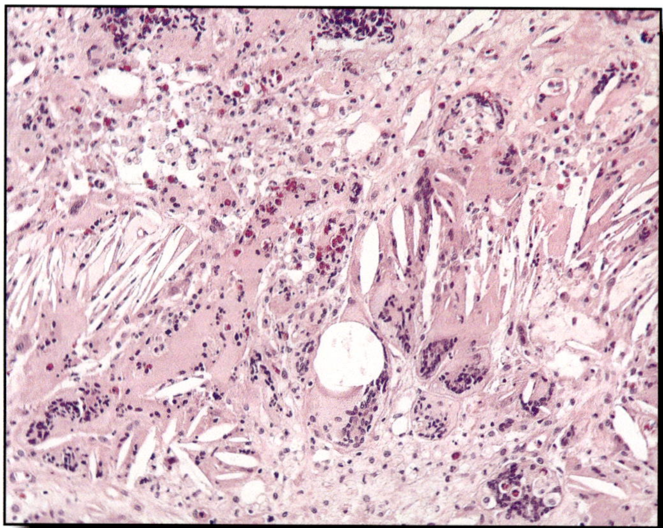

Figure 2.21 Sudan Plated Lizard, *Gerrhosaurus major.* Gerrhosauridae. Xanthomatosis. Inflammation surrounding cholesterol clefts is comprised of multinucleate giant cells, macrophages, and heterophils. H&E stain.

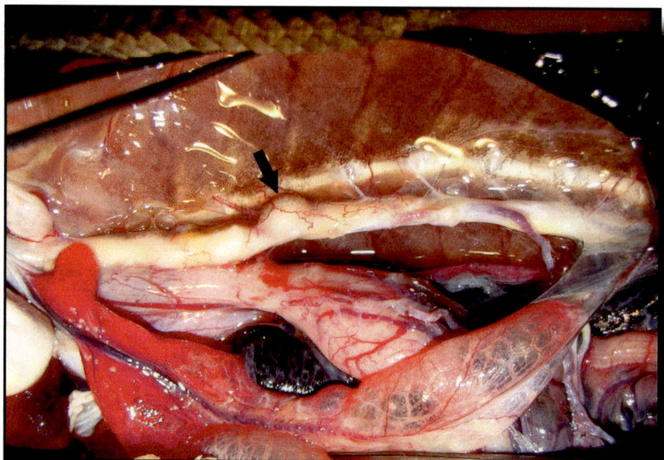

Figure 2.22 Central Bearded Dragon, *Pogona vitticeps*. Agamidae, Atheromatosis. Note tortuosity with multifocal yellowish discoloration of the aorta and segmental dilatation (arrow) resembling an aneurysm. (Photo courtesy of Guillermo Piñeiro.)

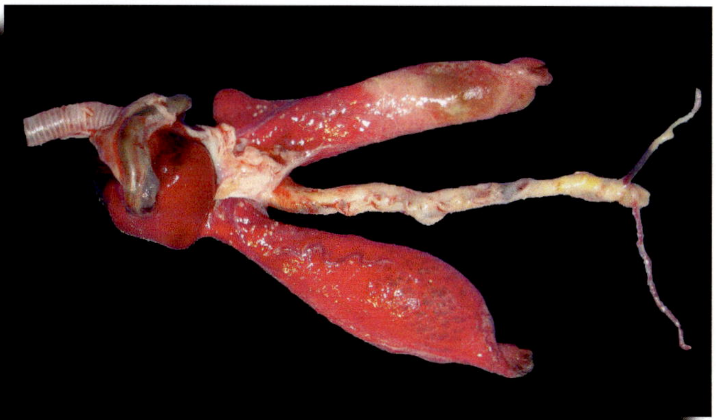

Figure 2.23 Central Bearded Dragon, *Pogona vitticeps*. Agamidae. Atheromatosis. Higher magnification of the aortic lesions in Figure 2.22 with the cardiovascular and respiratory tract removed from coelomic cavity. Note that tortuosity with yellow plaques also involve the iliac arteries. (Photo courtesy of Guillermo Piñeiro.)

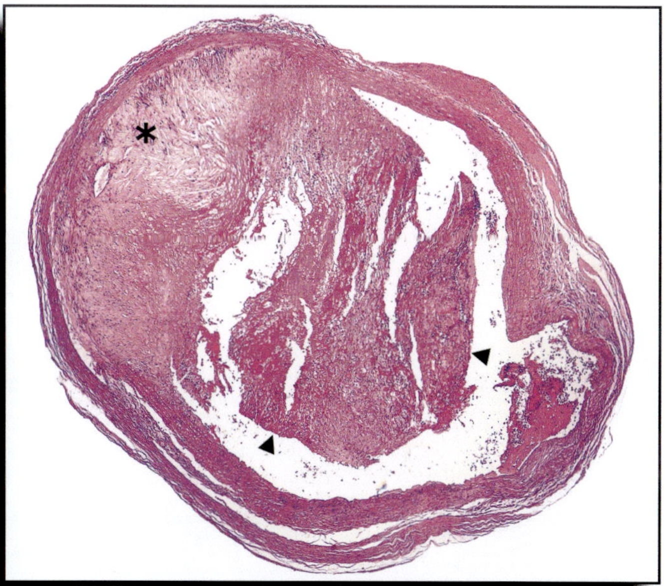

Figure 2.24 Central Bearded Dragon, *Pogona vitticeps*. Agamidae. Atheromatosis. Note thickening of the aortic wall by an atheromatous lesion (asterisk) to which a fibrin thrombus is attached (arrowheads). H&E stain.

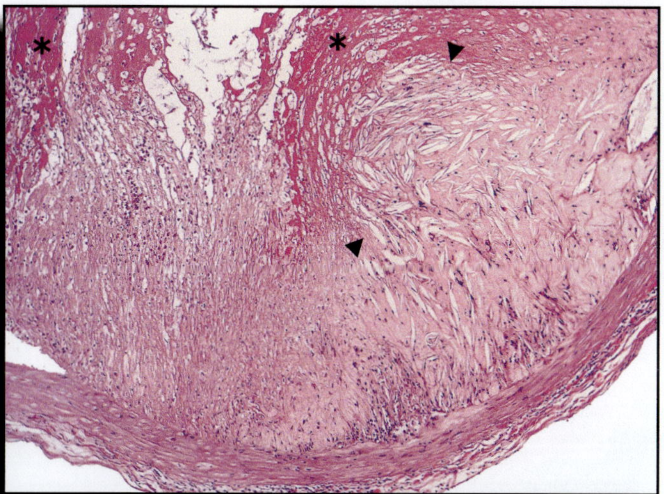

Figure 2.25 Central Bearded Dragon, *Pogona vitticeps*. Agamidae. Atheromatosis. A higher magnification of Figure 2.24 demonstrates an atheromatous plaque with clear acicular (cholesterol) clefts (arrowheads) surrounded by pale eosinophilic material, and an adherent fibrin thrombus (asterisks). H&E stain.

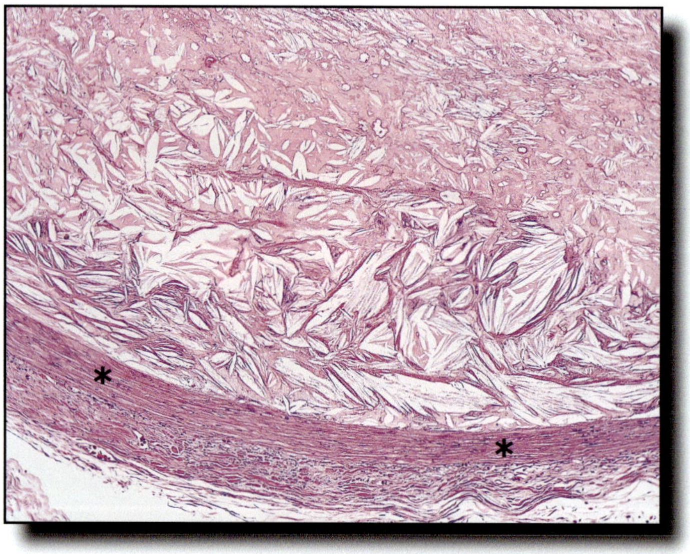

Figure 2.26 Central Bearded Dragon, *Pogona vitticeps*. Agamidae. Atheromatosis. Note a cholesterol cleft-rich focus in an atheromatous plaque, which compresses the medial tunic (asterisks). H&E stain.

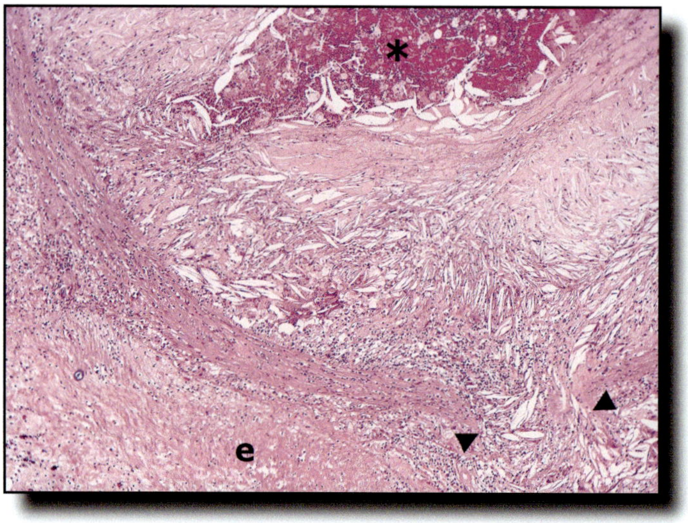

Figure 2.27 Central Bearded Dragon, *Pogona vitticeps*. Agamidae. Atheromatosis. This atheromatous plaque contains a focus of necrosis (asterisk) with focal discontinuity of the medial tunic (arrowheads). Infiltrates of lymphocytes and granulocytes are present in the atheromatous plaque, medial tunic and adventitia, which is thickened due to proteinaceous edema (e). H&E stain.

Figure 2.28 Boa Constrictor, *Boa constrictor*. Boidae. Atrophy of adipose tissue. Note marked atrophy of adipose tissue (arrows) (isolated red plaques) in this boa with inclusion body disease and retention of degenerate eggs (asterisks).

Figure 2.29 Leopard Gecko, *Eublepharis macularius*. Eublepharidae. Atrophy of adipose tissue. Atrophy of adipose tissue is striking in the tail due to the normally pronounced fat storage in gecko tails. Note generalized muscular atrophy as well, particularly in the limbs and tail. (Photo courtesy of Steve Barten.)

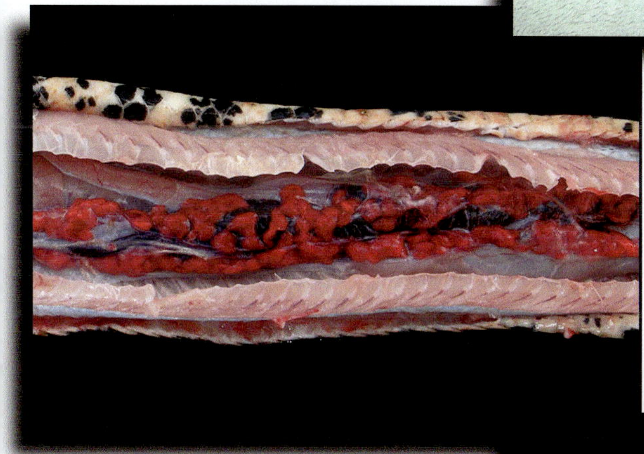

Figure 2.30 Boa Constrictor, *Boa constrictor*. Boidae. Atrophy of adipose tissue. The adipose tissue is atrophied and reddened.

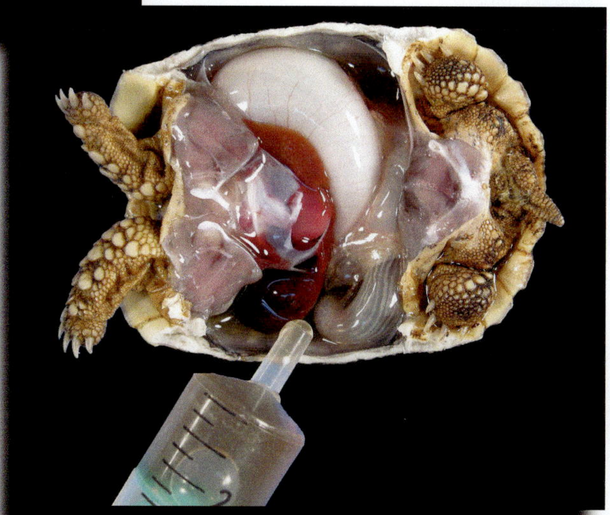

Figure 2.31 Leopard Tortoise, *Stigmochelys pardalis*. Testudinidae. Cachexia. Note the presence of straw-colored fluid in the coelomic cavity (hydrocoelom) and atrophy of the liver.

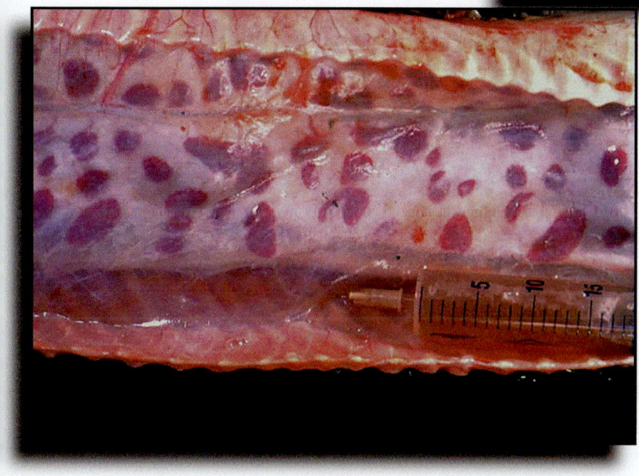

Figure 2.32 Reticulated Python, *Malayopython reticulatus*. Pythonidae. Hydrocoelom and atrophy of adipose tissue. The coelomic cavity contains straw-colored fluid (hydrocoelom). The adipose tissue is markedly atrophic and consists of isolated, flattened red plaques.

Figure 2.33 Chinese Water Dragon, *Physignathus cocincinus*. Agamidae. Cachexia. Note appearance of advanced cachexia with marked atrophy of skeletal muscles.

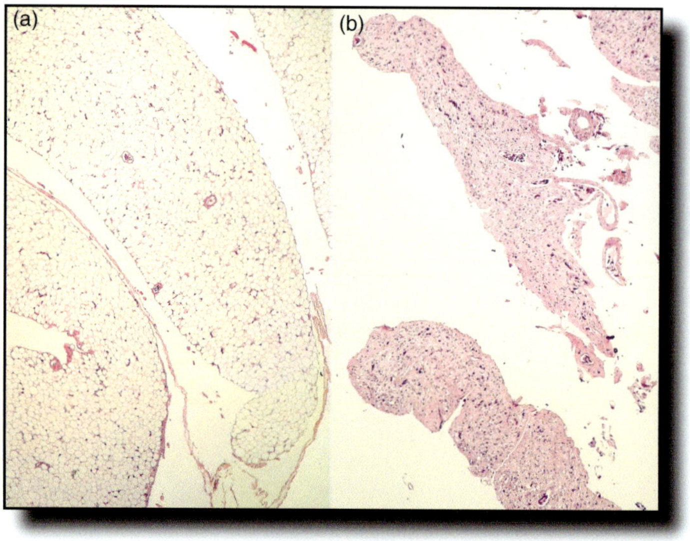

Figure 2.34 (a) Green Tree Python, *Morelia viridis*. Pythonidae. Normal adipose tissue. Note adjacent normal fat bodies. (b) Amazon Tree Boa, *Corallus hortulanus*. Boidae. Atrophy of adipose tissue. Marked atrophy of two fat bodies, which consist of severely atrophied adipocytes dispersed in fibrovascular stroma. Atrophy has resulted in accentuated separation of lobules of adipose tissue. H&E stain.

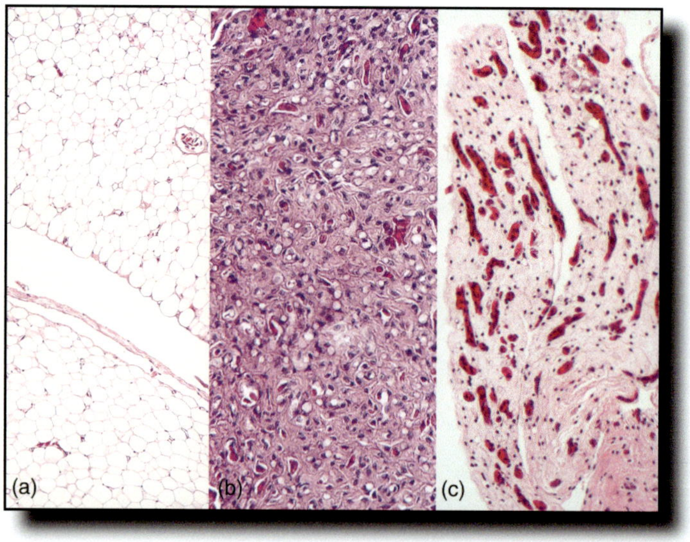

Figure 2.35 (a) Green Tree Python, *Morelia viridis*. Pythonidae. Normal adipose tissue. Adipocytes have normal lipid contents. (b) Corn Snake, *Pantheropis guttatus*. Colubridae. Atrophy of adipose tissue. Adipocytes are severely atrophied as a result of chronic biliary and pancreatic disease (cryptosporidiosis). Some adipocytes still contain small intracytoplasmic clear vacuoles. Atrophy of adipocytes has resulted in condensation of the fibrovascular stroma. (c) Western Ground Snake, *Sonora semiannulata*. Colubridae. Atrophy of adipose tissue. Adipocytes are markedly atrophied, with a shrunken nucleus. Capillaries are congested. H&E stain. (Photo courtesy of Michael M. Garner.)

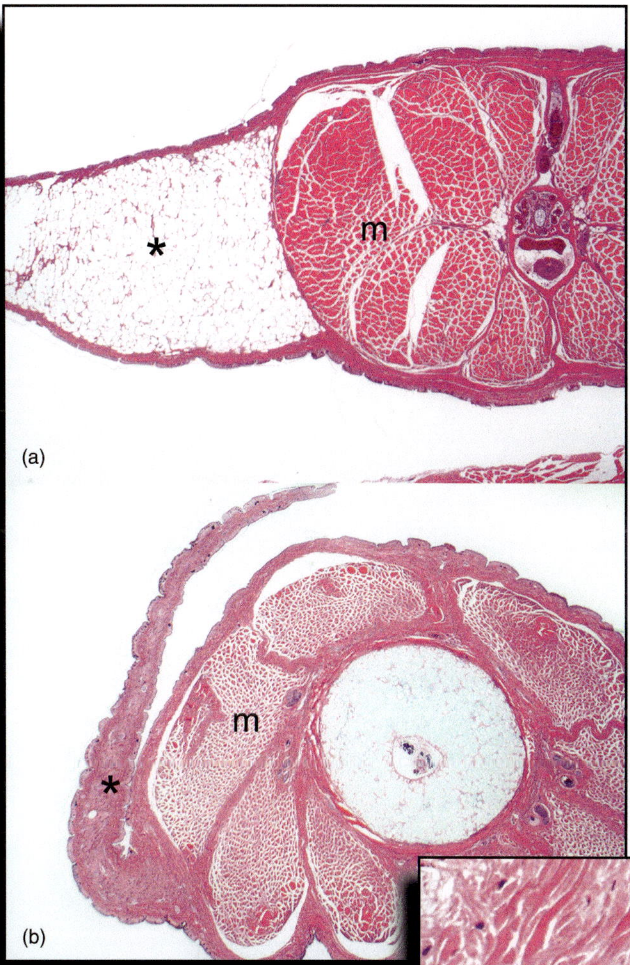

Figure 2.36 (a) Henkel's Leaf-Tailed Gecko, *Uroplatus hankeli*. Gekkonidae. Normal adipose tissue. Normal accumulation of adipocytes with lipid storage (*), adjacent to muscle bundles (m) in the tail. (b) Mossy Leaf-Tailed Gecko, *Uroplatus sikorae*. Gekkonidae. Atrophy of adipose tissue in the tail. Note marked atrophy of adipose tissue (*) adjacent to atrophied muscle bundles (m) in the tail as compared to control in (a). H&E stain. (Photo courtesy of Michael M. Garner.)

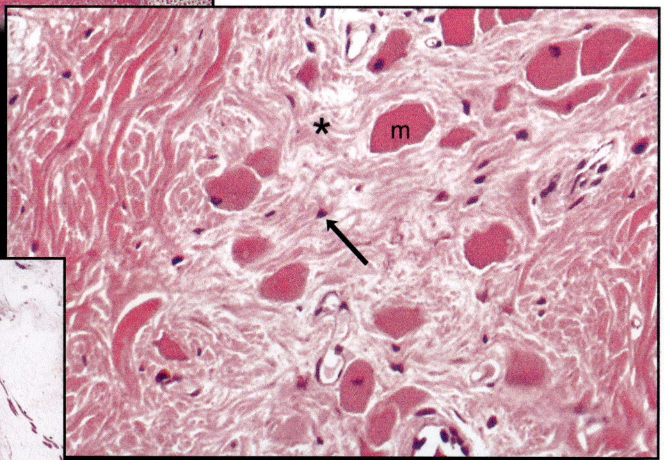

Figure 2.37 Mossy Leaf-Tailed Gecko, *Uroplatus sikorae*. Gekkonidae. Atrophy of tail's adipose tissue. A higher magnification of Figure 2.36(b) demonstrates the marked atrophy of adipocytes (arrow), consolidation of collagen in zone of atrophy (*) and variable atrophy of muscle fibers (m). H&E stain. (Photo courtesy of Michael M. Garner.)

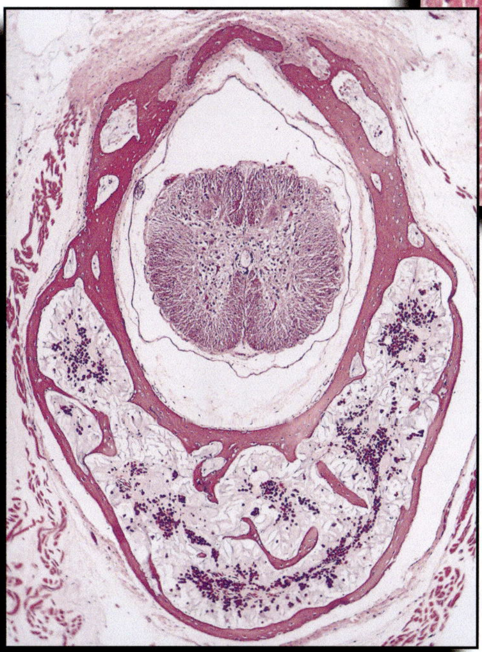

Figure 2.38 Hermann's Tortoise, *Testudo hermanni*. Testudinidae. Atrophy of adipose tissue. The adipose tissue in the bone marrow of this vertebral body is atrophied and replaced by abundant mucin-rich edema. H&E stain.

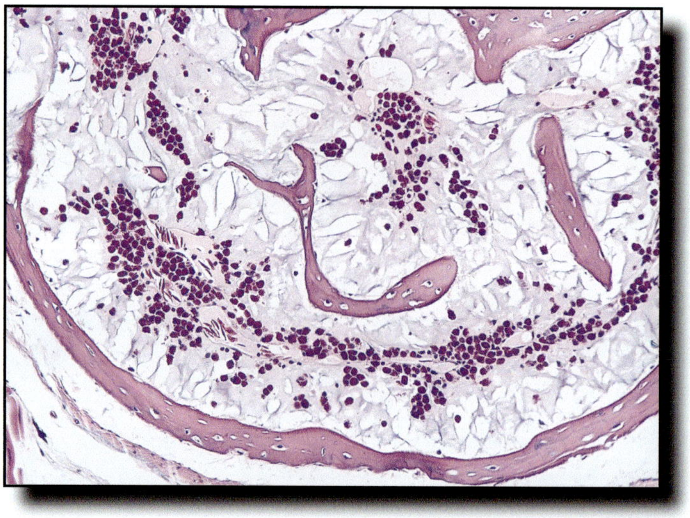

Figure 2.39 Hermann's Tortoise, *Testudo hermanni*. Testudinidae. Atrophy of adipose tissue. Higher magnification of the vertebral body shown in Figure 2.38 demonstrating the mucin-rich edema and severe atrophy of adipocytes. H&E stain.

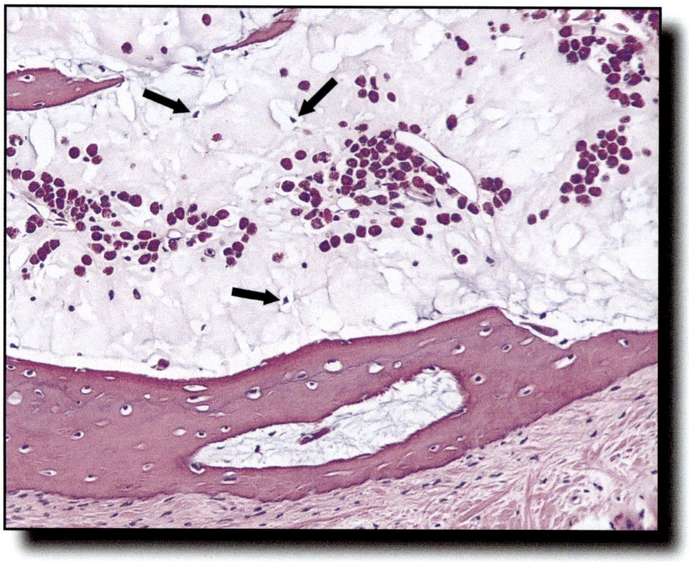

Figure 2.40 Hermann's Tortoise, *Testudo hermanni*. Testudinidae. Atrophy of adipose tissue. In this higher magnification of Figures 2.37 and 2.38, markedly atrophic marrow adipocytes (arrows) are scattered within the mucin-rich edema. H&E stain.

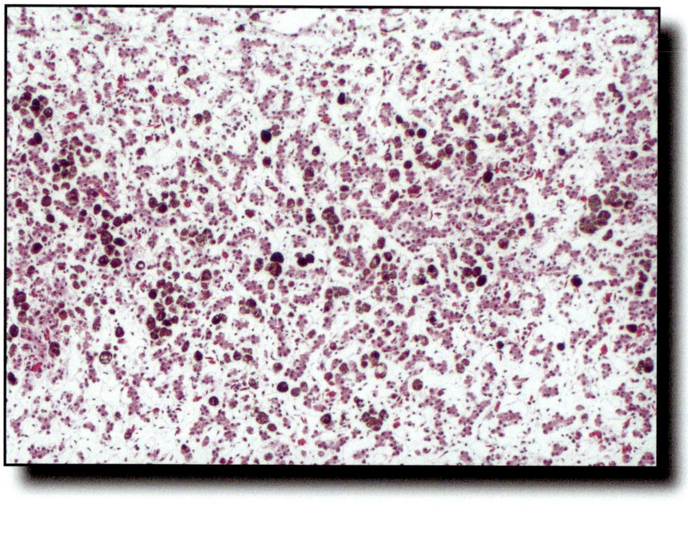

Figure 2.41 Loggerhead Sea Turtle, *Caretta caretta*. Cheloniidae. Hepatocellular atrophy. The hepatocellular cords are severely atrophic and in some foci are dissociated. H&E stain. (Photo courtesy of Michael M. Garner.)

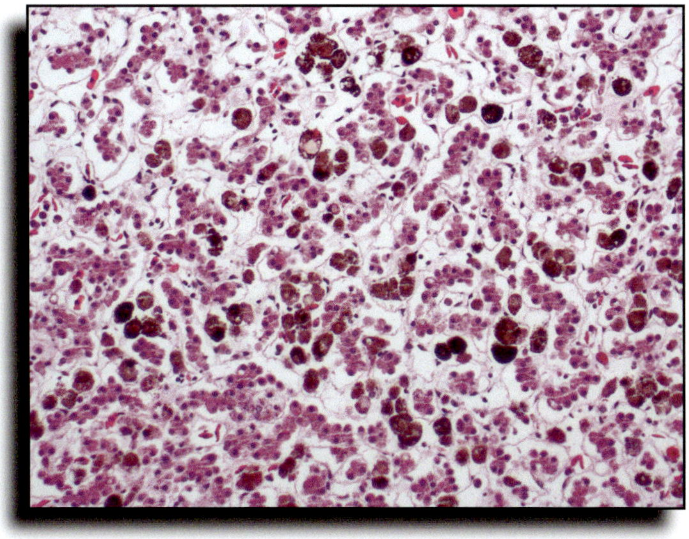

Figure 2.42 Loggerhead Sea Turtle, *Caretta caretta*. Cheloniidae. Hepatocellular atrophy. Higher magnification of Figure 2.41 illustrating severe atrophy of hepatocytes. Sinusoids and the space of Disse are secondarily dilated. H&E stain. (Photo courtesy of Michael M. Garner.)

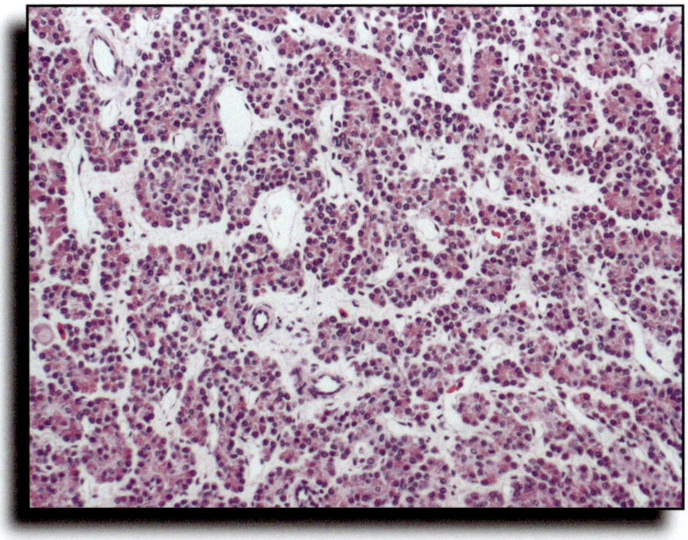

Figure 2.43 Loggerhead Sea Turtle, *Caretta caretta*. Cheloniidae. Pancreatic acinar cell atrophy. The acinar tissue is atrophied. H&E stain. (Photo courtesy of Michael M. Garner.)

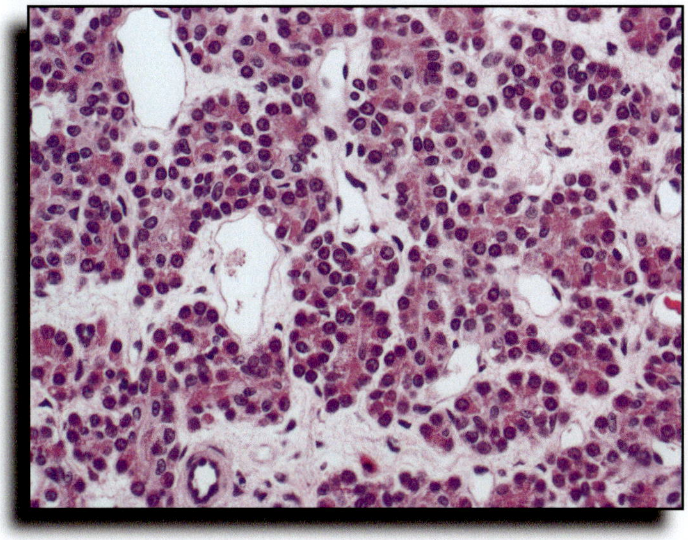

Figure 2.44 Loggerhead Sea Turtle, *Caretta caretta*. Cheloniidae. Pancreatic acinar cell atrophy. A higher magnification of Figure 2.43 demonstrates prominent atrophy of acinar cells, which have an increased nuclear to cytoplasmic ratio, and depleted zymogen granularity. The interstitium is expanded by clear spaces, attributed to edema and cellular atrophy. H&E stain. (Photo courtesy of Michael M. Garner.)

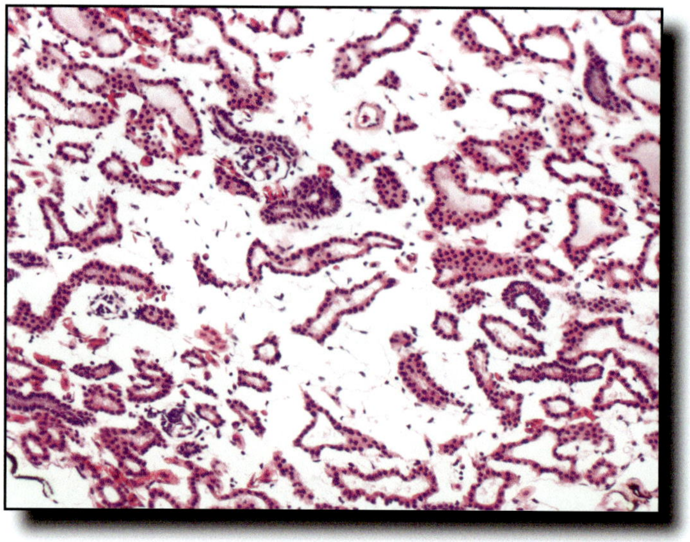

Figure 2.45 Texas Horned Lizard, *Phrynosoma cornutum*. Phrynosomatidae. Note marked atrophy of renal tubular epithelium with interstitial edema. The nucleus to cytoplasm ratio of tubular epithelial cells is prominently increased. H&E stain. (Photo courtesy of Michael M. Garner.)

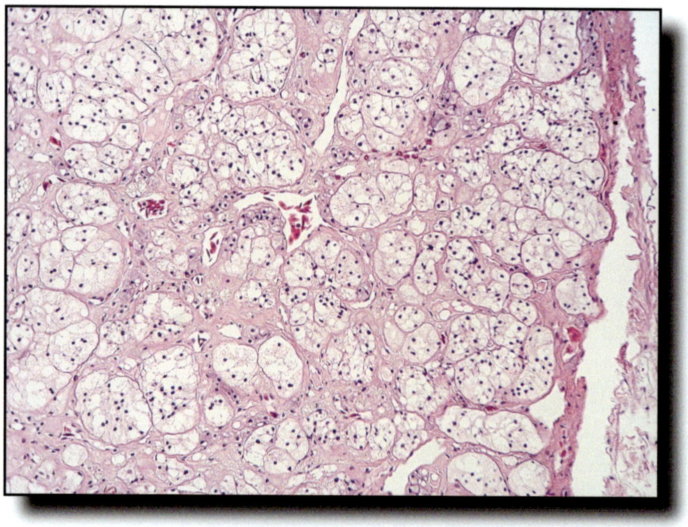

Figure 2.46 Green Tree Python, *Morelia viridis*. Pythonidae. Chronic stress response in adrenal glands. Interrenal cells in the adrenal gland are markedly hypertrophied due to the presence of clear intracytoplasmic vacuoles. H&E stain.

Figure 2.47 Red-Footed Tortoise, *Chelonoidis carbonaria*. Testudinidae. Pyramiding. Compared to a normal carapace for this species (a), note moderate pyramidal elevation of scutes, most notably in the dorsum (b). (Photos courtesy of Xavier Valls.)

Figure 2.48 Leopard Tortoise *Stigmochelys pardalis*. Testudinidae. Pyramiding. Pyramidal elevation of scutes is marked in this tortoise. (Photo courtesy of Xavier Valls.)

Figure 2.49 Indian Star Tortoise, *Geochelone elegans*. Testudinidae. Pyramiding. Prominent pyramidal elevation of scutes is noted. (Photo courtesy of Xavier Valls.)

Figure 2.50 Leopard Tortoise, *Stigmochelys pardalis*. Testudinidae. Metabolic bone disease. Note increased flexibility of carapace in this juvenile leopard tortoise with gastroduodenal cryptosporidiosis and retarded growth.

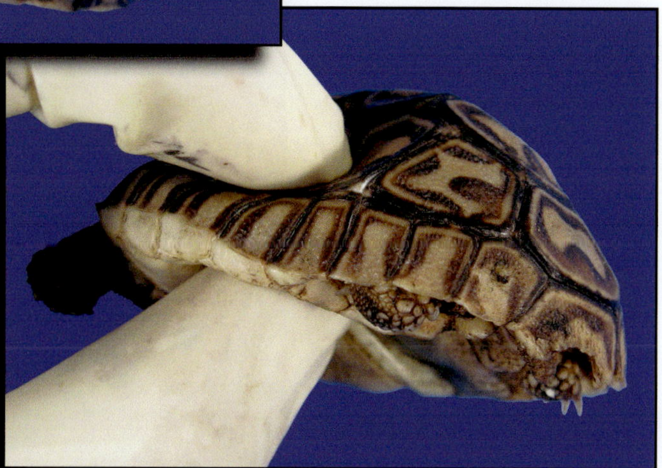

Figure 2.51 Leopard Tortoise, *Stigmochelys pardalis*. Testudinidae. Metabolic bone disease. Lateral view of the same tortoise as in Figure 2.50 showing both carapace and plastron involvement.

Figure 2.52 African Agama, *Agama africana*. Agamidae. Normal basking behavior. Basking under natural sunlight as in this wild African Agama in Sierra Leone is optimal for the prevention of metabolic bone disease in reptiles, but it may not be possible in some captive settings.

Figure 2.53 Leopard Gecko, *Eublepharis macularius*. Eublepharidae. Dietary metabolic bone disease. As compared to a normal posture (upper gecko with the sternum raised off of the substrate), the lower gecko is thinner and is sprawled on its sternum. (Photo courtesy of Steve Barten.)

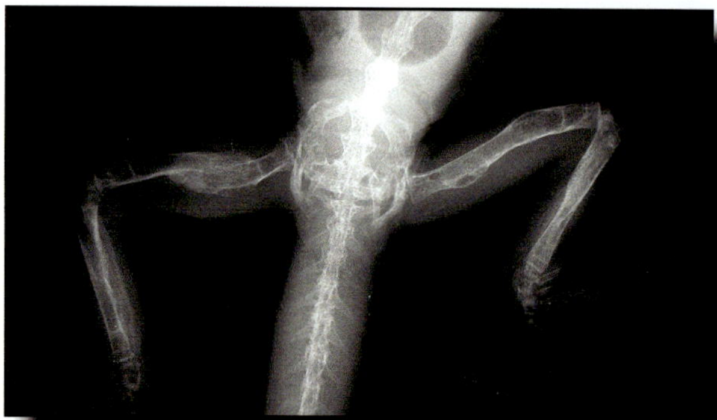

Figure 2.54 Green Iguana, *Iguana iguana*, Iguanidae. Dietary metabolic bone disease. This radiograph reveals deformities of long bones in the hindlimbs with multifocal to diffuse decreased density of cortical and trabecular bone, and a pathologic fracture of a femoral diaphysis with callus formation. Fibular diaphysis is slightly flared. (Photo courtesy of Xavier Valls.)

Figure 2.55 Veiled Chameleon, *Chamaeleo calyptratus*. Chamaeleonidae. Dietary metabolic bone disease. Note swelling of limbs and deformity of left tibial bone and casque as a result of metabolic bone disease. (Photo courtesy of Xavier Valls.)

Figure 2.56 Green Iguana, *Iguana iguana*. Iguanidae. Dietary metabolic bone disease. Note bowing of mandibular bones with shortening of the lower jaw. (Photo courtesy of Mireia Máinez.)

Figure 2.57 Veiled Chameleon, *Chamaeleo calyptratus*. Chamaeleonidae. Dietary metabolic bone disease. The abnormal posture of the right front limb is due to a pathological fracture of the right humerus. (Photo courtesy of Xavier Valls.)

Figure 2.58 Green Iguana, *Iguana iguana*. Iguanidae. Dietary metabolic bone disease. Marked kyphosis is evident in this juvenile iguana. (Photo courtesy of Xavier Valls.)

Figure 2.59 Central Bearded Dragon, *Pogona vitticeps*. Agamidae. Dietary metabolic bone disease. The upper maxillary and nasal bones are deformed and shortened, while the mandibular symphysis is elevated. (Photo courtesy of Andrés Montesinos.)

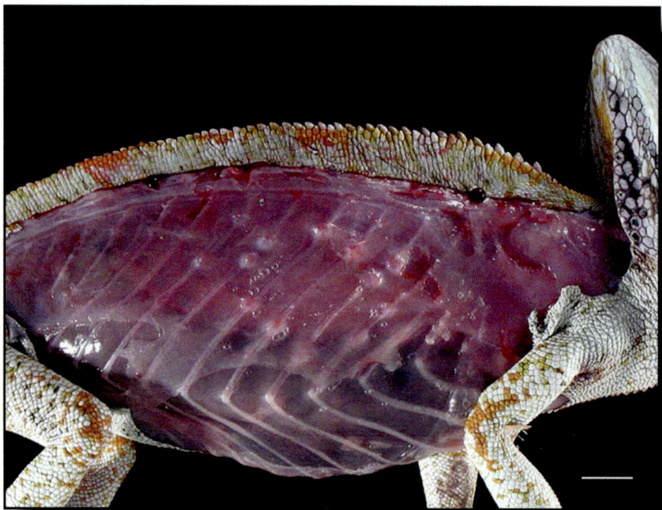

Figure 2.60 Veiled Chameleon, *Chamaeleo calyptratus*. Chamaeleonidae. Dietary metabolic bone disease. Note multifocal nodular swelling of the ribs. Bar = 1 cm.

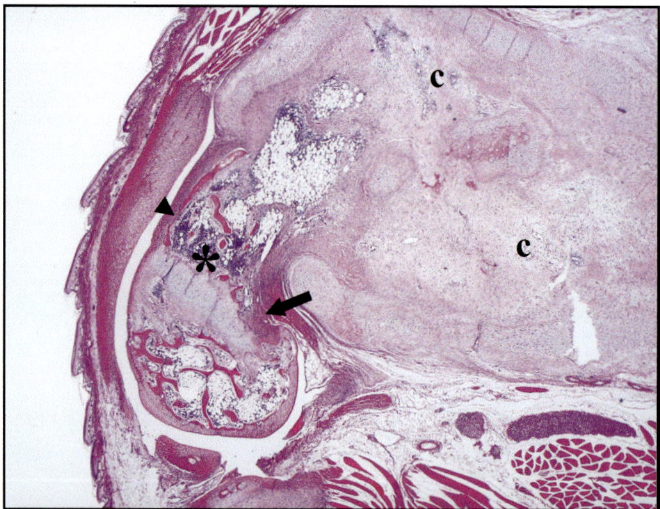

Figure 2.61 Madagascar Day Gecko, *Phelsuma madagascariensis*. Gekkonidae. Dietary metabolic bone disease. Note marked decreased bone density in metaphyseal growth plate (asterisk) and cortical bone of distal diaphysis (arrowhead). Focal collapse of the metaphysis (arrow), bowing of the affected bone and fracture with replacement by a large callus (c). H&E stain. (Photo courtesy of Michael M. Garner.)

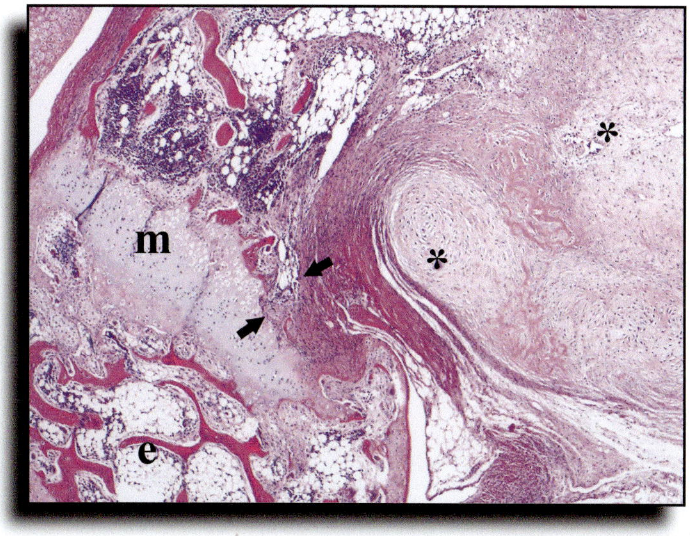

Figure 2.62 Madagascar Day Gecko, *Phelsuma madagascariensis*. Gekkonidae. Dietary metabolic bone disease. higher magnification of Figure 2.61 demonstrates markedly decreased bone density in the metaphyseal growth plate. Metaphyseal cartilage is focally collapsed (arrows). Note exuberant callus (*) formation. m, metaphyseal cartilage; e, epiphysis. H&E stain. (Photo courtesy of Michael M. Garner.)

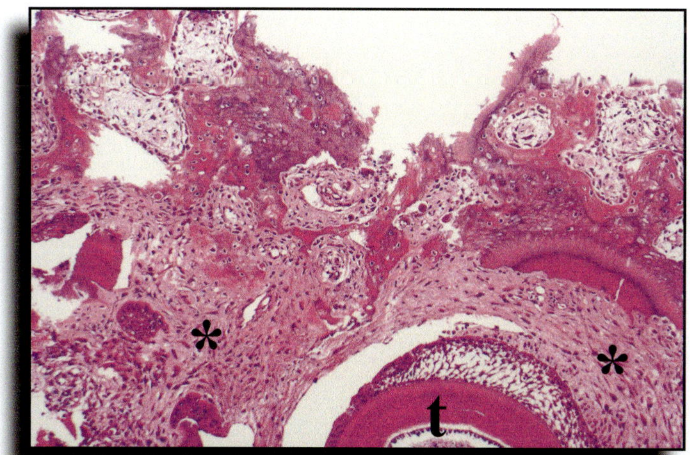

Figure 2.63 Green Iguana, *Iguana iguana*. Iguanidae. Dietary metabolic bone disease. Maxillary bone is partly replaced with fibrous tissue (asterisks). Note a fragment of a tooth (t). H&E stain. (Photo courtesy of Michael M. Garner.)

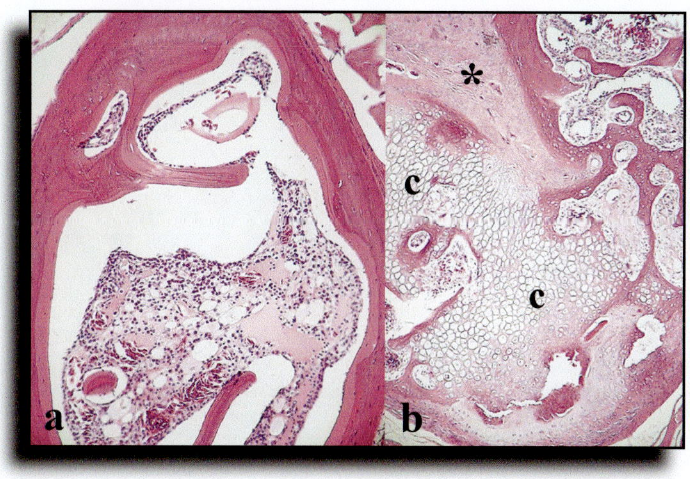

Figure 2.64 Veiled Chameleon, *Chamaeleo calyptratus*. Chamaeleonidae. Dietary metabolic bone disease. As compared to a normal rib section of the same chameleon (a), the rib lesions depicted in Figure 2.60 consist of replacement of trabecular and cortical bone with fibrous tissue (asterisk) and metaplastic cartilage (c, b). H&E stain.

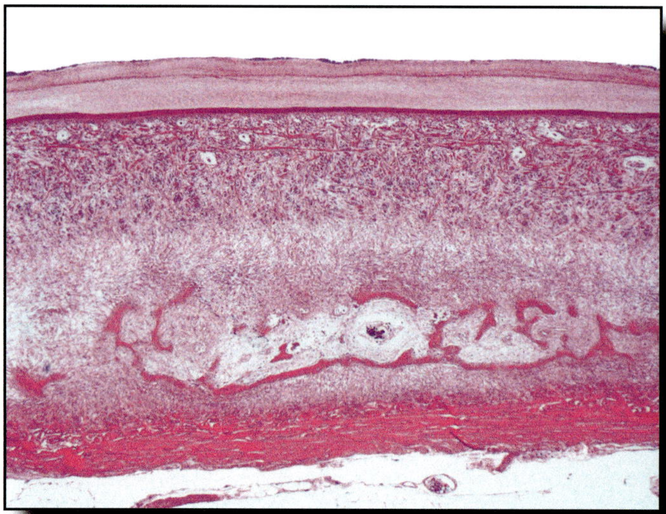

Figure 2.65 Box Turtle, *Terrapene sp.* Emydidae. Metabolic bone disease. Insufficient trabecular and cortical bone is observed in the carapace. Comparison with age-matched controls is necessary for a definitive diagnosis. H&E stain. (Photo courtesy of Michael M. Garner.)

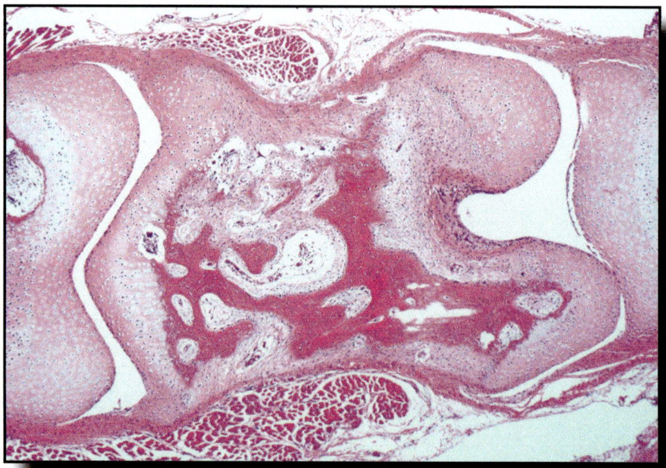

Figure 2.66 Box Turtle, *Terrapene sp.* Emydidae. Dietary metabolic bone disease. Lysis with increased osteoclastic resorption of upper cortical, subchondral and trabecular bone is noted in this phalangic bone. Mild fibrosis accompanies bone lysis. H&E stain. (Photo courtesy of Michael M. Garner.)

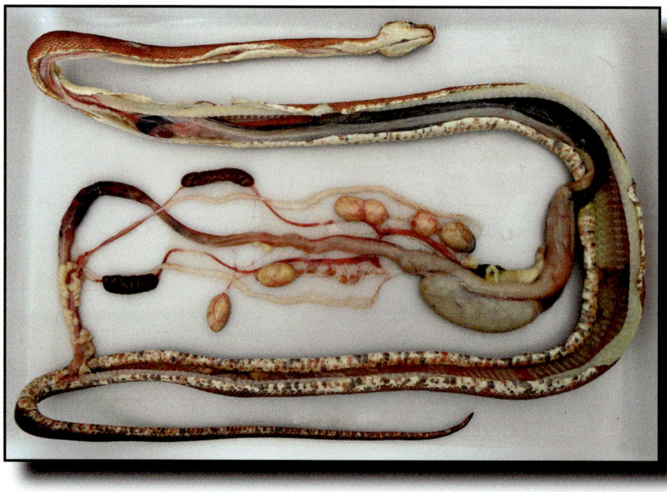

Figure 2.67 Amazon Tree Boa, *Corallus hortulanus.* Boidae. Ovarian follicular stasis. Note follicular stasis and degeneration of ovarian follicles, which was associated with microscopic evidence of soft tissue mineralization.

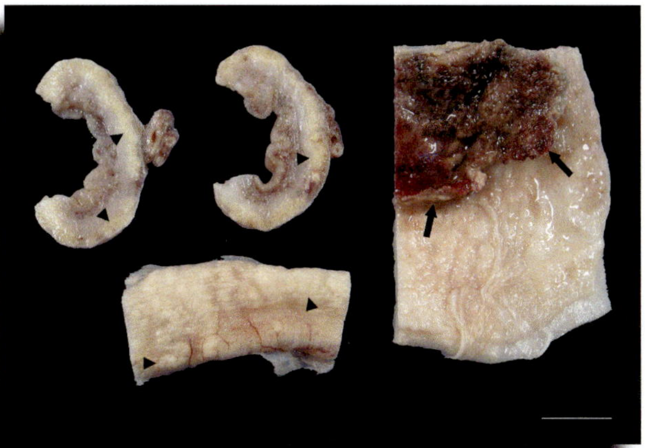

Figure 2.68 Common Snapping Turtle, *Chelydra serpentine*. Chelydridae. Enteric soft tissue mineralization. Note foci of thickened, discolored muscular tunics (arrowheads) due to mineralization. Associated with this deposition disorder are foci of necrotizing gastritis (arrows). Formalin-fixed specimens. Bar = 1 cm.

Figure 2.69 Green Iguana, *Iguana iguana*. Iguanidae. Cutaneous mineralization. The well-circumscribed green-gray to red cutaneous foci are due to mineralization. (Photo courtesy of Andrés Montesinos.)

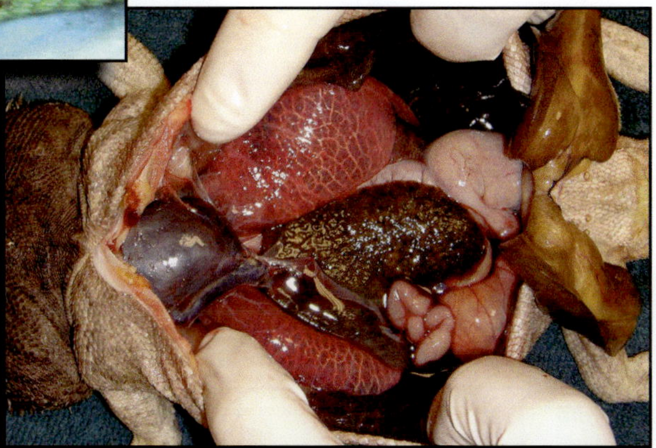

Figure 2.70 Central Bearded Dragon, *Pogona vitticeps*. Agamidae. Pulmonary mineralization. Note accentuated white reticular pattern in the interstitium, due to mineralization of muscular trabeculae. Hepatic cirrhosis is also present. (Photo courtesy of Max Conn.)

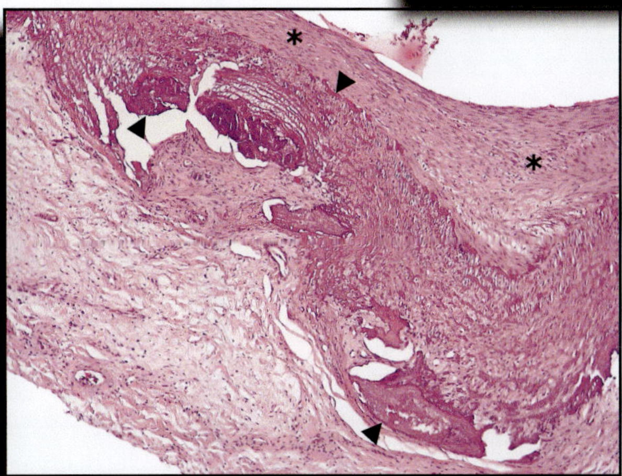

Figure 2.71 Green Iguana, *Iguana iguana*. Iguanidae. Arterial mineralization. Note marked mineralization and ossification (arrowheads) with disruption of the medial tunic and intimal (asterisks) and to a lesser extent, medial fibrosis. H&E stain.

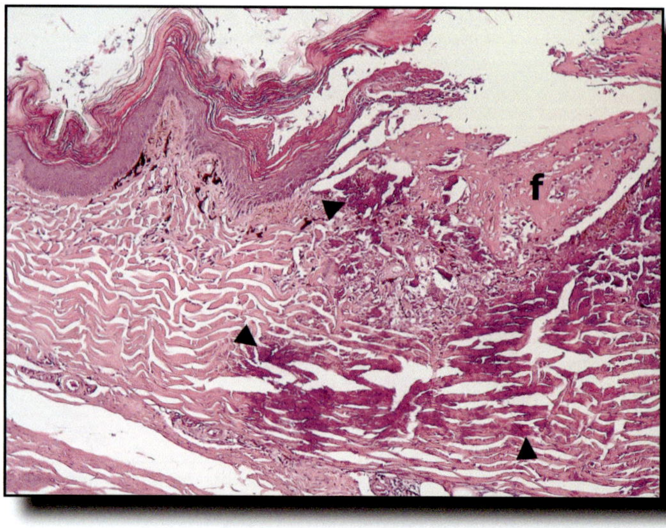

Figure 2.72 Green Iguana, *Iguana iguana*. Iguanidae. Cutaneous mineralization. The dermis contains marked deposits of mineral (arrowheads), which are associated with ulceration with fibrin deposition (f). Note epidermal hyperplasia and hyperkeratosis of non-mineralized skin (left). H&E stain.

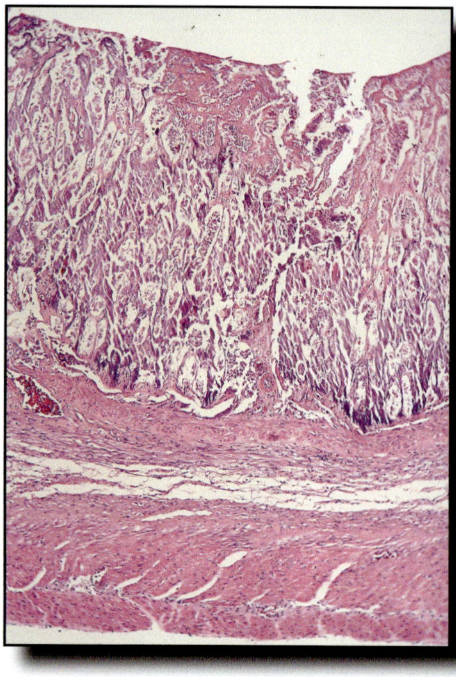

Figure 2.73 Panther Chameleon, *Furcifer pardalis*. Chamaeleonidae. Gastric mineralization. Marked diffuse mineral deposits efface the gastric mucosa. H&E stain.

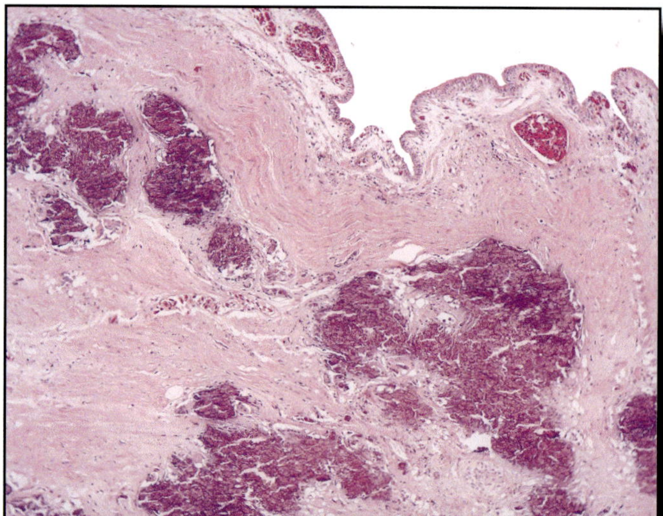

Figure 2.74 Common Snapping Turtle, *Chelydra serpentine*. Chelydridae. Pulmonary mineralization. Coarse mineral deposits are observed in the pulmonary smooth muscle. H&E stain.

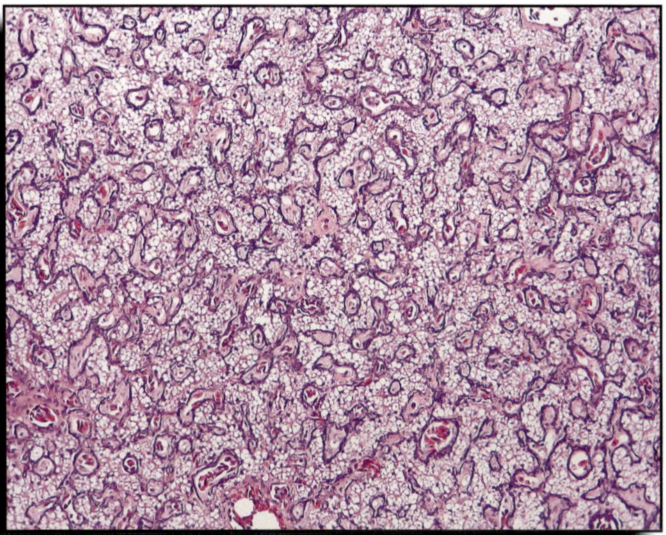

Figure 2.75 Panther Chameleon, *Furcifer pardalis*. Chamaeleonidae. Hepatic mineralization. Diffuse thickening of the wall of hepatic sinusoids is due to mineralization. Prominent hepatocellular lipidosis is noted as well. H&E stain.

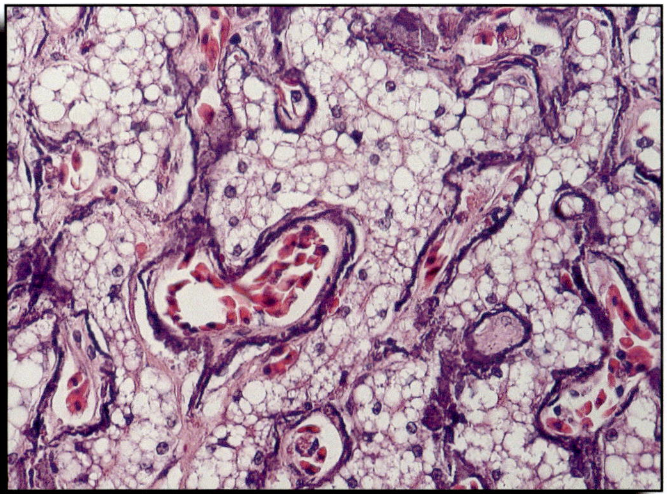

Figure 2.76 Panther Chameleon, *Furcifer pardalis*. Chamaeleonidae. Hepatic mineralization. Mineral deposits cause a thickening of the wall of hepatic sinusoids. Hepatocytes contain clear intracytoplasmic vacuoles (lipidosis). H&E stain.

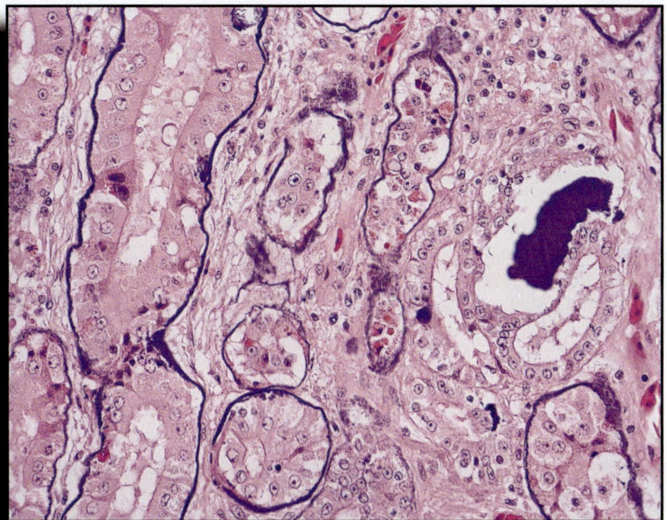

Figure 2.77 Madagascar Boa, *Acrantophis madagascariensis*. Boidae. Renal mineralization. The basement membrane of renal tubules is thickened due to mineral deposition (arrowheads); the lumen or epithelial layer of some tubules also contains mineral deposits. H&E stain.

Figure 2.78 Ouachita Map Turtle, *Graptemys ouachitensis*. Emydidae. Hypovitaminosis A. Note swelling of eyelids. (Photo courtesy of Xavier Valls. Specific identification of turtle in this figure by Cris Hsgen.)

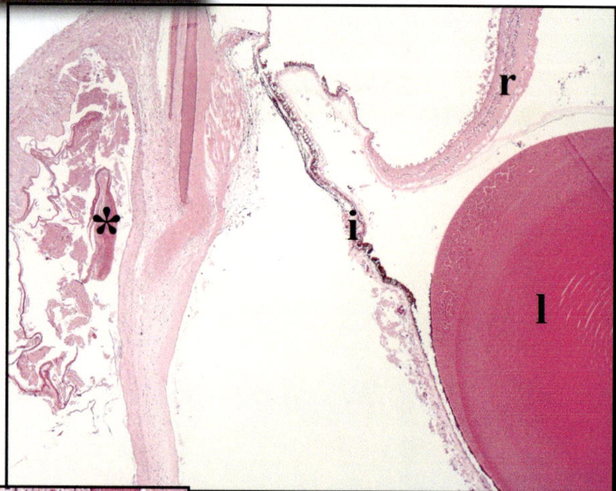

Figure 2.79 Leopard Gecko, *Eublepharis macularius*. Eublepharidae. Hypovitaminosis A. The conjunctival sac is filled with keratin (asterisk). l, lens; i, iris; r, retina. H&E stain.

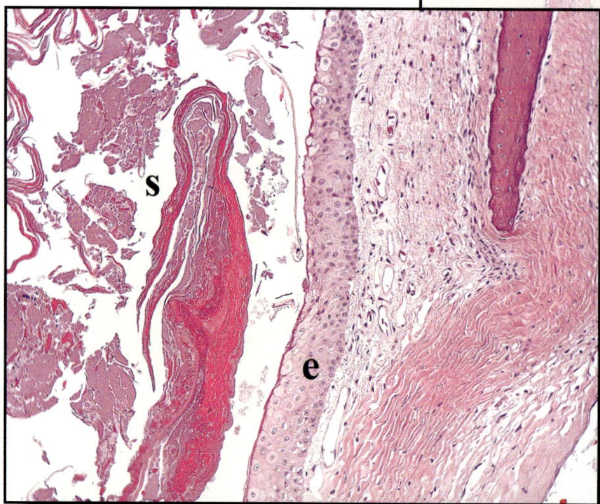

Figure 2.80 Leopard Gecko, *Eublepharis macularius*. Eublepharidae. Hypovitaminosis A. A higher magnification of the same eye depicted in Figure 2.79 reveals squamous metaplasia of the conjunctival epithelium (e) with accumulation of keratin in the conjunctival sac (s). H&E stain.

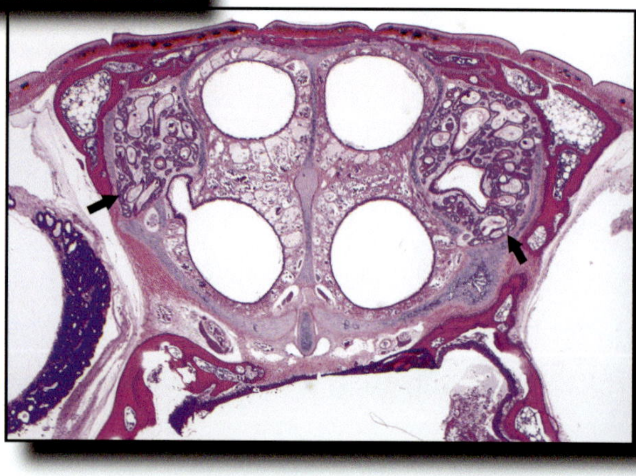

Figure 2.81 Desert Grassland Whiptail Lizard, *Aspidoscelis uniparens*. Teiidae. Note dilatation of nasal glands and ducts (arrows) on both sides of the nasal cavity due to luminal accumulation of keratin. H&E stain. (Photo courtesy of Michael M. Garner.)

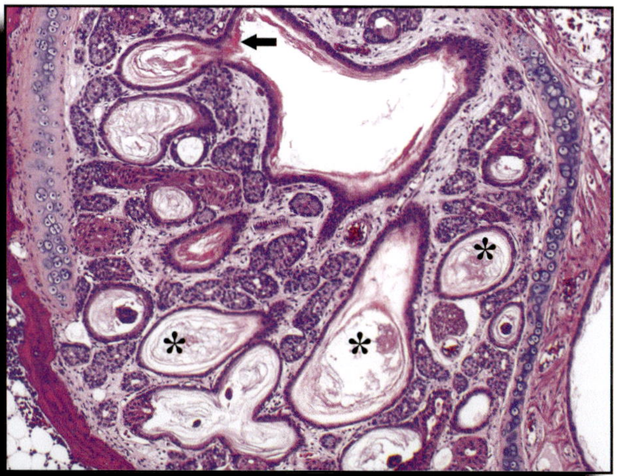

Figure 2.82 Desert Grassland Whiptail Lizard, *Aspidoscelis uniparens*. Teiidae. A higher magnification of Figure 2.81 demonstrates that dilated glands (asterisks) are lined by a squamous epithelium with hyperkeratosis (arrow). H&E stain. (Photo courtesy of Michael M. Garner.)

Figure 2.83 Hermann's Tortoise, *Testudo hermanni*. Testudinidae. Hypervitaminosis A. Note extensive skin ulceration in both hindlimbs.

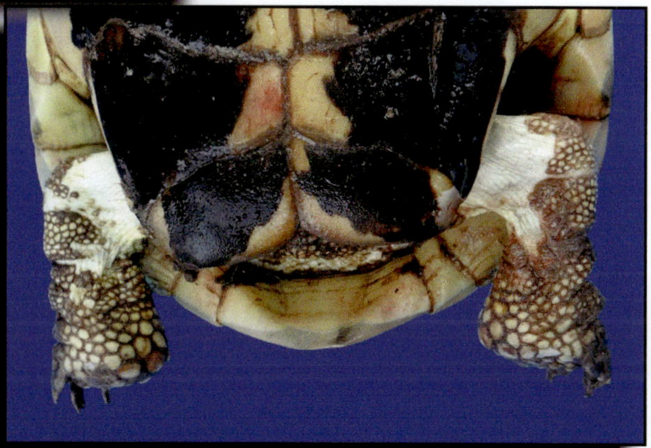

Figure 2.84 Hermann's Tortoise, *Testudo hermanni*. Testudinidae. Hypervitaminosis A. Close-up view of affected limbs in Figure 2.83 shows whitish areas of cutaneous ulceration.

Figure 2.85 Sulcatta Tortoise, *Centrochelys sulcata*. Testudinidae. Hypervitaminosis A. Note severe skin slough involving the front legs. (Photo courtesy of Doug Mader.)

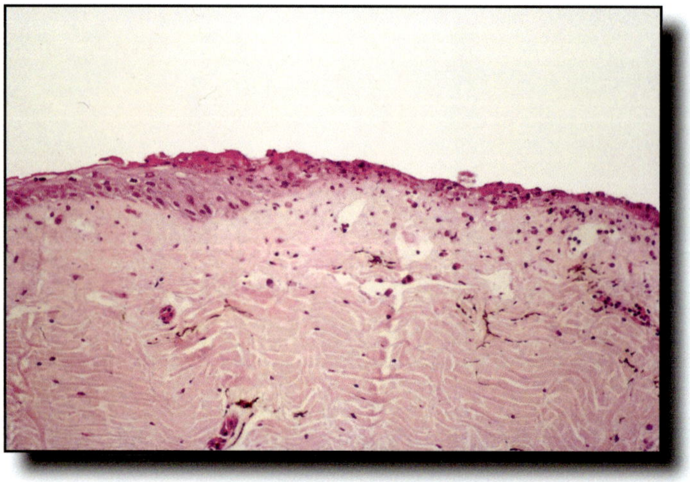

Figure 2.86 Eastern Box Turtle, *Terrapene c. carolina*. Emydidae. Hypervitaminosis A. Note epidermal ulceration and necrosis with mild infiltration of inflammatory cells in the underlying dermis. H&E stain. (Photo courtesy of Michael M. Garner.)

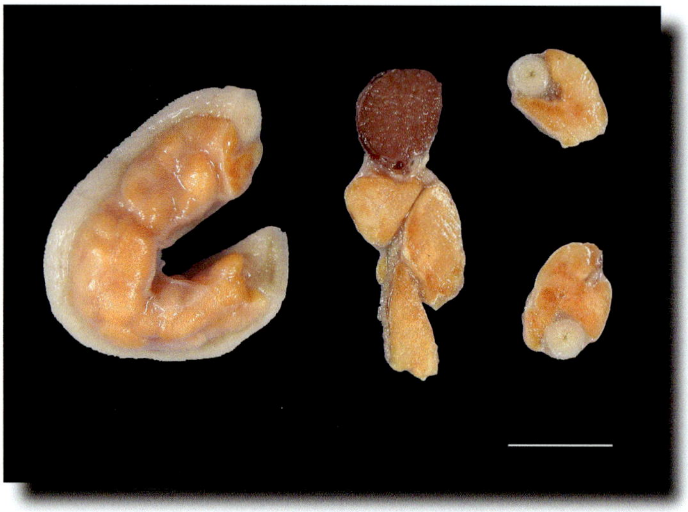

Figure 2.87 Caiman, *Caiman* sp. Alligatoridae. Pansteatitis. The visceral adipose tissue of the spleen and intestine is discolored orange to brown. Formalin-fixed specimens. Bar = 1 cm.

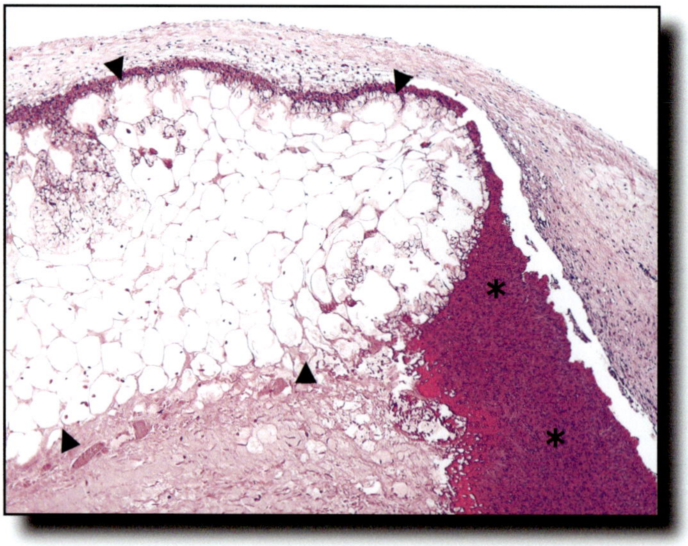

Figure 2.88 Caiman, *Caiman* sp. Alligatoridae. Pansteatitis. Note necrotic adipose tissue (arrowheads) surrounded by necrotic exudate (asterisks) and covered with fibrous tissue. H&E stain.

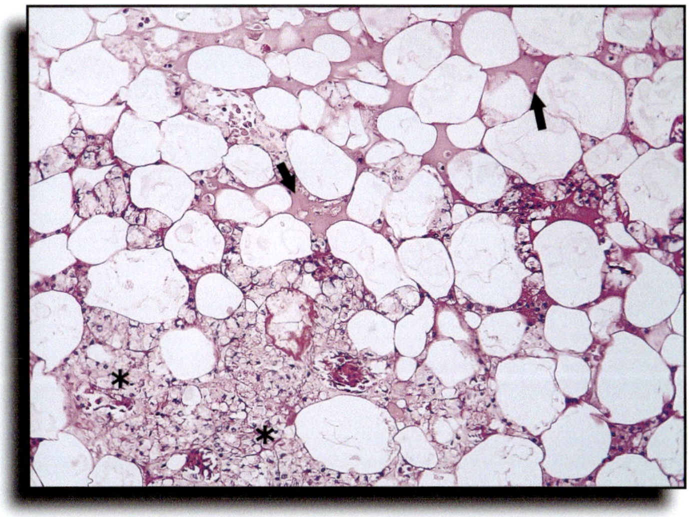

Figure 2.89 Caiman, *Caiman* sp. Alligatoridae. Pansteatitis. Necrotic adipose tissue with interstitial accumulation of proteinaceous edema (arrows) and infiltrates of vacuolated macrophages (asterisks) in the interstitium. H&E stain.

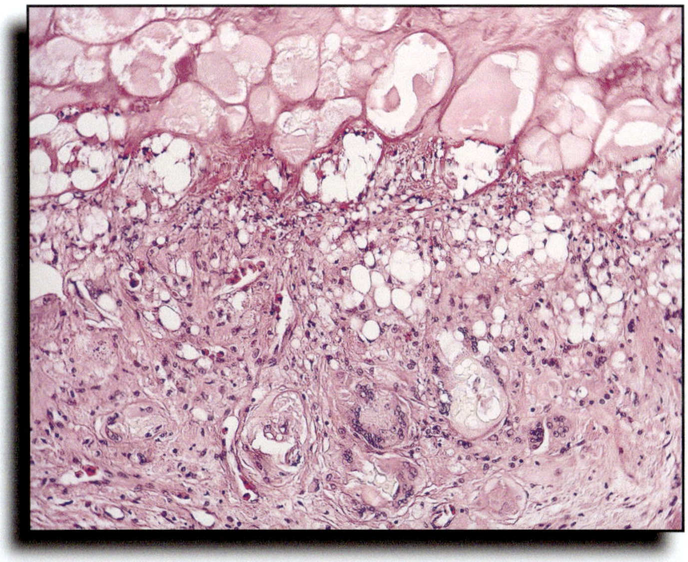

Figure 2.90 Caiman, *Caiman* sp. Alligatoridae. Pansteatitis. Necrotic adipose tissue (top) is surrounded by granulomatous inflammation consisting of multinucleated giant cells and vacuolated macrophages. H&E stain.

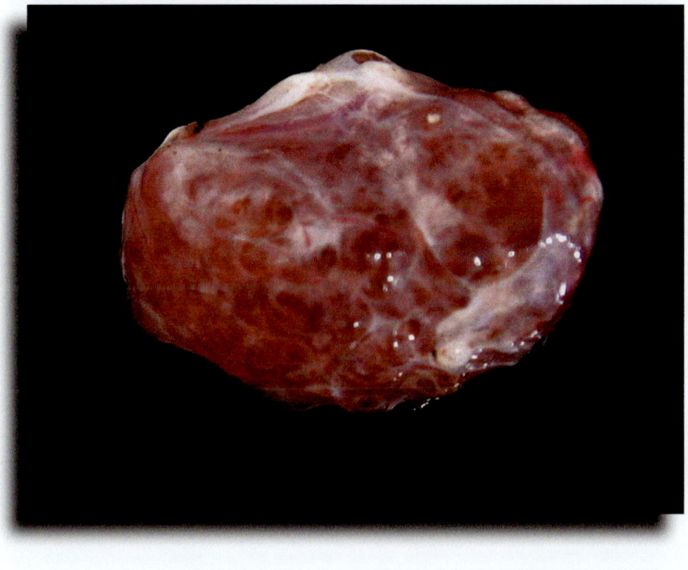

Figure 2.91 Boa Constrictor, *Boa constrictor*. Boidae. Goiter. The thyroid gland is enlarged and has an irregular surface due to dilatation of numerous follicles. Bar = 1 cm.

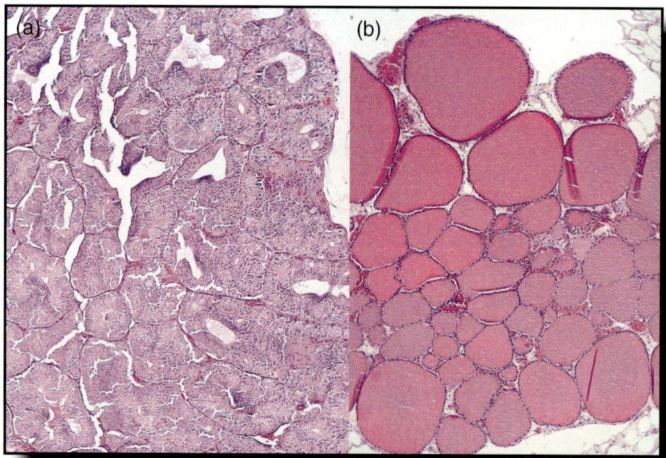

Figure 2.92 (a) Corn snake, *Pantherophis guttatus*. Colubridae. Hyperplastic goiter. Diffusely, follicles contain no colloid and are lined by a hyperplastic and hypertrophied epithelium resulting in an adenomatous appearance. H&E stain. (b) Central Bearded Dragon, *Pogona vitticeps*. Agamidae. Normal thyroid gland. Note follicles filled with colloid. H&E stain.

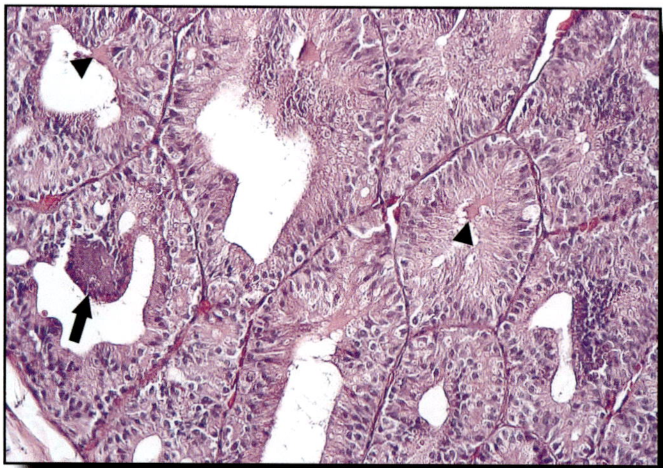

Figure 2.93 Corn snake, *Pantherophis guttatus*. Colubridae. Hyperplastic goiter. A higher magnification of Figure 2.92(a) reveals minimal amounts of colloid (arrowheads) in few follicles. Follicular epithelium is hyperplastic and tall, and micropapillary projections into the lumen are present (arrow). H&E stain.

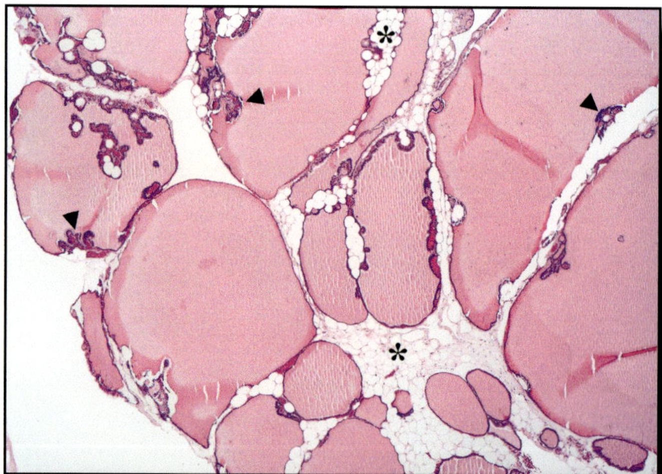

Figure 2.94 Western Bearded Anole, *Xiphosurus* (*Anolis*, *Chamaeleolis*) *barbatus*. Dactyloidae. Colloid goiter. Diffusely, follicles are markedly enlarged due to accumulation of excessive colloid. Note micropapillary luminal projections of hyperplastic and hypertrophied follicular epithelium (arrowheads). The interstitium is expanded with areas of lipomatosis (asterisks). H&E stain.

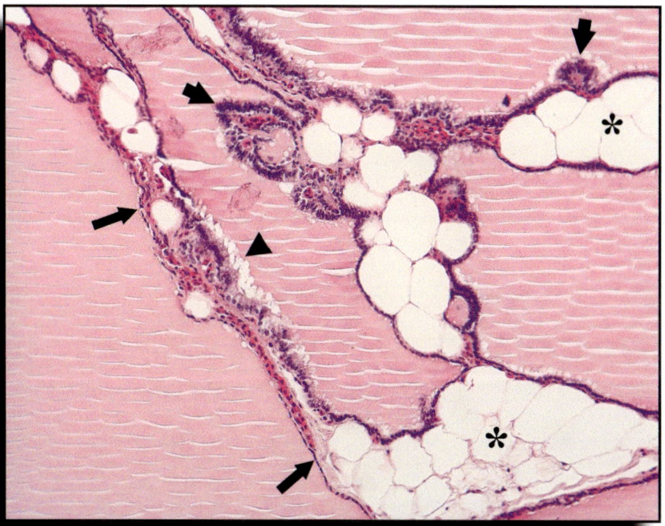

Figure 2.95 Western Bearded Anole, *Xiphosurus* (*Anolis*, *Chamaleolis*) *barbatus*. Dactyloidae. Colloid goiter. Higher magnification of Figure 2.94. Follicles are lined by a single layer of flattened epithelium (long arrows) with luminal micropapillary projections of hyperplastic and hypertrophied follicular epithelium (arrowheads). Note endocytosis of colloid in areas of hypertrophied epithelium (arrowhead). Interstitial lipomatosis (asterisks) is also present. H&E stain.

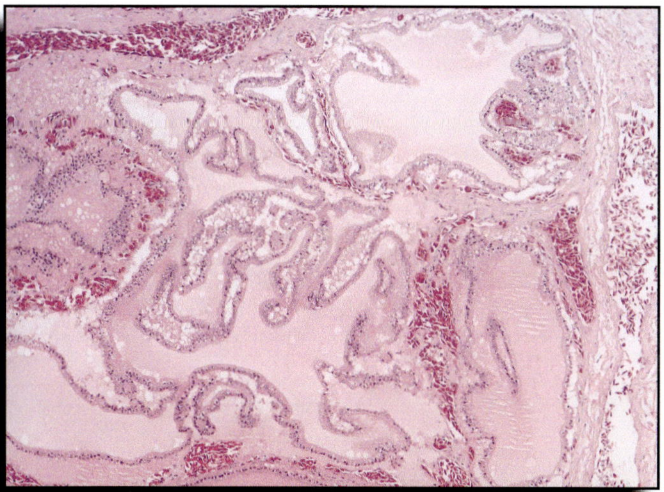

Figure 2.96 Common Snapping Turtle, *Chelydra serpentine*. Chelydridae. Goiter. Note hyperplastic goiter transitioning into colloid goiter. Numerous papillary projections of the follicular epithelium are still present, and the follicular epithelium is still cylindrical. H&E stain.

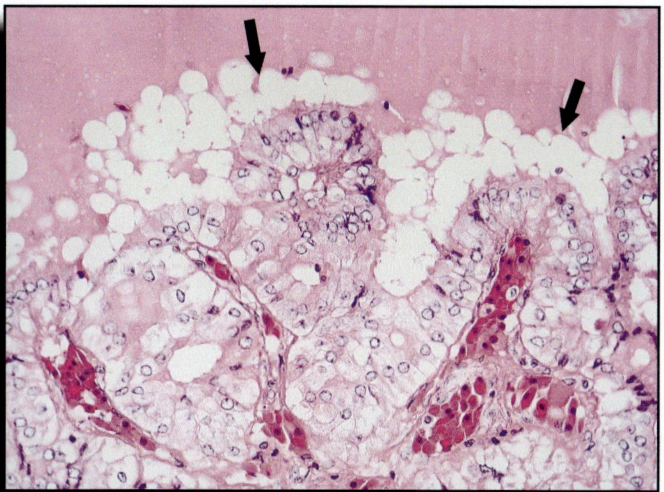

Figure 2.97 Common Snapping Turtle, *Chelydra serpentine*. Chelydridae. Goiter. High magnification of Figure 2.96 with hyperplastic goiter still not fully transitioned into colloid goiter, which is characterized by marked vacuolation of colloid (arrows) in areas adjacent to the remaining hyperplastic and hypertrophied follicular epithelium. H&E stain.

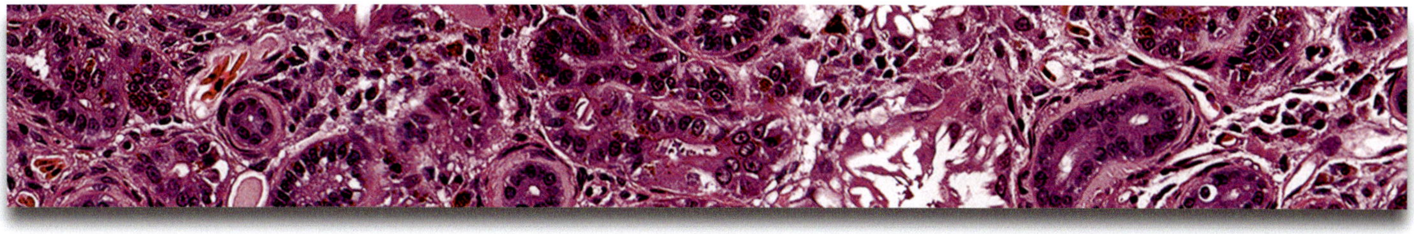

3

DEPOSITIONAL DISEASES

ERIN A. GRAHAM, RACHEL E. BURNS, AND ROBERT J. OSSIBOFF

Contents

3.1 Introduction

Depositional diseases, defined as the abnormal intra- and/or extracellular accumulation of various organic and inorganic compounds within the parenchyma or potential spaces of tissues and body cavities, can be a common finding in captive, and to a lesser degree, free-ranging reptiles. Such disorders can be the end point of a broad range of hereditary, metabolic, nutritional, and degenerative conditions. Some depositional diseases, such as gout, have important clinical and physiologic manifestations. Others, such as certain pigment accumulations, are commonly encountered as background lesions in the histologic examination of reptile tissues. This chapter will detail the gross and histologic manifestations of depositional diseases in reptiles and, when possible, provide information regarding the underlying pathophysiologic mechanisms and clinical sequelae.

3.2 Gout

Gout refers to deposition of uric acid in tissues. The term has been adopted from human medicine, in which the condition has been recognized for centuries, despite substantial differences in the disease between primates and reptiles. Gout in its various forms (renal, visceral, articular, and periarticular) is a common condition in all reptiles that secrete uric acid as their primary nitrogenous waste product, i.e. uricotelic species. When the concentration of uric acid becomes increased in the blood or synovial fluid, uric acid can precipitate in tissues as monosodium urate causing damage and inflammation. While uric acid is the most common nitrogenous waste product in reptiles, in some species, particularly crocodilians and certain chelonians, ammonia and urea can be important routes of nitrogen excretion (Divers and Innis, 2019). However, gout can occur even in species where uric acid is not the primary nitrogenous waste.

3.2.1 Causes

Though the basic mechanism of gout is relatively straightforward, the underlying predispositions are considerably less so. Uric acid is produced as a byproduct of protein and nucleotide metabolism and excreted by the renal tubules. Excessive concentrations of uric acid in the body therefore occur with increased production or decreased excretion. A relative increase in uric acid levels can also occur with dehydration, with a decrease in the total body fluid volume leading to increased uric acid concentration.

Increased production of uric acid in reptiles can be attributed to improper diet, particularly in regard to the type and amount of dietary protein. In animals that lack uricase, uric acid is the end stage of metabolism of purines (adenine and guanine) by xanthine oxidase. High levels of purines are found in animal proteins, and an association between increased intake of meat and seafood and increased serum levels of uric acid has been reported in humans (Choi et al., 2005). Dietary protein requirements for reptiles vary among herbivorous, carnivorous, and omnivorous species, with carnivorous reptiles requiring the highest protein levels from primarily animal-based sources and herbivorous reptiles utilizing substantially lower protein levels from plant-based sources (Boyer and Scott, 2019). Therefore, herbivorous reptiles (tortoises and some lizards, e.g., iguanas) are presumed to be at the greatest risk for a nutritional cause of gout when fed a diet high in animal-based proteins, such as dog or cat food (Driggers, 1997; Miller, 1998; Reavill and Schmidt, 2010; Orós, 2019). The excess protein in these diets results in rapid increases in blood uric acid levels (Miller, 1998). Experimentally, serum uric acid levels significantly increase in Green Iguanas fed a high protein diet compared to a normal vegetarian diet (Hernandez-Divers et al., 2008).

While carnivorous reptiles are considered less susceptible to excess dietary protein, protein levels higher than the recommended 25%–60% protein by dry matter in these species could still be a factor in development of gout (Donoghue, 2006). In a group of crocodile hatchlings, a diet of 69% protein, compared to the recommended level of 45%, could have contributed to a high prevalence of gout in this group, although concurrent renal damage from hypovitaminosis A was also involved (Ariel et al., 1997). An outbreak of gout in another large group of young Siamese Crocodiles (*Crocodylus siamensis*) was associated with feeding a diet of primarily porcine viscera (liver, lung, and spleen) for 4 months. These animals had hyperuricemia, and gout was attributed to the high protein diet (Ousavaplangchai, 1990).

Besides diet, other causes of increased production of uric acid in humans which could apply to reptiles include myeloproliferative and hemolytic diseases (Yü, 1965; Mader, 2006). These disorders accelerate metabolism of nucleic acids, potentially increasing purine metabolites, though this has not been demonstrated in reptiles. Starvation and breakdown of endogenous proteins is another potential source of elevated uric acid levels (Cowan, 1968).

Increased concentrations of uric acid also occur secondary to decreased excretion by the kidneys. In reptiles, the proximal renal tubules are responsible for secreting uric acid into the urine for elimination (Mader, 2006). Impaired renal function from dehydration or cold stress, damage to renal tubules, and obstruction of urinary outflow are all potential causes of gout. Nephrotoxic drugs, such as the aminoglycoside antibiotic gentamicin, which cause degeneration and necrosis of proximal renal tubular epithelium, have been shown to result in gout in snakes (Jacobson, 1976; Hodge, 1978; Montali et al., 1979). Other causes of renal compromise or failure predisposing to development of gout include bacterial and other infections of the kidney (Huchzermeyer, 2003; Reavill and Schmidt, 2010), chronic renal disease, renal neoplasia, and vitamin A deficiency. The latter results in squamous metaplasia of collecting ducts, which can lead to secondary ascending bacterial infection or obstruction by accumulations of keratin. This was considered the major factor, in addition to high dietary protein levels, in the development of gout in a group of young Australian Freshwater (*Crocodylus johnstoni*) and Saltwater (*Crocodylus porosus*) Crocodiles (Ariel et al., 1997). Renal gout is also reported secondary to renal adenocarcinoma in Red Spitting Cobras (*Naja pallida*) (Jacobson et al., 1986; Belasco-Zeitz et al., 2013).

Besides overproduction and decreased excretion, dehydration has the most considerable effect on fluid uric acid concentrations in reptiles. Uric acid and its circulating salt, monosodium urate, are poorly soluble, and normal hydration is required to prevent its precipitation in the body, though the exact concentration at which such precipitation occurs is unknown (Mader, 2006). Diuretic drugs, such as furosemide, predispose to development of gout (Boyer and Scott, 2019). Dehydration is often presumed to be a consistent requirement for gout in reptiles even when other factors, such as high protein diet, might also be involved (Appleby and Siller, 1960; Wallach and Hoessle, 1967; Cowan, 1968; Cooper and Jackson, 1981). The effect of dehydration on renal function is another consideration. Whether dehydration leads to decreased renal uric acid excretion is questionable (Mader, 2006), but chronic low-grade dehydration could contribute to nonspecific chronic renal disease (Reavill and Schmidt, 2010). In either case, it is important to recognize that dehydration, renal disease, and renal gout are often inseparable, and in advanced cases it can be difficult or impossible to determine the initial disease process. Renal damage occurs early in gout, and the resulting renal insufficiency perpetuates further dehydration, gout, and renal injury.

3.2.2 Prevalence of Gout

Gout has been reported in all types of reptiles, namely snakes, lizards, terrestrial chelonians (tortoises), and crocodilians. Among groups of captive reptiles, rates of gout are often relatively high. The overall prevalence in a review of reptile necropsy records from one large institution identified gout as a diagnosis in over 7.5% of cases (261 of 3412) (Burns, unpublished). The rates were highest in lizards, which constituted 70% of the total cases, with a prevalence of 10.7% (183 of 1707). Iguanids were overrepresented with 22% (43 of 196) of cases having a diagnosis of gout. Snakes and chelonians had lower rates at approximately 5% (65 of 1234) and 2% (8 of 446), respectively. Cases of gout were also identified in crocodilians (2 of 19) and tuatara (3 of 6), but overall numbers of these species were small. In other studies, in varanids, gout was the cause of death in 10% of these lizards from one institution (Mendyk et al., 2013) and was present in 33% of varanid necropsies in a broader review (Garner, 2008). European tortoises had a 16% occurrence of renal gout at necropsy in another retrospective analysis (Kölle and Hoffman, 2002). Gout (renal and articular) has also rarely been observed in free-ranging Desert Tortoises (*Gopherus agassizii*) (Homer et al., 1998; Jacobson, personal communication).

3.2.3 Common Pathologic Features of Gout

Precipitation and deposition of monosodium urate in tissues causes tissue damage and incites an inflammatory response. Gout is most commonly classified by the affected tissue: renal (**Figures 3.1–3.7**), visceral (**Figures 3.8–3.9**), and articular/periarticular (**Figures 3.10–3.12**). In all sites, precipitated uric acid appears grossly as white to off-white deposits that may cause fine stippling or streaking within visceral parenchyma, an opaque coating on serosal surfaces, or large nodular accumulations of pasty material within joints or periarticular tissues. Uric acid deposits dissolve in aqueous solutions and with routine processing for histologic examination. Preservation of uric acid deposits requires fixation in 70% ethanol; visualization of crystals can be achieved with eosin staining in absolute alcohol or special histochemical stains, including deGalantha and Gomori methenamine silver (deGalantha, 1935; Stevens, 1990; Shidham and Shidham, 2000). Without special processing, only pale, radiating, wispy basophilic or fibrillar material that disrupts tissue architecture may be evident microscopically (**Figure 3.13**). Acutely, there is associated granulocytic inflammation and necrosis which progresses to the formation of discrete tophi (**Figures 3.14–3.15**). Classical urate tophi are composed of a starburst arrangement of monosodium urate crystals surrounded by granulomatous inflammation in the form of heterophils, macrophages, and multinucleated giant cells (**Figures 3.16–3.17**) (Driggers, 1997; Knafo, 2019). Cytologically, monosodium urate crystals are fine, needle-like, and strongly birefringent under polarized light (**Figures 3.18a,b–3.19**; Volume 1, **Figure 6.20a,b**).

3.2.4 Renal and Visceral Gout

Renal gout is characterized by the disruption of the renal parenchyma by aggregates of crystallized monosodium urate and inflammation with generalized renal edema, swelling, and pallor (**Figures 3.1–3.7**). Histologically, crystal formation causes tubular rupture with heterophilic and granulomatous inflammation in tubules that can extend into the interstitium (**Figures 3.13, 3.17**). In the chronic stages of renal gout, there may be interstitial fibrosis, regeneration, and hyperplasia of renal tubular epithelium, and mineralization (**Figures 3.17b, 3.20**). Renal gout may begin with urate stasis, in which the intratubular accumulation of urates causes distension of tubules and gives a gross appearance of fine white streaking (**Figure 3.21**). Progression may lead to tubular obstruction and rupture with the formation of gouty tophi. However, urate stasis is commonly observed in reptiles without gout and care must be taken to not mistake urate stasis for renal gout on gross examination (**Figure 3.22**). Reproductively active male squamates develop hypertrophied, granular tubules (sexual segment; Volume 1, **Figures 1.220, 1.233–1.235**) that grossly can be misconstrued as renal gout.

Visceral gout affects organs in addition to the kidneys or sometimes in the absence of apparent renal gout. The most common gross presentation is precipitation of uric acid on serosal surfaces, especially epicardial and pericardial surfaces of the heart and capsular surfaces of the liver and spleen (**Figures 3.8a,b–3.9**), although oral mucosa can also be affected. Urate tophi within the parenchyma of liver, spleen, and lung can sometimes be seen grossly as fine white stippling on cut surface (**Figure 3.9**). In freshly euthanized animals, serosal monosodium urate deposition must be distinguished from barbiturate precipitate. For quick differentiation between the two during gross examination, uric acid is insoluble in ≥70% ethanol, while barbiturate precipitates will dissolve completely and rapidly in ≥70% ethanol. Histologically, parenchymal deposits often involve vessel walls and, in acute stages, consist of foci of necrosis, heterophilic inflammation, and variably visible outlines of monosodium urate crystalline clefts in a radiating, starburst pattern. These may progress to mature tophi with characteristic granulomatous inflammation (**Figures 3.23–3.24**). Precipitated uric acid on serosal surfaces is often poorly preserved in histologic sections due to its tendency to dissolve during processing. Wispy material and inflammation may remain. Vascular deposition can lead to thrombosis (**Figure 3.25**).

3.2.5 Articular and Periarticular Gout

Gout is the most common and best characterized cause of nodular deposits of crystalline material in and around the joints of reptiles. In avian gout, the articular and visceral forms are usually separate manifestations considered to have different causes (Crespo and Shivaprasad, 2013; Crespo et al., 2018). In reptiles, articular and periarticular gout can develop concurrently to or independent of the renal and visceral forms. Nodular swellings around joints may be the earliest clinical sign of articular gout (**Figures 3.10–3.11**). Cytologic preparations of joint aspirates should be examined prior to staining under polarized light to visualize the strongly birefringent, needle-like crystals, as some of the crystals will dissolve during staining (**Figure 3.19**).

Deposition of monosodium urate and associated inflammation within articular spaces can lead to destruction of cartilage and bone (**Figures 3.26–3.27**). Lytic lesions may be visible radiographically, but uric acid deposits are radiolucent unless they become mineralized with chronicity. Large, coalescing accumulations of crystalline uric acid may extend from joint spaces into the surrounding connective tissue, and along tendons and fascial planes (**Figures 3.28–3.29**). Urate tophi are commonly found in the subarticular trabecular bone of affected joints. Chronic lesions are surrounded by granulomatous inflammation and fibrosis.

3.2.6 Gout-Like Lesions

Nodular periarticular deposits of crystalline material other than urates also occur in reptiles and must be differentiated from gout. Variably termed pseudogout, false gout, tumoral calcinosis, and hydroxyapatite deposition disease, the material deposited in some of these cases has been shown by crystallographic analysis to be calcium phosphate in the form of hydroxyapatite [$Ca_{10}(PO_4)_6(OH)_2$], the predominant mineral found in bone (Frye and Dutra, 1976; Cooper and Jackson, 1981; Wenker et al., 1999; Jones and Fitzgerald, 2009). These cases therefore represent a type of mineralization disorder in the form of tumoral calcinosis, which is addressed in Section 3.4.3. Tumoral calcinosis or hydroxyapatite deposition disease (HADD) are preferred over the term pseudogout for these cases, since pseudogout in other species (humans, nonhuman primates, and dogs) refers to deposition of calcium pyrophosphate dihydrate ($Ca_2P_2O_7{\cdot}2H_2O$) in articular tissues (Roberts et al., 1984; Woodard, 1997; Thompson, 2007; Rosenberg, 2010). Deposition of calcium pyrophosphate dihydrate has not been confirmed in reptiles.

3.3 Urolithiasis

The formation of calculi within the reptilian urinary system occurs most commonly in reptiles with a urinary bladder, such as chelonians (**Figures 3.30–3.32**), some lizards (**Figure 3.33**), and tuatara, though it can also occur in species lacking a discrete urinary bladder, including snakes (**Figure 3.5**), crocodilians, and other lizards (Bennett and Lock, 2000; Reavill and Schmidt, 2013; Hedley and Eatwell, 2014). Calculi are macroscopic mineral precipitates or concretions within the urinary tract, and are termed "uroliths" or "cystic calculi" when found within the urinary bladder (Amat et al., 2012; Reavill and Schmidt, 2013). Calculi can also develop in the ureters (**Figure 3.34**) and cloaca (Innis and Kincaid, 1999; Keller et al., 2015). In snakes, calculi may form in the distal ureters where urates are stored, and can be a sequela to neoplasia (Frye, 1991; Reavill and Schmidt, 2013).

3.3.1 Prevalence and Predisposing Factors

Captive reptiles appear more susceptible to the formation of calculi due to dietary factors and hydration status (Ryer, 1983; Johnson et al., 2001). A common historical finding in reptilians with cystic calculi is dehydration with restricted access to water (Frye, 1991; Bennett and Lock, 2000; Amat et al., 2012; Reavill and Schmidt, 2013). High levels of dietary protein, especially in herbivorous reptiles, can also promote the formulation of urinary calculi due to the increased production of uric acid (Boyer, 1991; Frye, 1991; Amat et al., 2012; Kwantes, 1992). Other predisposing factors include vitamin A and D deficiency, high oxalate diets (such as spinach), and the presence of material in the urinary system that can act as a nidus for calculus formation, such as suture or retropulsed eggs (Thomas et al., 2002; Reavill and Schmidt, 2013).

Alterations in urine pH and supersaturation of urinary calculogenic salts may further contribute to calculus formation (Homer et al., 1998). In desert tortoises, as much as 60% of nitrogenous waste is excreted as uric acid, and high concentrations of relatively insoluble urate salts, especially when exacerbated by dehydration or high protein diets, predispose to the development of calculi (Cowan, 1968; Frye, 1991). Bacterial cystitis may also predispose reptiles to urolith formation. Alteration of urinary pH due to bacterial infections can decrease the solubility of certain calculogenic salts. For example, calcium phosphate is considerably less soluble in alkaline urine than in acidic urine (Innis and Kincaid, 1999). Necrotic or inflammatory cells, which may be increased with cystitis, can also serve as a nidus for stone formation (Kölle et al., 2001; Amat et al., 2012).

Urinary calculi are most common in captive chelonians and iguanas (Kölle et al., 2001; Reavill and Schmidt, 2013). In North America, *Gopherus* tortoises are overrepresented (Frye, 1983; Homer et al., 1998; Mader et al., 1999; Bennett and Lock, 2000; Johnson et al., 2001; Keller et al., 2015), and in Europe *Testudo* and *Geochelone* are the most frequently reported (Keymer, 1978; Kölle et al., 2001). Aquatic turtles appear less predisposed to urolith formation, though reports exist for Red-Eared Sliders (*Trachemys scripta elegans*), Yellow-Bellied Sliders (*T. scripta scripta*), and Western Spiny Softshell Turtles (*Apalone spinifera*) (McKown, 1998; Innis and Kincaid, 1999; Keller et al., 2015). Freshwater turtles excrete up to 60% of nitrogenous waste as highly water-soluble urea, with smaller amounts of ammonia. In marine turtles, nitrogenous wastes are excreted exclusively as ammonia and urea (Cowan, 1968). There are rare reports of calculi in wild reptiles, including three Desert Tortoises and a Western Spiny Softshell Turtle (*Apalone spinifera*) (Homer et al., 1998; McKown, 1998). While the majority of calculi occur in herbivorous species, there is a single report of a sodium urate cystic calculus in a wild-caught, insectivorous Green Anole (*Anolis carolinensis*) (Birke et al., 2017). It is unknown if the stone was present at the time of capture.

3.3.2 Composition

The majority of calculi in reptiles are composed of urate salts (Mader et al., 1999; Bennett and Lock, 2000; Johnson et al., 2001; Wolf et al., 2008; Amat et al., 2012; Keller et al., 2015; Birke et al., 2017). While urates are typically radiolucent, they are often complexed with minerals such as calcium phosphate, calcium carbonate, or struvite (magnesium ammonium phosphate) resulting in increased radiodensity (Bennett and Lock, 2000; Keller et al., 2015). Quantitative analysis of 246 uroliths from reptiles including iguanas, geckos, lizards (not otherwise specified), and chelonians submitted to the Minnesota Urolith Center reported 93.5% of uroliths analyzed were composed of purines; the remaining 6.5% were reported as being of mixed, compound, calcium carbonate, or other composition (Osborne et al., 2009). In an evaluation of 49 bladder stones and concretions from lizards and chelonians, 51% of the samples were composed of one substance and 49% consisted of two or more substances (Kölle et al., 2001). Grossly, calculi may be round to oval with annular lamellations resulting from the repeated deposition of supersaturated material around a central nidus (Frye, 1983; Keller et al., 2015).

3.3.3 Signs and Associated Lesions

Clinical signs of urinary calculi vary depending on the number, location, and size of calculi. In chelonians, many cystic calculi are detected incidentally on physical examination or radiographs. Palpation is an important diagnostic tool for stones with higher urate concentration that may not be evident on radiographs (Bennett and Lock, 2000). As the right lobe of the bilobed chelonian urinary bladder is closely associated with the liver, larger calculi are often displaced to the center of the bladder or to the left lobe (Bennett and Lock, 2000; Keller et al., 2015). In a study of urolithiasis in 100 California Desert Tortoises, calculi were located in the left lobe of the urinary bladder 47% of the time (Mader et al., 1999). Single calculi were present in 73 cases with an average size of 5.4 cm diameter. In the study by Mader et al. (1999), 60% of affected tortoises were male, although other studies do not report an apparent sex predilection (Keller et al., 2015). When calculi are large or present in high numbers, they may produce nonspecific clinical signs due to obstruction, displacement of coelomic viscera, or secondary infection (Frye, 1991; Innis and Kincaid, 1999; Denardo et al., 2000; Reavill and Schmidt, 2013). Clinical signs commonly include anorexia, lethargy, paresis of the pelvic limbs, dystocia (post-ovulatory egg retention), obstipation, tenesmus, cloacal prolapse, weight loss, and poor growth (Mader et al., 1999; Bennett and Lock, 2000; Denardo et al., 2000; Johnson et al., 2001; Amat et al., 2012; Mans and Sladky, 2012; Keller et al., 2015). Hematuria is a common clinical finding in reptiles with urinary calculi (Reavill and Schmidt, 2013).

The reptilian urinary bladder, when present, is normally a thin walled, semi-transparent structure (Minter et al., 2010). Chronic irritation of the mucosal epithelium by uroliths may result in the bladder wall becoming thickened, opaque, or hemorrhagic (**Figure 3.32b**) (Ryer, 1983; Bennett and Lock, 2000). Histologic lesions of the urinary bladder include pressure necrosis (especially with large, heavy uroliths), mucosal epithelial hypertrophy and hyperplasia, granulocytic or granulomatous inflammation, and mural and mucosal mucinosis (**Figure 3.32c**) (Homer et al., 1998; Minter et al., 2010; Reavill and Schmidt, 2013; Keller et al., 2015). Goblet cell metaplasia of the mucosal epithelium may represent a reaction to the presence of uroliths or a response to dehydration, as similar findings can be noted in dehydrated tortoises without urolithiasis (Homer et al., 1998). Some calculi, however, are fragile and even when large may induce only minimal inflammatory changes (Frye, 1991). The formation of uroliths in other parts of the urinary tract, including the ureters and collecting ducts, is often associated with tissue necrosis, granulocytic to granulomatous inflammation, and in chronic cases, fibrosis with epithelial hyperplasia (**Figure 3.34**).

3.4 Mineralization

Mineralization disorders are an important subset of depositional diseases in reptiles. Mineralization refers to the pathologic deposition of mineral salts in soft tissue and is interchangeable with the term "calcification" when referring specifically to the deposition of calcium salts. Mineral deposits are most commonly composed of calcium phosphate salts in the form of hydroxyapatite (Frye, 1991). There are two broad mechanisms of tissue mineralization: dystrophic and metastatic.

3.4.1 Causes and Types

Dystrophic mineralization occurs following tissue injury and necrosis despite normal calcium metabolism and circulating blood calcium levels. Mechanisms of tissue injury are vast but commonly include trauma (such thermal, chemical, and physical injury), parasitic migration, and inflammatory diseases. As such, deposition of calcium salts remains localized to injured and inflamed tissue and manifests as mineral deposits against a background of necrosis, inflammation, and eventual fibrosis as lesions become more chronic.

Periarticular mineralization (also known as "pseudogout," tumoral calcinosis, or hydroxyapatite deposition disease) is a type of dystrophic mineralization disorder in reptiles characterized by nodular deposits of mineral around joints which will be discussed later. In a few reports, the formation of discrete, "tumor-like" calcium deposits in the skin and underlying subcuticular tissues is referred to as calcinosis circumscripta (Frye, 1991; Reavill and Schmidt, 2002; Yanai et al., 2002). This condition appears morphologically identical to other forms of tumoral calcinosis and may represent various manifestations of the same disease. In one instance of an iguana with mineralization of the ventral skin, the inciting cause was attributed to thermal injury from an unregulated heating unit (hot rock) (Frye, 1991). In a study of aged captive Alligators (*Alligator mississippiensis*), opacification of the normally transparent third eyelid (nictitating membrane) was presumed to be calcium salt deposition secondary to inadequate environmental hygiene, although this was not confirmed histologically (Millichamp et al., 1983).

Metastatic mineralization refers to the deposition of calcium in soft tissues secondary to derangements in calcium and phosphorus metabolism. Pathologic processes associated with altered calcium metabolism include primary or secondary hyperparathyroidism, hypervitaminosis D_3, excessive exposure to ultraviolet radiation, and osteolytic bone disease (Frye, 1991). In reptiles, these conditions may be the result of dietary imbalances, metabolic disease, or chronic renal disease (Wallach and Hoessle, 1966; Rosenthal et al., 2000; Garner, 2001). In contrast to the localized distribution of dystrophic mineralization, metastatic lesions occur anywhere in the body and may be locally extensive or widespread. Metastatic mineralization is common in reptiles, and the lesions can be quite extensive (**Figures 3.35–3.40**). The most commonly affected anatomic locations include fibroelastic tissue (such as blood and lymphatic vessels) (**Figures 3.36a, 3.38, 3.40a,b**), alimentary mucosa (**Figure 3.37a**), respiratory tissues (**Figures 3.36d, 3.37b,c, 3.39d**), urogenital tissues (**Figure 3.36b**), the hepatic parenchyma (**Figures 3.35a,b, 3.39b,c**), and cardiac, smooth, and skeletal muscle (**Figures 3.36c, 3.39a**) (Wallach and Hoessle, 1966; Frye, 1991; Richman et al., 1995; Rosenthal et al., 2000; Raiti and Garner, 2006; Eatwell, 2008). Cutaneous and corneal epidermal basement membranes may also be affected.

Captive herbivorous reptiles appear most susceptible to the development of metastatic mineralization due to dietary imbalances of vitamin D, calcium, and phosphorus (Wallach and Hoessle, 1966; Richman et al., 1995). This is supported by the paucity of reports documenting pathologic mineralization in carnivorous or recently acquired wild-caught reptiles (Frye, 1991). Extensive deposition of calcium salts in elastic arteries including the aorta, pulmonary, renal, and iliac vessels, as well as throughout the renal parenchyma in captive Green Iguanas (*Iguana iguana*) is reported in a study by Wallach and Hoessle (1966). While they hypothesized the lesion was a result of excess dietary supplementation of vitamin D, evidence for absorption of vitamin D from the diet is lacking, and Green Iguanas are extremely susceptible to vitamin D deficiency in the absence of UVB radiation even with appropriate dietary supplementation (Allan and Oftedal, 2003).

Paradoxical metastatic mineralization has also been described in reptiles with normal or low vitamin D and calcium levels (Richman et al., 1995; Raphael et al., 1999; Raiti and Garner, 2006; Eatwell, 2008). Vascular mineralization was present in 20 Green Iguanas with normal calcium and low vitamin D_3 levels despite receiving supplementation with dietary vitamin D_3 (Richman et al., 1995). A similar report involving a captive Uromastyx (*Uromastyx* spp.) with mineralization of the skin and periarticular regions had lower levels of vitamin D than unaffected conspecifics (Raphael et al., 1999). In a Geoffrey's Side-Necked Turtle (*Phrynops geoffroanus*) with widespread metastatic mineralization due to renal secondary hyperparathyroidism, peripheral blood calcium levels were normal (Raiti and Garner, 2006). The pathogenesis for this paradoxical type of metastatic mineralization is unclear. Possible mechanisms include an exaggerated parathyroid hormone response leading to excessive calcium mobilization from bone or alterations in the ratio of calcium to phosphorus that encourage the precipitation of mineral in tissue by increasing the solubility index of calcium and phosphorus (Richman et al., 1995; Hernandez-Divers, 2005).

Oviparous females may be especially susceptible to tissue mineralization due to fluctuations in calcium during folliculogenesis and egg development (Rosenthal et al., 2000). Blood calcium levels rise dramatically during folliculogenesis. This transient physiologic hypercalcemia is well tolerated when the majority of calcium is present in its bound (inactive) form to albumin. However, metabolic disruptions by processes such as metabolic acidosis or dehydration can cause drastic shifts in calcium to its active ionized form, resulting in soft tissue mineralization (Rosenthal et al., 2000).

Conversely, increased demands for calcium during egg development may induce or further exacerbate hypocalcemia. Prolonged hypocalcemia can predispose for soft tissue mineralization due to the development of secondary hyperparathyroidism. A Tunisian Tortoise (*Furculachelys nabeulensis*) with egg stasis and hypocalcemia had concurrent parathyroid gland hyperplasia with mineralization of renal tubules, vessels, and cardiac muscle. Mineralization was attributed to secondary hyperparathyroidism due to increased calcium demand for eggshell production (Eatwell, 2008).

Nephrocalcinosis is an important disease characterized by accumulation of mineral within renal tubules, with concurrent mineralization of the renal parenchyma. Aggregates of granular to homogenous basophilic material within ectatic tubules can be associated with tubular epithelial damage, degeneration, and tubular rupture (**Figure 3.41**). This condition represents a gray area in the spectrum of reptile mineralization disorders, as renal mineralization may stem from both dystrophic and metastatic causes (Reavill and Schmidt, 2013). Nephrocalcinosis is considered the most common secondary renal lesion in captive Green Iguanas with systemic mineralization (Rosenthal et al., 2000).

Additional consideration should be given to reptiles possessing endolymphatic sacs such as geckos. Endolymphatic sacs are calcium reservoirs and are important for calcium homeostasis (see Volume 1, **Figure 1.410a,b**). As such, neoplasia or other diseases affecting these sacs may cause disruptions in calcium regulation (**Figure 3.42**). The mechanism by which this occurs is unknown but may involve compromise of the endolymphatic sac barrier or alterations in glandular activity (Sander et al., 2015).

Not all mechanisms resulting in metastatic mineralization that for mammals have well-described pathophysiologic mechanisms translate to reptilian disease conditions. Hypercalcemia secondary to chronic granulomatous inflammation is well documented in mammals due to the unregulated production of 1,25-dihydroxycholecalciferol (calcitriol) by activated macrophages. While reptile granulomas may contain areas of dystrophic mineralization, widespread metastatic mineralization has not been documented (Soldati et al., 2004). Humoral hypercalcemia of malignancy is a paraneoplastic syndrome caused by the production of parathyroid hormone-related protein (PTHrP) in neoplastic cells. It is a common process in many mammalian cancers (such as apocrine gland anal sac adenocarcinoma in dogs). Paraneoplastic syndrome in reptiles has not been officially documented, but similarities suggest that it may occur (Hernandez-Divers and Garner, 2003). Hypercalcemia was documented in a Western Hognose Snake (*Heterodon nasicus*) with metastatic ductal adenocarcinoma (Stern et al., 2010). The underlying cause of hypercalcemia was not determined, although paraneoplastic causes could not be ruled out. Soft tissue mineralization was not observed in this snake.

With few established blood reference ranges for various reptile species, especially uncommon species, interpretation of blood calcium, phosphorous, and vitamin D measurements can be challenging. Evaluation of serum biochemistry levels in 15 captive Indigo Snakes (*Drymarchon couperi* [*D. corais couperi*]) documented a pattern of persistent hypercalcemia and hyperphosphatemia in 13 of the snakes. Whole body radiographs revealed mild metastatic calcification in only one of the snakes (Drew, 1994). In reptiles, physiologic elevations of blood calcium in otherwise healthy individuals may be related to egg production, seasonal changes, or hormonal fluctuations. In a study of 68 Prehensile-Tailed Skinks (*Corucia zebrata*), extreme hypercalcemia was periodically documented only in females, suggesting a possible reproductive component (Bradford et al., 2009). Interestingly and somewhat paradoxically, Prehensile-Tailed Skinks are viviparous, and increased calcium is not necessary for shelling of eggs.

3.4.2 Visceral Mineralization

The effects of visceral mineralization vary based on the severity of mineral deposition, duration of disease, and the organs involved. Grossly, calcium salts may be visible as white to tan plaques or streaks within vessels, skin, and other soft tissue. Aggregates of calcium salts are firm and difficult to cut, with a chalky or gritty consistency. Deposition of calcium salts in organs may result in organ enlargement, pallor, or increased firmness. Histologically with hematoxylin and eosin staining, calcium deposits appear as deeply basophilic amorphous densities (**Figures 3.34–3.38**) or, less commonly, as distinct round collections of mineral known as psammoma bodies (**Figure 3.39e**). Special staining for calcium salts can be performed with the Von Kossa method that highlights areas of calcium salt deposition as black against a background of non-mineralized red to pink tissue (**Figures 3.35b, 3.37c, 3.39c**).

In metastatic mineralization disorders, calcium deposits demonstrate a predilection for fibroelastic tissue components (Wallach and Hoessle, 1966; Frye, 1991). In a study of arterial disease in reptiles, medial calcification represented the dominant vascular lesion and was most often secondary to parasitic arteritis in snakes and metabolic bone disease in lizards (Finlayson and Woods, 1977). Grossly, mineralized vessels appear thickened and tortuous with decreased pliability, white to tan plaques, or dimpling of the intimal surfaces (**Figure 3.40a**) (Wallach and Hoessle, 1966). The vessels may grate when cut or require the use of saws or decalcification prior to tissue sectioning for histology (Wallach and Hoessle, 1966; Frye, 1991). Calcium salt deposition occurs predominately in the tunica media and may be associated with disruption of the elastic lamina, necrosis of myocytes, and/or intimal thickening (**Figures 3.38, 3.40b**) (Wallach and Hoessle, 1966; Finlayson and Woods, 1977). In severe cases, vascular mineralization may be circumferential, with the formation of complete rings of mineral (Wallach and Hoessle, 1966). Possible sequelae of vascular mineralization, especially of the great vessels and lung, include aneurysm and/or fatal hemorrhage (Kik and Mitchell, 2005). Rupture of mineralized pulmonary capillaries can lead to hemoptysis, epistaxis, hypoxia, or fatal pulmonary hemorrhage (Rosenthal et al., 2000; Barten, 2006). In the heart, deleterious effects of cardiac muscle mineralization occur when the deposition of mineral salts interferes with electrical conduction pathways or muscular contraction (**Figure 3.39a**) (Frye, 1991).

Primary or secondary renal disease is often linked with systemic mineralization (Rosenthal et al., 2000; Raiti and Garner, 2006; Reavill and Schmidt, 2013). Two Green Iguanas with tissue mineralization secondary to hypervitaminosis D also had concurrent terminal nephritis and mineralization characterized by enlargement of the kidneys and a hyaline or "ground glass" appearance on cut surface (Wallach and Hoessle, 1966). Histologically, mineral may accumulate anywhere throughout the renal parenchyma including the glomerular mesangium, basement membranes, tubular epithelium and lumina, interstitium, and renal vessels (**Figure 3.36b**) (Reavill and Schmidt, 2013).

3.4.3 Articular and Periarticular Mineral Deposition

Nodular articular and periarticular deposits of mineral, specifically hydroxyapatite [$Ca_{10}(PO_4)_6(OH)_2$], have been reported in reptiles (Frye, 1976; Cooper and Jackson, 1981; Wenker et al., 1999; Burns et al., 2013). Considered to represent a type of tumoral calcinosis, the clinical and gross appearance of these lesions must be differentiated primarily from articular gout. As such, the condition has been variably termed pseudogout, false gout, tumoral calcinosis, and hydroxyapatite deposition disease, the latter having been proposed (Wenker et al., 1999) because the term pseudogout refers to deposition of a different material in other species, namely calcium pyrophosphate dihydrate ($Ca_2P_2O_7{\cdot}2H_2O$) (Roberts et al., 1984; Woodard, 1997; Thompson, 2007; Rosenberg, 2010).

Most reports of this condition have been in turtles, including 10 related juvenile Red-Bellied Short-Necked Turtles (*Emydura subglobosa*) (**Figure 3.43a–d**) (Burns et al., 2013), two other juvenile Red-Bellied Short-Necked Turtles (Wenker et al., 1999), a Red-Eared Slider Turtle (*Trachemys scripta elegans*) (Frye, 1976), and clutches of newly hatched turtles of unspecified species (Woodard, 1997). There is also a reported case in a uromastyx (Cooper and Jackson, 1981), and is possibly common in this species (Ed's comment: M. Garner). A couple of additional cases of described pseudogout for which crystallographic analysis was not performed include a Painted Turtle (*Chrysemys picta*) (Chambers et al., 2009) and a Basilisk (*Basilicus plumifrons*) (Jones and Fitzgerald, 2009); though not published, chameleons also seem to be predisposed to pseudogout-like disease. However, diagnosis of true reptilian pseudogout is questionable, as none of the deposits have been confirmed as calcium pyrophosphate by crystallographic analysis, using instead histologic appearance, staining qualities, and the presence or absence of birefringence, which may be unreliable for this determination (Burns et al., 2013).

In Red-Bellied Short-Necked Turtles, articular mineral deposits were exclusively or predominantly in the feet of juveniles (**Figures 3.43a–d, 3.44**) (Wenker et al., 1999; Burns et al., 2013). In another turtle, larger joints of limbs were affected (Frye, 1976). In chameleons, lesions can be found in the cervical region, distal limbs, and the tail (**Figures 3.45–3.46a,b**). The mineral deposits result in non-ulcerated, firm, multinodular swellings that are intensely radiopaque and may compromise mobility when extensive (**Figures 3.43a, 3.45**). No regression of these lesions has been reported. Grossly, the nodules consist of sharply demarcated accumulations of white to yellowish, pasty to chalky material within periarticular tissues (**Figure 3.43b**). Unstained cytologic preparations of the material show finely granular to coarsely clumped, acellular, refractile debris (**Figure 3.47**); when examined under polarized light the material has weak, glowing white birefringence (Burns et al., 2013). This is in sharp contrast to the bright, fine, spiculate crystals of gout (**Figure 3.18**). Stained cytologic impressions of tumoral calcinosis lesions consist of pleomorphic aggregates of round, refractile, pale eosinophilic, Von Kossa-positive material with associated histiocytic inflammation (**Figure 3.48a,b**).

Histologically, subcutaneous and periarticular connective tissues are expanded by multiple irregular, nodular accumulations of granular to amorphous, basophilic to eosinophilic material which may appear variably mineralized depending on the degree of tissue decalcification. The material is non-birefringent under polarized light. Surrounding and separating the mineral deposits are thin bands of fibrous tissue and variable numbers of epithelioid macrophages, multinucleated giant cells, and other inflammatory cells (**Figures 3.43c,d, 3.46b**) (Frye, 1976; Wenker et al., 1999; Burns et al., 2013). There may be associated erosion of articular cartilage or bone lysis (**Figure 3.43d**) (Frye, 1976; Burns et al., 2013).

The cause of articular tumoral calcinosis in reptiles is unknown. In addition to primary hereditary forms, tumoral calcinosis (or calcinosis circumscripta) in humans and other species can occur secondary to trauma, renal failure, primary hyperparathyroidism, hypervitaminosis D, connective tissue disorders, and diseases causing bone lysis (e.g., infection or neoplasia) (Slavin et al., 1993; Tafti et al., 2005; Choi et al., 2006; Ginn et al., 2007; Laskin et al., 2007; Radi and Sato, 2010). The mineralization associated with these diseases can therefore be metastatic or dystrophic. Evidence for any of these underlying conditions has been lacking in reptile reports to date. While Red-Bellied Short-Necked Turtles are overrepresented in the few reported cases, and the largest study involved 10 affected clutch mates, a hereditary basis for this condition has not been proven (Wenker et al., 1999; Burns et al., 2013). No evidence of renal disease or consistently elevated serum calcium and phosphorus levels has been found in affected reptiles based on reports to date. Vitamin D levels have yet to be investigated. No infectious organisms have been identified within the lesions; however, tissue damage from a prior infection is a consideration.

3.5 Amyloidosis

Amyloidosis constitutes a group of protein misfolding diseases in which abnormal fibrils are deposited extracellularly in tissues. In this well-recognized disorder in mammals and birds, accumulation of amyloid in tissues can eventually lead to organ dysfunction (Woldemeskel, 2012; Sipe et al., 2014). Amyloidosis has been reported in reptiles (Trautwein and Pruksaraj, 1967; Burns et al., 2017), but confirmed cases are rare.

3.5.1 Causes and Types

Amyloid diseases are named for the precursor proteins of the amyloid fibril, with 10 distinct amyloid diseases identified to date in animals (Sipe et al., 2014). Identification of the amyloid type is based on its microanatomic location, diagnosis of the associated condition, immunohistochemical typing and, more recently, protein sequence analysis (Sipe et al., 2014). Amyloid A (AA) amyloidosis, derived from the acute phase protein serum amyloid A (SAA), is the most common systemic amyloid disease of animals and the only one known to occur in birds (Woldemeskel, 2012). It is also the suspected type in most of the reported reptile cases, although this has never been proven by immunohistochemistry or protein sequence analysis (Burns et al., 2017). Other types of amyloidosis in mammals include AL amyloidosis derived from immunoglobulin light chains and associated with plasma cell tumors, Aβ amyloidosis derived from Aβ precursor protein and deposited in the brain as an age-related condition, and AIAPP amyloidosis derived from islet amyloid polypeptide and deposited in the pancreatic islets (Sipe et al., 2014).

Amyloid A amyloidosis is a systemic form that occurs secondary to inflammatory disease and prolonged elevation of SAA (Woldemeskel, 2012). Besides increased levels of SAA, accumulation of AA amyloid in tissues also requires a failure of normal enzymatic degradation, either as a result of an abnormal protein resistant to degradation or because of a defect in the enzyme systems responsible for removal (Kumar et al., 2010).

The SAA gene and gene family, as well as the elevation of SAA in response to inflammatory stimuli, are highly conserved across taxa (Uhlar and Whitehead, 1999). A gene for a SAA homologue has been identified in Chinese Softshell Turtles (*Trionyx sinensis*) and Tuataras (*Sphenodon punctatus*), and its expression was shown to be up-regulated in the former in response to a bacterial infection, typical of an acute phase protein (Zhou et al., 2008, 2011). The amino acid sequences of the corresponding proteins are most similar to avian SAA and contain hydrophobic N-terminal portions, which are important in determining the molecule's amyloidogenic potential (Zhou et al., 2011). Therefore, it is likely that a similar pathogenesis could lead to AA amyloidosis in reptiles as in birds and mammals.

3.5.2 Pathologic Features and Diagnosis

Macroscopically, severe amyloid deposition can cause enlargement, increased firmness, and a pale, waxy appearance to tissues. The starch-like properties that give amyloid its name can be demonstrated grossly by application of iodine and sulfuric acid, resulting in blue-black staining of the tissue (Kumar et al., 2010). More commonly, the diagnosis is made histologically. Regardless of the precursor protein, amyloid aggregates share a similar microscopic appearance and histochemical staining characteristics (Kumar et al., 2010; Toyama and Weissman, 2011; Woldemeskel, 2012). These features result from misfolding of the precursor proteins into a cross-β-pleated sheet conformation in long, non-branching, approximately 5–15 nm diameter fibrils as seen by electron microscopy (Cohen and Calkins, 1959; Kumar et al., 2010; Toyama and Weissman, 2011; Sipe et al., 2014). With routine H&E staining, amyloid deposits are extracellular and appear as amorphous accumulations of hyaline to finely fibrillar, eosinophilic to pale basophilic material (**Figure 3.49a**). Amyloid is often deposited along basement membranes and within vessel walls, eventually leading to atrophy and other degenerative changes through compression of adjacent tissue and/or restriction of blood flow (Kumar et al., 2010).

As other deposits like fibrin and collagen (**Figures 3.50–3.53**) can have a similar appearance with H&E staining, accurate diagnosis requires additional testing. Classically, amyloid is pink, red, or orange with a Congo red stain and has green birefringence with polarized light (**Figure 3.49b**). It also exhibits fluorescence when stained with thioflavin T or S (Kumar et al., 2010). A newer method of diagnosis utilizes luminescent conjugated oligothiophenes, such as h-FTAA, which are fluorescent molecules that specifically bind amyloid. In contrast to rigid Congo red and thioflavin-T molecules, these molecules can swivel and intercalate into amyloid protofibrils and fibrils in a conformationally specific way, resulting in increased sensitivity (Sigurdson et al., 2007; Sjölander et al., 2016). Though not widely available, this technique was used in the diagnosis of amyloidosis in an African Tiger Snake (*Telescopus semiannulatus*) (Burns et al., 2017). Immunohistochemistry has been used to identify and type amyloid in mammals and birds, but this technique has not been reported to be successful in reptiles. Possibilities for this finding include differences in tertiary protein structure and misfolding resulting in a buried epitope, or decreased sequence conservation of reptilian SAA in comparison with mammals (Burns et al., 2017). The Chinese Softshell Turtle SAA protein shares only 62%–72% amino acid identity with mammalian sequences of SAA (Zhou et al., 2011).

Reports of amyloidosis in reptiles are rare and often lack confirmation of the diagnosis beyond a description of an amyloid-like appearance with H&E staining. However, at least two reports clearly demonstrate that amyloidosis can occur in reptiles (Trautwein and Pruksaraj, 1967; Burns et al., 2017). Trautwein and Pruksaraj described vascular and splenic amyloidosis in 30 Hermann's Tortoises (*Testudo hermanni*) out of a group of 52 (Trautwein and Pruksaraj, 1967). The amyloid deposition in the tortoises started in the subintima and media of arteries and progressed to effacement of vessel walls and splenic architecture. Eight of the cases were subjected to various histochemical staining methods and demonstrated positive Congo red staining and birefringence and weak thioflavin T fluorescence. A link to *Salmonella* infection was investigated, but the type of amyloidosis could not be determined (Trautwein and Pruksaraj, 1967).

In the African Tiger Snake (*Telescopus semiannulatus*), amyloidosis similarly affected splenic parenchyma and blood vessels in various tissues (**Figure 3.49a–c**) (testes, kidney, adrenal gland) (Burns et al., 2017). The diagnosis of amyloidosis in that case was based on positive Congo red staining and birefringence (**Figure 3.49b**), characteristic fibrils by electron microscopy (**Figure 3.49c**), and fluorescence with the amyloid binding probe, heptameric oligothiophene (h-FTAA). Reactive AA amyloidosis was strongly suspected due to concurrent mycobacteriosis and an inflamed biliary cystadenocarcinoma, but attempted typing by immunohistochemistry using mammalian antibodies was unsuccessful.

Additional reported cases include renal glomerular amyloidosis in a Boa Constrictor (*Boa constrictor imperator*) and splenic amyloidosis in a Tree Boa (*Corallus enydris enydris*) (Cowan, 1968). Although affinity for Congo red was reported, the staining quality was "not identical" to that of mammalian amyloid. Other reports that had a typical appearance with H&E staining but no further reported diagnostic confirmation include splenic amyloidosis in a Spiny-Tailed Iguana (*Ctenosaura acanthura*) and glomerular amyloidosis in an American Alligator (*Alligator mississippiensis*) (Cosgrove and Anderson, 1984).

3.5.3 Paramyloid

Other deposits of non-congophilic material morphologically similar to amyloid have been seen in reptiles and termed *paramyloid*. At least one of these was demonstrated to be composed of collagen and not amyloid by transmission electron microscopy (**Figure 3.50a,b**) (B. Stacy, unpublished). An amyloid-like deposition disorder has been reported in female Komodo Dragons (*Varanus komodoensis*) (**Figure 3.53a–c**) (Garner et al., 2003). Four of five Komodo dragons, all females, had deposits of an amphophilic, amorphous, and homogeneous material in blood vessel walls in the ovary, kidney, and spleen, which was also present in the renal interstitium and glomeruli in two animals (**Figure 3.53a**). This material was amphophilic and not birefringent with a Congo red stain (**Figure 3.53b**), and ultrastructural findings were more consistent with collagen rather than amyloid (**Figure 3.53c**). Compromised ovarian vasculature from deposition of this amyloid-like material was important, as acute yolk embolism was considered to be the cause of death in three of the individuals (**Figure 3.53a**).

3.6 Glomerular Basement Membrane Diseases

Information regarding the pathogenic mechanisms and ultrastructural changes of glomerular basement membrane diseases in reptiles is limited. While the glomerulus can be a common site for some of the previously mentioned deposits, such as amyloid and mineral, other glomerular depositional diseases for which there are detailed pathogenic mechanisms recognized in mammals are less well described in reptiles.

Glomerular lesions are classified by the location of deposits in the glomerular tuft and by alterations in the character and cellularity of the mesangium. Generally, histologic features include deposits of hyaline material within glomerular capillary basement membranes and also within the mesangial connective tissue (Zwart, 2006). The deposits are often periodic acid-Schiff positive, indicating that they are of glycoprotein or mucopolysaccharide origin. These changes are best resolved beyond a light microscopic level by techniques such as transmission electron microscopy. However, ultrastructural examination of glomerular disorders is lacking in reptiles, and thus the pathophysiology and classification of these deposits is not clearly understood. Additional histologic findings may include enlargement or shrinking of the glomerular tuft, thickening of the basal lamina of Bowman's capsule, glomerular synechiae, and tubular changes.

There is considerable variability in the distribution and severity of glomerular lesions documented in reptiles, which suggests that these changes are the manifestation of a complex of pathophysiologic processes. Possible causes include age-related, immunological, inflammatory, infectious, and arteriosclerotic conditions (Cosgrove and Anderson, 1984). In a classic study of renal lesions in reptiles by Zwart, glomerular lesions were classified into four broad categories: (1) membranous glomerulonephrosis, (2) acute glomerulonephritis, (3) chronic intracapillary glomerulonephritis, and (4) chronic extracapillary glomerulonephritis (Zwart, 1964).

Membranous glomerulonephrosis is characterized by the thickening of the basal lamina of glomerular capillaries by hyaline deposits without an increase in mesangial cellularity. This lesion has been described in Pythons (*Python reticulatus* and *P. molurus*) with concurrent *Pseudomonas* infection at distant sites. While the etiology is not clear, the concurrent bacterial infection common to both of these snakes suggests that bacterial toxins may play a role (Zwart, 1964). Membranous glomerular changes have also been described in reptiles with systemic lupus erythematosus (Frye, 1976).

Lesions of acute glomerulonephritis are predominately proliferative. In the early phase, there is enlargement of the glomerular tuft resulting from proliferation and/or swelling of parietal and visceral podocytes and mesangial cells followed by hyaline deposits in the capillary basement membranes and protein precipitates in Bowman's space in later stages (Cowan, 1968; Zwart, 2006). A definitive cause has not been identified. Most cases have been identified in boas and colubrids (Cowan, 1968). Additional renal changes may include tubular dilation, hyaline casts, and heterophilic interstitial inflammation (Zwart, 2006).

Chronic glomerulonephritis can be separated into two categories: intracapillary and extracapillary. Lesions of chronic intracapillary glomerulonephritis include small, ischemic capillaries with narrowed lumina leading to eventual sclerosis of the tufts. In two tortoise cases, glomerular lesions were accompanied by the deposition of urate tophi within renal tubules and eventually within the glomerular tufts (Zwart, 1964, 2006). In contrast, chronic extracapillary glomerulonephritis, also referred to as "basement membrane disease," is a condition that appears limited to chelonians and monitors characterized by massive accumulations of hyaline material in glomeruli, tubules, and vessels (Cowan, 1968). These changes may be accompanied by mild glomerular hypercellularity, interstitial inflammation, and prominent periglomerular and interstitial fibrosis (Zwart, 2006). The etiology of chronic glomerulonephritis is unknown, although possible dietary imbalances involving deficiencies of calcium and vitamin A and D have been suggested (Zwart, 1964).

Glomerulosclerosis is a distinct glomerular lesion characterized by the deposition of collagen in the mesangial tuft with thickening and eventual obliteration of capillary loops in the absence of inflammation (**Figures 3.54–3.56a,b**) (Cowan, 1968). Glomerulosclerosis may result from progressive or end-stage acute or chronic glomerulonephritis. It has also been described as an age-related change in snakes and crocodilians. Histologically, sclerotic glomeruli appear small and acellular with absent capillary loops. Application of stains such as Masson's trichrome (**Figure 3.56b**) or hematoxylin-phloxine-saffron (HPS) highlight the increased amounts of collagen within the mesangial tuft.

3.7 Pigment Deposition

The mechanisms for the production and deposition of pigments are less clearly defined in reptiles than in teleosts and some amphibians. Pigments are most often found within cytoplasmic vacuoles, and there are several pigment types, including iron, lipofuscin, and melanin. Some of these pigments, such as prominent melanin deposition in the liver, can be detected macroscopically as dark brown to black foci in the hepatic parenchyma, often distinctly contrasted with the pale parenchyma of hepatic lipidosis. The normal melanosis on parietal surfaces of the body wall and visceral surfaces of the viscera in some lizard species is also a prominent macroscopic observation. However, most pigments are best characterized by light microscopic examination of routine histologic sections and with special histochemical stains and/or electron microscopy.

In reptiles, as well as amphibians and fish, a commonly encountered pigment-containing cell is the pigmented macrophage, also known as the "melanomacrophage." Pigmented macrophages are mononuclear phagocytic cells that accumulate large amounts of various pigments, including lipofuscin, melanin, and hemosiderin (Christiansen et al., 1996). Given the variability of the pigment content in these cells and the variable appearance of the cells with routine staining, the term "pigmented macrophage" is favored over the classical term "melanomacrophage." These cells can be found individually or within clusters (pigmented macrophage centers) that are present in highest numbers in the liver (**Figures 3.57–3.58**; Volume 1, **Figures 1.78–1.79, 1.81–1.84, 5.11–5.13**), but also can be seen in other organs such as the spleen, kidney, lung, gonad, and heart. The size and number of these macrophage centers in the hepatic parenchyma increases linearly with age (Christiansen et al., 1996). Other factors that result in the increase in size and number of pigmented macrophage centers include immune stimulation, chronic inflammatory disease, or chronic negative energy balance (Agius and Roberts, 2003). In states of chronic negative energy balance, a relative increase in the number of pigmented macrophage centers due to hepatic atrophy should be differentiated from absolute hyperplasia that is often accompanied by an increase in the size of the centers (hypertrophy).

In cases of severe pigmented macrophage hyperplasia, organ function could possibly be impaired, especially hepatic function. An adult Crocodile Lizard (*Shinisaurus crocodilurus*) with a darkly colored liver had effacement of the hepatic parenchyma by severe pigmented macrophage hyperplasia histologically, which likely impaired hepatic function (Gyimesi and Howerth, 2004). The cause of the hyperplasia was unclear but possible contributing factors included seasonal and nutritional factors. This form of hyperplasia seems common in certain species, especially geckos, where pigmented macrophage centers can replace nearly the entire hepatic parenchyma (M. Garner, personal communication).

3.7.1 Iron

Iron is a complex molecule that is found in a variety forms. It can exist as a divalent cation (ferrous; Fe^{2+}) or trivalent cation (ferric; Fe^{3+}) and may be bound to a variety of protein carriers to form iron-storage compounds such as hemosiderin and ferritin. Histologically, one of the most commonly encountered iron pigments is hemosiderin, which is composed of protein complexed with ferric iron (Fe^{3+}) and is derived from hemoglobin breakdown products during normal erythrocyte turnover or during times of increased erythrocyte destruction. By light microscopy, hemosiderin appears golden-brown to brown-black and forms coarse granules inside phagocytic cells and less often free within the extracellular matrix (**Figures 3.59a, 3.60a**). Special staining with Perl's Prussian blue will react with ferric iron to highlight these deposits in tissue section as bright blue granules (**Figures 3.59b, 3.60b**; Volume 1, **Figure 1.183**).

Iron capture and storage appears to be one of the primary functions of pigmented macrophages. Iron pigment concentrations may rise in times of increased erythrocyte turnover (**Figure 3.61**), such as seen in hemolytic disease processes (Agius and Roberts, 2003). Iron storage disease, which is often caused in part by increased dietary iron intake in mammals and birds, has not been documented in reptiles (Kik et al., 2003). However, the accumulation of cytoplasmic iron in hepatocytes (**Figures 3.57, 3.59a,b**), renal tubular epithelial cells (**Figure 3.62**), and enterocytes (**Figure 3.60a,b**) not associated with tissue damage is a common finding in all reptiles.

3.7.2 Lipofuscin

Lipofuscin is a pigment associated with cellular "wear and tear" derived from the oxidative polymerization of polyunsaturated fatty acids from plasma membranes and senescent organelles, and lipofuscin levels may rise dramatically during times of cachexia or chronic disease (Agius and Roberts, 2003). Lipofuscin pigments are present in highest concentrations in pigmented macrophage centers (**Figures 3.57–3.59**) but can also be widely distributed through other tissues including neurons (**Figure 3.63**), renal tubular epithelium, and cardiac myocytes. By light microscopy, lipofuscin appears as globular to globoid, yellow-brown intracytoplasmic granules that exhibit positive periodic acid-Schiff staining, although electron microscopy is the preferred method for resolving lipofuscin pigment.

In studies on the effect of aging on lipid peroxides and lipofuscin in the lungs and brain of male Garden Lizards (*Calotes versicolor*), lipid peroxides were shown to increase while lipofuscin concentrations decreased significantly in older lizards (Manibabu and Patnaik, 1997; Majhi et al., 2000). This finding suggests that lipid peroxidation products may play less of a role in lipofuscinogenesis in reptiles than in mammals (Reichel, 1968; Majhi et al., 2000).

3.7.3 Melanin

In general, melanin is synthesized by melanocytes, which are embryologically derived from the neural crest. Melanin granules are stored within these cells as melanosomes and can be phagocytosed and sequestered inside vacuoles by pigmented macrophages and other phagocytic cells. It has also been documented that some melanin is synthesized directly by pigmented macrophages in teleosts, amphibians, and reptiles (Scalia et al., 1988; Christiansen et al., 1996; Agius and Roberts, 2003). Melanin pigments function as antioxidants and can neutralize free radicals and participate in the production of bactericidal compounds (Wolke et al., 1985). Melanin concentrations may increase in colder temperatures due to its increased efficacy over other enzyme-driven bactericidal systems in poikilotherms (Wolke et al., 1985).

In histologic section, melanin appears as dark brown to black, round, well demarcated, cytoplasmic granules. These granules can also be released in the extracellular matrix. Fontana Masson stain is used to highlight melanin granules histologically. Conversely, a melanin bleach can also be applied in highly melanized tissue to clear the pigment in order to observe cellular detail. Melanocytes are commonly found within the visceral serosa of many lizards (agamids, chameleons, varanids) and chelonians, which imparts a black-brown appearance to the walls of the coelomic cavity and to some tissues (kidneys, testicles, coelomic fat bodies, gastrointestinal tract) (**Figures 3.2, 3.6, 3.64**).

3.8 Fatty Deposits

The accumulation of lipid within organs and tissues is a frequent finding in reptiles. Captive reptiles may be especially prone to lipid deposition disorders, as many of the predisposing factors leading to these derangements may stem from aspects of captive husbandry (Simpson, 2006).

Careful evaluation of lipid deposits is necessary to determine normal physiologic processes from pathologic conditions. The amount of lipid present within visceral and subcutaneous tissue can vary drastically depending on species and gender differences, reproductive status, hibernation, diet, and activity levels (Hernandez-Divers and Cooper, 2001; Dutra, 2014). Differences in the mechanism of lipid metabolism may determine the type of lipid deposited. Within the liver, fatty acids are esterified to triglycerides whereas cholesterol predominates in the deposition of lipids at peripheral anatomic locations such as the brain and cornea.

3.8.1 Lipid

One of the most important organs for the interpretation of lipid deposits is the reptilian liver. Hepatic lipid deposits can occur due to physiologic mechanisms of fat metabolism or as a pathologic process. The lipid content of reptilian hepatocytes rises dramatically in preparation for hibernation or during vitellogenesis in females (Dutra, 2014; Boyer and Scott, 2019). Causes of pathologic accumulation of lipid within hepatocytes (hepatic lipidosis) include toxins such as ivermectin, follicular stasis, high fat diets, and obesity (Teare and Bush, 1983; Dutra, 2014). Nonbreeding female reptiles may be especially predisposed to hepatic lipidosis due to prolonged accumulation of lipids in preparation of vitellogenesis (Simpson, 2006). Clinical signs are usually nonspecific and may include lethargy, anorexia, and weight loss.

In animals with hepatic lipidosis, there is an increase in liver weight which may be associated with diffuse or regional rounding of lobe margins, increased pallor, and friability of the hepatic parenchyma (**Figures 3.65–3.66a**). Samples of liver with high lipid content will often float in formalin solution. Histologically, hepatocytes contain colorless to pale yellow, well-defined vacuoles within the cytoplasm that displace or compress the nucleus (**Figures 3.66b–3.68**). Frequently, fusion of lipid vacuoles of adjacent hepatocytes is noted with often hypereosinophilic surrounding cytoplasm (hepatocellular degeneration and necrosis) (**Figure 3.66b**). Observation of the zonal patterns of lipid distribution in hepatocytes is also important, as metabolic disturbances are usually generalized rather than zonal, which may suggest an underlying toxic etiology (Hernandez-Divers and Cooper, 2001). Oil red O and periodic acid-Schiff histochemical staining may be applied to differentiate positively staining lipid vacuoles from other cytoplasmic products, such as glycogen or water. Other secondary changes, such as inflammation, fibrosis, and cholestasis may accompany chronic or severe hepatic lipidosis (**Figure 3.68**). Long-term hepatic lipidosis may progress to cirrhosis over time, especially in tortoises and bearded dragons (M. Garner, personal communication). Hepatic biopsies from a female Black-Headed Python (*Aspidites melanocephalus*) with hepatic lipidosis revealed generalized hepatocellular lipid accumulation with portal fibrosis, mononuclear portal hepatitis, and cholestasis as evidenced by the accumulation of bile within macrophages (Simpson, 2006).

While serum biochemistry values may also serve as a reliable marker for hepatic lipidosis or severe hepatic disease in some species (Dutra, 2014), hepatic biopsies are an important antemortem technique for the diagnosis of clinical lipidosis in reptile patients (Hernandez-Divers and Cooper, 2001). Accurate differentiation between the presence of hepatocellular lipid (**Figure 3.69**) and hepatic lipidosis is important. In some cases, interpretation of histopathologic findings and patient clinical history and signs are necessary for an accurate diagnosis. Cytologic evaluation of hepatic lipidosis in reptiles is insensitive. Lipid-laden hepatocytes of reptiles seem especially fragile and readily lyse during aspiration and slide preparation (M. Garner, personal communication).

Obesity is both a common finding in captive reptiles and a predisposing factor for hepatic lipidosis. In reptiles, obesity is characterized predominately by the deposition of fat within coelomic and caudal (tail) fat bodies as well as certain organs, such as the liver and spleen. Abnormal lipid accumulation can also be seen in other tissues, including renal tubular epithelial cells, that while often innocuous (**Figure 3.70**), can sometimes greatly distort renal architecture (**Figure 3.71**). The storage of subcutaneous fat in reptiles is less common as to avoid disruption of thermoregulation (Simpson, 2006). Larger-bodied reptiles, such as pythons, boas, vipers, crocodilians, monitor lizards, and snapping turtles, appear especially susceptible to increased sedentary behavior and obesity (Bays and de Souza Dantas, 2019). For additional review of hepatic lipidosis, see Chapter 2.

3.8.2 Cholesterol

Cholesterol deposition disorders are well documented in reptiles. While the cause of cholesterol deposition in tissue is not well understood, females are overrepresented, suggesting that changes in lipid metabolism associated with vitellogenesis may be a contributing factor (Cooper et al., 1998; Garner et al., 1999; Gibbons and Garner, 2001; Gyimesi et al., 2002; Reed et al., 2007; Kummrow et al., 2010). Other considerations include the feeding of diets high in cholesterol (e.g., neonatal mice), age, trauma, mineral imbalances, and genetic factors (Cooper et al., 1998; Garner et al., 1999; Gibbons and Garner, 2001; Gyimesi et al., 2002; Reed et al., 2007; Kummrow et al., 2010). Obesity may also be a contributing factor (Cooper et al., 1998; Vargas-Soto et al., 2017).

Accumulations of cholesterol within the tissue may be innocuous, with little to no clinical impact. This is particularly true for corneal and spectacular cholesterol deposition in reptiles (**Figure 3.72**). In older chelonians and varanids it can be observed as bilateral blue to silvery white discolorations of the cornea near the limbus (arcus senilis) that do not appear to be associated with the impairment of vision (Frye, 1991). Band keratopathies are considered rare in reptiles, but the condition has been described by Frye in a Red-Tailed Ratsnake (*Gonyosoma oxycephalum*) as an oblique band of opacification along the cornea with visual impairment (Frye, 1991).

Tissue deposition of cholesterol crystals, however, frequently incites a granulomatous inflammatory response (xanthoma or xanthogranuloma) with the potential for substantial clinical disease. Xanthomas can occur as solitary or widespread foci in tissues but appear to have a predilection for the choroid plexus and ventricles (**Figures 3.73–3.75**). Other commonly reported anatomic locations include the ovary, lung (**Figure 3.76a,b**), coelomic surfaces (**Figure 3.77**), fat body, cornea, pericardium, liver, and subcutis. The cholesterol crystals dissolve during processing leaving the empty, sharp, angular (acicular) spaces (**Figures 3.72, 3.73b, 3.74b, 3.76b**). Grossly, these granulomas appear as tan serosal plaques on coelomic organs or as granular masses that with increased size cause compression of adjacent tissues (**Figures 3.73a, 3.76a, 3.77**) (Gibbons and Garner, 2001; Gyimesi et al., 2002). Histologically, xanthomas are composed of central cores of extracellular acicular clefts surrounded by a collar of epithelioid macrophages, multinucleated giant cells, and fewer lymphocytes and heterophils (**Figures 3.73b–3.74a,b, 3.75–3.76b**) (Garner et al., 1999; Gibbons and Garner, 2001; Gyimesi et al., 2002; Anderson et al., 2010). When xanthomas occur in the brain, they often arise from the choroid plexus and extend into the lateral and third ventricles which can result in hydrocephalus, compression, and rarefaction of the adjacent neuroparenchyma. Large, compressive xanthomas may result in neurologic signs, such as stargazing, head tilt, torticollis, abnormal recumbency, diminished righting reflexes, seizures, circling, opisthotonus, head bobbing, and incoordination (Garner et al., 1999; Gibbons and Garner, 2001; Anderson et al., 2010; Kummrow et al., 2010; Vargas-Soto et al., 2017).

Xanthomas have been reported in multiple squamate species (Cooper et al., 1998; Gibbons and Garner, 2001; Gyimesi et al., 2002; Reed et al., 2007; Anderson et al., 2010; Kummrow et al., 2010; Vargas-Soto et al., 2017). Commonly reported concurrent diseases in affected lizards and snakes include hepatic lipidosis, lipid keratopathy, and atherosclerosis (Garner et al., 1999; Gibbons and Garner, 2001; Anderson et al., 2010; Kummrow et al., 2010; Vargas-Soto et al., 2017). Many of these affected reptiles also had documented hypercholesterolemia (Gyimesi et al., 2002; Anderson et al., 2010; Kummrow et al., 2010; Vargas-Soto et al., 2017). The importance of hypercholesterolemia in these animals is unclear. In three captive female Water Dragons with varying degrees of intracranial xanthomatosis, there was no apparent difference in cholesterol levels among other unaffected animals or male Water Dragons (Kummrow et al., 2010). Coelomic xanthomas may be treated successfully by surgical removal; however, treatment is limited for invasive or intracranial xanthomas (Reed et al., 2007).

3.9 Miscellaneous Deposits

Other deposition conditions are sporadically encountered in reptiles. Choleliths and inspissated biliary secretory products have been reported multiple times in bearded dragons

(**Figure 3.78**) (Ritzman and Garner, 2009; Gimmel et al., 2017). Analysis of the biliary material by Gimmel et al. (2017) identified a single cholelith composed of 100% calcium carbonate and other inspissated material being composed of protein-based material, suggesting that potentially high protein diets may predispose bearded dragons to the formation of biliary sludge and choleliths. Choleliths are also sporadically observed in other lizards, including a Green Iguana (*Iguana iguana*) and a Giant Ameiva (*Ameiva ameiva*) (**Figure 3.79**) (T. Viner, personal communication). Damage to the biliary tree by parasites may also predispose individuals to cholelithiasis (**Figure 3.79**). Aggregates of proteinaceous material resulting in biliary dilatation, cholecystitis, hepatitis, and enteritis have also been observed in a Mata Mata (*Chelus fimbriatus*) (R. Montali, personal communication).

Aggregates of amphophilic to basophilic material can occasionally be seen within the lumen of faveoli in squamates (**Figures 3.80–3.83**). In most instances, the material is seen in conjunction with the presence of protein-rich fluid and likely represents inspissated pulmonary edema (**Figures 3.80–3.82**). Morphologically similar material can also be seen in individuals with no evidence of altered hemodynamics and likely represents an incidental finding (**Figure 3.83**).

References

Agius C, and Roberts RJ. 2003. Melano-macrophage centres and their role in fish pathology. *J Fish Dis* 26:499–509.

Allan ME, and Oftedal OT. 2003. Nutrition in captivity, in *Biology, Husbandry, and Medicine of the Green Iguana*. Jacobson ER (Ed.), Kreiger Publishing Co, Malabar, Fl. Chapter 4, Nutrition in captivity. Pg. 58.

Amat AC, Gabriel B, and Chee NW. 2012. Cystic calculi removal in African spurred Tortoise (*Geochelone sulcata*) using transplstron coeliotomy. *Veterinary World* 5:489–492.

Anderson ET, Troan BV, Stringer EM, Chinnadurai SK, and DeVoe RS. 2010. Cerebral xanthoma in a long-nosed snake (*Rhinocheilus lecontei*). *J Herpetol Med Surg* 20:58–60.

Appleby EC, and Siller WG. 1960. Some cases of gout in reptiles. *J Path Bact* 80:427–430.

Ariel E, Ladds PW, and Buenviaje GN. 1997. Concurrent gout and suspected hypovitaminosis A in crocodile hatchlings. *Aust Vet J* 75:247–249.

Barten SL. 2006. Lizards, in *Reptile Medicine and Surgery*. Mader DR (Ed.), Saunders Elsevier, Philadelphia, p. 690.

Bays TB, and de Souza Dantas LM. 2019. Clinical behavioral medicine, in *Reptile and Amphibian Medicine and Surgery*. Divers SJ and Stahl SJ (Eds.), Elsevier, St. Louis, Missouri, pp. 927–928.

Belasco-Zeitz M, Pye GW, Burns RE, and Pessier AP. 2013. Clinical challenge. *J Zoo Wildl Med* 44:807–810.

Bennett RA, and Lock BA. 2000. Nonreproductive surgery in reptiles. *Vet Clin Exot Anim* 3:715–732.

Birke LL, Cespedes AM, Schachner ER, and Lailvaux SP. 2017. Cystic calculus in a laboratory-housed Green Anole (*Anolis carolinensis*). *Comp Med* 67(2):112–115.

Boyer TH. 1991. Common problems and treatment of green iguanas (*Iguana iguana*). *Bull Assoc Reptilian Amphibian Veterinarians* 1:8–11.

Boyer TH, and Scott PW. 2019. Nutrition, in *Reptile and Amphibian Medicine and Surgery*. Divers SJ and Stahl SJ (Eds.), Elsevier, St. Louis, Missouri, pp. 201–223.

Bradford CM, Denver MC, Bronson E, Bradford J, and Martin S. 2009. Hypercalcemia in Prehensile-Tailed Skinks (*Corucia zebrata*). *J Herpetol Med Surg* 19:57–59.

Burns RE, Bicknese EJ, Westropp JL, Shiraki R, and Stalis IH. 2013. Tumoral calcinosis form of hydroxyapatite deposition disease in related Red-Bellied Short-Necked Turtles, *Emydura subglobosa*. *Vet Pathol* 50:443–450.

Burns RE, Gaffney PM, Nilsson KPR, Armién AG, and Pessier AP. 2017. Systemic amyloidosis in an African tiger snake (*Telescopus semiannulatus*). *J Comp Path* 157:136–140.

Chambers JK, Suzuki T, and Une Y. 2009. Tophaceous pseudogout of the femorotibial joint in a painted turtle (*Chrysemys picta*). *J Vet Med Sci* 71:693–695.

Choi HK, Liu S, and Curhan G. 2005. Intake of purine-rich foods, protein, and dairy products and relationship to serum levels of uric acid: The Third National Health and Nutrition Examination Survey. *Arthritis Rheum* 52:283–289.

Choi MH, MacKenzie JD, and Dalinka MK. 2006. Imaging features of crystal-induced arthropathy. *Rheum Dis Clin North Am* 32:247–446.

Christiansen JL, Grzybowski JM, and Kodama RM. 1996. Melanomacrophage aggregations and their age relationships in the yellow mud turtle, *Kinosternon favescens* (Kinosternidae). *Pigment Cell Res* 9:185–190.

Cohen AS, and Calkins E. 1959. Electron microscopic observations on a fibrous component in amyloid of diverse origins. *Nature* 183:1202–1203.

Cooper JE, and Jackson OF. 1981. Nutritional diseases, in *Diseases of the Reptilia*, Vol 2. Academic Press, London, UK, pp. 409–428.

Cooper JE, Bloxam QMC, and Tonge SJ. 1998. Pathology of Round Island geckos *Phelsuma guentheri*: Some unexpected findings. *Dodo* 34:153–158.

Cosgrove GE, and Anderson MP. 1984. Aging and degenerative diseases, in *Diseases of Amphibians and Reptiles*. Hoff GL, Frye FL and Jacobson ER (Eds.), Plenum Press, New York.

Cowan DF. 1968. Diseases of captive reptiles. *J Am Vet Med Assoc* 153:848–859.

Crespo R, and Shivaprasad HL. 2013. Developmental, metabolic, and other noninfectious disorders, in *Diseases of Poultry*, 13th ed. Swayne DE (Ed.), John Wiley & Sons, Inc. Ames, Iowa, pp. 1254–1256.

Crespo R, Franca MS, Fenton H, and Shivaprasad HL. 2018. Galliformes and columbiformes, in *Pathology of Wildlife and Zoo Animals*. Terio KA, McAloose D, St. Leger J (Eds.), Academic Press, London, UK, pp. 743–744.

DeGalantha E. 1935. Technic for preservation and microscopic demonstration of nodules in gout. *Am J Clin Path* 5:165–166.

DeNardo D, Barten SL, Rosenthal KL, Raiti P, and Nathan R. 2000. Dystocia. *J Herpetol Med Surg* 10:8–17.

Divers SJ, and Innis CJ. 2019. Urology, in *Mader's Reptile and Amphibian Medicine and Surgery.* Divers SJ and Stahl SJ (Eds.), Elsevier, St. Louis, Missouri, pp. 624–648.

Donoghue S. 2006. Nutrition, in *Reptile Medicine and Surgery.* Mader DR (Ed.), Saunders Elsevier, Philadelphia, pp. 251–298.

Drew ML. 1994. Hypercalcemia and hyperphosphatemia in indigo snakes (*Drymarchon corais*) and serum biochemical reference values. *J Zoo Wildl Med* 25:48–52.

Driggers T. 1997. Internal medicine, in *The Biology, Husbandry and Health Care of Reptiles*, Vol 3. Ackerman L (Ed.), T.F.H. Publications, Inc, Neptune City, NJ, pp. 574–592.

Dutra GHP. 2014. Diagnostic value of hepatic enzymes, triglycerides and serum proteins for the detection of hepatic lipidosis in *Chelonoidis carbonaria* in captivity. *J Life Sci* 8:633–639.

Eatwell K. 2008. Hypocalcemia in a Tunisian Tortoise (*Furculachelys nabeulensis*). *J Herpetol Med Surg* 18:117–122.

Finlayson R, and Woods SJ. 1977. Arterial disease of reptiles. *J Zool* 183:397–410.

Frye FL. 1976. Observations on systemic lupus erythematosus accompanying lymphoreticular neoplasia in poikilothermic species. *Theriogenology* 6(2–3):127–131.

Frye FL. 1983. Urinary calculosis and cystotomy in a lizard. *Vet Med Small Anim Clin* 78:431–433.

Frye FL. 1991. Common pathologic lesions and disease processes, in *Reptile Care: An Atlas of Diseases and Treatments*, Vol 2. Frye FL (Ed.), TFH, Neptune City, NJ, pp. 533–557.

Frye FL, and Dutra FR. 1976. Articular pseudogout in a turtle (*Chrysemys s. elegans*). *Vet Med Small Anim Clin* 71:655.

Garner MM, Lung NP, and Murray S. 1999. Xanthomatosis in geckos: Five cases. *J Zoo Wildl Med* 30:443–447.

Garner M. 2001. Regenerative erythroid response without anemia in Iguanidae: Associated with soft tissue mineralization. *Proceedings of the Association of Reptilian and Amphibian Veterinarians*, Orlando, Florida, pp. 213–216.

Garner MM. 2008. A retrospective study of disease in monitor lizards (*Varanus* spp.). *Proceedings of the Association of Reptilian and Amphibian Veterinarians*, Los Angeles, California, pp. 5–6.

Garner MM, Gyimesi ZS, Rasmussen JM, Joslin JO, Aguilar RF, Raymond JT, Gallo G, and Nordhausen RW. 2003. An amyloid-like deposition disorder in Komodo dragons. *Proceedings of the Association of Reptilian and Amphibian Veterinarians*, Minneapolis, Minnesota, pp. 50–52.

Gibbons PM, and Garner MM. 2001. Disseminated xanthomatosis in a female green water dragon, *Physignathus cocincinus*. *Proceedings of the Association of Reptilian and Amphibian Veterinarians*, Orlando, Florida, pp. 157–160.

Gimmel A, Kempf H, Öfner S, Müller D, and Liesegang A. 2017. Cholelithiasis in adult bearded dragons: Retrospective study of nine adult bearded dragons (*Pogona vitticeps*) with cholelithiasis between 2013 and 2015 in southern Germany. *J Anim Physiol Anim Nutr* 101:122–126.

Ginn PE, Mansell JEKL, and Rakich PM. 2007. Skin and appendages, in *Jubb, Kennedy and Palmer's Pathology of Domestic Animals*, Vol 1, 5th ed. Maxie MG (Ed.), Elsevier Limited, Edinburgh, UK, pp. 608–609.

Gyimesi ZS, and Howerth EW. 2004. Severe melanomacrophage hyperplasia in a crocodile lizard, *Shinisaurus crocodilurus:* A review of melanomacrophages in ectotherms. *J Herpetol Med Surg* 14:19–23.

Gyimesi ZS, Stedman NL, and Crossett, VR. 2002. Cholesterol granulomas in a great plated lizard, *Gerrhosaurus major. J Herpetol Med Surg* 12:36–39.

Hedley J, and Eatwell K. 2014. Cloacal prolapses in reptiles: A retrospective study of 56 cases. *J Small Anim Prac* 55:265–268.

Hernandez-Divers SJ. 2005. Reptile Non-Infectious Diseases. *Proceedings of the World Small Animal Veterinary Association*, Mexico City, Mexico, pp. 1–5.

Hernandez-Divers SJ, and Cooper JE. 2001. Reptile hepatic lipidosis. *Proceedings of the Association of Reptilian and Amphibian Veterinarians*, Orlando, Florida, pp. 193–200.

Hernandez-Divers SM, and Garner MM. 2003. Neoplasia of reptiles with an emphasis on lizards. *Vet Clin Exot Anim* 6:251–273.

Hernandez-Divers SJ, Martinez-Jiminez D, Bush S, Latimer KS, Zwart P, and Veldhuis-Kroeze EJ. 2008. Effects of allopurinol on plasma uric acid levels in normouricaemic and hyperuricaemic green iguanas (*Iguana iguana*). *Veterinary Record* 162:112–115.

Hodge MK. 1978. The effect of acclimation temperature on gentamicin nephrotoxicity in the Florida broad banded water snake (Natrix fasciata). *Annual Proceedings American Association of Zoo Veterinarians*, Knoxville, Tennessee, p. 226.

Homer BL, Berry KH, Brown MB, Ellis G, and Jacobson ER. 1998. Pathology of diseases in wild desert tortoises from California. *J Wildl Dis* 34:508–523.

Huchzermeyer FW. 2003. Non-transmissible diseases, in *Crocodiles: Biology, Husbandry and Diseases.* CAB International, Cambridge, MA, pp. 230–233.

Innis CJ, and Kincaid AL. 1999. Bilateral calcium phosphate ureteroliths and spirorchid trematode infection in a Red-eared Slider Turtle, *Trachemys scripta elegans*, with a review of the pathology of spirorchiasis. *Bull Assoc Reptilian Amphibian Veterinarians* 9:32–35.

Jacobson ER. 1976. Gentamicin-related visceral gout in two boid snakes. *Vet Med Small Anim Clin* 71:361–363.

Jacobson ER, Long PH, Miller RE, Kramer LW, Beehler B, Harshbarger JC, and Buergelt CD. 1986. Renal neoplasia of snakes. *J Amer Vet Med Assoc* 189:1134–1136.

Johnson JD, Averill-Murray RC, and Jarchow JL. 2001. Captive care of the desert tortoise, *Gopherus agassizii*. *J Herpetol Med Surg* 11:8–15.

Jones YL, and Fitzgerald SD. 2009. Articular gout and suspected pseudogout in a basilisk lizard (*Basilicus plumifrons*). *J Zoo Wildl Med* 40:576–578.

Keller KA, Hawkins MG, Weber III ES, Ruby AL, Guzman DSM, and Westropp JL. 2015. Diagnosis and treatment of urolithiasis in client-owned chelonians: 40 cases (1987–2012). *J Am Vet Med Assoc* 247:650–658.

Keymer IF. 1978. Diseases of chelonians: (1) necropsy survey of tortoises. *Vet Rec* 103:548–552.

Kik MJ, and Mitchell MA. 2005. Reptile cardiology: A review of anatomy and physiology, diagnostic approaches, and clinical disease. *Semin Avian Exotic Pet Med* 14:52–60.

Kik MJL, Dorrestein GM, and Beynen AC. 2003. Evaluation of 15 commercial diets and their possible relation to metabolic diseases in different species of reptiles. Proc Sym. Erkrank. *Zoo- und Wildtiere* 41:87–90.

Knafo SE. 2019. Musculoskeletal system, in *Reptile and Amphibian Medicine and Surgery*. Divers SJ and Stahl SJ (Eds.), Elsevier, St. Louis, Missouri, pp. 910–912.

Kölle P, and Hoffman R. 2002. Incidence of nephropathies in European Tortoises. *Proceedings of the Association of Reptilian and Amphibian Veterinarians*, Reno, Nevada, pp. 33–35.

Kölle P, Hoffman R, Wolters M, and Hesse A. 2001. Cystic calculi in reptiles. *Proceedings of the Association of Reptilian and Amphibian Veterinarians*, Orlando, Florida, pp. 191–192.

Kumar V, Abbas AK, Fauso N, and Aster JC. 2010. *Diseases of the immune system in Robbins' and Cotran's Pathologic Basis of Disease*, 8th ed. Saunders Elsevier, Philadelphia, pp. 249–255.

Kummrow MS, Berkvens CN, Paré JA, and Smith DA. 2010. Cerebral xanthomatosis in three green water dragons (*Physignathus cocincinus*). *J Zoo Wildl Med* 41:128–132.

Kwantes LJ. 1992. Surgical correction of cystic urolithiasis in an iguana. *Can Vet J* 33(11):752.

Laskin WB, Miettinen M, Fetsch JF. 2007. Calcareous lesions of the distal extremities resembling tumoral calcinosis (tumoral calcinosis-like lesions): Clinicopathologic study of 43 cases emphasizing a pathogenesis-based approach to classification. *Am J Surg Pathol* 31:15–25.

Mader DR. 2006. Gout, in *Reptile Medicine and Surgery*. Mader DR (Ed.), Saunders Elsevier, Philadelphia, pp. 793–800.

Mader DR, Ling GV, and Ruby AL. 1999. Cystic calculi in the California desert tortoise, *Gopherus agassizii*: Evaluation of 100 cases. *Proceedings of the Association of Reptilian and Amphibian Veterinarians*, Columbus, Ohio, pp. 81–82.

Majhi S, Jena BS, and Patnaik BK. 2000. Effect of age on lipid peroxides, lipofuscin and ascorbic acid contents of the lungs of male garden lizard. *Comp Biochem Physiol C* 126:293–298.

Manibabu PV, and Patnaik BK. 1997. Lipofuscin concentration of the brain shows a reduction with age in male garden lizard. *Comp Biochem Physiol C* 117:229–232.

Mans C, and Sladky KK. 2012. Endoscopically guided removal of cloacal calculi in three African spurred tortoises (*Geochelone sulcata*). *J Am Vet Med Assoc* 240:869–875.

McKown RD. 1998. A cystic calculus from a wild western spiny softshell turtle (*Apalone [Trionyx] spiniferus hartwegi*). *J Zoo Wildl Med* 29:347–347.

Mendyk RW, Newton AL, and Baumer M. 2013. A retrospective study of mortality in varanid lizards (Reptilia:Squamata:Varanidae) at the Bronx Zoo: Implications for husbandry and reproductive management in zoos. *Zoo Biol* 32:152–162.

Miller HA. 1998. Urinary diseases of reptiles: Pathophysiology and diagnosis. *Semin Avian Exotic Pet Med* 7:93–103.

Millichamp NJ, Jacobson ER, and Wolf ED. 1983. Diseases of the eye and ocular adnexae in reptiles. *J Am Vet Med Assoc* 183:1205–1212.

Minter LJ, Wood MW, Hill TL, and Lewbart GA. 2010. Cystoscopic guided removal of ectopic eggs from the urinary bladder of the Florida cooter turtle (*Pseudemys floridana floridana*). *J Zoo Wildl Med* 41:503–509.

Montali RJ, Bush M, and Smeller JM. 1979. The pathology of nephrotoxicity of gentamicin in snakes. *Vet Pathol* 16:108–115.

Orós J. 2019. Gout, in *Reptile and Amphibian Medicine and Surgery*. Divers SJ and Stahl SJ (Eds.), Elsevier, St. Louis, Missouri, pp. 1308–1309.

Osborne CA, Albasan H, Lulich JP, Nwaokorie E, Koehler LA, and Ulrich LK. 2009. Quantitative analysis of 4468 uroliths retrieved from farm animals, exotic species, and wildlife submitted to the Minnesota Urolith Center: 1981 to 2007. *Vet Clin North Am Small Anim Pract* 39(1):65–78.

Ousavaplangchai L. 1990. Gout in freshwater crocodiles (Crocodylus siamensis). *Proceedings of the Federation of Asian Veterinary Associations*, Pattaya, Thailand.

Radi ZA, and Sato K. 2010. Bilateral dystrophic calcinosis circumscripta in a cynomolgus macaque (*Macaca fascicularis*). *Tox Pathol* 38:637–641.

Raiti P, and Garner MM. 2006. Metastatic mineralization in a Geoffrey's Side-Necked Turtle, *Phrynops geoffroanus*. *J Herpetol Med Surg* 16:135–139.

Raphael BL, James SB, and Cook RA. 1999. Evaluation of vitamin D concentrations in *Uromastyx* spp. with and without radiographic evidence of dystrophic mineralization. *Proceedings of the Association of Reptilian and Amphibian Veterinarians*, Puerto Vallarta, Mexico, pp. 20–22.

Reavill DR, and Schmidt RE. 2002. Mineralized skin lesions in lizards. *Proceedings of the Association of Reptilian and Amphibian Veterinarians*, Reno, Nevada, pp. 77–78.

Reavill DR, and Schmidt RE. 2010. Urinary tract diseases of reptiles. *J Exot Pet Med* 1:280–289.

Reavill DR, and Schmidt RE. 2013. Urogenital tract diseases of reptiles. *Proceedings of the Association of Reptilian and Amphibian Veterinarians*, Wiesbaden, Germany, p. 84.

Reed SD, Reed FM, and Castleman WL. 2007. Successful surgical management of advanced xanthomatosis in a leopard gecko, *Eublepharis macularius*. *J Herpetol Med Surg* 17:19–21.

Reichel W. 1968. Lipofuscin pigment accumulation in five rat organs as a function of age. *J Geront* 23:145–153.

Richman LK, Montali RJ, Allen ME, and Oftedal OT. 1995. Paradoxical pathologic changes in vitamin D deficient green iguanas (*Iguana iguana*). *Proceedings of the Association of Reptilian and Amphibian Veterinarians*, Sacramento, California, pp. 231–232.

Ritzman T, and Garner M. 2009. Cholelithiasis and surgical cholelith removal in a bearded dragon (*Pogona vitticeps*). *Proceedings of the Association of Reptilian and Amphibian Veterinarians*, Milwaukee, Wisconsi, p. 117.

Roberts ED, Baskin GB, Watson E, Henk WG, Shelton TC, and Bowen MS. 1984. Calcium pyrophosphate deposition in non-human primates. *Vet Pathol* 21:592–596.

Rosenberg AE. 2010. Bones, joints and soft-tissue tumors, in *Robbins and Cotran Pathologic Basis of Disease*, 8th ed. Kumar V, Abbas AK, Fausto N et al. (Eds.), Saunders Elsevier, Philadelphia, PA, p. 1246.

Rosenthal K, Divers SJ, Donoghue S, Gamer M, Klingenberg RJ, and Nathan R. 2000. Renal disease. *J Herpetol Med Surg* 10:34–43.

Ryer KA. 1983. Urinary calculi in a green iguana. *Vet Med Small Anim Clin* 78:607–607.

Sander SJ, Ossiboff RJ, Stokol T, Steeil JC, and Neiffer DL. 2015. Endolymphatic sac carcinoma *in situ* in a Tokay gecko (*Gekko gecko*). *J Herpetol Med Surg* 25:82–86.

Scalia M, Geremia E, Corsaro C, Santoro C, Sciuto S, and Sichel G. 1988. The extracutaneous pigmentary system: Evidence for the melanosynthesis in amphibian and reptilian liver. *Comp Biochem Physiol B* 89:715–717.

Shidham V, and Shidham G. 2000. Staining method to demonstrate urate crystals in formalin-fixed, paraffin-embedded tissue sections. *Arch Pathol Lab Med* 124:774–776.

Sigurdson CJ, Nilsson KP, Hornemann S, Manco G, Polymenidou M, Schwarz P, Leclerc M, Hammarström P, Wüthrich K, and Aguzzi A. 2007. Prion strain discrimination using luminescent conjugated polymers. *Nat Methods* 4:1023–1030.

Simpson M. 2006. Hepatic lipidosis in a Black-Headed python (*Aspidites melanocephalus*). *Vet Clin Exot Anim* 9:589–598.

Sipe JD, Benson MD, Buxbaum JN, Ikeda SI, Merlini G, Saraiva MJ, and Westermark P. 2014. Nomenclature 2014: Amyloid fibril proteins and clinical classification of the amyloidosis. *Amyloid* 21:221–224.

Sjölander D, Röcken C, Westermark P, Westermark GT, Nilsson, KPR, and Hammarström P. 2016. Establishing the fluorescent amyloid ligand h-FTAA for studying human tissues with systemic and localized amyloid. *Amyloid* 23:98–108.

Slavin RE, Wen J, Kumar D, and Burke Evans E. 1993. Familial tumoral calcinosis: A clinical, histopathologic, and ultrastructural study with analysis of its calcifying process and pathogenesis. *Am J Surg Pathol* 17:788–802.

Soldati G, Lu ZH, Vaughan L, Polkinghorne A, Zimmermann DR, Huder JB, and Pospischil A. 2004. Detection of mycobacteria and chlamydiae in granulomatous inflammation of reptiles: A retrospective study. *Vet Pathol* 41:388–397.

Stern AW, Velguth KE, and D'Agostino J. 2010. Metastatic ductal adenocarcinoma in a western hognose snake (*Heterodon nasicus*). *J Zoo Wildl Med* 41:320–324.

Stevens A. 1990. Pigment and minerals, in *Theory and Practice of Histological Techniques*. Bancroft JD and Stevens AD (Eds.), Churchill Livingstone, New York, NY, p. 262.

Tafti AK, Hanna P, and Bourque AC. 2005. Calcinosis circumscripta in the dog: A retrospective pathological study. *J Vet Med A Physiol Pathol Clin Med* 52:13–17.

Teare JA, and Bush M. 1983. Toxicity and efficacy of ivermectin in chelonians. *J Am Vet Med Assoc* 183:1195–1197.

Thomas HL, Wilier CJ, Wosar MA, Spaulding KA, and Lewbart GA. 2002. Egg-retention in the urinary bladder of a Florida cooter turtle, *Pseudemys floridana floridana*. *J Herpetol Med Surg* 12:4–6.

Thompson K. 2007. Bones and joints, in *Jubb, Kennedy and Palmer's Pathology of Domestic Animals*, Vol 1, 5th ed. Maxie MG (Ed.), Elsevier Limited, Edinburgh, UK, pp. 1–184.

Toyama BH, and Weissman JS. 2011. Amyloid structure: Conformational diversity and consequences. *Annu Rev Biochem* 80:557–585.

Trautwein G, and Pruksaraj D. 1967. On amyloidosis in turtles. *Deutsche Tierarztliche Wochenschrift* 74:184–186.

Uhlar CM, and Whitehead AS. 1999. Serum amyloid A, the major vertebrate acute-phase reactant. *Eur J Biochem* 265:501–523.

Vargas-Soto LM, Vázquez-Briones DB, Núñez-Ochoa L, and Reyes-Matute A. 2017. Xanthogranuloma and hepatic lipidosis in a Mexican moccasin snake (*Akistrodon bilineatus*) with hypercholesterolemia: A case report. *J Vet Med Surg* 1:15.

Wallach JD, and Hoessle C. 1966. Hypervitaminosis D in green iguanas. *J Am Vet Med Assoc* 149:912–914.

Wallach JD, and Hoessle C. 1967. Visceral gout in captive reptiles. *J Am Vet Med Assoc* 151:897–899.

Wenker CJ, Bart M, Guscetti F, Hatt J-M, and Isenbügel E. 1999. Periatricular hydroxyapatite deposition disease in two Red-Bellied Short-Necked Turtles (Emydura albertisii). *Proceedings of the American Association of Zoo Veterinarians*, Columbus, Ohio, pp. 23–26.

Woldemeskel M. 2012. A concise review of amyloidosis in animals. *Vet Med Int* 2012:1–11.

Wolf KN, Troan BV, and DeVoe R. 2008. Chronic urolithiasis and subsequent cystectomy in a San Esteban Island Chuckwalla, *Sauromalus varius*. *J Herpetol Med Surg* 18:106–112.

Wolke RE, George CJ, and Blazer, VS. 1985. Pigmented macrophage accumulations (MMC; PMP): Possible monitors of fish health, in *Parasitology and Pathology of Marine Organisms of the World Ocean*. Hargis Jr. WG (Ed.), pp. 93–97.

Woodard JC. 1997. Skeletal system, in *Veterinary Pathology*, 6th ed. Jones TC, Hunt RD and King NW (Eds.), Williams & Wilkins, Baltimore, MD, pp. 899–946.

Yanai T, Noda A, Hasegawa K, Sakai H, and Masegi T. 2002. Calcinosis circumscripta in a Malayan box turtle (*Cuora amboinensis kamaroma*). *Jpn J Zoo Wildlife Med* 7:149–152.

Yü TA. 1965. Secondary gout associated with myeloproliferative diseases. *Arthritis Rheum* 8:765–771.

Zhou X, Guo Q, and Dai H. 2008. Identification of differentially expressed immune-relevant genes in Chinese Softshell Turtle (*Trionyx sinesis*) infected with *Aeromonas hydrophila*. *Vet Immunol Immunopathol* 125:82–91.

Zhou X, Wang L, Feng H, Guo Q, and Dai H. 2011. Acute phase response in Chinese Softshell Turtle (*Trionyx sinensis*) with *Aeromonas hydrophila* infection. *Dev Comp Immunol* 35:441–451.

Zwart P. 1964. Studies on renal pathology in reptiles. *Pathologia Veterinaria* 1:542–556.

Zwart P. 2006. Renal pathology in reptiles. *Vet Clin Exot Anim* 9:129–159.

Figure 3.1 Parson's Chameleon, *Calumma parsonii*. Chamaeleonidae. Renal gout. Numerous punctate, white urate tophi are distributed throughout the pale tan and swollen renal parenchyma. (Courtesy of Disease Investigations, San Diego Zoo Global.)

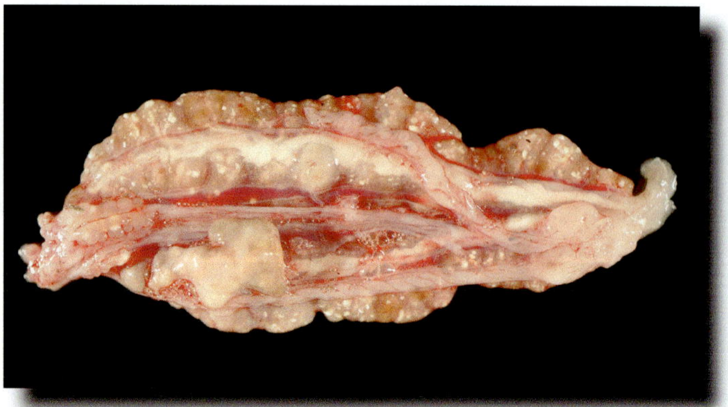

Figure 3.2 Panther Chameleon, *Furcifer pardalis*. Chamaeleonidae. Renal gout. Multiple, punctate, white-yellow foci are scattered throughout the renal parenchyma (arrows). Also note the white, serpiginous sexual segments of the nephron (arrowheads), as well as the pigmented serosa of the testicles (T). Serosal pigmentation is a common finding in many lacertilians. (Courtesy of Rachel Burns, William R. Pritchard Veterinary Medical Teaching Hospital, UC Davis, CA.)

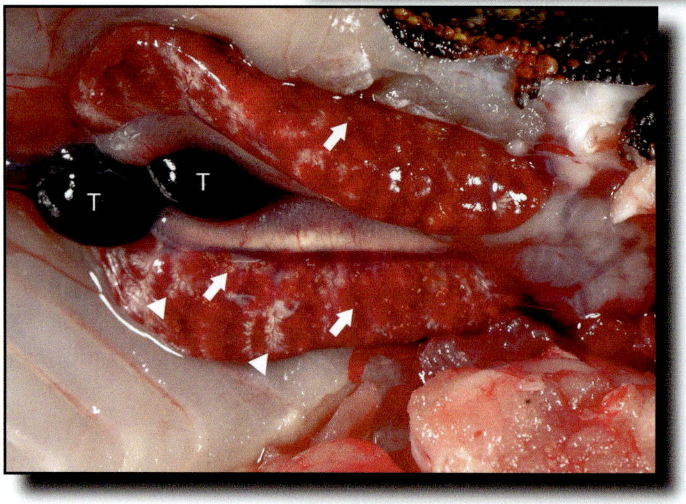

Figure 3.3 Bearded Dragon, *Pogona vitticeps*. Agamidae. Renal gout. The cranial pole (arrow) of the left kidney is regionally expanded by urate tophi. Urate-producing renal tubular adenocarcinoma can have a similar appearance.

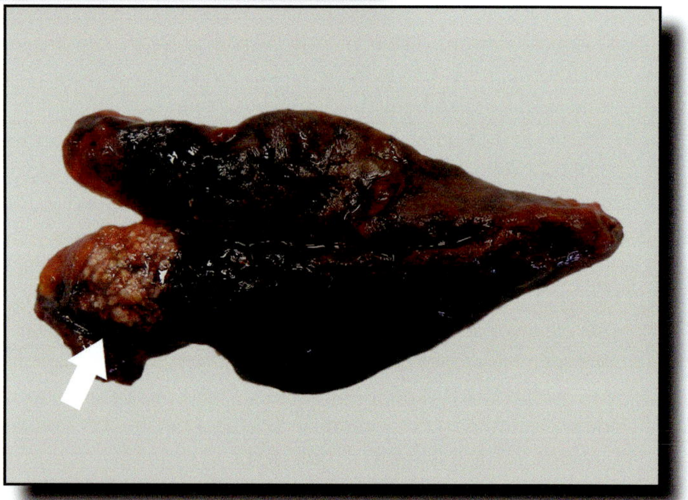

Figure 3.4 Corn Snake, *Pantherophis guttatus*. Colubridae. Renal gout. Renal lobules are markedly and extensively expanded by multiple coalescing urate tophi.

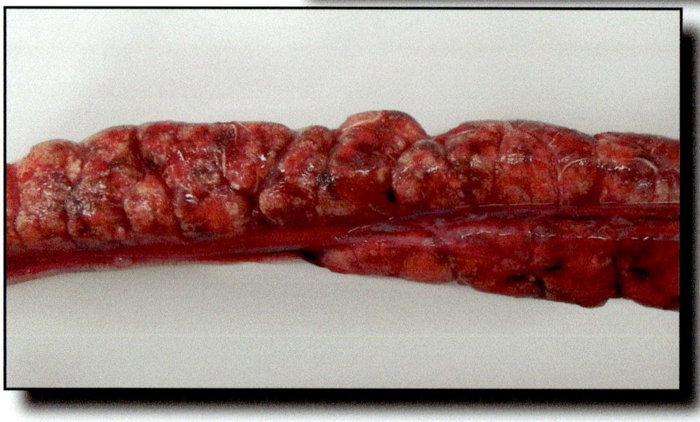

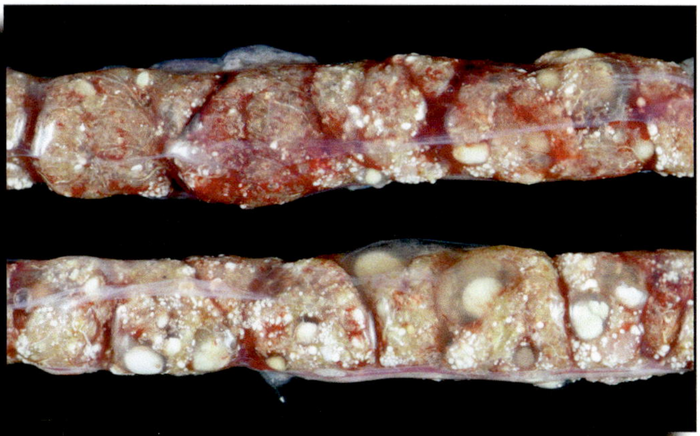

Figure 3.5 Caatinga Lancehead, *Bothrops erythromelas*. Viperidae. Renal gout and nephrolithiasis. Punctate white foci (urate tophi) are intermixed with large nephroliths contained within ectatic ducts/tubules. (Courtesy of Disease Investigations, San Diego Zoo Global.)

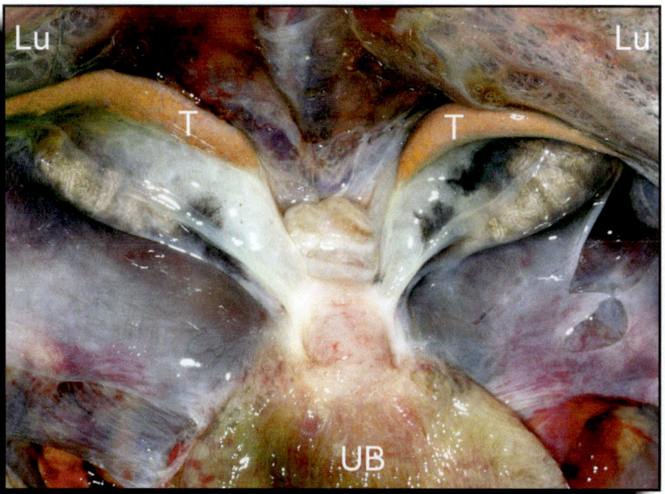

Figure 3.6 Desert Tortoise, *Gopherus* species. Testudinidae. Renal gout. Urate tophi are present throughout the renal parenchyma. Note the presence of pigment within the serosa of the renal capsule, which is also a common finding in many chelonians. UB = urinary bladder; Lu = lung; T = testicle. (Courtesy of Michael Garner.)

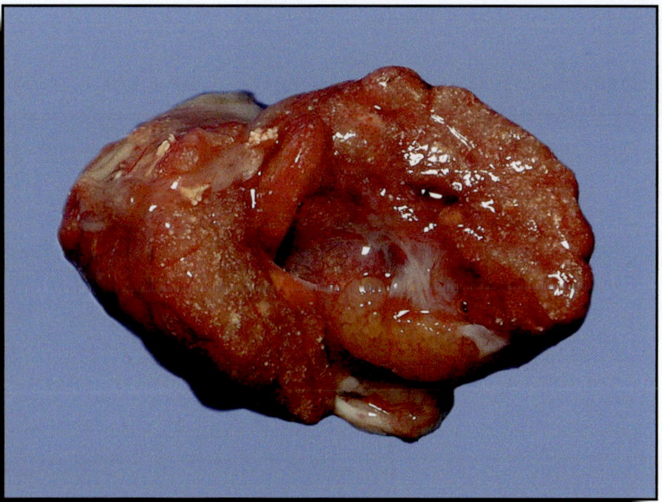

Figure 3.7 Burmese Star Tortoise, *Geochelone platynota*. Testudinidae. Renal gout. (Courtesy of Paul Gibbons, Turtle Conservancy.)

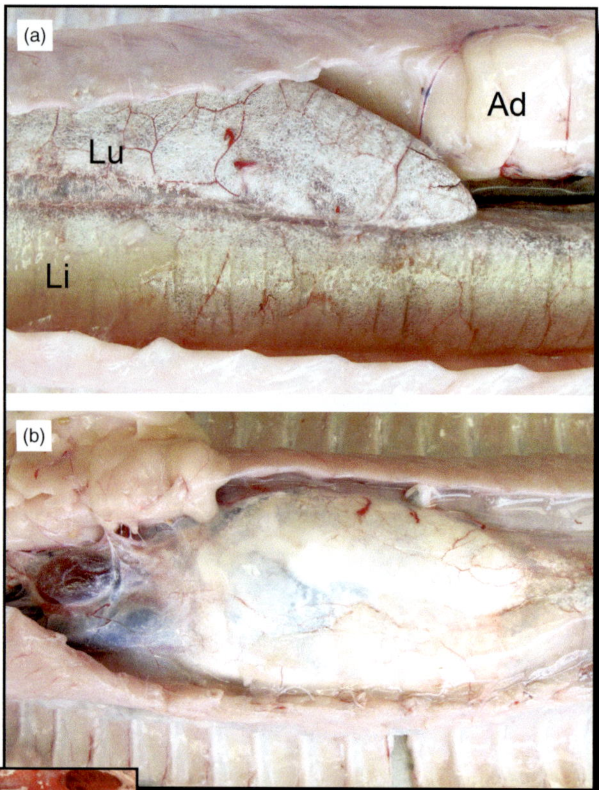

Figure 3.8 King Cobra, *Ophiophagus hannahi*. Elapidae. Visceral gout. There is extensive precipitation of white, granular crystalline uric acid along the hepatic capsule (a) and the surface of the epicardium, and along the serosa of the great vessels and thyroid gland (b). Lu = lung, cranial; Li = liver; Ad = adipose, caudal. (Courtesy of Disease Investigations, San Diego Zoo Global.)

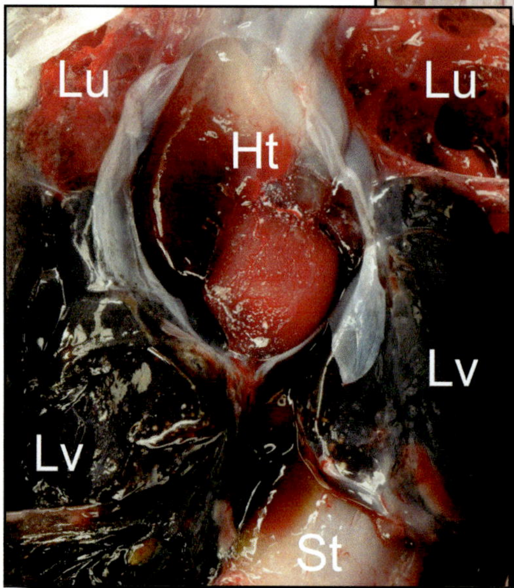

Figure 3.9 Broad-Snouted Caiman, *Caiman latirostris*. Alligatoridae. Visceral gout. Numerous punctate white foci of precipitated uric acid are present on the ventricular epicardial surface. Note also the presence of multifocal, white urate tophi present within the liver. Ht = heart; Lu = lung; Lv = liver; St = stomach.

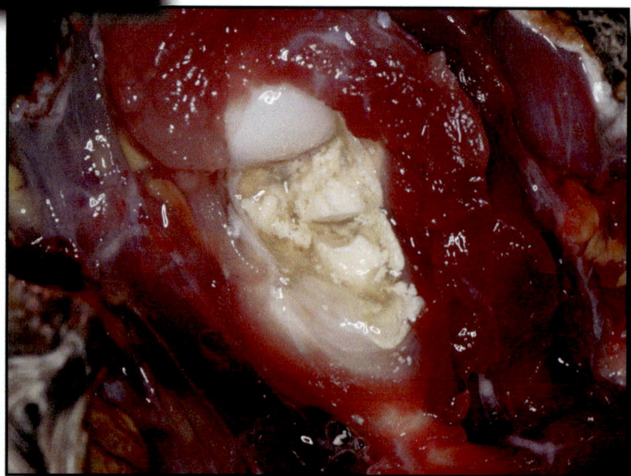

Figure 3.10 Desert Tortoise, *Gopherus* species. Testudinidae. Articular gout. The synovial fluid contains abundant, granular, crystalline uric acid. (Courtesy of Elliot Jacobson.)

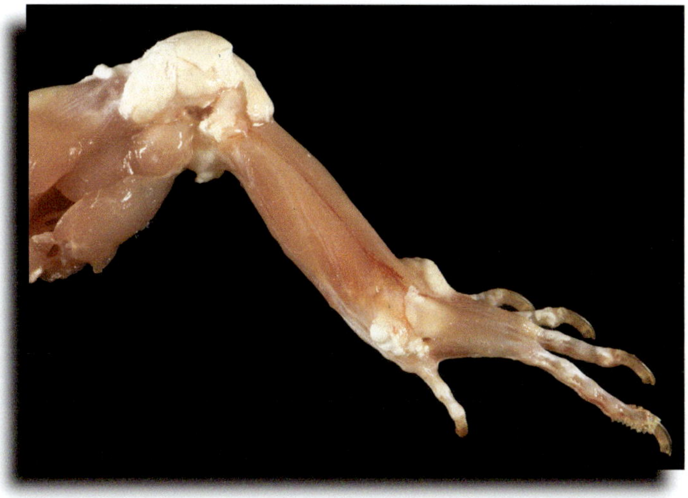

Figure 3.11 Bearded Dragon, *Pogona vitticeps*. Agamidae. Articular and periarticular gout. The stifle and tarsal joints are expanded and surrounded by precipitated uric acid. (Courtesy of Brian Stacy.)

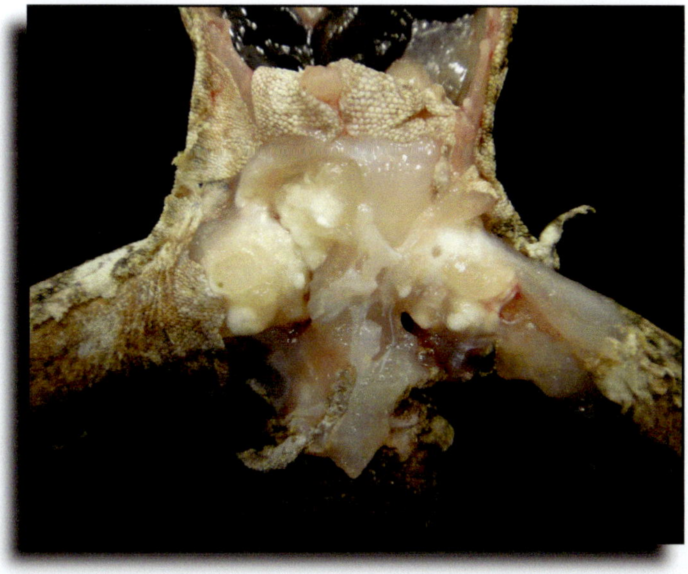

Figure 3.12 Henkel's Leaf-Tailed Gecko, *Uroplatus henkeli*. Gekkonidae. Articular and periarticular gout. The coxofemoral joints are expanded and surrounded by abundant crystalized uric acid that extends into the surrounding connective tissue and skeletal muscle. (Courtesy of Robert Ossiboff/Wildlife Conservation Society.)

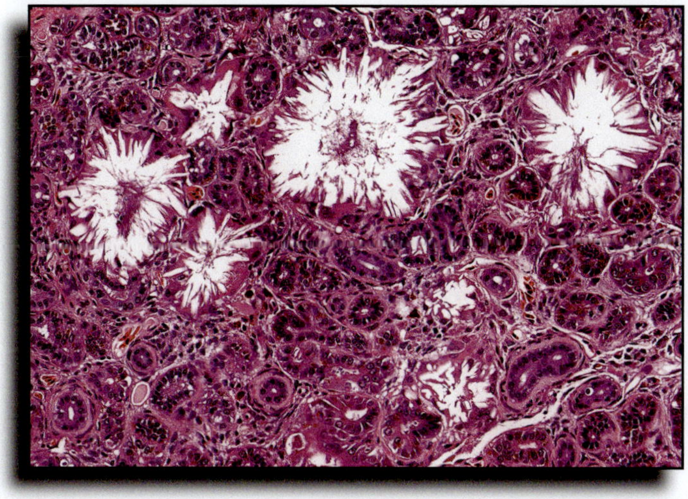

Figure 3.13 Western Hognose Snake, *Heterodon nasicus*. Colubridae. Photomicrograph of urate tophi within the kidney. Crystallized uric acid often dissolves during routine processing for histology, leaving colorless acicular clefts surrounded by inflammation and associated tissue damage. Note the presence of granulomatous inflammation surrounding the acicular clefts, as well as generalized interstitial inflammation and fibrosis that can be seen with chronicity in cases of renal gout. Hematoxylin and eosin.

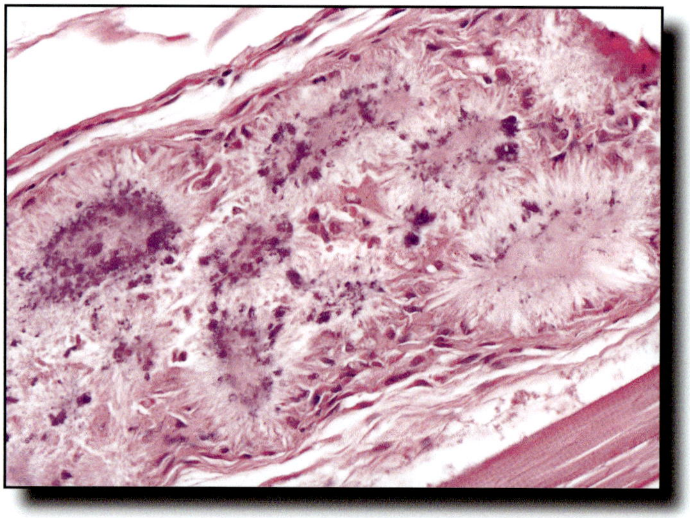

Figure 3.14 Frilled Lizard, *Chlamydosaurus kingii.* Agamidae. Photomicrograph of periarticular gout. Note the presence of granular basophilic material (mineral) and radiating acicular clefts and granulomatous inflammation. Hematoxylin and eosin. (Courtesy of Disease Investigations, San Diego Zoo Global.)

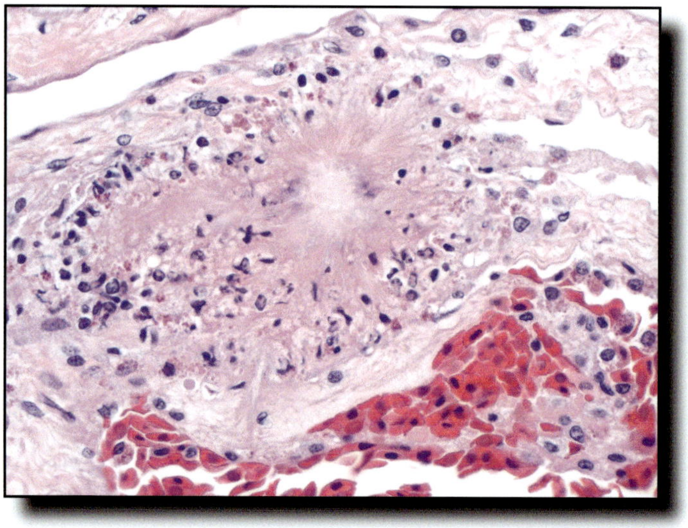

Figure 3.15 Broad-Snouted Caiman, *Caiman latirostris*. Alligatoridae. Photomicrograph of an acute perivascular urate tophus. Note the acicular, radiating clefts and pale basophilic material surrounded by intact and degenerate granulocytes and cellular debris. Hematoxylin and eosin.

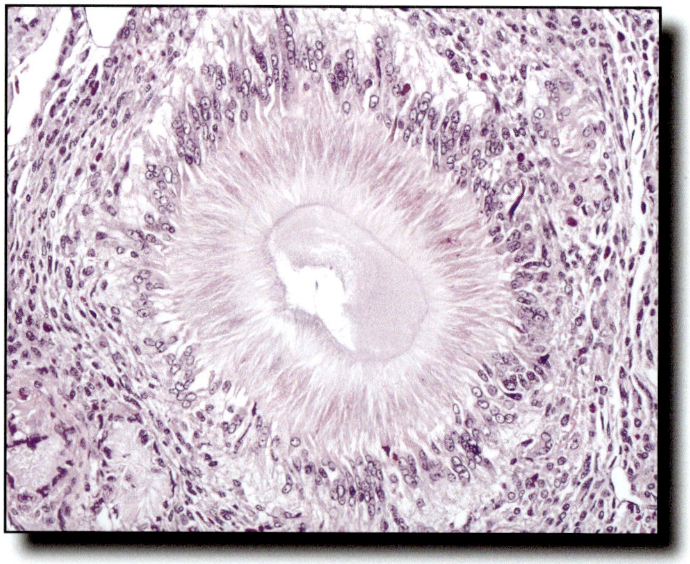

Figure 3.16 Tuatara, *Sphenodon punctatus*. Sphenodontidae. Photomicrograph of a chronic urate tophus with a band of epithelioid macrophages and multinucleated giant cells forming the tophus surrounded by mixed inflammation. Hematoxylin and eosin. (Courtesy of Disease Investigations, San Diego Zoo Global.)

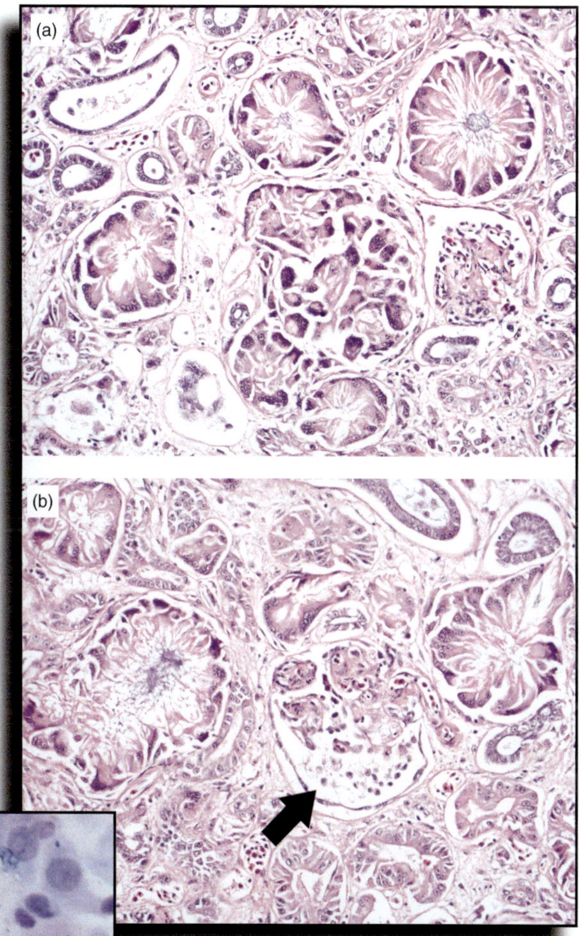

Figure 3.17 Anegada Ground Iguana, *Cyclura pinguis*. Iguanidae. Photomicrograph of chronic renal gout characterized by urate tophi with prominent multinucleated giant cells that efface the renal parenchyma. Note the tubular degeneration, tubular epithelial necrosis, and tubular regeneration in the surrounding parenchyma (a). Surrounding the urate tophi, note the presence of renal interstitial fibrosis, renal tubular regeneration, and the presence of low numbers of inflammatory cells within Bowman's space (arrow; b). Hematoxylin and eosin. (Courtesy of Disease Investigations, San Diego Zoo Global.)

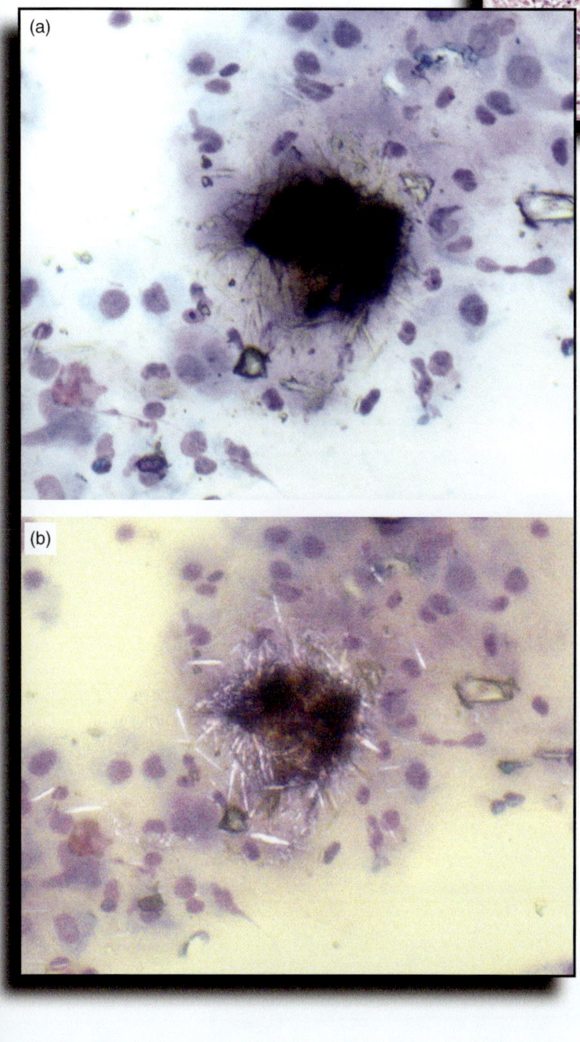

Figure 3.18 Gecko, species NOS. Gekkota. Photomicrograph of cytologic impressions of gout from a gecko. Fine, needle-like crystals, consistent with monosodium urate are surrounded by histiocytic and granulocytic inflammation (a). With polarized light, the crystals are birefringent (b). Wright-Giemsa stain.

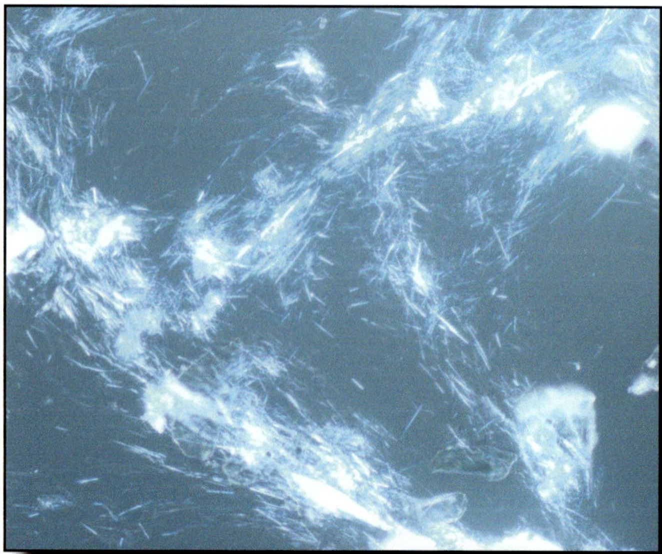

Figure 3.19 Green Crested Basilisk, *Basiliscus plumifrons*. Corytophanidae. Abundant birefringent needle-like crystals are present in an unstained joint aspirate with polarized light. (Courtesy of Disease Investigations, San Diego Zoo Global.)

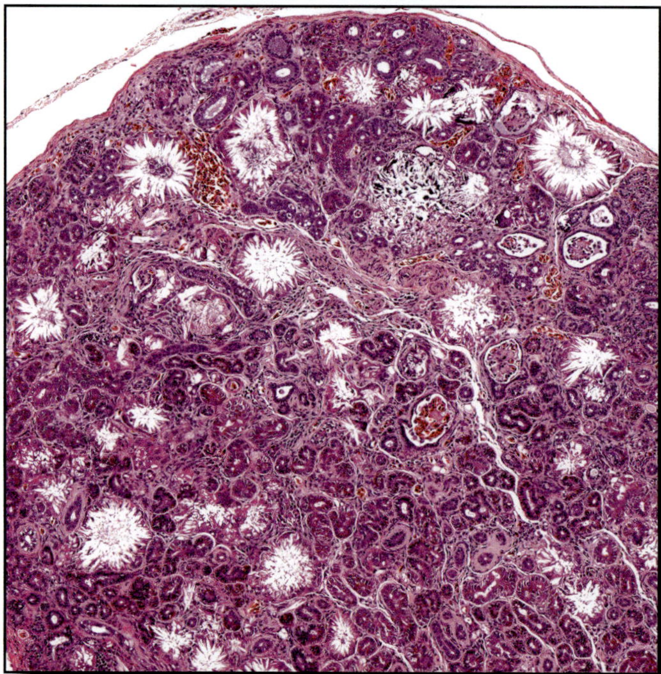

Figure 3.20 Western Hognose Snake, *Heterodon nasicus*. Colubridae. Photomicrograph of chronic renal gout. Note the prominent interstitial fibrosis dissecting between urate tophi, renal tubules, and glomeruli. Hematoxylin and eosin.

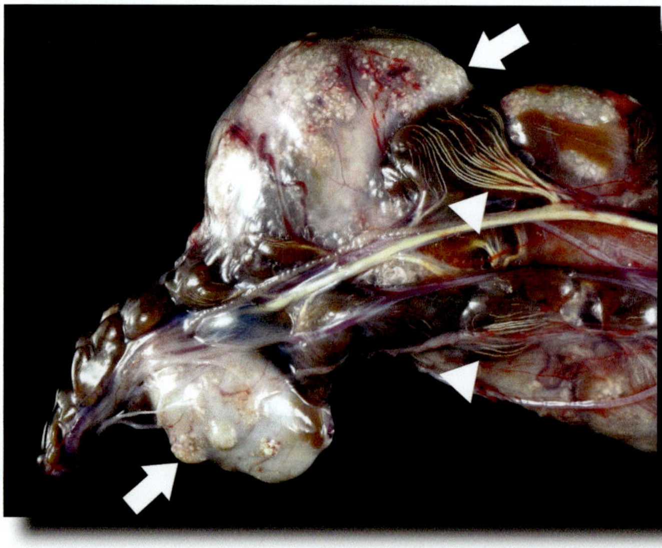

Figure 3.21 Red Spitting Cobra, *Naja pallida*. Elapidae. Renal carcinoma with concurrent renal gout and urate stasis. Note the linear accumulation of white material (urate stasis; arrowheads) within the ureter and collecting ducts and contrast the appearance of urate stasis to the punctate accumulation of urate tophi within the neoplastic lobules (arrows). (Courtesy of Disease Investigations, San Diego Zoo Global.)

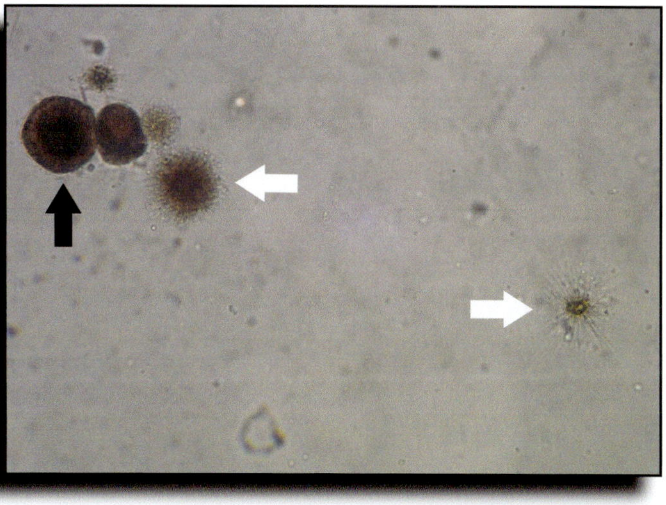

Figure 3.22 Henkel's Leaf-Tailed Gecko, *Uroplatus henkeli*. Gekkonidae. Unstained renal squash preps of an individual with gout, demonstrating aggregates of radiating, needle-like monosodium urate crystals (white arrows) as well as round, well demarcated urates (black arrow). (Courtesy of Robert Ossiboff/ Wildlife Conservation Society.)

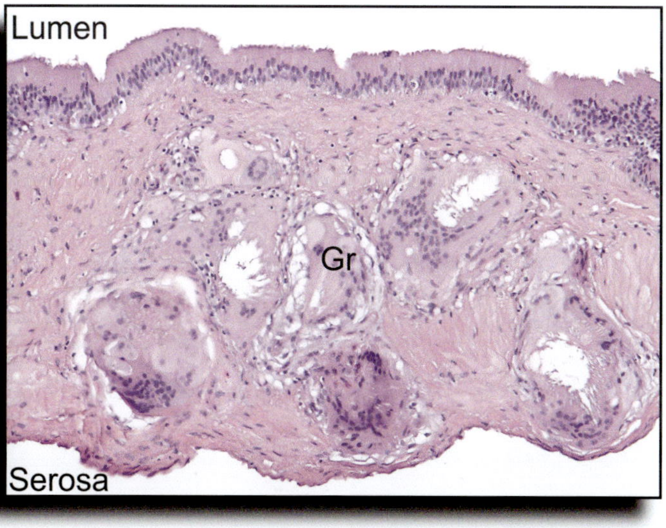

Figure 3.23 Geyr's Spiny-Tailed Lizard, *Uromastyx geyri*. Agamidae. Photomicrograph of visceral gout associated with the urinary bladder. Foci of granulomatous inflammation (Gr) surrounding colorless, acicular clefts expand the wall of the urinary bladder. The luminal and serosal surfaces of the urinary bladder are indicated. Hematoxylin and eosin.

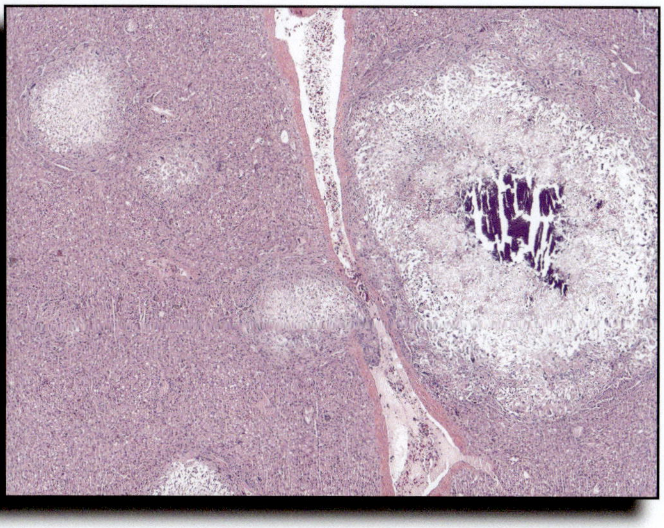

Figure 3.24 Western Hognose Snake, *Heterodon nasicus*. Colubridae. Photomicrograph of visceral (hepatic) gout. Urate tophi efface regions of the hepatic parenchyma and focally the wall of a hepatic artery. Within the center of one large chronic tophus, abundant granular basophilic material (mineral) is present. Hematoxylin and eosin.

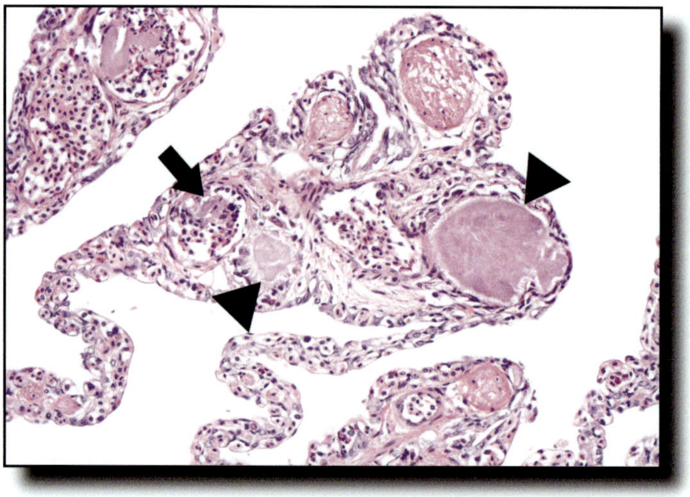

Figure 3.25 Fiji Banded Iguana, *Brachylophus fasciatus*. Iguanidae. Photomicrograph of uric acid precipitation in vessels (arrow) and interstitium (arrowhead) of the lungs, associated with thrombosis and granulocytic and granulomatous inflammation. Hematoxylin and eosin. (Courtesy of Disease Investigations, San Diego Zoo Global.)

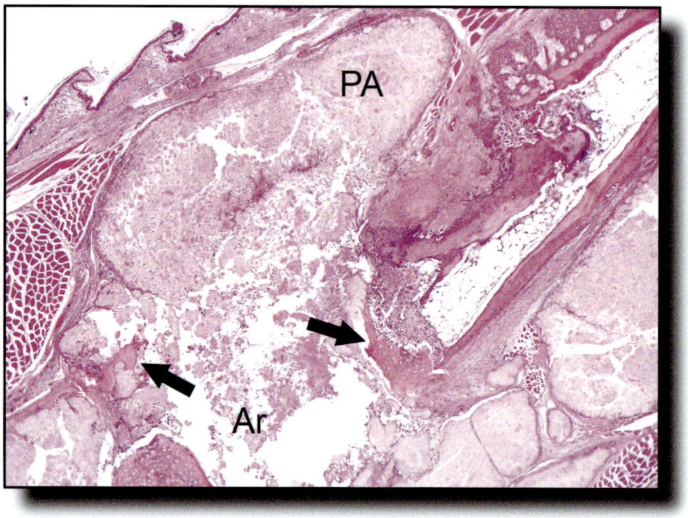

Figure 3.26 Frilled Lizard, *Chlamydosaurus kingii*. Agamidae. Photomicrograph of articular (Ar) and periarticular (PA) gout. Large aggregates of precipitated uric acid expand the joint space and extend into the surrounding periarticular tissue. There is associated granulomatous inflammation and lysis of articular cartilage and bone (arrows). Hematoxylin and eosin. (Courtesy of Disease Investigations, San Diego Zoo Global.)

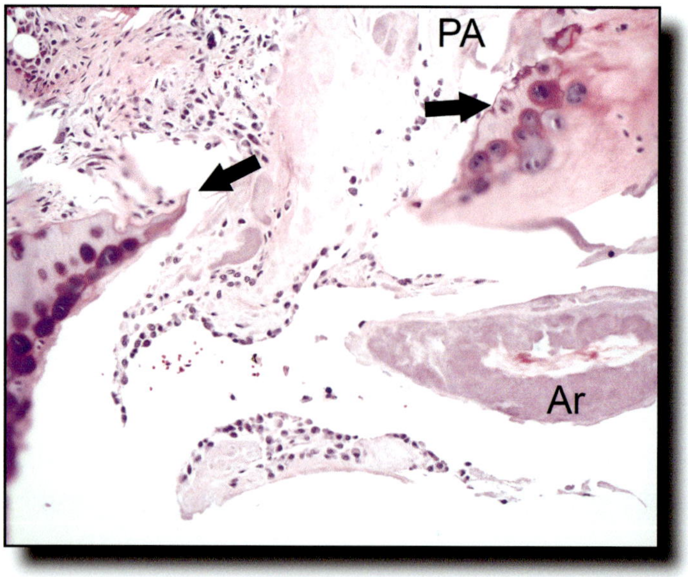

Figure 3.27 Fiji Banded Iguana, *Brachylophus fasciatus*. Iguanidae. Photomicrograph of articular (Ar) and periarticular (PA) gout. Granulomatous and lymphoplasmacytic inflammation surrounds amphophilic urate material. Note the lysis and necrosis of articular cartilage (arrows). Hematoxylin and eosin. (Courtesy of Disease Investigations, San Diego Zoo Global.)

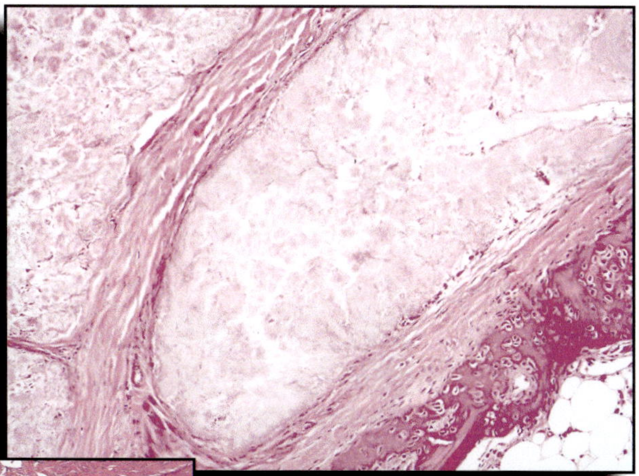

Figure 3.28 Frilled Lizard, *Chlamydosaurus kingii*. Agamidae. Photomicrograph of periarticular gout that extends throughout the connective tissue adjacent to the joint and associated bones. Hematoxylin and eosin. (Courtesy of Disease Investigations, San Diego Zoo Global.)

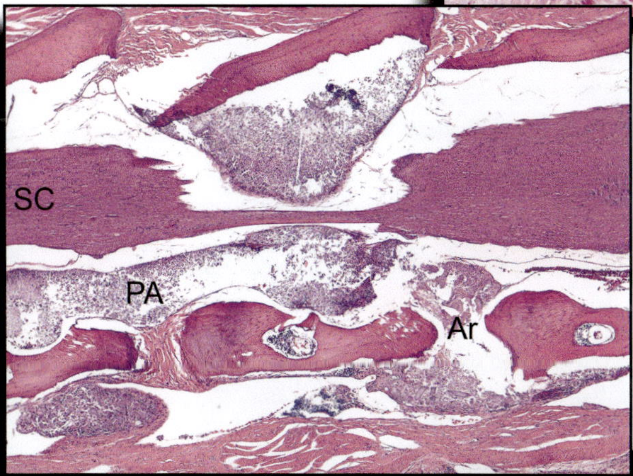

Figure 3.29 Western Hognose Snake, *Heterodon nasicus*. Colubridae. Photomicrograph of articular (Ar) and periarticular (PA) gout of the spinal column. Urate tophi extend from articular spaces between adjacent vertebral bodies into the spinal canal and into the surrounding connective tissue. SC = spinal cord. Hematoxylin and eosin.

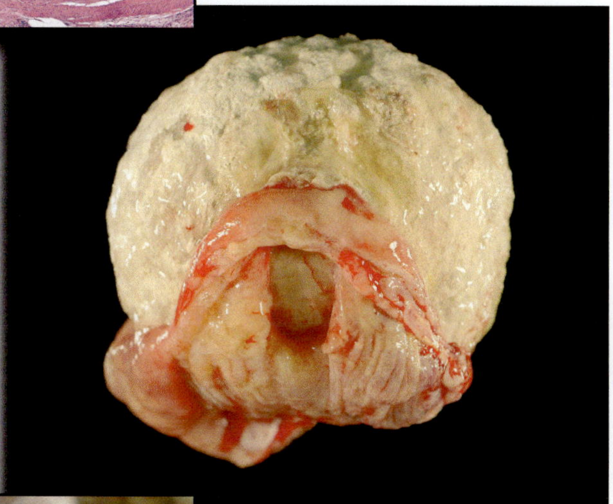

Figure 3.30 Desert Tortoise, *Gopherus agassizii*. Testudinidae. A large urolith is present within the lumen of the urinary bladder. (Courtesy of Disease Investigations, San Diego Zoo Global.)

Figure 3.31 Desert Tortoise, *Gopherus agassizii*. Testudinidae. Uroliths and granular urates are visible through the wall of one lobe of the urinary bladder. (Courtesy of Disease Investigations, San Diego Zoo Global.)

Figure 3.32 Sulcata Tortoise, *Centrochelys sulcate*. Testudinidae. Urolithiasis. A large urolith is present within the lumen of the urinary bladder (a). The wall of the urinary bladder is diffusely thickened with prominent vasculature (a and b). A photomicrograph of the urinary bladder wall shows a band of granulocytic inflammation, cellular debris, and fibrin along the surface of the exposed submucosa with complete mucosal ulceration. Hematoxylin and eosin (c).

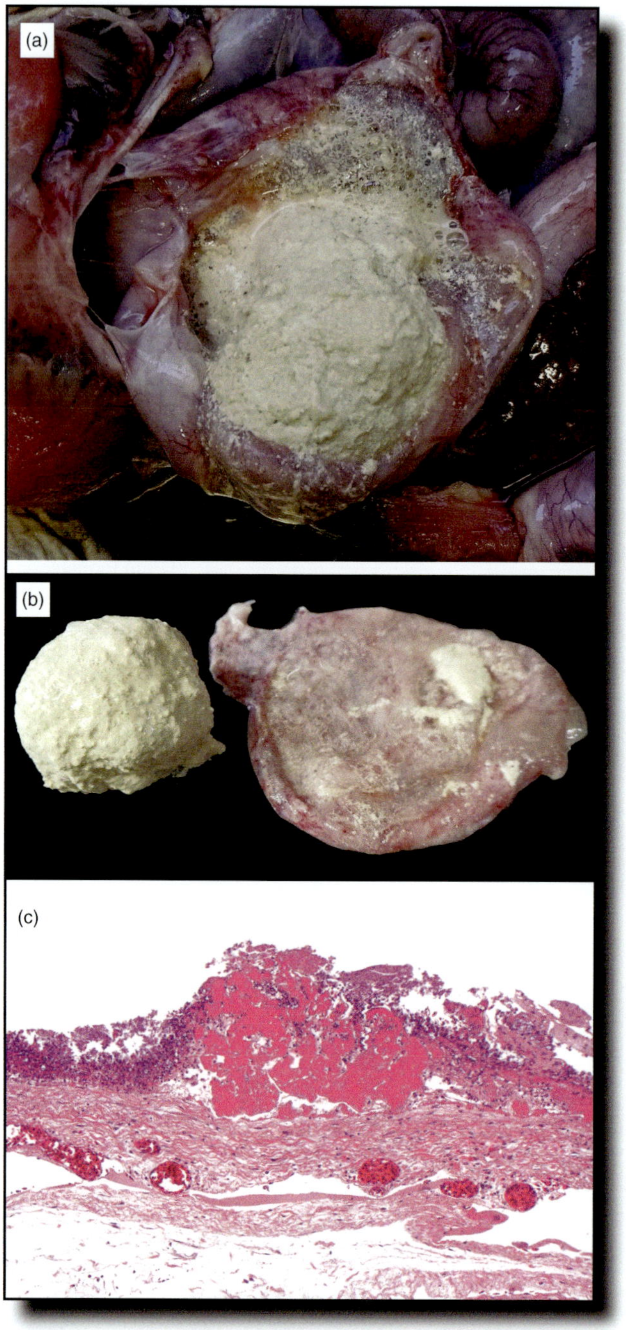

Figure 3.33 Fiji Banded Iguana, *Brachylophus fasciatus*. Iguanidae. A urolith is visible through the wall of the urinary bladder. (Courtesy of Disease Investigations, San Diego Zoo Global.)

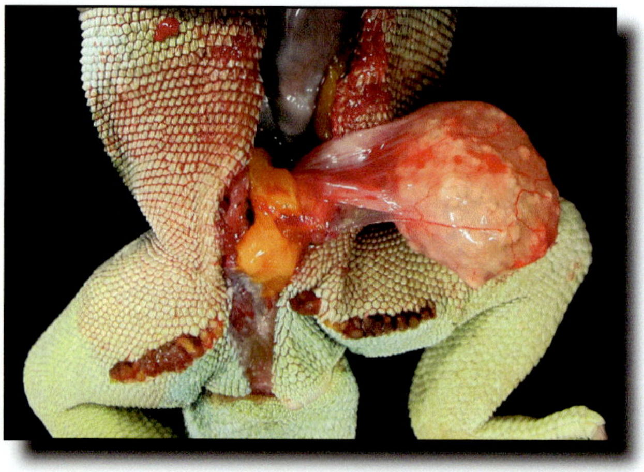

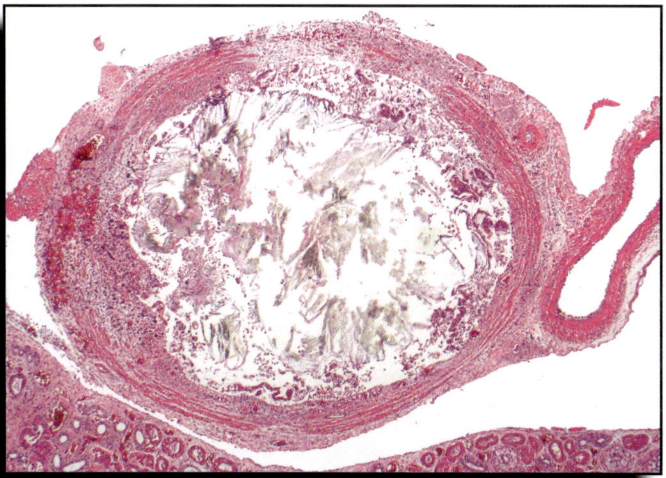

Figure 3.34 Tiger Rat Snake, *Spilotes pullatus*. Colubridae. A photomicrograph of a ureterolith associated with pressure necrosis and inflammation of the ureteral wall. Note the translucent, refractile material of the ureterolith and the surrounding cellular and proteinaceous debris and inflammation. Hematoxylin and eosin. (Courtesy of Michael Garner.)

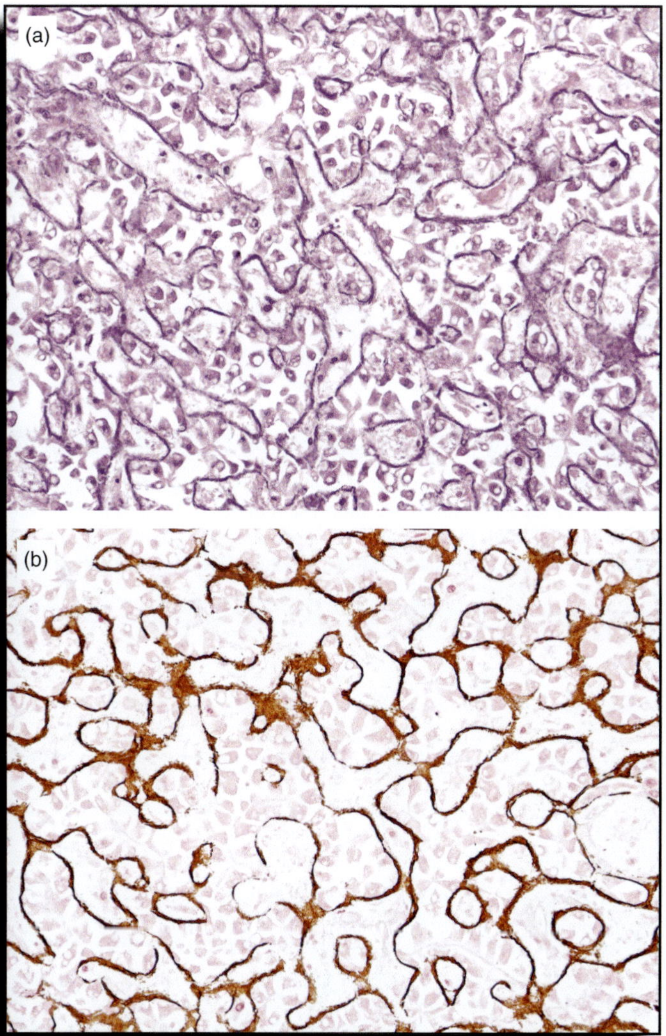

Figure 3.35 Blacktail Cribo, *Drymarchon melanurus*. Colubridae. A photomicrograph of generalized perisinusoidal basement membrane mineralization. Note the highlighting of the sinusoidal space by bands of hyperbasophilic material (a). Hematoxylin and eosin. The presence of calcium is highlighted by brown-black staining (b). Von Kossa. (Courtesy of Brian Stacy.)

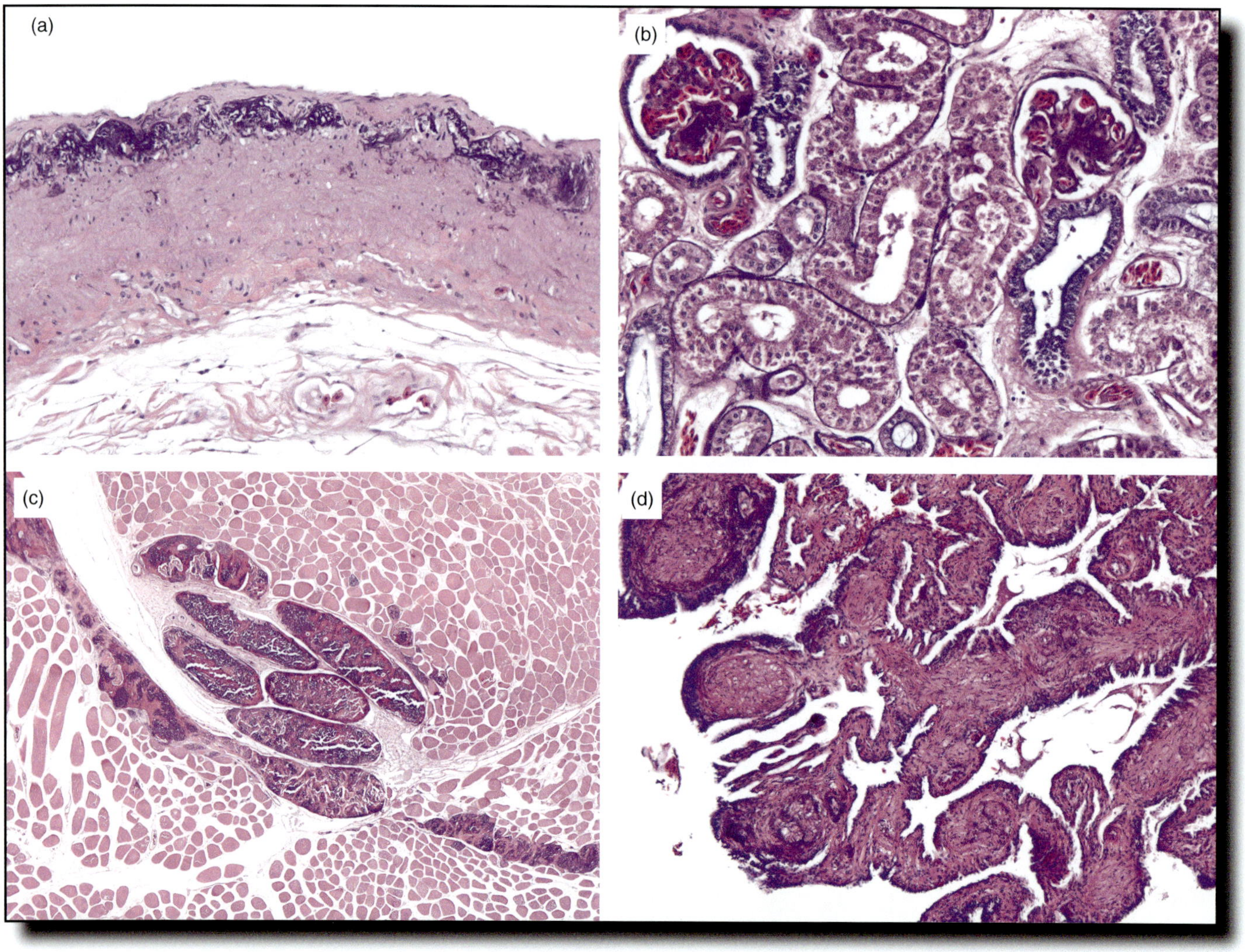

Figure 3.36 Black-Tailed Rattlesnake, *Crotalus molossus*. Viperidae. Photomicrographs of systemic metastatic mineralization of undetermined etiology. Mineral disrupts the subintima and media of a large elastic artery (a). Deposition of basophilic mineral is present in the basement membrane of renal tubules as well as the mesangium and basement membranes of glomeruli (b). Scattered, individual myofibers are swollen and hypereosinophilic to basophilic, progressing to ossification (myodegeneration and mineralization; c). Similar mineralization is prominent within the pulmonary interstitium and deep to the respiratory epithelium (d). Hematoxylin and eosin.

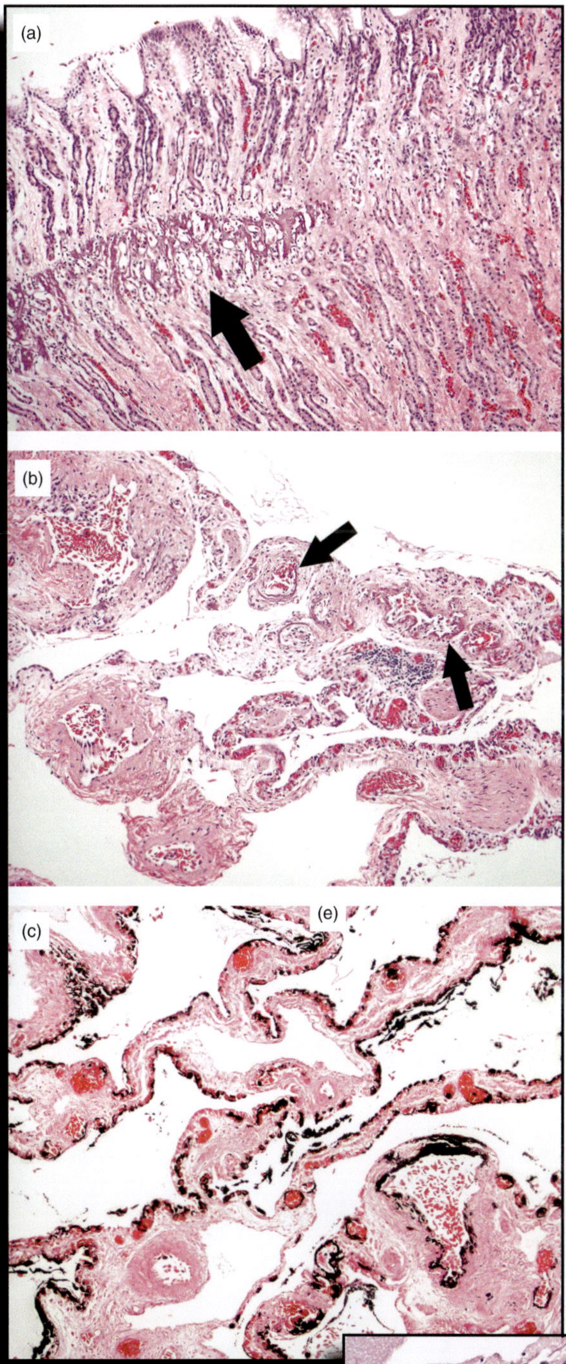

Figure 3.37 Gharial, *Gavialis gangeticus*. Gavialidae. Photomicrographs of metastatic mineralization. A mid-mucosal band of amphophilic to basophilic mineral disrupts the gastric mucosa (arrow; a). Similar mineral is present within the pulmonary interstitium and the wall of vessels (arrows; b). Hematoxylin and eosin. The interstitial and vascular mineral deposition is highlighted by positive (black) Von Kossa histochemical reactivity (c). (Courtesy of Disease Investigations, San Diego Zoo Global.)

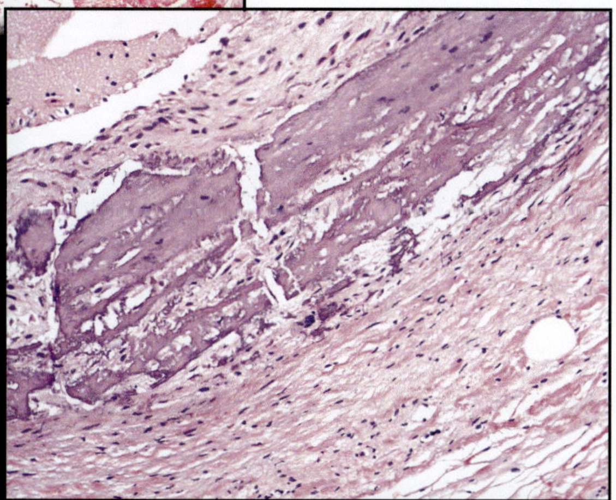

Figure 3.38 Anegada Ground Iguana, *Cyclura pinguis*. Iguanidae. Photomicrograph of arteriosclerosis. The tunica media is disrupted and expanded by lamellar and granular mineral deposition. Hematoxylin and eosin. (Courtesy of Disease Investigations, San Diego Zoo Global.)

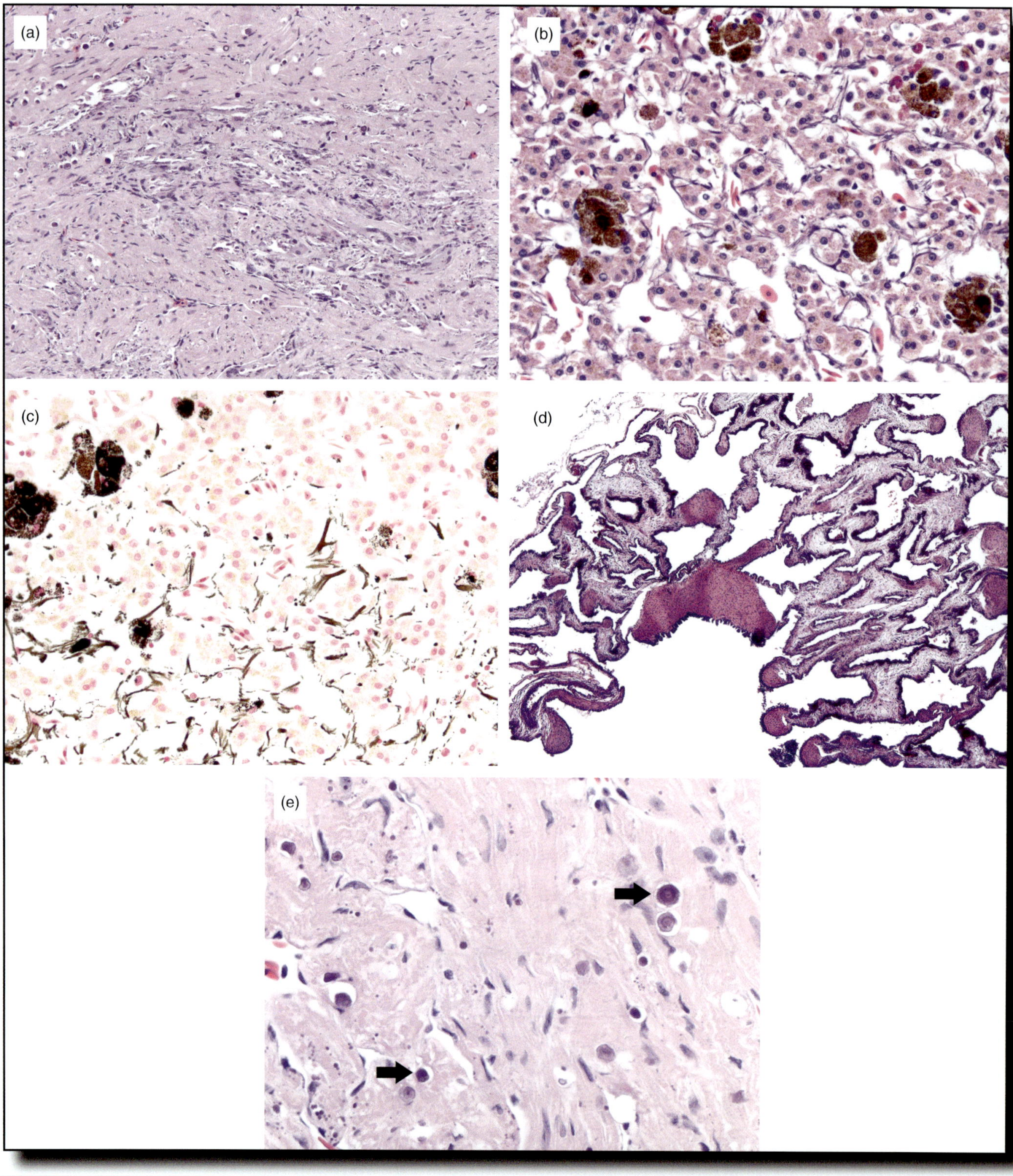

Figure 3.39 Diamondback Terrapin, *Malaclemys terrapin*. Emydidae. Photomicrographs of systemic metastatic mineralization of undetermined etiology. The myocardium is disrupted by the deposition of granular basophilic material and granulomatous inflammation with associated myocardial degeneration and necrosis. Hematoxylin and eosin (a). Linear bands of basophilic material are present in hepatic basement membrane, outlining sinusoids. Hematoxylin and eosin (b). Perisinusoidal mineral deposits exhibit positive Von Kossa histochemical reactivity; note the pigmented macrophage centers that also retain their black staining with this method (c). Dense bands of hyperbasophilic mineral are present throughout the interstitium, most commonly just deep to the respiratory epithelium and faveolar capillaries along basement membranes (d). Note the small round foci of lamellated mineral (psammoma bodies; black arrows) present in this section of myocardium (e).

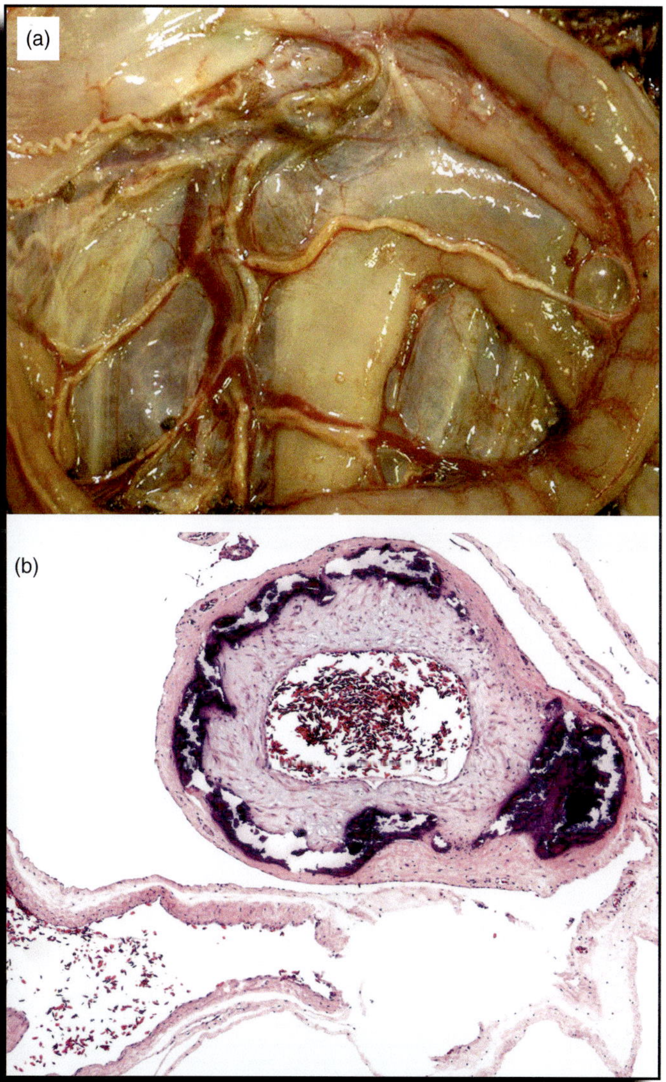

Figure 3.40 Desert Tortoise, *Gopherus agassizii*. Testudinidae. Mesenteric vessels are thickened, prominent, and tan white (a). Photomicrograph of basophilic aggregates of mineral within the wall of a mesenteric artery (b). (Courtesy of John Roberts.)

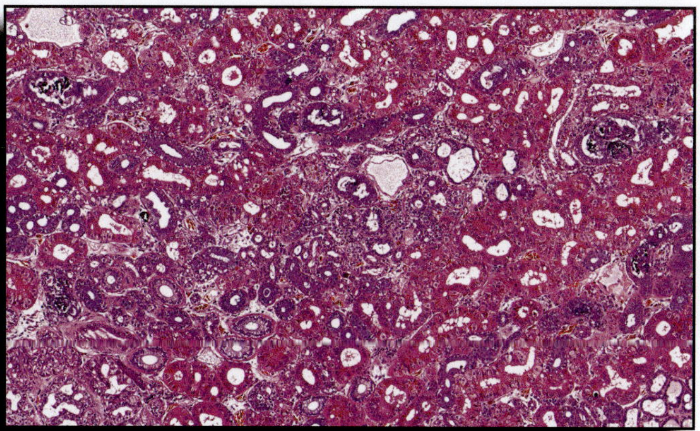

Figure 3.41 Corn Snake, *Pantherophis guttatus*. Colubridae. Nephrocalcinosis. Aggregates of hyperbasophilic mineral are present within the lumen of renal tubules, renal tubular epithelial cells, and the renal interstitium. There is associated renal tubular ectasia and concurrent mild interstitial nephritis and interstitial fibrosis. Hematoxylin and eosin.

Figure 3.42 Tokay Gecko (*Gekko gecko*). Gekkonidae. The left endolymphatic sac is markedly more distended than the right, due to presumed excess calcium retention. (Courtesy of Elliot Jacobson.)

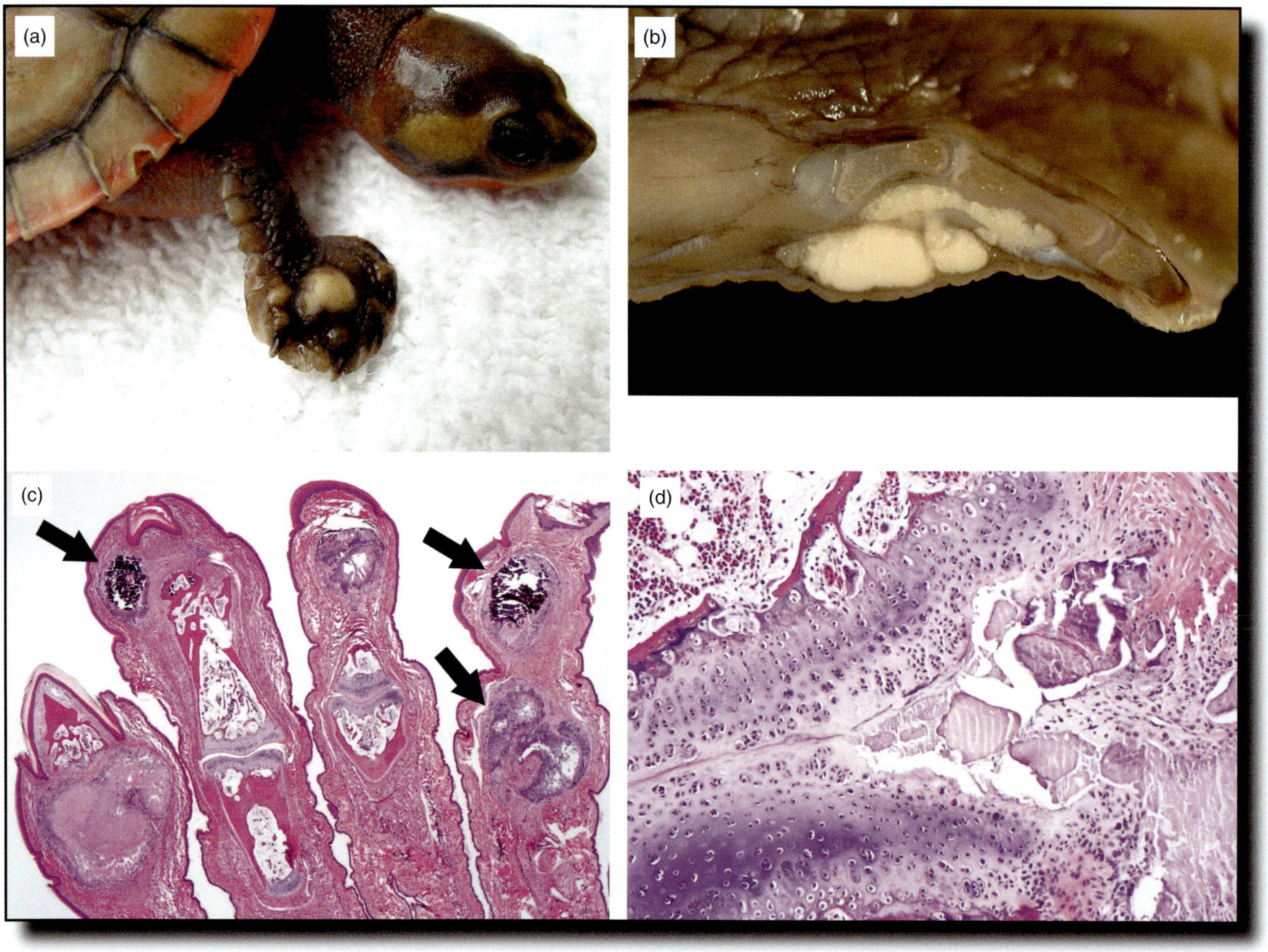

Figure 3.43 Red-Bellied Short-Necked Turtle, *Emydura subglobosa.* Chelidae. Nodular articular and periarticular mineral deposition. Firm nodular swellings are associated with the joints of the foot (a). On section, white, pasty to chalky material is present in the periarticular tissue (b). Photomicrographs of affected digits demonstrate the accumulation of intraarticular and periarticular mineral (partially decalcified) that is associated with bone and cartilage loss. Hematoxylin and eosin (c and d). (Courtesy of Vet Services and Disease Investigations, San Diego Zoo Global.)

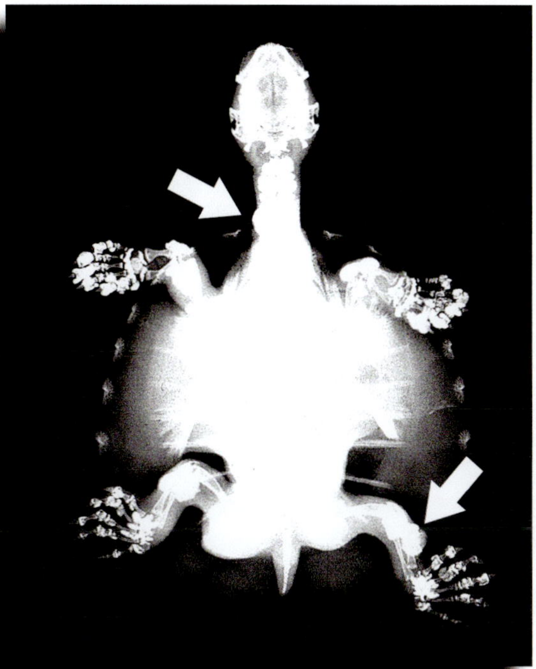

Figure 3.44 Red-Bellied Short-Necked Turtle, *Emydura subglobosa*. Chelidae. Radiograph demonstrating the accumulation of radiopaque mineral, most prominently in the cervical and stifle regions (arrows). (Courtesy of Elliot Jacobson.)

Figure 3.45 Veiled Chameleon, *Chamaeleo calyptratus*. Chamaeleonidae. Nodular accumulations of periarticular mineral (white arrows) are present in the distal limbs and the cervical region. (Courtesy of John Roberts.)

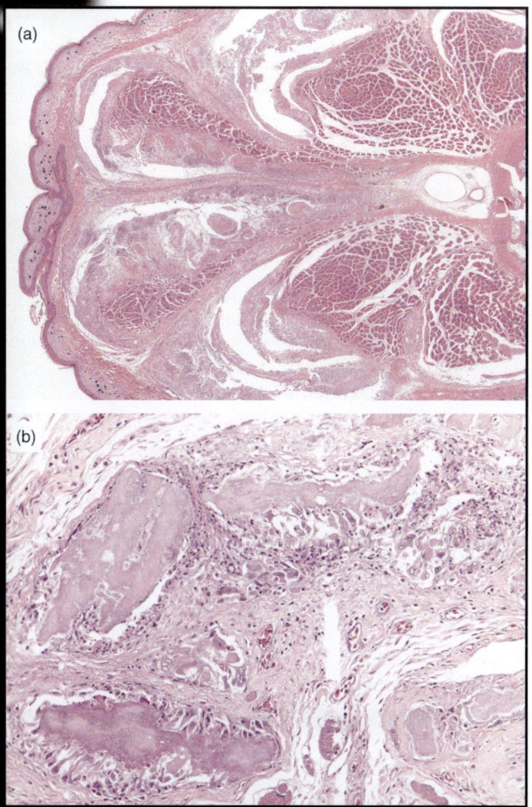

Figure 3.46 Oustalet's Chameleon, *Furcifer oustaleti*. Chamaeleonidae. Photomicrographs of mineral accumulation within the muscle bodies and tendon sheaths of the ventral portion of the prehensile tail. Abundant basophilic mineral and associated granulomatous inflammation disrupts the ventral muscle bundles (a). Aggregates of amphophilic to basophilic mineral are surrounded by epithelioid macrophages and multinucleated giant cells (b). Hematoxylin and eosin.

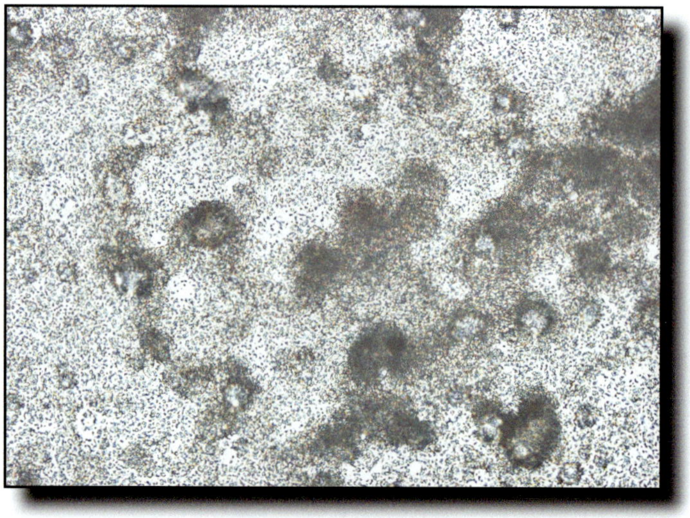

Figure 3.47 Red-Bellied Short-Necked Turtle, *Emydura subglobosa*. Chelidae. Photomicrograph of an unstained cytologic preparation of periarticular mineral, characterized by granular to coarsely clumped, acellular, refractile material. (Courtesy of Disease Investigations, San Diego Zoo Global.)

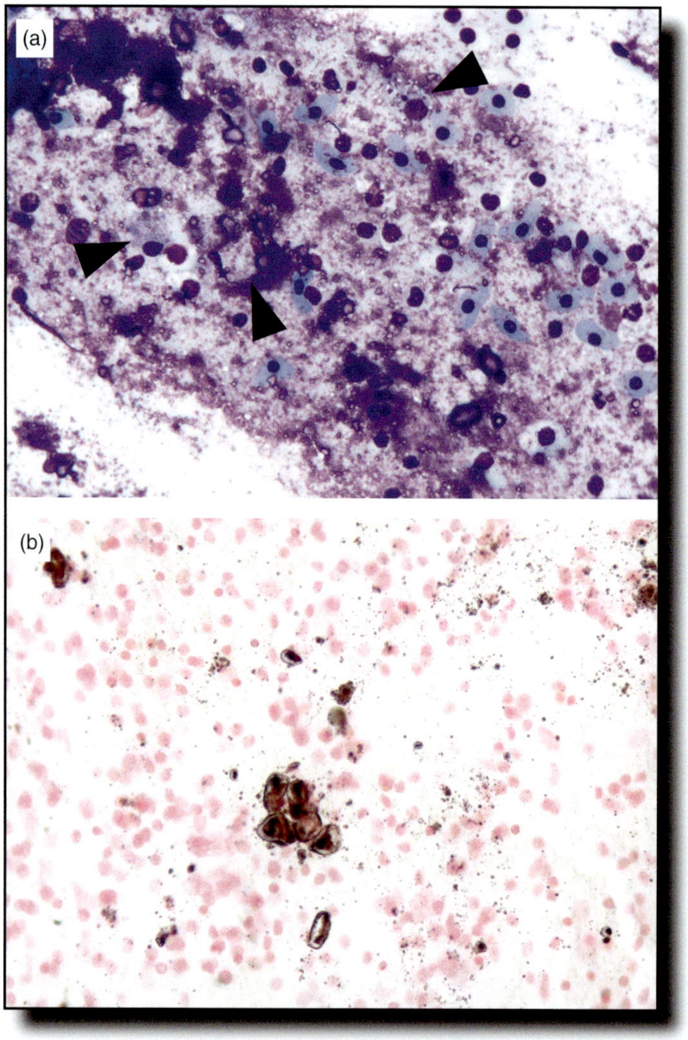

Figure 3.48 Chinese Box Turtle, *Cuora flavomarginata*. Geomydidae. Photomicrograph of tissue imprints of a soft tissue mass close to the right knee composed of mild histiocytic inflammation (arrowheads) and abundant variably sized, pale pink, refractile, pleomorphic material suggestive of calcium deposition. Wright Giemsa (a). The refractile material exhibits positive cytochemical reactivity with Von Kossa staining, consistent with calcium (b). (Courtesy of Nicole Stacy.)

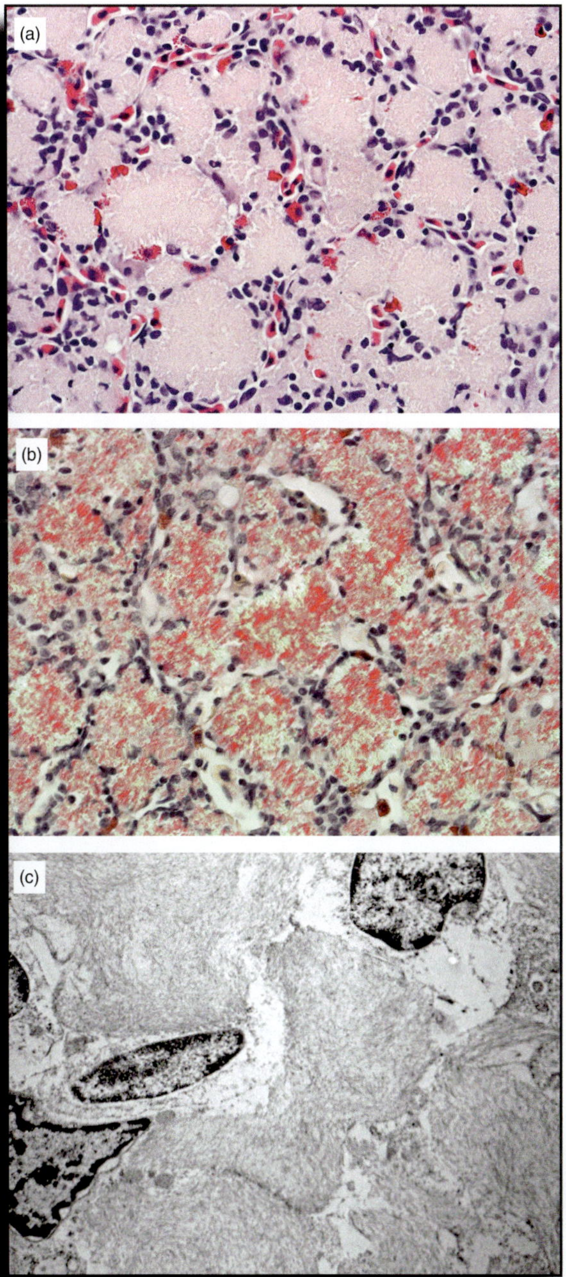

Figure 3.49 African Tiger Snake, *Telescopus semiannulatus*. Colubridae. Splenic amyloidosis. Photomicrograph of splenic deposits of amorphous, extracellular, eosinophilic material expanding the splenic parenchyma. Hematoxylin and eosin (a). Photomicrograph of the congophilic nature of the material and apple-green birefringence under polarized light. Congo red (b). (Courtesy of Disease Investigations, San Diego Zoo Global.) Transmission electron photomicrograph extracellular splenic deposits composed of haphazardly arranged, non-branching, deposits of 4–8 nm wide fibrils of indeterminate length, consistent with amyloid. Uranyl acetate and lead citrate stain (c). (Courtesy of Anibal Armien.)

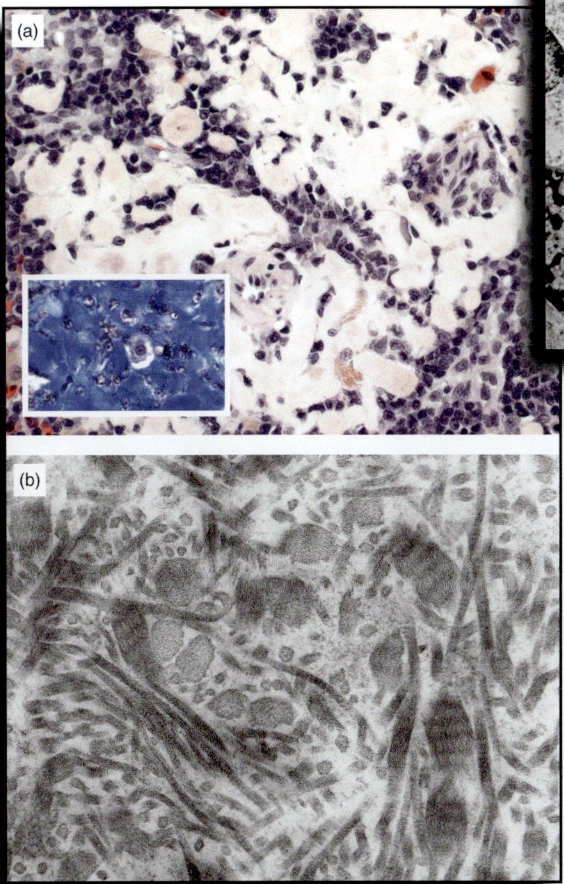

Figure 3.50 Green Anaconda, *Eunectes murinus*. Boidae. Photomicrographs of atypical splenic collagen deposition (amyloid-like appearance). The spleen contains coalescing deposits of homogenous eosinophilic material. Hematoxylin and eosin (a). The deposits stain intensely blue with Masson's trichrome stain, which is consistent with collagen (a, inset) (Courtesy of Brian Stacy.) Transmission electron microscopy reveals that the deposits are composed of atypical collagen fibers, and inconsistent with amyloid. Uranyl acetate and lead citrate stain (b). (Courtesy of Dalen Agnew.)

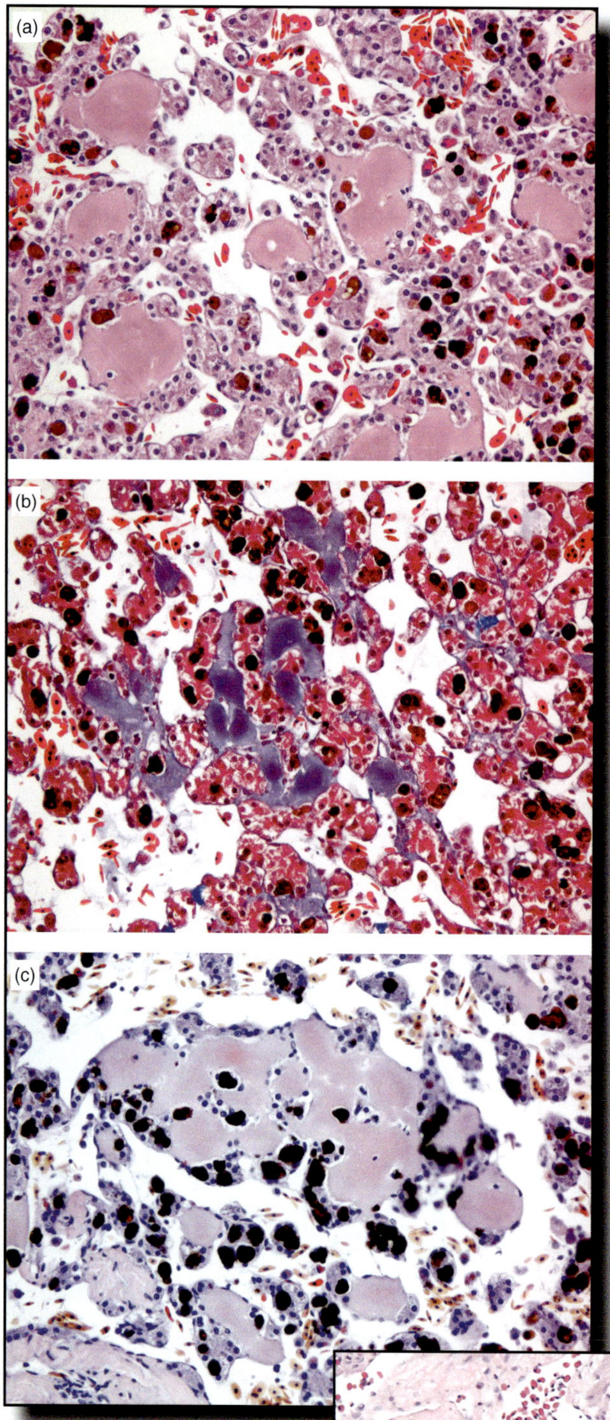

Figure 3.51 American Alligator, *Alligator mississippiensis*. Alligatoridae. Photomicrographs of atypical hepatic collagen deposition (amyloid-like appearance). Hepatic perisinusoidal spaces contain multifocal deposits of homogenous eosinophilic material. Hematoxylin and eosin (a). The deposits stain intensely blue with Masson's trichrome stain (b) but diffusely negative with Congo red staining (c). These results are consistent with collagen. (Courtesy of Brian Stacy.)

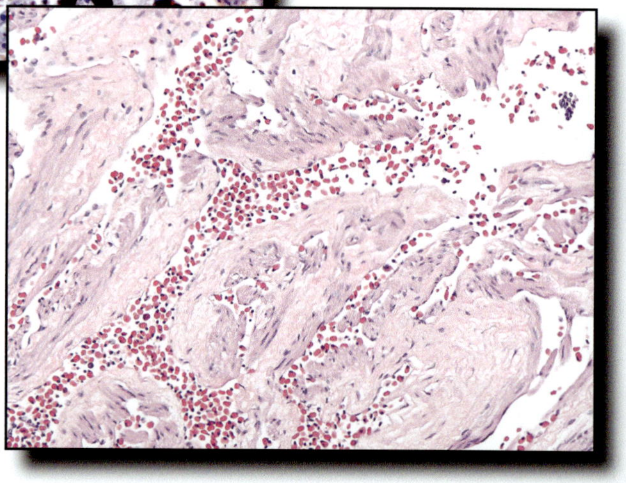

Figure 3.52 Pig-Nosed Turtle, *Carettochelys insculpta*. Carettochelyidae. Photomicrograph of atypical atrial subendocardial deposition of fibrillar, pale-eosinophilic material, most consistent with collagen. Hematoxylin and eosin.

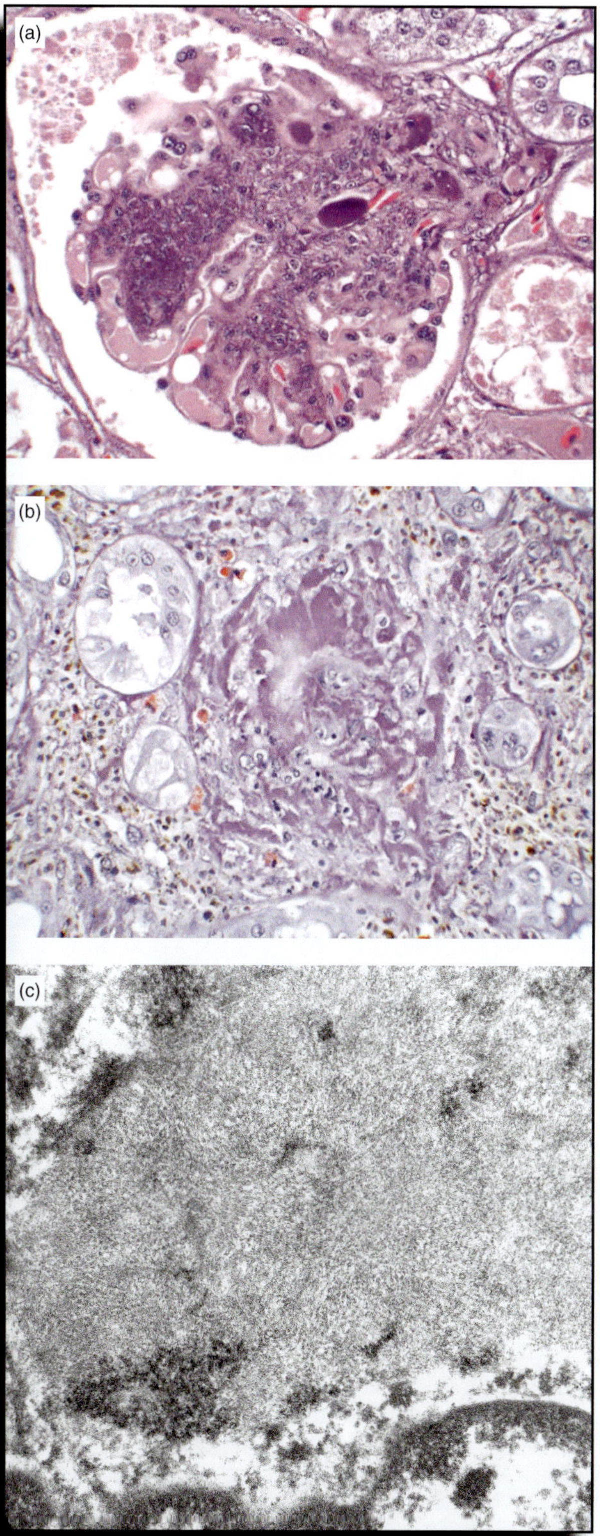

Figure 3.53 Komodo Dragon, *Varanus komodoensis*. Glomerular paramyloidosis and yolk embolism. The glomerular mesangium is expanded by abundant amphophilic to basophilic material. Note the presence of aggregates of basophilic material within glomerular vessels, consistent with yolk embolization. Hematoxylin and eosin (a). The deposited mesangial material is non-congophilic. Congo Red (b). Transmission electron microscopy reveals that the deposits are composed of fibrillar to amorphous material, inconsistent with amyloid. Uranyl acetate and lead citrate stain (c). (Courtesy of Michael Garner.)

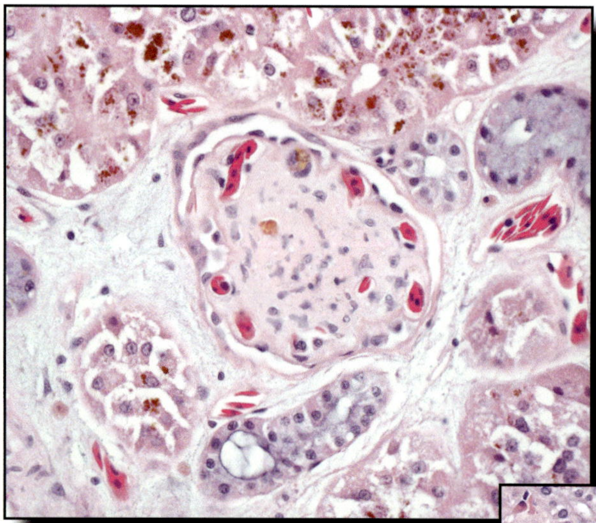

Figure 3.54 Boa Constrictor, *Boa constrictor.* Boidae. Photomicrograph of the accumulation of abundant, pale eosinophilic material within glomerular mesangium that also expands the basement membranes of glomerular capillaries. The mesangium is also hypercellular. Note the presence of pigmented macrophages within the glomerular tuft as well as the presence of abundant cytoplasmic pigment in surrounding tubular epithelial cells. Hematoxylin and eosin. (Courtesy of Disease Investigations, San Diego Zoo Global.)

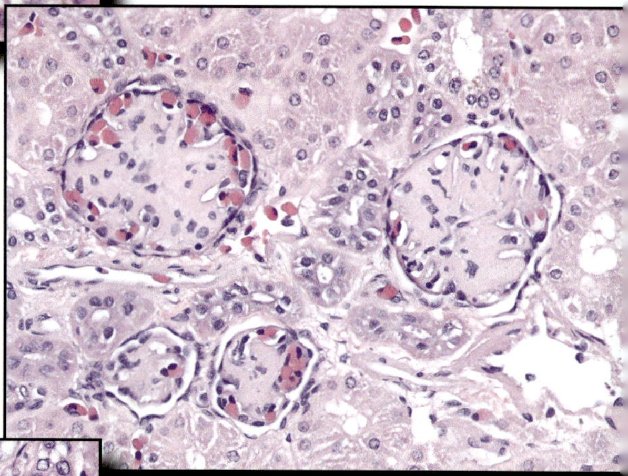

Figure 3.55 Pig-Nosed Turtle, *Carettochelys insculpta*. Carettochelyidae. Photomicrograph of glomerular mesangial expansion by abundant pale, eosinophilic material. Hematoxylin and eosin.

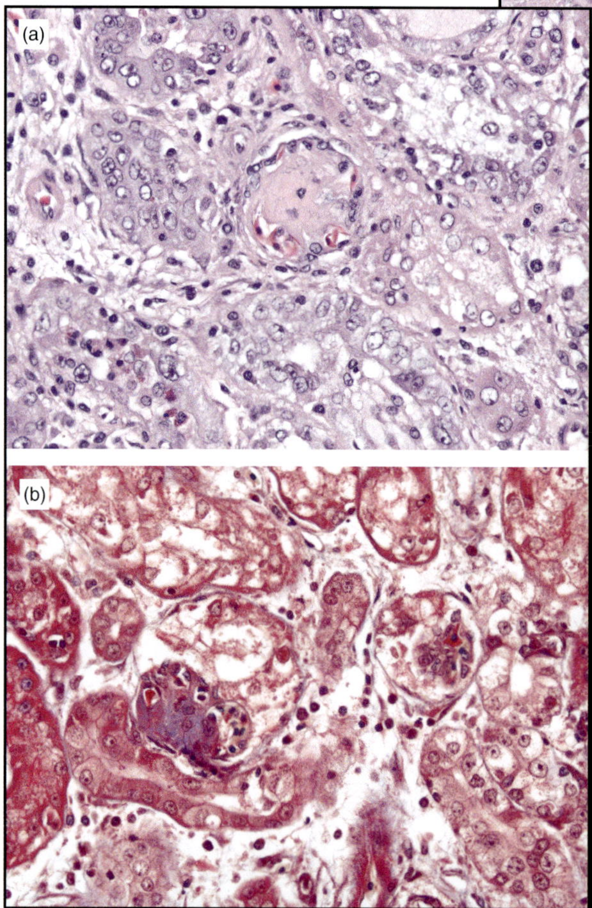

Figure 3.56 Indian Black Turtle, *Melanochelys trijuga*. Geomydidae. Photomicrographs of glomerulosclerosis. Abundant pale, eosinophilic material expands the mesangium of a sclerotic glomerulus. Hematoxylin and eosin (a). The mesangial material exhibits positive blue cytochemical reactivity, consistent with collagen. Note the unaffected glomerulus present in the photomicrograph with minimal appreciable positive staining. Masson's Trichrome (b).

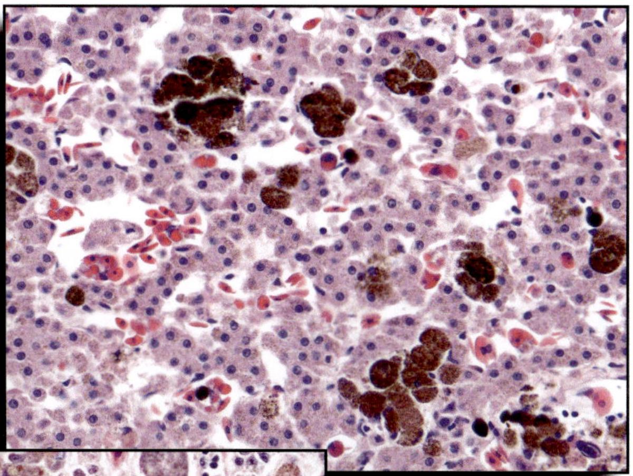

Figure 3.57 Diamondback Terrapin, *Malaclemys terrapin*. Emydidae. Photomicrograph of pigmented macrophage center hyperplasia. Pigmented macrophage aggregates are in close relative apposition to neighboring centers and are composed of 2–12 cells. Also note the diminutive appearance of surrounding hepatocytes (consistent with hepatocellular atrophy) and the deposition of abundant, yellow-brown cytoplasmic pigment. Hematoxylin and eosin.

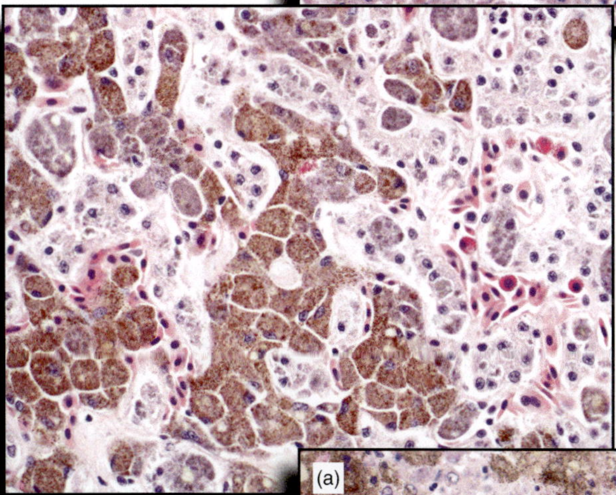

Figure 3.58 Chinese Crocodile Lizard, *Shinisaurus crocodilurus*. Shinisauridae. Photomicrograph of pigmented macrophage center hyperplasia. Pigmented macrophage centers are hyperplastic and coalesce. Note the presence of brown, brown-gray, and yellow-brown pigment within pigmented macrophages, likely representative of melanin, hemosiderin, and lipofuscin. Also note the accumulation of brown-gray cytoplasmic pigment within surrounding hepatocytes. Hematoxylin and eosin. (Courtesy of Disease Investigations, San Diego Zoo Global.)

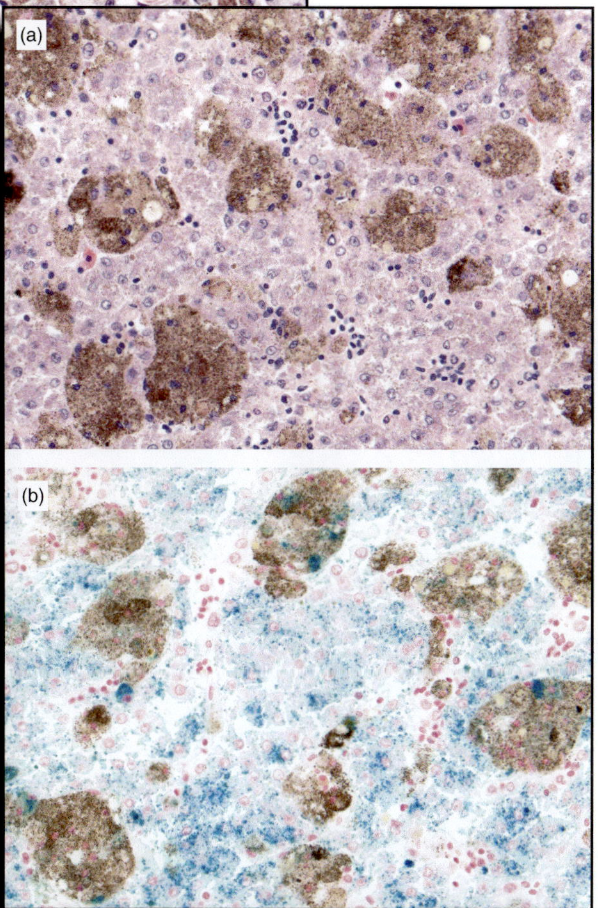

Figure 3.59 Bog Turtle, *Glyptemys muhlenbergii*. Emydidae. Photomicrographs of pigmented macrophage center hyperplasia and hepatocellular iron deposition. Pigmented macrophage centers are hyperplastic and coalesce, and contain macrophages swollen with brown, yellow-brown, and translucent pale yellow pigment, likely representative of melanin, hemosiderin, and lipofuscin/ceroid. Hepatocytes also contain abundant yellow-brown cytoplasmic pigment. Hematoxylin and eosin (a). Hepatocellular pigment and a subset of the pigmented macrophage pigment exhibits strong positive Prussian blue reactivity, consistent with iron (b).

Figure 3.60 Henkel's Leaf-Tailed Gecko, *Uroplatus henkeli*. Gekkonidae. Photomicrographs of iron deposition in small intestinal enterocytes. Note the presence of abundant, brown-yellow, globular cytoplasmic pigment in many enterocytes (a). The pigment exhibits strong positive Prussian Blue reactivity, consistent with iron (b). (Courtesy of Michael Garner.)

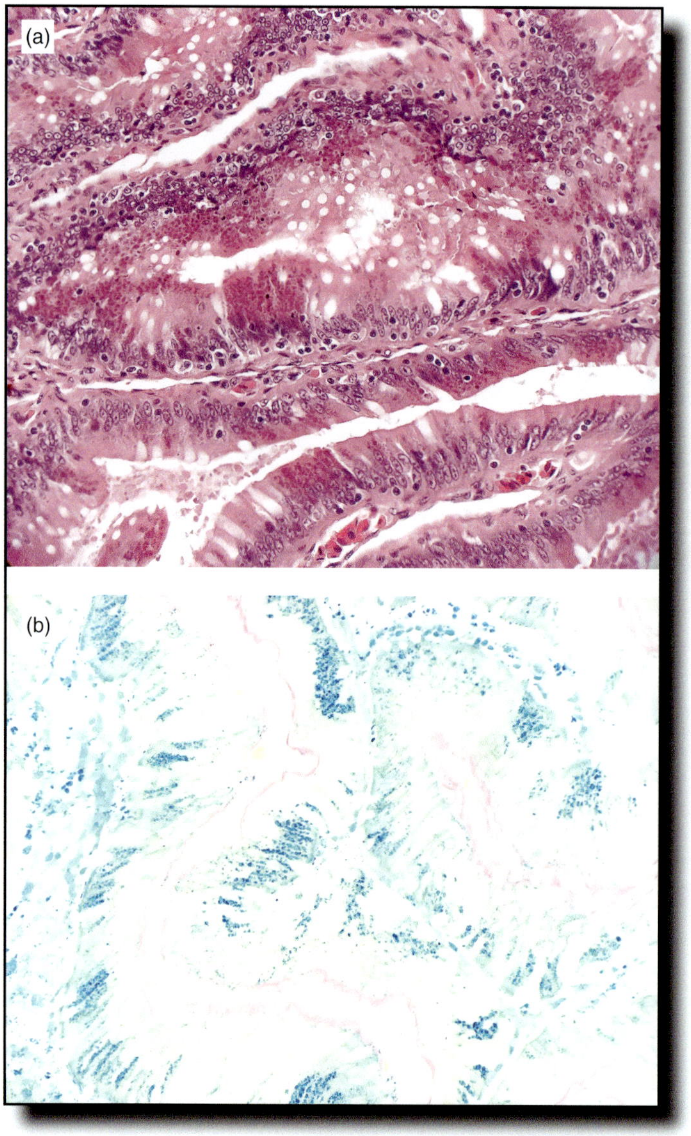

Figure 3.61 Gharial, *Gavialis gangeticus*. Gavialidae. Photomicrograph of splenic hemosiderosis and erythrophagocytosis. Note the presence of free and intracellular brown-yellow pigment, consistent with hemosiderin and the phagocytosis and degradation of erythrocytes by macrophages. Hematoxylin and eosin. (Courtesy of Disease Investigations, San Diego Zoo Global.)

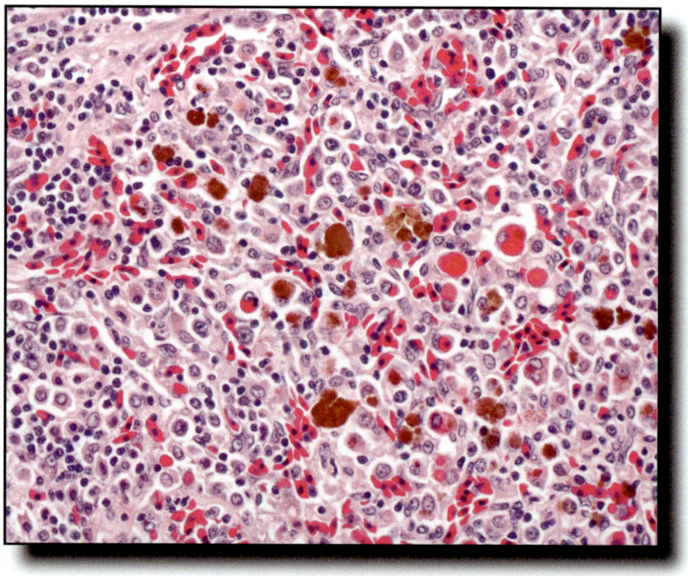

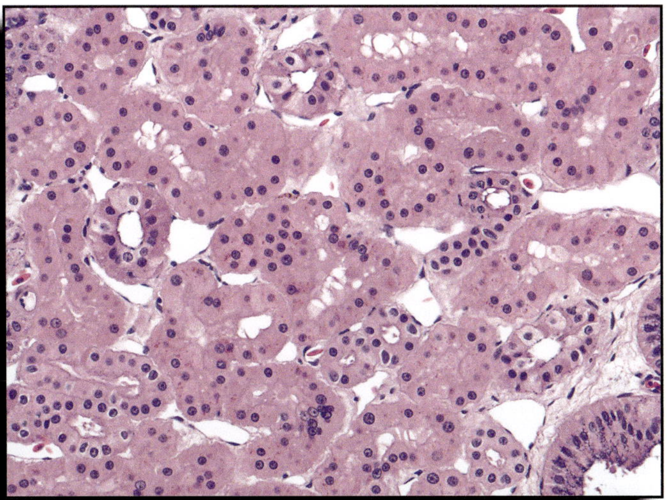

Figure 3.62 Chinese Water Dragon, *Physignathus cocincinus*. Agamidae. Photomicrograph of mild renal tubular epithelial cell pigment accumulation. Note the presence of small amounts of granular, brown-yellow pigment within the cytoplasm of tubular epithelial cells. Hematoxylin and eosin.

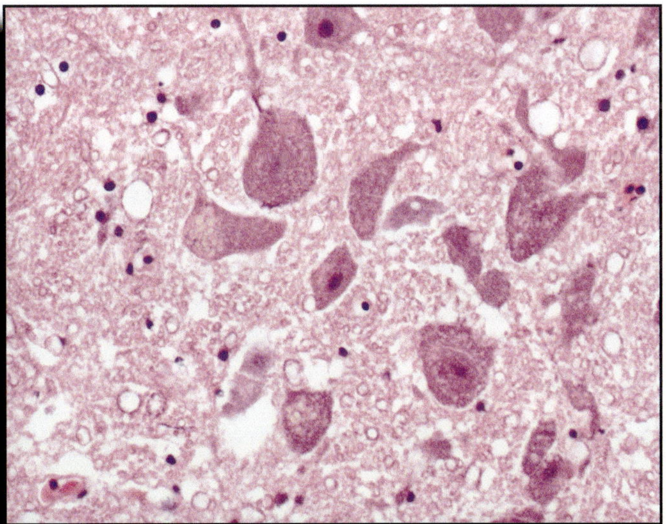

Figure 3.63 Rat Snake, unknown species. Colubridae. Photomicrograph of neuronal lipofuscinosis. Note the accumulation of pale yellow, slightly translucent material, consistent with lipofuscin, within the cytoplasm of neuronal cell bodies. Hematoxylin and eosin. (Courtesy of Michael Garner.)

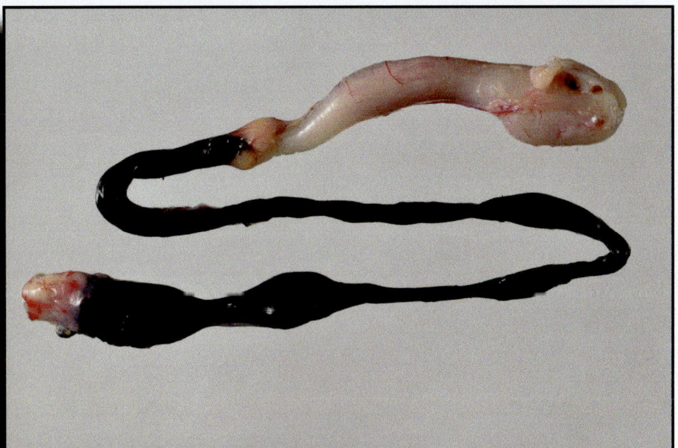

Figure 3.64 Henkel's Leaf-Tailed Gecko, *Uroplatus henkeli*. Gekkonidae. Pigmentation of the intestinal serosa. The presence of melanocytes is a common finding in the serosal lining of many coelomic tissues in some reptiles, particularly lizards. (Courtesy of Disease Investigations, San Diego Zoo Global.)

Figure 3.65 Radiated Tortoise, *Astrochelys radiata*. Testudinidae. Hepatic lipidosis. Note the diffuse pale yellow-white appearance to the hepatic parenchyma and the presence of milky-white fluid on cut surface. (Courtesy of Randy Junge.)

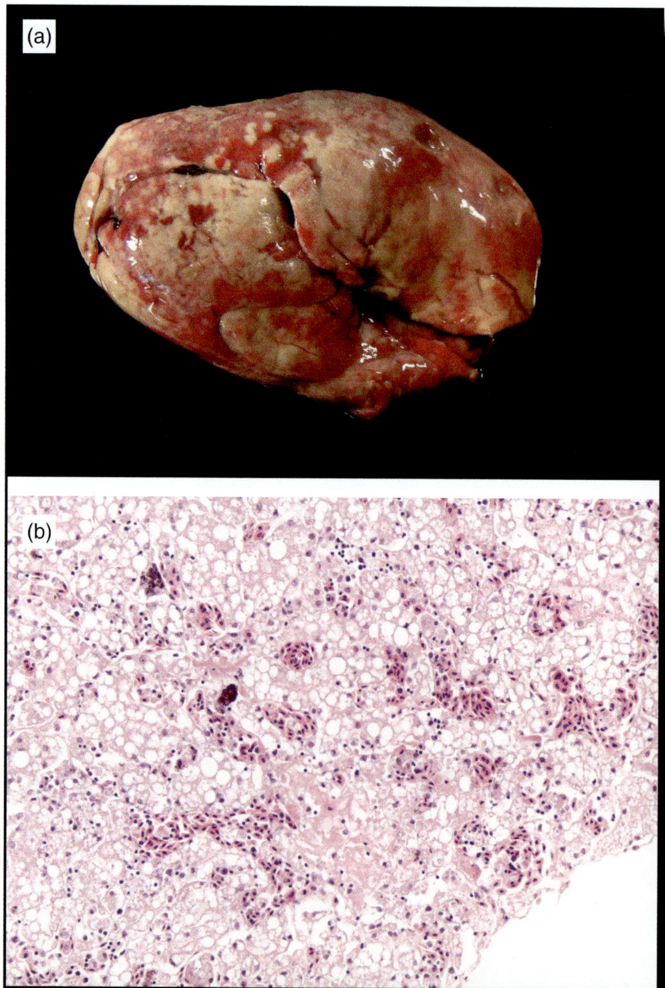

Figure 3.66 Horned Lizard, *Phrynosoma* species. Phrynosomatidae. Hepatic lipidosis and necrosis. The hepatic parenchyma is mottled yellow-tan to red-pink with multiple foci of hemorrhage and congestion (a). A photomicrograph of the hepatic parenchyma demonstrates the accumulation of colorless, well-demarcated hepatocellular vacuoles (consistent with lipid accumulation) of variable size. Note the fusion of lipid vacuoles within and between adjacent hepatocytes as well as foci of necrosis characterized by loss of cellular detail and lakes of eosinophilic material. Hematoxylin and eosin (b).

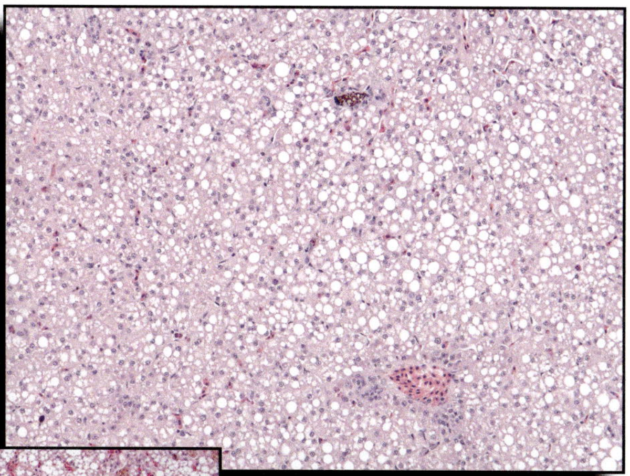

Figure 3.67 Chinese Water Dragon, *Physignathus cocincinus*. Agamidae. Photomicrograph of inconsistent hepatocellular lipid accumulation. Note the variable appearance of hepatocellular lipid accumulation and the fusion of lipid vacuoles within and between adjacent hepatocytes. Though the lipid accumulation in this photomicrograph is abnormal, it is not yet associated with hepatocellular degeneration and necrosis. Hematoxylin and eosin.

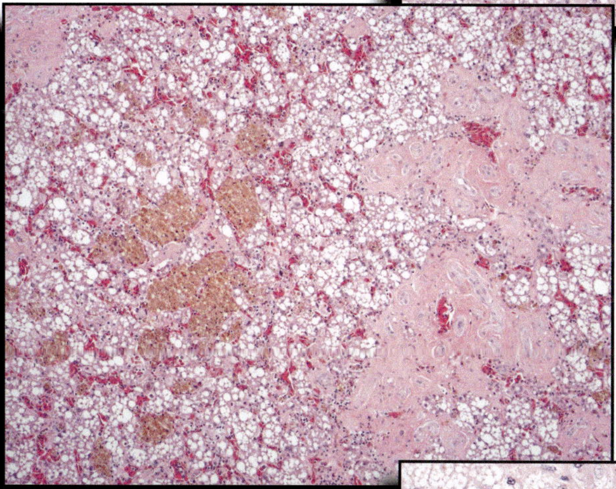

Figure 3.68 Boa Constrictor, *Boa constrictor.* Boidae. Photomicrograph of chronic hepatic lipidosis associated with parenchymal loss and fibrosis. Note the irregular size and distribution of hepatocellular lipid vacuoles and the presence of large areas of pale eosinophilic connective tissue, consistent with fibrosis, that disrupt the hepatic architecture. Also note the presence of large, coalescing pigmented macrophage centers, consistent with hyperplasia and hypertrophy. Hematoxylin and eosin. (Courtesy of Disease Investigations, San Diego Zoo Global.)

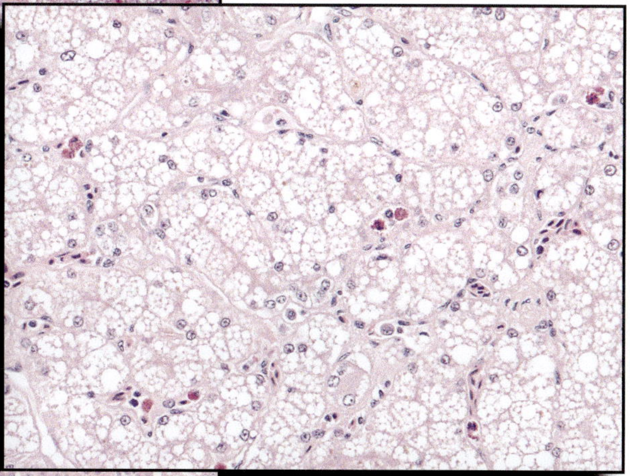

Figure 3.69 Black-Tailed Monitor, *Varanus tristis*. Varanidae. Photomicrograph of hepatocellular lipid accumulation. Note the presence of abundant round, well demarcated lipid vacuoles in hepatocytes in the absence of extensive vacuolar fusion or hepatocellular degeneration. Hematoxylin and eosin.

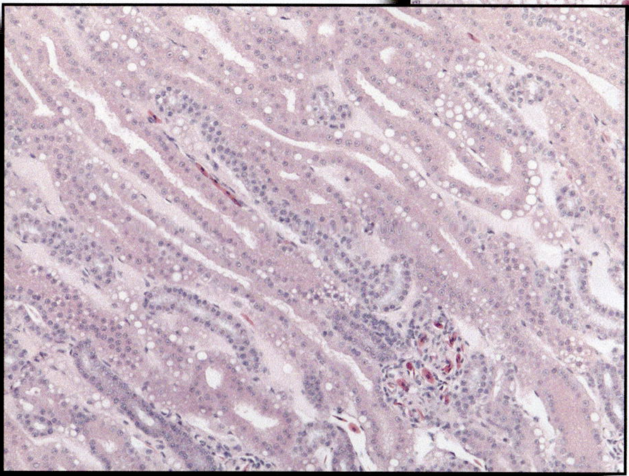

Figure 3.70 Rhinoceros Viper, *Bitis nasicornis.* Viperidae. Photomicrograph of lipid accumulation in renal tubular epithelial cells. Note the presence of well-demarcated, colorless vacuoles, consistent with lipid, in a subset of renal tubular epithelial cells. Also, note the presence of pale, yellow-tan, apical cytoplasmic pigment present in renal tubular epithelial cells (presumptive lipofuscin). Hematoxylin and eosin.

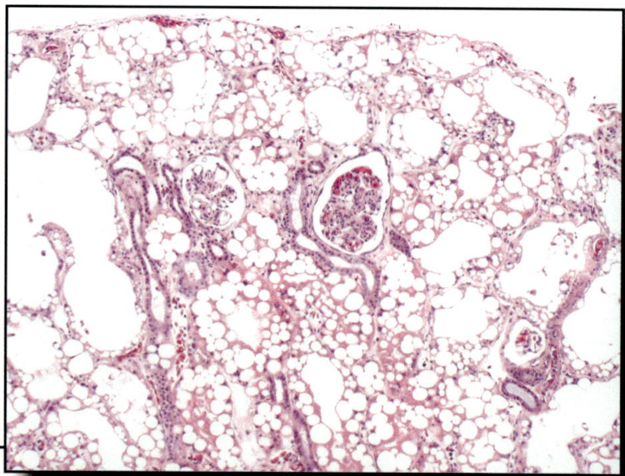

Figure 3.71 Henkel's Leaf-Tailed Gecko, *Uroplatus henkeli*. Gekkonidae. Photomicrograph of renal lipidosis. Large, coalescing cytoplasmic lipid vacuoles expand renal tubular epithelial cells and are multifocally associated with tubular degeneration and necrosis. Note the presence of ectatic tubules lined by attenuated and vacuolated tubular epithelial cells. (Courtesy of Michael Garner.)

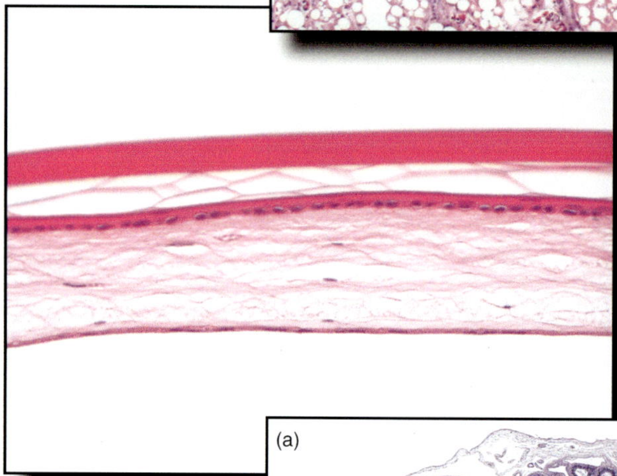

Figure 3.72 Corn Snake, *Pantherophis guttatus*. Colubridae. Spectacular xanthomatosis. Note the presence of colorless, acicular clefts within the stroma of the spectacle, consistent with cholesterol deposition. Cholesterol crystals dissolve during routine processing and leave the colorless clefts visible histologically. (Courtesy of Michael Garner.)

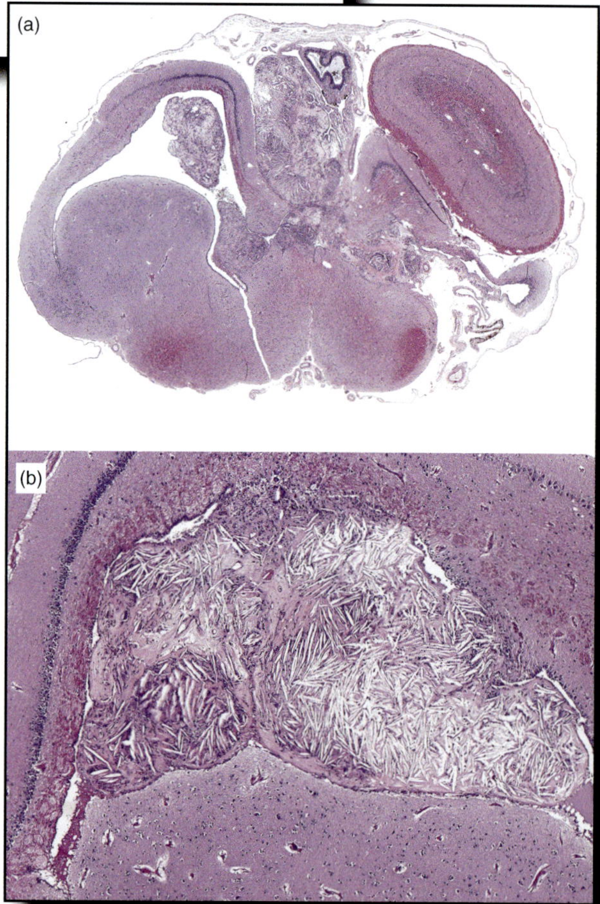

Figure 3.73 Chinese Water Dragon, *Physignathus cocincinus*. Agamidae. Photomicrographs of cerebral xanthomatosis. Nodular, coalescing aggregates of granulomatous inflammation and colorless acicular clefts (xanthomas) are present within a lateral ventricle and in the meninges of the median sulcus (a). Needle-like, colorless acicular clefts are surrounded by epithelioid macrophages, multinucleated giant cells, and fewer granulocytes expand the lateral ventricle and extend into and disrupt the surrounding neuroparenchyma (b). Hematoxylin and eosin.

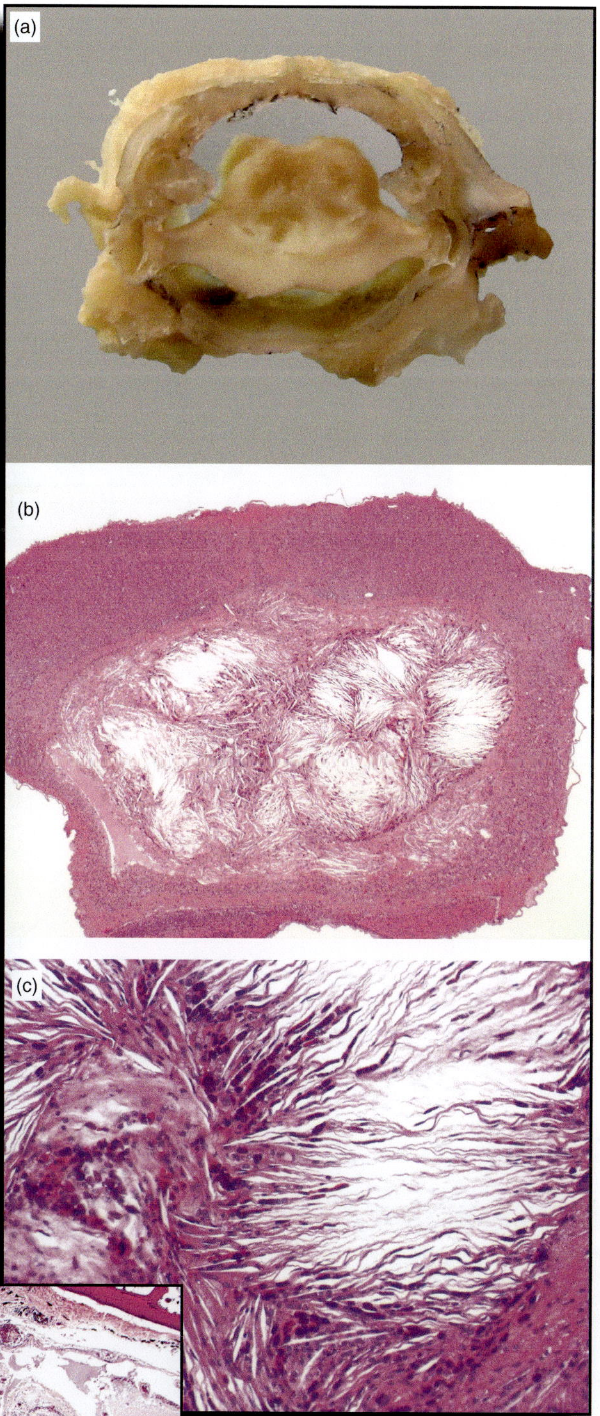

Figure 3.74 Snake, unknown species. Cerebral xanthomatosis. Cross-section of the formalin-fixed skull and brain at the level of the vestibulocochlear nerve. The neuropil is expanded and effaced by a locally extensive, pale yellow mass (a). The mass is composed of acicular clefts and granulomatous and granulocytic inflammation that effaces the parenchyma (b). Colorless, needle-like, acicular clefts are surrounded by macrophages and granulocytes (c). Hematoxylin and eosin. (Courtesy of Michael Garner.)

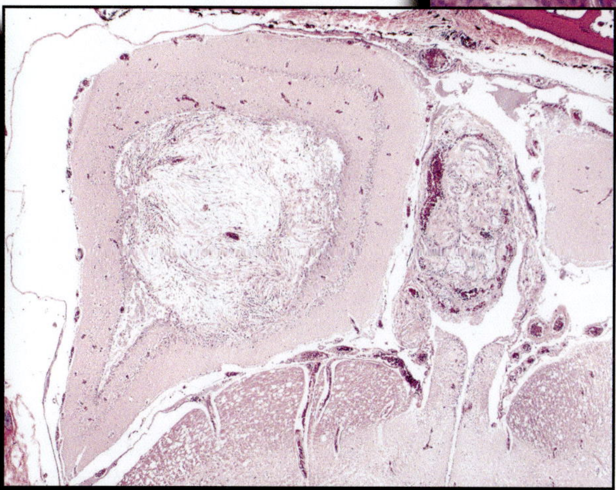

Figure 3.75 Tropical Girdled Lizard, *Cordylus tropidosternum*. Cordylidae. Photomicrograph of a xanthoma within the lateral ventricle of the cerebral hemisphere that expands the lumen of the ventricle and extends into the surrounding neuropil. Hematoxylin and eosin. (Courtesy of Disease Investigations, San Diego Zoo Global.)

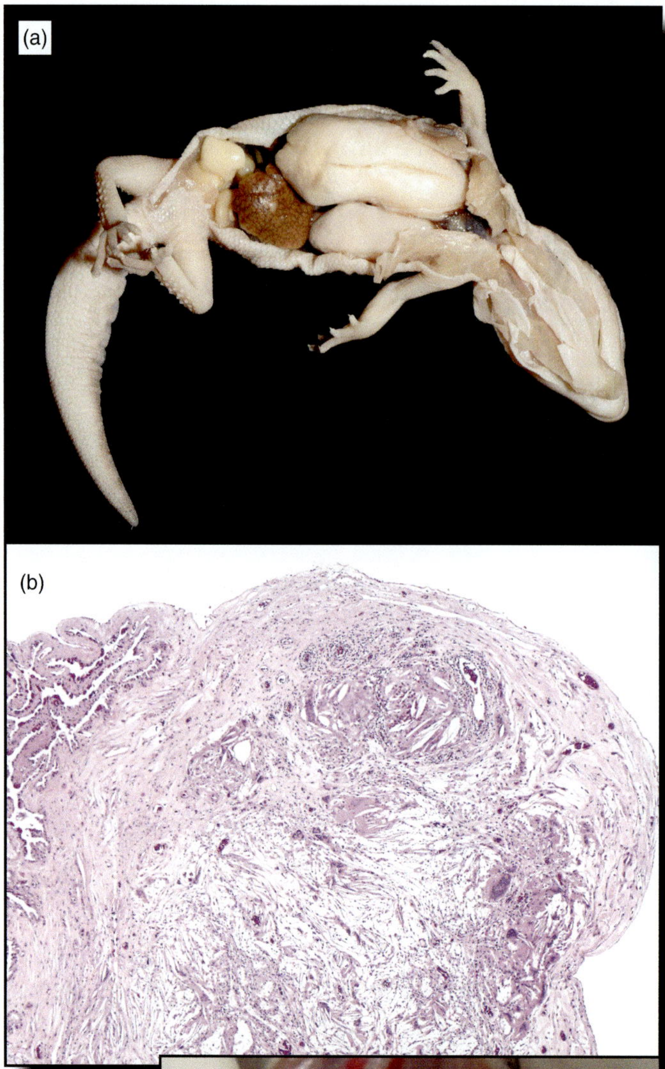

Figure 3.76 Leopard Gecko, *Eublepharis macularius*. Eublepharidae. Pulmonary xanthomatosis. The lungs are diffusely and markedly enlarged in the formalin-fixed carcass (a). A photomicrograph of granulomatous inflammation associated with colorless, acicular clefts, consistent with cholesterol clefts adjacent to faveolar spaces. Note the associated perivascular lymphocytic inflammation surrounding the xanthoma. Hematoxylin and eosin (b). (Courtesy of John Roberts.)

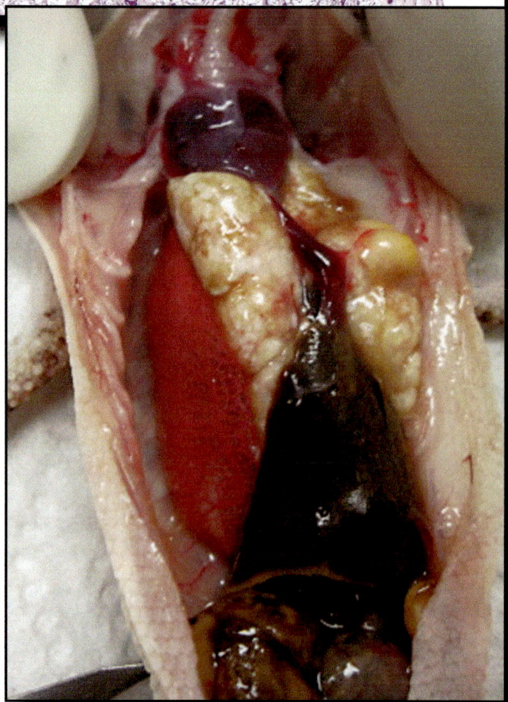

Figure 3.77 Leopard Gecko, *Eublepharis macularius*. Eublepharidae. The coelomic cavity contains multifocal to coalescing, yellow-white nodules (xanthomas). (Courtesy of Alexandra Goe.)

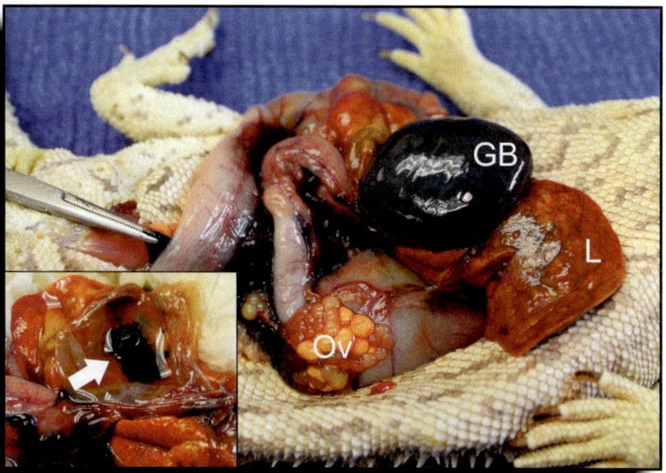

Figure 3.78 Bearded Dragon, *Pogona vitticeps*. Agamidae. The gallbladder (GB) is markedly distended and the hepatic parenchyma (Lv) is orange-tan. Within the lumen of the gall bladder (inset) is a focal cholelith (arrow). (Courtesy of Scott Stahl.)

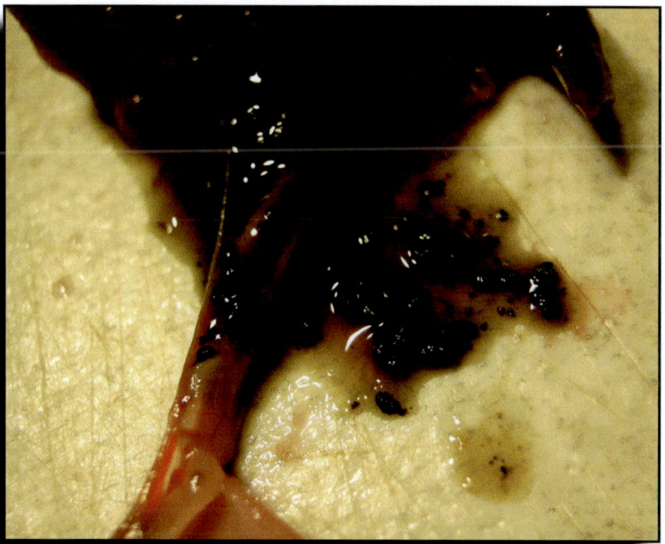

Figure 3.79 Giant Amieva, *Ameiva ameiva*. Teiidae. Cholelithiasis. Numerous dark brown-green concretions are present along the bile duct (opened). (Courtesy of Chriss Miller.)

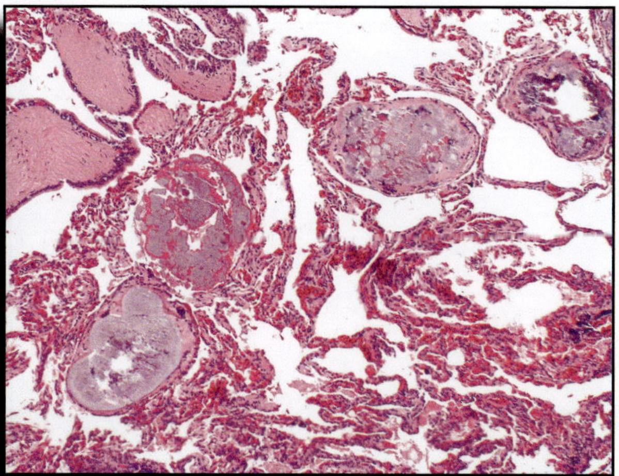

Figure 3.80 Agama, *Agama* species. Agamidae. Photomicrograph of the accumulation of amphophilic to eosinophilic, variably mineralized material (inspissated edema) within faveolar spaces. Note the presence of homogenous, eosinophilic material (protein-rich fluid) in surrounding faveoli. Hematoxylin and eosin. (Courtesy of Michael Garner.)

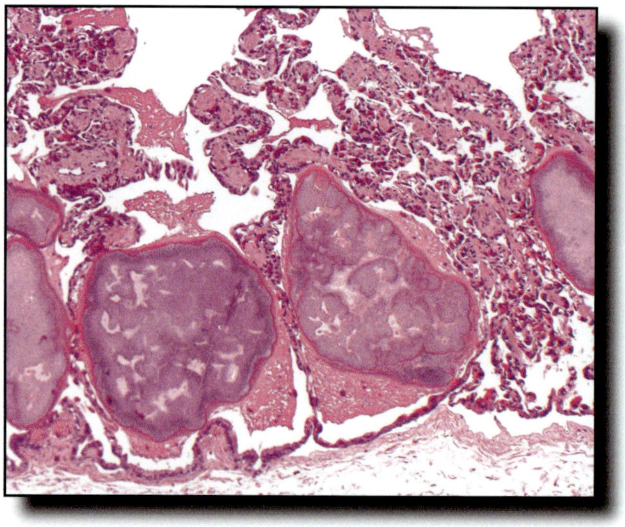

Figure 3.81 Gopher Snake, *Pituophis* species. Colubridae. Photomicrograph of the accumulation of amphophilic to eosinophilic material (inspissated edema) within faveolar spaces. Note the presence of homogenous, eosinophilic material (protein-rich fluid) surrounding the inspissated aggregates. Hematoxylin and eosin. (Courtesy of Michael Garner.)

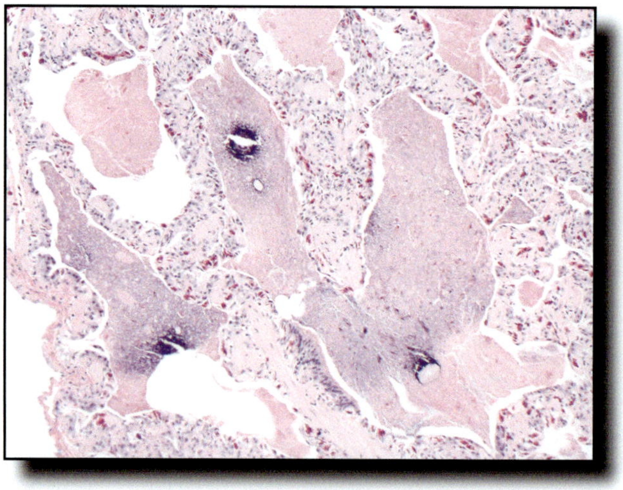

Figure 3.82 Corn Snake, *Pantherophis guttatus*. Colubridae. Photomicrograph of pulmonary edema with inspissation and mineralization. Hematoxylin and eosin. (Courtesy of Michael Garner.)

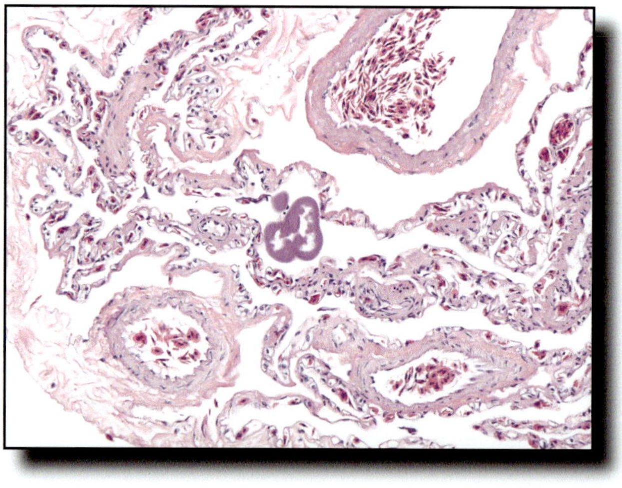

Figure 3.83 Bearded Dragon, *Pogona vitticeps*. Agamidae. Photomicrograph of faveolar basophilic material in the absence of pulmonary edema; etiology undetermined. Hematoxylin and eosin.

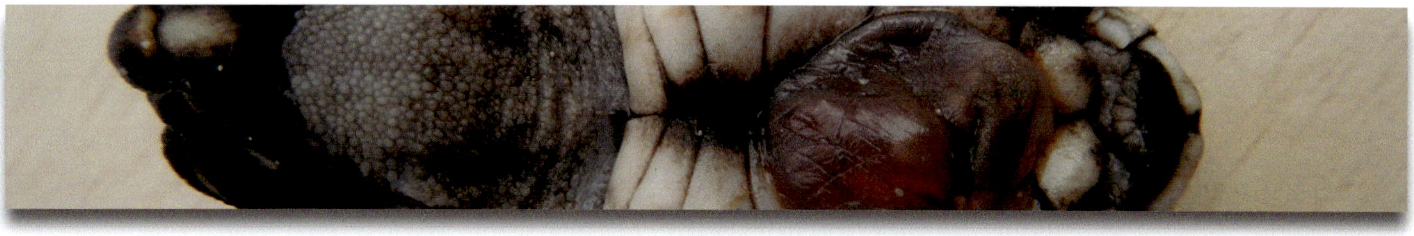

4

NORMAL AND ABNORMAL REPTILE DEVELOPMENT

MARCELO P.N. DE CARVALHO, GREGORY A. LEWBART, JAMES R. STEWART, AND JEANETTE WYNEKEN

Contents

4.1 Introduction and General Background

The development of reptiles has been studied in a just a few species. A substantial amount of the literature focuses on vivipary and temperature effects during incubation and sex determination. This chapter provides an overview of normal development and developmental anomalies in reptilian taxa. Recent use of several taxa as model species has increased understanding of development in several species of turtles (Testudines), tuatara, lizards, and snakes (Lepidosauria), and crocodilians. Somewhat limited information is available for the tuatara (Sphenodontia), primarily because of conservation concerns. There has been very little study of development in worm lizards (Amphisbaena), likely because most species are fossorial and not observed.

4.1.1 Reproductive Cycles

Reproductive cycles in reptiles can be quite variable and hence knowledge of specific taxa should be sought from those that work with the species of interest in the field or in captivity. Briefly, reproduction tends to be seasonal. Many reptile species from tropical regions tend to breed prior to or during the rainy season in the warmer spring and summer months. In temperate species, mating may be in the spring and oviposition may follow in weeks or months.

The development of reptiles reflects diversity. Not only do major clades differ in egg structure and embryonic tempo, but also how embryos develop in a nest or within the adult (herein referred to as the dam). This diversity is summarized in this chapter along with references to aid the reader seeking further detail.

In some temperate species that nest at higher latitudes, such as Map Turtles (*Graptemys geographica*), Common Snapping Turtles (*Chelydra serpentina*), and Painted Turtles (*Chrysemys picta*), nest emergences may be delayed until spring. Hatchlings overwintering in their nests or layers below the nest may develop hyperglycemia and hyperlactemia. These conditions are thought to provide cold-hardiness that will allow overwintering in the nest by preventing freezing and allowing recovery (Yntema, 1978; Packard et al., 1997; Costanzo et al., 2004; Nagle et al., 2004; Ultsch, 2006).

4.1.2 Ovipary and Vivipary

There are two major characteristics that classify embryonic development or reproductive mode: (1) where the embryo develops and (2) how it is nourished (Blackburn and Starck, 2015). The two components can impact the reproductive health of the dam and the success of reproduction.

Ovipary refers to egg-laying reproduction. It occurs in all reptilian orders: Testudinata, Squamata, Crocodylia, and Sphenodontia (= Rhynchocephalia). Vivipary refers to live-bearing reproduction. "Ovo-vivipary" is an antiquated term that confounds where the embryo develops and the extent to which it is nourished by yolk or nutrition from the dam across the vitelline membranes.

Blackburn (2015) reports that vivipary has arisen independently in more than 150 vertebrate lineages, including a minimum of 115 clades of extant squamates. Approximately 20% of the lizards and snakes are viviparous. There are no viviparous chelonians, crocodilians, or the monotypic tuatara, *Sphenodon punctatus*.

In oviparous development, the embryo receives nutrients via "lecithotrophy," in which yolk from the ovum provides nutrients for development. In viviparous species, embryonic nutrition is "matrotrophy," such that maternal provisioning occurs during gestation in the reproductive tract (Blackburn and Starck, 2015). Matrotrophy arose independently in vertebrates at least 33 times so there can be considerable diversity. Matrotrophy has multiple independent origins and diverse forms, many of which are not fully distinct because embryos can transition from one form to another. Matrotrophy includes *placentotrophy* (in which nutrition and gas exchange cross a placenta formed by fetal membranes in contact with uterine wall) and is found a number of skink species (Scincidae) including several species of *Chalcides*, *Eumecia*, and *Mabuya* (see Blackburn, 2015). Histotrophy is a form of matrotrophy in which embryos ingest nutrients, typically liquids of maternal origin, that diffuse across tissue spaces to absorptive sites of the embryo. Embryos that develop in the reproductive tract and are nourished by ingesting sibling yolks or yolk breakdown products is termed oophagy. Embryos that actively eat maternal tissues such as the oviductal mucosa engage in matrophagy. Embryophagy is defined embryos that obtain nutrition by cannibalizing siblings (Blackburn, 2015).

4.2 Review of Normal Development of Lizards and Snakes

Squamates are a diverse clade of Reptilia with a broad geographic distribution. They have the highest diversity of reproductive patterns of any amniote lineage and have evolved a wide range of body sizes and shapes. Viviparity, giving birth to free-living offspring (Blackburn, 1993a,b), is known for Cretaceous lizards (*Yabeinosaurus* sp.) (Wang and Evans, 2011). Mosasauroids (reviewed in Blackburn and Sidor, 2014) and approximately 20% of extant species of lizards and snakes are viviparous (Blackburn, 1982, 1985; Shine 1985). Unisexual reproduction in the absence of sperm (obligate parthenogenesis) has evolved in at least 40 lineages of squamates yet does not occur in any other vertebrates (Kearney et al., 2009). Body sizes range from diminutive lizards and snakes (Hedges and Thomas, 2001; Hedges, 2008; Glaw et al., 2012) to the largest of varanid lizards and large pythons of the modern world (Jessop et al., 2006; O'Shea, 2011) and ancient (giant pythons) (Head et al., 2009) and varanid and mosasauroid lizards (Lingham-Soliar, 1995; Molnar, 2004; Grigoriev, 2014) of a former world. In contrast to other amniotes (mammals, turtles, crocodilians, birds), squamates have evolved elongate body forms and reduced limbs in numerous lineages (Wiens et al., 2006). These patterns of diversity pose intriguing questions for evolutionary and developmental biologists.

4.2.1 Early Embryogenesis

Modern reptiles inherited an egg with a substantial amount of yolk and a novel system to metabolize and transport nutrients to developing embryos. The eggs produce relatively large offspring and the large yolk mass has a profound influence on embryonic development. The yolk component of the zygote does not undergo mitotic divisions. Cells destined to contribute tissue to the embryo are located at the edge of the yolk, a condition termed telolecithal (Balfour, 1880), and early cell divisions, i.e., cleavage divisions, occur only in these cells. This was first studied by Peter (1904) in the Sand Lizard (*Lacerta agilis*) (**Figures 4.1, 1–5**). This developmental pattern, in which only a portion of the egg cell divides, is termed *meroblastic cleavage*, and the early embryo, or blastodisk, forms on the surface of the mass of yolk (Hubert, 1985). Establishment of embryonic germ layers and the initial organization of the embryonic body plan occur during gastrulation (**Figure 4.1**, 5–7; *Lacerta agilis*, 1904). A series of classical studies of early embryogenesis of reptiles established that gastrulation in turtles, lizards, and snakes proceeds by movement of cells through a blastopore which is associated with a thickening on the blastodisk termed the *blastoporal plaque* or *plate* (reviewed in Bertocchini et al., 2013). Amphibian gastrulation also features a structure similar to the reptilian blastopore. In contrast, birds and eutherian mammals lack a blastopore, and gastrulation occurs by ingression along an elongate thickening termed the primitive streak. Recent studies of development of the Chinese Softshell Turtle (*Pelodiscus sinensis*), the Red-Eared Slider (*Trachemys scripta*), and the Veiled Chameleon (*Chamaeleo calyptratus*) (Bertocchini et al., 2013; Stower et al., 2015) suggest that gastrulation of turtles, lizards, and snakes contains shared components with amphibians as well as with birds and eutherian mammals. Thus, the unique early embryology of non-avian reptiles may be a useful model for gastrulation in the earliest amniotes (Bertocchini et al., 2013; Stower et al., 2015).

4.2.2 Extraembryonic Membranes

In addition to contributing ectoderm, mesoderm, and endoderm to embryonic tissues, the differentiation and growth of germ layers during gastrulation also gives rise to cell lineages that contribute to the extraembryonic membranes. These tissues, which extend outward from the embryo, support and protect the embryo and mediate physiological exchange between egg compartments and between the egg and the external environment. Extraembryonic tissues include the yolk sac, chorion, amnion, and allantois. Although there are lineage-specific differences, the general pattern of development of the extraembryonic membranes is similar among Reptilia (Luckett, 1977; Stewart, 1997). A prominent feature is the development of a system that provides embryonic access to the large yolk mass, which is accomplished by an elaborate yolk sac with a dedicated blood vascular system. The amnion and chorion are protective membranes that surround the embryo and egg respectively, while the allantois is a large reservoir with an extensive vascular network with multiple functions including respiration. The primordia of the yolk sac, amnion, and chorion are all apparent during neurogenesis, while the allantois is recognizable as a small outgrowth of the hind gut in older embryos with 16–18 somites (Dufaure and Hubert, 1961; Hubert and Dufaure, 1968).

Amniogenesis proceeds in all Reptilia as extraembryonic ectoderm grows upward over the embryo (Luckett, 1977). During neurogenesis in most lizards and snakes, the ectoderm extends to enclose the head region of the embryo and, as the fold continues caudally, it is joined by mesoderm. Fusion of the leading edge of the folds results in an outer and inner membrane, the chorion and amnion, respectively (reviewed in Luckett, 1977; Hubert, 1985). Amniogenesis has been described in relatively few lizards and snakes and thus the pattern of phylogenetic diversity is uncertain. However, some variation is apparent because the development of the amnion in some species of chameleons is unusual in that a circular ridge extends over the blastodisk prior to gastrulation and the amnion encloses the embryo prior to neural tube formation (Blanc, 1974; reviewed in Fisk and Tribe, 1949; Luckett, 1977).

The chorion is a nonvascular tissue that forms the outermost living tissue of the egg, i.e., is adjacent to the eggshell. Extraembryonic ectoderm contributes the outer epithelium of the chorion, which extends over the embryo during amniogenesis and is continuous with the ectodermal epithelium that contributes to the yolk sac. As the neural tube is forming, the growth of extraembryonic endoderm and ectoderm over the surface of the yolk mass signals the initial phases of yolk sac development. The growth of extraembryonic mesoderm between the layers of endoderm and ectoderm results in a trilaminar yolk sac. The earliest indication of vascular development is the presence of blood islands within the mesoderm of the trilaminar yolk sac. Blood islands are visible in neurula stage embryos with four pairs of somites in the Viviparous Lizard (*Zootoca* [*Lacerta*] *vivipara*) (Dufaure and Hubert, 1961) and the Asp Viper (*Vipera aspis*) (Hubert and Dufaure, 1968). As it extends over the yolk, the vascular trilaminar yolk sac splits to form an outer nonvascular tissue, the chorion, and an inner vascular tissue the definitive yolk sac. The chorion (ectoderm, mesoderm) adjacent to the yolk sac is continuous with the chorion adjacent to the embryo. The definitive yolk sac, which is initially continuous with the embryonic gut, consists of an inner epithelium of large endodermal cells and an outer epithelium derived from mesoderm. Yolk sac blood vessels are associated with the mesoderm.

In contrast to other amniotes, the vascular extraembryonic mesoderm of squamates dips into the yolk and grows parallel to the surface of the yolk mass as it approaches the portion of the yolk mass opposite the embryo (abembryonic region). As a result, there is an absence of blood vessels in the abembryonic region of the yolk sac and a small amount of yolk, the isolated yolk mass, remains between the outer nonvascular yolk sac and the vascular yolk sac. In oviparous squamates, the vascular yolk sac ultimately surrounds the entire yolk mass, but in viviparous species the region of the isolated yolk mass contributes to a yolk sac placenta (reviewed in Blackburn and Stewart, 2011; Stewart and Blackburn, 2014).

There is additional penetration of the yolk mass by blood vessels in squamates associated with the mechanism of yolk metabolism. The yolk sac of reptiles consists of an inner epithelium of large endodermal cells with blood vessels in the outer layer. This system has been studied most extensively in birds, in which yolk platelets and/or yolk nutrients are taken up by the endodermal cells, where they are processed and transported to blood vessels that lie adjacent to the basal membrane of the cells (Mobbs and McMillan, 1979; Yoshizaki et al., 2004; Nakazawa et al., 2011; Bauer et al., 2013). Fluid yolk is taken up and processed by the endodermal cells throughout development. Lizards and snakes have a similar yolk sac surrounding the perimeter of the yolk mass, but endodermal cells also proliferate within the yolk sac cavity and take up yolk throughout the cavity (Virchow, 1892; Elinson and Stewart, 2014; Elinson et al., 2014; Powers and Blackburn, 2017a,b; Stewart and Thompson, 2017). These yolk-filled cells subsequently associate with blood vessels that extend inward from the perimeter of the yolk mass. This process continues until the entire yolk mass is a network of blood vessels surrounded by yolk-filled endodermal cells, i.e., all yolk is within cells. Thus, yolk sac development and yolk metabolism differ in squamates compared to birds.

The allantois, like the amnion, is a shared, derived character for amniotes. The allantois develops as an outgrowth of the embryonic hindgut and eventually fills the cavity between the nonvascular chorion and the vascular yolk sac. The allantoic blood vascular system is separate from that of the yolk sac. Fusion of the allantois to the chorion to form the chorioallantoic membrane positions the allantoic blood vessels near the outer perimeter of the egg and contributes the primary embryonic respiratory system (Blackburn, 1993a,b). In addition to a respiratory function, the chorioallantoic membrane plays a nutritional role in mobilization and transport of calcium from the eggshell to the developing embryo (Packard, 1994; Stewart and Ecay, 2010). These functions are exaptations for viviparity and are important functional attributes of the chorioallantoic placenta of viviparous species (Guillette, 1982; Blackburn, 1993a,b; Stewart, 2013).

4.2.3 Neurogenesis, Somitogenesis

Organization of the embryonic body plan continues as the extraembryonic membranes are forming. In the Sand Lizard (Peter, 1904), a major phase of this process following gastrulation is the development of the neural tube and formation of somites (body segments) (**Figure 4.1**, 8–19). Development of the neural tube (neurogenesis) commences during the latest phases of gastrulation as the embryo takes on a more elongate form (Dufaure and Hubert, 1961; Hubert and Dufaure, 1968; Blanc, 1974). The neural tube develops by growth of ectoderm on either side of a neural groove and these neural folds extend over and enclose the neural groove.

Somites are blocks of tissue that become organized from the paraxial (presomitic) mesoderm located on either side of the developing neural tube. They are produced by an oscillating mechanism of cyclic gene expression ("segmentation clock") that travels along the presomitic mesoderm (Gomez et al., 2008; Gomez and Pourquie, 2009). Somites, which give rise to skeletal and muscular tissues, are responsible for the segmental organization of the axial skeleton of vertebrates and thus the number of somites determines the number of vertebrae. One of the notable characteristics of development of squamates is an evolutionary change in somitogenesis resulting in an increase in the number of body segments (Muller et al., 2010). Among extant amniotes, squamates exhibit the greatest diversity in number of body segments. Some lizards have as few as 20 vertebrae, legless lizards may have more than 100, and snakes often have more than 170 (reviewed in Lindell, 1994; Muller et al., 2010). Reduction or loss of limbs is associated with extreme body elongation, and this body plan has evolved independently in more than 20 lineages of squamates (Wiens et al., 2006).

In addition to a wide diversity in number of vertebrae, the axial skeleton of squamates contrasts with that of archosaurs and mammals in that the vertebral column is not distinctly regionalized (Muller et al., 2010; Head and Polly, 2015). Regional specialization within the vertebral column is regulated by *Hox* gene expression, raising questions about the role of *Hox* genes in the evolution of body elongation and limb reduction in lizards and snakes (Cohn and Tickle, 1999; Woltering et al., 2009). Comparative vertebral morphometrics of a wide diversity of amniotes suggests that squamates retain an ancestral pattern of subtle vertebral column regionalization and that the evolution of limb reduction is independent of vertebral evolution (Head and Polly, 2015). Recent advances in the study of the genetics of limb development in snakes point to mutations in a critical enhancer of vertebrate limb development resulting in loss of function and limb reduction (Leal and Cohn, 2016; Kvon et al., 2016). These studies suggest a central role for modification of regulatory genes in the evolution of limb reduction and that an increase in body elongation is a response to variation in somitogenesis.

4.2.4 External Morphology

Embryonic staging systems describe the sequential appearance of anatomical features with emphasis on structures that are visible externally. Presented as normal tables, or normal plates, of development, these embryological series are aimed at providing a reference for a typical developmental sequence of a selected species (**Figure 4.1**). Conceptually, normal tables represent a pattern of embryological development produced by interrelated processes. Operationally they are a static series of snapshots extracted from a continuous process, and their reproducibility is influenced by the interdependence of the processes as well as the sampling interval. Nonetheless, they provide an important reference system for both intra-specific (Stewart et al., 2004) and interspecific comparative embryology (Werner, 1971; Guerra-Fuentes et al., 2014) and can be usefully employed in detecting developmental variation (Andrews, 2004; Andrews and Greene, 2011) and abnormalities (Simoniello et al., 2011). Normal tables are particularly relevant in comparative analyses of development of reptiles because of the varied embryonic response to differences in incubation conditions. However, there is sufficient variation among squamate lineages in the sequence of developmental events that comparisons between distantly related taxa are likely to lack rigor (Andrews et al., 2013). Fortunately, embryonic sequences, portrayed as stages of development, have been described for a diversity of lizards and snakes, although only a small percentage of taxa have been studied (**Table 4.1**).

4.2.5 Intrauterine Gestation

The length of intrauterine gestation distinguishes squamate reproductive patterns from those of other lineages of reptiles. Squamates are the only lineage of reptiles with viviparous species, and most oviparous species retain eggs in the uterus for a considerable portion of the incubation interval. Although there is a wide range of variation in embryonic stage at oviposition among oviparous squamates, ranging from the gastrula stage (Blanc, 1974; Andrews and Donoghue, 2004; Andrews et al., 2008) to the latest embryonic stages (Qualls et al., 1995; Smith and Shine, 1997), most species oviposit eggs with embryos that have completed approximately 40%–50% of developmental time to hatching (Shine, 1983; Blackburn, 1995; Andrews and Mathes, 2000). At oviposition, lizard embryos typically exhibit early stages of limb formation, i.e., limb buds, or approximately stage 30–31 in the Dufaure and Hubert (1961) staging system. Because of the difficulty in obtaining earlier embryos, many staging systems begin at oviposition. Viviparity, has evolved independently more than 100 times among modern lizards and snakes, and approximately 20% of species are viviparous (reviewed in Blackburn, 2006; Blackburn and Stewart, 2011; Stewart and Blackburn, 2014). Thus, for both oviparous and viviparous species, the potential for postovulatory maternal effects on embryonic development of lizards and snakes is greater than for other reptiles.

4.3 Developmental Anomalies in Lizards and Snakes

4.3.1 Introduction

The embryonic development of vertebrates consists of a highly complex biological process, and due to the delicate synchronization of these events, aberrations or errors may occur during gametogenesis, fertilization, blastogenesis, embryogenesis, or fetogenesis, giving rise to developmental anomalies (Szabo, 1989a). Some of these malformations are primarily genetic mutations, whereas others are more strongly impacted by environmental factors (Bellairs, 1981; Szabo, 1989a; Sant'Anna et al., 2013). However, there are unknown ways in which genetic and environmental causes interact, making it difficult to attribute the cause of developmental anomalies to a single source (Szabo, 1989a).

In humans, only 30%–35% of congenital malformations have clearly defined causes: about 20% have a genetic origin, 5%–6% are related to chromosomal defects, 4%–5% occur due to maternal infections and noninfectious diseases, and 2%–3% are attributed to the actions of drugs and environmental pollutants (Brent, 1976; Kalter and Warkany, 1983). Most (65%–70%) human anomalies still present unknown causes or are attributed to complex multifactorial situations and factors (Szabo, 1989a; Corsello and Giufrè, 2012). Such etiological complexity is probably also present in the developmental malformation processes involving domestic, laboratory, and wild animals (Szabo, 1989a). Thus, while definitive causal evidence has not been discovered, it is conceivable that interactions between genetic and environmental factors are important in the incidence of anomalies in animals (Szabo, 1989a).

The nature and frequency of anomalies in animals differ mainly according to their species, geographic location, occurrence of maternal prenatal infections, and maternal nutritional condition (Leipold et al., 1972; Leipold, 1980). Regarding the form of presentation, developmental anomalies can be macroscopic or microscopic and be present in any type of tissue or organ (Warkany, 1947). However, in all animal species studied, the most commonly affected systems include skeletal, muscular, and central nervous systems (Leipold et al., 1972; Leipold, 1980), and often with little or no detectable abnormality in the histogenesis of the affected tissue.

A wide variety of developmental anomalies have been observed in reptiles, especially in species that comprise the Squamata order (Bellairs, 1981). Malformations in these reptiles, as well as in other vertebrates, may arise due to genetic and/or environmental factors (Bellairs, 1981; Frye, 1991).

The genetic pool of some wild reptile populations is severely restricted or concentrated due to geographic limitations, such as islands, or to the isolation resulting from habitat fragmentation caused by anthropic action (Frye, 1991; Tóth et al., 2005), which contribute to inbreeding processes and consequent expression of genes responsible for the manifestation of developmental anomalies. Furthermore, some Squamata reproduce by parthenogenesis, giving rise to genetically similar individuals, which also contributes to the expression of recessive genes, promoting developmental malformations (Billy, 1986; Frye, 1991). Malformed embryos produced by parthenogenetic Rock Lizard (*Darevskia* [*Lacerta*] *saxicola*) are good examples of genetic effect in the expression of developmental anomalies in reptiles (Darevsky, 1966; Bellairs, 1981).

Table 4.1 Squamate Embryonic Staging Systems

Taxon	Developmental Interval	References
Gekkonidae		
Eublepharis macularius	Post-oviposition	Wise et al. (2009)
Paroedura pictus	Post-oviposition	Noro et al. (2009)
Tarentola annularis	Post-oviposition	Khannoon (2015)
Lacertidae		
Lacerta agilis	Post-zygote	Peter (1904)
Lacerta muralis	Post-oviposition	Dhouailly, Saxod (1974)
Zootoca vivipara	Post-zygote	Dufaure, Hubert (1961)
Gymnothalmidae		
Calyptommatus sinebrachiatus	Post-oviposition	Roscito, Rodrigues (2012)
Nothobachia ablephara	Post-oviposition	Roscito, Rodrigues (2012)
Teiidae		
Cnemidophorus uniparens	Post-neurula	Billy (1988)
Varanidae		
Varanus indicis	Post-oviposition	Gregorovicova et al. (2012)
Varanus rosenbergi	Post-oviposition	Andrews et al. (2017)
Chamaeleonidae		
Furcifer lateralis	Post-zygote	Blanc (1974)
Chamaeleo calyptratus	Post-gastrula	Andrews (2007)
Agamidae		
Agama impalearis	Post-zygote	El Mouden et al. (2000)
Calotes versicolor	Post-oviposition	Muthukkaruppan et al. (1970)
Calotes versicolor	Post-zygote to oviposition	Thapliyal et al. (1973)
Phrynosomatidae		
Uta stansburiana	Post-oviposition	Andrews, Greene (2011)
Dactyloidae		
Anolis sagrei	20 somites to hatching	Sanger et al. (2008)
Tropiduridae		
Tropidurus torquatus	Post-oviposition	Py-Daniel et al. (2017)
Liolaemidae		
Liolaemus gravenhorstii	Post-zygotic	Lemus (1967)
Liolaemus tenuis	Post-zygotic to oviposition	Lemus, Duvauchelle (1966)
Liolaemus tenuis	Post-oviposition	Lemus et al. (1981)
Pythonidae		
Python sebae	Post-oviposition	Boughner et al. (2007)
Viperidae		
Vipera aspis	Post-zygote	Hubert, Dufaure (1968)
Lamprophiidae		
Boaedon fuliginosus	Post-oviposition	Boback et al. (2012)
Psammophis sibilans	Post-oviposition	Khannoon, Zahradnicek (2017)
Elapidae		
Naja kaouthia	Post-oviposition	Jackson (2002)
Naja h. haje	Post-oviposition	Khannoon, Evans (2014)
Colubridae		
Thamnophis sirtalis	Post-oviposition	Zehr (1962)

Moreover, reptiles reproduce by oviparity or viviparity, both employing a simple form of placentation which favors the influence of external environmental (environment or maternal body) agents on embryonic development (Frye, 1991). Of environmental causes, temperature and humidity are the main factors related to the induction of developmental anomalies in Squamates, although contaminants (mainly agricultural pesticides) and infections are also cited as important environmental causative factors (Bellairs, 1981; Wallach, 2007; Johnson et al., 2010).

In the viviparous Multi-Ocellated Racerunner (*Eremias multiocellata*), environmental temperature in suboptimal and supraoptimal bands during the initial period of embryogenesis promoted significant changes in several biometric traits of the offspring, including changes in tail length, anterior and posterior limb deformities, and head malformations (Xue-Feng et al., 2001). Predictably, high temperatures appear to compromise development by accelerating embryo ossification processes, while low temperatures tend to retard it, both resulting in malformations in the embryo (Bellairs, 1981). Most gross or bizarre phenotypic changes appear to be more closely correlated with environmental causes, while more discrete abnormalities, such as altered coloring patterns, are related to the manifestation of recessive autosomal genes (Bechtel and Bechtel, 1989; Frye, 1991).

In snakes, some anomalies are described as albinism, anophthalmia, axial bifurcation, bicephaly or dicephaly, brachygnathia, buphthalmia, coiled tail, cyclopia, dwarfism, fusion and heterotopia of scales, head malformations, hydrocephalus, incomplete tail, labial clefts, micrognathia, microphthalmia, schistosomus reflexus syndrome, and spine malformations (kyphosis, scoliosis and lordosis) (Cunningham, 1937; Pendlebury, 1976; Bellairs, 1981; Federsoni, 1981; Andrade and Abe, 1993; Wallach, 2007; Sant'Anna et al., 2013).

In a study published by Sant'Anna et al. (2013), developmental anomalies were present in about 2.3% and 11.1% of the newborns from the Jararaca (*Bothrops jararaca*) and the Tropical Rattlesnake (*Crotalus durissus*), respectively. Spinal abnormalities in this study were the most common developmental anomaly in both species, followed by the fusion of ventral scales. In lizards, malformations such as short upper jaw, microphthalmia, abnormal forebrain, cleft palate, absence or deficiency of the pituitary, tail defects, and partial Siamese twinning are also described (Bellairs, 1965).

Less frequently, the presence of visceral malformations in Squamatas has been reported including deformities in heart, kidneys, stomach, esophagus, liver, gallbladder, and trachea (Scott, 1925; Óros et al., 1997; Jensen and Wang, 2009).

4.3.2 Pigmentation Anomalies

Pigmentation anomalies in reptiles range from total absence of melanophores (true albinism) to extreme accumulation of melanin (melanism) (Villa, 1981; Smith, 1983). Snakes lacking a part of pigment cells are called leucistic (see Volume 1, Chapter 1 for more details in types of chromatophores). These anomalies are found in most species of Squamates, most frequently in snakes (Pérez-Higareda, 1980; Krecsák, 2008). Although pigmentation anomalies are considered to have low biological interference, alterations of pigmentation substantially influence the survival of individuals in the wild (Frye, 1991). As in any other animal group, pigmentation patterns in reptiles are strongly influenced by the camouflage capacity in front of their predators and preys, thus, changes in coloration patterns can compromise camouflage, making affected individuals more susceptible to predation, and risking the hunting strategy of predatory species (Jackson et al., 1976).

In young Jararaca, a lighter tail tip, together with the caudal bait maneuver (sinuous movements and tail-tip curls that mimic the movements of a larva or caudal luring), work as an attractant for small prey (Hartmann et al., 2003). Thus, pigmentation anomalies in this species could negatively influence the success of seizure of food in the wild. Moreover, pigmentation is cited as a physical property that directly influences the body temperature of ectothermal vertebrates (Tanaka, 2005). In a study with Japanese Four-Line Ratsnake (*Elaphe quadrivirgata*), melanic patterns reached an optimum body temperature in a shorter time when compared to animals that presented striated pigmentation patterns, which demonstrates the importance of pigmentation in the thermoregulation process (Tanaka, 2005).

Due to the influence on survival in the natural environment, reptiles with pigmentation anomalies are rarely documented in nature (Frye, 1991). In the last 20 years, the Butantan Institute in Brazil received a wild adult Tropical Rattlesnake (**Figure 4.2**), and a juvenile Jararaca (**Figure 4.3**) with xanthism (coloration predominantly yellow). In neotropical snake species, xanthism was already reported in a Patagonia Green Racer (*Philodryas patagoniensis*) (**Figure 4.4**), Brazilian White Collared Slug Eater (*Dipsas* (*Sibynomorphus*) *mikania*) (Amaral, 1934), Cotiara (*Bothrops cotiara*) (Hoge, 1952), Rainbow Boa (*Epicrates cenchria*) (Hoge and Belluomini, 1957), Tropical Rattlesnake (Sazima and Di-Bernardo, 1991; Travaglia-Cardoso and Parpinelli, 2006), and Jararaca (Sueiro et al., 2010). Most pigmentation abnormalities in reptiles present a genetic cause (Bechtel and Bechtel, 1989; Krecsák, 2008).

4.3.3 Malformations and Anomalies of Scales

Changes in number, form, and location of scales are commonly described in snakes (Frye, 1991). The exact mechanism by which these anomalies develop is still not well defined, but some hypotheses are proposed. Fox (1948) observed that snakes of the Western Terrestrial Garter Snake (*Thamnophis elegans*) born from dams maintained at low temperatures during the gestation period had fewer rows of longitudinal scales when compared to individuals that were incubated at optimal temperatures for the species. However, genetic causes were described for heterotopia of scales in inbred Red Rat (Corn) Snakes (*Elaphe guttata*) (Frye, 1991).

In most of the cases described in snakes and lizards, scale anomalies are discrete and do not present great damage to the affected individual (Frye, 1991). Exceptions are when these malformations affect the ventral scales of serpents, compromising their locomotion (**Figure 4.5**). Ventral scale fusion in association with spinal anomalies were found in 22.1% of the malformed snakes analyzed by Sant'Anna et al. (2013).

4.3.4 Bicephaly

The term *bicephaly* refers to a congenital malformation characterized by the presence of two heads in a single individual. This type of anomaly has been described in mammals, including humans (Wu et al., 2002) and reptiles, being frequent in snakes (**Figure 4.6**) (Belluomini, 1959; Lema, 1982; Maryan, 2001; Hoser and Gibbons, 2003). Bicephaly may be the result of occasional periods of anoxia or abrupt changes in temperature during the embryonic development of reptiles, resulting in a temporary cessation or a radical retardation of its development, directly interfering with the embryo's cleavage and gastrulation stage (Newman, 1917; Bellairs, 1965).

Rates of bicephaly in captive snakes appear to be ten times higher than that estimated for wild populations (Wallach, 2003). In the Butantan Institute, Brazil, between 2007 and 2012, the incidence of bicephaly in snakes born from wild-caught pregnant Jararaca represented 3.2% (3/95) of the malformations identified in this species (Sant'Anna et al., 2013). Complementary to this work, in a study published by Carvalho et al. (2017), three bicephalic Jararaca (differing from each other in relation to the location of bifurcation) were analyzed by x-ray microtomography. Snake 1 was characterized externally by the presence of four nostrils, but with the remaining cephalic morphology within normal patterns (**Figure 4.7**), which classifies this serpent as a *rinodymus* (double nose), according to Smith and Pérez-Higareda (1987). The external characteristics of this snake were complemented by x-ray microtomographic examination, demonstrating the duplication of the nasal, premaxillary, and mandibular bones (**Figure 4.8**). An unexpected finding was the presence of duplicate mandibular bones, with the secondary mandible being positioned transversely to the main mandibular axis (**Figure 4.9**). Furthermore, presence of morphological alterations in the prefrontal and frontal bones was identified (Carvalho et al., 2017).

The second snake presented the bifurcation point positioned more caudally, when compared to snake 1. In this second case, four ocular globes were present, and the skulls were fused by the surangular, supraoccipital, occipital, basioccipital, and basisphenoid bones (**Figure 4.10**) (Carvalho et al., 2017). A similar condition of bicephaly was described in *C. durissus* by Wiley (1930), but without mention of the bones involved in the malformation.

In the third snake, the bifurcation point was located in the proximal region of the spine (**Figures 4.11–4.12**). In this case, each column was composed of 18 vertebral bodies. The Jararacussu (*Bothrops jararacussu*) and the Fer-de-Lance (*Bothrops atrox*) were described presenting bicephaly arising from double columns, with 20 and 27 vertebral bodies respectively in each duplicate segment (Nakamura, 1938; Belluomini, 1959; Carvalho et al., 2017). Also, based on radiographic exams, Wallach (2003) states that the number of vertebrae that comprise the duplicate columns in bicephalic snakes can range from 1 to 77.

Brongersma (1952) compiled the anatomical features of 17 bicephalic serpents of various species. Seven of them had two hearts, sometimes partially fused or with communications between the great vessels. Most also had duplicate esophagus and stomach. Anatomical transposition of viscera and internal organs is relatively common in bicephalic animals. Usually, the anterior organs such as esophagus and trachea are duplicated; two hearts and two sets of lungs are often observed (Wallach, 2003). No correlation seems to occur between the duplication of the spine and that of the viscera, an example of this being a craniodichotomous Western Diamondback Rattlesnake (*Crotalus atrox*), with duplicated tracheae, tracheal lungs, right lungs, left and right thymus, hearts, livers, gallbladders, pancreas, spleens, esophagi, stomachs, and anterior small intestines (McAllister and Wallach, 2007).

There are reports of bicephalic specimens surviving for several years in captivity. As an example, the literature cites a Pacific Gopher Snake (*Pituophis melanoleucus catenifer*) that was held in captivity for two decades in a zoo (Frye, 1991). Reports of wild reptiles with bicephaly are rare, probably because of their greater susceptibility to predation and the greater difficulty presented to these reptiles for prehension of food (Bellairs, 1981).

4.3.5 Spine Deformities

Spinal deformities in Squamates, as well as in other vertebrates, are classified as scoliosis (lateral deviation of the vertebral column), kyphosis (abnormal increase of the posterior convexity of the vertebral column), and lordosis (abnormal concave curvature of the vertebral column), and their interactions as kyphoscoliosis (**Figure 4.13**) and kypholordosis (Sant'Anna et al., 2013).

Spine abnormalities may be present soon after birth and may become more severe as the animal ages or become evident throughout the individual's life (Frye, 1991). For Ahboucha and Gamrani (2001), neonatal axial malformations in lizards are related to dysfunctions of the subcommissural organ, a neuronal gland located at the top of the diencephalon, important in the control of skeletal morphogenesis during the stage of embryonic organogenesis. Late lesions are probably due to environmental changes (sudden changes in temperature, physical trauma, or diet), or as a result of late manifestations of sublethal genes (Frye, 1991).

Spinal malformations were the most frequent developmental anomaly present in malformed newborns from the Jararaca and Tropical Rattlesnake (Sant'Anna et al., 2013). In this study, the incidence of kyphosis, scoliosis, and lordosis represented, respectively, 67.4%, 33.7%, and 12.6% of the malformations found in the Jararaca and 75%, 30.6%, and 11.1% of those found in Tropical Rattlesnakes. According to Sant'Anna et al. (2013), spinal malformations compromise the hatchability and locomotor function of newborns, negatively influencing their survival in the wild.

In a study published by Carvalho et al. (2017), 50 Jararaca and 26 Tropical Rattlesnakes with spinal malformations were morphometrically analyzed. Data analysis shows that snakes of *Bothrops jararaca* species had malformed sites evenly distributed throughout the spine, whereas South American Rattlesnakes (*Crotalus durissus terrificus*) had a significantly higher prevalence of lesions in the caudal segment. It is possible that malformations located in the caudal segment of the spine may not directly interfere with apprehension or ingestion of prey, thereby interfering less with the snakes' development in the wild (Sant'Anna et al., 2013), but might increase their reproductive difficulties (Carvalho et al., 2017). This hypothesis is supported by the rare observation of adult wild snakes with discrete areas of kyphosis or scoliosis in the caudal segment of the body (Sant'Anna et al., 2013).

Spinal malformations such as kyphosis, kyphoscoliosis, scoliosis, and lordosis were characterized by x-ray microtomography, and alterations in vertebral bodies such as hemivertebra, spina bifida, duplication of ribs and fusion of the head of ribs were identified in association with spine malformations in South American Rattlesnakes (**Figures 4.14–4.15**) (Carvalho et al., 2017). Hemivertebra associated to ankylosis of vertebral bodies and duplication of ribs were also present in other reptiles such as the African Rock Python (*Python sebae*), Indian Python (*Python molurus*), Yellowbelly Sea Snake (*Hydrophis platurus* [*Pelamis platura*]), Ring Neck Snake (*Diadophis punctatus*), European Grass Snake (*Natrix natrix*), unidentified Ratsnake (*Elaphe*), Unidentified King Snake (*Lampropeltis*), unidentified Brown Snake (*Rhadinaea*), and unidentified Garter Snake (*Thamnophis*) (Baur, 1891; King, 1959).

The hemivertebra is a malformation that can be caused by hemimetameric displacement of the somites, or by erroneous vascularization and ossification of the vertebra, resulting in scoliosis and absence of closure of vertebral bodies (Braund, 1987). Effects of teratogenic substances are well documented as causes of neural tube malformations (Marques et al., 2012); examples are valproic acid in rats and humans, griseofulvin in felines, and trypan blue in rats (Marques et al., 2012). Even though there is no information reported about the participation of different substances in the induction of this process in snakes, environmental contaminants could possibly play a role in the appearance of spina bifida in wild snakes (Carvalho et al., 2017).

Most snakes with spinal anomalies presented similar histopathological alterations, such as periaxial musculature derangement and fusion of the vertebral bodies (Carvalho et al., 2017). The structural disarrangement of the spine may promote, concomitantly, a disorganization of the periaxial muscle fibers, which are presented histologically in a disordered manner (Carvalho et al., 2017). In some situations, the pressure exerted by the unformed anatomy leads to alterations in local blood circulation, causing coagulative necrosis at muscular points, edema, and noninfectious myositis (Valentine and McGavin, 2009; Carvalho et al., 2017).

Concurrently with spine anomalies, most of the snakes studied by Riches (1967), Pendlebury (1976), and Sant'Anna et al. (2013) also presented with fusion of ventral scales (**Figure 4.16**), periaxial musculature dystrophy, and spinal cord inflammation (Frye, 1991).

4.3.6 Malformations or Agenesis of the Tail

Reptiles can exhibit different types of tail malformations, such as the presence of permanent lateral folds, sharp twisting, curling, shortening (**Figure 4.17**), or complete absence of the tail (Kar and Bustard, 1982). Among the forms of presentation, tail agenesis is the rarest occurrence described in Squamata (Mathur and Goel, 1974; Frye, 1991).

The percentage of coiled tails in the Jararaca (**Figure 4.18**) was 13.7% in the individuals studied by Sant'Anna et al. (2013). This kind of abnormality was analyzed by x-ray microtomography, in which the presence of spiral kyphosis and scoliosis with absence of deformities in the vertebral bodies (**Figure 4.19**) were verified (Carvalho et al., 2017).

Frye (1991) reports the existence of a population of Pacific Rattlesnakes (*Crotalus oreganus*) that shows no rattle, attributed to genetic causes. In addition to the possible genetic etiology, there are studies that point out the abnormal incubation temperature in the early stages of embryonic development as a causal possibility for these tail malformations (Mathur and Goel, 1974).

4.3.7 Dwarfism

The incidence of dwarfism in Squamates is rarely described, with only one case in a newborn Jararaca (Sant'Anna et al., 2013). In this dwarf snake, the snout vent length was about half the normal value without change in the overall mass of the snake (Sant'Anna et al., 2013). Achondroplastic dwarfism in reptile species seems to present a similar form of dwarfism to that seen in humans. It is thought that this form of achondroplasia in reptiles is genetically linked (Frye and Carney, 1974).

4.3.8 Schistosomus Reflexus Syndrome

Schistosomous reflexus syndrome (SRS) comprises the most severe type of developmental anomaly that affects the coelomic cavity. This anomaly is characterized by improper closure of the coelomatous wall, which may lead to the exposure of viscera (Szabo, 1989b). Anomalies of this nature appear to be caused by the interruption or interference in the process of collecting the viscera into the embryo cavity during organogenesis. The reasons for this failure are still not clearly defined, although changes in fetal gut maturation and peritoneum development are suggestive factors (Szabo, 1989b). Factors that interfere with fibrosis and fusion of coelomic muscles may also be involved in the cause of SRS (Szabo, 1989b).

Rare cases of SRS are described in Squamata. Frye (1991) describes a juvenile of the Eastern Hognose Snake (*Heterodon platyrhinos*) that had a large opening in the anterior portion of the body from which the heart and the great vessels protruded. Barten (1988, personal communication; Frye, 1991, p. 408) reports the case of successive litters that presented with SRS and came from the same pair of Red Rat Snakes.

Sant'Anna et al. (2013) reported the presence of SRS in 21 of 95 (22.1%) Jararaca hatchlings (**Figures 4.20–4.21**) born from recent wild-caught mothers. In this study, SRS was one of the most frequent alterations. The incomplete closure of the body of snakes during their formation and, in many cases, the extrusion of internal organs, caused death within 24 hours of birth (Sant'Anna et al., 2013). Due to its deleterious characteristics, there are no reports of lizards and snakes with SRS in the wild.

4.3.9 Ocular Malformations

Ocular malformations in reptiles may be present as a discrete reduction in the size of the eyeball, with no impairment of vision, or in the form of severe alterations in which vision is absent (Bellairs, 1981; Frye, 1991). For Squamata, ocular malformations described include anophthalmia, microphthalmia (which can be differentiated only by histological examination), buphthalmia, exophthalmia, and cyclopia.

A group of Northern Pine Snakes (*Pituophis melanoleucus melanoleucus*) with absence of eyes (globes), was kept in the San Diego Zoo reptile collection (Enseley, 1978) between the years 1940 and 1950. All individuals had some degree of kinship, which supported the presence of genes carrying these anomalies in snakes (Enseley, 1978). For snakes, total absence of vision does not seem to interfere with the localization, seizure, and ingestion of prey. An example cited by Frye (1991) comprises an Indian (or Burmese) Python exhibiting bilateral anophthalmia, which fed normally and showed no other behavioral changes due to lack of vision.

Of malformed snakes analyzed by Sant'Anna et al. (2013), ocular malformations such as anophthalmia and buphthalmia were present in 3.2% and 5.3% of Jararacas and 8.3% and 2.8% of Tropical Rattlesnakes, respectively. In the same study, microphthalmia (**Figure 4.22**) was present in 2.8% of Tropical Rattlesnakes. An isolated condition of exophthalmos has been described in a Rhinoceros Viper (*Bitis nasicornis*) by Cooper (1975); one of the eyes of this snake was distended due to the presence of structures similar to cysts that protruded from the ocular conjunctiva. No signs of infection were identified in this individual and the lesion was considered a true congenital malformation.

A Jararaca with buphthalmia was submitted for x-ray microtomographic evaluation of the skull; from the exam, the presence of a larger and shallower orbital cavity ipsilateral to the protruded eye was evidenced, when compared to the contralateral orbital cavity (Carvalho et al., 2017). Furthermore, alterations in the conformation of adjacent cranial bones, such as the reduction of the prefrontal, frontal, and post-frontal bones, and the enlargement of the frontoparietal sutures were also identified (Carvalho et al., 2017).

The incidence of cyclopia in Squamates is rarely reported (Frye, 1991). Bellairs (1965) describes a case of a cyclops in the Jeweled Lizard (*Timon lepidus* [*Lacerta lepida*]), in which globes were presented together and located in the rostral region of the skull, altering the conformation of the nostrils with their opening in the ventral surface of the choana. This form of anomaly satisfies the criteria of arhinencephaly, characterized by shortening of the maxilla and absence of external nostrils (Frye, 1991). Millichamp et al. (1983) also described cases of a cyclops in certain individuals in a clutch of the Burmese Pythons. Morphological alterations were identified by x-ray microtomography exam in cranial bones of snakes with anophthalmia and cyclopia (**Figures 4.23–4.24**). In these animals, there was a decrease in the orbital cavities and morphological changes in the ectopterygoid, nasal, frontal, parietal, premaxillary, and mandibular bones (Carvalho et al., 2017).

Studies involving children and domestic animals show that the volumetric increase in orbital pressure in cases of buphthalmia and exophthalmia, and decreased pressure in anophthalmias and cyclopias, promote an increase and decrease, respectively, in the size of the orbital cavities, and consequent morphological derangement of adjacent cranial bones (Lo et al., 1990; Cepela et al., 1992; Eppley et al., 1993; Heinz et al., 1998). This occurs because the presence and size of the globes function as the main parameters in the development of the orbital cavity and skull bones (Lo et al., 1990; Cepela et al., 1992; Eppley et al., 1993; Heinz et al., 1998). Although studies on the development of cranial bones in snakes are rare in the literature, similarities between the alterations found in snakes, cats, and humans with ocular anomalies suggest that the process of organogenesis and origin of ophthalmic morphological changes are similar (Carvalho et al., 2017).

Some factors known to induce cyclopia in reptiles include environmental contamination by magnesium salt, lithium and cadmium, ionizable radiation, changes in the incubation temperature of the embryo, and genetic defects (Aldeman, 1936; Bellairs, 1965). In mammals, the ingestion of some plants of the genus *Veratrum* sp. presents a high correlation with the incidence of cyclopia in offspring (Binns et al., 1963). Most ocular abnormalities of reptiles likely are the result of a combination of genetic and unfavorable environmental conditions (Szabo, 1989c; Frye, 1991).

4.3.10 Cephalic Malformations

The most commonly described cephalic malformations in Squamates include brachygnathia (shortening of the jaw), prognathism (mandibular prominence), agnathia (absence of mandible), and cleft palate (Bellairs and Gamble, 1960; Bellairs, 1965). The anatomy of cleft palate in reptiles is similar to that found in human anomalies (Bellairs, 1981). The Jacobson's organ, located medially in the cleft palate of some individuals described, presents a slightly lower position than the unaffected contralateral organ (Bellairs, 1981).

The formation of palatine and labial clefts is described in lizard embryos of the species Viviparous Lizard and the Jeweled Lizard (*Timon lepidus* [*Lacerta lepida*]), and in the Green Anaconda (*Eunectes murinus*), Grass Snake (*Natrix natrix*), and Northern Viper (*Vipera berus*) (Bellairs and Boyd, 1957; Bellairs and Gamble, 1960; Bellairs, 1965). In these citations, unilateral and bilateral lesions were described. In some cases, this anomaly appears together with brachygnathia and microphthalmia (Bellairs, 1981).

Lip clefts were described by Fox et al. (1961) in young snakes of the Western Terrestrial Garter Snake born from pregnant females kept at low temperatures during the gestational period. Such description suggests the influence of environmental temperature on the incidence of cephalic malformations. Snakes with palatine clefts in captivity appear to have no impairment in their hunting and seizure abilities (Bellairs, 1981).

Dean et al. (1980) report that six of seven newborns from the same litter of Prairie Rattlesnakes (*Crotalus viridis*) had bilateral palatal and labial clefts. Curiously, all the individuals affected were females.

Cephalic malformations represented 5.3% of malformations present in Jararacas born from wild-caught dams at the Instituto Butantan between 2007 and 2012 (Sant'Anna et al., 2013). Specific cephalic malformations, such as hydrocephalus, brachygnathia, cleft lip, and shortening of the maxilla, represented respectively 2.1%, 9.5%, 2.1%, and 1.1% of the malformations identified in newborn Jararacas. In the same study, brachygnathia and cleft lip were present in 5.6% and 2.8%, respectively, of the anomalies found in newborns of Tropical Rattlesnakes (Sant'Anna et al., 2013).

The prevalence of encephalic malformations in Squamates, such as hydrocephalus, is low and rarely reported. Sant'Anna et al. (2013) describe hydrocephalus in snakes as a dilation of the skull, characterized by the development of a bulging structure in the upper part of the head (**Figure 4.25**). In addition to this description, x-ray microtomographic analysis of snakes presenting hydrocephalus characterized the bones directly involved in this type of malformation, such as the thinning of parietal bones and the enlargement of the cranial sutures (**Figure 4.26**) (Carvalho et al., 2017). Morphological changes in the orbital cavities were also observed, demonstrating that snakes with hydrocephalus present more complex morphological alterations than externally evidenced (Carvalho et al., 2017).

Similarly, in domestic animals affected by hydrocephalus, the skull bones show enlargement of the cranial sutures and hypoplasia (thinning) of the bones that constitute the skull cap (Szabo, 1989d). Congenital hydrocephalus in domestic animals is frequently accompanied by brain malformations such as corpus callosum agenesis and cerebellar defects, and less frequently by cardiac, facial, palatal, urogenital, and vertebral anomalies (Szabo, 1989d).

Hydrocephalus can be caused by genetic, infectious, nutritional, and environmental factors (Szabo, 1989d). Some environmental pollutants such as mercury and organosulfur derived from fossil fuels are indicated as experimental inducers of hydrocephalus in rats and mice (Murakami, 1972; Khera and Iverson, 1980). In this context, the participation of environmental pollutants in the incidence of hydrocephalus in wild snakes should be considered possible.

Malformations such as meningoencephalocele were first described in reptiles by Hibberd (1996) and Webb and Manolis (1998) in an unidentified crocodile. A similar condition was recently described in a Jararaca, characterized macroscopically by the presence of prognathism and mandibular asymmetry (**Figure 4.27**) (Carvalho et al., 2017). An unexpected deformity, identified in this snake by x-ray microtomographic analysis, was the presence of a cranial fissure characterized by agenesis of the frontal, parietal, and occipital bones (**Figure 4.28**) (Carvalho et al., 2017). In rats, the incidence of these lesions is related to genetic factors and spontaneous mutations, but experimental inductions were performed with fetal exposure to ionizable radiation (Szabo, 1989d).

4.3.11 Malformations of Internal Organs

Internal organ malformations are not often described in reptiles; however, some isolated cases are reported. Developmental malformations involving deformities in heart, kidneys, stomach, liver, and trachea are described in snakes and chelonians (Scott, 1925; Jensen and Wang, 2009). Jensen and Wang (2009) describe two occurrences of cardiac malformations, with presence of bifid ventricles, in young Ball Pythons (*Python regius*). In both individuals, all cardiac chambers were enlarged and abnormal in shape, causing changes in blood systemic and pulmonary pressures. After euthanasia, internal access to the ventricles revealed a profound alteration of the cardiac septum. This anomaly triggered the presence of other alterations, such as the lateral projection of the atrioventricular orifice to the left and right, with concomitant distortion of the atrioventricular valves and thickening of the basal portion of the atrial septum (Jensen and Wang, 2009).

A condition of unilateral congenital renal cystic disease was reported in the European Grass Snake by Scott (1925). In this snake, the right kidney was comprised of a series of cysts. Other findings included perirenal hemorrhage, probably due to ruptured cysts (Scott, 1925). A case of polycystic liver associated with cystic kidney in a White-Throated Monitor (*Varanus albigularis*) is described by Scott (1925). At necropsy, both kidneys presented an irregular surface with cystic dilations of varied size with the largest being equal to "that of a pea" (Scott, 1925). These cysts appeared to be limited to the cortical region. The liver showed small cysts projecting through the surface; some of the bile ducts were dilated with numerous cysts in the parenchyma (Scott, 1925). Cysts in this monitor lizard seem to have been congenital (Scott, 1925). In jararacas, renal glomerulocystic and tubular disease were significantly correlated with snakes presenting with brachygnathia (2/3; 66.6%) (Carvalho et al., 2019).

According to Frye (1991), solitary or multiple renal cysts have been found in many species of reptiles. The etiology of renal cysts in these animals is unknown, but they appear to be due to hereditary defects mediated by autosomal allele genes. However, several types of chemical substances have been associated with the presence of renal cysts in experimental animals, among them polychlorinated biphenyls (PCBs) (Maxie and Newman, 2007). Paraovarian cysts have been seen in an American Alligator (*Alligator mississippiensis*) (Elliott Jacobson, personal communication). Paraovarian cysts are remnants of Wolffian duct in the mesosalpinx that do not arise from the ovary.

4.4 Review of Normal Testudines Development

4.4.1 Reproductive Biology and Natural History

Reproduction by Testudines (turtles and tortoises) is characterized by seasonality and by oviparity. Many more studies focus upon the female reproductive cycles than those of males, perhaps because females may be more accessible when aquatic species come up on land to lay eggs and terrestrial species are exposed during oviposition. However, some have very unique reproductive biology, such as the Central American River Turtle (*Dermatemys mawii*), locally known as the Hicatee, or the Australian Pig-Nosed Turtle (*Carettochelys insculpta*), which lay eggs in terrestrial chambers near flood-prone areas. Eggs incubate into the rainy season and hatch when nests become inundated by rainwater or flooding (including under water). Many taxa have distinct nesting sites that are used across multiple years. The reproductive cycles of tropical species may be characterized by production of multiple clutches across weeks to months. Temperate species tend to produce few clutches per breeding season. There are exceptions: Some turtle species reproduce annually, although it is common for iteroparous species (those with multiple reproductive cycles over a lifetime) to skip annual reproduction either on a regular schedule or occasionally, and instead assimilate nutrients for reproduction in subsequent years. In sea turtles, which are species that expend considerable energy migrating long distances from feeding grounds to breeding and nesting grounds, biannual or triannual nesting is common. Turtles are termed *capital breeders* because they sequester essential resources prior to reproducing (Perrault and Stacy, 2018). Consequently, feeding is rare during nesting season for gravid dams (however, maintaining hydration is essential).

Hatchlings and neonate females have ovaries that have multicellular layers of cortical cells overlying a disorganized medulla (stroma). Occasionally, primordial germ cells are present and appear as individual large cells on the surface of the ovary (**Figure 4.29**). The cortical cells of juvenile females slowly hypertrophy until puberty. Follicular growth is associated with puberty. Upon reaching puberty, follicular maturation occurs each breeding season. It is common for a season's follicles to mature simultaneously. Maturing oocytes gain yolk as yolk platelets and lipid droplets increase in size, density, and proximity to the periphery of the oocyte. As ovaries mature and function, they contain pre-vitellogenic follicles, maturing follicles, corpora lutea, and atretic follicles (corpora atretica) supported by the stroma (**Figure 4.30**) (after Miller and Dinkelacker, 2007). For successful reproduction, sufficient nutritional status must be reached (often manifest as minimum weight by body size) for vitellogenesis to occur and for successful breeding.

In general, spermatogenesis is associated with hypertrophy of the testes and the epididymis. Moll (1973) reports that engorgement and pale color of the epididymis is a good indicator of the presence of mature sperm. Spermatogenesis, mating, and egg production may be temporally decoupled. Spermatogenesis may occur after a season's ovulation and oviposition are complete so that each season's clutches are fertilized by a previous season's sperm. Several taxa that undergo spermatogenesis prior to mating in the same season, including all sea turtles, the Agassiz's Desert Tortoise (*Gopherus agassizii*), and the Central American River Turtle (*Dermatemys mawii*) (Rostal et al., 1994; Meylan et al., 2002; Legler and Vogt, 2013). Few species are acyclic, having year-round spermatogenesis (reviewed by Legler and Vogt, 2013).

4.4.2 Sperm Storage

Sperm storage by females has been documented in several squamates and turtles (Saint-Girons, 1975; Sakaoka et al., 2013). Sperm are sequestered in tissues of the oviduct (Sever and Hamlett, 2002). Gist and Jones (1989) identified sperm storage tubules in the albumin-secreting tissues of distal oviducts of 11 turtle species. The tubules are located in a small portion of the female's reproductive tract along the posterior portion of the albumin-secreting section of the oviduct and located between the infundibulum and the uterus (**Figure 4.31**). The ducts to sperm storage tubules connect to the oviduct lumen and allow entrance of sperm to the tubules. In their review of sperm storage, Gist and Jones (1989) noted that sperm were identified in tubules of female turtles that were isolated from males for more than a year (423 days). Previous reports of live sperm in the oviduct after more than 4 years were not found in their study. Reptiles are exceptional in sperm storage, compared with amphibians, birds, and mammals; up to 7 years of sperm storage is documented in several snake species (Magnusson, 1979), but in most turtles storage appears to be on the order of months (Holt and Fazeli, 2016). Chen et al. (2015) reported differences in the locations of sperm storage structures in several species (Chinese Softshells [*Pelodiscus sinensis*], Painted Turtles [*Chrysemys picta*], Red-Ear Sliders [*Trachemys scripta*]). Sperm storage is a key aspect of multiple paternity of clutches in many species (Sarkar et al., 2003; Rostal, 2007; Bu et al., 2016).

Chelonian clutches may be large. In marine turtles, females lay multiple clutches in a breeding season. Clutches of relatively small eggs often range from 50 to 150 and females may produce several hundred eggs per season, depending upon species. Conversely, the Pancake Tortoise (*Malacochersus tornieri*) may lay one to two large egg clutches, up to five times per season in captivity (Darlington and Davis, 1990), and the smallest of the tortoises, the Namaqualand Speckled Padloper Tortoise (*Chersobius* [*Homopus*] *signatus*) lays single egg clutches and more than one clutch per season (Loehr et al., 2004). The small tortoises produce just a few relatively large eggs per season. Clutches are deposited in underground nests dug by the dam (**Figure 4.32**).

4.4.3 Eggshells

Eggs vary in shell texture, organization, and rigidity. Two major types of eggshells are found in the testudines: rigid and pliable. Eggs with a thick mineralized calcareous layer overlying the shell membrane are rigid. Eggs with a thin or loosely organized calcareous layer over the shell membrane are pliable (also termed flexible or parchment shelled). Ewert (1985) further separates non-brittle eggs into pliable and hard-expansible eggs. He defines hard-expansible eggs and those that feel brittle when slightly dehydrated and pliable when slightly expanded after absorbing water. Pliable shelled eggs differ in their turgidity with water absorption. These eggs will collapse partially and fold if insufficient moisture is available and rebound when moisture is available unless dehydration is extreme. Rigid-shelled eggs remain rigid regardless of moisture during incubation. If insufficient water is available, the embryo and its membranes will collapse within the eggshell and not recover.

The eggshell mineral layer is primarily calcium carbonate as aragonite crystals. It both protects the developing embryo and is its major calcium source. It is common for the mineral layer to crack and crumble toward the end of development (**Figure 4.33**). Ewert (1985) describes the mineral layer and its organization in detail. Kusuda et al. (2013) surveyed eggshells from 56 species and further categorized eggs into 5 categories from the thinnest parchment and pliable eggs to rigid-shelled eggs with layered mineralization. Further discussion of the eggshell is beyond the scope of this chapter. Additional information can be found in Volume 1, Chapter 1.

4.4.4 Fertilization and Development

There is abundant literature on developmental diversity in reptiles. Hereafter we focus on each major clade.

In all turtles and tortoises, fertilization is thought to occur within the upper part of the reproductive tract at or just cranial to the albumin-producing gland (termed *uterine tube* by Girling, 2002). There is no literature on fertilization and early zygote formation in turtles. Development starts within the oviduct and proceeds so that embryos undergo cell division and blastulation. The embryos then cease development at the blastula or gastrula stage until oviposition (Ewert, 1985; Miller, 1985; Rafferty and Reina 2014). There are increasing numbers of studies on developmental arrest in turtles, which demonstrate that hypoxia is responsible for preovipositional developmental arrest (Rafferty and Reina, 2012, 2014; Rafferty et al., 2013; Rings et al., 2015; Williamson et al., 2017). Emerging from developmental arrest, eggs visibly change color via the progressive formation of a chalky white area, termed chalking (or white spot formation). The embryo and yolk are separated from the albumen by a vitelline membrane (Ewert, 1985). Resumption of development is associated with segregation so that as the yolk moves toward the egg's bottom (relative to gravity), the vitelline membrane expands rapidly and displaces albumen as it rises to occupy the upper portion of the egg. Where the vitelline membrane contacts and adheres to the eggshell membrane, chalking occurs (Yntema, 1964; Moll and Legler, 1971; Ewert, 1985; Miller, 1985).

Some species' embryos undergo post-ovipositional arrest due to cold torpor, delayed hatching, or diapause. Torpor and delayed hatching are often associated with species that nest in late summer or in mountain areas where cold temperatures, oxygen availability, and rainfall can influence final developmental stages and hatching (Rafferty and Reina, 2012). Development ceases during cold periods. Diapause is normal cessation of development prior to white spot formation and under conditions that would normally support incubation; normal development resumes in response to a stimulus such as day length, temperature, or hydric conditions (Ewert, 1991; Rafferty and Reina, 2012).

Embryonic developmental rates of turtles and tortoises is temperature dependent. The duration of incubation tends to be inversely related to temperature (Yntema, 1979). When embryos are developing (not undergoing cold torpor, diapause, or developmental arrest), development tends to be faster as temperatures increase, within the range of incubation temperatures that are normal for the species. At temperatures that are warmer than normal but still allow successful embryonic growth and hatching, development tends to slow (Bowden et al., 2014). At incubation above the critical thermal maximum (CTM) for a species, the embryo will die. The CTM can vary within a species and with latitude.

Ewert (1985) partitions turtles into two groups by normal incubation period: rapid developers and slow developers. Rapid developers include many emydid and chelid species, marine turtles, and soft-shelled turtles. They tend to incubate in 60 days or less. Slow developers can have highly variable incubation duration and include several tropical and South African tortoises, tropical kinosternids, and several New- and Old-World pleurodires. Slow developers incubate over months to more than a year (summarized in Ewert, 1985). For example, emergence within clutch from might span 6 weeks from the first to the last eggs to hatch in a captive group of Leopard Tortoises (Jacobson, unpublished data).

Postovipositional embryonic development follows typical vertebrate embryonic patterns such that the earliest embryos undergo basic differentiation, with little growth. Neurulation (neural tube formation) follows, during which a primitive brain and primitive circulatory components arise, as do a clear head and tail. Organogenesis and histogenesis follow for approximately the middle third of development, during which overall embryo growth is relatively slow; it is during organogenesis that the embryo's sex is determined. The embryos are smaller than their yolks during this middle phase of development (**Figure 4.34**). Over approximately the last third of development, growth predominates and the embryos become larger than their yolk sacs as yolk is absorbed (**Figure 4.35**). Due to variation in egg sizes as well as nest environmental effects (e.g., temperature, moisture) and incubation duration, the morphological characteristics of embryos and their extra-embryonic structures contribute to developmental stages. Standard developmental stages (staging tables) are species-specific in detail but share common features. Describing embryos by stages "eliminates" variations due to incubation conditions so that ontogenetic changes can define embryonic development somewhat independently of incubation duration. Perhaps, not surprisingly, all standardized embryonic stages are of rapid-developing species. Yntema (1968) established a staging series for the Common Snapping Turtle. This series serves as the standard for many subsequent species' stages. Mahmoud et al. (1973) developed a standard set of stages for the Western Painted Turtle (*Chrysemys picta bellii*). The staging series was refined and expanded by Cordero and Janzen (2014) to include traits that are now accessible with contemporary technology. Greenbaum (2002) developed a detailed staging series for the Red-Eared Slider. The stages of the Olive Ridley Turtle (*Lepidochelys olivacea*), a species that tends to incubate eggs in nests that get warmer than those of other marine turtles, are detailed by Crastz (1982). All marine species' developmental stages are described using similar criteria by Miller (1985); Tokita and Kuratani (2001) developed a standard staging series for the Chinese Softshell Turtle (*Pelodiscus sinensis*). An embryonic staging series for the Spiny Softshell (*Apalone spinifera*) was described by Greenbaum and Carr (2002). Beggs et al. (2000) created an egg aging and embryo staging system for the Pig-Nose Turtle (*Carettochelys insculpta*). Werneburg et al. (2009) developed a staging series for the Australian pleurodire, the Red-Bellied Short-Necked Turtle (*Emydura subglobosa*) and compared embryonic ossification between pleurodire and cryptodire species. Magalhães et al. (2017) developed a staging series for a New World pleurodire, the South American River Turtle (*Podocnemis expansa*). Miller et al. (2017) created a generalized post-ovipositional staging system for use in the field across all marine turtle species. **Table 4.2** summarizes characteristics common to the embryonic turtle staging systems.

Most turtle species have environmentally determined sex (temperature-dependent sex determination [TSD]) (Janzen and Phillips, 2006). In species with TSD, the embryo's sex is directed by the incubation environment. Turtles have one of two types of TSD: warm female–cool male or hot and cool male with intermediate temperatures producing females. Those with chromosomal sex determination have independently evolved heteromorphic (XY) sex chromosomes in several distinct lineages (South American chelids [*Acanthochely* spp.], Mexican Giant Musk Turtles [*Saurotypus* spp.], and *Siebenrockiella* spp., the Geoemydid from Southeast Asia and the Phillipines). Several other lineages have homomorphic sex chromosomes (Australian pleurodires in the genera *Emydura* and *Chelodina*, softshelled turtles in the genera *Pelodiscus* and *Apalone*, the Narrow-Bridged Musk Turtle [*Claudius angustatus*] from Central America, the Wood [*Glyptemys {Clemmys} insculpta*] and Bog Turtle [*Glyptemys{Clemmys} muhlenbergii*] of North America) (Janzen and Phillips, 2006).

4.5 Review of Normal Crocodilian Development

Ferguson (1985a,b) provides a detailed and thoroughly referenced review of crocodilian reproduction and development. This review includes information on general behavior, courtship, copulation, and maternal care. Detailed tables and figures provide species-specific information. For the purpose of this chapter only normal crocodilian embryology will be included in this section.

All crocodilians lay eggs. The eggshells are composed of an organic matrix/calcite crystal complex and are generally about 1 mm thick (Ferguson, 1985a,b). It appears that for all crocodilians, embryonic development begins prior to egg laying (Ferguson, 1984). Detailed early embryonic development of *Crocodylus niloticus* has been meticulously illustrated in manuscripts dating back over 100 years (Voeltzkow, 1899, 1901). These drawings have been modified and/or included in Ferguson (1985a,b). While various workers have attempted to stage crocodilian embryology over time, a gold standard is

Table 4.2 Some Developmental Staging Criteria for Testudines

Description	Stage Characteristics
Yolk with light blastoderm surrounded by albumen, eggshell membrane, and forming eggshell.	Pre-oviposition. Zygote-gastrula. Eggs in oviduct.
Eggshell cream-colored with shell becoming bright white. Embryo resumes development.	Post-oviposition. Embryonic disc has shaped as a transverse slit or crescent.
Embryo is small but elongated with a forming head with head flexing. Neural tube forms. Head of embryo has as neural groove closes as neural tube. Eyes start to form.	Approximately 18%–25% of development.
Forelimb ridge forms that becomes a limb bud. Hind limb ridge and bud formation follows; optic cup pigmented. Embryo increases in length, flexes; tail obvious.	Approximately 25%–30% of incubation.
Embryo increases in size, carapace formation obvious; digital plate forms.	Approximately 30% of development is complete. Embryo looks like a turtle.
Pigmentation forming; digital ridge clear; eyelids, scutes, and plastron form, integument pigmentation appears.	Approximately 50%–60% of development is complete.
Embryo as large as yolk.	Approximately 70% incubation.
Lower jaw closure is complete, intestinal loops herniated. Plastral crease increasing. Embryo larger than yolk.	Approximately 75%–85% of incubation.
Embryo much larger than yolk, pigmentation near completion, to pipping.	Approximately 85%–100% of incubation.

Note: Embryonic stage is frequently assessed to identify when embryonic death occurred. This table summarizes some morphologic benchmarks for staging. Early embryonic death is challenging to identify. For detailed assessment of embryonic stages, please refer to references in the main text. The major characters described here focus on head and eye, limbs, carapace, and relative size of the embryo to the yolk. Species differ in detail; consequently, this table is a general guide. Werneburg (2009) developed a standardized developmental event system for assessing presence or absence of staging vertebrate embryos that allows definitive staging based upon suites of existing characteristics.

not possible, as species differences and temperature-induced variations complicate these efforts. Ferguson (1985a,b) compiled a table that establishes 28 stages for three species: Nile Crocodile (*C. niloticus*), Freshwater Crocodile (*C. johnsoni*), and American Alligator. These embryonic stages occur over a period of 64–90 days depending on the species (the American Alligator has the shortest and the Freshwater Crocodile the longest developmental window). The table is accompanied by detailed descriptions of each stage with numerous photographs and drawings.

In 2008, Iungman et al. published a detailed account of the Broad-Snouted Caiman's (*Caiman latirostris*) embryological development. The authors used Ferguson's (1985a,b, 1987) 28 stages to qualitatively and qualitatively describe and illustrate the caiman's development. There were some subtle differences between the staging dynamics, and these are evident in **Table 4.3** (Iungman et al., 2008). Still, it appears that crocodilian development is quite similar among the handful of taxa that have been studied.

4.6 Developmental Anomalies in Testudines and Crocodylia

4.6.1 Introduction

Numerous congenital and developmental diseases and anomalies have been reported in the orders Testudines and Crocodylia. Only in some cases is the etiologic agent or pathogenesis known.

4.6.2 Testudines

There is a relatively large body of literature of developmental anomalies in the Testudines. Most of these are reported from aquatic turtles (Hildebrand, 1930; Bell et al., 2006; de Solla et al., 2008; Davy and Murphy, 2009; Palmieri et al., 2013; Bárcenas-Ibarra et al., 2015; Sönmez et al., 2017). In many cases malformed turtles possess more than one anomaly. For example, a turtle with an abnormal flipper or spinal deformity may also display a misshapen carapace or tail.

One area of research focus has been with Common Snapping Turtles as a sentinel species for the effects of environmental pollution on embryonic development (Bishop et al., 1998; Bell et al., 2006; de Solla et al., 2008). In one study of a suburban Philadelphia wildlife refuge, deformities were common in common snapping turtles and Eastern Painted Turtles (*Chrysemys picta picta*) (Bell et al., 2006).

4.6.2.1 Bicephaly Two-headed turtles have been reported from Green Turtles (*Chelonia mydas*) (Sönmez et al., 2017), Olive Ridley Turtles (Bárcenas-Ibarra et al., 2015), Diamondback Terrapins, (*Malaclemys terrapin*) (Hildebrand, 1930), Hermann's Tortoises (*Testudo hermanni boettgeri*) (Sailer et al., 1997), and an African Spurred Tortoise (*Centrochelys sulcate*) and Mediterranean Spur-Thighed Tortoise (*Testudo graeca ibera*) (Palmieri et al., 2013). In the latter case, the hatchling survived for 22 days, despite being anorexic, lethargic, and without fecal production. This animal's heads functioned independently, the hearts beat asynchronously, and

Table 4.3 Detailed List of Specific Characteristics for Aging Caiman Latirostris Embryos That Match, and Those That Do not Match Ferguson (1985a,b, 1987) Stages

Age (Days)	Ferguson Stages	Features Do not Match Ferguson Stages	Features Match Ferguson Stages
2	2(E)	**Three primary brain vesicles.** Primary optic vesicles are not invaginated, and lens placodes are not formed yet. Visceral arches are not noticeable, visceral cleft absent.	**Optic and auditory placodes are present. Twenty pairs of somites.** Cranial and cervical flexure indicated. Paired primordia of heart fused.
3	3(E)	Amnion absent. Heart S-shaped, bent to right. Tail bud absent.	**Twenty-six pairs of somites.**
4	4 (E)	Optic vesicles are invaginating, lens placodes are forming. **Tail bud forms and bent.**	Head fold of amnion present. **Allantois bud formed. Thirty-two pairs of somites. Head turned to the right.**
5	5 (L)	**Forelimb-primordia present.**	**Thirty-six pairs of somites.**
6	6 (L)	Auditory pits still open, endolymphatic duct present. **Maxillary processes are distinct.**	Lens placodes are formed. Telencephalon indicated. **Indication of nasal placodes. Embryo rotation is completed. Hindlimb-primordia present.** Tail-bud curved.
7	7 (L)	**Nasal placodes are distinct pits**, nasal processes are conspicuous.	**Optic cups complete**, choroid fissure visible. Auditory pits closed. **Three visceral arches, two grooves, two clefts and** branchial sinus are present. **Distinct hindlimb-bud.**
8	8 (L)	Allantois and chorion fused. **Four visceral arches.**	**Distinct forelimb-bud.** AER is preset on hindlimb-buds. Genital primordial appears.
9	9 (L)	**Eyes pigmented** in the iris. AER is present on forelimb-buds.	Liver is visible.
10	10 (E)	**Limb-buds symmetrical**, digital plate absent.	**Five visceral arches are present.**
11	12 (E)	Note constriction for proximal and distal elements is marked in the limbs.	Retina is dark grey. **Visceral arches are reduced.** Maxillary and lateral nasal processes are fused, mandibular process is inconspicuous. **Limbs enlarged.**
12	12		**Trunk flexure well developed.**
15	15	Eyelid development: nictitating membrane is present.	**Visceral arches are not visible. Lower jaw conspicuous. Hand plate and footplate form**, but smooth. Trunk-flexure less bent.
18	17 (L)-18 (E)	**Lower jaw extends to the end of the upper jaw.** Scales are absent.[a]	Upper eyelids development. **Jaw is hook-shaped. Cervical flexure straightens. Limbs digital condensations.**
21	18		Lower eyelids development. **Limbs interdigit tissue receding.**
24	19	**Lower jaw is straight.**	Upper and lower lids are distinct.
27	20 (L)	**Interdigital tissue disappears**, fingers and toes get lengthened. Sensory papillae are present.	**Jaw formed. Nail anlagen on 1st, 2nd, and 3rd toes**, but not on fingers. **Scales evident on dorsal and ventral aspect, nuchal scutes beginning to appear.**
30	21 (L)	**First indication of pigmentation.** [a]Nuchal scutes are not established.	Circumference of eyelids oval. **Nail anlagen are now present on 1st, 2nd, and 3rd fingers.**
33	22 (L)	Eyelids are not able to close. **Heart is not longer visible. Claws are enclosed in "sheaths," curved** and pigmented. Sensory papillae clearly evident.	Scales absent on the proximal and distal elements of the limbs. Pigmentation is still faint.
36–39	23 (L)	**Lower eyelid covers less than half of the pupil.** Claws formation is complete.	**Pigmentation more extensive, but embryos are a light brown color.** Scales present on the elements of the limbs, nuchal scutes established.
42–51	24	Midbrain is not enclosed by bone.	**Eyelids are able to close. Embryos are a dark brown color.**
54–57	25		**Brain is completely enclosed by bone.** Musk glands visible.
	26	This stage does not occur in Caiman latirostris, which is characterized by eruption of egg teeth.	
60–63	27		Withdrawal of remaining yolk into abdominal cavity.
66	28		Yolk scar diminishing.

Bold font represents features of *C. latirostris* embryos that are useful for aging embryos.

Abbreviation: AER, apical ectodermal ridge.

[a] Represents specific characteristics.

the esophagus, gallbladder, liver, and trachea were all duplicated. Radiographic and ultrasonographic imaging findings were confirmed at necropsy (Palmieri et al., 2013). **Figure 4.36** is a radiographic image of a dicephalic Loggerhead Sea Turtle (*Caretta caretta*) that survived for about 1 week.

4.6.2.2 Spine Deformities Spinal deformities have been reported in sea turtles with a prevalence of 0.11%, based on fieldwork and a retrospective review of the literature (Rhodin et al., 1984). However, the authors noted that Indonesian Green Turtles displayed almost 10 times this rate with an overall combined incidence of 1.0% for cases of lordosis and kyphosis (Rhodin et al., 1984). Spinal deformities are also reported in freshwater and terrestrial chelonia. An extreme case of kyphosis was described in a Florida Softshell Turtle (*Apalone ferox*), although no radiographs were obtained (Taylor and Mendyk, 2017). This turtle, which was found in a Fleming Island, Florida parking lot, was described as though its carapace was draped over a "large grapefruit." In 1978, Plymale et al. presented a detailed description of kyphosis in a Mexican Yaqyi Slider (*Trachemys* [*Chrysemys scripta*] *yaquia*]). This specimen was a mature female that had been stuffed and prepared for the curio/souvenir trade. The authors radiographed the specimen and include the image in their paper. They also present a detailed table of kyphosis in turtles published to that point. Their table lists 21 species belonging to 5 different families. In **Table 4.4** their work has been expanded to include those reports published after 1978.

4.6.2.3 Tail Deformities Tail deformities are widely reported in a number of species including Alligator Snapping Turtles (*Macroclemys temminickii*) (McCallum and Trauth, 2000), Blanding's Turtles, (*Emydoidea blandingii*) (Ryan, 1986), Common Snapping Turtles (Bishop et al., 1998; de Solla et al., 2008; Davy and Murphy, 2009), and Diamondback Terrapins (Hildebrand, 1930). In many cases tail deformities, duplication, or even missing tails accompany other types of physical deformities. **Figure 4.37** shows the split tail of a mature common snapping turtle from North Carolina. It is unclear whether the animal hatched this way or if the condition resulted from trauma or some other disease process. Minor tail deformities, when unaccompanied by other problems, may not significantly impact survival. However, a condition of Alligator Snapping Turtles known as curly tail may cause hatchlings to drown or be washed away, as their normally prehensile tail is not functional (McCallum and Trauth, 2000).

4.6.2.4 Limb Deformities A number of limb deformities have been reported in the Testudines. Bárcenas-Ibarra et al. (2015) reported a large variety of limb deformities in hatchling Green, Hawksbill (*Eretmochelys imbricata*), and Olive Ridley Turtles from the Mexican Pacific. These included adactyly, amelia, aplastic claws, brachydactyly, diplopodia, dysmelia, hemimeliamacrodactyly, micromelia, phocomelia, polydactyl, and syndactyly.

Table 4.4 Comprehensive Listing of Published Cases of Spinal Deformities in Turtles

Family	Species	References
Chelonidae	*Caretta caretta*	Coker (1910)
	Chelonia mydas	Rhodin et al. (1984)
	Eretmochelys imbricata	Bárcenas-Ibarra et al. (2015)
	Lepidochelys olivacea	Bárcenas-Ibarra et al. (2015)
Chelydridae	*Chelydra serpentina*	Cahn (1937)
Emydidae	*Chrysemys picta*	Ernst (1971)
	C. p. marginata	Werner (1959)
	C. p. bellii X	
	C. p. marginata	Necker (1940)
	C.scripta scripta	Carr (1952)
	C. scripta yaquai	Plymale et al. (1978)
	Graptemys pseudogeographica ouachitensis	Carpenter (1958)
	Malaclemys terrapin centrata	Hildebrand (1930)
	Melanochelys trijuga thermalis	Deraniyagala (1939)
	Pseudemys floridana	Hildebrand (1938)
	Terrapene carolina triunguis	Black (1976)
	Terrapene ornate	Fox (1941)
Kinosternidae	*Kinosternum subrubrum*	Mertens (1940)
	Sternotherus odoratus	Nixon and Smith (1949)
Testudinidae	*Testudo hermanni*	Wandolleck (1904)
Trionychidae	*Apalone ferox*	Nixon and Smith (1949)
		Taylor and Mendyk (2017)
	Apalone mutica	Smith (1947)
	Apalone spinifera	Cahn (1937); Smith (1947)
	Apalone spinifera emoryi	Smith (1947)
	Palea steindachneri	Gressitt (1936)
	Pelodiscus sinensis	Vogt (1922); Gressitt (1936)
	Trionyx triunguis	Mertens (1940)

Figure 4.38 shows complete amelia in a Loggerhead Sea Turtle hatchling and **Figure 4.39** shows partial amelia in an Olive Ridley hatchling. Other species with limb deformities include common snapping turtles (Bishop et al., 1998; Bell et al., 2006; de Solla et al., 2008) and Eastern Painted Turtles (Bell et al., 2006). Bell et al. (2006) reported adactyly, amelia, apodia, and dysmelia in snapping turtle hatchlings exposed to abnormally high levels of polycyclic aromatic hydrocarbons (PAH).

4.6.2.5 Schistosomus Reflexus Syndrome Schistosomus reflexus syndrome (SRS) is a fatal condition where parts of the viscera are exposed via a large herniation of the abdominal wall. This condition has been reported with some frequency in ruminants but only once in the chelonia, and that was in Olive Ridley Sea Turtles (Bárcenas-Ibarra et al., 2017). In this report the authors describe a relatively low prevalence of SRS (0.6%) while examining over 20,000 embryos. Most of the 124 cases were associated with other abnormalities, including shell, neck, tail, and flipper deformities. **Figure 4.40a,b** shows the dorsal and ventral aspects of a hatchling Olive Ridley with SRS.

4.6.2.6 Ocular Malformations Ocular malformations are well documented in turtles (Bishop et al., 1998; Bell et al., 2006; de Solla et al., 2008; Bárcenas-Ibarra et al., 2015; Lanteri et al., 2017). In some cases, an eye or both eyes are missing, in other cases the eyes are present but malformed, and cyclopia (**Figure 4.41**) has been reported in some hatchlings. One paper (Bárcenas-Ibarra et al., 2015) reported seven different types of ocular deformities in sea turtle hatchlings (anophthlamia, cyclopia, eyelid hyperplasia, macrophthalmia, microphthalmia, paroxistic cyclops, and synophthalmia).

4.6.2.7 Cephalic Malformations Cephalic malformations are well reported, especially in hatchling sea turtles. Bárcenas-Ibarra et al. (2015) reported eight different types of cephalic anomalies (acephaly, anencephaly, bicephaly, encephalocele, exencephaly, hydrocephaly, microcephaly, and skull dysostosis) from three species of sea turtles (Hawksbills, Loggerheads, and Olive Ridleys). Other species where general cephalic malformations have been reported include common snapping turtles (Bell et al., 2006; de Solla et al., 2008), Diamondback Terrapins (Hildebrand, 1930), Green Sea Turtles (Türkozan and Durmus, 2001), and a Red-Eared Slider (Lanteri et al., 2017). **Figure 4.42** shows a case of diprosopus, or craniofacial duplication, in a Loggerhead Sea Turtle hatchling. This condition is extremely rare in vertebrates, and affected animals and humans rarely survive (Borzabadi-Farahani et al., 2012).

4.6.2.8 Albinism Albinism is an anomaly that has been reported with some frequency in turtles. The following species are represented in the literature: *Malaclemys terrapin* (Hildebrand, 1930), *Caretta caretta* (Marcovaldi et al., 1995; Türkozan and Durmus, 2001), and *Chelonia mydas* (Harrison, 1963; Fletemeyer, 1977; Türkozan and Durmus, 2001; Sönmez and Özdilek, 2011). In at least one case the incidence was quite high, with 22 out of 98 viable *Caretta caretta* on a Brazil beach being albino (Marcovaldi et al., 1995). **Figures 4.43–4.44** illustrate albinism in the Common Snapping Turtle and Leatherback Sea Turtle, respectively, and **Figure 4.45** shows amelanism in a Loggerhead Sea Turtle.

4.6.2.9 Twinning A single egg containing two embryos appears to be relatively uncommon in most chelonian species (Yntema, 1970). Several species producing twins have been documented in the literature and these have been compiled in **Table 4.5**. There are undoubtedly numerous natural occurrences of twinning in chelonia, including that of the Eastern River Cooter (*Pseudemys concinna*) (**Figure 4.46a,b**). In this case, the smaller twin was not viable and was surgically separated from the larger twin. In some cases, the twinning is conjoined and the distinction between bicephaly and twinning becomes imprecise.

Table 4.5 Comprehensive Listing of Published Cases of Twinning in Turtles

Family	Species	References
Chelonidae	*Chelonia mydas*	Glaesner (1924)
Chelydridae	*Chelydra serpentina*	Yntema (1970, 1971)
Emydidae	*Malaclemys terrapin*	Hildebrand (1938)
	Terrapene carolina	Crooks and Smith (1958)
	Pseudemys concinna	Lewbart, unpublished
Testudinidae	*Gopherus polyphemus*	Hunsaker (1968)
	Geochelone pardalis	Cooper (2009)
	Testudo hermanni	Harkewicz (2002)

4.6.3 Crocodylia

The literature for developmental anomalies in crocodilians is mostly limited to case reports. Ferguson (1985a,b) summarized the literature to that time and included a table that lists referenced, personal observation, and personal communication data for a wide variety of deformities and anomalies. Ferguson used the American Alligator as a model for craniofacial development, and as part of this research, investigated the effects of the teratogens 5-fluoro-2-desoxyuridine and hydrocortisone (Ferguson, 1981a,b,c).

One of the most extensive works is a color atlas of crocodile diseases that highlights a number of deformities and malformations (Youngprapakorn et al., 1994). While much of this information is not in the peer-reviewed literature, this is a valuable resource for those working with Old World crocodiles. Singh and Bustard (1982) describe over a dozen different congenital deformities of hatchling gharials (*Gavialis gangeticus*). It is thought these abnormalities were a result of egg desiccation during incubation.

4.6.3.1 Spine Deformities Spinal deformities have been reported in the Gharial (Singh and Bustard, 1982), Mugger Crocodiles (*Crocodylus palustris*) (Singh et al., 2001), and the Siamese Crocodile (*Crocodylus siamensis*) (Youngprapakorn et al., 1994). Singh and Bustard identified 34 cases of spinal deformity out of 476 fertile eggs that were incubated in captivity. Interestingly, of the 21 animals with "bent necks," all but four were normal 1 month after hatching. **Figure 4.47** illustrates abnormal cervical flexion in juvenile gharials. In crocodiles it appears that a condition known as "hunch back" is most likely related to nutrition and husbandry of the young animals and not a congenital problem (Singh et al., 2001). A very low number of spinal deformities were noted in American Alligator hatchlings (Sepúlveda et al., 2006) exposed to high levels of organochlorines. In this study, just 3% of 236 hatchlings (eight animals) displayed some type of deformity. Scoliosis and tail kinking accounted for almost a third of these cases. Similarly, a low percentage (<5%) of spine and tail deformities were noted in cultured American and Orinoco (*Crocodylus intermedius*) Crocodiles on farms in Venezuela (Boede and Sogbe, 2000).

4.6.3.2 Tail Deformities Singh and Bustard (1982) report a condition known as "bent tail" in gharial hatchlings. The condition persists as the animals grow but does not appear to be a fatal condition. Affected animals may have compromised swimming ability. Youngprapakorn et al. (1994) illustrate a condition referred to as "docked tail." In such cases the crocodiles hatch with a partial or completely absent tail. The animals eat and grow but have difficulty navigating deep water. Clubbed tail has also been seen in Orinoco Crocodile hatchlings.

4.6.3.3 Limb Deformities A number of limb deformities have been reported in captive-reared Siamese Crocodiles (Youngprapakorn et al., 1994). These include amelia, supernumerary limb, and polydactyly. Extra digits were reported in American Alligators at a very low prevalence (Sepúlveda et al., 2006).

4.6.3.4 Ocular Malformations A number of ocular deformities have been reported in captive-reared Siamese Crocodiles (Youngprapakorn et al., 1994). These include aniridia, anophthalmia, exophthalmia, microphthalmia, and sealed nictitans. **Figure 4.48** illustrates microphthalmia in a crocodile. Ferguson (1981b, 1985a,b) reported cyclopia in the American Alligator as both the result of a teratogen and spontaneously.

4.6.3.5 Cephalic Malformations A number of cephalic deformities have been reported in captive-reared Siamese Crocodiles (Youngprapakorn et al., 1994). These include meningoencephalocele, cleft chin, cleft lip, and cleft palate. Ferguson (1981a,c) reviews the development of the crocodilian palate and reports cleft palate in hatchling American Alligators.

4.6.3.6 Twinning Sepúlveda et al. (2006) reported one case of conjoined American Alligator embryos. These embryos died before hatching. Tryon (1980) reported twinning in hatchling West African Dwarf Crocodiles (*Osteolaemus tetraspis*).

4.7 Review of Normal Tuatara (*Sphenodon punctatus*) Development

4.7.1 Reproductive Biology and Natural History

The tuatara is a sexually dimorphic reptile that is closely related to lizards but is the only extant member of its ancient linage, the Rhynchocephalia (also termed the Sphenodontia). Much of the biology and health of Tuataras are described in detail by Cree (2014), and that reference provided much of the foundation information briefly summarized here. The current taxonomic status is that the sole species is *Sphenodon punctatus*. It is a New Zealand endemic formed of populations or assemblages with distinctive and important geographic variants (Hay et al., 2010). Hatchling and small juvenile Tuataras generally are not sexually dimorphic (Cree, 2014); however, at least some partially grown juvenile males develop dimorphic features such as throat coloration, dorsal spines, and neck elongation.

Following puberty, males reproduce annually. Females reproduce interannually (ranging from 2–5 years between clutches, with 4–5 years between clutches being average). Females lay a single clutch when they nest (Cree, 2014).

Tuataras have TSD and is the only taxon known to have females develop from cool incubated eggs. Laboratory studies showed that eggs incubated at constant temperatures produced females 18 to ~21.5°C and males develop from eggs incubated between 21.6–24°C (Cree et al., 1995; Nelson et al., 2004; Mitchell et al., 2006). Hatchlings have gonads and ducts that are differentiated and can be distinguished laparoscopically and histologically but lack secondary sex characteristics until at least half grown (males develop a more predominant dorsal crest and larger head than females; Cree, 2014). The gonads and ducts of hatchlings and very young juveniles are positioned in close association with the kidneys (Cree, 2014). Gonads are paired and located to either side of the vertebral column in the posterior half of the abdomen. Adults lack intromittent organs; however, reproductively active males partially evert the cloaca and mating is by cloacal apposition.

Ovaries are suspended from the dorsal coelom by mesovarium, and each oviduct (that will form the infundibulum, uterine tube, uterus, and vagina) is suspended by mesotubarium. The mesovarium also suspends the adrenal grand on each side.

Testes are ovoid, cream-colored structures. Each testis and the ipsilateral adrenal and vas deferens are suspended from the dorsal coelom by the mesorchium.

Mating occurs in the Austral summer (January–March) and egg deposition occurs the following November–December (Cree, 2014). Cree et al. (1992) documented that tuatara eggs remain in the oviduct for 6–8 months prior to oviposition.

4.7.2 Nests and Incubation

The eggs and nests of *Sphenodon punctatus* are well described. Eggs are parchment shelled (flexible shelled). The eggshell is described as having two distinct layers: an outer calcium carbonate later with calcium "pillars" that penetrate through an inner fibrous eggshell membrane. The eggs are oval and contain a relatively small about of albumen compared to turtle or squamate eggs (**Figure 4.49**). Clutches average 9–10 eggs (range 1–18) for Stephens Island Tuataras with mean egg masses of 4.4–4.9 g and lengths and widths averaging 2.7 cm long and 1.9 cm wide for naturally laid eggs (Cree, 2014). Females migrate up to several hundred meters from vegetated areas to nest in sunny open areas of soil (Thompson et al., 1996). Females compete for nest sites. Nesting tuataras dig underground nests (**Figure 4.50**) and sometimes dig more than one nest cavity per day (Cree, 2014). Oviposition typically occurs in a single night in just one nest hole, and that hole is refilled with soil and grass. Nest guarding may occur for several nights following oviposition.

Incubation is unattended after the first few days and extends for ~11-months (McIntyre, 1997). The embryos develop under a warm male: cool female TSD pattern (Cree et al., 1995; Nelson et al., 2002). Hatching asynchrony occurs such that eggs may vary in hatch date by as much 25 days (Thompson, 1989; Thompson et al., 1996).

4.7.3 Embryogenesis

Embryogenesis appears to follow gross patterns of amniote developmental events while development proceeds. The earliest embryos undergo basic differentiation to form a blastoderm. There is very little growth and early development within the oviduct in spite of the prolonged period in which eggs are oviducal. Post oviposition, embryonic development remains exceptionally slow (Dendy, 1899; Cree, 2014). Embryonic stages of tuatara eggs, collected on Stephens Island over several years, were described by Dendy (1899). Cree (2014) summarizes his staging system that is lettered (rather than more common numbering) and stages are demarked by the appearances of several unique (or at least rare) structures as well as major structural events common in vertebrate development. For example, the primitive streak is present in stage D, the amnionic cavity forms a posterior extension in stage F, the parietal eye appears at stage K, the allantois forms at stage L, and limb buds appear at stage M (Dendy, 1899).

4.7.4 Hatchling to Adult

Hatchlings average 5.3 cm in snout–vent length and ~4.5–4.6 g (**Figure 4.51**). Initial egg size significantly affects hatchling size, and growth rate of posthatchlings (**Figure 4.52**) is affected by egg size and by incubation temperatures during the first 10 months (Nelson et al., 2004). Body size appears to be poorly correlated with age in *Sphenodon* (as well as other reptiles) and age to maturity is likely variable, with males growing until they are 25–35 years old and females until approximately 20–25 years old (Castanet et al., 1988). Estimates from skeletochronology and reanalysis of mark-recapture size data from the 1940s suggest an age to maturity of 11–13 years for males and about 17 years for females (Castanet et al., 1988; Cree, 2014).

4.8 Developmental Anomalies in the Tuatara (*Sphenodon punctatus*)

There are no published reports of developmental anomalies in *Sphenodon punctatus*. It would be highly unlikely that these vertebrates lack developmental anomalies that are seen in all other reptiles. The absence of a literature likely reflects a research opportunity. The sole mention of *Sphenodon* developmental pathology is that embryonic death is associated with rapid collapse of the egg (Blanchard, 2002).

Acknowledgments

The authors thank Drs. Sávio Stefanini Sant'Anna (Laboratory of Herpetology, Butantan Institute, Brazil), Steve Barten (Vernon Hills Animal Hospital, Mundelin, Illinois), Nicola Mitchell (University of Western Australia), Anna Carter (Victoria University and Iowa State University), Adrienne Breaux (Wright Veterinary Medical Center Bethlehem, Pennsylvania), Anne Bárcenas-Ibarra (Centro de Investigación en Alimentación y Desarrollo, Mazatlán, Sinaloa, Mexico), and Elliot Jacobson (College of Veterinary Medicine, University of Florida) for the photos ceded to this chapter. We also thank Dr. Michael M. Garner and Dr. Elliot Jacobson for reviewing this chapter and providing comments.

References

Ahboucha S, and Gamrani H. 2001. Differences in protein expression in the subcomissural organ of normal and lordotic lizard (*Agama impalearis*). *Metab Brain Dis* 16:219–226.

Aldeman HB. 1936. The problem of cyclopia. Parts I and II. *Q Rev Biol* 11:284–304.

Amaral A. 1934. Notas sobre xantismo de ophidios. III. Um caso de xanthismo e um novo caso de albinismo, observados no Brasil. *Mem Inst Butantan* 8:151–153.

Andrade DV, and Abe AS. 1993. Malformações em ninhadas de caiçaca, *Bothrops moojeni* (Serpentes: Viperidae). *Mem Inst Butantan* 54:61–67.

Andrews RM. 2004. Patterns of embryonic development, in *Reptilian Incubation. Environment, Evolution and Behavior.* Deeming DC (Ed.), Nottingham University Press, Nottingham, United Kingdom, pp. 75–102.

Andrews RM. 2007. Effects of temperature on embryonic development of the veiled chameleon, *Chamaeleo calyptratus*. *Comp Biochem Physiol* 148A:698–706.

Andrews RM, Brandley MC, and Greene VW. 2013. Developmental sequences of squamate reptiles are taxon specific. *Evol Dev* 15:326–343.

Andrews RM, Díaz-Paniagua C, Marco A, and Portheault A. 2008. Developmental arrest during embryonic development of the Common Chameleon (*Chamaeleo chamaeleon*) in Spain. *Physiol Biochem Zool* 81:336–344.

Andrews RM, and Donoghue S. 2004. Effects of temperature and moisture on embryonic diapause of the veiled chameleon (*Chamaeleo calyptratus*). *J Exp Zool* 301A:629–635.

Andrews RM, and Greene VW. 2011. Character variation in embryonic development of the lizard *Uta stansburiana*. *J Herpetol* 45:451–454.

Andrews RM, and Mathes T. 2000. Natural history of reptilian development: Constraits on the evolution of viviparity. *Bioscience* 50:227–238.

Andrews RM, Pezaro N, Doody JS, Guarino F, and Green B. 2017. Oviposition to hatching: Development of *Varanus rosenbergi*. *J Herpetol* 51:396–401.

Balfour FM. 1880. *A Treatise on Comparative Embryology*, Vol 1. Macmillan Publishers Limited, London.

Bárcenas-Ibarra A, de la Cueva H, Rojas-Lleonart I, Abreu-Grobois FA, Lozano-Guzmán RI, Buevas E, and García-Gásca A. 2015. First approximation to congenital malformation rates in embryos and hatchlings of sea turtles. *Birth Defects Res A* 103:203–224.

Bárcenas-Ibarra A, Rojas-Lleonart I, Lozano-Guzmán RI, and García-Gasca A. 2017. Schistosomus reflexus syndrome in olive ridley sea turtles (*Lepidochelys olivacea*). *Vet Path* 54(1):171–177.

Bauer R, Plieschnig JA, Finkes T, Riegler B, Hermann M, and Schneider WJ. 2013. The developing chicken yolk sac acquires nutrient transport competence by an orchestrated differentiation process of its endodermal epithelial cells. *J Biol Chem* 288:1088–1098.

Baur G. 1891. On intercalation of vertebrae. *J Morphol* 4:331–336.

Bechtel HB, and Bechtel E. 1989. Color mutation in the corn snake (*Elaphe guttate guttata*): Review and additional breeding data. *J Hered* 4:272–276.

Beggs K, Young J, Georges A, and West P. 2000. Ageing the eggs and embryos of the pig-nosed turtle, *Carettochelys insculpta* Chelonia: Carettochelydidae, from northern Australia. *Can J Zool* 783:373–392.

Bell B, Spotila JR, and Congdon J. 2006. High incidence of deformity in aquatic turtles in the John Heinz National Wildlife Refuge. *Environ Poll* 142:457–465.

Bellairs A d'A. 1965. Cleft palate, microphthalmia and other malformations in embryos of lizards and snakes. *J Zool* 2:239–252.

Bellairs A d'A. 1981. Congenital and developmental diseases, in *Diseases of the Reptilia*, Vol 2. Cooper JE and Jackson OF (Eds.), Academic Press, London, pp. 469–485.

Bellairs A d'A, and Boyd JD. 1957. Anomalous cleft palate in snake embryos. *Proc Zool Soc* 129:525–539.

Bellairs A d'A, and Gamble HJ. 1960. Cleft palate, microphthalmia and other malformations in an embryo lizard (*Lacerta vivipara*). *Brit J Herpetol* 2:171–176.

Belluomini HE. 1959. Bicefalia em *Xenodon merremii* (Wagier 1824) (Serpentes). *Mem Inst Butantan* 28:85–89.

Bertocchini F, Alev C, Nakaya Y, and Sheng G. 2013. A little winning streak: The reptilian-eye view of gastrulation in birds. *Dev Growth Differentiation* 55:52–59.

Billy AJ. 1986. Developmental deformities in the parthenogenetic lizard *Cnemidophorus uniparens* (Teiidae) and the "anomalous male" phenomenon. *Can J Zool* 64:2418–2424.

Billy AJ. 1988. Observations on the embryology of the unisexual lizard *Cnemidophorus uniparens* (Teiidae). *J Zool* 215:55–81.

Binns W, James LF, Shupe JL, and Everett GA. 1963. Congenital cyclopian-type malformation in lambs induced by maternal ingestion of a range plant, *Veratrum californicum*. *Amer J Vet Res* 24:1164–1175.

Bishop CA, Ng P, Pettit KE, Kennedy SW, Stebeman JJ, Norstrom RJ, and Brooks RJ. 1998. Environmental contamination and developmental abnormalities in eggs and hatchlings of the common snapping turtle (*Chelydra serpentina serpentina*) from the Great Lakes-St Lawrence River basin (1989–1991). *Environ Poll* 101:143–156.

Black JH. 1976. Kyphosis in the three-toed box turtle. *Chelonia* 2:2–3.

Blackburn DG. 1982. Evolutionary origins of viviparity in the Reptilia. I. Sauria. *Amphib-Reptilia* 3:185–205.

Blackburn DG. 1985. Evolutionary origins of viviparity in the Reptilia. II. Serpentes, Amphisbaenia, and Ichthyosauria. *Amphib-Reptilia* 5:259–291.

Blackburn DG. 1993a. Chorioallantoic placentation in squamate reptiles: Structure, function, development, and evolution. *Jour Exp Zool* 266:414–430.

Blackburn DG. 1993b. Standardized criteria for the recognition of reproductive modes in squamate reptiles. *Herpetologica* 49:118–132.

Blackburn DG. 1995. Saltationist and punctuated equilibrium models for the evolution of viviparity and placentation. *J Theor Biol* 174:199–216.

Blackburn DG. 2006. Squamate reptiles as model organisms for the evolution of viviparity. *Herp Monogr* 20:131–146.

Blackburn DG. 2015. Evolution of vertebrate viviparity and specializations for fetal nutrition: A quantitative and qualitative analysis. *J Morphol* 276(8):961–990.

Blackburn DG, and Sidor CA. 2014. Evolution of viviparous reproduction in Paleozoic and Mesozoic reptiles. *Int J Dev Biol* 58:935–948.

Blackburn DG, and Starck JM. 2015. Morphological specializations for fetal maintenance in viviparous vertebrates: An introduction and historical retrospective. *J Morphol* 276(11):E1–E16.

Blackburn DG, and Stewart JR. 2011. Viviparity and placentation in snakes, in *Reproductive Biology and Phylogeny of Snakes*. Aldridge RD, and Sever DM (Eds.), CRC Press, Boca Raton, FL, pp. 119–181.

Blanchard B. 2002. *Tuatara captive management plan and husbandry manual. Threatened Species Occasional Publication 21.* Department of Conservation, Wellington, New Zealand.

Blanc F. 1974. Table de développement de *Chameleo lateralis* Gray, 1831. *Annales D'Embryologie et de Morphogenese* 7:99–115.

Boback SM, Dichter EK, and Mistry HL. 2012. A developmental staging series for the African house snake, *Boaedon (Lamprophis) fuliginosus*. *Zool* 115:38–46.

Boede EO, and Sogbe AE. 2000. Diseases in Orinoco crocodile (*Crocodylus intermedius*) and American crocodile (*Crocodylus acutus*) kept in Venezuelan farms. *Revista Cientifica FCV-LUZ* 4:328–338.

Borzabadi-Farahani A, Yen SLK, Yamashita DD, and Sanchez-Lara PA. 2012. Bilateral maxillary duplication: Case report and literature review. *Oral Surg Oral Med Oral Pahtol Pral Radiol* 113:e29–e32.

Boughner JC, Buchtová M, Fu K, Diewert V, Hallgrímsson B, and Richman JM. 2007. Embryonic development of *Python sebae* - I: Staging criteria and macroscopic skeletal morphogenesis of the head and limbs. *Zool* 110:212–230.

Bowden RM, Carter AW, and Paitz RT. 2014. Constancy in an inconstant world: Moving beyond constant temperatures in the study of reptilian incubation. *Integr Comp Biol* 54(5):830–840.

Braund KG. 1987. Degenerative and developmental diseases, in *Veterinary Neurology*. Oliver JE, Hoerlein BF, Mayhew ZG (Eds.), Saunders, Philadelphia, p. 99.

Brent RL. 1976. Environmental factors: Miscellaneous, in *Prevention of Embryonic, Fetal, and Perinatal Disease*, Vol 3. Harris MI, and Brent RL (Eds.), Dhew Publication, Bethesda, pp. 1–18.

Brongersma LD. 1952. Notes upon the arteries of the lungs in *Python reticulatus*. *Proc K ned Akad Wet* 55:49–61.

Bu X, Jiang Y, Nie L, and Xia X. 2016. High frequency of multiple paternity in the Chinese softshelled turtle. *Aquac Res* 1–4, doi: 10.1111/are.13210

Cahn AR. 1937. The turtles of Illinois. *Illinois Biol Mongr* 16:1–218.

Carpenter CC. 1958. An unusual Ouachita map turtle. *Herpetologica* 14:116.

Carr A. 1952. *Handbook of Turtles*. Comstock Pub Co., Ithaca NY, 542.

Carvalho MPN. 2014. Comparative pathology of congenital malformations in *Bothrops jararaca* and *Crotalus durissus* from Southeast Brazil. Masters thesis. School of Veterinary Medicine and Animal Science, University of São Paulo, Brazil.

Carvalho MPN, Sant'Anna SS, Díaz-Delgado J, Grego KF, Gogone CVP, Queiroz-Hazarbassanov NGT, and Catão-Dias JL. 2019. Histopathological findings in the viscera of malformed neotropical viperids. *J Comp Pathol* 167:73–81.

Carvalho MPN, Sant'Anna SS, Grego KF, Fonseca-Pinto ACBC, Lorigados CAB, Queiroz-Hazarbassanov NGT, and Catão-Dias JL. 2017. Microcomputed tomographic, morphometric, and histopathologic assessment of congenital bone malformations in two neotropical viperids. *J Wildl Dis* 4:804–815.

Castanet J, Newman DG, and Girons HS. 1988. Skeletochronological data on the growth, age, and population structure of the tuatara, *Sphenodon punctatus*, on Stephens and Lady Alice Islands, New Zealand. *Herpetologica* 1:25–37.

Cepela MA, Nunery WR, and Martin RT. 1992. Stimulation of orbital growth by the use of expandable implants in the anophthalmic cat orbit. *Ophthal Plast Reconstr Surg* 8:157–167.

Chen S, Zhang L, Le Y et al. 2015. Sperm storage and spermatozoa interaction with epithelial cells in oviduct of Chinese Softshell Turtle, *Pelodiscus sinensis*. *Ecol Evol* 5(15):3023–3030.

Cohn MJ, and Tickle C. 1999. Developmental basis of limblessness and axial patterning in snakes. *Nature* 399:474–479.

Coker RE. 1910. Diversity in scutes of chelonia. *J Morph* 21:1–75.

Cooper JE. 1975. Exophthalmia in a rhinoceros viper (*Bitis nasicornis*). *Vet Rec* 16:130–131.

Cooper JE. 2009. Conjoined ("Siamese") twins of the leopard tortoise (*Geochelone pardalis*), with a plea for documentation for such abnormalities in reptiles. *J Herp Med Surg* 3:69–71.

Cordero GA, and Janzen FJ. 2014. An enhanced developmental staging table for the painted turtle, *Chrysemys picta* (Testudines: Emydidae). *J Morphol* 275(4):442–455.

Corsello G, and Giufrè M. 2012. Congenital malformations. *J Matern Neonatal Med* 25:25–29.

Costanzo JP, Dinkelacker SA, Iverson JB, and Lee Jr. RE. 2004. Physiological ecology of overwintering in the hatchling painted turtle: Multiple-scale variation in response to environmental stress. *Physiol Biochem Zool* 77(1):74–99.

Crastz F. 1982. Embryological stages of the marine turtle *Lepidochelys olivacea* (Eschscholtz). *Rev Biol Trop* 30(2):113–120.

Cree A. 2014. *Tuatara: Biology and Conservation of a Venerable Survivor*. Canterbury University Press, Christchurch.

Cree A, Cockrem JF, and Guillette LJ. 1992. Reproductive cycles of male and female tuatara (*Sphenodon punctatus*) on Stephens Island, New Zealand. *J Zool* 226(2):199–217.

Cree A, Thompson MB, and Daugherty CH. 1995. Tuatara Sex Determination. *Nature* 6532:543.

Crooks FD, and Smith PW. 1958. An instance of twinning in the box turtle. *Herpet* 14:170–171.

Cunningham B. 1937. *Axial Bifurcation in Serpents*. Duke University Press, Durham.

Darevsky IS. 1966. Natural parthenogenesis in a polymorphic group of Caucasian rock lizards related to *Lacerta saxicola* eversman. *J Ohio Herpet Soc* 5:115–152.

Darlington AR, and Davis RB. 1990. Reproduction of the pancake tortoise, *Malacochersus tornieri*, in captive collections. *Herpetol Rev* 21:16–18.

Davy DM, and Murphy RW. 2009. Explaining patterns of deformity in freshwater turtles using MacCulloch's hypothesis. *Canadian J Zool* 87:433–439.

Dean JN, Glenn JL, and Straight RC. 1980. Bilateral cleft labial and palate in the progeny of a *Crotalus virids virids*. *Herpetol Rev* 11:91–92.

Dendy A. 1899. Memoirs: Outlines of the Development of the Tuatara, *Sphenodon (Hatteria) punctatus*. *J Cell Sci* 2(165):1–87.

Deraniyagala PEP. 1939. *The Tetrapod Reptiles of Ceylon, Vol. 1: Testudinates and Crocodilians*. Colombo Museum, Sri Lanka. Z9: 1–412. Reprint edition (1994).

de Solla SR, Fernie KJ, and Ashpole S. 2008. Snapping turtles (*Chelydra serpentina*) as bioindicators in Canadian areas of concern in the Great Lakes Basin. II. Changes in hatching success and hatchling deformities in relation to persistent organic pollutants. *Environ Poll* 153:529–536.

Dhouailly D, and Saxod R. 1974. Les stades du developpement de *Lacerta muralis* Laur. entre la ponte dt l'eclosion. *Bull Soc Zool France* 99:489–494.

Dufaure JP, and Hubert J. 1961. Table de developpement du lezard vivipare: *Lacerta (Zootoca) vivipara* Jacquin. *Arch Anat Microsc Morphol Exp* 50:309–328.

Elinson RP, Stewart JR, Bonneau LJ, and Blackburn DG. 2014. Amniote yolk sacs: Diversity in reptiles and a hypothesis on their origin. *Int J Dev Biol* 58:889–894.

Elinson RP, and Stewart JR. 2014. The corn snake yolk sac becomes a solid tissue filled with blood vessels and yolk-rich endodermal cells. *Biol Lett* 10:20130870.

Enseley PK. 1978. Ophthalmic disorders in three snakes. *J Zoo Wildl Med* 9:57–58.

Eppley BL, Holley S, and Sadove AM. 1993. Experimental effects of intraorbital tissue expansion on orbitomaxillary growth in anophthalmos. *Ann Plast Surg* 31:19–26.

Ernst CH. 1971. Observations of the painted turtle, *Chrysemys picta*. *J Herpet* 4:216–220.

Ewert MA. 1985. Embryology of turtles, in *Biology of the Reptilia*. Gans C, Billet F, and Maderson P (Eds.), John Wiley & Sons, New York, pp. 75–267.

Ewert MA. 1991. Cold torpor, diapause, delayed hatching and aestivation in reptiles and birds, in *Egg Incubation: Its Effects on Embryonic Development in Birds and Reptiles*. Deeming C, and Ferguson MWJ (Eds.), Cambridge University Press, Cambridge, pp. 173–191.

Federsoni PAJ. 1981. Casos teratogênicos em *Bothrops atrox* (Serpentes: Viperidae: Crotalinae). *Mem Inst Butantan* 43:1978–1979.

Ferguson MWJ. 1981a. The structure and development of the palate in *Alligator mississippiensis*. *Arch Oral Biol* 26(5):427–443.

Ferguson MWJ. 1981b. Developmental mechanisms in normal and abnormal palate formation with particular reference to the aetiology, pathogenesis and prevention of cleft palate. *Brit J Orthodontics* 8(3):115–137.

Ferguson MWJ. 1981c. Review: The value of the American alligator (*Alligator mississippiensis*) as a model for research in craniofacial development. *J Craniofacial Gen Develop Biol* 1(1):123–144.

Ferguson MWJ. 1984. Craniofacial development in *Alligator mississippiensis*, in *The Structure, Development and Evolution of Reptiles*. Ferguson MWJ (Ed.), Academic Press, London, pp. 223–273.

Ferguson MWJ. 1985a. Reproductive biology and embryology of the crocodilians, in *Biology of the Reptilia*. Gans C, Billet F, and Maderson PFA (Eds.), John Wiley & Sons, Inc., New York, pp. 329–491.

Ferguson MWJ. 1985b. Reproductive biology and embryology of the crocodilians, in *Biology of the Reptilia*. Gans C, Billet F, and Maderson PFA (Eds.), John Wiley & Sons, New York, pp. 451–460.

Ferguson MWJ. 1987. Post-laying stages of embryonic development in crocodilians, in *Wildlife Management: Crocodiles and Alligators*. Webb GJW, Manolis SC, and Whitehead PJ (Eds.), Surrey Beatty & Sons Chipping Norton, NSW, pp. 427–444.

Fisk A, and Tribe M. 1949. The development of the amnion and chorion of reptiles. *Proc Zool Soc London* 24:83–114.

Fletemeyer J. 1977. Rare albino turtle. *Sea Frontiers* 23:233.

Fox H. 1941. Kyphoscoliosis in a painted tortoise. *Report Penrose Res Lab* 1941:16.

Fox W. 1948. Effect of temperature on development of scutellation in the garter snake, *Thamnophis elegans atratus*. *Copeia* 4:252–262.

Fox W, Gordon C, and Fox MH. 1961. Morphological effects of low temperature during embryonic development of the garter snake *Tamnophis elegans*. *Zoologica* 46:57–71.

Frye FL. 1991. Developmental anomalies, in *Reptile Care: An Atlas of Diseases and Treatments*, Vol 2. Frye FL (Ed.), T.H.F. Publications Inc, Neptune City, pp. 393–419.

Frye FL, and Carney JD. 1974. Achondroplastic dwarfism in a turtle. *Vet Med/Small Anim Clin* 69:299–301.

Girling JE. 2002. The reptilian oviduct: A review of structure and function and directions for future research. *J Exp Zool A: Ecol Gen Physiol* 293(2):141–170.

Gist DH, and Jones JM. 1989. Sperm storage within the oviduct of turtles. *J Morphol* 199:379–384.

Glaesner L. 1924. Ueber drei doppelbildungen von *Chelonia mydas*. *Zool. Anz* 60:185–194.

Glaw F, Kohler J, Townsend TM, and Vences M. 2012. Rivaling the world's smallest reptiles: Discovery of miniaturized and micro-endemic new species of leaf chameleons (*Brookesia*) from Northern Madagascar. *PLoS ONE* 7:e31314.

Gomez C, Özbudak EM, Wunderlich J, Baumann D, Lewis J, and Pourquié O. 2008. Control of segment number in vertebrate embryos. *Nature* 454:335–339.

Gomez C, and Pourquie O. 2009. Developmental control of segment numbers in vertebrates. *J Exp Zool B: Mol Dev Evol* 312:533–544.

Greenbaum E. 2002. A standardized series of embryonic stages for the emydid turtle *Trachemys scripta*. *Can J Zool* 80(8):1350–1370.

Greenbaum E, and Carr JL. 2002. Staging criteria for embryos of the spiny softshell turtle, *Apalone spinifera* (Testudines: Trionychidae). *J Morph* 254(3):272–291.

Gregorovicova M, Zahradnicek O, Tucker AS, Velensky P, and Horacek I. 2012. Embryonic development of the monitor lizard, *Varanus indicus*. *Amphib-Reptilia* 33:451–468.

Gressitt JL. 1936. Camel-back turtle. *Zoologica* 39:246.

Grigoriev DV. 2014. Giant *Mosasaurus hoffmanni* (Squamata, Mosasauridae) from the Late Cretaceous (Maastrichtian) of Penza, Russia. *Proceedings of the Zoological Institute RAS*, 318, pp. 148–167.

Guerra-Fuentes RA, Daza JD, and Bauer AM. 2014. The embryology of the retinal pigmented epithelium in dwarf geckos (Gekkota: Sphaerodactylinae): A unique developmental pattern. *BMC Devel Biol* 14:29.

Guillette LJ Jr. 1982. The evolution of viviparity and placentation in the high altitude Mexican lizard *Sceloporus aeneus*. *Herpetologica* 38:94–103.

Harkewicz KA. 2002. Prolapse of the penis in a pair of co-joined twin male Hermann's tortoises (*Testudo hermanni* spp.). *Proc Assoc Rept Amph Vet*, Reno, Nevada, pp. 11–12.

Harrison T. 1963. Notes on albino turtles: 14. Albino green sea turtles and sacred ones. *Sarawak Museum J* 11:304–306.

Hartmann PA, Hartmann MT, and Giasson LOM. 2003. Uso do hábitat e alimentação em juvenis de *Bothrops jararaca* (Serpentes, Viperidae) na Mata Atlântica do sudeste do Brasil. *Phyllomedusa* 2:35–41.

Hay JM, Sarre SD, Lambert DM, Allendorf FW, and Daugherty CH. 2010. Genetic diversity and taxonomy: A reassessment of species designation in tuatara (*Sphenodon*: Reptilia). *Conserv Genet* 11:1063–1081.

Head JJ, Bloch JI, Hastings AK, Bourque JR, Cadena EA, Herrera FA, Polly PD, and Jaramillo CA. 2009. Giant boid snake from the Palaeocene neotropics reveals hotter past equatorial temperatures. *Nature* 457:715–717.

Head JJ, and Polly PD. 2015. Evolution of the snake body form reveals homoplasy in amniote Hox gene function. *Nature* 520:86–89.

Hedges SB. 2008. At the lower size limit in snakes: Two new species of threadsnakes (Squamata: Leptotyphlopidae: *Leptotyphlops*) from the Lesser Antilles. *Zootaxa* 1841:1–30.

Hedges SB, and Thomas R. 2001. At the lower size limit in amniote vertebrates: A new diminutive lizard from the West Indies. *Caribb J Sci* 37:168–173.

Heinz GW, Clunie DA, and Mullaney PB. 1998. The effect of buphthalmos on orbital growth in early childhood: Increased orbital soft tissue volume strongly correlates with increased orbital volume. *J AAPOS* 2:39–42.

Hibberd EM. 1996. Abnormalities in eggs, embryos and juvenile farmed esturarine crocodiles *Crocodylus porosus*. *Proceedings for the 13th Working Meeting of the Crocodile Specialist Group, IUCN – The World Conservation Union, Gland,* Switzerland, pp. 290–295.

Hildebrand SF. 1930. Duplicity and other abnormalities in diamondback terrapins. *J Elisha Mitchell Sci Soc* 46:41–53.

Hildebrand SF. 1938. Twinning in turtles. *J Hered* 29:243–253.

Hoge AR. 1952. Notas herpetológicas. *Mem Inst Butantan* 24:237–240.

Hoge AR, and Belluomini HE. 1957. Aberrações cromáticas em serpentes brasileiras. *Mem Inst Butantan* 28:95–98.

Holt WV, and Fazeli A. 2016. Sperm storage in the female reproductive tract. *Ann Rev Anim Biosc* 4:291–310.

Hoser R, and Gibbons D. 2003. A bicephalic coastal Queensland carpet snake, *Morelia spilota mcdowelli* (Serpentes: Pythonidae). *Herpetofauna* 33:111.

Hubert J. 1985. Embryology of the Squamata, in *Biology of the Reptilia*, Vol 15, Development B. Gans C, and Billett F. (Eds.), John Wiley & Sons, New York, pp. 1–34.

Hubert J, and Dufaure JP. 1968. Table de developpement de la vipere aspic: *Vipera aspis*. *Bull Soc Zool France* 93:135–148.

Hunsaker D II. 1968. Twinning in tortoises. *Turtle and Tortoise Soc J* 2(6):38.

Iungman J, Piña CI, and Siroski P. 2008. Embryological development of *Caiman latirostris* (Crocodylia: Alligatoridae). *Genesis* 46:401–417.

Jackson K. 2002. Post-ovipositional development of the monocled cobra, *Naja kaouthia* (Serpentes: Elapidae). *Zool* 105:203–214.

Jackson JF, Ingram W III, and Campbell HW. 1976. The dorsal pigmentation pattern of snakes as an antipredator strategy: A multivariate approach. *Am Nat* 110:1029–1053.

Janzen FJ, and Phillips PC. 2006. Exploring the evolution of environmental sex determination, especially in reptiles. *J Evol Biol* 19(6):1775–1784.

Jensen B, and Wang T. 2009. Hemodynamic consequences of cardiac malformations in two juvenile ball pythons (*Python regius*). *J Zoo Widl Med* 40:752–756.

Jessop TS, Madsen T, Sumner J, Rudiharto H, Phillips JA, and Ciofi C. 2006. Maximum body size among insular Komodo dragon populations covaries with large prey density. *Oikos* 112:422–429.

Johnson PTJ, Reeves MK, Krest SK, and Pinkey AE. 2010. A decade of deformities: Advances in our understanding of amphibian malformations and their implications, in *Ecotoxicology of Amphibians and Reptiles*. Sparlig DW, Linder G, Bishop CA, and Krest SK (Eds.), SETAC and CRC Press, Pensacola, pp. 524–549.

Kalter H, and Warkany J. 1983. Congenital malformations. Etiologic factors and their role in prevention. *N Engl J Med* 308:424–431.

Kar SK, and Bustard HR. 1982. Embryonic tail deformation in the saltwater crocodile (*Crocodylus porosus*, Schneider) in Orissa, India. *Brit J Herpetol* 6:221–222.

Kearney M, Fujita MK, and Ridenour J. 2009. Lost sex in reptiles: Constraints and correlations, in *Lost Sex*. Schon I, Martens K, and van Dijk P (Eds.), Springer, London, pp. 447–474.

Khannoon ER. 2015. Developmental stages of the climbing gecko *Tarentola annularis* with special reference to the claws, pad lamellae, and subdigital setae. *J Exp Z B: Mol Dev Evol* 324:450–464.

Khannoon ER, and Evans SE. 2014. The embryonic development of the Egyptian cobra *Naja h. haje* (Squamata: Serpentes: Elapidae). *Acta Zool* 95:472–483.

Khannoon ER, and Zahradnicek O. 2017. Postovipositional development of the sand snake *Psammophis sibilans* (Serpentes:Lamprophiidae) in comparison with other snake species. *Acta Zool* 98:144–153.

Khera KS, and Iverson F. 1980. Hydrocephalus induced by N-nitrosoethylenethiourea in the progeny of rats treated during gestation. *Teratology* 21:367–370.

King W. 1959. Vertebra duplication, an osteological anomaly widespread in snakes. *Herpetologica* 15:87–88.

Krecsák L. 2008. Albinism and leucism among European viperianae: A review. *Russ J Herpetol* 15:97–102.

Kusuda S, Yasukawa Y, Shibata H, Saito T, Doi O, Ohya Y, and Yoshizaki N. 2013. Diversity in the matrix structure of eggshells in the Testudines (Reptilia). *Zool Sci* 30(5):366–374.

Kvon EZ, Kamneva OK, Melo US et al. 2016. Progressive loss of function in a limb enhancer during snake evolution. *Cell* 167:633–642.e11.

Lanteri G, Macri F, Passantino A, Monteverde V, Marino F, and Mazzullo G. 2017. A case of diprosopiasis in *Trachemys scripta scripta*. *Vet Ital* 53(2):171–174.

Leal F, and Cohn MJ. 2016. Loss and re-emergence of legs in snakes by modular evolution of sonic hedgehog and HOXD enhancers. *Curr Biol* 26:2966–2973.

Legler JM, and Vogt RC. 2013. *The Turtles of Mexico: Land and Freshwater Forms*. University of California Press, Berkeley.

Leipold HW. 1980. Congenital defects in zoo and wild animals: A review, in *The Comparative Pathology of Zoo Animals*. Montalli R, and Migaki G (Eds.), Smithsonian Institution Press, Washington, pp. 457–470.

Leipold HW, Dennis SM, and Huston K. 1972. Polydactyly in cattle. *Cornell Vet* 62:337–345.

Lema T. 1982. Descrição de dois espécimes bicéfalos de *Liophis miliaris* (Linnaeus, 1758) (Serpentes, Colubridae). *Inheringia Zool* 61:67–79.

Lemus D. 1967. Contribucion al estudio de la embriologia de reptiles Chilenos. II. tabla de desarollo de la lagartija vivipara *Liolaemus gravenhorsti*. *Biol Chili* 40:39–61.

Lemus D, and Duvauchelle CR. 1966. Desarrollo intrauterino de *Liolaemus tenuis tenuis* (Dumeril Y Bibron). Contribucion al estudio del desarrollo embriologico de reptiles Chilenos. *Biol Chili* 39:80–98.

Lemus D, Illanes J, Fuenzalida M, Paz del la Vega Y, and Garcia M. 1981. Comparative analysis of the development of the lizard, *Liolaemus tenuis*. II. A series of normal postlaying stages in embryonic development. *J Morphol* 169:337–349.

Lindell LE. 1994. The evolution of vertebral number and body size in snakes. *Func Ecol* 8:708–719.

Lingham-Soliar T. 1995. Anatomy and functional morphology of the largest marine reptile known, *Mosasaurus hoffmanni* (Mosasauridae, Reptilia) from the Upper Cretaceous, Upper Maastrichtian of the Netherlands. *Philos Trans Royal Soc London* 347:155–180.

Lo AK, Colcleugh RG, Allen L, Van Wyck L, and Bite U. 1990. The role of tissue expanders in anophthalmic animal model. *Plast Reconstr Surg* 86:399–408.

Loehr VJT, Henen B, and Hofmeyr M. 2004. Reproduction of the smallest tortoise, the Namaqualand speckled padloper, *Homopus signatus signatus*. *Herpetologica* 60(4):444–454.

Luckett WP. 1977. Ontogeny of amniote membranes and their application to phylogeny, in *Major Patterns in Vertebrate Evolution*. Hecht MK, Goody PC, and Hecht BM (Eds.), Plenum Press, New York, pp. 439–516.

Magalhães, MS, Vogot RC, Sebben A, Dias LC, Oliveira MF, and Moura CEB. 2017. Embryonic development of the Giant South American River Turtle, *Podocnemis expansa* (Testudines: Podocnemididae). *Zoomorphology* 136:523.

Magnusson W. 1979. Production of an embryo by an Acrochordus javanicus isolated for seven years. *Copeia* 1979(4):744–745. doi:10.2307/1443886.

Mahmoud IY, Klicka J, and Hess GL. 1973. Normal embryonic stages of the western painted turtle, *Chrysemys picta bellii*. *J Morph* 141(3):269–279.

Marcovaldi M, Lima EP, and Penteado R. 1995. Albino sea turtle hatchlings in Brazil. *Marine Turtle Newsletter* 69:10.

Marques DSM, Ramos CH, Freitas AB, and Souza Júnior P. 2012. Espinha bífida oculta em um sagui-de-tufos-pretos (*Callithrix penicillata*). *Acta Sci Vet* 40:1093.

Maryan B. 2001. A note on bicephalic death adder. *Herpetofauna* 31:73.

Mathur JK, and Goel SC. 1974. A note on a tailless embryo if the lizard *Calotes versicolor*. *Brit J Herpetol* 5:420–422.

Maxie MG, and Newman SJ. 2007. Urinary system, in *Pathology of Domestic Animals*. Jubb KVF, Kennedy PC, Palmer NC (Eds.), Elsevier, Philadelphia, pp. 425–522.

McAllister CT, and Wallach V. 2007. Discovery of dicephalic western diamond rattlesnake, *Crotalus atrox* (Serpents: Viperidae), from Texas with a summary of dicephalism among members of the genus *Crotalus*. *J Arkansas Acad Sci* 60:67–73.

McCallum ML, and Trauth SE. 2000. Curly-tail malformity in hatchlings of the alligator snapping turtle, *Macroclemys temminickii* (Testudines: Chelydridae), from Northeastern Arkansas. *J Ark Acad Sci* 54:150–152.

McIntyre M. 1997. *Conservation of the Tuatara*. Wellington: Victoria University Press.

Mertens R. 1940. Der knochenpanzer einter kyphotischen weichschildkröte. *Senckenbergiana* 22:236–243.

Meylan PA, Schuler R, and Moler P. 2002. Spermatogenic cycle of the Florida softshell turtle, *Apalone ferox*. *Copeia* 3:779–786.

Miller JD. 1985. Embryology of marine turtles, in *Biology of the Reptilia*. Gans C, Billet F, and Maderson P (Eds.), John Wiley & Sons Inc., New York, pp. 269–328.

Miller JD, and Dinkelacker SA. 2007. Reproductive structures and strategies of turtles, in *Biology of Turtles: From Structures to Strategies of Life*. Wyneken J, Godfrey M, and Bels V (Eds.), CRC Press, Boca Raton, pp. 225–278.

Miller JD, and Limpus CJ. 2003. Ontogeny of marine turtle gonads, in *The Biology of Sea Turtles*, Vol. II. Lutz PL, Musick JA, and Wyneken J (Eds.), CRC Press, Boca Raton, pp. 199–224.

Miller JD, Mortimer JA, and Limpus CJ. 2017. A field key to the developmental stages of marine turtles (Cheloniidae) with notes on the development of *Dermochelys*. *Chelonian Conserv Biol* 16(2):111–122.

Millichamp N, Jacobson ER, and Wolf ED. 1983. Diseases of the eye and ocular adnexae in reptiles. *J Amer Vet Med Assoc* 183:1205–1212.

Mitchell NJ, Nelson NJ, Cree A, Pledger S, Keall SN, and Daugherty CH. 2006. Support for a rare pattern of temperature-dependent sex determination in archaic reptiles: Evidence from two species of tuatara (*Sphenodon*). *Front Zool* 3(1):9–20.

Mobbs IG, and McMillan DB. 1979. Structure of the endodermal epithelium of the chick yolk sac during early stages of development. *Am J Anat* 155:287–309.

Moll E. 1973. Latitudinal and intersubspecific variation in reproduction of the painted turtle, *Chrysemys picta*. *Herpetologica* 29:307–318.

Moll EO, and Legler JM. 1971. The life history of a neotropical slider turtle, *Pseudemys script* a (Schoepff), in Panama. *Bull Los Angeles County Mus Nat Hist Sci* 11:1–102.

Molnar RE. 2004. *Dragons in the Dust: The Paleobiology of the Giant Monitor Lizard Megalania*. Indiana University Press, Bloomington.

Mouden E, Bons J, Pieau C, Renous S, Znari M, and Boumezzough A. 2000. Table de developpement embryonnaire d'un lizard agamide, *Agama impaleuris* Boettger, 1874. *Annales des Sciences Naturelles* 21:93–115.

Muller J, Scheyer TM, Head JJ, Barrett PM, Werneburg I, Ericson PGP, Pol D, and Sanchez-Villagra MR. 2010. Homeotic effects, somitogenesis and the evolution of vertebral numbers in recent and fossil amniotes. *Proc Nat Acad Sci* 107:2118–2123.

Murakami U. 1972. The effect of organic mercury on intrauterine life. *Adv Exp Med Biol* 27:301–336.

Muthukkaruppan V, Kanakambika P, Manickavel V, and Veeraraghavan K. 1970. Analysis of the development of the lizard, *Calotes versicolor* I. A series of normal stages in the embryonic development. *J Morphol* 130:479–490.

Nagle RD, Lutz CL, and Pyle AL. 2004. Overwintering in the nest by hatchling map turtles (*Graptemys geographica*). *Can J Zool* 82(8):1211–1218.

Nakamura K. 1938. Studies on some specimens of double monsters of snakes and tortoises. *Mem Coll Sci Kyoto Imperial University (B)* 4:171.

Nakazawa F, Alev C, Jakt LM, and Sheng G. 2011. Yolk sac endoderm is the major source of serum proteins and lipids and is involved in the regulation of vascular integrity in early chick development. *Dev Dyn* 240:2002–2010.

Necker WL. 1940. Hump-backed turtles. *Chicago Nat* 3:62.

Newman HH. 1917. *The Biology of Twins (Mammals)*. University Chicago Press, Chicago IL, 186.

Nixon CW and Smith HM. 1949. The occurrence of kyphosis in turtles. *Turtox News* 27:1–2.

Nelson NJ, Keall SN, Pledger S, and Daugherty CH. 2002. Male-biased sex ratio in a small tuatara population. *J Biogeogr* 29(5–6):633–640.

Nelson NJ, Thompson MB, Pledger S, Keall SN, and Daugherty CH. 2004. Egg mass determines hatchling size, and incubation temperature influences post-hatching growth, of tuatara *Sphenodon punctatus*. *J Zool* 263(1):77–87.

Noro M, Uejima A, Abe G, Manabe M, and Tamura K. 2009. Normal developmental stages of the Madagascar ground gecko *Paroedura pictus* with special reference to limb morphogenesis. *Dev Dyn* 238:100–109.

O'Shea M. 2011. *Boas and Pythons of the World*. Princeton Univ. Press, Princeton.

Óros J, Rodríguez JL, Monteros AE, Rodríguez F, Herráez P, and Fernández A. 1997. Tracheal malformation in a bicephalic honduran milk snake (*Lampropeltis hondurensis*) and subsequente fatal *Salmonella arizonae* infection. *J Zoo Wildl Med* 28:331–335.

Packard MJ. 1994. Patterns of mobilization and deposition of calcium in embryos of oviparous amniotic vertebrates. *Isr J Zool* 40:481–492.

Packard GC, Fasano SL, Attaway MB, Lohmiller LD, and Lynch TL. 1997. Thermal environment for overwintering hatchlings of the painted turtle (*Chrysemys picta*). *Can J Zool* 75(3):401–406.

Palmieri C, Selleri P, Di Girolamo N, Montani A, and Della Salda L. 2013. Multiple congenital malformations in a dicephalic spur-thighed tortoise (*Testudo graeca ibera*). *J Comp Path* 149(2–3):368–371.

Pendlebury GB. 1976. Congenital defects in the brood of a Prairie rattlesnake. *Can J Zool* 54:2023–2025.

Pérez-Higareda G. 1980. Albinism in *Corytophanes hernandezi* (Lacertilia: Iguanidae). *Bull Md Herp Soc* 16:97–98.

Perrault JR, and Stacy NI. 2018. Note on the unique physiologic state of loggerhead sea turtles (Caretta caretta) during nesting season as evidenced by a suite of health variables. *Mar Biol* 165(4):71–76.

Peter K. 1904. Normentafeln Zur Entwicklungsgeschichte Der Zauneidechse (*Lacerta agilis*), in *Normentafeln Zur Entwicklungsgeschichte Der Wirbeltiere*. Keibel F (Ed.), Verlag, Von Gustav Fischer, Jena, pp. 1–165.

Plymale HH, Jackson Jr. CG, and Collier G. 1978. Kyphosis in *Chrysemys scripta yaquia* (Testudines: Emydidae) and other turtles. *Southwest Nat* 23(3):457–461.

Powers KG, and Blackburn DG. 2017a. A novel pattern of yolk processing in developing snake eggs (Colubridae: Lampropeltini) and its functional and evolutionary implications. *J Exp Zool B: Mol Dev Evol* 328:462–475.

Powers KG, and Blackburn DG. 2017b. Morphological specializations of the yolk sac for yolk processing in embryonic corn snakes (*Pantherophis guttatus*: Colubridae). *J Morphol* 278:768–779.

Py-Daniel TR, De-Lima AKS, Lima FC, Pic-Taylor A, Junior ORP, and Sebben A. 2017. A staging table of post-ovipositional development for the South American collared lizard *Tropidurus torquatus* (Squamata: Tropiduridae). *Anat Rec* 300:277–290.

Qualls CP, Shine R, Donnellan S, and Hutchinson M. 1995. The evolution of viviparity within the Australian scincid lizard *Lerista bouguinuillii*. *J Zool (Lond)* 237:13–26.

Rafferty AR, and Reina RD. 2012. Arrested embryonic development: A review of strategies to delay hatching in egg-laying reptiles. *Proceedings of the Royal Society of London B: Biological Sciences*. doi: 10.1098/rspb.2012.0100

Rafferty AR, and Reina RD. 2014. The influence of temperature on embryonic developmental arrest in marine and freshwater turtles. *J Exp Mar Biol Ecol* 450:91–97.

Rafferty AR, Evans RG, Scheelings TF, and Reina RD. 2013. Limited oxygen availability in utero may constrain the evolution of live birth in reptiles. *Am Nat* 181(2):245–253.

Rhodin AGJ, Pritchard PCH, and Mittermeier RA. 1984. The incidence of spinal deformities in marine turtles, with notes on the prevalence of kyphosis in Indonesian *Chelonia mydas*. *British J Herp* 6:369–373.

Riches RJ. 1967. Notes on a clutch of eggs of the viperinae snake (*Natrix maura*). *Brit J Herpetol* 4:14–16.

Rings CC, Rafferty AR, Guinea ML, and Reina RD. 2015. The impact of extended preovipositional arrest on embryonic development and hatchling fitness in the flatback sea turtle. *Physiol Biochem Zool* 88(2):116–127.

Roscito JG, and Rodrigues MT. 2012. Embryonic development of the fossorial gymnophthalmid lizards *Nothobachia ablephara* and *Calyptommatus sinebrachiatus*. *Zoology* 115:302–318.

Rostal DC. 2007. Reproductive physiology of the ridley sea turtle, in *Biology and Conservation of Ridley Sea Turtles*. Plotkin P (Ed.), Johns Hopkins University Press, Baltimore, pp. 151–165.

Rostal DC, Lance VA, Grumbles JS, and Alberts AC. 1994. Seasonal reproductive cycle of the desert tortoise (*Gopherus agassizii*) in the eastern Mojave Desert. *Herpetological Monogr* 8:72–82.

Ryan JJ. 1986. Congenital tail's abnormality in the Blanding's turtle *Emyoidea blandingi*. *Bull Soc Herp France* 38:1–5.

Sailer A, Pyczak C, and Hartman UK. 1997. Siamese twins of *Testudo hermanni boettgeri*. *Herpetofauna* 19:12–13.

Saint-Girons H. 1975. Sperm survival and transport in the female genital tract of reptiles, in *The Biology of Spermatozoa*. Hafez ESE, Thibault CG (Eds.), Karger, Basel, pp. 105–113.

Sakaoka K, Sakai F, Yoshii M, Okamoto H, and Nagasawa K. 2013. Estimation of sperm storage duration in captive loggerhead turtles (*Caretta caretta*). *J Exp Mar Biol Ecol* 439:136–142.

Sanger TJ, Losos JB, and Gibson-Brown JJ. 2008. A developmental staging series for the lizard genus *Anolis*: A new system for the integration of evolution, development, and ecology. *J Morphol* 269:129–137.

Sant'Anna SS, Grego KF, Lorigados CAB, Fonseca-Pinto ACBC, Fernandes W, Sá-Rocha LC, and Catão-Dias JL. 2013. Malformations in neotropical viperids: Qualitative and quantitative analysis. *J Comp Pathol* 149:503–508.

Sarkar S, Sarkar NK, and Maiti BR. 2003. Oviductal sperm storage structure and their changes during the seasonal (dissociated) reproductive cycle in the soft-shelled turtle *Lissemys punctata punctata*. *J Exp Zool A Comp Exp Biol* 295(1):83–91.

Sazima I, and Di-Bernardo M. 1991. Albinismo em serpentes neotropicais. *Mem Inst Butantan* 53:167–173.

Scott HH. 1925. Congenital malformations of the kidney in reptiles, birds, and mammals. *J Zool* 4:1259–1269.

Sepúlveda MS, Del Piero F, Wiebe JJ, Rauschenberger HR, and Gross TS. 2006. Embryos and hatchlings from north central Florida lakes contaminated with organochlorines pesticides. *J Wild Dis* 42(1):56–73.

Sever DM, and Hamlett WC. 2002. Female sperm storage in reptiles. *J Exp Zool A Ecol Genet Physiol* 292(2):187–199.

Shine R. 1983. Reptilian reproductive modes: The oviparity-viviparity continuum. *Herpetologica* 39:1–8.

Shine R. 1985. The evolution of viviparity in reptiles. An ecological analysis, in *Biology of the Reptilia*, Vol 4. Gans C and Billett F (Eds.), Wiley, New York, pp. 605–694.

Simoniello P, Motta CM, Scudiero R, Trinchella F, and Filosa S. 2011. Cadmium-induced teratogenicity in lizard embryos: Correlation with metallothionein gene expression. *Comp Biochem Physiol C Toxicol Pharmacol* 153:119–127.

Singh LAK, and Bustard HR. 1982. Congenital defects in the gharial *Gavialis gangeticus* (Gmelin). *Brit J Herp* 6:215–219.

Singh LAK, Srivastava SS, Mohanty AP, and Rout SD. 2001. Prevention of rubbery-snout and x-ray revelations on hunchback condition in mugger crocodile (*Crocodylus palustris*). *Zoo's Print J* 16(4):465–466.

Smith DB. 1983. Aberrant pattern/coloration in *Elaphe subocularis*. *Herpetol Rev* 14:47.

Smith HM. 1947. Kyphosis and other variations in soft-shelled turtles. *Univ Kansas Publ Nat Hist* 1:117–124.

Smith HM, and Pérez-Higareda G. 1987. The literature on somatodichotomy in snakes. *Bull Md Herpetol Soc* 23:139–153.

Smith SA, and Shine R. 1997. Intraspecific variation in reproductive mode within the scincid lizard *Saiphos equalis*. *Aust J Zool* 45:435–445.

Sönmez B, and Özdelik Y. 2011. Morphologic characters of albino green turtle (*Chelonia mydas*) hatchlings on Samandag Beach in Turkey. *Mar Turt News* 131:46–47.

Sönmez B, Sert M, Kayikci S, Bedirci O, and Ozdilek Y. 2017. A two-headed green sea turtle (*Chelonia mydas*) hatchling on Samandag Beach, Turkey. *Russian J Herp* 24(2):155–157.

Stewart JR. 1997. Morphology and evolution of the egg of oviparous amniotes, in *Amniote Origins*. Sumida S, and Martin K (Eds.), Academic Press, New York, pp. 291–326.

Stewart JR. 2013. Fetal nutrition in lecithotrophic squamate reptiles: Toward a comprehensive model for evolution of viviparity and placentation. *J Morphol* 274:824–843.

Stewart JR, and Blackburn DG. 2014. Viviparity and placentation in lizards, in *Reproductive Biology and Phylogeny of Lizards and Tuatara*. Rheubert J, Seigel D, and Trauth S (Eds.), CRC Press, Florida, pp. 448–563.

Stewart JR, and Ecay TW. 2010. Patterns of maternal provision and embryonic mobilization of calcium in oviparous and viviparous squamate reptiles. *Herp Cons Biol* 5:341–359.

Stewart JR, and Thompson MB. 2017. Yolk sac development in lizards (Lacertilia: Scincidae): New perspectives on the egg of amniotes. *J Morphol* 278:574–591.

Stewart JR, Heulin B, and Surget-Groba Y. 2004. Extraembryonic membrane development in a reproductively bimodal lizard, *Lacerta (Zootoca) vivipara*. *Zool* 107:289–314.

Stower MJ, Diaz RE, Fernandez LC, Crother MW, Crother B, Marco A, Trainor PA, Srinivas S, and Bertocchini F. 2015. Bi-modal strategy of gastrulation in reptiles. *Dev Dyn* 244:1144–1157.

Sueiro LR, Rojas CA, Risk JY, França FOS, and Almeida-Santos SM. 2010. Anomalias cromáticas em *Bothrops jararaca* (Serpentes, Viperidae): Xantismo interfere na sobrevivência? *Biotemas* 23:155–160.

Szabo KT. 1989a. Etiology of congenital malformations, in *Congenital Malformations in Laboratory and Farm Animals*. Szabo KT (Ed.), Academic Press, Inc., San Diego, pp. 1–2.

Szabo KT. 1989b. Alimentary and digestive systems, in *Congenital Malformations in Laboratory and Farm Animals*. Szabo KT (Ed.), Academic.

Szabo KT. 1989c. Sensory systems, in *Congenital Malformations in Laboratory and Farm Animals*. Szabo KT (Ed.), Academic Press, Inc., San Diego, pp. 245–268.

Szabo KT. 1989d. Central nervous system, in *Congenital Malformations in Laboratory and Farm Animals*. Szabo KT (Ed.), Academic Press, Inc., San Diego, pp. 95–143.

Tanaka K. 2005. Thermal aspects of melanistic and striped morphs of the snake *Elaphe quadrivirgata*. *Zool Sci* 22:1173–1179.

Taylor M, and Mendyk RW. 2017. Apalone ferox (Florida Softshell Turtle); Kyphosis. *Herp Rev* 28(2):418–419.

Thapliyal JP, Singh KS, and Chandola A. 1973. Prelaying stages in the development of Indian Garden Lizards *Calotes versicolor*. *Ann Embr Morph* 6:253–259.

Tokita M, and Kuratani S. 2001. Normal embryonic stages of the Chinese softshelled turtle *Pelodiscus sinensis* (Trionychidae). *Zool Sci* 18(5):705–715.

Thompson MB. 1989. Patterns of metabolism in embryonic reptiles. *Respir Physiol* 76(1989):243–256.

Thompson MB, Packard GC, Packard MJ, and Rose B. 1996. Analysis of the nest environment of tuatara *Sphenodon punctatus*. *J Zool* 238(2):239–251.

Tóth T, Gál J, Újvari B, and Madsen T. 2005. Severe malformation in neonate *Vipera ursinii rakosiensis*. *Amphib-reptil* 26:388–390.

Travaglia-Cardoso SRT, and Parpinelli C. 2006. *Crotalus durissus terrificus* (rattlesnake): A case of xanthism. *Herpetol Bull* 97:39–40.

Tryon B. 1980. Observations on reproduction in the West African dwarf crocodile with a description of parental behavior. J.

Türkozan O, and Durmus SH. 2001. Albino loggerhead and green turtle (*Caretta caretta* and *Chelonia mydas*) hatchlings in Turkey. *Zool Middle East* 24:133–136.

Ultsch GR. 2006. The ecology of overwintering among turtles: Where turtles overwinter and its consequences. *Biol Rev* 81(3):339–367.

Valentine BA, and McGavin MD. 2009. Músculo esquelético, in *Bases da Patologia Veterinária*, 4th ed. McGavin MD, Zachary JF (Eds.), Elsevier, Rio de Janeiro, pp. 973–1040.

Villa J. 1981. Three albino snakes from Nicaragua: An update. *Herpetol Rev* 12:81.

Virchow H. 1892. Das dotterorgan der wilbelthiere. *Archiv Für Mikroskopische Anatomie* 40:39–101.

Voeltzkow A. 1899. Beiträge zur entwicklungsgeschichte der reptilien. I. Biologie und entwicklung der äusseren körperform von *Crocodilus madagascariensis* Grand. *Abhandl Swnchenb Naturf Ges* 26(1):1–150.

Voeltzkow A. 1901. Beiträge zur entwicklungsgeschichte der reptilien. IV. Keimblätter, dottersack und erste anlage des blutes un der gefässe bei *Crocodilus madagascariensis* Grand. *Abhandl Swnchenb Naturf Ges* 26(3):337–418.

Vogt T. 1922. Zur reptilian und amphibienfauna Südchinas. *Arch Naturgesch* 10:135–146.

Wallach V. 2003. See live mutant snakes right here. *Natural New England* 16:50–55.

Wallach V. 2007. Axial bifurcation and duplication in snakes. Part I. A synopsis of authentic and anecdotal cases. *Bull Md Herpetol Soc* 43:57–95.

Wandolleck B. 1904. Eine bucklige *Testudo graeca* L. *Zool Jahrb Syst* 20:151–166.

Wang Y, and Evans SE. 2011. A gravid lizard from the Cretaceous of China and the early history of squamate viviparity. *Naturwissenschaften* 98:739–743.

Warkany J. 1947. Etiology of congenital malformations. *Advc Pediatr* 2:1–63.

Webb GJW, and Manolis C. 1998. *Australian Crocodiles: A Natural History.* New Holland Publishers, Sydney, 160.

Werner WE Jr. 1959. Amphibians and reptiles of the Thousand Islands region, New York. *Copeia* 2:170–172.

Werner YL. 1971. The ontogenetic development of the vertebrae in some gekkonoid lizards. *J Morphol* 133:41–92.

Werneburg I. 2009. A standard system to study vertebrate embryos. *PLOS ONE* 4(6):e5887.

Werneburg I, Hugi J, Muller J, and Sánchez-Villagra MR. 2009. Embryogenesis and ossification of *Emydura subglobosa* (Testudines, Pleurodira, Chelidae) and patterns of turtle development. *Dev Dyn* 238:2770–2786.

Wiens JJ, Brandley MC, and Reeder TW. 2006. Why does a trait evolve multiple times within a clade? Repeated evolution of snakelike body form in squamate reptiles. *Evol* 60:123–141.

Wiley GO. 1930. Notes on the neotropical rattlesnake (*Crotalus terrificus brasiliensis*) in captivity. *Bull Antivenin Inst Am* 3:100–103.

Williamson SA, Evans RG, and Reina RD. 2017. When is embryonic arrest broken in turtle eggs? *Physiol Biochem Zool* 90(5):523–532.

Wise PAD, Vickaryous MK, and Russell AP. 2009. An embryonic staging table for in ovo development of *Eublepharis macularius*, the leopard gecko. *Anat Rec* 292:1198–1212.

Woltering JM, Vonk FJ, Müller H, Bardine N, Tuduce IL, de Bakker MAG, Knöchel W, Sirbu IO, Durston AJ, and Richardson MK. 2009. Axial patterning in snakes and caecilians: Evidence for an alternative interpretation of the Hox code. *Dev Biol* 332:82–89.

Wu J, Staffenberg DA, Mulliken JB, and Shanske AL. 2002. Disprosopus: A unique case and review of the literature. *Teratology* 66:282–287.

Xue-Feng Y, Xiao-Long T, Feng Y, De-Jiu Z, Ying X, Cui W, and Qiang C. 2001. Influence of ambient temperature on maternal thermoregulation and neonate phenotypes in a viviparous lizard, *Eremias multiocellata*, during the gestation period. *J Therm Biol* 36:187–192.

Yntema CL. 1964. Procurement and use of turtle embryos for experimental procedures. *The Anatomical Record* 149(4):577–585.

Yntema CL. 1968. A series of stages in the embryonic development of *Chelydra serpentina*. *J Morphol* 125(2):219–251.

Yntema CL. 1970. Twinning in the common snapping turtle, *Chelydra serpentina*. *Anat Rec* 166:491–498.

Yntema CL. 1971. Incidence and survival of twin embryos of the common snapping turtle, *Chelydra serpentina*. *Copeia* 4:755–758.

Yntema CL. 1976. Effects of incubation temperatures on sexual differentiation in the turtle, *Chelydra serpentina*. *J Morphol* 150(2):453–461.

Yntema CL. 1978. Incubation times for eggs of the turtle *Chelydra serpentina* (Testudines: Chelydridae) at various temperatures. *Herpetologica* 274–277.

Yntema CL. 1979. Temperature levels and periods of sex determination during incubation of eggs of *Chelydra serpentina*. *J Morphol* 159(1):17–27.

Yoshizaki N, Soga M, Ito Y, Kun MM, Sultana F, and Yonezawa S. 2004. Two-step consumption of yolk granules during the development of quail embryos. *Dev Growth Diff* 46:229–238.

Youngprapakorn P, Ousavaplangchai L, and Kanchanapangka S. 1994. *A Color Atlas of the Diseases of the Crocodile.* Style Creative House, Thailand, 76.

Zehr DR. 1962. Stages in the normal development of the common garter snake, *Thamnophis sirtalis sirtalis*. *Copeia* 1962:322–329.

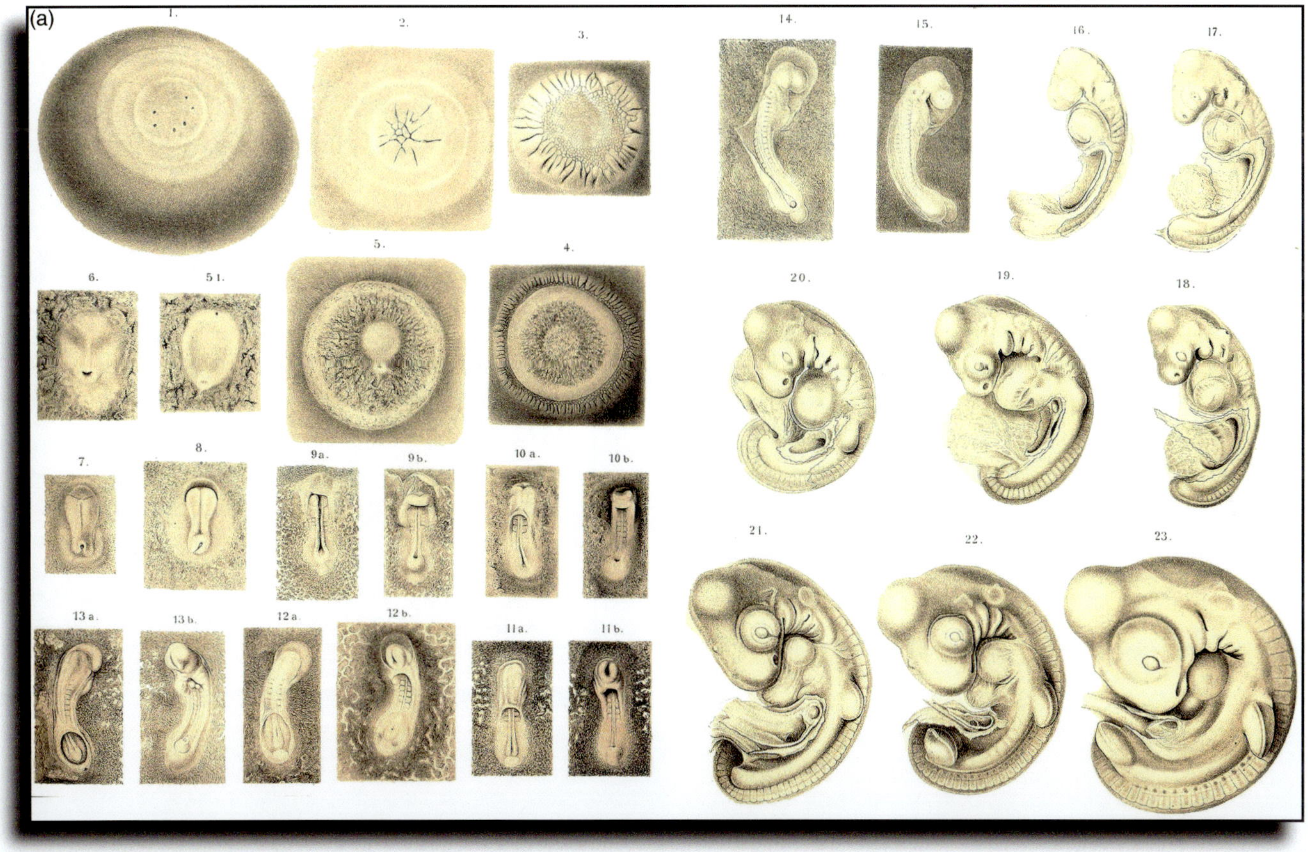

Figure 4.1 Normal plate (Normentafel) of embryonic development of the Sand Lizard, *Lacerta agilis* by Karl Peter (1904). The illustrations, which appeared in a monograph as Volume 4 in the classic series of vertebrate normal tables edited by Franz Keibel, are the first detailed descriptions of normal development of a squamate. (a) Embryonic stages 1–23.

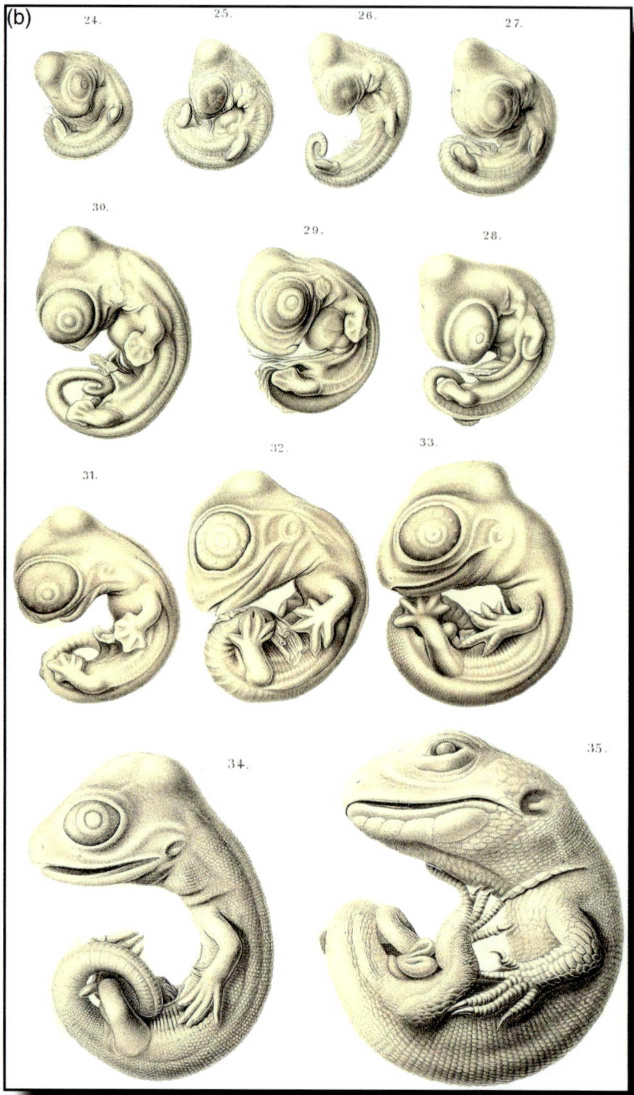

Figure 4.1 (Continued) Normal plate (Normentafel) of embryonic development of the Sand Lizard, *Lacerta agilis* by Karl Peter (1904). The illustrations, which appeared in a monograph as Volume 4 in the classic series of vertebrate normal tables edited by Franz Keibel, are the first detailed descriptions of normal development of a squamate. (b) Embryonic stages 24–35.

Figure 4.2 Rattlesnake, *Crotalus durissus*. Viperidae. A juvenile with xanthism. (Courtesy of Dr. Sávio Stefanini Sant'Anna.)

Figure 4.3 Jararaca, *Bothrops jararaca*. Viperidae. A juvenile with xanthism. (Courtesy of Dr. Sávio Stefanini Sant'Anna.)

Figure 4.4 Patagonia Green Racer, *Phylodrias patagoniensis*. Colubridae. A juvenile with xanthism. (Courtesy of Dr. Sávio Stefanini Sant'Anna.)

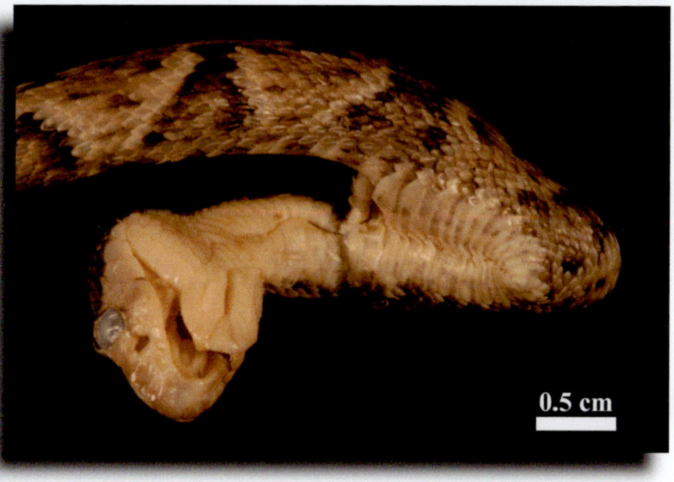

Figure 4.5 Jararaca, *Bothrops jararaca*. Viperidae. A newborn with fused ventral scales. (From Carvalho MPN. 2014. Comparative pathology of congenital malformations in *Bothrops jararaca* and *Crotalus durissus* from Southeast Brazil. Masters thesis. School of Veterinary Medicine and Animal Science, University of São Paulo, Brazil.)

Figure 4.6 Tropical Rattlesnake, *Crotalus durissus*. Viperidae. A juvenile bicephalous rattlesnake. (Courtesy of Dr. Sávio Stefanini Sant'Anna.)

Figure 4.7 Jararaca, *Bothrops jararaca*. Viperidae. A newborn with rostral bifurcation. (From Carvalho MPN et al. 2017. *J Wildl Dis* 53(4):804–815. With permission.)

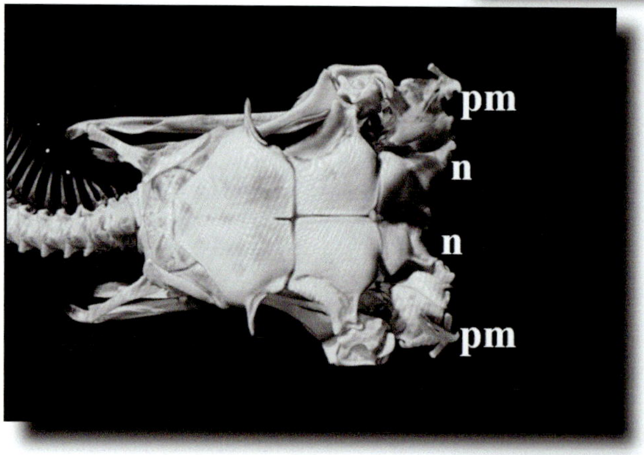

Figure 4.8 Jararaca, *Bothrops jararaca*. Viperidae. 3D microtomographic reconstruction of rostral bifurcation (dorsal view). The image exhibits duplication of premaxillae (pm) and nasal (n) bones. (From Carvalho MPN et al. 2017. *J Wildl Dis* 53(4):804–815. With permission.)

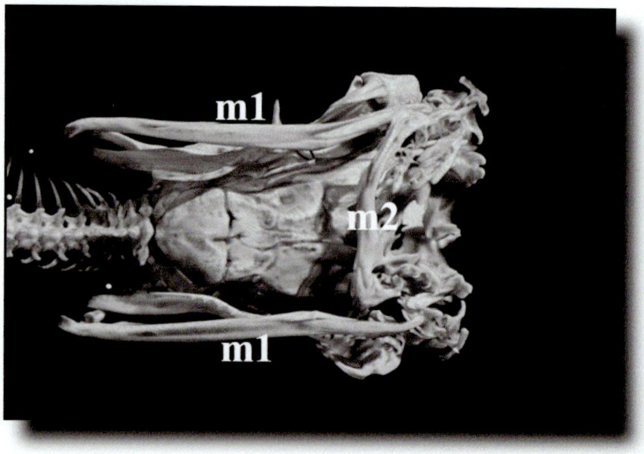

Figure 4.9 Jararaca, *Bothrops jararaca*. Viperidae. 3D microtomographic reconstruction of rostral bifurcation (ventral view). Image shows the secondary jaw (m2), transversal to the axis of the main jaw (m1). (From Carvalho MPN et al. 2017. *J Wildl Dis* 53(4):804–815. With permission.)

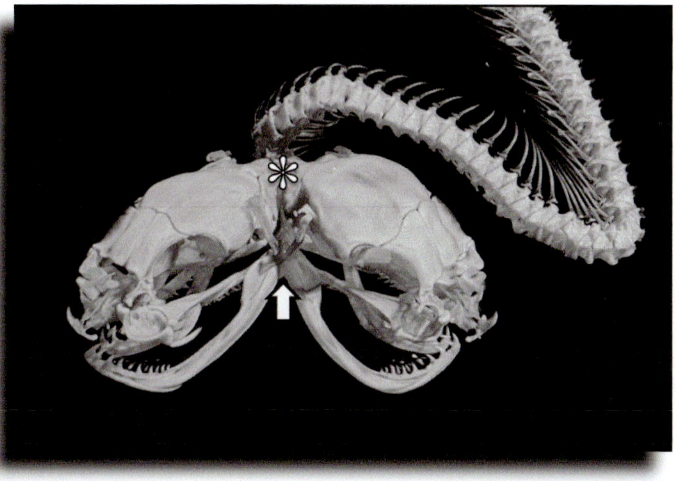

Figure 4.10 Jararaca, *Bothrops jararaca*. Viperidae. 3D microtomographic reconstruction of bicephaly. Image depicts fusion of the surangular bones (arrow) and supraocciptal bones (asterisk). (From Carvalho MPN et al. 2017. *J Wildl Dis* 53(4):804–815. With permission.)

Figure 4.11 Jararaca, *Bothrops jararaca*. Viperidae. Newborn with bicephaly. (From Carvalho MPN et al. 2017. *J Wildl Dis* 53(4):804–815. With permission.)

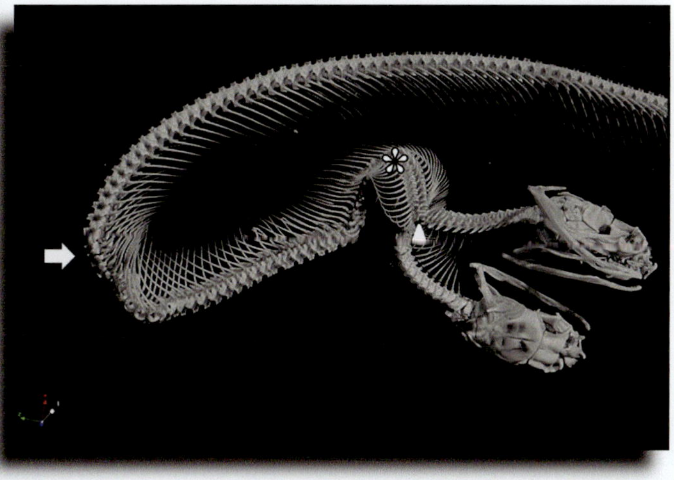

Figure 4.12 Jararaca, *Bothrops jararaca*. Viperidae. 3D microtomographic reconstruction of a bicephalic jararaca with axial bifurcation (triangle), lordosis (asterisk), and kyphosis (arrow). (From Carvalho MPN et al. 2017. *J Wildl Dis* 53(4):804–815. With permission.)

Figure 4.13 Jararaca, *Bothrops jararaca*. Viperidae. Newborn with marked kyphoscoliotic vertebral column. (Photo by MPN Carvalho.)

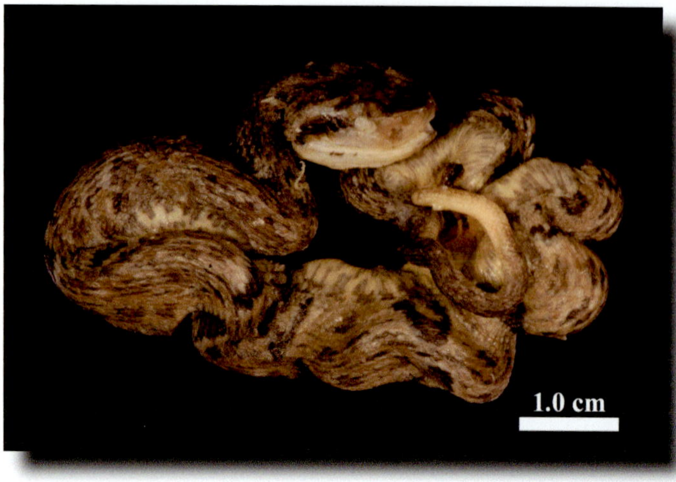

Figure 4.14 Tropical Rattlesnake, *Crotalus durissus*. Viperidae. 3D microtomographic reconstruction of hemivertebrae (white triangle), fusion of the head of ribs (black arrow), and duplication of ribs (white arrow). (From Carvalho MPN et al. 2017. *J Wildl Dis* 53(4):804–815. With permission.)

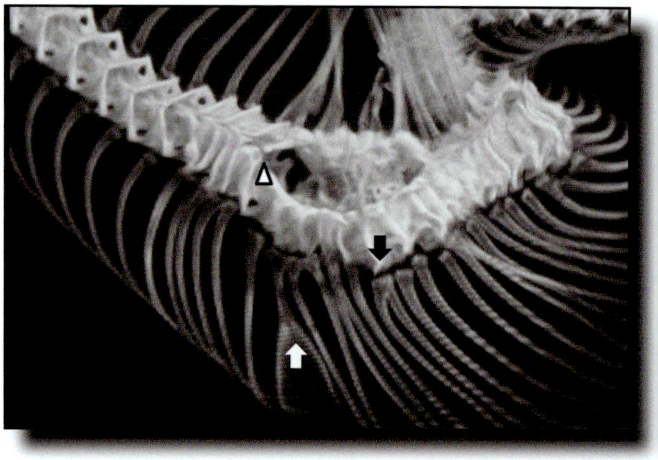

Figure 4.15 Tropical Rattlesnake, *Crotalus durissus*. Viperidae. 3D microtomographic reconstruction of spina bifida (arrows). (From Carvalho MPN et al. 2017. *J Wildl Dis* 53(4):804–815. With permission.)

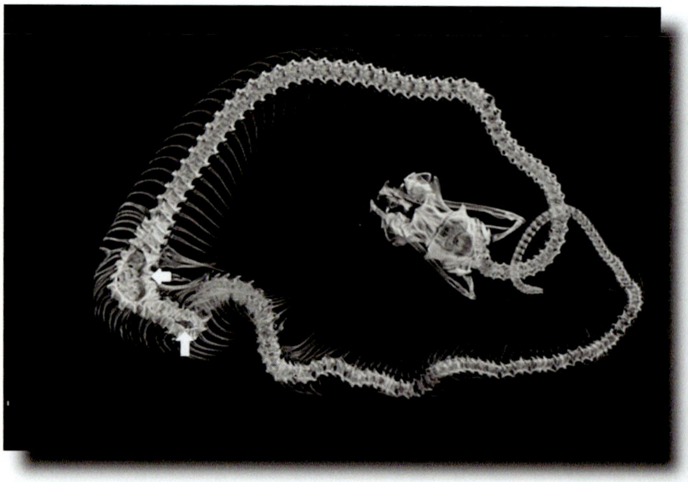

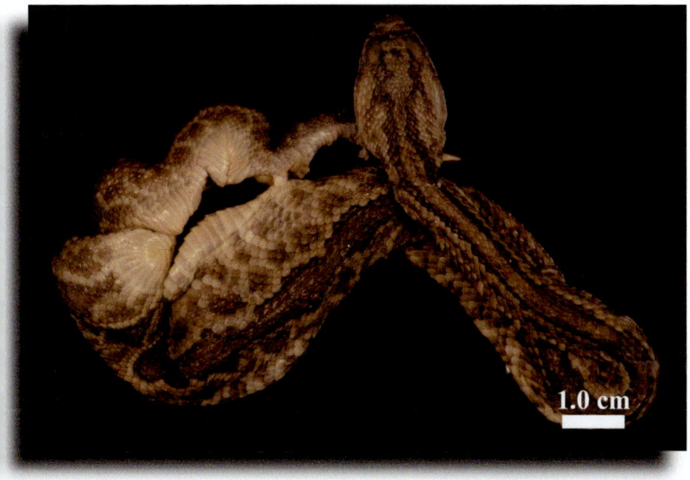

Figure 4.16 Tropical Rattlesnake, *Crotalus durissus*. Viperidae. A newborn Tropical Rattlesnake with kyphoscoliosis associated with fusion of ventral scales. (Photo by MPN Carvalho.)

Figure 4.17 Jararaca, *Bothrops jararaca*. Viperidae. A newborn with shortened tail. (From Carvalho MPN. 2014. Comparative pathology of congenital malformations in *Bothrops jararaca* and *Crotalus durissus* from Southeast Brazil. Masters thesis. School of Veterinary Medicine and Animal Science, University of São Paulo, Brazil.)

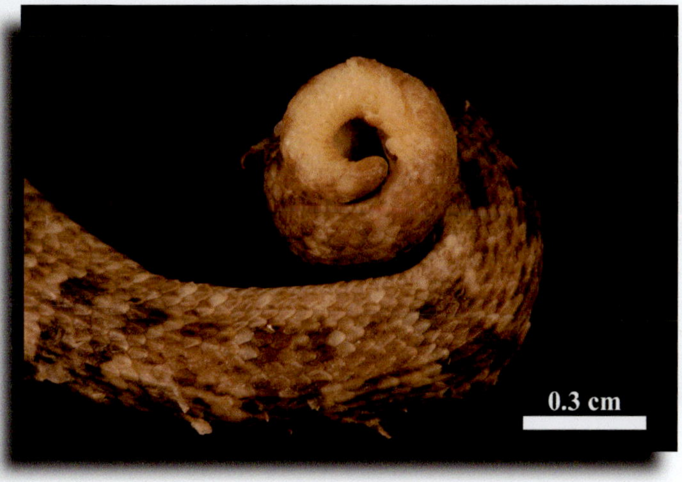

Figure 4.18 Jararaca, *Bothrops jararaca*. Viperidae. A newborn with coiled tail. (From Carvalho MPN. 2014. Comparative pathology of congenital malformations in *Bothrops jararaca* and *Crotalus durissus* from Southeast Brazil. Masters thesis. School of Veterinary Medicine and Animal Science, University of São Paulo, Brazil.)

Figure 4.19 Jararaca, *Bothrops jararaca*. Viperidae. 3D microtomographic reconstruction of coiled tail. (From Carvalho MPN. 2014. Comparative pathology of congenital malformations in *Bothrops jararaca* and *Crotalus durissus* from Southeast Brazil. Masters thesis. School of Veterinary Medicine and Animal Science, University of São Paulo, Brazil.)

Figure 4.20 Jararaca, *Bothrops jararaca*. Viperidae. A newborn with schistosomus reflexus syndrome in the cranial region to the cloaca. (From Carvalho MPN. 2014. Comparative pathology of congenital malformations in *Bothrops jararaca* and *Crotalus durissus* from Southeast Brazil. Masters thesis. School of Veterinary Medicine and Animal Science, University of São Paulo, Brazil.)

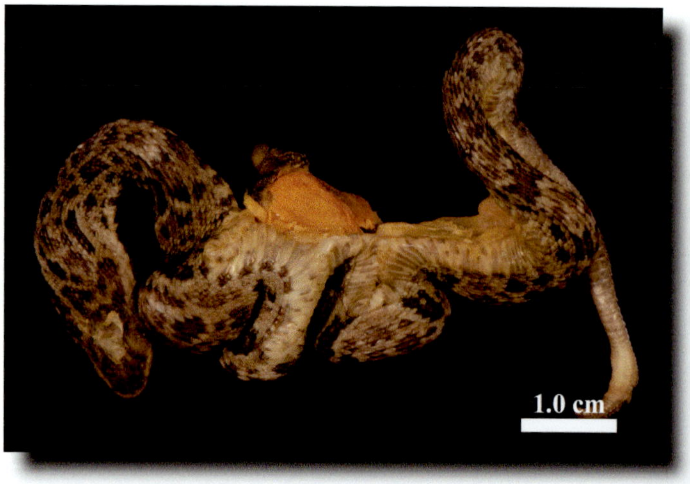

Figure 4.21 Jararaca, *Bothrops jararaca*. Viperidae. A newborn with schistosomus reflexus syndrome in the middle third of the body. (Photo by MPN Carvalho.)

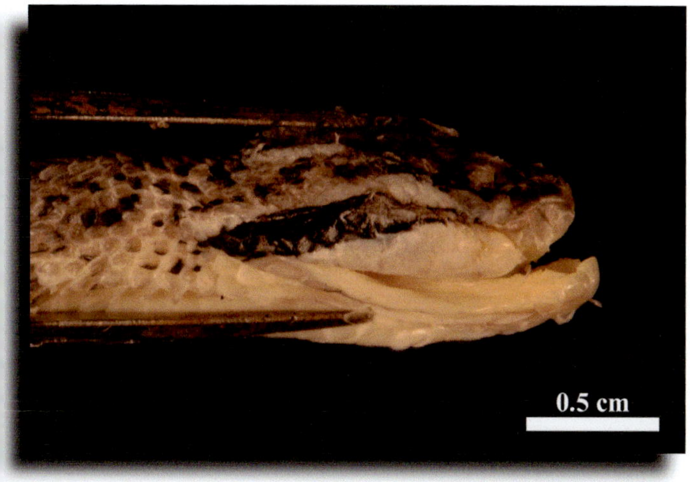

Figure 4.22 Jararaca, *Bothrops jararaca*. Viperidae. Newborn with microphthalmia. (From Carvalho MPN. 2014. Comparative pathology of congenital malformations in *Bothrops jararaca* and *Crotalus durissus* from Southeast Brazil. Masters thesis. School of Veterinary Medicine and Animal Science, University of São Paulo, Brazil.)

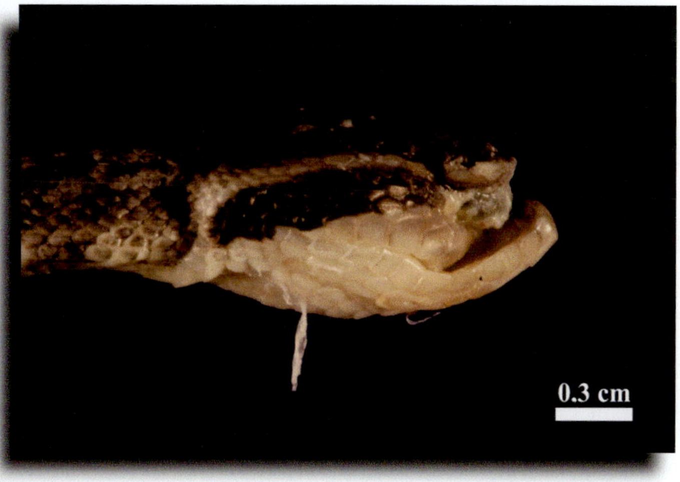

Figure 4.23 Jararaca, *Bothrops jararaca*. Viperidae. Cyclops in a newborn Jararaca. (From Carvalho MPN et al. 2017. *J Wildl Dis* 53(4):804–815. With permission.)

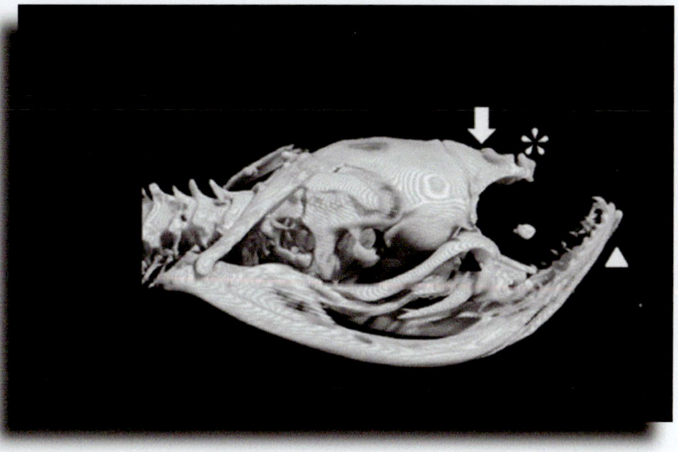

Figure 4.24 Jararaca, *Bothrops jararaca*. Viperidae. 3D microtomographic reconstruction of cyclops. The image depicts absence of nasal bones and premaxillae; prefrontal (asterisk) and frontal (arrow) bones have small dimensions; abnormal curvature of ectopterygoid bone shifted toward orbit (black triangle), and prognathism (white triangle). (From Carvalho MPN et al. 2017. *J Wildl Dis* 53(4):804–815. With permission.)

Figure 4.25 Jararaca, *Bothrops jararaca*. Viperidae. Hydrocephalus in a newborn Jararaca. (From Carvalho MPN et al. 2017. *J Wildl Dis* 53(4):804–815. With permission.)

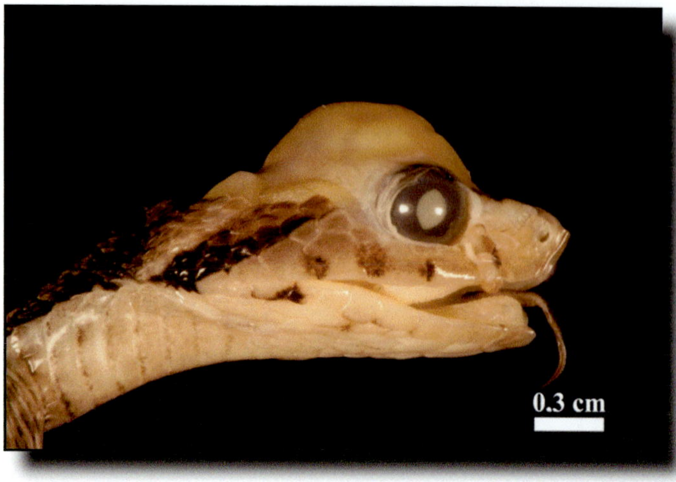

Figure 4.26 Jararaca, *Bothrops jararaca*. Viperidae. 3D microtomographic reconstruction of hydrocephalus. The image shows thinning of the parietal bone (arrow). (From Carvalho MPN et al. 2017. *J Wildl Dis* 53(4):804–815. With permission.)

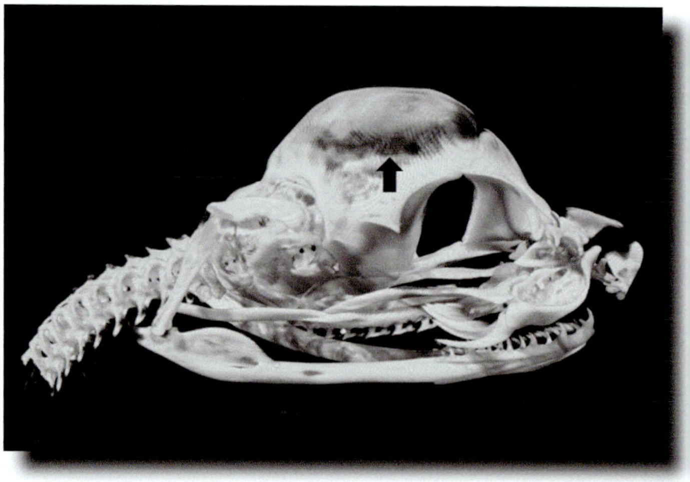

Figure 4.27 Jararaca, *Bothrops jararaca*. Viperidae. Head malformation and kyphoscoliosis in a newborn Jararaca. (From Carvalho MPN et al. 2017. *J Wildl Dis* 53(4):804–815. With permission.)

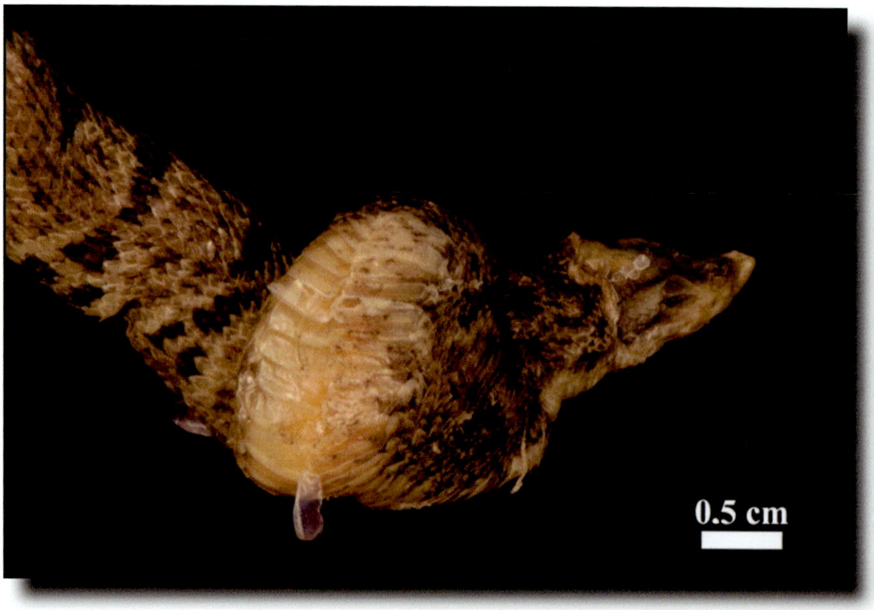

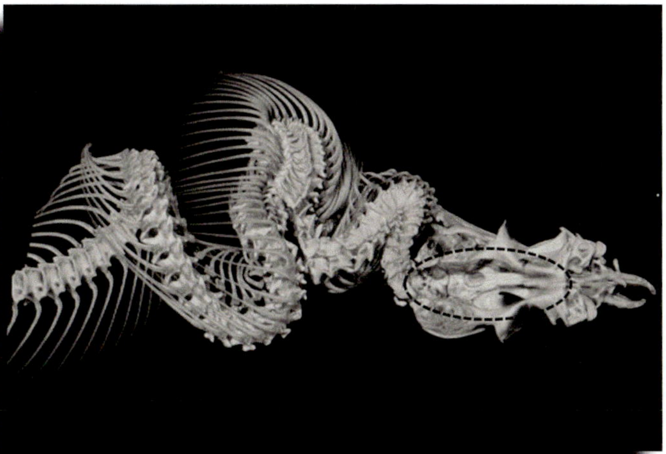

Figure 4.28 Jararaca, *Bothrops jararaca*. Viperidae. 3D microtomographic reconstruction image depicts jaw asymmetry and agenesis of frontal, parietal, and supraoccipital bones (dotted ellipse). (From Carvalho MPN et al. 2017. *J Wildl Dis* 53(4):804–815. With permission.)

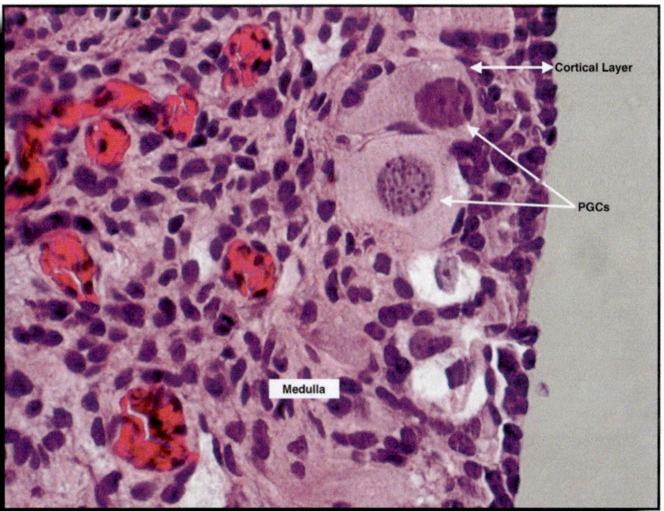

Figure 4.29 Loggerhead Sea Turtle, *Caretta caretta*. Cheloniidae. H&E-stained immature ovary of a Loggerhead Sea Turtle. The ovary is characterized by a multicellular cortical layer. Occasionally, primordial germ cells (PGCs) are found in the cortex of neonatal ovaries, appearing as hypertrophied cells. The medullary portion of the ovary is loosely organized. PGCs have been observed only in the cortex of neonate females. (Photo: J. Wynokon. With permission.)

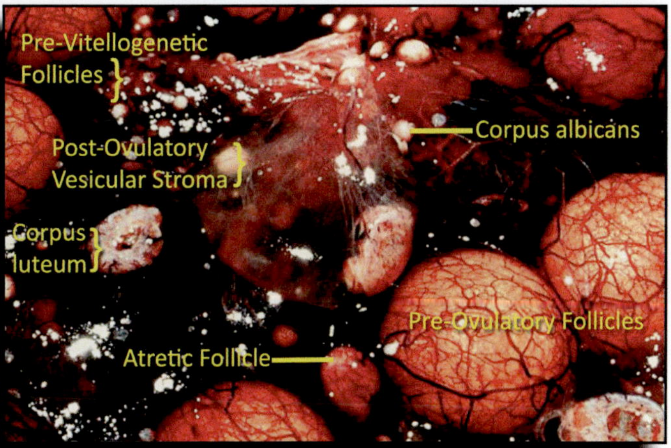

Figure 4.30 Green Turtle, *Chelonia mydas*. Cheloniidae. Mature ovary with several stages of oocytes and follicles, in ventral view. Several stages of oocytes are present as pre-vitellogenic follicles, mature preovulatory follicles and somewhat flattened, atretic follicles. The scars of ovulated follicles, the corpora lutea (recently ovulated) and corpora albicans, are present within the ovarian stroma. (Reproduced from Miller JD, and Limpus CJ. 2003. Ontogeny of marine turtle gonads. *Biology of Sea Turtles*. Volume II. CRC Press, Boca Raton, pp. 199–224. With permission.)

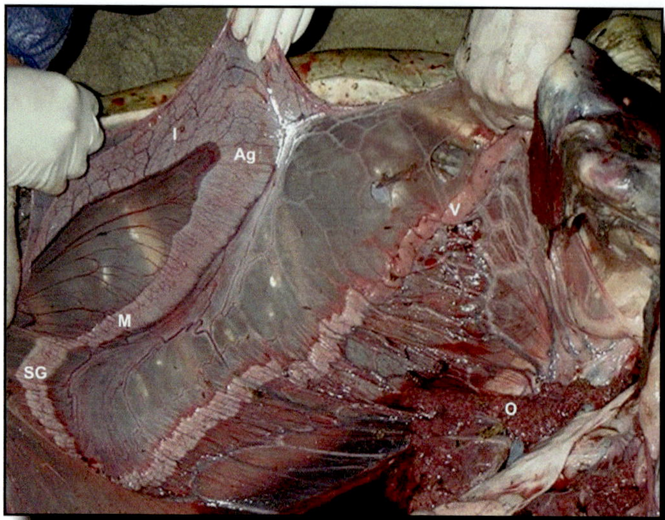

Figure 4.31 Green Turtle, *Chelonia mydas*. Cheloniidae. Oviduct of a Green Turtle. This turtle was not reproductively active based upon the small size of the ovarian follicles across the ovary (O). The oviduct appears to be mature; all segments are enlarged and have the plicate form of a mature turtle's oviduct. The infundibulum (I) is held up in the upper left. Oocytes collected by the infundibulum pass through the funnel-like opening toward an aglandular segment (AG). Fertilization occurs between the infundibulum and the magnum (M) segments where albumen is produced and envelops the oocyte. The yolk and its embryo, surrounded by albumen, pass through the long shell gland (SG) where the eggshell membrane and mineral layer form. Eggs pass through the vagina (V) and may be stored in this region prior to oviposition. (From Miller JD, and Limpus CJ. 2003. Ontogeny of marine turtle gonads, in The Biology of Sea Turtles, Vol. II. CRC Press, Boca Raton, pp. 199–224.)

Figure 4.32 Red-Eared Slider (*Trachemys scripta*). Emydidae. The Red-Eared Slider is a daytime nesting species. It is common for sliders to urinate on the soil, softening and hydrating it where they dig their nests. Wet soil from an abandoned nest is seen beyond the turtle's head. (Photo: J. Wyneken. With permission.)

Figure 4.33 Green Turtle, *Chelonia mydas*. Cheloniidae. These Green Turtle eggs reflect embryo-associated calcium depletion. At pipping, the shell crumbles, releasing a substantial amount of fluid as the turtles emerge from the eggs. (Photo: J. Wyneken. With permission.)

Figure 4.34 Eastern Box Turtle, *Terrapene carolina*. Emydidae. Hatchling size relative to yolk size is a characteristic used to identify embryonic stage. This late-stage Box Turtle (*Terrepene carolina*) embryo is pigmented, has scales and claws, but is smaller than its yolk sac indicating it was not a full-term embryo. (Photo: J. Wyneken. With permission.)

Figure 4.35 Green Turtle, *Chelonia mydas*. Cheloniidae. Late-stage embryos are larger than their yolk sac. This Green Turtle neonate died after pipping. The turtle disrupted the chorioallantoic membrane, seen between yolk sac and the eggshell. The remains of the blood vessels are present still. (Photo: J. Wyneken. With permission.)

Figure 4.36 Loggerhead Sea Turtle, *Caretta caretta*. Cheloniidae. Dorsoventral radiograph of a hatchling Loggerhead Sea Turtle with axial bifurcation and bicephaly. (Photo by G.A. Lewbart.)

Figure 4.37 Common Snapping Turtle, *Chelydra serpentine*. Chekydridae. Malformed tail of a mature common snapping turtle. It is possible that trauma contributed to this condition. (Photo by G.A. Lewbart.)

Figure 4.38 Loggerhead Sea Turtle, *Caretta caretta*. Cheloniidae. Amelia (lack of limbs) in a Loggerhead Sea Turtle hatchling. (Photo courtesy of A. Bárcenas-Ibarra.)

Figure 4.39 Olive Ridley Sea Turtle, *Lepidochelys olivacea*. Cheloniidae. Partial amelia (lack of one or some limbs) in an Olive Ridley Sea Turtle hatchling. (Photo courtesy of A. Bárcenas-Ibarra.)

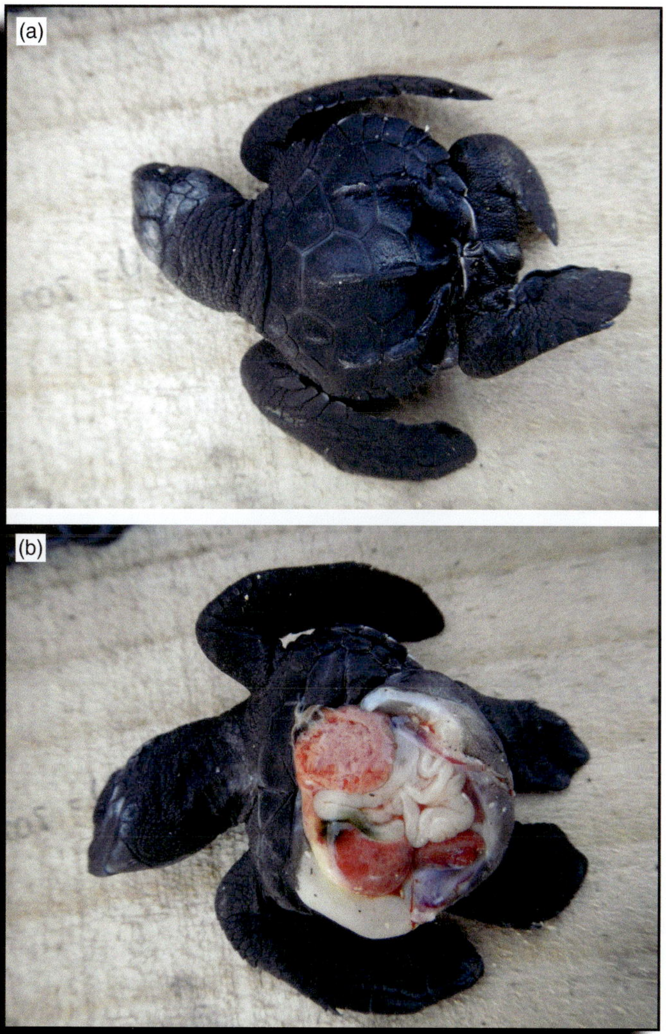

Figure 4.40 Olive Ridley Sea Turtle, *Lepidochelys olivacea*. Cheloniidae. Hatchling Olive Ridley Sea Turtle with schistosomus reflexus syndrome. (a) Dorsal view of the affected animal. (b) Ventral view of the affected animal. (Photos courtesy of A. Bárcenas-Ibarra.)

Figure 4.41 Olive Ridley Sea Turtle, *Lepidochelys olivacea*. Cheloniidae. Cyclopia in a hatchling Olive Ridley Turtle. (Photo courtesy of A. Bárcenas-Ibarra.)

Figure 4.42 Loggerhead Sea Turtle, *Caretta caretta*. Cheloniidae. Diprosopus, or craniofacial duplication, in a Loggerhead Sea Turtle hatchling. (Photo courtesy of A. Bárcenas-Ibarra.)

Figure 4.43 Common Snapping Turtle. *Chelydra serpentine*. Chelydridae. True albino. This individual weighs approximately 7 kg. (Photo by G.A. Lewbart.)

Figure 4.44 Leatherback Sea Turtle, *Dermochelys coriacea*. Dermochelyidae. Albino Leatherback Sea Turtle hatchling. (Photo by E.R. Jacobson.)

Figure 4.45 Loggerhead Sea Turtle, *Caretta caretta*. Cheloniidae. Subadult amelanistic. (Photo by E.R. Jacobson.)

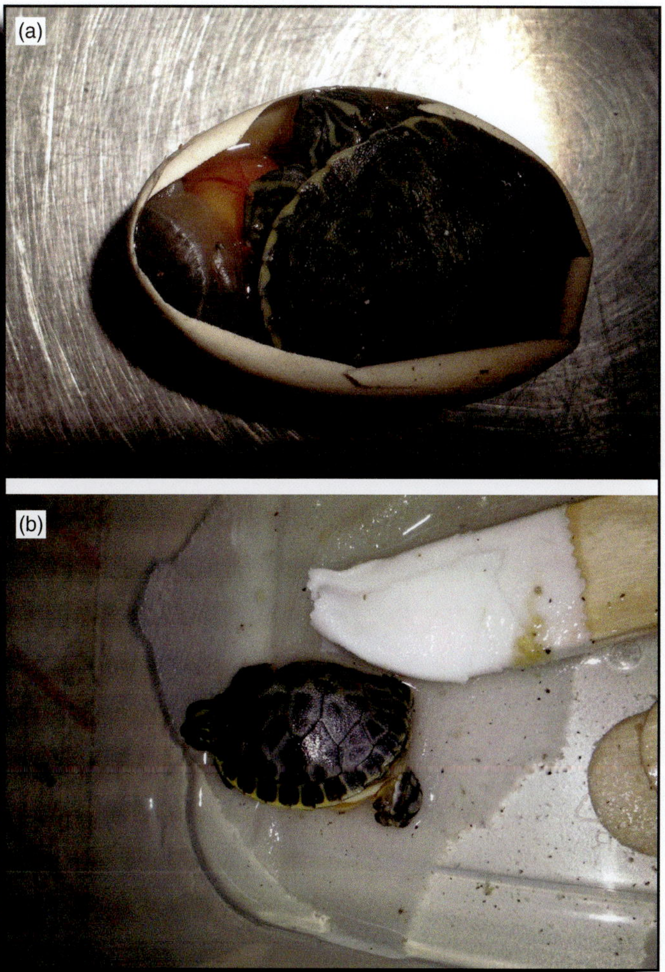

Figure 4.46 River Cooter, *Pseudemys concinna*. Emydidae. Twinning in the River Cooter. (a) Twins in situ before removal from the egg. (b) Twins after removal from the egg. In this case the smaller twin was not viable and was surgically separated from the larger twin. (Photos courtesy of A. Breaux.)

Figure 4.47 Gharial, *Gavialis gangeticus*. Gavialidae. Cervical flexion (hemivertebrae), or "bent neck," in a Gharial. (Photo courtesy of E.R. Jacobson.)

Figure 4.48 Siamese Crocodile, *Crocodylus siamensis*. Crocodylidae. Microphthalmia in a hatchling Siamese Crocodile. (Photo by E.R. Jacobson.)

Figure 4.49 Tuatara, *Sphenodon punctatus*. Sphenodontidae. Eggs excavated from an underground nest. The oval, white eggs are stained with soil. Eggs are small, 4–5 g and less than 3 cm in length. This clutch is slightly larger than average. (Photo courtesy of Anna Carter.)

Figure 4.50 Tuatara, *Sphenodon punctatus*. Sphenodontidae. Tuatara nest opened to reveal a clutch of eggs in soil. Nests are often under stones and in open areas that promote successful incubation. Clutches of parchment-shelled eggs are buried with soil and vegetation to close the neck of the nest. (Photo courtesy of Nicola Nelson.)

Figure 4.51 Tuatara, *Sphenodon punctatus*. Sphenodontidae. Tuatara embryos are slow to develop, requiring 10–11 months from fertilization, to oviposition, to hatching in natural nests. In laboratory conditions, without over-wintering temperatures, development can occur much faster. As eggs approach hatching, it is common for fluids to weep from one end. This full-term embryo has torn its parchment eggshell along the long axis. The remnant of the chorioallantoic membrane with blood vessels crossing the lateral face and neck is still present. (Photo courtesy of Nicola Mitchell.)

Figure 4.52 Tuatara, *Sphenodon punctatus*. Sphenodontidae. Tuatara hatchlings can emerge from their eggs asynchronously. Hatchlings are small (4–5 g), cryptically colored, and have large claws. (Photo courtesy of Nicola Mitchell.)

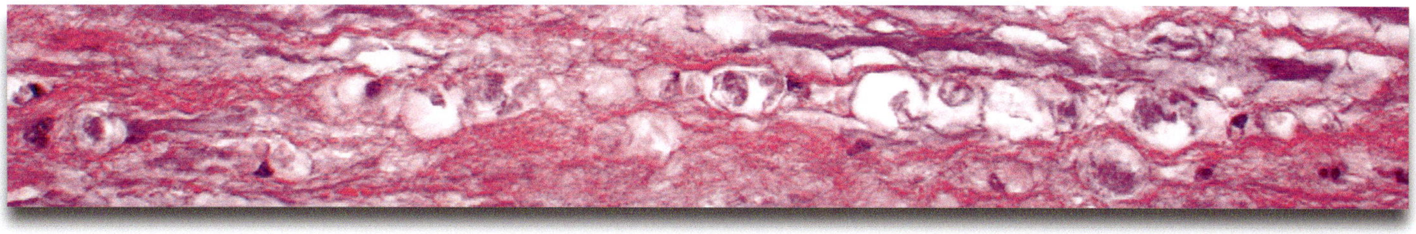

5

DEGENERATIVE DISEASES

NANCY L. STEDMAN

Contents

5.1 Introduction

The diverse spectrum of degenerative diseases includes those caused by deterioration of cells and tissue over time. Although some degenerative diseases have a genetic basis, others are related to aging and repeated use, possibly exacerbated by lifestyle factors, exposure to toxins, and other concurrent diseases and conditions. Although data are deficient for many reptile species, reptiles generally appear to have slower aging compared with mammals and birds (Hoekstra et al., 2019). Consequently, reptiles also appear to be more resistant to some degenerative diseases compared with mammals and birds. Some reptile species are known for longevity with minimal degenerative changes, and some freshwater turtle species have even been suggested to have "negligible senescence" (Krivoruchko and Storey, 2010). Factors that may promote resistance to degenerative diseases in reptiles include low metabolic rate relative to endotherms, adaptations for indeterminate growth or continued growth after reproductive maturity, and adaptations for survival during extreme conditions and stress. For example, the metabolic rate of the Western Fence Lizard (*Sceloporus occidentalis*) is only 3%–4% of the metabolic rate of a bird or mammal of similar size (Bennett and Nagy, 1977). Lower metabolic rate results in generation of fewer harmful products of metabolism, such as reactive oxygen species, resulting in less cellular damage. Adaptations for indeterminate growth and survival under extreme conditions and stress that have been demonstrated in some reptile species include the ability to suppress metabolic rate, resistance to shortening of telomeres in somatic cell DNA with aging, enhanced ability for DNA repair, high constitutive and inducible antioxidant mechanisms, high constitutive heat shock protein expression, anoxia-inducible molecular chaperones, and ability to suppress apoptosis (Bronikowski, 2008; Gomes et al., 2010; Krivoruchko and Storey, 2010). These same factors also contribute to extreme longevity in some reptile species, particularly the Testudines, and are responsible for the ability of some reptiles, such as turtles of the genera *Trachemys* and *Chrysemys*, to hibernate without breathing oxygen for weeks (Milton and Prentice, 2007; Krivoruchko and Storey, 2010). In addition, some reptiles also have remarkable tissue regenerative capacity. Perhaps the best examples of tissue regeneration are those that occur following caudal (tail) autotomy in most species of lizards and tuataras (*Sphenodon* spp.) and regional avulsion of the skin to evade capture in some species of lizards (Jacyniak et al., 2017). Despite these adaptations, reptiles clearly age and die, and may develop degenerative diseases described in this chapter. In addition to covering these degenerative diseases, this chapter will also cover normal findings and physiologic variations that should not be mistaken for degeneration.

5.2 Skin

Ultraviolet (UV) light is required by many reptile species for normal growth and metabolism. However, reptiles are also susceptible to the harmful effects of exposure to short wavelength photoreactive UVB radiation (Adkins et al., 2003; Gardiner et al., 2009). In one report, a male Ball Python (*Python regius*) and a female Blue-Tongue Skink (*Tiliqua* sp.), representative of a collection of multiple species of snakes and lizards, developed clinical signs and gross lesions that included anorexia, lethargy, decreased motor activity, skin thickening, increased skin pigmentation, excessive shedding, overly moist skin, skin erosions and ulcers, eye closure, and corneal opacity. The affected reptiles were exposed to UV lamps 12–24 inches from the reptiles, with most UVB output from the lamps below 300 nm and within the range known to cause harmful skin and eye effects in humans and other mammals (Gardiner et al., 2009). Clinical signs and lesions resolved following removal of the lamps. UV light-associated skin lesions in reptiles are often confined to the head and/or dorsum and occur over bony prominences. Microscopically, necrotic keratinocytes are present in the stratum basale and spinosum, reminiscent of sunburn cells in mammalian skin (**Figure 5.1**). Necrosis may progress to segmental epidermal erosion and ulceration with secondary infection. The superficial dermis has perivascular infiltrates of heterophils and lymphocytes. To date, exposure to UV radiation has not been associated with cutaneous neoplasia in reptiles, although more investigation of the potential for an association is needed. Exposure to short wavelength photoreactive UVB radiation may also be associated with poor reproduction and death.

Although changes in the frequency and completeness of ecdysis in reptiles are not truly a degenerative condition of the skin, they may be indicative of other underlying conditions. Ecdysis frequency in reptiles is dependent on numerous factors including temperature, humidity, diet, growth rate, age, and underlying disease. For normal stages of ecdysis in snakes, see Volume 1, Chapter 1. Nonspecific changes in ecdysis in reptiles with underlying disease are decreased frequency of shedding and incomplete shedding with retention of sloughing skin fragments. These changes may be difficult to distinguish from dysecdysis associated with improper husbandry conditions. Ecdysis in reptiles is also closely associated with function of the pituitary–thyroid axis, and dysecdysis may be an indication of underlying abnormalities in this axis, specifically with thyroid gland function. Experimental thyroidectomy performed in a small number of lizard and snake species resulted in reduced ecdysis or complete cessation of ecdysis in lizards and increased frequency of ecdysis in snakes (Boyer and Steffes, 2011). Therefore, in addition to considering husbandry-related dysecdysis, primary diseases of the thyroid gland or involvement of the thyroid gland in systemic disease should also be considered. Dysecdysis and retained shed often are associated with opportunistic bacterial and mycotic epidermitis which can lead to sepsis. Retained shed often causes constriction of toes and tail, with subsequent ischemic necrosis.

Reptiles in very poor nutritional condition may have connective tissue fragility that manifests as separation of the epidermis from the dermis (Jacobson, 1990). Early lesions may manifest as fluid accumulation between the epidermis and dermis. The skin becomes fragile and may tear following manual restraint or other trauma. Ulcerated dermis may quickly develop secondary bacterial and/or fungal infection. For more information see Chapter 2.

5.3 Nervous System

The reptilian nervous system is more resistant to degeneration compared with the mammalian and avian nervous systems. This phenomenon is likely related to some of the same factors described previously. In many reptiles, there is continuous growth of the brain and spinal cord throughout life and constitutive neurogenesis. Growth includes an increase in mass of the brain and spinal cord, spinal cord length, ocular volume, and numbers of cortical neurons (Kaslin et al., 2008; Ngwenya et al., 2013). In adult mammals, constitutive neurogenesis is predominantly limited to the subventricular zone of the lateral ventricle and the subgranular zone of the dentate gyrus, whereas in some reptile species, constitutive adult neurogenesis occurs additionally in the olfactory bulbs, other areas of the telencephalon, and cerebellum (Kaslin et al., 2008). The network of radial glial cells that guide migration of new neurons and serve as progenitors for neurons and glial cells is widespread throughout the brain of adult reptiles, whereas this network does not persist in the adult mammalian brain (**Figure 5.2**) (Font et al., 2001; Shao et al., 2012). Thus, adult reptiles have a greater capacity to generate neurons and incorporate these neurons into existing neural networks compared with adult mammals.

In addition to adaptations for constitutive growth, the brains of some freshwater turtles in the family *Emydidae*, such as the Western Painted Turtle (*Chrysemys picta bellii*) and Red-Eared Slider (*Trachemys scripta elegans*), show resistance to anoxic conditions that would rapidly result in neuronal necrosis in mammals (Krivoruchko and Storey, 2010). Resistance to anoxia in the turtle brain relative to mammals is mediated by numerous constitutive and inducible factors including high glycogen stores to support anaerobic glycolysis, lower aerobic enzyme activity, lower ion channel density, high antioxidant production, transcriptional arrest, reduction in ATP use, ion channel permeability, excitatory neurotransmitter release, and increased expression of heat shock proteins (Lutz and Milton, 2004; Keenan et al., 2015). These same factors likely also provide resistance to degenerative changes in the turtle brain. Healing in the reptilian brain also differs from mammals and may account for some resistance to degenerative changes. Astrocytes are minimal or nonexistent in the reptile brain, and formation of glial scarring following brain injury in mammals that can interfere with axon regrowth does not seem to occur in reptiles. In a model of neuronal necrosis in the medial cortex induced by 3-acetylpyridine administration, there is a transient disappearance of microglia and increased thickening of processes of radial glia that are responsible for removal of cell debris (Lopez-Garcia et al., 1994; Nacher et al., 1999). It is likely that the radial glia have multiple functions including removal of debris, creating a scaffold to guide the migration of new neurons, and possibly also serving as progenitor cells for neurons and glia. The remarkable capability of the reptilian brain to undergo neurogenesis results in generation of a new medial cortex with normal histological features in this model.

In lizard species that undergo caudal autotomy and regeneration, the regenerated tail segment develops a spinal cord consisting of a central canal originating from ependymal cells at the site of autotomy, and axons organized into nerve tracts that originate from the white matter of the original spinal cord and dorsal root ganglia (**Figure 5.3a,b**) (Jacyniak et al., 2017). The new spinal cord in the regenerated tail segment does not contain gray matter or dorsal root ganglia; these findings should not be mistaken for degenerative changes.

Degenerative changes in the brain and spinal cord of reptiles may be encountered in geriatric animals and consist of small numbers of pigment-laden phagocytic cells around vessels, neuronal lipofuscin accumulation, and corpora amylacea (**Figure 5.4**). These findings are of no clinical significance. Degenerative changes of the spinal cord may be observed in reptiles secondary to spinal lesions, such as spinal osteoarthrosis or osteoarthritis or traumatic spinal injuries. Clinical manifestations of degenerative changes in the spinal cord of reptiles may be minimally apparent despite extensive changes, and may include abnormal posture, difficult ambulation or movement, and altered intestinal motility. Grossly, the affected segment may show compression of the spinal cord, or in extreme cases, severance of the cord (**Figure 5.5**). Microscopically, lesions are similar to those in mammals. Myelin sheaths are enlarged, axons show fragmentation and spheroid formation, and phagocytic cells may be present within myelin sheaths (**Figure 5.6**). In severe cases, gitter cells may infiltrate or even efface segments of the spinal cord. Lesions predominantly affect white matter unless compression is severe enough to extend into gray matter. If lesions extend to the central canal, ependymal cells may show attenuation, hypertrophy, and/or hyperplasia.

5.4 Special Sense Organs

The reptile cornea is susceptible to the harmful effects of exposure to short wavelength photoreactive UVB radiation. Clinical signs and gross findings are eye closure and corneal opacity that is usually bilateral. Microscopically, lesions within the affected eyes of a Ball Python were severe bilateral ulceration with neovascularization, heterophilic and lymphocytic keratoconjunctivitis and anterior uveitis, and secondary bacterial colonization (Gardiner et al., 2009). Lesions are nonspecific and difficult to distinguish grossly and microscopically from other causes of ulcerative keratoconjunctivitis, and the diagnosis of UVB-induced corneal damage should be supported by concurrent skin lesions described previously, and ruling out other causes such as trauma, xanthomatosis, mineralization, hypovitaminosis A, and mycotic infection.

Cataracts are common in old reptiles (**Figure 5.7**). Their causes are uncertain but speculated to be nutritional deficiencies, ocular inflammation or trauma, aging, or possibly genetic factors. They have not been associated with exposure to harmful UV radiation, as in some mammals. In tortoises, cataracts have been associated with exposure to freezing temperatures, and may spontaneously improve over time (Lawton, 2000). It is important to be familiar with the normal microscopic anatomy of the reptilian lens. Most reptiles have equatorial annular pads where lens epithelial cells are elongated and oriented in a radial manner. In snakes, elongation of lens epithelium is not equatorial, but occurs in the anterior aspect of the lens, forming the anterior pad. Degenerative change may occur anywhere in the reptilian lens and are easiest to appreciate microscopically in the annular and anterior pads. Pathologic changes in reptilian cataracts are generally similar to those in mammals. Grossly, unilateral or bilateral lenticular opacity is present. Microscopically, degenerate lens fibers form round to oval eosinophilic Morgagnian globules (**Figure 5.8**). As the lesion progresses, proteinaceous fluid accumulates between degenerate lens fibers. Lens epithelial cells, particularly at the annular and anterior pads, may also degenerate and be replaced by proteinaceous fluid. Reactive lens epithelial cells may also form bladder cells similar to mammals. These cells have a rounded shape and abundant eosinophilic and sometimes foamy or vacuolated cytoplasm. As lens fibers continue to degenerate and proteinaceous fluid is lost through the lens capsule, a hypermature cataract forms that consists of a collapsed lens capsule that may surround small numbers of bladder cells and proteinaceous fluid (**Figure 5.9**).

5.5 Coelom

Although widespread steatonecrosis may have a nutritional basis in reptiles similar to other species, spontaneous necrosis of the intracoelomic adipose is a common incidental finding in older reptiles that are in good nutritional condition at the time of death, particularly in snakes and lizards that have excessive intracoelomic adipose. It is similar to nodular fat necrosis that occurs in the abdominal fat of some geriatric mammals. The change may be related to local tissue ischemia due to vascular compromise, as might occur with age-related vascular degeneration, trauma, parasite migration, mineralization, etc. Grossly, foci of necrotic fat appear as well demarcated, yellow or tan nodules within the intracoelomic adipose. Microscopically, lesions are a mixture of acute necrosis of adipocytes and chronic necrosis with infiltrates of epithelioid and giant cell macrophages, lipophages, and ceroid (**Figure 5.10**). Small numbers of lymphocytes may also be present. Although mineralization of necrotic fat is common in mammals, mineralization is not a common feature of this lesion in reptiles. Chronic localized steatitis can resemble involuted ovarian follicles, so gender and location of the finding should be considered in this regard.

Size of the coelomic adipose stores varies considerably with season and stage of the reproductive cycle in many reptiles, and the absence of substantial adipose stores may not indicate poor nutritional condition. For example, in many male reptiles, size of the coelomic adipose stores is greatest prior to spermatogenesis, and gradually decreases during reproductive activity (Trauth, 1979; Guillette and Sullivan, 1985; Ramirez-Bautista et al., 2002; Ortiz et al., 2014). In females, coelomic adipose stores are used for vitellogenesis, and are often largest prior to reproductive activity, gradually decrease in size during vitellogenesis, and are smallest prior to and during ovulation (Guillette and Sullivan, 1985; Flemming and van Wyk, 1992; Pinilla, 1995; Ramirez-Bautista et al., 2002; Ortiz et al., 2014). For viviparous species, coelomic adipose stores may increase during early gestation but decrease again during late gestation (van Wyk, 1989). These findings reflect seasonal conditions such as temperature and humidity and food availability, and generally apply to free-ranging reptiles.

5.6 Respiratory System

Degenerative diseases of the reptilian respiratory system are rare. In lizards, incidental accumulation of mucus and cell debris in the lumen of peribronchiolar faveolae may be seen as an age-related change. This finding is not associated with clinical signs. Rarely, enough material may accumulate to cause mild enlargement of the parenchyma grossly that may mimic pneumonia, but without the gross appearance of consolidation. Microscopically, affected faveolae contain luminal masses of inspissated mucus and cell debris without associated inflammation or with minimal associated infiltrates of macrophages in the faveolar lumina or around the inspissated mucus (**Figure 5.11**). Degenerative changes in the cartilage matrix of the trachea, most commonly manifesting histopathologically as matrix eosinophilia, may also be seen in geriatric reptiles, and may be associated with flattening of the trachea.

5.7 Cardiovascular System

The reptilian heart is relatively resistant to degenerative diseases for many of the same reasons that reptiles in general are resistant to degenerative diseases. Because of the lower reptilian metabolic rate relative to mammals and birds, reptilian cardiac mass relative to body weight, cardiac output, and cardiac work are lower compared with values for mammals and birds (Jensen et al., 2014). Lower cardiac output is related to lower heart rate in reptiles, and lower cardiac work is related to both lower heart rate and mean arterial blood pressure. Reptiles also have some remarkable physiologic abilities that also may render resistance to degenerative diseases in the heart and other viscera. Because most reptiles have a single ventricle, they are able to control shunting of blood from the pulmonary to systemic circulation, and vice-versa, presumably to accommodate physiologic needs, facilitate warming, control myocardial oxygenation, and assist digestion, although the hypothesized benefits from this ability have yet to be definitively demonstrated (Jensen et al., 2014). Crocodilians, with a complete ventricular septum and four-chambered heart, have the ability to shunt blood from the pulmonary to systemic circulation through the left aorta. Some pythons have a unique ability to undergo cardiac myocyte hypertrophy within 2–3 days following ingestion of a large prey item that can increase heart weight by up to 40% (Riquelme et al., 2011). Some pythons can also increase stroke volume in the postprandial period, without substantial cardiac hypertrophy (Enok et al., 2016). These changes result in increased cardiac output and increased blood flow to the digestive tract to facilitate digestion. The python heart may be able to accommodate these changes because of the presence of greater myocardial connective tissue stroma to provide structural support (Riquelme et al., 2011). Some freshwater turtles in the family *Emydidae* have also evolved mechanisms to protect the heart during periods of anoxia, similar to adaptive mechanisms to protect the brain. For example, anoxic Western Painted Turtles have lowered heart rate and contractility, up-regulation of genes involved in suppression of cardiac metabolism, and overall reduction of RNA resulting in translational arrest in the ventricle (Keenan et al., 2015).

Dilated cardiomyopathy and congestive heart failure are rarely reported in reptiles, and most reports are in snakes. Some predisposing factors include atrioventricular valvular insufficiency and aortic stenosis; however, many cases are idiopathic and may occur in juvenile animals (Jacobson et al., 1979; Barten, 1980; Clippinger, 1993; Rishniw and Carmel,

1999). Gross lesions are cardiomegaly and transudative pericardial effusion. Visceral and great vessel congestion, coelomic effusion, pulmonary edema, gular edema, and skin and mucus membrane discoloration or pallor may be present. Affected reptiles are often in poor nutritional condition. In snakes, cardiomegaly can be visible prior to dissection as focal enlargement of the body one-quarter of the length from the head. Microscopically, there may be no apparent lesions in the myocardium, or myocardial interstitial edema, myofiber hypertrophy, myofiber necrosis, and mineralization, and replacement of myofibers by fibrosis at the junction of the ventricular inner and outer muscular tunics may be present. A single case of congestive heart failure secondary to idiopathic circumferential ventricular fibrosis and restrictive cardiomyopathy has been described in a McDowell's Carpet Python (*Morelia spilota mcdowelli*) (Schilliger et al., 2016).

Degenerative vascular lesions are also rare in reptiles. Arteriosclerosis with mineralization and atherosclerosis of the great vessels are reported to occur in reptiles, but likely have underlying nutritional or other husbandry-related causes and may not be true degenerative diseases. Great vessel aneurysms occasionally occur at the heart base or in the descending aorta of snakes and other reptiles and often present as sudden death following consumption or attempted consumption of a prey item in snakes. Death is most often related to pericardial hemorrhage and cardiac tamponade, but fatal cardiac arrest secondary to thromboembolism and myocardial or visceral ischemia without pericardial hemorrhage may also occur (Rush et al., 2001). These aneurysms may develop as a result of transient increases in blood pressure within the great vessels associated with passage of a prey item through the alimentary tract, or transient hypertension associated with attacking and constricting prey. Because systolic pressures in the pulmonary and systemic circulation are similar in reptiles, heart base aneurysms could likely arise in the left or right aorta or the pulmonary artery. Most reports of great vessel aneurysms at the heart base of reptiles do not specify the vessel involved, and in chronic lesions the affected vessel can be difficult to identify. In reptiles that die from ruptured aneurysms of the great vessels, massive hemopericardium or hemocoelom may be present. The aneurysm may be difficult to find grossly except in very large reptiles, but adventitial accumulation of pigment may be visible around the great vessels at the heart base related to chronic hemorrhage prior to fatal rupture of the aneurysm. Microscopically, separation of the intima from the media is present at the aneurysm. Medial necrosis with dissection of the aneurysm into and sometimes through the media may also be present. The dissection may be filled with hemorrhage, fibrin thrombi, or both. Fibrin thrombi may also be present in the vessel lumen at the site. Depending on chronicity, fibrosis and infiltrates of siderophages may be present around the dissection. The author has observed three emaciated and anemic wild Loggerhead Sea Turtles (*Caretta caretta*) with ruptured aneurysms of the left aorta or its vaso vasorum at the origin of the left aorta, resulting in fatal pericardial hemorrhage and cardiac tamponade similar to other reptiles with ruptured great vessel aneurysms (**Figure 5.12a,b**). Affected animals have concurrent spirorchid trematodiasis, but the aneurysms do not appear to be a direct consequence of vasculitis caused by parasites or ova at the site. The left aorta shows medial degenerative changes consisting of increased loose collagenous or mucinous stroma around vessels and increased tortuous vessels that progress to transmural medial necrosis in the aorta and its vaso vasorum resulting in mural, adventitial, and pericardial hemorrhage, suggesting the aneurysms may be consequence of altered hemodynamics secondary to anemia, downstream left aortitis related to spirorchid trematodiasis, or both. Because animals are emaciated, lesions may also be exacerbated by concurrent nutritional deficiencies. Aneurysms presumably from vascular degenerative changes may also occur in other vessels, such as aneurysms of the internal carotid artery that result in hemorrhage in the neck of Bearded Dragons (*Pogona vitticeps*) (Barten et al., 2006).

Rarely, increased mucinous stroma may be seen in the atrioventricular valves of geriatric reptiles, similar to valvular endocardiosis commonly encountered in geriatric mammals (**Figure 5.13**). This finding should be interpreted cautiously because these valves normally contain greater amounts of stroma relative to mammals. Increased mucinous stroma with a disorganized appearance, and/or increased mucinous stroma that causes gross enlargement of the valve leaflet and partial or complete occlusion of the valve is likely pathologic. However, similar to mammals, valvular endocardiosis in reptiles does not appear to be clinically significant.

For more information on cardiovascular disease see Chapter 9.

5.8 Digestive System

The digestive tract is an organ system with a high mucosal turnover rate and great capacity for regeneration, and degenerative diseases of the digestive tract are generally rare in all species including reptiles. One exception is tooth wear, which is commonly encountered in mammals as a progressive change associated with aging. Except for mammals, most toothed vertebrates, including most toothed reptiles, are polyphyodonts with the ability to regenerate teeth, and therefore tooth wear is not commonly encountered, although the ability to regenerate teeth may decline with age. Reptiles with the ability to regenerate teeth include crocodilians and most squamates; these animals have thecodont teeth (crocodilians) or pleurodont teeth (some squamates) that have connective tissue attachments to the medial aspects of the bones of the jaw (**Figure 5.14**). In contrast, the cheek teeth of agamid lizards and all teeth of adult chameleons are acrodont and fused to the bone of the jaw with no connective tissue attachments (**Figure 5.15**). Acrodont teeth cannot be regenerated and are subject to wear. Acrodont tooth wear and age-related changes have been studied in the Egyptian Spiny-Tailed Lizard (*Uromastyx aegyptius*) and in chameleons (Throckmorton, 1979; Dosedelova et al., 2016). In the Egyptian Spiny-Tailed Lizard, tooth wear occurs predominantly on the labial side of the cheek teeth. As dentine wears away, odontoblasts in the pulp cavity produce secondary dentine that undergoes progressive mineralization and eventually fills the pulp cavity (**Figure 5.16**). Subjacent bone of the jaw gradually changes from cancellous to compact. Eventually, the secondary dentine is worn away. When the entire tooth has been worn away, the bone of the jaw will serve the function of the missing tooth (**Figure 5.17**). Because newer acrodont teeth are added at the caudal aspect of the arcade as the jaw grows, the most severe tooth wear in agamid lizards is present on the rostral cheek teeth.

For more information on the dentition of reptiles, see Volume 1, Chapter 1.

5.9 Hepatobiliary System and Pancreas

There are few degenerative diseases of the reptilian hepatobiliary system or pancreas. It is important to be aware of the normal anatomy of the liver of reptiles to avoid misinterpretation of normal findings. Pigmented macrophages containing hemosiderin, melanin, and lipofuscin are a normal finding in the hepatic sinusoids of reptiles. Although increased numbers of pigmented macrophages can be an indication of disease, the number and size of aggregates of pigmented macrophages increases with age in reptiles and may constitute up to 20% of the hepatic volume in a geriatric animal (Christiansen et al., 1996). The American Alligator (*Alligator mississippiensis*) and Spectacled Caiman (*Caiman crocodilus*), and likely all adult crocodilians, have fibrous trabeculae in their livers as part of the normal anatomy (Beresford, 1993). These trabeculae extend from portal triads and the hepatic capsule and should not be mistaken for fibrosis. Hepatic fibrosis is occasionally seen in reptiles, often around portal triads and vessels with extension into the surrounding parenchyma and bridging of triads and vessels (**Figure 5.18a,b**). The fibrosis may be extensive and replace most of the hepatic parenchyma (cirrhosis). However, it is unknown if hepatic fibrosis is a true degenerative change or a sequel to other hepatic insults, such as exposure to a hepatotoxin, chronic hepatic lipidosis, biliary obstruction (choleliths, tumor, etc.), aging, or combinations thereof. The author has seen one case of mildly increased immature connective tissue within multifocal hepatic sinusoids in a geriatric Honduran Milksnake (*Lampropeltis triangulum hondurensis*) that underwent repeated radiographs to monitor the progression of spinal arthritis. Lesions were suggestive of early sinusoidal obstruction syndrome (formerly hepatic veno-occlusive disease) related to repeated radiation exposure and adverse effects on sinusoidal endothelial cells. Connective tissue deposition in the space of Disse and extending into sinusoids has also been documented in Southern Watersnakes (*Nerodia fasciata fasciata*) fed a diet of fish with arsenic, cadmium, selenium, strontium, and vanadium contamination from coal ash exposure (Ganser et al., 2003).

5.10 Immune System

Reptilian lymphoid organs are modulated by seasonal and hormonal effects, and changes in morphology related to these effects should not be mistaken for degenerative changes. Seasonal changes in the cellularity of lymphoid organs have been documented for multiple reptile species, and the morphologic appearance of lymphoid tissues depends on the optimum and adverse seasons for that species (Hussein et al., 1979; el Ridi et al., 1988). Reptiles may undergo one or two cycles of involution followed by increasing cellularity of lymphoid organs depending on the species and its reproductive patterns. These changes appear to be mediated by varying levels of endogenous corticosteroids and reproductive hormones. For example, in the male Caspian Turtle (*Mauremys caspica*), thymic involution correlates with peaks in circulating testosterone, and exogenous testosterone administration can cause thymic involution and reduced lymphoid cellularity in the spleen (Saad et al., 1991; Varas et al., 1992). In the viviparous Ocellated Skink (*Chalcides ocellatus*), numbers of leukocytes in the spleen in mid and late gestation decline to approximately half of the corresponding numbers in a nongravid lizard (Saad and el Deeb, 1990).

Thymic atrophy is associated with starvation in reptiles and can be severe (**Figure 5.19**). The atrophied thymus is difficult to find grossly. Histologically, the thymus is very poorly cellular without a distinct cortex and medulla. Changes may be difficult to distinguish from involution, and history and signalment should be considered when interpreting reptilian thymus.

For more information on the immune system of reptiles, see Volume 1, Chapter 2.

5.11 Urinary System

Unlike mammals, degenerative diseases of the kidneys of clinical relevance are uncommon in reptiles. The resistance of the reptilian kidney to degenerative changes is likely related to several features of reptilian anatomy and physiology, including a dual renal blood supply through both the renal arteries and renal portal veins, low metabolic rate, low blood pressure, low glomerular filtration rate, short nephrons, the ability to grow new nephrons throughout postembryonic life, and use of postrenal modification of urine osmolality in the urinary bladder (except for crocodilians and snakes that lack a bladder), cloaca, and posterior colon (Zwart, 1964; Camarata et al., 2016). Most reptiles rely on tubular secretion of uric acid by terrestrial species and ammonia by aquatic species, thereby reducing the need for glomerular filtration of nitrogenous wastes and increasing the need for tubular blood supply. Some aquatic species also excrete urea; the aquatic Chinese Softshell Turtle (*Pelodiscus sinensis*) excretes the majority of its urea through the oral cavity, further reducing the need for renal excretion (Ip et al., 2012). Glomerular morphology is also influenced by environment and water availability. Aquatic chelonians in the family *Emydidae* have a high density of capillaries in the glomerular tuft (**Figure 5.20**). Tortoises that reside in arid environments have fewer capillaries and greater mesangial stroma that should not be mistaken for a pathologic change (**Figure 5.21**) (Zwart, 2006). Nephrogenic zones may be visible microscopically and sometimes grossly beneath the renal capsule in chelonians and should not be mistaken for a pathologic change (**Figure 5.22**). Interstitial connective tissue is generally minimal in reptiles, but can be found surrounding the ureteral branches, especially in sea turtles. Male snakes and some lizards have a modified segment of the distal renal tubules designated the sexual segment (see Volume 1, **Figures 1.220, 1.233–1.235**). During periods of spermatogenesis, this segment undergoes epithelial hypertrophy due to accumulation of cytoplasmic eosinophilic granular material that is often visible grossly as pale, enlarged foci within the renal parenchyma. Grossly, this seasonal alteration can be mistaken for gout or other renal lesions.

The most common degenerative change in the kidneys of reptiles is the accumulation of granular brown pigment in tubular epithelial cells, most commonly in the proximal segments (**Figure 5.23a,b**). This pigment may stain with periodic acid-Schiff or Fontana-Masson stains, and does not stain with Prussian blue stains for ferric iron. It is likely lipofuscin, melanin, or both. The pigment can be abundant in geriatric reptiles and is of no clinical consequence.

Glomerulosclerosis and interstitial fibrosis are occasionally seen in reptiles and need to be interpreted carefully considering the variations in normal species anatomy (**Figures 5.24a,b–5.25a,b**). Because of low blood pressure and low glomerular filtration rate in reptiles, it is uncertain if the pathophysiology of glomerulosclerosis in reptiles is similar to that in mammals. It is possible that glomerulosclerosis in reptiles occurs as a sequel to chronic glomerulonephritis, and not a true degenerative change. Nevertheless, glomerulosclerosis is occasionally seen in some geriatric reptiles without clear evidence of previous or chronic inflammatory disease, suggesting glomerulosclerosis in some individuals may be a degenerative, age-related change. Increased renal interstitial connective tissue in reptiles is more likely to be a consequence of previous tubulointerstitial nephritis or nephrotoxicity affecting the renal tubules. Similar to mammals, identifying the etiopathogenesis in cases of chronic glomerulosclerosis or interstitial fibrosis in reptiles is difficult or impossible. Because of the renal portal system, glomerulosclerosis in reptiles is not typically associated with tubular atrophy or other tubular changes. Interstitial fibrosis, however, is often associated with a spectrum of tubular changes including tubular dilation, tubular atrophy, tubular epithelial hypertrophy or regeneration, or loss of tubular epithelial differentiation resulting in a monomorphic appearance to all renal tubules, and accumulation of proteinaceous material within the tubular lumina (**Figure 5.26**) (Zwart, 2006). Most likely, these changes are a consequence of the initial insult to the tubules and loss of tubular blood supply with ensuing interstitial fibrosis.

Degenerative diseases of the ureters and urinary bladder (in those reptiles that possess a bladder) have not been reported.

5.12 Reproductive System

Almost all reptiles are seasonal breeders, and seasonal variation exists in the morphology of reproductive tissues that should not be mistaken for degenerative change, similar to lymphoid tissues. Some tropical reptiles that can breed throughout the year may show seasonal changes although not to the same degree as reptiles from other climates.

In male reptiles, the size of seminiferous tubules and overall testicular size are small during the nonbreeding season. Seminiferous tubules are lined only by small numbers of Sertoli cells and spermatogonia in one or two cell layers. As the reptile enters the reproductive season, spermatogonia undergo division, and sequential appearance of spermatocytes, spermatids, and luminal mature spermatozoa occurs, resulting in marked increase in the size and cellularity of seminiferous tubules that may obscure the interstitium. Following cessation of the breeding season, cellularity of the tubules decreases, and cell debris and lipofuscin may be seen in the lumen of seminiferous tubules (**Figure 5.27**). Tubules decrease in size and the interstitium is visible. Interstitial Leydig cells may also show hypertrophy during the breeding season and atrophy during the nonbreeding season. In some male reptiles, the epithelium of epididymal ductules also may vary in appearance from stratified in the nonbreeding season to a single layer of columnar secretory cells during the breeding season, coinciding with the presence of luminal spermatozoa in the epididymal ductules. Some male reptiles are able to store spermatozoa for many months in the epididymis and therefore these seasonal changes in the appearance of the epididymis may not occur in all species (Goldberg and Parker, 1975). In species that are able to store spermatozoa, it is not uncommon to see mature spermatozoa in the epididymis in an animal with completely regressed testes.

True degenerative changes of the testes are uncommon. In geriatric male reptiles, lipofuscin accumulation in germinal epithelium and Sertoli cells and within sloughed cell debris in the lumen of tubules can be excessive. Seminiferous tubules may also become cystic with luminal accumulation of cell debris and mucinous material with attenuation of the epithelium, and the change may be severe enough to see grossly (**Figure 5.28**). Spermatogenic arrest or formation of multinucleated spermatids may also be encountered (**Figure 5.29**). The author has seen one case of testicular edema in a Jeweled Lizard (*Timon lepidus*) following whole body radiography. Gross lesions were not apparent, but microscopically, abundant proteinaceous fluid separating seminiferous tubules was observed. Findings were similar to those produced experimentally in rats following radiation exposure (Porter et al., 2006).

In female reptiles, follicles at any stage of development may undergo atresia and should not be mistaken for abnormal follicles. In species that are seasonal breeders, atretic previtellogenic follicles can be found in the nonbreeding season. Grossly, in reproductively active females, atretic follicles are variable in size and appear dark orange, pale yellow, or white depending on the stage of development. Atretic follicles in seasonal breeders in the nonbreeding season are small and white. Microscopically, follicular atresia begins with folding and fragmentation of the zona pellucida, vacuolation and fragmentation of the ooplasm, hyperplasia of granulosa cells, vascularization of the theca, and invasion and phagocytosis of the ooplasm by granulosa cells (**Figures 5.30–5.31**). As atresia progresses, macrophages and fibroblasts also invade the ooplasm. The atretic follicle is eventually replaced by a small mass of pigmented and vacuolated macrophages and connective tissue, a finding that can be confused with inflamed or necrotic fat bodies (**Figures 5.32–5.33**). Because follicular atresia is a method of controlling clutch and litter size in reptiles, numbers of atretic follicles are variable in female reptiles of the same species but exposed to different environmental conditions that would affect breeding success.

Degenerative changes of the female reproductive tract in reptiles have not been reported. Occasional oviductal acini may show ectasia in geriatric female reptiles, but this finding is of no clinical significance.

For more information on the reproductive system of reptiles, see Volume 1, Chapter 1 and Volume 2, Chapter 8.

5.13 Endocrine System

Endocrine organs of reptiles include the thyroid gland, parathyroid gland, ultimobranchial body, pituitary gland, adrenal gland, pineal gland, and pancreatic islets. A pineal complex or "third eye" is present in some species of lizards. Interpretation of degenerative changes in the endocrine organs of reptiles should be done cautiously because of seasonal and anatomic variation in some endocrine organs, particularly the thyroid gland. Height of the thyroid follicular epithelium is greatest during highest metabolic activity, such as breeding season, exposure to warm ambient temperatures, and impending or ongoing ecdysis. Reptiles with low metabolic activity, such as nonbreeding, ill, and geriatric animals, have thyroid follicles lined by low cuboidal to attenuated cells with abundant colloid that should not be mistaken for colloid goiter. In most reptiles, connective tissue is present originating from the capsule of the gland and accompanying and surrounding vessels that extend into the gland. This connective tissue can be prominent in some species, such as American Alligators, and should not be mistaken for fibrosis.

Accumulation of lipofuscin is a common age-related change in multiple endocrine organs in reptiles. In the adrenal gland, lipofuscin can accumulate in both interrenal and chromaffin cells. In the thyroid gland, accumulation of sloughed lipofuscin-laden cells and occasional mineral can also be entrapped within colloid in geriatric reptiles. Pituitary acini in geriatric reptiles may accumulate luminal proteinaceous product. These changes are of no known clinical importance.

The author has seen two cases of abundant adipose tissue within the thyroid gland of two geriatric male Jeweled Lizards. The adipose tissue was separating follicles, and in one animal was associated with infiltrates of lymphocytes and fewer macrophages. These findings were suggestive of thyroid atrophy and/or lymphocytic thyroiditis (**Figure 5.34**).

For more information on the endocrine system of reptiles, see Volume 1, Chapter 1.

5.14 Musculoskeletal System

Degenerative diseases of skeletal muscle are rare in reptiles, likely related to the same factors that render other reptile tissues resistant to degenerative diseases. For example, external gastrocnemius muscle from the Red-Eared Slider is resistant to disuse atrophy in experimental models that induce atrophy in muscle from endothermic mammals (McDonagh et al., 2004). It is likely that this resistance is an adaptation to retain muscle strength and composition during prolonged periods of inactivity and hypometabolic states, such as hibernation. There is a single report of temporal muscle myopathy in an adult male Veiled Chameleon (*Chamaeleo calyptratus*) presenting as progressive inability to open the mouth (Rowland, 2011). Microscopically, affected muscle had multifocal myofiber atrophy with internalization of nuclei and cytoplasmic vacuolation in myofibers. Dense collagenous connective tissue without inflammatory cells infiltrated and separated myofibers. Other muscles were not affected, and the cause for the condition was not determined.

Compared with other degenerative diseases, degenerative joint disease (osteoarthritis) is relatively common in geriatric reptiles, despite the tremendous capacity of some lizard species to regenerate cartilage following caudal autotomy and in experimental models of articular cartilage damage in the appendicular skeleton. Chondrocyte progenitors in reptiles are located in the superficial articular cartilage and metaphyseal growth plate of the epiphysis, and likely contribute to continuous growth of the skeleton in some reptiles (Alibardi, 2015). However, the growth plate may not persist in all squamates of the clade Pleurodonta (Frydlova et al., 2019). The absence of the growth plate in some mature squamates suggests some species have determinant growth and may also have less regenerative capacity. In reptiles with degenerative joint disease, stem cells in the superficial articular cartilage may be lost following cartilage erosion and eburnation, thereby limiting regenerative capacity. In the author's experience, in reptiles with both an axial and appendicular skeleton (excluding chelonians), the axial skeleton develops degenerative joint disease more commonly than the appendicular skeleton, but this observation may be biased by body shapes and sizes in the patient population, and more extensive studies are needed to investigate the effects of body size and shape on the prevalence and distribution of degenerative joint disease in reptiles. In the axial skeleton, degenerative joint disease appears to be more common in the atlantooccipital joint and cervical, dorsal (thoracolumbar), and sacral spine compared with the caudal spine. Geriatric chelonians may develop degenerative joint disease at the atlantooccipital joint and at the articulation of the seventh and eighth cervical vertebra in the axial skeleton, possibly because the former is freely movable and the latter is fused to the carapace in almost all chelonian species. In the appendicular skeleton, the more proximal joints (shoulder, cubital, coxofemoral, and stifle) are more commonly affected by degenerative changes than the distal joints. Presumably, degenerative joint disease in reptiles is caused by factors similar to those in endothermic vertebrates, including repetitive use and age-related changes in the composition of articular cartilage. There are scant data on the prevalence of primary degenerative joint disease compared with secondary joint disease in reptiles and the factors that predispose to secondary degenerative joint disease in reptiles. Because nutritional disorders such as metabolic bone disease and hypovitaminosis A are relatively common in some reptiles, it is possible their effects on skeletal development may predispose to secondary degenerative joint disease, but data are lacking in this regard.

In most cases, degenerative joint disease is relatively straightforward to diagnose in reptiles and is similar to that in mammals. Because the reptile spine generally lacks intervertebral discs (geckos are an exception), degenerative joint disease of the spine is similar to that in other joints. Grossly, joint enlargement may be visible if degenerative joint disease is severe enough (**Figure 5.35**). Gross lesions may include thickening of the joint capsule, a roughened articular surface, cartilage eburnation, and osteophyte formation at the margins of the articular cartilage (**Figure 5.36–5.37**). In the appendicular skeleton, secondary synovial chondromatosis may be present as hard nodules of cartilage in the thickened synovium, and in severe cases, these nodules may break off and appear as smooth, free floating, white firm masses within the joint space (**Figure 5.38**). In the spine, joint capsule thickening may also undergo cartilaginous or osseous metaplasia, resulting in grossly visible dorsal and/or ventral spondylosis. Microscopically, articular cartilage shows matrix eosinophilia, fibrillation, chondrocyte necrosis, erosion, and eburnation similar to mammals (**Figure 5.39**). Inflammatory cell infiltrates are absent. In the author's experience, reparative attempts, such as multicellular chondrocyte clusters, are rarely seen. Increased subchondral bone may be present, and there may be herniation of small fragments of articular cartilage into the medullary spaces of subchondral bone. In severe cases of spinal degenerative joint disease with dorsal spondylosis, spinal cord compression with associated degenerative changes may be present.

A potential challenge of diagnosing degenerative joint disease in reptiles is ruling out other causes for osteoarthritis, particularly infection. Generally, if inflammatory cell infiltrates are absent and ancillary testing such as culture fails to identify any infectious agents, a diagnosis of degenerative joint disease is reasonable. However, spinal osteoarthritis in snakes can be more challenging to interpret. Snakes commonly develop marked bony proliferation as both subchondral bone and large exostoses extending from the vertebral epiphyses and other sites along the vertebral cortices (**Figures 5.40–5.41**). Proliferating bone is often severe enough to be palpable and visible grossly in the live patient. Bony proliferation can result in effacement of the joint space and ankylosis, sometimes involving multiple vertebrae and the costovertebral articulations. Microfractures can develop in these exostoses, further complicating interpretation of the lesions. These lesions in snakes are often associated with bacterial osteomyelitis and osteoarthritis of the spine, often involving salmonellae (Isaza et al., 2000). Identification of inflammation and bacterial microcolonies confirms the diagnosis, but in chronic cases, inflammation and microorganisms may be difficult to demonstrate microscopically; examining multiple sections of the spine may be helpful in this regard, as early lesions with bacteria may not be grossly apparent. Ancillary testing such as culture is important in these cases to rule out bacterial infection of the spine. It is likely that geriatric snakes develop sterile degenerative joint disease of the atlantooccipital joint and spine that result in similar lesions. Snakes are unique among vertebrates in having five points of articulation at each vertebral junction and locomotion via lateral undulation (in terrestrial species) that involves propagation of bending movement along most of the length of the spine; this method would seemingly predispose to degenerative joint disease of the spine in aged animals. If degenerative changes develop, it seems likely that continued locomotion, and possibly other activities such as striking and prey capture, could exacerbate the lesions and result in these severe changes. It is also possible that degenerative changes in the spine create a microenvironment conducive to bacterial colonization, resulting in secondary septic osteoarthritis and osteomyelitis in some snakes with degenerative joint disease.

In lizard species that undergo caudal autotomy and regeneration, the regenerated tail segment develops a hollow, cone-shaped, unsegmented cartilaginous structure that does not differentiate further into coccygeal vertebrae (Jacyniak et al., 2017). The periosteum of the vertebrae at the site of autotomy will produce new bone that extends for several millimeters into the cartilaginous structure to anchor it to the preexisting spine, but the newly formed cartilage does not ossify (**Figure 5.42–5.43**). The newly formed cartilage also contains larger chondrocytes at a higher density compared with preexisting cartilage at other sites of the skeleton. These findings should not be mistaken for degenerative changes. For more information see also Chapter 4.

References

Adkins ET, Driggers GW, Ferguson WH et al. 2003. Ultraviolet light and reptiles, amphibians. *J Herpetol Med Surg* 13:27–37.

Alibardi L. 2015. Regeneration of articular cartilage in lizard knee from resident stem/progenitor cells. *Int J Mol Sci* 16:20731–20747.

Barten SL. 1980. Cardiomyopathy in a Kingsnake (*Lampropeltis calligaster rhombomaculata*). *Vet Med Small Anim Clin* 75:125–129.

Barten S, Wyneken J, Mader D, and Garner M. 2006. Aneurysm in the dorsolateral neck of two Bearded Dragons, in *Proceedings of the Association of Reptilian and Amphibian Veterinarians*, Baltimore, Maryland, pp. 43–44.

Bennett AF, and Nagy KA. 1977. Energy expenditure in free-ranging lizards. *Ecology* 58:697–700.

Beresford WA. 1993. Fibrous trabeculae in the liver of Alligator (*Alligator mississippiensis*). *Ann Anat* 175:357–359.

Boyer TH, and Steffes ZJ. 2011. Reptilian thyroid anatomy, physiology, and disease, in *Proceedings of the Association of Reptilian and Amphibian Veterinarians*, Seattle, Washington, pp. 18–35.

Bronikowski AM. 2008. The evolution of aging phenotypes in snakes: A review and synthesis with new data. *Age* 30:169–176.

Camarata T, Howard A, Elsey RM et al. 2016. Postembryonic nephrogenesis and persistence of six2-expressing nephron progenitor cells in the reptilian kidney. *PLOS ONE* 11:e0153422.

Christiansen JL, Grzybowski JN, and Kodama RM. 1996. Melanomacrophage aggregations and their age relationships in the Yellow Mud Turtle, *Kinosternon flavescens* (Kinosternidae). *Pigm Cell Res* 9:185–190.

Clippinger TL. 1993. Aortic stenosis and atrioventricular dilatation in a Green Iguana (*Iguana iguana*), in *Proceedings of the American Association of Zoo Veterinarians*, St. Louis, Missouri, pp. 390–393.

Dosedelova, H, Stepankova K, Zikmund T et al. 2016. Age-related changes in the tooth-bone interface area of acrodont dentition in the chameleon. *J Anat* 229:356–368.

el Ridi R, Zada S, Afifi A et al. 1988. Cyclic changes in the differentiation of lymphoid cells in reptiles. *Cell Differentiation* 24:1–8.

Enok S, Leite GS, Leite CA et al. 2016. Improved cardiac filling facilitates the postprandial elevation of stroke volume in *Python regius*. *J Exp Biol* 219:3009–3018.

Flemming AF, and van Wyk JH. 1992. The female reproductive cycle of the lizard *Cordylus p. polyzonus* (Sauria: Cordylidae) in the Southwestern Cape Province, South Africa. *J Herpetol* 26:121–127.

Font E, Desfilis E, Pérez-Cañellas MM et al. 2001. Neurogenesis and neuronal regeneration in the adult reptilian brain. *Brain Behav Evol* 58:276–295.

Frydlova P, Mrzílkova J, Seremeta M et al. 2019. Universality of indeterminate growth in lizards rejected: The micro-CT reveals contrasting timing of growth cartilage persistence in iguanas, agamas, and chameleons. *Sci Rep* 9:18913.

Ganser LR, Hopkins WA, O'Neil L et al. 2003. Liver histopathology of the Southern Watersnake, *Nerodia fasciata fasciata*, following chronic exposure to trace element-contaminated prey from a coal ash disposal site. *J Herpetol* 37:219–226.

Gardiner DW, Baines FM, and Pandher K. 2009. Photodermatitis and photokeratoconjunctivitis in a Ball Python (*Python regius*) and a Blue-tongue Skink (*Tiliqua* spp.). *J Zoo Wildl Med* 40:757–766.

Goldberg SR, and Parker WS. 1975. Seasonal testicular histology of the colubrid snakes, *Masticophis taeniatus* and *Pituophis nelanoleucus*. *Herpetologica* 31:317–322.

Gomes, NMV, Shay JW, and Wright WE. 2010. Telomere biology in metazoa. *FEBS Lett* 584:3741–3751.

Guillette Jr LJ, and Sullivan WP. 1985. The reproductive and fat body cycles of the lizard, *Sceloporus formosus*. *J Herpetol* 19:474–480.

Hoekstra LA, Schwartz TS, Sparkman AM et al. 2019. The untapped potential of reptile biodiversity for understanding how and why animals age. *Funct Ecol* 00:1–17.

Hussein MF, Badir N, el Ridi R et al. 1979. Lymphoid tissues of the snake, *Spalerosophis diadema*, in the different seasons. *Dev Comp Immunol* 3:77–88.

Ip YK, Loong AM, Lee SML et al. 2012. The Chinese Soft-shelled Turtle, *Pelodiscus sinensis*, excretes urea mainly through the mouth instead of the kidney. *J Exp Biol* 215:3723–3733.

Isaza R, Garner M, and Jacobson E. 2000. Proliferative osteoarthritis and osteoarthrosis in 15 snakes. *J Zoo Wildl Med* 31:20–27.

Jacobson ER. 1990. Diseases of the integumentary system of reptiles, in *Small Animal Dermatology*. Nesbitt G, and Ackerman L (Eds.), Veterinary Learning Systems, NJ, pp. 225–239.

Jacobson ER, Seely JC, Novilla MN et al. 1979. Heart failure associated with unusual hepatic inclusions in a Deckert's Rat Snake. *J Wildl Dis* 15:75–81.

Jacyniak K, McDonald RP, and Vickaryous MK. 2017. Tail regeneration and other phenomena of wound healing and tissue restoration in lizards. *J Exp Biol* 220:2858–2869.

Jensen B, Moorman AFM, and Wang T. 2014. Structure and function of the hearts of lizards and snakes. *Biol Rev* 89:302–336.

Kaslin J, Ganz J, and Brand M. 2008. Proliferation, neurogenesis and regeneration in the non-mammalian vertebrate brain. *Philos Trans R Soc Lond B* 363:101–122.

Keenan SW, Hill CA, Kandoth C et al. 2015. Transcriptomic responses of the heart and brain to anoxia in the Western Painted Turtle. *PLOS ONE* 10(7):e0131669.

Krivoruchko A, and Storey KB. 2010. Forever young: Mechanisms of natural anoxia tolerance and potential links to longevity. *Oxid Med Cell Longev* 3:186–198.

Lawton MPC. 2000. Reptilian ophthalmology, in *Proceedings of the Association of Reptilian and Amphibian Veterinarians*, Reno, Nevada, pp. 197–202.

Lopez-Garcia C, Nacher J, Castellano B et al. 1994. Transitory disappearance of microglia during the regeneration of the lizard medial cortex. *Glia* 12:52–61.

Lutz PL, and Milton SL. 2004. Negotiating brain anoxia survival in the turtle. *J Exp Biol* 207:3141–3147.

McDonagh JC, Callister RJ, Favron ML et al. 2004. Resistance to disuse atrophy in a turtle hindlimb muscle. *J Comp Physiol A Neuroethol Sens Neural Behav Physiol* 190:321–329.

Milton SL, and Prentice HM. 2007. Beyond anoxia: The physiology of metabolic downregulation and recovery in the anoxia-tolerant turtle. *Comp Biochem Physiol A Mol Integr Physiol* 147:277–290.

Nacher J, Ramírez C, Palop JJ et al. 1999. Radial glia and cell debris removal during lesion-regeneration of the lizard medial cortex. *Histol Histopathol* 14:89–101.

Ngwenya A, Patzke N, Spocter MA et al. 2013. The continuously growing central nervous system of the Nile Crocodile (*Crocodylus niloticus*). *Anat Rec* 296:1489–1500.

Ortiz MA, Boretto JM, Piantoni C et al. 2014. Reproductive biology of the Amazon Lava Lizard (*Tropidurus torquatus*) from the Wet Chaco of Corrientes (Argentina): Congeneric comparisons of ecotypic and interspecific variations. *Can J Zool* 92:643–655.

Pinilla MPR. 1995. Reproductive and fat body cycles of the oviparous lizard *Liolaemus bitaeniatus* (Sauria: Tropiduridae). *J Herpetol* 29:256–260.

Porter KL, Gunapala S, and Meistrich ML. 2006. Testicular edema is associated with spermatogonial arrest in irradiated rats. *Endocrinology* 147:1297 1305.

Ramirez-Bautista A, Ramos-Flores O, and Sites Jr JW. 2002. Reproductive cycle of the Spiny Lizard *Sceloporus jarrovii* (Sauria: Phrynosomatidae) from North-Central Mexico. *J Herpetol* 36:225–233.

Riquelme CA, Magida JA, Harrison BC et al. 2011. Fatty acids identified in the Burmese Python promote beneficial cardiac growth. *Science* 334:528–531.

Rishniw M, and Carmel BP. 1999. Atrioventricular valvular insufficiency and congestive heart failure in a Carpet Python. *Aust Vet J* 77:580–583.

Rowland MN. 2011. Fibrosing myopathy of the temporal muscles causing lockjaw in a Veiled Chameleon (*Chamaeleo calyptratus*). *Vet Rec* 169:527.

Rush EM, Donnelly TM, and Walberg J. 2001. What's your diagnosis? Cardiopulmonary arrest in a Burmese Python. *Lab Anim (NY)* 30:24–27.

Saad AH, and el Deeb S. 1990. Immunological changes during pregnancy in the viviparous lizard, *Chalcides ocellatus*. *Vet Immunol Immunopathol* 25:279–286.

Saad AH, Torroba M, Varas A et al. 1991. Testosterone induces lymphopenia in turtles. *Vet Immunol Immunopathol* 28:173–180.

Schilliger L, Chetboul V, Damoiseaux C et al. 2016. Restrictive cardiomyopathy and secondary congestive heart failure in a Mcdowell's Carpet Python (*Morelia spilota mcdowelli*). *J Zoo Wildl Med* 47:1101–1104.

Shao H, Fan L, Xu XJ et al. 2012. Characterization of adult neurogenesis in lizard *Phrynocephalus vlangalii* (Agamidae: Reptilia). *Ital J Zool* 79:547–558.

Throckmorton GS. 1979. The effect of wear on the cheek teeth and associated dental tissues of the lizard *Uromastix aegyptius* (Agamidae). *J Morphol* 160:195–208.

Trauth SE. 1979. Testicular cycle and timing of reproduction in the Collared Lizard (*Crotaphytus collaris*) in Arkansas. *Herpetologica* 35:184–192.

van Wyk JH. 1989. The female reproductive cycle of the lizard, *Cordylus polyzonus polyzonus* (Sauria: Cordylidae) in the Orange Free State. *S Afr J Zool* 24:263–269.

Varas A, Torroba M, and Zapata AG. 1992. Changes in the thymus and spleen of the turtle *Mauremys caspica* after testosterone injection: A morphometric study. *Dev Comp Immunol* 16:165–174.

Zwart P. 1964. Studies on renal pathology in reptiles. *Vet Pathol* 1:542–556.

Zwart P. 2006. Renal pathology in reptiles. *Vet Clin North Am Exot Anim Pract* 9:129–159.

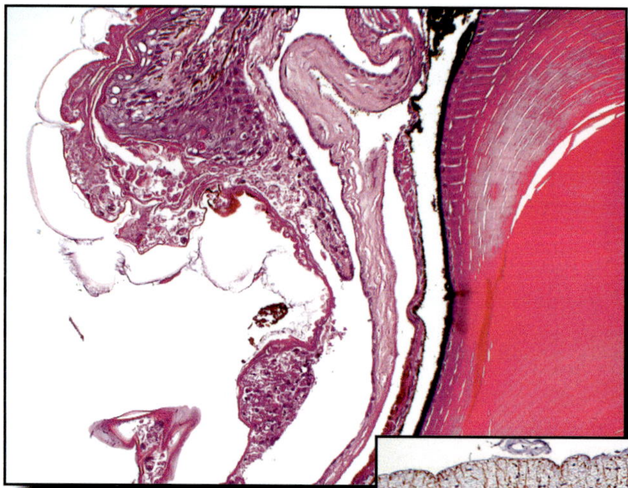

Figure 5.1 Day Gecko, *Phelsuma* spp. Gekkonidae. Photomicrograph of ultraviolet light damage. The eyelid epithelium has single-cell necrosis and is covered by abundant cell debris that extends along the deep aspect of the spectacle. Hematoxylin and eosin.

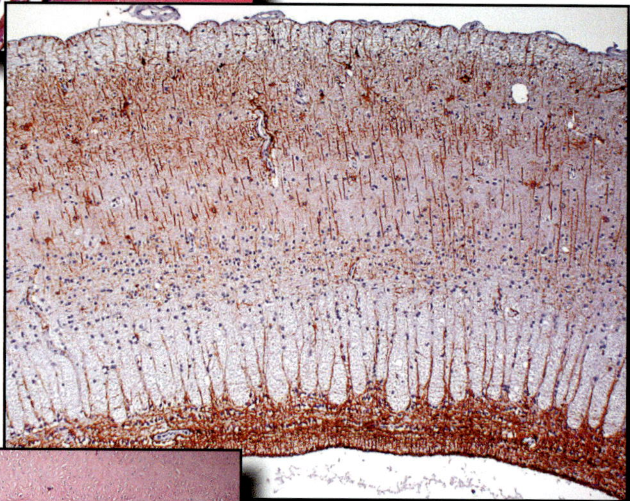

Figure 5.2 Green Basilisk, *Basiliscus plumifrons*. Corytophanidae. Photomicrograph of optic tectum with radial glia. Radial glial cells and ependyma of the optic ventricle stain strongly with anti-glial fibrillary acidic protein (GFAP) antibodies. Radial glial cell processes extend from the ependyma to the meninges. Immunohistochemical stain with anti-GFAP antibodies, diaminobenzidine chromogen, and hematoxylin counterstain.

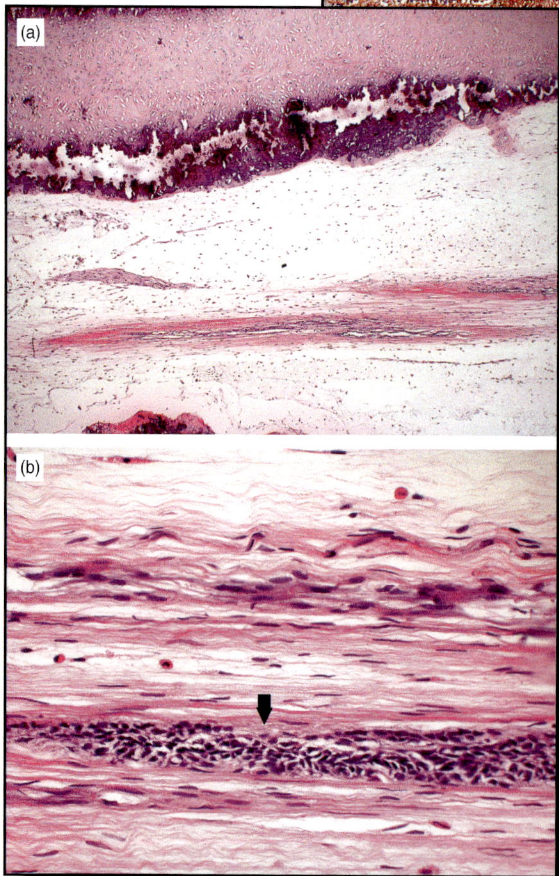

Figure 5.3 Basilisk, *Basiliscus plumifrons*. Corytophanidae. Photomicrograph of tail regrowth following autotomy. Cartilage surrounds loose stroma containing ependymal and nerve tissue (a). Higher magnification of the regrowth shows ependymal cells forming an elongated structure resembling a rudimentary central canal (arrow; b) Tracts of nerve fibers are present around the ependymal cells, but gray matter is not present. Hematoxylin and eosin.

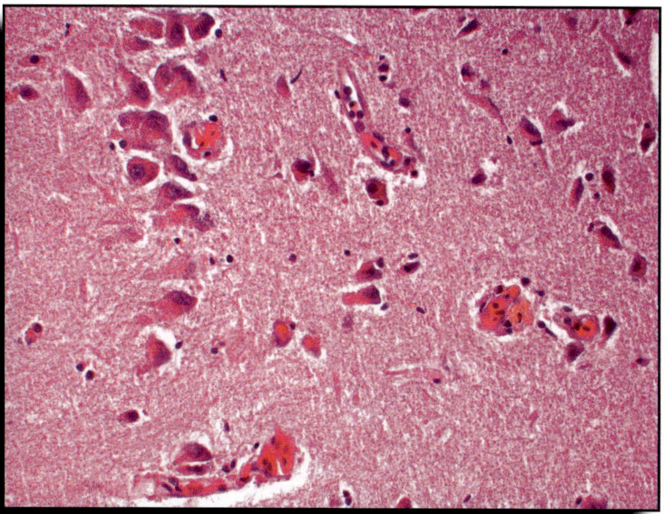

Figure 5.4 Ball Python, *Python regius*. Pythonidae. Photomicrograph of the telencephalon from a geriatric animal. Perivascular pigment-laden phagocytic cells and neuronal lipofuscin accumulation are present. Hematoxylin and eosin.

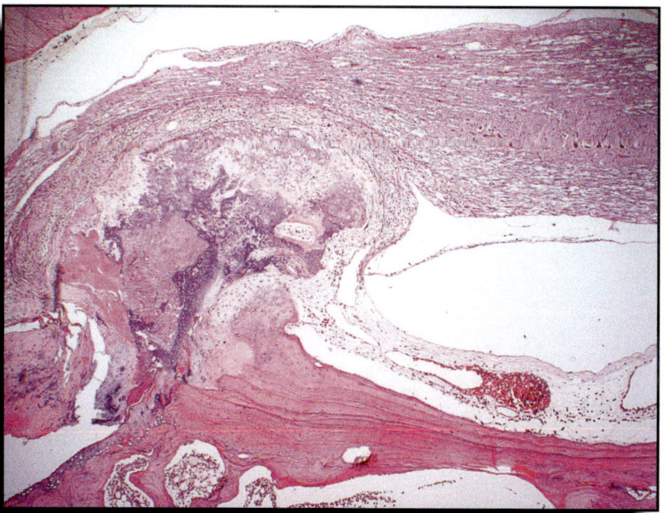

Figure 5.5 Central Bearded Dragon, *Pogona vitticeps*. Agamidae. Photomicrograph of spinal cord compression secondary to spinal arthritis. A large epiphyseal osteophyte extends dorsally into the lumbar spinal canal resulting in marked spinal cord compression. Hematoxylin and eosin.

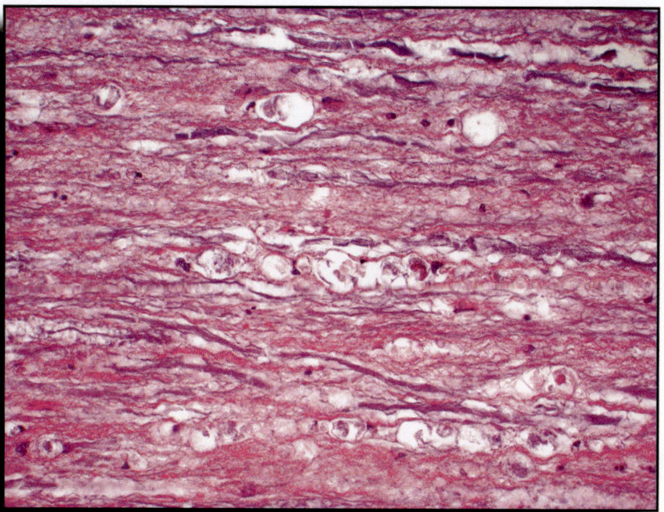

Figure 5.6 Ball Python, *Python regius*. Pythonidae. Photomicrograph of degenerative leukomyelopathy. Multiple axons and myelin sheaths are enlarged. Hematoxylin and eosin.

Figure 5.7 Corn Snake, *Pantherophis guttatus*. Colubridae. Cataract. Diffuse lens opacity is present. (Courtesy of Sara and Estevan Rojas.)

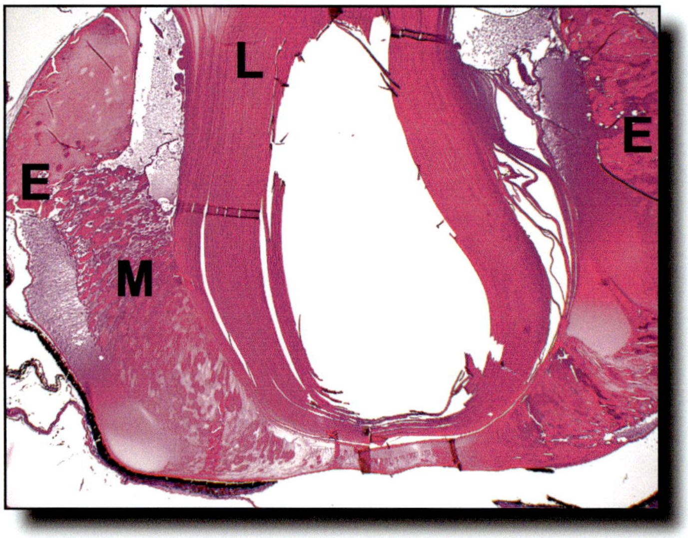

Figure 5.8 Veiled Chameleon, *Chamaeleo calyptratus*. Chamaeleonidae. Photomicrograph of a cataract. Lens fiber degeneration is present at the equatorial regions. Lens fibers form Morgagnian globules separated by flocculent proteinaceous material. Intact lens fibers are present in the center of the lens. The clear space in the center of lens is artifact from sectioning. E = equatorial regions; M = Morgagnian globules; L = intact lens fibers. Hematoxylin and eosin.

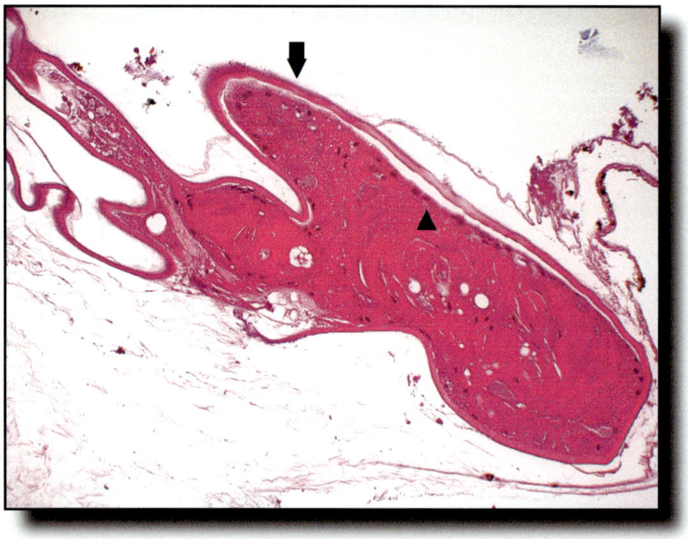

Figure 5.9 Corn Snake, *Pantherophis guttatus*. Colubridae. Photomicrograph of a hypermature cataract. A small lens remnant is composed of degenerate fiber material and peripheral lens epithelial cells (arrowhead) surrounded by an undulating capsule (arrow). Hematoxylin and eosin.

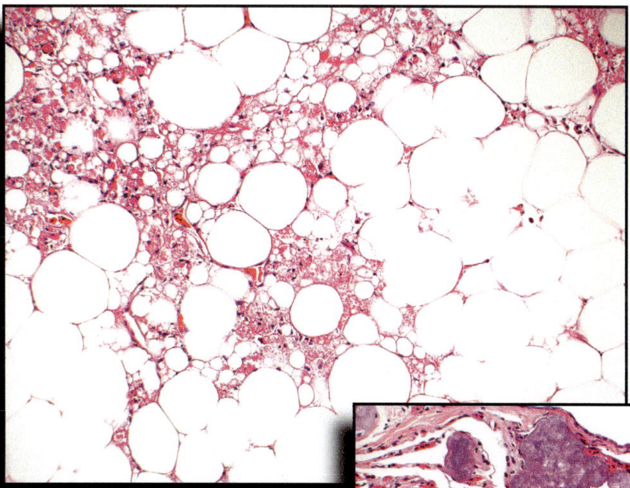

Figure 5.10 Ball Python, *Python regius*. Pythonidae. Photomicrograph of steatonecrosis. Infiltrates of lipophages and ceroid-laden macrophages are present among intact adipocytes of the coelomic adipose. Hematoxylin and eosin.

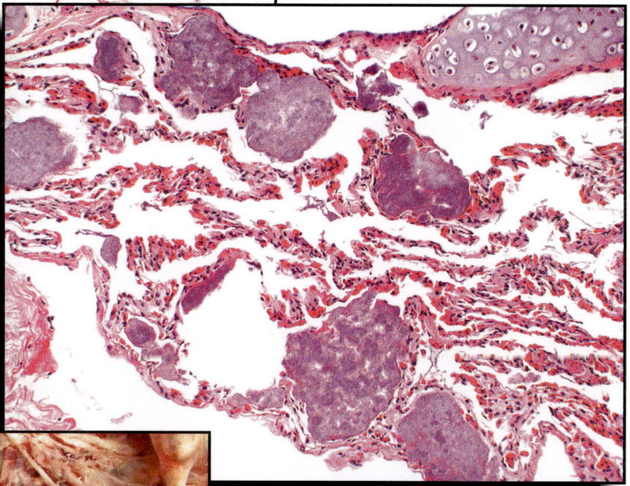

Figure 5.11 Egyptian Spiny-Tailed Lizard (*Uromastyx aegyptius*). Agamidae. Photomicrograph of faveolar accumulations of inspissated material. Mucoproteinaceous material forms concretions in multiple faveolae around an airway. Hematoxylin and eosin.

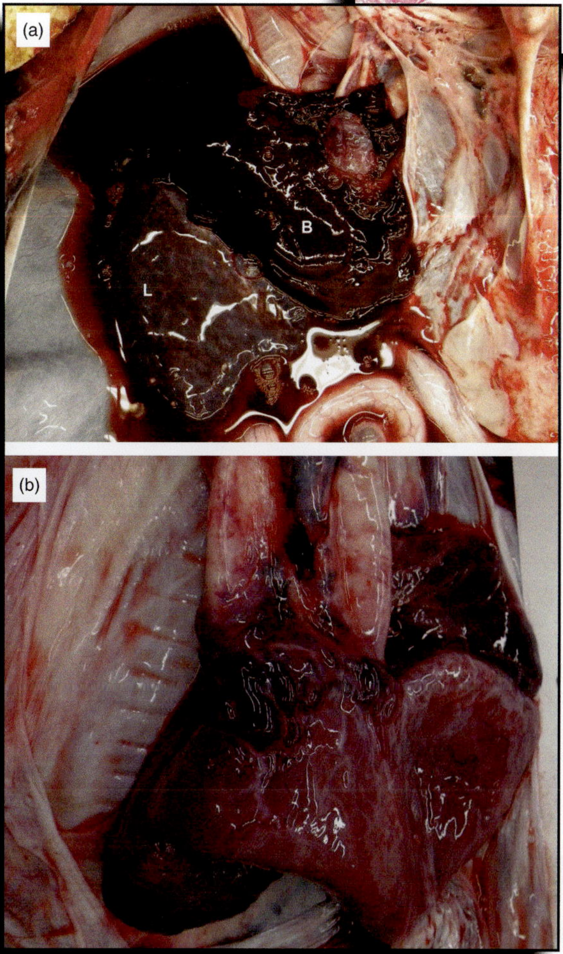

Figure 5.12 Loggerhead Sea Turtle, *Caretta caretta*. Cheloniidae. Hemopericardium secondary to ruptured aneurysm in the left aorta. The pericardial sac has been opened to show abundant clotted blood that obscures the heart. (a) The animal is emaciated, and a small dark brown liver is present caudal to the opened pericardial sac B = blood in pericardial sac; L = liver. (b) The cause for hemopericardium is ruptured aneurysm in the lateral wall of the left aorta at its origin.

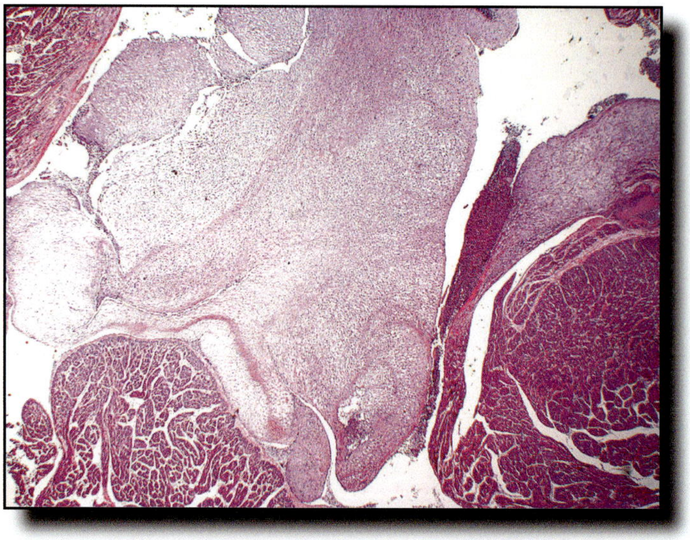

Figure 5.13 Honduran milk snake, *Lampropeltis triangulum hondurensis*. Colubridae. Photomicrograph of endocardiosis. One atrioventricular valve leaflet contains nodular, poorly organized accumulations of mucinous stroma resulting in marked enlargement. Hematoxylin and eosin.

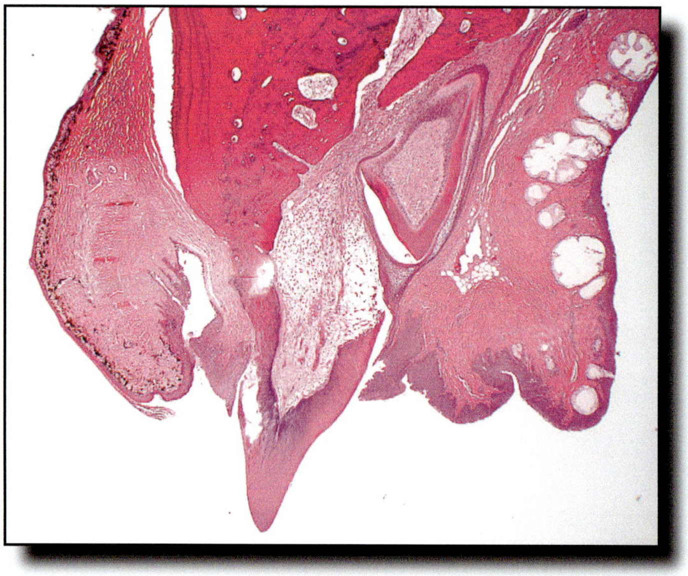

Figure 5.14 Green Basilisk, *Basiliscus plumifrons*. Corytophanidae. Photomicrograph of pleurodont teeth. A mature tooth is attached to maxillary bone by fibrous tissue. A developing tooth is present on the medial aspect of the bone. Hematoxylin and eosin.

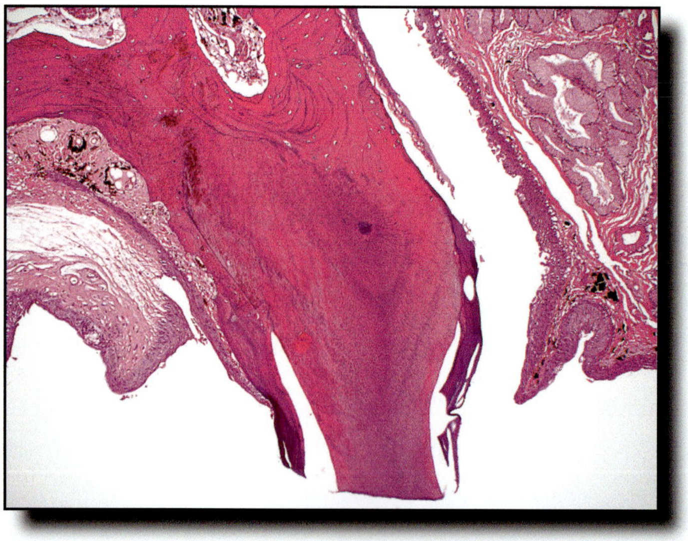

Figure 5.15 Chameleon Forest Dragon, *Gonocephalus chamaeleontinus*. Agamidae. Photomicrograph of an acrodont tooth. Dentine is attached to maxillary bone. Hematoxylin and eosin.

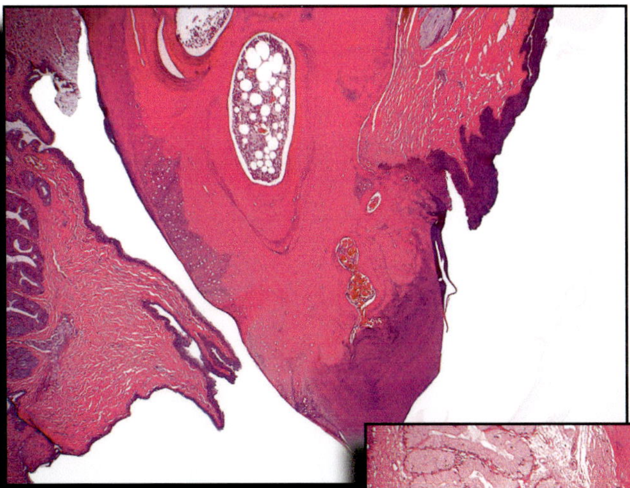

Figure 5.16 Central Bearded Dragon, *Pogona vitticeps*. Agamidae. Photomicrograph of a worn acrodont tooth. Worn dentine is present upon maxillary bone that is proliferating and replacing the dentine of the deep aspect of the tooth. Hematoxylin and eosin.

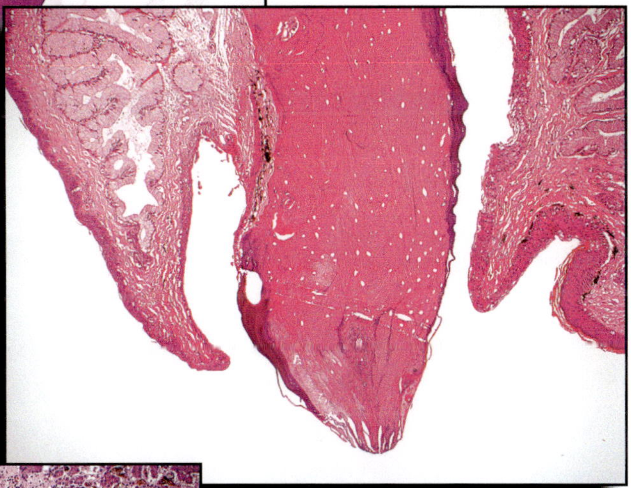

Figure 5.17 Chameleon Forest Dragon, *Gonocephalus chamaeleontinus*. Agamidae. Photomicrograph of a worn acrodont tooth. Dentine is worn away and maxillary bone is functioning as the tooth. Hematoxylin and eosin.

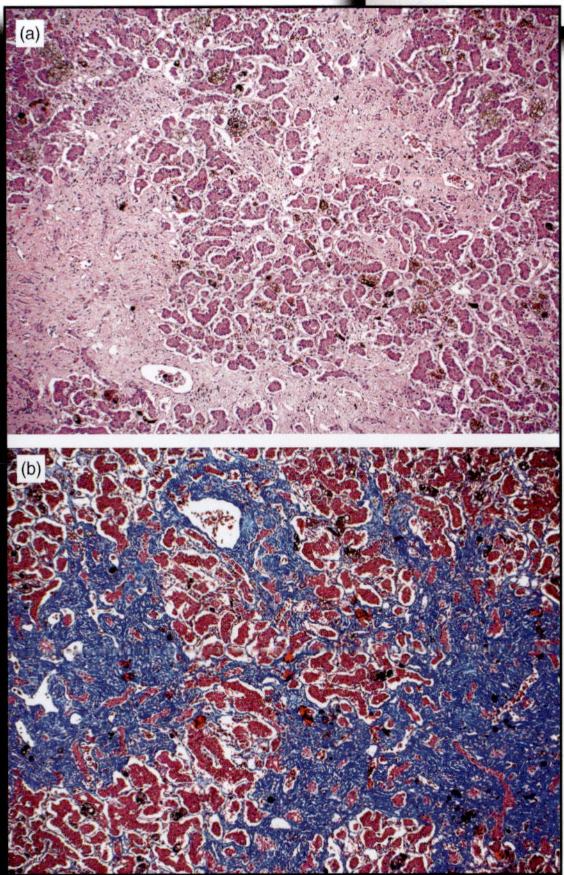

Figure 5.18 Panther Chameleon, *Furcifer pardalis*. Chamaeleonidae. Photomicrograph of hepatic fibrosis. (a) Bridging streams of fibrosis separate and truncate hepatic cords. Pigmented macrophage centers are also hyperplastic. Hematoxylin and eosin. (b) A trichrome stain confirms the presence of mature collagen. Masson's trichrome.

Figure 5.19 Loggerhead Sea Turtle, *Caretta caretta*. Cheloniidae. Photomicrograph of thymic hypocellularity. Thymic tissue is very poorly cellular and lacks a distinct cortex and medulla. Thymic hypocellularity is associated with starvation in this animal. Hematoxylin and eosin.

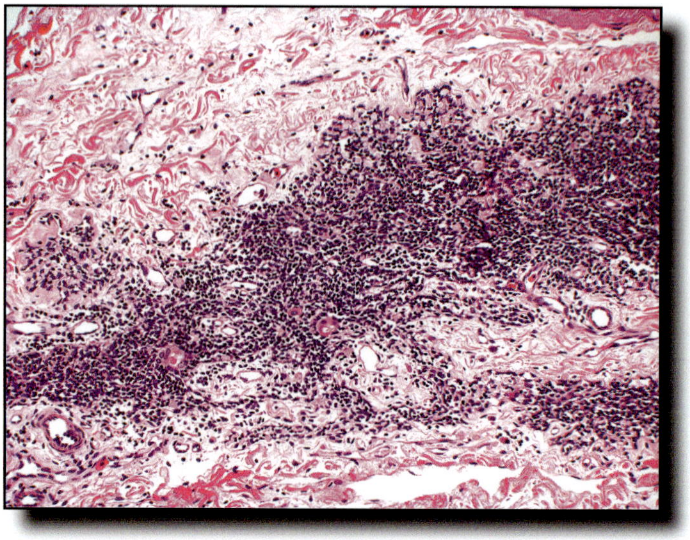

Figure 5.20 Wood Turtle, *Glyptemys insculpta*. Emydidae. Photomicrograph of renal glomeruli. This species is an aquatic turtle with less connective tissue and more prominent capillaries within glomerular tufts. Hematoxylin and eosin.

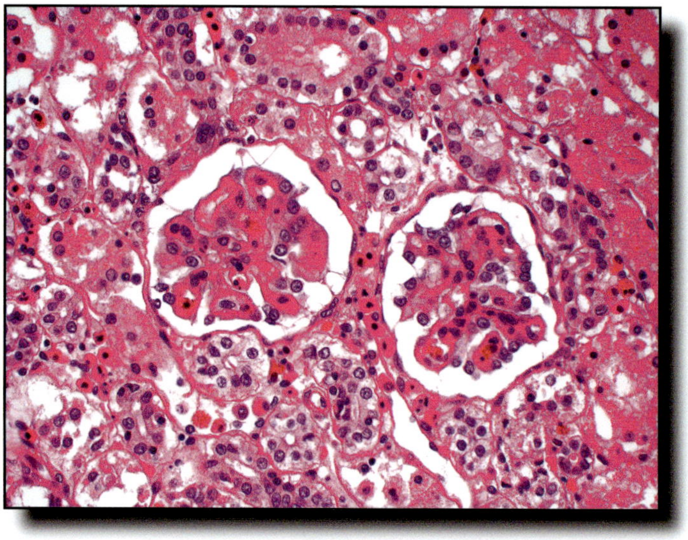

Figure 5.21 African Spurred Tortoise, *Centrochelys sulcata*. Testudinidae. Photomicrograph of renal glomeruli. This species is found in an arid environment and has more mesangial connective tissue within glomerular tufts. Hematoxylin and eosin.

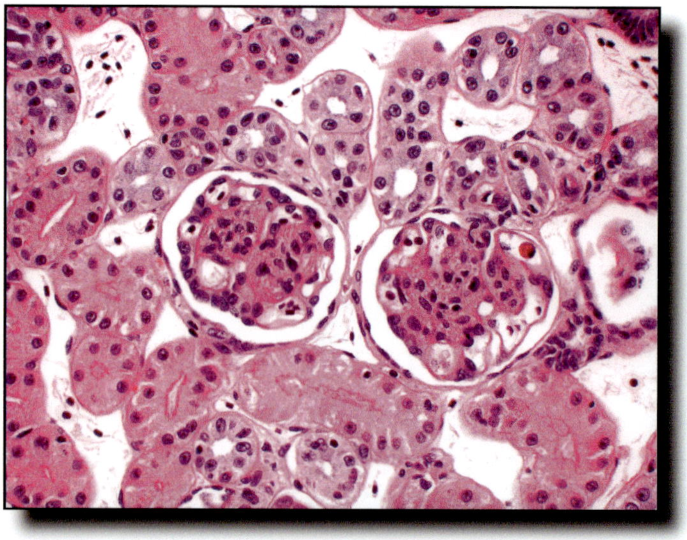

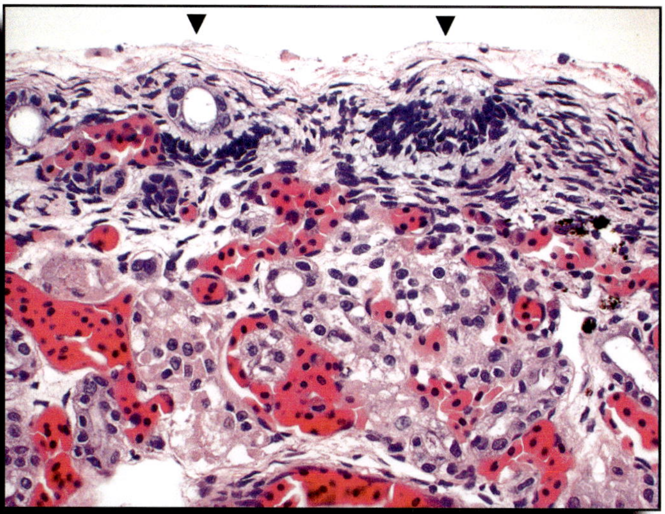

Figure 5.22 Green Turtle, *Chelonia mydas*. Cheloniidae. Photomicrograph of a nephrogenic zone. A subcapsular focus of embryonal nephrogenic tissue is present in the kidney (arrowheads). Hematoxylin and eosin.

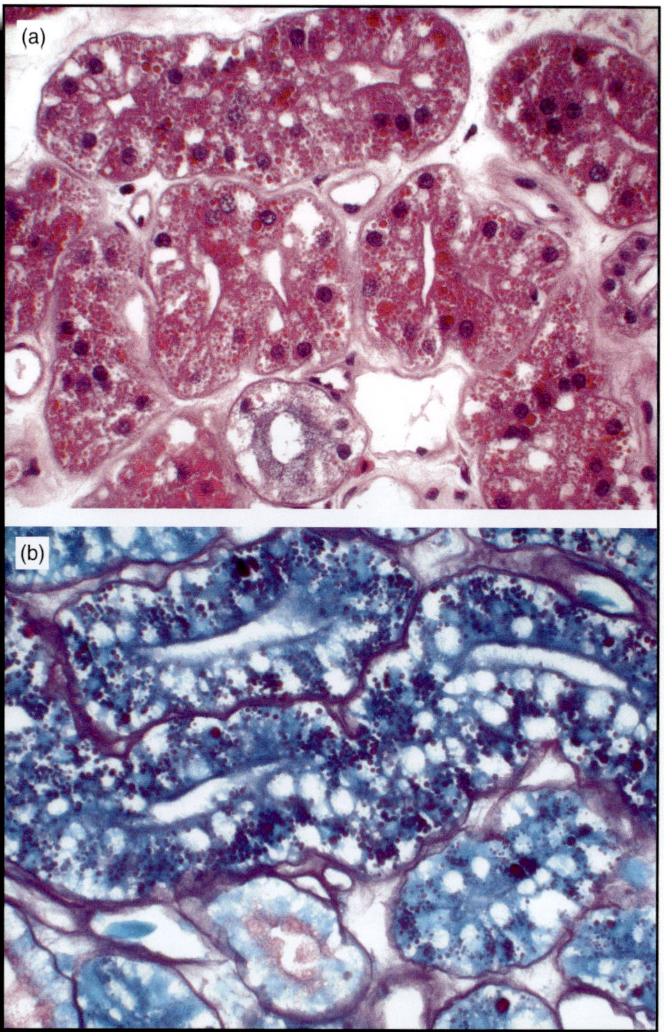

Figure 5.23 Southern Copperhead, *Agkistrodon contortrix contortrix*. Viperidae. Photomicrograph of renal tubular pigment. (a) Abundant granular brown pigment is present in proximal tubular epithelial cells. Hematoxylin and eosin. (b) Pigment stains dark pink with periodic acid-Schiff and is likely lipofuscin.

Figure 5.24 Panther Chameleon, *Furcifer pardalis*. Chamaeleonidae. (a) Photomicrograph of glomerulosclerosis. Glomeruli have increased mesangial collagenous connective tissue, and two are adhered to Bowman's capsule (arrowheads). Surrounding tubular parenchyma has pigment accumulation in epithelial cells but is otherwise relatively unremarkable. Sexual segment tubules are also present. S=sexual segment tubules. Hematoxylin and eosin. (b) A Masson's trichrome stain confirms deposition of collagenous connective tissue within glomeruli.

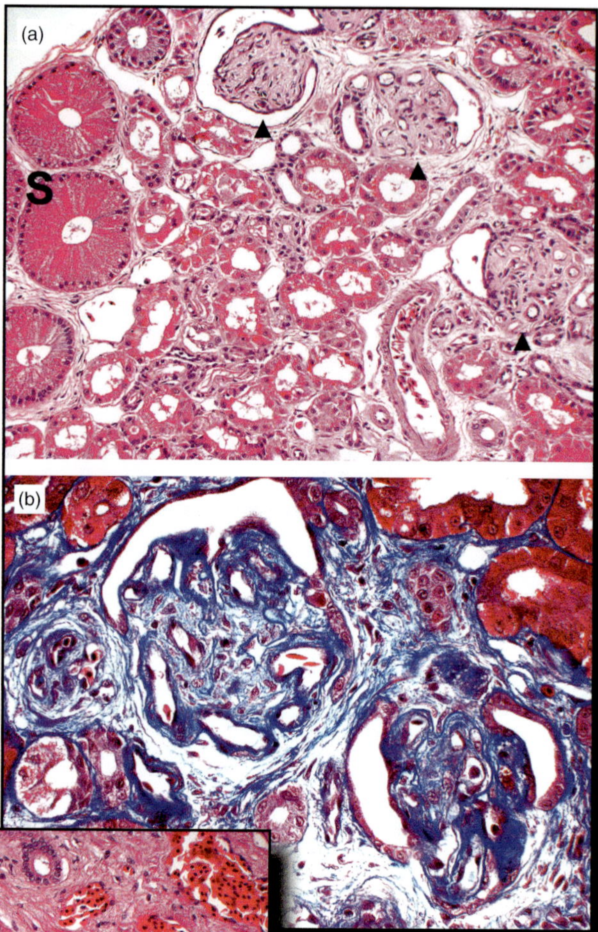

Figure 5.25 Green Turtle, *Chelonia mydas*. Cheloniidae. Photomicrograph of renal interstitial fibrosis. (a) Marked interstitial deposition of collagenous connective tissue is present. A glomerulus is not affected. Hematoxylin and eosin. (b) A Masson's trichrome stain confirms deposition of collagenous connective tissue in the interstitium, sparing the glomerulus.

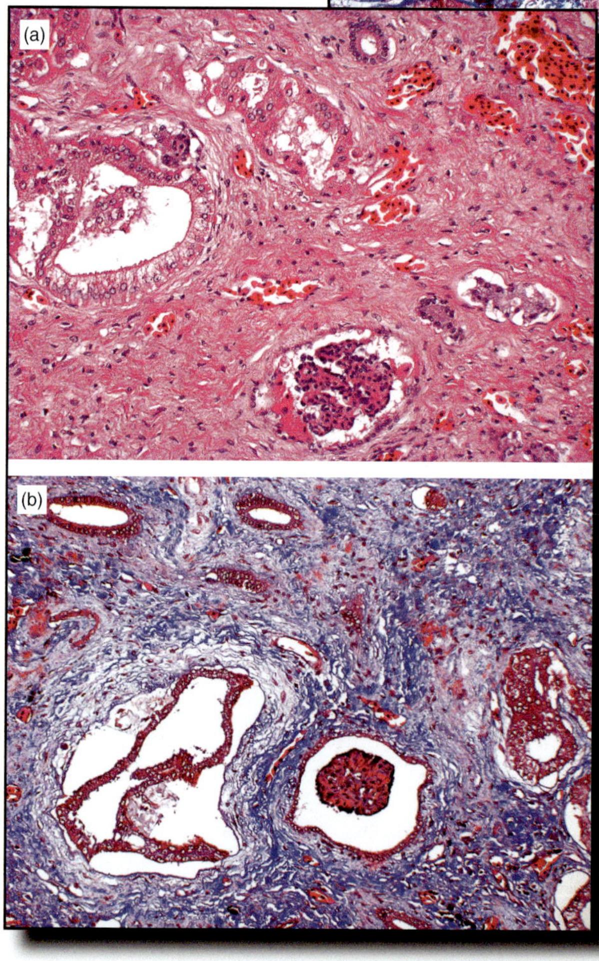

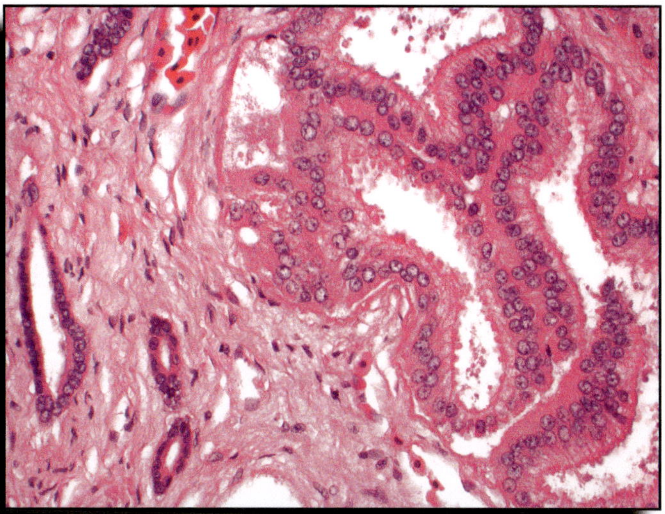

Figure 5.26 Green Turtle, *Chelonia mydas*. Cheloniidae. Photomicrograph of tubular changes associated with renal interstitial fibrosis. Tubules show marked atrophy or hyperplasia. Hematoxylin and eosin.

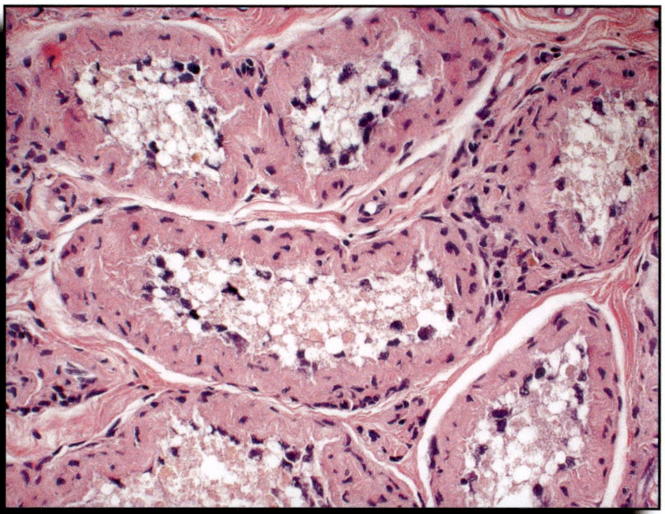

Figure 5.27 Corn snake, *Pantherophis guttatus*. Colubridae. Photomicrograph of a nonbreeding testis. Seminiferous tubules are lined predominantly by Sertoli cells. Small numbers of spermatogenic precursors are present, and some are necrotic and sloughing. Lipofuscin is present in the lumen of the tubule. Hematoxylin and eosin.

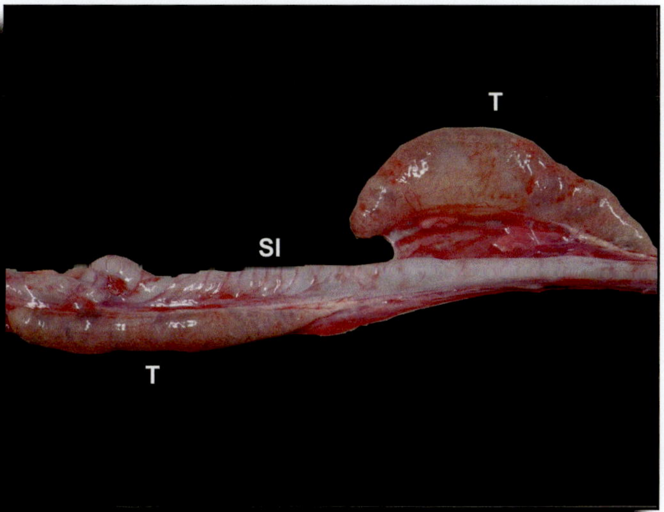

Figure 5.28 California Kingsnake, *Lampropeltis californiae*. Colubridae. Cystic testes. Testes are shown with the small intestine. Multiple cystic seminiferous tubules are present in both testes. Lipofuscin accumulation is also present resulting in an orange appearance to the tissue. T = testes. SI = small intestine.

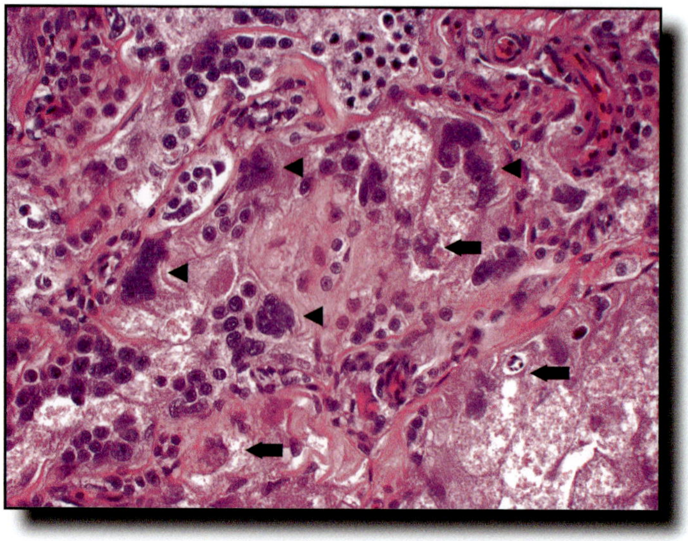

Figure 5.29 Veiled Chameleon, *Chamaeleo calyptratus*. Chamaeleonidae. Photomicrograph of testicular degeneration. Seminiferous tubules are lined by variable numbers of spermatogenic precursors with necrosis and giant cell formation. Hematoxylin and eosin.

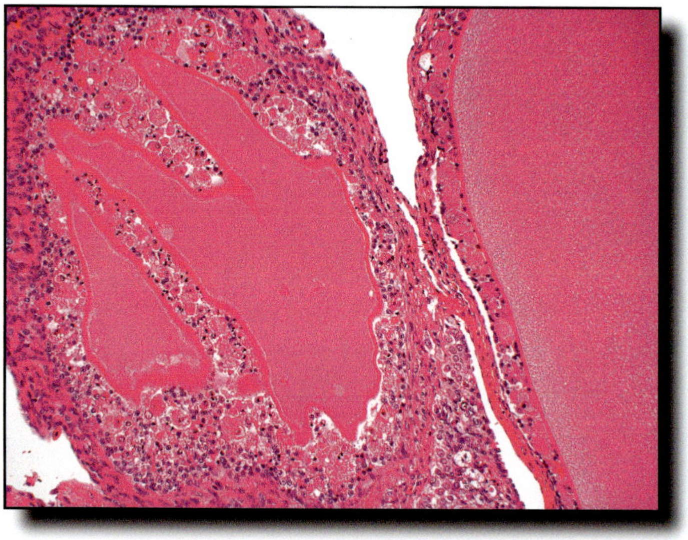

Figure 5.30 Alligator lizard, *Elgaria* spp. Anguidae. Photomicrograph of ovarian follicular atresia. An early atretic previtellogenic follicle has an undulating zona pellucida around the remaining ooplasm. Proliferating granulosa cells are present around the periphery. An adjacent intact previtellogenic follicle is also shown. Hematoxylin and eosin.

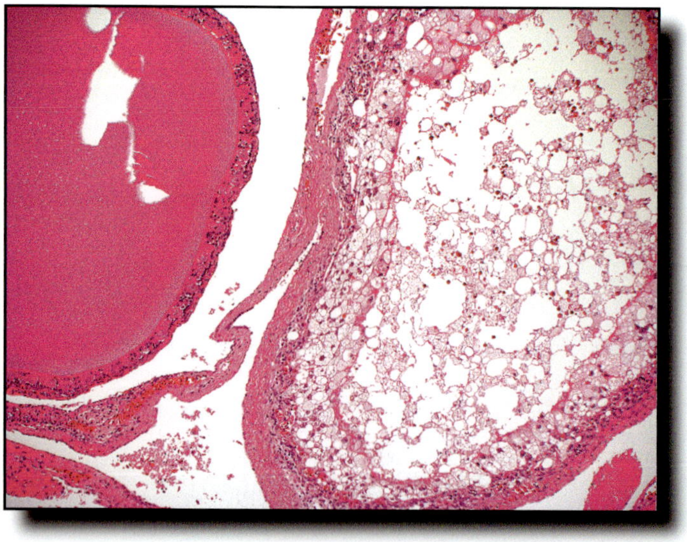

Figure 5.31 Alligator lizard, *Elgaria* sp. Anguidae. Photomicrograph of ovarian follicular atresia. An atretic vitellogenic follicle consists of lipid droplets and yolk granules surrounded by theca and granulosa cells with abundant foamy cytoplasm. Hematoxylin and eosin.

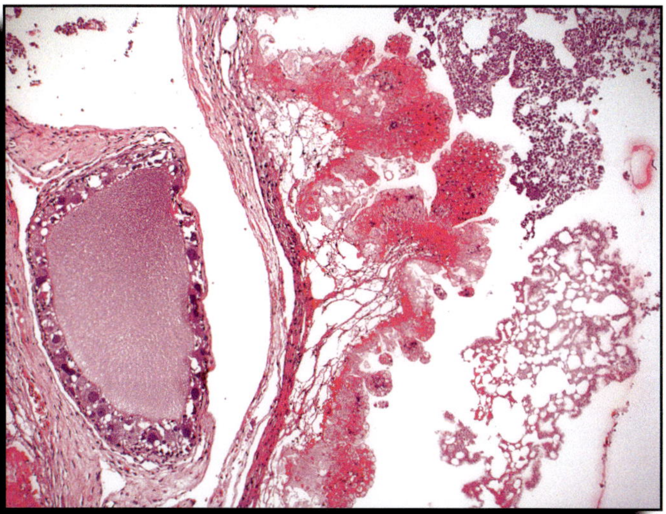

Figure 5.32 Chinese Crocodile Lizard, *Shinisaurus crocodilurus*. Shinisauridae. Photomicrograph of ovarian follicular atresia. An atretic vitellogenic follicle consists of peripheral lipid and cell debris with some remaining central yolk granules. The presence of yolk granules and proximity to an atretic previtellogenic follicle help distinguish the atretic vitellogenic follicle from necrotic adipose. Hematoxylin and eosin.

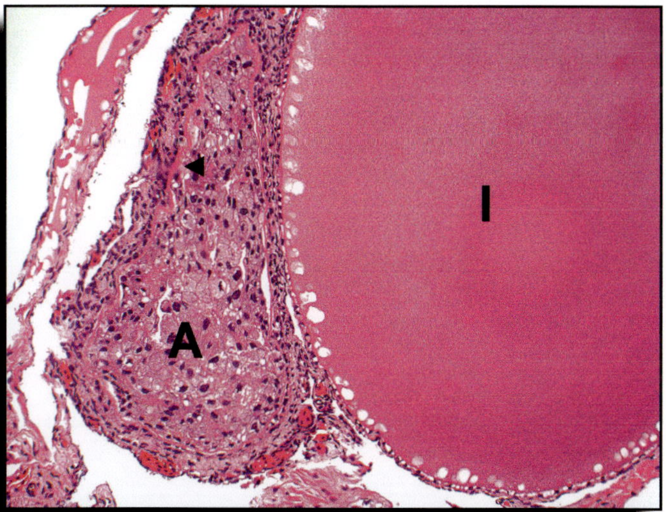

Figure 5.33 Loggerhead Sea Turtle, *Caretta caretta*. Cheloniidae. Photomicrograph of ovarian follicular atresia in a subadult. Atretic follicles can be found in animals that are not yet reproductively mature. This atretic follicle adjacent to an intact immature follicle resembles a granuloma but remnants of the zona pellucida (arrowhead) identify it as an atretic follicle. A = atretic follicle; I = immature follicle. Hematoxylin and eosin.

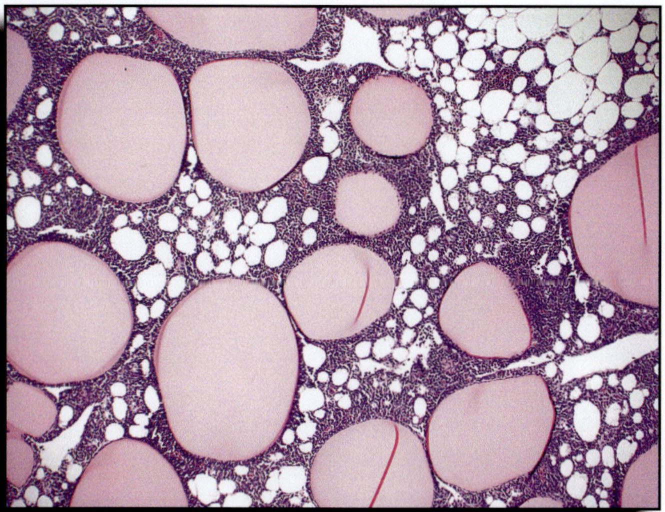

Figure 5.34 Jeweled Lizard, *Timon lepidus*. Lacertidae. Photomicrograph of the thyroid gland. The interstitium contains abundant lipid and mature lymphocytes. Hematoxylin and eosin.

Figure 5.35 Central Bearded Dragon, *Pogona vitticeps.* Agamidae. Spinal degenerative arthritis. Marked enlargement of one sacral vertebral body articulation with ventral proliferation of bone and cartilage (spondylosis) is present.

Figure 5.36 American Alligator, *Alligator mississippiensis*. Alligatoridae. Bilateral stifle arthritis. The stifle joints of this geriatric animal are opened and shown side by side. In both, there is prominent soft tissue proliferation around the joints and cartilage erosion of the lateral femoral condyles.

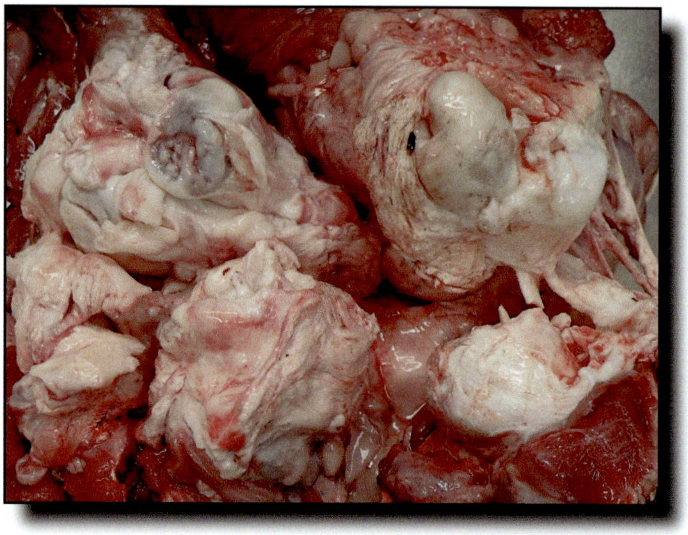

Figure 5.37 Burmese Python, *Python bivittatus*. Pythonidae. Atlantooccipital arthritis. The caudal aspect of the head is shown. The occipital condyle has cartilage erosion and marked periarticular proliferation of soft tissue (arrow).

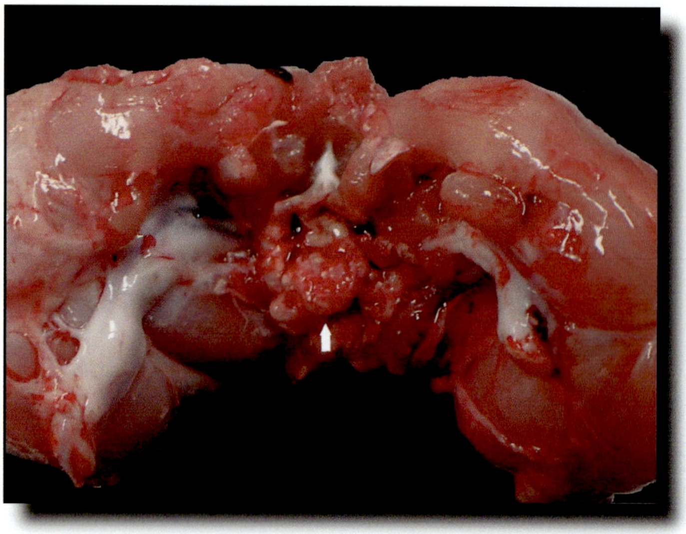

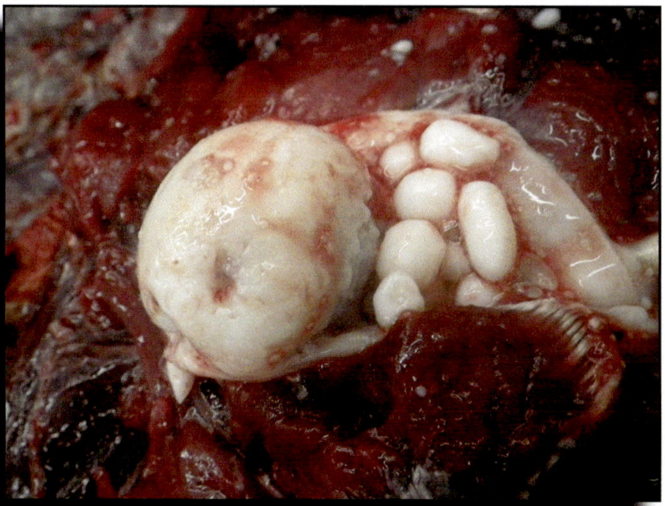

Figure 5.38 Hawksbill Sea Turtle. *Eretmochelys imbricate*. Cheloniidae. Synovial chondromatosis. One femoral head and coxofemoral joint capsule of this geriatric animal are shown. The femoral head has multifocal cartilage erosion, and the joint space contains numerous cartilage nodules.

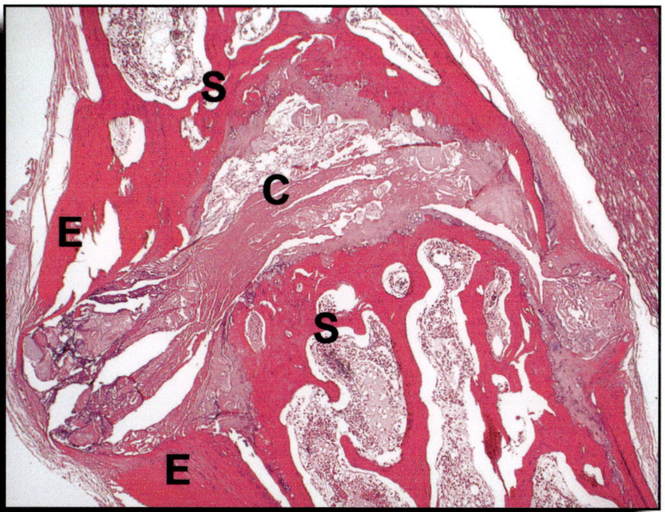

Figure 5.39 Chameleon Forest Dragon, *Gonocephalus chamaeleontinus*. Agamidae. Photomicrograph of spinal arthritis. A cervical vertebral body articulation has diffuse cartilage matrix eosinophilia and erosion with subchondral sclerosis and accumulation of degenerate cartilage in the joint space, including the dorsal and ventral aspects with associated protrusion of the joint space. Prominent epiphyseal osteophytes are also present around the material within the enlarged ventral joint space. C = degenerate cartilage in joint space; S = subchondral sclerosis; E = ventral epiphyseal osteophytes. Hematoxylin and eosin.

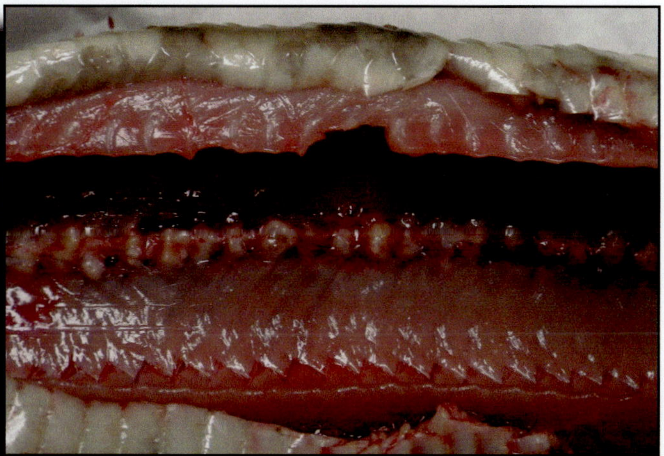

Figure 5.40 Ball Python, *Python regius*. Pythonidae. Spinal arthritis with ankylosing spondylosis. Multiple vertebral body articulations are enlarged and nodular.

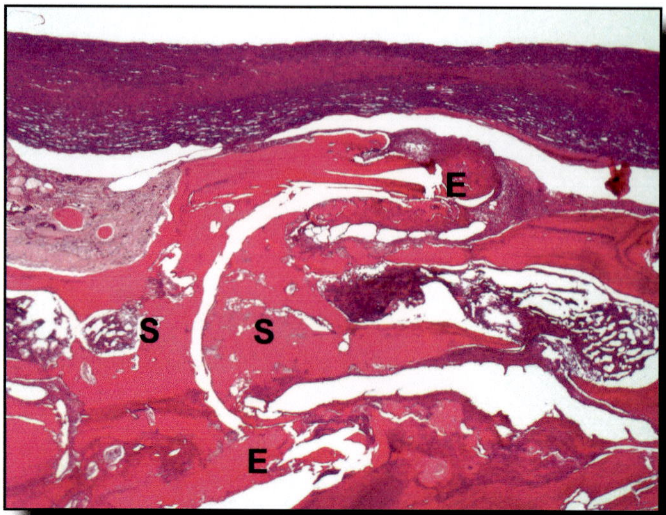

Figure 5.41 Green Mamba, *Dendroaspis angusticeps*. Elapidae. Photomicrograph of spinal arthritis with ankylosing spondylosis. No articular cartilage remains on the vertebral body epiphyses. Vertebral subchondral bone is sclerotic. Epiphyseal osteophytes are bridging (ankylosing spondylosis). S = subchondral sclerosis; E = bridging epiphyseal osteophytes. Hematoxylin and eosin. (Courtesy of Michael Garner.)

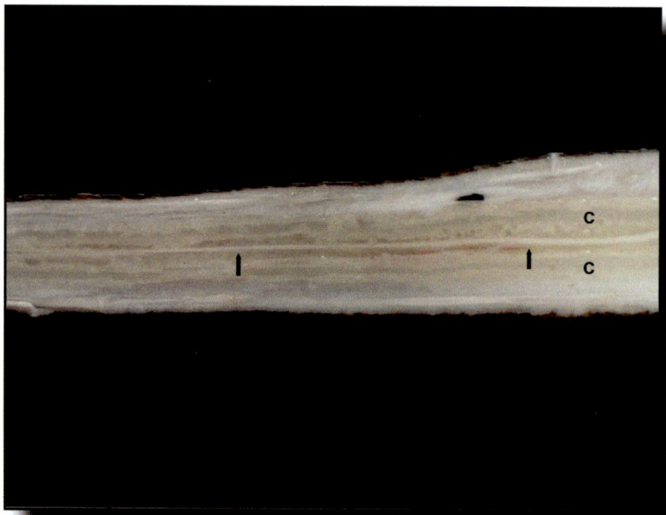

Figure 5.42 Green Basilisk, *Basiliscus plumifrons*. Corytophanidae. Tail autotomy. The regrown tail contains a cylinder of cartilage around nerve tissue (arrows). The cartilage does not have segmentation. C = cartilage.

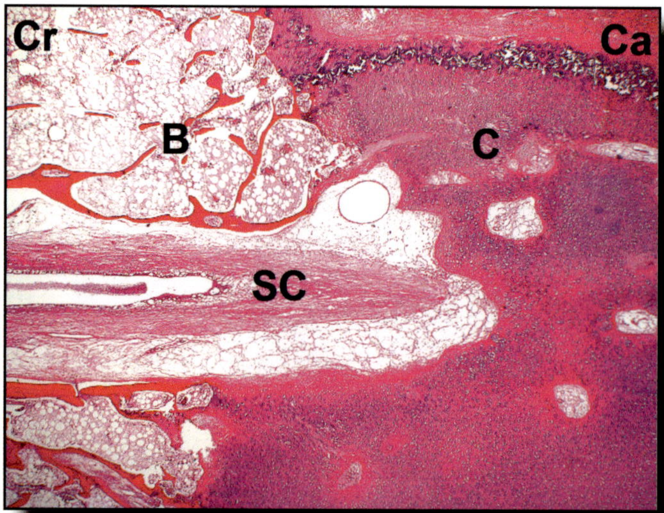

Figure 5.43 Green Basilisk, *Basiliscus plumifrons*. Corytophanidae. Photomicrograph of tail autotomy. At the fracture plane, the regrown tail cartilage is originating from the remaining vertebral bone. The remaining truncated spinal cord has dilation of the central canal. Cr = cranial; Ca = caudal; B = vertebral bone; C = regrown cartilage; SC = spinal cord. Hematoxylin and eosin.

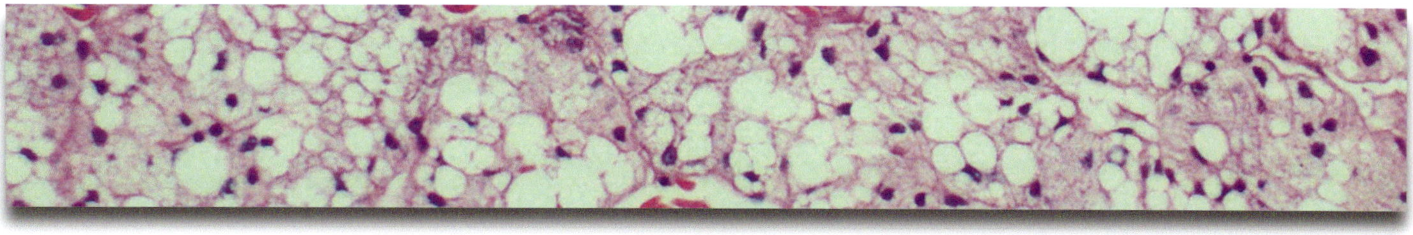

6

TRAUMA AND PHYSICAL DISEASES

CHRISTINE L. MILLER

Contents

6.1 General Comments

Trauma and other physical injuries may affect free-roaming native and non-native reptiles as well as captive individuals. These injuries may be the result of naturally occurring or anthropogenic factors in the physical environment or may be influenced by social interactions. Generally, for the purposes of this chapter, trauma is defined as a physical injury caused by a violent force applied to the body. The extent and severity of injury depends on the type of force or energy applied and the organs or tissues involved (Cooper, 1996). Some published surveys of morbidity and mortality in various reptile species presented to wildlife rehabilitation facilities, or veterinarians providing services for captive specimens, reported trauma in 21%–78% of presentations (Homer et al., 1998; Brown and Sleeman, 2002; Schrader et al., 2008; Frolich et al., 2010; Rivas et al., 2014; Scheelings, 2015). For captive reptiles, a majority of traumatic and physical disorders are directly influenced by enclosure design and husbandry issues (Frye, 1981). Inadequacies in providing for the environmental and social needs of captive reptiles may specifically cause physical injuries or induce stress.

Stress may be defined as an individual's physical and psychological response to any adverse forces or events. Stress may compromise health if it exceeds the individual's ability to compensate or adapt. Substantial or prolonged stress may result in physiological disorders or behavioral changes leading to trauma or disease in reptiles (Denardo, 2006; Nevarez,

2007). Evidence of stress can be identified histologically in reptiles. The adrenal glands may develop hypertrophic interrenal cells and intracytoplasmic vacuolar change in the interrenal and chromaffin cells, resulting in organ enlargement (**Figure 6.1a**). The lymphoid tissues such as spleen and thymus can have lymphocytolysis or lymphoid depletion, resulting in organ shrinkage. Skeletal muscle may have rhabdomyolysis and edema, resulting in swelling and pallor (**Figure 6.1b**). Erosion and hemorrhage of the gastric or enteric mucosa may also occur.

6.2 Conspecific or Cagemate Trauma

6.2.1 Male–Male Aggression

Males of various species in all of the orders of reptiles exhibit territoriality and ritualized or aggressive competition for breeding access to mates, especially during the breeding season. Crocodilians (Frye, 1981), lizards (Boyer, 1998; Barten, 2006a), tuataras (Ramstad et al., 2012), turtles (Frye, 1981), tortoises and snakes (Olsson and Madsen, 1998; Fearn et al., 2005; Funk, 2006a) can all deliver bite wound injuries to a competing male if warning displays fail, or if competitors cannot escape or hide due to enclosure constraints. These bite wounds frequently involve the head and neck or the feet, legs, or tail, and may include shell margins and penis in turtles. Bite wounds can vary in severity and include multiple punctures, lacerations or avulsed skin, crush injuries, and fractures or amputations of digits, limbs, rostrums, or tails (**Figure 6.2a,b**). Bite wounds may result in opportunistic bacterial, mycotic, or parasitic infection at the wound sites, sepsis, or vascular thrombosis and necrosis of traumatized extremities. Especially with snakes or lizard species that have caudally recurved teeth, the individual inflicting the bite may also suffer injury to the mandible or the dentition and oral cavity if the bite site (such as a leg) is withdrawn violently by the victim rather than released by the aggressor (**Figure 6.3**) (Boyer, 1998, p. 123). Some tortoises also engage in ramming behavior and will attempt to overturn rival males (Greenberg and Crews, 2013). These behaviors may lead to blunt force trauma, penetration from the aggressor's long or sharp gular scutes, and suffocation or predation in the overturned tortoise (**Figure 6.4a,b**). If the tortoise is in an outdoor enclosure and overturned in a sunny area, it potentially could die of hyperthermia.

Smaller or weaker subordinate males that are not directly attacked but are unable to escape from the dominant males may experience chronic intimidation and stress. Submissive or fearful behavior may limit access to resources (food, water, warmth) and lead to stunted growth or weight loss, and stress-related diseases (**Figure 6.5a,b**) (Boyer, 1998, p. 77; Denardo, 2006).

6.2.2 Mate/Breeding Trauma

During the breeding season, the actual period of receptivity to a male may be very limited for individual females in many species. These females may then be faced with aggressive advances from males trying to force breeding encounters, especially in captive situations where escape is impeded or impossible. During an act of sexual submission or in situations where escape is impossible, the female may be injured by the male. Bite wound trauma from males during breeding is common among turtles, lizards, and snakes. Female tortoises may sustain mild or substantial abrasions to their shells from persistent male mounting attempts (**Figure 6.6a,b**). Submissive male tortoises may also sustain shell damage from repeated mounting by other males.

In a turnabout, defensive or aggressive females may bite the attending male. Female tortoises who are not receptive may fail to lift the caudal portion of their carapace to give cloacal access to the male, or if startled or interrupted during intromission, they may suddenly shift their position or drop their carapace to the ground, resulting in potential injury to the male's penis (**Figure 6.7a–c**).

6.2.3 Overcrowding or Mixed-Species Encounters

Especially in captivity, overcrowding or limitations of useable space can lead to competition for resources such as basking sites, resting areas, water, or sheltered retreats. For example, an enclosure with abundant square footage that houses an arboreal lizard species may have insufficient height or perching structures or cover to allow all individuals to rest or bask without intimidation or aggression from dominant lizards in the group. Many reptile species are solitary through much of their lifetimes, and some are intolerant of conspecifics in their vicinity (Boyer and Boyer, 2006). True chameleons are particularly prone to stress from conspecifics and should be separated until being paired for breeding. Other species need appropriate social groupings, as they may not tolerate other males or juveniles but may be content cohabitating with a few mature females, for instance. Forced encounters and inability to escape or hide may lead to injuries or death from aggression, or will result in varying degrees of chronic stress for all individuals involved (**Figure 6.8a,b**) (Griner, 1983; Leong et al., 1989; Shilton et al., 2014). Juveniles of some species may be cannibalized by larger, older conspecifics (lizards, crocodilians, turtles) (**Figure 6.9**) (Boyer, 1998; Boyer and Boyer, 2006; Lane, 2006). Young crocodilians and hatchling turtles or tortoises tend to pile up on each other in crowded pens with limited hiding or basking spaces or poorly designed facilities, and this may result in suffocation or drowning or other injuries (**Figure 6.10a,b**) (Lane, 2006; Warwick et al., 2013).

During the nesting season, some female crocodilians and lizards, such as varanids, may vigorously defend their nesting site and aggressively attack conspecifics of either sex that venture too close. Insufficient nesting materials or lack of multiple appropriate nest sites may enhance competition between females, leading to aggressive encounters.

Mixed species of reptiles in an enclosure likewise may not be tolerated. Errors of mixing predators with nonpredatory species may lead to trauma or fatalities. Snake species which may be naturally ophidiophagus or prey on lizards may cause chronic stress to potential prey, even when housed next to those individuals. Some turtle species are intolerant of other turtles in their space, such as snapping turtles, musk or mud turtles, and some soft-shelled turtles (Boyer, 1998, p. 47).

Mixing some reptiles with mammals or birds may lead to interactions that can cause traumatic injuries (predatory and nonpredatory). For instance, birds may mob and harass a snake or lizard or tortoise in an area where the birds are nesting or feeding, and can cause pecking or scratching injuries that may include damage to the eyes (**Figure 6.11a,b**). Some large mammals, such as giraffe or antelopes and cervid species, or primates, may find lizards or tortoises and turtles in their enclosure to be objects of curiosity and may stomp, gore, hit, toss, or flip these animals as a part of their investigatory or play behavior (**Figure 6.12a,b**). A giraffe stomping even a large tortoise into the mud may cause severe trauma, including shell fractures. Alternatively, the mammal may find the tortoise to be "useful furniture," and in the author's experience, at least one large tortoise sustained multiple gore wounds through the neck and axillary area from a muntjac deer rubbing on the animal's shell while shedding his antler velvet.

6.2.4 "Feeding Frenzy"

In situations where food is delivered within a limited space and where the reptiles in an enclosure are motivated to eat or competition for food is high, such as with overcrowding, individuals may mistakenly seize one another rather than a food item. Crocodilians and turtles may bite or scratch each other aggressively in competition for food held above, or thrown into, the water (Frye, 1981). Snakes may strike the same prey item or each other mistakenly, potentially leading to bite wounds or unintentional cannibalism (Funk, 2006a). Snakes may even envenomate each other when competing for the same prey item. When venomous snakes are of considerably different size, a large venom injection may result in the death of a small snake whereas a large snake of the same species may survive. There is little published data regarding snake to snake envenomation; however, it seems that some species may be relatively "immune" to the effects of their own species' venom, while others are not. Hoser describes multiple envenomating-bite incidents between various Australian snakes held together in enclosures from which the victims suffered no apparent toxic effects from the venom (Hoser, 1996), whereas others have published observations of various rattlesnake species that bit conspecifics and the victims died, although it is not certain in every case if this was a result of trauma or envenomation (Nichol et al., 1933; Griner, 1983). Takacs et al. (2001) report a molecular basis for protection of neurotoxin receptors in cobras granting them resistance to their own species' venom.

6.3 Self-Induced Trauma

6.3.1 Escape Behaviors and Enclosure-Related Trauma

Reptiles held in captivity, especially those recently acquired from the wild, have a natural tendency to explore the boundaries of their enclosures and attempt to escape from them. This behavior is enhanced by failures to observe appropriate social groupings or in overcrowded situations, or when an enclosure fails to provide for a specific need to hide or escape from cagemates or other environmental stressors (such as excess light or noise, human activity, or inadequate thermal ranges) (Warwick et al., 2013). A natural tendency to hunt for prey, or for mates during the breeding season, will also lead to escape attempts. During their natural breeding period some reptiles will wander their cage for weeks to months. This activity commonly results in self-induced trauma such as rostral and labial abrasions from rubbing along screens or other rough surfaces, or repeated attempts to go through transparent glass or plexiglass walls (**Figure 6.13a–c**). Shell abrasions may occur in tortoises or turtles from rough surfaces or bumping along overhanging structures (**Figure 6.14a,b**). Frye (1991) described a case of fiberglass fragments from a cage resulting in a granulomatous response in the skin of a Cuban Boa (*Epicrates angulifer*). The same has been seen in another snake (E. Jacobson, personal communication). The toes and nails of individuals trying to climb mesh-sided enclosures may incur cracks, lacerations, or avulsions. Foot ulcers, or chronic abrasions of the ventrum or mandibles, may be seen in animals pacing, swimming along pool margins, or moving repeatedly in and out of pools or along areas that are surfaced with rough substrates (such as concrete or gunnite) (**Figure 6.15a,b**). In addition, the tail-whipping behavior of some lizards when threatened may lead to wounds, autotomy, fractures, or avascular necrosis of the distal tail if they repeatedly strike walls or other structures in their enclosure (**Figure 6.16a,b**) (Funk, 2006b).

The Komodo Dragon (*Varanus komodoensis*) has specifically been noted to be predisposed to compressive spinal injuries due to luxations, subluxations, or fractures of the vertebrae—especially the cervical vertebral column (Zimmerman et al., 2009). These injuries and other fractures have often occurred secondary to traumatic events such as falls from enclosure furnishings or walls, or possibly running into the glass barriers in viewing areas (**Figure 6.17a,b**). A vertebral malformation or some genetic predisposition may exist in the captive population, but this has not been confirmed at this time. A nutritional predisposition to skeletal problems stemming from vitamin D deficiencies may also play a role in some of these captive Komodo Dragons (Gyimesi and Burns, 2002).

6.3.2 Other Stress-Related Behaviors

Abnormal stereotypic behaviors in reptiles are occasionally seen, such as snakes circling compulsively in their cages (E. Jacobson, personal communication), but are not widely documented. Other stress-related abnormal behaviors have been reviewed (Warwick et al., 2013). Self-biting in snakes or other species has generally been attributed to internal pain, such as intestinal obstruction or intracoelomic infections (**Figure 6.18**). Self-mutilation due to stress or "boredom" in reptiles seems to be rarely reported (Burghardt et al., 1996).

6.4 Other Trauma

6.4.1 Entrapments

Reptiles of varying species may become entrapped in a multitude of situations. Lizards and snakes have been caught on glue traps set in the environment for "pest control" (**Figure 6.19a,b**). Once mired in the glue, the reptile can lose chunks of skin, or toes and tails, or injure eyes and tongue in the struggle to be freed, or during attempts by handlers to extricate the animal. If left trapped for a prolonged period, the victim may die from shock, exhaustion, or dehydration (or possibly asphyxia). Occasionally a reptile may also be injured in a spring-loaded mouse or rat trap, while hunting in a rodent-infested building or lot. This sudden crushing injury is often lethal, but this is dependent on the body part caught when the trap is sprung.

Sea turtles or estuarine chelonians (such as Diamond Back Terrapins [*Malaclemys* spp.]) may be found with heads or flippers caught up in discarded trash, especially macroplastics such as six-pack rings for beverages, fishing lines, or pieces of netting (Norton, 2005; Mader, 2009). These may produce chronic lesions, including deep cutting wounds, vascular injuries, and thrombosis or necrosis from constriction of extremities, and may interfere with ability to swim, see, feed, or breathe freely (**Figure 6.20a–c**).

In the wild, large lizards (such as varanids), large snakes, or tortoises may be caught accidentally or on purpose in snares set out by trappers. Wild reptiles may also become entangled in manmade barriers of other sorts (**Figure 6.21a–c**). Occasional captive animals may become ensnared in their enclosures by stringy plant fibers among foliage or floor substrates, or by electric cords, or wires, or cables used for remote door manipulation.

Other unintentional hazards that may cause entrapment in captive situations include narrow spaces or "holes" which reptiles may try to squeeze through during exploration, or attempts to hide or escape (**Figure 6.22a,b**). Snakes may get caught between the sliding glass fronts of their enclosure if the space between panes is too large for the size of snake being held. Other "narrow spaces" may include piping, and in the case of aquatic enclosures or those with substantial pools, these may include water pumps, filters, or drains. Entrapment in these situations may include the inability of a small or weakened reptile to escape from the suction of water flowing through the site. Injury from these types of entrapments may range from minimal bruising and edema due to physical restriction and struggling, to rhabdomyolysis, shock, asphyxiation, or drowning. These animals may also be vulnerable to tankmate or live prey predation.

6.4.2 Crush Injuries

Unsecured enclosure "props," such as large perches or rocks, can dislodge and fall onto reptiles, with traumatic consequences including death, dependent on the body part that is struck or crushed and the weight and form of the fallen object. Toe, tongue, and tail amputations may occur from being caught in enclosure lids or in sliding or guillotine doors (**Figure 6.23a,b**).

6.4.3 Handling Injuries and Human Cruelty

Improper use of restraining or handling devices such as hooks or tongs can cause spinal injuries in snakes and some lizards. Heavy-bodied species lifted with most of their body weight supported at a single site or held by the neck may overextend or subluxate their vertebral columns (Frye, 1991, p. 172), especially if they struggle or have preexisting vertebral disease. Individuals that writhe or twist in tongs or under hooks while pinned to a surface may strain spinal muscles and ligaments or dislocate their cervical vertebrae, especially at the occipital attachment, with resulting hemorrhage or direct spinal cord trauma (**Figure 6.24**). Tongs used with an overzealous grip can break ribs and cause crush injuries or tear skin. Snakes provoked by humans can incur injuries to the face and mouth after striking transparent walls (such as exhibit glass), or when biting or accidentally hitting instruments used for handling or to offer food items.

Tears or other damage to the skin may occur with handling accidents. Minor injury from displacement of the edematous "blue" superficial skin during an early ecdysis cycle is common and usually does not result in any long-term consequences. Some species, such as Day Geckos (*Phelsuma* spp.), readily tear full-thickness layers of skin, generally considered part of a defense-escape mechanism. These species will often survive and heal rapidly, although skin loss may occur over an extensive portion of the body and result in scars. Likewise, many lizard species will autotomize (auto-amputate) their tails as an escape tactic (see Volume 1, Chapter 1) (**Figure 6.25**). Specific injury to the eyes of snakes, and some lizard species, may occur from ill-informed or rough attempts by handlers to remove retained spectacles (**Figure 6.26**).

Traumatic injuries from being dropped during handling or from a fall from a table can vary in severity and in what part of the body struck during the fall. Head injuries, leg fractures, or shell fractures are possible, as well as internal hemorrhage.

Some reptiles, captive and wild, are injured or stressed as a result of direct human cruelty, including beating the animal with objects, hacking with sharp instruments, shooting, stomping or otherwise crushing, or burning (**Figure 6.27**). Reptiles in shipping containers have arrived at their destinations with penetrating injuries, or missing tongues that were presumably maliciously amputated when the animals investigated the environment by protruding them through an opening in the crate. These injurious actions by humans sometimes stem from societal beliefs with regard to reptiles having a connotation of "evil," disbelief that reptiles can feel pain, or from fear of the animals—particularly snakes.

6.4.4 Vehicle Strikes

Free-roaming reptiles, especially those in areas near human habitation, are subjected to trauma from vehicle strikes (Sack et al., 2017). Reptiles crossing roads while seeking nesting sites or mates, ranging across their territory to hunt, or seeking the warmth of a paved surface are frequent hit-by-car (or -truck) victims (**Figure 6.28a–c**). Likewise, individuals using other pathways, such as off-road vehicle trails or waterways, may suffer similar injuries. Boat strikes affect sea turtles and some other aquatic reptiles (crocodilians and freshwater turtles) and can result in shell fractures and severe soft tissue injuries after collision with the hull or the propellers (**Figure 6.29a–d**) (Norton, 2005; Mader, 2009). Commonly, rear-end trauma results in transection of the spinal cord with turtles becoming permanent floaters (rear end up).

Tortoises and snakes that are caught unaware or cannot otherwise escape from the path of horticultural equipment, such as mowers, weed-eaters, or other farm equipment, can incur severe blunt and sharp trauma (**Figure 6.30**) (Applegate et al., 2016).

6.4.5 Predator/Prey Bites

Large numbers of both adult and juvenile free-ranging reptiles are killed or injured by natural predators (such as sea turtle hatchlings taken by gulls, or young Desert Tortoises [*Gopherus agassizii*] preyed upon by ravens [*Corvus corax*] or coyotes [*Canis latrans*], or snakes killed by birds of prey or other wetland species). Natural predators notwithstanding, domestic dogs and cats are often responsible for predatory trauma to a variety of reptile species. Cats eagerly seek small lizards to entertain their predatory instincts. Tortoises are commonly chewed on by dogs, sustaining shell damage, limb injuries or amputations, or coelomic perforations (**Figure 6.31a–g**).

Small, weakened, or otherwise immobilized individuals, as well as hatchlings and clutches of eggs, may be attacked by fire ants (*Solenopsis invicta*), and may die from intoxication/envenomation and shock (**Figure 6.32a,b**) (Parris et al., 2002; Allen et al., 2004).

Live animals intended to be prey items for captive reptiles are a common source of bite wounds. Rodents, such as mice and rats, left unattended in cages with snakes will bite in defense if a snake fails to kill it or render it unconscious quickly by constriction, bite, or envenomation. Damage to the neck and face of the reptile, including the eyes, may occur under these circumstances. If the reptile does not choose to feed at all, for whatever reason, an unattended live rodent may choose to bite and partially consume the predator—resulting in tissue loss (skin and muscle), frequently over the dorsum of the caudal portion of the body (**Figure 6.33a–c**).

Similarly, insect prey items, such as crickets, left in enclosures with lizards may bite and partially consume body, head, or extremities of the lizard. This risk may be lessened by ensuring that the insects have a source of food and water in the enclosure, and that they are not left in with an ill or weak reptile.

6.4.6 Foreign Body Ingestion and Inappropriate Food Items

Ingestion of nonfood items by reptiles is common and may be accidental or intentional. Geophagy is a natural behavior in some species (such as tortoises) as a method to acquire calcium or for other nutrient benefits (Jones and Weeks, 1996). Wild crocodilians are frequently found with stones in their stomachs, commonly believed to assist in the breakup of food that is swallowed in large whole chunks (Lane, 2006). Both in the wild and in captivity, many foreign objects have been found in crocodilian stomachs. Coins and other objects are often tossed by people into the open mouths of basking crocodilians on public display, and subsequently swallowed (**Figure 6.34**). A cachectic American Alligator (*Alligator mississippiensis*) in Everglades National Park was found to have a lens cap of a camera over the pyloric outflow tract of the stomach (E. Jacobson, personal communication). Polished glass is common in the stomach of wild alligators. Lead or zinc toxicosis from swallowed coins or fishing sinkers is possible, as is chemical corrosion and injury from alkaline batteries. Gastritis and chronic irritation of the mucosa by any persistent object, even if otherwise inert, may occur. Ingested foreign bodies may be vomited at some point (Andrews et al., 2000), or may remain indefinitely in the stomach if not digested or passed. Crocodilians and aquatic chelonians may attempt to swallow bait or hooked fish from fishing lines. The hook and a portion of line may pass through the gastrointestinal tract without consequence or could lead to plication and laceration of the intestines with subsequent septic ceolomitis (**Figure 6.35a–c**) (Mader,

2009; Franchini et al., 2018). In other instances, especially if the fishing line is pulled, the hook may penetrate the oral cavity, esophagus, or stomach (**Figure 6.36a,b**). This may cause mucosal or mural tears and lead to disseminated infection or sepsis, or will lead to localized infections that become walled off as chronic abscesses or granulomas. Rocks, sand, wood chips, or other natural or artificial substrates may be ingested by lizards, tortoises, turtles, or snakes, especially if adherent to or mixed with food. Other foreign matter, such as floating plastic bags or other discarded objects, especially if brightly colored or located in a feeding zone, may be mistaken for food and ingested by turtles, tortoises, and less commonly by lizards or snakes. In some cases, the ingested material may cause ileus and enteritis, or may cause pyloric obstruction, intestinal impactions, perforations, or constipation (**Figure 6.37a–h**) (Norton, 2005; Reavill and Schmidt, 2011; Romeijer et al., 2016; Ilayaraja et al., 2018). A recent study linked the quantity of ingested plastic marine debris with increasing risk of mortality in sea turtles (Wilcox et al., 2018). Polyacrylamide gel, used as a hydration source for plants in some commercial potting soils and by some cricket farms, forms a small dark grain-like pellet when fully dehydrated, but expands markedly when rehydrated, making it a potential obstructive foreign body if ingested by a small reptile (**Figure 6.38**) (Miller et al., 2009).

Inappropriately large-sized prey fed to captive snakes may result in injury to the mandible or quadrate during strenuous attempts to swallow the item (Waffa, 2015), or distention and mural tears of the esophagus (De Oliveira Nogueira et al., 2013). "Power feeding" an inappropriate volume of food to achieve accelerated growth may cause obstipation (**Figure 6.39a–c**) (Griffin, 2006). Gorging on abundant, possibly oversized or abrasive food items, has been reported to be a probable cause of gastroduodenal impactions and perforations with subsequent mortality in juvenile turtles (Couture et al., 2017). Excessive hair may accumulate in the stomachs of some crocodilians, carnivorous lizards, or snakes fed primarily furred prey items (such as rabbits and guinea pigs) and can cause gastric obstruction with matted fur or development of trichobezoars.

6.5 Environmental and Physical Diseases

6.5.1 Hypothermia

The ectothermic physiology of reptiles makes them susceptible to injury (physical and metabolic) when environmental temperatures extend outside of their critical thermal ranges. Free-ranging reptiles are often found in torporous or comatose states when sudden localized climate fluctuations result in extreme low air and water temperatures. These animals may develop systemic illnesses, dysbiosis due to gut stasis, and a suppressed immune system that remains depressed despite restoration of optimal body temperatures. Sea turtles commonly develop aspiration pneumonia weeks after the event (Stockman et al., 2010). "Cold stunning" in sea turtles trapped in suddenly cooled waters (below 50°F/10°C) during wintertime regularly deploys rescue efforts (Norton, 2005; Mettee, 2010). Clinical presentations and secondary diseases are well recognized for sea turtles but may apply to any hypothermic reptile. The individual's responsiveness, including reflexes, may be severely reduced depending on the length of time and severity of the temperature decrease. This may cause an inability to maintain hydration and avoid trauma, leading to corneal and skin lesions (**Figure 6.40**). Ischemic necrosis of the appendages due vasoconstriction and frostbite may also occur (Norton, 2005).

Reptiles may also suffer from acute hypothermia in captive situations, when thermal accidents occur during transportation, or with malfunctions of environmental temperature control units. Induced hypothermia is still sometimes practiced for reduction of reptile mobility and ease of restraint and is considered an unacceptable means of euthanasia at this time (Leary et al., 2013; Lillywhite et al., 2016). Induced hibernation of ill reptiles can lead to death during hibernation or potentially irreversible disorders following emergence. Steep or rapid temperature changes during hibernation, or maintenance temperatures too close to freezing, may result in death. Dysbiosis and mixed mycotic/bacterial gastroenteritis may be seen in these animals.

Chronic hypothermia in captive reptiles is a common problem when ambient temperatures are kept below, or at the low end of, the preferred optimum temperature range for the species (as in air-conditioned homes) and an opportunity for basking is not provided. Offering only a hot rock for a large lizard or snake, or heating under the ground substrate rather than providing a basking light for an arboreal species are examples of inadequate warming opportunities for captive reptiles. Chronic hypothermia affects the metabolism and behavior of the captive reptile, and may lead to anorexia, vomiting, weight loss, dysecdysis or retained sheds, and vulnerability to infections and other disorders associated with chronic stress (Boyer, 1998, p. 8; Fitzgerald and Vera, 2006; Funk, 2006c; Rossi, 2006).

6.5.2 Hyperthermia

As with hypothermia, reptiles are vulnerable to temperatures above their critical thermal maximums that may quickly cause coma and death. Animals in shipping containers exposed to excessive heat may be found in a moribund state or dead upon arrival at their destination. Pet reptiles in aquaria or small plastic containers set in a window or in the direct sunlight outdoors may succumb to hyperthermia after only 10–15 minutes (Boyer, 1998, p. 9). Weak or immobilized animals that cannot remove themselves from heat sources may also succumb to hyperthermia, and these carcasses are often found in an advanced state of autolysis due to persistently high postmortem body temperature. The physiological mechanisms of thermoregulation in reptiles have been reviewed (Seebacher and Franklin, 2005), but the pathophysiology of hyperthermia-induced morbidity and mortality in reptiles is not as thoroughly documented in the literature. Hyperthermic injuries in reptiles most likely have many similarities to that of endothermic mammals and birds. Hyperthermia in these animals leads to initial redistribution of blood flow in attempts to cool the body, followed by increased permeability of blood vessels with damage to multiple organ systems including the CNS, skeletal and cardiac muscles, kidneys, liver, and the gastrointestinal tract. Disseminated intravascular coagulopathy and systemic inflammatory response syndrome (or sepsis) are involved in organ damage and mortality in the period following a hyperthermia episode in endotherms (Leon and Kenefick, 2012), and it seems reasonable to believe that this is also true for reptiles. A syndrome resembling hyperviscosity has been associated with abrupt temperature fluctuations in snakes (LaDouceur and Garner, 2017; also see Chapter 9).

6.5.3 Burns

Captive reptiles with thermal burns requiring medical care are well covered by Mader (2006c). These burns are usually caused by inappropriate access to a heat source (such as an incandescent bulb or heat lamp) or by a malfunctioning heating element (such as a hot-rock or heating pad) or thermostat. The severity of burns varies from superficial skin damage and erythema (first degree), to partial thickness skin injury with edema and hemorrhage of the underlying tissue (second degree) to full-thickness skin sloughing (third degree), and potential injury to underlying muscles, bone, or organs (fourth degree). Accordingly, the rate and type of healing, and the risks for scarring, secondary infections, and sepsis vary with the depth of the burn (**Figure 6.41a–e**).

Reptiles may occasionally be caught in a fire, from an accident in a captive situation (Zehnder et al., 2010) or from wildfires in free-ranging situations (**Figure 6.42a–g**) (Cooper, 1996; Smith et al., 2012). The types of thermal burns to the integument in these instances may be classified as above, but thermal injury and smoke inhalation damage to the respiratory system may also occur. Respiratory damage from fire and smoke may include rhinitis, tracheitis, and pulmonary edema with hemorrhage, a risk for secondary infections of both upper and lower respiratory tract, and sepsis.

Solar radiation injuries (sunburn) are not common but may be seen in reptiles not adapted for such exposure and trapped in direct sunlight. Individuals with albinism or other atypical color variations due to dermal chromatophore alterations have been found to develop burns when not adequately shaded or kept indoors, such as "white" alligators or leucistic snakes and lizards. These are generally manifested as superficial epidermal injuries and erythema with or without secondary infections. Potentially deeper burns would likely be superseded by hyperthermia-related systemic injuries. Similarly, these individuals may develop burns from inescapable exposure to ultraviolet-emitting lamps in their enclosures (**Figure 6.43a,b**).

Chemical burns to reptiles may be caused by exposures to substances that have markedly alkaline or acidic properties. These may occur in accidents with cleaning agent spills, or in aquatic environments with marked pH-disruptions due to failure to properly treat pool walls (such as unsealed gunnite or uncured new paint). Other abrupt changes in water chemistry including ozone spikes, pump failures, or marked changes in municipal chlorination may lead to similar contact injuries. Damage to soft skin, oral mucous membranes, and the eyes are most likely to occur under these circumstances, or to skin directly in contact with a spilled chemical.

6.5.4 Electrocution

Electrocution is not common in reptiles but might occasionally occur if an individual were to bite into an exposed electric cord or there is a faulty circuit in a heating appliance (Frye, 1981). Lightning strike near a wire fence or pool might also be an uncommon cause of electrocution of a reptile in the vicinity. As for mammals or birds with such injuries, tissue charring and burns of the tissues in contact with the electricity could be expected, as well as possible myocardial damage and pulmonary and other hemorrhages (Cooper, 1996).

6.5.5 Dehydration

A species-appropriate source of potable drinking water is essential for all but the most fastidious of animals. Some captive reptiles, such as some chameleons, geckos, and other small arboreal lizards, will only drink from water droplets (mimicking dew or rainfall) on enclosure walls and foliage (**Figure 6.44a,b**). Some snakes and tortoises also prefer to soak in standing water for a time to fulfill their hydration needs and maintain good skin health (**Figure 6.45a,b**). Environments that are chronically too arid for a species may lead to dehydration and deterioration in some metabolic activities and digestion efficiency, and to retained shed.

Retained shed, sometimes referred to as dysecdysis, is commonly associated with lack of access to adequate humidity or sufficient water for soaking during the period immediately preceding shedding (Frye, 1981; Boyer, 1998; Fitzgerald and Vera, 2006). (True dysecdysis is probably a more complex metabolic disorder in which the ecdysis cycle is not appropriately synchronized, but with similar clinical presentations and outcomes [see Chapter 2]). Retained shed may be observed as disrupted patches of unshed superficial epidermal layers over large portions of the body but is most problematic if skin is retained on digits or tail tips and forms constricting bands that may cause ischemic necrosis and loss of the extremity. Retained shed on the spectacles may impair vision (or result in direct trauma to the eyes, if removal is performed too forcefully). Retained shed may also obstruct nares and lead to open-mouthed breathing or to face-rubbing and abrasions in the affected individual (**Figure 6.46a–d**). Retained shed notoriously predisposes to opportunistic fungal infections, which can invade the underlying dermis and lead to sepsis.

Acute or severe dehydration may lead to uric acid precipitation within the kidneys, followed by gout. Gout may be observed as urate precipitates and tophus formation in renal tubules, the serosa of the liver and other viscera, pericardial sac, submucosal and subcutaneous tissues, interstitial tissues of muscles, and in and around joints (Mader, 2006b). In some instances, chronic dehydration is likely to be involved in the formation of uroliths in the bladders of tortoises (Mader, 2006a; Highfield, 2015). (See Chapter 3.)

6.5.6 Excessive Humidity/Wet Conditions

A captive environment that has excessive humidity or overly moist substrate is often coincident with a poorly ventilated situation. Constant exposure to wet conditions can predispose individuals to respiratory disease and dermatitis with secondary cutaneous and visceral infections, especially when there is an overgrowth of pathogens in the substrate or there is poor water quality (Frye, 1981; Fitzgerald and Vera, 2006; Highfield, 2015). Tortoises kept in excessively humid or damp environments may develop "shell rot"—keratolytic bacterial and fungal infections of the shells—particularly of the plastron (**Figure 6.47a–c**).

6.5.7 Toenail/Beak Overgrowth or Deformity

Lack of sufficient exercise or an inadequately abrasive substrate for wear of toenails can lead to nail overgrowth and deformity (**Figure 6.48**). Likewise, insufficiently fibrous or abrasive dietary items (along with unbalanced nutrition) can lead to beak (rhamphotheca) overgrowth and deformity (**Figure 6.49a–c**) (Frye, 1981). In chelonians, severe beak problems can impair prehension of food and predispose to infections in the oral cavity and affected mouth parts.

6.5.8 Lighting and Other Physical Disturbances

In the captive environment, artificial lighting without external photoperiod cues can control reptiles' diurnal and seasonal cyclicity. Inattention to the regularity of the light cycle or frequent disruptions of lighting for these animals may theoretically disturb a normal awake–rest cycle and confuse seasonal urges to hibernate, thus also disrupting seasonal reproductive and appetite/metabolism cycles. Likewise, frequent and unpredictable physical disturbances, such as excessive or irregular handling, visual disturbances caused by extraneous activity outside of an enclosure, or noise or vibrational disturbances, may interfere with regular rest intervals and cause fear-related responses. Insufficient hiding opportunities may also interfere with normal rest, vigilance, and activity intervals. Nocturnal reptiles might be expected to be more seriously affected by these sorts of stressors. Although these physical and photoperiod disturbances in themselves are unlikely to cause grossly obvious injuries or other pathological changes, the chronic stress accompanying lack of regular rest and diurnal cyclicity may lead to increased vulnerability to disease and impair welfare (Warwick et al., 2013).

6.5.9 Irradiation

Information on gamma irradiation of reptiles is based mostly on chronic low-level environmental exposures to nuclear incident sites and suggests some gametogenic damage may occur (Pearson et al., 1978), but no indicators of other physiological disturbances were found in lizards (Nagy and Medica, 1985). An older study indicated population demographics remained similar between exposed versus non-exposed populations of lizards (Turner et al., 1969). Brisbin et al. (1974) suggested that free-ranging snakes could be used as a sentinel species to measure radionuclide contamination of environmental sites.

X-ray exposure investigations suggest some snakes' susceptibility to radiation (LD50 300–400R for rat snakes and racers) may be comparable to that of humans, while some turtles and tortoises (LD50 approximately 1000R for box turtles, painted turtles, and snapping turtles) may be somewhat more radioresistant (Cosgrove, 1971). Clinical overdoses may be manifested as localized skin damage and color changes, subcutaneous and pulmonary hemorrhage, enteric necrosis and consequent enteritis and septicemia, and necrosis of regenerative cell lines in bone marrow, spleen, and gonads, potentially leading to death of the individual (Cosgrove, 1971; Kothari and Patil, 1975; Cooper and Jackson, 1981). Clinical radiation treatment applied to neoplasia or other disease entities in reptiles will likely provide further understanding of its role and effects in these animals in the future (Diaz-Figueroa and Mitchell, 2006).

Ultraviolet irradiation from the sun or from UVA–UVB-emitting lamps may induce some pathologic changes in the eyes (corneas, lenses, and retinas), skin (see discussion of burns in this chapter and **Figure 6.52**), or metabolism (vitamins, renal and hepatic) of reptiles; however, research is still needed to verify and further elucidate these risks (Adkins et al., 2003; Baines, 2008; Gardiner et al., 2009).

Artificial sources of infrared radiation (basking lamps) can result in thermal burns, as previously noted, as well as localized and systemic dehydration, and potential ocular injuries. A good discussion of these infrared effects and critical differences between natural solar and artificial sources is provided by Highfield (2015).

6.5.10 Acoustic Injuries and Underwater Explosion Injuries

The effects of anthropogenic underwater noise have been studied in many marine organisms, although study of its effects on aquatic reptiles is still very limited at this time (Williams et al., 2015). Some research has been performed regarding hearing sensitivity and the potential effects of anthropogenic noise on sea turtles (Dow Piniak et al., 2012). Documented effects in sea turtles so far seem to be predominantly modifications in behavior (diving and avoidance) in response to sudden noise of air guns (Peng et al., 2015), although high-intensity sounds have caused hearing losses, other physiological effects, and even death in some aquatic species including mammals, fish, and invertebrates. Specific ocular and acoustic injuries or strandings in aquatic turtles following high-powered sonar use, like those observed in some marine mammals, have not been documented as yet; however, behavioral responses, strandings, and injuries (lethal and non-lethal), have been reported in sea turtles found in the vicinity of underwater explosions such as underwater demolitions (**Figure 6.50**). These injuries may result from direct shock-wave trauma or be combined with cavitation-induced injuries in some instances. Reported effects ranged from erythema around eyes and flippers, to unconsciousness, internal hemorrhages especially affecting the lungs and gastrointestinal tract, shell fractures, and acute fatality (Viada et al., 2008).

6.6 Other "Accidents"

6.6.1 Intestinal or Other Visceral Accidents

Foreign body ingestion and the possible consequences of impaction, obstruction, and perforation have been discussed earlier in this chapter. Gastrointestinal accidents causing acute or chronic illness may also include intussusception, torsion or volvulus, visceral entrapments, or ulcer-related perforations. While these may have infectious predisposing factors including parasitism or bacterial enteritis (or other pathogenic agents), in some instances they may be related to traumatic incidents, physical displacements (rolling/overturning), neoplasia, or may be idiopathic (Diaz-Figueroa and Mitchell, 2006; Applegate et al., 2016; Wills et al., 2016). Neurological damage from trauma or disease may secondarily alter visceral motility, leading to ileus and impaction or displacements (**Figure 6.51a–d**) (Di Girolamo and Selleri, 2016). In addition to the gastrointestinal tract, other viscera such as ovaries, oviducts, or a portion of the liver may become entrapped in mesenteric tears, adhesions, or may develop a torsion or volvulus.

Prolapses of the cloaca, colon, oviduct, penis/phallus, or hemipenes are also frequent presentations in reptiles. Intussusceptions may be related to altered motility, dystocia, or infection (see Chapter 8). A prolapsed or extended hemipene or phallus may become traumatized during sexual arousal or as part of defensive or distress behavior (**Figure 6.52a–c**) (Barten, 2006b). Cloacal scent gland impactions may also be seen in snakes, either secondary to infections, neoplasia, trauma, or potentially husbandry or age-related factors (**Figure 6.53**) (Couture et al., 2018).

6.6.2 Overturned Tortoises

Most normal healthy tortoises are capable of righting themselves when overturned by accident or from interaction with an aggressive male or predator. However, nearby obstacles or suboptimal substrates (see Section 6.6.3), or abnormal shell development secondary to captive management issues (such as dietary imbalances), may prevent an individual from returning to normal position and mobility (**Figure 6.54**). In this event, weight of the viscera on the lungs and major blood vessels may eventually result in respiratory distress and circulatory collapse. These gravity-dependent problems may be combined with exposure to direct sunlight or other ambient factors, enhancing distress and contributing to more rapid morbidity and mortality.

6.6.3 Drowning

While many reptile species are adapted to live in or around water, they are nonetheless all air breathers, and thus subject to drowning. Many species of lizards and snakes that are excellent swimmers may suffer exhaustion and drown if they are unable to escape from deep water. Flooding caused by natural disasters or by the construction of dams may cause widescale drowning of wildlife, including reptiles. Accidental falls into other structures such as swimming pools, steep-sided ditches, moats, or canals seem to be special risks for free-ranging or pet tortoises. Relatively shallow water that does not allow complete submersion of a turtle creates a significant drowning hazard if the animal is accidentally overturned, as when trying to escape from an aquarium or other container by attempting to climb the sides. The physical ability of a chelonian to right itself in these circumstances may be seriously compromised and can lead to exhaustion and drowning in only a few inches of water. Any reptile in captivity, left to "soak" in a covered container in water that is too deep, may succumb to exhaustion and drowning, especially if sick or weak.

Highly adapted reptilian diving species, such as turtles, survive long periods of breath-holding with diving reflexes somewhat different from those of diving mammals, increasing blood flow to lungs and thereby increasing arterial oxygen exchange from pulmonary gases (Burggren, 1988). Sea turtles may remain submerged for prolonged periods by means of additional physiologic responses to hypoxia, including bradycardia, anaerobic metabolism, and marked slowing of metabolism (Berkson, 1966). Crocodilians have unique hemoglobin molecules that increase oxygen transfer to tissues in states of hypoxia (Perutz et al., 1981). Regardless, if trapped while submerged in water, all will eventually drown. Sea turtles and some other chelonians are unfortunately frequent bycatch victims and drown when trapped in fishing nets or crab pots, or caught on long lines and dragged underwater for a prolonged period (Roosenburg et al., 1997).

The pathogenesis of drowning involves the aspiration of water into the respiratory tract, as well as swallowing of large volumes of water. Differentiation of drowning from postmortem saturation of tissues can be difficult. Flooding of the lungs may occur with either scenario, and trauma may be found in both cases, especially when an animal was knocked into or fell into a body of water. Evidence for drowning includes an acute response to water in the lungs such as edema and hemorrhages. Whether drowning occurs in fresh water or saltwater may result in changes noted due to osmotic imbalances, such as hemolysis due to hypotonicity of fresh water, or hemoconcentration and dehydration due to hypertonicity of saltwater (Frye, 1981; Cooper, 1996). In addition to pulmonary edema and hemorrhages, specific columnar epithelial cell sloughing was noted in drowned sea turtles versus those that died of other causes while submerged (Work and Balazs, 2010). Laryngospasm and "dry-drowning" sometimes noted in other vertebrates may occur in some instances, but if the reptile dies and the glottis relaxes, then the lungs may flood perimortem, making this specific diagnosis elusive at best. Reptiles may sometimes be found on land around an aquatic enclosure with histologic changes in the lung that support drowning, such as congestion, hemorrhage, and water in the faveolae, suggesting that they may be capable of briefly surviving a fatal influx of water into the pulmonary system. If an individual survives a near-drowning incident, aspiration pneumonia may develop. The other consequences of a long period of anoxia in the brain and other organs may or may not resolve with time.

6.6.4 Decompression Sickness and Gas Bubble Disease

Decompression sickness (the "bends") has been described in sea turtles examined after fisheries bycatch accidents (Garcia-Parraga et al., 2014). Flipper spastic rigidity, behavioral disturbances, and coma were followed by death in many of the turtles, although some mild cases recovered after a few days, and a few more moderate cases responded to decompression in hyperbaric chambers and survived. Premortem and postmortem diagnoses were based on findings of gas embolism in major blood vessels and heart chambers (especially the right atrium) observed grossly and by radiography and ultrasonography. Gas was also found in organs such as the kidney, spleen, and liver. Collapsed lungs with and without emphysema were also noted. The gases in the intravascular bubbles were analyzed for one turtle fatality and were reported to be characteristic for decompression, rather than off-gassing or bacterial decomposition.

Gas bubble disease has also been reported in a group of turtles living in an exhibit pool that had developed supersaturation of gasses from a malfunctioning filter valve, and later had a heater malfunction leading to a drop in water temperature (Lung and Garner, 2000). The turtles spent much of their time at the bottom of the 9-foot deep pool where the water was warmer, and all of them developed varying amounts of cutaneous emphysema, which gradually resolved over 8 to 10 weeks. Two of the turtles required hospitalization due to secondary infections in ruptured sites of the emphysematous skin.

Acknowledgments

I extend my thanks to all friends and colleagues who generously provided images for this chapter and who are individually recognized in the figure legends. I also wish to acknowledge the following institutions responsible for the care or rescue of some of the animals in those photographs: Currumbin Wildlife Hospital (Brisbane, Australia), Massachusetts Audubon's Wellfleet Bay Wildlife Sanctuary, New England Aquarium (Boston, MA), Pittsburgh Zoo, Rattlesnake Solutions LLC (Phoenix AZ), and Zoo Miami.

References

Adkins E, Driggers T, Ferguson G et al. 2003. Roundtable—Ultraviolet light and reptiles, amphibians. *J Herp Med Surg* 13:27–37.

Allen CR, Epperson DM, and Garmestani AS. 2004. Red imported fire ant impacts on wildlife: A decade of research. *Am Midl Nat* 152:88–103.

Andrews PLR, Axelsson M, Franklin C, and Holmgren S. 2000. The emetic reflex in a reptile (*Crocodylus porosus*). *J Exp Biol* 203:1625–1632.

Applegate Jr JR, Drapp RL, and Lewbart GA. 2016. Nonfatal traumatic gastric evisceration in two box turtles (*Terrapene carolina carolina*). *J Herp Med Surg* 26:80–84.

Baines F. 2008. Photo-kerato-conjunctivitis in reptiles, in *British Veterinary Zoological Society Proceedings*, Stretton, Cheshire, p. 43.

Barten SL. 2006a. Lizards, in *Reptile Medicine and Surgery*, 2nd ed. Mader DR (Ed.), Saunders/Elsevier Inc, St. Louis, pp. 59–77.

Barten SL. 2006b. Penile prolapse, in *Reptile Medicine and Surgery*, 2nd ed. Mader DR (Ed.), Saunders/Elsevier Inc, St. Louis, pp. 862–864.

Berkson H. 1966. Physiological adjustments to prolonged diving in the pacific green turtle (*Chelonia mydas agassizii*). *Comp Biochem Physiol* 18:101–119.

Boyer TH. 1998. *Essentials of Reptiles, A Guide for Practitioners*. AAHA Press, San Diego.

Boyer TH, and Boyer DM. 2006. Turtles, tortoises, and terrapins, in *Reptile Medicine and Surgery*, 2nd ed. Mader DR (Ed.), Saunders/Elsevier Inc, St. Louis, pp. 78–99.

Brisbin Jr IL, Staton MA, Pinder JE, and Geiger RA. 1974. Radiocesium concentrations of snakes from contaminated and non-contaminated habitats of the AEC Savannah River Plant. *Copeia* 1974:501–506.

Brown JD, and Sleeman JM. 2002. Morbidity and mortality of reptiles admitted to the Wildlife Center of Virginia, 1991 to 2000. *J Wildl Dis* 38:699–705.

Burggren W. 1988. Cardiovascular responses to diving and their relation to lung and blood oxygen stores in vertebrates. *Can J Zool* 66:20–28.

Burghardt GM, Ward B, and Rosscoe R. 1996. Problem of reptile play: Environmental enrichment and play behavior in a captive nile soft-shelled turtle, *Trionyx triunguis*. *Zoo Biol* 15:223–238.

Cooper JE. 1996. Physical injury, in *Noninfectious Diseases of Wildlife*. Fairbrother A, Locke LN and Hoff GL (Eds.), Iowa State University Press, Ames, pp. 157–172.

Cooper JE, and Jackson OF. 1981. Miscellaneous diseases, in *Diseases of the Reptilia*, Vol 2. Cooper JE and Jackson OF (Eds.), Academic Press, London, pp. 487–504.

Cosgrove GE. 1971. Reptilian radiobiology. *J Am Vet Med Assoc* 159:1678–1684.

Couture EL, Ferrell ST, Desmarchelier M, Lazure L, and Lair S. 2017. Gastro-duodenal perforations and coelomitis in a group of juvenile spiny softshell turtles (*Apalone spinifera*). *J Herp Med Surg* 27:36–43.

Couture EL, Lung NP, Lair S, and Ferrell ST. 2018. Ophidian cloacal scent gland impaction and surgical excision: A case series in six different snake species. *J Herp Med Surg* 28:81–88.

Denardo D. 2006. Stress in captive reptiles, in *Reptile Medicine and Surgery*, 2nd ed. Mader DR (Ed.), Saunders/Elsevier Inc, St. Louis, pp. 119–123.

De Oliveira Nogueira CH, Figueiredo-de-Andrade CA, and de Freitas NN. 2013. Death of a juvenile snake *Oxyrhopus petolarius* (Linnaeus, 1758) after eating an adult house Gecko *Hemidactylus mabouia* (Moreau de Jonnes, 1818). *Herpetology Notes* 6:39–43.

Diaz-Figueroa O, and Mitchell MA. 2006. Gastrointestinal anatomy and physiology. In *Reptile Medicine and Surgery*, 2nd ed. Mader DR (Ed.), Saunders/Elsevier Inc, St. Louis, pp. 145–162.

Di Girolamo N, and Selleri P. 2016. What is your diagnosis? *J Am Vet Med Assoc* 249:271–273.

Dow Piniak WE, Eckert SA, Harms CA, and Stringer EM. 2012. Underwater hearing sensitivity of the leatherback sea turtle (*Dermochelys coriacea*): Assessing the potential effect of anthropogenic noise. US Dept of the Interior, Bureau of Ocean Energy Management, Headquarters, Herndon, VA. OCS Study BOEM 2012–01156. pp. 35.

Fearn S, Schwarzkopf L, and Shine R. 2005. Giant snakes in tropical forests: A field study of the Australian Scrub Python, *Morelia kinghorni*. *Wildl Res* 32:193–201.

Fitzgerald KT, and Vera R. 2006. Dysecdysis, in *Reptile Medicine and Surgery*, 2nd ed. Mader DR (Ed.). Saunders/Elsevier Inc, St. Louis, pp. 778–786.

Franchini D, Valastro C, Ciccarelli S, Caprio F, Lenoci D, and Di Bello A. 2018. Ultrasonographic detection of ingested fishing lines in loggerheads (*Caretta caretta*). *J Wildl Dis* 54:680–690.

Frolich J, and Sutherland-Smith M. 2010. Morbidity of Fijian banded Iguanas (*Brachylophus bulabula*) in the San Diego Zoo collection from 2000–2010, in *Proceedings of the Association of Reptilian and Amphibian Veterinarians*, South Padre Island, TX, p. 130.

Frye FL. 1981. Traumatic and physical diseases, in *Diseases of the Reptilia*, Vol 2. Cooper JE and Jackson OF (Eds.), Academic Press, London, pp. 387–407.

Frye FL. 1991. *Biomedical and Surgical Aspects of Captive Reptile Husbandry*, Vol 1. Krieger Publishing Co, Malabar (Florida).

Funk RS. 2006a. Snakes, in *Reptile Medicine and Surgery*, 2nd ed. Mader DR (Ed.), Saunders/Elsevier Inc, St. Louis, pp. 42–58.

Funk RS. 2006b. Tail damage, in *Reptile Medicine and Surgery*, 2nd ed. Mader DR (Ed.), Saunders/Elsevier Inc, St. Louis, pp. 913–915.

Funk RS. 2006c. Vomiting and regurgitation, in *Reptile Medicine and Surgery*, 2nd ed. Mader DR (Ed.), Saunders/Elsevier Inc, St. Louis, pp. 939–940.

Garcia-Parraga D, Crespo-Picazo JL, Bernaldo de Quiros Y et al. 2014. Decompression sickness ("the bends") in sea turtles. *Dis Aquat Org* 111:191–205.

Gardiner DW, Baines FM, and Pandher K. 2009. Photodermatitis and photokeratoconjunctivitis in a Ball Python (*Python regius*) and a blue-tongue skink (*Tiliqua* spp.). *J Zoo Wildl Med* 40:757–766.

Greenberg N, and Crews D. 2013. Physiological ethology of aggression in amphibians and reptiles, in *Hormones and Aggressive Behavior*. Svare BB (Ed.), Springer Science and Business Media, pp. 469–506.

Griffin C. 2006. Severe obstipation as a result of power feeding in a red tail boa (*Boa constrictor*), in *Proceedings of the Association of Reptilian and Amphibian Veterinarians*, Baltimore, MD, pp. 112–113.

Griner LA. 1983. *Pathology of Zoo Animals, A Review of Necropsies Conducted over a Fourteen-Year Period at the San Diego Zoo and San Diego Wild Animal Park*. Zoological Society of San Diego, San Diego.

Gyimesi ZS, and Burns RB. 2002. Monitoring of plasma 25-hydroxyvitamin D concentrations in two Komodo Dragons, *Varanus komodoensis*: A Case Study. *J Herp Med Surg* 12:4–9.

Highfield AC. 2015. The Effect of Basking Lamps on the Health of Captive Tortoises and Other Reptiles. http://www.tortoisetrust.org/articles/baskinghealth.html (accessed February 27, 2017).

Homer BL, Berry KH, Brown MB, Ellis G, and Jacobson ER. 1998. Pathology of diseases in wild desert tortoises from California. *J Wildl Dis* 34:508–523.

Hoser RT. 1996. Interspecific immunity to venom in snakes. *Herpetofauna (Journal of the Australasian Herpetological Societies)* 26:26–27.

Ilayaraja S, Mukul CG, Acharya PR, Praqdeep R, and Sha Arun A. 2018. Ingested foreign body (thorn): Primary cause of death in a wild python (*Python molurusmolurus*). *J Wildl Res*, www.jakraya.com. https://www.researchgate.net/publication/323695181_Ingested_Foreign_Body_Thorn_Primary_Cause_of_Death_in_a_Wild_Python_Python_molurusmolurus (accessed January 13, 2019).

Jones RL, and Weeks Jr HP. 1996. Geophagy, in *Noninfectious Diseases of Wildlife*. Fairbrother A, Locke LN, and Hoff GL (Eds.), Iowa State University Press, Ames, 2nd edition, pp. 19–23.

Kothari RM, and Patil SF. 1975. Effect of gamma irradiation on the common garden lizard, *Calotes versicolor* (Daud). *J Herpetology* 9:103–105.

LaDouceur EEB, and Garner MM. 2017. Hyperviscosity-like syndrome in reptiles, in *Proc 49th AAZV Ann Conf*, Dallas, TX, pp. 79–80.

Lane T. 2006. Crocodilians, in *Reptile Medicine and Surgery*, 2nd ed. Mader DR (Ed.), Saunders/Elsevier Inc, St. Louis, pp. 100–117.

Leary S, Underwood W, Anthony R et al. 2013. AVMA Guidelines for the Euthanasia of Animals: 2013 Edition. http://www.avma.org/KB/Policies/Pages/Euthanasia-Guidelines.aspx (accessed February 19, 2017).

Leon LR, and Kenefick RW. 2012. Pathophysiology of heat-related illnesses, in *Wilderness Medicine Textbook*, 6th ed. Auerbach P (Ed.), Mosby, pp. 113–129.

Leong JK, Smith DL, Revera DB et al. 1989. Health care and diseases of captive-reared Loggerhead and Kemp's Ridley sea turtles, in *Proceedings of the First International Symposium on the Kemp's Ridley Sea Turtle Biology, Conservation, and Management*. Caillouet CW, and Landry AM (Eds.), Texas A & M Sea Grant College Program: Galveston, TX, pp. 178–201.

Lillywhite HB, Shine R, Jacobson E et al. 2017. Anesthesia and euthanasia of amphibians and reptiles used in scientific research: Should hypothermia and freezing be prohibited? *Bioscience* 67:53–61.

Lung N, and Garner M. 2000. Gas bubble disease in Malaysian painted terrapins, in *Proceedings of the Association of Reptilian and Amphibian Veterinarians*, Reno, NV, pp. 119–120.

Mader D. 2009. Master class: Sea turtles part B. Medical care of sea turtles, in *Proceedings of the Association of Reptilian and Amphibian Veterinarians*, Milwaukee, WI, pp. 37–50.

Mader DR. 2006a. Calculi: Urinary, in *Reptile Medicine and Surgery*, 2nd ed. Mader DR (Ed.), Saunders/Elsevier Inc, St. Louis, pp. 763–771.

Mader DR. 2006b. Gout, in *Reptile Medicine and Surgery*, 2nd ed. Mader DR (Ed.), Saunders/Elsevier Inc, St. Louis, pp. 793–800.

Mader DR. 2006c. Thermal burns, in *Reptile Medicine and Surgery*, 2nd ed. Mader DR (Ed.), Saunders/Elsevier Inc, St. Louis, pp. 916–923.

Mettee NS. 2010. Florida's 2010 cold stunned sea turtles, in *Proceedings of the Association of Reptilian and Amphibian Veterinarians*, South Padre Island, TX, p. 43.

Miller CL, Bischoff KL, and Hoff B. 2009. Polyacrylamide gel ingestion leading to fatal intestinal obstruction in two birds in a zoological collection. *J Avian Med Surg* 23:286–289.

Nagy KA, and Medica PA. 1985. Altered energy metabolism in an irradiated population of lizards at the Nevada test site. *Radiat Res* 103:98–104.

Nevarez JG. 2007. Observations from the field: Common disease presentations in Louisiana alligator ranches, in *Proceedings of the Association of Reptilian and Amphibian Veterinarians*, New Orleans, LA, pp. 30–33.

Nichol A, Douglas V, and Peck L. 1933. On the immunity of rattlesnakes to their venom. *Copeia* 1933:211–213.

Norton TM. 2005. Chelonian emergency and critical care. *Semin Avian Exot Pet Med* 14:106–130.

Olsson M, and Madsen T. 1998. 13: Sexual selection and sperm competition in reptiles, in *Sperm Competition and Sexual Selection*. Birkhead TR and Moller AP (Eds.), Academic Press, pp. 503–577.

Parris LB, Lamont MM, and Carthy RR. 2002. Increased incidence of red imported fire ant (*Hymenoptera: Formicidae*) presence in loggerhead sea turtle (*Testudines: Cheloniidae*) nests and observations of hatchling mortality. *Fla Entomol* 85:514–417.

Pearson AK, Licht P, Nagy KA, and Medica PA. 1978. Endocrine function and reproductive impairment in an irradiated population of the lizard *Uta stansburiana*. *Radiat Res* 76:610–623.

Peng C, Zhao X, and Liu G. 2015. Noise in the sea and its impacts on marine organisms. *Int J Environ Res Public Health* 12:12304–12323.

Perutz MF, Bauer C, Gros G et al. 1981. Allosteric regulation of crocodilian haemoglobin. *Nature* 291:682–684.

Ramstad KM, Moore JA, and Refsnider JM. 2012. Intrasexual aggression in tuatara: Males and females respond differently to same-sex intruders. *Herpetol Rev* 43:19–21.

Reavill DR, and Schmidt RE. 2011. Pathology of reptilian gastrointestinal tract, in *Proceedings of the Association of Reptilian and Amphibian Veterinarians*, Seattle, WA, pp. 72–88.

Rivas AE, Allender MC, Mitchell M, and Whittington JK. 2014. Morbidity and mortality in reptiles presented to a wildlife care facility in Central Illinois. *Human-Wildlife Interactions*. 8:78–87.

Romeijer C, Beaufrere H, Laniesse D et al. 2016. Vomiting and gastrointestinal obstruction in a red-footed tortoise (*Chelonoidis carbonaria*). *J Herp Med Surg* 26:32–35.

Roosenburg WM, Cresko W, Modesitte M, and Robbins MB. 1997. Diamondback terrapin (*Malaclemys terrapin*) mortality in crab pots. *Conserv Biol* 11:1166–1172. https://www.jstor.org/stable/2387399 (accessed May 13, 2019)

Rossi JV. 2006. General husbandry and management, in *Reptile Medicine and Surgery*, 2nd ed. Mader DR (Ed.), Saunders/Elsevier Inc, St. Louis, pp. 25–41.

Sack A, Butler E, Cowen P, and Lewbart GA. 2017. Morbidity and mortality of wild turtles at a North Carolina Wildlife Clinic: A 10-year retrospective. *J Zoo Wildl Med* 48:716–724.

Scheelings TF. 2015. Morbidity and mortality of reptiles admitted to the Australian Wildlife Health Centre, Healesville Sanctuary, Australia, 2000–13. *J Wildl Dis* 51:712–718.

Schrader G, Allender MC, and Odoi A. 2008. A retrospective study of initial complaints and diagnoses of box turtles (*Terrapene carolina carolina*) presented to a wildlife clinic in Tennessee from 1995–2007, in *Proceedings of the Association of Reptilian and Amphibian Veterinarians*, Los Angeles, CA, p. 82.

Seebacher F, and Franklin CE. 2005. Physiological mechanisms of thermoregulation in reptiles: A review. *J Comp Physiol B* 175:533–541.

Shilton C, Brown GP, Chambers L et al. 2014. Pathology of Runting in Farmed Saltwater Crocodiles (*Crocodylus porosus*) in Australia. *Vet Pathol* 51:1022–1034.

Smith A, Meulders B, Bull CM, and Driscoll D. 2012. Wildfire-induced mortality of Australian reptiles. *Herpetology Notes* 5:233–235.

Stockman J, Weber ES, Innis C, O'Sullivan-Brisson J, and Solano M. 2010. Prevalance and distribution of lung radiographic abnormalities in cold stunned Kemp's Ridley sea turtles (*Lepidochelys kempii*): 102 cases, in *Proceedings of the Association of Reptilian and Amphibian Veterinarians*, South Padre Island, TX, p. 42.

Takacs Z, Wilhelmsen KC, and Sorota S. 2001. Snake alpha-neurotoxin binding site on the Egyptian Cobra (*Naja haje*) nicotinic acetylcholine receptor is conserved. *Mol Biol Evol* 18:1800–1809.

Turner FB, Medica PA, Lannom JR, and Hoddenbach GA. 1969. A demographic analysis of continuously irradiated and nonirradiated populations of the lizard, *Uta stansburiana*. *Radiat Res* 38:349–356.

Viada ST, Hammer RM, Racca R et al. 2008. Review of potential impacts to sea turtles from underwater explosive removal of offshore structures. *Environmental Impact Assessment Review* 28:267–285. http://www.seaturtle.org/PDF/Viada_2008_EnvironImpactAssessRev.pdf (accessed February 4, 2017).

Waffa BJ. 2015. Management of quadrate bone trauma in a green tree Python (*Morelia viridis*), in *Proceedings of the Association of Reptilian and Amphibian Veterinarians*, San Antonio, TX, pp. 81–84.

Warwick C, Arena P, Lindley S, Jessop M, and Steedman C. 2013. Assessing reptile welfare using behavioral criteria. *In Practice* 35:123–131.

Wilcox C, Puckridge M, Schuyler QA, Townsend K, and Hardesty BD. 2018. *A Quantitative Analysis Linking Sea Turtle Mortality and Plastic Debris Ingestion*. Scientific Reports 8. #12356. https://www.nature.com/articles/s41598-018-30038-z (accessed on February 27, 2019).

Williams R, Wright AJ, Ashe E et al. 2015. Impacts of anthropogenic noise on marine life: Publication patterns, new discoveries, and future directions in research and management. *Ocean and Coastal Management* 115:17–24. http://www.dx.doi.org/10.1016/j.ocecoaman.2015.05.021 (accessed on February 28, 2017).

Wills S, Beaufrere H, Watrous G, Oblak ML, and Smith DA. 2016. Proximal duodenoileal anastomosis for treatment of small intestinal obstruction and volvulus in a green Iguana (*Iguana iguana*). *J Am Vet Med Assoc* 249:1061–1066.

Work TM, and Balazs GH. 2010. Pathology and distribution of sea turtles landed as bycatch in the Hawaii-based North Pacific pelagic longline fishery. *J Wildl Dis* 46:422–432.

Zehnder A, Koski M, Paul-Murphy J et al. 2010. Clinical treatment and pathology of severe burns in *Geochelone sulcata* and *Geochelone pardalis*: A case series, in *Proceedings of the Association of Reptilian and Amphibian Veterinarians*, South Padre Island, TX, p. 81.

Zimmerman DM, Douglass M, Sutherland-Smith M, Aguilar R, Schaftenaar W, and Shores A. 2009. Compressive myelopathy of the cervical spine in Komodo Dragons (*Varanus komodensis*). *J Zoo Wildl Med* 40:207–210.

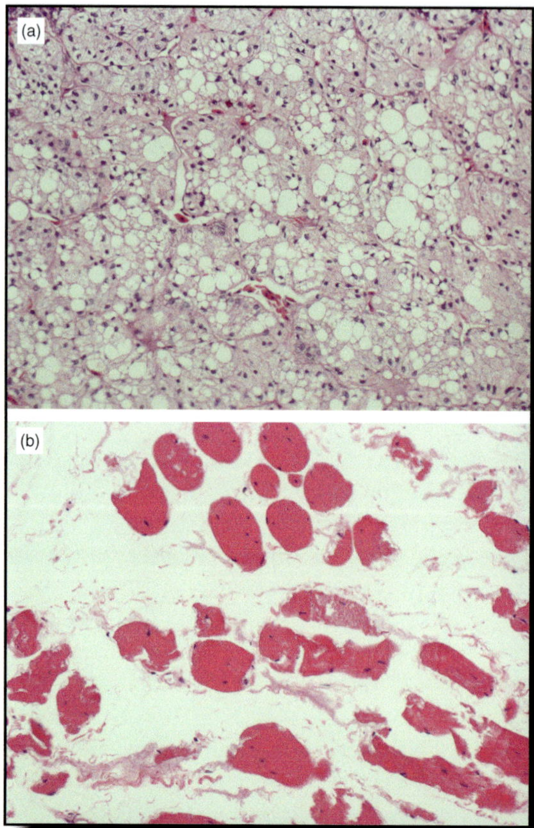

Figure 6.1 Histologic images of stress. (a) Green Anole, *Anolis carolinensis*, Dactyloidea. Interrenal cell hypertrophy and vacuolar change typical of a stress response. (b) Eastern Musk Turtle, *Sternotherus odoratus*, Kinosternidae. Myofiber swelling, loss of cross striations, and fragmentation with associated edema, typical of stress-related or exertional myopathy. (Courtesy of Michael Garner.)

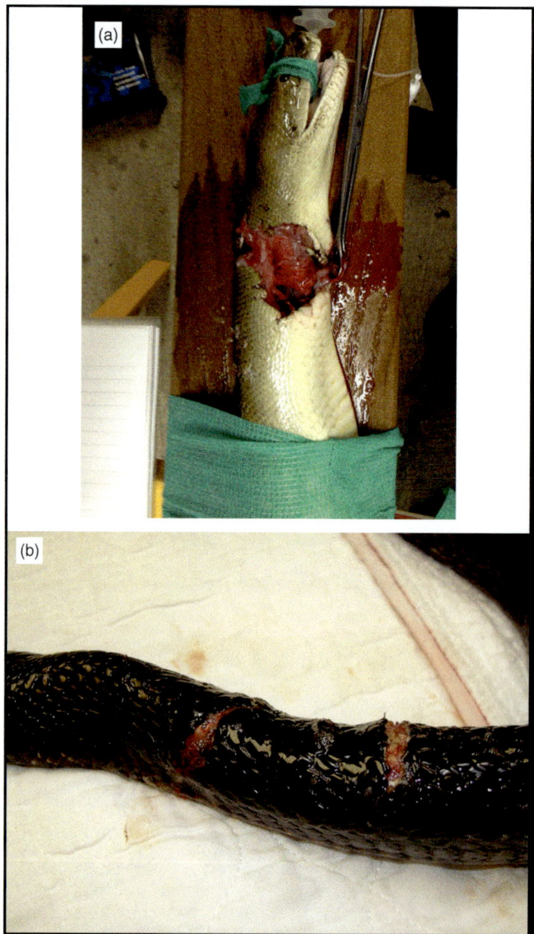

Figure 6.2 (a) Burmese Python, *Python bivittatus*. Pythonidae. Severe bite wounds with a lacerated trachea caused by a Reticulated Python, *Python reticulatus*. Pythonidae. (Courtesy of James Bogan.) (b) Indigo Snake, *Drymarchon melanurus rubidus*. Colubridae. Multiple bite wounds from a male conspecific. (Courtesy of Elliott Jacobson.)

Figure 6.3 Cuban Crocodile, *Crocodylus rhombifer*. Crocodylidae. Juvenile with mandibular fracture subsequent to biting a conspecific. (Courtesy of Adam Stern.)

Figure 6.4 (a) Male Desert Tortoise, *Gopherus agassizii*. Testudinidae. Substantially developed gular projections for overturning opponent tortoises. (Courtesy of Adam Stern.) (b) Yellow-Footed Tortoise, *Chelonoidis denticulata*. Testudinidae. Male ramming and pushing another male in un-returned combat effort. (Courtesy of Adam Stern.)

Figure 6.5 (a) Hermann's Tortoises, *Testudo hermanni*. Testudinidae. Three young tortoises from the same clutch raised on the same diet in the same enclosure with obviously different growth rates. (Courtesy of Doug Mader.) (b) Emydidae. Competition for food among numerous turtles in a pond can result in bite and scratch wounds. (Courtesy of Doug Mader.)

Figure 6.6 (a) Galapagos Tortoises, *Chelonoidis nigra*. Testudinidae. Male mounting a female (note front foot placement). (Courtesy of Ron Magill.) (b) Chronic eburnation and thinning of female's carapace from persistent mounting and friction from male's feet. (Courtesy of Adam Stern.)

Figure 6.7 (a) Desert Box Turtle, *Terrapene ornata luteola*. Emydidae. Breeding trauma—potential injury to this male's phallus after being overturned during intromission. (Courtesy of John Roberts.) (b) African Spurred Tortoise, *Centrochelys sulcata*. Testudinidae. Breeding trauma to this male's phallus. (Courtesy of Heather Bjornebo.) (c) Radiated Tortoise, *Astrochelys radiata*. Testudinidae. Phallus prolapse and trauma. (Courtesy of Doug Mader.)

Figure 6.8 (a) Fly River Turtle, *Carettochelys insculpta*. Carettochelyidae. Turtle with bacterial keratitis, but note cervical skin and marginal carapace bite wounds from conspecifics. (b) Mountain Horned Lizard, *Acanthosaura capra*. Agamidae. Several very deep bite wounds from conspecific due to overcrowding. (Courtesy of Tim Tristan.)

Figure 6.9 American Alligator, *Alligator mississippiensis*. Alligatoridae. Individual in a captive group cannibalizing a smaller individual. (Courtesy of James Bogan.)

Figure 6.10 (a) Yellow-Bellied Sliders, *Trachemys scripta scripta*. Emydidae. Markedly overcrowded container with hatchlings. Note individuals piled on top of each other at the far end where some at the bottom could be in danger of drowning if the water were too deep. (Courtesy of Tim Tristan.) (b) Indian Star Tortoises, *Geochelone elegans*. Testudinidae. Overcrowded hatchlings piled up on each other. (Courtesy of Paul Gibbons.)

Figure 6.11 (a) Eastern Diamond-Backed Rattlesnake, *Crotalus adamanteus*. Viperidae. Wild birds mobbing a snake in Everglades National Park. (Courtesy of Kresimir Golenja.) (b) Kemp's Ridley Sea Turtle, *Lepidochelys kempii*. Cheloniidae. Cold-stunned turtle with eye damage from a bird or birds. (Courtesy of Charles Innis.)

Figure 6.12 (a) Leopard Tortoise, *Stigmochelys pardalis*. Testudinidae. Wild tortoise in Okavango region of Botswana found with crush injury to carapace after being stomped on by a large ungulate—presumed to have been an African Buffalo, *Syncerus caffer*. Bovidae. (Courtesy of Rebeccah Mulder.) (b) Green Iguana, *Iguana iguana*. Iguanidae. Captive Spider Monkey, *Ateles geoffroyi*. Atelidae. "Petting" a free-ranging iguana in her enclosure. (Courtesy of Dolora Batchelor.)

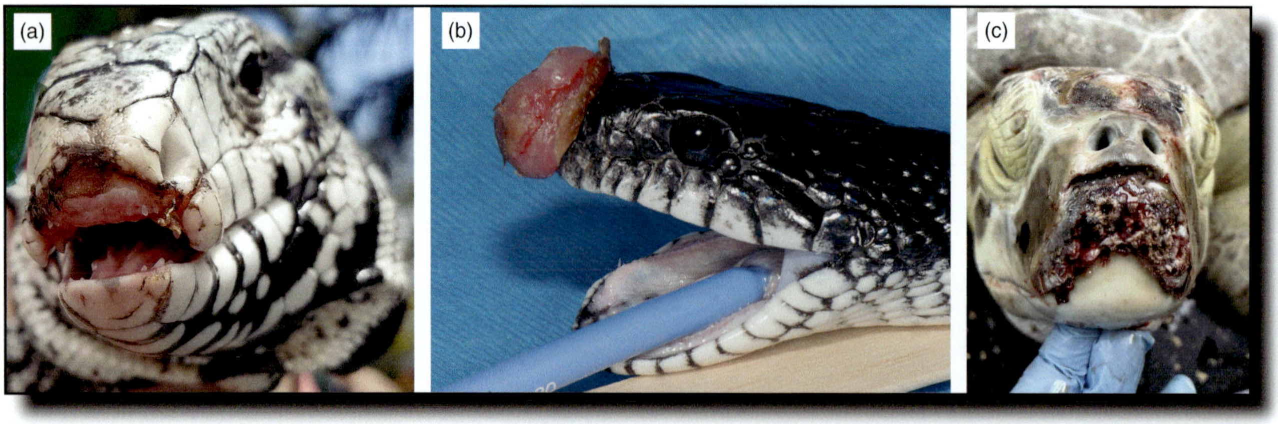

Figure 6.13 (a) Black and White Tegu, *Tupinambis merianae*. Teiidae. Marked chronic rostral damage from repeated attempts to find an escape from enclosure. (Courtesy of Adam Stern.) (b) Gray Ratsnake, *Pantherophis spiloides*. Colubridae. Chronic rostral trauma with proliferative response. (Courtesy of Stephen Barten.) (c) Green Turtle, *Chelonia mydas*. Cheloniidae. Marked chronic trauma and necrosis of the rhamphotheca and dorsal head from rubbing and bumping along tank walls. (Courtesy of Brooke Burkhalter.)

Figure 6.14 (a) Fly River Turtle, *Carettochelys insculpta*. Carettochelyidae. Abrasion of the skin of the carapace from bumping along an overhanging ledge in the pool. (b) Red-Footed Tortoise, *Chelonoidis carbonaria*. Testudinidae. Multifocal, superficial scratch trauma to the shell from rubbing along enclosure furniture. (Courtesy of Doug Mader.)

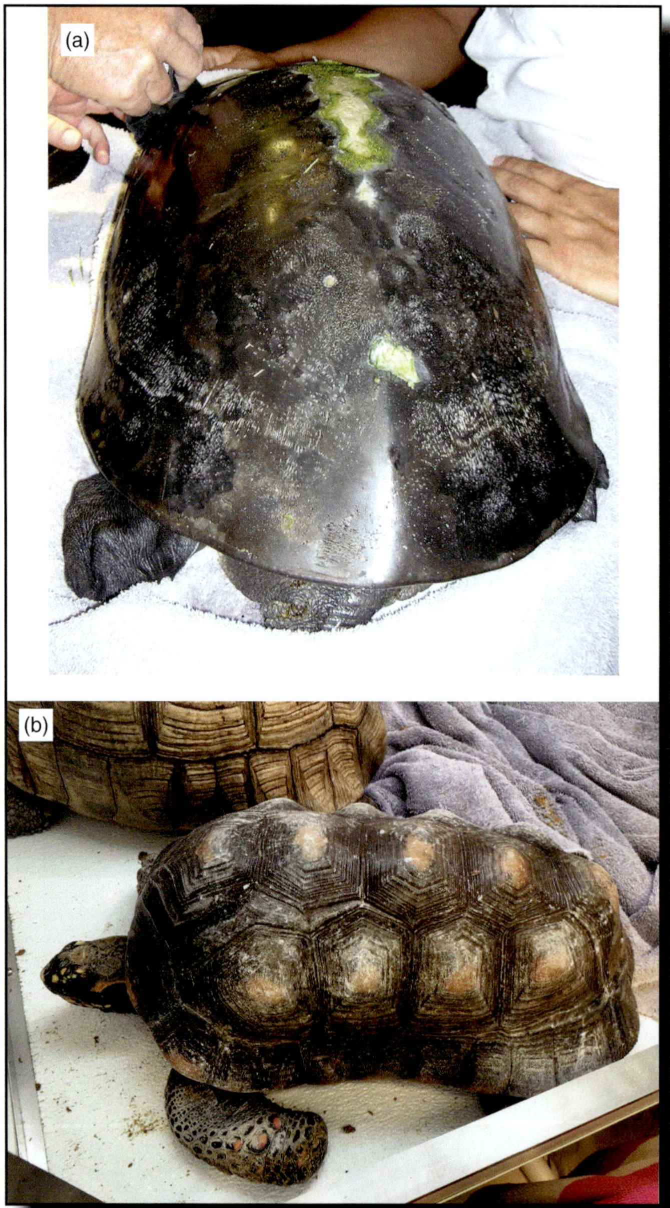

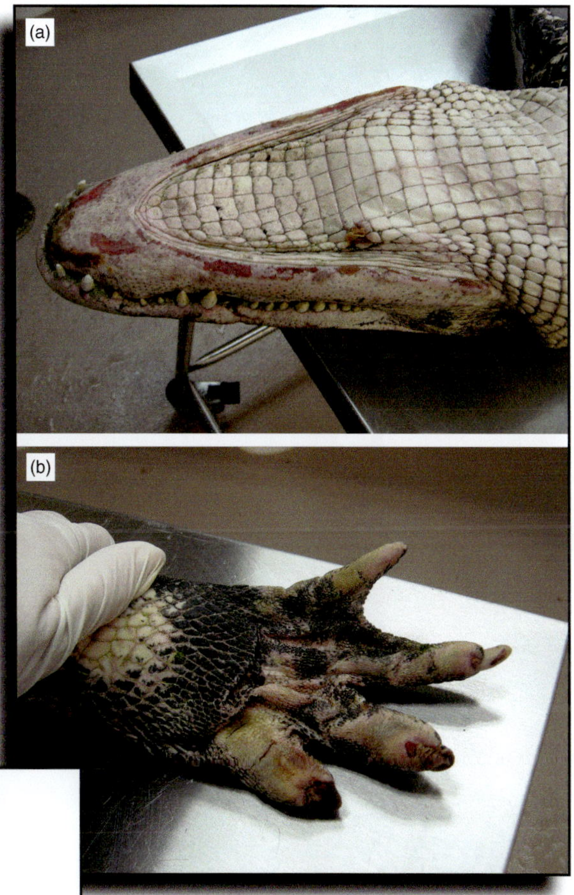

Figure 6.15 (a) Black Caiman, *Melanosuchus niger*. Alligatoridae. Ventral mandibular abrasions from rubbing along the concrete pool and flooring. (b) Digit abrasions in the same individual.

Figure 6.16 (a) Green Basilisk, *Basiliscus plumifrons*. Corytophanidae. Tail injury with necrosis from striking objects or walls within enclosure. (Courtesy of Adam Stern.) (b) Leopard Gecko, *Eublepharis macularius*. Eublepharidae. Recent tail autotomy. (Courtesy of James Bogan.)

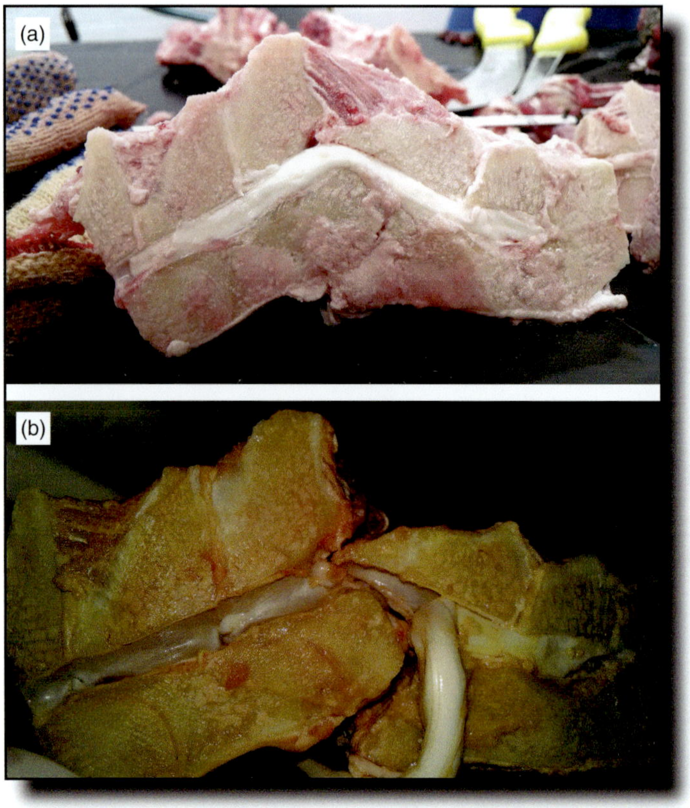

Figure 6.17 Komodo Dragon, *Varanus komodoensis*. Varanidae. Vertebral fracture with consequent spinal cord compression demonstrated on necropsy—spinal cord in place (a), and spinal cord partially removed to reveal narrowing of spinal canal (b). (Courtesy of Kim Cook.)

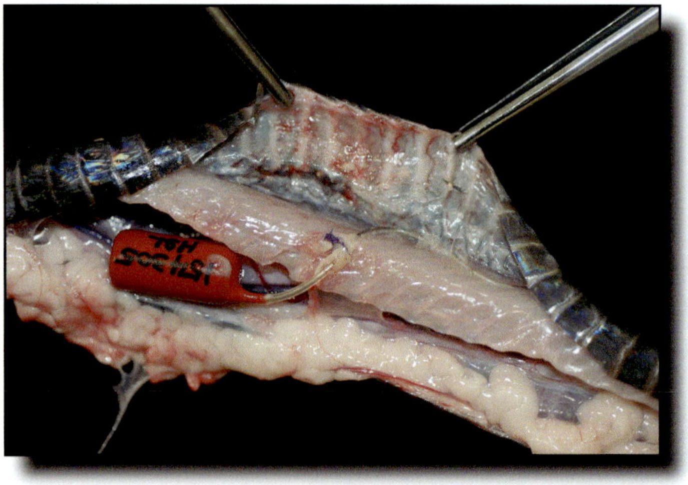

Figure 6.18 Indigo Snake, *Drymarchon couperi*. Colubridae. Localized infection such as this one from an implanted transmitter may lead to self-biting or self-mutilation behavior. (Courtesy of John Roberts.)

Figure 6.19 (a) Western Diamondback Rattlesnake, *Crotalus atrox*. Viperidae. Mature snake inescapably adhered to the walls in a glue trap. (b) California Kingsnake, *Lampropeltis getula californiae*, Colubridae; and two unspecified Whiptails, *Aspidoscelis* spp., Teiidae; and an unidentified Gecko. Glue traps indiscriminately trap animals, as seen here where several reptiles and invertebrates have become victims. (Photographs courtesy of Bryan Hughes.)

Figure 6.20 (a) Green Turtle, *Chelonia mydas*. Cheloniidae. Young turtle entangled in a beverage six-pack ring. (Courtesy of Doug Mader.) (b) Green Turtle, *Chelonia mydas*. Cheloniidae. Fishing line entanglement. (Courtesy of Doug Mader.) (c) Leatherback Sea Turtle, *Dermochelys coriacea*. Dermochelyidae. Fatal entanglement in fixed-pot fishing gear. (Courtesy of Charles Innis.)

Figure 6.21 (a) Gray Ratsnake, *Pantherophis spiloides*. Colubridae. Wild snake entangled in garden netting. (b) Indigo Snake, *Drymarchon couperi*. Colubridae. Wild snake entrapped in sod netting that has been exposed by a roadside. (Courtesy of Brad Tornwall.) (c) Green Iguana, *Iguana iguana*. Iguanidae. Free-roaming, non-indigenous individual trapped in a chainlink fence. (Courtesy of Adam Stern.)

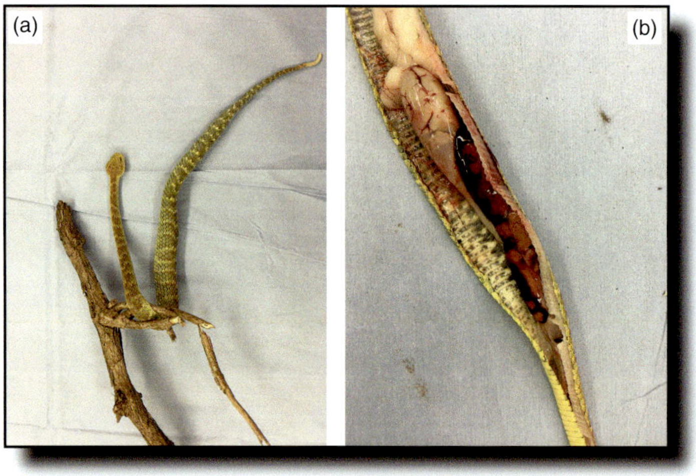

Figure 6.22 African Bush Viper, *Atheris squamigera*. Viperidae. (a) Captive snake trapped in a narrow space between branches of enclosure "furniture," with subsequent mortality. (b) Gross postmortem photograph of the same snake shows hemorrhage and swelling in body and organs caudal to branch entrapment site. (Photographs courtesy of James Bogan.)

Figure 6.23 (a) Komodo Dragon, *Varanus komodoensis*. Varanidae. Tongue injury of unknown cause. (Courtesy of Ginger Sturgeon.) (b) Leopard Gecko, *Eublepharis macularius*. Eublepharidae. Toe and tail damage from being caught in an enclosure lid. (Courtesy of Elizabeth Hyde.)

Figure 6.24 Tiger Ratsnake, *Spilotes pullatus*. Colubridae. Skin and muscle injury to the neck from being aggressively pinned with a snake hook. (Courtesy of Adam Stern.)

Figure 6.25 Green Iguana, *Iguana iguana*. Iguanidae. Fresh tail autotomy. (Courtesy of Doug Mader.)

Figure 6.26 Burmese Python, *Python bivittatus*. Pythonidae. Chronic scarring of injured spectacle. (Courtesy of Lorraine Karpinski.)

Figure 6.27 Gopher Tortoise, *Gopherus polyphemus*. Testudinidae. Individual that has been shot with an arrow, the shaft of which has subsequently been broken off. (Courtesy of Elliott Jacobson.)

Figure 6.28 (a) Eastern Diamondback Rattlesnake, *Crotalus adamanteus*. Viperidae. Hit-by-car mortality. (b) American Crocodile, *Crocodylus acutus*. Crocodylidae. Multiple skull fractures following vehicular collision. (c) Florida Snapping Turtle, *Chelydra serpentina osceola*. Chelydridae. Hit-by-car trauma with shell fractures, and shell eburnation from the tire. (Courtesy of Adam Stern.)

Figure 6.29 (a) Loggerhead Sea Turtle, *Caretta caretta*. Cheloniidae. Severe boat-strike injury. (Courtesy of Doug Mader.) (b) Leatherback Sea Turtle, *Dermochelys coriacea*. Dermochelyidae. Propeller lacerations and mortality from a boat strike. (Courtesy of Karen Dourdeville.) (c) Olive Ridley Sea Turtle, *Lepidochelys olivacea*. Cheloniidae. Wild female with healed front limb amputation, of unknown cause. (Courtesy of Carles Juan Salles.) (d) Olive Ridley Sea Turtle, *Lepidochelys olivacea*. Cheloniidae. Wild female leaving La Escobilla beach in Oaxaca, Mexico, after nesting, with healed partial carapace amputations of undetermined cause. (Courtesy of Carles Juan Salles.)

Figure 6.30 Burmese Python, *Python bivittatus*. Pythonidae. Free-roaming, non-indigenous snake killed by mower strike in tall grass near Miami, Florida. (Courtesy of Rebeccah Mulder.)

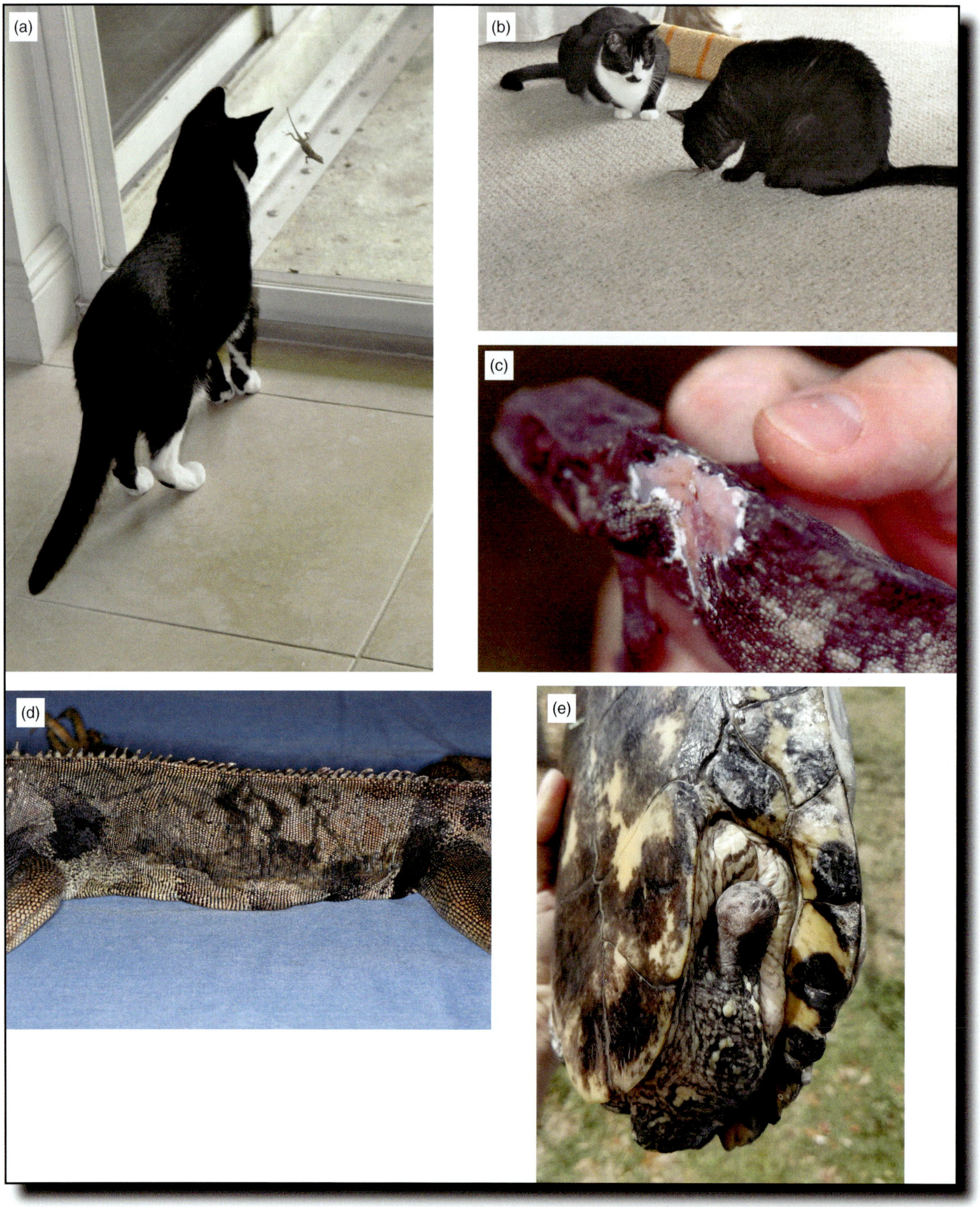

Figure 6.31 (a) Brown Anole, *Anolis sagrei*. Dactyloidae. Housecat (*Felis catus*. Felidae) prepared to pounce on a wild anole in the window. (b) Housecats "playing" with an anole. (c) Panther Chameleon, *Furcifer pardalis*. Chamaeleonidae. Cat bite injury over the side of this individual. (Courtesy of Rob Coke.) (d) Green Iguana, *Iguana iguana*. Iguanidae. Healed dog bite injuries. (Courtesy of Elliott Jacobson.) (e) Red-Eared Slider, *Trachemys scripta elegans*. Emydidae. Healed hindlimb amputation from dog predation. (Courtesy of Adam Stern.) *(Continued)*

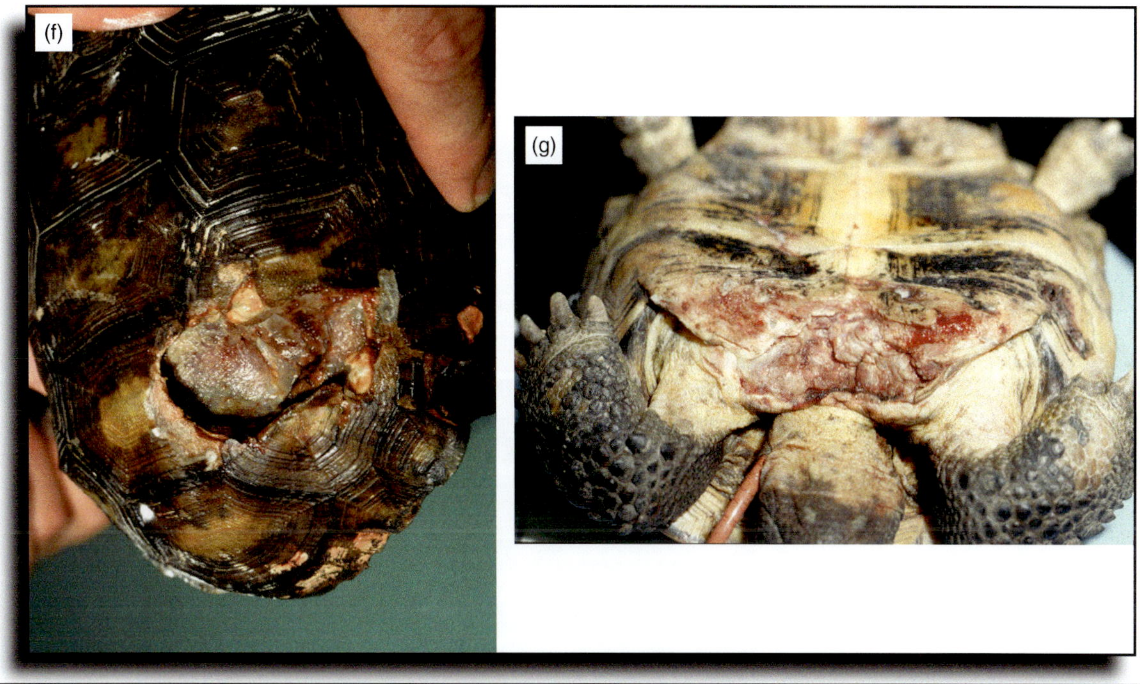

Figure 6.31 (Continued) (f) Gopher Tortoise, *Gopherus polyphemus*. Testudinidae. Young tortoise with recent dog bite injuries to the carapace. (Courtesy of Elliott Jacobson.) (g) Gopher Tortoise, *Gopherus polyphemus*. Testudinidae. Tortoise with recent dog bite trauma to the plastron, with mangled cranial bones and soft tissues. (Courtesy of Elliott Jacobson.)

Figure 6.32 (a) Red Imported Fire Ants, *Solenopsis invicta*. Formicidae. Ants at nest entrance. (b) Gopher Tortoise, *Gopherus polyphemus*. Testudinidae. Juvenile with facial, neck, and eyelid swelling due to fire ant bites. (Courtesy of Elliott Jacobson.)

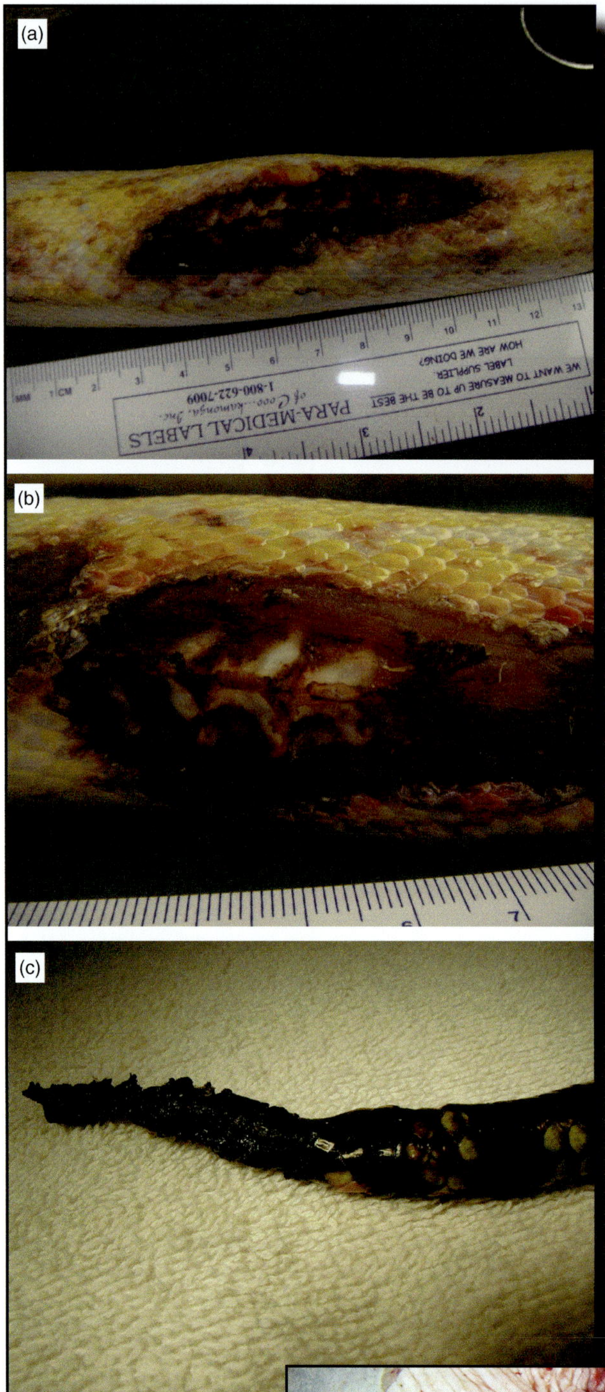

Figure 6.33 (a,b) Burmese Python, *Python bivittatus*. Pythonidae. Rat predation trauma over the dorsum. (Courtesy of Charles Innis.) (c) Carpet Python, *Morelia spilota*. Pythonidae. Rodent bite injury to the tail. (Courtesy of Charles Innis.)

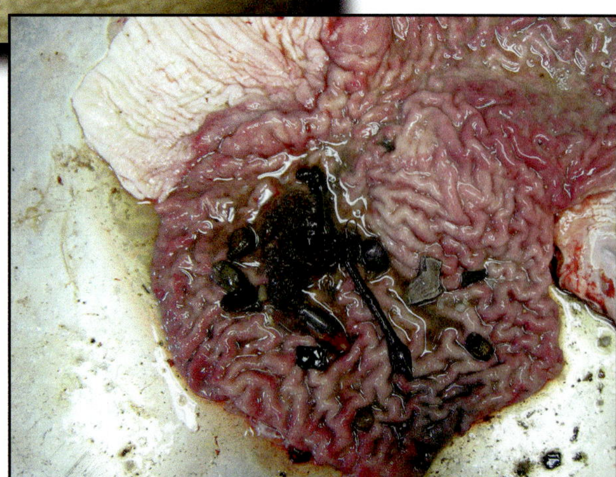

Figure 6.34 American Crocodile, *Crocodylus acutus*. Crocodylidae. Foreign materials found in the stomach during a necropsy exam.

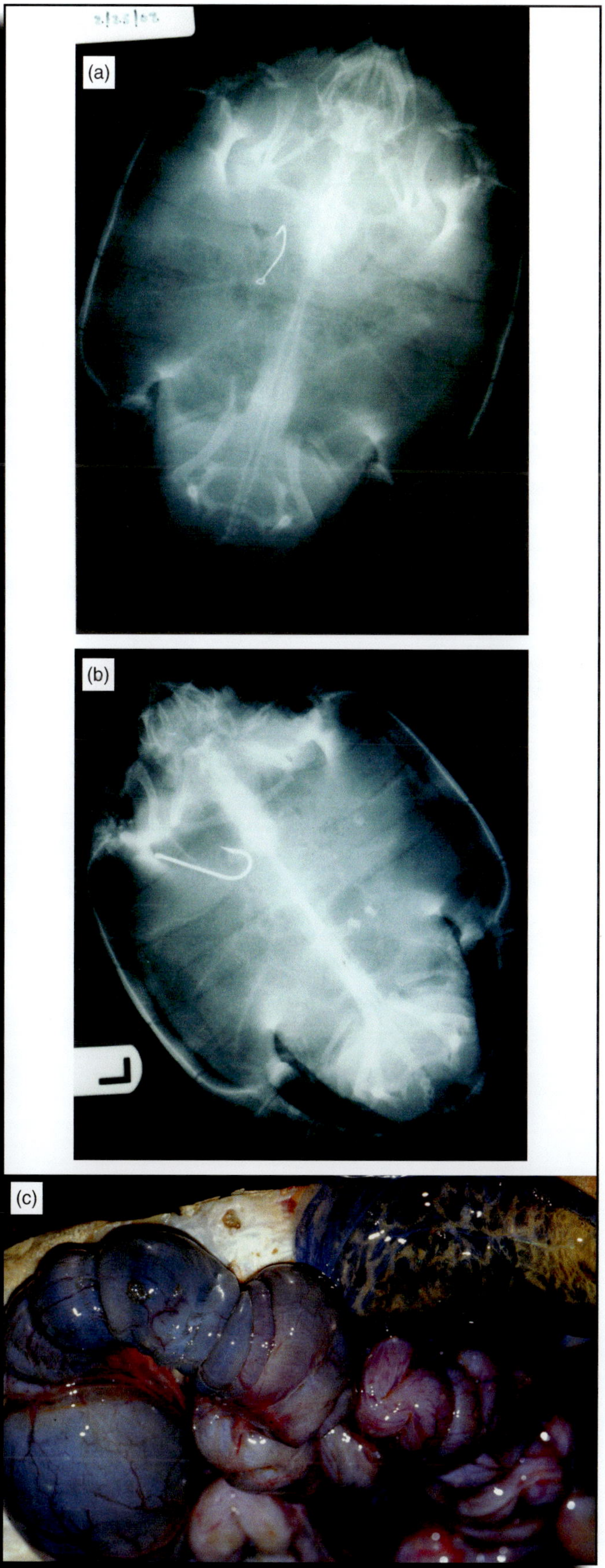

Figure 6.35 (a,b) Malayan Giant Turtles, *Orlitia borneensis*. Geoemydidae. Two radiographs reveal large fishhooks swallowed by these turtles. (c) Loggerhead Sea Turtle, *Caretta caretta*. Cheloniidae. Plicated intestines due to ingested fishing line. (Courtesy of Brian Stacy.)

Figure 6.36 (a) Loggerhead Sea Turtle, *Caretta caretta*. Cheloniidae. Fish hook chronically embedded in the mouth. (Courtesy of Doug Mader.) (b) Florida Softshell Turtle, *Apalone ferox*. Trionychidae. Individual with fish hook lodged in palate. (Courtesy of Elliott Jacobson.)

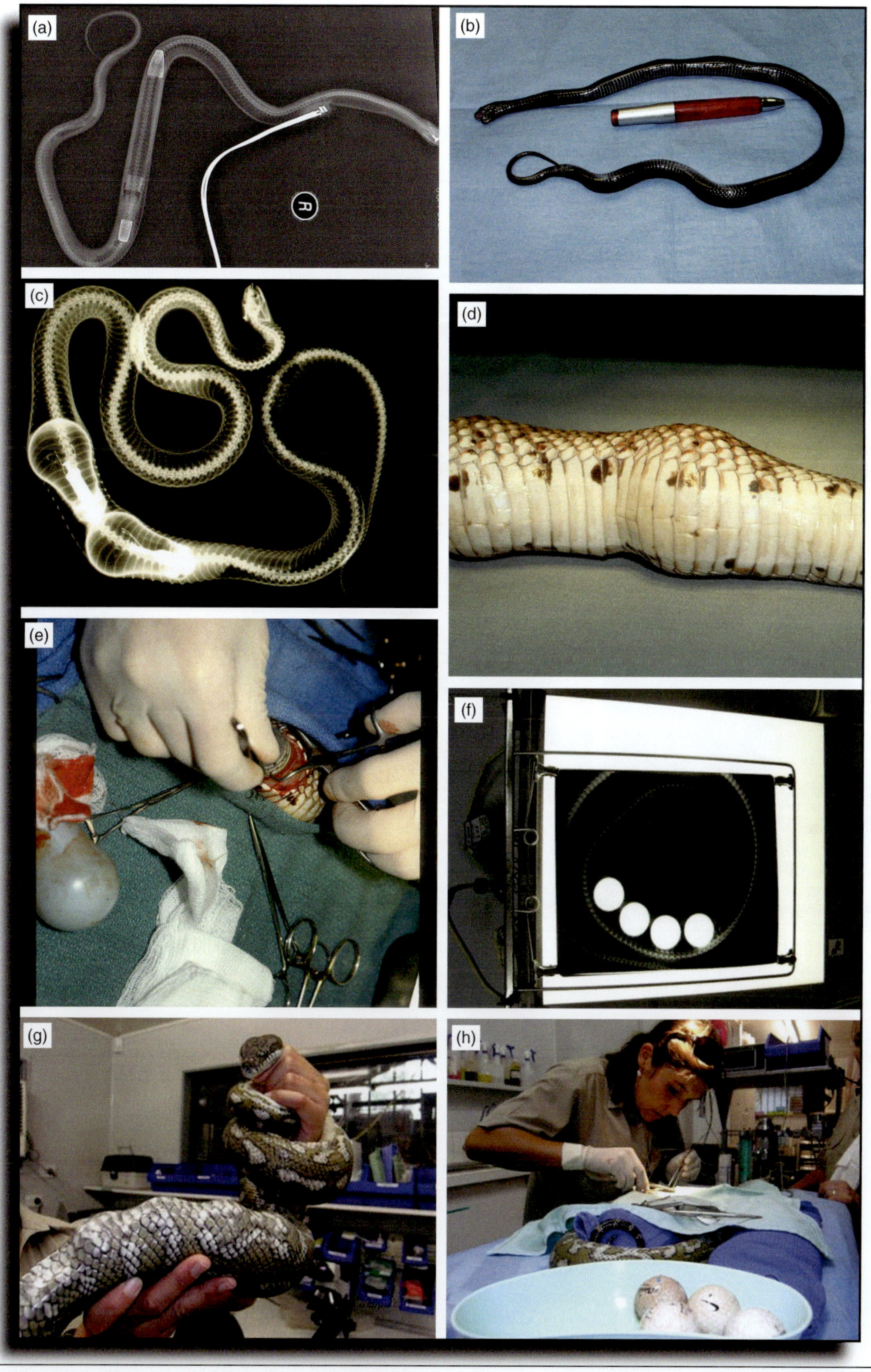

Figure 6.37 Ingested foreign bodies causing obstruction or perforation of the gastrointestinal tracts in snakes. (a,b) Black Kingsnake, *Lampropeltis nigra*. Colubridae. Radiograph of the whole snake with an ingested pen, and image of the same snake with the extracted pen beside it. (Courtesy of Elliott Jacobson.) (c–e) Florida Pine Snake, *Pituophis melanoleucus mugitus*. Colubridae. Light bulb ingestion. Radiograph of snake with ingested light bulbs, expanded body segment over bulbs, and image showing surgical removal of the foreign bodies. (Courtesy of Elliott Jacobson.) (f–h) Carpet Python, *Morelia spilota*. Pythonidae. Golf ball ingestion. Radiograph of snake with four ingested golf balls, enlarged body segment overlying the ingested foreign bodies, and image showing surgical removal of the golf balls. (Courtesy of Michael Pyne.)

Figure 6.38 Polyacrylamide gel particles in dehydrated and hydrated state demonstrating the potential for gastrointestinal obstruction if a dried particle is ingested by a small reptile or other animal and rehydrated within the tract.

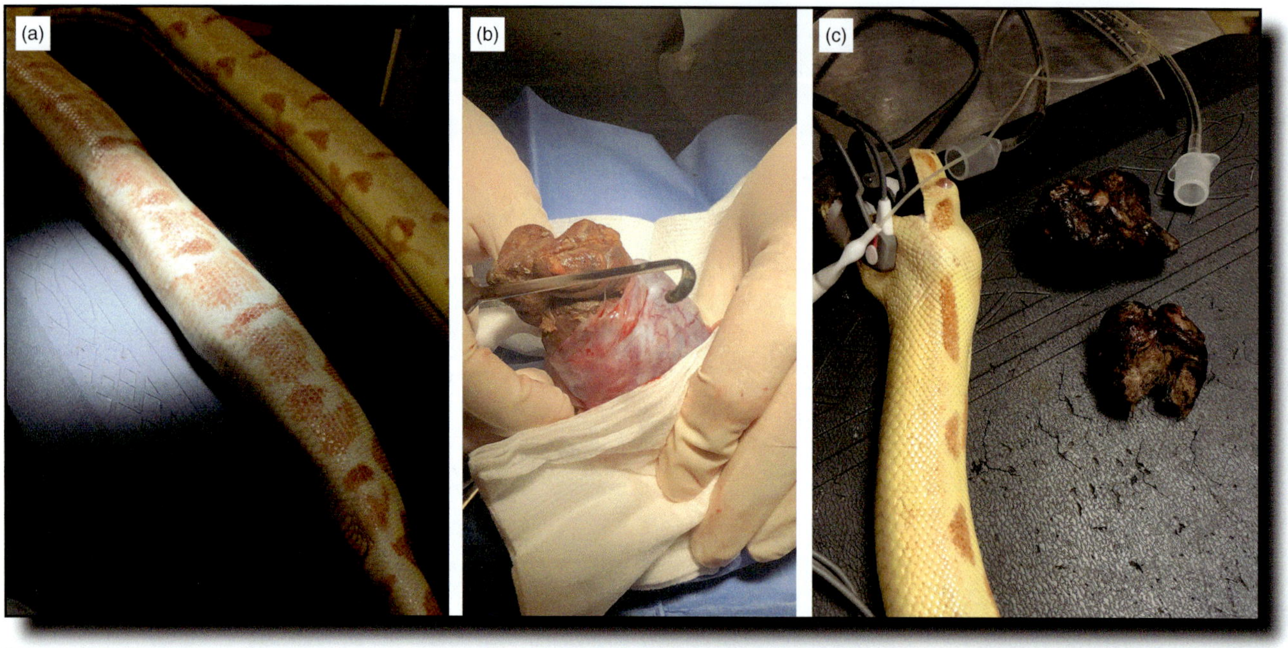

Figure 6.39 (a–c) Boa constrictor, *Boa constrictor*. Boidae. This chronically obstipated snake failed to move its bowels in response to medical treatment but recovered well following surgical evacuation of the colon. (Courtesy of Rob Coke.)

Figure 6.40 Kemp's Ridley Sea Turtle, *Lepidochelys kempii*. Cheloniidae. Corneal ulcer in a cold-stunned turtle. (Courtesy of Tim Tristan.)

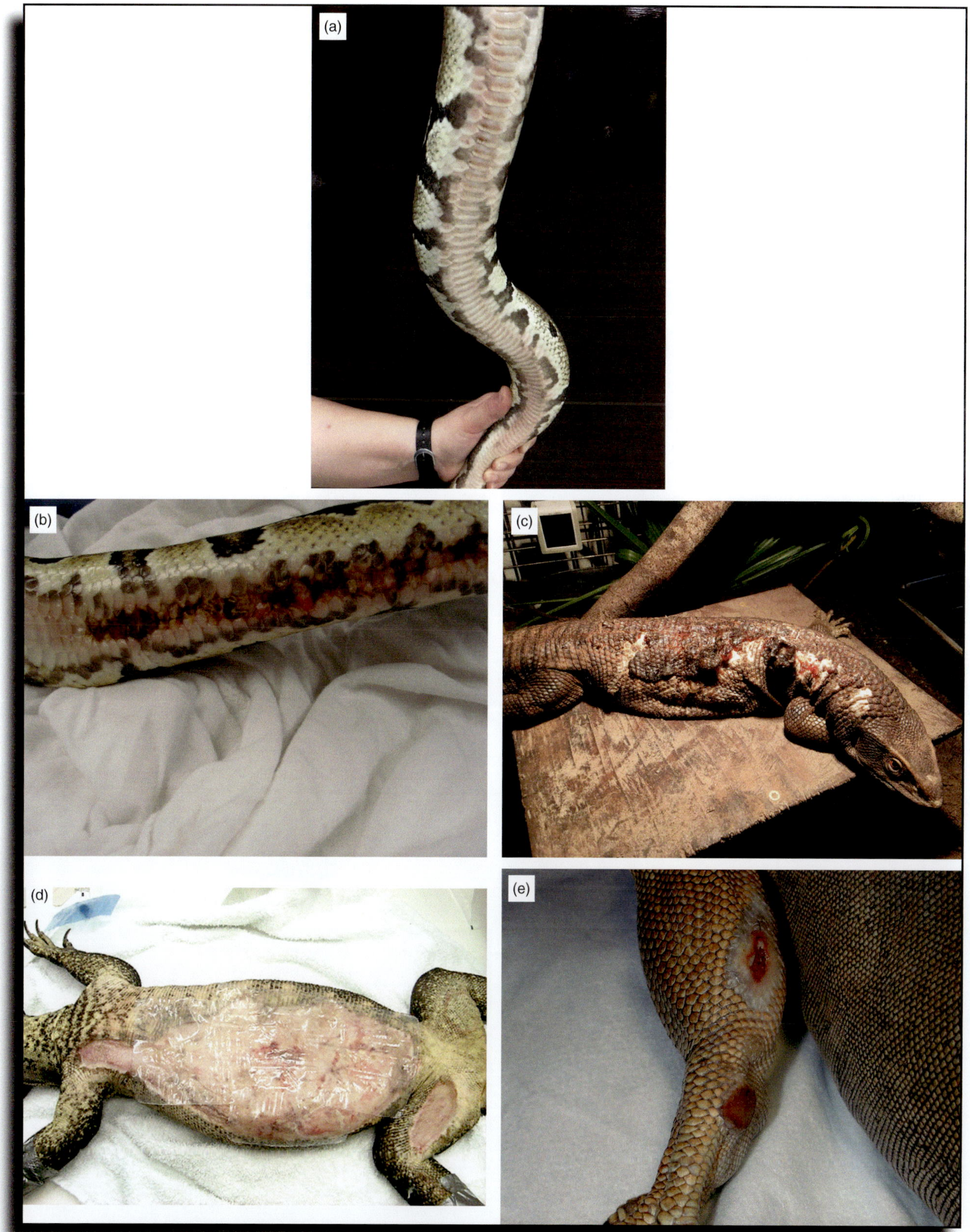

Figure 6.41 (a) Ball Python, *Python regius*. Pythonidae. First-degree burns along the ventrum of this snake. Note erythema of the injured scales. (Courtesy of James Bogan.) (b) Ball Python, *Python regius*. Pythonidae. Second- and third-degree burns along the ventrum. Note partial to full-thickness necrosis of the more severely injured scales. (Courtesy of Charles Innis.) (c) Savannah Monitor, *Varanus exanthematicus*. Varanidae. Third-degree burns over the side and back. (Courtesy of James Bogan.) (d) Komodo Dragon, *Varanus komodensis*. Varanidae. Third-degree burns on ventral trunk and legs—debrided. (Courtesy of Susan Clubb.) (e) Green Iguana, *Iguana iguana*. Iguanidae. Third-degree burns on the forelimb. (Courtesy of Paul Gibbons.)

Figure 6.42 (a,b) Desert Tortoise, *Gopherus agassizii*. Testudinidae. Injuries to the shell and hindlimbs sustained during a wildfire. (Courtesy of John Roberts.) (c) Painted Turtle, *Chrysemys picta*. Emydidae. Subacute injury to the carapace from a fire, showing edema and lifting of carapace scutes shortly before sloughing. (Courtesy of Charles Innis.) (d–f) Blanding's Turtle, *Emydoidea blandingii*. Emydidae. Sequential photographs over the course of healing of a large area of carapace that was burned showing recent scute sloughing, followed by granulation and re-epithelialization. (Courtesy of Steve Barten.) (g) Desert Tortoise, *Gopherus agassizii*. Testudinidae. Chronic scarring of the carapace from old wildfire injury. (Courtesy of Elliott Jacobson.)

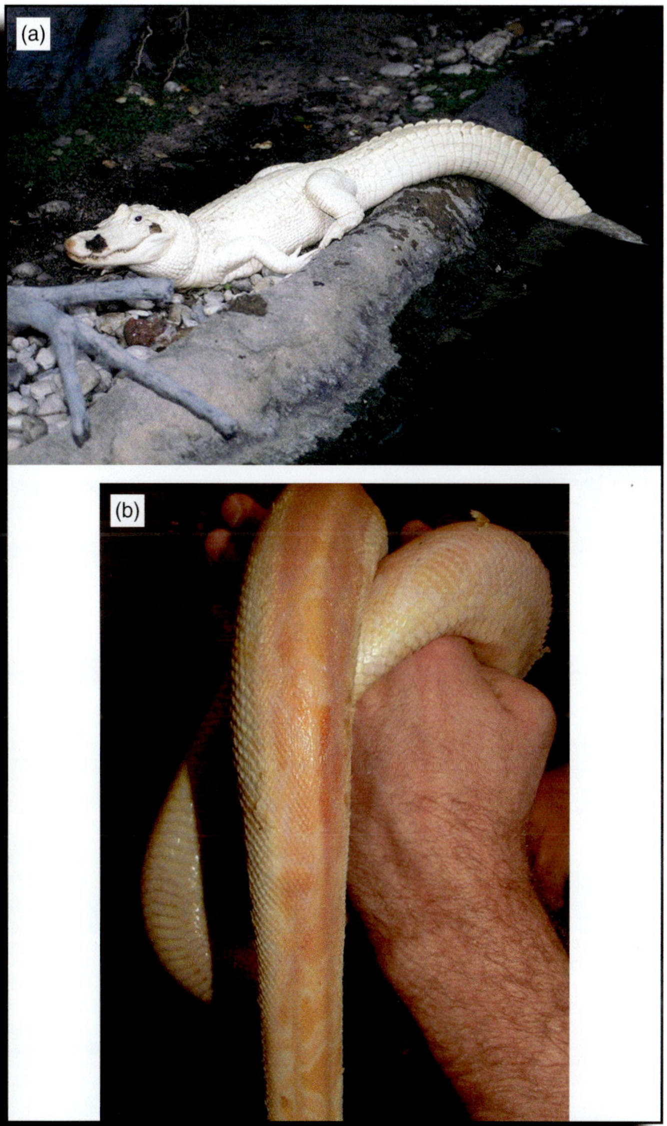

Figure 6.43 (a) American Alligator, *Alligator mississippiensis*. Alligatoridae. Leucistic individual housed in a shaded exhibit to avoid sunburn. (Courtesy of Ron Magill.) (b) Burmese Python, *Python bivittatus*. Pythonidae. Albino individual with ultraviolet exposure burns from an enclosure UVB bulb. (Courtesy of John Roberts, from Kevin Wright).)

Figure 6.44 (a) Childreni Python, *Antaresia children*. Pythonidae. Young snake drinking from water drops on glass-walled enclosure. (Courtesy of Adam Stern.) (b) Four-Horned Chameleon, *Trioceros quadricornis*. Chamaeleonidae. This female is reaching for a water droplet on the artificial foliage in her enclosure. (Courtesy of Paul Gibbons.)

Figure 6.45 (a) Sinaloan Milksnake, *Lampropeltis triangulum sinaloae*. Colubridae. Individual soaking in a water dish in its enclosure prior to ecdysis. (Courtesy of Adam Stern.) (b) Galapagos Tortoise, *Chelonoidis nigra*. Testudinidae. Wild tortoise soaking in a small pool on ranchland on Santa Cruz Island.

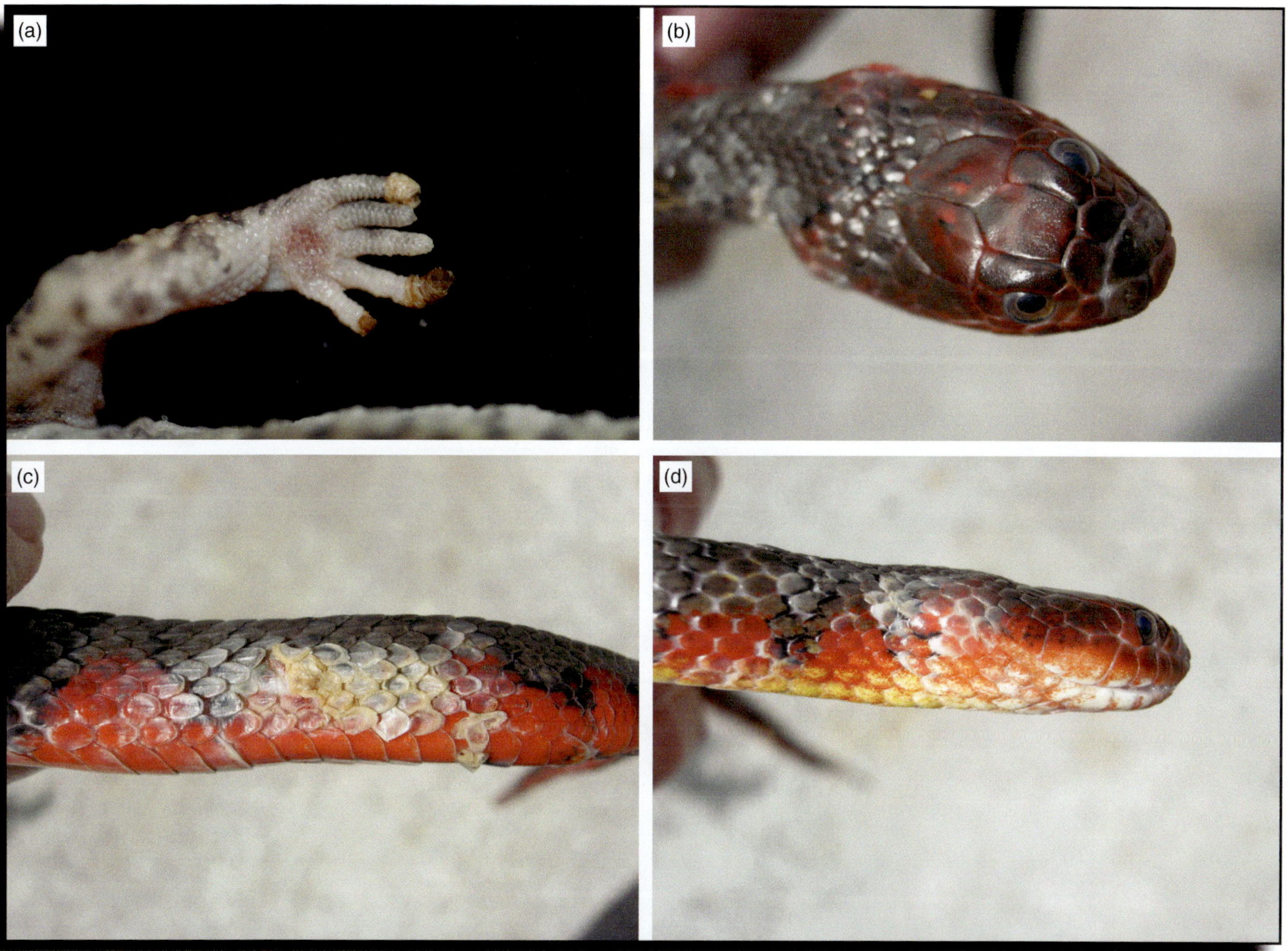

Figure 6.46 (a) Leopard Gecko, *Eublepharis macularius*. Eublepharidae. Retained shed causing constriction around toes. Note other toe tips have already been lost and have healed. (Courtesy of John Roberts.) (b–d) Mountain Keelback Watersnake, *Helicops angulatus*. Colubridae. Dysecdysis and partially retained shed causing obstructed nares and layers of redundant skin in patches over the body. Note the bubbles of saliva between the upper and lower labial scales (d).

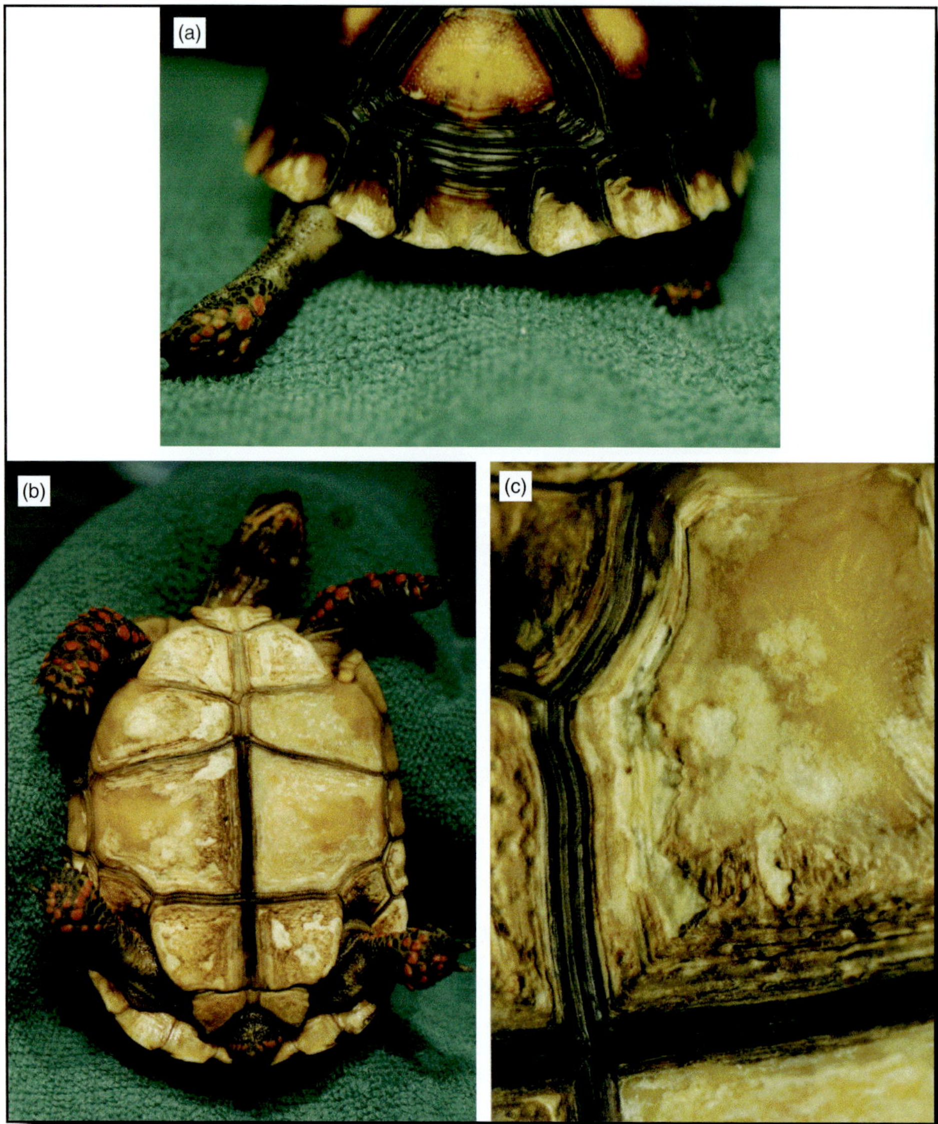

Figure 6.47 (a–c) Red-Footed Tortoise, *Chelonoidis carbonaria*. Testudinidae. "Shell rot" (keratolytic degenerative changes) in the marginal scutes and plastron of a juvenile tortoise—a portion of the plastron has just been debrided in image (b).

Figure 6.48 Russian Tortoise, *Testudo horsfieldii*. Testudinidae. Individual with overgrown toenails. (Courtesy of Adam Stern.)

Figure 6.49 (a,b) Box Turtles, *Terrapene carolina*. Emydidae. Overgrown rhamphotheca (beak). (c) Russian Tortoise, *Testudo horsfieldii*. Testudinidae. Overgrown rhamphotheca (beak). (Photographs courtesy of Charles Innis.)

Figure 6.50 Galapagos Green Turtle, *Chelonia agassizii*. Cheloniidae. Fatal stranding of undetermined cause.

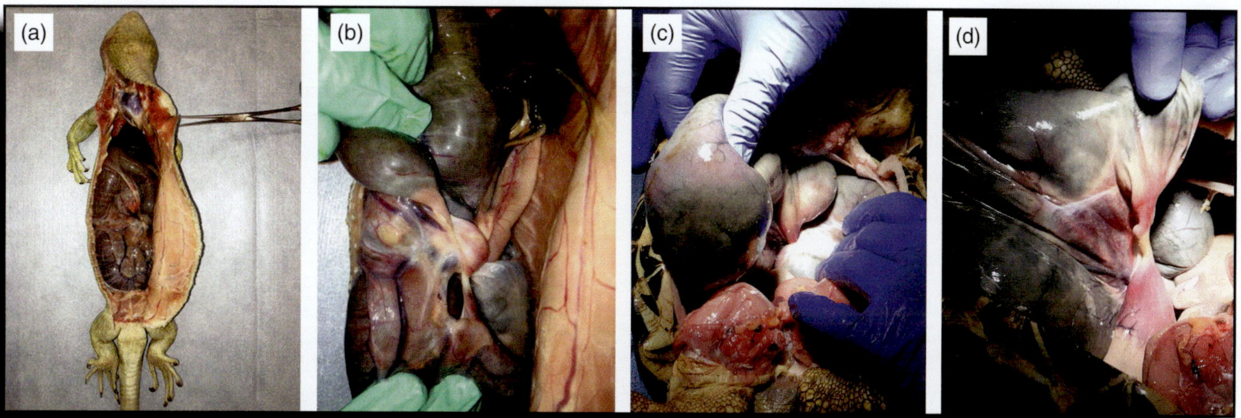

Figure 6.51 (a,b) Prehensile Tailed Skink, *Corucia zebrata*. Scincidae. Mesenteric volvulus. (Courtesy of James Bogan.) (c,d) Radiated Tortoise, *Astrochelys radiata*. Testudinidae. Mesenteric torsion with severe intestinal dilatation and congestion. (Courtesy of Cord Offerman.)

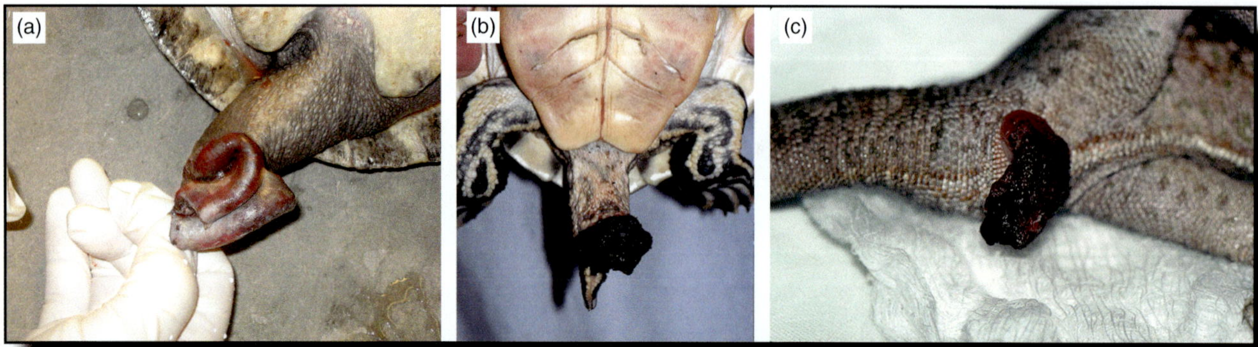

Figure 6.52 (a) Yellow Spotted Amazon River Turtle, *Podonemis unifilis*. Podocnemididae. Injury to a prolapsed penis from fish bites. (Courtesy of Charles Innis.) (b) Yellow-Bellied Slider, *Trachemys scripta scripta*. Emydidae. Penile prolapse with subsequent exposure injury. (Courtesy of Elliott Jacobson.) (c) Veiled Chameleon, *Chamaeleo calyptratus*. Chamaeleonidae. Hemipenile prolapse with exposure damage. (Courtesy of Elliott Jacobson.)

Figure 6.53 California Kingsnake, *Lampropeltis getula californiae*. Colubridae. Impacted cloacal scent glands. (Courtesy of Richard Funk.)

Figure 6.54 Red-Footed Tortoise, *Chelonoidis carbonaria*, Testudinidae. Overturned male potentially stuck in soft substrate or from nearby obstacles such as tree roots or branches. (Courtesy of Adam Stern.)

7

ENVIRONMENTAL AND MISCELLANEOUS TOXICOSES IN REPTILES

JORGE ORÓS, MARÍA CAMACHO, AND OCTAVIO P. LUZARDO

Contents

7.1 General Comments

Until the 1990s, there were very few reports of environmental pollutants measured in reptiles. Since then, studies on contaminants in reptiles have increased, but to a lesser extent than attention to contaminants in mammals, birds, fish, and amphibians. Most research has focused on the study of pollutant levels (reviewed, among others, by de Solla, 2010; Grillitsch and Schiesari, 2010; Pauli et al., 2010, and Keller, 2013a), with fewer studies of their biological effects on reptiles (reviewed, among others, by Burger, 2006; Keller et al., 2006b; McClellan-Green et al., 2006; Novillo et al., 2006; Willingham, 2006; Grillitsch and Schiesari, 2010; McConnell and Sparling, 2010, and Keller, 2013a). It is difficult to investigate the effects of environmental pollutants on reptiles for several reasons. Free-living reptiles often harbor various pollutants from the environment in which they live, the synergistic effect of these

Table 7.1 Chemicals Included in the United Nations Stockholm Convention on Persistent Organic Pollutants (2020)

	Pesticides	Industrial Chemicals	Unintentional Production
The 12 initial POPs	Aldrin	Hexachlorobenzene	Hexachlorobenzene
	Chlordane	Polychlorinated biphenyls	Polychlorinated biphenyls
	DDT		Polychlorinated dibenzo-*p*-dioxins
	Dieldrin		Polychlorinated dibenzofurans
	Endrin		
	Heptachlor		
	Hexachlorobenzene		
	Mirex		
	Toxaphene		
The new POPs	α-hexachlorocyclohexane	Hexabromobiphenyl	
	β-hexachlorocyclohexane	Hexabromocyclododecane	
	Chlordecone	Hexabromobiphenyl ether and heptabromodiphenyl ether (commercial octabromodiphenyl ether)	
	Lindane	Hexachlorobutadiene	Hexachlorobutadiene
	Pentachlorobenzene	Pentachlorobenzene	Pentachlorobenzene
	Pentachlorophenol and its salts and esters	Polychlorinated naphthalenes	Polychlorinated naphthalenes
	Technical endosulfan and its related isomers	Tetrabromodiphenyl ether and Pentabromodiphenyl ether	
		Perfluoroctane sulfonic acid (PFOS), its salts and perfluoroctane sulfonyl fluoride (PFOSF)	
		Decabromodiphenyl ether (commercial mixture, c-DecaBDE)	
		Short-chain chlorinated paraffins (SCCPs)	

Orange box: Chemical included in the Annex A (to eliminate its production and use).
Yellow box: Chemical included in the Annex B (to restrict its production and use).
Blue box: Chemical included in the Annex C (to reduce its unintentional release).

contaminants is unknown in most cases, biological effects of pollutants are a function of the concentration or contaminant burden, and experimental studies are especially difficult to carry out in protected reptiles.

This chapter reviews the information available in the scientific literature on the effects on the different groups of reptiles of (a) major pollutants of anthropogenic origin, (b) main pesticides causing acute poisoning, (c) biotoxins, and (d) main drug and plant toxicoses (reported in captive reptiles). In addition, the chapter provides several tables summarizing the effects of pollutants of anthropogenic origin on organs and health parameters (**Tables 7.2–7.6**).

7.2 Persistent Organic Pollutants

Persistent organic pollutants (POPs) are defined by the United Nations Stockholm Convention on Persistent Organic Pollutants (2015) as organic chemical (carbon based) substances that remain intact for exceptionally long periods of time, become widely distributed throughout the environment, accumulate in the adipose- or lipid-containing tissue of living organisms, and are toxic to both humans and wildlife. The Stockholm Convention originally recognized 12 POPs in 2001, placed in three categories: pesticides, industrial chemicals, and byproducts. Subsequent additions increased the list of POPs in the Stockholm Convention up to 26 chemicals or chemical classes (**Table 7.1**). In addition, perfluorohexane sulfonic acid (PFHxS), its salts and PFHxS-related compunds, dechlorane plus, and methoxychlor have been proposed for listing under the Stockholm Convention and currently are under review.

7.2.1 Chelonia

7.2.1.1 Organochlorine Pesticides in Freshwater Turtles

Several in vitro experiments were carried out in the 1970s to study the effect of different organochlorine pesticides (OCPs) on active transport across cellular membranes of freshwater turtles. In vitro treatment with aldrin and dieldrin produced inhibition of intestinal mucosa and urinary bladder adenosine triphosphatase (ATPase) activity in Map Turtles (*Graptemys geographica*) (Wells et al., 1974). Phillips and Wells (1974) demonstrated in vitro inhibition of the total dependent ATPases by DDT in liver, intestinal mucosa, urinary bladder, and kidney of five freshwater turtle species, suggesting that active transport is severely affected by the action of DDT upon these ATPase systems. Significant inhibition of the ATPase systems in cellular fractions of tissues from the Red-Eared Slider (*Trachemys* [formerly *Chrysemys*] *scripta elegans*) after in vitro treatment with DDT, DDD, and DDE, was also reported (Witherspoon and Wells, 1975).

The effects of OCPs on embryo development have also been investigated. Bishop et al. (1994) did not find significant correlations between body size (body mass, carapace length, carapace width, plastron length) and concentrations of 4,4′-DDE, mirex, and dieldrin, concluding that body size and clutch characteristics are not reliable predictors of the level of contaminants in Snapping Turtle (*Chelydra serpentina*) eggs. Other authors found no evidence that the reproductive success of Eastern Spiny Softshell Turtles (*Apalone spiniferus spiniferus*) was compromised due to OCP contamination (de Solla et al., 2003).

The potential of some OCPs to act as endocrine disruptors has also been investigated. In an experiment with hormone-binding proteins from oviduct cytosol of Yellow-Bellied Turtle (*Trachemys scripta scripta*) in competitive binding assays with [^{3}H]17β-estradiol, 2,4′-DDT, and toxaphene exhibited the greatest affinity for the binding proteins, concluding that cytosolic and circulating binding proteins bind many environmental contaminants with much less affinity than native steroids (Crain et al., 1998).

While the sex ratio of young Florida Red-Bellied Cooters (*Pseudemys nelsoni*) in Lake Apopka was highly skewed toward females, several had ovotestes, indicative of the feminization abilities of these compounds (Guillette et al., 1994). Portelli et al. (1999) demonstrated experimentally that 4,4′-DDE, at levels that existed in the environment in the Great Lakes did not cause feminization of snapping turtles during embryonic development.

When Willingham and Crews (1999) tested the ability of the pesticides toxaphene, dieldrin, 4,4′-DDD, *cis*-nonachlor, *trans*-nonachlor, 4,4′-DDE, and chlordane to override a male-producing incubation temperature in Red-Eared Sliders, compounds assayed alone that caused significant sex reversal were *trans*-nonachlor, *cis*-nonachlor, 4,4′-DDE, and chlordane. When administered with estradiol, only chlordane caused sex reversal at significant levels. In addition, chlordane did not affect aromatase activity in either the brain or the adrenal-kidney-gonad complex, indirectly suggesting that it is an antiandrogen and exerts its effects by inhibiting the male developmental pathway, rather than interfering with the female developmental pathway (Willingham and Crews, 2000).

Male Red-Eared Slider hatchlings treated in ovo with chlordane exhibited significantly lower testosterone concentrations than controls, whereas chlordane-treated females had significantly lower progesterone, testosterone, and 5α-dihydrotestosterone concentrations than controls. Thus, the effects of these endocrine disruptors extend beyond embryonic development, altering sex steroid physiology in exposed animals (Willingham et al., 2000). In addition, hatchlings of this species orally exposed to low doses of *trans*-nonachlor and chlordane grew significantly more than controls, suggesting a role for pesticides in endocrine disruption that extends beyond sex determination and sex development (Willingham, 2001). Willingham (2004) showed that lower doses of 4,4′-DDE caused more sex reversal than higher doses in this species, suggesting that the effect follows a hermetic dose–response curve (Keller, 2013a).

The disease-causing effects of OCPs on freshwater turtles have not been fully proven. Immunosuppressive effects of low-level exposure to chlordane and endosulfan could have been involved in the pathogenesis of ocular, nasal, or otic infections in Eastern Box Turtles (*Terrapene carolina carolina*) (Tangredi and Evans, 1997). Three OCPs (α-benzene hexachloride, heptachlor epoxide, and oxychlordane) and total OCP concentrations were significantly higher in wild-caught box turtles with aural abscesses compared with turtles without abscesses (Holladay et al., 2001); in addition, reduced serum and hepatic vitamin A levels were observed in the turtles with abscesses, suggesting a possible environmental chemical effect on retinoid metabolism or utilization in wild box turtles, resulting in hypovitaminosis A. In a more recent study on 68 box turtles (40 with aural abscesses and 28 without), there was no correlation between pathologic score and total hepatic OCP concentration. However, pathologic score was positively correlated with 2,4′-DDT. Authors also found an initial increase in vitamin A as the OCP burden increased, followed by a decline as OCP levels increased further (Sleeman et al., 2008). Another study demonstrated that 6 months of exposure to a selected mixture of OCPs (chlordane, aroclor, and lindane) did not influence the formation of squamous metaplasia and aural abscesses in Red-Eared Slider Turtles (Kroenlein et al., 2008).

The identification of sensitive and specific biomarkers of exposure to OCPs in freshwater turtles is a challenge for scientists. The use of plasma vitellogenin as a biomarker for estrogenic xenobiotics in freshwater turtles was first investigated in adult male Red-Eared Slider Turtles (Palmer and Palmer, 1995). The estrogenic induction of the lipoprotein vitellogenin is expressed in all oviparous and ovoviviparous vertebrates. In females, vitellogenin is transported in the circulatory system to the ovaries, where it is incorporated into the developing ovarian follicles as yolk. Males normally have no detectable production of vitellogenin due to their normally low levels of endogenous estrogens. However, their liver is capable of synthesizing and secreting vitellogenin into the blood in response to exogenous estrogen stimulation. While intracelomic injections of 17β-estradiol and diethylstilbestrol induced the most vitellogenin, DDT treatments induced smaller amounts of vitellogenin in a dose-dependent fashion (Palmer and Palmer, 1995). These data indicate that induction of plasma vitellogenin in males may be a useful biomarker of xenobiotic estrogen activity in this freshwater turtle species.

Micronuclei in red blood cells (RBCs) (a common assay for genotoxicity) of European Pond Turtles (*Emys orbicularis*) from an industrial area of Azerbaian were statistically correlated with tissue levels of heptachlor, DDD, hexachlorobenzene, and *trans*-nonachlor (Swartz et al., 2003).

More recently, a significant negative relationship was found between *cis*-chlordane concentration and plasma total thyroxine levels in adult male common snapping turtles (Letcher et al., 2015).

7.2.1.2 Polychlorinated Biphenyls in Freshwater Turtles

Regarding the effects of polychlorinated biphenyls (PCBs) on embryo development, hatchling deformities occurred in common snapping turtle populations with the highest PCB concentrations in eggs (Bishop et al., 1991). Bishop et al. (1994) did not find significant correlations between body size (body mass, carapace length, carapace width, plastron length) and concentrations of PCBs (52, 105, 118, 138, 153, 180, 194, and ∑PCBs). Hebert et al. (1993) identified differences in contaminant composition in adult snapping turtles from sites in the Great Lakes, where differences in reproductive success had been previously reported, suggesting that PCB 105 could be an important contributor to the toxic burden of these turtle populations. However, after evaluating hatching success and deformities in snapping turtle hatchlings using eggs collected from 14 sites in the Great Lakes Basin, the authors did not find any direct association between deformities and higher organochlorine body burden (de Solla et al., 2008).

Bergeron et al. (1994) demonstrated that two PCBs (2′,4′,6′-trichloro-4-biphenylol and 2′,3′,4′,5′-tetrachloro-4-biphenylol) significantly reversed sex of Red-Eared Slider embryos at a male-producing temperature. In the same turtle species, when Aroclor 1242 was assayed on eggs using environmentally relevant concentrations detected in alligator eggs from Lake Apopka, it also resulted in significant sex reversal (Willingham and Crews, 1999). Male Red-Eared Slider hatchlings treated in ovo with Aroclor 1242 had significantly lower testosterone concentrations than controls (Willingham et al., 2000). Aroclor 1242 significantly altered aromatase activity levels in the brain within 24 h of application to the eggshell and during the temperature-sensitive window, but not in the adrenal-kidney-gonad complex; after this crucial period, Aroclor 1242 caused an increase in aromatase activity in the adrenal-kidney-gonad complex of embryos just prior to hatch (Willingham and Crews, 2000). Because aromatase is the enzyme that converts testosterone to estradiol, lower testosterone levels and higher estradiol levels are an expected outcome of increased aromatase activity. It was suggested that Aroclor 1242 effects are via the aromatase pathway, causing an increase in activity at crucial time periods, and finally exerting its effects even beyond embryogenesis (Willingham and Crews, 2000). However, when PCB 126 was applied on Red-Eared Slider eggs in a temperature-sensitive turtle sex determination assay, no sex reversal was observed (Gale et al., 2002). More recently, Matsumoto et al. (2014) examined the genetic and epigenetic effect of the embryonic exposure to PCBs on gonad differentiation in this turtle species. They found that the embryonic exposure to specific PCBs both redirected the gonadal trajectory against the ambient egg incubation temperature and produced a subset of individuals with malformations of the reproductive tracts. Because the regression of the Müllerian duct is regulated by Müllerian-inhibiting substance produced from embryonic testes, a possible PCB mechanism may be interruption of the normal Müllerian-inhibiting substance expression in differentiating gonads.

Conversely, in the field studies to assess body burdens, maternal transfer of PCBs, and indices of sexual dimorphism in snapping turtles from the upper Hudson River, no discernible expression of abnormal sexual development or reproduction was demonstrated (Kelly et al., 2008).

Acute exposure of male 8-month-old Diamondback Terrapins (*Malaclemys terrapin*) from the Chesapeake Bay to PCB 126 significantly decreased growth and metabolic rate and significantly increased ventilation frequency over time. Six months after PCB 126 exposure, terrapins remained unable to physiologically recover from the PCB 126 and maintained growth and metabolic rates significantly lower than those of uncontaminated individuals. Whether this growth effect is the result of PCB 126-mediated alterations on the ability of the gut to extract nutrients from food, a disruption of growth hormones, or some other mechanism remains to be demonstrated (Holliday et al., 2009). PCB exposure also reduced hematocrit; because lower hematocrit means a reduced ability to transport oxygen to and metabolic waste products from tissues to support cellular metabolism, a reduced ability to transport blood gases could lead to a lower metabolic state (Holliday et al., 2009).

Eisenreich et al. (2009) compared sublethal and lethal responses of juvenile snapping turtles collected from the upper Hudson River exposed maternally and/or through the diet to PCBs over 14 months post hatching. Maternal exposure did not affect embryonic development or hatching success, but dietary PCB exposure reduced metabolic rates of juveniles. Approximately 8 months after hatching, high rates of mortality began to emerge in individuals having been exposed maternally to PCBs, and mortality rate correlated with PCB total in eggs. By 14 months posthatching, only 40% of juveniles derived from females in the contaminated area had survived, compared to 90% from the reference area. Aside from growth rate, metabolic rate was the only sublethal endpoint altered during PCB exposure, and it appeared to only be altered by proximate, dietary exposure rather than maternal exposure. The authors were unable to suggest a putative mode of toxicosis responsible for the latent onset of mortality of turtles exposed maternally to high concentrations of PCBs (Eisenreich et al., 2009).

Juvenile Diamondback Terrapins experimentally exposed to PCB 126 exhibited reduced apparent bone mineral density and altered bone microstructure in skeletal elements (**Figures 7.1–7.2**) (Ford and Holliday, 2004, 2005; Holliday and Holliday, 2012). Skulls of exposed animals had a skeletal phenotype more typical of younger animals (larger frontal-parietal fontanels, weakly developed nuchal crests, less-pronounced tomial ridges of the maxillae and dentaries, smaller jaw muscle attachments, and generally less-dense elements of the masticatory apparatus), and the femora had significantly more cortical area occupied by tissue other than bone. An increase of osteoclastic activity or a failure of osteogenesis caused by PCB exposure was suggested (Holliday and Holliday, 2012).

Regarding the identification of biomarkers, Yawetz et al. (1993) demonstrated the induction of P450 in liver microsomes of the Marsh Turtle (*Mauremys caspica rivulata*) after treatment with Aroclor 1254. Although the content of P450IA1 in liver microsomes of turtles collected from their natural habitat was relatively high, treatment with Aroclor 1254 increased the mean content of P450IA1 in the turtle livers up to 5 times (Yawetz et al., 1993).

Total white blood cell (WBC) counts and number of heterophils were negatively correlated with concentrations of total PCBs and Aroclor 1260, respectively, measured in liver and fat of Red-Eared Slider Turtles collected from ponds near or on the Paducah Gaseous Diffusion Plant in Kentucky. However, disease and other contaminants in the study area could have influenced the results (Yu et al., 2012). More recently, Letcher et al. (2015) concluded that levels of circulating total thyroxine (without additional assessments of the thyroid axis) could not be a suitable biomarker of PCB exposure in snapping turtles.

7.2.1.3 Other POPs in Freshwater Turtles Bishop et al. (1998) found a significant increase in developmental abnormalities in common snapping turtle embryos and hatchlings with increasing polychlorinated dibenzodioxin (PCDD) and polychlorinated dibenzofuran (PCDF) concentrations. Turtles with the highest PCDD/PCDF concentrations also had elevated 7-ethoxyresorufin-*O*-deethylase activity and cytochrome P4501A concentrations in liver microsomes.

A small percentage (10%) of male Yellow-Blotched Map Turtles (*Graptemys flavimaculata*) from a historically 2,3,7,8-tetrachlorodibenzo-*p*-dioxin (TCDD) contaminated site had high levels of 17β-estradiol (equivalent to levels found in females) and low plasma testosterone concentrations. However, no males had detectable plasma vitellogenin (Shelby and Mendonça, 2001). When TCDD was applied on Red-Eared Slider eggs in a temperature-sensitive turtle sex determination assay, no sex reversal was observed (Gale et al., 2002). A study examining Painted Turtles (*Chrysemys picta*) from a contaminated site found hepatic biotransformation enzymes to be elevated, notably CYP1A1 protein, suggesting contamination with dioxin-like compounds (Rie et al., 2000). In addition, juvenile turtles exposed to this environment had elevated gonadal aryl hydrocarbon receptor-1, an orphan receptor that is activated by environmental contaminants (Márquez et al., 2011).

Recently, a significant positive correlation was found between plasma brominated diphenyl ether (BDE-99) concentration and plasma total thyroxine levels in adult male common snapping turtles (Letcher et al., 2015).

7.2.1.4 Organochlorine Pesticides in Sea Turtles A significant correlation between increasing Green Turtle (*Chelonia mydas*) egg POP (including OCP, PCB, and polybrominated diphenyl ether [PBDE]) concentrations and decreasing hatchling mass/length ratio has been reported (van de Merwe et al., 2010). Egg mass of Loggerhead Turtle (*Caretta caretta*) eggs correlated significantly and negatively with *cis*-nonachlor concentrations (Keller, 2013b).

Podreka et al. (1998) investigated the potential effects of DDE on sexual differentiation of green turtles. After treating eggs with 4,4′-DDE at the beginning of the thermosensitive period for sexual determination, the sex ratio did not differ from what would be expected for temperature alone. In addition, incubation time, hatching success, incidence of body deformities, hatchling size, and weight were within the limits of healthy developed hatchlings.

In an in vitro study using an immortal testis cell line from a green turtle, 4,4′-DDE inhibited the cytochrome P450 aromatase activity, the key enzyme responsible for the conversion of testosterone to 17β-estradiol, but only at cytotoxic concentrations (Keller and McClellan-Green, 2004). Also, in vitro studies using plasma of nesting green turtles demonstrated that dieldrin competed with estradiol for binding sites; testosterone-binding affinity was affected at various DDT concentrations and it was hypothesized that DDT in vivo could act to inhibit steroid–protein interactions in nesting green turtles. The authors suggested that DDT could have an effect upon the protein conformation, thus affecting testosterone binding (Ikonomopoulou et al., 2009).

Several adverse effects of OCPs on sea turtle immunity have also been reported (Keller et al., 2006a). In loggerhead turtles, lysozyme activity was significantly and negatively correlated with whole blood concentrations of 4,4′-DDE and ∑chlordanes; and mitogen-induced lymphocyte proliferation was also significantly and positively correlated with 4,4′-DDE blood concentrations. After in vitro exposure of peripheral blood leukocytes from 16 green turtles to 4,4′-DDE, this contaminant increased mitogen-induced proliferation at concentrations below those that affected cell viability (Keller et al., 2006a). More recently, blood 4,4′-DDE concentrations in green turtles resident to an urbanized estuary in San Diego exceeded no-effects levels for lymphocyte proliferation, suggesting that immunological effects could be a concern at current concentrations of this pollutant (Komoroske et al., 2011).

The first study conducted to determine the possible involvement of POPs (including OCPs and PCBs) in a sea turtle's disease was reported by Aguirre et al. (1994). After testing the presence of several POPs in tissue samples of juvenile green turtles afflicted with fibropapillomatosis, no correlation was found between the contaminants and fibropapillomatosis because of the low detected levels. Since then, the hypothesis that exposure to OCPs could be a cofactor for fibropapillomatosis occurrence has not been supported by several studies (Keller et al., 2014; Sánchez-Sarmiento et al., 2016). Keller et al. (2014) suggested that the rise of POPs in late-stage, severely tumored, emaciated turtles might contribute to progression of the disease through immunosuppression or toxic effects to other organs or systems.

Keller et al. (2004) compared the OCP concentrations in loggerhead turtles with their clinical health parameters. Blood concentrations of ∑chlordanes were negatively correlated with RBC counts, hemoglobin, and hematocrit, indicative of anemia. Positive correlations were observed between most classes of OCPs and WBC counts and between mirex and the heterophil:lymphocyte ratio, suggesting modulation of the immune system. All classes of OCPs in the blood except dieldrin were correlated positively with aspartate aminotransferase activity (AST), indicating possible hepatocellular damage. Mirex blood concentration was negatively correlated with alkaline phosphatase (ALP) activity. Glucose concentrations were negatively correlated with adipose concentrations of dieldrin and ∑DDTs. The ratio of albumin:globulin was negatively correlated with concentrations of ∑chlordanes in blood and fat. Blood urea nitrogen (BUN) concentrations were positively correlated with concentrations of most OCPs measured in blood. Sodium concentrations were positively correlated with blood concentrations of mirex. Magnesium concentrations were negatively correlated with ∑DDTs in the blood. Body condition was negatively correlated with dieldrin in the blood.

When OCP baseline levels were determined in loggerhead turtles from the Canary Islands, emaciated turtles showed higher average ∑DDT concentrations in the liver than in the fat. Histologically, the authors detected hepatic lesions in four loggerhead turtles with high ∑DDT concentrations (severe multifocal necrotic hepatitis and diffuse vacuolar hepatic degeneration), but these were not exclusively attributed to acute effects of DDTs (Monagas et al., 2008). In a survey of OCPs in blood of Kemp's Ridley Turtles (*Lepidochelys kempii*) from the Gulf of Mexico (Swarthout et al., 2010), significant negative correlations were observed between ∑DDTs and T-lymphocyte proliferation, and between dieldrin and testosterone concentrations in female turtles. In addition, BUN concentrations were positively correlated with ∑chlordane concentrations. However, in adult loggerheads from a contaminated site, no significant correlations were detected between ∑DDTs or ∑chlordanes and any hematological or plasma chemistry value (Ragland et al., 2011). In green turtles from San Diego Bay 4,4′-DDE, lindane, and *trans*-chlordane plasma concentrations were also correlated with a few health parameters (Komoroske et al., 2011). In green turtles the activity of the antioxidant enzymes glutathione-S-transferase (GST) and catalase were positively correlated with the plasmatic concentrations of many OCPs (Labrada-Martagón et al., 2011). A significant positive correlation was also found between catalase activity and hexachlorohexane in Hawksbill Sea Turtles (*Eretmochelys imbricata*) (Tremblay et al., 2017); they also found a negative relationship between butyrylcholinesterase (BChE) activity and ΣHeptachlors families, suggesting a disruption of the nerve impulse, as BChE is inhibited by high concentrations of this component. More recently, a negative relationship was found between OCP concentrations in eggs and BChE activity in the plasma of female green turtles laying the eggs. Since OCPs are not cholinesterase inhibitors, an antagonistic effect between OCPs and organophosphate pesticides and the mobilization of OCPs from fatty tissues to eggs was suggested (Rivas-Hernández et al., 2018).

The effects of OCPs on the health status of adult nesting loggerheads from Cape Verde were studied using hematological and biochemical parameters (Camacho et al., 2013b). A negative association between some OCPs and WBC and thrombocyte estimates was noted. Higher concentrations of 4,4′-DDE were associated with lower heterophil values, and with higher eosinophil values. 4,4′-DDE was also negatively correlated with the Na:K ratio, in agreement with the results found by McConnell (1985) reporting that kidneys are sensitive to POPs. The authors observed a negative correlation between blood concentrations of 4,4′-DDE and OCPs and γ-glutamyl transferase (GGT). High levels of GGT could indicate liver damage. Keller et al. (2004) also observed a negative correlation between GGT and the pesticide dieldrin.

In a study on the influence of rehabilitation of injured loggerhead turtles on their blood levels of environmental organic pollutants (Camacho et al., 2014), all turtles that died during hospitalization had higher OCP concentrations than turtles that survived, and all of them were cachectic, concluding that mobilization of lipid stores may cause variations in the plasma concentrations of these contaminants.

7.2.1.5 Polychlorinated Biphenyls in Sea Turtles Fresh egg mass from loggerhead turtles correlated significantly and negatively with PCB 138 concentrations (Keller, 2013b).

Keller et al. (2006a) reported mitogen-induced lymphocyte proliferation significantly and positively correlated with ∑PCB blood concentrations in loggerhead turtles. In addition, in vitro exposure of peripheral blood leukocytes from loggerheads to Aroclor 1254 increased mitogen-induced proliferation at concentrations below those that affected cell viability (Keller et al., 2006a). More recently, functional immune assays (phagocytosis and natural killer cell activity) were used to quantify the direct effects of PCB congeners 105, 138, and 169 on innate immune functions upon in vitro exposure of loggerhead turtle cells; PCB 105 and 138 significantly increased eosinophil phagocytosis and significantly decreased natural killer cell activity (Rousselet et al., 2017).

Keller et al. (2004) also compared the PCB concentrations in loggerhead turtles with their clinical health parameters. Positive correlations were observed between ∑PCBs and WBC counts; ∑PCBs in adipose tissue positively correlated with an increase in circulating eosinophils. Increasing concentrations of some PCBs correlated with an elevation in the heterophil:lymphocyte ratio, and were negatively correlated with hematocrit, hemoglobin, and RBC counts. While significant positive correlations were detected between PCBs and AST and plasma sodium concentrations, significant negative correlations were observed for albumin:globulin ratio, ALP, and magnesium. In a survey of POPs in blood of Kemp's Ridley Turtles from the Gulf of Mexico (Swarthout et al., 2010), significant negative correlations were observed between ∑PCBs and creatine phosphokinase (CPK). However, in adult loggerheads from a contaminated site, no significant correlations were detected between ∑PCBs and any hematological or plasma chemistry value (Ragland et al., 2011).

When studying the effects of PCBs on the health status of adult nesting loggerheads from Cape Verde, packed cell volume decreased with higher concentrations of PCBs, which suggests that PCB exposure could result in anemia in sea turtles (Camacho et al., 2013b). Additionally, PCB 52 was negatively associated with WBC counts. Correlations between PCBs and biochemical parameters were also found in this study: turtles showing the highest levels of PCB 52 had the lowest concentrations of glucose, and higher phosphorus concentrations were associated with higher levels of PCB 52.

Richardson et al. (2010) measured PCB concentrations as well as the expression patterns of cytochrome P450 enzymes and GST activities in livers of three sea turtle species. GST activity and cytochrome P450 enzyme expression did not correlate significantly with PCB concentrations.

In a recent study on the influence of the rehabilitation of injured loggerhead turtles on their blood levels of environmental organic pollutants (Camacho et al., 2014), ∑PCBs and ∑marker-PCBs were significantly reduced at the end of the rehabilitation period. This lowering effect was much more prominent in the subset of turtles admitted with signs of severe emaciation. The authors attributed such reduction to a redistribution of these contaminants into the superficial peripheral compartment (adipose tissue) after receiving veterinary assistance and a brief period of normal feeding.

The effects of PCBs on diseases affecting sea turtles have not been demonstrated. Very high levels of PCB 138, 153, 180, 209 were detected in the celomic fat of a loggerhead turtle with severe pansteatitis (**Figure 7.3**) (Orós et al., 2013). Although a nutritional cause could not be ruled out, the authors suggested that the high levels of PCBs detected in the celomic fat could have induced lipid peroxidation in adipocytes, resulting in cell damage, deposition of ceroid pigment, and inflammatory response. In a study on stranded loggerhead turtles in the Canary Islands, a positive correlation was detected between ∑PCB concentration and cachexia, probably due to a remobilization process. It was also remarkable that almost all turtles with septicemia showed very high levels of PCBs, suggesting a possible immunosuppression (Orós et al., 2009a).

In a recent study on Hawaiian green turtles, PCBs concentration in turtles with moderate or severe fibropapillomatosis affliction was much higher than in healthy turtles, suggesting that PCBs may play a role in fibropapillomatosis (Yan et al., 2018).

7.2.1.6 Other POPs in Sea Turtles Keller et al. (2012) compared blood plasma concentrations of 13 perfluoroalkyl contaminants (PFCs) in five species of sea turtles with different trophic levels. The measured perfluorooctane sulfonate (PFOS) concentrations were compared with concentrations known to cause toxic effects in laboratory animals, and estimated margins of safety (EMOS) were calculated. Small EMOS (<100), suggestive of potential risk of adverse health effects (immunosuppression), were observed for all five sea turtle species. EMOS <100 were also observed for liver, thyroid, and neurobehavioral effects for the more highly exposed species.

Significant and positive correlations between plasma PFOS concentrations and T-lymphocyte proliferation, AST activity, and globulin, glucose, total protein, BUN, and potassium were observed in loggerhead turtles (Peden-Adams et al., 2005). In addition, PFOS significantly and negatively correlated with plasma lysozyme activity. Guerranti et al. (2014) detected a positive statistical correlation between hepatic levels of PFOS and uroporphyrins. Porphyrins, intermediate metabolites of heme byosinthesis or oxidative byproducts of metabolites, are produced and accumulated in trace amounts in erythropoietic tissues, liver, and kidneys and are excreted via urine or feces. As heme byosinthesis may be altered by several environmental contaminants, leading to changes in accumulation or excretion profiles, porphyrins have been proposed as sensitive biomarkers of exposure to contaminants (Casini et al., 2006). There are two hypotheses on the mechanism of action of some organic contaminants on the accumulation of uroporphyrins and both involve the cytochrome P-450. The cytochrome P-450 can transform these compounds in metabolites that can directly inhibit uroporphyrinogen decarboxylase. These compounds can also act indirectly, causing a mechanism of oxidative stress (Guerranti et al., 2014).

The cytotoxicity of PFOA to primary skin fibroblast cell cultures derived from biopsies of multiple healthy Loggerhead Sea Turtles has been confirmed (Webb et al., 2014).

In adult male loggerheads captured near Cape Canaveral, Florida, no significant correlations were detected between ∑PBDEs and any hematological or plasma chemistry value (Ragland et al., 2011). De Andrés et al. (2016) provided additional baseline data on PCB and PBDE concentrations in Leatherback Turtle (*Dermochelys coriacea*) eggs and investigated whether any of these compounds was correlated to viability, fertility, hatching success, or hatchling morphometrics. ∑PBDEs showed a negative correlation to the hatching success, suggesting potentially harmful effects of these contaminants on the reproduction of leatherbacks.

7.2.2 Crocodylia

7.2.2.1 Organochlorine Pesticides After a major pesticide spill in 1980, composed mainly of dicofol, DDT, and DDT metabolites, into Lake Apopka, Florida, the American Alligator (*Alligator mississippiensis*) population inhabiting the lake declined dramatically between 1980 and 1987. The principal contaminant found in alligator eggs and juveniles collected from Lake Apopka was 4,4′-DDE, a major persistent metabolite of DDT (Heinz et al., 1991). When embryo and neonate alligators were studied, evidence of reduced hatchability, reduced viability of offspring, and endocrine demasculinization of males or supermasculinization of females was seen (Guillette et al., 1994). Gonadal abnormalities included ovaries with multinuclear oocytes and polyovular follicles, and testes with aberrant germ cells (Guillette et al., 1995). Several hypotheses were also suggested to explain the reduced penis size observed in juvenile male alligators from Lake Apopka (Guillette et al., 1996, 1999): reduced plasma androgen concentrations due to modifications in liver degradation of androgens or modifications in the synthesis of plasma proteins responsible for binding steroid (Guillette et al., 1995; Guillete and Crain, 1996), alteration of the ratio tes tosterone:dihydrotestosterone due to the suppression of the enzyme 5-reductase; reduced numbers of androgen receptors on phallic tissue, presence of an androgen antagonist competing for the androgen receptor, and presence of a xenobiotic estrogen inducing a female phenotype poorly responsive to androgens later in life.

Since then, the potential of OCPs to act as endocrine disruptors has been investigated in several species of crocodilians. Vonier et al. (1996) provided the first evidence that some OCPs bind the alligator estrogen receptor (aER) (4,4′-DDD, trans-nonachlor, endosulfan) and alligator progesterone receptor (endosulfan, chlordecone, alachlor), supporting the hypothesis that the reproductive abnormalities may be related to the modulation of endocrine-related responses. Other OCPs, mainly heptachlor epoxide and heptachlor, showed affinity for the aER (Guillette et al., 2002). In addition, combinations of chlordane, dieldrin, and toxaphene inhibited [^{3}H]17β-estradiol binding to the aER in a synergistic manner (Arnold et al., 1997).

Experimental studies demonstrated sex reversal in alligator eggs treated with 4,4′-DDE and incubated at male-induced temperatures (Matter et al., 1998; Milnes et al., 2005). However, toxaphene failed to affect sexual differentiation, or in vitro thyroxin, testosterone, and estradiol production (Milnes et al., 2004). In ovo exposure to endosulfan did not cause estrogen-like effects on sex determination of Broad-Snouted Caiman (*Caiman latirostris*) hatchlings but caused a reduction in hatchling fractional weight (Beldoménico et al., 2007). Neonatal female caimans exposed in ovo to endosulfan had decreased testosterone levels (Stoker et al., 2008). In another study, the testes of endosulfan-exposed caiman showed tortuous seminiferous tubules with increased perimeter, disrupted distribution of peritubular myoid cells, and emptied tubular lumens; in addition, an imbalance between proliferative activity and cell death was observed in the testes of caiman exposed to the higher doses of endosulfan, mainly due to a high frequency of apoptosis in intratubular cells. This altered turnover was associated with decreased testosterone levels (Rey et al., 2009). An increased mRNA expression of sex-determining genes was also observed in this species after in-ovo exposure to endosulfan (Durando et al., 2013). It was also suggested that the higher mRNA expression of *sf-1* gen in caiman males exposed to endosulfan could modify the steroidogenic pathway, since these animals showed decreased levels of testosterone, compared to untreated males (Rey et al., 2009). Gunderson et al. (2006) suggested that exposure to OCPs at concentrations present in Lake Apopka does not lead to variation in the expression of CYP3A77 or steroid-xenobiotic receptor genes in alligators. The regulation of CYP3A77 by OCPs is of interest because of the role it plays in the biotransformation of testosterone and xenobiotics. The induction of CYP3A77 activity by OCPs could potentially lead to changes in plasma testosterone concentrations as well as lead to activational alterations in sexually dimorphic patterns of hepatic enzyme expression. They demonstrated that toxaphene up-regulates CYP3A77 in yearling female alligators and thus could activationally modulate the expression of this enzyme in alligators inhabiting contaminated sites.

The effects of OCPs on embryo viability have also been investigated. Histologic examination of alligator embryos and neonates from several north central Florida lakes contaminated with OCPs revealed lesions in 35% of the animals, with pneumonia, pulmonary edema, and atelectasis in over half of the cases; authors also reported an increase in prevalence of histologic lesions (heterophilic pneumonia and atelectasis) in clutches with high OCPs (Sepúlveda et al., 2006). Experimental maternal exposure to a mixture of 4,4′-DDE, toxaphene, dieldrin, and chlordane was associated with decreased egg/embryo viability in American Alligators (Rauschenberger et al., 2004, 2007). Exposure of alligator eggs to a mixture of OCPs via injection or topical application caused embryonic mortality (Muller et al., 2007). Thiamine deficiency could influence alligator hatching success at the lakes with high OCP exposure (Sepúlveda et al., 2004). When evaluating embryo mortality in alligators at reference and OCP-contaminated sites, Rauschenberger et al. (2009) found that the total thiamine levels in the eggs were positively associated with clutch viability and negatively associated with the lipid content and OCPs measured in egg yolks. Stoker et al. (2013) suggested that exposure of Broad-Snouted Caiman eggs to OCPs exerts a direct effect on the mother's oviductal functions, evidenced by decreased eggshell porosity, and an indirect effect on hatchling survival as a consequence of the chronic hypoxia experienced by embryos with a reduced shell pore density during development.

Studies reporting adverse effects of OCPs on juvenile or adult crocodilians are scarce. Alligators captured from Lake Apopka had decreased thymic ratio (medulla:cortex), and decreased size of malpighian bodies and lymphocyte sheaths in spleen, compared with alligators from two reference lakes (Rooney et al., 2003). Female alligators from this contaminated lake also had greater trabecular bone mineral density (BMD), total BMD, and trabecular mineral content, suggesting that contaminants present in the lake could inhibit the natural and continuous resorption of bone tissue, resulting in increased bone mass (**Figure 7.4**) (Lind et al., 2004). The very high levels of 4,4′-DDE and toxaphene found in Australian Freshwater Crocodiles (*Crocodylus johnstoni*) from a historically contaminated site did not result in obvious effects on blood chemistry or gonadal microanatomy, although the limited number of samples and the variability of the breeding state of the animals examined could have masked possible effects (Yoshikane et al., 2006). More recently, body condition of 14 adult Spectacled Caimans (*Caiman crocodilus*) from Costa Rica was negatively correlated with total pesticide concentrations (Grant et al., 2013).

7.2.2.2 Polychlorinated Biphenyls Juvenile alligators from Lake Apopka had significantly elevated serum concentrations of ∑PCBs compared to juveniles from the other lakes, but no relationships between phallus size and serum contaminant levels were detected (Guillette et al., 1999).

Gonzalez-Jauregui et al. (2012) observed that ∑PCB concentrations in Morelet's Crocodiles (*Crocodylus moreletii*) were related to testosterone levels among female crocodiles. This androgenic effect of PCBs had not been reported previously. Because testosterone promotes aggressive behavior in vertebrates, the authors suggested that excessive aggression during the estrous season, or when female crocodiles should be caring for their young, could result in reproductive failure in Morelet's Crocodiles and potential long-term decline of the population.

7.2.2.3 Other POPs Matter et al. (1998) reported sex reversal in alligator eggs treated with TCDD and incubated at male-induced temperatures.

7.2.3 Sauria

Although some species have been used as environmental sentinels (White and Krynitsky, 1986; Lambert, 1993; Berny et al., 2006; Gómara et al., 2007), detailed studies on the acute and chronic effects of POPs in this group are scarce. Kihara and Yamashita (1978) reported the lethal effects of several pesticides, including POPs, on Ousima Skinks (*Plestiodon* [formerly *Eumeces*] *marginatus*). Everts et al. (1983) studied side effects following a spraying campaign to control *Glossina* spp. (Tsetse flies) with endosulfan, but no measurable changes in lizard populations were observed.

When a spillage of organochlorines was studied, authors found that lizard species avoided soil contamination levels above 1 ppm total pesticides, and were absent above 10 ppm (Lambert, 1997). Ciliberti et al. (2012) assessed the environmental contamination caused by OCPs in Mali using the Nile Monitor (*Varanus niloticus*) as a sentinel species, but they did not find any negative influence of pollutant concentrations on fat somatic index or body condition index. In other experiments, Savannah Monitors (*Varanus exanthematicus*) were orally exposed during a 6-month interval to a selected set of pollutants (lead + 4,4′-DDT + chlorpyrifos-ethyl). Some critically intoxicated monitors showed typical signs of cholinesterasic activity inhibition, probably due to the organophosphate compound; however, the individuals that did not die recovered completely, had extraordinary high 4,4′-DDT, 4,4′-DDD or 4,4′-DDE body burden, and their hematological and biochemical parameters were normal (Ciliberti et al., 2013).

7.2.4 Ophidia

Snakes have also been used as environmental sentinels, but the significance of tissue POP residues to reproduction, survival, growth/development, and population dynamics in contaminated habitats have remained unknown (Winger et al., 1984; Santos et al., 1999; Bishop and Rouse, 2000; Fontenot et al., 2000; Rainwater et al., 2005; Wu et al., 2008, 2009, 2011; Zhou et al., 2016).

Stafford et al. (1977) measured pesticide detoxifying enzymes and residues of 4,4′-DDE and PCBs in Water Moccasins (*Agkistrodon piscivorus*), Copperheads (*Agkistrodon contortrix*), and several water snake species (*Natrix erythrogaster*, *Natrix fasciata*, and *Natrix rhombifera*) from three aquatic ecosystems in central Texas, concluding that snakes with high levels of detoxifying enzymes, particularly microsomal oxidases, are more likely to occur in contaminated ecosystems. Bishop and Rouse (2006) found no significant correlations among body mass, snout–vent length, number of young per female, or pergram body mass of Lake Erie Water Snakes (*Nerodia sipedon insularum*) and organochlorine concentrations in plasma.

Experimental studies have also been scarce. Corn Snakes (*Elaphe guttata*) experimentally fed with mice contaminated with a mixture of alpha-chlordane, Aroclor 1254, and lindane, did not show signs of suggested toxicosis as indicated by feeding behavior, weight gain, shedding frequency, or peripheral leukocyte counts (Jones et al., 2005). Exposure of hepatocytes of African Brown House Snakes (*Lamprophis fuliginosus*) to TCDD and PCB 126 resulted in a dose-dependent increase in ethoxyresorufin-*O*-deethylase activity, this activity directly associated with the induction of hepatic cytochrome

Table 7.2 Summary of the Main Effects of Several Organochlorine Pesticides on Reptiles

Compound	Species	Environment	Effect	Reference
trans-nonachlor, *cis*-nonachlor, 4,4′-DDE, chlordane	*Trachemys scripta elegans*	Laboratory	Sex reversal toward female	Willingham and Crews (1999)
4,4′-DDE	*Trachemys scripta elegans*	Laboratory	Sex reversal toward female	Willingham (2004)
Heptachlor, DDD, hexachlorobenzene, *trans*-nonachlor	*Emys orbicularis*	Field	Micronuclei in RBCs	Swartz et al. (2003)
cis-chlordane	*Chelydra serpentina*	Field	Decreased plasma total thyroxine levels	Letcher et al. (2015)
4,4′-DDE and ∑chlordanes	*Caretta caretta*	Field	Decreased lysozyme activity	Keller et al. (2006a)
4,4′-DDE	*Caretta caretta*	Field and laboratory	Increased mitogen-induced lymphocyte proliferation	Keller et al. (2006a)
∑chlordanes	*Caretta caretta*	Field	Decreased RBC counts, hemoglobin, and hematocrit Increased WBC counts	Keller et al. (2004)
Mirex, ∑DDTs	*Caretta caretta*	Field	Increased WBC counts	Keller et al. (2004)
∑chlordanes, Mirex, ∑DDTs	*Caretta caretta*	Field	Increased AST levels	Keller et al. (2004)
4,4′-DDE	*Caretta caretta*	Field	Decreased Na:K ratio and GGT levels	Camacho et al. (2013b)
∑DDTs	*Lepidochelys kempii*	Field	Decreased mitogen-induced T-lymphocyte proliferation	Swarthout et al. (2010)
Hexachlorohexane	*Eretmochelys imbricata*	Field	Increased catalase activity	Tremblay et al. (2017)
Dicofol, DDT, and DDT metabolites	*Alligator mississippiensis*	Field	Gonadal abnormalities	Guillette et al. (1995)
4,4′-DDE	*Alligator mississippiensis*	Laboratory	Sex reversal toward female	Milnes et al. (2005)
Mixture of OCPs	*Alligator mississippiensis*	Field	Pneumonia, pulmonary edema, and atelectasis	Sepúlveda et al. (2006)
4,4′-DDE, toxaphene, dieldrin, and chlordane	*Alligator mississippiensis*	Laboratory	Decreased egg/embryo viability	Rauschenberger et al., 2004, 2007)
Mixture of OCPs	*Alligator mississippiensis*	Field	Altered lymphoid tissue architecture	Rooney et al. (2003)
Mixture of OCPs	*Alligator mississippiensis*	Field	Increased bone mass	Lind et al. (2004)
Endosulfan	*Caiman latirostris*	Laboratory	Altered testicular architecture Decreased testosterone levels	Rey et al. (2009)

Table 7.3 Summary of the Main Effects of Several Polychlorinated Biphenyls (PCBs) on Reptiles

Compound	Species	Environment	Effect	Reference
2′,4′,6′-trichloro-4-biphenylol 2′,3′,4′,5′-tetrachloro-4-biphenylol	*Trachemys scripta elegans*	Laboratory	Sex reversal toward female	Bergeron et al. (1994)
Aroclor 1242	*Trachemys scripta elegans*	Laboratory	Sex reversal toward female	Willingham and Crews (1999)
Aroclor 1242	*Trachemys scripta elegans*	Laboratory	Decreased testosterone levels	Willingham et al. (2000)
4-hydroxy-2′,4′,6′-trichlorobiphenyl 4-hydroxy-2′,3′,4′,5′-tetrachlorobiphenyl	*Trachemys scripta elegans*	Laboratory	Reproductive tract malformations	Matsumoto et al. (2014)

(*Continued*)

Table 7.3 (*Continued*) Summary of the Main Effects of Several Polychlorinated Biphenyls (PCBs) on Reptiles

Compound	Species	Environment	Effect	Reference
PCB 126	*Malaclemys terrapin*	Laboratory	Decreased growth and metabolic rate Reduced hematocrit	Holliday et al. (2009)
PCB126	*Malaclemys terrapin*	Laboratory	Reduced bone mineral density Altered bone microstructure	Ford and Holliday (2004, 2005); Holliday and Holliday (2012)
ΣPCBs	*Chelydra serpentina*	Laboratory	Reduced metabolic rate Increased mortality	Eisenreich et al. (2009)
ΣPCBs	*Caretta caretta*	Field	Increased mitogen-induced lymphocyte proliferation	Keller et al. (2006a)
Aroclor 1254	*Caretta caretta*	Laboratory	Increased mitogen-induced lymphocyte proliferation	Keller et al. (2006a)
ΣPCBs	*Caretta caretta*	Field	Increased WBC counts Increased eosinophil counts Increased AST levels	Keller et al. (2004)
ΣTCDD-like PCBs	*Caretta caretta*	Field	Increased heterophil:lymphocyte ratio Reduced albumin:globulin ratio	Keller et al. (2004)
ΣPCBs	*Caretta caretta*	Field	Decreased hematocrit	Camacho et al. (2013b)
PCB 52	*Caretta caretta*	Field	Reduced WBC counts	Camacho et al. (2013b)
ΣPCBs	*Lepidochelys kempii*	Field	Decreased CPK levels	Swarthout et al. (2010)
ΣPCBs	*Crocodylus moreletii*	Field	Increased testosterone levels in females	Gonzalez-Jauregui et al. (2012)

Table 7.4 Summary of the Main Effects of Other Persistent Organic Pollutants (POPs) on Reptiles

Compound	Species	Environment	Effect	Reference
Polychlorinated dibenzodioxins Polychlorinated dibenzofurans	*Chelydra serpentina*	Field	Developmental abnormalities	Bishop et al. (1998)
Brominated diphenyl ether (BDE-99)	*Chelydra serpentina*	Field	Increased plasma total thyroxine levels	Letcher et al. (2015)
2,3,7,8-tetrachlorodibenzo-*p*-dioxin	*Graptemys flavimaculata*	Field	Increased 17β-estradiol levels	Shelby and Mendonça (2001)
Perfluorooctane sulfonate	*Caretta caretta*	Field	Increased T-lymphocyte proliferation, AST activity, globulin, glucose, total protein, BUN, and potassium levels Reduced plasma lysozyme activity	Peden-Adams et al. (2005)
Perfluorooctane sulfonate	*Caretta caretta*	Field	Increased uroporphyrins	Guerranti et al. (2014)
Σpolybrominated diphenyl ethers (PBDEs)	*Dermochelys coriacea*	Field	Decreased hatching success	De Andrés et al. (2016)
2,3,7,8-tetrachlorodibenzo-*p*-dioxin	*Alligator mississippiensis*	Laboratory	Sex reversal toward female	Matter et al. (1998)
Brominated diphenyl ether (BDE-47)	*Lamprophis fuliginosus*	Laboratory	Reduced corticosterone levels Increased size and higher thyroid follicular epithelial height	Neuman-Lee et al. (2015)

P4501A (Hecker et al., 2006). Neuman-Lee et al. (2015) tested how short-term, repeated exposure to PBDE congener BDE 47 during pregnancy affected physiological processes in pregnant female Garter Snakes (*Thamnophis elegans*) and their resulting offspring. The baseline corticosterone levels during the stress-reactivity test provided some evidence that BDE 47 may affect the hypothalamic–pituitary–adrenal axis; treated females had lower corticosterone concentrations than the control females, potentially indicating energy limitations after the chronic BDE 47 exposure. They also found increased size and higher thyroid follicular epithelial height in dams and increased size in their offsprings. There is evidence in other species that the thyroid gland and its hormonal products are impacted by exposure to PBDEs probably due to structural similarity of PBDEs to thyroxine (Neuman-Lee et al., 2015).

7.3 Polycyclic Aromatic Hydrocarbons and Crude Oil

Polycyclic aromatic hydrocarbons (PAHs) are ubiquitous organic contaminants found in air, water, sediment, and soil. Most environmental sources of PAHs are from anthropogenic activity, and they include petrogenic PAHs (fossil fuels) and pyrogenic PAHs (from incomplete combustion of fossil fuels), petrogenic PAHs being more bioavailable than pyrogenic PAHs (de Solla, 2010). Studies investigating PAHs have primarily focused on aquatic ecosystems. Crude oil spills are rare events, but usually they are devastating to vulnerable species, including reptiles (Camacho et al., 2012).

7.3.1 Chelonia

7.3.1.1 PAHs and Crude Oil in Freshwater Turtles Bell et al. (2006) suggested that PAHs were involved in the high occurrence of deformities in Painted Turtle (*Chrysemys picta*) and Snapping Turtle (*Chelydra serpentina*) embryos from a contaminated site. Mean annual deformity rate of pooled Painted Turtle clutches ranged from 45% to 71%, while that of snapping turtle clutches ranged from 13% to 19%. The most common deformities among Painted Turtle embryos and hatchlings were dwarfism and malformations of the shell, followed by skull abnormalities. In snapping turtle embryos, tail and carapace or plastron malformations were the most common deformities, followed by kyphosis or scoliosis; facial deformities often consisted of unfused maxilla and premaxilla bones, and occasionaly missing eyes and short upper and/or lower jaws. Authors attributed the greater occurrence and severity of embryonic deformities in Painted Turtles to a high susceptibility to pollution. Significant compromised embryonic development and malformations associated with total petroleum hydrocarbons, PAHs, and trace metals were also observed in Diamondback Terrapin (*Malaclemys terrapin*) eggs from Bermuda (**Figure 7.5**) (Outerbridge et al., 2016). A significant positive correlation was documented between three-ring PAH sediment concentrations and flow cytometry method estimates of chromosomal damage in European pond turtles (*Emys orbicularis*) inhabiting wetlands in Azerbaijan, suggesting genotoxicity (Matson et al., 2005). Saba and Spotila (2003) examined survival and post-rehabilitation behavior of Painted Turtles, Snapping Turtles, Red-Eared Slider Turtles (*Trachemys scripta*), and Red-Bellied Turtles (*Pseudemys rubriventris*) exposed to a crude oil spill in February 2000 at the John Heinz National Wildlife Refuge in southeastern Pennsylvania; freshwater turtle behavior, home range size, and water temperature preference were not affected by crude oil exposure and rehabilitation, but reproductive biology and endocrinology were not studied.

Experimental exposure of snapping turtle eggs to benzo[*a*]pyrene or 7,12-dimethylbenz[*a*]anthracene had significant effects on survival rates in embryos, and there was a positive linear relationship between level of exposure to PAHs and severity of deformities in embryos (Van Meter et al., 2006); however, neither righting response nor upper temperature tolerance of hatchlings with no or minor deformities were significantly affected by exposure to PAHs. In addition, exposure to crude oil did appear to cause more damage to a developing embryo in comparison to a single PAH (Van Meter et al., 2006). However, in another experiment, in which snapping turtle eggs were exposed to naturally and chemically dispersed oil, no effects on hatching success or hatchling/juvenile traits (DNA integrity, survival, growth, metabolism, energy storage, or behavior) were observed (Rowe et al., 2009).

More recently, the biochemical changes in liver, heart, lung, pancreas, brain, and small and large intestine of Geoffroy's Side-Necked Turtle (*Phrynops geoffroanus*) exposed to benzo[a]pyrene for 7 days have been described (Silva et al., 2016); this freshwater turtle species demonstrated exposure resistance and effective defense mechanism against contamination, since there was no increase in malondialdehyde levels in the organs. In addition, antioxidant enzymes (7-ethoxyresorufin-*O*-deethylase, 7-benzyloxyresorufin-*O*-debenzylase, 7-pentoxyresorufin-*O*-depentylase) were also responsive to benzo[a]pyrene (500–1000 μg/kg) exposure, suggesting the role of these enzymes in the metabolism of this compound. The analysis of GGT activity and comet assay demonstrated a significant and dose-dependent increase in turtles treated with benzo[a]pyrene doses of 500 and 1000 μg/kg, indicating an association between exposure to the contaminant and an increase in hepatic and blood cell DNA damages (Silva et al., 2016).

7.3.1.2 PAHs and Crude Oil in Sea Turtles After measuring PAHs in plasma samples of 162 juvenile Loggerheads (*Caretta caretta*) from the Canary Islands and 205 adult nesting loggerheads from Cape Verde, authors did not find bioaccumulation, and concluded that PAH concentrations in loggerhead turtles were directly related to recent exposure to contaminated waters or food (Camacho et al., 2012). In addition, no differences were detected in PAH levels in injured loggerheads during their hospitalization period, indicating that there was no mobilization (Camacho et al., 2014).

Several correlations between certain PAHs and electrolyte balances were found in loggerheads, suggesting that exposure to these environmental contaminants could affect the kidneys and salt glands; in addition, correlations between PAHs and enzyme activity (alanine aminotransferase [ALT], ALP, and amylase) and serum protein levels were also observed, pointing to the possibility that these contaminants could induce adverse metabolic effects in sea turtles (Camacho et al., 2013b).

A significant cytochrome CYP1A5 induction was observed in loggerhead turtle skin fibroblasts following an experimental 72-h exposure to benzo[*a*]pyrene, demonstrating that cell cultures can be used to obtain toxicity information on contaminants of current environmental concerns, such as crude oil exposure in marine turtles (Webb et al., 2014). More recently, a significant correlation was found between whole blood cell expression of three biomarkers (HSP60, CYP1A, and ERα) and plasma levels of some PAH congeners in Loggerhead Sea Turtles (Cocci et al., 2018). The global DNA methylation levels were significantly and positively correlated with the concentrations of most of the PAHs and only one of the PCB congeners (PCB52); in contrast, no significant association between PCB burden and gene expression was observed (Cocci et al., 2018). A statistically significant positive correlation between carcinogenic PAHs in blood and erythrocytic DNA fragmentation was also recently observed in Mediterranean Loggerhead Turtles (Casini et al., 2018).

Saturation of turtle nests with oil is known to induce mortality in embryos (Fritts and McGehee, 1981; Yender and Mearns, 2003). Reported sublethal effects of oil ingestion have included metabolic production of potential carcinogenic compounds, immunosuppression, diminished salt gland function, and hormonal and behavioral abnormalities (Hall et al., 1983; Lutz et al., 1986).

Vargo et al. (1986) found that acute (0.5 cm oil layer for 48 h) and chronic (0.05 cm oil layer for 96 h) crude oil exposure adversely affected respiratory and salt gland function, blood chemistry composition, and integument surfaces in loggerheads. In laboratory studies using juvenile loggerheads, almost all the major physiological systems were adversely affected by short exposure to crude oil (Lutcavage et al., 1995). The presence of oil had no consistent effect on breathing frequency, although during recovery there was some indication of reduced ventilation rate. Oiled turtles had up to fourfold increase in WBC counts, a 50% reduction in RBC counts, and RBC polychromasia. Acute inflammatory cell infiltrates and moderate to extensive multifocal epithelial cell necrosis was observed in the skin and mucosal surface of oiled turtles. In addition, there was moderate dysplasia of epidermal epithelium and a loss of cellular architectural organization of the skin layers. The superficial dermis was characterized by moderate to severe diffuse edema with associated diffuse infiltrates of heterophils and occasional plasma cells, histiocytic cells, and small lymphoid cells. Authors incidentally sampled salt gland fluid, observing that salt gland failed to function properly in two oiled turtles for a period of 7 days. The failure of the salt gland to produce fluids several days after exposure to crude oil pointed to attenuated or delayed effects in salt transport.

Loggerheads stranded by crude oil ingestion in the Canary Islands had esophageal impaction with balls of crude oil (**Figure 7.6**), necrotizing gastritis (**Figure 7.7**), necrotizing enteritis, necrotizing hepatitis, and tubulonephrosis (Orós et al., 2004, 2005; Camacho et al., 2013a). They had significantly lower RBC and thrombocyte counts than those observed in clinically normal turtles. In addition, they had significantly higher values for BUN, total cholesterol, ALT, and ALP, and significantly lower values for triglycerides, glucose, and calcium than the clinically normal turtles (Casal and Orós, 2009). Torrent et al. (2002) reported presence of crude oil balls in an esophageal diverticulum in a Loggerhead Sea Turtle (**Figure 7.8**).

Mignucci-Giannoni (1998) found that rehabilitated Green Turtles (*Chelonia mydas*) and Hawksbill Turtles (*Eretmochelys imbricata*) had the highest chances of survival compared to seabirds after being exposed to an oil spill in Puerto Rico in 1994. In a long-term retrospective study (1998–2014) on causes of stranding of 1860 Loggerhead Sea Turtles, the highest release rate was achieved in the crude oil category (93.87%), whereas the global survival rate for all causes of admission was 86.29%. Although ingestion of crude oil and subsequent internal lesions can threaten sea turtle survival, lesions in the skin, carapace, and plastron are not fatal in the majority of cases (Orós et al., 2016).

7.3.2 Crocodylia

Arukwe et al. (2015) investigated the biotransformation and oxidative stress responses in relation to chemical burden in the liver of male and female Nile Crocodiles (*Crocodylus niloticus*) from a commercial crocodile farm passively exposed to various anthropogenic aquatic pollutants. Male crocodiles consistently produced higher biotransformation and oxidative stress responses compared to females. Authors observed relationships between these responses and concentrations of aliphatic hydrocarbons and PAHs. More recently, Dzul-Caamal et al. (2016) assessed the effects of 11 PAHs and several metals from a panel of biomarkers evaluated in the snout scraping, serum, and peripheral blood cells of Morelet's Crocodiles (*Crocodylus moreletii*) from two sites (contaminated and noncontaminated). In male crocodiles from the reference site, only H_2O_2 in peripheral blood cells was related to levels of fluoranthene on the keel of caudal scales, but in females no association was detected. Levels of benzo[*a*]pyrene, benzo[*a*]anthracene, chrysene, pyrene, phenanthrene, and acenaphthene in keel of caudal scales of the female crocodiles were related to the oxidative stress biomarkers on peripheral blood cells, including the total CYP450 activity and levels of H_2O_2 in serum. However, in male crocodiles, the oxidative stress in snout scraping and in the serum and superoxide dismutase in peripheral blood cells was related to benzo[*a*]pyrene water concentrations and to the burdens of indeno[1,2,3 cd]pyrene in keel of caudal scales. Authors concluded that in this contaminated site, PAHs are responsible for the oxidative stress response in females, in contrast with male crocodiles, that suffer from the oxidative stress elicited by metals.

Ekpubeni and Ekundayo (2002) studied the toxicity of petroleum waste drilling fluid on Short-Nosed Crocodiles (*Osteolemus tetraspis*) and Nile Crocodiles. Both crocodile species exhibited a high resistance to the undiluted waste and the different dilutions, with mortality rates <1.6%. Because the majority of this low number of deaths occurred at the end of the study period, the authors suggested that there was a delayed toxicant-induced mortality effect.

7.3.3 Sauria

When Santos et al. (1997) tested Puerto Rican Ground Lizards (*Ameiva exsul*) from a contaminated site, the elevated levels of ethoxyresorufin-*O*-deethylase and CYP450 strongly suggested that some of these lizards were exposed to a variety of compounds related to PAHs. In a study using Italian Wall Lizards (*Podarcis sicula*) as sentinels, the most sensitive tests for the evaluation of toxicological impact of an on-shore oil center in the lizards were found to be Cd, Hg, total and carcinogenic PAH levels, and PAH metabolites in bile (Marsili et al., 2009).

Results from a study on Desert Lizards (*Acanthodactylus scutellatus*) and their ant prey in an oil field in Kuwait suggested that total PAHs in the tar mat sites for both lizards and their prey could be responsible for reducing the lizard's survival, and could be responsible for histopathological features that link tissue damage to prey contamination (Al-Hashem et al., 2007). The hepatocytes of lizards collected from the tar mat sites showed swelling, ballooning degeneration of the hepatic cytoplasm, and cell death (Al-Hashem, 2011).

Wikelski et al. (2002) studied a massive 62% mortality of Marine Iguanas (*Amblyrhynchus cristatus*) in the year after an oil spill in the Galapagos, due to a small amount of residual oil contamination in the sea. Authors concluded that fermenting endosymbionts in the iguanas' hindgut must be very sensitive to low-level environmental disturbance or contamination. This sensitivity probably compromises the digestive efficiency of affected iguanas, causing their corticosterone concentrations to rise and triggering an increase in mortality.

7.3.4 Ophidia

Concentrations of PAHs in several tissues of Annulated Sea Snakes (*Hydrophis cyanocinctus*), Short Sea Snakes (*Lapemis curtus*), and Beaked Sea Snakes (*Enhydrina schistose*) have been reported, but no pathologic studies were performed (Sereshk and Bakhtiari, 2014; Mote et al., 2015).

Jones et al. (2009) detected PAH in the shed skin of Corn Snakes (*Elaphe guttata*) experimentally fed with mice contaminated with benzo[*a*]pyrene and 3-methylcholanthrene. According to authors, because normal shedding of skins in snakes is a process that involves lymph diffusing into the space between old and new skins, this may be a route through which the PAHs are deposited in the shed skins.

Table 7.5 Summary of the Main Effects of Several Polycyclic Aromatic Hydrocarbons (PAHs) and Crude Oil on Reptiles

Compound	Species	Environment	Effect	Reference
PAHs	*Chrysemys picta* *Chelydra serpentina*	Field	Developmental abnormalities	Bell et al. (2006)
PAHs	*Malaclemys terrapin*	Field	Malformations	Outerbridge et al. (2016)
Benzo[*a*]pyrene 7,12-dimethylbenz[*a*]anthracene	*Chelydra serpentina*	Laboratory	Decreased embryo survival Deformities in embryos	Van Meter et al. (2006)
Benzo[*a*]pyrene	*Phrynops geoffroanus*	Laboratory	Increased GGT levels	Silva et al. (2016)
Phenantrene	*Caretta caretta*	Field	Increased plasma Na:K ratio Increased ALT levels	Camacho et al. (2013b)
Carcinogenic PAHs	*Caretta caretta*	Field	Erytrhocytic DNA damage	Casini et al. (2018)
Crude oil	*Caretta caretta*	Laboratory	Increased WBC counts Decreased RBC counts Necrotizing dermatitis	Lutcavage et al. (1995)
Crude oil	*Caretta caretta*	Field	Esophageal impaction Necrotizing gastritis/enteritis Necrotizing hepatitis Tubulonephrosis	Orós et al., 2004, 2005); Camacho et al. (2013a)
Crude oil	*Caretta caretta*	Field	Reduced RBC and thrombocyte counts Increased ALT and ALP levels	Casal and Orós (2009)
PAHs	*Acanthodactylus scutellatus*	Field	Hepatocyte damage	Al-Hashem (2011)

7.4 Inorganic Pollutants: Metals

Metals such as cobalt (Co), copper (Cu), chromium (Cr), iron (Fe), magnesium (Mg), manganese (Mn), molybdenum (Mo), nickel (Ni), selenium (Se), and zinc (Zn) are essential nutrients that are required for various biochemical and physiological functions (WHO, 1996); although they are necessary, they become toxic at high concentrations.

Other metals such as aluminum (Al), antimony (Sb), arsenic (As), barium (Ba), beryllium (Be), bismuth (Bi), cadmium (Cd), gallium (Ga), germanium (Ge), gold (Au), indium (In), lead (Pb), lithium (Li), mercury (Hg), nickel (Ni), platinum (Pt), silver (Ag), strontium (Sr), tellurium (Te), thallium (Tl), tin (Sn), titanium (Ti), vanadium (V), and uranium (U) have no established biological functions and are considered as non-essential metals (Tchounwou et al., 2012)

7.4.1 Chelonia

7.4.1.1 Metals in Freshwater Turtles and Tortoises

Mercury. Experimental studies on the effects of $HgCl_2$ on ion transport in the urinary bladder of slider turtles (*Trachemys* [formerly *Pseudemys*] *scripta*) demonstrated that Hg inhibits Na^+ transport in the urinary bladder without altering passive ion fluxes (Schwartz and Flamenbaum, 1976).

Jacobson et al. (1991) observed that desert tortoises (*Gopherus* [formerly *Xerobates*] *agassizii*) with upper respiratory tract disease had higher concentrations of Hg and Fe than healthy tortoises. While Fe probably accumulated in the livers as a consequence of red blood breakdown and inability to reutilize iron released from hemoglobin, it was not possible to determine if Hg was responsible for the thymic atrophy observed in the ill tortoises.

Hg concentration in Common Snapping Turtle (*Chelydra serpentina*) eggs was negatively correlated with hatching success, and this effect was driven by both increased egg infertility and embryonic mortality (Hopkins et al., 2013b). The authors suggested that maternally transferred Hg is more embryotoxic than teratogenic, as malformation frequencies were uniformly low and were not correlated with total Hg concentration in eggs. Recently, high levels of maternally transferred Hg disrupted magnetic responses of snapping turtle hatchlings, suggesting that Hg may compromise the performance of turtles in a wide variety of spatial tasks (Landler et al., 2017). Micronuclei in RBCs of European Pond Turtles (*Emys orbicularis*) from an industrial area of Azerbaian were statistically correlated with tissue levels of Hg, suggesting genotoxicity; no other lesions were studied (Swartz et al., 2003).

Meyer et al. (2014) found evidence that Hg exposure in Western Pond Turtles (*Emys marmorata*) was negatively correlated with plasma triiodothyronine and positively correlated with thyroxine, suggesting that methylmercury may influence the endocrine system in turtles, specifically altering the thyroid hormone response; no lesions were studied.

Lead. Snapping turtles from Pb-contaminated sites showed lower δ-aminolevulinic acid dehydratase (δ-ALAD) (an enzyme of the heme synthesis pathway) activity levels (Overmann and Krajicek, 1995). A negative and significant correlation was also found between δ-ALAD activity and blood Pb levels in Spur-Thighed Tortoises (*Testudo graeca*), indicating the suitability of this enzyme as biomarker for Pb in this species (Martínez-López et al., 2010).

Burger et al. (1998) examined the effects of Pb on behavioral development of hatchling slider turtles, concluding that Pb at $\geq$1 mg/g had a major effect on survival, a lesser effect on growth, and a small but significant effect on the righting response. Bishop et al. (2010) did not find positive correlation between Pb concentration and shell disease in emydid turtles (32 Slider Turtles and 10 Northern Redbelly Turtles [*Pseudemys rubriventris*]).

A negative correlation between total WBC counts and liver Pb concentration was found in Red-Eared Slider Turtles (*Trachemys scripta elegans*) inhabiting aquatic ecosystems associated with a gaseous diffusion plant in Kentucky, suggesting that Pb concentrations may have an immunosuppressive effect in this species (Yu et al., 2011); however, hematocrit values were not correlated with metal concentrations in tissues. A trend toward a decrease in WBC with increasing Pb was also observed in free-ranging Eastern Box Turtles (*Terrapene carolina carolina*); no morphological changes in the leukocytes were studied (Allender et al., 2015).

Arsenic. Mortality of Red-Eared Sliders and common snapping turtles showing symptoms linked to chronic exposure to As was reported in Finfeather Lake, Texas in 1973 (Cearley, 1973). More recently, Allender et al. (2015) found a trend for decreasing mass and total solids with increasing As concentration in free-ranging Eastern Box Turtles.

Cadmium. Experimental studies on the distribution of Cd in tissues of Painted Turtles (*Chrysemys picta*) showed that Cd was accumulated preferentially in liver, followed by kidney, pancreas, and gastrointestinal tract. In addition, after Cd injection, the concentration of metallothionein-like protein was highest in liver, followed by pancreas and kidney; no lesions were studied (Rie et al., 2001). Authors also suggested the potential for interference by Cd with steroidogenesis and steroid action (i.e., vitellogenin induction and oviduct function) (Rie et al., 2005; Kitana et al., 2006). More recently, liver cells from Chinese Pond Turtles (*Mauremys reevesii*) experimentally exposed to Cd chloride by intraperitoneal injection, showed swollen, degeneration and necrosis depending on the administered dose. Under electronic microscope, nucleus, mitochondria, and rough endoplasmic reticulum presented various degrees of lesions depending on the dose (Huo et al., 2017). The activities of superoxide dismutase and catalase in plasma of these turtles after Cd exposure were lower than those in the controls; in addition, the treated animals had a higher content of malonyldialdehyde (secondary product of lipid peroxidation, used as an indicator for membrane damage) and lower content of glutathione in plasma, suggesting that Cd causes oxidative stress and damage in these turtles under experimental conditions (Huo et al., 2018).

Kitana and Callard (2008) investigated the effect of an environmentally relevant Cd dose on gonadal development during the end of the germ cell migration phase and postnatal gonadal maturation in lab-reared slider turtles. The total number of germ cells in Cd-treated embryos was less than half that found in control embryos, suggesting that Cd may reduce proliferation and/or delay migration of germ cells to the genital ridge. In addition, the effects of Cd on turtle gonadal development were found to extend into 3 months post hatch. Proliferation of oocytes was not influenced by exposure to Cd in ovo; in contrast, apoptosis of oocytes was significantly increased in Cd-exposed turtles. Overall, the data provided evidence that environmentally relevant doses of Cd may affect gonadal development processes of freshwater turtles during embryonic and postnatal stages and may result in disruption of reproductive processes later in life.

A significantly higher frequency of aneuploidy was observed in RBCs of Red-Eared Slider Turtles from the Tar Creek Superfund Site (a site contaminated with Pb, Zn, and Cd) when compared with reference sites. While blood and carapace Pb levels were not significantly different between different site populations, blood Cd was significantly higher in animals at the contaminated site (Hays and McBee, 2007). Results of the study carried out by Yu et al. (2011) (see -Lead-) suggested that Cd exposure could increase heterophil:lymphocyte ratios because Cd is a highly toxic element.

Selenium. Incubation in coal ash–contaminated soil was associated with reduced embryo survivorship in Sliders (*Trachemys scripta*). Adult females from a contaminated site accumulated high levels of As, Cd, Cr, and Se in the liver, yet Se was the only element transferred maternally to hatchlings at relatively high levels. Hatchlings from females at the polluted site exhibited reduced O_2 consumption rates compared to hatchlings from reference sites. The authors concluded that relatively high levels of Se transferred to hatchlings by females at the ash-polluted site might contribute to the observed differences in hatchling physiology (Nagle et al., 2001). In a more recent study, the potential immunological effects of coal combustion residues (with high levels of As, Cd, and Se) in Yellow-Bellied Sliders (*Trachemys scripta scripta*) were evaluated via bacterial killing assays and phytohemagglutinin assays; turtles from reference wetlands exhibited lower bacterial killing abilities than individuals from contaminated sites (Haskins et al., 2017). More recently, Haskins et al. (2018) reported histopathologic changes in Yellow-Bellied Sliders experimentally exposed to selenomethionine; lesions included acute tubular degeneration, necrosis and regeneration in the kidney, with hyaline droplets in the high-dose animals (**Figure 7.9**), acute degenerative changes in the pectoralis muscle, and changes in the claws ranging from epidermal hyperplasia with disorganization and intercellular edema to ulceration, and accumulation of seroheterophilic exudate between the epidermis and cornified layer (**Figure 7.10**).

7.4.1.2 Metals in Sea Turtles

Mercury. When Hg toxicity was studied in Loggerhead Turtles (*Caretta caretta*) (Day et al., 2007), blood Hg concentrations were positively correlated with hematocrit and CPK activity, and negatively correlated with lymphocyte cell counts and AST; ex vivo negative correlations between blood Hg concentrations and B-cell proliferation were also observed. According to the authors, the positive correlations between blood Hg concentration and hematocrit reflects the higher affinity of Hg species for erythrocytes than plasma and demonstrates the importance of measuring hematocrit when analyzing whole blood for Hg. In vitro exposure of peripheral blood leukocytes to methylmercury resulted in suppression of proliferative responses for B cells and T cells. Peden-Adams et al. (2003) found no correlation between blood Hg levels and total WBC counts or WBC differential counts in Kemp's Ridley Turtles (*Lepidochelys kempii*), but they did report a negative relationship between Hg levels in scutes and the number of circulating eosinophils, and between blood Hg levels and T-cell proliferation. A positive correlation between blood Hg concentration and creatine kinase was also found in cold-stunned Kemp's Ridley Turtles (Innis et al., 2008).

Mercury concentrations in green turtles (*Chelonia mydas*) inhabiting an urbanized estuary in San Diego exceeded immune function no-effects thresholds, and increased carapace metal loads were correlated with higher levels of multiple health markers (ALP, CPK, glucose, and albumin:globulin ratio); however, interpretation of these relationships was confounded by covariance with turtle size (Komoroske et al., 2011). When Hg and Se concentrations were measured in nesting female Leatherbacks (*Dermochelys coriacea*) and their hatchlings from Florida (Perrault et al., 2011), both liver Se and the Se:Hg ratio positively correlated with Leatherback hatching and emergence success. Selenium deficiency can result from elevated concentrations of bodily Hg, resulting in degeneration of heart and skeletal muscle of hatchlings (Miller et al., 2009). More recently, concentrations of Hg and Pb in RBCs from green turtles negatively correlated with body condition index, while Hg positively correlated with total tumor score of green turtles afflicted with fibropapillomatosis, suggesting a possible association with Hg and increased tumor growth (Perrault et al., 2017a).

Wang et al. (2013) tested comparatively several cell lines established from different organs and tissues from Green Turtles for their cytotoxic response to $HgCl_2$. Turtle liver cells appeared to be most tolerant to Hg exposure, while turtle lung cells were most sensitive to the compound. More recently, Tong et al. (2016) evaluated the cytotoxicity of $HgCl_2$ and methylmercury in five neural cell lines established from different animal species. Experimental results indicated that both forms of Hg were toxic to all neural cells, and that organic Hg is much more toxic to these neural cells than inorganic Hg; among these cell lines, two cell lines derived from green turtles showed a significantly high tolerance to $HgCl_2$ as compared to the three mammalian neural cells.

Lead. Recently, the reduced concentration of serum cholesterol observed in Green Turtles from Brazil heavily afflicted with fibropapillomatosis was related to a Cu- and Pb-induced inhibition of 3-hydroxy-3-methylglutaryl-CoA reductase activity paralleled by a higher lipid peroxidaton rate induced by increased Fe and Pb concentrations. Because oxidative stress is implicated in the pathogenesis of viral infections, the authors suggested that metal contamination, especially by Cu, Fe, and Pb, could be contributing to the pathogenesis of fibropapillomatosis through oxidative stress generation (da Silva et al., 2016). However, no correlation was found between metal concentrations and fibropapillomatosis in green turtles from Hawaii because of the low detected levels (Aguirre et al., 1994). Perrault et al. (2017b) observed negative relationships between whole blood Pb and albumin and α_2-globulins in nesting loggerheads, suggesting that Pb exposure may interfere with the humoral immune response. More recently, Cortés-Gómez et al. (2018b) found negative correlations between several metals (Pb, Ti, As, Cr, and Se) and p-nitrophenyl acetate esterase activity in the serum of nesting Olive Ridley Turtles (*Lepidochelys olivacea*).

Arsenic. Torrent et al. (2004) detected severe diffuse hepatocellular vacuolar degeneration and multifocal necrotizing hepatitis in three juvenile loggerhead turtles with high As hepatic concentrations (**Figure 7.11**). Ley-Quiñónez et al. (2017) observed significant associations among blood As and Cd levels with ALP enzyme in loggerheads, suggesting liver damage. In other study, concentrations of As, Cd, and Pb in RBCs from Kemp's Ridley Turtles positively correlated with superoxide dismutase, suggesting oxidative stress (Perrault et al., 2017a). In a recent survey on nesting loggerheads from Florida, a negative relationship between whole blood As and γ-globulins was observed, suggesting altered humoral immunity (Perrault et al., 2017b).

Cadmium. Andreani et al. (2008) found positive correlations between Cd concentration and Cd-metallothionein (MT) in liver and kidney of green turtles from Costa Rica and loggerheads from the Mediterranean Sea, suggesting a role for MT in metal storage and detoxification. Although MT has been found in a wide range of animal species, only Anan et al. (2002) isolated MT in hepatic cytosols of green and Hawksbill Turtles from Japan. Lipid peroxidation levels were positively correlated with Cd concentrations in green turtles from Baja California (Labrada-Martagón et al., 2011). However, green turtles stranded in the Arabian Sea with high Cd hepatic concentrations showed low antioxidant enzyme activity and lipid peroxidation levels (Bicho et al., 2006). More recently, Cortés-Gómez et al. (2018c) found significant negative correlations between Cd and Pb and p-nitrophenyl acetate esterase activity in the serum of Olive Ridley Turtles (*Lepidochelys olivacea*) from a Mexican Pacific population. Cadmium was also negatively correlated with creatinine and glucose (Cortés-Gómez et al., 2018b). Several positive relationships of superoxide dismutase, catalase, and glutathione reductase gene expression in the different tissues and very high Cd concentrations were found in this population of Olive Ridley Turtles, suggesting that these turtles are adapting to the metal production of reactive oxygen species (ROS) and damage through a high transcription of these antioxidants (Cortés-Gómez et al., 2018a).

In the largest published study investigating the levels of inorganic pollutants and their potential adverse effects on the hematological and biochemical parameters of sea turtles, decreased RBC counts were associated with higher concentrations of Cd, indicating that Cd might contribute to anemia in loggerheads (Camacho et al., 2013c). Ley-Quiñónez et al. (2017) observed significant associations among blood Cd levels with ALP enzyme in loggerheads, suggesting liver damage. A statistically significant positive correlation between Cd in carapace and GGT in plasma was also recently observed in Mediterranean Loggerhead Turtles (Casini et al., 2018). Perrault et al. (2017b) observed negative relationships between whole blood Cd and γ-globulins and total globulins in nesting loggerheads, suggesting a potential immunosuppressive effect.

Tan et al. (2010) evaluated the cytotoxicity of Cd, Cr, Cu, and Zn in 10 cell lines established from juvenile green turtles. All these metals were cytotoxic to the cell lines at varied concentrations, but the cytotoxicities of Cd and Cr were significantly more potent than the other two metals.

Selenium. Camacho et al. (2013c) detected a negative association between blood Se levels and WBC, suggesting that this essential element may induce immunosuppressive effects on nesting loggerhead turtles. More recently, negative relationships between whole blood Se and total γ-globulins, and positive relationships between whole blood Se and lymphocyte and monocyte counts were observed in nesting loggerheads, suggesting that the humoral immune system may be suppressed/more sensitive to Se exposure, while the cell-mediated immune system may be modulated (Perrault et al., 2017b).

Copper. In vitro studies showed that high concentrations of Cu reduced testosterone and estradiol binding activities and demonstrated a competition for binding sites with both sex steroids in the plasma of green turtles; however, Pb and Zn had no effect on the binding activities of testosterone (Ikonomopoulou et al., 2009).

Manganese. A positive association between blood Mn levels and the heterophil:lymphocyte ratio was reported in nesting loggerheads (Camacho et al., 2013c). Because the numbers of heterophils and lymphocytes are differently affected by stress, the elevated ratio of heterophils to lymphocytes in reptiles may reflect a composite measure of the stress response.

Chromium. Wise et al. (2014) demonstrated the cytotoxicity and genotoxicity of particulate and soluble Cr(VI) compounds in skin cells developed from a Hawksbill Sea Turtle (*Eretmochelys imbricata*). Because the induction of chromosome aberrations is a standard accepted short-term test for cancer, authors concluded that Cr(VI) is likely a carcinogen in sea turtles, as it is in humans and rodents. More recently, the cytotoxicity and genotoxicity of particulate and soluble Cr (VI) have also been proved in leatherback sea turtle lung cells (Speer et al., 2018).

7.4.2 Crocodylia

Mercury. Although American Alligators (*Alligator mississippiensis*) from the Everglades had significantly higher Hg concentrations than alligators from reference sites, no clinical signs of neurologic, hepatic, or renal toxicosis were detected; in addition, Hg accumulation was not associated with gross or light microscopic lesions in any organs (Heaton-Jones et al., 1997). Recently, Nilsen et al. (2016) demonstrated that global measures of genomic DNA methylation in RBCs of alligators were negatively associated with high Hg exposure. DNA methylation, consisting of the covalent addition of a methyl-group to the 5′ carbon of cytosine, is an epigenetic modification functioning in roles that include regulating gene expression, promoting chromosome stability, and silencing the transcription of transposons. They also identified a possible link connecting environmental exposures to the age-associated loss of global DNA methylation. In a recent study, when Hg and metallothioneins (MTs) were studied in blood fractions and tissues of captive Morelet's Crocodiles (*Crocodylus moreletii*), a significant negative relationship between Hg and hemoglobin was observed, probably due to disorders in the oxidation process and even denature of this protein; a positive trend was observed between MTs and hemoglobin, confirming the detoxifing functions of the MTs in blood fractions providing protection against oxidative damage (Buenfil-Rojas et al., 2018). More recently, erythrocytic DNA damage associated with high Hg levels in Spectacled Caimans (*Caiman crocodilus*) from tropical ecosystems impacted by gold mining has been reported (Marrugo-Negrete et al., 2019).

Lead. Camus et al. (1998) reported high Pb levels in farmed American Alligators consuming nutria meat contaminated by lead bullets; clinical signs included anorexia, weight loss, poor growth, lethargy, and death, but there were no hematologic abnormalities. Karyomegaly with margination of chromatin was seen in renal tubular epithelial cells, but intranuclear lead inclusions were not detected. The authors suggested that alligators were relatively resistant to the toxic effects of Pb. Another report described two False Gharials (*Tomistoma schlegelii*) and a Cuban Crocodile (*Crocodylus rhombifer*) with high blood Pb levels that showed no signs of Pb toxicosis (Cook et al., 1988). When juvenile Saltwater Crocodiles (*Crocodylus porosus*) were fed several Pb bullets, blood Pb concentrations increased by an order of magnitude within a week, and the crocodiles remained in apparent good physical condition and displayed no clinical signs of Pb toxicosis throughout the experiment, also suggesting that crocodilians may possess a relatively high degree of resistance to Pb toxicosis (Hammerton et al., 2003). It was postulated that the metabolism of Pb in crocodiles differs markedly from that described in birds and mammals and includes an enhanced mechanism for preventing or alleviating the effects of this toxic metal. A similar conclusion was drawn from a recent study in which no clinical effects of Pb toxicosis were observed in free-ranging Nile Crocodiles (*Crocodylus niloticus*) in South Africa, even though the highest concentration detected (960 µg/dL) represents the most elevated blood Pb concentration recorded to date for a free-ranging vertebrate (Warner et al., 2016).

However, high-yolk Pb was suggested as a probable cause for early embryonic death in captive alligators fed with nutria meat contaminated with Pb shot (Lance et al., 2006). Other studies suggested that altered deposition of yolk steroids, and possibly differential utilization by the embryo, could be a contributory mechanism in the reproductive and developmental abnormalities seen in alligators from contaminated sites (Hamlin et al., 2010).

Cadmium. Metallothioneins (MTs) are metal-binding proteins that play an important role in the transport of metals in internal tissues and have been related to heavy metal detoxification in mammals. Juvenile alligators from a contaminated site had higher MT concentrations and lower GST activity compared to those from other organochlorine-polluted site and a reference site (Gunderson et al., 2016). Mean blood concentrations of Cd and MTs in Morelet's Crocodiles (*Crocodylus moreletii*) also suggested that MTs may be related to Cd exposure (Buenfil-Rojas et al., 2015). More recently, when Finger et al. (2016) investigated the possible effects of chronic dietary ingestion of prey contaminated with coal combustion wastes (rich in Cd, As, and Se) on juvenile alligators, no negative effects on the immune and hematological system were detected.

Selenium. Although Roe et al. (2004) reported lower American Alligator clutch viability (30%–54%) at a Se-contaminated site compared to reference nests (67%–74% viability), other factors, such as dietary nutrient deficiencies, genetic abnormalities, female size and age, population density, and the presence of other contaminants, could affect clutch viability in crocodilians. More recently, dietary selenomethionine administration in American Alligators significantly reduced body condition; accumulation also was associated with mortality, suggesting toxicity, but no lesions were investigated (Finger et al., 2017).

Chromium. Wise et al. (2016) demonstrated that particulate and soluble Cr(VI) are both cytotoxic and genotoxic to alligator cells derived from scute tissue in a concentration-dependent manner.

Iron. In a recent study on metallic elements in Nile Crocodile eggs from the Kruger National Park, authors found Fe possibly contributing toward thicker eggshells, which could act as barrier to gas and water exchange, as well as increased effort required for the hatchling to emerge (du Preez et al., 2018).

7.4.3 Sauria

Lead. Thirty percent of Western Fence Lizards (*Sceloporus occidentalis*) orally exposed to 1000 mg/kg Pb died within

24 hours, and 50% of the surviving lizards exhibited a significant increase in skin pigmentation (darkened) but showed no other overt signs of poisoning. Changes in skin coloration are a classic sign of stress in lizards and are primarily mediated by chromoactive hormones. Despite showing these signs of stress at near-lethal doses of Pb, sprint velocity of these lizards was not impaired (Holem et al., 2006). In a sub-chronic study (daily dose for 60 days) on this species (Salice et al., 2009), signs of suspected Pb-induced anemia were observed, including hemoglobin, hematocrit, and serum total solids; calcium and glucose were reduced. More recently, Simonyan et al. (2018) observed a positive correlation between DNA damage in erythrocytes of Azerbaidzhan Lizards (*Darevskia raddei*) and Armenian Lizards (*Darevskia armeniaca*) and concentration of Pb, Cu, As, and other metals in soil.

Arsenic. When Iberian Rock Lizards (*Lacerta monticola cyrenni*) were exposed to an artificial substrate watered with As concentrations ranging from 50 to 500 ppb, no effect on embryo survival, incubation duration, and hatchling size was observed; however, there was a strong negative relationship between the average hatchling running speed and the As concentration in the substrate water (Marco et al., 2004).

Cadmium. Brasfield et al. (2004) reported acute mortality when high Cd doses (1480 and 14800 μg Cd/g) were applied to the substrate in which Eastern Fence Lizard (*Sceloporus undulatus*) eggs were incubated. There was a decrease in triiodothyronine:thyroxine ratio at the highest surviving dose (148 μg Cd/g), being the only indication that thyroid function may be compromised in these animals. However, adult Carbonell Wall Lizards (*Podarcis carbonelli*) exposed to an environmentally realistic concentration of Cd in a 21-week dietary study appeared to be in good physical condition at the end of the experiment (Mann et al., 2006, 2007).

The effects of Cd on the Italian Wall Lizard (*Podarcis sicula*) have been widely studied under experimental conditions. Regarding its effects on the glial architecture in the brain, Favorito et al. (2007) demonstrated that a chronic Cd exposure involved a maximum reduction of the glial cells at 60 days, whereas at 120 days there was a return of occurrence of glial fibrillary acidic protein-structures similar to that of control lizards. More recently, Favorito et al. (2017) demonstrated that an acute exposure to Cd resulted in diffuse edema in the brain and decrease of the expression of glial fibrillary acidic protein marker, with possible consequent damage of glial cell functions.

Ferrandino et al. (2009) demonstrated that acute exposure to $CdCl_2$ induces apoptosis in the pituitary gland cells of Italian Wall Lizards, and this effect appears irreversible. According to the authors, Cd activates caspase-3 and increases DNA fragmentation in the entire gland, but these effects are more evident in the rostral pars distalis where prolactin cells are particularly located; these cells could therefore be those most involved in apoptosis. Intraperitoneal injection of a single and massive dose of $CdCl_2$ produced an inhibitory effect on the normal hormonal secretion of adrenocorticotropic (ACTH), prolactin (PRL), and somatotropin (gonadotropic, GTH) cells (Favorito et al., 2010); however, a chronic treatment for 120 days showed that the GTH cells, whose secretion was initially inhibited, returned indeed at normal values (Ferrandino et al., 2007). After a chronic oral $CdCl_2$ treatment (Ferrandino et al., 2010), PRL and ACTH cells showed an increase in occurrence and immunostaining intensity, indicative of a hormonal accumulation in these cells due to the inhibiting effect induced by Cd. The authors suggested that Cd may compete with calcium at the pituitary level through membrane channels or by changing intracellular calcium mobilization, as postulated for mammals, and inhibit hormone secretion; cadmium could also cause alterations in receptor binding and secretory mechanisms of pituitary hormones as reported in rats. Unlike tissue alterations, which persisted over time, the inhibitory action both on ACTH- and PRL-secreting cells diminished. The authors suggested a probable adaptation to the toxic action of Cd over time or the activation of defense mechanisms such as metallothionein.

De Falco et al. (2010) demonstrated that acute and chronic Cd treatments were able to influence the morphophysiology of the adrenal gland of the Italian Wall Lizard, inducing in a time-dependent manner steroidogenic cord hyperplasia, disorganization of steroidogenic parenchyma until necrotic degeneration that in turn evokes macrophage infiltration (**Figures 7.12–7.13**).

Experimental studies demonstrated that Cd induced hepatic edema, activated metallothionein expression in Kupffer cells and extracellular matrix production in fat storing cells, and caused swelling and alteration in lipid and sugar metabolism in hepatocytes (**Figures 7.14–7.15**) (Simoniello et al., 2010a). In addition, the response was not strictly dose and time dependent and almost no recovery occurred over short time periods. In adults, the accumulation in liver was accompanied by severe damage at multiple levels (hepatic congestion, hydropic swelling of hepatocytes, edema, altered sugar metabolism) interfering with tissue structure and function (Scudiero et al., 2011).

In ovary Cd, accumulation was accompanied by stimulation of oogonial proliferation, oocyte recruitment, and follicular atresia (**Figures 7.16–7.17**). The effects exerted by Cd in the ovary of this species seem to be not typically estrogenic but rather FSH-like. Collectively, these effects strongly reduce fecundity and reproductive performance (**Figures 7.18–7.20**) (Simoniello et al., 2010b; Scudiero et al., 2011; Simoniello et al., 2013). The effects of the oral dose were time independent, persisting essentially unchanged regardless of single or multiple administration, so the authors concluded that even occasional sublethal Cd contamination may significantly impair reproductive performance in these animals (Simoniello et al., 2013).

Incubation of eggs in Cd-contaminated soil had no effect on embryo survival but strongly affected cranial morphogenesis, showing abnormalities in the development of proencephalic vesicles, mesencephalon, and eyes. Authors suggested a possible correlation between inability to induce MT expression and abnormalities observed in the head region (**Figure 7.21**) (Simoniello et al., 2011). Trinchella et al. (2010) also demonstrated that Cd modifies the expression of many genes, particularly genes involved in molecular pathways associated with membrane trafficking, signal transduction, neuronal transmission, and regulation of gene transcription. The disregulation in the expression of these genes could explain many of the morphological alterations displayed by lizard embryos (anencephaly, telencephalon, diencephalon, and mesencephalon abnormalities, eye abnormalities, etc.) (Scudiero et al., 2011).

Cadmium is highly toxic for the retina of *Podarcis* embryos where it causes substantial developmental anomalies, such as anophthalmia, microphthalmia, and retinal dysplasia, probably due to a loss of control over the proliferative processes (**Figure 7.22**) (Simoniello et al., 2009, 2014). Authors concluded that the increased expression of the genes Pax6 and Otx2 could be ascribed to the hyperproliferation of retinal cells rather than to an effective Cd-induced gene overexpression.

Trinchella et al. (2006) suggested that, although MT mRNA induction in gut is a clear indicator of dietary Cd intoxication in this lizard species, in other tissues the MT mRNA level is not a suitable tool for the detection of low levels of Cd contamination in the environment.

Ultrastructural findings in the lungs of Egyptian Spiny-Tailed Lizards (*Uromastyx aegyptius*) orally exposed to Cd for 7 weeks included desquamated pneumocytes with degenerated surfactant materials, thickened alveolar wall, and thickening of alveolar septum due to proliferation of endothelial cells lining the pulmonary capillaries as a result of an active transmigration. Cadmium caused a significant elevation in blood leukocytes, lymphocytes, neutrophils, and monocytes, suggesting chemotactic activity and enhanced transmigration from pulmonary microcirculation into inflammed tissue. Lung tissue free radical production, lactate dehydrogenase, and glutathione peroxidase activities increased significantly from the baseline control (Al-Johany and Haffor, 2009).

Recently, Georgiev et al. (2014) characterized the first metal-responsive transcription factor-1 (MTF-1) from a reptile, the Slow Worm (*Anguis fragilis*). The best characterized target genes of MTF-1 encode the metallothioneins. They demonstrated that its inductibility upon zinc and cadmium load is as high as that of human MTF-1, and that it clearly outperforms the induction of mouse MTF-1.

Selenium. In an experiment on trophic transfer of Se in Western Fence Lizards, female gonad concentrations approached the highest of thresholds for reproductive toxicity in oviparous vertebrates; however, lizards appeared healthy, survived until the end of the study, and reached sexual maturity (Hopkins et al., 2005). Rich and Talent (2009) reported significant lower overall growth in body mass in Leopard Geckos (*Eublepharis macularius*) that ingested high doses of Se compared to controls.

7.4.4 Ophidia

Mercury. Hopkins et al. (2013a) measured wound healing rates, a highly integrative and biologically relevant measure of innate immunity, of female Northern Water Snakes (*Nerodia sipedon*) and their offspring collected from Hg-contaminated sites, but they detected no influence of environmental pollution on maternal or offspring healing rates.

In the same snake species, Chin et al. (2013b) demonstrated maternal transfer of Hg, but they found little evidence of adverse effects of Hg on reproductive output and embryonic survival. However, Hg concentrations in offspring negatively correlated with motivation to feed and strike efficiency, and offspring from contaminated areas maintained consistently lower efficiencies in these regards than in reference individuals (Chin et al., 2013a). More recently, Cusaac et al. (2016) tested the effects of maternally transferred methylmercury on the stress physiology of this snake species, but they observed no effects on leukocyte differentials or corticosterone levels.

Arsenic. Banded Water Snakes (*Nerodia fasciata*) inhabiting a coal ash disposal site had extremely high tissue burdens of As, Se, and Cd. Snakes with high tissue burdens exhibited increased standard metabolic rates (oxygen consumption), possibly indicating that a significant physiological cost was associated with contaminant exposure (Hopkins et al., 1999).

Cadmium. Despite the high levels of Cd, Hg, and Pb found in the shed skins of Corn Snakes (*Elaphe guttata*) experimentally injected with these three metals, all snakes appeared to be healthy with no visible signs of toxicity (Jones and Holladay, 2006). Orós et al. (2009b) described a metastatic fibrosarcoma, with metastasis to the lung, right atrium, both ovaries, and the celomic wall, in a captive Saharan Horned Viper (*Cerastes cerastes*) (**Figure 7.23**), in which high hepatic Cd levels (223 mg/kg) were detected; ultrastructural examination of the tumor cells did not reveal the presence of retroviral particles. Absence of hematology and blood chemistry in this case impeded knowledge of how Cd could affect these parameters.

Aluminum. When blood samples of Dice Snakes (*Natrix tessellate*) caught in a mixed metals–contaminated area of Serbia were examined, the activities of catalase, glutathione peroxidase, glutathione reductase, and acetylcholinesterase and the concentration of lipid peroxides were significantly decreased, while GST activity and total glutathione concentration were significantly increased compared to specimens from the control area (Gavrić et al., 2015).

Selenium. When juvenile Banded Water Snakes were fed prey items collected from a coal ash-contaminated site, Se accumulation was most notable, greatly exceeding established toxicity thresholds for other vertebrates. However, the snakes survived through the study and exhibited normal food consumption, growth, condition factors, overwinter survival and mass loss, metabolic rate, and gonadosomatic index (Hopkins et al., 2002). The majority (71%) of snakes did not exhibit any differences in liver histology when compared to control

Table 7.6 Summary of the Main Effects of Several Metals on Reptiles

Compound	Species	Environment	Effect	Reference
Mercury	*Chelydra serpentina*	Laboratory	Increased egg infertility and embryonic mortality	Hopkins et al. (2013b)
	Emys orbicularis	Field	Micronuclei in RBCs	Swartz et al. (2003)
	Emys marmorata	Field	Reduced plasma triiodothyronine levels Increased plasma thyroxine levels	Meyer et al. (2014)
	Caretta caretta	Field	Increased hematocrit levels and CPK activity Decreased lymphocyte cell counts and AST levels	Day et al. (2007)
	Lepidochelys kempii	Field	Decreased T-cell proliferation	Peden-Adams et al. (2003)
	Lepidochelys kempii	Field	Increased CPK levels	Innis et al. (2008)
	Chelonia mydas	Field	Increased ALP, CPK, glucose, and albumin:globulin ratio	Komoroske et al. (2011)
	Caiman crocodilus	Field	Erythrocytic DNA damage	Marrugo-Negrete et al. (2019)
Lead	*Trachemys scripta elegans*	Field	Decreased WBC counts	Yu et al. (2011)
	Terrapene carolina carolina	Field	Decreased WBC counts	Allender et al. (2015)
	Chelonia mydas	Field	Decreased serum cholesterol levels	da Silva et al. (2016)
	Caretta caretta	Field	Decreased albumin and α_2-globulin levels	Perrault et al. (2017b)
	Alligator mississippiensis	Farm	Damage in renal tubular epithelial cells	Camus et al. (1998)
	Alligator mississippiensis	Farm	Early embryonic death	Lance et al. (2006)
	Sceloporus occidentalis	Laboratory	Increased skin pigmentation	Holem et al. (2006)
	Sceloporus occidentalis	Laboratory	Decreased hemoglobin, hematocrit, serum total solids, calcium, and glucose values	Salice et al. (2009)
Arsenic	*Terrapene carolina carolina*	Field	Decreased plasma total solids values	Allender et al. (2015)
	Caretta caretta	Field	Diffuse hepatocellular vacuolar degeneration and multifocal necrotizing hepatitis	Torrent et al. (2004)
	Caretta caretta	Field	Increased ALP levels	Ley-Quiñónez et al. (2017)
	Caretta caretta	Field	Decreased γ-globulins levels	Perrault et al. (2017b)
Cadmium	*Mauremys reevesii*	Laboratory	Swollen, degeneration and necrosis of hepatocytes Altered nucleus, mitochondria and rough endoplasmic reticulum (hepatocytes)	Huo et al. (2017)
	Mauremys reevesii	Laboratory	Decreased plasmatic superoxide dismutase and catalase activities Increased malonyldialdehyde levels Decreased glutathione levels	Huo et al. (2018)
	Trachemys scripta	Laboratory	Altered gonadal development	Kitana and Callard (2008)
	Trachemys scripta elegans	Field	Aneuploidy in RBCs	Hays and McBee (2007)
	Trachemys scripta elegans	Field	Increased heterophil:lymphocyte ratio	Yu et al. (2011)
	Caretta caretta	Field	Decreased RBC counts	Camacho et al. (2013c)
	Caretta caretta	Field	Increased ALP levels	Ley-Quiñónez et al. (2017)
	Caretta caretta	Field	Decreased γ-globulins and total globulins levels	Perrault et al. (2017b)

(Continued)

Table 7.6 (*Continued*) Summary of the Main Effects of Several Metals on Reptiles

	Compound	Species	Environment	Effect
	Caretta caretta	Field	Increased GGT	Casini et al. (2018)
	Sceloporus undulatus	Laboratory	Acute mortality Decreased triiodothyronine:thyroxine ratio	Brasfield et al. (2004)
	Podarcis sicula	Laboratory	Damage of glial cells	Favorito et al., 2007, (2017)
	Podarcis sicula	Laboratory	Apoptosis of the pituitary gland cells	Ferrandino et al. (2009)
	Podarcis sicula	Laboratory	Disorganization of steroidogenic parenchyma and necrotic degeneration (adrenal gland)	De Falco et al. (2010)
	Podarcis sicula	Laboratory	Hepatic damage	Simoniello et al. (2010a); Scudiero et al. (2011)
	Podarcis sicula	Laboratory	Oocyte recruitment, follicular atresia Reduced fecundity	Simoniello et al., 2010b, (2013); Scudiero et al. (2011)
	Podarcis sicula	Laboratory	Developmental abnormalities (proencephalic vesicles, mesencephalon and eye)	Simoniello et al. (2011); Scudiero et al. (2011)
	Podarcis sicula	Laboratory	Anophthalmia, microphthalmia, and retinal dysplasia	Simoniello et al. (2009, 2014)
	Uromastyx aegyptius	Laboratory	Damage of pulmonary cells Increased WBC counts	Al-Johany and Haffor (2009)
Selenium	*Trachemys scripta scripta*	Laboratory	Renal tubular damage Lesions in claws	Haskins et al. (2018)
	Caretta caretta	Field	Decreased WBC counts	Camacho et al. (2013c)
	Caretta caretta	Field	Decreased γ-globulins Increased lymphocyte and monocyte counts	Perrault et al. (2017b)
	Alligator mississippiensis	Laboratory	Reduced body condition	Finger et al. (2017)
	Eublepharis macularius	Laboratory	Reduced overall growth in body mass	Rich and Talent (2009)
Chromium	*Eretmochelys imbricata*	Laboratory	Cytotoxicity and genotoxicity in skin cells	Wise et al. (2014)
	Dermochelys coriacea	Laboratory	Cytotoxicity and genotoxicity in lung cells	Speer et al. (2018)
	Alligator mississippiensis	Laboratory	Cytotoxicity and genotoxicity in skin cells	Wise et al. (2016)
Manganese	*Caretta caretta*	Field	Increased heterophil:lymphocyte ratio	Camacho et al. (2013c)

snakes fed an uncontaminated diet. The remaining contaminant-exposed snakes had hepatic intrasinusoidal fibrosis, with proliferation of collagen fibers that resulted in narrowing or occlusion. A substantial individual variability in response to trace element exposure was suggested (Ganser et al., 2003).

Brown House Snakes (*Lampropeltis fuliginosus*) experimentally exposed to excessive Se accumulated statistically significant concentrations of Se in kidney, liver, and ovarian tissue compared to controls, but accumulation had no effect on female survival, food consumption, growth, or body condition (Hopkins et al., 2004).

7.5 Emerging Pollutants

According to Sauvé and Desrosiers (2014), we can extend the focus on emerging contaminants (contaminants which have appeared only recently) to contaminants of emerging concern (contaminants which have been in the environment for a while but for which concerns have been raised much more recently). Some of the pollutants considered emerging a few years ago are persistent pollutants, and readers can find the information in the POPs section. Most recent emerging contaminants include anionic surfactants (nonylphenol, octylphenol), antibacterial products (triclosan, triclocarban), fungicides (strobilurins, triazoles, imidazoles, dimethomorph, boscalid, famoxadone, etc.), and pharmaceuticals. Studies on the effects of these emerging pollutants on reptiles are scarce.

7.5.1 Chelonia

7.5.1.1 Emerging Pollutants in Freshwater Turtles The endocrine disruption by bisphenol-A (BPA) has been investigated in freshwater turtles. Most turtle sexual development begins with a bipotential gonad, which under the influence of temperature and chemicals will develop into an ovary or testis. Initially, the sex-determining region box 9 (*sox9*) is expressed in the bipotential gonad. However, with the development of Sertoli cells in the male testis, *sox9* is up-regulated. If the gonad is to develop into an ovary, the existing Sertoli cells become lacunae, medullary *sox9* expression is down-regulated, and cortical beta-catenin expression is up-regulated (Jandegian et al., 2015).

Clairardin et al. (2013) observed that BPA interfered with estradiol metabolism in Slider Turtles (*Trachemys scripta*). Eggs treated with 40 μg/egg BPA early in development prior to the temperature-sensitive period had greater concentrations of yolk estradiol and estrone and lesser concentrations of estrone sulfate than untreated control eggs. They suggested that developmental exposure to BPA alters metabolism of maternal estrogens and possibly makes them bioavailable during later times in development, including the temperature-sensitive period.

When three doses of BPA were applied to eggs of Painted Turtles (*Chrysemys picta*) at the male-producing temperature, the gonads had varying amounts of ovarian-like cortical tissue and disorganized testicular tubules in the medulla. In all three BPA treatments, histologically and immonohistochemically *sox9* patterns revealed disorganized medullary testicular tubules and beta-catenin expression in thickened cortex, suggesting that developmental exposure to BPA disrupts sexual differentiation in Painted Turtles (Jandegian et al., 2015).

7.5.1.2 Emerging Pollutants in Sea Turtles Antibiotic-resistant bacteria isolated from sea turtles are causing deep concern regarding the dissemination of antibiotic resistance in marine wildlife, suggesting that antibiotic resistances are mainly caused by widespread use of antimicrobial agents for treatment of human and animal infections and as growth promoters in agriculture and aquaculture (Al-Bahry et al., 2009, 2011, 2012; Foti et al., 2009; Zavala-Norzagaray et al., 2015; Prichula et al., 2016).

Recently, the use of primary erythrocyte cell cultures of loggerheads as in vitro models for evaluating the effects of 4-nonylphenol and diisodecyl phthalate led to a significant impairment of cellular metabolic activity (Cocci et al., 2017).

7.5.2 Crocodylia

Broad-Snouted Caiman (*Caiman latirostris*) eggs incubated at male-induced temperature and exposed to high doses of bisphenol A experienced sex reversal (Stoker et al., 2003, 2008). Neonatal females produced by sex reversion lacked type III follicles, while females prenatally exposed to the lowest doses of bisphenol A showed an increase in type III follicles. Juvenile caiman prenatally exposed to bisphenol A showed an augmented incidence of multioocyte follicles. Even though the mechanism by which multioocyte follicles develop was unknown, it was postulated to be a consequence of oocyte clusters that did not separate and became enclosed individually in a follicle. In addition, neonatal female caiman exposed in ovo to 17β-estradiol o bisphenol A had higher estrogen serum levels (Stoker et al., 2008).

Although a significant disruption of the gonadal histoarchitecture (**Figures 7.24–7.26**) of Broad-Snouted Caiman was observed after in-ovo exposure to bisphenol A, the changes in *amh* and *sf-1* mRNA expression were not significant (Durando et al., 2013, 2016).

7.5.3 Sauria

Several effects of nonylphenol on the Italian Wall Lizard (*Podarcis sicula*) have been investigated. Acute exposure through intracelomic injection induced a significant decrease of thyroxine and triiodothyronine plasma levels, correlating with a decrease in epithelial cell height, retracted colloid, and few reabsorption vacuoles (**Figure 7.27**) (Sciarrillo et al., 2010). Intracelomic injections in this lizard species caused effects on adrenal gland morphology: increase of steroidogenic cord diameter and vascularization, a strong escalation of adrenaline cell number, and a decrease of noradrenaline cells. The notably elevated levels of adrenal hormones suggested a permanent activation of hypothalamic corticotropin-releasing factor secretion and impaired negative feedback of the hypothalamic–pituitary–adrenal gland axis (**Figures 7.28–7.29**) (De Falco et al., 2014). Verderame et al. (2011) demonstrated that an experimentally nonylphenol-polluted diet (food and water) induced the presence of vitellogenin in the plasma of males and its expression in the liver where it appears related to the expression of the α type of estrogen receptors (ERα). Verderame and Limatola (2015) demonstrated that nonylphenol administered during the mating period causes in this lizard a reduction of spermatogenesis and affects the testicular and epididymal structure, making it similar to that of the nonreproductive period (**Figures 7.30–7.32**). They hypothesized that the lack of androgen-receptor and estrogen-receptor expression in spermatogonia and spermatocytes I could be responsible for the reduction of spermatogenesis.

7.5.4 Ophidia

Zhou et al. (2016) examined several species from the Yangtze River Delta (China) for organohalogen contaminants and detected high concentrations of chlorinated paraffins in the terrestrial snake Short-Tailed Mamushi (*Gloydius brevicaudus*), but no pathologic effects were measured.

7.6 Macroplastics and Microplastics

Plastics are synthetic organic polymers derived from the polymerization of monomers extracted from oil or gas; they show high resistance to aging and minimal biological degradation (Cole et al., 2011). The total annual global production of plastics grew from 1.5 million tons in 1950 to 299 million tons in 2015 and 335 million tons in 2016, resulting in an important environmental concern (Nelms et al., 2016; Alimba and Faggio, 2019). In general, freshwater environments have received less attention regarding plastic pollution than the marine environment (Costa and Barietta, 2015). On a global scale, plastic items are consistently among the most numerically abundant types of marine debris (Thompson et al., 2009; Gall and Thompson, 2015), and are divided into two categories: macroplastics (>5 mm) and microplastics (<5 mm). Plastics that are manufactured to be of a microscopic size are defined as primary microplastics (typically used in facial cleansers and cosmetics), whereas secondary microplastics are plastic fragments derived from the breakdown of larger plastic debris, both at sea and on land (Cole et al., 2011); when exposed to short-wave ultraviolet B radiation in sunlight, the oxidative properties of the atmosphere, and the hydrolytic properties of seawater, plastics break into smaller pieces, eventually becoming individual polymer molecules, which must undergo further degradation before becoming bioavailable (Moore, 2008; Andrady, 2015).

Long retention times of plastic debris in the intestine may cause the releasing of toxic chemicals (e.g., plasticisers, POPs) that may act as endocrine disruptors and therefore can compromise the fitness of individuals (Teuten et al., 2009). In a recent survey on three species of Pacific Pelagic Sea Turtles, the majority of POP accumulation most likely came from their natural prey; however, within the group of green turtles, a significant positive correlation was observed between concentrations of ΣPCBs and the number of plastic pieces ingested (Clukey et al., 2018).

In a recent study in which gut contents from 102 stranded sea turtles belonging to 7 species over 3 ocean basins (Atlantic, Mediterranean, and Pacific) were analyzed, synthetic particles including microplastics were present in every turtle (Duncan et al., 2018). Ingestion of microplastics can cause toxic effects. Microplastics have a large surface area to volume ratio, and consequently they can transport contaminants incorporated in the plastic as part of the manufacturing process (monomers, oligomers, bisphenol-A, phthalate plasticizers, flame retardants, and antimicrobials) and persistent bioaccumulative and toxic substances adsorbed from the marine environment (Gall and Thompson, 2015). The biological repercussions depend on to the size of microplastics ingested, with smaller sizes having greater effects on organisms at the cellular level. According to Lusher (2015), nanometer-size plastics can pass through cell membranes. The effects of microplastics at the cellular level have not been documented in reptiles.

7.7 Pesticide Poisoning

Pesticides are the main chemicals that can cause acute poisoning in reptiles. While reports of poisonings in wild reptiles are scarce, the increasing incidence of reptiles as pets has increased the information available about it. It is estimated that less than 2% of pesticide application is directed to eliminate vertebrates, yet it is well documented that these are often affected after application of pest control treatments. Logically, reptiles are no exception, and though there is scarce literature documenting this, different species of reptiles also appear to be commonly affected. Pesticides can exert their effect on reptiles, either directly because of their intrinsic toxicity, after the direct ingestion of the formulated product, by eating contaminated prey, or through the skin after fumigations. However, pesticides can also affect populations of reptiles indirectly by reducing their food supply, by a loss in quantity or quality of prey, or even by habitat modification produced by pesticides (i.e., herbicides). Depending on the exposure dose, the effects of pesticides can result in death of the reptiles, or low and longer doses over time may exert sublethal effects, which are more difficult to evaluate, such as impaired growth or reproduction, modifications of behavior and skills for getting food, alterations of the ability of regulating body temperature, etc. (Gibbons et al., 2015).

The United States Environmental Protection Agency establishes five categories for the ecological risk assessment of pesticides, according to their lethal dose 50 (LD50) for the different species. For reptiles, this classification is: practically nontoxic (LD50 > 2000 mg/kg), slightly toxic (LD50 = 501–2000 mg/kg), moderately toxic (LD50 = 51–500 mg/kg), highly toxic (LD50 = 10–50 mg/kg), and very highly toxic (LD50 < 10 mg/kg) (Gibbons et al., 2015).

7.7.1 Acetylcholinesterase Inhibitors

Organophosphates (OPs) and carbamates of N-methyl (CMs) are two chemical groups that have had, and still have, wide use as pesticides. While today these compounds face severe restrictions so that many applications are being replaced by less toxic compounds, many are still widely used at home, gardens, agriculture, and veterinary practice. There are many OP and CM compounds currently registered, and all incur the risk of causing acute, subacute, and chronic toxicity in vertebrates, including reptiles.

Apparently, all OPs and CMs share a common mechanism of inhibition of acetylcholinesterase and can cause similar symptoms, but affinities for the enzymes are different. Because they share this mechanism, mixed exposure to compounds of these groups could lead to serious additive.

Possible impacts on reptiles (especially lizards) include acute poisoning by direct eating (or licking) poisoned bait, and secondary poisoning from eating dead, dying, or disorientated insects containing relevant amounts of these pesticides (Spurr, 1993). Given the lack of data it is difficult to draw any strong conclusions about the relationship between toxicity to reptiles and other terrestrial vertebrates (Fryday and Thompson, 2009), but research seems to indicate that sensitivity and patterns of mortality from exposure of reptiles to these pesticides resemble those of birds and mammals (Hall and Clark, 1982). These toxic chemicals have a very high potency because they react rapidly with the active-site serine of cholinesterases (ChEs). Non-degraded acetylcholine leads to its accumulation and therefore affects its physiological functions, causing muscle paralysis, seizures and apnea, seizures and apnea followed by death (Poyot et al., 2006). Several reports of wildlife poisonings throughout the world include acute poisoning with the result of death of different species of reptiles (Fryday and Thompson, 2009; Rattner, 2009; Buckmaster et al., 2014; Ruiz-Suárez et al., 2015).

Interestingly, some authors have also studied the effects of sublethal chronic exposure of reptiles to OPs and CMs on their basic physiological and behavioral performance, since at environmentally relevant sublethal doses, these neurotoxic chemicals can potentially interfere with centrally based essential functions in reptiles such as thermoregulation, locomotion, feeding behaviors, and egg hatchability (Bain et al., 2004; Holem et al., 2006; Fryday and Thompson, 2009; Amaral et al., 2012; Valdes et al., 2015). This research has yielded inconsistent results and, while some authors found that the exposure to non-lethal doses of OPs do not pose a clear adverse effect on these basic functions of reptiles, and even at the highest doses assayed an increase of sprint velocity was observed (Holem et al., 2006), others reported the opposite (Fryday and Thompson, 2009; Amaral et al., 2012; Valdes et al., 2015). However, what is actually common in all these studies is that biomarkers of exposure (activities of GST and enzymes involved in the glutathione redox cycle, glutathione concentrations, esterase activities, or liver and testes histopathology) were substantially affected (Amaral et al., 2012), and some, such as ChE levels, remained substantially depressed up to 21 days after dosing (Bain et al., 2004). Therefore, regardless of those observational studies where no adverse effects were observed, these results suggest that chronic exposure can cause adverse health effects in reptiles (Amaral et al., 2012; Valdes et al., 2015). Undoubtedly, more research is needed in this field.

7.7.2 Pyrethrins and Pyrethroids

Pyrethroids are synthetic insecticides, which are chemically similar to natural pyrethrins but whose basic structures have been modified to increase stability in the natural environment. Pyrethrins were widely used as insecticides until the introduction of cheaper synthetically produced compounds such as organochlorine compounds and acetylcholinesterase inhibitors. After the ban of the vast majority of these, pyrethrin insecticides reemerged as pyrethroids (chemically modified pyrethrins), which are currently widely used as insecticides in agriculture, home, and garden, mainly as insecticides, and for the treatment of ectoparasites of humans and animals. Pyrethrins and pyrethroids mainly act by slowing the opening and closing of neural sodium channels, and thus cause repetitive neural discharge and membrane depolarization. Sensory nerves are more sensible, and thus the most frequently observed clinical sign in all animal species is paresthesia. However, other neurologic signs are frequently observed in acutely poisoned animals, such as tremors and seizures, ataxia, or proprioceptive deficit (Gwaltney-Brant, 2003). Currently there are a number of formulations and commercial preparations with registered uses worldwide, since they are highly effective and are generally considered of middle to low toxicity to mammals. One of the major reasons for this decreased sensitivity of mammals to this class of compounds compared to insects is the fact that they are more potent on the sodium channel at low temperatures than at high temperatures (Gwaltney-Brant, 2003). Therefore, cold-blooded vertebrates are more sensitive probably for the same reason. This is the case for fish, which are very sensitive to the toxic effects of pyrethrins and pyrethroids (Maund et al., 2012). There are a few studies that indicate that the toxicity of pyrethrins and pyrethroids to reptiles would also be very high (Brooks et al., 1998; Fryday and Thompson, 2009; Weir et al., 2015), especially in those inhabiting colder places. Thus, a study done in Green Anole Lizards (*Anolis carolinensis*) demonstrated that the mortality rate at the same dose of pyrethrins was much higher in animals kept at 15°–20°C than in those kept at 35°–38°C (Talent, 2005). However, other studies reported that some pyrethroids like permethrin, fenvalerate, tetramethrin, or phenothrin produced little or no mortality in Brown Tree Snakes (*Boiga irregularis*) (Brooks et al., 1998).

Several studies demonstrated that chronic exposure to pyrethroids at doses as low as 1% of their LC50 value is capable of rendering toxicity in fish (Maund et al., 2012; Sapana-Devi and Gupta, 2014). However, the chronic effects of pyrethrins and pyrethroids in other cold-blooded vertebrates such as reptiles have been scarcely studied, and only a few studies can be found in the scientific literature to date. The most relevant of those focused on the effect of these chemicals on eggs. As many reptiles oviposit in soils associated with agricultural landscapes, the exposure of eggs to agricultural chemical residues in soil may be relevant. In a controlled study, the authors demonstrated that at typical field application rates, pyrethroids may adversely affect hatching success, and increase both mortality and the frequency of deformities in snapping turtles (de Solla et al., 2011).

7.7.3 Anticoagulant Rodenticides

Rodenticides are a large group of chemical substances that have in common the aim of eliminating rodent pests. Most currently used rodenticides are anticoagulant compounds that interfere with blood clotting, and as a result cause death due to blood loss and tissue hypoxia, which occurs 2–4 days after being consumed. However, there are a number of rodenticides that are not anticoagulants, such as zinc phosphide, bromethalin, cholecalciferol, and strychnine, whose toxicities rely on different mechanisms of action. Rodenticides are widely used in agricultural lands to control rodents that destroy crops. In addition, as invasive mammals are a great threat to biodiversity, especially on islands, and invasive rodents are likely responsible for the greatest number of extinctions and ecosystem changes, rodenticides are also widely used in eradication campaigns (Howald et al., 2007). However, these compounds are nonselective and have proved to be toxic to many nontarget species. They are perhaps the class of agrochemicals that most frequently affect populations of reptiles, at least according to the number of reports found in the scientific literature (Brooks et al., 1998; Howald et al., 2007; Witmer et al., 2007; Sánchez-Barbudo et al., 2012; Pitt et al., 2015; Weir et al., 2015; Rueda et al., 2016).

Anticoagulant rodenticides are perhaps the group of pesticides that have a longer and successful history of use, given their high effectiveness. However, this also motivates numerous accidental poisonings of domestic animals, wildlife, and even humans, and for this reason many of these products have been banned in recent years. Interestingly, it has recently been reported that reptiles seem to be less susceptible to the toxicity of anticoagulant rodenticides than mammals or birds (Weir et al., 2015). In a recent invasive rodent eradication campaign in the Galapagos Islands, no population-level poisoning mortality of Pinzon Island Lava Lizards (*Microlophus duncanensis*) was observed. However, they accumulated significant amounts of the rodenticide employed, brodifacoum, which was detected in lizards' tissues several months after the campaign. Thus, the lizards constituted a living reservoir of the rodenticide, and this was a significant contributor to unplanned mortality of predatory birds in these islands (Rueda et al., 2016). This does not mean, however, that the reptile species are resistant to poisoning by anticoagulant rodenticides, especially when exposed to massive doses through direct ingestion of baits, and it has been shown that mortality rates exposed to bait prepared with diphacinone or warfarin reached levels close to 100% (Brooks et al., 1998), and several reports described wild reptile species poisoning by these chemicals (Sánchez-Barbudo et al., 2012; Pitt et al., 2015).

7.7.4 Cholecalciferol (and Calcium)

After international regulations have removed most second-generation anticoagulant rodenticides from the over-the-counter market and severely restricted the open-field use of some of them, other types of rodenticides have become more popular. Among them, cholecalciferol is one of the most used. Chlolecalciferol is marketed as rodenticide, either alone or in combination with permitted anticoagulant rodenticides such as coumatetralyl. Rodenticide baits prepared with vitamin D_3 as the active ingredient represent the primary source of exposure to this chemical in domestic and wild animals (Peterson and Fluegeman, 2013). There is considerable species variation in susceptibility to cholecalciferol among mammals and birds, but it is considered that the risk of secondary poisoning is low, as multiple exposures over several days would be required to cause toxicosis (O'Connor and Eason, 2000). The data in reptiles are virtually nonexistent, but in a controlled assay about the toxicity of cholecalciferol in Brown Tree Snakes the authors found no mortality after oral administration of high doses (40 mg/kg) (Brooks et al., 1998).

Apart from exposure to this type of rodenticide, another very important source of exposure of reptiles to toxic doses of vitamin D_3 are calcium supplements in captive reptiles, as nutritional metabolic bone disease (NMBD) is one of the most frequently observed diseases in captive reptiles, especially herbivores (Hoby et al., 2010). However, since the minimum daily requirements of calcium and vitamin D_3 for most species of reptiles have not been established, owners, breeders, and veterinarians often overdose them inadvertently, and this may expose these animals to doses that can be several hundred times the physiological requirements of calcium and vitamin D_3.

Whatever the source of exposure—rodenticides or supplements—vitamin D toxicosis arises from the induction of hypercalcemia. Hypercalcemia may be prolonged, especially when due to supplementation, and causes metastatic calcification of blood vessels, gastrointestinal tissues, kidneys, lungs, blood vessels, and joints. In particular, the destruction of the glomerular and tubular basement membrane is responsible for the production of acute renal failure, which further exacerbates widespread tissue mineralization, leading to death. (Wallach, 1966). Calcification of soft tissues was observed at levels of only three times the recommended supplement dose (Fledelius et al., 2005). Once calcification of soft tissues occurs, therapies are often ineffective and the damage is permanent, so prophylaxis is the best approach.

7.7.5 Sodium Monofluoroacetate (Compound 1080)

Sodium monofluoroacetate (MFAS), or compound 1080, is an extremely toxic product of natural origin, which was introduced in the market in the 1940s for use by professionals. MFAS toxicity relates to an interference in the Krebs cycle, that is inhibited. The MFAS combines with acetyl coenzyme A to form fluoroacetyl coenzyme A, which quickly becomes fluorocitric acid. This conversion is called "lethal synthesis" since the fluorocitric acid inhibits the enzyme aconitase, preventing the passage of citrate to isocitrate, and ultimately breaking the cycle of tricarboxylic acids. This compound has been demonstrated to be lethal to all classes of vertebrates after a single dose, including reptiles (McIlroy et al., 1985; Fryday and Thompson, 2009). However, lethal doses described for different species of reptiles are around 4 times higher than those described in birds, and around 40–50 times higher than those in mammals (McIlroy et al., 1985). Compound 1080 has been widely used to control rodents and other invasive mammal species worldwide, but has been discontinued in many countries, including the United States and the European Union; however, in Australia and other countries it is still approved for the control of invasive species, especially fox control.

Despite its high toxicity, certain zones of Australia have many native fauna, including reptiles, that have an increased tolerance to 1080 as a result of coevolution with *Gastrolobium* and *Oxylobium* plant species that contain a natural form of MFAS (Dundas et al., 2014). It has been described that after a baiting campaign for the control of foxes a relevant amount of baits were removed by Lace Monitors (*Varanus varius*) but no mortality of these reptiles was registered, which would support the lower toxicity of compound 1080 for reptiles, or at least in this species (Woodford et al., 2012).

7.7.6 Zinc Phosphide

Zinc phosphide (ZP) is another chemical that has been widely used as a rodenticide and for the control of invasive species. ZP is a highly toxic compound for all vertebrates. After ingestion, zinc phosphide is rapidly converted to phosphine gas, which is easily absorbed and distributed to the stomach and intestines and is captured by the liver and lungs, where it interferes with several essential metabolic pathways. Clinical signs are circulatory collapse, hypotension, shock symptoms, myocarditis, pericarditis, acute pulmonary edema, and congestive heart failure. All this usually causes death within minutes (Ruder et al., 2011). The use of zinc phosphide is still permitted in most countries for licensed applicators. Its toxicity for reptiles has been demonstrated after campaigns for the control of invasive lizards, where it was estimated that amounts as low as 25 mg/lizard (in 0.5% baits) produced a mortality rate close to 100% (Avery et al., 2011). However, it has been reported that with strict control of the doses used for the preparation of baits, the risk of poisoning to non-target reptiles is very low (Engeman et al., 2006; Witmer et al., 2010), so it can be deduced that they are presumably more resistant to the effects of this chemical, as is the case with other rodenticides discussed in this chapter.

7.7.7 Glyphosate

Glyphosate-based herbicides (GBH) are the most widely used pesticides worldwide. Their uses have increased exponentially since their introduction on the market, basically linked to the parallel use of genetically modified crops engineered to tolerate GBH. Although initially glyphosate and GBH were marketed as virtually nontoxic pesticides, the evidence of their toxic effects on living organisms, ecosystems, and humans has been increasing steadily in the scientific literature (Myers et al., 2016). Furthermore, the World Health Organization's International Agency for Research on Cancer recently has classified glyphosate and GBH as potentially carcinogenic to humans (Guyton et al., 2015). Glyphosate has been clearly demonstrated to be toxic for amphibians (mainly) and reptiles, particularly those living in aquatic environments (Carpenter et al., 2016). As also described for other species of vertebrates, the toxicity of glyphosate and GBH to reptiles seems to be highly potentiated by additives such as surfactants within certain formulations (Carpenter et al., 2016). Several studies demonstrated the effects of glyphosate or GBH exposure on genotoxicity, thyroid disrupting effects, mortality, cellular immune response, body temperature (fever), body size, and morphological abnormalities in several reptile species (Poletta et al., 2011; Bicho et al., 2013; Latorre et al., 2013; López-González et al., 2013; Carpenter et al., 2016; Schaumburg et al., 2016; Siroski et al., 2016). It seems clear that strict regulation of the conditions for application of agrochemicals is in order for those areas where wild populations of terrestrial and aquatic species live, to minimize the adverse effects on these ecosystems.

7.7.8 Atrazine

Atrazine is the second most widely used herbicide after glyphosate. Although it has been banned since 2004 in the European Union, atrazine is sill massively used in many countries around the globe including the United States and Australia, as its environmental safety remains controversial. Although there are many studies in cold-blooded vertebrates (fish, amphibians, and reptiles) assessing the toxicity of atrazine, no clear acute effects have been reported. The majority of studies in reptiles exposed to atrazine (eight species investigated to date) find alterations in biomarkers (expression of genes and/or associated proteins, concentrations of hormones, and biochemical processes), but a recent meta-analysis has concluded that these alterations were not translated to adverse outcomes in terms of apical endpoints (Van Der Kraak et al., 2014).

7.7.9 Neonicotinoids and Fipronil

Neonicotinoids and fipronil are insecticides that have recently become commercially available and have gained popularity quickly. There are several neonicotinoids registered worldwide (imidacloprid, clothianidin, acetamiprid, thiamethoxam), which are widely used mainly for crop protection. The synthetic analogs of nicotine, namely neonicotinoids, and the phenyl pyrazole fipronil act as systemic insecticides, sharing a similar mechanism of action. These insecticides act as agonists at the nicotinic acetylcholine receptors, causing paralysis and death. The high specificity for receptors in insects was considered to have highly selective toxicity to insects and relatively low toxicity to vertebrates; however, in recent years increasing numbers of cases of acute poisoning and damage to ecosystems and wildlife derived from the intensive use of these chemicals have been reported (Gibbons et al., 2015).

Although more than 150 studies to date have focused on the acute toxic effects of neonicotinoids on non-target vertebrates, none of them have studied the effects of neonicotinoids in reptiles, and only three have studied fipronil in these vertebrates. One of those studies found that fipronil can be considered highly toxic to Fringe-Toed Lizards (*Acanthodactylus dumerili*), where after the administration of a single dose of 30 mg/kg via contaminated prey ingestion, the authors found a mortality rate of 62% and survivors showed significantly reduced feeding activity, food consumption, body weight, and organ-to-body-weight ratios (liver and/or fat body) (Peveling and Demba, 2003). In other reptile species, such as Madagascar Iguana (*Chaladoron madagascariensis*) and the skink *Mabuy elegans*, the same authors described a reduction of the size of the population attributable to an indirect effect of the use of fipronil in the region, which resulted in a decrease in the availability of termites to these lizards (Peveling et al., 2003). These important studies reveal the acute toxicity of fipronil in reptiles and are the only studies that have shown indirect adverse effects of these systemic insecticides in vertebrates with consequences at the population level.

While similar data for neonicotinoids are not available, these compounds could possibly exert indirect effects on populations of reptiles (and other vertebrates) similar to those of fipronil. These indirect effects may be as important as the direct toxic effects on vertebrates due to the efficacy of these drugs for killing the vertebrate prey base. It seems clear that a reevaluation of the ecological risks of neonicotinoids and fipronil is needed, and that this evaluation should take into account direct toxicity as well as that derived from the alteration of the food chain that these pesticides may produce (Peveling and Demba, 2003; Peveling et al., 2003; Gibbons et al., 2015).

7.8 Biotoxins

7.8.1 Saxitoxins

Saxitoxins are neurotoxins produced by several species of microalgae and cyanobacteria. In a recent mortality event affecting hundreds of Diamondback Terrapins (*Malaclemys terrapin*), postmortem analyses revealed the presence of measurable levels of saxitoxin analogs (paralytic shellfish toxins [PST]) within various tissues; in addition, high-throughput sequencing revealed that in the Ribbed Mussel (*Geukensia demissa*), a PST vector was present in the gastrointestinal lumen of all terrapin samples tested. Concurrent with the mortality event, elevated densities of the PST-producing dinoflagellate *Alexandrium fundyense* and high levels of PST in bivalves were observed in the region. No consistent significant gross or histologic changes were present in the examined tissues of the turtles (Hattenrath-Lehmann et al., 2017).

Most sea turtle mortality events involving saxitoxins have been associated with blooms of the dinoflagellate *Pyrodinium* spp. (Licea et al., 2008; Amaya et al., 2018). Flacid paralysis and incoordination have been reported in a stranded green turtle with high levels of saxitoxin (Manire et al., 2017). No histological lesions were detected in sea turtles that died from saxitoxicosis (Manire et al., 2017).

7.8.2 Brevetoxins

Brevetoxins are neurotoxins produced by the dinoflagellate *Karenia brevis*, frequently involved in sea turtle stranding events (Fauquier et al., 2013; Perrault et al., 2017a; Walker et al., 2018). Manire et al. (2013) reported neurologic signs, such as the inability to control the head (head bobbing) and nervous twitching in Green Turtles and Kemp's Ridley Turtles with brevetoxicosis; Loggerheads had more extreme neurologic symptoms, including coma and other symptoms that included generalized edema, conjunctival edema, and cloacal or penile prolapse.

In a red tide bloom event in the Gulf of Mexico, positive correlations were observed between brevetoxins and α_2-globulins in Kemp's Ridleys and α_2- and γ-globulins in green turtles, indicating potential immunostimulation (Perrault et al., 2017a). Plasma lysozyme activity was positively correlated with brevetoxin concentrations measured in the blood of Loggerhead Sea Turtles rescued from red tide toxin exposure (Walsh et al., 2010). Effects on loggerheads long after a bloom had dissipated were also reported. Brevetoxin concentrations in nesting females positively correlated with γ-globulins, indicating a potential for immunomodulation, and also correlated with liver brevetoxin concentrations of dead-in-nest hatchlings, suggesting a decreased reproductive success in this species (Perrault et al., 2016). No specific histological lesions have been described in sea turtles that died from brevetoxicosis (Fauquier et al., 2013).

In a recent experimental study, Slider Turtles (*Trachemys scripta*) and Diamondback Terrapins were administered brevetoxin-3 via esophageal tube or by intratracheal instillation, exhibiting muscular and neurological symptoms consistent with brevetoxicosis including head bobbing, muscle twitching, ataxia, swimming in circles, partial to complete paralysis, penile prolapse in males, edema, and in some cases apparent coma. Lesions included inflammatory changes of the trachea and lungs, and to a lesser extent the small intestine, meninges, and endocardium/myocardium. Importantly, similar lesions were found in some control animals (Cocilova et al., 2017).

7.8.3 Microcystins

Microcystins are toxins produced by several freshwater cyanobacteria (also known as blue-green algae). Most microcystins are hepatotoxins and are associated with *Microcystis aeruginosa*. Deaths of European Pond Turtles (*Emys orbicularis*) and Mediterranean Turtles (*Mauremys leprosa*) from Lake Oubeira (Algeria) were observed during a bloom of *Microcystis* spp. High levels of microcystins were detected in the liver and viscera tissues, and necropsy revealed liver crumbling; no histological study was performed (Nasri et al., 2008).

Microcystis blooms are also a concern to zoos and farms. Doster et al. (2014) reported several Yellow-Bellied Slider deaths in a zoo moat with recurring blooms of toxic *Microcystis aeruginosa*; an extremely high and potentially lethal concentration of the hepatotoxin microcystin (166 ng/g) was found in the liver of a necropsied turtle, and microscopic examination revealed moderate hepatic atrophy. Singo et al. (2017) demonstrated vertical transmission of microcystins to Nile Crocodile (*Crocodylus niloticus*) eggs, but it was not possible to determine whether microcystin affected the hatchability of Nile Crocodile eggs.

7.8.4 Other Toxins from Cyanobacteria

Aetokthonos hydrillicola is a newly described cyanobacteria that grows in abundance on *Hydrilla verticillata*, a widespread invasive aquatic plant, and is associated with vacuolar myelinopathy and mortality of birds (Mercurio et al., 2014; Maerz et al., 2019). Mercurio et al. (2014) conducted a feeding trial to determine whether Painted Turtles (*Chrysemys picta*) would develop vacuolar myelinopathy after feeding on *Hydrilla verticillata* colonized by this cyanobacterium. All turtles fed toxic *Hydrilla* displayed physical and/or neurologic impairment; histologic examination of the brain and spinal cord revealed vacuolations in all treatment turtles (Mercurio et al., 2014).

7.8.5 Tetrodotoxin

Tetrodotoxin (TTX) is an extremely potent neurotoxin found in some fish, such as puffer fish, globefish, and toadfish (order Tetraodontiformes) and in some amphibian, octopus, and shellfish species. Townsend et al. (2012) reported two cases of envenomation in Green Sea Turtles caused by accidental ingestion of Blue-Lined Octopus (*Hapalochlaena fasciata*), whose saliva contains TTX; the octopus was encased within the seagrass bolus in each of the two turtles, and the esophageal tissue directly surrounding the octopus was red and inflammed in both cases. TTX was detected in tissues of both turtles.

Williams et al. (2016) reported a fatal case in a captive Cuvier's Dwarf Caiman (*Paleosuchus palpebrosus*) that had been fed with a Green Spotted Pufferfish (*Tetraodon nigroviridis*) (**Figure 7.33**). On initial presentation, the caiman was unresponsive and limp, and both eyes exhibited severe mydriasis and were unresponsive to light (**Figure 7.34**). Microscopic findings included acute ulcerative tracheitis and aspiration pneumonia attributed to paralysis while still confined to the water-filled aquarium, and acute gastric mucosal erosions and necrosis with bacterial overgrowth. TTX was positively identified and quantified in both the pufferfish and caiman.

In some populations of Common Garter Snakes (*Thamnophis sirtalis*) syntopic with Rough-Skinned Newts (*Taricha granulosa*) (which contain TTX), snake individuals have evolved resistance to TTX, and may benefit from accumulated TTX stores as protection from their own predators (Williams et al., 2012).

7.9 Drug Toxicoses

7.9.1 Ivermectin

Teare and Bush (1983) reported paralysis and death in Red-Footed Tortoises (*Geochelone carbonaria*) treated with a single intramuscular injection of ivermectin (0.4 mg/kg). Experimental studies also demonstrated ivermectin toxicosis at similar dosages (0.1 mg/kg or less) in Leopard Tortoises (*Geochelone pardalis*), Eastern Box Turtles (*Terrapene carolina carolina*), and Red-Eared Turtles (*Trachemys scripta elegans*). The Leopard Tortoise appeared to be the most susceptible of the species tested, consistently developing mild paresis with a dosage of 0.025 mg/kg; death occurred with dosages as low as 0.3 mg/kg (Teare and Bush, 1983).

Ivermectin toxicosis was also described in a Senegalese Chameleon (*Chamaeleo senegalensis*) infected with *Foleyella furcata* after a single subcutaneous dose (0.2 mg/kg). Complete inertia developed within 24 hours after injection and lasted for 7 days (Széll et al., 2001). Neurologic disorders were also reported in Ball Pythons (*Python regius*) treated with ivermectin (Wozniak et al., 2000).

7.9.2 Benzimidizole Anthelmintics

Fenbendazole toxicosis due to overdose was reported in Fea's Vipers (*Azemiops feae*); at necropsy, gross lesions were limited to segmental thickening and reddening of the small intestine. Severe transmural intestinal necrosis was histologically diagnosed (**Figure 7.35**) (Alvarado et al., 2001). In an experimental study, six male Hermann's Tortoises (*Testudo hermanni*) were treated orally with two 5-day courses of fenbendazole 2 weeks apart at a dosage of 50 mg/kg; although the tortoises remained healthy, blood sampling indicated an extended heteropenia with transient hypoglycemia, hyperuricemia, hyperphosphatemia, and equivocal hyperproteinemia/hyperglobulinemia, which were considered to be in response to fenbendazole administration (Neiffer et al., 2005).

7.9.3 Metronidazole

Doses >100 mg/kg of metronidazole were associated with fatal toxicosis in Indigo Snakes (*Drymarchon corais*), California Mountain Kingsnakes (*Lampropeltis zonata*), and Arizona Mountain Kingsnakes (*Lampropeltis pyromelena*) (Funk, 1988). Severe neurological disease and increase of AST levels were also reported in several reptiles after treatment with high doses (>200 mg/kg) of metronidazole (Mitchell, 2006).

7.9.4 Gentamicin

Renal lesions caused by high dose of gentamicin were experimentally demonstrated in Gopher Snakes (*Pituophis catenifer*), consisting of cloudy swelling, hydropic degeneration, and tubular necrosis of the proximal tubules; the snakes subsequently also developed visceral gout (Montali et al., 1979).

Bagger-Sjöbäck and Wersäll (1976) also studied the toxic effects of gentamicin on the basilar papilla in the Oriental Garden Lizard (*Calotes versicolor*). In a more recent study on the saccule of the Italian Wall Lizard (*Podarcis sicula*), gentamicin induced damage in the hair cells (**Figure 7.36**), and the damage increased with increasing duration of treatment (Avallone et al., 2008). In addition, the authors provided morphological evidence of hair cell recovery and regeneration after damage induced by gentamicin.

7.9.5 Dioctyl Sodium Sulfosuccinate

Paul-Murphy et al. (1987) reported necrosis of the esophageal and gastric mucosa in gophersnakes after administration of dioctyl sodium sulfosuccinate (DSS) via oral route (250 mg/kg).

7.9.6 Ibuprofen

Gladden (2006) reported a potential ibuprofen toxicosis in a South American Red-Footed Tortoise (*Geochelone carbonaria*) that accidentally ingested a 200-mg solubilized ibuprofen gelatin capsule. The total dose of 256 mg/kg ingested by the tortoise raised concerns of potential complications based on mammalian toxicity profiles, but no clear evidence of early renal/hepatic damage from ibuprofen toxicity was provided; despite this, decontamination was considered the safest approach.

7.10 Plant Toxicoses

Although herbivorous reptiles can be especially susceptible to accidental ingestion of toxic plants, clinical reports are very scarce. Fitzgerald and Vera (2006) reported depression and death in iguanas associated with ingestion of azalea plants, poisoning in an iguana associated with ingestion of Eastern lily, death in a tortoise and an iguana associated with ingestion of cigarette butts, and stupor in an iguana associated with the ingestion of marijuana. Rotstein et al. (2003) reported necrosis of the oral cavity, esophagus, stomach, and kidneys in an African Spurred Tortoise (*Geochelone sulcata*) that died after ingesting oak *Quercus* leaves.

7.11 Envenomation and Self-Envenomation

Snakebite envenomation is a huge topic, and most scientific literature refers to human envenomation (Gutiérrez et al., 2017). The World Health Organization considers snakebite envenoming as a neglected public health issue (WHO, 2007). Snake venoms have a high diversity of proteins and peptids, displaying many pharmacological and toxic actions (Gutiérrez et al., 2017). Cardiovascular shock, systemic hemorrhage, coagulopathy, paralysis involving respiratory muscles, generalized rhabdomyolysis, and acure renal failure are common life-threatening effects of snakebite envenoming (WHO, 2007; Serino-Silva et al., 2018).

Scientific literature reporting the effects of snakebite envenoming in other reptiles is scarce. Berry et al. (2016) reported the case of a Desert Tortoise (*Gopherus agassizii*) found in distress, with drying blood extending from the left ocular area to the maxillary rhamphotheca (**Figure 7.37**); at necropsy, subcutaneous hemorrhage extending from the left periorbital area to the base of the cervical subcutaneous connective tissue was observed (**Figure 7.38**). Tissue damage was compatible with rattlesnake envenomation (Jacobson and Berry, 2012).

Although not frequently reported, cases of reptile envenomation by other reptiles, even cagemates of the same species, or by ants, scorpions, etc., may occur. While several proteins in the plasma of some venomous and nonvenomous snakes have been found capable of inhibiting the snake venom's action (Serino-Silva et al., 2018), these analytical determinations have not been made in all the venomous snake species.

Johnston and Mulley (2014) reported a possible in-ovo self-envenomation in an Egyptian Saw-Scaled Viper (*Echis pyramidum*); whether death was a direct result of a bite involving penetration of the fangs or an indirect result of a non-penetrative bite (e.g., drowning) was unclear. More recently, Hill and McKillop (2017) reported a fatal self-envenomation in an adult Brown Tree Snake (*Boiga irregularis*). During handling, the snake chewed its own tail with its back teeth. Clinical signs included muscle twitching, ataxia, and heat-seeking behavior which progressed to partial paralysis by 6 hours, generalized paralysis and respiratory arrest at 10 hours, and cardiac arrest by 12 hours post envenomation. Clinical signs were suggestive of potent neurotoxicity for this snake species by its own venom (Hill and McKillop, 2017).

Acknowledgments

The authors thank many colleagues for providing images and comments; those providing images are also acknowledged in the figure legends for this chapter. We also thank the collaboration of the Library of the University of Las Palmas de Gran Canaria, through its Document Access Service (SOD), and M. Carmen Iglesias.

References

Aguirre AA, Balazs GH, and Zimmerman B, and Galey FD. 1994. Organic contaminants and trace metals in the tissues of green turtles (*Chelonia mydas*) afflicted with fibropapillomas in the Hawaiian Islands. *Mar Pollut Bull* 28:109–114.

Al-Bahry SN, Al-Zadjali MA, Mahmoud IY, and Elshafie AE. 2012. Biomonitoring marine habitats in reference to antibiotic resistant bacteria and ampicillin resistance determinants from oviductal fluid of the nesting green sea turtle, *Chelonia mydas*. *Chemosphere* 87:1308–1315.

Al-Bahry SN, Mahmoud IY, Al-Zadjali M, Elshafie A, Al-Harthy A, and Al-Alawi W. 2011. Antibiotic resistant bacteria as bio-indicator of polluted effluent in the green turtles, *Chelonia mydas* in Oman. *Mar Environ Res* 71:139–144.

Al-Bahry SN, Mahmoud IY, Elshafie A et al. 2009. Bacterial flora and antibiotic resistance from eggs of green turtles *Chelonia mydas*: An indication of polluted effluents. *Mar Pollut Bull* 58:720–725.

Al-Hashem MA. 2011. Evidence of hepatotoxicity in the sand lizard *Acanthodactylus scutellatus* from Kuwait's Greater Al-Burgan oil field. *Ecotoxicol Environ Saf* 74:1391–1395.

Al-Hashem MA, Brain PF, and Omar SA. 2007. Effects of oil pollution at Kuwait's greater Al-Burgan oil field on polycyclic aromatic hydrocarbon concentrations in the tissues of the desert lizard *Acanthodactylus scutellatus* and their ant prey. *Ecotoxicology* 16:551–555.

Alimba CG, and Faggio C. 2019. Microplastics in the marine environment: Current trends in environmental pollution and mechanisms of toxicological profile. *Environ Toxicol Pharmacol* 68:61–74.

Al-Johany AM, and Haffor AS. 2009. Effects of cadmium exposure on the ultrastructural pathology of different pulmonary cells, leukocyte count, and activity of glutathione peroxidase and lactate dehydrogenase in relation to free radical production in *Uromastyx aegyptius*. *Ultrastruct Pathol* 33:39–47.

Allender MC, Dreslik MJ, Patel B et al. 2015. Select metal and metalloid surveillance of free-ranging Eastern box turtles from Illinois and Tennesse (*Terrapene carolina carolina*). *Ecotoxicology* 24:1269–1278.

Alvarado TP, Garner MM, Gamble K, Levens GP, Raymond JT, and Nordhausen R. 2001. Fenbendazole overdose in four Fea's vipers (Azemiops feae). In *Proc Joint Conf Am Assoc Zoo Vet, Am Assoc Wildl Vet, Assoc Reptile Amphibian Vet, and Nat Assoc Zoo Wildl Vet.* Orlando, 28–29.

Amaral MJ, Sánchez-Hernández JC, Bicho RC et al. 2012. Biomarkers of exposure and effect in a lacertid lizard (*Podarcis bocagei* Seoane) exposed to chlorpyrifos. *Environ Toxicol Chem* 31:2345–2353.

Amaya O, Quintanilla R, Stacy BA et al. 2018. Large-scale sea turtle mortality events in El Salvador attributed to paralytic shellfish toxin-producing algae blooms. *Front Mar Sci* 5:411.

Anan T, Kunito T, Sakai H, and Tanabe S. 2002. Subcellular distribution of trace elements in the liver of sea turtles. *Mar Pollut Bull* 45:224–229.

Andrady AL. 2015. Persistence of plastic litter in the oceans, in *Marine anthropogenic litter.* Bergmann M, Gutow L and Klages M (Eds.), Springer, Berlin, pp. 57–72.

Andreani G, Santoro M, Cottignoli S, Fabbri M, Carpenè E, and Isani G. 2008. Metal distribution and metallothionein in loggerhead (*Caretta caretta*) and Green (*Chelonia mydas*) sea turtles. *Sci Total Environ* 390:287–294.

Arnold SF, Vonier PM, Collins BM, Klotz DM, Guillette, LJ Jr, and McLachlan JA. 1997. In vitro synergistic interaction of alligator and human estrogen receptors with combinations of environmental chemicals. *Environ Health Perspect* 105:615–618.

Arukwe A, Røsbak R, Adeogun AO et al. 2015. Biotransformation and oxidative stress responses in captive Nile Crocodile (*Crocodylus niloticus*) exposed to organic contaminants from the natural environment in South Africa. *PLOS ONE* 10:e0130002.

Avallone B, Fascio U, Balsamo G, and Marmo F. 2008. Gentamicin ototoxicity in the saccule of the lizard *Podarcis sicula* induces hair cell recovery and regeneration. *Hear Res* 235:15–22.

Avery ML, Eisemann JD, Keacher KL, and Savarie PJ. 2011. Acetaminophen and zinc phosphide for lethal management of invasive lizards *Ctenosaura similis*. *Current Zoology* 57:625–629.

Bagger-Sjöbäck D, and Wersäll J. 1976. Toxic effects of gentamicin on the basilar papilla in the lizard *Calotes versicolor*: A surface study. *Acta Otolaryngol* 81:57–65.

Bain D, Buttemer WA, Astheimer L, Fildes K, and Hooper MJ. 2004. Effects of sublethal fenitrothion ingestion on cholinesterase inhibition, standard metabolism, thermal preference, and prey-capture ability in the Australian central bearded dragon (*Pogona vitticeps*, Agamidae). *Environ Toxicol Chem* 23:109–116.

Beldoménico PM, Rey F, Prado WS, Villareal JC, Muñoz-de-Toro M, and Luque EH. 2007. In ovum exposure to pesticides increases the egg weight loss and decreases hatchlings weight of *Caiman latirostris* (Crocodylia: Alligatoridae). *Ecotoxicol Environ Saf* 68:246–251.

Bell B, Spotila JR, and Congdon J. 2006. High incidence of deformity in aquatic turtles in the John Heinz National Wildlife Refuge. *Environ Pollut* 142:457–465.

Bergeron JM, Crews D, and McLachlan JA. 1994. PCB as environmental estrogens: Turtle sex determination as a biomarker of environmental contamination. *Environ Health Perspect* 102:780–781.

Berny PJ, de Buffrénil V, and Hémery G. 2006. Use of the Nile monitor, *Varanus niloticus L* (Reptilia: Varanidae), as a bioindicator of organochlorine pollution in African wetlands. *Bull Environ Contam Toxicol* 77:359–366.

Berry KH, Shields T, and Jacobson ER. 2016. *Gopherus agassizii* (Mohave desert tortoise). Probable rattlesnake envenomation. *Herpetol Rev* 47:652–653.

Bicho RC, Amaral MJ, Faustino AM et al. 2013. Thyroid disruption in the lizard *Podarcis bocagei* exposed to a mixture of herbicides: A field study. *Ecotoxicology* 22:156–165.

Bicho R, Joaquim N, Mendonça V, Kiyumi AA, Mahmoud IY, and Kindi AA. 2006. Levels of heavy metals and antioxidant enzymes in the green turtle (*Chelonia mydas*) in the Arabian Sea, Sultanate of Oman, in *Proc 6th An Symp Sea Turtle Biol Conserv*. Frick M, Panagopoulou A, Rees AF, and Williams K (Eds.), International Sea Turtle Society, Greece, pp. 49–50.

Bishop CA, Brooks RJ, Carey JH, Ng P, Norstrom RJ, and Lean DRS. 1991. The case for a cause-effect linkage between environmental contamination and development in eggs of the common snapping turtle (*Chelydra s. serpentina*) from Ontario, Canada. *J Toxicol Environ Health* 33:521–547.

Bishop CA, Brown GP, Brooks RJ, Lean DRS, and Carey JH. 1994. Organochlorine contaminant concentrations in eggs and their relationship to body size, and clutch characteristics of the female common snapping turtle (*Chelydra serpentina serpentina*) in Lake Ontario, Canada. *Arch Environ Contam Toxicol* 27:82–87.

Bishop CA, Ng P, Pettit KE et al. 1998. Environmental contamination and developmental abnormalities in eggs and hatchlings of the common snapping turtle (*Chelydra serpentina serpentina*) from the Great Lakes–St Lawrence River basin (1989–91). *Environ Pollut* 101:143–156.

Bishop CA, and Rouse JD. 2000. Chlorinated hydrocarbon concentrations in plasma of the Lake Erie water snake (*Nerodia sipedon insularum*) and northern water snake (*Nerodia sipedon sipedon*) from the Great Lakes basin in 1998. *Arch Environ Contam Toxicol* 39:500–505.

Bishop CA, and Rouse JD. 2006. Polychlorinated biphenyls and organochlorine pesticides in plasma and the embryonic development in Lake Erie water snakes (*Nerodia sipedon insularum*) from Pelee Island, Ontario, Canada (1999). *Arch Environ Contam Toxicol* 51:452–457.

Bishop BE, Savitzky BA, and Abdel-Fattah T. 2010. Lead bioaccumulation in emydid turtles of an urban lake and its relationship to shell disease. *Ecotoxicol Environ Saf* 73:565–571.

Brasfield SM, Bradham K, Wells JB, Talent LG, Lanno RP, and Janz DM. 2004. Development of a terrestrial vertebrate model for assessing bioavailability of cadmium in the fence lizard (*Sceloporus undulatus*) and in ovo effects on hatchling size and thyroid function. *Chemosphere* 54:1643–1651.

Brooks JE, Savarie PJ, and Johnston JJ. 1998. The oral and dermal toxicity of selected chemicals to brown tree snakes (*Boiga irregularis*). *Wildl Res* 25:427–435.

Buckmaster T, Dickman CR, and Johnston MJ. 2014. Assessing risks to non-target species during poison baiting programs for feral cats. *PLOS ONE* 9:e107788.

Buenfil-Rojas AM, Álvarez-Legorreta T, and Cedeño-Vázquez JR. 2015. Metals and metallothioneins in Morelet's Crocodile (*Crocodylus moreletii*) from a transboundary river between Mexico and Belice. *Arch Environ Contam Toxicol* 68:265–273.

Buenfil-Rojas AM, Álvarez-Legorreta T, and Cedeño-Vázquez JR. 2018. Mercury and metallothioneins in blood fractions and tissues of captive Morelet's Crocodiles in Quintana Roo, Mexico. *Chemosphere* 199:630–636.

Burger J. 2006. Neurotoxicology and behavioral effects in reptiles, in *Toxicology of reptiles*. Gardner SC, and Oberdörster E (Eds.), CRC Press, Pensacola, FL, pp. 173–197.

Burger J, Carruth-Hinchey C, Ondroff J, and McMahon M. 1998. Effects of lead on behavior, growth, and survival of hatchling slider turtles. *J Toxicol Environ Health Part A* 55:495–502.

Camacho M, Boada LD, Orós J, Calabuig P, Zumbado M, and Luzardo OP. 2012. Comparative study of polycyclic aromatic hydrocarbons (PAHs) in plasma of eastern Atlantic juvenile and adult nesting Loggerhead Sea Turtles (*Caretta caretta*). *Mar Pollut Bull* 64:1974–1980.

Camacho M, Calabuig P, Luzardo OP, Boada LD, Zumbado M, and Orós J. 2013a. Crude oil as stranding cause among Loggerhead Sea Turtles (*Caretta caretta*) in the Canary Islands, Spain (1998–2011). *J Wildl Dis* 49:637–640.

Camacho M, Luzardo OP, Boada LD et al. 2013b. Potential adverse health effects of persistent organic pollutants on sea turtles: Evidences from a cross-sectional study on Cape Verde Loggerhead Sea Turtles. *Sci Total Environ* 458–460:283–289.

Camacho M, Orós J, Boada LD et al. 2013c. Potential adverse effects of inorganic pollutants on clinical parameters of Loggerhead Sea Turtles (*Caretta caretta*): Results from a nesting colony from Cape Verde, West Africa. *Mar Environ Res* 92:15–22.

Camacho M, Orós J, Henríquez-Hernández LA et al. 2014. Influence of the rehabilitation of injured loggerhead turtles (*Caretta caretta*) on their blood levels of environmental organic pollutants and elements. *Sci Total Environ* 487:436–442.

Camus AC, Mitchell MM, Williams JF, and Jowett PLH. 1998. Elevated lead levels in farmed American alligators *Alligator mississippiensis* consuming nutria *Myocastor coypus* meat contaminated by lead bullets. *J World Aquacult Soc* 29:370376.

Carpenter JK, Monks JM, and Nelson N. 2016. The effect of two glyphosate formulations on a small, diurnal lizard (*Oligosoma polychroma*). *Ecotoxicology* 25:548–554.

Casal AB, and Orós J. 2009. Plasma biochemistry and haematological values in juvenile Loggerhead Sea Turtles undergoing rehabilitation. *Veterinary Record* 164:663–665.

Casini S, Caliani I, Giannetti M et al. 2018. First ecotoxicological assessment of *Caretta caretta* (Linnaeus, 1758) in the Mediterranean Sea using an integrated nondestructive protocol. *Sci Total Environ* 631–632:1221–1233.

Casini S, Marsili L, Fossi MC et al. 2006. Use of biomarkers to investigate toxicological effects of produced water treated with conventional and innovative methods. *Mar Environ Res* 62:S347–S351.

Cearley JE. 1973. *Arsenic Contamination of Finfeather and Municipal Lakes in the City of Bryan*. Texas Water Quality Board, Austin, TX.

Chin SY, Willson JD, Cristol DA, Drewett DVV, and Hopkins WA. 2013a. Altered behavior of neonatal norther watersnakes (*Nerodia sipedon*) exposed to maternally transferred mercury. *Environ Pollut* 176:144–150.

Chin SY, Willson JD, Cristol DA, Drewett DVV, and Hopkins WA. 2013b. High levels of maternally transferred mercury do not affect reproductive output or embryonic survival of northern watersnakes (*Nerodia sipedon*). *Environ Toxicol Chem* 32:619–626.

Ciliberti A, Berny P, Vey D, and de Buffrénil V. 2012. Assessing environmental contamination around obsolete pesticide stockpiles in West Africa: Using the Nile monitor (*Varanus niloticus*) as a sentinel species. *Environ Toxicol Chem* 31:387–394.

Ciliberti A, Martin S, Ferrandez E et al. 2013. Experimental exposure of juvenile savannah monitors (*Varanus exanthematicus*) to an environmentally relevant mixture of three contaminants: Effects and accumulation in tissues. *Environ Sci Pollut Res* 20:3107–3114.

Clairardin SG, Paitz RT, and Bowden RM. 2013. In ovo inhibition of steroid metabolism by bisphenol-A as a potential mechanism of endocrine disruption. *Proc Biol Sci* 280:20131773.

Clukey KE, Lepczyk CA, Balazs GH et al. 2018. Persistent organic pollutants in fat of three species of Pacific pelagic longline caught sea turtles: Accumulation in relation to ingested plastic marine debris. *Sci Total Environ* 610–611:402–411.

Cocci P, Capriotti M, Mosconi G, and Palermo FA. 2017. Effects on endocrine disrupting chemicals on estrogen receptor alpha and heat shock protein 60 gene expression in primary cultures of loggerhead sea turtle (*Caretta caretta*) erythrocytes. *Environ Res* 158:616–624.

Cocci P, Mosconi G, Bracchetti L et al. 2018. Investigating the potential impact of polycyclic aromatic hydrocarbons (PAHs) and polychlorinated biphenyls (PCBs) on gene biomarker expression and global DNA methylation in Loggerhead Sea Turtles (*Caretta caretta*) from the Adriatic Sea. *Sci Total Environ* 619–620:49–57.

Cocilova CC, Flewelling LJ, Bossart GD, Granholm AA, and Milton SL. 2017. Tissue uptake, distribution and excretion of brevetoxin-3 after oral and intratracheal exposure in the freshwater turtle *Trachemys scripta* and the diamondback terrapin *Malaclemys terrapin*. *Aquat Toxicol* 187:29–37.

Cole M, Lindeque P, Halsband C, and Galloway TS. 2011. Microplastics as contaminants in the marine environment: A review. *Mar Pollut Bull* 62:2588–2597.

Cook R, Behler J, Braezitis P, and Dolensek E. 1988. A survey of blood lead levels in crocodilians, in *Int Assoc Aquat Anim Med Proc.* Stoskopf MK (Ed.), pp. 149–150.

Cortés-Gómez AA, Morcillo P, Guardiola FA et al. 2018a. Molecular oxidative stress markers in Olive Ridley Turtles (*Lepidochelys olivacea*) and their relation to metal concentrations in wild populations. *Environ Pollut* 233:156–167.

Cortés-Gómez AA, Tvarijonaviciute A, Girondot M, Tecles F, and Romero D. 2018b. Relationship between plasma biochemistry values and metal concentrations in nesting olive ridley sea turtles. *Environ Sci Pollut Res* 25:36671–9.

Cortés-Gómez AA, Tvarijonaviciute A, Teles M, Cuenca R, Fuentes-Mascorro G, and Romero D. 2018c. *p*-nitrophenyl acetate esterase activity and cortisol as biomarkers of metal pollution in blood of Olive Ridley Turtles (*Lepidochelys olivacea*). *Arch Environ Contam Toxicol* 75:25–36.

Costa MF, and Barietta M. 2015. Microplastics in coastal and marine environments of the western tropical and sub-tropical Atlantic Ocean. *Environ Sci Process Impacts* 17:1868–1879.

Crain DA, Noriega N, Vonier PM, Arnold SF, Mclachlan JA, and Guillette L.J.Jr. 1998. Cellular bioavailability of natural hormones and environmental contaminants as a function of serum and cytosolic binding factors. *Toxicol Ind Health* 14:261–273.

Cusaac JPW, Kremer V, Wright R, Henry C, Otter RR, and Bailey FC. 2016. Effects of maternally-transferred methylmercury on stress physiology in northern water snake (*Nerodia sipedon*) neonates. *Bull Environ Contam Toxicol* 96:725–731.

da Silva CC, Klein RD, Barcarolli IF, and Bianchini A. 2016. Metal contamination as a possible etiology of fibropapillomatosis in juvenile female green sea turtles (*Chelonia mydas*) from the southern Atlantic Ocean. *Aquat Toxicol* 170:42–51.

Day RD, Segars AL, Arendt MD, Lee AM, and Peden-Adams MM. 2007. Relationship of blood mercury levels to health parameters in the loggerhead sea turtle (*Caretta caretta*). *Environ Health Perspect* 115:1421–1428.

De Andrés E, Gómara B, González-Paredes D, and Ruiz-Martín J. 2016. Persistent organic pollutant levels in eggs of leatherback turtles (*Dermochelys coriacea*) point to a decrease in hatching success. *Chemosphere* 146:354–361.

De Falco M, Sciarrillo R, Valiante S et al. 2010. A preliminary study of cadmium effects on the adrenal gland of the lizard *Podarcis sicula*. *Open Zool J* 3:23–29.

De Falco M, Sellitti A, Sciarrillo R et al. 2014. Nonylphenol effects on the HPA axis of the bioindicator vertebrate, *Podarcis sicula* lizard. *Chemosphere* 104:190–196.

de Solla SR. 2010. Organic contaminants in reptiles, in *Ecotoxicology of Amphibians and Reptiles*, 2nd ed. Sparling DW, Linder G, Bishop CA, and Krest SK (Eds.), SETAC Press and CRC Press, Pensacola, FL, pp. 289–324.

de Solla SR, Fernie KJ, and Ashpole S. 2008. Snapping turtles (*Chelydra serpentina*) as bioindicators in Canadian areas of concern in the Great Lakes Basin. II. Changes in hatching success and hatchling deformities in relation to persistent organic pollutants. *Environ Pollut* 153:529–536.

de Solla SR, Fletcher ML, and Bishop CA. 2003. Relative contributions of organochlorine contaminants, parasitism, and predation to reproductive success of eastern spiny softshell turtles (*Apalone spiniferus spiniferus*) from Southern Ontario, Canada. *Ecotoxicology* 12:261–270.

de Solla SR, Martin PA, and Mikoda P. 2011. Toxicity of pesticide and fertilizer mixtures simulating corn production to eggs of snapping turtles (*Chelydra serpentina*). *Sci Total Environ* 409:4306–4311.

Doster E, Chislock MF, Roberts JF, Kottwitz JJ, and Wilson AE. 2014. Recognition of an important water quality issue at zoos: Prevalence and potential threat of toxic cyanobacteria. *J Zoo Wildl Med* 45:174–177.

Duncan EM, Broderick AC, Fuller WJ et al. 2018. Microplastic ingestion ubiquitous in marine turtles. *Glob Change Biol*, 1–9. DOI: 10.1111/gcb.14519.

Dundas S, Flemming PA, and Adams PJ. 2014. First in, first served: Uptake of 1080 poison fox baits in south-west Western Australia. *Wildl Res* 41:117–126.

du Preez M, Govender D, Kylin H, and Bouwman H. 2018. Metallic elements in Nile Crocodile eggs from the Kruger national Park, South Africa. *Ecotoxicol Environ Saf* 148:930–941.

Durando M, Canesini G, Cocito LL et al. 2016. Histomorphological changes in testes of Broad-Snouted Caimans (*Caiman latirostris*) associated with in ovo exposure to endocrine-disrupting chemicals. *J Exp Zool A Ecol Genet Physiol* 325:84–96.

Durando M, Cocito L, Rodríguez HA et al. 2013. Neonatal expression of *amh*, *sox9* and *sf-1* mRNA in *Caiman latirostris* and effects of in ovo exposure to endocrine disrupting chemicals. *Gen Comp Endocrinol* 191:31–38.

Dzul-Caamal R, Hernández-López A, González-Jaúregui M, Padilla SE, Girón-Pérez MI, and Vega-López A. 2016. Usefulness of oxidative stress biomarkers evaluated in the snout scraping, serum and peripheral blood cells of *Crocodylus moreletii* from Southeast Campeche for assessment of the toxic impact of PAHs, metals and total phenols. *Comp Biochem Physiol A: Mol Integr Physiol* 200:35–46.

Eisenreich KM, Kelly AM, and Rowe CL. 2009. Latent mortality of juvenile snapping turtles from the upper Hudson River, New York, exposed maternally and via the diet to polychlorinated biphenyls (PCBs). *Environ Sci Technol* 43:6052–6057.

Ekpubeni FA, and Ekundayo EO. 2002. Effects of exposure of crocodiles to sublethal concentrations of petroleum waste drilling fluid in the Niger Delta basin of Midwestern Nigeria. *Environ Monit Assess* 76:291–298.

Engeman R, Woolard JW, Perry ND et al. 2006. Rapid assessment for a new invasive species threat: The case of the Gambian giant pouched rat in Florida. *Wildl Res* 33:439–448.

Everts JW, Van Frankenhuyzen K, Román B, Cullen J, Copplestone J, and Koeman JH. 1983. Observations on side effects of endosulfan used to control tsetse in a settlement area in connection with a campaign against human sleeping sickness in Ivory Coast. *Trop Pest Manag* 29:177–182.

Fauquier DA, Flewelling LJ, Maucher J et al. 2013. Brevetoxin in blood, biological fluids, and tissues of sea turtles naturally exposed to *Karenia brevis* blooms in central West Florida. *J Zoo Wildl Med* 44:364–375.

Favorito R, Grimaldi MC, Coppola M, and Ferrandino I. 2010. Effects of acute cadmium exposure on the pituitary gland of *Podarcis sicula*. *Open Zool J* 3:30–36.

Favorito R, Grimaldi MC, and Ferrandino I. 2007. Effects of a chronic exposure to cadmium on the glial architecture in the lizard brain: An immunocytochemical study. *Eur J Histochem* 51 suppl 1:21.

Favorito R, Monaco A, Grimaldi MC, and Ferrandino I. 2017. Effects of cadmium on the glial architecture in lizard brain. *Eur J Histochem* 61:14–19.

Ferrandino I, Favorito R, Annunziata M, and Grimaldi MC. 2009. Cadmium induces apoptosis in the pituitary gland of *Podarcis sicula*. *Ann N Y Acad Sci* 1163:386–388.

Ferrandino I, Favorito R, and Grimaldi MC. 2007. Cadmium effects on GTH cells in pituitary gland of *Podarcis sicula*. *Atti del 6° Congresso Nazionale Della Societas Herpetologica Italica*, pp. 151–158.

Ferrandino I, Favorito R, and Grimaldi MC. 2010. Cadmium induces changes in corticotropic and prolactin cells in the *Podarcis sicula* lizard pituitary gland. *Eur J Histochem* 54:200–204.

Finger JW Jr., Hamilton MT, Glenn TC, and Tuberville TD. 2017. Dietary selenomethionine administration in the American alligator (*Alligator mississippiensis*): Hepatic and renal Se accumulation and its effects on growth and body condition. *Arch Environ Contam Toxicol* 72:439–448.

Finger JW Jr, Hamilton MT, Metts BS, Glenn TC, and Tuberville TD. 2016. Chronic ingestion of coal fly-ash contaminated prey and its effects on health and immune parameters in juvenile American alligators (*Alligator mississippiensis*). *Arch Environ Contam Toxicol* 71:347–358.

Fitzgerald KT, and Vera R. 2006. *Reported toxicities in reptiles. In Reptile Medicine and Surgery* 2nd ed. Mader DR (Ed.), Saunders-Elsevier, St. Louis, MO, pp. 1068–1080.

Fledelius B, Jorgensen GW, Jensen HE, and Brimer L. 2005. Influence of the calcium content of the diet offered to leopard tortoises (*Geochelone pardalis*). *Vet Rec* 156:831–835.

Fontenot LW, Noblet GP, Akins JM, Stephens MD, and Cobb GP. 2000. Bioaccumulation of polychlorinated biphenyls in ranid frogs and northern water snakes from a hazardous waste site and a contaminated watershed. *Chemosphere* 40:803–809.

Ford DK, and Holliday CM. 2004. PCB 126 has detrimental effects on bone density in juvenile diamondback terrapins. *Integr Comp Biol* 44:695.

Ford DK, and Holliday CM. 2005. An investigation into the effects of PCB-126 on bone density in turtles. *Integr Comp Biol* 45:997.

Foti M, Giacopello C, Bottari T, Fisichella V, Rinaldo D, and Mammina C. 2009. Antibiotic resistance of gram negatives isolates from Loggerhead Sea Turtles (*Caretta caretta*) in the central Mediterranean Sea. *Mar Pollut Bull* 58:1363–1366.

Fritts TH, and McGehee MA. 1981. *Effects of Petroleum on the Development and Survival of Marine Turtle Embryos*. Contract No. 14-16-0009-870-946, FWS/OBS-81/37. US Fish and Wildlife Service, US Dep. of Interior, Washington DC.

Fryday S, and Thompson H. 2009. *Compared Toxicity of Chemicals to Reptiles and Other Vertebrates*. Contract No. CFT/EFSA/PPR/2008/01. Food and Environment Research Agency, United Kingdom.

Funk RS. 1988. Herp health hints and husbandry: parasiticide dosages for captive amphibians and reptiles. *Chicago Herp Soc Newsletter* 23:30.

Gale RW, Bergeron JM, Willingham EJ, and Crews D. 2002. Turtle sex determination assay: Mass balance and responses to 2,3,7,8-tetrachlorodibenzo-p-dioxin and 3,3′,4,4′,5-pentachlorobiphenyl. *Environ Toxicol Chem* 21:2477–2482.

Gall SC, and Thompson RC. 2015. The impact of debris on marine life. *Mar Pollut Bull* 92:170–179.

Ganser LR, Hopkins WA, O'Neil L, Hasse S, Roe JH, and Sever DM. 2003. Liver histopathology of the southern watersnake, *Nerodia fasciata fasciata*, following chronic exposure to trace element-contaminated prey from a coal ash disposal site. *J Herpetol* 37:219–226.

Gavrić JP, Prokić MD, Anđelković MZ et al. 2015. Effects of metals on blood oxidative stress biomarkers and acetylcholinesterase activity in dice snakes (*Natrix tessellate*) from Serbia. *Arch Biol Sci* 67:303–315.

Georgiev O, Günther V, Steiner K, Schönrath K, and Schaffner W. 2014. The legless lizard *Anguis fragilis* (slow worm) has a potent metal-responsive transcription factor 1 (MTF-1). *Biol Chem* 395:425–431.

Gibbons D, Morrissey C, and Mineau P. 2015. A review of the direct and indirect effects of neonicotinoids and fipronil on vertebrate wildlife. *Environ Sci Pollut Res Int* 22:103–118.

Gladden JN 2006. Clinical management of potential ibuprofen toxicosis in a South American Red-Footed tortoise (*Geochelone carbonaria*). *Vet Clin North Am Exot Anim Pract* 9:599–607.

Gómara B, Gómez G, Díaz-Paniagua C, Marco A, and González MJ. 2007. PCB, DDT, arsenic, and heavy metal (Cd, Cu, Pb, and Zn) concentrations in chameleon (*Chamaeleo chamaeleon*) eggs from Southwest Spain. *Chemosphere* 68:25–31.

Gonzalez-Jauregui M, Valdespino C, Salame-Méndez A, Aguirre-León G, and Rendón-vonOsten J. 2012. Persistent organic contaminants and steroid hormone levels in Morelet's Crocodiles from the Southern Gulf of Mexico. *Arch Environ Contam Toxicol* 62:445–454.

Grant PBC, Woudneh MM, and Ross PS. 2013. Pesticides in blood from spectacled caiman (*Caiman crocodilus*) downstream of banana plantations in Costa Rica. *Environ Toxicol Chem* 32:2576–2583.

Grillitsch B, and Schiesari L. 2010. The ecotoxicology of metals in reptiles, in *Ecotoxicology of Amphibians and Reptiles*, 2nd ed. Sparling DW, Linder G, Bishop CA, and Krest SK (Eds.), SETAC Press and CRC Press, Pensacola, FL, pp. 337–448.

Guerranti C, Baini M, Casini S et al. 2014. Pilot study on levels of chemical contaminants and porphyrins in *Caretta caretta* from the Mediterranean Sea. *Mar Environ Res* 100:33–37.

Guillette, Jr. LJ, Brock JW, Rooney AA, and Woodward AR. 1999. Serum concentrations of various environmental contaminants and their relationship to sex steroid concentrations and phallus size in juvenile American alligators. *Arch Environ Contam Toxicol* 36:447–455.

Guillette Jr LJ, and Crain DA. 1996. Endocrine-disrupting contaminants and reproductive abnormalities in reptiles. *Comments Toxicol* 5:381–398.

Guillette Jr. LJ, Gross TS, Gross D, Rooney AA, and Percival HF. 1995. Gonadal steroidogenesis in vitro from juvenile alligators obtained from contaminated and control lakes. *Environ Health Perspect* 103(Suppl. 4):31–36.

Guillette Jr LJ, Gross TS, Masson GR, Matter JM, Percival HF, and Woodward AR. 1994. Developmental abnormalities of the gonad and abnormal sex hormone concentrations in juvenile alligators from contaminated and control lakes in Florida. *Environ Health Perspect* 102:680–688.

Guillette Jr LJ, Pickford DB, Crain DA, Rooney AA, and Percival HF. 1996. Reduction in penis size and plasma testosterone concentrations in juvenile alligators living in a contaminated environment. *Gen Comp Endocrinol* 101:32–42.

Guillette Jr LJ, Vonier PM, and McLachlan JA. 2002. Affinity of the alligator estrogen receptor for serum pesticide contaminants. *Toxicology* 181–182:151–154.

Gunderson MP, Kohno S, Blumberg B, Iguchi T, and Guillette LJ Jr. 2006. Up-regulation of the alligator CYP3A77 gene by toxaphene and dexamethasone and its short term effect on plasma testosterone concentrations. *Aquat Toxicol* 78:272–283.

Gunderson MP, Pickett MA, Martin JT et al. 2016. Variations in hepatic biomarkers in American alligators (*Alligator mississippiensis*) from three sites in Florida, USA. *Chemosphere* 155:180–187.

Gutiérrez JM, Calvete JJ, Habib AG, Harrison RA, Williams DJ, and Warrell DA. 2017. Snakebite envenoming. *Nat Rev Dis Primers* 3:17063.

Guyton KZ, Loomis D, Grosse Y et al. 2015. Carcinogenicity of tetrachlorvinphos, parathion, malathion, diazinon, and glyphosate. *Lancet Oncol* 16:490–491.

Gwaltney-Brant S. 2003. Insecticides and molluscicides, in *Clinical Veterinary Toxicology*. Plumlee KH (Ed.), Mosby, St. Louis, MO, pp. 177–192.

Hall RJ, Belisle AA, and Sileo L. 1983. Residues of petroleum hydrocarbons in tissue of sea turtles exposed to the Ixtoc I oil spill. *J Wildl Dis* 19:106–109.

Hall RJ, and Clark DR. 1982. Responses of the iguanid lizard *Anolis carolinensis* to four organophosphorus pesticides. *Environ Pollut A, Ecol Biol* 28:45–52.

Hamlin HJ, Lowers RH, Albergotti LC, McCoy MW, Mutz J, and Guillette LJ Jr. 2010. Environmental influence on yolk steroids in American alligators (*Alligator mississippiensis*). *Biol Reprod* 83:736–741.

Hammerton KM, Jayasinghe N, Jeffree RA, and Lim RP. 2003. Experimental study of blood lead kinetics in estuarine crocodiles (*Crocodylus porosus*) exposed to ingested lead shot. *Arch Environ Contam Toxicol* 45:390–398.

Haskins DL, Hamilton MT, Jones AL, Finger Jr. JW, Bringolf RB, and Tuberville TD. 2017. Accumulation of coal combustion residues and their immunological effects in the Yellow-Bellied Slider (*Trachemys scripta scripta*). *Environ Pollut* 224:810–819.

Haskins DL, Howerth EW, and Tuberville TD. 2018. Experimentally induced selenosis in Yellow-Bellied Slider durtles (*Trachemys scripta scripta*). *Vet Pathol* 55:473–477.

Hattenrath-Lehmann TK, Ossiboff RJ, Burnell CA et al. 2017. The role of a PSP-producing *Alexandrium* bloom in an unprecedented diamondback terrapin (*Malaclemys terrapin*) mortality event in Flanders Bay, New York, USA. *Toxicon* 129:36–43.

Hays KA, and McBee K. 2007. Flow cytometric analysis of Red-Eared Slider Turtles (*Trachemys scripta*) from Tar Creek Superfund Site. *Ecotoxicology* 16:353–361.

Heaton-Jones TG, Homer BL, Heaton-Jones DL, and Sundlof SF. 1997. Mercury distribution in American alligators (Alligator *mississippiensis*) in Florida. *J Zoo Wildl Med* 28:62–70.

Hebert CE, Glooschenko V, Haffner GD, and Lazar R. 1993. Organic contaminants in snapping turtle (*Chelydra serpentina*) populations from southern Ontario, Canada. *Arch Environ Contam Toxicol* 24:35–43.

Hecker M, Murphy MB, Giesy JP, and Hopkins WA. 2006. Induction of cytochrome P4501A in African Brown House Snake (*Lamprophis fuliginosus*) primary hepatocytes. *Environ Toxicol Chem* 25:496–502.

Heinz GH, Percival HF, and Jennings ML. 1991. Contaminants in American alligator eggs from Lake Apopka, Lake Griffin, and Lake Okeechobee, Florida. *Environ Monit Assess* 16:277–285.

Hill AG, and McKillop L. 2017. Fatal self-envenomation in a brown tree snake, *Boiga irregularis*, from south-east Queensland. *Toxicon* 126:1–3.

Hoby S, Wenker C, Robert N et al. 2010. Nutritional metabolic bone disease in juvenile veiled chameleons (*Chamaeleo calyptratus*) and its prevention. *J Nutr* 140:1923–1931.

Holem RR, Hopkins WA, and Talent LG. 2006. Effect of acute exposure to malathion and lead on sprint performance of the Western Fence Lizards (*Sceloporus occidentalis*). *Arch Environ Contam Toxicol* 51:111–116.

Holladay SD, Wolf JC, Smith SA, Jones DE, and Robertson JL. 2001. Aural abscesses in wild-caught box turtles (*Terrapene carolina*): Possible role of organochlorine-induced hypovitaminosis A. *Ecotoxicol Environ Saf* 48:99–106.

Holliday DK, Elskus AA, and Roosenburg WM. 2009. Impacts of multiple stressors on growth and metabolic rate of *Malaclemys terrapin*. *Environ Toxicol Chem* 28:338–345.

Holliday DK, and Holliday CM. 2012. The effects of the organopollutant PCB 126 on bone density in juvenile diamondback terrapins (*Malaclemys terrapin*). *Aquat Toxicol* 109:228–233.

Hopkins BC, Chin SY, Willson JD, and Hopkins WA. 2013a. Like mother, like offspring: Maternal and offspring wound healing correlate in snakes. *J Exp Biol* 216:2545–2547.

Hopkins BC, Willson JD, and Hopkins WA. 2013b. Mercury exposure is associated with negative effects on turtle reproduction. *Environ Sci Technol* 47:2416–2422.

Hopkins WA, Roe JH, Snodgrass JW, Staub BP, Jackson BP, and Congdon JD. 2002. Effects of chronic dietary exposure to trace elements on banded water snakes (*Nerodia fasciata*). *Environ Toxicol Chem* 21:906–913.

Hopkins WA, Rowe CL, and Congdon JD. 1999. Elevated trace element concentrations and standard metabolic rate in banded water snakes (*Nerodia fasciata*) exposed to coal combustion wastes. *Environ Toxicol Chem* 18:1258–1263.

Hopkins WA, Staub BP, Baionno JA, Jackson BP, Roe JH, and Ford NB. 2004. Trophic and maternal transfer of selenium in Brown House Snakes (*Lampropeltis fuliginosus*). *Ecotoxicol Environ Saf* 58:285–293.

Hopkins WA, Staub BP, Baionno JA, Jackson BP, and Talent LG. 2005. Transfer of selenium from prey to predators in a simulated terrestrial food chain. *Environ Pollut* 134:447–456.

Howald G, Donlan CJ, Galvan JP et al. 2007. Invasive rodent eradication on islands. *Conserv Biol* 21:1258–1268.

Huo J, Dong A, Niu X et al. 2018. Effects of cadmium on oxidative stress activities in plasma of freshwater turtle *Chinemys reevesii*. *Environ Sci Pollut Res* 25:8027–8034.

Huo J, Dong A, Wang, Y, Lee S, Ma C, and Wang L. 2017. Cadmium induces histopathological injuries and ultrastructural changes in the liver of freshwater turtle (*Chinemys reevesii*). *Chemosphere* 186:459–465.

Ikonomopoulou MP, Olszowy H, Hodge H, and Bradley AJ. 2009. The effect of organochlorine and heavy metals on sex steroid-binding proteins in vitro in the plasma of nesting green turtles, *Chelonia mydas*. *J Comp Physiol B: Biochem, Syst Environ Physiol* 179:653–662.

Innis C, Tlusty M, Perkins C, Holladay S, Merigo C, and Weber ES III. 2008. Trace metal and organochlorine pesticide concentrations in cold-stunned juvenile Kemp's Ridley Turtles (*Lepidochelys kempii*) from Cape Cod, Massachusetts. *Chelonian Conserv Biol* 7:230–239.

Jacobson ER, and Berry KH. 2012. *Mycoplasma testudineum* in free-ranging desert tortoises, *Gopherus agassizii*. *J Wildl Dis* 48:1063–8.

Jacobson ER, Gaskin JM, Brown MB et al. 1991. Chronic upper respiratory tract disease of free-ranging desert tortoises (*Xerobates agassizii*). *J Wildl Dis* 27:296–316.

Jandegian CM, Deem SL, Bhandari RJ et al. 2015. Developmental exposure to bisphenol A (BPA) alters sexual differentiation in Painted Turtles (*Chrysemys picta*). *Gen Comp Endocrinol* 216:77–85.

Johnston RE, and Mulley J. 2014. The biter bit? Investigation of possible *in-ovo* self-envenomation in an Egyptian saw-scaled viper using region of interest X-ray microtomography. *PeerJ Preprints* 2:e2624v1.

Jones DE, Gogal RM Jr, Nader PB, and Holladay SD. 2005. Organochlorine detection in the shed skin of snakes. *Ecotoxicol Environ Saf* 60:282–287.

Jones DE, and Holladay SD. 2006. Excretion of three heavy metals in the shed skin of exposed Corn Snakes (*Elaphe guttata*). *Ecotoxicol Environ Saf* 64:221–225.

Jones DE, Magnin-Bissel G, and Holladay SD. 2009. Detection of polycyclic aromatic hydrocarbons in the shed skins of Corn Snakes (*Elaphe guttata*). *Ecotoxicol Environ Saf* 72:2033–2035.

Keller JM. 2013a. Exposure to and effects of persistent organic pollutants, in *The Biology of Sea Turtles*. Wyneken J, Lohmann KJ, and Musick JA (Eds.), CRC Press, Boca Raton, FL, pp. 285–328.

Keller JM. 2013b. Forty-seven days of decay does not change persistent organic pollutant levels in loggerhead sea turtle eggs. *Environ Toxicol Chem* 32:747–756.

Keller JM, Balazs GH, Nilsen F, Rice M, Work TM, and Jensen BA. 2014. Investigating the potential role of persistent organic pollutants in Hawaiian green sea turtle fibropapillomatosis. *Environ Sci Technol* 48:7807–7816.

Keller JM, Kucklick JR, Stamper MA, Harms CA, and McClellan-Green PD. 2004. Associations between organochlorine contaminant concentrations and clinical health parameters in Loggerhead Sea Turtles from North Carolina, USA. *Environ Health Perspect* 112:1074–1079.

Keller JM, and McClellan-Green P. 2004. Effects of organochlorine compounds on cytochrome P450 aromatase activity in an inmortal sea turtle cell line. *Mar Environ Res* 58:347–351.

Keller JM, McClellan-Green PD, Kucklick JR, Keil DE, and Peden-Adams MM. 2006a. Effects of organochlorine contaminants on loggerhead sea turtle immunity: Comparison of a correlative field study and in vitro exposure experiments. *Environ Health Perspect* 114:70–76.

Keller JM, Peden-Adams MM, and Aguirre AA. 2006b. Immunotoxicology and implications for reptilian health, in *Toxicology of Reptiles*. Gardner SC, and Oberdörster E (Eds.), CRC Press, Pensacola, FL, pp. 199–240.

Keller JM, Ngai L, Braun McNeill J et al. 2012. Perfluoroalkyl contaminants in plasma of five sea turtle species: Comparisons in concentration and potential health risks. *Environ Toxicol Chem* 31:1223–1230.

Kelly SM, Eisenreich KM, Baker JE, and Rowe CL. 2008. Accumulation and maternal transfer of polychlorinated biphenyls in snapping turtles of the upper Hudson River, New York, USA. *Environ Toxicol Chem* 27:2565–2574.

Kihara H, and Yamashita H. 1978. Lethal effects of various agricultural chemicals against reptiles. *Snake* 10:10–15.

Kitana N, and Callard IP. 2008. Effect of cadmium on gonadal development in freshwater turtle (*Trachemys scripta, Chrysemys picta*) embryos. *J Environ Sci Health A Tox Hazard Subst Environ Eng* 43:262–271.

Kitana N, Khonsue W, Won SJ, Lance VA, and Callard IP. 2006. Gonadotropin and estrogen responses in freshwater turtle (*Chrysemys picta*) from Cape Cod, Massachussetts. *Gen Comp Endocrinol* 149:49–57.

Komoroske LM, Lewison RL, Seminoff JA, Deheyn DD, and Dutton PH. 2011. Pollutants and the health of green sea turtles resident to an urbanized estuary in San Diego, CA. *Chemosphere* 84:544–552.

Kroenlein KR, Sleeman JM, Holladay SD et al. 2008. Inability to induce tympanic squamous metaplasia using organochlorine compounds in vitamin A-deficient Red-Eared Sliders (*Trachemys scripta elegans*). *J Wildl Dis* 44:664–669.

Labrada-Martagón V, Tenorio-Rodríguez PA, Méndez-Rodríguez LC, and Zenteno-Savín T. 2011. Oxidative stress indicators and chemical contaminants in East Pacific green turtles (*Chelonia mydas*) inhabiting two foraging coastal lagoons in the Baja California peninsula. *Comp Biochem Physiol C: Toxicol Pharmacol* 154:65–75.

Lambert MRK. 1993. Effects of DDT ground-spraying against tsetse flyes on lizards in NW Zimbabwe. *Environ Pollut* 82:231–237.

Lambert MRK. 1997. Environmental effects of heavy spillage from a destroyed pesticide store near Hargeisa (Somaliland) assessed during the dry season, using reptiles and amphibians as bioindicators. *Arch Environ Contam Toxicol* 32:80–93.

Lance VA, Horn TR, Elsey RM, and de Peyster A. 2006. Chronic incidental lead ingestion in a group of captive-reared alligators (*Alligator mississippiensis*): Possible contribution to reproductive failure. *Comp Biochem Physiol C: Toxicol Pharmacol* 142:30–35.

Landler L, Painter MS, Hopkins Coe B, Youmans PW, Hopkins WA, and Phillips JB. 2017. High levels of maternally transferred mercury disrupt magnetic responses of snapping turtle hatchlings (*Chelydra serpentina*). *Environ Pollut* 228:19–25.

Latorre MA, López-González EC, Larriera A, Poletta GL, and Siroski PA. 2013. Effects of in vivo exposure to Roundup® on immune system of *Caiman latirostris*. *J Immunotoxicol* 10:349–354.

Letcher RJ, Lu Z, de Solla SR, Sandau CD, and Fernie KJ. 2015. Snapping turtles (*Chelydra serpentina*) from Canadian areas of concern across the southern Laurentian Great Lakes: Chlorinated and brominated hydrocarbon contaminants and metabolites in relation to circulating concentrations of thyroxine and vitamin A. *Environ Res* 143:266–278.

Ley-Quiñónez CP, Rossi-Lafferriere NA, Espinoza-Carreon TL et al. 2017. Associations between trace elements and clinical health parameters in the North Pacific loggerhead sea turtle (*Caretta caretta*) from Baja California Sur, *Mexico. Environ Sci Pollut Res* 24:9530–9537.

Licea A, Navarrete J, Bustillos J, and Martínez B. 2008. Monitoring bloom of Pyrodinium bahamense var. compressum in El Salvador and the southern coast of Mexico (November 2005–March 2006), in *Proceedings of the 12th International Conference on Harmful Algae*. Moestrup Ø, Doucette G, Enevoldsen H et al. Copenhagen: International Society for the Study of Harmful Algae and Intergovernmental Oceanographic Commission of UNESCO, 86–89.

Lind PM, Milnes MR, Lundberg R, Bermudez D, Örberg J, and Guillette LJ Jr. 2004. Abnormal bone composition in female juvenile American alligators from a pesticide-pollutant lake (Lake Apopka, Florida). *Environ Health Perspect* 112:359–362.

López-González EC, Latorre MA, Larriera A, Siroski PA, and Poletta GL. 2013. Induction of micronuclei in broad snouted caiman (*Caiman latirostris*) hatchlings exposed in vivo to Roundup® (glyphosate) concentrations used in agriculture. *Pestic Biochem Physiol* 105:131–134.

Lusher A. 2015. Microplastics in the marine environment: Distribution, interactions and effects, in *Marine anthropogenic litter*. Bergmann M, Gutow L, and Klages M (Eds.), Springer, Berlin, pp. 245–307.

Lutcavage ME, Lutz PL, Bossart GD, and Hudson DM. 1995. Physiologic and clinicopathologic effects of crude oil on Loggerhead Sea Turtles. *Arch Environ Contam Toxicol* 28:417–422.

Lutz PL, Lutcavage M, and Hudson D. 1986. Physiological effects, in *Final Report: Study of the Effect of Oil on Marine Turtles*. Varge S, Lutz PL, Odell DK, Van Vleet T, and Bossart G (Eds.). Minerals Management Service Contract No.14-12-0001-30063.

Maerz JC, Wilde SB, Terrell VK et al. 2019. Seasonal and plant specific vulnerability of amphibian tadpoles to the invasion of a novel cyanobacteria. *Biol Invasions* 21:821–831.

Manire CA, Anderson ET, Byrd L, and Fauquier DA. 2013. Dehydration as an effective treatment for brevetoxicosis in Loggerhead Sea Turtles (*Caretta caretta*). *J Zoo Wildl Med* 44:447–452.

Manire CA, Flewelling LJ, and Fauquier DA. 2017. Harmful algae and biotoxins, in *Sea Turtle Health and Rehabilitation*. Manire CA, Norton TM, Stacy BA, Innis CJ, and Harms CA (Eds.), J. Ross Publishing, Plantation, FL, 791–804.

Mann RM, Sánchez-Hernández JC, Serra EA, and Soares AMVM. 2007. Bioaccumulation of Cd by European lacertid lizard after chronic exposure to Cd-contaminated food. *Chemosphere* 68:1525–1534.

Mann RM, Serra EA, and Soares AMVM. 2006. Assimilation of cadmium in a European lacertid lizard: Is trophic transfer important? *Environ Toxicol Chem* 25:3199–3203.

Marco A, López-Vicente M, and Pérez-Mellado V. 2004. Arsenic uptake by reptile flexible-shelled eggs from contaminated nest substrates and toxic effect on embryos. *Bull Environ Contam Toxicol* 72:983–990.

Marquez EC, Traylor-Knowles N, Novillo-Villajos A, and Callard IP. 2011. Novel cDNA sequences of aryl hydrocarbon receptors and gene expression in turtles (*Chrysemys picta* and *Pseudemys scripta*) exposed to different environments. *Comp Biochem Physiol C: Toxicol Pharmacol* 154:305–317.

Marrugo-Negrete J, Durango-Hernández J, Calao-Ramos C, Urango-Cárdenas I, and Díez S. 2019. Mercury levels and genotoxic effect in caimans from tropical ecosystems impacted by gold mining. *Sci Total Environ* 664:899–907.

Marsili L, Casini S, Mori G et al. 2009. The Italian Wall Lizard (*Podarcis sicula*) as a bioindicator of oil field activity. *Sci Total Environ* 407:3597–3604.

Martínez-López E, Sousa AR, María-Mojica P, Gómez-Ramírez P, Guilhermino L, and García-Fernández AJ. 2010. Blood δ-ALAD, lead and cadmium concentrations in spur-thighed tortoises (Testudo graeca) from southeastern Spain and northern Africa. *Ecotoxicology* 19:670–677.

Matson CW, Palatnikov G, Islamzadeh A et al. 2005. Chromosomal damage in two species of aquatic turtles (*Emys orbicularis* and *Mauremys caspica*) inhabiting contaminated sites in Azerbaijan. *Ecotoxicology* 14:513–525.

Matsumoto Y, Hannigan B, and Crews D. 2014. Embryonic PCB exposure alters phenotypic, genetic, and epigenetic profiles in turtle sex determination, a biomarker of environmental contamination. *Endocrinology* 155:4168–4177.

Matter JM, McMurry CS, Anthony AB, and Dickerson RL. 1998. Development and implementation of endocrine biomarkers of exposure and effects in American alligators (*Alligator mississippiensis*). *Chemosphere* 37:1905–1914.

Maund SJ, Campbell PJ, Giddings JM et al. 2012. Ecotoxicology of synthetic pyrethroids. *Top Curr Chem* 314:137–165.

McClellan-Green P, Celander M, and Oberdörster E. 2006. Hepatic, renal, and adrenal toxicology, in *Toxicology of Reptiles*. Gardner SC, and Oberdörster E (Eds.), CRC Press, Pensacola, FL, pp. 123–148.

McConnell EE. 1985. Comparative toxicity of PCBs and related compounds in various species of animals. *Environ Health Perspect* 60:29–33.

McConnell LL, and Sparling DW. 2010. Emerging contaminants and their potential effects on amphibians and reptiles. In *Ecotoxicology of Amphibians and Reptiles*, 2nd ed. Sparling DW, Linder G, Bishop CA, and Krest SK (Eds.), SETAC Press and CRC Press, Pensacola, FL, pp. 487–509.

McIlroy JC, King DR, and Oliver AJ. 1985. The sensitivity of Australian animals to 1080 poison. VIII. Amphibians and reptiles. *Aust Wildl Res* 12:113–118.

Mercurio AD, Hernandez SM, Maerz JC et al. 2014. Experimental feeding of *Hydrilla verticillata* colonized by Stigonematales cyanobacteria induces vacuolar myelinopathy in paited turtles (*Chrysemys picta*). *PLOS ONE* 9(4):e93295.

Meyer E, Eagles-Smith CA, Sparling D, and Blumenshine S. 2014. Mercury exposure associated with altered plasma thyroid hormones in the declining western pond turtle (*Emys marmorata*) from California mountain streams. *Environ Sci Technol* 48:2989–2996.

Mignucci-Giannoni AA. 1998. Assessment and rehabilitation of wildlife affected by an oil spill in Puerto Rico. *Environ Pollut* 104:323–333.

Miller DL, Wyneken J, Rajeev S et al. 2009. Pathological findings in hatchling and post-hatchling leatherback sea turtles (*Dermochelys coriacea*) from Florida. *J Wildl Dis* 45:962–971.

Milnes MR, Allen D, Bryan TA, Sedacca CD, and Guillette LJ Jr. 2004. Development effects of embryonic exposure to toxaphene in the American alligator (*Alligator mississippiensis*). *Comp Biochem Physiol C: Toxicol Pharmacol* 138:81–87.

Milnes MR, Bryan TA, Medina JG, Gunderson MP, and Guillette LJ Jr. 2005. Developmental alterations as a result of in ovo exposure to the pesticide metabolite p,p'-DDE in *Alligator mississippiensis*. *Gen Comp Endocrinol* 144:257–263.

Mitchell MA. 2006. Therapeutics, in *Reptile Medicine and Surgery*, 2nd ed. Mader DR (Ed.), Saunders-Elsevier, St. Louis, MO, pp. 631–664.

Monagas P, Orós J, Araña J, and González-Díaz OM. 2008. Organochlorine pesticide levels in loggerhead turtles (*Caretta caretta*) stranded in the Canary Islands, Spain. *Mar Pollut Bull* 56:1949–1956.

Montali RJ, Bush M, and Smeller JM. 1979. The pathology of nephrotoxicity of gentamicin in snakes. *Vet Pathol* 16:108–115.

Moore CJ. 2008. Synthetic polymers in the marine environment: A rapidly increasing, long-term threat. *Environ Res* 108:131–139.

Mote S, Kumar R, Naik BG, and Ingole BS. 2015. Polycyclic aromatic hydrocarbons (PAHs) and n-alkanes in beaked sea snake *Enhydrina schistose* (Daudin, 1803) from the Mandovy Estuary, Goa. *Bull Environ Contam Toxicol* 94:171–177.

Muller JK, Scarborough JE, Sepúlveda MS, Casella G, Gross TS, and Borgert CJ. 2007. Dose verification after topical treatment of alligator (*Alligator mississippiensis*) eggs. *Environ Toxicol Chem* 26:908–913.

Myers JP, Antoniou MN, Blumberg B et al. 2016. Concerns over use of glyphosate-based herbicides and risks associated with exposures: A consensus statement. *Environ Health* 15:19.

Nagle RD, Rowe CL, and Congdon JD. 2001. Accumulation and selective maternal transfer of contaminants in the turtle *Trachemys scripta* associated with coal ash deposition. *Arch Environ Contam Toxicol* 40:531–536.

Nasri H, El Herry S, and Bouaïcha N. 2008. First reported case of turtle deaths during a toxic *Microcystis* spp. Bloom in Lake Oubeira, Algeria. *Ecotoxicol Environ Saf* 71:535–544.

Neiffer DL, Lydick D, Burks K, and Doherty D. 2005. Hematologic and plasma biochemical changes associated with fenbendazole administration in Hermann's tortoises (*Testudo hermanni*). *J Zoo Wildl Med* 36:661–672.

Nelms SE, Duncan EM, Broderick AC et al. 2016. Plastic and marine turtles: A review and call for research. *ICES J Mar Sci* 73:165–181.

Neuman-Lee LA, Carr J, Vaughn K, and French SS. 2015. Physiological effects of polybrominated diphenyl ether (PBDE-47) on pregnant gartersnakes and resulting offdpring. *Gen Comp Endocrinol* 219:143–151.

Nilsen FM, Parrott BB, Bowden JA et al. 2016. Global DNA methylation loss associated with mercury contamination and aging in the American alligator (*Alligator mississippiensis*). *Sci Total Environ* 545–546:389-397.

Novillo A, Kitana N, Marquez E, and Callard IP. 2006. Reptilian genotoxicity, in *Toxicology of reptiles*. Gardner SC and Oberdörster E, CRC Press, Pensacola, FL, pp. 241–265.

O'Connor CE, and Eason CT. 2000. Rodent baits and delivery systems for island protection. *Sci Conserv* 150:25p.

Orós J, Calabuig P, and Déniz S. 2004. Digestive pathology of sea turtles stranded in the Canary Islands between 1993 and 2001. *Vet Rec* 155:169–174.

Orós J, González-Díaz OM, and Monagas P. 2009a. High levels of polychlorinated biphenyls in tissues of Atlantic turtles stranded in the Canary Islands, Spain. *Chemosphere* 74:473–478.

Orós J, Monagas P, Andrada M, Calabuig P, and Pether J. 2009b. Metastatic fibrosarcoma in a captive Saharan Horned Viper (*Cerastes cerastes*) with high hepatic levels of cadmium. *Vet Rec* 164:690–692.

Orós J, Monagas P, Calabuig P, Luzardo OP, and Camacho M. 2013. Pansteatitis associated with high levels of polychlorinated biphenyls in a wild loggerhead sea turtle *Caretta caretta*. *Dis Aquat Org* 102:237–242.

Orós J, Montesdeoca N, Camacho M, Arencibia A, and Calabuig P. 2016. Causes of stranding and mortality, and final disposition of Loggerhead Sea Turtles (*Caretta caretta*) admitted to a wildlife rehabilitation center in Gran Canaria Island, Spain (1998–2014). A long-term retrospective study. *PLOS ONE* 11:e0149398.

Orós J, Torrent A, Calabuig P, and Déniz S. 2005. Diseases and causes of mortality among sea turtles stranded in the Canaray Islands, Spain (1998–2001). *Dis Aquat Org* 63:13-24.

Outerbridge ME, O'Riordan R, Fort DJ, and Davenport J. 2016. Ecotoxicological assessment of diamondback terrapin (*Malaclemys terrapin*) pond habitat, prey and eggs in Bermuda. *Mar Pollut Bull* 102:36–43.

Overmann SR, and Krajicek JJ. 1995. Snapping turtles (*Chelydra serpentina*) as biomonitors of lead contamination of the big river in Missouri's old lead belt. *Environ Toxicol Chem* 14:689–695.

Palmer BD, and Palmer SK. 1995. Vitellogenin induction by xenobiotic estrogens in the red-eared turtle and African clawed frog. *Environ Health Perspect* 103:19–25.

Pauli BD, Money S, and Sparling DW. 2010. Ecotoxicology of pesticides in reptiles, in *Ecotoxicology of amphibians and reptiles*, 2nd ed. Sparling DW, Linder G, Bishop CA, and Krest SK (Eds.), SETAC Press and CRC Press, Pensacola, FL, 203–224.

Paul-Murphy J, Mader DR, Kock N, and Frye FL. 1987. Necrosis of esophageal and gastric mucosa in snakes given oral dioctyl sodium sulfosuccinate, in *Proceedings of the American Association of Zoo Veterinarians*, Hawaii, pp. 474–477.

Peden-Adams MM, Kannan K, Lee AM et al. 2005. Perfluorinated contaminants measured in sea turtle blood correlate to modulations in plasma chemistry values and immune function measurements, in *Proc 25th Symp Sea Turtle Biol Conserv*, Savannah, GA, p. 37.

Peden-Adams MM, Wang A, Johnson A, EuDaly J, Smythe J, and Heesemann L. 2003. Relationship of lymphoproliferation and clinical blood parameters to heavy metals in Kemp's ridley sea turtles, in *Proc SETAC 24th Annual Meeting in North America*, Society of Environmental Toxicology and Chemistry, Pensacola, FL, p. 240.

Perrault JR, Bauman KD, Greenan TM, Blum PC, Henry MS, and Walsh CJ. 2016. Maternal transfer and sublethal immune system effects of brevetoxin exposure in nesting Loggerhead Sea Turtles (*Caretta caretta*) from western Florida. *Aquat Toxicol* 180:131–140.

Perrault JR, Stacy NI, Lehner AF et al. 2017a. Potential effects of brevetoxins and toxic elements on various health variables in Kemp's ridley (*Lepidochelys kempii*) and green (*Chelonia mydas*) sea turtles after a red tide bloom event. *Sci Total Environ* 605–606:967–979.

Perrault JR, Stacy NI, Lehner AF et al. 2017b. Toxic elements and associations with hematology, plasma biochemistry, and protein electrophoresis in nesting Loggerhead Sea Turtles (*Caretta caretta*) from Casey Key, Florida. *Environ Pollut* 231:1398–1411.

Perrault JR, Wyneken J, Thompson LJ, Johnson C, and Miller DL. 2011. Why are hatching and emergence success low? Mercury and selenium concentrations in nesting leatherback sea turtles (*Dermochelys coriacea*) and their young in Florida. *Mar Pollut Bull* 62:1671–1682.

Peterson ME, and Fluegeman K. 2013. Cholecalciferol. *Top Companion Anim Med* 28:24–27.

Peveling R, and Demba SA. 2003. Toxicity and pathogenicity of *Metarhizium anisopliae* var. *acridum* (Deuteromycotina, Hyphomycetes) and fipronil to the fringe-toed lizard *Acanthodactylus dumerili* (Squamata: Lacertidae). *Environ Toxicol Chem* 22:1437–1447.

Peveling R, McWilliam AN, Nagel P et al. 2003. Impact of locust control on harvester termites and endemic vertebrate predators in Madagascar. *J Appl Ecol* 40:729–741.

Phillips JB, and Wells MR. 1974. Adenosine triphosphatase activity in liver, intestinal mucosa, cloacal bladder, and kidney tissue of five turtle species following in vitro treatment with 1,1,1-trichloro-2,2-bis(p-chlorophenyl)ethane (DDT). *J Agric Food Chem* 22:404-407.

Pitt WC, Berentsen AR, Shiels AB et al. 2015. Non-target species mortality and the measurement of brodifacoum rodenticide residues after a rat (*Rattus rattus*) eradication on Palmyra Atoll, tropical Pacific. *Biol Conserv* 185:36–46.

Podreka S, Georges A, Maher B, and Limpus CJ. 1998. The environmental contaminant DDE fails to influence the outcome of sexual differentiation in the marine turtle *Chelonia mydas*. *Environ Health Perspect* 106:185–188.

Poletta GL, Kleinsorge E, Paonessa A, Mudry MD, Larriera A, and Siroski PA. 2011. Genetic, enzymatic and developmental alterations observed in *Caiman latirostris* exposed in ovo to pesticide formulations and mixtures in an experiment simulating environmental exposure. *Ecotoxicol Environ Saf* 74:852–859.

Portelli MJ, de Solla SR, Brooks RJ, and Bishop CA. 1999. Effect of dichlorodiphenyltrichloroethane on sex determination of the common snapping turtle (*Chelydra serpentina serpentina*). *Ecotoxicol Environ Saf* 43:284–291.

Poyot T, Nachon F, Froment MT et al. 2006. Mutant of *Bungarus fasciatus* acetylcholinesterase with low affinity and low hydrolase activity toward organophosphorus esters. *Biochim Biophys Acta* 1764:1470–1478.

Prichula J, Inhoque-Pereira R, Raffo-Wachholz G et al. 2016. Resistance to antimicrobial agents among enterococci isolated from fecal samples of wild marine species in the southern coast of Brazil. *Mar Pollut Bull* 105:51–57.

Ragland JM, Arendt MD, Kucklick JR, and Keller JM. 2011. Persistent organic pollutants in blood plasma of satellite-tracked adult male Loggerhead Sea Turtles (*Caretta caretta*). *Environ Toxicol Chem* 30:1549–1556.

Rainwater TR, Reynolds KD, Cañas JE et al. 2005. Organochlorine pesticides and mercury in cottonmouths (*Agkistrodon piscivorus*) from northeastern Texas, USA. *Environ Toxicol Chem* 24:665–673.

Rattner BA. 2009. History of wildlife toxicology. *Ecotoxicology* 18:773–783.

Rauschenberger RH, Sepúlveda MS, Wiebe JJ, Wiebe JE, Honeyfield DC, and Gross TS. 2009. Nutrient and organochlorine pesticide concentrations in American alligator eggs and their association with clutch viability. *J Aquat Anim Health* 21:249–261.

Rauschenberger RH, Wiebe JJ, Buckland JE, Smith JT, Sepúlveda MS, and Gross TS. 2004. Achieving environmentally relevant organochlorine pesticide concentrations in eggs through maternal exposure in *Alligator mississippiensis*. *Mar Environ Res* 58:851–856.

Rauschenberger RH, Wiebe JJ, Sepúlveda MS, Scarborough J, and Gross TS. 2007. Parental exposure to pesticides and poor clutch viability in American alligators. *Environ Sci Technol* 41:5559–5563.

Rey F, González M, Zayas MA et al. 2009. Prenatal exposure to pesticides disrupts testicular architecture and alters testosterone levels in male *Caiman latirostris*. *Gen Comp Endocrinol* 162:286–292.

Rich CN, and Talent LG. 2009. Soil ingestion may be an important route for the uptake of contaminants by some reptiles. *Environ Toxicol Chem* 28:311–315.

Richardson KL, López-Castro M, Gardner SC, and Schlenk D. 2010. Polychlorinated biphenyls and biotransformation enzymes in three species of sea turtles from the Baja California peninsula of Mexico. *Arch Environ Contam Toxicol* 58:183–193.

Rie MT, Kitana N, Lendas KA, Won SJ, and Callard IP. 2005. Reproductive endocrine disruption in a sentinel species (*Chrysemys picta*) on Cape Cod, Massachusetts. *Arch Environ Contam Toxicol* 48:217–224.

Rie MT, Lendas KA, and Callard IP. 2001. Cadmium: Tissue distribution and binding protein induction in the Painted Turtle, *Chrysemys picta*. *Comp Biochem Physiol C: Toxicol Pharmacol* 130:41–51.

Rie MT, Lendas KA, Woodin BR, Stegeman JJ, and Callard IP. 2000. Hepatic biotransformation enzymes in a sentinel species, the Painted Turtle (*Chrysemys picta*), from Cape Cod, Massachusetts: Seasonal-, sex- and location-related differences. *Biomarkers* 5:382–394.

Rivas-Hernández G, May-Uc Y, Noreña-Barroso E, Cobos-Gasca V, and Rodríguez-Fuentes G. 2018. Plasma cholinesterase activity in female green turtles *Chelonia mydas* nesting in Laguna de términos, Mexico, related to organochlorine pesticide in their eggs. *Bull Environ Contam Toxicol* 100:101–105.

Roe JH, Hopkins WA, Baionno JA, Staub BP, Rowe CL, and Jackson BP. 2004. Maternal transfer of selenium in *Alligator mississippiensis* nesting downstream from a coal burning power plant. *Environ Toxicol Chem* 23:1969–1972.

Rooney AA, Bermudez DS, and Guillette LJ Jr. 2003. Altered histology of the thymus and spleen in contaminant-exposed juvenile American alligators. *J Morphol* 256:349–359.

Rotstein D, Lewbart G, Hobbie K et al. 2003. Suspected oak *Quercus* toxicity in an African spurred tortoise, *Geochelone sulcata*. *J Herpetol Med Surg* 13:20–21.

Rousselet E, Levin M, Gebhard E, Higgins BM, DeGuise S, and Godard-Codding CAJ. 2017. Polychlorinated biphenyls (PCBs) modulate both phagocytosis and NK cell activity in vitro in juvenile Loggerhead Sea Turtles (*Caretta caretta*). *J Toxicol Environ Health A* 80:556–561.

Rowe CL, Mitchelmore CL, and Baker JE. 2009. Lack of biological effects on water accommodated fractions of chemically- and physically-dispersed oil on molecular, physiological, and behavioral traits of juvenile snapping turtles following embryonic exposure. *Sci Total Environ* 407:5344–5355.

Ruder MG, Poppenga RH, Bryan JA 2nd, Bain M, Pitman J, and Keel MK. 2011. Intoxication of nontarget wildlife with rodenticides in northwestern Kansas. *J Wildl Dis* 47:212–216.

Rueda D, Campbell KJ, Fisher P, Cunninghame F, and Ponder JB. 2016. Biologically significant residual persistence of brodifacoum in reptiles following invasive rodent eradication, Galapagos Islands, Ecuador. *Conserv Evid* 13:38.

Ruiz-Suárez N, Boada LD, Henríquez-Hernández LA et al. 2015. Continued implication of the banned pesticides carbofuran and aldicarb in the poisoning of domestic and wild animals of the Canary Islands (Spain). *Sci Total Environ* 505:1093–1099.

Saba VS, and Spotila JR. 2003. Survival and behavior of freshwater turtles after rehabilitation from an oil spill. *Environ Pollut* 126:213–223.

Salice CJ, Suski JG, Bazar MA, and Talent LG. 2009. Effects of inorganic lead on Western Fence Lizards (*Sceloporus occidentalis*). *Environ Pollut* 157:3457–3464.

Sánchez-Barbudo IS, Camarero PR, and Mateo R. 2012. Primary and secondary poisoning by anticoagulant rodenticides of non-target animals in Spain. *Sci Total Environ* 420:280–288.

Sánchez-Sarmiento AM, Rossi S, Vilca FZ et al. 2016. Organochlorine pesticides in green sea turtles (*Chelonia mydas*) with and without fibropapillomatosis caught at three feeding areas off Brazil. *J Mar Biol Assoc U K* 97:215–223.

Santos L, Hupka A, and Winston G. 1997. *Ameiva Exsul*: A Potential Sentinel Model of Environmental Contamination. *FASEB J* 11:A1341.

Santos X, Pastor D, Llorente GA, and Albaigés J. 1999. Organochlorine levels in viperine snake *Natrix maura* carcasses from the Ebro Delta (NE Spain): Sexual and size-related differences. *Chemosphere* 39:2641–2650.

Sapana-Devi M, and Gupta A. 2014. Sublethal toxicity of commercial formulations of deltamethrin and permethrin on selected biochemical constituents and enzyme activities in liver and muscle tissues of *Anabas testudineus*. *Pestic Biochem Physiol* 115:48–52.

Sauvé S, and Desrosiers M. 2014. A review of what is an emerging contaminant. *Chem Cent J* 8:15.

Schaumburg LG, Siroski PA, Poletta GL, and Mudry MD. 2016. Genotoxicity induced by Roundup® (Glyphosate) in tegu lizard (*Salvator merianae*) embryos. *Pestic Biochem Physiol* 130:71–78.

Schwartz JH, and Flamenbaum W. 1976. Heavy metal-induced alterations in ion transport by turtle urinary bladder. *Am J Physiol* 230:1582–1589.

Sciarrillo R, Capaldo A, Valiante S et al. 2010. Thyroid hormones as potential early biomarkers of exposure to nonylphenol in adult male lizard (*Podarcis sicula*). *Open Zool J* 3:17–22.

Scudiero R, Filosa S, Motta CM, Simoniello P, and Trinchella F. 2011. Cadmium in the wall lizard Podarcis sicula: Morphological and molecular effects on embryonic and adult tissues, in *Reptiles: Biology, behavior and conservation*. Baker KJ (Ed.), Nova Science Publishers, New York, pp. 147–162.

Sepúlveda MS, Del Piero F, Wiebe JJ, Rauschenberger HR, and Gross TS. 2006. Necropsy findings in American alligator late-stage embryos and hatchlings from northcentral Florida lakes contaminated with organochlorine pesticides. *J Wildl Dis* 42:56–73.

Sepúlveda MS, Wiebe JJ, Honeyfield DC et al. 2004. Organochlorine pesticides and thiamine in eggs of largemouth bass and American alligators and their relationship with early life-stage mortality. *J Wildl Dis* 40:782–786.

Sereshk ZH, and Bakhtiari AR. 2014. Distribution patterns of PAHs in different tissues of annulated sea snake (*Hydrophis cyanocinctus*) and short sea snake (*Lapemis curtus*) from the Hara Protected Area on the north coast of the Persian Gulf, Iran. *Ecotoxicol Environ Saf* 109:116–123

Serino-Silva C, Morais-Zani K, Hikari-Toyama M et al. 2018. Purification and characterization of the first γ-phospholipase inhibitor (γPLI) from Bothrops jararaca snake serum. *PLOS ONE* 13:e0193105.

Shelby JA, and Mendonça MT. 2001. Comparison of reproductive parameters in male yellow-blotched map turtles (*Graptemys flavimaculata*) from a historically contaminated site and a reference site. *Comp Biochem Physiol C: Toxicol Pharmacol* 129:233–242.

Silva MIA, Silva TL, and Almeida EA. 2016. Study of biochemical changes in blood and various organs of *Phrynops geoffroanus* (Schweigger, 1812) (Testudines: Chelidae) collected in contaminated environment or exposed to venzo[*a*]pyrene. *Genet Mol Res* 15:1–2.

Simoniello P, Filosa S, Riggio M et al. 2010a. Responses to cadmium intoxication in the liver of the wall lizard *Podarcis sicula*. *Comp Biochem Physiol C: Toxicol Pharmacol* 151:194–203.

Simoniello P, Filosa S, Scudiero R, Trinchella F, and Motta CM. 2013. Cadmium impairment of reproduction in the female wall lizard *Podarcis sicula*. *Environ Toxicol* 28:553–562.

Simoniello P, Motta CM, Scudiero R, Trinchella F, and Filosa S. 2011. Cadmium-induced teratogenicity in lizard embryos: Correlation with metallothionein gene expression. *Comp Biochem Physiol C: Toxicol Pharmacol* 153:119–127.

Simoniello P, Trinchella F, Borrelli L et al. 2009. Effects of cadmium on retinal development in lizard embryo: A molecular and morphological study. *Comp Biochem Physiol A: Mol Integr Physiol* 154:S21–S22.

Simoniello P, Trinchella F, Filosa S et al. 2014. Cadmium contaminated soil affects retinogenesis in lizard embryos. *J Exp Zool* 321A:207-219.

Simoniello P, Trinchella F, Scudiero R, Filosa S, and Motta CM. 2010b. Cadmium in *Podarcis sicula* disrupts prefollicular oocyte recruitment by mimicking FSH action. *Open Zool J* 3:37–41.

Simonyan A, Hovhannisyan G, Sargsyan A, Arakelyan M, Minasyan S, and Aroutiounian R. 2018. DNA damage and micronuclei in parthenogenetic and bisexual *Darevskia* rock lizards from the areas with different levels of soil pollution. *Ecotoxicol Environ Saf* 154:13–18.

Singo A, Myburgh JG, Laver PN et al. 2017. Vertical transmission of microcystins to Nile Crocodile (*Crocodylus niloticus*) eggs. *Toxicon* 134:50–56.

Siroski PA, Poletta GL, Latorre MA, Merchant ME, Ortega HH, and Mudry MD. 2016. Immunotoxicity of commercial-mixed glyphosate in broad snouted caiman (*Caiman latirostris*). *Chem-Biol Interact* 244:64–70.

Sleeman JM, Brown J, Steffen D, Jones D, Robertson J, and Holladay S. 2008. Relationships among aural abscesses, organochlorine compounds, and vitamin A in free-ranging eastern box turtles (*Terrapene carolina carolina*). *J Wildl Dis* 44:922–929.

Speer RM, Wise CF, Young JL et al. 2018. The cytotoxicity and genotoxicity of particulate and soluble hexavalent chromium in leatherback sea turtle lung cells. *Aquat Toxicol* 198:149–157.

Spurr EB. 1993. A review of the effects of pesticides on lizards. *Conserv Advisory Sci Notes* 3:1–6.

Stafford DP, Plapp, Jr. FW, and Fleet RR. 1977. Snakes as indicators of environmental contamination: Relation of detoxifying enzymes and pesticide residues to species occurrence in three aquatic ecosystems. *Arch Environ Contam Toxicol* 5:15–27.

Stockholm Convention on Persistent Organic Pollutants. 2015. http://chm.pops.int (accessed May 5, 2020).

Stoker C, Beldoménico PM, Bosquiazzo VL et al. 2008. Developmental exposure to endocrine disruptor chemicals alters follicular dynamics and steroid levels in *Caiman latirostris*. *Gen Comp Endocrinol* 156:603–612.

Stoker C, Rey F, Rodríguez H et al. 2003. Sex reversal effects on *Caiman latirostris* exposed to environmentally relevant doses of the xenoestrogen bisphenol A. *Gen Comp Endocrinol* 133:287–296.

Stoker C, Zayas MA, Ferreira MA et al. 2013. The eggshell features and clutch viability of the Broad-Snouted Caiman (*Caiman latirostris*) are associated with the egg burden of organochlorine compounds. *Ecotoxicol Environ Saf* 98:191–195.

Swarthout RF, Keller JM, Peden-Adams M, Landry AM, Fair PA, and Kucklick JR. 2010. Organohalogen contaminants in blood of Kemp's ridley (*Lepidochelys kempii*) and green sea turtles (*Chelonia mydas*) from the Gulf of Mexico. *Chemosphere* 78:731–741.

Swartz CD, Donnelly KC, Islamzadeh A et al. 2003. Chemical contaminants and their effects in fish and wildlife from the industrial zone of Sumgayit, Republic of Azerbaijan. *Ecotoxicology* 12:509–521.

Széll Z, Sréter T, and Varga I. 2001. Ivermectin toxicosis in a chameleon (*Chamaeleo senegalensis*) infected with *Foleyella furcata*. *J Zoo Wildl Med* 32:115–117.

Talent LG. 2005. Effect of temperature on toxicity of a natural pyrethrin pesticide to green anole lizards (*Anolis carolinensis*). *Environ Toxicol Chem* 24:3113–3116.

Tan F, Wang M, Wang W, Aguirre AA, and Lu Y. 2010. Validation of an in vitro cytotoxicity test for four heavy metals using cell lines derived from a green sea turtle (*Chelonia mydas*). *Cell Biol Toxicol* 26:255–263.

Tangredi BP, and Evans RH. 1997. Organochlorine pesticides associated with ocular, nasal or otic infection in the eastern box turtle (*Terrapene carolina carolina*). *J Zoo Wildl Med* 28:97–100.

Tchounwou PB, Yedjou CG, Patlolla AK, and Sutton DJ. 2012. Heavy metals toxicity and the environment. *Experientia Suppl* 101: 133–164.

Teare JA, and Bush M. 1983. Toxicity and efficacy of ivermectin in chelonians. *J Am Vet Med Assoc* 183:1195–1197.

Teuten EL, Saquing JM, Knappe DRU et al. 2009. Transport and release of chemicals from plastics to the environment and to wildlife. *Philos Trans R Soc Lond B: Biol Sci* 364:2027–2045.

Thompson RC, Moore CJ, vom Saal FS, and Swan SH. 2009. Plastics, the environment and human health: Current consensus and future trends. *Philos Trans R Soc Lond B: Biol Sci* 364:2153–2166.

Tong J, Wang Y, and Lu Y. 2016. In vitro evaluation of inorganic and methyl mercury mediated cytotoxic effect on neural cells derived from different animal species. *J Environ Sci* 41:138–145.

Torrent A, Déniz S, Ruiz A, Calabuig P, Sicilia J, and Orós J. 2002. Esophageal diverticulum associated with *Aerococcus viridans* infection in a loggerhead sea turtle (*Caretta caretta*). *J Wildl Dis* 38:221–223.

Torrent A, González-Díaz OM, Monagas P, and Orós J. 2004. Tissue distribution of metals in loggerhead turtles (*Caretta caretta*) stranded in the Canary Islands, Spain. *Mar Pollut Bull* 49:854–860.

Townsend KA, Altvater J, Thomas MC, Schuyler QA and Nette GW. 2012. Death in the octopus' garden: Fatal blue-lined octopus envenomations of adult green sea turtles. *Mar Biol* 159:689–695.

Tremblay N, Otiz-Arana A, González-Jaúregui M, and Rendón-von Osten J. 2017. Relationship between organochlorine pesticides and stress indicators in hawksbill sea turtle (*Eretmochelys imbricata*) nesting at Punta Xen (Campeche), Southern Gulf of Mexico. *Ecotoxicology* 26:173–183.

Trinchella F, Cannetiello M, Simoniello P, Filosa S, and Scudiero R. 2010. Differential gene expression profiles in embryos of the lizard *Podarcis sicula* under in ovo exposure to cadmium. *Comp Biochem Physiol C: Toxicol Pharmacol* 151:33–39.

Trinchella F, Riggio M, Filosa S, Volpe MG, Parisi E, and Scudiero R. 2006. Cadmium distribution and metallothionein expression in lizard tissues following acute and chronic cadmium intoxication. *Comp Biochem Physiol C: Toxicol Pharmacol* 144:272–278.

Valdes SA, Vieira LG, Ferreira CH et al. 2015. Effects of exposure to methyl parathion on egg hatchability and eggshell chemical composition in *Podocnemis expansa* (Testudines, Podocnemididae). *Zoolog Sci* 32:135–140.

van de Merwe JP, Hodge M, Whittier JM, Ibrahim K, and Lee SY. 2010. Persistent organic pollutants in the green sea turtle *Chelonia mydas*: Nesting population variation, maternal transfer, and effects on development. *Mar Ecol Prog Ser* 403:269–278.

Van Der Kraak GJ, Hosmer AJ, Hanson ML, Kloas W, and Solomon KR. 2014. Effects of atrazine in fish, amphibians, and reptiles: An analysis based on quantitative weight of evidence. *Crit Rev Toxicol* 44 Suppl 5:1–66.

Van Meter RJ, Spotila JR, and Avery HW 2006. Polycyclic aromatic hydrocarbons affect survival and development of common snapping turtle (*Chelydra serpentina*) embryos and hatchlings. *Environ Pollut* 142:466–475.

Vargo S, Lutz P, Odell D, Vleet, EV, and Bossart G. 1986. *Study of the effects of oil on marine turtles. Final Report-US Department of the Interior, MMS 86-0070.* Florida Institute of Oceanography.

Verderame M, and Limatola E. 2015. Interferences of an environmental pollutant with estrogen-like action in the male reproductive system of the terrestrial vertebrate *Podarcis sicula*. *Gen Comp Endocrinol* 213:9–15.

Verderame M, Prisco M, Andreuccetti P, Aniello F, and Limatola E. 2011. Experimentally nonylphenol-polluted diet induces the expression of silent genes VTG and ERα in the liver of male lizard Podarcis sicula. *Environ Pollut* 159:1101–1107.

Vonier PM, Crain DA, McLachlan JA, Guillette, Jr. LJ, and Arnold SF. 1996. Interaction of environmental chemicals with the estrogen and progesterone receptors from the oviduct of the American alligator. *Environ Health Perspect* 104:1318–1322.

Walker JS, Shaver DJ, Stacy BA, Flewelling LJ, Broadwater MH, and Wang Z. 2018. Brevetoxin exposure in sea turtles in south Texas (USA) during *Karenia brevis* red tide. *Dis Aquat Org* 127:145–150.

Wallach JD. 1966. Hypervitaminosis D in green iguanas. *J Am Vet Med Assoc* 149:912–914.

Walsh CJ, Legget SR, Carter BJ, and Colle C. 2010. Effects of brevetoxin exposure on the immune system of Loggerhead Sea Turtles. *Aquat Toxicol* 97:293–303.

Wang H, Tong J, Bi Y, Wang C, Guo L, and Lu Y. 2013. Evaluation of mercury mediated in vitro cytotoxicity among cell lines established from green sea turtles. *Toxicol In vitro* 27:1025–1030.

Warner JK, Combrink X, Myburgh JG, and Downs CT. 2016. Blood lead concentrations in free-*ranging Nile Crocodiles (Crocodylus niloticus)* from South Africa. *Ecotoxicology* 25:950–958.

Webb SJ, Zychowski GV, Bauman SW et al. 2014. Establishment, characterization, and toxicological application of loggerhead sea turtle (*Caretta caretta*) primary skin fibroblast cell cultures. *Environ Sci Technol* 48:14728–14737.

Weir SM, Yu S, Talent LG, Maul JD, Anderson TA, and Salice CJ. 2015. Improving reptile ecological risk assessment: Oral and dermal toxicity of pesticides to a common lizard species (*Sceloporus occidentalis*). *Environ Toxicol Chem* 34:1778–1786.

Wells MR, Phillips JB, and Murphy GG. 1974. ATPase activity in tissues of the map turtle, *Graptemys geographica* following in vitro treatment with aldrin and dieldrin. *Bull Environ Contam Toxicol* 11:572–576.

White DH, and Krynitsky AJ. 1986. Wildlife in some areas of New Mexico and Texas accumulate elevated DDE residues, 1983. *Arch Environ Contam Toxicol* 15:149–157.

WHO/FAO/IAEA. 1996. *Trace Elements in Human Nutrition and Health*. World Health Organization, Geneva, Switzerland, p. 361.

WHO. 2007. *Rabies and Envenomings: A Neglected Public Health Issue*. World Health Organization, Geneva, Switzerland, p. 32.

Wikelski M, Wong V, Chevalier B, Rattenborg N, and Snell HL. 2002. Marine iguanas die from trace oil pollution. *Nature* 417:607–608.

Williams BL, Hanifin CT, Brodie, Jr. ED, and Brodie III ED. 2012. Predators usurp prey defenses? Toxicokinetics of tetrodotoxin in common garter snakes after consumption of rough-skinned newts. *Chemoecology* 22:179–185.

Williams BL, Powers LV, and Garner MM. 2016. A pufferfish (*Tetradon nigroviridis*) available in the common pet trade harbors lethal concentrations of tetrodotoxin: A case study of poisoning in a Cuvier's dwarf caiman (*Paleosuchus palpebrosus*). *J Zoo Wildl Med* 47:676–680.

Willingham E. 2001. Embryonic exposure to low-dose pesticides: Effects on growth rate in the hatchling Red-Eared Slider Turtle. *J Toxicol Environ Health Part A* 64:257–272.

Willingham E. 2004. Endocrine-disrupting compounds and mixtures: Unexpected dose–response. *Arch Environ Contam Toxicol* 46:265–269.

Willingham E. 2006. Developmental and reproductive effects, in *Toxicology of reptiles*. Gardner SC, and Oberdörster E (Eds.), CRC Press, Pensacola, FL, 149–171.

Willingham E, and Crews D. 1999. Sex reversal effects of environmentally relevant xenobiotic concentrations on the Red-Eared Slider Turtle, a species with temperature-dependent sex determination. *Gen Comp Endocrinol* 113:429–435.

Willingham E, and Crews D. 2000. The Red-Eared Slider Turtle: An animal model for the study of low doses and mixtures. *Am Zool* 40:421–428.

Willingham E, Rhen T, Sakata JT, and Crews D. 2000. Embryonic treatment with xenobiotics disrupts steroid hormone profiles in hatchling Red-Eared Slider Turtles (*Trachemys scripta elegans*). *Environ Health Perspect* 108:329–332.

Winger PV, Sieckman C, May TW, and Johnson WW. 1984. Residues of organochlorine insecticides, polychlorinated biphenyls, and heavy metals in biota from Apalachicola River, *Florida*, 1978. *J-Assoc Off Anal Chem* 67:325–333.

Wise SS, Wise C, Xie H et al. 2016. Hexavalent chromium is cytotoxic and genotoxic to American alligator cells. *Aquat Toxicol* 171:30–36.

Wise SS, Xie H, Fukuda T, Thompson WD, and Wise JP. 2014. Hexavalent chromium is cytotoxic and genotoxic to hawksbill sea turtle cells. *Toxicol Appl Pharmacol* 279: 113–118.

Witherspoon, Jr. FG, and Wells MR. 1975. Adenosine triphosphatase activity in brain, intestinal mucosa, kidney, and liver cellular fractions of the red-eared turtle following in vitro treatment with DDT, DDD, and DDE. *Bull Environ Contam Toxicol* 14:537–544.

Witmer GW, Boyd F, and Hillis-Starr Z. 2007. The successful eradication of introduced roof rats (*Rattus rattus*) from Buck Island using diphacinone, followed by an irruption of house mice (*Mus musculus*). *Wildl Res* 34:108–115.

Witmer GW, Eisemann JD, Primus TM et al. 2010. Assessing potential risk to alligators, *Alligator mississippiensis*, from nutria control with zinc phosphide rodenticide baits. *Bull Environ Contam Toxicol* 84:698–702.

Woodford LP, Robley A, Maloney P, and Reside J. 2012. The impact of 1080 bait removal by lace monitors (*Varanus varius*) on a red fox (*Vulpes vulpes*) control programme. *Ecol Manage Restor* 13:306–308.

Wozniak EJ, DeNardo D, Burridge MJ, Walker DH, Funk RS, and Nathan R. 2000. Ectoparasites. *J Herpetol Med Surg* 10:15–21.

Wu JP, Guan YT, Zhang Y et al. 2011. Several current-use, non-PBDE brominated flame retardants are highly accumulative: Evidence from field determined bioaccumulation factors. *Environ Int* 37:210–215.

Wu JP, Luo XJ, Zhang Y et al. 2008. Bioaccumulation of polybrominated diphenyl ethers (PBDEs) and polychlorinated biphenyls (PCBs) in wild aquatic species from an electronic waste (e-waste) recycling site in South China. *Environ Int* 34:1109–1113.

Wu JP, Luo XJ, Zhang Y et al. 2009. Biomagnification of polybrominated diphenyl ethers (PBDEs) and polychlorinated biphenyls in a highly contaminated freshwater food web from South China. *Environ Pollut* 157:904–909.

Yan M, Nie H, Wang W, Huang Y, Li QX, and Wang J. 2018. The risk of polychlorinated biphenyls facilitating tumors in Hawaiian green sea turtles (*Chelonia mydas*). *Int J Environ Res Public Health* 15:1243.

Yawetz A, Goldman D, Stegeman JJ, and Woodin B. 1993. The effect of PCB on the induction of cytochrome P450 enzyme, detected by monoclonal antibody against the P450IA1 gene products, in barn owl and marsh turtle hepatic microsomes. *Water Sci Technol* 27:457–464.

Yender RA, and Mearns AJ. 2003. Case studies of spills that threaten sea turtles, in *Oil and Sea Turtles: Biology, Planning, and Response*. Shigenaka G (Ed.), NOAA National Ocean Service, Office of Response and Restoration, Seattle, pp. 69–84.

Yoshikane M, Kay WR, Shibata Y et al. 2006. Very high concentrations of DDE and toxaphene residues in crocodiles from the Ord River, Western Australia: An investigation into possible endocrine disruption. *J Environ Monit* 8:649–661.

Yu S, Halbrook RS, and Sparling DW. 2012. Accumulation of polychlorinated biphenyls (PCBs) and evaluation of hematological and immunological effects of PCB exposure on turtles. *Bull Environ Contam Toxicol* 88:823–827.

Yu S, Halbrook RS, Sparling DW, and Colombo R. 2011. Metal accumulation and evaluation of effects in a freshwater turtle. *Ecotoxicology* 20:1801–1812.

Zavala-Norzagaray AA, Aguirre AA, Velazquez-Roman J et al. 2015. Isolation, characterization, and antibiotic resistance of *Vibrio* spp. in sea turtles from northwestern Mexico. *Front Microbiol* 6:635.

Zhou Y, Asplund L, Yin G et al. 2016. Extensive organohalogen contamination in wildlife from a site in the Yangtze River Delta. *Sci Total Environ* 554–555:320–328.

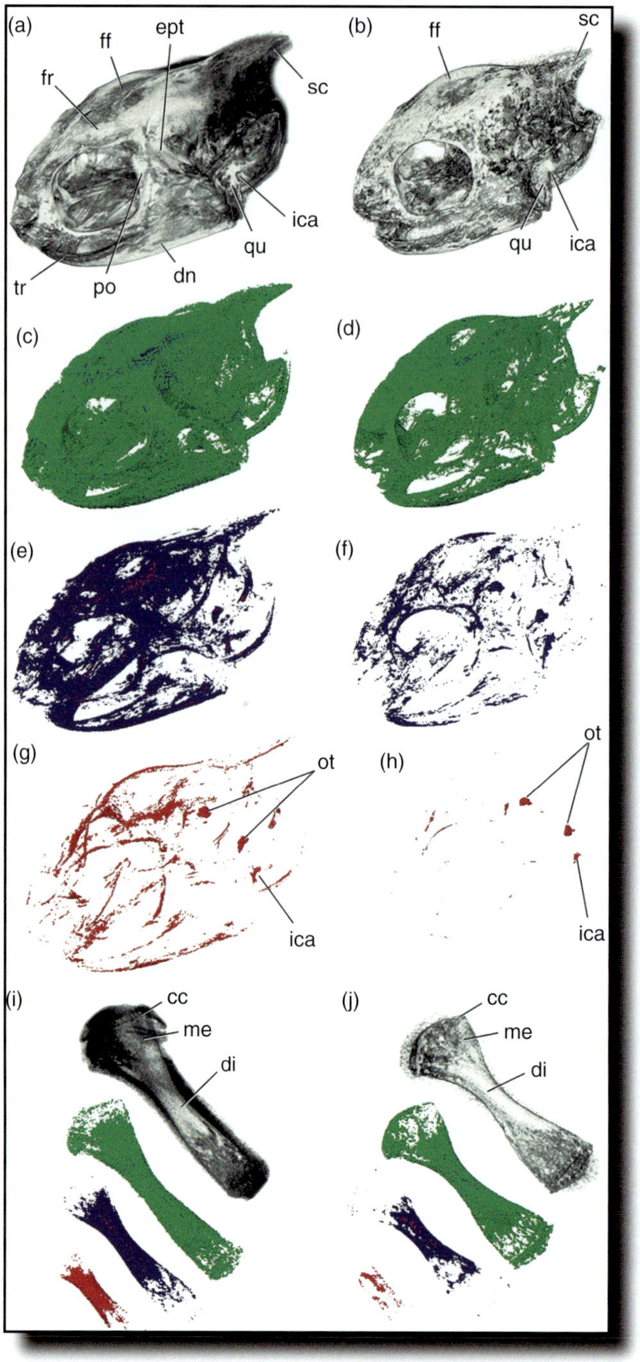

Figure 7.1 Diamondback Terrapin, *Malaclemys terrapin*. Emydidae. Patterns of bone mineral density in the skulls and femora of two specimens scaled to the same size. (a) Rendered visualization of the skull of a control specimen in left, oblique view; (b) specimen treated with PCB 126; (c,d) segmented volume of low-density bin; (e,f) segmented volume of mid-density bin; (g,h) segmented volume of high-density bin; (i) rendering and segmented bone density bins of the left femur of the control specimen, colors as in (a–h); (j) left femur of the specimen treated with PCB 126. *Abbreviations*: cc, calcified cartilage; di, diaphysis; dn, dentary; ept, epipterygoid; ff, frontal fontanel; fr, frontal; ica, incisura columellae auris; me, metaphysis; ot, otolith; po, postorbital; qu, quadrate; sc, sagittal crest; tr, tomial ridge. (Photo courtesy of Dawn Holliday, from Holliday and Holliday. 2012. *Aquat Toxicol* 109:228–233. With permission.)

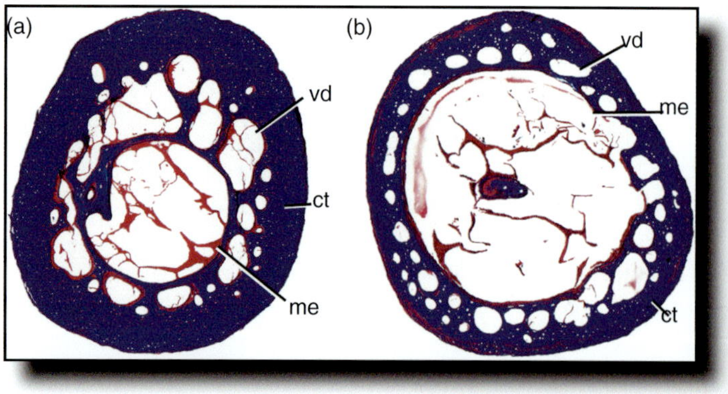

Figure 7.2 Diamondback Terrapin, *Malaclemys terrapin*. Emydidae. Midshaft cross-sections stained with Mallory's trichrome, scaled to similar sizes. (a) Specimen from the control group; (b) specimen treated with PCB 126. Femora of terrapins exposed to PCB 126 had significantly more area occupied by void than similarly aged unexposed individuals. *Abbreviations:* ct, cortical bone; me, medullary cavity; vd, void space. (Image courtesy of Dawn Holliday, from Holliday and Holliday. 2012. *Aquat Toxicol* 109:228–233. With permission.)

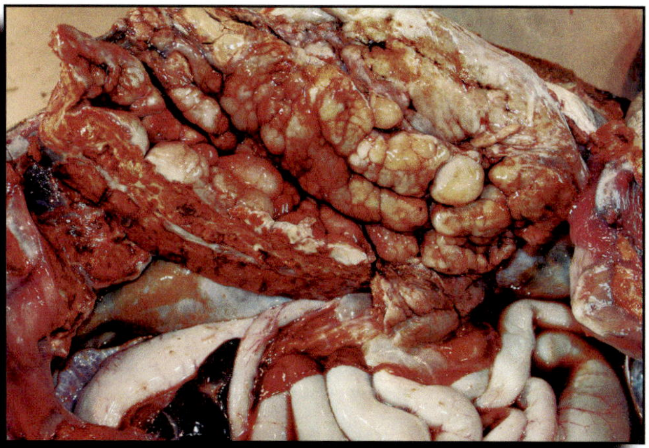

Figure 7.3 Loggerhead Sea Turtle, *Caretta caretta*. Cheloniidae. Large amount of adipose tissue showing very firm irregular formations in the left part of the celomic cavity. Celomic fat was analyzed for PCBs and DDTs, resulting in very high levels of PCB 138, 153, 180, 209. (Photo courtesy of Jorge Orós, from Orós et al. 2013. *Dis Aquat Org* 102:237–242. With permission.)

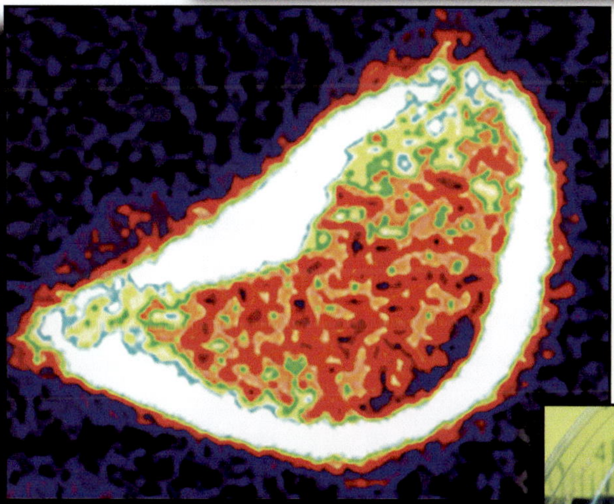

Figure 7.4 American Alligator, *Alligator missisippiensis*. Alligatoridae. Peripheral quantitative computed tomography scan of the proximal tibial metaphysis of a female specimen from the Lake Apopka (pesticide-contaminated lake). The white outer shell is cortical bone with a high bone mineral density (BMD), and the inner spongy structure is trabecular bone with a much lower BMD. The blue areas contain bone tissue with a lower mineral density compared with the yellow and red areas. (Image courtesy of Monica Lind, from Lind et al. 2004. *Environ Health Perspect* 112:359–362. With permission.)

Figure 7.5 Diamondback Terrapin, *Malaclemys terrapin*. Emydidae. Significant compromised embryonic development and malformations associated with total petroleum hydrocarbons, PAHs, and trace metals in terrapin eggs in Bermuda. (Photo courtesy of Mark Outerbridge.)

Figure 7.6 Loggerhead Sea Turtle, *Caretta caretta*. Cheloniidae. Severe esophageal impaction with crude oil. (Photo courtesy of Jorge Orós, from Orós et al. 2004. *Vet Rec* 155:169–174. With permission.)

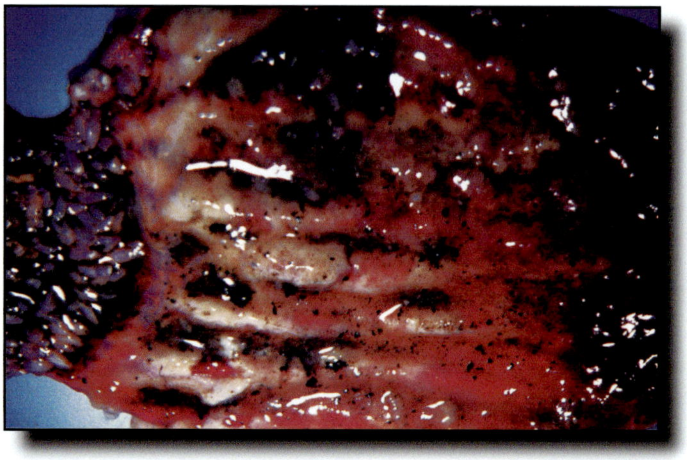

Figure 7.7 Loggerhead Sea Turtle, *Caretta caretta*. Cheloniidae. Necrotizing gastritis associated with the presence of small particles of crude oil. (Photo courtesy of Jorge Orós, from Camacho et al. 2013. *J Wildl Dis* 49:637–640. With permission.)

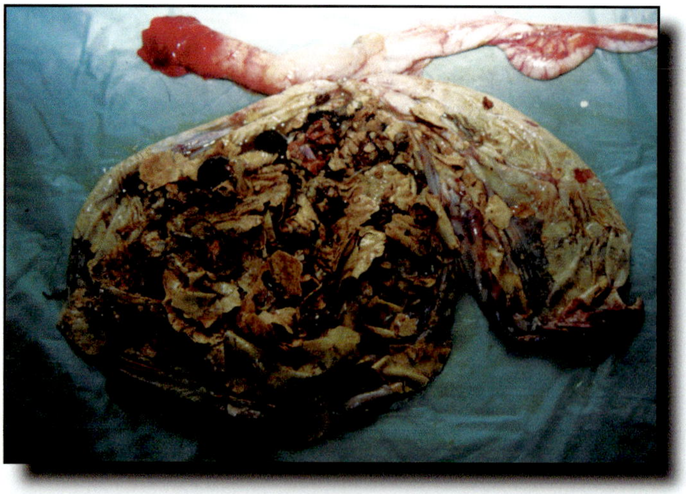

Figure 7.8 Loggerhead Sea Turtle, *Caretta caretta*. Cheloniidae. Esophageal diverticulum with crude oil balls. (Photo courtesy of Jorge Orós.)

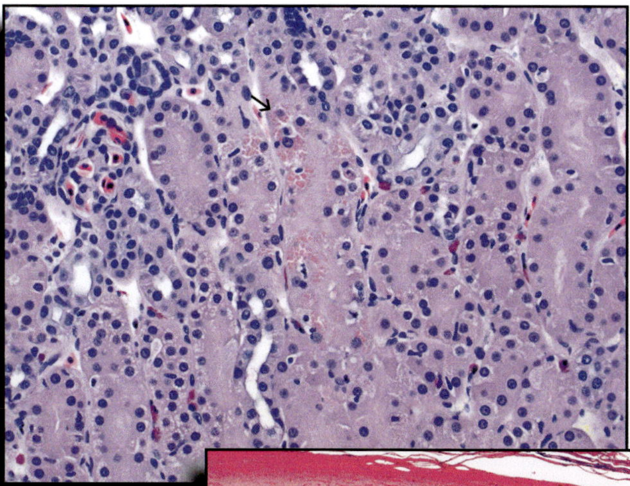

Figure 7.9 Yellow-Bellied Slider Turtle, *Trachemys scripta scripta*. Emydidae. Experimentally induced selenosis. Renal tubules are necrotic with loss of nuclei and hyaline droplets (arrow), or regenerative and lined by flattened tubular epithelium. (Image courtesy of Elizabeth W. Howerth, from Haskins et al. 2018. *Vet Pathol* 55:473–477. With permission.)

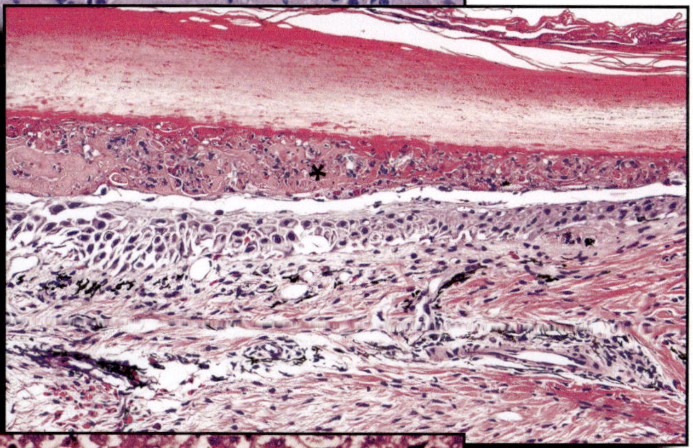

Figure 7.10 Yellow-Bellied Slider Turtle, *Trachemys scripta scripta*. Emydidae. Experimentally induced selenosis. Necrosis of the palmar epidermis of the claw, which is extensively hyperplastic and disorganized with intercellular edema. The epidermis is separated from the stratum corneum by a thick layer of serocellular exudate (*). (Image courtesy of Elizabeth W. Howerth, from Haskins et al. 2018. *Vet Pathol* 55:473–477. With permission.)

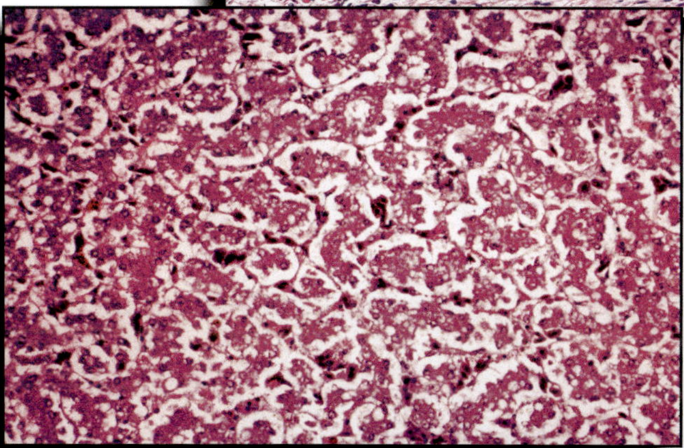

Figure 7.11 Loggerhead Sea Turtle, *Caretta caretta*. Cheloniidae. Severe diffuse vacuolar hepatic degeneration associated with high arsenic hepatic concentration. H&E stain. (Image courtesy of Jorge Orós.)

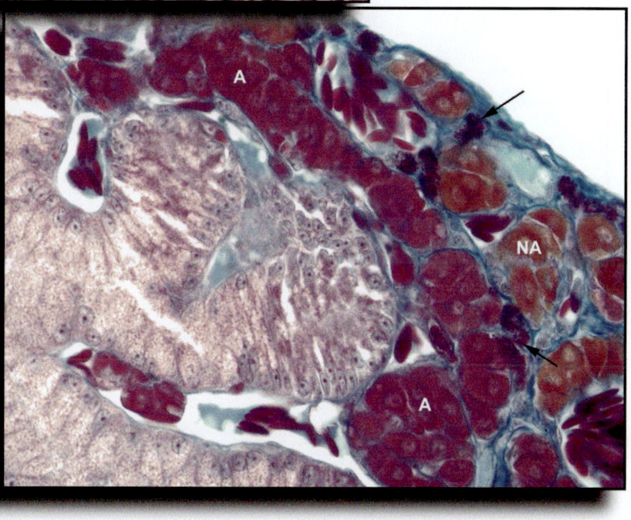

Figure 7.12 Italian Wall Lizard, *Podarcis sicula*. Lacertidae. Chromaffin tissue of lizard adrenal glands forms a dorsal ribbon, with noradrenaline (NA, gold colored) cells in the outer layers, and adrenaline (A, red colored) cells in the inner layers and islets interspersed between steroidogenic tissue. In the picture an adrenal gland of specimen treated with acute cadmium treatment for 2 days: macrophage infiltration (arrows) is localized among chromaffin cells of dorsal ribbon; a disorganization of steroidogenic cords close to chromaffin ribbon is already visible. Mallory trichromic stain. (Image courtesy of Maria de Falco.)

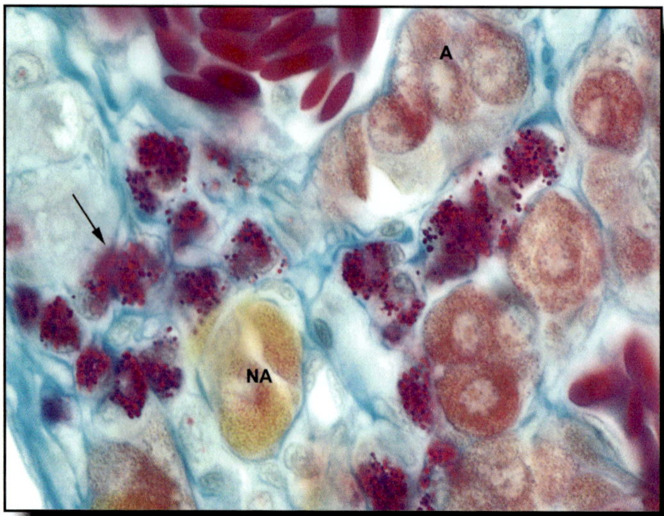

Figure 7.13 Italian Wall Lizard, *Podarcis sicula*. Lacertidae. Adrenal gland of specimen treated with chronic cadmium treatment for ten days. Abundant macrophage infiltration distributed among chromaffin cells. NA cells are gold colored and A cells are red colored. Mallory trichromic stain. (Image courtesy of Maria de Falco.)

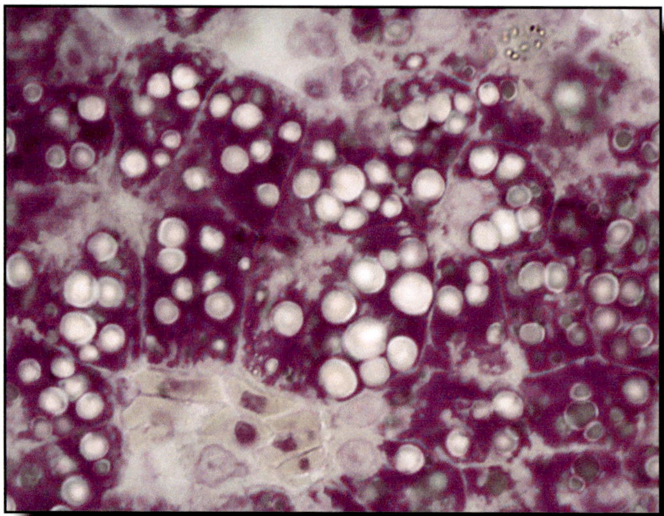

Figure 7.14 Italian Wall Lizard, *Podarcis sicula*. Lacertidae. Liver of specimen fed with cadmium-contaminated insect larvae for 4 weeks (chronic treatment). Note the intensely stained hepatocyte cytoplasm filled with large PAS-negative vacuoles. (Image courtesy of Rosaria Scudiero and Chiara M. Motta, from Simoniello et al. 2010. *Comp Biochem Physiol C Toxicol Pharmacol* 151:194–203. With permission.)

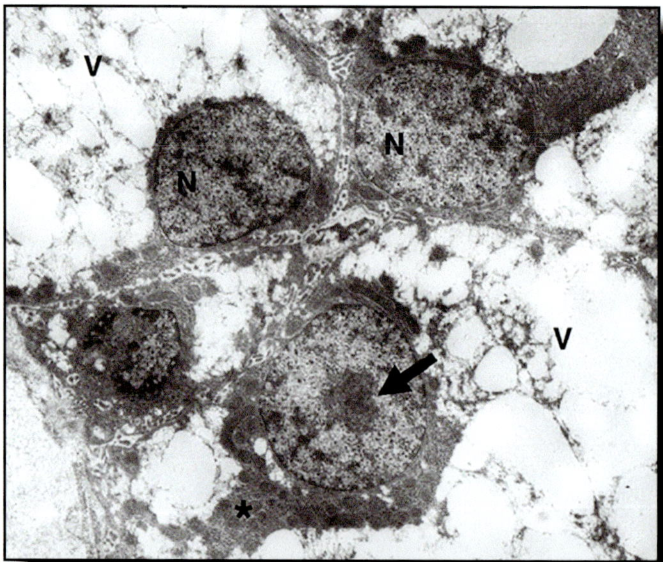

Figure 7.15 Italian Wall Lizard, *Podarcis sicula*. Lacertidae. Liver of specimen fed with cadmium-contaminated insect larvae (chronic treatment). Group of hepatocytes with dense cytoplasm (*) containing several lipid vesicles (V). Nuclei (N) have dense nucleoplasm, chromatin in patches, and a single large nucleolus (arrow). Uranyl acetate and lead citrate staining. (Image courtesy of Rosaria Scudiero and Chiara M. Motta, from Simoniello et al. 2010. *Comp Biochem Physiol C Toxicol Pharmacol* 151:194–203. With permission.)

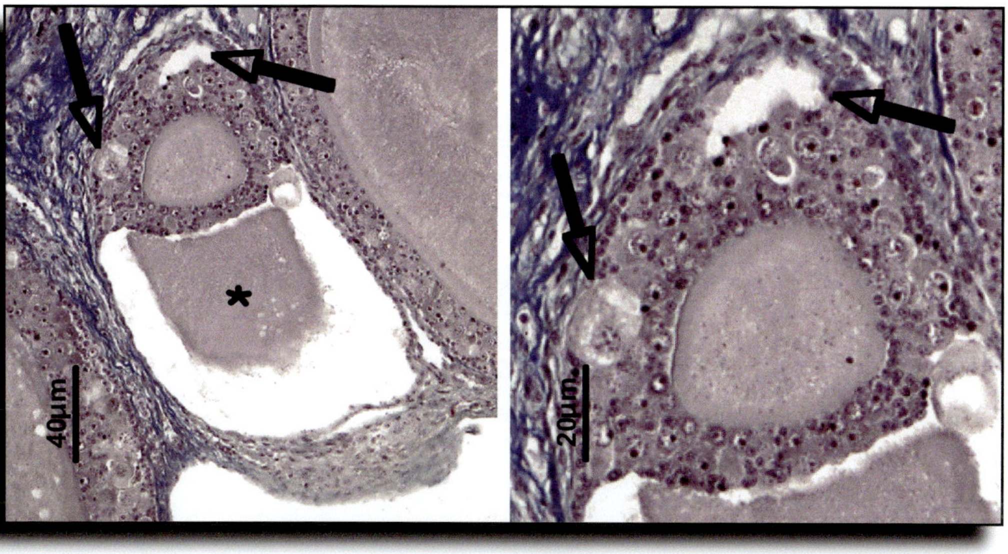

Figure 7.16 Italian Wall Lizard, *Podarcis sicula*. Lacertidae. Atretic early previtellogenic follicles in a specimen that received a single intraperitoneal injection of a cadmium solution. Note the cytoplasmic remnants (*) of an atretic oocyte and the presence of alterations in the epithelium (arrows) of a intact follicle. (Image courtesy of Rosaria Scudiero and Chiara M. Motta, from Simoniello et al. 2010. *Open Zool J* 3:37–41. With permission.)

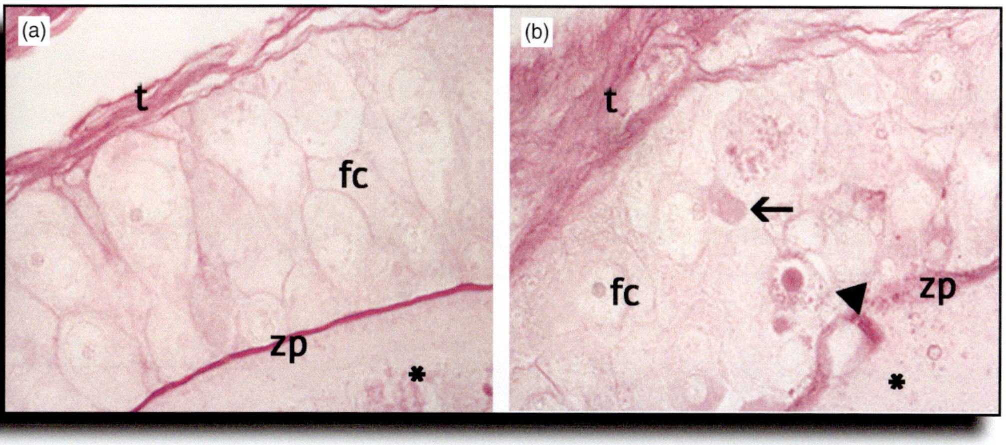

Figure 7.17 Italian Wall Lizard, *Podarcis sicula*. Lacertidae. Effects of cadmium on follicular epithelium. PAS staining. (a) Intensely stained theca (t) and zona pellucida (zp). Follicle cells (fc) and oocyte cytoplasm (*) are moderately stained. (b) Theca (t) is thicker and intensely stained; the zona pellucida (zp) is interrupted and severely orrugated (arrowhead). Follicle cell cytoplasm (fc) often contains dense PAS positive bodies (arrows). (Image courtesy of Chiara M. Motta.)

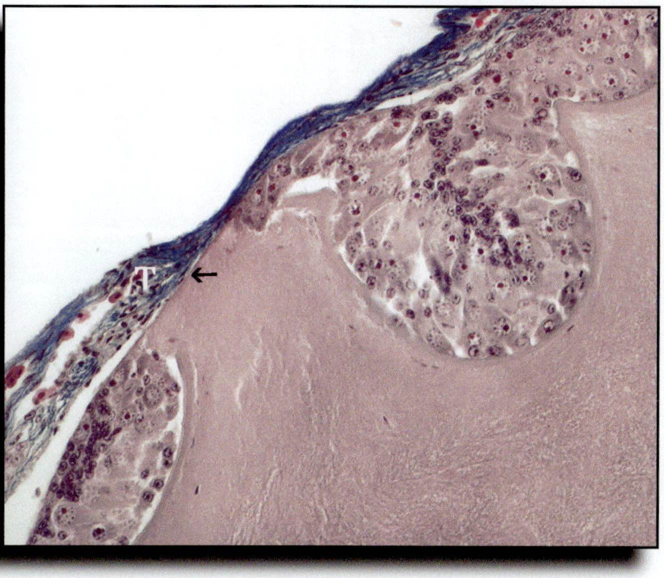

Figure 7.18 Italian Wall Lizard, *Podarcis sicula*. Lacertidae. Effects of cadmium on ovarian follicle morphology. Absence of follicular epithelium: the oocyte membrane (arrow) is in direct contact with the theca (T). H&E staining. (Image courtesy of Rosaria Scudiero and Chiara M. Motta, from Simoniello et al. 2013. *Environ Toxicol* 28:553–562. With permission.)

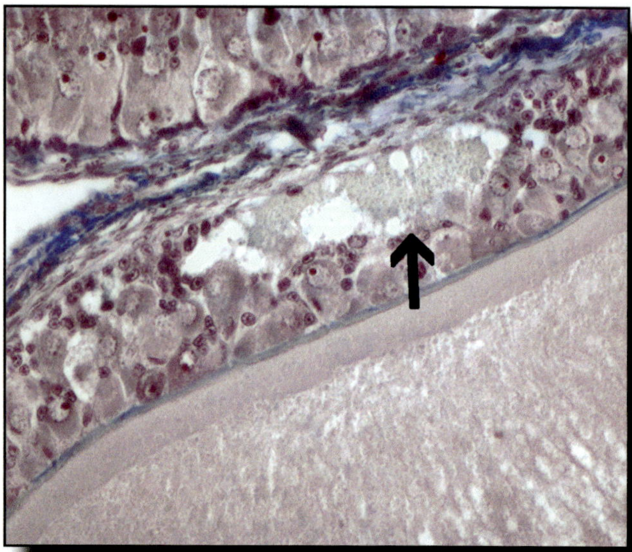

Figure 7.19 Italian Wall Lizard, *Podarcis sicula. Lacertidae*. Effects of cadmium on ovarian follicle cell proliferation. Extensive cell regression causes the appearance of disorganized areas (arrow) in the epithelium. H&E staining. (Image courtesy of Rosaria Scudiero and Chiara M. Motta, from Simoniello et al. 2013. *Environ Toxicol* 28:553–562. With permission.)

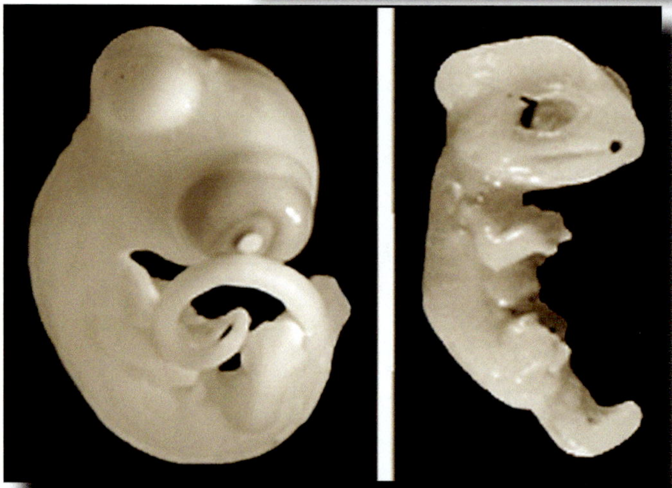

Figure 7.20 Italian Wall Lizard, *Podarcis sicula. Lacertidae*. (Left) Regularly developed 15 day-embryo obtained from a control female. (Right) Embryo recovered from a female fed with cadmium-contaminated insect larvae showing anophthalmia and defects in limb and tail development. (Photo courtesy of Rosaria Scudiero and Chiara M. Motta, from Simoniello et al. 2013. *Environ Toxicol* 28:553–562. With permission.)

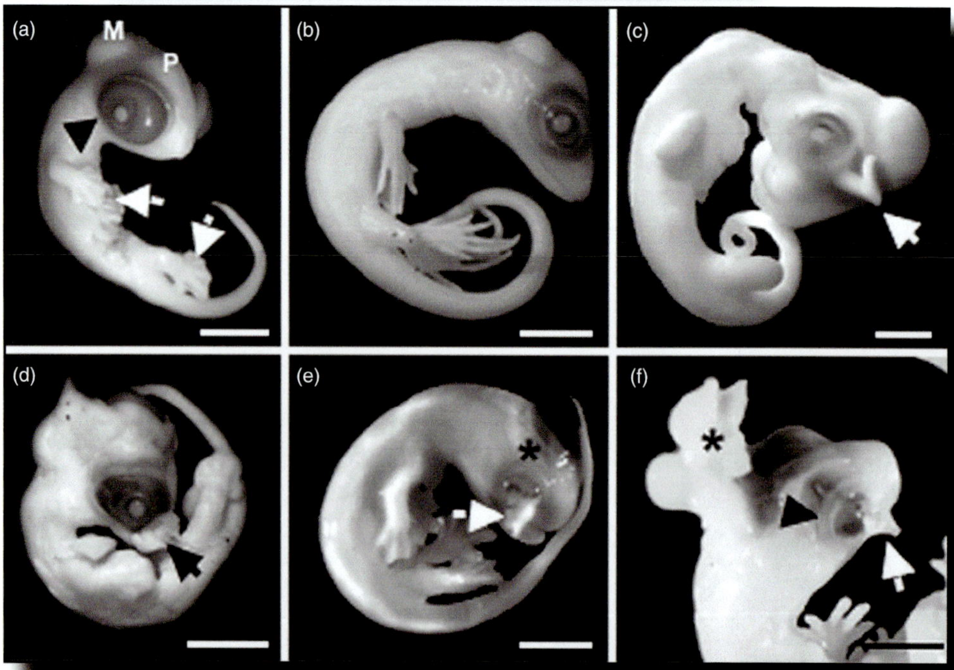

Figure 7.21 Italian Wall Lizard, *Podarcis sicula. Lacertidae*. Morphological alterations in embryos incubated in cadmium-contaminated soil. (a) Control embryo at 25 days post deposition (pd). The eye (arrowhead) is well developed, pigmented, with lens differentiated; mesencephalon (M) and proencephalon (P) are prominent; limb paddles (arrows) show developing fingers. (b) Control embryo at 40 days pd. The eye has completely developed eyelid, the encephalon has become less prominent, digits are fully separated. (c) Cadmium-treated embryo at 15 and (d) 20 days pd. Malformed eye with fleshy growths (arrow). (e) Cadmium-treated embryo, 25 days pd. The encephalon (*) and the eye are missing, a large fleshy growth (arrow) is present. (f) Cadmium-treated embryo, 35 days pd. The encephalon is extruded (*); the eye is poorly developed (arrowhead) and shows a fleshy growth (arrow). (Photo courtesy of Rosaria Scudiero and Chiara M. Motta, from Simoniello et al. 2011. *Comp Biochem Physiol C Toxicol Pharmacol* 153:119–127. With permission.)

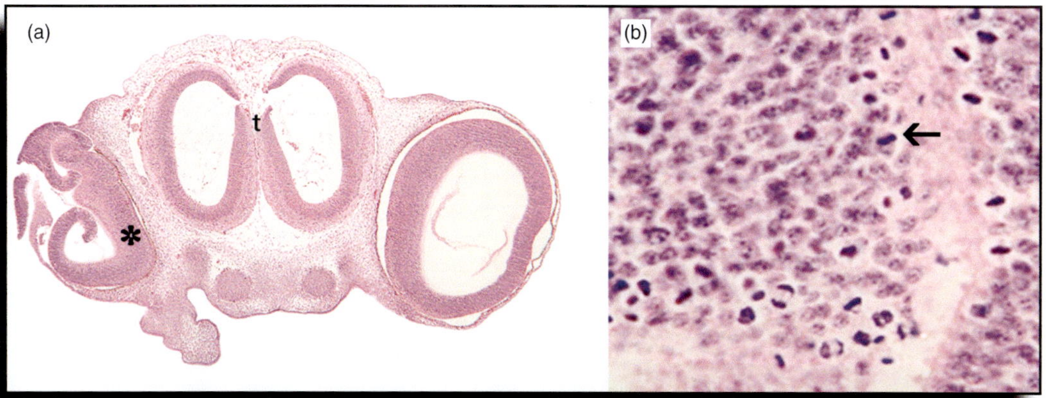

Figure 7.22 Italian Wall Lizard, *Podarcis sicula. Lacertidae.* Retinal development in cadmium-treated embryos. (a) Embryo with unilateral folding of the retina (*). Note the asymmetric buccal vault. Telencephalon (t). (b) Note increase in proliferation (mitotic cells, arrow) in a portion of folded, undifferentiated distal retina. H&E staining. (Image courtesy of Rosaria Scudiero and Chiara M. Motta, from Simoniello et al. 2014. *J Exp Zool* 321A:207–219. With permission.)

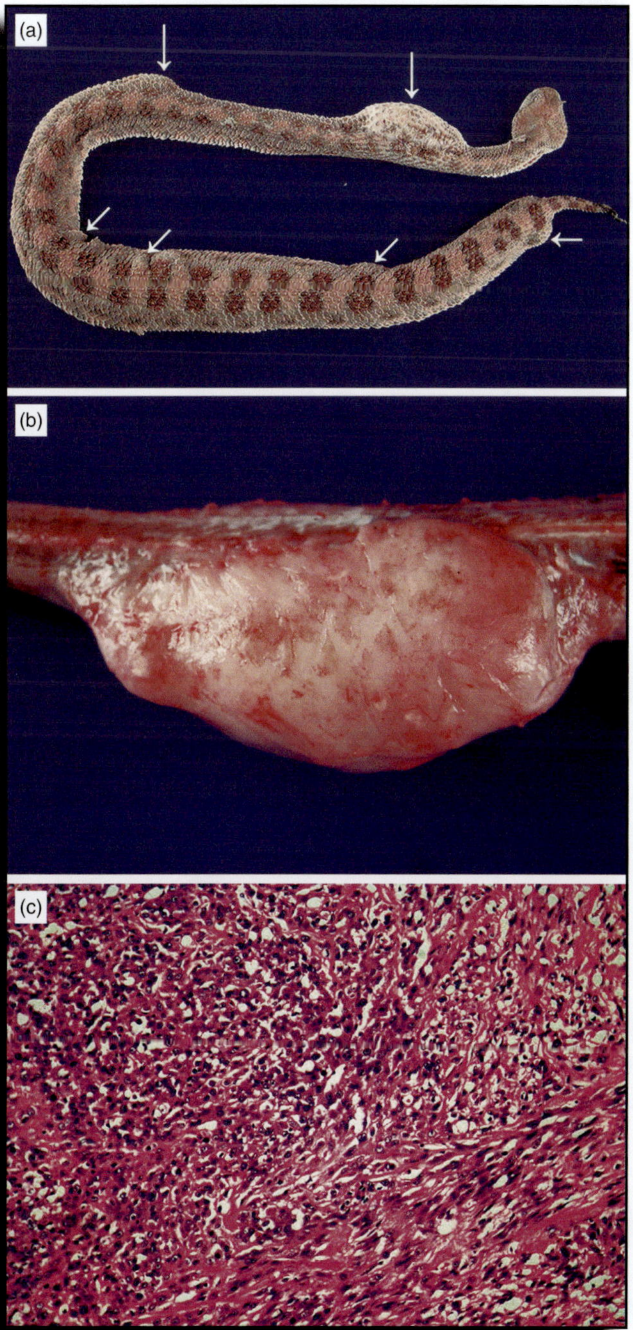

Figure 7.23 Saharan Horned Viper, *Cerastes cerastes*. Viperidae. (a) Several subcutaneous tumors (arrows) on the dorsal and lateral surfaces of the body. (b) Gray-white surface of the largest tumor, identified as a fibrosarcoma. (c) Proliferation of pleomorphic populations of spindle-shaped or round tumor cells arranged in interwoven patterns. H&E stain. High hepatic Cd levels were detected in this snake. (Photo courtesy of Jorge Orós, from Orós et al. 2009. *Vet Rec* 164:690–692. With permission.)

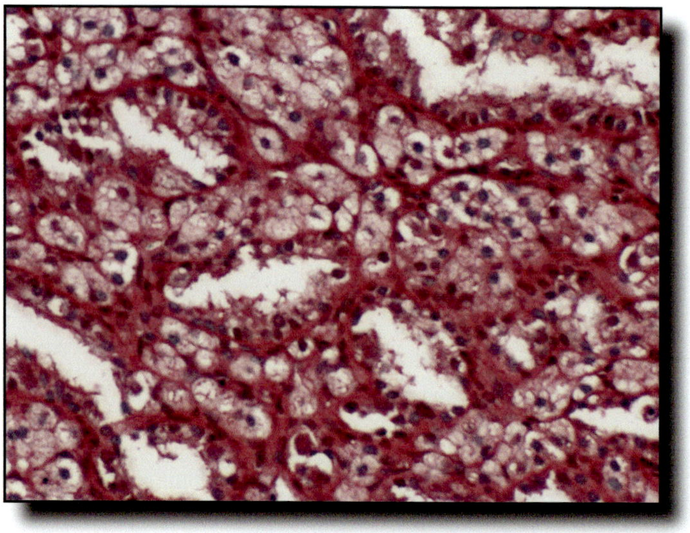

Figure 7.24 Broad-Snouted Caiman, *Caiman latirostris*. Alligatoridae. Disorganized seminiferous tubules in a 10-day-old caiman in ovo exposed to bisphenol A. (Image courtesy of Mónica Muñoz de Toro, from Durando et al. 2013. *Gen Comp Endocrinol* 191:31–38. With permission.)

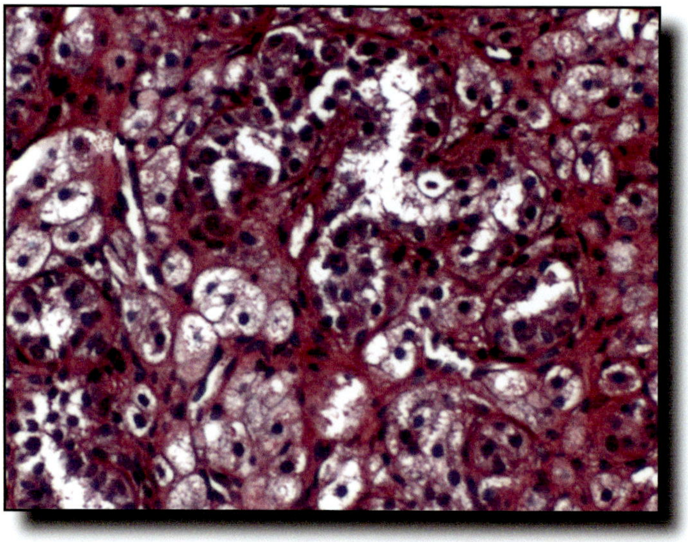

Figure 7.25 Broad-Snouted Caiman, *Caiman latirostris*. Alligatoridae. Disorganized seminiferous tubules in a 90-day-old caiman in ovo exposed to bisphenol A. (Image courtesy of Mónica Muñoz de Toro, from Durando et al. 2013. *Gen Comp Endocrinol* 191:31–38. With permission.)

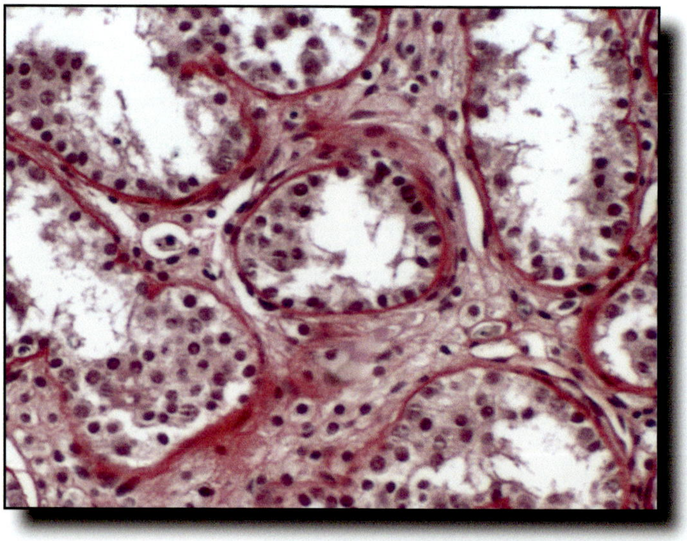

Figure 7.26 Broad-Snouted Caiman, *Caiman latirostris*. Alligatoridae. Disorganized seminiferous tubules and a thickening of interstitial spaces in a juvenile caiman in ovo exposed to bisphenol A. (Image courtesy of Mónica Muñoz de Toro, from Durando et al. 2013. *Gen Comp Endocrinol* 191:31–38. With permission.)

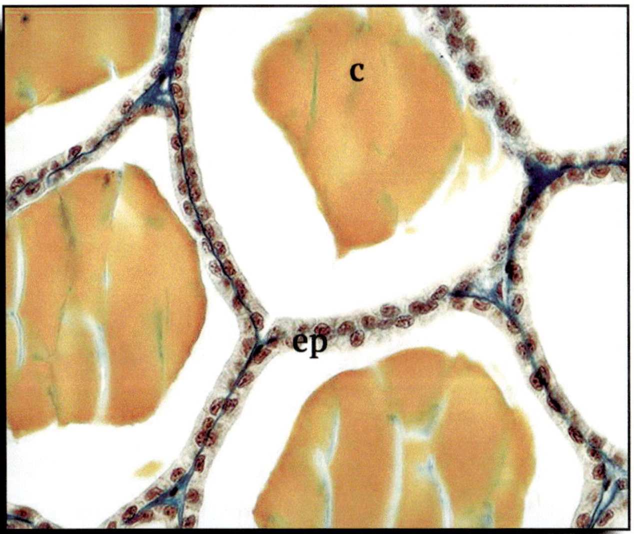

Figure 7.27 Italian Wall Lizard, *Podarcis sicula. Lacertidae*. Thyroid gland from a specimen treated with nonylphenol for 14 days; the follicular epithelium (ep) is low cuboidal. (Image courtesy of Maria de Falco, from Sciarrillo et al. 2010. *Open Zool J* 3:17–22. With permission.)

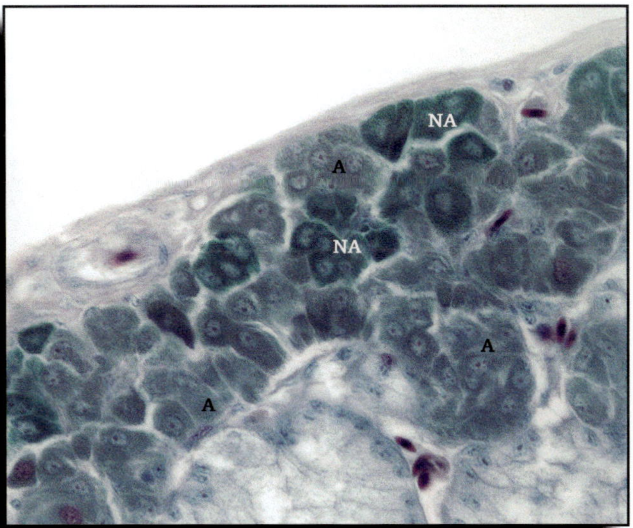

Figure 7.28 Italian Wall Lizard, *Podarcis sicula. Lacertidae*. Adrenal gland. Animal treated with 12 intraperitoneal injections of nonylphenol (the treatment lasted 24 days). Chromaffin ribbon showed an increase of adrenaline cells (A) that occupied the outer layers of the dorsal ribbon. Moreover, steroidogenic cord was increased in diameter. Giemsa stain. (Image courtesy of Maria de Falco, from De Falco et al. 2014. *Chemosphere* 104:190–196. With permission).

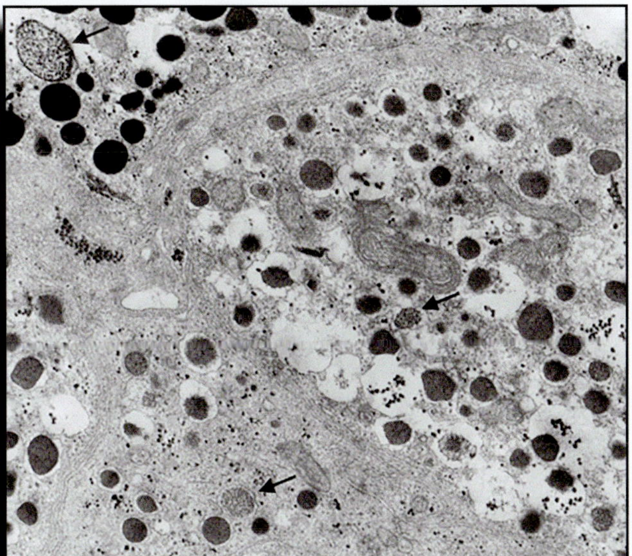

Figure 7.29 Italian Wall Lizard, *Podarcis sicula. Lacertidae*. Adrenal gland. Animal treated with 38 intraperitoneal injections of nonylphenol (the treatment lasted 76 days). Adrenaline granules were smaller than controls; some granules showed a re-synthesis phase (arrows). (Image courtesy of Maria de Falco, from De Falco et al. 2014. *Chemosphere* 104:190–196. With permission.)

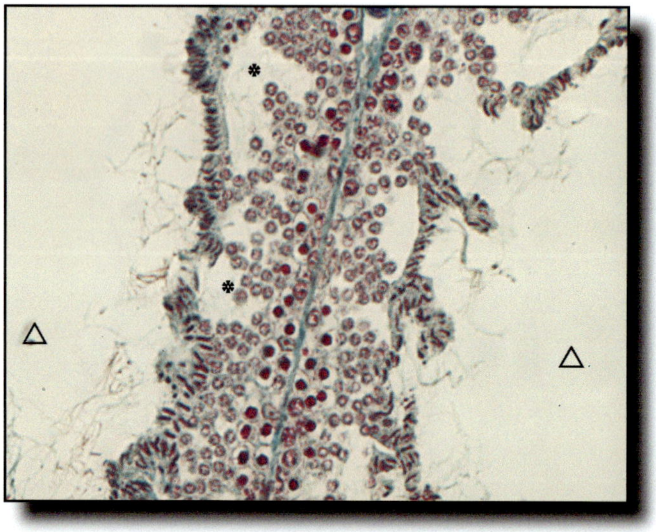

Figure 7.30 Italian Wall Lizard, *Podarcis sicula. Lacertidae*. Seminiferous tubules. Animal treated with nonylphenol administered via food and water. Several empty spaces (*) are evident in the seminiferous epithelium. Note the enlarged lumen of the tubule (△). Mallory's trichrome staining. (Image courtesy of Ilaria Verderame, from Verderame and Limatola 2015. *Gen Comp Endocrinol* 213:9–15. With permission.)

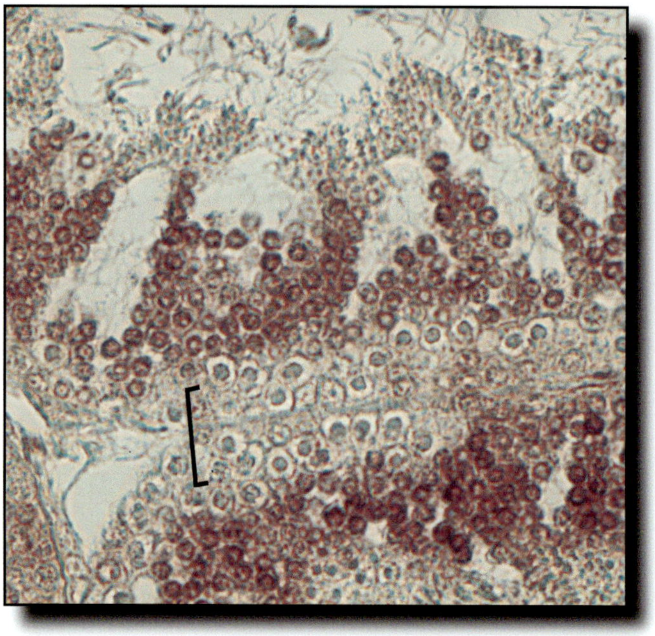

Figure 7.31 Italian Wall Lizard, *Podarcis sicula. Lacertidae*. Seminiferous tubules. Animal treated with nonylphenol administered via food and water. In situ hybridization with ERβ homologous probe. ERβ mRNA is unexpressed in spermatogonia and primary spermatocytes ([) and detected toward the lumen in secondary spermatocytes, spermatids, and spermatozoa. (Image courtesy of Ilaria Verderame, from Verderame and Limatola 2015. *Gen Comp Endocrinol* 213:9–15. With permission.)

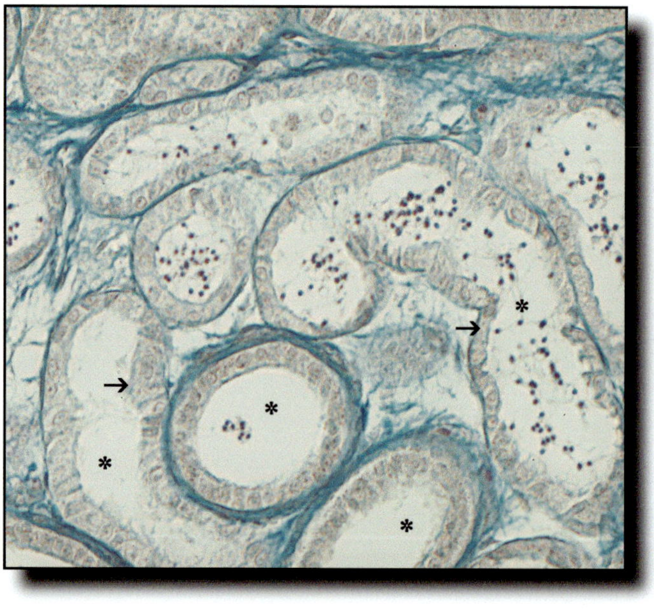

Figure 7.32 Italian Wall Lizard, *Podarcis sicula. Lacertidae*. Epididymis. Animal treated with nonylphenol administered via food and water. The epithelium of the corpus (arrows) is regressed and few granules and sperm are present in the lumen (*). Mallory's trichrome staining. (Image courtesy of Ilaria Verderame, from Verderame and Limatola 2015. *Gen Comp Endocrinol* 213:9–15. With permission.)

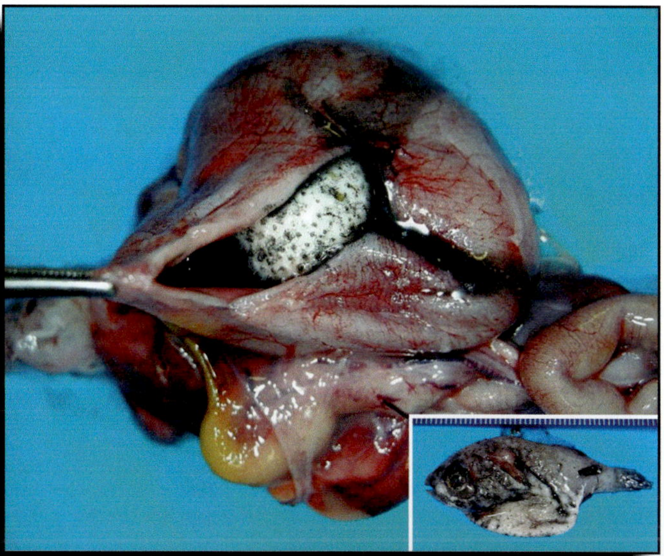

Figure 7.33 Cuvier's Dwarf Caiman, *Paleosuchus palpebrosus*. Alligatoridae. This captive caiman had been fed a green spotted pufferfish (*Tetraodon nigroviridis*) (inset), source of tetrodotoxin. (Courtesy of Lauren Powers.)

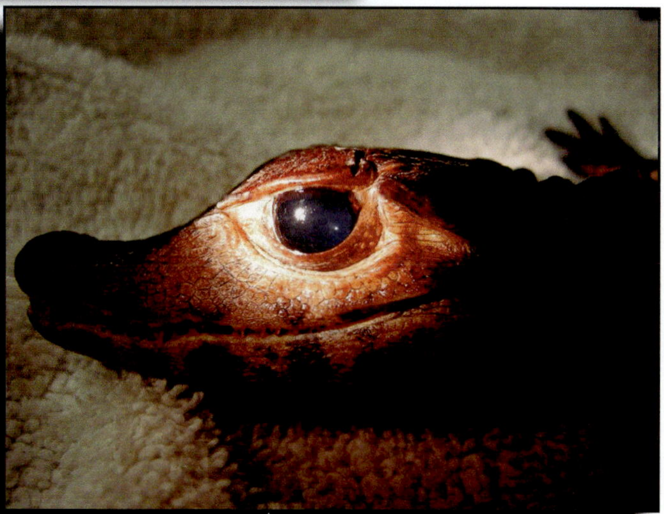

Figure 7.34 Cuvier's Dwarf Caiman, *Paleosuchus palpebrosus*. Alligatoridae. Mydriasis associated with tetrodotoxicosis after being fed a Green Spotted Pufferfish. (Courtesy of Lauren Powers.)

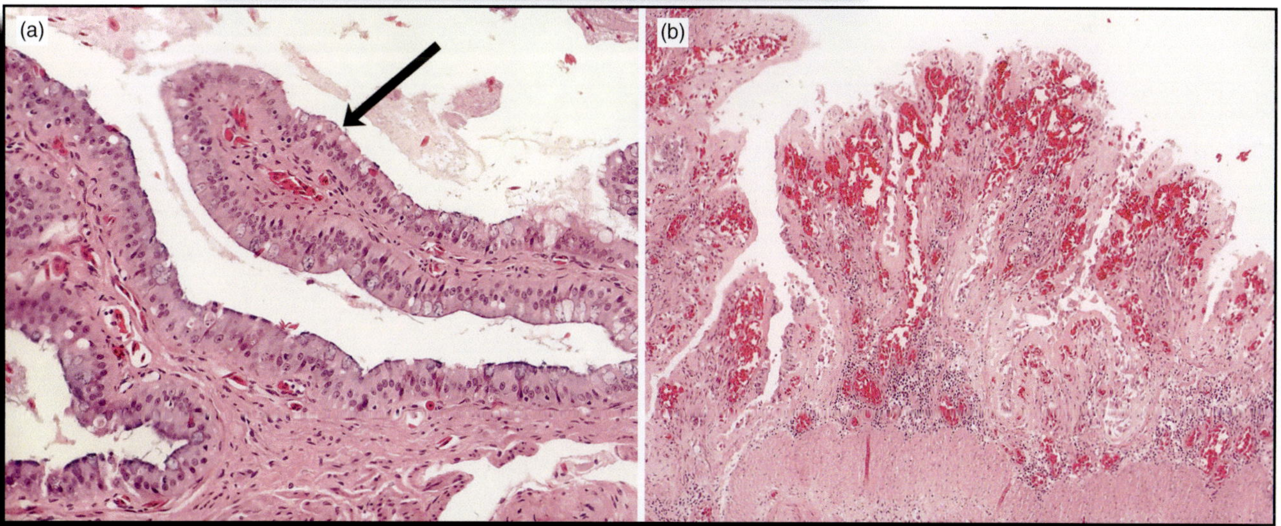

Figure 7.35 Fea's Viper, *Azemiops feae*. Viperidae. Fenbendazole toxicosis. (a) small intestine of Fea's Viper with normal villi, lined by one or two layers of elongated mucosal epithelium (arrow). (b) Acute fenbendazole toxicosis resulting in necrosis and exfoliation of villar epithelium, collapse and fusion of the lamina propria, and acute hemorrhage in the lamina propria. H&E. (Courtesy of Michael Garner.)

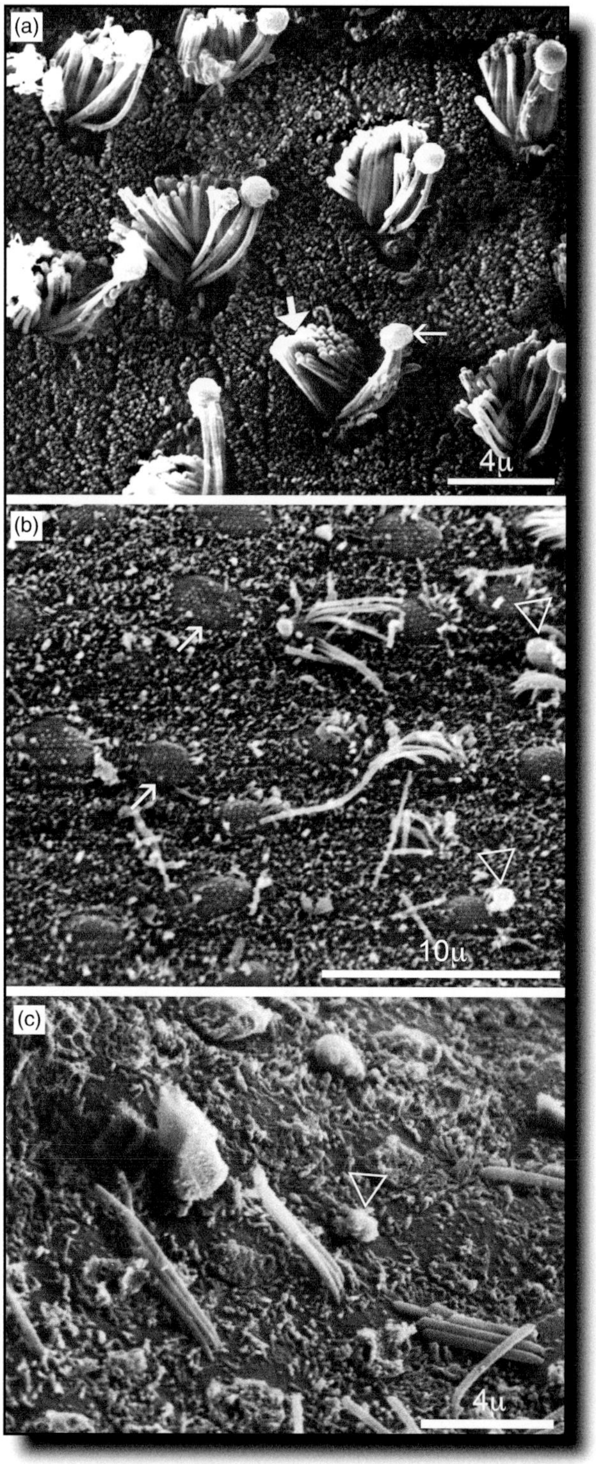

Figure 7.36 Italian Wall Lizard, *Podarcis sicula. Lacertidae.* Saccule. (a) The otic saccule sensory epithelium of untreated lizards has hair cells with an enlarghed, bulb-like kinocilium (thin arrow) and stereovilli in an organ pipe configuration (bold arrow). (b) Lizard euthanized 3 days after the third gentamicin injection: widespread loss of hair bundles (arrows) from the hair cells in the striolar region, apical blebs are present (arrowheads). (c) Lizard euthanized 8 days after the third gentamicin injection: in extrastriolar region numerous hair cells are missing and apical blebs are still present (arrowhead). (Image courtesy of Bice Avallone, [a,b] from Avallone et al. 2008. *Hearing Res* 235:15–22. With permission).

Figure 7.37 Mohave Desert Tortoise, *Gopherus agassizii*. Testudinidae. Probable envenomation from a rattlesnake. The tortoise had dried blood extending from the left ocular area to the maxillary rhamphotheca. (Photo courtesy of Elliot R. Jacobson, from Berry et al. 2016. *Herpetol Rev* 47:652–653. With permission).

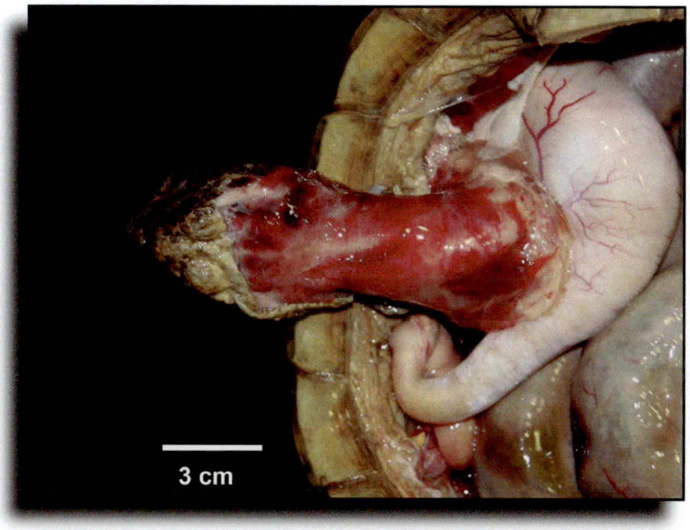

Figure 7.38 Mohave Desert Tortoise, *Gopherus agassizii*. Testudinidae. Probable envenomation from a rattlesnake. The tortoise had subcutaneous hemorrhage that extended from the left periorbital area to the base of the cervical subcutaneous connective tissue. (Photo courtesy of Elliot R. Jacobson, from Berry et al. 2016. *Herpetol Rev* 47:652–653. With permission.)

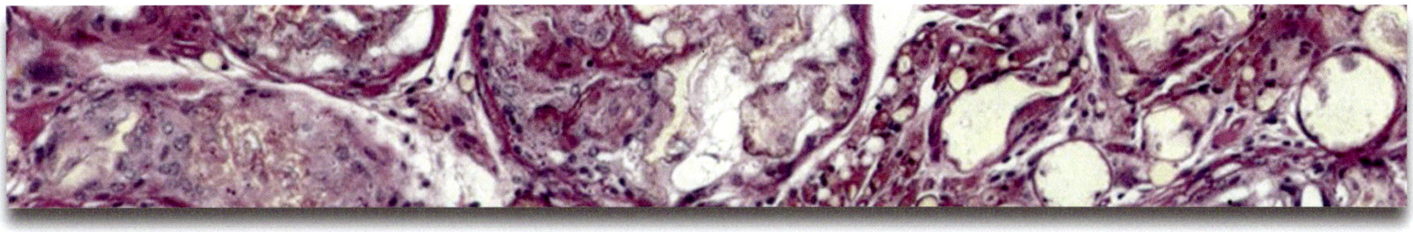

8

PHYSIOLOGY AND DISEASES OF THE REPRODUCTIVE SYSTEM

JOHN F. ROBERTS AND DANIEL A. WARNER

Contents

8.1 Introduction

Reproduction is an inherently broad and complex topic that incorporates many components of biology (e.g., life history, behavior, physiology, ecology, anatomy, sex determination, and development). In reptiles, reproductive biology is extremely diverse, which reflects the long evolutionary history and species diversity of this group (>10,450 species) (Pough et al., 2016; Uetz, 2016). Phylogenetically, the modern Class Reptilia includes two major clades of extant taxa: the Lepidosauria and its sister clade Archosauromorpha. Lepidosauria includes Squamata (lizards and snakes) and Rhynchocephalia (Tuatara), and Archosauromorpha includes Testudines (Chelonia; turtles and tortoises) and Archosauria (crocodilians and birds) (Lee, 2013). Classification of birds as reptiles is taxonomically correct (Benton, 2014), but traditionally birds have been treated as very different taxa due to several derived features (e.g., endothermy, feathers). Nevertheless, references to avian reproductive physiology and disease may be useful when literature on "non-avian reptiles" is incomplete, especially for their sister group the Crocodilia (Johnson, 2014). Although hormonal control of reproduction is well conserved across amniotes, reptile reproductive strategies are diverse, even between closely related taxa. For example, multidirectional evolutionary transitions between oviparity and viviparity have occurred within squamates (Blackburn, 1992; Albergotti and Guillette, 2011), and delivery of nutrients to embyros varies from lecithotrophic oviparity (yolk-derived nutrition) to matrotrophic viviparity (placental-derived nutrition) (Blackburn, 1992). Sex-determining mechanisms also vary considerably, sometimes even within the same genus or species (Bull, 2004; Pen et al., 2010).

Diversity in native habitats and geographic ranges has also produced enormous seasonal variation in male and female reproductive strategies (Cunningham et al., 2018), as well as generated variation of sensitivities of developing embryos to environmental conditions (Warner et al., 2018; While et al., 2018). For example, reptile reproduction tends to be cyclic (i.e., varying with seasonal changes in environmental conditions) and depends greatly on energy homeostasis despite fluctuating metabolic demands. Annual reproductive cycles are primarily influenced by day length and temperature in temperate climates and by rainfall in tropical climates (Van Dyke, 2015). Energy maintenance requirements and nutrient intake, which vary seasonally, also influence the production of viable ova, sperm, and offspring (Brown and Shine, 2006; Warner et al., 2008; Hall et al., 2018). In addition, humidity and temperature linked to local and seasonal climates greatly influence the viability of eggs.

Simulating native environmental conditions and nutrient requirements facilitates captive propagation of reptiles (Rivera, 2008). When nutrient imbalances arise, infertility and disease can manifest during gamete production, fertilization, sexual differentiation, gonad maturation, pregnancy, parturition/oviposition, incubation, or aging. Compounding influences of multiple factors (e.g., nutrition, toxins, and environmental stressors) on reproductive tissues can influence the development of reproductive diseases. Since reproductive disease is usually a response to systemic challenge, accompanying clinical signs may include changes in behavior, neurologic function, appetite, thermoregulation, and even skin coloration (Kummrow et al., 2010; Johnson and Wade, 2011). Chronic infectious diseases in digestive, urinary, neurologic, endocrine, cardiovascular, and respiratory systems will negatively affect energy balance, reduce the ability to compensate, and produce physiologic regression and/or disease in reproductive organs (**Figure 8.1**). When nutrient resources are limited or adipose reserves are depleted, reproductive demands may also suppress immune function and result in susceptibility to pathogens (French et al., 2007). The aim of this chapter is to provide a general overview of reproductive physiologic processes and noninfectious diseases in reproductive organs, eggs, and neonates.

8.2 Sex-Determining Mechanisms

Reptile sex determination has been historically classified as either genotypic sex determination (GSD) whereby an individual's sex is determined by inherited genes on sex chromosomes, or as temperature-dependent sex determination (TSD) whereby sex is determined by the temperature experienced by embryos during development (Valenzuela et al., 2003). Strong evidence has accumulated, however, that these forms of sex determination are more properly viewed as ends of a continuum since elements of both mechanisms co-occur within many taxa (Sarre et al., 2004; Warner, 2011; Holleley et al., 2016). For example, the Australian Three-lined Skink (*Bassiana duperreyi*) and Central (Inland) Bearded Dragon (*Pogona vitticeps*) have heteromorphic sex chromosomes, yet autosomal sex-determining genes may be activated by temperature during egg incubation (Shine et al., 2002; Holleley et al., 2015). Factors other than temperature (e.g., yolk allocation, maternal nutrition, incubation moisture and stress) also impact sex ratios of offspring, but the role of these factors in sex determination remains poorly studied (Warner et al., 2007, 2009; Radder et al., 2009; Sifuentes-Romero et al., 2018). Thus, sex determination can be influenced by multiple factors, including inherited genes, captive husbandry practices, and global change (e.g., environmental toxins, climate change) (Pezaro et al., 2016).

8.2.1 Genotypic Sex Determination (GSD)

Genotypic sex determination (GSD) may be determined by heteromorphic sex chromosomes, influenced by multiple alleles on several genes or can be dependent on dosage effects of specific genes (Yao and Capel, 2005). When sex chromosomes vary in size and shape this is termed heterogametic and this is often, but not always, true of chromosome pairs in GSD. In mammals, individuals that inherit a Y chromosome from their father develop into males (due to the SRY gene that triggers testes development), and those that inherit an X chromosome develop as females (i.e., the absence of the SRY gene leads to ovarian development); both sexes always inherit an X chromosome from their mother. Thus, males (XY) are termed heterogametic and females with two X sex chromosomes (XX) are termed homogametic. This same chromosomal arrangement occurs in many reptiles, but it is not universal. For instance, males of numerous species contain two Z sex chromosomes (ZZ) and thus are the homogametic sex while females, with one Z and one W chromosome (ZW), are heterogametic (Warner, 2011). GSD may also occur in species that have sex chromosomes that are morphologically indistinguishable (Warner, 2011). Heteromorphic sex chromosomes are the minority in lizards and are present in <200 of the approximately 1000 species that have been karyotyped (Ezaz et al., 2009). Yet, most snakes have a female heterogametic system (ZZ/ZW sex chromosomes) (Vicoso et al., 2013; Gamble et al., 2017).

In some reptile species ploidy of sex-determining genes is variable. Diploid-triploid mosaicism is associated with sex ratio in the Twist-Neck Turtle *Platemys platycephala* (Bickham and Hanks, 2009). Triploid (XXY) female lizards and male (ZWW) snakes also occur (Olmo, 2005; Ezaz et al., 2009; Georges et al., 2010). Occurrence of aberrant ploidy in species that are normally diploid may cause gonad changes and defective sperm in birds (Lin et al., 1995). Sex in reptiles may also be determined by interspecific hybridization in obligate parthenogenesis (Neaves and Baumann, 2011) and facultative parthenogenesis (Booth et al., 2012). Parthenogenesis was first described in 1958 in the Iberian Rock Lizard (*Darevskia saxicola*) and has since been documented in several agamid, teiid, gekkonid, and varanid lizard families and fewer snakes including the Brahminy Blind Snake (*Indotyphlops braminus*), Papuan Taipan (*Oxyuranus scutellatus*) and Southern (or Common) Death Adder (*Acanthophis antarcticus*) (Nussbaum, 1980; Vitt, 2015; Allen et al., 2018). Parthenogenetic species have increased diversity within the individual genome and less diversity in the whole population while species with sexual reproduction have less individual diversity and greater diversity in the population (Vitt, 2015).

8.2.2 Temperature-Dependent Sex Determination

Temperature-dependent sex determination (TSD) was first described in the Common Agama (Rainbow Lizard, *Agama agama*) (Charnier, 1966; Steele et al., 2018), but TSD was not fully accepted by the scientific community until after additional discoveries in turtles and crocodilians in the early 1970s (Bull, 1980). TSD is triggered by thermodynamic regulation of autosomal gene expression in the developing embryo (Pieau, 1975). This usually involves an absence of heteromorphic sex-determining chromosomes (XY/XX or ZW/ZZ). TSD is often considered a major driver of skewed sex ratios due to certain weather patterns or selected incubator settings. TSD is the most commonly deployed mechanism of sex determination in orders Testudines, Crocodilia, and Rhynchocephalia, but also occurs in several squamates, most notably the families Gekkonidae and Agamidae (Gamble, 2010; Warner, 2011). TSD has not been documented in snakes (Warner, 2011), but sex-biased embryo mortality in relation to incubation temperature could give the impression of TSD in some taxa (Burger and Zappalorti, 1988; Rood and Freedberg, 2016; Kato et al., 2017). Typically, TSD is characterized by a narrow range of constant incubation temperatures (1°–2°C) that produces mixed sex ratios (called the transitional range of temperatures), but temperatures above or below that range produce exclusively one or the other sex depending on the species. In addition, sex determination typically occurs during the middle third of embryonic development, which corresponds to the timing of gonadal differentiation (Shine et al., 2007). Importantly, most studies of TSD involve incubation of eggs at constant temperatures that do not simulate the complex thermal conditions of natural nests. Several studies that incubate eggs under fluctuating temperatures demonstrate strong effects on sex determination (Georges et al. 1994; Neuwald and Valenzuela, 2011; Warner and Shine, 2011).

Prior to morphologic differentiation of gonads, temperature affects shifts in transcriptome expression in liver, kidney, and most prominently in brain tissue. Changes in neurologic gene expression initiate steroidogenic shifts (Czerwinski et al., 2016). In addition, temperature can stimulate aromatase activity and activate expression of the aromatase gene. In turn, aromatase converts androgens to estrogens during development and plays a key role in sex determination for both TSD and GSD species (Crews et al., 2001). Once this cascade of genes and enzymes is activated, TSD is irreversible and directed gonadogenesis follows in the middle third of embryonic development (Yao and Capel, 2005; Shoemaker-Daly et al., 2010).

Sex determination in species with heteromorphic sex chromosomes can also be sensitive to incubation temperature. That is, the effects of temperature can override the effects of sex chromosomes, particularly toward thermal extremes (Shine et al., 2002; Quinn et al., 2011; Holleley et al., 2015). In the Central Bearded Dragon, sex reversal in genotypic female embryos (ZZ) occurs at warmer incubation temperatures, resulting in a mismatch between genotypic and phenotypic sex (ZZ male). When these homozygous offspring are mated (ZZ_{female} + ZZ_{male}) they produce ZZ homozygous progeny with a temperature-sensitive sex-determining mechanism (Holleley et al., 2015). The Indian Garden Lizard (*Calotes versicolor*) has been shown to have a sex determination mechanism that is neither strictly GSD or TSD, and development of male embryos is dependent on exposure to male hormones (Chakraborty and Raman, 2010). Similar effects of steroids on sex determination has been shown in Painted Turtles (*Chrysemys picta*) (Bowden et al., 2001) and Green Anoles (*Anolis carolinensis*) (Lovern and Wade, 2003a).

8.3 Gonad and Sex Duct Differentiation

Bipotential primordial cells in reptile embryos have potential to develop into male or female gonads. The gonadal sex cords in both sexes originate from migration of epithelized somatic cells out of the germinal epithelium (Pieau et al., 1999; Capel, 2017). The Red-Eared Slider (*Trachemys scripta elegans*) has served as a reptile model for studies of gene expression during gonadal development. Conserved genes, steroidogenic factor 1 (Sf1) and Wilms tumor 1 (Wt1) verified direct formation of the bipotential gonad. Loss of either gene results in gonadal agenesis in both sexes (Yao and Capel, 2005; Ge et al., 2017). In the Red-Eared Slider, Dmrt1 (double sex and mab-3-related transcription factor 1) is the master switch gene regulating testes determination in TSD (Ge et al., 2017). Expression of Dmrt1 occurs after temperature-determining events and prior to expression of anti-Müllerian hormone (AMH). Unlike mammals, expression of Sox 9 in the Red-Eared Slider and the American Alligator (*Alligator mississippiensis*) occurs after up-regulation of AMH (Yao and Capel, 2005). In Red-Eared Slider eggs incubated at the male-producing temperature, Dmrt1 up-regulates first, followed by Sox9 and AMH. In eggs incubated at the female-producing temperature, Foxl2 is up-regulated after the thermosensitive period (Bieser and Wibbels, 2014).

The nongonadal tubular portion of the reproductive tract has embryologic origin in the mesonephric (Wolffian) duct that becomes the oviduct or the paramesonephric (Müllerian) duct that develops into the spermatic duct. Sertoli cells secrete AMH that causes Müllerian duct regression and Leydig cells secrete testosterone (T) that halts regression of the Wolffian duct (Agnew and MacLacklan, 2017). Mutations in genes controlling these processes may result in intersex individuals. Glandular development of the oviduct (adenogenesis) in the Broad-Snouted Caiman (*Caiman latirostris*) and other reptiles occurs postnatally and is regulated by gonad origin steroids including estrogen and progestins. Aberrant development of glands in the time between neonatal to the juvenile life stages may result in eggshell malformations when these females reach maturity (Galoppo et al., 2016). Altered morphology of the oviduct gland epithelium occurs simultaneously with changes in the sub-epithelium including collagen remodeling and muscle patterns of α-actin and desmin distribution (Galoppo et al., 2016).

8.4 Endocrine Regulation of Reproduction

Reproductive hormones are well conserved in most reptile species and there is considerable overlap with other classes of vertebrates (Lovern, 2011; Milnes, 2011). The hypothalamus releases gonadotropin releasing hormone (GnRH) that stimulates pituitary release of reproductive gonadotropins or glycoprotein hormones, luteinizing hormone (LH), and follicular-stimulating hormone (FSH) or FSH-like hormones. FSH and LH activate G protein-coupled receptors (GPCRs) on granulosa cells in the ovary and Sertoli cell in the testicle (Dubois and Callard, 1985; Zhang et al., 2016). FSH receptors in alligators are phylogenetically closest to birds among examined classes (Zhang et al., 2015). FSH influences folliculogenesis, clutch size, offspring growth, and regulation of reproductive energy needs in reptiles (Sinervo and Licht, 1991; Sinervo and Miles, 2011; Zhang et al., 2015). The gonads feed back to the hypothalamus through plasma estrogen and T. This feedback mechanism is termed the hypothalamic-pituitary-gonadal (HPG) axis and is influenced by photoperiod, temperature, and moisture parameters (Van Dyke, 2015).

Notably, peptide structural differences in reptile GnRH and FSH are unique, and commercially available assays for mammal gonadotropins are invalid in reptiles (Van Dyke, 2015). However, commercially available assays for mammal- and avian-derived estradiol (E2), T, and progesterone (P) are used with success across reptile taxa (Van Dyke, 2015). GnRH has multiple isoforms that likely vary in activity in Chelonia (Tsai and Licht, 1993). Although LH has not been found in Squamata, this hormone has been identified in Testudines and Crocodilia (Van Dyke, 2015). FSH may be more conserved among amniotes as ovulation in Green Anoles has been stimulated with porcine origin FSH (Jones et al., 1988).

In many reptiles, cyclic HPG axis activity is driven by photo stimulation of the parietal eye atop the head or lateral eye which communicates to the pineal gland for secretion of melatonin (Van Dyke, 2015). Trauma or disease in photoreceptors could affect photodynamic cyclicity. Melatonin influences the pineal gland to activate stimulatory neuropathways regulating the hypothalamic release of GnRH (Vivien-Roels et al., 1979; Blanvillain et al., 2011). Release of hypothalamic hormones is also regulated in the brain by kisspeptin and its receptor (KissR) and through the feedback mechanism within the HPG axis (Pasquier et al., 2014). Mutations resulting in underexpression of kisspeptin may cause hypogonadism, and overexpression of kisspeptin may cause precocious puberty (Ball and Wade, 2013; Pasquier et al., 2014). All dynamic events in the reproductive tract such as spermatogenesis, folliculogenesis, maintenance of pregnancy, and ovipostion are also influenced at some point by hormones (E2, T, P) that originate in the gonad.

8.5 Seasonal Cyclicity

Reptiles exhibit considerable variation in hormonally timed gonadal recrudescence, gametogenesis, and sexual behavior. Coordinated cyclic gonad response between testes and ovaries may be prenuptial (associated), postnuptial (dissociated), or a variation of both types of cycles ("mixed"). These cycles refer primarily to the degree of synchronicity between maximum sex hormone secretion and gametogenesis before courtship and mating. A prenuptial gonadal cycle is more common in Squamata and Crocodilia, while a post-nuptial cycle occurs more frequently in Testudines. The prenuptial cycle is typical of tropical and subtropical species while the postnuptial cycle is characteristic of temperate species (Blanvillain et al., 2011). The ovaries of tropical chelonians may have small vitellogenic follicles in waiting year-round that undergo yolking shortly before ovulation (Duval et al., 1982). Geography may influence seasonal cycling within a species. For example, Painted Turtles have seasonal patterns in the northern extent of their range and tropical patterns of folliculogenesis in the southern populations (Moll, 1979). During the prenuptial cycle, both sexes are coordinated so that peak sperm production and maximum testicular weight occurs concurrently with rapid ovary follicle growth immediately prior to the breeding season (Blanvillain et al., 2011; Van Dyke, 2015). Postnuptial cycles are seasonally split so that males and females will initiate sperm production and follicle growth before hibernation and resume gametogenesis prior to mating in the spring (Van Dyke, 2015). Patterns of reproduction can also be disassociated between males and females over different seasons (Crews, 1984). The Tuatara (*Sphenodon punctatus*) may initiate ovary follicular growth three years prior to breeding, while sperm production initiates immediately prior to mating (Brown et al., 1994; Cree, 2015). Extended sperm storage prior to fertilization is also associated with postnuptial cycles (Gribbins, and Rheubert, 2011). In some parthenogenic species, male-typical sexual behaviors performed by females (i.e., pseudocopulation) can directly trigger gonadal activity (e.g., ovulation) (Crews and Moore, 1993). Apoptosis, or programmed cell death, is essential for post-breeding season atrophy of testes and ovaries. During periods of seasonal regression, the epithelium of the oviduct and spermatic duct also undergoes apoptosis (Assisi et al., 2011). Toxins disrupting apoptosis can potentially affect cyclicity (Simoniello et al., 2013). Knowledge of follicular cycling and expected time of ovipostion is often the key to diagnosis of maternal reproductive failure. Depending on species and local environments, oviparous reptiles may produce multiple clutches per year or skip years between clutches (Vitousek et al., 2010). For example, lizards from temperate climates mate in spring and produce offspring during the summer, but patterns can vary considerably depending on seasonality associated with latitude, local climatic conditions (rainfall), or reproductive mode (e.g., viviparity vs. oviparity) (Vitt, 2015; Pough et al., 2016).

8.6 Spermatogenesis

Sperm production (spermatogenesis) occurs within seminiferous tubules of paired testes. Testes may have dynamic variation in size and shape depending on stage of maturity and sexual activity. Spermatogenesis has three general stages: (1) mitosis, proliferation of spermatogonia along the basement membrane (2) meiosis, formation of spermatocytes, and (3) spermiogenesis, maturation of spermatids to spermatozoa (Gribbins, 2011; Foster, 2012). During mitosis, a large population of clonal germ cells is produced. During meiosis, segregation and recombination produces haploid gametes (spermatocytes and spermatids). During spermiogenesis, spermatids differentiate into motile spermatozoa in a seven- or eight-step process (**Figure 8.2**) (Gribbins, 2011). Mature spermatozoa have wide variation among reptile taxa (**Figure 8.3**). Spermiogenesis can be subdivided into (1) acrosome development, (2) elongation, and (3) maturation (Gribbins, 2011). Gamete production in the seminiferous epithelium during spermatogenesis can be affected by toxic and endocrine disruptive chemicals (Gribbins, 2011).

Immature testes are small and may be mistaken for adipose tissue during gross examination. Immature testes are adjacent to and approximate the size of the adrenal gland (**Figure 8.4**). Interstitial tissue separating seminiferous tubules of immature testes are primarily composed of mixed mesenchymal, myoepithelial, and immunologic cells (Gribbins and Rheubert, 2011). Testosterone-producing Leydig cells (interstitial cells) in the interstitium are not prominent prior to puberty. In prepubescent reptiles, a single layer of spermatogonia and Sertoli cells (sustentacular cells) line the basement membrane (**Figure 8.5**), and a similar configuration may be observed in turtles during seasonal quiescence (Gribbins et al., 2003). Detached immature spermatogenic cells may be present in the lumen of the prepubescent epididymis (**Figure 8.6**) (Barnes et al., 2016).

After puberty, and during seasonal spermatogenesis, seminiferous tubules have a progressive increase of cellular density in the seminiferous epithelium and testicular interstitium (**Figures 8.7–8.8**) (Barnes et al., 2016). In temperate climates, Leydig cells hypertrophy in early spring to coincide with elevated breeding blood androgen level and then atrophy in mid-summer when spermatogenesis is decreasing (Dubois et al., 1988). FSH and LH regulate spermatogenesis and testicular steroidogenesis (Bluhm et al., 2004; Kumar et al., 2011). LH acts on Leydig cells to produce T while FSH stimulates Sertoli cells to produce androgen binding protein (ABP) and aromatase (Foster, 2012). In the Northern House Gecko (*Hemidactylus flaviviridis*), interstitial macrophage numbers increase with Leydig cell hypertrophy and decrease as Leydig cells regress (Khan and Rai, 2008).

The width of the seminiferous epithelium is greatest during peak spermatogenesis, corresponding with increased mitosis and meiosis. The width of the entire seminiferous tubule and efferent ductules is greatest post spermiogenesis when the lumina of seminiferous tubules are dilated with spermatozoa (**Figure 8.9**) (Gribbins, 2011). After the breeding season, the width of the seminiferous epithelium decreases as germ cell apoptosis increases (**Figure 8.10**). Estrogen functions in post-spermatogenic regression of testes

(Kumar et al., 2011). The regression of spermatogenesis is also controlled through variation in spermatogonia apoptosis (Foster, 2012). Molecular profiles expressed in testicular tissue of the Northern House Gecko during the active phase include genes for cytoskeleton organization, cell growth, GTPase regulator activity, transcription, and apoptosis. In the regressing phase, genes that promote maintenance of stem cells are up-regulated (Gautam et al., 2013). In times of quiescence excess spermatocytes may be removed from seminiferous epithelium through the process of entosis and apoptosis (**Figure 8.11**) (Zhang et al., 2008; Ahmed et al., 2017). Sperm may be phagocytized within the lumen of seminiferous tubules by Sertoli cells or in the epididymis by mononuclear phagocytic cells (**Figure 8.12**) (Gribbins, 2011; Da Silva and Smith, 2015; Panneerdoss et al., 2017).

Viability of sperm is dependent on paracellular sealing of the "blood–testes barrier" created by tight junctions between Sertoli cells. The term "Sertoli cell barrier" has been proposed as a more anatomically correct term and has been used by some authors. Tight junctions separate basal spermatogonia and luminal spermatocytes. Although most reptiles have a permanent population of Sertoli cells, the osmotic integrity of the blood–testis barrier is variable depending on seasonal cycling (Gribbins, 2011). Examination of Claudin 11, a tight junction protein in Korean Soft-Shelled Turtles (*Pelodiscus maackii*) demonstrated that tight junction integrity increases during seasonal spermatogenesis and decreases during quiescent periods (Park et al., 2015).

Efferent spermatic ducts (rete tubules in chelonians and epididymis in other reptiles) are lined by ciliated epithelial cells that maintain the blood–testes barrier, aid sperm transit with cilia, secrete products supporting sperm, and when needed, phagocytize spermatozoa (Foster, 2012; Barnes et al., 2016). Increased levels of testosterone stimulate hypertrophy and filling of intracytoplasmic secretory granules in the luminal epithelium of the epididymis (**Figure 8.13**) (Gist, 2011). Lysosome enzymes including β-hexosaminidase are present in the epididymis epithelium of the Keeled Indian Mabuya (*Eutropis carinata*), and contribute to endocytosis of proteins, degradation of residual bodies and abnormal sperm, and mediation of carbohydrate-based transient sperm–egg binding (**Figure 8.14**) (Koyanagi and Honegger, 2003; Medini et al., 2018). Secretory granules produce lipocalin proteins that function in transport of small lipophilic molecules. These molecules include lizard epididymal secretory protein (LESP) that bind to the head of sperm, resulting in increased sperm motility (Gist, 2011). At peak seasonal spermatogenesis, the epididymis and vas deferens serve as the primary storage site, which can swell to a size larger than the testes in some species (**Figures 8.15–8.16**) (Gist, 2011). In a postnuptial cycle, there is testicular hyperplasia after breeding, followed by sperm storage (**Figure 8.17**) (Kumar et al., 2011). Direct influence of T produced by Leydig cells, without aromatization to E2 or 5α-dihydrotestosterone, is associated with territorial size, sexual and aggressive behavior, skin color, endurance, and external ornamentation (Johnson and Wade, 2011). Seasonal variation in testicular hypertrophy is also influenced by paracrine factors secreted from Leydig cells, Sertoli cells, and spermatogonia (Kumar et al., 2011).

Many snakes have androgen-dependent renal sex segments that secrete products necessary for viable semen (**Figures 8.18–8.19**) (Gist, 2011). Renal epithelial secretory granule size peaks at the time of mating and regresses in concurrence with post spermatogenesis regression. Grossly, at peak hypertrophy sexual segments may be mistaken for renal disease, especially renal gout, by inexperienced prosectors (Volume 1, **Figures 1.220, 4.7, 4.31**). Hemipenes may be hypertrophied in response to testosterone (**Figure 8.20**) (Gist, 2011). In some species secretions of the renal sex segment form a copulatory plug that exerts pheromones involved in discouragement of further breeding (**Figure 8.21**) (Ross and Crews, 1977; Shine et al., 2000; Gist, 2011).

8.7 Ovarian Physiology

8.7.1 Oogenesis

Oogenesis depends on a pool of viable oogonia. Ovum formation is initiated by migration of primary germ cells (PGCs) from the extraembryonic area into the developing ovary during the middle third of fetal development. PGC migration occurs by interstitial ameboid movements and vascular migration followed by mitosis to form diploid primary and secondary oogonia (Ramírez-Pinilla et al., 2015). Oogonia mature and congregate within single or multiple germinal beds (or nests) located in the mesovarium near the ovarian hilum (dorsal aspect of the ovary) (**Figures 8.22–8.23**) (Klosterman, 1983; Guraya, 1989; Ramírez-Pinilla et al., 2015). In reptiles, mitosis may continue from neonates into adulthood, but in birds this process is completed prior to hatching (Guraya, 1989). Oogonia undergo meiosis to become haploid primary oocytes.

8.7.2 Folliculogenesis

Folliculogenesis from primary oocytes to maturation of ovarian follicles can be divided into the previtellogenic and vitellogenic stages, the latter characterized by yolk deposition in the follicles. The progression of folliculogenesis in lizards in the genus *Sceloporus* has been categorized into nine stages (Lozano et al., 2014). These stages include formation of primary oocytes (stage I), maturation of primary follicles (stages II–III), stable previtellogenic follicles (stage IV), and maturation of vitellogenic follicles (stages V–IX) (Lozano et al., 2014). These stages are defined by oocyte morphologic changes in the nucleus, ooplasm, and stored yolk. Granulosa cell morphology varies greatly between taxa, and morphologic changes are variable between orders.

8.7.2.1 Previtellogenesis Initially, primordial follicles are formed from primary oocytes that have left the germinal beds. During maturation, primordial oocytes may increase in size by approximately five times and the nucleus becomes eccentric, developing chromosomes with "lamp-brush" morphology and intranuclear Balbiani bodies. After primordial follicles are recruited from germinal beds and populate the cortex, they are surrounded by capillaries that deliver nutrients for follicular growth (Ramírez-Pinilla et al., 2015). A single layer of cuboidal granulosa cells will initially surround primordial follicles. The zona pellucida becomes clearly visible during stage IV in previtellogenic follicles. In stage IV the zona pellucida becomes surrounded by multiple layers of granulosa cells including small, intermediate, and outer pyriform cells (**Figures 8.24–8.25**) (Laughran et al., 1981). Testudines (Callebaut et al., 1997) and Crocodilia (Uribe and Guillette, 2000) have a single layer of granulosa cells at this stage. Toxins can cause genetic miscoding that leads to defective oogenesis in vertebrates (Lai et al., 2017). Cadmium exposure has been associated with defective progression of primary follicles (Simoniello et al., 2013). Congenital or environmental induced mutations in genes encoding formation of the zona pellucida also result in oocyte morphologic abnormalities (Yang et al., 2017). Static previtellogenic follicles may be milky white (**Figure 8.26**).

8.7.2.2 Vitellogenesis During vitellogenesis, milky white previtellogenic follicles increase in volume by 10- to 100-fold to become yellow (yolked) vitellogenic follicles (**Figures 8.27–8.28**) (Tinkle, 1967; DeNardo, 2006a). At the initial stages of vitellogenesis (stage V), the zona pellucida develops dual banding, producing an inner hyaline layer and outer zona radiata (**Figure 8.29**). The zona radiata is associated with increased uptake of vitellogenin and rapid expansion of the oocyte cytoplasm. Microvilli from the zona radiata penetrate the granulosa cell layer and aid the transport of nutrients into the ooplasm (Guraya, 1989). During vitellogenesis, estrogen drives the conversion of stored adipose and consumed nutrients into vitellogenin within hepatocytes (Taylor and DeNardo, 2011). The liver will be enlarged and yellow during this time and microscopically hepatocytes will contain vacuoles (**Figure 8.30**). Vitellogenin is transported from the liver in the blood to the developing ovarian follicles. During progression of vitellogenesis (stage V and VIII follicles) yolk platelets form at the periphery of the ooplasm, increase in size, and fuse to form yolk. Expression of FHS receptor is seasonally variable in Chinese Alligators (*Alligator sinensis*) with peak activity during folliculogenesis and low activity during hibernation (Zhang et al., 2015). Serum T, triglyceride, cholesterol, and calcium may also be elevated during vitellogenesis (Sykes, 2010). Yolking follicles contain about double the E2 found in plasma, and also contain higher concentrations of T than observed in the blood. In the Green Anole, levels of T increase and estrogen remains steady in the follicles as they mature (Lovern and Wade, 2003b).

As the ooplasm accumulates homogenous yolk and lipid vacuoles, the granulosa cells flatten, along with increased vascularity in the surrounding theca. Developing follicles are suspended on fibrovascular stalks (pedicles) (Barnes et al., 2016). At maturation and prior to ovulation, all mature yolk follicles should be of similar size (**Figure 8.31**) and hypoechoic or dark (**Figure 8.32**) with ultrasound, but once ova enter the oviduct and begin to develop a shell the ultrasound image of the ova will become hyperechoic or whitish. Environmental factors that affect vitellogenesis can vary. For example, when female Ball Pythons undergoing vitellogenesis are placed with non-vitellogenic females, the active female may resorb her follicles or the inactive female may be stimulated to initiate vitellogenesis (Barker and Barker, 2006); the precise stimuli of these reproductive events are unknown.

8.7.3 Ovulation

Follicle recruitment for ovulation is regulated by GnRH secreted from the hypothalamus. Just prior to ovulation, the volume of ooplasm in oocytes peaks (stage IX). Stage IX follicles have an "animal pole" attracting migration of the oocyte nucleus and an opposing "vegetative pole." Release of the oocyte during ovulation occurs through a disc-like zone in the follicular wall called the stigma. In the Green Anole the stigma has decreased theca externa vascularity and the diameter is about half that of the entire follicle (Jones et al., 1988). Rupture of the granulosa cell layer and theca occurs simultaneously due to hormone-induced hypoxia and ischemic necrosis in the stigma (Jones et al., 1988). Inflammation at the stigma post ovulation is not observed in Green Anoles (Jones et al., 1988). At ovulation the anterior end of the oviduct, or infundibulum, captures the released oocyte in a funnel-shaped structure called the ostium. Mature oocytes captured by the infundibulum are transported into the proximal oviduct. It is not possible for the right ostium to capture an oocyte from the left ovary, as oviducts are separated by a medial septum (Barker and Barker, 2006). Although Crocodilians have two ovaries, one may be dominant and produce a larger number of ovulated follicles (Johnson, 2014). In crocodilians, E2 levels increase gradually and peak in late vitellogenesis and subsequently fall around the time of ovulation. At the onset of ovulation, there is a sharp decrease in plasma E2 and a rise in P (Milnes, 2011). In tortoises, induced ovulation from either pheromone influence or male courtship behavior is suspected (Kuchling and Razandrimamilafiniarivo, 1999). In alligators, serum calcium also increases prior to ovulation (Guillette et al., 1997). Some changes in female morphology and physiology during ovulation are notable. For example, egg-laying snakes may have visible swelling in the midsection prior to ovulation, which could be mistaken for pathogenic gastric enlargement of undigested prey (Barker and Barker, 2006). At ovulation, ova enter the ostium of the oviduct and fertilization occurs before the oviductal secretions are added to the ovum.

8.7.4 Corpus Luteum Development

The corpus luteum (CL) is formed from follicular lutinization of postovulatory follicles. This transitory endocrine organ produces P prior to ovulation and during pregnancy. Progesterone usually peaks during mid gravidity. For some species, P regulates timing of ovipostion. The structure of the CL, level of persistence, and endocrine function are variable between reptile species (Xavier, 1987; Guillette and Cree, 1997; Martínez-Torres et al., 2003). After ovulation, the central cavity is filled with a luteal cell mass and the ovulation fossa at the stigma may be filled in or persist. The site of theca rupture is also termed the *ovulation aperture* (**Figures 8.33–8.34**) (Guillette and Cree, 1997). At least two classification schemes have been proposed for CLs in squamates. Martínez-Torres et al. (2003) describes four stages of development and three stages of regression for the Mexican Imbricate Alligator Lizard (*Barisia imbricata imbricata*). In *Barisia* species, the CL begins the degenerative stage in the second third of pregnancy. Lozano et al. (2014) describes four stages (I to IV) of development and atrophy for the viviparous Mexican Graphic Spiny Lizard (*Sceloporus grammicus*). During stage I, the theca of the ruptured follicle forms an irregular central cavity that is soon filled with proliferating monomorphic cells containing a central nucleus (**Figure 8.35**). The collapsed follicle is invaded by theca that slowly transforms into mesenchymal septa. As the CL matures (stage II), luteinized granulosa cells are exposed to increased vascularization and fibrosis (Lozano et al., 2014). Older CL accumulate brown pigment that is likely lipofuscin, and may also have dystrophic mineralization (**Figure 8.36**). At luteolysis (stage III), nuclei of luteal cells become pyknotic. During stage IV, prior to complete regression, luteal cells become homogenous and merge with the theca. In species that produce multiple clutches in a season, the pattern of vitellogenesis will be reinitiated when the CL regresses at oviposition (Tinkle, 1967). In viviparous reptiles parturition is less dependent on luteal regression (Xavier, 1987). Surgical removal of CL in gravid Mexican Imbricate Alligator Lizard indicates the conceptus participates in the maintence of late-term P (**Figure 8.37**) (Martínez-Torres et al., 2014). Progesterone secretion by luteal structures is required to maintain oviduct egg development and placenta-derived fetal development (Graham et al., 2011).

8.7.5 Follicular Atresia

During atresia, yolk is either phagocytized by follicular mesothelial cells and replaced by invading cells (obliterative atresia) or released into the coelomic cavity (bursting atresia) (Cornax et al., 2013; Barnes et al., 2016). Obliterative follicular atresia, which has been described in the Chinese Softshell Turtle (*Pelodiscus simensis*), occurs throughout follicular cycling and can be broken into four stages: (1) disorder of follicular cell structure; loss of a distinct zona pellucida distinguishes viable from degenerated follicles (**Figure 8.38**); (2) elimination of yolk granules secondary to cell proliferation (**Figure 8.39**); (3) influx of erythrocytes into the ooplasm (**Figure 8.40**); (4) gradual degeneration of atretic follicles (**Figures 8.41–8.42**) (Wang et al., 2013; Lozano et al., 2014).

8.8 Copulation and Fertilization

Delivery of sperm to the external terminal portion of the female reproductive tract is a characteristic of amniotes. Disruption of this process should be ruled out when investigating cases of infertility. Ineffective copulation may result from loss or paralysis of the tail in snakes (Barker and Barker, 2006), or lack of sperm transfer as shown in some lizards (Warner et al., 2013). Copulation may precede ovulation by days to months, as viable sperm are stored within oviduct pouches in many reptile species (Fox, 1963; Shankar et al., 2015). In squamates, sperm are stored in tubuloalveolar glands at the junction between the infundibulum and uterine tube (Fox, 1963), and storage is regulated by estrogen (Gist, 2011). Sperm stored from different males may result in clutches with multiple paternity (Pearse et al., 2002; Perry and Mitchell, 2017). Circulating estrogen associated with late vitellogenic follicles can affect female receptivity to mating by changing pheromone secretion by the skin (Rhen et al., 1999; Vitt, 2015).

Intromittent organ structure varies among reptiles. The chelonian phallus (penis) originates from the ventral aspect of the cloaca. The outer surface of the phallus contains a seminal groove between two corpora cavernosa (spongiosa) that directs semen flow. During erection, the corpora cavernosa are engorged from blood originating from the internal iliac artery (Innis and Boyer, 2002). In Squamata, paired open sac-like hemipenes enter the caudal wall of the cloaca and evert through the vent.

Fertilization usually occurs shortly following ovulation of the vitellogenic follicle and entry into the oviduct (Jones, 2011). In squamates sperm must be present in the oviduct near the infundibulum prior to ovulation for fertilization to occur (Barker and Barker, 2006). As with birds, each ovum may be penetrated by multiple sperm (polyspermy) but only one sperm nucleus fuses with one haploid ovum nucleus. Supernumerary sperm migrate to the periphery of the ooplasm and undergo degeneration (Mizushima, 2017). Aberrant physiology associated with degeneration of supernumerary sperm or creation of polyspermic zygotes may result in polyploidy, abnormal development, and early embryonic death (Stoker et al., 2003). Once the ovum and sperm merge, a diploid zygote is formed.

8.9 Embryo Development and Birth

8.9.1 Extraembryonic Membrane Development

Extraembryonic membranes (homologous with placental structures) exist in both oviparous and viviparous species, and are composed of the yolk sac, chorion, allantois, amnion, and umbilical cord. The umbilical cord contains the vitelline (omphalomesenteric) duct that transfers yolk into the intestine of the embryo. Occlusion of this duct may result in fetal death. Fetal membranes make up a larger percentage of the embryo in oviparous species than viviparous species (Bonnet et al., 2017). Yolk deposition into eggs is greatly influenced by maternal nutrient and energy supply (Warner and Lovern, 2014). The primary source of energy, vitamins, maternal hormones, fatty acids, proteins, and minerals for the developing embryo is the yolk. Gas and water exchange to the embryo must also occur through the yolk sac membrane (Albergotti and Guillette, 2011). Water is stored within the amnion, allantois, and chorion and this reserve may serve as a buffer for desiccation in eggs (Bonnet et al., 2017). Maternal antibodies are transferred through the yolk sac (Keller, 2017). The yolk sac can also be the route of entry for exogenous toxins (Perrault et al., 2016).

The poorly vascularized chorion and the highly vascularized allantois fuse to form the chorioallantoic membrane (CAM). In shelled eggs, the chorion lines the internal porous surface of the shell and facilitates gas exchange with the environment. In viviparous species the CAM and yolk sac have a direct epitheliochorial attachment to the endometrium of the oviduct and form a chorioallantoic placenta. In reptiles, epitheliochorial placentation does not erode or invade the maternal endometrium (Albergotti and Guillette, 2011). Nitrogenous waste from the embryo is excreted as urea or relatively insoluble uric acid. These end products are secreted into the allantoic sac during late fetal development (Sartori et al., 2012). Pregnancy may influence cellular immunity, as a decrease in plasma bactericidal capacity was observed in gravid vs non-gravid snakes (Graham et al., 2011).

8.9.2 Oviparous Development in the Oviduct

After fertilization, progressive development of self-contained calcified eggs occurs within the oviduct of oviparous reptiles. The oviduct of oviparous squamates is composed of three different regions: the infundibulum, which opens into the coelomic cavity and receives the oocyte at ovulation; the uterine tube, which is responsible for the eggshell production; and the vagina, which empties into the cloaca (Nogueira et al., 2011; Siegel et al., 2015). Grossly, the muscular wall of the gravid oviduct is thin compared to the intestine (**Figure 8.43**). Oviduct disease may disrupt mechanisms required for successful reproduction, including sperm transport and storage, fertilization, transport of ovulated follicles, secretion of albumen, production of the eggshell, environmental maintenance of eggs, and oviposition (Stewart and Blackburn, 2015).

The fertilized ovum becomes an egg with the addition of albumen and shell that are derived from the oviduct epithelium (Albergotti and Guillette, 2011). As the fertilized zygote passes through the uterus, the albumin-producing glandular segments secrete albumin around the zygote. The albumen provides structural internal support, contributes antimicrobial peptides and maintains water storage for the embryo. Unlike chelonians and crocodilians, squamates do not secrete a prominent layer of albumen around the yolk. Many reptiles are megalecithal, in that the yolk occupies the largest percentage of the egg mass (**Figure 8.44**) (Gist, 2011); thus, much of the water necessary for proper development is not present in the egg at oviposition. Consequently, nest or incubation environments must provide enough moisture for the embryos via absorption through the eggshell (Packard and Packard, 1988).

When the zygote enters the uterine tube (or isthmus section of the uterus), the uterine shell gland secretes proteinaceous fibers along columns of calcium carbonate to form a shell (Cree, 2015). Unlike birds, squamates do not have segmental shell glands, and calcium is deposited into eggs throughout the length of the oviductal uterus (Barker and Barker, 2006). Depending on calcium density, the general structure of eggshells varies between species of reptiles from flexible to rigid (Packard and Packard, 1988). Calcium density observed by radiology can be used to evaluate the progression of gravidity and health of eggs before and after oviposition (DeNardo, 2006a). The length of time the fully calcified egg is retained in the oviduct varies among species and environmental conditions (Buhlmann et al., 1995; Blanvillain et al., 2011; Jones, 2011). Egg retention is correlated with luteinization and retention of the CL (Albergotti and Guillette, 2011). Some species will hold their eggs for as little as 9 days, while others may retain eggs for as long as 4 to 6 months (**Figure 8.45**) (McArthur, 2004; Kummrow et al., 2010; Vitt, 2015).

Stage of development at time of oviposition varies from very early embryonic development (e.g., primitive streak or gastrula stage as in Tuatara and Testudines) to about 30% or more (e.g., at limb bud stage as in most squamates). Some lizard and snake species lay eggs at very advanced embryo stages (Stewart and Blackburn, 2015). Diapause refers to eggs laid at early embryonic stage that may undergo cessation of development after parturition (e.g., chameleons), and this cessation of growth may be influenced by temperature and oxygen availability (Andrews and Donoghue, 2004). Ambient temperature may also affect the length of time between ovulation and oviposition in snakes, cooler temperatures prolonging this period (Barker and Barker, 2006). Oviduct secretions of chelonians can limit oxygen diffusion and arrest development of embryos (Rafferty et al., 2013), which may provide a reason why oviposition occurs at an early embryonic stage in turtles.

The vagina enters the cloaca at the genital papillae. At the time of oviposition, an increase in elasticity of the closed vagina allows egg passage (Barker and Barker, 2006). Oviposition is initiated by P, the neurohypophseal hormone arginine vasotocin (AVT), and PGF2α (Gist, 2011). Circulating T may also increase at oviposition (Kummrow et al., 2010). Near the time of oviposition, prostaglandins induce luteolysis of the CL (Guillette et al., 1997) causing the P level return to baseline (Kummrow et al., 2010), which reduces contractility of the oviduct (Gist, 2011).

8.9.3 Egg Physiology

Embryos of most reptiles enter a period of incubation within an external amniotic egg. The components of the external egg are the shell, albumen, yolk and yolk sac, extraembryonic membranes, and the embryo (Keller, 2017). The shell provides protection from the environment, is a source of calcium for developing embryos, and allows for gas and water exchange (Keller, 2017). Flexible eggs are more frequently observed in species from humid environments, and more rigidly shelled eggs are typical of reptiles from dryer environments (Keller, 2017), although there are many exceptions to this general pattern. Species with softer shells contain less calcium in the shell and generally have less albumen. These species are highly dependent on diffusion of external moisture during incubation. Moisture uptake by eggs from the nest environment is less important for species with rigid, highly calcified eggshells (e.g., crocodilians, many geckos). In species with soft eggshells, added water absorption can increase egg weight gain over 60% during incubation (Packard and Packard, 1988). Acidity of the nesting environment also likely affects calcification of the shell, with acidic environments favoring thick shells (Deeming, 2018). The texture and porosity of the shell is important for oxygen exchange and may be influenced by toxins such as organochlorine compounds (Stoker et al., 2013). Some species (e.g., *Drymacharon* spp.) normally have small struvite crystals on the surface of eggs (**Figure 8.46**; Volume 1, **Figures 1.305–1.306**).

The yolk is the primary source of nutrients for the embryo. The presence of a blastodisc, a whitish spot on the surface of the yolk, can be assessed for fertility in birds (Barnes et al., 2016), but this is not well documented in reptiles. Eggs of many reptiles develop an opaque white spot within hours to days following oviposition, which indicates fusion of extraembryonic membranes to the shell membrane (Williamson et al., 2017a,b). This white spot expands as embryo metabolic demands increase during incubation and eventually encompasses the entire egg. This membrane contact functions as the respiratory surface for developing embryos, and if the attachments rupture (due to vigorous movement) it can result in embryo mortality (**Figures 8.47–8.48**). Chalazae attached to the yolk function to keep the yolk and embryo properly aligned in the crocodilian egg but are not visible in squamate eggs. Albumen in squamates tends to be clear (Keller, 2107). Reduced albumen may indicate dehydration and result in decreased shock absorbing properties for the embryo.

Biotic (e.g., clutch size, microbial respiration, embryo characteristics) and abiotic factors (e.g., moisture, oxygen, and temperature) have strong effects on patterns of incubation (Cordero et al., 2017). Oxygen requirements are particularly important in the last third of embryo development and are dependent on amniotic vascular development (**Figure 8.49**). Turtles are resistant to periods of intermittent hypoxia but less tolerant as the embryo matures (Cordero et al., 2017). In the natural environment, periods of high rainfall and flooding have the most impact on oxygen tension in nests (Cordero et al., 2017). Variation in eggshell thickness often correlates with the microhabitat that females use for nesting. For example, species that lay eggs in loose leaf litter or in places where eggs are directly exposed to air (i.e., many geckos) have thicker shells than species that bury their eggs under substrate. Intriguingly, experimental exposure to light during incubation negatively influences survival of skink eggs, whereas gecko eggs are largely unaffected (Zhang et al., 2016); thus, exposure to light (or lack thereof) should be considered in incubator design.

The process of hatching can be very slow to extremely rapid. In some cases, hatching occurs in response to some environmental cue, such as rainfall (Pig-Nose Turtles [*Carettochelys insculpta*]) or movement from neighboring eggs. In some species, cues from neighboring eggs facilitate synchronous hatching among clutch mates (Loudon et al., 2013). Some species have a prominent egg tooth at hatching (**Figure 8.50**) that aids in tearing the eggshell, whereas others use their claws to emerge from the egg (Warner et al., 2006).

8.9.4 Viviparous Development

Viviparity is defined by retention of the embryo in utero until fully developed and subsequent birth of a full-term neonate at parturition. Viviparity has evolved >100 independent times in squamates and occurs in both temperate and tropical species (Ma et al., 2018). The evolution of viviparity is generally associated with cold climates (e.g., high latitudes or elevations), and has advantages over oviparity such as eliminating the vulnerable egg-stage of reproduction from predation and desiccation and providing females with an enhanced ability to facilitate embryo growth and development through thermoregulatory behaviors (Albergotti and Guillette, 2011; Vitt, 2015). These effects are particularly important at high latitude or elevation environments, as they ensure that parturition occurs before winter. Viviparous reptiles may undergo egg maturation within the oviduct (lecithotrophy dependent) or depend entirely on placental exchange of nutrients (matrotrophy dependent) (DeNardo, 2006a). The oviductal uterus of viviparous reptiles has a reduced shell gland, increased vascularity, and decreased epithelial thickness (Stewart and Blackburn, 2015). In general, viviparous species have a longer period of energy/nutrient intake near the time of parturition than do oviparous species (DeNardo, 2006a).

Matrotrophic species have respiratory gas and nutrient (calcium, sodium, and amino acid) exchange through a placenta rather than through the yolk and eggshell (**Figure 8.51**). Four types of chorioallantoic placentas (CAP [Type I CAP–IV CAP]) have been identified in viviparous reptiles (Blackburn,

1993). Type I CAP occurs in species with reduction of the uterine shell gland. Type I CAP has indirect apposition of the outer chorion to the uterine epithelium and functions primarily only for gas exchange. Gas exchange occurs when fetal blood oxygen affinity is higher than that of maternal blood. Oxygen affinity gradient is influenced by variation in hemoglobin or altered maternal nucleoside triphosphate level (Ragsdale and Ingermann, 1993; Ragsdale et al., 1993). Viviparous reptiles with Type I CAP receive nutrients from the yolk and are thus lecithotrophic (formally termed ovoviviparity). Most viviparous squamates have Type II through IV chorioallantoic placentas receiving nutrients from the uterus and are matrotrophic (Stewart and Blackburn, 2015). Type II and III CAPs achieves direct apposition with the mucosal epithelium of the oviduct through formation of a placentome (Albergotti and Guillette, 2011). At the placentome, chorionic trophoblasts have absorptive properties and maternal uterine epithelium has secretory properties (Blackburn et al., 2017). Type IV CAP are described in the genera *Mabuya*, *Chalcides*, and *Pseudemoia* having placentomes with interdigitating, hypertrophied uterine and chorioallantoic interfaces that more closely resemble the placenta of eutherian mammals (Blackburn, 1993; Jerez and Ramírez-Pinilla, 2003). For additional information on the development of the chorioallantoic placenta, see Volume 1, Chapter 1.

Gestation periods of most viviparous lizards range from 2 to 4 months, but those from colder climates can extend up to 14 months (Stewart and Blackburn, 2015). Pregnancy may be compromised by decreased locomotion, constraints on feeding, greater metabolic burden, and abdominal space availability (Stewart and Blackburn, 2015). The role of P secreted by functional CL for maintenance of pregnancy is debated in reptiles (Stewart and Blackburn, 2015). Experiments have demonstrated fetal loss after surgical removal of corpora luteum (Albergotti and Guillette, 2011). Chorionic membranes in both oviparous and viviparous reptiles produce P and have steroid signaling capacity (Cruze et al., 2013). Uterine P receptors were shown to be essential for maintenance of early pregnancy in the viviparous Southern Grass Skink (*Pseudemoia entrecasteauxii*) but were not needed to maintain pregnancy in mid and late gestation (Biazik et al., 2012).

8.9.5 Embryo Development

Reptilian embryos are not classified as "embryos" or "fetuses" based on age of gestation; rather, they are termed "embryos" throughout all stages of pre-hatching or pre-birth development in biological focused literature. In clinical and husbandry settings, however, veterinarians, pathologists, and caretakers will use the term "fetus" for late term embryos. The basis for the embryo grading utilized by embryologists and developmental biologists is the 1–40 stage table for development established for the European Common Lizard (Zootoca [*Lacerta*] *vivipara*) (Dufaure and Hubert, 1961). This grading system has been modified slightly depending on species (Common Leopard Gecko [*Eublepharis macularius*], Wise et al., 2009; Brown Anole [*Anolis sagrei*], Sanger et al., 2008). Stages 1–33 are characterized by neurulation, organogenesis, and early growth (**Figure 8.52**). In some species, embryos undergo embryonic diapause after oviposition, prior to somite and vitelline membrane formation, which greatly influences the length of egg incubation (Andrews and Donoghue, 2004; Rafferty and Reina, 2012). Arrested egg development may occur at any stage from oviposition to hatching as a result of cold torpor, reduced temporary suspension due to decreased temperature (Rafferty and Reina, 2012), or low oxygen levels (e.g., in turtles that nest underwater; Kennet et al., 1998). The majority of embryo growth occurs during late development during stages 34–40 (**Figure 8.53**) (Andrews and Donoghue, 2004). Eggs of some species of tortoises need to be cooled for a period of time before incubation at the ideal temperatures for males and females.

8.9.6 Neonatal Reptiles

Neonatal reptiles have an umbilical cord (**Figure 8.54**). Inflammation and epidermal crusting at the junction of the umbilical cord with the body wall in neonates is typical and a normal component of involution (**Figure 8.55**). The umbilicus attaches to the mid-segment of the small intestines (**Figure 8.56**). At the time of hatching, the yolk sac inverts through the umbilicus to occupy the coelomic cavity. Most reptiles retain a substantial yolk sac in the coelomic cavity after birth (**Figure 8.57**). The amnion contains smooth muscle that retracts, promoting yolk internalization through the umbilicus prior to hatching. This produces an umbilical scar that heals rapidly in healthy reptiles (**Figure 8.58**). The amount of residual yolk internalized in neonatal reptiles affects their ability to conduct rigorous activity during the first few weeks of life (Pezaro et al., 2013); however, the importance of residual yolk for neonatal growth varies among species (**Figure 8.59**) (Troyer, 1987; Radder et al., 2007). Bacterial infection of the yolk may occur and result in an omphalitis (Volume 1, **Figures 11.161–11.162**).

8.10 Nutrient Metabolism Effect on Reproduction

Nutrient uptake, lipid metabolism, and body stores of energy are closely linked to reproductive physiology (Price, 2016). For example, offspring from female Leatherback Sea Turtles (*Dermochelys coriacea*) with low plasma selenium levels had lower hatchability (Perrault et al., 2011). Especially for females, meeting the energy demand for the reproductive processes is often the tipping point between health and disease. In captivity, sexual maturity may be achieved considerably faster and clutch sizes may be larger than in natural populations due to more frequent energy intake and less energy expenditure associated with foraging (DeNardo, 2006a; Kummrow et al., 2010).

Maternal placental nutrient supplementation in viviparous species varies with forage availability and conditioning. Likewise, available adipose storage in oviparous species highly influences vitellogenesis and yolk deposition in ovum (Derickson, 1976; Etheridge et al., 1986; Blackburn, 1992; Price, 2016). The majority of adipose tissue in females is composed of energy-dense triglycerides (Tinkle, 1967). Triglycerides are about 8–10 times more efficient at storing energy than carbohydrates (McWilliams et al., 2004; Cornax et al., 2013). Subcutaneous adipose stores are uncommon in most reptile species, with exceptions such as the Mediterranean House Gecko (*Hemidactylus turcicus*) (Greene, 1969). In Squamata, the most common anatomical locations of adipose storage are intracoelomic adipose foci or fat bodies (also corpora adiposa) and to a lesser extent, the tail (Tinkle, 1967; Greene, 1969). These two locations may hold up to 70% of the energy reserve in many species (Santos et al., 2007). Although lipid is also stored in the liver, it may only account for about 5% of the energy stores (Santos and Llorente, 2004). Adipose stores are also found in the bone marrow, and although a small component of total adipose stores, are the last to atrophy in cases of advanced inanition. The most common source of energy for lecithotrophic and matrotrophic growth of embryos originates from mobilization of stored lipids (Blackburn, 1992; Price, 2016). In female snakes, 40% of body mass may be associated with gravidity (DeNardo, 2006a), and energy allocated for egg production ranges from 10% to 44% in lizards (Lovern, 2011). Energy cost for oviparous species is greater during initial vitellogenesis than during the gravid period of egg formation. Energy expenditure for viviparous species is greater later in gestation in response to fetal development (Bleu et al., 2013b).

The source of energy that fuels reproduction varies considerably among reptiles. "Capital breeders" obtain the energy for reproduction from stored resources, primarily coelomic adipose tissue, whereas "income breeders" use recently acquired nutrition during gestation (Bleu et al., 2013a). Some species (e.g., Jacky Dragon [*Amphibolurus muricatus*]) use a combination of capital and income to fuel reproduction (Warner et al., 2008), and the Indian Garden Lizard shifts from capital use of lipid in the spring to dietary income as the summer progresses (Shanbhag, 2003). Reproductive output is strongly influenced by immediate food intake in Brown Anoles (Hall et al., 2018). Viviparous reptiles are usually capital breeders and have a longer gestation than oviparous species. Capital breeders usually do not consume food during pregnancy (DeNardo, 2006a). Some species, such as the Viperine Water Snake (*Natrix maura*), are capital breeders for lipids but income breeders for needed proteins (Santos et al., 2007). In Chelonia, Aldabra Giant Tortoises (*Aldabrachelys* [*Geochelone*] *gigantean*) are income breeders, while the Green Turtle (*Chelonia mydas*) is a capital breeder (Kuchling and Razandrimamilafiniarivo, 1999). Testicular weight fluctuates dramatically in turtles that are income breeders and may be linked to foraging (Blanvillain et al., 2011). Gestation costs in viviparous squamates may also be managed by shifts in thermal preference and decreased movement (Bleu et al., 2013b). Reptiles with depleted caloric reserves often fall victim to morbidity and mortality from reproductive failure. Energy expenditure for reproduction is an important factor for both sexes, but of greater importance for females (**Figure 8.60**). Despite the fact that male Green Anoles are 40% larger than females, the females expended 46% more energy toward reproduction (**Figure 8.61**) (Orrell et al., 2004). Studies in wild European Common Lizards indicate that accumulation of lipid energy stores may be sufficient to support vitellogenesis if food is scarce during follicle yolking (Bleu et al., 2013b). In the Ornate Tree Lizard (*Urosaurus ornatus*), leptin is an endocrine signal that mediates the tradeoff between reproductive activity and immune function (French et al., 2011). Induction of stress also alters energy demands in reptiles and is discussed later in this chapter (Josserand et al., 2017).

Mobilized adipose tissue and gut-absorbed nutrients are synthesized in the liver before they become available for use in follicular growth (Price, 2016). Transport of lipids in reptiles largely occurs via very-low-density lipoproteins (VLDLs) (Janeiro-Cinquini et al., 1999). Accumulation of hepatic-synthesized yolk triglycerides is a limiting factor in reproduction, and dysfunction of this mechanism is a common pathway for reproductive failure and disease. Yolk triglycerides reach the liver and are converted via four pathways: (1) direct import from dietary nutritional intake without affecting adipose stores (i.e., income breeders), (2) import from dietary nutrients while stored triglycerides are simultaneously used for maintenance metabolism, (3) conversion of stored liver triglycerides to VLDLs, or (5) mobilization of adipose stores (Price, 2016).

8.11 Stress Effect on Reproduction

Disruption of hormonal and metabolic homeostasis by noxious stimuli generates stress. For the purposes of this chapter, stress is defined as an overstimulation of the adrenal gland resulting in fluctuations of baseline plasma glucocorticoids (primarily corticosterone [CORT]). Reproductive biologists more broadly define stress as exposure to external conditions that disrupt homeostasis and provoke a physiologic and behavioral response (Pottinger, 1997). Steroids influence gene expression patterns, directly affecting phenotype (sex-biased gene expression). Body condition and nutrient intake directly affect steroid allegation in the developing embryo (Cox et al., 2016). Release of CORT from the adrenal cortex is primarily controlled within the feedback mechanism of the hypothalamic-pituitary-adrenal (HPA) axis. Hormonal activity of the HPA axis influences regulation of reproductive activity when it intersects with the hypothalamic-pituitary-gonadal (HPG) axis.

Some level of normal CORT secretion is required for reproductive activity, but large fluctuations in plasma glucocorticoids can adversely affect gonad physiology, reproductive behavior, pregnancy, and viability of offspring (Moore and Jessop, 2003; Tokarz and Summers, 2011). Baseline corticosterone increases in Asp Vipers (*Vipera aspis*), a capital breeder, throughout pregnancy and influences available energy and structural protein mobilization for embryos (Lorioux et al., 2016).

Stress is induced by factors including nutritional challenges, infectious agents, population dynamics, predator density, inadequate husbandry, and other anthropomorphic factors. Stress can be categorized as acute or chronic (Tokarz and Summers, 2011). An example of acute stress with an immediate behavior response would be threat display in a chameleon, which is often accompanied by color change (Tokarz and Summers, 2011). The concept of "allostasis" entails reactive fluctuation in hormonal and metabolic activity to maintain a physiologically stable interaction between the HPA and HPG axes (McEwen and Wingfield, 2010). When allostasis can no longer be maintained because of severe acute or chronic exposure to stressors, the physiologic balance, or "allostatic load," is tipped and pathophysiologic processes ensue (Tokarz and Summers, 2011; Josserand et al., 2017). Reproductive success is often dependent on overriding the stress response to provide a "coping mechanism." Coping mechanisms that counteract effects of increased plasma CORT may include mechanisms to block stimulus input in the central nervous system, blockade of a pathway in the HPA or HPG axes, and changes in gonad hormone production (Wingfield and Sapolsky, 2003). Chronic stress may alter external sexual characteristics, behavior, and breeding activity. This is observed in stress-induced changes with dewlap coloration in the Ornate Tree Lizard, and this affects mate selection (Moore et al., 1998). Numerous studies show that chronically elevated plasma CORT is associated with decreased T (Tokarz and Summers, 2011). Increased CORT inhibits oogonial mitosis and spermatogenesis in squamates (Nijagal and Yajurvedi, 1999; Yajurvedi and Nijagal, 2000). In vertebrates, CORT disrupts hypothalamic GnRH release and thus LH secretion from the pituitary, and this impact on gonadal hormone activity is most important in females; however, CORT also reduces gonadal responsiveness to LH and concentrations of LH receptor, which is more detrimental in males (Sapolsky et al., 2000). Resorption of follicles prior to the copulatory period in Galápagos Marine Iguanas (*Amblyrhynchus cristatus*) is associated with increases in T and P as well as CORT. It is thought that increases in CORT initiate follicular atrophy for females that have insufficient body reserves (Vitousek et al., 2010). When Josserand et al. (2017) manipulated habitat to increase allostatic load in the viviparous European Common Lizard, fecundity predictably decreased (Josserand et al., 2017). Eggs produced by CORT-treated birds tend to be lighter and contain lower concentrations of T and P in the yolk (Henriksen et al., 2011).

8.12 Noninfectious Causes of Reproductive Disease

8.12.1 Aging (Degenerative Diseases)

Declines in physiological and reproductive function with age are not well studied in reptiles. However, some important findings from captive reptiles and long-term field studies provide insights into this aspect of reptile biology. Several extrinsic (e.g., nutrition, stress) and intrinsic factors (e.g., growth rate, reproductive rate) may influence the age at sexual maturation and start of gonad regression (DeNardo, 2006a). These factors can also influence lifespan and rates of aging (i.e., senescence), as predicted by life-history theory (Charlesworth, 2000; Roach and Carey, 2014). Although reproductive senescence and age-related declines in survival rates are traditionally viewed as negligible in reptiles (Olsson and Shine, 1996; Congdon et al., 2001, 2003; Schwartz and Bronikowski, 2011), some evidence from a long-term field study of Painted Turtles suggests that survival and egg hatching success declines with female age (Warner et al., 2016a). In addition, an experimental field study of Brown Anoles demonstrates that survival rates of ovariectimized females increases substantially compared to control individuals (i.e., those capable of reproduction); this pattern supports the hypothesis that reproduction is costly, thereby decreasing survival and lifespan (possibly due to reduced allocation of energy for maintenance) (Cox and Calsbeek, 2010).

Reptiles are also prone to several age-related diseases, which include cystic degeneration and atrophy of gonads. In addition, melanin often accumulates in tissues of older reptiles (**Figures 8.62–8.63**). Disease of nonspecific primary etiology, such as egg yolk peritonitis, may also increase with age (Innis and Boyer, 2002). Several studies also show that telomere length is associated with survival and lifespan in reptiles. For example, telomere length decreases with age in male Red-Sided Garter Snakes (*Thamnophis sirtalis parietalis*) but remains stable in female snakes (Rollings et al., 2017). Telomere length also differs between the sexes in Sand Lizards (*Lacerta agilis*) and positively correlates with lifespan and reproductive success more so in females than in males (Olsson et al., 2011).

8.12.2 Neoplasia of the Reproductive Tract

Neoplasia originating in reproductive organs is most often reported in squamates and has less documentation in Testudines and Crocodilia (Perry and Mitchell, 2017). Reproductive neoplasia may be of germ cell, mesenchymal (sex cord-stromal), or epithelial origins (Agnew and MacLacklan, 2017). During embryogenesis and sexual differentiation, germ cells originating from the embryonic yolk sac merge with somatic mesenchymal cells originating from the urogenital ridge to form testicular seminiferous tubules and ovarian follicles. Gonadal tumors of germ cell origin include seminoma in males, dysgerminoma in females, and teratomas in both sexes. Teratomas originate from totipotential cells, which lead to multiple distinct cell layers (Tocidlowski et al., 2001). Sex cord stromal tumors originate from gonadal precursor cells, derived from coelomic epithelium (Piprek et al., 2016). Sex cord stromal tumors include granulosa cell tumors in ovaries and Sertoli cell tumors and Leydig (intersitial) cell tumors in testicles. Epithelial origin tumors are not reported from the testes of amniotes but do occur in the ovaries of reptiles. Epithelial origin tumors have been reported in female oviducts but reports of epithelial origin tumors from the male spermatic duct or renal accessory sexual segments are absent. As observed in hens, coelomic disseminated carcinoma in female reptiles is often suspected to be of reproductive origin but not always verifiable (Garner et al., 2004).

Reproductive neoplasia is inherently linked to embryologic and endocrine reproductive physiology. Disruption of mechanisms involved in gender differentiation can lead to development of neoplasia. Exposure to environmental endocrine disrupters such as dichlorodiphnytrichlorooethane (DDT) and bisphenol will initiate neoplasia in vertebrates, but no reports of neoplasia caused by a toxin are verified in wild reptiles. Epigenetic alteration in utero may lead to vertical transmission of tumor susceptibility and transgenerational reproductive abnormalities (Crews and McLachlan, 2006; Schug et al., 2016; Evans, 2018), and altered endocrine function may predispose to endometrial cancer (Mallozzi et al., 2017) in amniotes. Selective estrogen receptor modulators (SERMs) contribute to the pathogenesis of estrogen-responsive neoplasia (Evans, 2018).

Our knowledge of reproductive neoplasia in reptiles reflects a bias toward geriatric species commonly kept in captivity (Bel et al., 2016). However, ovarian teratoma was documented in a garter snake soon after capture from the wild (Hertha et al., 2000). Often, a fluid-filled and distended coelomic cavity is the initial clinical sign of female reproductive neoplasia (Anderson et al., 1996; Tocidlowski et al., 2001; Stacy et al., 2004; Bel et al., 2016). Secondary lesions or signs that may accompany reproductive neoplasia include anorexia or partial inappetence, colonic impaction, egg yolk and/or bacterial coelomitis, follicular stasis, penile, cloacal, and/or oviduct prolapse, bacterial salpingitis, and septicemia (Pees et al., 2014; Bel et al., 2016). For additional information on neoplasia in reptiles, see Chapter 1.

8.12.2.1 Male Reproductive Tract Seminomas have been occasionally reported in testes of squamates, chelonians, and crocodilians (Garner et al., 2004). Seminomas are common in birds, usually benign and localized to one testicle. A lower percentage of avian seminomas are bilateral and malignant (Wernick et al., 2015). A unilateral benign testicular seminoma was diagnosed in a Spur-Thighed Tortoise (*Testudo graeca*) (Pees et al., 2014), an American Alligator, (Huchzermeyer, 2003) and a Chinese Alligator (Conley and Shilton, 2018). Similar to seminomas in mammals and birds, individual cells in seminomas have large nuclei, multiple nucleoli, high mitotic rate, anisocytosis, and anisokaryosis (Garner et al., 2004; Martorell et al., 2009). Sertoli cell tumors have been reported from squamates (Rivera, 2008) and a Chinese Alligator (Conley and Shilton, 2018). Leydig cell (interstitial cell) tumors have been described in squamates and chelonians and are usually composed of nodules of well-differentiated Leydig-like cells displacing seminiferous tubules (Frye et al., 1988; Frye, 1994; Garner et al., 2004). For additional information regarding tumors of the male reproductive tract, see Chapter 1.

8.12.2.2 Female Reproductive Tract Dysgerminomas are uncommonly reported in reptile ovaries; a dysgerminoma was reported in a young adult Common Snapping Turtle (*Chelydra serpentine*) (**Figures 8.64–8.65**) (Machotka et al., 1992). Dysgerminomas are also infrequently reported in birds and usually are benign (Strunk et al., 2011; Shimonohara et al., 2013). Teratomas are commonly reported from ovaries of reptiles. Teratomas may contain ectoderm, mesoderm, and endoderm origin tissues that originate from primordial germ cells in the blastocyst. Well-differentiated tissues occurring in teratomas include stratified squamous epithelium with melanin-laden dendritic cells, cuboidal or columnar mucosal epithelium, teeth, fibrous tissue, adipose tissue, smooth muscle, hyaline cartilage, nervous tissue, and woven bone (**Figure 8.66**) (Newman et al., 2003; Bel et al., 2016). Two ovarian teratomas were diagnosed in parthenogenic Desert Grassland Whiptail Lizards (*Cnemidophorus uniparens*) from one facility, suggesting a predisposition in this species (Tocidlowski et al., 2001). Teratoma has been reported in the ovaries of Green Iguanas (*Iguana iguana*) (Anderson et al., 1996; Levine, 2004; Wenger et al. 2010), Garter Snakes (*Thamnophis* sp.) (Hertha et al., 2000), Red-Eared Sliders (Newman et al., 2003) and Mediterranean Tortoises (*Testudo* spp.) (Martorell et al., 2009). If a teratoma contains an immature, poorly differentiated embryonal cellular component, it will have greater potential for malignancy. Tumors with immature ectodermal origin cells may be termed a teratocarcinoma, but those with immature endothelial and mesothelial components are named "malignant teratomas" (Bel et al., 2016). Granulosa cell tumors are well represented in squamates, particularly in Garter Snakes (**Figures 8.67–8.68**) (Onderka and Zwart, 1978; Garner et al., 2004; Pirson et al., 2018) as well as Green Iguanas, (Cruz-Cardona et al., 2011) and Monitor Lizards (species not reported) (Christman et al., 2017). Malignant granulosa cell tumors occur in squamates, as reported in the Mangrove Snake (*Boiga dendrophila*) (**Figure 8.69**) (Pirson et al., 2018). Adenocarcinomas are most often reported from squamates, and have tubulopapillary morphology (Garner et al., 2004). Ovarian adenocarcinoma has been reported in several snake species, including the Cottonmouth (*Agkistrodon piscivorus*) and Boa Constrictor (*Boa constrictor*) (Michaels and Saneck, 1988; Gibbons and Schiller, 2000; Page-Karjian et al., 2017), and cystadenocarcinoma has been reported in the Green

Iguana (Stacy et al., 2004). Oviduct carcinoma has been documented in numerous snake species, including the Gaboon Viper (*Bitis gabonica*) and Corn Snake (*Pantherophis guttatu*) (**Figure 8.70**) (Pereira and Viner, 2008; Page-Karjian et al., 2017). Oviduct tumors including leiomyoma and leiomyosarcoma have been reported in chelonians (Frye, 1994).

8.12.3 Environmental Factors

Multiple interrelated environmental factors contribute to reproductive failure in wild and captive reptile populations. The incidence of reproductive failure, however, is greater in captivity because environmental requirements of individual species are difficult to replicate (Highfield, 1996). Wild turtles require suitable nesting devoid of human activity, excessive competition from other turtles, or flooding (Booth and Dunstan, 2018). Captive aquatic turtles and tortoises need appropriate space and substrate to lay eggs. Deep substrates for nesting should be equal to double the carapace length, and large areas over five times the carapace length in diameter are needed for some species (Boyer and Boyer, 1994). Failure to provide a proper captive environment may induce egg retention.

8.12.3.1 Light Sunlight contains electromagnetic radiation classified by wavelength as ultraviolet (285–320 nm), visible (400–700 nm), and infrared light (800 nm–1.0 mm). Deficiency or excess of light exposure may be the underlying factor in multiple types of reproductive disease. The influence of light effects early embryotic development and gonadal cycling in adults. Insufficient ultraviolet light (UVB) may disrupt calcium homeostasis, affecting embryo viability, growth of fetus, vitellogenesis, and eggshell formation (Ferguson et al., 2002; Burger et al., 2007; Zhang et al., 2016). Visible UVA light is detected by photoreceptors associated with the pineal gland and regulates reproductive cycles or circannual (yearly) rhythms. Disrupted photoperiod or blocked visible light in captive reptiles may disrupt seasonal folliculogenesis and spermatogenesis (Burger et al., 2007; Ooninex and van Leeuwen, 2017). Garter snakes held indoors with light restriction demonstrated stasis of follicular cycling and static follicular atresia (**Figures 8.71–8.72**) (Jenison and Nolte, 1980). Wild Green Turtles will have impaired ability to nest if beach sites are disturbed by excessive lighting from human development (Demetropoulos and Hadjichristophorou, 1995). Light and temperature are closely linked environmental parameters.

Infrared light generates heat. Failure to absorb infrared light results in low metabolic rate and incubation failure. In the northern hemisphere, wild turtles often use south-facing slopes with sun exposure for nesting (McArthur, 2004); south-facing slopes receive relatively high solar radiation and typically provide suitable thermal conditions for egg incubation. The incidence of abortion and final litter sizes are influenced by ability of lizards to bask (Hare and Cree, 2010).

8.12.3.2 Temperature Failure to maintain the preferred optimal temperature zone (POTZ) initiates metabolic influences that greatly effect reproductive cycling in adults and retards development of embryos. Regulation of POTZ though behavior is variable between genders in tegus (*Tupinambis* sp.) (Cecchetto and Naretto, 2015). During incubation, temperature influences reproductive organ structure, growth rate, endocrine profiles, neuronal pathways, neuroanatomy, and a variety of other traits (Gutzke and Crews, 1988; While et al., 2018). When Soft Shelled Turtles (*Lissemys punctata punctata*) were exposed to a 3°C variation above and below the normal mean incubation temperature during preparatory, recrudescent, and quiescent phases of the seasonal reproductive cycle, high temperatures were found to enhance ovarian growth and estrogen secretion early in vitellogenesis, but had a regressive effect on follicles in late vitellogenesis (Sarkar et al., 1996). Temperature influence during incubation and embryogenesis affects early gene expression and this influence may be carried into adulthood (Rhen et al., 2000). High or low fluctuation in temperature will cause variations or arrested embryo development, such as increased periods of diapause and abortion (Rock and Cree, 2003; Rafferty and Reina, 2012). Low incubation temperatures cause prolonged hatching in the Common Wall Lizard (*Podarcis muralis*) (Van Damme et al., 1992), and this is a common observation in most species. Although increased temperature during incubation speeds hatching, mild deviation above the optimal range will produce heat shock. Brown Anoles, exposed to brief heat shock during incubation experienced decreased neurologic function and slower growth rate (Sanger et al., 2018), and embryos had reduced survival (Hall and Warner, 2018). Temperature may affect the viability of hatchling reptiles by reducing the amount of residual yolk internalized prior to hatching (Pezaro et al., 2013).

8.12.3.3 Water Balance and Humidity Dehydration is a predisposing or resultant state in numerous reproductive disorders including dystocia and yolk coelomitis. Tropical species may need variation in humidity to initiate vitellogenesis (Sykes, 2010). Dehydrated individuals may have an elevated packed cell volume, blood urea nitrogen, and total protein (Rivera, 2008). Optimal humidity range is required for yolk metabolism of embryos during egg incubation (Santidrián Tomillo et al., 2012). Water essential to maintain embryonic viability is maintained in extraembryotic membranes (Bonnet et al., 2017). Eggs maintained at low humidity may crinkle and collapse, and excessive moisture will result in embryonic death and opportunistic fungal and bacterial overgrowth on the shell (Keller, 2017).

8.12.3.4 Social Structure The social structure of reptile populations varies widely, and no overarching general statements about reptile social structures will be meaningful. Thus, recommendations of whether reptiles are housed in groups or as individuals depends on the species of interest.

Nevertheless, cohabitation for long periods may suppress mating behaviors and produce perceived infertility; separation of individuals and subsequent reintroduction have been used to initiate successful matings (Sykes, 2010; Perry and Mitchell, 2017). Multiple mating by different males in polyandrous European common lizard females produced larger clutches than monandrous females, possibly selecting males with more heterozygosity in subsequent matings (Laloi et al., 2011).

8.12.4 Reproductive Toxicants and Endocrine Disruption

Reptiles are poikilotherms and they do not clear contaminants as fast as homoeothermic mammals and birds. Free-ranging, long-lived predatory reptiles in aquatic environments, such as chelonians and crocodilians, are excellent populations for monitoring the effects of industrial and agriculture chemicals that exhibit bioaccumulation (Selcer, 2006). Reproductive toxins can affect multiple mechanisms of reptile reproduction including spermiogenesis, oogenesis, release of gametes, formation of the zygote, blastocyst development, organogenesis, viability of newborn offspring, and proper pubertal development (Evans, 2018). The yolk of reptile eggs is a good site to sample for lipophilic compounds (Bhandari et al., 2015). Toxicants classified as teratogens can cause abnormalities in the conceptus any time between conception and birth (Evans, 2018). Toxicosis can be induced by oxidative damage, interference with normal enzymatic reactions, and dysregulation of hormone signaling (endocrine disruption). Exposure to multiple toxins simultaneously results in compounding effects (Boggs et al., 2011; Murray et al., 2017). Several pharmaceuticals administered clinically are potential reproductive toxicants, but the side effects of these agents for reptiles are largely unknown (Evans, 2018). Determination of specific reproductive diseases from toxic dumps or industrial wastewater discharge sites is difficult since these sites contain multiple classes of reproductive toxicants present in spatial gradients (Swartz et al., 2003; Shelby-Walker et al., 2009). For additional information on reptile toxicoses, see Chapter 7.

8.12.4.1 Endocrine-Disrupting Chemicals Affecting Reproduction Endocrine-disrupting chemicals (EDCs) cause agonistic or antagonistic modulation of endocrine, paracrine, or autocrine hormone receptor function by either direct (receptor-ligand complex) or indirect mechanisms. EDCs may include insecticides, herbicides, industrial dispersants, fungicides, heavy metals, and pharmaceuticals (Boggs et al., 2011). Specific EDCs of concern to humans and other vertebrates include cadmium, organochlorine pesticides, polychlorinated biphenyls (PCBs), bisphenol A (BPA), phthalates, dioxins, and furans (Rich et al., 2016; Mallozzi et al., 2017). The environmental presence of these toxins is reflected in tissue of free-ranging reptiles (Ashpole et al., 2004; White and Birnbaum, 2009; Bhandari et al., 2015; Brock et al., 2016). Hormones that affect reproduction in the reptiles include steroids (estrogens), eicosanoids (prostaglandin), the glycoproteins FSH and LH, the peptide GnRH, amino acid derivatives (melatonin), and proteins (Evans, 2018). The most common type of receptors involved in EDC-induced endocrine disruption belong to the nuclear receptor superfamily. Nuclear receptors affected by EDC binding include estrogen receptors (ERα and ERβ), peroxisome-proliferator activated receptor (PPAR), androgen receptor (AR), liver X receptors (LXR), and small heterodimer partner (SHP) (Rouiller-Fabre et al., 2015). EDCs may also activate transcription factors that may increase or decrease in hormone synthesis. In addition, EDCs can function by dysregulation of circulating hormone levels through interaction with hormone transport proteins, blockage of endogenous hormone clearance, disruption of hepatic biotransformation, and modification of hormone receptor sites. Levels of circulating hormones can also be altered by toxic nonspecific destruction of endocrine or reproductive tissues (Evans, 2018). Effects of EDCs may be permanent or transient, depending on the mechanism activated by the hormone (Boggs et al., 2011).

"Classic" endocrine disruption involves mimicry (xenoestrogens) or blockage (antiestrogens) at the receptor site. The terminology can be confusing, as androgenic toxicants may also be termed "antiestrogens" (Evans, 2018). Xenoestrogens are chemicals similar to 17β-estradiol in regard to their actions at estrogen receptors. Xenoestrogens have variable actions, including antiestrogenic activity, depending on tissue types. Actions of xenoestrogens may be dependent on binding affinities for either ERα- or ERβ-type receptors (Safe et al., 2018). Structural classes within categories of xenoestrogens will have wide variations in binding affinity (Matsumoto et al., 2014). Plant-derived phytoestrogens, such as isoflavones from soybeans, are not discussed, as they are largely absent from reptile literature. New synthetic xenoestrogen compounds are manufactured yearly, and the list of potential xenoestrogens continues to grow (Safe et al., 2018). Antiestrogenic human-use pharmaceuticals, such as diethylstilbesterol (DES) and tamoxifen, are also potential reptile EDCs, but exposure to reptiles is not reported.

EDCs are most notable for disruption of gonad organogenesis. This was observed in American Alligator hatchlings from Lake Apopka in Florida where increased exposure to EDCs was correlated with development of multi-oocytic follicles (MOFs) (**Figures 8.73–8.74**) (Guillette et al., 1994; Guillette and Moore, 2006). MOFs have also been generated in ovaries of Broad-Snouted Caimans exposed to increased levels of estrogen. This experiment mimics overstimulation of estrogen receptors by exogenous xenoestrogens (Stoker et al., 2008). Altered patterns of gene expression and hormone signaling during early development can be delayed and expressed later at sexual maturation resulting in loss of reproductive function (**Figure 8.75**) (Boggs et al., 2011). EDCs can also transiently alter spermatogenesis in adults (Boggs et al., 2011). Altered levels of circulating T will affect sexual behavior, and this may have adverse effects on reproduction (Bhandari et al., 2015).

EDC exposure can impact sexual differentiation in reptiles that have TSD by altering aromatase synthesis (Crews et al., 1995). Aromatase is needed early in embryonic sexual differentiation to convert androgens to estrogens (Crain et al., 1997). Estrogen or xenoestrogenic compounds are able to inhibit expression of SOX9 gene, the predominant male determining gene in the Red-Eared Slider (Barske and Capel, 2010). Sex reversal in TSD species has been demonstrated with exposure to Chlorodane, p.p.-dichlorodiphenyldichloro ethylens (p.p.DDE), Aroclor, and PCBs (Bergeron et al., 1994; Willingham and Crews, 1999). In heavily contaminated environments, combinations of multiple EDCs have demonstrated synergistic action and caused sex reversal at lower individual concentrations (Willingham and Crews, 1999; Boggs et al., 2011).

Oviduct development in Broad-Snouted Caiman can be altered by EDC binding to both estrogen receptor-α (ERα) and P receptors (Galoppo et al., 2016). Postnatally, female Broad-Snouted Caimans exposed to estrogenic compounds have abnormal development of oviduct glands resulting in malformed eggshells and neonatal mortality (Galoppo et al., 2017). Eggshells from caimans with disrupted oviduct development may have impaired pore development (**Figure 8.76**) (Stoker et al., 2013). Binding of EDC to ARs may lead to morphologic disorders of the testis and phallus in American Alligators (Miyagawa et al., 2015). The ERα receptor in liver initiates hepatic vitellogenesis and zona pellucida formation. EDCs can adversely affect hepatic induction of vitellogenin, zona pellucida protein production, hydroxylation of steroids, and systemic androgen-to-estrogen ratios (Arukwe et al., 2016).

8.12.5 Industrial and Petroleum Compounds or Byproducts

8.12.5.1 Bisphenol A Bisphenol A (BPA) is an estrogenic organic synthetic compound that leaches from polycarbonate plastics, the linings of metal food cans, thermal receipt paper, and numerous other household packaging products. These compounds are a component in microplastic particles, and microplastic ingestion is rapidly becoming ubiquitous in aquatic species worldwide (Duncan et al., 2019). BPA has been found in sewage and surface and drinking water, and therefore poses a risk for humans and wildlife (Stoker et al., 2003; Bhandari et al., 2015; Zychowski and Godard-Codding, 2017). When farm-raised turtle eggs were treated with BPA at a male-producing temperature (26°C), 89% of the male offspring had gonads with ovarian-like cortical tissue and disorganized testicular medullary tubules (Jandegian et al., 2015). When Blunt-Snouted Caimans were exposed to BPA early in life, aberrant activation of androgen receptors altered oviduct morphology (Galoppo et al., 2017). Alligators experimentally exposed to BPA hatch earlier and grow quicker in the short term, but their growth eventually slows compared to control alligators (Cruze et al., 2015). When 10-day-old caimans were exposed to BPA, the testicular tubular area was decreased and seminiferous tubules became disorganized (Stoker et al., 2008; Durando et al., 2016). Red-Eared Slider embryos exposed to BPA following oviposition had suppressed conversion of estrone to estrone sulfate, leading to early depletion of maternally derived steroids (Paitz and Bowden, 2015).

8.12.5.2 Halogenated Aromatic Hydrocarbons These compounds were produced by coolant, transformer oil, and flame-retardant industries creating potential toxicant zones that persist for long periods in the environment. Halogenated aromatic hydrocarbons (HAHs) are now banned in the United States, but their persistence in sediments allow bioaccumulation and material transfer within the food chains (Ming-Ch'eng Adams et al., 2016; Kodavanti et al., 2018). HAHs include dioxins and dioxin-like compounds such as polychlorinated dibenzo-p-dioxins (PCDDs). They also include dibenzofuanrans (PCDFs), PCBs, and polybrominated biphenyls (PBBs) (Ming-Cheng Adams et al. 2016; Kodavanti et al., 2018). The aryl hydrocarbon (AhR) type receptor is a ligand-activated transcription factor regulating the expression of cytochrome P450 enzymes (Hale et al., 2017). AhR signaling mediates hepatic clearance of xenobiotic chemicals, induces enzymes in sex hormone metabolism, alters neuroendocrine development, interferes with sex hormone biosynthesis, and disrupts testosterone signal transduction (Hale et al., 2017; Evans, 2018). Most of the effects from these chemicals occur during embryo development (Kodavanti et al., 2018). The ability of PCBs to elicit sex reversal in turtles was hypothesized in the 1990s, and further studies have confirmed changes in gene expression profiles during sexual differentiation (Bergeron et al., 1994; Matsumoto et al., 2014). Slider turtles experimentally exposed to PCB during incubation at male-producing temperatures exhibit up-regulation of feminizing cyp19a1 (aromatase), forkhead box protein L2 (FoxL2) and R-spondin 1 (Rspo1) genes, and down-regulation of male-determining Doublesex mab3-related transcription factor 1 (Dmrt1) and sex-determining region on Y chromosome-box 9 (Sox9) genes (Matsumoto et al., 2014). Snapping Turtles from a PCB-contaminated site on the Hudson River, New York, have clutches that hatch normally but are followed by substantial neonate mortality (Eisenreich et al., 2009).

8.12.5.3 Polycyclic Aromatic Hydrocarbons Polycyclic aromatic hydrocarbons (PAHs) are created through the coal and oil industries, and excessive environmental accumulation results from discharge of industrial wastewater and exposure to crude oil. Review of literature associated with PAH exposure suggests that deformities and developmental abnormalities, neoplasia, general reproductive toxicity, and mortalities near time of hatching deserve further investigation (Zychowski and Godard-Codding, 2017). Snapping Turtles (*Chelydra serpentina*) exposed to crude oil, Benzo[a]pyrene, and 7,12-Dimethylbenz[a]anthracene all had tetrogenic deformities and decreased embryo survival (Van Meter et al., 2006).

8.12.5.4 Dimethoxyethane-Type Compounds These chemicals are used as industrial solvents and also found in lithium batteries. The petroleum product ethane 1,2-dimethane sulfonate was toxic to testicular interstitial cells in Italian Wall Lizard (*Podarcis sicula*) (Minucci et al., 1995).

8.12.5.5 2,4,6-Trinitrotoluene 2,4,6-trinitrotoluene (TNT) is an explosive chemical that may be found in environments previously used by the military and causes testicular hypertrophy and decreased sperm count in the Western Fence Lizard (*Sceloporus occidentalis*) (McFarland et al., 2008).

8.12.5.6 Corexit 9500 Corexit is a chemical dispersant used on oil spills. When administered to alligator eggs during the temperature-sensitive period for male sex determination, the gonads of embryos developed into ovaries (Williams et al., 2018).

8.12.6 Pharmaceutical Contaminants in the Environment

8.12.6.1 Ethinylestradiol Ethinylestradiol (EE2) is an estrogenic human pharmaceutical that is used widely in birth control and readily leaches into the environment. Experimentally, EE2 has a similar effect to that of BPA (Bhandari et al., 2015). In-ovo administration of EE2 to alligator eggs at male-producing temperatures during the incubation period shifted the sex ratio to a female bias (Matter et al., 1998). Although EE2 is a xenoestrogen, it has a neuroendocrine disrupting ability. Neuroendocrine-disrupting (NED) chemicals disrupt pathways in the brain, hypothalamus, and pituitary that effect sex steroid feedback loops (Rosenfeld et al., 2017).

8.12.6.2 Methyltestosterone 17α-methyltestosterone (MT) is a synthetic androgen and has been documented in tissue of the American Crocodile (*Crocodylus acutus*), possibly due to industrial spillage from the aquaculture industry. Persistent exposure to methyltestosterone has resulted in a male-bias sex ratio in this population (Murray et al., 2017).

8.12.7 Pesticides

8.12.7.1 Organochlorine Compounds (Chlorinated Hydrocarbons) The exposure of reptiles to organochlorine pesticides (OCP), known reproductive toxicants, has decreased as a result of regulation in some countries, but still remains prevalent in many areas of the world (Buah-Kwofie et al., 2018). Dichlorodiphenyltrichloroethane (DDT) remains persistent in the environment after 30 years of discontinued use. These compounds bioaccumulate in the food chain and are often present in combination with other organochlorines. American Alligators exposed to multiple OCP in contaminated lakes have experienced high rates of embryonic mortality (Rauschenberger et al., 2004, 2007). Relevant OCP include aldrin, alpha-benzene hexachloride (a-BHC), b-BHC, lindane, d-BHC, p,p0–dichlorodiphenyldichloroethane (p,p0-DDD), p,p0-dichlorodiphenyldichloroethylene (p,p0-DDE), dichlorodiphenyltrichloroethane (p,p0-DDT), dieldrin, endosulfan compounds, endrin compounds, heptachlor compounds, hexachlorobenzene, kepone, methoxychlor, mirex, cis-nonachlor, and trans-chlordane compounds, 525, 525.1 polychlorinated biphenyl and oxychlordane and toxaphene (Rauschenberger et al., 2009).

Red-Eared Sliders incubated at male-producing incubation temperatures exhibited significant sex reversal when exposed to chlordane (Willingham and Crews, 1999). DDT metabolites, p,p-DDE and p,p-DDD were some of the primary EDC compounds identified in American Alligator eggs from the often-cited research conducted at Lake Apopka in Florida (Semenza et al., 1997). The levels of the organochlorine pesticides, Dieldrin and p.p′-DDE in the adipose tissue of alligators from Lake Apopka were comparable to that of Freshwater Crocodiles (*Crocodylus johnstoni*) in the lower Ord River, Australia (Yoshikane et al., 2006) and 30 times greater than that in crocodilians sampled at two other contaminated sites in South Africa and Costa Rica (Buah-Kwofie et al., 2018). When DDT was given to male Red-Eared Sliders, plasma vitellogenin increased presumably from estrogen receptor stimulation in the liver (Palmer and Palmer, 1995). Caimans exposed to endosulfan during development developed tortuous seminiferous tubules devoid of sperm (Durando et al., 2016).

8.12.7.2 Triazine Class Compounds Triazine class herbicides are commonly used worldwide. Atrazine has generated much attention from experiments demonstrating demasculinization of anurans (Hayes et al., 2002). Atrazine administered at low dose to eggs during the thermosensitive period enhanced female differentiation (Willingham, 2005). Caimans exposed in ovum to atrazine developed disorganized seminiferous tubules with decreased spermatozoa (Rey et al., 2009). Atrazine has also been shown to alter gene regulation in the HPG and HPA axes of snapping turtles (Russart and Rhen, 2016).

8.12.8 Heavy Metals—Trace Elements

In-ovo uptake of a number of heavy metals has been documented. These include Cadmium (Cd), copper (Cu), Iron (Fe), Mercury (Hg), lead (Pb), and zinc (Zn) (Lance et al., 2006). Large-scale release of trace elements into aquatic environments has occurred by fly ash discharge or disturbance originating from coal-burning power plants. (Van Dyke et al., 2017). Mineral composition of the chelonian shell correlates well with mineral levels in the fetus (Roe et al., 2004).

8.12.8.1 Cadmium Cadmium (Cd) is an EDC. Increased cadmium exposure in vertebrates has been linked to oxidative stress and disruption of the blood–testis barrier, altered spermiation, gonadal steroidogenesis, disruption of hypothalamus-controlled endocrine pathways, alteration of epididymis pH, increased follicular atresia, increase in the incidence of chromosomal abnormalities, reduced gonadotropin binding to granulosa cells, decreased production of P, compromise of embryo implantation into the placenta, and disrupted embryo development (Rie et al., 2005; Thompson and Bannigan, 2008; Simoniello et al., 2013; Evans, 2018). Cd directly affects endothelium of testicular and epididymal vasculature, predisposing to ischemic necrosis of the testicle (Prozialeck et al., 2006). Experimental exposure of Italian Wall Lizards to Cd stimulates oocytes into meiotic prophase and increases oocyte recruitment from reserves (Simoniello et al., 2013). Premature exhaustion of germ cell nests leads to infertility. Cd also increases the number of atretic and deformed follicles in the Italian Wall Lizard. The Cd-binding protein metallothionein is located in the membrane of the Golgi apparatus in small stem cells, pyriform cells, and yolk platelets of ovarian follicles. Pyriform cells exposed to exogenous Cd undergo translocation and functional change (Scudiero et al., 2017). Cd-treated lizards have dysregulation of folliculogenesis, leading to deformed follicles with large protrusions, increased embryonic mortality, fetal deformity, and impaired fecundity (**Figure 8.77–8.80**) (Simoniello et al., 2013). In early vitelline follicles, Cd induces increased vascularization of the theca, disrupted or adhered outer epithelium, disruption of the granulosa cell layer, and disruption of the zona pellucida (Simoniello et al., 2013). Disruption of the zona pellucida could result in hydropic swelling of follicles and increased vesicle formation in the cytoplasm. In late vitelline follicles, increased apoptosis of pyriform cells is associated with increased follicular atresia. During experimental exposure to cadmium, apoptosis can also be observed in primary and early previteogenic follicles, while under normal development apoptosis is absent from this stage (Simoniello et al., 2013).

8.12.8.2 Mercury Environmental mercury contamination often occurs near gold mines. The water-soluble form methylmercury is released into the atmosphere by coal-powered electrical generators and may contaminate pristine areas via rainwater (Eisler, 2004). Mercury can be transmitted maternally to eggs. Common Snapping Turtles from a contaminated environment had accumulation of mercury in maternal tissues. Females with higher mercury concentrations had increased incidence of infertility and embryo mortality (Hopkins et al., 2013; Thompson et al., 2018). Offspring from Leatherback Sea Turtles with elevated blood mercury had lower hatchability (Perrault et al., 2011). Mercury may also decrease sperm quality in other amniotes (Evans, 2018).

8.12.8.3 Lead Lead may accumulate in crocodilians through ingestion of lead shot or fishing sinkers retained as gastroliths (Warner et al., 2016a,b). Lead is directly toxic to spermatogonia, causing decreased sperm count and decreased fertilization rates (Evans, 2018). Lead decreases pituitary secretion of LH and FSH. Lead ingestion by captive American Alligators was correlated with lipid peroxidation using the thiobarbituric acid (TBA) test, and this was hypothesized to be the likely cause of early embryonic death in this population (Lance et al., 2006).

8.12.9 Algal Toxin

Blooms of *Karenia brevis* (also called red tides) occur often in the Gulf of Mexico, and increased concentrations of fetal hepatic brevetoxin are correlated with "dead-in-nest" mortality in Loggerhead Turtle (*Caretta caretta*) hatchlings (Perrault et al., 2016).

8.13 Observations on Intersex

The term "intersex" refers to chromosomal, gonadal, or anatomical gender characteristics that are atypical (Rich et al., 2016). There is minimal published information on intersex in reptiles, and the pathogenesis is difficult to trace back to the initiating cause, but the incidence of intersex is likely much greater than reported. The frequency of intersex individuals is greater in species with TSD (Yao and Capel, 2005). Aberrant temperature fluctuation during egg incubation has been associated with increased incidence of intersex (Pieau et al., 1998). Some reports of intersex in Testudines have not been accompanied by microscopic description of gonads (Limpus, 1982; Highfield, 1996), and more work is needed in this area.

At a cellular level, bipotential well-differentiated sex cord cells forming granulosa cells and Sertoli cells may revert to the opposite sex. Dysregulation of steroidogenic factor-1 (Sf-1) may occur from exposure to EDCs and cause female-to-male sex reversal in mammals and fish, but this potential cause of intersex has not been clinically verified (Durando et al., 2013; Xie et al., 2016).

8.13.1 True Hermaphrodites (Ovotestes)

True hermaphrodites have ovotestes. In marine turtles observed with ovotestes, seminiferous tubules were present in the medulla of the gonad and previtelline follicles in the cortex (Limpus et al., 2009). There are a mixture of genetic, hormonal, and TSD-related explanations for the formation of ovotestes. In European Pond Turtles (*Emys orbicularis*) with male differentiation, slight increases in estrogen could induce the transient formation of an ovarian-like cortex but does not inhibit the development of testicular cords (Pieau et al., 1998). Eggs incubated at the TSD pivotal temperature (i.e., the temperature that produces balanced sex ratio) have a tendency to produce intersex individuals, and this condition has been suggested for production of ovotestes in the African Dwarf Crocodile (*Osteolaemus tetraspis*) (Langer et al., 2014). Ovotestes in Red-Eared Sliders were also observed in sex-reversal experiments using organochlorine pesticides (Willingham and Crews, 1999). In the European Pond Turtles, mild increases in estrogen exposure during the transient temperature period resulted in a slight increase in number of ovotestes. These gonads have prefollicular structures in the ovary, but ovum will degenerate around the time of hatching and the ovary will take on a testicular phenotype (Pieau et al., 1998). Hermaphroditism has been reported in a Green Iguana with concurrent ectopic eggs (Miller et al., 2003).

8.13.2 Pseudohermaphroditism

Partial sex reversal occurs when normal genetic regulation of sexual organ development is redirected by an incomplete hormonal influence or genetic signaling. Partial sex reversal can occur in both XY and ZW systems, and within oviparous and viviparous species. Reversal of gonadal phenotype leads to downstream genetic products that alter sexual phenotype, create gonadal dysgenesis, and induce reproductive dysfunction (Holleley et al., 2016). The action of steroidogenic enzyme aromatase, converting T to estrogen, is pivotal in ovarian differentiation and when inhibited will cause female-to-male sex reversal (Lance and Bogart, 1992; Barske and Capel, 2010). Effects of aromatase on sex reversal are dose dependent and may change in accordance with timing or level of inhibition (Wibbels and Crews, 1994). Experimental knockdown of gene Dmrt1 in male Chinese Softshell Turtle results in feminization of embryonic gonads, reduced expression of testicular markers Amh and Sox9, and up-regulation of the ovarian regulators Cyp19a1 and Foxl2 (Sun et al., 2017). In lizards, a high diversity of sex chromosomes and variation in TSD suggest mechanisms of sex determination may be susceptible to rapid evolution of hormone function (Ezaz et al., 2009). Sex reversal over entire populations is particularly overrepresented in species laying eggs in or near an aquatic environment (Barske and Capel, 2010). Two Loggerhead Turtles with precocious male external characteristics but female gonad differentiation were diagnosed as pseudohermaphrodites by endoscopy and histology (Crespo et al., 2013).

8.13.3 Dysplasia and Aplasia of the Tubular Tract

Red-Eared Slider Turtles experimentally exposed to PCBs during periods of sexual differentiation had complete or segmental aplasia of the uterus. This change can be caused by interruption of Müllerian-inhibiting substance expression in vertebrates (Matsumoto et al., 2014).

8.14 Diseases of the Male Reproductive Tract

The male reproductive tract has testes, efferent tubules (rete testis, epididymis, and ductus deferens), hemipenes or penis (phallus), and cloaca. In most reptile species, testes are located adjacent to the adrenal gland in the coelomic cavity cranial to right and left kidneys, respectively. Congenital diseases of the avian and mammalian testes include fusion of the cranial poles, spermatocele formation, hypoplasia, and agenesis (Schmidt et al., 2015). These congenital diseases have the potential to occur in Reptilia. Congenital testicular cysts have been observed grossly in juvenile Nile Crocodiles (*Crocodylus niloticus*) (Huchzermeyer, 2003).

The cloacal scent gland is included in this discussion of male genitalia, as diseases in this structure must be differentiated from hemipene abnormalities, and secretions of this gland are associated with courtship behavior (de Haan, 2003).

8.14.1 Testicular Degeneration

Pathologic testicular degeneration and atrophy must be distinguished from normal seasonal regression. Testicular degeneration is often the result of inadequate nutrition or disease associated cachexia. In birds, testicular degeneration has been reported in lead, selenium, cadmium, insecticide, and herbicide toxicosis (Barnes et al., 2016). During pathologic degeneration, testes may become pale and fibrous, remaining small and unresponsive despite appropriate seasonal cues and sufficient adipose reserve (**Figure 8.81**) (Barnes et al., 2016). During seasonal atrophy and disease, injury or immunosuppression of the interstitial cell population will increase apoptosis of spermatocytes and alter maturation of spermatozoa (Foster, 2012). Degeneration of germ cells usually follows degeneration of Sertoli cells and starts when germ cells are disassociation from Sertoli cells (Foster, 2012). In the proliferative phase of spermatogenesis, groups of spermatids should have their heads attached to Sertoli cells. Increased numbers of spermatids attached to Sertoli cells may indicate sperm retention or disrupted spermatogenesis. Failure to release spermatogonia is a characteristic of early degeneration and can result in multinucleated spermatids within seminiferous tubules (Foster, 2012). Spermatic giant cells with condensed nuclei are described in birds with testicular degeneration (**Figure 8.82**). Spermiophagy, the process of sperm phagocytosis, occurs in seminiferous tubules and the epididymis (Carrell et al., 2003). Sertoli cells phagocytize degenerated germ cells creating cytoplasmic residual bodies (**Figure 8.83**). In late stages of degeneration, one or two layers of Sertoli cells and spermatogonia will line the basement membrane of tubules, and in severe cases only Sertoli cells will persist (**Figure 8.84**) (Barnes et al., 2016). Advanced testicular degeneration is accompanied by interstitial fibrosis and loss of seminiferous tubule basement membrane integrity (**Figure 8.85**). Testes of some species of avian and non-avian reptiles contain a greater number of melanoblasts in the testicular interstitium (Schmidt et al., 2015). In reptiles with pigmented testes, as in birds (Barnes et al., 2016), the capsule may be a darker shade of brown or black during degeneration. Unilateral size discrepancy of testes may indicate a vascular anomaly or infectious process, rather than physiologic degeneration (**Figure 8.86**). The testes of geriatric reptiles may undergo cystic degeneration yet remain functional (**Figure 8.87**). Incidental sperm granulomas often are associated with small tubular cysts. Large cysts may represent congenital defects or dilatation of structures closely associated with the testicle such as the efferent ductules (epididymis), as might occur with obstruction downstream from the testis (**Figure 8.88**).

8.14.2 Breach of Blood–Testis Barrier

Disruption of the blood–testis barrier, as might occur with testicular trauma, infection, or neoplasia, will elicit a foreign body response in leukocytes exposed to antisperm antibody. Interstitial infiltration by lymphocytes will signal a break in the blood–testis barrier. In severe cases a granulomatous response will result, with phagocytosis of the sperm and formation of sperm granuloma (Barnes et al., 2016). Granuloma formation and local scarring may result in regional obstruction of tubules and development to cysts (Foster, 2012).

8.14.3 Orchitis

Orchitis is not widely reported in avian or non-avian reptiles (Schmidt et al., 2015). Cellular degeneration, toxins, and infectious agents may induce orchitis. Bacterial orchitis usually occurs with concurrent multiorgan infection by the implicated pathogen (i.e., *Mycobacterium* spp. or *Salmonella* spp.).

8.14.4 Disease of Epididymis and Vas Deferens

Morphologic features of the proximal efferent tubules (epididymis) are variable between species. These structures communicate with the paired ductus deferens. The ductus deferens terminate in the urodeum of the cloaca. Damage to the epididymis may result in blockage and rupture of the tubules or formation of an extratesticular sperm granuloma (Foster, 2012). Interstitial fibrosis may also be present between efferent tubules of epididymis in geriatric reptiles (**Figure 8.89**). A spermatocele is a fluid-filled cyst that forms adjacent to the efferent tubules or epididymis (**Figure 8.90**).

8.14.5 Defective Spermatogenesis

Sperm defects in reptiles are similar to those reported for domestic mammals and can be documented through examination of fresh spermatozoa. Abnormalities reported in reptiles include cytoplasmic droplets attached near the proximal midpiece, teratoid sperm, sperm heads without flagella, and macrocephalic and microcephalic sperm (**Figure 8.91**) (Johnston et al., 2014). Double-tailed (biflagellate) spermatozoa are also frequently observed in reptiles (**Figure 8.92**) (Trauth and Buchanan, 1997). The quality of semen is dependent on seasonality and nutritional status (Johnston et al., 2014). Sperm abnormalities increase during testicular degeneration (Foster, 2012). Bent tail formation may occur in Green Iguanas undergoing electro ejaculation, and this may be an iatrogenic consequence of this technique (Zimmerman et al., 2013).

8.14.6 Disease of Male External Genitalia and Cloaca

Since reptiles reproduce by internal fertilization, all males with exception of Tuatara have copulatory organs. These copulatory organs are usually retracted within the pericloacal region. In Testudines and Crocodilia, the male has a single median penis (phallus), while squamates, including snakes and lizards, possess a pair of hemipenes, only one of which is typically used during intromission. Reproduction in Tuatara is similar to birds and involves cloacal kissing (Lutz, 2005).

8.14.6.1 Prolapsed Penis Penile prolapse in Testudines and Crocodilia often occurs in association with breeding trauma or bite wounds. A study of pet reptiles submitted to a university teaching hospital revealed prolapsed copulatory organs to be more frequent in Testudines than Squamata (**Figure 8.93**) (Hedley and Eatwell, 2014). The chelonian penis is attached to the ischia by the ligamentum rami and rotates out from the cloaca by contraction of paired levator cloacae muscles. The elastic recoil of the ligamentum rami retracts the penis post copulation (Kelley, 2103). Injury to the ligamentum rami may result in prolapse. Disruption of the internal iliac vessels supplying the penis may alter hemodynamic properties of erection (Perry and Mitchell, 2017). Substrate adhered to a prolapsed hemipene or penis exacerbates desiccation, prevents retraction of the affected organ, and may result in ulceration, cellular edema, and bacterial or mycotic balanitis (**Figures 8.94–8.96**) (Raiti, 1995; Perry and Mitchell, 2017). Shell deformities associated with nutritional secondary hyperparathyroidism will predispose to penile prolapse in Testudines (Innis and Boyer, 2002). Other predisposing conditions of penile prolapse in Testudines include obstipation, lower gastrointestinal straining, masturbation, and neurologic deficits. Gastrointestinal or urogenital conditions resulting in straining or coelomic distention may also predispose to penis prolapse (Sykes, 2010).

8.14.6.2 Prolapsed Hemipene Predisposing factors for hemipene prolapse in Squamata include breeding trauma and age. Sexing of juvenile snakes and lizards with "popping technique" as well as improper use of sexing probes can result in injury to hemipenes (DeNardo, 2006a). Damage or atrophy of the retractor penis muscle associated transverse penis muscle or cloacal sphincter muscle may result in hemipene eversion (Wade, 2005; Barker and Barker, 2006). Usually only one of the paired hemipenes is prolapsed. If the prolapse is prolonged, edema, congestion, and desiccation of hemipene will follow (**Figures 8.97–8.98**). Snakes undergoing amputation of one hemipene may still breed successfully.

8.14.6.3 Hemipenal Cast or Plugs In Squamata the hemipene sulcus can become clogged with inspissated semen and desquamated keratin, resulting in a hemipenal plug and inflamed hemipenal sulcus (**Figures 8.99–8.101**). Hemipenal plugs may appear caseous and can be infected with opportunistic bacteria or fungi (Knotek et al., 2017). Hemipenal plugs may also be associated with unsuccessful copulatory attempts (Raiti, 1995). Infected and distended caudal scent glands may mimic the gross appearance of hemipenal plugs (**Figure 8.102**).

8.15 Diseases of Female Reproductive Tract

The inner (medulla) or vascular zone of the ovary is composed of compact mesenchymal tissue and vessels. The outer zone (cortex) or parenchymous zone contains the large developing follicles. Removal or injury of one ovary will result in compensatory follicular development in the non-injured ovary (Johnson and Wade, 2011).

8.15.1 Atresia and Congenital Ovarian Disease

Congenital ovarian disease in reptiles is not well documented. Mechanisms preventing migration of primordial cells from the germinal ridge, failure of oocyte nest formation during embryologic development, and viability of primary oogonia would limit the number of ova available to enter folliculogenesis. These processes could initiate during embryologic development. Epigenetic influence of the HOX pathway can cause altered expression of genes controlling development of the gonad and tubular structures (Martín-Del-Campo et al., 2019). Estrogen and therefore EDCs may influence Hox genes expression (Garcia-Gasca and Spyropuolos, 2000). Congenital ovarian cysts lined by granulosa cells are observed in birds and are likely to occur in some reptiles (Schmidt et al., 2015). Diffuse polycystic dysplastic folliculogenesis has been observed in a wild Gopher Tortoise (*Gopherus Polyphemus*) (**Figures 8.103–8.104**).

8.15.2 Follicular Atresia and Degeneration

Ovary cycling can be disrupted by environmental factors, nutritional deficiencies, toxin exposure, and infectious disease. Normal physiologic ovarian atresia must be distinguished from pathologic degenerative atresia (Barnes et al., 2016). Follicular degeneration may occur secondary to preovulatory follicular stasis, often leading to rupture and release of yolk into the coelomic cavity (chronic bursting atresia), with subsequent coelomitis (**Figures 8.105–8.106**). Activated mesothelial cells on the serosa surface of the ovary and other visceral organs phagocytize free yolk in the coelomic cavity. In birds, acute bursting atresia with mesothelial uptake of yolk occurs in clinically normal individuals. Yolk phagocytized by mesothelial cells is transported via a lipoprotein-dependent mechanism and reenters circulation with properties similar to that of hepatic derived plasma triglycerides (Cornax et al., 2013). In chickens (*Gallus gallus domesticus*), when small amounts of yolk were experimentally introduced into the coelomic cavity, circulating inflammatory cytokines did not increase, and it was hypothesized that cells absorbing yolk are able to regulate the inflammatory response (Cornax et al., 2013).

Sustained ovarian follicular degeneration can be associated with chronic metabolic and nutritional disruption. Geriatric and obese reptiles often have larger numbers of atretic follicles (**Figure 8.107**) (Cree, 2015). Follicles that undergo chronic delayed atresia may contain central cholesterol clefts and granulomatous or xanthogranulomatous inflammation (**Figures 8.108–8.109**). Obese captive reptiles that become ill are susceptible to degeneration of multiple vitelline follicles (**Figures 8.110–8.111**). Metabolic disease resulting in soft tissue and vascular mineralization may compromise perfusion of follicles leading to degeneration (**Figures 8.112–8.113**). Vascular compromise associated with an amyloid-like deposition disorder in Komodo Dragons (*Varanus komodoensis*) resulted in fatal acute yolk embolization (Garner et al., 2003).

The pathogenesis of acquired ovarian cyst is not well defined for any species, and also occurs in reptiles (**Figures 8.114–8.116**). Large cysts in ovary follicles of reptiles likely have the same pathogenesis as birds and can be caused by twisting of the ovarian pedicle (Barnes et al., 2016).

8.15.3 Ovarian Hemorrhage

Fatal hemorrhage has been observed from the ovary in Mugger Crocodiles near the time of ovulation. This was proposed by Huchzermeyer to be related to increased blood vessel proliferation in response to late folliculogenesis (Huchzermeyer, 2003). Clotting functions associated with folliculogenesis and gravidity are not well studied in reptiles, and this should also be considered. In birds, some bleeding from the follicular wall is expected at the time of ovulation (Barnes et al., 2016).

8.15.4 Oophoritis

Follicles may have mild inflammatory cell infiltration during atresia and degeneration. Yolk within follicles in vitellogenesis is a ready culture medium for circulating bacteria. Bacterial colonization and oophoritis is most often secondary to preovulatory stasis or systemic disease and septicemia. Oophoritis in late vitellogenesis and preovulatory stasis can result in coelomitis and adherence between follicles, especially if affected follicles rupture (**Figures 8.116–8.118**). Oophoritis occurring in postovulatory egg stasis

(POES) can result from vascular compromise and cystic degeneration of non-ovulated follicles. Thrombosis of vasculature draining the ovaries may result in follicle congestion, degeneration, and secondary bacterial colonization (**Figures 8.119–8.120**).

8.15.5 Preovulatory Disease

Preovulatory follicular stasis (POFS) is a disease of captivity resulting in failure of mature follicles to ovulate or undergo atresia (Rivera, 2008; Knotek et al., 2017). POFS is also termed "preovulatory egg binding and retained follicles," and may accumulate over multiple cycles. This syndrome occurs in captive Testudines and Squamata and often requires surgical intervention (**Figures 8.121–8.124**) (Innis and Boyer, 2002; McArthur, 2004; Sykes, 2010). Ovulation is dependent on multiple factors including visual stimulation between males and females, group dynamics, copulatory behavior, photoperiod, hormonal cycling, correct wavelength of light, and metabolic status. Captive Veiled Chameleons (*Chamaeleo calyptratus*) experience longer photoperiods with high temperatures and higher levels of dietary protein than occur naturally. This contradicts the naturally occurring breeding season and exceeds physiologic requirements (Knotek et al., 2017). Breeding group cohabitation in small enclosures may increase serum cortisol, and this could also contribute to the development of POFS. Static follicles will become necrotic and produce inspissated yolk, friable follicles, and eventually chronic coelomitis (Rivera, 2008). Antemortem diagnosis of POFS requires the ability to monitor vitellogenic progression and time of ovulation. Vitellogenic follicle numbers, sizes, location, and clustered appearance can be monitored by palpation, transillumination, endoscopy, ultrasound, and magnetic resonance imaging, depending on the species (Rhen et al., 2000; Kummrow et al., 2010). Vitellogenic follicles that have not entered the oviduct and not progressed to eggs are spherical, clustered, and noncalcified (**Figures 8.125–8.126**) Ovaries with chronic POFS will become adhered to each other and develop heterophilic and granulomatous oophoritis (**Figures 8.127–8.128**) (Knotek et al., 2017). Reptiles with retained follicles may be anorexic, with hypercalcemia, hyperphosphatemia, hyperalbunemia, hyperproteinemia, and elevated serum alkaline phosphatase. Hemograms may demonstrate decreased circulating lymphocytes and heterophils (leucopenia) and anemia (Innis and Boyer, 2002; Rivera, 2008).

8.15.6 Periovulatory Disease

Oviduct fimbria may fail to capture ovulated follicles or expel follicles because adhesions, obstructions, retrograde oviduct peristalsis, rough handling, or other trauma may result in ectopic follicles free in the coelomic cavity (**Figure 8.129**). Chronic ectopic follicles are composed of layered inspissated yolk, inflammation, and dystrophic mineralization (**Figures 8.130–8.133**). Fertilized eggs or sterile shelled eggs ("slugs") may enter the coelomic cavity through rupture of the oviduct or retrograde oviduct peristalsis (Mans and Sladky, 2012).

Premature ovulation in captive snakes may be associated with obesity, rough handling, trauma, or a large meal. Ovulated immature follicles will produce an infertile ovum known by reptile breeders as "slugs." Slugs may be laid outside the normal time frame of oviposition or simultaneously with fertile eggs. The passage of slugs may be correlated with defecation in captive snakes (**Figure 8.134**) (Barker and Barker, 2006). Early embryonic death is a proposed pathogenesis for slug formation. Early embryonic death rate in wild lizards has been estimated at necropsy by counting corpora luteum and subtracting the number of viable ova or neonates. Data compiled from multiple independent studies utilizing 15 lizard species under variable conditions indicated that the average pregnancy failure for wild lizards was estimated at 15% (Hare and Cree, 2010).

8.15.7 Postovulatory Disease

8.15.7.1 Salpingitis, Infertile Follicles, and Diminished Egg Viability Remnants of previous infertile eggs and free albumin are observed in lumina of oviducts, and if colonized by bacteria will result in salpingitis (**Figure 8.135**). Static eggs in the oviduct will develop adhesions and initiate an inflammatory response in the oviduct mucosa (**Figures 8.136–8.137**). Metabolic compromise may lead to devitalization of eggs in oviparous species. Viviparous species may have arrested development and autolysis within the oviduct, leading to salpingitis and placentitis (**Figures 8.138–8.139**).

POES is one of the most common reproductive diseases encountered in captive reptiles. Developing eggs within the oviduct may occupy most of the coelomic cavity. Clinically, lizards with POES will have a distended coelom, anorexia, and dehydration (**Figure 8.140**). In most Squamates, follicles entering the oviduct lose the spherical shape and assume an oval shape. Individual calcified eggs may be visible by radiograph (**Figures 8.141–8.142**). Affected captive lizards may pass a few eggs and then stop or lay eggs in inappropriate areas such as a water dish. Wild reptiles can experience postovulatory stasis as a result of chronic disease (**Figure 8.143**). Failure of eggs to calcify may be associated with nutritional deficiency, toxins, and viral pathogens in birds (Schmidt et al., 2015). Static noncalcified eggs have potential to form adhesions to oviduct mucosa, and chronic cases may progress to coelomitis.

8.15.8 Yolk Coelomitis

Yolk coelomitis is a major cause of death in all classes of reptiles as an acute or chronic disease. Yolk coelomitis may occur in combination with oophoritis, salpingitis, failure of fimbria to catch ova, preovulatory follicular stasis, rupture of the oviduct, coelomic trauma, POES, obstructive and physiologic dystocia, and other conditions. Transmural tears of the oviduct can occur from overzealous palpation. Yolk released during excessive bursting atresia can exceed the phagocytic capacity of activated mesothelial cells and result in coelomitis (**Figures 8.144–8.145**) (Cornax et al., 2013). Vessels within the subserosal tunic may undergo reactive change during yolk uptake by activated mesothelial cells (**Figure 8.146**). Follicles in postovulatory follicular stasis often degenerate or rupture, resulting in release of free yolk into the coelomic cavity (**Figure 8.147**). Captive Fiji Banded Iguanas (*Brachylophus fasciatus*) were shown to be predisposed to yolk coelomitis (Stacy et al., 2008). This report described chronic yolk leakage from large vitellogenic follicles undergoing atresia and also noted the difficulty in microscopically differentiating activated mesothelial cells from diseased follicular granulosa cells (**Figure 8.148**). Yolk is a good medium for bacterial growth, and if the intracoelomic yolk becomes colonized with bacteria from blood or intestines, a more severe and exudative inflammatory reaction will ensue (**Figure 8.149**). Chronic yolk coelomitis with bacterial infection produces a thick, whitish-yellow, malodorous exudate (**Figures 8.150–8.151**). Clinical signs of yolk coelomitis may include weight loss, inactivity, loose stool, and a decrease in fecal and urine volume (Innis and Boyer, 2002). Reptiles with yolk coelomitis usually will have leukocytosis with heterophilia and azurophilia. Segmented white blood cells may have toxic changes (Rivera, 2008; Stacy et al., 2008). Hyperuricemia has also been observed with egg yolk coelomitis in tortoises (Sykes, 2010).

Yolk embolism occurs when yolk aberrantly enters the bloodstream through vessels located near the ovary. Several species, especially Monitor Lizards, seem to have an increased incidence of yolk embolization (**Figure 8.152**). Sea Turtles may develop yolk embolism associated with trauma, including falls, entrapment, burial, or obstruction (Stacy et al., 2013).

8.15.9 Dystocia

Dystocia in mammals most often refers to obstructive parturition. In reptile medicine the term has been used nonspecifically to reference parturition failure ranging from preovulatory stasis to failed oviposition (Raiti, 1995). For this chapter, "dystocia" is not used to indicate POES. Dystocia indicates failure to complete ovipostion or parturition from mechanical (obstruction) or physiologic (uterine inertia) factors.

Obstructive dystocia may occur from fetal or maternal causes. Maternal causes of obstructive dystocia include uterine compression (cloacoliths, fecal impaction, uroliths, intestinal foreign body, and neoplasia), physically immature individuals, oviduct strictures, oviduct torsion, pelvic fractures, and inflammatory diseases (salpingitis, cloacitis) (Raiti, 1995; Funk, 2002; DeNardo, 2006b; Rivera, 2008; Stahl and DeNardo, 2019). Egg or embryo causes of obstructive dystocia include embryologic abnormalities, dead embryos, retained slugs, mummified products from previous pregnancies, large embryos in viviparous species, chronically retained eggs, fractured eggs, and excessively large, deformed, or fused eggs, (**Figure 8.153**) (Raiti, 1995; Funk, 2002; DeNardo, 2006b). In chelonians, fractured eggs may be caused by sexually aggressive males. Radiographs are useful for antemortem diagnosis of obstructive dystocia. Excessive circumferential calcareous lamellae indicate prolonged egg retention in chelonians (**Figures 8.154–8.155**) (Raiti, 1995). Females that produce smaller than expected clutches may produce larger or oversized eggs (Sinervo and Licht, 1991). The size of the egg is determined by available resources rather than the amount of time spent in the oviduct (Sinervo and Licht, 1991). Chronically retained eggs in squamates may undergo increased mineralization of the shell (**Figures 8.156–8.157**). Obstruction of the pelvic canal in chelonians by misshapen or large eggs may result in compression of the obturator nerve (**Figures 8.158–8.159**) (Raiti, 1995). Death of late-term embryos will result in dystocia in viviparous species (**Figures 8.160–8.161**). As a result of dystocia, secondary salpingitis accompanies the presence of degenerated eggs or embryos in the oviduct and eventually can lead to bacterial overgrowth, coelomitis, and septicemia (**Figures 8.162–8.163**).

Physiologic dystocia is more difficult to diagnose and may involve multiple factors in captive reptiles (DeNardo, 2006b). Predisposing factors for physiologic dystocia in captive reptiles are malnutrition, lack of nesting site, and inappropriate temperature and substrate (Lloyd, 1990; DeNardo, 2006b). Environmental moisture is a key initiator of oviposition, particularly in many tropical squamates (Van Dyke, 2015). Low environmental humidity may prolong oviposition and result in dystocia (DeNardo, 2006a). Associated metabolic causes of dystocia may include secondary hyperparathyroidism and general nutritional deficiency (Sykes, 2010). Failure to pass an egg through the vaginal stricture of the oviduct in birds is termed "egg binding" and is associated with hypocalcemia, vitamin E/selenium deficiency, obesity, and stress (Schmidt et al., 2015). These factors likely also affect reptiles. Calcium sequestration for shell formation may result in hypocalcemic myopathy and impairment of oviposition. Nutritional secondary hypoparathyroidism may also cause hypocalcemia and affect oviductal contractility. Other nutritional and husbandry factors associated with dystocia in captive reptiles are "power-fed" young, small females at breeding, inactivity, and poor conditioning (Funk, 2002; Barker and Barker, 2006). Obesity in sedentary captive reptiles also predisposes to dystocia (DeNardo, 2006b). Squamates with a prior history of infertility will often develop dystocia in subsequent pregnancies (DeNardo, 2006b), and in severe cases ectopic eggs and yolk coelomitis accompany dystocias.

Dystocia in snakes is more common in oviparous than live-bearing species (**Figures 8.164–8.165**) (Stahl, 2015). Viviparous snakes, however, are more likely to retain the entire clutch (DeNardo, 2006b). It is common for captive snakes to lay an otherwise normal clutch and retain the final few eggs or embryos (DeNardo, 2006b). Physically immature and first-time gravid captive snakes may have a greater incidence of dystocia (Raiti, 1995; Di Giolamo and Selleri, 2017). When captive animals are double clutched, the second batch of eggs may be prone to dystocia (Raiti, 1995). Pythons experiencing dystocia may lay eggs in random locations over an extended period (Barker and Barker, 2006).

Complications of dystocia include salpingitis, rupture of the oviduct, and coelomitis (Sykes, 2010). In snakes, rupture of the oviduct and coelomitis may be a serious complication if manual manipulation of static eggs or slugs is attempted (**Figures 8.166–8.167**). Deposition of calcified eggs into the coelomic cavity (internal laying) may result from reverse peristalsis of the oviduct or rupture during dystocia (**Figure 8.168**). Clinical signs of dystocia are excessive straining, anorexia, lethargy, fecal obstruction, urinary obstruction, prolapse of the oviduct or urinary bladder, cloacal prolapse, and paresis of the hindlimbs (**Figure 8.169**) (Sykes, 2010).

8.15.10 Female Cloaca and Oviduct Prolapse

Distal oviductal prolapse occurs with frequency in chelonians (**Figure 8.170**). This condition may be associated with dystocia or any condition that causes straining, such as parasitism or obstipation (Raiti, 1995; Divers, 1996). A prolapsed oviduct may be difficult to distinguish from prolapsed cloaca, urinary bladder, rectum, or colon (**Figure 8.171**) (De Voe, 2002).

8.15.11 Disease of Non-Gravid Oviduct

Cystic endometrial hyperplasia, a disease generally attributed to abnormal hormonal influence on mucosa epithelium, is reported in birds but there are no reports in non-avian reptiles (Schmidt et al., 2015). The oviduct contains lymphatic tracts that may become distended with lymph in compromised individuals (**Figure 8.172**). Acquired or congenital strictures of the oviduct may affect fertilization, development, birth, and oviposition. Ectatic oviductal mucosal glands filled with inspissated secretions associated with aging or dehydration should not be overinterpreted as cystic endometrial hyperplasia.

8.16 Egg Viability

Numerous factors including fertilization status, temperature, humidity, oxygen, carbon dioxide, and hydrodynamic characteristics of substrate may contribute to egg infertility (Innis and Boyer, 2002). Investigation of fertility problems in captivity requires knowledge of the species' natural history, repticulture and incubation techniques, and review of reproductive records. Eggs that are laid outside the expected time period of oviposition are often nonviable (Barker and Barker, 2006). The reproductive efficiency of wild reptile populations can be affected by fluctuations in temperature, humidity, and rainfall. Complete fertility and hatchability are usually not achieved in the wild or captivity, so knowledge of the normal hatching percentage is necessary to judge the clinical importance of mortality events (Rideout, 2012). Nutritional factors associated with infertility in birds includes vitamin A, vitamin D, niacin, biotin, and iodine deficiencies (St. Leger, 2012), and could also be significant in reptiles, but these factors have not been widely studied.

One of the primary causes for infertile eggs is the failure of ova to contact sperm. Sperm oocyte interaction can be verified with various methods of previtelline membrane-bound sperm detection (**Figure 8.173**) (Croyle et al., 2016; Augustine, 2017). In birds, excessive sperm penetration of the germinal disc can cause early embryonic mortality (Bakst and Howarth, 1977), but this has not been documented in reptiles. Vessels of the chorioalllantois in the fertile egg are often visible when examined with bright backlighting, or "candling" (**Figure 8.174**) (Barker and Barker, 2006; Keller, 2017). Chorionic morphologic features are variable between species; for instance, candled crocodile eggs have distinct banding (**Figures 8.175–8.177**).

Egg necropsy should be performed for cases of increased or unexpected infertility. Mineralization of the shell must be assessed with knowledge of species variability. Extraembryonic membranes evaluated at necropsy include the vitelline or yolk sac membrane, amnion, allantois, and chorion. Knowledge of reptile embryo staging is needed to determine age of embryos. Examination of the entire normal clutch may reveal discrepancy in egg size (even small, pearl-shaped slugs) (**Figure 8.178**), or discoloration of the shell. Discolored eggs are often misshapen and/or collapsed (**Figures 8.179–8.180**). Secondary bacterial and fungal overgrowth can be observed by candling (**Figure 8.181**). Fungal and bacterial organisms are best identified to genus and species by a microbiologist experienced in reptile pathogens, using a combination of culture and molecular techniques (**Figure 8.182**). Autolysis, often advanced by incubation temperature and interval between death and detection of dead embryo, negatively affects diagnostic capability in egg necropsy and this process can be difficult to distinguish from true necrosis or bacterial-driven liquefaction. Determination of fertility or cause of infertility may not be possible in cases of severe bacterial overgrowth (**Figure 8.183**).

8.17 Early Embryonic Death

Early embryonic death generally includes mortality occurring during the first trimester of pregnancy in viviparous reptiles, or prior to or near the time of oviposition in oviparous species. This time period is often difficult to estimate because of embryonic diapause. Most reptiles undergo growth arrest at some time during development. Time periods when arrest may occur include (1) pre-oviposition arrest, (2) delayed hatching and estivation, (3) embryonic diapause after oviposition, or (4) cold torpor encountered at any stage of development (Rafferty and Reina, 2012). Marine and freshwater turtle embryos typically arrest during early development stages—stage 6 of 31 for marine embryos (Miller, 1985) and stage 1 of 26 for freshwater embryos (Ewert, 1985), whereas lizard embryos typically arrest at around stage 30 of a 40-stage developmental chronology (Andrews and Mathies, 2000). Eggs of some species require a cooling period before embryo development will initiate (Sykes, 2010). Embryonic inflammatory response is absent during the early stages of development. Parental health fitness is often the primary factor leading to infertility or early embryonic death. In nonparthenogenic species, embryos contain paternal and maternal genomes; thus the embryo is genetically foreign to the female and must evade the maternal immune system (Hendrawan et al., 2017). Egg development abnormalities, oversized eggs, and preovipositional arrest may occur from decreased maternal oviductal oxygen tension and inappropriate environmental temperature (Rafferty and Reina, 2012). Changes in phenotype can be influenced by nest moisture content in Painted Turtles (*Chrysemys picta*) (Bodensteiner et al., 2015). Hydrodynamics of flexible-shelled eggs, such as the Chinese Skink (*Eumeces chinensis*), may also influence the phenotype of hatchlings (Du, 2004). Decreased environmental oxygen may affect the growth of the embryo during incubation, but aquatic chelonians can tolerate periods of anoxia (Cordero et al., 2017). These conditions may be experienced in subterranean nests or in conditions where eggs become submerged by rain or flood. Hypoxia may also induce diapause (Williamson et al., 2017a,b). Both temperature and oxygen have significant effects on survivability, growth, and embrydifferentiation in the Flatback Sea Turtle (*Natator depressus*) and Ground Skink (*Scandella laterals*) (Flewelling and Parker, 2015; Rings et al., 2015). Toxin exposure may arrest development and is more likely near aquatic environments. High levels of lead in yolk were suggested as a probable cause for early embryonic death in American Alligators (Lance et al., 2006). Abnormalities observed in the eggs of Amazon River Turtle (*Podocnemis expansa*) exposed to oil included eggs without embryos and eggs with increased diameter and abnormal shape (Alves-Júnior et al., 2012). Husbandry practices such as cleaning the outer cuticle of the egg or using cleaning products or fumigants should be considered in epidemiologic investigations of captive infertility or embryonic death. Spatial clutch disorientation—turning the eggs after embryo attachment—will result in early mortality (Grahame et al., 1987; Aubret et al., 2015).

8.18 Late Embryonic Death

Late embryonic death includes mortalities that are most likely to be diagnosed in the final third of development. Radiographs can be used to evaluate viability of late-term reptile embryos in-ovo (**Figure 8.184**). Late embryonic mortality coincides with development of the immune system and ability for the embryo to mount an autogenous inflammatory response. Microscopic and gross lesions observed in late-term neonatal deaths are usually limited to necrosis, heterophilic infiltration, hemorrhage, congestion, and edema (Rideout, 2012). Renal tubular mineralization is a common finding associated with late-term embryonic and perinatal death in birds and reptiles, suggesting dehydration during incubation (M. Garner, personal communication). In oviparous species, late embryonic death is often observed near the time of hatching. Failure to hatch, delayed hatching, and estivation may be influenced by temperature, humidity, substrate, ambient oxygen, and carbon dioxide levels (Innis, 2004; Rafferty and Reina, 2012). Overgrowth by opportunistic organisms secondary to environment disruption may contribute to late fetal death (**Figures 8.185–8.186**) (St. Leger, 2012). Yolk cytology is a rapid method to evaluate presence of pathogenic or opportunistic organisms in the egg (Rideout, 2012). Birds utilize an air chamber within the egg prior to hatching, and malposition may inhibit the neonate's ability to initiate pulmonary respiration (Rideout, 2012). Death from late fetal malposition as seen in birds is not documented in reptiles. Reptiles do, however, transition from chorioallantoic respiration to pulmonary respiration at some point in the hatching process, prompting closure of the ductus arteriosus and release from chorioallantoic circulation (Rideout, 2012). Reptiles can be susceptible to membrane entrapments during hatching. Many congenital defects are grossly obvious, such as amelia or dicephaly, but others may involve internal organs and present a diagnostic challenge (Palmieri et al., 2013; Charruau and Niño-Torres, 2014). Congenital defects account for a large percentage of neonatal deaths (**Figure 8.187**). For additional information regarding congenital defects, see Chapter 4.

8.19 Perinatal Disease

Reptile perinatology refers to diseases occurring in near-term embryos and neonates (Keller, 2017). Partially unretracted yolk sacs and umbilical hernias result from failure to internalize the yolk sac around the time of hatching or live birth. Yolk internalization is dependent on involuntary contraction of smooth muscle in the amnion (**Figure 8.188**) (Pezaro et al., 2013) and may be affected by incubation temperature and humidity (Nechaeva et al., 2005; St. Leger, 2012). Neonates may develop a patent urachus if yolk internalization is incomplete, and this may be associated with omphalophlebitis (**Figures 8.189–8.190**) (Goe et al., 2015). Neonates born with umbilical hernias or open umbilicus may survive but later succumb to bacterial or fungal infections (**Figures 8.191–8.192**). Maternally derived antibodies are transferred to the yolk of the developing egg (Barnes et al., 2016). As a result of metabolic or infectious etiologies, neonates may fail to resorb the yolk from the intracoelomic yolk sac, resulting in yolk sacculitis and predisposing to septicemia (**Figures 8.193–8.194**). In birds, the yolk sac should be completely involuted by 2 weeks, and non-absorption of the yolk sac often results in colonization by opportunistic bacteria (St. Leger, 2012). These time intervals are no doubt variable between reptile species. Birds are susceptible to epigenetic regulation caused by maternal stress. Females with experimentally elevated plasma corticosterone levels produced offspring that are smaller at hatching, less competitive, less fearful, and have impaired spacial orientation and altered hormone levels (Henriksen et al., 2011).

The complex genetic signaling influenced by environmental factors plays a role in all aspects of reptile reproduction. Given the enormous diversity of the Reptilia, the study of reproductive biology and pathogenesis in this class will be greatly expanded in coming years. Declining reptile populations are occurring as a result of global climate change (Jensen et al., 2018), while pollution and habitat destruction underscore the importance of further study in the field of reptile reproductive pathology.

Acknowledgments

The authors would like to thank the biologists, reptile breeders, and veterinarians who graciously answered unsolicited requests and contributed images to this chapter. Those who contributed images are acknowledged in figure legends. John Roberts is also grateful to Auburn University, College of Veterinary Medicine, Thompson Bishop Sparks State Diagnostic Laboratory, Alabama Department of Agriculture and Industries, Smithsonian Institution, National Zoological Park and University of Florida, College of Veterinary Medicine for support.

References

Agnew DW, and MacLacklan NJ. 2017. Tumors of the genital system, in *Tumors in Domestic Animals*, 5th ed. Meuten DJ (Ed.), John Wiley & Sons Inc, Ames, Iowa, pp. 689–722.

Ahmed, N, Yang P, Huang Y et al. 2017. Entosis Acts as a novel way within sertoli cells to eliminate spermatozoa in seminiferous tubule. *Front Physiol* 8:361.

Albergotti LC, and Guillette LJ. 2011. Viviparity in reptiles: Evolution and reproductive endocrinology, in *Hormones and Reproduction of Vertebrates, Reptiles*, Vol 3. Norris DO, and Lopez KH (Eds.), Academic Press: Elsevier, San Diego, California, pp. 247–275.

Allen L, Sanders KL, and Thomson VA. 2018. Molecular evidence for the first records of facultative parthenogenesis in elapid snakes. *R Soc Open Sci* 5(2). doi: 10.1098/rsos.171901. eCollection.

Alves-Júnior JRF, Lustosa APG, Bosso ACS et al. 2012. Reproductive indices in natural nests of giant Amazon river turtles *Podocnemis expansa* (Schweigger, 1812) (Testudines, Podocnemididae) in the Environmental Protection Area Meanders of the Araguaia river. *Braz J Biol* 72:199–203.

Anderson NL, Williams J, Sagartz JE, and Barneswall R. 1996. Ovarian teratoma in a green iguana (*Iguana iguana*). *J Zoo Wildl Med* 27:90–95.

Andrews RM, and Donoghue S. 2004. Effects of temperature and moisture on embryonic diapause of the veiled chameleon (*Chamaeleo calyptratus*). *J Exp Zool A* 301:629–635.

Andrews RM, and Mathies TM. 2000. Natural history of reptilian development: Constraints on the evolution of viviparity. *Bioscience* 50:227–238.

Arukwe A, Myburgh J, Langberg HA et al. 2016. Developmental alterations and endocrine-disruptive responses in farmed Nile crocodiles (*Crocodylus niloticus*) exposed to contaminants from the Crocodile River. *South Africa Aquat Toxicol* 173:83–93.

Ashpole SL, Bishop CA, and Brooks RJ. 2004. Contaminant residues in snapping turtle (*Chelydra s. serpentina*) eggs from the Great Lakes-St. Lawrence River basin (1999 to 2000). *Arch Environ Contam Toxicol* 47:240–252.

Assisi L, Raucci F, and Di Fiore MM. 2011. Seasonal study of apoptotic markers in lizard oviduct. *J Exp Zool A* 315:291–301.

Aubret F, Blanvillain G, and Kok PJ. 2015. Myth busting? Effects of embryo positioning and egg turning on hatching success in the water snake *Natrix maura*. *Sci Rep* doi: 10.1038/srep13385.

Augustine L. 2017. Crocodilian perivitelline membrane-bound sperm detection. *Zoo Biol* 36:213–219.

Bakst MR, and Howarth B Jr. 1977. Hydrolysis of the hen's previteline layer by cock sperm *in vitro*. *Biol Reprod* 17:370–379.

Ball GF, and Wade J. 2013. The value of comparative approaches to our understanding of puberty as illustrated by investigations in birds and reptiles. *Horm Behav* 64:211–214.

Barker DG, and Barker TM. 2006. *Ball Pythons: Their History, Natural History, Care and Breeding*. VPI Library, Boerne, Texas.

Barnes HJ, Fletcher OJ, and Abdul-Aziz T. 2016. Reproductive system, in *Avian Histopathology*. Abdul-Aziz T, Fletcher OJ, and Barnes HJ (Eds.), American Association of Avian Pathologists, Inc, Omnipress, Madison, Wisconsin, pp. 581–614.

Barske LA, and Capel B. 2010. Estrogen represses SOX9 during sex determination in the red-eared slider turtle Trachemys scripta. *Dev Biol* 341:305–314.

Bel L, Tecilla M, Borza G et al. 2016. Diagnosis and surgical management of malignant ovarian teratoma in a green iguana (*Iguana iguana*). *BMC Vet Res* 12(1):144.

Benton MJ. 2014. *Vertebrate Paleontology*, 4th ed. Oxford: Blackwell Science Ltd., Hoboken, New Jersey.

Bergeron JM, Crews D, and McLachlan JA. 1994. PCBs as environmental estrogens: Turtle sex determination as a biomarker of environmental contamination. *Environ Health Perspect* 102:780–781.

Bhandari RK, Deem SL, Holliday DK et al. 2015. Effects of the environmental estrogenic contaminants bisphenol A and 17α-ethinyl estradiol on sexual development and adult behaviors in aquatic wildlife species. *Gen Comp Endocrinol* 214:195–219.

Biazik JM, Parker SL, Murphy CR, and Thompson MB. 2012. Uterine epithelial morphology and progesterone receptors in a mifepristone-treated viviparous lizard *Pseudemoia entrecasteauxii* (Squamata: Scincidae) during gestation. *J Exp Zool B* 318:148–158.

Bickham JW, and Hanks G. 2009. Diploid-triploid mosaicism and tissue ploidy diversity within *Platemys platycephala* from Suriname. *Cytogenet Genome Res* 127:280–286.

Bieser KL, and Wibbels T. 2014. Chronology, magnitude and duration of expression of putative sex-determining/differentiation genes in a turtle with temperature-dependent sex determination. *Sex Dev* 8:364–375.

Blackburn DG. 1992. Convergent evolution of viviparity, matrotrophy, and specialization for fetal nutrition in reptiles and other vertebrates. *Amer Zool* 32:313–321.

Blackburn DG. 1993. Chorioallantoic placentation in squamate reptiles: Structure, function, development, and evolution. *J Exp Zool* 266:414–430.

Blackburn DG, Anderson KE, Aronson KW et al. 2017. Placentation in watersnakes I: Placental histology and development in North American Nerodia (Colubridae: Natricinae). *J Morphol* 278:665–674.

Blanvillain G, Owens DW, and Kuchling G. 2011. Hormones and reproductive cycles in turtles, in *Hormones and Reproduction of Vertebrates, Reptiles*, Vol 3. Norris DO, and Lopez KH (Eds.), Academic Press: Elsevier, San Diego, California, pp. 277–303.

Bleu J, Le Galliard JF, Fitze PS, Meylan S, Clobert J, and Massot M. 2013a. Reproductive allocation strategies: A long-term study on proximate factors and temporal adjustments in a viviparous lizard. *Oecologia* 171:141–151.

Bleu J, Massot M, Haussy C, and Meylan S. 2013b. An experimental study of the gestation costs in a viviparous lizard: A hormonal manipulation. *Physiol Biochem Zool* 86:690–701.

Bluhm AP, Toledo RA, Mesquita FM et al. 2004. Molecular cloning, sequence analysis and expression of the snake follicle-stimulating hormone receptor. *Gen Comp Endocrinol* 137:300–311.

Bodensteiner BL, Mitchell S, Strickland JT, and Janzen FJ. 2015. Hydric conditions during incubation influence phenotypes of neonatal reptiles in the field. *Funct Ecol* 29:1365–2435.

Boggs SP, Botteri NL, Hamlin HJ, and Guillette LJ Jr. 2011. Endocrine disruption of reproduction in reptiles, in *Hormones and Reproduction of Vertebrates, Reptiles*, Vol 3. Norris DO, and Lopez KH (Eds.), Academic Press: Elsevier, San Diego, California, pp. 373–396.

Bonnet X, Naulleau G, and Shine R. 2017. The evolutionary economics of embryonic-sac fluids in squamate reptiles. *Am Nat* 189:333–344.

Booth DT, and Dunstan A. 2018. A preliminary investigation into the early embryo death syndrome (EEDS) at the world's largest green turtle rookery. *PLOS ONE* 13(4):e0195462. eCollection.

Booth W, Smith CF, Eskridge PH, Hoss SK, Mendelson JR, and Schuett GW. 2012. Facultative parthenogenesis discovered in wild vertebrates. *Biol Lett* 23:983–985.

Bowden RM, Ewert MA, Lipar JL, and Nelson CE. 2001. Concentrations of steroid hormones in layers and biopsies of chelonian egg yolks. *Gen Comp Endocrinol* 121(1):95–103.

Boyer DM, and Boyer TH. 1994. Tortoise care. *Bull Assoc Rept Amphib Vet* 4:16–28.

Brock JW, Bell JM, and Guillette LJ Jr. 2016. Urinary phthalate metabolites in American alligators (*Alligator mississippiensis*) from selected Florida wetlands. *Arch Environ Contam Toxicol* 71:1–6.

Brown GP, and Shine R. 2006. Why do most tropical animals reproduce seasonally? Testing hypotheses on an Australian snake. *Ecology* 87:133–143.

Brown MA, Cree A, Daugherty CH, Dawkins BP, and Chambers GK. 1994. Plasma concentrations of vitellogenin and sex steroids in female Tuatara (*Sphenodon punctatus punctatus*) from northern New Zealand. *Gen Comp Endocrinol* 95:201–212.

Buah-Kwofie A, Humphries MS, Combrink X, and Myburgh JG. 2018. Accumulation of organochlorine pesticides in fat tissue of wild Nile crocodiles (*Crocodylus niloticus*) from iSimangaliso Wetland Park, South Africa. *Chemosphere* 195:463–471.

Buhlmann K, Lynch T, Gibbons JW, and Greene J. 1995. Prolonged egg retention in the turtle *Deirochelys reticularia* in South Carolina. *Herpetologica* 51:457–462.

Bull JJ. 1980. Sex determination in reptiles. *Q Rev Biol* 55:3–21.

Bull JJ. 2004. Perspective on sex determination: Past and future, in *Temperature-Dependent Sex Determination in Vertebrates*. Valenzuela N, and Lance VA (Eds.), Smithsonian Institution Press, Washington, District of Columbia, pp. 5–8.

Burger J, and Zappalorti RT. 1988. Effects of incubation temperature on sex ratios in pine snakes: Differential vulnerability of males and females. *Am Nat* 132:492–505.

Burger M, Gehrmann WH, and Ferguson GW. 2007. Evaluation of UVB reduction by materials commonly used in reptile husbandry. *Zoo Biol* 26:417–423.

Callebaut M, Van Nassauw L, and Harrisson F. 1997. Comparison between oogenesis and related ovarian structures in a reptile, *Pseudemys scripta elegans* (turtle) and in a bird *Coturnix coturnix japonica* (quail). *Reprod Nutr Dev* 37:233–252.

Capel B. 2017. Vertebrate sex determination: Evolutionary plasticity of a fundamental switch. *Nat Rev Genet* 18:675–689.

Carrell DT, Emery BR, and Hamilton B. 2003. Seminal infection with *Ralstonia picketti* and cytolysosomal spermophagy in a previously fertile man. *Fertil Steril* 79(Suppl 3):1665–1667.

Cecchetto NR, and Naretto S. 2015. Do sex, body size and reproductive condition influence the thermal preferences of a large lizard? A study in *Tupinambis merianae*. *J Therm Biol* 53:198–204.

Chakraborty A, and Raman R. 2010. Modulation of gene activity in androgen-induced sex reversal in the garden lizard, *Calotes versicolor*. *Sex Dev* 4:162–169.

Charlesworth B. 2000. Fisher, Medawar, Hamilton and the evolution of aging. *Genetics* 156:927–931.

Charnier M. 1966. Action de la temperature sur la sex-ratio chez l'embryon d' *Agama agama* (Agamidae, *Lacertilien*). *C R Seances Soc Biol Fil* 160:620–622.

Charruau P, and Niño-Torres A. 2014. A third case of amelia in Morelet's crocodile from the Yucatan Peninsula. *Dis Aquat Organ* 109:263–267.

Christman J, Devau M, Wilson-Robles H et al. 2017. Oncology of reptiles: Diseases, diagnosis, and treatment. *Vet Clin North Am Exot Anim Pract* 20:87–110.

Congdon JD, Nagle RD, Kinney OM, and van Loben Sels RC. 2001. Hypotheses of aging in a long lived vertebrate, Blanding's turtle (*Emydoidea blandingii*). *Exp Gerontol* 36:813–827.

Congdon JD, Nagle RD, Kinney OM, van Loben Sels RC, Quinter T, and Tinkle DW. 2003. Testing hypotheses of aging in long-lived painted turtles (*Chrysemys picta*). *Exp Gerontol* 38:765–772.

Conley KJ, and Shilton CM. 2018. Crocodilia, in *Pathology of Wildlife and Zoo Animals*. Terio KA, McAloose D, and St. Ledger J (Eds.), Academic Press, Elsevier, Cambridge, Massachusetts, pp. 849–864.

Cordero GA, Karnatz ML, Svendsen JC, and Gangloff EJ. 2017. Effects of low-oxygen conditions on embryo growth in the painted turtle, *Chrysemys picta*. *Integrative Zoology* 12:148–156.

Cornax I, Walzem RL, Larner C, Macfarlane RD, and Klasing KC. 2013. Mobilization of ectopic yolk in *Gallus gallus domesticus*: A novel reverse lipid transport process. *J Exp Biol* 216:1949–1958.

Cox RM, and Calsbeek R. 2010. Severe costs of reproduction persist in Anolis lizards despite the evolution of a single-egg clutch. *Evolution* 64:1321–1330.

Cox RM, McGlothlin JW, and Bonier F. 2016. Hormones as mediators of phenotypic and genetic integration: An evolutionary genetics approach. *Integr Comp Biol* 56:126–137.

Crain DA, Guillette LJ Jr, Rooney AA, and Pickford DB. 1997. Alterations in steroidogenesis in alligators (*Alligator mississippiensis*) exposed naturally and experimentally to environmental contaminants. *Environ Health Perspect* 105:528–533.

Cree A. 2015. Reproductive anatomy and cycles of Tuatara (*Sphenodon punctatus*), an intriguing non-squamate lepidosaur, in *Reproduction Biology and Phylogeny of Lizards and Tuatara*. Rheubert JL, Siegal DS, Trauth SE, and Jamieson BGM (Eds.), CRC Press-Taylor and Francis Group, Boca Raton, Florida, pp. 621–646.

Crespo JL, García-Párraga D, Giménez I et al. 2013. Two cases of pseudohermaphroditism in loggerhead sea turtles *Caretta caretta*. *Dis Aquat Organ* 105:183–191.

Crews D. 1984. Gamete production, sex hormone secretion, and mating behavior uncoupled. *Horm Behav* 18:22–28.

Crews D, Bergeron JM, and McLachlan JA. 1995. The role of estrogen in turtle sex determination and the effect of PCBs. *Environ Health Perspect* 103(Suppl 7):73–77.

Crews D, Fleming A, Willingham E, Baldwin R, and Skipper JK. 2001. Role of steroidogenic factor 1 and aromatase in temperature-dependent sex determination in the red-eared slider turtle. *J Exp Zool* 290:597–606.

Crews D, and McLachlan JA 2006. Epigenetics, evolution, endocrine disruption, health, and disease. *Endocrinology* 147(6 Suppl):S4–S10.

Crews D, and Moore MC. 1993. Psychobiology of reproduction of parthenogenetic whiptail lizards, in *Biology of Whiptail Lizards (Genus Cnemidophorus)*. Wright JW, and Vitt LJ (Eds.), Herpetologist's League Special Publication No. 3, Oklahoma Museum of Natural History, Norman, Oklahoma, pp. 257–282.

Croyle K, Gibbons P, Light C, Goode E, Durrant B, and Jensen T. 2016. Chelonian perivitelline membrane-bound sperm detection: A new breeding management tool. *Zoo Biol* 35:95–103.

Cruz-Cardona JA, Conley KJ, Wellehan JF, Farina LL, Origgi FC, and Wamsley HL. 2011. Incomplete ovariosalpingectomy and subsequent malignant granulosa cell tumor in a female green iguana (*Iguana iguana*). *J Am Vet Med Assoc* 239:237–242.

Cruze L, Hamlin HJ, Kohno S, McCoy MW, and Guillette LJ Jr. 2013. Evidence of steroid hormone activity in the chorioallantoic membrane of a Turtle (*Pseudemys nelsoni*). *Gen Comp Endocrinol* 186:50–57.

Cruze L, Roark AM, Rolland G, Younas M, Stacy N, Guillette LJ Jr. 2015. Endogenous and exogenous estrogens during embryonic development affect timing of hatch and growth in the American alligator (*Alligator mississippiensis*). *Comp Biochem Physiol B Biochem Mol Biol.* 184:10–18.

Cunningham GD, Fitzpatrick LJ, While GM, and Wapstra E. 2018. Plastic rates of development and the effect of thermal extremes on offspring fitness in a cold-climate viviparous lizard. *J Exp Zool A* 329:262–270.

Czerwinski M, Natarajan A, Barske L, Looger L, and Capel BA. 2016. Timecourse analysis of systemic and gonadal effects of temperature on sexual development of the red-eared slider turtle. *Dev Biol* 420:166–177.

Da Silva N, and Smith TB. 2015. Exploring the role of mononuclear phagocytes in the epididymis. *Asian J Androl* 17:591–596.

Deeming DC. 2018. Nesting environment may drive variation in eggshell structure and egg characteristics in the Testudinata. *J Exp Zool A Ecol Integr Physiol* 329(6–7):331–342.

de Haan CC. 2003. Extrabuccal infralabial secretion outlets in *Dromophis*, *Mimophis* and *Psammophis* species (Serpentes, Colubridae, Psammophiini). A probable substitute for "self-rubbing" and cloacal scent gland functions, and a cue for a taxonomic account. *C R Biol* 326:275–286.

Demetropoulos A, and Hadjichristophorou M. 1995. *Manual on Marine Turtle Conservation in the Mediterranean*. UNEP, Coordinating Unit for the Mediterranean Action Plan (MAP) IUCN Cyprus, Ministry of Agriculture, Natural Resources and Environment, Cyprus Wildlife Society, Cyprus.

DeNardo D. 2006a. Reproductive biology, in *Reptile Medicine and Surgery*, Vol 2. Mader DR (Ed.), Saunders: Elsevier, Philadelphia, Pennsylvania, pp. 376–390.

DeNardo D. 2006b. Dystocias, in *Reptile Medicine and Surgery*, Vol 2. Mader DA (Ed.), Saunders: Elsevier, Philadelphia, Pennsylvania, pp. 787–792.

Derickson WK. 1976. Lipid storage and utilization in reptiles. *Am Zool* 16:711–723.

De Voe RS. 2002. The herptile cloaca: Anatomy, function and disease. *Proc Assoc Rep Amphib Vet*, Reno Nevada.

Di Girolamo N, and Selleri P. 2017. Reproductive disorders in snakes. *Vet Clin North Am Exot Anim Pract* 20:391–409.

Divers SJ. 1996. Medical and surgical treatment of pre-ovulatory ova stasis and post-ovulatory egg stasis in oviparous lizards. *Proc Assoc Amphib Rept Vet*, Tampa, Florida.

Du WG. 2004. Water exchange of flexible-shelled eggs and its effect on hatchling traits in the Chinese skink, *Eumeces chinensis*. *J Comp Physiol B* 174:489–493.

Dubois W, and Callard IP. 1985. Characterization of the testicular binding site for iodinated rat FSH in the turtle, *Chrysemys picta*. *Comp Biochem Physiol A* 82:891–898.

Dubois W, Pudney J, and Callard IP. 1988. Annual testicular cycle in the turtle, *Chrysemys picta*: A histochemical and electron microscopic study. *Gen Comp Endocrinol* 71:191–204.

Dufaure JP, and Hubert J. 1961. Table de développment du lézard Vivipare: *Lacerta (Zootoca) vivipara* Jacquin. *Archiv Anat Microscop Morphol Exp* 50:309–328.

Duncan EM, Broderick AC, and Fuller WJ. 2019. Microplastic ingestion ubiquitous in marine turtles. *Glob Chang Biol* 25(2):744–752.

Durando M, Canesini G, Cocito LL et al. 2016. Histomorphological changes in testes of broad-snouted caimans (*Caiman latirostris*) associated with in ovo exposure to endocrine-disrupting chemicals. *J Exp Zool A* 325:84–96.

Durando M, Cocito L, Rodríguez HA et al. 2013. Neonatal expression of amh, sox9 and sf-1 mRNA in *Caiman latirostris* and effects of in ovo exposure to endocrine disrupting chemicals. *Gen Comp Endocrinol* 191:31–38.

Duval D, Guillette LJ Jr, and Jones RE. 1982. Environmental control of reptilian reproduction cycles, in *The Biology of the Reptilia*, Vol 13. Gans C, and Pough FH (Eds.), Wiley and Sons, New York, New York, pp. 201–223.

Eisenreich KM, Kelly SM, and Rowe CL. 2009. Latent mortality of juvenile snapping turtles from the Upper Hudson River, New York, exposed maternally and via the diet to polychlorinated biphenyls (PCBs). *Environ Sci Techno* 43:6052–6057.

Eisler R. 2004. Mercury hazards from gold mining to humans, plants, and animals. *Rev Environ Contam Toxicol* 181:139–198.

Etheridge K, Wit LC, Sellers JC, and Trauth SE. 1986. Seasonal changes in reproductive condition and energy stores in *Cnemidophorus sexlineatus*. *J Herpetol* 20:554–559.

Evans TJ. 2018. Reproductive toxicology and endocrine disruption, in *Veterinary Toxicology*. Gupta RC (Ed.), Elsevier, Academic Press, Cambridge, Massachusetts, pp. 273–316.

Ewert MA. 1985. Embryology of turtles, in *Biology of the Reptilia*. Gans C, Billett F, and Maderson PFA (Eds.), John Wiley & Sons, New York, New York, pp. 75–268.

Ezaz T, Sarre SD, O'Meally D, Graves JA, and Georges A. 2009. Sex chromosome evolution in lizards: Independent origins and rapid transitions. *Cytogenet Genome Res* 127:249–260.

Ferguson GW, Gehrmann WH, Chen TC, Dierenfeld ES, and Holick MF. 2002. Effects of artificial ultraviolet light exposure on reproductive success of the female panther chameleon (*Furcifer pardalis*) in captivity. *Zoo Biol* 21:525–537.

Flewelling S, and Parker SL. 2015. Effects of temperature and oxygen on growth and differentiation of embryos of the ground skink, *Scincella lateralis*. *J Exp Zool A* 323:445–455.

Foster RA. 2012. Male reproductive system, in *Pathologic Basis of Veterinary Disease*. Zachary JF, and McGavin MD (Eds.), Mosby: Elsevier, Maryland Heights, Missouri, pp. 1127–1151.

Fox W. 1963. Special tubules for sperm storage in female lizards. *Nature* 198:500–501.

French SS, Dearing MD, and Demas GE. 2011. Leptin as a physiological mediator of energetic trade-offs in ecoimmunology: Implications for disease. *Integr Comp Biol* 51:505–513.

French SS, DeNardo DF, and Moore MC. 2007. Trade-offs between the reproductive and immune systems: Facultative responses to resources or obligate responses to reproduction? *Am Nat* 170:79–89.

Frye FL. 1994. The diagnosis and surgical treatment of reptilian neoplasia with a compilation of cases 1966–1993. *In Vivo* 8:885–892.

Frye FL, Dybdal NO, and Harshberger JC. 1988. Testicular interstitial cell tumor in a desert tortoise (*Gopherus agassizii*). *J Zoo Anim Med* 19:55–58.

Funk RS. 2002. Reptile reproduction. *Proceedings of Western Veterinary Conference*, Las Vegas.

Galoppo GH, Canesini G, Tavalieri YE et al. 2017. Bisphenol A disrupts the temporal pattern of histofunctional changes in the female reproductive tract of *Caiman latirostris*. *Gen Comp Endocrinol* 254:75–85.

Galoppo GH, Stoker C, Canesini G et al. 2016. Postnatal development and histofunctional differentiation of the oviduct in the broad-snouted caiman (*Caiman latirostris*). *Gen Comp Endocrinol* 236:42–53.

Gamble T. 2010. A review of sex determining mechanisms in geckos (Gekkota, Squamata). *Sex Dev* 4:88–103.

Gamble T, Castoe TA, Nielsen SV et al. 2017. The discovery of XY sex chromosomes in a boa and python. *Curr Biol* 27:2148–2153.

Garcia-Gasca A, and Spyropoulos DD. 2000. Differential mammary morphogenesis along the anteroposterior axis in Hoxc6 gene targeted mice. *Dev Dyn* 219:261–276.

Garner MM, Gyimesi ZS, Rasmussen JM et al. 2003. An amyloid-like deposition disorder in Komodo dragons. *Proceedings, Association of Reptilian and Amphibian Veterinarians*, Minneapolis, Minnesota.

Garner MM, Hernandez-Divers SM, and Raymond JT. 2004. Reptile neoplasia: A retrospective study of case submissions to a specialty diagnostic service. *Vet Clin North Am Exotic Anim Pract* 7:653–671.

Gautam M, Mathur A, Khan MA, Majumdar S, and Rai U. 2013. Transcriptome analysis of spermatogenically regressed, recrudescent an active phase testis of seasonally breeding wall lizards *Hemidactylus flaviviridis*. *PLOS ONE* 8(3):e58276.

Ge C, Ye J, Zhang H et al. 2017. Dmrt1 induces the male pathway in a turtle species with temperature-dependent sex determination. *Development* 144:2222–2233.

Georges A, Limpus C, and Stoutjeskijk R. 1994. Hatchling sex in the marine turtle *Caretta caretta* is determined by proportion of development at a temperature, not daily duration of exposure. *J Exp Zool* 270:432–444.

Georges A, Ezaz T, Quinn AE, and Sarre SD. 2010. Are reptiles predisposed to temperature-dependent sex determination? *Sex Dev* 4:7–15.

Gibbons P, and Schiller C. 2000. What's your diagnosis: Ovarian adenocarcinoma in a green iguana (*Iguana iguana*)? *J Herpetol Med Surg* 10:34–38.

Gist DH. 2011. Hormones and the sex ducts and sex accessory structures of reptiles, in *Hormones and Reproduction of Vertebrates, Reptiles*, Vol 3. Norris DO, and Lopez KH (Eds.), Academic Press: Elsevier, San Diego, California, pp. 117–139.

Goe A, Heard DJ, Fredholm DV, Stacy NI, McCoy L, and Wellehan JF. 2015. Management of a patent urachus and yolk coelomitis in a prehensile-tailed skink (*Corucia zebrata*). *J Zoo Wildl Med* 46:909–912.

Graham SP, Earley RL, Guyer C, and Mendonça MT. 2011. Innate immune performance and steroid hormone profiles of pregnant versus nonpregnant cottonmouth snakes (*Agkistrodon piscivorus*). *Gen Comp Endocrinol* 174:348–353.

Grahame JW, Webb SC, Manolis P, Whitehead J, and Dempsey K. 1987. The possible relationship between embryo orientation opaque banding and the dehydration of albumen in crocodile eggs. *Copeia* 1987:252–257.

Greene HW. 1969. Fat storage in females of an introduced lizard, *Hemidactylus trucicus*, from *Texas*. *Texas J Sci* 21:233–235.

Gribbins KM. 2011. Reptilian spermatogenesis: A histological and ultrastructural perspective. *Spermatogenesis* 1:250–269.

Gribbins KM, Gist DH, and Congdon JD. 2003. Cytological evaluation of spermatogenesis and organization of the germinal epithelium in the male slider turtle, *Trachemys scripta*. *J Morphol* 255:337–346.

Gribbins KM, and Rheubert JL. 2011. The ophidian testis, spermatogenesis, and mature spermatozoa, in *Reproductive Biology and Phylogeny of Snakes*, Vol 9. Aldridge RD, and Sever DM (Eds.), CRC Press, Taylor & Francis Group, Boca Raton, Florida, pp. 183–264.

Guillette LJ Jr, and Cree A. 1997. Morphological changes in the corpus luteum of Tuatara (*Sphenodon punctatus*) during gravidity. *J Morphol* 232:79–91.

Guillette LJ Jr, Gross TS, Masson GR, Matter JM, Percival HF, and Woodward AR. 1994. Developmental abnormalities of the gonad and abnormal sex hormone concentrations in juvenile alligators from contaminated and control lakes in Florida. *Environ Health Perspect* 102:680–688.

Guillette LJ Jr, and Moore BC. 2006. Environmental contaminants, fertility, and multioocytic follicles: A lesson from wildlife? *Semin Reprod Med* 24:134–141.

Guillette LJ Jr, Woodward AR, Crain DA et al. 1997. The reproductive cycle of the female American alligator (*Alligator mississippiensis*). *Gen Comp Endocrinol* 108:87–101.

Guraya SS. 1989. Ovarian follicles in reptiles and birds. *Zoophysiol* 24:1–285.

Gutzke WH, and Crews D. 1988. Embryonic temperature determines adult sexuality in a reptile. *Nature* 332:832–834.

Hale MD, Galligan TM, Rainwater TR et al. 2017. AHR and CYP1A expression link historical contamination events to modern day developmental effects in the American alligator. *Environ Pollut* 230:1050–1061.

Hall JM, Buckelew A, Lovern M, Secor S, and Warner DA. 2018. Seasonal shifts in reproduction depend on prey availability for an income breeder. *Physiol Biochem Zool* 91:1129–1147.

Hall JM, and Warner DA. 2018. Thermal spikes from the urban heat island increase mortality and alter physiology of lizard embryos. *J Exp Biol* doi: 10.1242/jeb.181552.

Hare KM, and Cree A. 2010. Incidence, causes and consequences of pregnancy failure in viviparous lizards: Implications for research and conservation settings. *Reprod Fertil Dev* 22:761–770.

Hayes TB, Collins A, Lee M et al. 2002. Hermaphroditic, demasculinized frogs after exposure to the herbicide atrazine at low ecologically relevant doses. *Proc Natl Acad Sci USA* 99:5476–5480.

Hedley J, and Eatwell K. 2014. Cloacal prolapses in reptiles: A retrospective study of 56 cases. *J Small Anim Pract* 55:265–268.

Hendrawan K, Whittington CM, Brandley MC, Belov K, and Thompson MB. 2017. The Regulation of Uterine Proinflammatory Gene Expression during Pregnancy in the Live-Bearing Lizard, *Pseudemoia entrecasteauxii*. *J Exp Zool B Mol Dev Evol* 328(4):334–346.

Henriksen R, Rettenbacher S, and Groothuis T. 2011. Maternal corticosterone elevation during egg formation in chickens (*Gallus gallus domesticus*) influences offspring traits, partly via prenatal undernutrition. *Gen Comp Endocrinol* 191:83–91.

Hertha DW, Giles HD, and Frye FL. 2000. Ovarian teratoma in a garter snake, *Thamnophis sirtalis*. *J Herp Med Surg* 10:22–23.

Highfield AC. 1996. *Practical Encyclopedia of Keeping and Breeding Tortoises and Freshwater Turtles*. Carpace Press, Carmarthen, United Kingdom.

Holleley CE, O'Meally D, Sarre SD et al. 2015. Sex reversal triggers the rapid transition from genetic to temperature-dependent sex. *Nature* 523:79–82.

Holleley CE, Sarre SD, O'Meally D, and Georges A. 2016. Sex reversal in reptiles: Reproductive oddity or powerful driver of evolutionary change? *Sex Dev* 10:279–287.

Hopkins BC, Willson JD, and Hopkins WA. 2013. Mercury exposure is associated with negative effects on turtle reproduction. *Environ Sci Technol* 47:2416–2422.

Huchzermeyer FW. 2003. *Crocodiles, Biology, Husbandry and Disease*. CABI Publishing/CAB International, Wallingford, Oxfordshire, England.

Innis CJ, and Boyer TH. 2002. Chelonian reproductive disorders. *Vet Clin Exot Anim* 5:555–578.

Innis CJ. 2004. Infertility and embryonic death, in *Medicine and Surgery of Tortoises and Turtles*. McArthur S, Wilkinson R, and Meyer J (Eds.), Oxford: Blackwell Publishing, Hoboken, New Jersey, pp. 63–68.

Jandegian CM, Deem SL, Bhandari RK et al. 2015. Developmental exposure to bisphenol A (BPA) alters sexual differentiation in painted turtles (*Chrysemys picta*). *Gen Comp Endocrinol* 216:77–85.

Janeiro-Cinquini TR, Ribolla PE, Capurro Mde L, and Winter CE. 1999. Vitellogenic and yolk protein processing in *Bothrops jararaca* (Wied), a tropical venomous snake. *Comp Biochem Physiol B* 122:189–198.

Jenison G, and Nolte J. 1980. An ultraviolet-sensitive mechanism in the reptilian parietal eye. *Brain Res* 194:506–510.

Jensen MP, Allen CD, Eguchi T et al. 2018. Environmental Warming and Feminization of One of the Largest Sea Turtle Populations in the World. *Curr Biol* 28(1):154–159.

Jerez A, and Ramírez-Pinilla MP. 2003. Morphogenesis of extraembryonic membranes and placentation in *Mabuya mabouya* (Squamata, Scincidae). *J Morphol* 258:158–178.

Johnson AL. 2014. The avian ovary and follicle development: Some comparative and practical insights. *Turk J Vet Anim Sci* 38:660–669.

Johnson MA, and Wade J. 2011. Neuroendocrinology of reptilian reproductive behavior, in *Hormones and Reproduction of Vertebrates, Reptiles*, Vol 3. Norris DO, and Lopez KH (Eds.), Academic Press: Elsevier, San Diego, California, pp. 39–61.

Johnston SD, Lever J, McLeod R et al. 2014. Semen collection and seminal characteristics of the Australian salt water crocodile (*Crocodylus porosus*). *Aquaculture* 423:25–35.

Jones SM. 2011. Hormone regulation of ovarian function in reptiles, in *Hormones and Reproduction of Vertebrates, Reptiles*, Vol 3. Norris DO, and Lopez KH (Eds.) Academic Press: Elsevier, San Diego, California, pp. 89–115.

Jones RE, Austin HB, Lopez KH, Rand MS, and Summers CH. 1988. Gonadotropin-induced ovulation in a reptile (*Anolis carolinensis*): Histological observations. *Gen Comp Endocrinol* 72:312–322.

Josserand R, Dupoué A, Agostini S, Haussy C, Le Galliard JF, and Meylan S. 2017. Habitat degradation increases stress-hormone levels during the breeding season, and decreases survival and reproduction in adult common lizards. *Oecologia* 184:75–86.

Kato T, Matsui S, Terai Y et al. 2017. Male-specific mortality biases secondary sex ratio in Eurasian tree sparrows *Passer montanus*. *Ecol Evol* 7:10675–10682.

Keller KA. 2017. Reptile perinatology. *Vet Clin Exot Anim* 20:439–454.

Kennet R, Christian K, and Bedford G. 1998. Underwater nesting by the Australian freshwater turtle *Chelonia rugosa*: Effect of prolonged immersion and eggshell thickness on incubation period, egg survivorship, and hatchling size. *Can J Zool* 76:1019–1023.

Khan UW, and Rai U. 2008. Paracrine role of testicular macrophages in control of Leydig cell activities in the wall lizard, *Hemidactylus flaviviridis*. *Gen Comp Endocrinol* 156:44–50.

Klosterman LL. 1983. The ultrastructure of germinal beds in the ovary of *Gerrhonotus coeruleus* (Reptilia: Anguidae). *J Morphol* 178:247–265.

Knotek Z, Cermakova E, and Oliveri M. 2017. Reproductive medicine in lizards. *Vet Clin North Am Exot Anim Prat* 20:411–438.

Kodavanti PRS, Valdez MC, Yang J, and Curras-Collazo M. 2018. Polycholorinated biphenyls, polybrominated biphenyls, polycholorinated dibenzo-p-dioxins and polychlorinated dibenzofurans, in *Veterinary Toxicology*. Gupta RC (Ed.), Elsevier, Academic Press, Cambridge, Massachusetts, pp. 675–690.

Koyanagi R, and Honegger TG. 2003. Molecular cloning and sequence analysis of an ascidian egg beta-N-acetylhexosaminidase with a potential role in fertilization. *Dev Growth Differ* 45:209–218.

Kuchling G, and Razandrimamilafiniarivo OC. 1999. The use of ultrasound scanning to study the relationship of vitellogenesis, mating, egg production and follicular atresia in captive ploughshare tortoises, *Geochelone yniphora*. *Dodo* 35:109–115.

Kumar S, Brototi R, and Rai U. 2011. Hormonal regulation of testicular function in reptiles, in *Hormones and Reproduction of Vertebrates, Reptiles*, Vol 3. Norris DO, and Lopez KH (Eds.), Academic Press, Elsevier, San Diego, California, pp. 63–84.

Kummrow MS, Smith DA, Crawshaw G, and Mastromonaco GF. 2010. Characterization of fecal hormone patterns associated with the reproductive cycle in female veiled chameleons (*Chamaeleo calyptratus*). *Gen Comp Endocrinol* 168:340–348.

Lai YH, Ding YJ, Moses D, and Chen YH. 2017. Teratogenic Effects of Topiramate in a Zebrafish Model. *Int J Mol Sci* doi: 10.3390/ijms18081721.

Laloi D, Eizaguirre C, Fédérici P, and Massot M. 2011. Female choice for heterozygous mates changes along successive matings in a lizard. *Behav Processes* 88:149–154.

Lance VA, and Bogart MH. 1992. Disruption of ovarian development in alligator embryos treated with an aromatase inhibitor. *Gen Comp Endocrinol* 86:59–71.

Lance VA, Horn TR, Elsey RM, and de Peyster A. 2006. Chronic incidental lead ingestion in a group of captive-reared alligators (*Alligator mississippiensis*): Possible contribution to reproductive failure. *Comp Biochem Physiol C* 142:30–35.

Langer S, Ternes K, Widmer D, and Mutschmann F. 2014. The first case of intersexuality in an African dwarf crocodile (*Osteolaemus tetraspis*). *Zoo Biol* 33:459–462.

Laughran LJ, Larsen JH Jr, and Schroeder PC. 1981. Ultrastructure of developing ovarian follicles and ovulation in the lizard *Anolis carolinensis* (Reptilia). *Zoomorphol* 98:191–208.

Lee MSY. 2013. Turtle origins: Insights from phylogenetic retrofitting and molecular scaffolds. *J Evol Biol* 26:2729–2738.

Levine B. 2004. Treatment of a malignant ovarian teratoma in a green iguana (*Iguana iguana*). *Exot DVM* 6:12–14.

Limpus CJ. 1982. Intersexuality in a loggerhead turtle *Caretta caretta*. *Herpetol Rev* 13:323–333.

Limpus CJ, Limpus DJ, Read MA, and Fitz Simmons NN. 2009. When is a male turtle not a male? Observations on intersex turtles. *Chel Conserv Biol* 8:102–105.

Lin M, Thorne MH, Martin IC, Sheldon BL, and Jones RC. 1995. Development of the gonads in the triploid (ZZW and ZZZ) fowl, *Gallus domesticus*, and comparison with normal diploid males (ZZ) and females (ZW). *Reprod Fertil Dev* 7:1185–1197.

Lloyd ML. 1990. Reptile dystocias review—causes, prevention, management and comments on the synthetic hormone vasotocin. *Proc Assoc Zoo Vet, South Padre Island*, Texas, pp. 290–294.

Lorioux S, Angelier F, and Lourdais O. 2016. Are glucocorticoids good indicators of pregnancy constraints in a capital breeder? *Gen Comp Endocrinol* 232:125–133.

Loudon FK, Spencer RJ, Strassmeyer A, and Harland K. 2013. Metabolic circadian rhythms in embryonic turtles. *Integ Comp Biol* 53:175–182.

Lovern MB. 2011. Hormones and reproductive cycles in lizards, in *Hormones and Reproduction of Vertebrates, Reptiles*, Vol 3. Norris DO, and Lopez KH (Eds.), Elsevier, Academic Press, San Diego, California, pp. 321–355.

Lovern MB, and Wade J. 2003a. Yolk testosterone varies with sex in eggs of the lizard, *Anolis carolinensis*. *J Exp Zool* 295A:206–210.

Lovern MB, and Wade J. 2003b. Sex steroids in green anoles (*Anolis carolinensis*): Uncoupled maternal plasma and yolking follicle concentrations, potential embryonic steroidogenesis, and evolutionary implications. *Gen Comp Endocrinol* 134:109–115.

Lozano A, Ramírez-Bautista A, and Uribe MC. 2014. Oogenesis and ovarian histology in two populations of the viviparous lizard *Sceloporus grammicus* (Squamata: Phrynosomatidae) from the central Mexican Plateau. *J Morphol* 275:949–960.

Lutz D. 2005. *Tuatara: A Living Fossil.* DIMI Press, Salem, Oregon.

Ma L, Buckley LB, Huey RB, and Du WG. 2018. A global test of the cold-climate hypothesis for the evolution of viviparity of squamate reptiles. *Glob Ecol Biogeogr* 27:679–689.

Machotka SV, Wisser J, Ippen R, and Nawab E. 1992. Report of dysgerminoma in the ovaries of a snapping turtle (*Chelydra serpentina*) with discussion of ovarian neoplasms reported in reptilians and women. *In Vivo* 6:349–354.

Mallozzi M, Leone C, Manurita F, Bellati F, and Caserta D. 2017. Endocrine disrupting chemicals and endometrial cancer: An overview of recent laboratory evidence and epidemiological studies. *Int J Environ Res Public Health* 14(3). pii: E334. doi: 10.3390/ijerph14030334.

Mans C, and Sladky KK. 2012. Diagnosis and management of oviductal disease in three red-eared slider turtles (*Trachemys scripta elegans*). *J Small Anim Pract* 53:234–239.

Martín-Del-Campo R, Sifuentes-Romero I, García-Gasca A. 2019. Hox Genes in Reptile Development, Epigenetic Regulation, and Teratogenesis. *Cytogenet Genome Res* 157(1-2):34–45.

Martínez-Torres M, Elena M, Hernández-Caballero M, Alvarez-Rodriguez C, Alba Luis-Díaz J, and Ortíz-López G. 2003. Luteal development and progesterone levels during pregnancy of the viviparous temperate lizard *Barisia imbricata imbricata* (Reptilia: Anguidae). *Gen Comp Endocrinol* 132:55–65.

Martínez-Torres M, Salcedo-Álvarez M, Alvarez-Rodríguez C, Cárdenas-León M, Luis J, and Moreno-Fierros L. 2014. Does the conceptus of the viviparous lizard *Barisia imbricata imbricata* participates in the regulation of progesterone production and the control of luteolysis? *Anim Reprod Sci* 8(3–4):212–220.

Martorell J, Soto S, Barrera S, and Ramis A. 2009. Case report: Ovarian teratoma in a Mediterranean tortoise. *Compend Contin Educ Vet* 31:193–196.

Matsumoto Y, Hannigan B, and Crews D. 2014. Embryonic PCB exposure alters phenotypic, genetic, and epigenetic profiles in turtle sex determination, a biomarker of environmental contamination. *Endocrinol* 155:4168–4177.

Matter JM, Crain DA, Sills-McMurry C et al. 1998. Effects of endocrine-disrupting contaminants in reptiles: Alligators, in *Principles and Processes for Evaluating Endocrine Disruption in Wildlife.* Kendall RD, Giesy J, and Suk W (Eds.), SETA Press, Pensacola, Florida, pp. 267–289.

McArthur SDJ. 2004. Follicular stasis, in *Medicine and Surgery of Tortoises and Turtles.* McArthur S, Wilkinson R, and Meyer J (Eds.), Blackwell Publishing, Oxford, pp. 325–329.

McEwen BS, and Wingfield JC. 2010. What is in a name? Integrating homeostasis, allostasis and stress. *Horm Behav* 57:105–111.

McFarland CA, Quinn MJ Jr, Bazar MA, Remick AK, Talent LG, and Johnson MS. 2008. Toxicity of oral exposure to 2, 4, 6–trinitrotoluene in the western fence lizard (*Sceloporus occidentalis*). *Environ Toxicol Chem* 27:1102–1111.

McWilliams SR, Guglielmo C, Pierce B, and Klaassen M. 2004. Flying, fasting, and feeding in birds during migration: A nutritional and physiological ecology perspective. *J Avian Biol* 35:377–393.

Medini R, Bhagya M, and Ravindra PV. 2018. Expression of β-hexosaminidase in the male reproductive system of the lizard, Eutropis carinata. (Reptilia, Squamata) (Schneider, 1801). *Cell Tissue Res* 374:413–421.

Michaels SJ, and Saneck R. 1988. Undifferentiated carcinoma in the ovary of a boa constrictor (*Boa constrictor ortoni*). *J Zoo Wildl Med* 19:237–240.

Milnes MR. 2011. Hormones and reproductive cycles in crocodilians, in *Hormones and Reproduction of Vertebrates, Reptiles,* Vol 3. Norris DO, and Lopez KH (Eds.), Academic Press: Elsevier, San Diego, California, pp. 305–319.

Miller JD. 1985. Embryology of marine turtles, in *Biology of the Reptilia,* Vol 14. Gans C, Billett F, Maderson PFA (Eds.), John Wiley & Sons, New York, New York, pp. 269–328.

Miller SM, Serrano MA, and Garner MM. 2003. Ectopic ovum in a hermaphrodite common green iguana, *Iguana iguana. J Herpetol Med Surg* 13:23–26.

Ming-Ch'eng Adams CI, Baker JE, and Kjellerup BV. 2016. Toxicological effects of polychlorinated biphenyls (PCBs) on freshwater turtles in the United States. *Chemosphere* 154:148–154.

Minucci S, Fasano S, Marmorino C, Chieffi P, and Pierantoni R. 1995. Ethane 1,2-dimethane sulfonate effects on the testis of the lizard, *Podarcis s. sicula* Raf: Morphological and hormonal changes. *Gen Comp Endocrinol* 97:273–282.

Miyagawa S, Yatsu R, Kohno S et al. 2015. Identification and characterization of the androgen receptor from the American alligator, *Alligator mississippiensis. Endocrinol* 156:2795–2806.

Mizushima S. 2017. Fertilization 2: Polyspermic Fertilization. *Adv Exp Med Biol* 1001:105–123.

Moll EO. 1979. Reproductive cycles and adaptations in turtles, in *Perspectives and Research. In Turtles: Perspectives and Research.* Harless M, and Morlock H (Eds.), John Wiley & Sons Inc, New York, New York, pp. 305–331.

Moore IT, and Jessop TS. 2003. Stress, reproduction, and adrenocortical modulation in amphibians and reptiles. *Horm Behav* 43:39–47.

Moore MC, Hews DK, and Knapp R. 1998. Hormonal control and evolution of alternative male phenotypes. *Am Zool* 38:133–151.

Murray CM, Merchant M, Easter M et al. 2017. Detection of a synthetic sex steroid in the American crocodile (*Crocodylus acutus*): Evidence for a novel environmental androgen. *Chemosphere* 180:125–129.

Neaves WB, and Baumann P. 2011. Unisexual reproduction among vertebrates. *Trends Genet* 27:81–88.

Nechaeva MV, Makarenko IG, Tsitrin EB, and Zhdanova NP. 2005. Physiological and morphological characteristics of the rhythmic contractions of the amnion in veiled chameleon (*Chamaeleo calyptratus*) embryogenesis. *Comp Biochem Physiol A* 140:19–28.

Neuwald JL, and Valenzuela N. 2011. The lesser known challenge of climate change: Thermal variance and sex-reversal in vertebrates with temperature-dependent sex determination. *PloS ONE* 6(3):e18117.

Newman SJ, Brown CJ, and Patnaik AK. 2003. Malignant ovarian teratoma in a red-eared slider (*Trachemys scripta elegans*). *J Vet Diagn Invest* 15:77–81.

Nijagal BS, and Yajurvedi HN. 1999. Influence of corticosterone on FSH-induced ovarian recrudescence in the lizard *Mabuya carinata*. *Gen Comp Endocrinol* 115:364–369.

Nogueira Kde O, Rodrigues SS, Araújo VA, and Neves CA. 2011. Oviductal structure and ultrastructure of the oviparous gecko, *Hemidactylus mabouia* (Moreau De Jonnès, 1818). *Anat Rec (Hoboken)* 294:883–892.

Nussbaum RA. 1980. The brahminy blind snake (*Ramphotyphlops braminus*) in the Seychelles Archipelago: Distribution, variation, and further evidence for parthenogenesis. *Herpetologica* 36:215–221.

Olmo E. 2005. Rate of chromosome changes and speciation in reptiles. *Genetica* 125:185–203.

Olsson M, and Shine R. 1996. Does reproductive success increase with age or with size in species with indeterminate growth? A case study using sand lizards (*Lacerta agilis*). *Oecologia* 105:175–178.

Olsson M, Pauliny A, Wapstra E et al. 2011. Sexual differences in telomere selection in the wild. *Mol Ecol* 20:2085–2099.

Onderka DK, and Zwart P. 1978. Granulosa cell tumor in a garter snake (*Thamnophis sirtalis*). *J Wildl Dis* 14:218–221.

Oonincx D, and van Leeuwen J. 2017. Evidence-based reptile housing and nutrition. *Vet Clin North Am Exot Anim Pract* 20:885–898.

Orrell KS, Congdon JD, Jenssen TA, Michener RH, and Kunz TH. 2004. Intersexual differences in energy expenditure of *Anolis carolinensis* lizards during breeding and postbreeding seasons. *Physiol Biochem Zool* 77:50–64.

Packard GC, and Packard MJ. 1988. The physiological ecology of reptilian eggs and embryos, in *Biology of the Reptilia*, Vol 16. Gans C, and Huey BB (Eds.), Alan R. Liss, New York, New York, pp. 523–605.

Page-Karjian A, Hahne M, Leach K, Murphy H, Lock B, and Rivera S. 2017. Neoplasia in snakes at Zoo Atlanta during 1992–2012. *J Zoo Wildl Med* 48:521–524.

Paitz RT, and Bowden RM. 2015. The in ovo conversion of oestrone to oestrone sulfate is rapid and subject to inhibition by Bisphenol A. *Biol Lett* 11(4):20140946.

Palmer BD, and Palmer SK. 1995. Vitellogenin induction by xenobiotic estrogens in the red-eared turtle and African clawed frog. *Environ Health Perspect* 103(Suppl 4):19–25.

Palmieri C, Selleri P, Di Girolamo N, Montani A, and Della Salda L. 2013. Multiple congenital malformations in a dicephalic spur-thighed tortoise (*Testudo graeca ibera*). *J Comp Pathol* 149:368–371.

Panneerdoss S, Viswanadhapalli S, Abdelfattah N et al. 2017. Cross-talk between miR-471-5p and autophagy component proteins regulates LC3-associated phagocytosis (LAP) of apoptotic germ cells. *Nat Commun* 8:598.

Park CJ, Ha CM, Lee JE, and Gye MC. 2015. Inter-Sertoli tight junctions in the testis of the Korean soft-shelled turtle (*Pelodiscus maackii*). *Biol Reprod* 92:96.

Pasquier J, Lafont A, Rousseau K, Quérat B, Chemineau P, and Dufour S. 2014. Looking for the bird kiss: Evolutionary scenario in sauropsids. *BMC Evol Biol* 14:30.

Pearse DE, Janzen FJ, and Avise JC. 2002. Multiple paternity, sperm storage, and reproductive success of female and male painted turtles (*Chrysemys picta*) in nature. *Behav Ecol Sociobiol* 51:164–171.

Pees M, Ludewig E, Plenz EB et al. 2014. Imaging diagnosis: Seminoma causing liver compression in a spur-thighed tortoise (*Testudo graeca*). *Vet Radiol Ultrasound* 56:E21–E24.

Pen I, Uller T, Feldmeyer B, Harts A, While GM, and Wapstra E. 2010. Climate-driven population divergence in sex-determining systems. *Nature* 468:436–438.

Pereira ME, and Viner TC. 2008. Oviduct adenocarcinoma in some species of captive snakes. *Vet Pathol* 45:693–697.

Perrault JR, Bauman KD, Greenan TM, Blum PC, Henry MS, and Walsh CJ. 2016. Maternal transfer and sublethal immune system effects of brevetoxin exposure in nesting loggerhead sea turtles (*Caretta caretta*) from western Florida. *Aquat Toxicol* 180:131–140.

Perrault JR, Wyneken J, Thompson LJ, Johnson C, and Miller DL. 2011. Why are hatching and emergence success low? Mercury and selenium concentrations in nesting leatherback sea turtles (*Dermochelys coriacea*) and their young in Florida. *Mar Pollut Bull* 62:1671–1682.

Perry SM, and Mitchell MA. 2017. Reproductive medicine in freshwater turtles and land tortoises. *Vet Clin North Am Exot Anim Pract* 20:371–389.

Pezaro N, Doody JS, Green B, and Thompson MB. 2013. Hatching and residual yolk internalization in lizards: Evolution, function and fate of the amnion. *Evol Dev* 15(2):87–95.

Pezaro N, Doody JS, and Thompson MB. 2016. The ecology and evolution of temperature-dependent reaction norms for sex determination in reptiles: A mechanistic conceptual model. *Biol Rev Camb Philos Soc* 92(3):1348–1364.

Pieau C. 1975. Temperature and sex differentiation in embryos of two chelonians, *Emys orbicularis L.* and *Testudo graeca L*, in *Intersexuality in the Animal Kingdom*. Reinboth R (Ed.), New York: Springer-Verlag, New York, New York, pp. 332–339.

Pieau C, Dorizzi M, and Richard-Mercier N. 1999. Temperature-dependent sex determination and gonadal differentiation in reptiles. *Cell Mol Life Sci* 55:887–900.

Pieau C, Dorizzi M, Richard-Mercier N, and Desvages G. 1998. Sexual differentiation of gonads as a function of temperature in the turtle *Emys orbicularis*: Endocrine function, intersexuality and growth. *J Exp Zool* 281:400–408.

Piprek RP, Kloc M, and Kubiak JZ. 2016. Early Development of the Gonads: Origin and Differentiation of the Somatic Cells of the Genital Ridges. *Probl Cell Differ* 58:1–22.

Pirson S, Geerinchx L, Bosseler L, Simard J, and Hellebuyck T. 2018. Metastatic granulosa cell tumor in a mangrove snake (*Boiga dendrophila melanota*). *J Exotic Pet Med* 27:6–10.

Pough FH, Andrews RM, Crump ML, Savitzky AH, Wells KD, and Brandley MC. 2016. *Herpetology*, 4th ed. Sinauer Associates Inc, Sunderland, Massachusetts.

Pottinger TG. 1997. The impact of stress on animal reproductive activities, in *Stress Physiology in Animals*. Balm PHM (Ed.), Sheffield Academic Press, Ltd, Sheffield, pp. 130–177.

Price ER. 2016. The physiology of lipid storage and use in reptiles. *Biol Rev* doi: 10.1111/brv.12288.

Prozialeck WC, Edwards JR, and Woods JM. 2006. The vascular endothelium as a target of cadmium toxicity. *Life Sci* 79:1493–1506.

Quinn AE, Sarre SD, Ezaz T, Marshall Graves JA, and Georges A. 2011. Evolutionary transitions between mechanisms of sex determination in vertebrates. *Biol Lett* 7:443–448.

Radder RS, Pike DA, Quinn AE, and Shine R. 2009. Offspring sex in a lizard depends on egg size. *Curr Biol* 19:1102–1105.

Radder RS, Warner DA, Cuervo JJ, and R Shine R. 2007. The functional significance of residual yolk in hatchling lizards *Amphibolurus muricatus* (Agamidae). *Funct Ecol* 21:302–309.

Rafferty AR, Evans RG, Scheelings TF, and Reina RD. 2013. Limited oxygen availability in utero may constrain the evolution of live birth in reptiles. *Am Nat* 181:245–253.

Rafferty AR, and Reina RD. 2012. Arrested embryonic development: A review of strategies to delay hatching in egg-laying reptiles. *Proc R Soc B* 279:2299–2308.

Ragsdale FR, and Ingermann RL. 1993. Biochemical bases for difference in oxygen affinity of maternal and fetal red blood cells of rattlesnakes. *Am J Physiol* 264:481–486.

Ragsdale FR, Imel KM, Nilsson EE, and Ingermann RL. 1993. Pregnancy-associated factors affecting organic phosphate levels and oxygen affinity of garter snake red cells. *Gen Comp Endocrinol* 91:181–188.

Raiti P. 1995. Reproductive problems of reptiles. *Proc Assoc Rep Amphib Vet*, Tucson, Arizona.

Ramírez-Pinilla MP, de Pérez GR, and Sever DM. 2015. Oogenesis and the Ovarian Cycle, in *Reproductive Biology and Phylogeny of Lizards and Tuatara*. Rheubert JL, Siegel DS, and Trauth SE (Eds.), CRC Press, Taylor and Francis Group, Boca Raton, Florida, pp. 231–252.

Rauschenberger RH, Sepúlveda MS, Wiebe JJ, Wiebe JE, Honeyfield DC, and Gross TS. 2009. Nutrient and organochlorine pesticide concentrations in American alligator eggs and their associations with clutch viability. *J Aquat Anim Health* 21:249–261.

Rauschenberger RH, Wiebe JJ, Buckland JE, Smith JT, Sepúlveda MS, and Gross TS. 2004. Achieving environmentally relevant organochlorine pesticide concentrations in eggs through maternal exposure in *Alligator mississippiensis*. *Mar Environ Res* 58:851–856.

Rauschenberger RH, Wiebe JJ, Sepúlveda MS, Scarborough JE, and Gross TS. 2007. Parental exposure to pesticides and poor clutch viability in American alligators. *Environ Sci Technol* 41:5559–5563.

Rey F, González M, Zayas MA et al. 2009. Prenatal exposure to pesticides disrupts testicular histoarchitecture and alters testosterone levels in male *Caiman latirostris*. *Gen Comp Endocrinol* 162:286–292.

Rhen T, Ross J, and Crews D. 1999 Effects of testosterone on sexual behavior and morphology in adult female leopard geckos, *Eublepharis macularius*. *Horm Behav* 36(2):119–128.

Rhen TJ, Sakata T, Zeller M, and Crews D. 2000. Sex steroid levels across the reproductive cycle of female leopard geckos, *Eublepharis macularius*, from different incubation temperatures. *Gen Comp Endocrinol* 118:322–331.

Rich AL, Phipps LM, Tiwari S, Rudraraju H, and Dokpesi PO. 2016. The increasing prevalence in intersex variation from toxicological dysregulation in fetal reproductive tissue differentiation and development by endocrine-disrupting chemicals. *Environ Health Insights* 10:163–171.

Rideout BA. 2012. Investigating embryo deaths and hatching failure. *Vet Clin North Am Exot Anim Pract* 15:155–162.

Rie MT, Kitana N, Lendas KA, Won JS, and Callard IP. 2005. Reproductive endocrine disruption in a sentinel species (*Chrysemys picta*) on Cape Cod, Massachusetts. *Arch Environ Contam Toxicol* 48:217–224.

Rings CC, Rafferty AR, Guinea ML, and Reina RD. 2015. The impact of extended preovipositional arrest on embryonic development and hatchling fitness in the flatback sea turtle. *Physiol Biochem Zool* 88:116–127.

Rivera S, 2008. Health Assessment of the Reptilian Reproductive Tract. *Journal of Exotic Pet Medicine*, 17(4):259–266.

Roach DA, and Carey JR. 2014. Population biology of aging in the wild. *Annu Rev Ecol Evol Syst* 45:421–443.

Rock J, and Cree A. 2003. Intraspecific variation in the effect of temperature on pregnancy in the viviparous gecko *Hoplodactylus maculatus*. *Herpetologica* 59:8–22.

Roe JH, Hopkins WA, Baionno JA, Staub BP, Rowe CL, and Jackson BP. 2004. Maternal transfer of selenium in Alligator mississippiensis nesting downstream from a coal-burning power plant. *Environ Toxicol Chem* 23:1969–1972.

Rollings N, Uhrig EJ, Krohmer RW et al. 2017. Age-related sex differences in body condition and telomere dynamics of red-sided garter snakes. *Proc Biol Sci* 284(1852). pii: 20162146. doi: 10.1098/rspb.2016.2146.

Rood ES, and Freedberg S. 2016. Intragenomic conflict produces sex ratio dynamics that favor maternal sex ratio distorters. *Ecol Evol* 6:8085–8093.

Rosenfeld CS, Denslow ND, Orlando EF, Gutierrez-Villagomez JM, and Trudeau VL. 2017. Neuroendocrine disruption of organizational and activational hormone programming in poikilothermic vertebrates. *J Toxicol Environ Health B Crit Rev* 20:276–304.

Ross P Jr, and Crews D. 1977. Influence of the seminal plug on mating behavior in the garter snake. *Nature* 267(5609):344–345.

Rouiller-Fabre V, Guerquin MJ, N'Tumba-Byn T et al. 2015. Nuclear receptors and endocrine disruptors in fetal and neonatal testes: A gapped landscape. *Front Endocrinol* 6:58. eCollection 2015.

Russart KL, and Rhen T. 2016. Atrazine alters expression of reproductive and stress genes in the developing hypothalamus of the snapping turtle, *Chelydra serpentina*. *Toxicology* 29:366–367.

Safe S, Jin UH, and Li X. 2018. Chemical-induced estrogenicity, in *Veterinary Toxicology*. Gupta RC (Ed.), Elsevier, Academic Press, Cambridge, Massachusetts, pp. 805–816.

Sanger TJ, Kyrkos J, Lachance DJ, Czesny B, and Stroud JT. 2018. The effects of thermal stress on the early development of the lizard *Anolis sagrei*. *J Exp Zool A* 329:244–251.

Sanger TJ, Losos JB, and Gibson-Brown JJ. 2008. A developmental staging series for the lizard genus Anolis: A new system for the integration of evolution, development, and ecology. *J Morphol* 269:129–137.

Santidrián Tomillo P, Saba VS, Blanco GS, Stock CA, Paladino FV, and Spotila JR. 2012. Climate driven egg and hatchling mortality threatens survival of eastern Pacific leatherback turtles. *PLOS ONE* 7(5):e37602.

Santos X, and Llorente GA. 2004. Lipid dynamics in the viperine snake, *Natrix maura*, from the Ebro Delta (NE Spain). *Oikos* 105(1):132–140.

Santos X, Arenas C, Llorente GA, and Ruiz X. 2007. Exploring the origin of egg protein in an oviparous water snake (*Natrix maura*). *Comp Biochem Physiol A* 147:165–172.

Sapolsky RM, Romero LM, and Munck AU. 2000. How do glucocorticoids influence stress responses? Integrating permissive, suppressive, stimulatory, and preparative actions. *Endocr Rev* 21:55–89.

Sarkar S, Sarkar NK, Das P, and Maiti BR. 1996. Photothermal effects on ovarian growth and function in the soft-shelled turtle, *Lissemys punctata punctata*. *J Exp Zool* 274:41–55.

Sarre SD, Georges A, and Quinn A. 2004. The ends of a continuum: Genetic and temperature-dependent sex determination in reptiles. *BioEssays* 26:639–645.

Sartori MR, Taylor EW, and Abe AS. 2012. Nitrogen excretion during embryonic development of the green iguana, *Iguana iguana* (Reptilia; Squamata). *Comp Biochem Physiol A* 163:210–214.

Schmidt RE, Reavill DR, and Phalen DN. 2015. *Pathology of Pet and Aviary Birds*, 2nd ed. Wiley Blackwell, Hoboken, New Jersey.

Schug TT, Johnson AF, Birnbaum LS et al. 2016. Minireview: Endocrine disruptors: Past lessons and future directions. *Mol Endocrinol* 30:833–847.

Schwartz TS, and Bronikowski AM. 2011. Molecular stress pathways and the evolution of life histories in reptiles, in *Molecular Mechanisms of Life History Evolution: The Genetics and Physiology of Life History Traits and Trade-offs*. Flatt T, and Heyland A (Eds.), Oxford Univ Press, Oxford, pp. 193–209.

Scudiero R, Verderame M, Motta CM, and Simoniello P. 2017. Unravelling the role of metallothionein on development, reproduction and detoxification in the wall lizard *Podarcis sicula*. *Int J Mol Sci* 18(7). pii: E1569. doi: 10.3390/ijms18071569.

Selcer KW. 2006. Reptile ecotoxicology: Studying effects of contaminants on populations, in *Toxicology of Reptiles*. Gardner SC, and Oberdoster E (Eds.), Taylor and Francis Group, Boca Raton, pp. 267–298.

Semenza JC, Tolbert PE, Rubin CH, Guillette LJ Jr, and Jackson RJ. 1997. Reproductive toxins and alligator abnormalities at Lake Apopka, Florida. *Environ Health Perspect* 105:1030–1032.

Shanbhag BA. 2003. Reproductive strategies in the lizard, *Calotes versicolor*. *Curr Sci* 84:646–652.

Shankar G, Sagar BK, Kumari TR, and Marathe GK. 2015. Ultrastructural features of sperm storage tubules in the oviduct of the Indian garden lizard, *Calotes Versicolor*. *Anat Rec (Hoboken)*. 298:1932–1937.

Shelby-Walker JA, Ward CK, and Mendonça MT. 2009. Reproductive parameters in female yellow-blotched map turtles (*Graptemys flavimaculata*) from a historically contaminated site vs. a reference site. *Comp Biochem Physiol A* 154:401–408.

Shimonohara N, Holland CH, Lin TL, and Wigle WL. 2013. Naturally occurring neoplasms in pigeons in a research colony: A retrospective study. *Avian Dis* 57:133–139.

Shine R, Elphick MJ, and Donnellan S. 2002. Co-occurrence supposedly incompatible modes of sex determination in a lizard population. *Ecol Lett* 5:486–489.

Shine R, Olsson MM, and Mason RT. 2000. Chastity belts in gartersnakes: The functional significance of mating plugs. *Biol J Linnean Soc* 70:377–390.

Shine R, Warner DA, and Radder RS. 2007. Windows of embryonic sexual lability in two lizard species with environmental sex determination. *Ecology* 88:1781–1788.

Shoemaker-Daly CM, Jackson K, Yatsu R, Matsumoto Y, and Crews D. 2010. Genetic network underlying temperature-dependent sex determination is endogenously regulated by temperature in isolated cultured *Trachemys scripta* gonads. *Dev Dyn* 239:1061–1075.

Siegel DS, Miralles A, Rheubert JL, and Sever DM. 2015. Female reproductive anatomy: Cloaca, oviduct and sperm storage, in *Reproductive Biology and Phylogeny of Lizards and Tuatara*. Rheubert JL, Siegal DS, Trauth SE, and Jamieson BGM (Eds.), CRC Press, Taylor and Francis Group, Boca Raton, Florida, pp. 144–195.

Sifuentes-Romero IB, Tezak M, Milton SL, and Wyneken J. 2018. Hydric environment effects on turtle development and sex ratio. *Zoology* 126:89–97.

Simoniello P, Filosa S, Scudiero R, Trinchella F, and Motta CM. 2013. Cadmium impairment of reproduction in the female wall lizard *Podarcis sicula*. *Environ Toxicol* 28:553–562.

Sinervo B, and Licht P. 1991. Hormonal and physiological control of clutch size: Constraints on the evolution of lizard life histories egg size, and egg shape in side-blotched lizards (*Uta stansburiana*). *J Exp Zool* 257:252–264.

Sinervo B, and Miles DB. 2011. Hormones and behavior of reptiles, in *Hormones and Reproduction of Vertebrates, Reptiles*, Vol 3. Norris DO, and Lopez KH (Eds.), Academic Press: Elsevier, San Diego, California, pp. 215–246.

Stacy BA, Vidal JD, Osofsky A, Terio K, Koski M, and De Cock HEV 2004. Ovarian Papillary Cystadenocarcinomas in a Green Iguana (*Iguana iguana*). *J Comp Path* 130:223–228.

Stacy BA, Howard L, Kinkaid J, Vidal JD, and Papendick R. 2008. Yolk coelomitis in Fiji Island banded iguanas (*Brachylophus fasciatus*). *J Zoo Wildl Med* 39:161–169.

Stacy BA, Foley A, Garner MM, and N Mettee N. 2013. Yolk embolism associated with trauma in vitellogenic sea turtles in Florida (USA): A review of 11 cases. *J Zoo Wildl Med* 44:1043–1048.

Stahl SJ. 2015. Managing dystocia in snakes. *Pro Association of Reptile and Amphibian Veterinarians*, Seattle, Washington.

Stahl SJ, and DeNardo DF. 2019 Theriogenology, in *Mader's Reptile and Amphibian Medicine and Surgery*, 3rd ed. Divers SJ, and Stahl SJ (Eds.), Elsevier, St. Louis, Missouri, pp. 848–893.

Steele AL, Wibbels T, and Warner DA. 2018. Revisiting the first report of temperature-dependent sex determination in a vertebrate, the African redhead agama. *J Zool* 306:16–22.

Stewart JR, and Blackburn DG. 2015. Viviparity and placentation, in *Reproductive Biology and Phylogeny of Lizards and Tuatara*. Rheubert JL, Siegal DS, Trauth SE, and Jamieson BGM (Eds.), CRC Press, Taylor & Francis Group, Boca Raton, Florida, pp. 448–563.

St Leger J. 2012. Nondomestic avian pediatric pathology. *Vet Clin North Am Exot Anim Pract* 15:233–250.

Stoker C, Beldoménico PM, Bosquiazzo VL et al. 2008. Developmental exposure to endocrine disruptor chemicals alters follicular dynamics and steroid levels in *Caiman latirostris*. *Gen Comp Endocrinol* 156:603–612.

Stoker C, Rey F, and Rodriguez H. 2003. Sex reversal effects on *Caiman latirostris* exposed to environmentally relevant doses of the xenoestrogen bisphenol A. *Gen Comp Endocrinol* 133:287–296.

Stoker C, Zayas MA, Ferreira MA et al. 2013. The eggshell features and clutch viability of the broad-snouted caiman (*Caiman latirostris*) are associated with the egg burden of organochlorine compounds. *Ecotoxicol Environ Saf* 98:191–195.

Strunk A, Imai DM, Osofsky A, and Tell LA. 2011. Dysgerminoma in an eastern rosella (*Platycercus eximus eximus*). *Avian Dis* 55:133–138.

Sun W, Cai H, Zhang G et al. 2017. Dmrt1 is required for primary male sexual differentiation in Chinese soft-shelled turtle *Pelodiscus sinensis*. *Sci Rep* 7:4433.

Swartz CD, Donnelly KC, Islamzadeh A et al. 2003. Chemical contaminants and their effects in fish and wildlife from the industrial zone of Sumgayit, Republic of Azerbaijan. *Ecotoxicology* 12:509–521.

Sykes JM. 2010. Updates and practical approaches to reproductive disorders in reptiles. *Vet Clin North Am Exot Anim Pract* 13:349–373.

Taylor EN, and DeNardo DF. 2011. Hormone and reproductive cycles in snakes, in *Hormones and Reproduction of Vertebrates, Reptiles*, Vol 3. Norris DO, and Lopez KH (Eds.), Academic Press: Elsevier, San Diego, California, pp. 355–372.

Thompson J, and Bannigan J. 2008 Cadmium: Toxic effects on the reproductive system and the embryo. *Reprod Toxicol* 25:304–315.

Thompson M, Coe B, Andrews RM, Cristol DA, Crossley DA, and Hopkins WA. 2018. Agricultural land use creates evolutionary traps for nesting turtles and is exacerbated by mercury pollution. *J Exp Zool* 329:230–243.

Tinkle DW. 1967. The life and demography of the side-blotched lizard, *Uta stansburiana*. *Misc Publ Mus Zool Univ Mich* 132:1–182.

Tocidlowski ME, Merrill CL, Loomis MR, and Wright JF. 2001. Teratoma in desert grassland whiptail lizards (*Cnemidophorus uniparens*). *J Zoo Wildl Med* 32:257–259.

Tokarz RR, and Summers CH. 2011. Stress and reproduction in reptiles, in *Hormones and Reproduction of Vertebrates, Reptiles*, Vol 3. Norris DO, and Lopez KH (Eds.), Academic Press: Elsevier, San Diego, California, pp. 39–61.

Trauth SE, and Buchanan RA. 1997. Characterization of spermatozoa of the lizard *Cnemidophorus sexlineatus* using fluorescence microscopy. *J Herpetol* 31(1):98–102.

Troyer K. 1987. Posthatching yolk in a lizard: Internalization and contribution to growth. *J Herpetol* 21:102–106.

Tsai P, and Licht SP. 1993. In vivo GnRH responsiveness of LH secretion in the female turtle, Trachemys scripta, in relation to the reproductive stage. *Gen Comp Endocrinol* 3:328–337.

Uetz P. 2016. http://www.reptile-database.org/db-info/SpeciesStat.html.

Uribe MC, and Guillette LJ Jr. 2000. Oogenesis and ovarian histology of the American alligator, *Alligator mississippiensis*. *J Morphol* 245:225–240.

Valenzuela N, Adams DC, and Janzen FJ. 2003. Pattern does not equal process: Exactly when is sex environmentally determined? *Am Nat* 161:676–683.

Van Damme R, Bauwens D, Braña F, and Verheyen RF. 1992. Incubation temperature differentially affects embryo survival, development and hatchling performance in the lizard *Podarcis muralis*. *Herpetologica* 48:220–228.

Van Dyke JU. 2015. Cues for reproduction in squamate reptiles, in *Reproductive Biology and Phylogeny of Lizards and Tuatara*. Rheubert JL, Siegal DS, Trauth SE, and Jamieson BGM (Eds.), CRC Press, Taylor and Francis Group, Boca Raton, Florida, pp. 107–143.

Van Dyke JU, Jachowski CM, Steen DA, Jackson BP, and Hopkins WA. 2017. Spatial differences in trace element bioaccumulation in turtles exposed to a partially remediated coal fly ash spill. *Environ Toxicol Chem* 36:201–211.

Van Meter RJ, Spotila JR, and Avery HW. 2006. Olycyclic aromatic hydrocarbons affect survival and development of common snapping turtle (*Chelydra serpentina*) embryos and hatchlings. *Environ Pollut* 142:466–475.

Vicoso B, Emerson JJ, Zektser Y, Mahajan S, and Bachtrog D. 2013. Comparative sex chromosome genomics in snakes: Differentiation, evolutionary strata, and lack of global dosage compensation. *PLoS Biol* 11:e1001643.

Vitousek MN, Mitchell MA, Romero LM, Awerman J, and Wikelski M. 2010. To breed or not to breed: Physiological correlates of reproductive status in a facultatively biennial iguanid. *Horm Behav* 57:140–146.

Vitt LJ. 2015. Lizard reproduction: A history of discovery, in *Reproductive Biology and Phylogeny of Lizards and Tuatara*. Rheubert JL, Siegal DS, Trauth SE, and Jamieson BGM (Eds.), CRC Press, Taylor and Francis Group, Boca Raton, Florida, pp. 1–26.

Vivien-Roels B, Arendt J, and Bradtke J. 1979. Circadian and circannual fluctuations of pineal indolemines (serotonin and melatonin) in *Testudo hermanni*. *Gen Comp Endocrinol* 37:197–210.

Wade J. 2005. Current research on the behavioral neuroendocrinology of reptiles. *Horm Behav* 48:451–460.

Wang W, Li C, Yu J et al. 2013. Morphologic analysis of atretic follicles in the Chinese soft-shelled turtle, *Pelodiscus sinensis*. *Tissue Cell* 45:383–386.

Warner DA. 2011. Sex determination in reptiles, in *Hormones and Reproduction of Vertebrates, Reptiles*, Vol 3. Norris DO, and Lopez KH (Eds.), Academic Press: Elsevier, San Diego, California, pp. 1–38.

Warner DA, Bonnet X, Hobson KA, and Shine R. 2008. Lizards combine stored energy and recently acquired nutrients flexibly to fuel reproduction. *J Anim Ecol* 77:1242–1249.

Warner DA, Du WG, and Georges A. 2018. Introduction to the special issue—developmental plasticity in reptiles: Physiological mechanisms and ecological consequences. *J Exp Zool A* 329(4–5):149–270.

Warner DA, Kelly CD, and Lovern MB. 2013. Reproductive experience affects mating behavior, but does not impact parental allocation in a lizard. *Behav Ecol Sociobiol* 67:973–983.

Warner DA, and Lovern MB. 2014. The maternal environment affects offspring viability via an indirect effect of yolk investment on offspring size. *Physiol Biochem Zool* 87:276–228.

Warner DA, Lovern MB, and Shine R. 2007. Maternal nutrition affects reproductive output and sex allocation in a lizard with environmental sex determination. *Proc R Soc B* 274:883–890.

Warner DA, Miller DA, Bronikowski AM, and Janzen FJ. 2016a. Decades of field data reveal that turtles senesce in the wild. *Proc Natl Acad Sci USA* 113:6502–6507.

Warner JK, Combrink X, Myburgh JG, and Downs CT. 2016b. Blood lead concentrations in free-ranging Nile crocodiles (*Crocodylus niloticus*) from South Africa. *Ecotoxicol* 25:950–958.

Warner DA, Radder RS, and Shine R. 2009. Corticosterone exposure during embryonic development affects offspring growth and sex ratios in opposing directions in two lizard species with environmental sex determination. *Physiol Biochem Zool* 82:363–371.

Warner DA, and Shine R. 2011. Interactions among thermal parameters determine offspring sex under temperature-dependent sex determination. *Proc R Soc B* 278:256–265.

Warner DA, Tucker JK, Filoramo NI, and Towey JB. 2006. Claw function of hatchling and adult red-eared slider turtles (*Trachemys scripta elegans*). *Chel Conserv Biol* 5:317–320.

Wenger S, Simova-Curd S, and Great P. 2010. Ovarian teratoma in a Fiji island banded iguana (*Brachylopus fasciatus*) and a green iguana (*Iguana iguana*). *J Herp Med Surg* 20:20–24.

Wernick MB, Tobias JR, Moeller RB, Barnes J, Palmieri C, and Shivaprasad HL. 2015. Bilateral malignant seminomas in two unrelated, aged trumpeter hornbills (*Bycanistes buccinator*). *J Avian Med Surg* 29:125–129.

While GM, Noble DWA, Uller T et al. 2018. Patterns of developmental plasticity in response to incubation temperature in reptiles. *J Exp Zool* 329(4–5):162–176.

White SS, and Birnbaum LS. 2009. An overview of the effects of dioxins and dioxin-like compounds on vertebrates, as documented in human and ecological epidemiology. *J Environ Sci Health C Environ Carcinog Ecotoxicol Rev* 27:197–211.

Wibbels T, and Crews D. 1994. Putative aromatase inhibitor induces male sex determination in a female unisexual lizard and in a turtle with temperature-dependent sex determination. *J Endocrinol* 141(2):295–299.

Williams CE, McNabb NA, Brunell A et al. 2018. Feminizing effects of exposure to Corexit-enhanced water-accommodated fraction of crude oil *in vitro* on sex determination in *Alligator mississippiensis*. *Gen Comp Endocrinol* 265:46–55.

Williamson SA, Evans RG, Manolis SC, Webb GJ, and Reina RD. 2017a. Ecological and evolutionary significance of a lack of capacity for extended developmental arrest in crocodilian eggs. *R Soc Open Sci* 4(12):171439.

Williamson SA, Evans RG, and Reina RD. 2017b. When is embryonic arrest broken in turtle eggs? *Physiol Biochem Zool* 90:523–532.

Willingham EJ. 2005. The effects of atrazine and temperature on turtle hatchling size and sex ratios. *Front Ecol Environ* 3:309–313.

Willingham E, and Crews D. 1999. Sex reversal effects of environmentally relevant xenobiotic concentrations on the red-eared slider turtle, a species with temperature-dependent sex determination. *Gen Comp Endocrinol* 113:429–435.

Wingfield JC, and Sapolsky RM. 2003. Reproduction and resistance to stress: When and how. *J Neuroendocrinol* 15:711–724.

Wise PA, Vickaryous MK, and Russell AP. 2009. An embryonic staging table for in ovo development of *Eublepharis macularius*, the leopard gecko. *Anat Rec (Hoboken)* 292:1198–1212.

Xavier F. 1987. Functional morphology and regulation of the corpus luteum, in *Hormones and Reproduction in Fishes, Amphibians and Reptiles*. Norris DO, and Jones RE (Eds.), Plenum Press, New York, New York, pp. 241–282.

Xie QP, He X, Sui YN, Chen LL, Sun LN, and Wang DS. 2016. Haploinsufficiency of SF-1 causes female to male sex reversal in Nile tilapia, *Oreochromis niloticus*. *Endocrinol* 157:2500–2514.

Yajurvedi HN, and Nijagal BS. 2000. Corticosterone inhibits normal and FSH-induced testicular recrudescence in the lizard, *Mabuya carinata*. *Gen Comp Endocrinol* 120:283–288.

Yang P, Luan X, Peng Y et al. 2017. Novel zona pellucida gene variants identified in patients with oocyte anomalies. *Fertil Steril* 107:1364–1369.

Yao HH, and Capel B. 2005. Temperature, genes, and sex: A comparative view of sex determination in *Trachemys scripta* and *Mus musculus*. *J Biochem* 138:5–12.

Yoshikane M, Kay WR, Shibata Y et al. 2006. Very high concentrations of DDE and toxaphene residues in crocodiles from the Ord River, Western Australia: An investigation into possible endocrine disruption. *J Environ Monit* 8:649–661.

Zhang L, Han XK, Qi YY, Liu Y, and Chen QS. 2008. Seasonal effects on apoptosis and proliferation of germ cells in the testes of the Chinese soft-shelled turtle, *Pelodiscus sinensis*. *Theriogenology* 69:1148–1158.

Zhang R, Zhang S, Zhu X, Zhou Y, and Wu X. 2015. Follicle-stimulating hormone receptor (FSHR) in Chinese alligator, *Alligator sinensis*: Molecular characterization, tissue distribution and mRNA expression changes during the female reproductive cycle. *Anim Reprod Sci* 156:40–50.

Zhang R, Hu Y, Wang H et al. 2016. Molecular cloning, characterization, tissue distribution and mRNA expression changes during the hibernation and reproductive periods of estrogen receptor alpha (ESR1) in Chinese alligator, *Alligator sinensis*. *Comp Biochem Physiol B* 200:28–35.

Zimmerman DM, Mitchell MA, and Perry BH. 2013. Collection and characterization of semen from green iguanas (*Iguana iguana*). *Am J Vet Res* 74:1536–1541.

Zychowski GV, and Godard-Codding CA. 2017. Reptilian exposure to polycyclic aromatic hydrocarbons and associated effects. *Environ Toxicol Chem* 36(1):25–35.

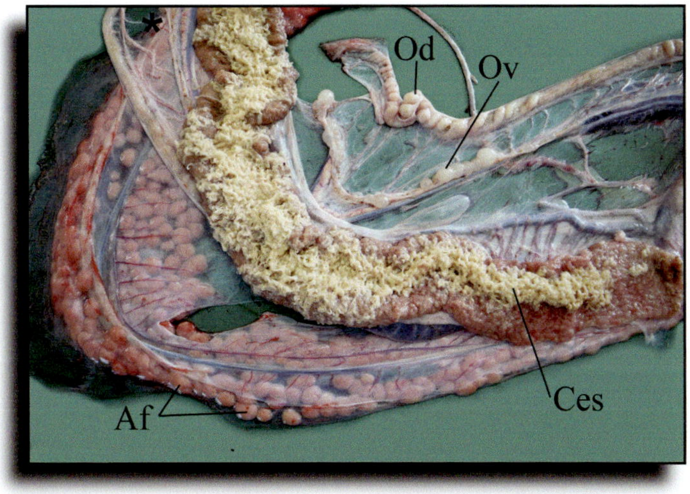

Figure 8.1 Burmese Python, *Python bivittatus*. Pythonidae. Wild-caught captive female exhibiting delayed folliculogenesis of the ovary (Ov) and severe parasitism by cestode (Ces) species. The oviduct (Od) is non-gravid. Spacing between coelomic adipose foci (Af) indicates mild atrophy and possible negative energy balance. Anterior (*).

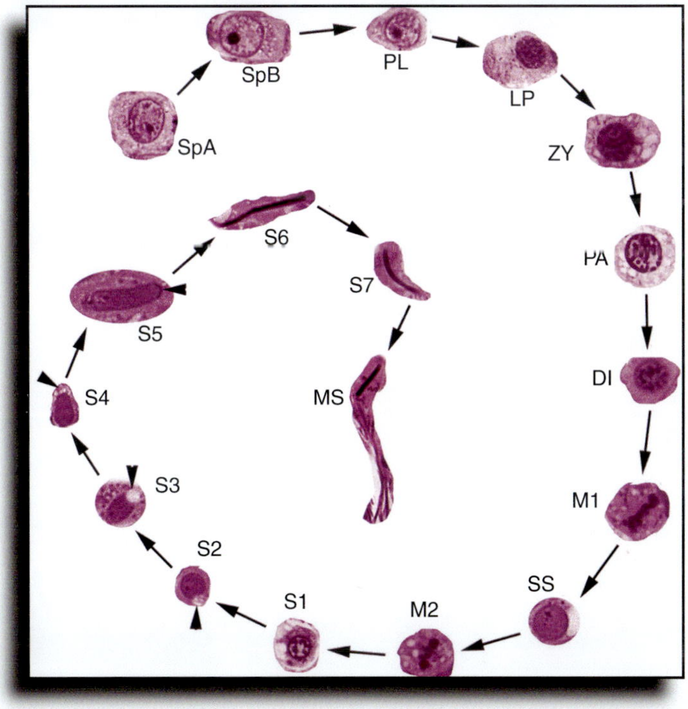

Figure 8.2 Map Turtle, *Graptemys* sp. Emydidae. Wild, male. The germ cells found in the mature seminiferous epithelium progress in order of maturity from Type A spermatogonia (SpA), type B spermatogonia (SpB), preleptotene spermatocytes (PL), leptotene spermatocytes (LP), zygotene spermatocytes (ZY), pachytene spermatocytes (PA), diplotene spermatocytes (DI), spermatocytes in meiosis I (M1), secondary spermatocytes (SS), spermatocytes in meiosis II (M2), step 1 spermatid (S1), step 2 spermatid (S2), step 3 spermatid (S3), step 4 spermatid (S4), step 5 spermatid (S5), step 6 spermatid (S6), step 7 spermatid (S7), mature sperm (MS). H&E stain. (Courtesy of Kevin Gribbins.)

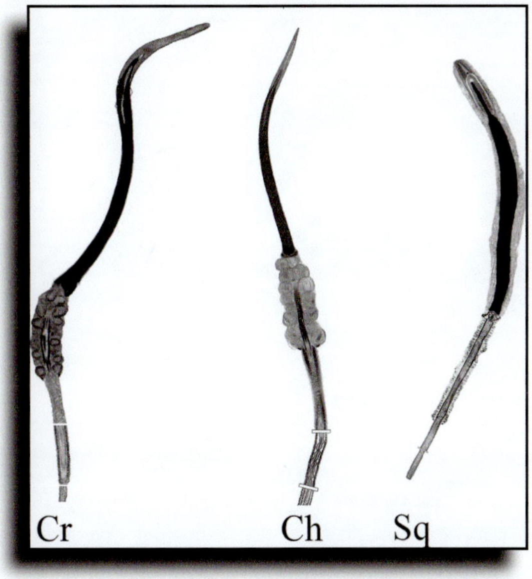

Figure 8.3 Comparison of spermatozoa from a Crocodilian (Cr). American Alligator, *Alligator mississippiensis*. Alligatoridae; Chelonian (Ch). Red Eared Slider, *Trachemys scripta elegans*. Emydidae; Squamata (Sq) Copperhead, *Agkistrodon contortrix*. Viperidae. Transmission electron microscope. Uranyl acetate and lead citrate stain (Courtesy of Kevin Gribbins.)

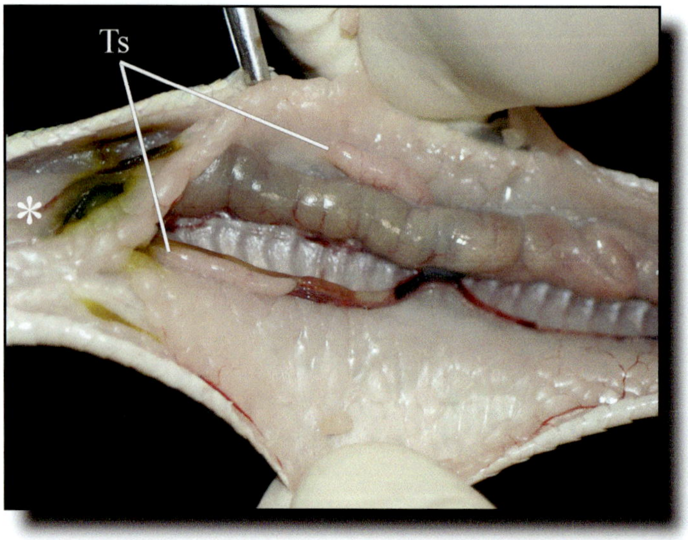

Figure 8.4 Green Tree Python, *Morelia viridis*. Pythonidae. Captive propagated male, 1-year-old. Normal white immature testicle (Ts). Green staining is postmortem autolysis from gallbladder bile leakage. Anterior (*).

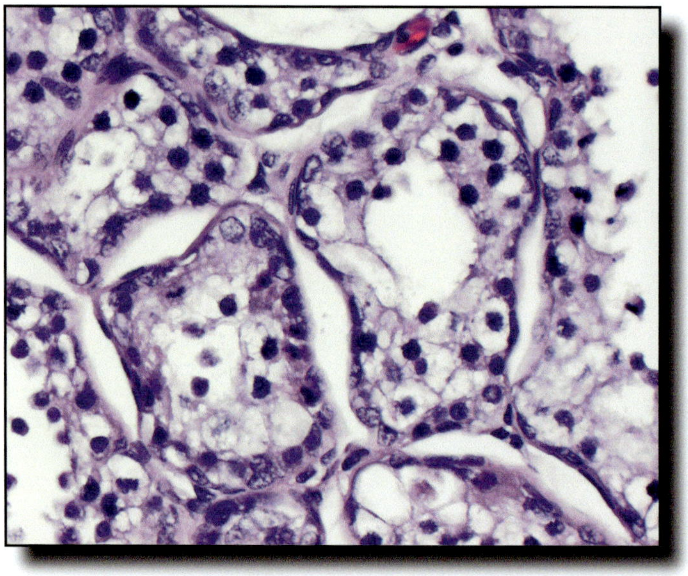

Figure 8.5 Eastern Indigo Snake, *Drymarchon couperi*. Colubridae. One-year-old captive propagated male. Immature seminiferous tubules. Spermiogenesis and spermatocyte progression to spermatids is absent. The lumina of seminiferous tubules are empty of spermatozoa. H&E stain.

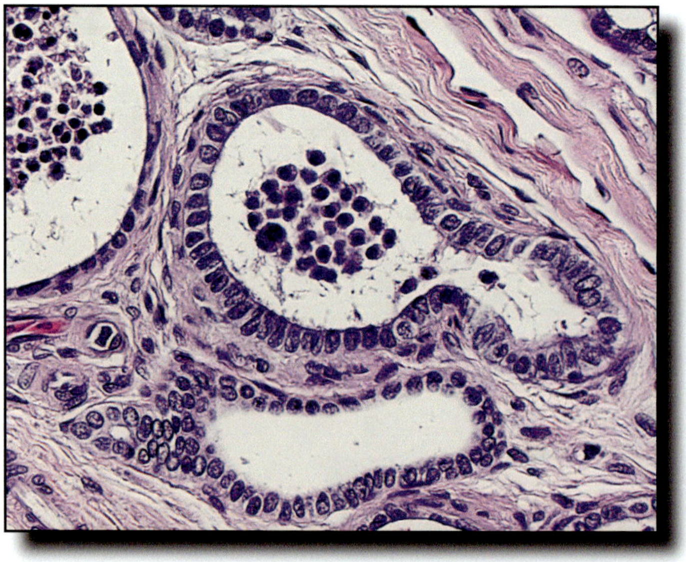

Figure 8.6 Eastern Indigo Snake, *Drymarchon couperi*. Colubridae. One-year-old captive propagated male. The lumen of the epididymis contains immature spermatogenic cells. H&E stain.

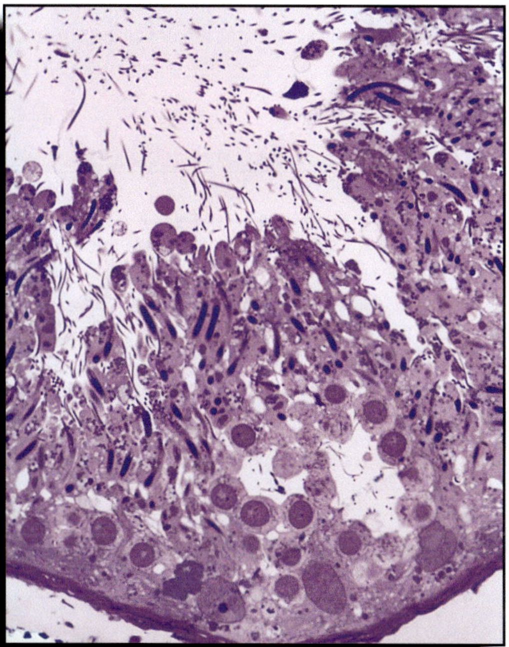

Figure 8.7 Map Turtle, *Graptemys geographica*. Emydidae. Wild male. The progression of germ cells found in the mature seminiferous epithelium during spermatogenesis can be seen. Spermatogonia in mitosis are intermingled with Sertoli cells along the basement membrane. Spermatocytes are in meiotic division. Spermatids can be seen maturing to spermatozoa. H&E stain. (Courtesy of Kevin Gribbins.)

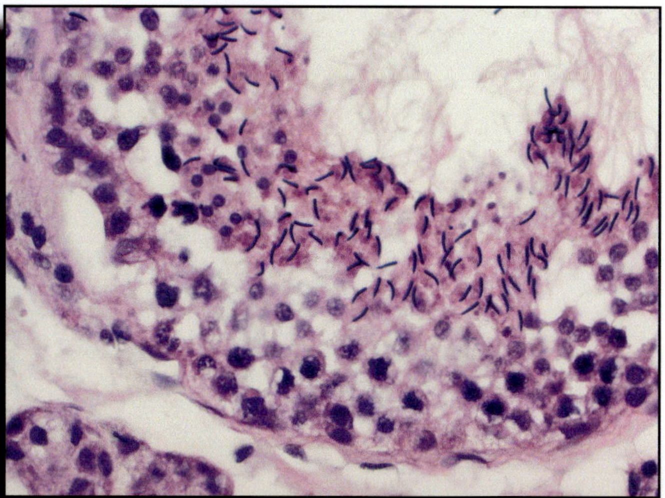

Figure 8.8 Canebrake Rattlesnake, *Crotalus horridus atricaudatus*. Viperidae. Captive male. Seminiferous tubule during spermatogenesis. H&E stain.

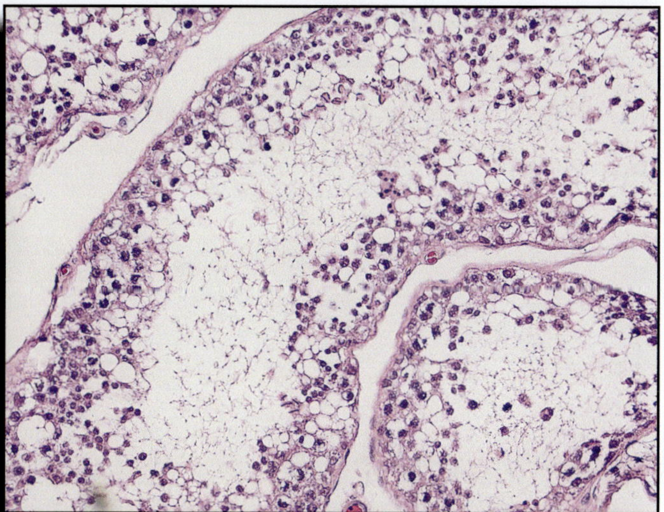

Figure 8.9 Long-Nosed Snake, *Rhinocheilus lecontei*. Colubridae. Wild male. Testes. Seminiferous tubules are distended, and interstitial space is decreased. Tubules have normal progression of spermatogenesis with mature spermatozoa and a few spermatids occupying the lumen of seminiferous tubules. H&E stain.

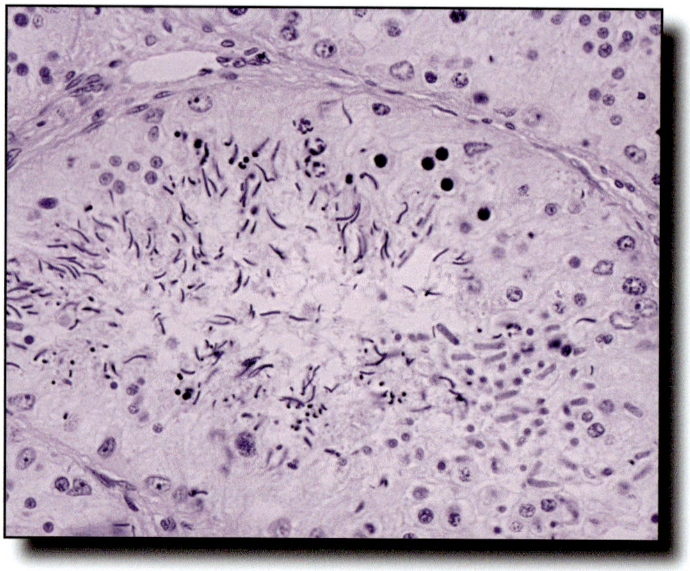

Figure 8.10 American Alligator. *Alligator mississippiensis,* Alligatoridae. Wild male. Post-breeding testes. The seminiferous tubules have decreased width and spermatogonia are depleted. H&E stain. (Courtesy of Kevin Gribbins.)

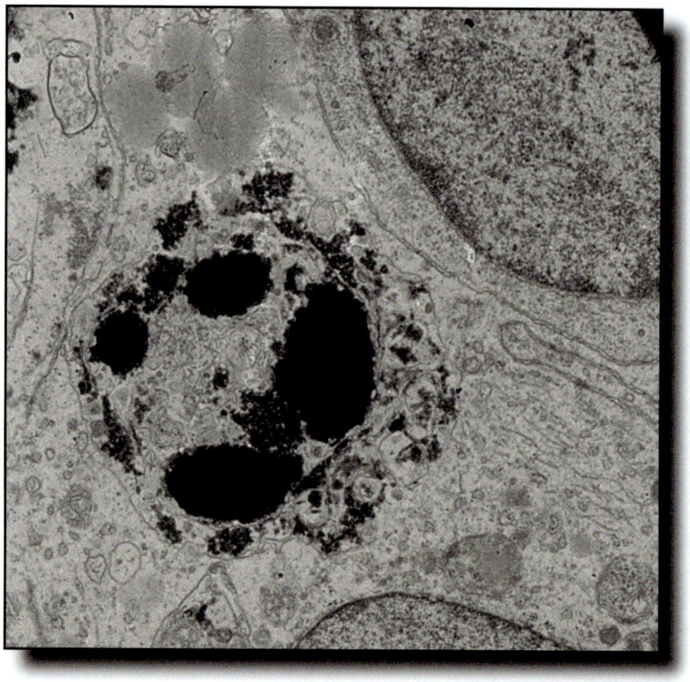

Figure 8.11 Chinese Softshell Turtle, *Pelodiscus sinensis.* Trionychidae. Captive male. Sperm are undergoing apoptosis in seminiferous tubules. H&E stain. (From Liu T, Wang L, Chen H, Huang Y, Yang P, Ahmed N, Wang T, Liu Y, and Chen Q. 2017. *Front Physiol* 8:188. Copyright (2017) Creative Commons Attribution License. With permission.)

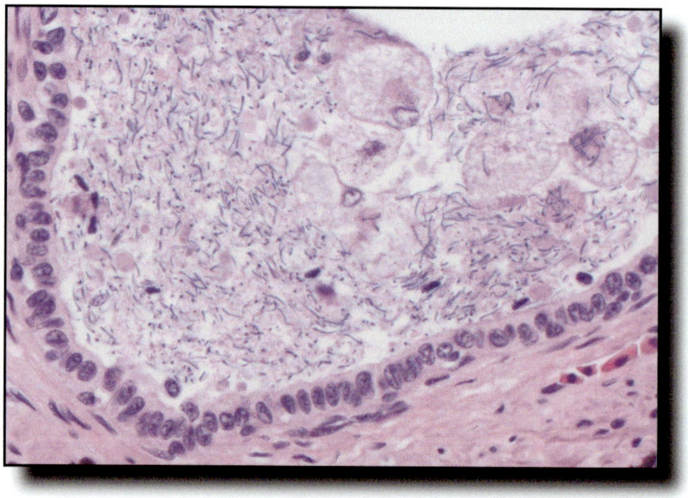

Figure 8.12 Russian Tortoise, *Agrionemys horsfieldii.* Testudinidae. Captive male. Testes. Phagocytosis of sperm in the epididymis by mononuclear phagocytic cells of undetermined origin. H&E stain.

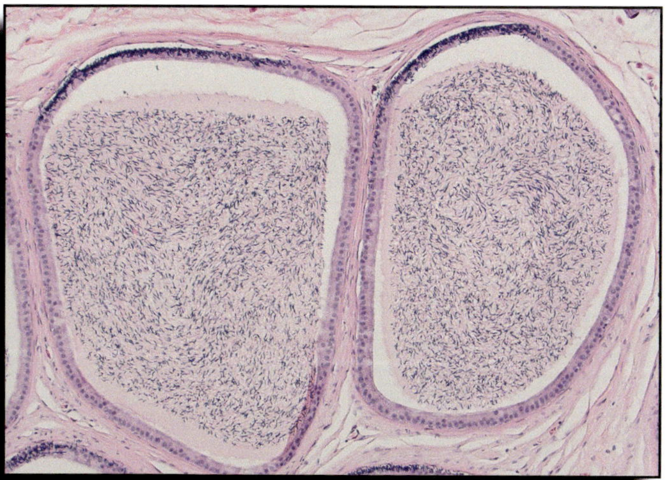

Figure 8.13 Canebrake Rattlesnake, *Crotalus horridus atricaudatus*. Viperidae. Captive male. Testes. Epididymal tubules are distended with spermatozoa. H&E stain.

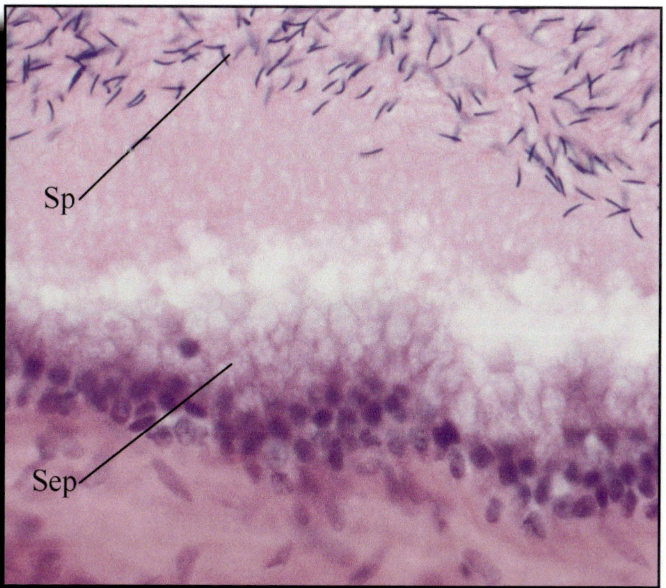

Figure 8.14 Canebrake Rattlesnake, *Crotalus horridus atricaudatus*. Viperidae. Captive male. The epididymal secretory epithelium (Sep) contains goblet-shaped cells with clear secretory granules. Sperm (Sp) are present in the lumen of the epididymis. H&E stain.

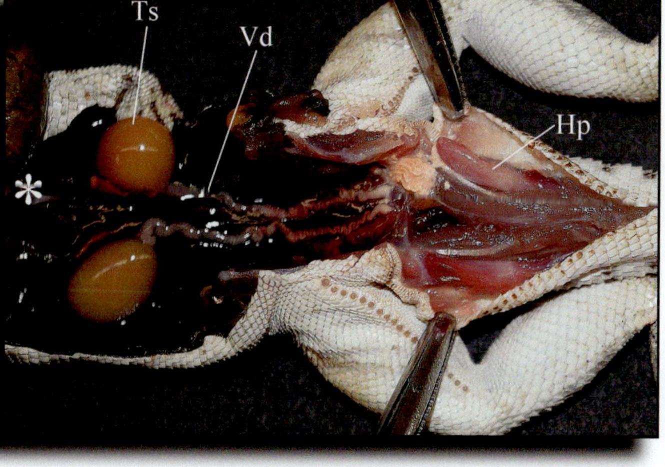

Figure 8.15 Desert Iguana, *Dipsosaurus dorsalis*. Iguanidae. Wild male. Seasonal physiologic hypertrophy of testes (Ts) with corresponding increased sperm production. The vas deferens (Vd) are distended with spermatozoa. Normal hemipenes (Hp) are present. Anterior (*).

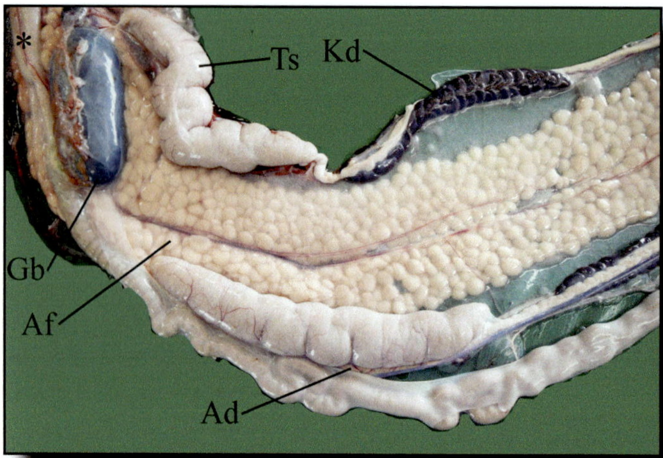

Figure 8.16 Burmese Python, *Python bivittatus*. Pythonidae. Captive male during the breeding season. There is seasonal hypertrophy of testes (Ts), and enlargement and apposition of adjacent coelomic adipose foci (Af) indicates sufficient energy storage. The gallbladder (Gb) is distended, the adrenal gland (Ad) abuts the testis, and kidneys (Kd) are adjacent to the caudal pole of the testes. Anterior (*).

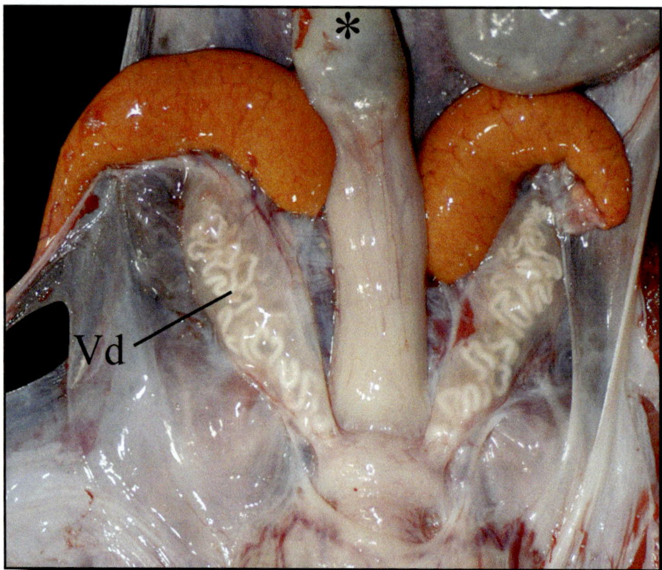

Figure 8.17 Agassiz's Desert Tortoise, *Gopherus agassizii.* Testudinidae. Wild male. There is seasonal hypertrophy of the testes. Vas deferens (Vd) are distended with spermatozoa. Anterior (*).

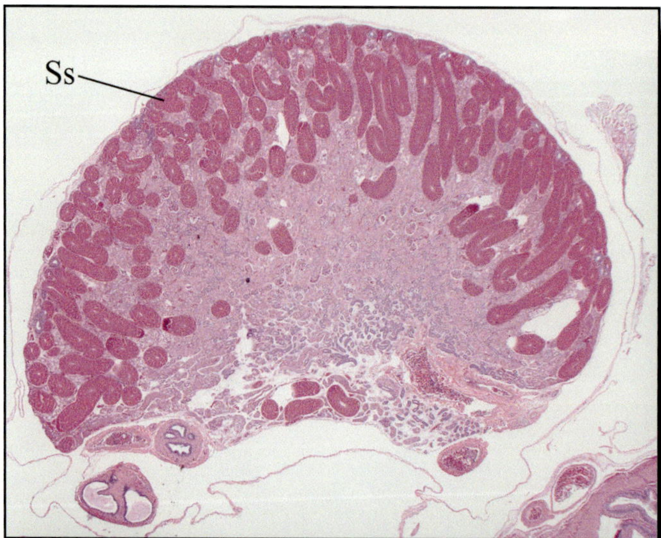

Figure 8.18 Long-Nosed Snake, *Rhinocheilus lecontei.* Colubridae. Wild male. Kidney. Renal tubules with prominent eosinophilic sexual segment (Ss) in the peak of breeding season. H&E stain.

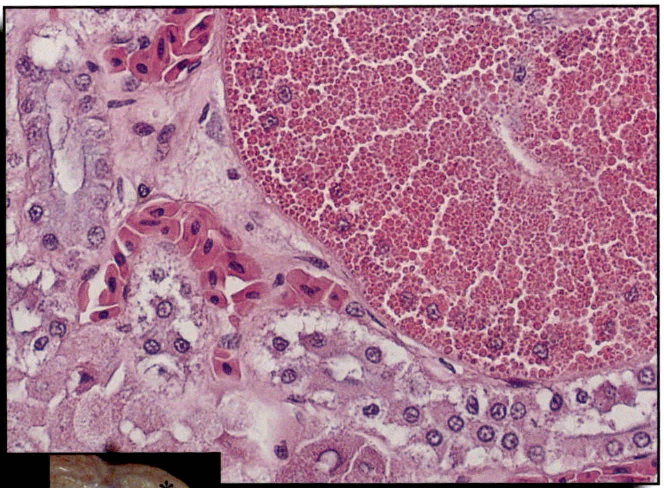

Figure 8.19 Long-Nosed Snake, *Rhinocheilus lecontei*. Colubridae, Wild male. Kidney. Renal tubular epithelial cells of the sexual segment contain prominent eosinophilic secretory granules during the peak of breeding season. H&E stain.

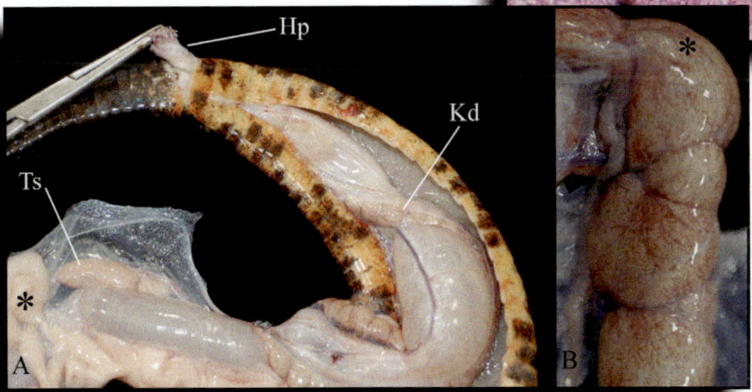

Figure 8.20 Canebrake Rattlesnake, *Crotalus horridus atricaudatus*. Viperidae. A. Complete male reproductive tract during time of seasonal spermatogenesis with hypertrophied testicle (Ts), kidneys (Kd), and an everted hemipene (Hp) with hormone-induced enlargement. B. Tan kidney and adjacent ductus deferens (arrowhead). Hypertrophy of the sexual segment during breeding season creates light tan streaks that grossly can resemble disease, especially gout. Anterior (*).

Figure 8.21 Red-Sided Garter Snake, *Thamnophis sirtalis parietalis*. Colubridae. Wild female. A seminal plug expands the opening to the vent. Anterior (*). (Courtesy of Christopher Friesen.)

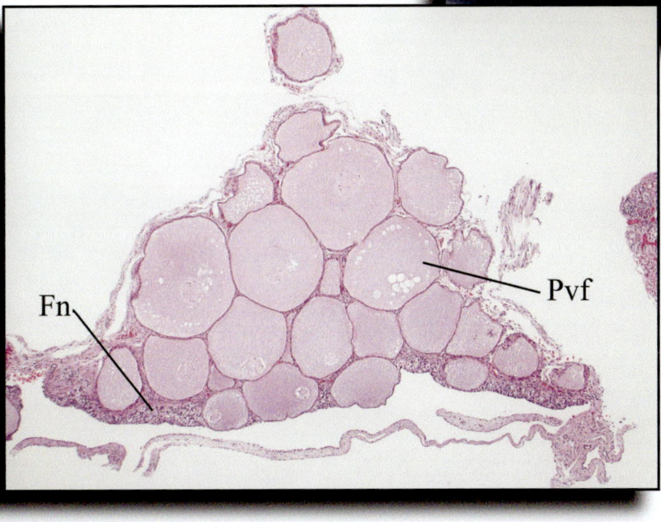

Figure 8.22 Brown Roofed Turtle, *Pangshura smithii*. Geoemydidae. Captive female. Multiple follicular nests (Fn) or germinal beds can be observed at the hilus of the ovary. Previtellogenic follicles (Pvf) in this ovary are quiescent. H&E stain.

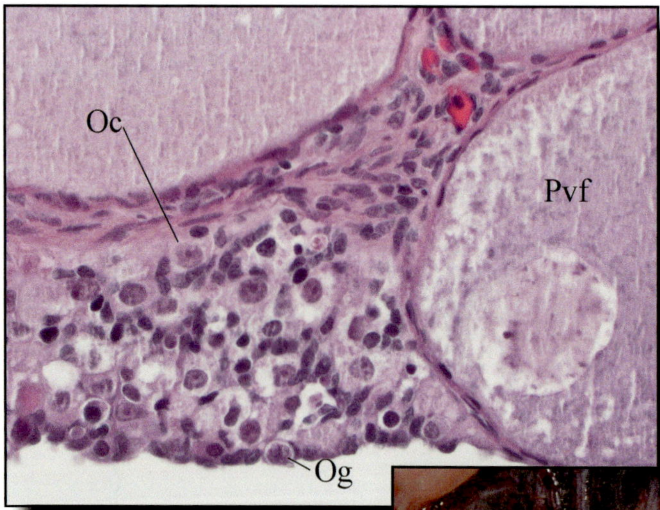

Figure 8.23 Brown Roofed Turtle, *Pangshura smithii*. Geoemydidae. Ovary. Captive female. A section of a follicular germinal bed (or nest) with oogonia (Og) and primary oocytes (Oc). The cytoplasm of the oocytes stains clear compared to other stromal cells. Previtellogenic follicles (Pvf) in this ovary are quiescent. H&E stain.

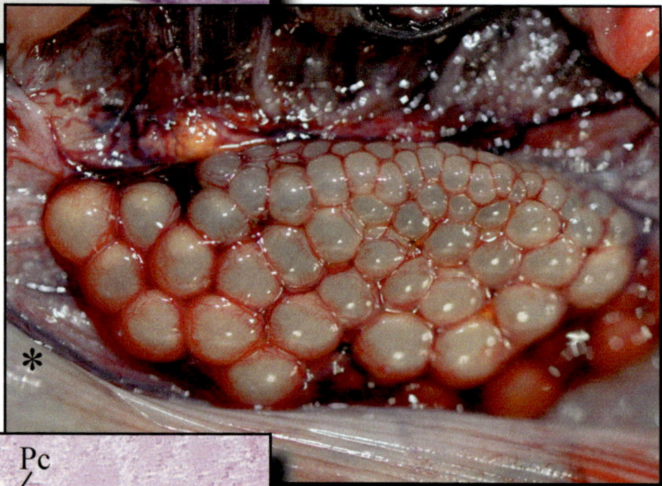

Figure 8.24 Argentine Black and White Tegu, *Salvator merianae*. Teiidae. Wild female. Previtellogenic stage follicles with congested blood supply during hibernation. Anterior (*).

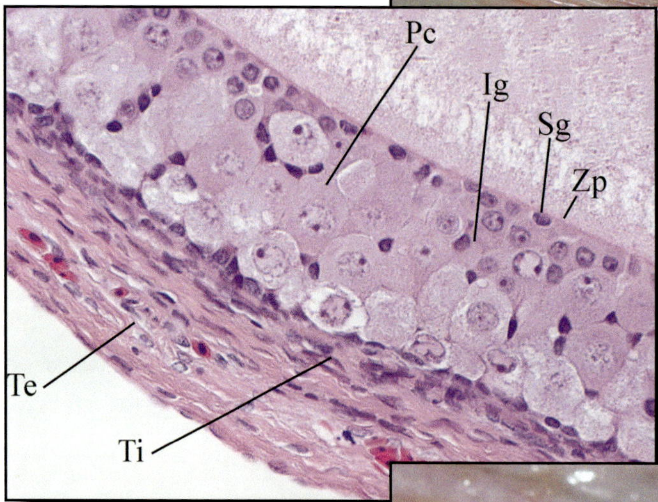

Figure 8.25 Argentine Black and White Tegu, *Salvator merianae*. Teiidae. Wild female. Ovary with previtellogenic stage follicles. Small granulosa (Sg) and intermediate granulosa (Ig) cells abut a thin zona pellucida (Zp). Large pyriform cells (Pc) are present adjacent to the theca interna (Ti) and developing theca extera (Te). H&E stain.

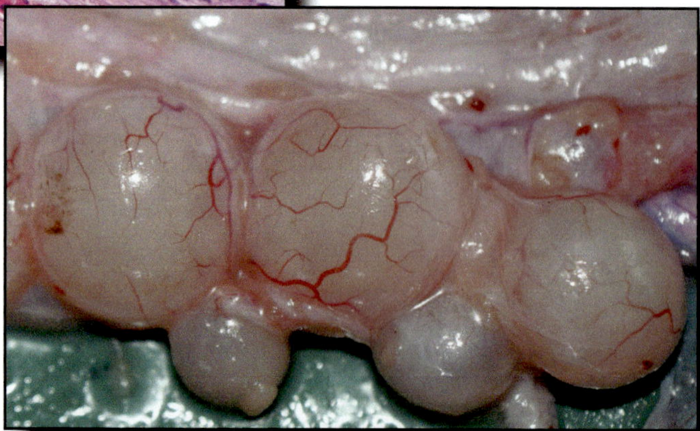

Figure 8.26 Burmese Python, *Python bivittatus*. Pythonidae. Wild female. Ovary with previtellogenic stage follicles, 1–2 months preceding vitellogenesis and peak breeding.

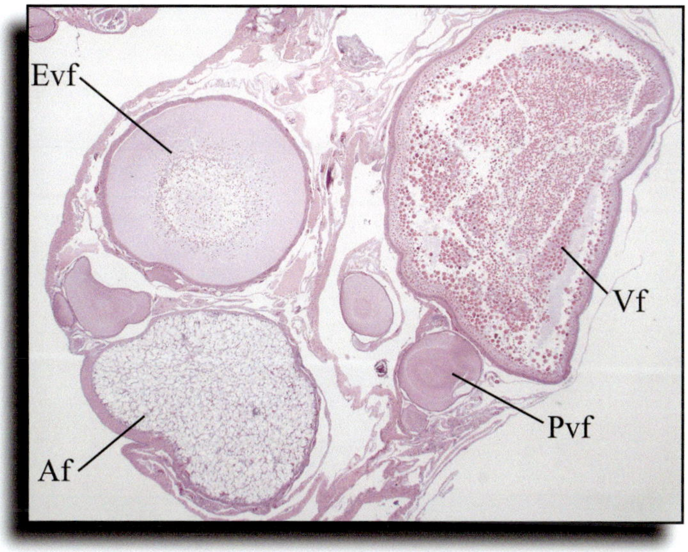

Figure 8.27 Red-Eared Slider, *Trachemys scripta elegans*. Emydidae. Wild female. Ovary with previtellogenic follicle (Pvf), transition to early vitellogenic follicle (Evf), vitellogenic follicle (Vf) and atretic follicles (Af). H&E stain.

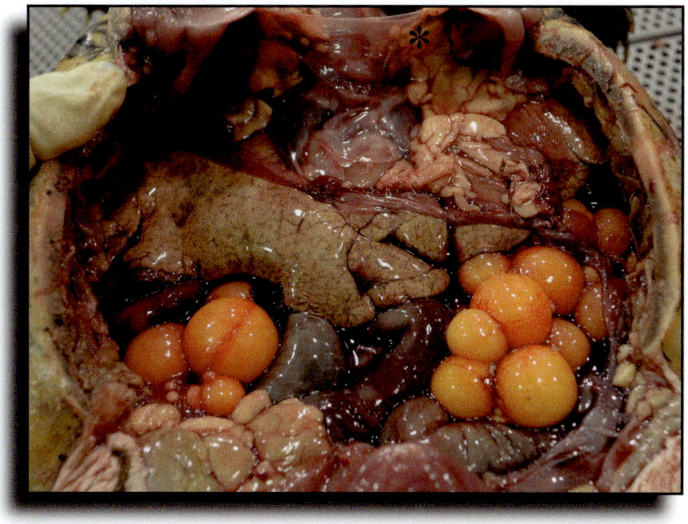

Figure 8.28 Red-Eared Slider, *Trachemys scripta elegans*. Emydidae. Wild female. Coelomic cavity with multiple variably sized yellow vitellogenic follicles. The vitellogenic follicles are preovulatory. The liver is pale due to production of vitellogenin. Prior to ovulation and egg production; spring season in the northern hemisphere. Anterior (*). (Courtesy of Kelley Steury.)

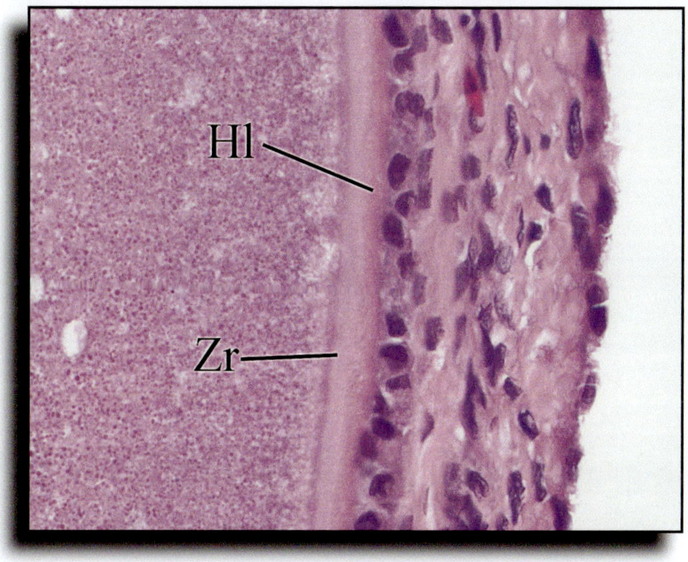

Figure 8.29 Eastern Box Turtle, *Terrapene carolina*. Emydidae. Captive female. Initial stages of vitellogenesis during which the zona pellucida develops duel bands, producing a hyaline layer (Hl) adjacent to granulosa cells and inner zona radiata (Zr). H&E stain.

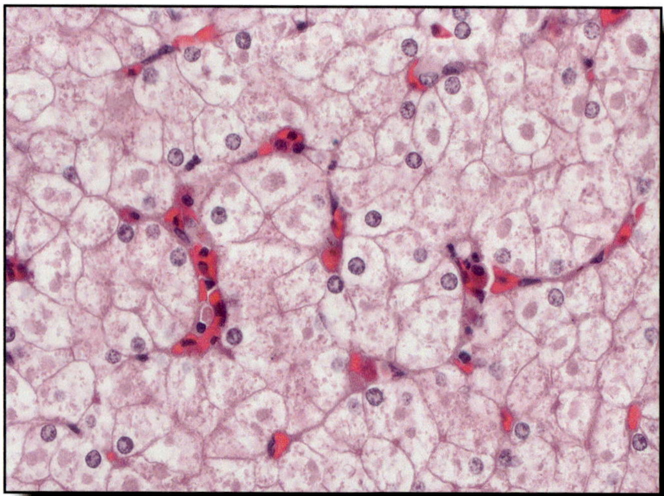

Figure 8.30 Brown Roofed Turtle, *Pangshura smithii*. Geoemydidae. Captive female. Liver during previtellogenic stage. Hepatocytes are about to initiate production and mobilization of vitellogenin. Matched with previtellogenic ovary in Figure 8.22. H&E stain.

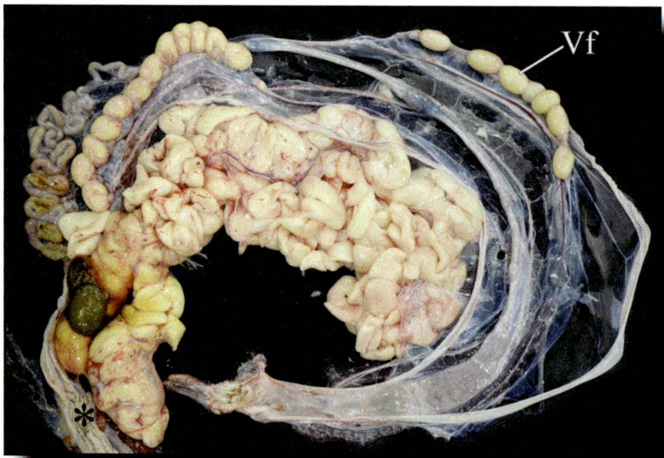

Figure 8.31 Bushmaster, *Lachesis muta*. Viperidae. Captive female. Ovary. Distribution of middle stage vitellogenic follicles (Vf) This snake is in the preovulatory stage. Follicles are oval. Anterior (*).

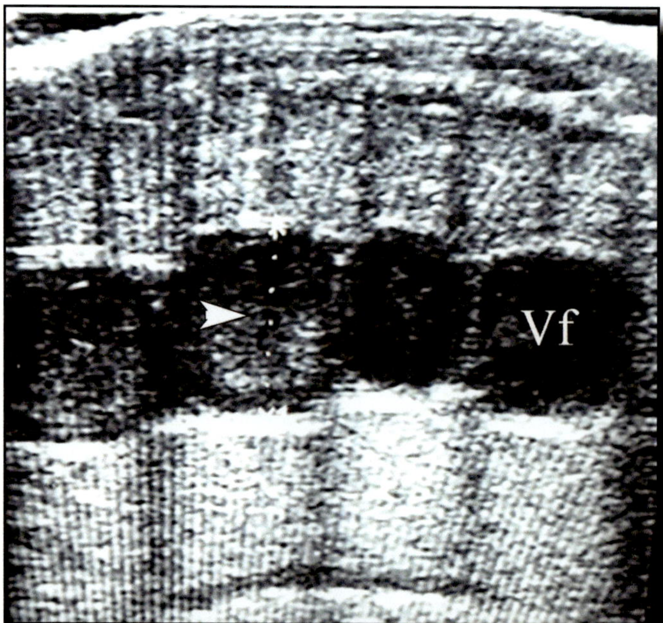

Figure 8.32 Ball Python, *Python regius*. Pythonidae. Captive female. Ultrasound image of late vitellogenic follicles (Vf). Dotted line indicates measurement of a 17 mm (arrow) diameter follicle. (Courtesy of Tracy Barker.)

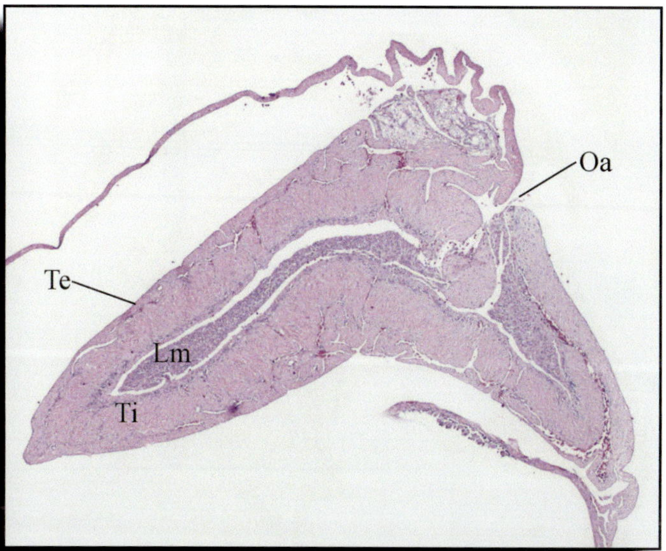

Figure 8.33 Crested Gecko, *Correlophus ciliatus*. Diplodactylidae. Captive female. Corpus luteum. The lumen is filled by the luteal mass (Lm). The theca externa (Te) and theca interna (Ti) from the previous follicle persist. An ovulation aperture (Oa) is present in the gap left by ovulation. H&E stain.

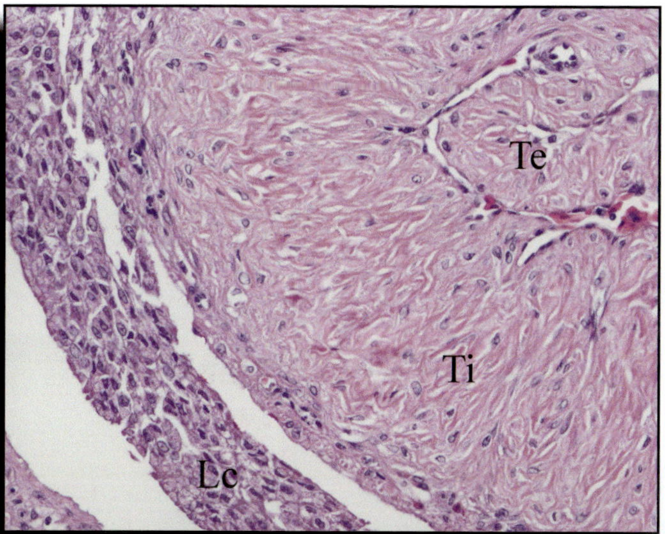

Figure 8.34 Crested Gecko, *Correlophus ciliates*. Diplodactylidae. Captive female. Corpus luteum. Monomorphic polygonal luteal cells (Lc) fill the central cavity. The theca externa (Te) and theca interna (Ti) contain eosinophilic homogenous extracellular matrix. H&E stain.

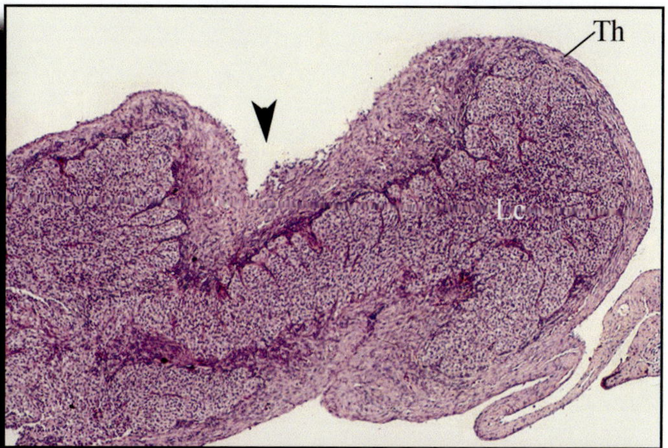

Figure 8.35 *Agama* sp. Agamidae. Captive female. Corpus luteum. Lumen filled with monomorphic cells with a central nucleus. This Lozano stage II CL has infolding (arrowhead) at the probable point of past ovulation. Luteal cells (Lc) completely occupy the central space and are surrounded by a theca (Th). (Lozano et al., 2014) H&E stain. (Courtesy of Peernel Zwart.)

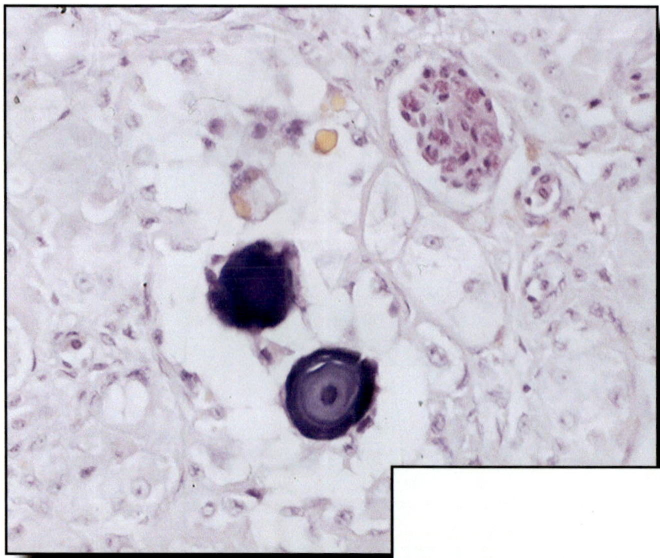

Figure 8.36 Green Iguana, *Iguana iguana*. Iguanidae. Captive female. Dystrophic mineralization in corpus luteum. H&E stain. (Courtesy of Peernel Zwart.)

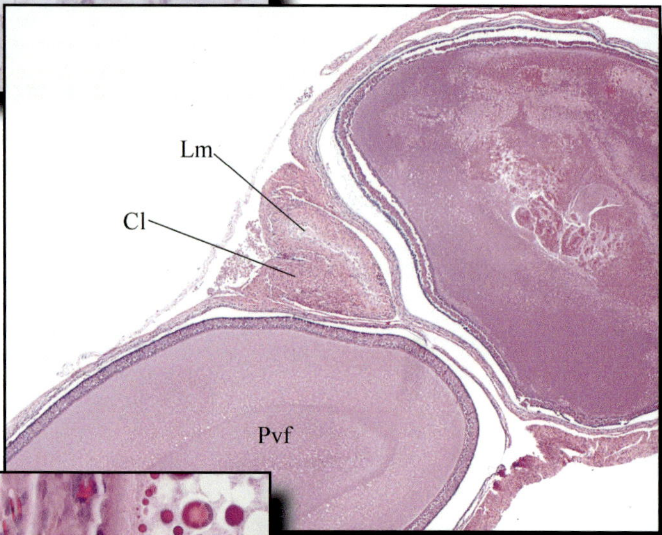

Figure 8.37 Midland Water Snake, *Nerodia sipedon pleuralis*. Colubroidea. Wild female. Late-term matrotrophy-dependent fetus near time of parturition. The ovary still contains numerous previtellogenic follicles (Pvf), but all CLs are in a regressive stage (Lozano Stage IV) with scant luteal mass (Lm) (Lozano et al., 2014). H&E stain.

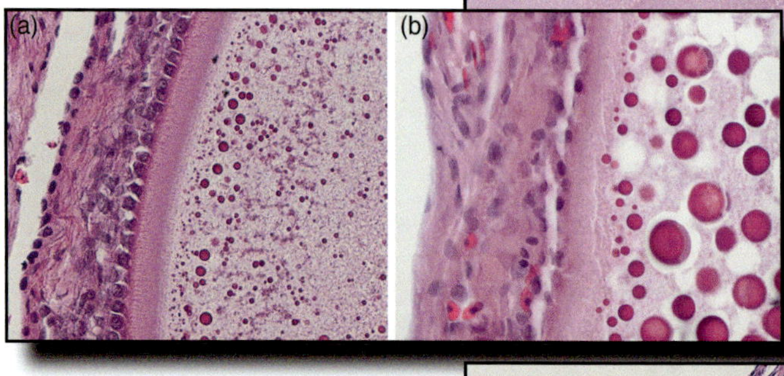

Figure 8.38 Red-Eared Slider, *Trachemys scripta elegans*. Emydidae. Wild female vitellogenic follicles. (a) Non-atretic follicle adjacent to (b) follicle in follicular atresia (Wang Stage I) demonstrating loss of the zona pellucida (Wang et al., 2013). H&E stain.

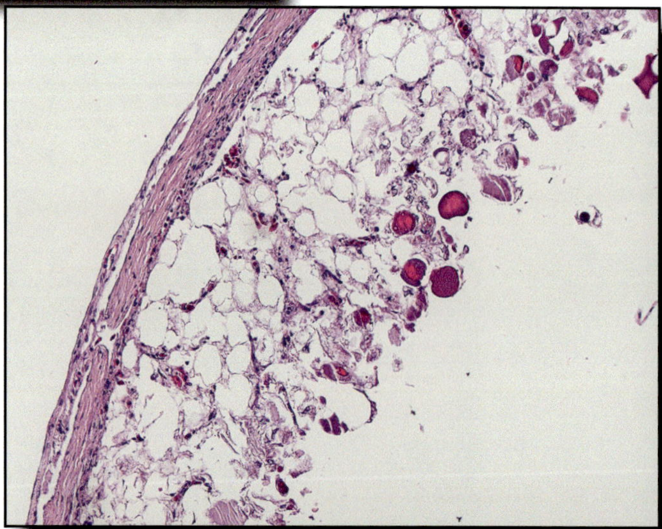

Figure 8.39 Red-Eared Slider, *Trachemys scripta elegans*. Emydidae. Wild female. In follicular atresia (Wang Stage II), yolk is replaced by cell proliferation (Wang et al., 2013). H&E stain.

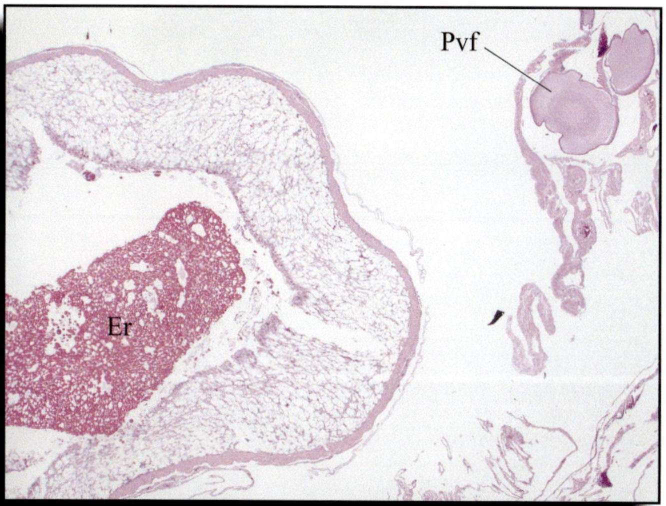

Figure 8.40 Red-Eared Slider, *Trachemys scripta elegans*. Emydidae. Wild female. In follicular atresia (Wang Stage III) , influx of erythrocytes (Er) occurs into the central ooplasm (Wang et al., 2013) H&E stain.

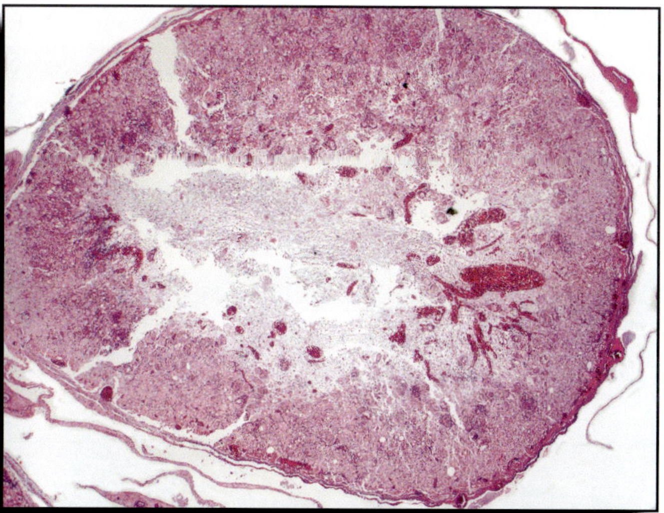

Figure 8.41 Agassiz's Desert Tortoise, *Gopherus agassizii*. Testudinidae. Wild female. Ovary. In follicular atresia (early Wang Stage IV) (Wang et al., 2013) H&E stain.

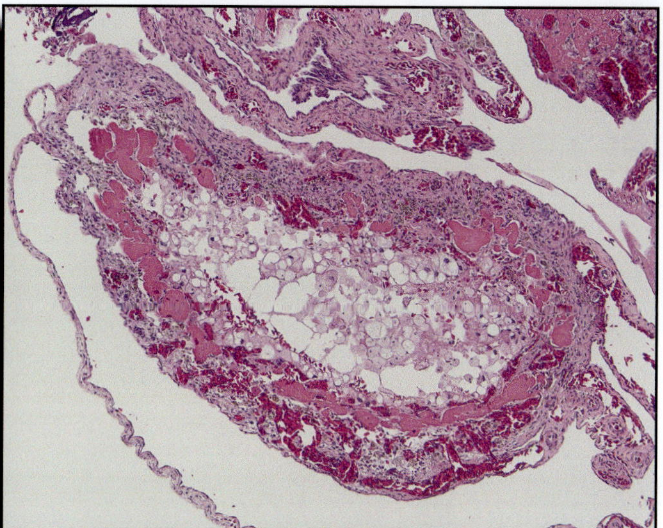

Figure 8.42 Central Bearded Dragon, *Pogona vitti* Agamidae. Captive female. Late stage 4 Atretic follicle (late Wang Stage IV). This lizard was housed alone and succumbed to dietary associated impaction (Wang et al., 2013). H&E stain.

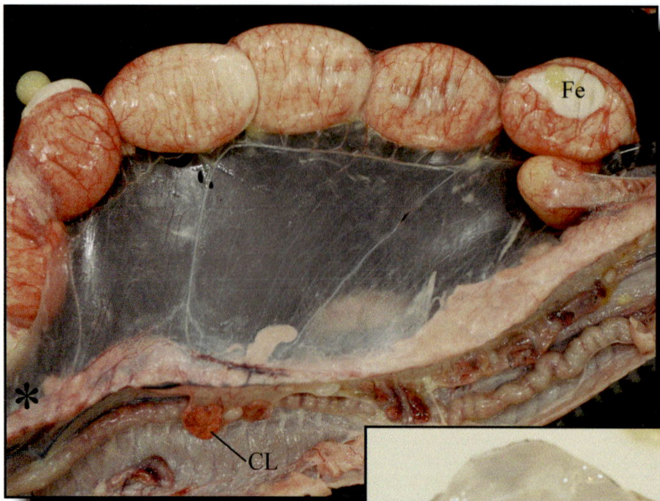

Figure 8.43 Eastern Indigo Snake, *Drymarchon couperi*. Colubridae. Wild female. Oogenesis with fertilized eggs (Fe) in the thin-walled oviduct. The ovary with multiple corpora luteum (Cl) is adjacent to the small intestine. Mineralization of eggshell has not yet occurred. Anterior (*).

Figure 8.44 Red-footed Tortoise, *Chelonoidis carbonarius*. Testudinidae. Captive female. Non-fertile egg. Eggs of chelonians are megalecithal (contain large volumes of yolk). (Courtesy of The Smithsonian Institution.)

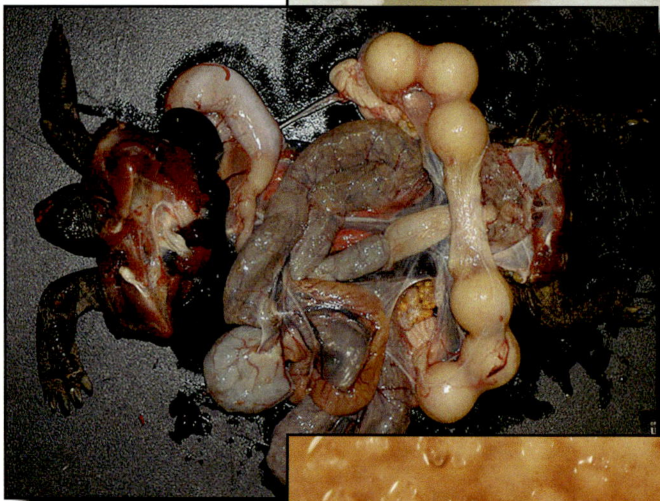

Figure 8.45 Gopher Tortoise, *Gopherus polyphemus*. Testudinidae. Wild female. Calcified fertilized eggs are present in the oviduct. This gopher tortoise was in pre-oviposition at time of death.

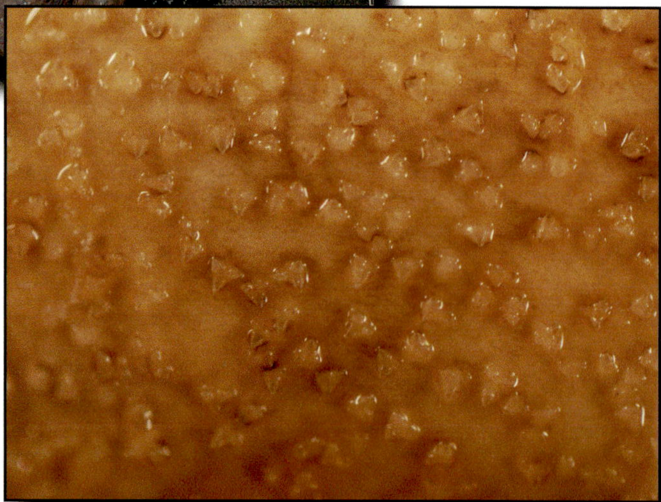

Figure 8.46 Eastern Indigo Snake, *Drymarchon couperi*. Colubridae. Egg from a wild female. Calcite crystals on surface of egg. This is a normal variation in this species. (See Volume 1, Figures 1.305, 1.306.)

Figure 8.47 Green Iguana, *Iguana iguana*. Iguanidae. Pre-ovipositional eggs from a captive female. Late-term fertilized eggs with the presence of pink spots representing the location of fusion between the fetal and allantoic surface of the shell. (Courtesy of Sid Knotek, from Knotek Z, Cermakova E, Oliveri M. 2017. *Vet Clin North Am Exot Anim Pract* 20:411–438. With permission.)

Figure 8.48 Green Iguana, *Iguana iguana*. Iguanidae. Opened pre-ovipositional eggs from Figure 8.47 demonstrating late-term embryos. (Courtesy of Sid Knotek, from Knotek Z, Cermakova E, Oliveri M. 2017. *Vet Clin North Am Exot Anim Pract* 20:411–438. With permission.)

Figure 8.49 Ball Python, *Python regius*. Pythonidae. Egg from captive female. Note normal placental vessels. (Courtesy David Barker and Tracy Barker, from Barker DG and Barker TM. *Ball Pythons*. Copyright 2006, VPI library, with permission.)

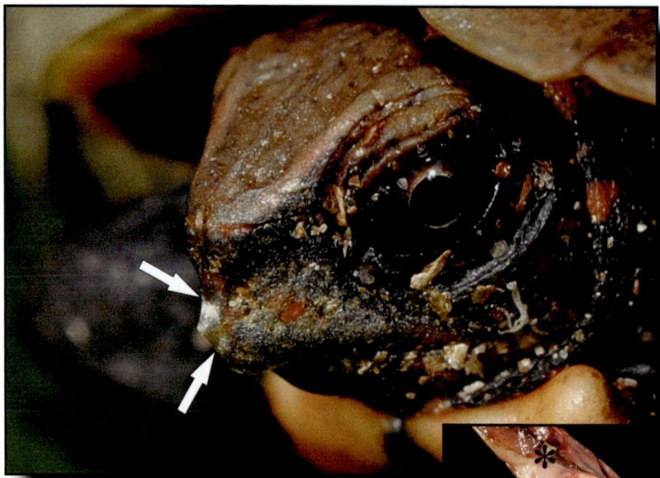

Figure 8.50 Ryukyu Black-Breasted Leaf Turtle, *Geoemyda japonica*. Geoemydidae. The small white egg tooth is at the tip of the rostrum in this recently hatched turtle (arrows). (Courtesy of Paul Gibbons, Turtle Conservancy.)

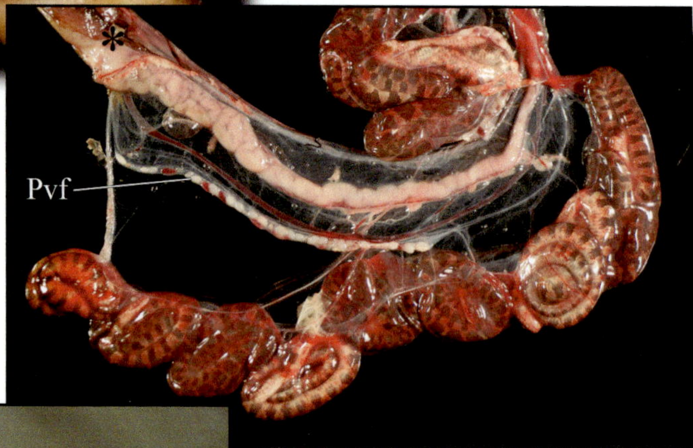

Figure 8.51 Midland Water Snake, *Nerodia sipedon pleuralis*. Colubridae. Wild female. Late-term matrotrophy dependent fetuses near time of parturition. The ovary still contains numerous previtellogenic follicles (Pvf). Regressed corpora lutea were identified microscopically between previtellogenic follicles, Figure 8.37. Anterior (*).

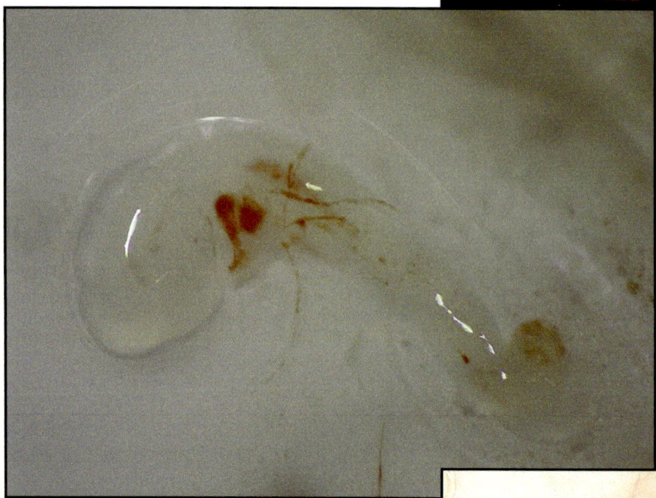

Figure 8.52 Cuban Crocodile, *Crocodylus rhombifer*. Crocodylidae. Egg from a captive female. Amnion and early embryo. (Courtesy of Lauren Augustine.)

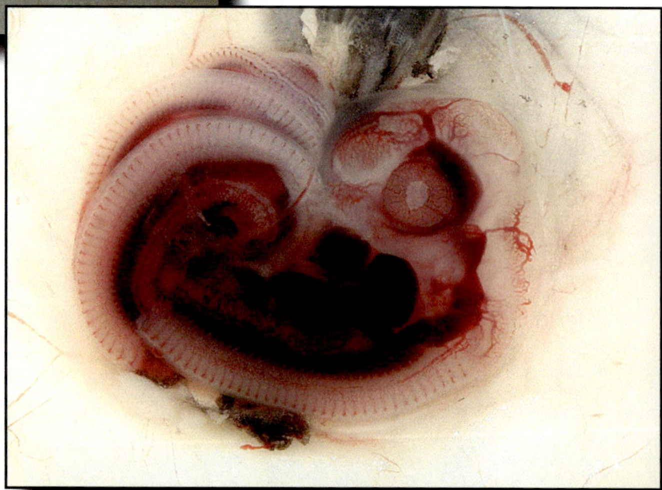

Figure 8.53 Boa Constrictor, *Boa constrictor*. Boidae. Mid-term embryo from captive female. (Courtesy of Tom Hellebuyck.)

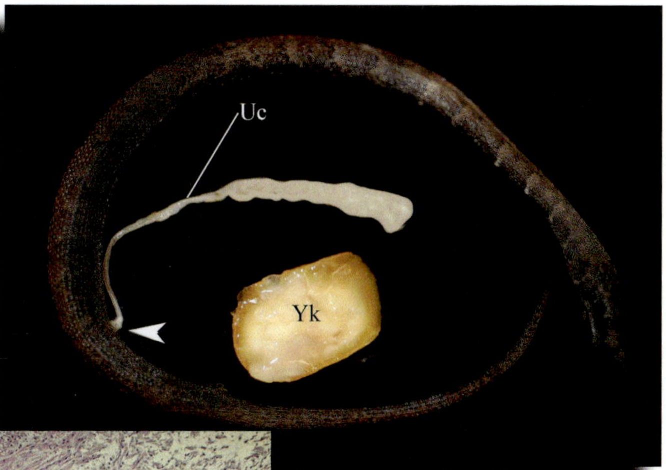

Figure 8.54 Tentacled Snake, *Erpeton tentaculatum*. Homalopsidae. Neonate from captive female. Attachment (arrowhead) of the umbilical cord (Uc) in the middle third of the body. The umbilicus is attached to the yolk (Yk) of this live-bearing species. Eggs are retained within the oviduct and this species does not depend on matrophy for late embryonic development.

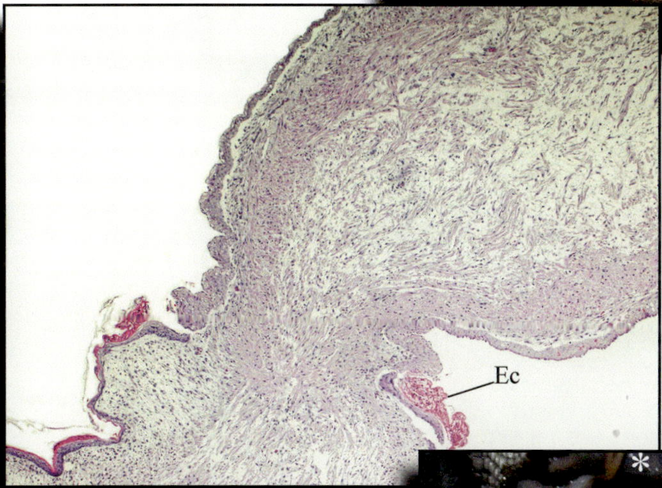

Figure 8.55 Rhinoceros Snake, *Gonyosoma boulengeri*. Colubridae. Neonate from captive female. Normal epidermal crusting (Ec) is present at the attachment of the resorbing umbilical cord and body wall. H&E stain.

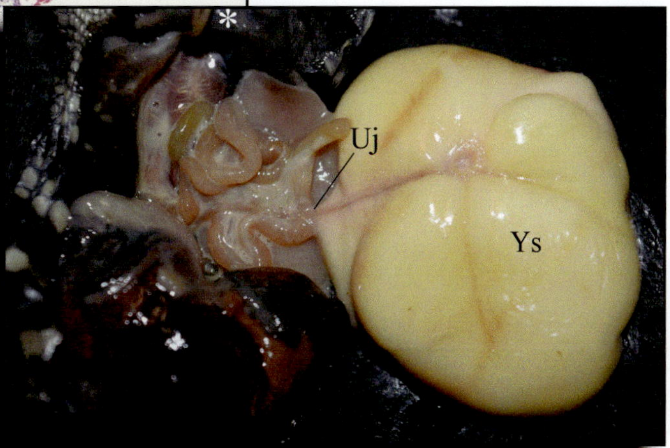

Figure 8.56 American Alligator, *Alligator mississippiensis*. Alligatoridae. Neonate from captive female. Umbilical junction (Uj) of yolk sac (Ys) with middle intestine. Anterior (*).

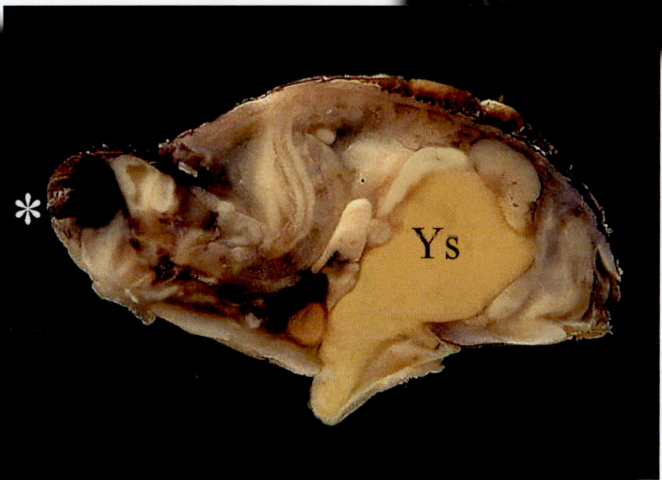

Figure 8.57 Eastern Box Turtle, *Terrapene carolina carolina*. Emydidae. Neonate from captive female. The yolk sac (Ys) occupies a large space in the coelom of recently hatched chelonians. Anterior (*). (Courtesy of Peernel Zwart.)

Figure 8.58 Painted Turtle, *Chrysemys picta*. Emydidae. Captive neonate. At pipping stage, these turtles have a prominent external yolk sac that is retracted into the coelom before and soon after emergence from the egg. The large umbilical scar (Us) normally heals rapidly. (Courtesy of Stephen Barten.)

Figure 8.59 Gopher Tortoise, *Gopherus polyphemus*. Testudinidae. Wild neonate. The urinary bladder contains urate deposits (Ud). A remnant yolk sac (Rys) is attached to the intestine. Anterior (*).

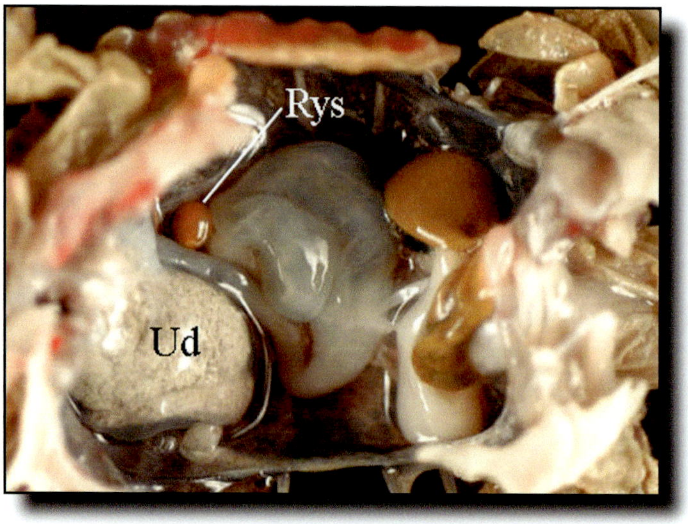

Figure 8.60 Reticulated Python, *Malayopython reticulatus*. Pythonidae. Captive emaciated female. A completely quiescent ovary with "pinpoint" corpora albicans (Ca). Only one visible previtelline follicle (Pvf) is present between the oviduct (Od) and fimbria (Fr) and small intestine (Si). Coelomic adipose foci (fat bodies) (Af) are markedly atrophic.

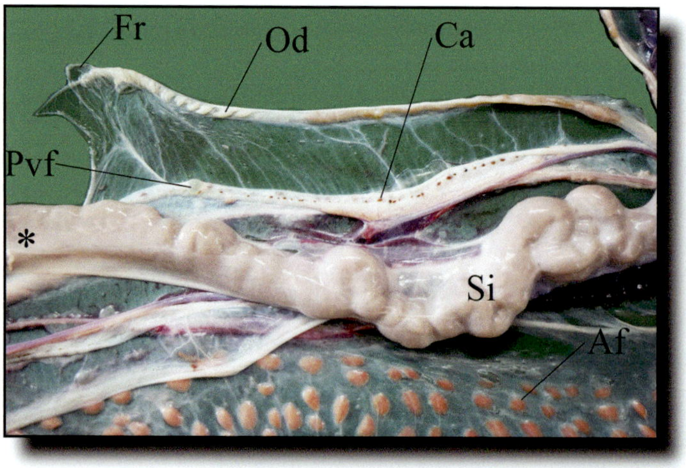

Figure 8.61 Green Anole, *Anolis carolinensis*. Dactyloidae. Wild male and female. The bright green male anole is larger (40% greater body mass on average) and practices courtship displays, but the female has a much higher energy requirement (approximately 46%) during breeding and post breeding seasons (Orrell et al., 2004).

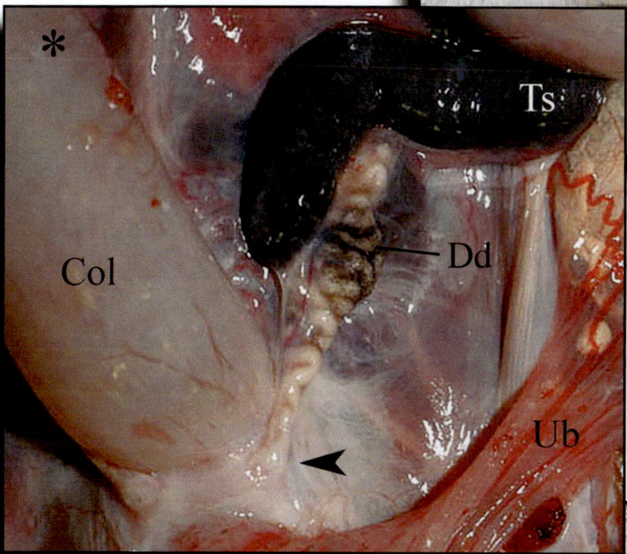

Figure 8.62 Gopher Tortoise, *Gopherus polyphemus*. Testudinidae. Wild male. Testis (Ts) and ductus deferens (Dd) adjacent to colon (Col), urinary bladder (Ub) and cloaca. The testis is atrophied due to decreased nutrient intake associated with mycoplasma induced morbidity. Melanin-laden cells are present in the serosa surrounding ductus deferens. A papilla (arrowhead) is present at the point where the vas deferens enters the cloaca. Anterior (*).

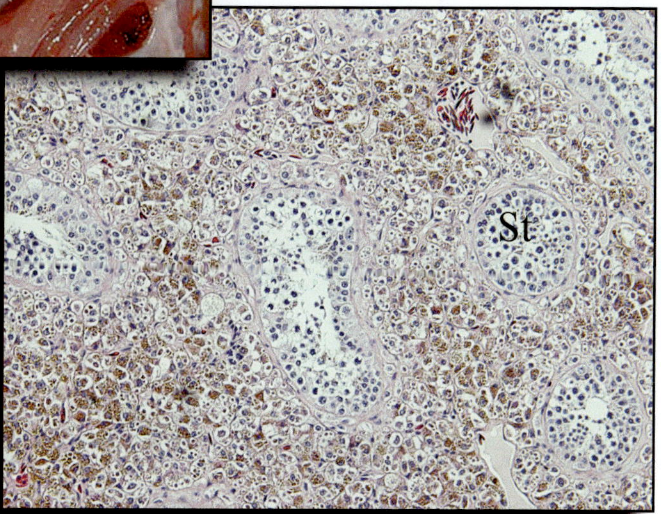

Figure 8.63 Gopher Tortoise, *Gopherus polyphemus*. Testudinidae. Wild male. Testis with melanin-laden cells in the interstitium between seminiferous tubules (St). H&E stain.

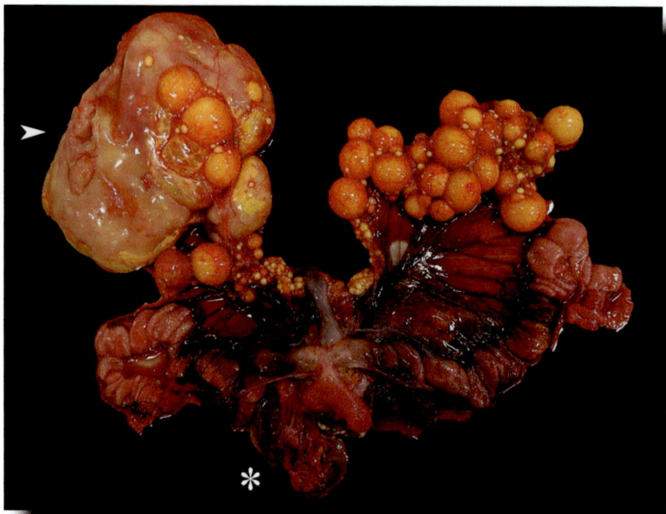

Figure 8.64 Red-Eared Slider, *Trachemys scripta elegans*. Emydidae. Captive female. Dysgerminoma (arrowhead). Reproductive tract including both ovaries and oviducts. The tumor is tan-yellow and envelopes vitellogenic follicles of the right ovary. Posterior (*). (Courtesy of Nicholas A. Robinson, Cummings School of Veterinary Medicine, Tufts University.)

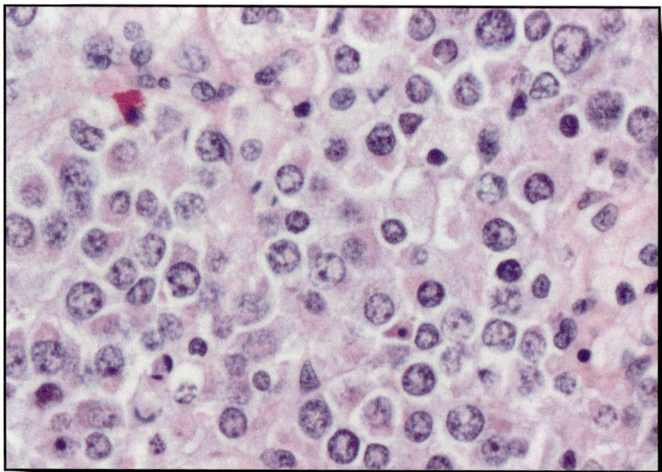

Figure 8.65 Red-Eared Slider, *Trachemys scripta elegans*. Emydidae. Captive female. Dysgerminoma. Sheets of neoplastic germ cells have moderate anisocytosis with distinct cell borders that contain moderate amounts of homogenous, pale eosinophilic cytoplasm. Nuclei are round to oval, centrally located with a vesiculate chromatin pattern, one to three nucleoli and moderate anisokaryosis. Nineteen mitotic figures were observed per 10 × high power fields (400 × magnification). H&E stain. (Courtesy of Nicholas A. Robinson, Cummings School of Veterinary Medicine, Tufts University.)

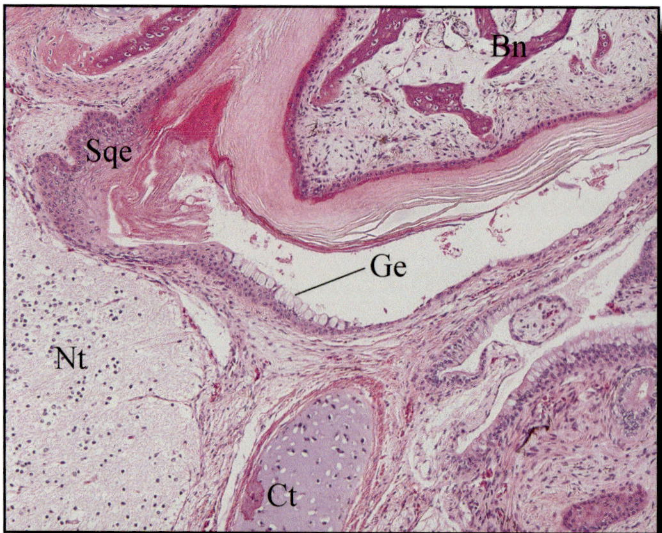

Figure 8.66 Red-Eared Slider, *Trachemys scripta elegans*. Emydidae. Captive female. Ovarian teratoma composed of three germlines of cells including ectoderm [squamous epithelium (Sqe)-skin, nervous tissue (Nt)-brain], mesoderm [cartilage (Ct) and bone (Bn)] and endoderm [glandular epithelium (Ge)]. (Courtesy of Shelley J. Newman.)

Figure 8.67 Mangrove Snake, *Boiga dendrophila*. Colubridae. Captive female. Granulosa cell tumor. Ovarian features are replaced by homogenous tan tissue. Anterior (*). (Courtesy of Tom Hellebuyck, from Pirson et al., 2018. *J Exotic Pet Med* 27:6–10. With permission.)

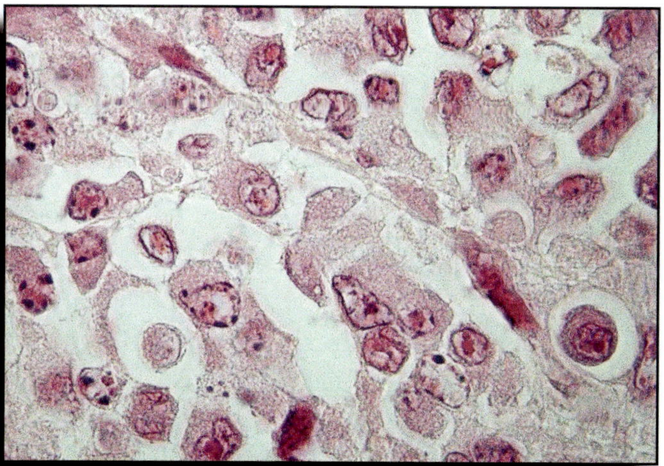

Figure 8.68 Mangrove Snake, *Boiga dendrophila*. Colubridae. Captive female. Granulosa cell tumor. Solid sheets of variably sized polyhedral neoplastic granulosa cells with abundant eosinophilic cytoplasm. H&E stain. (Courtesy of Tom Hellebuyck, from Pirson et al., 2018. *J Exotic Pet Med* 27:6–10. With permission.)

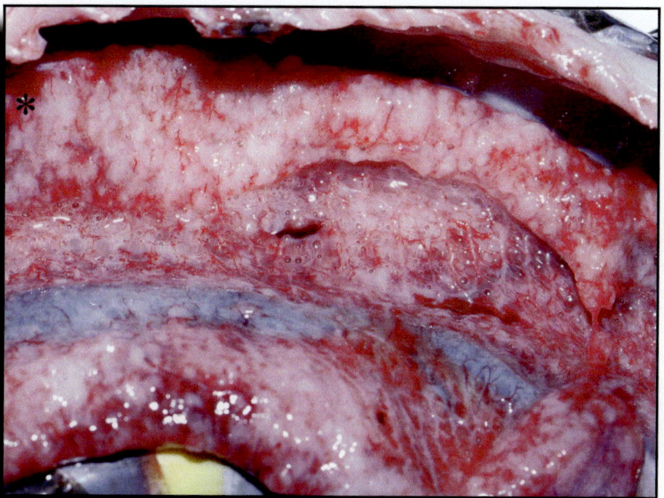

Figure 8.69 Mangrove Snake, *Boiga dendrophila*. Colubridae. Captive female. Granulosa cell tumor. Note extensive bilateral pulmonary metastasis. Anterior (*). (Courtesy of Tom Hellebuyck.)

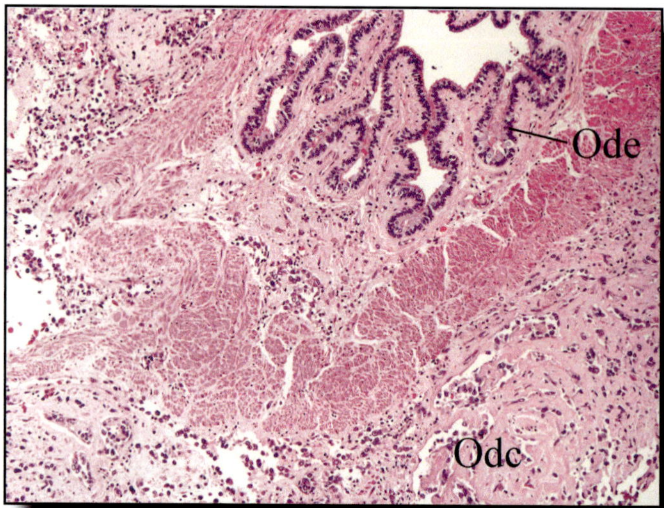

Figure 8.70 Corn Snake, *Pantherophis guttatus*. Colubridae. Captive female. Oviduct carcinoma. The normal oviduct and oviduct epithelium (Ode) is distinguishable from the oviduct carcinoma (Odc) that has invaded submucosa and subserosa. The tumor is comprised of small nests of neoplastic epithelium and is associated with a prominent scirrhous response.

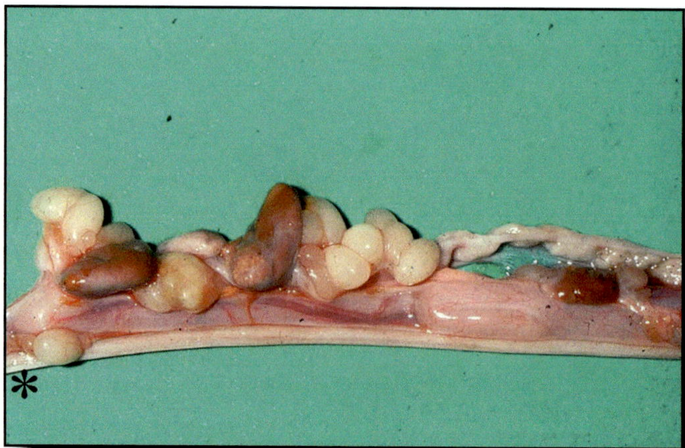

Figure 8.71 Garter Snake, *Thamnophis* sp. Colubridae. Captive, laboratory-housed female. This snake was housed under conditions of restricted light cycling. There was stasis of preovulatory follicles over multiple cycles resulting in vitellogenic oophoritis (fused and reddened follicles). Anterior (*). (Courtesy of Peernel Zwart.)

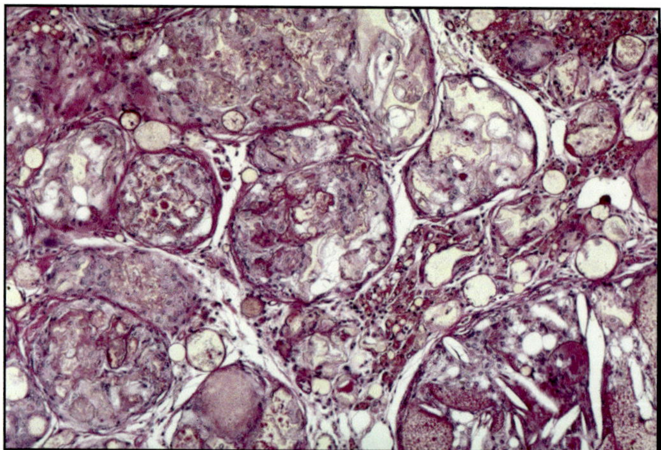

Figure 8.72 Garter Snake, *Thamnophis* sp. Colubridae. Captive, laboratory-housed female. This snake was housed under conditions of restricted light cycling. There was stasis of preovulatory follicles resulting in vitelogenic oophoritis. Macrophages contain phagocytized atretic follicular material. Periodic acid-Schiff (PAS) stain. (Courtesy of Peernel Zwart.)

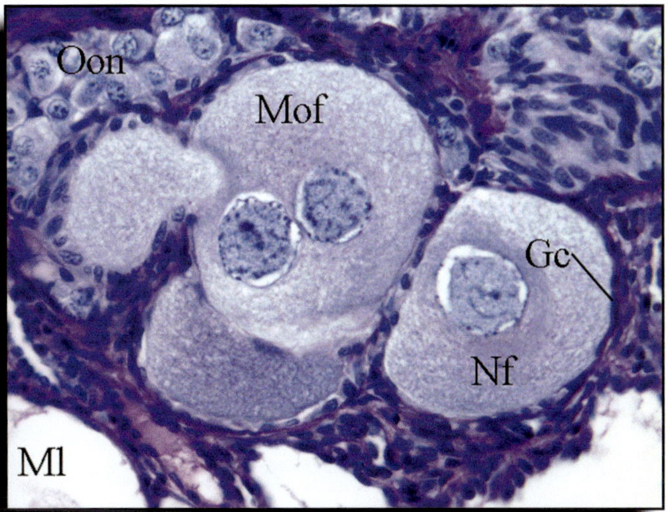

Figure 8.73 American Alligator, *Alligator mississippiensis*. Alligatoridae. Wild female. An ovary from a 5-month-old alligator that was exposed to an environmental endocrine disrupting chemical, (xenoestrogen) during multiple stages of development. A multi-oocytic follicle containing a multinucleated follicle (Mnf), surrounded by granulosa cells (Gc), an oocyte nest (Oon) with stage 1 and stage 2 nonfollicular oocytes containing meiotic chromatin, a normal stage 3 follicle (Nf) and medullary lacunae (Ml). PAS counterstained with hematoxylin. (Courtesy of Dr. Brandon Moore and Dr. Louis Guillette Jr.)

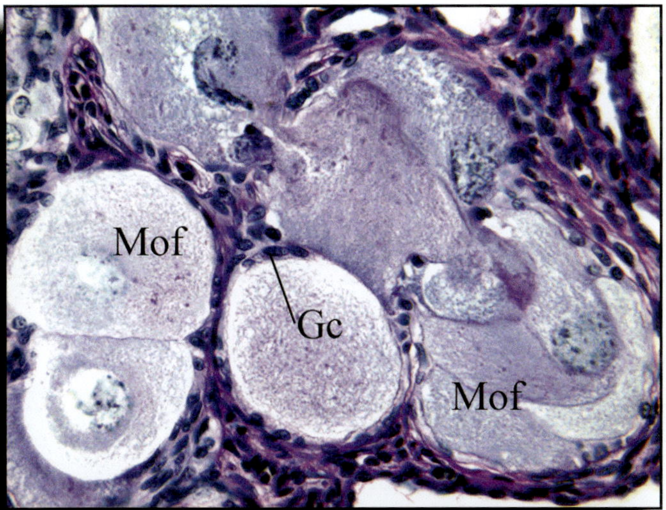

Figure 8.74 American Alligator, *Alligator mississippiensis*. Alligatoridae. Wild female. Ovary of a 5-month-old alligator that was exposed to an environmental endocrine disrupting chemical, (xenoestrogen) during multiple stages of development. Irregular multi-oocytic follicles (Mof) surrounded by well differentiated granulosa cells (Gc) that are contiguous between oocytes. PAS counterstained with hematoxylin. (Courtesy of Dr. Brandon Moore and Dr. Louis Guillette Jr.)

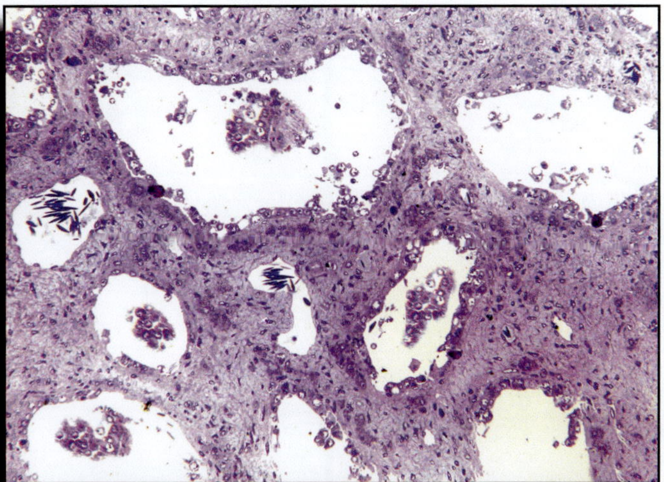

Figure 8.75 American Alligator, *Alligator mississippiensis*. Alligatoridae. Testes revealing loss of seminiferous epithelium. Suspected endocrine-disrupting chemical (EDC) exposure during development. PAS with Toluidine blue counterstain. (Courtesy of Kevin Gribbins.)

Figure 8.76 Broad-Snouted Caiman, *Caiman latirostris*. Alligatoridae. Egg from wild female. Scanning electron microscopic surface of eggshell demonstrating loss of pore structure due to endocrine disrupting substance in environment. (Courtesy of Mónica Muñoz-de-Toro.)

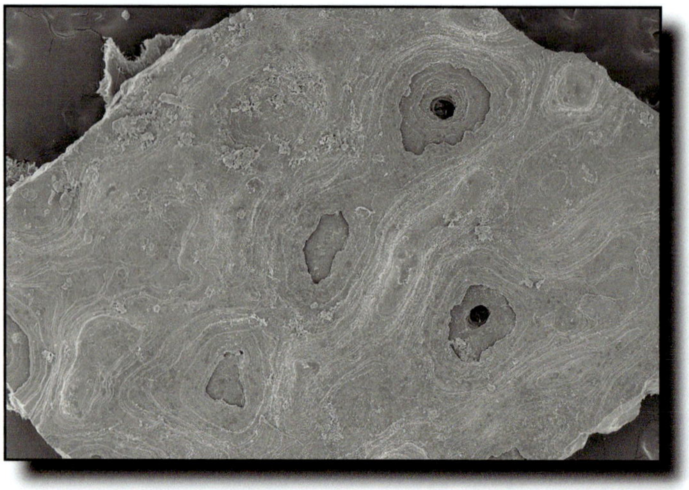

Figure 8.77 Italian Wall Lizard, *Podarcis sicula*. Lacertidae. Captive female from laboratory colony. Cadmium exposure. Single intraperitioneal cadmium chloride administration (1 µg / gr lizard body weight), 30 days. Large previtellogenic follicle with the granulosa cell (Gc) layer forming an irregular protrusion (*) into the oocyte cytoplasm (Occ). Also notice a regressing cell (Rc) and the thick zona pellucida (Zp). Oocyte nucleus (Oon). H&E stain. (Simoniello et al., 2013) (Courtesy of Chiara M. Motta.)

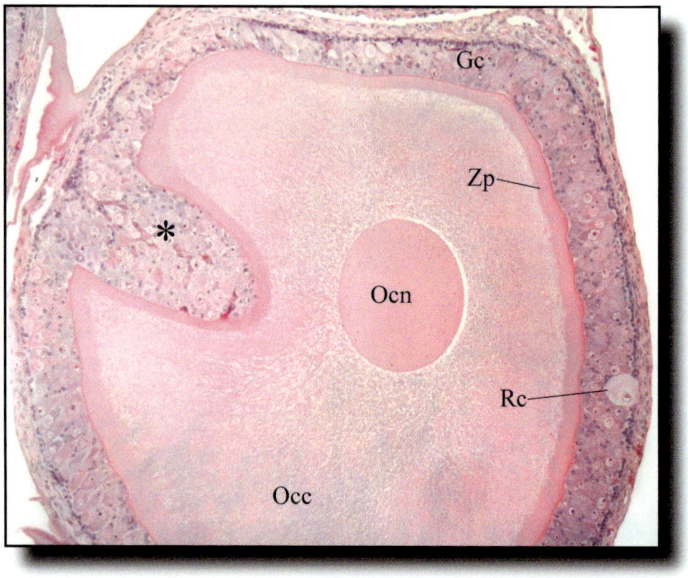

Figure 8.78 Italian Wall Lizard, *Podarcis sicula*. Lacertidae. Captive female from laboratory colony. Cadmium exposure. Following cadmium exposure, lesions are recognizable in the zona pellucida. In the control, the zona pellucida has a regular organization (a); after cadmium exposure, discontinuities or vesicles (arrowhead) are formed in the zona pellucida (b). PAS stain. (Simoniello et al., 2013) (Courtesy of Chiara M. Motta.)

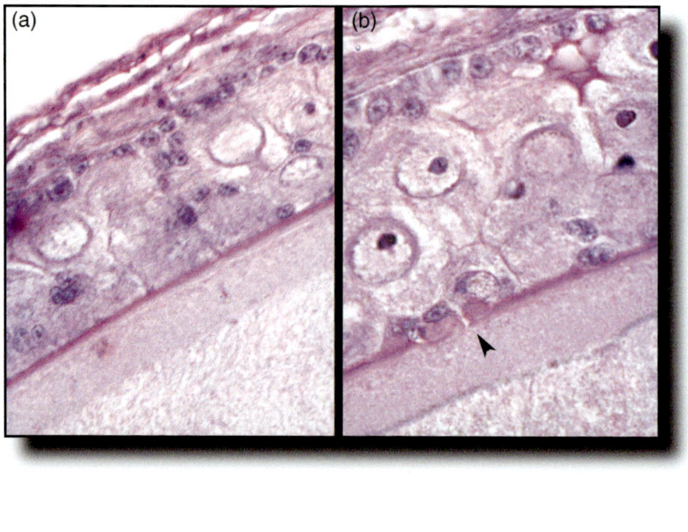

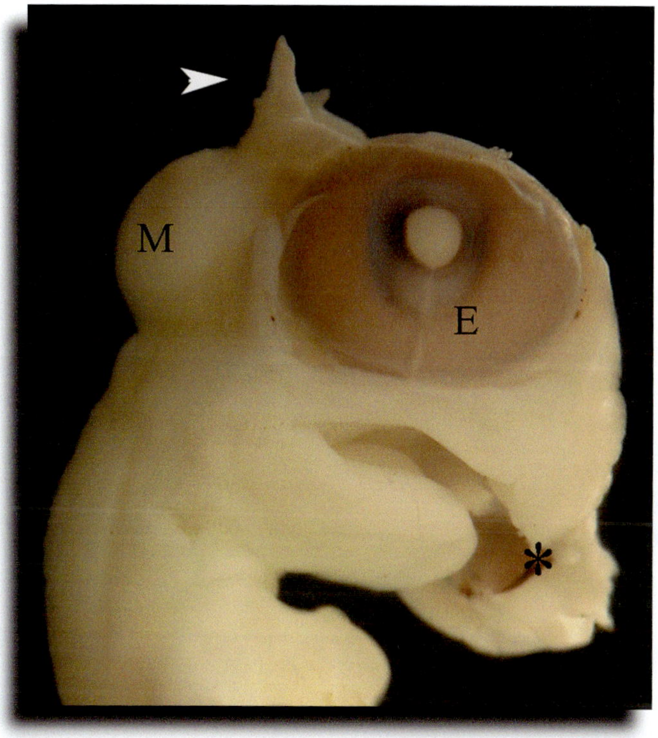

Figure 8.79 Italian Wall Lizard, *Podarcis sicula*. Lacertidae. Captive female from a laboratory colony. Cadmium treatment. Head of an embryo at 20 days post in ovo exposure of soil contaminated with 50 mg Cd/ Kg soil. Note the eye (E), the cephalic vesicle (M), a fleshy protrusion from the diencephalon (arrowhead) and the completely dislocated left eye (*). H&E stain. (Simoniello et al., 2013) (Courtesy of Chiara M. Motta.)

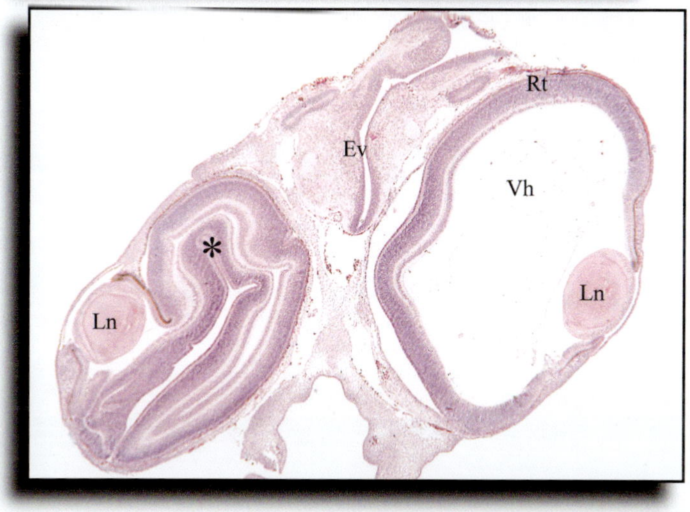

Figure 8.80 Italian Wall Lizard, *Podarcis sicula*. Lacertidae. Captive female form laboratory colony. Cadmium treatment. Cross section of the head of an embryo at twenty days post cadmium treatment showing the altered encephalic vesicle (Ev) and unilateral microphthalmia (*) with folded globe. Notice the contralateral eye has an organized retina (Rt) and large vitreous humor (Vh). Lens (L). H&E stain. (Simoniello et al., 2013) (Courtesy of Chiara M. Motta, from Simoniello et al., 2013.)

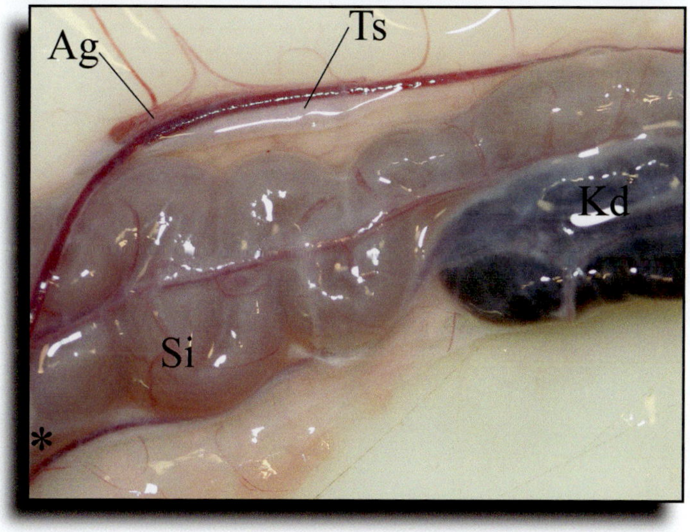

Figure 8.81 Emerald Tree Boa, *Corallus caninus*. Boidae. Emaciated captive adult male. The atrophied white testicle (Ts) is adjacent to a dark red adrenal gland (Ag) and near the dark red kidney (Kd) and small intestine (Si). Anterior (*). (Courtesy of The Smithsonian Institution.)

Figure 8.82 Gila monster, *Heloderma suspectum*. Helodermatidae. Captive geriatric male. Multinucleated germ cells (arrowheads) in a captive lizard with chronic debilitating disease. H&E stain.

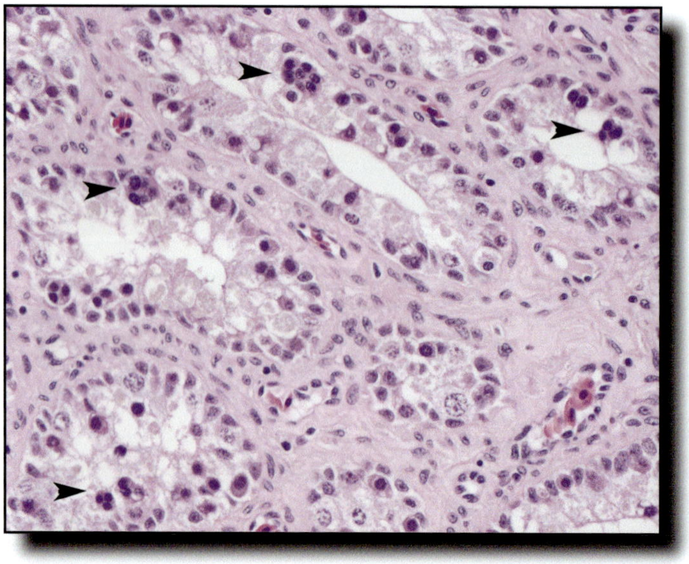

Figure 8.83 Amazon Tree Boa, *Corallus hortulanus*. Boidae. Captive geriatric male. Degeneration of seminiferous epithelium. Seminiferous tubules are devoid of spermatozoa and the lumen contains eosinophilic and golden droplets (residual bodies); the interstitium has increased fibrosis. H&E stain.

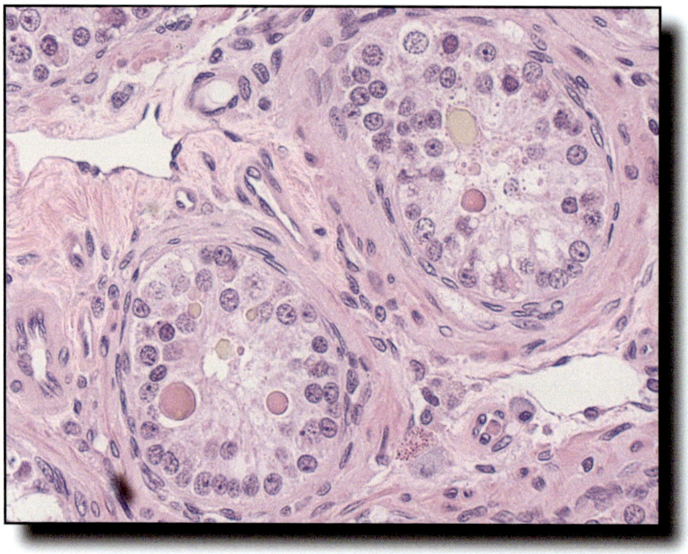

Figure 8.84 Blue-Tongued Skink, *Tiliqua scincoides*. Scincidae. Testis. Captive male, geriatric. There is overall degeneration of the testis. There is diminished spermatogenesis, with seminiferous tubules lined by a single layer of Sertoli cells containing cytoplasmic brown pigment and low numbers of spermatogonia. The basement membranes of the seminiferous tubules are thickened by eosinophilic extracellular matrix. H&E stain.

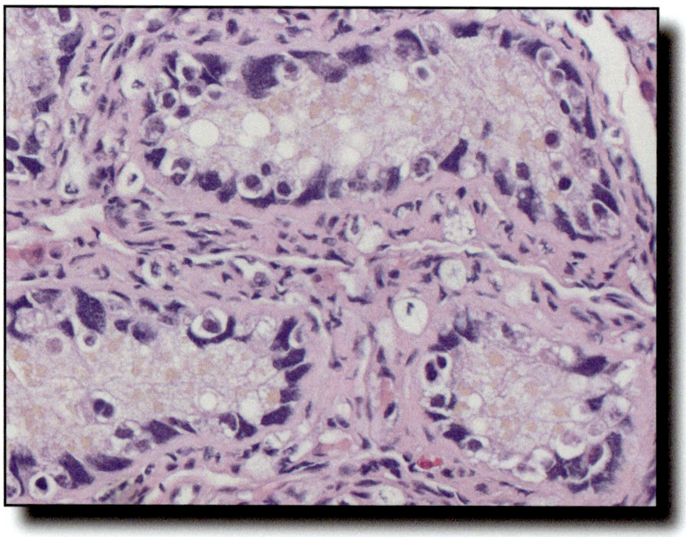

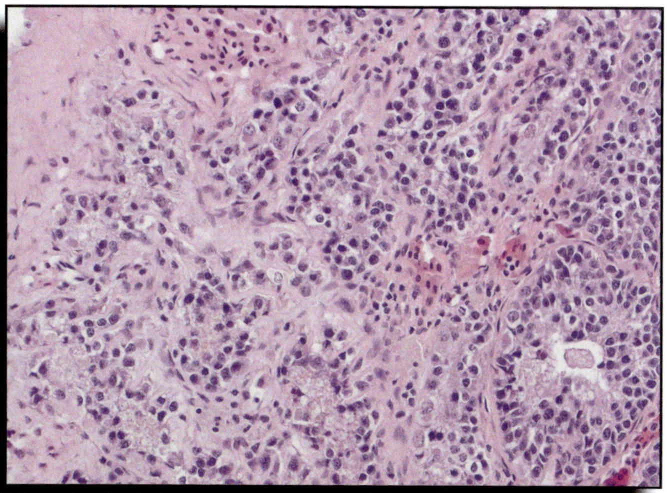

Figure 8.85 Leopard Gecko, *Eublepharis macularius*. Eublepharidae. Captive geriatric male. Testis. There is capsular and interstitial fibrosis, disassociation of Sertoli cells and disruption of the tubular basement membrane. H&E stain.

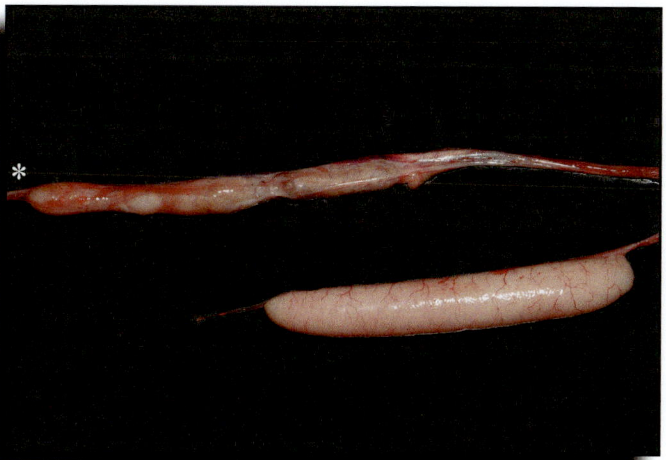

Figure 8.86 Corn Snake, *Pantherophis guttatus*. Colubridae. Captive male. Testis. Unilateral orchitis in a testis leading to asymmetry, degeneration and collapse of the tunica albuginea. *Salmonella arizonae* was cultured from the collapsed testis. Anterior (*). (Courtesy of Tom Hellebuyck.)

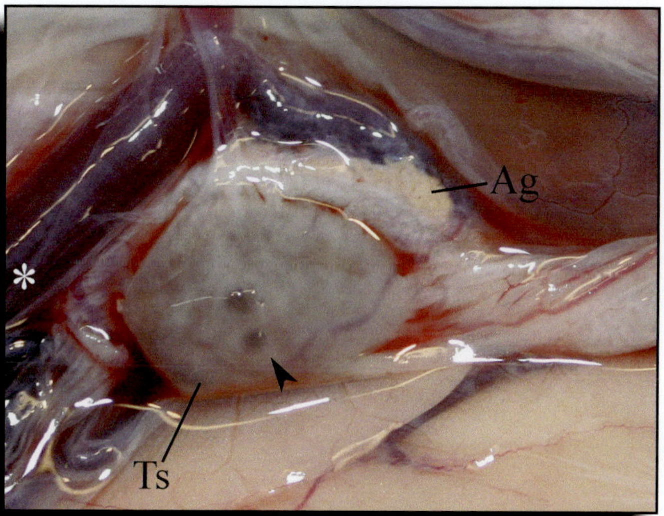

Figure 8.87 Gila Monster, *Heloderma suspectum*. Helodermatidae. Captive geriatric male. The gray testis (Ts) is adjacent to the epididymis and adrenal gland (Ag). The testis is contracted and contains cysts (arrowhead). The coiled epididymis is separated by edema. Anterior (*). (Courtesy of The Smithsonian Institution.)

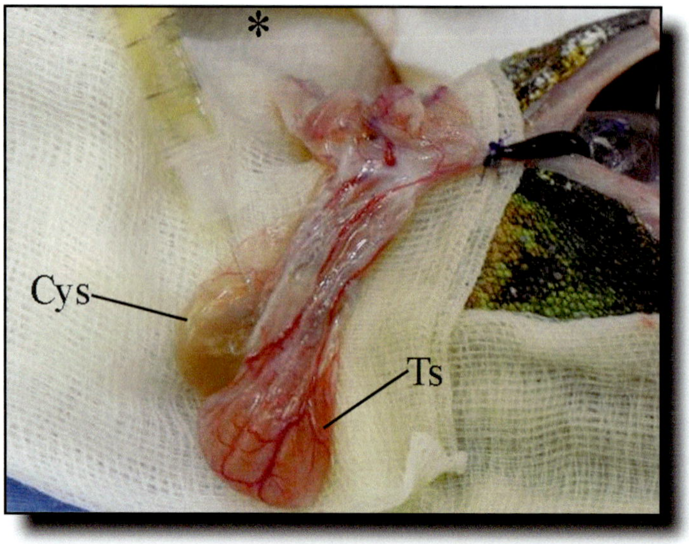

Figure 8.88 Veiled Chameleon, *Chamaeleo calyptratus*. Chamaeleonidae. Captive male. Testicular cyst (Tc) after fluid aspiration. Testicle (Ts). Anterior (*). (Courtesy of Sid Knotek. From Knotek et al., 2017. *Vet Clin North Am Exot Anim Pract* 20:411–438. With permission.)

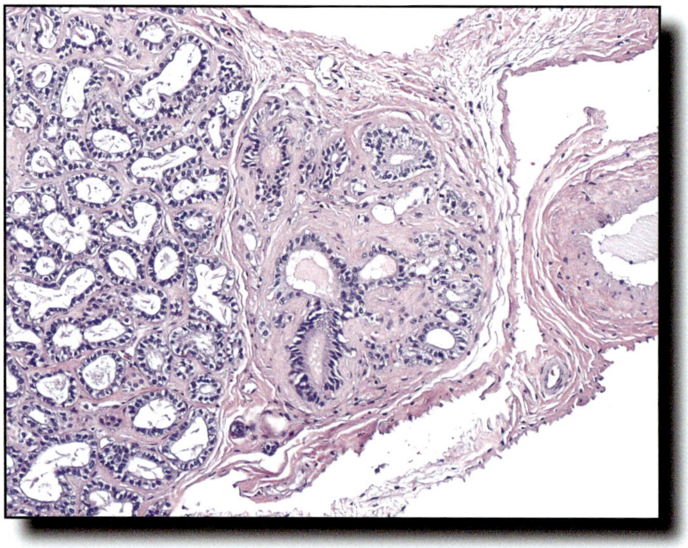

Figure 8.89 Gray Banded Kingsnake, *Lampropeltis alterna*. Colubridae. Captive geriatric male. Interstitial fibrosis between tubules at junction of testicle and epididymis. H&E stain.

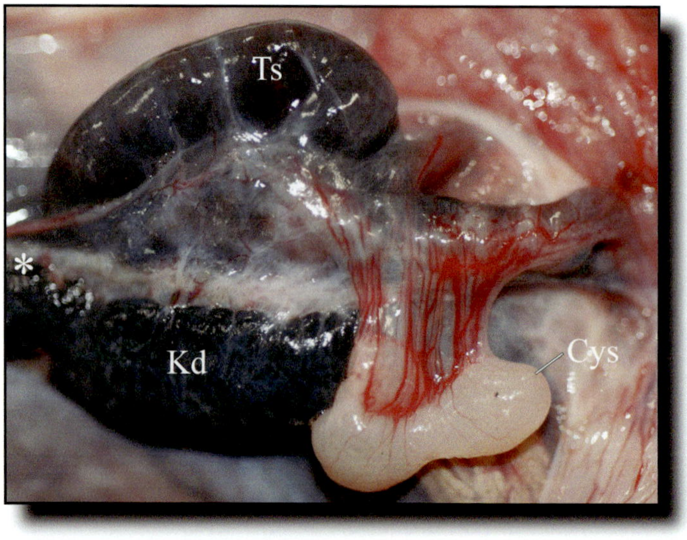

Figure 8.90 Gopher Tortoise, *Gopherus Polyphemus*. Testudinidae. Wild male. Spermatocele cyst (Cys), involving the vas deferens, distal to the testis (Ts). Both the testicle (Ts) and kidney (Kd) have prominent accumulation of melanin. Anterior (*).

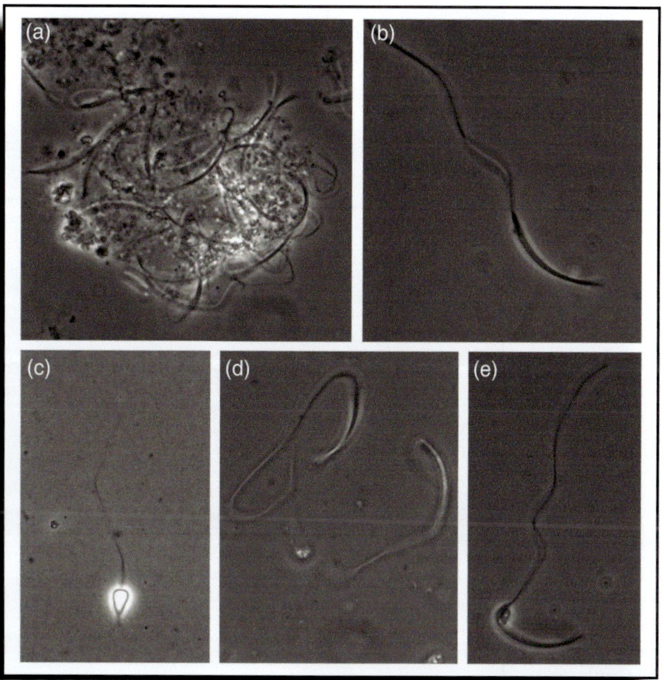

Figure 8.91 Grand Cayman Rock Iguana hybrid, *Cyclura lewisi x nubilia*. Iguanidae. Captive male. Spermatozoa. (a) Normal clumping of spermatozoa. (b) Normal sperm. (c) Pyriform sperm head. (d) Kinking of spermatozoa. (e) Spermatozoa with a cytoplasmic droplet. Dark field microscopy. (Courtesy of Sean Perry.)

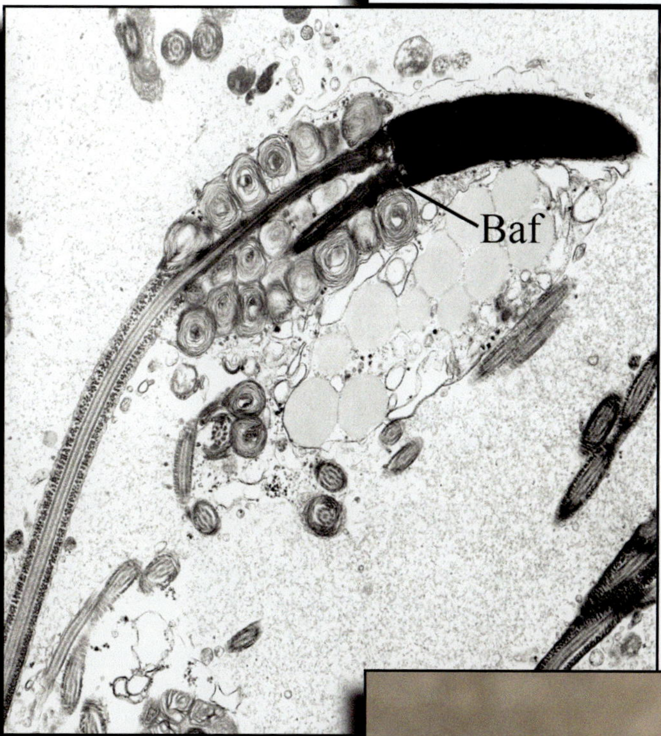

Figure 8.92 Chinese Softshell Turtle, *Pelodiscus sinensis*. Trionychidae. Captive male. Double-tailed (biflagellate) sperm. The point of bifurcation of the axial filament (Baf) is present at the junction of the mid piece and head. Transmission electron microscopy. (Courtesy of Qiusheng Chen.)

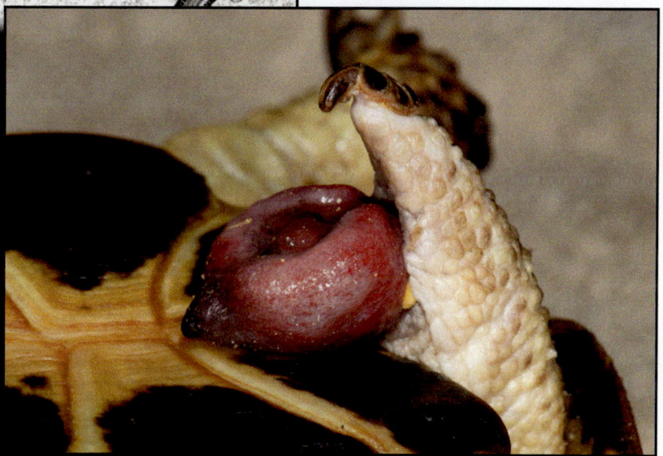

Figure 8.93 Russian Tortoise, *Agrionemys horsfieldii*. Testudinidae. Captive male. Prolapsed penis. (Courtesy of Stephen Barten.)

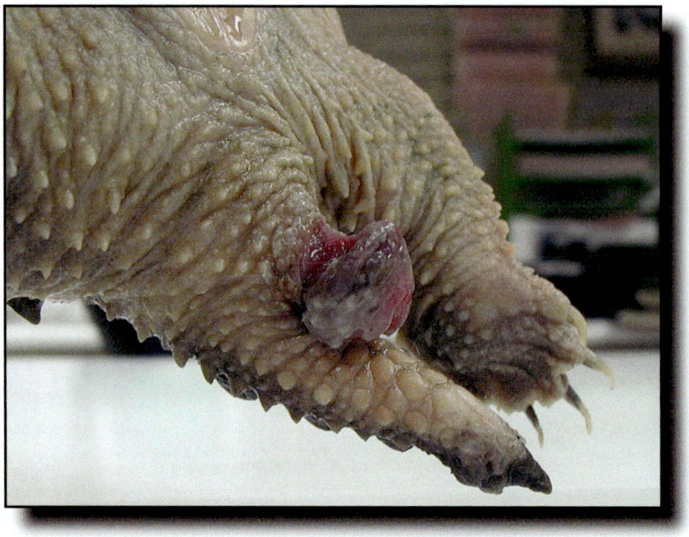

Figure 8.94 Common Snapping Turtle, *Chelydra serpentina*. Chelydridae. Prolapsed penis and exudative balanitis. (Courtesy of Charles Innis.)

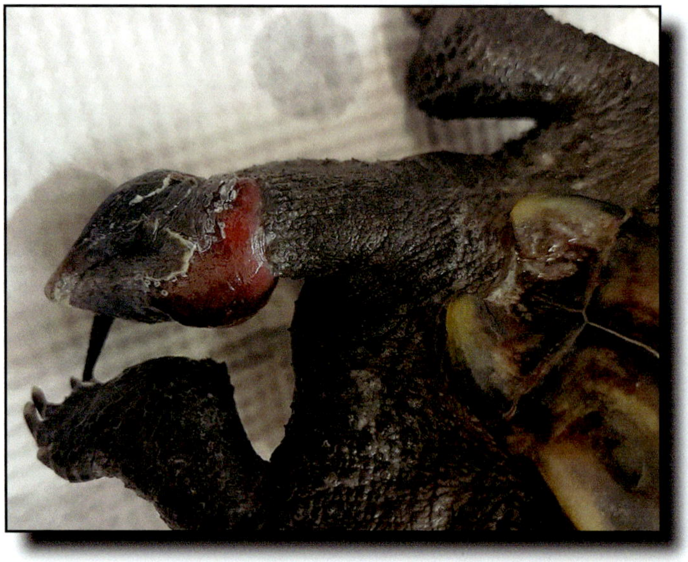

Figure 8.95 Chinese Pond Turtle (Reeves turtle), *Mauremys reevesii*. Geoemydidae. Prolapsed penis with chronic erosive balanitis. (Courtesy Nicola Di Girolamo.)

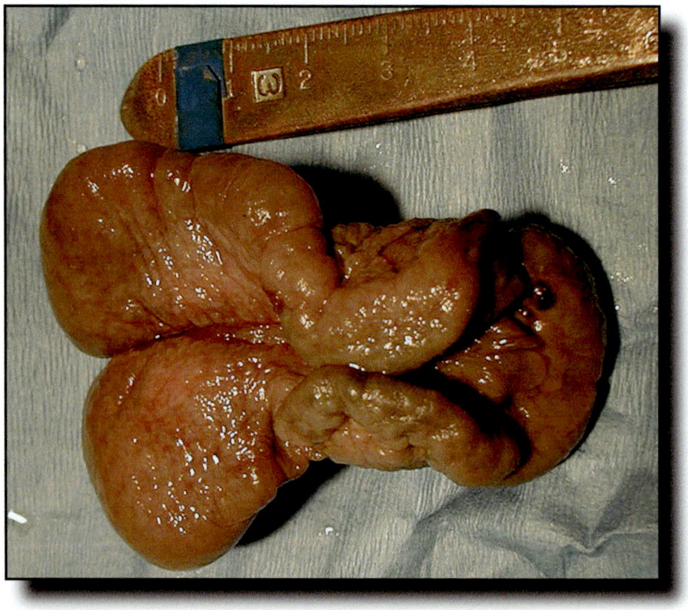

Figure 8.96 African Spurred Tortoise, *Centrochelys sulcata*. Testudinidae. Prolapsed and edematous penis prior to surgical resection. The central groove (sulcus spermaticus) is prominent in the edematous penis. In good health the groove transfers sperm to the female cloaca during copulation. (Courtesy of Charles Innis, reprinted from Innis CJ1, Boyer TH, *Vet Clin North Am Exot Anim Pract* 5(3):555–78. Chelonian reproductive disorders. Copyright (2002), with permission from Elsevier.)

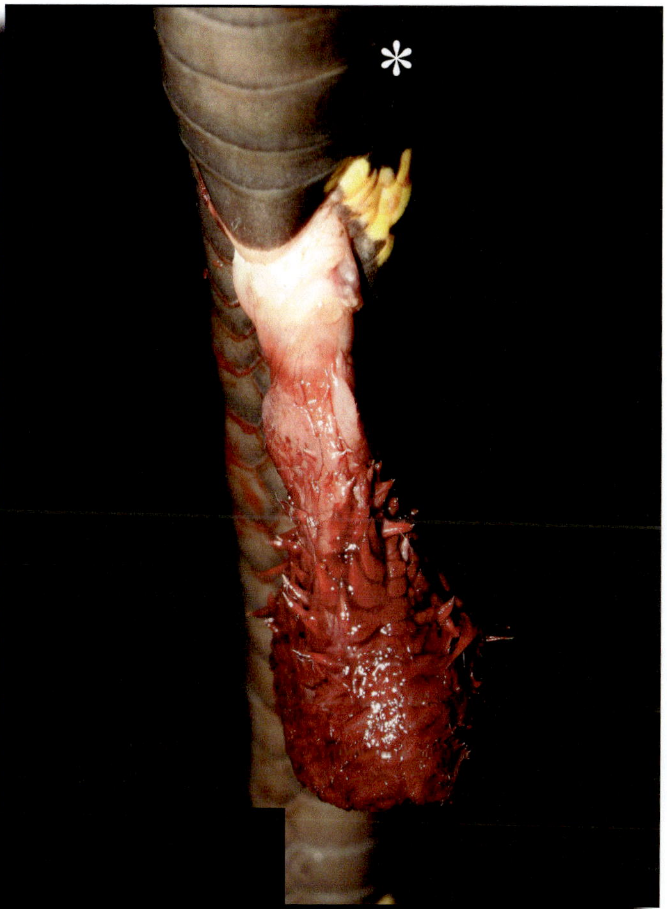

Figure 8.97 Mangrove Snake, *Boiga dendrophila*. Colubridae. Captive male. Hemipene prolapse. The prolapsed hemipene is congested and edematous. Anterior (*). (Courtesy of Tom Hellebuyck.)

Figure 8.98 Boa Constrictor, *Boa constrictor.* Boidae. Captive male. Chronic hemipene prolapse. The tissue has become desiccated and necrotic. Anterior (*). (Courtesy of Paul Raiti.)

Figure 8.99 Leopard Gecko, *Eublepharis macularius*. Eublepharidae. Captive male. Severely inflamed hemipenal sulcus with bilateral retained hemipenal plugs. (Courtesy of Paul Raiti.)

Figure 8.100 Veiled Chameleon, *Chamaeleo calyptratus*. Chamaeleonidae. Captive male. Seminal plug. (Courtesy of Stephen Barten.)

Figure 8.101 Green Iguana, *Iguana iguana*. Iguanidae. Captive male. Note prominent femoral pores (Fp) and seminal plugs (Sp), one of which has been removed. (Courtesy of Stephen Barten.)

Figure 8.102 Sinaloan Milk Snake, *Lampropeltis triangulum sinaloa.* Colubridae. Captive female. Scent gland adenitis. (a) Bilateral swelling distal to the vent (V). (b) Dilated caudal scent glands (Sg) or "anal sacs." Based on gross appearance and location, swelling in scent glands can be mistaken for a hemipene plug. Anterior (*). (Courtesy of The Smithsonian Institution.)

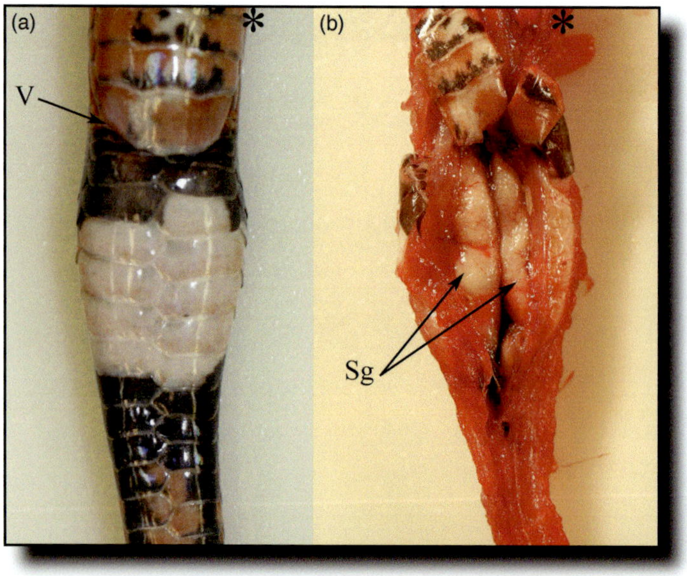

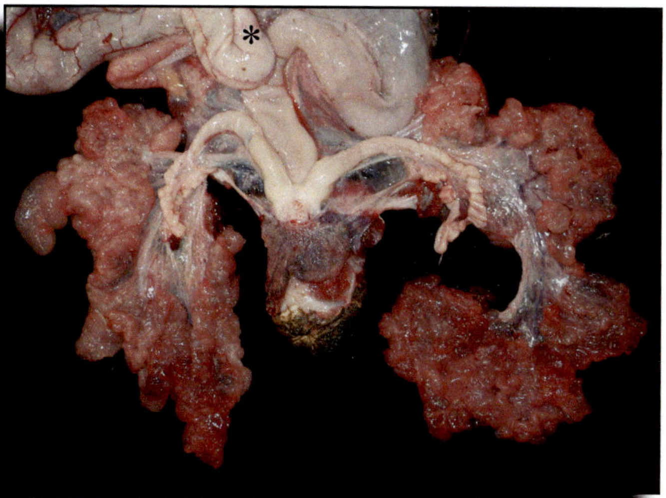

Figure 8.103 Gopher Tortoise, *Gopherus Polyphemus*. Testudinidae. Wild female. There is polycystic dysplastic folliculogenesis. Both ovaries are devoid of previtellogenic or vitellogenic follicles. Anterior (*).

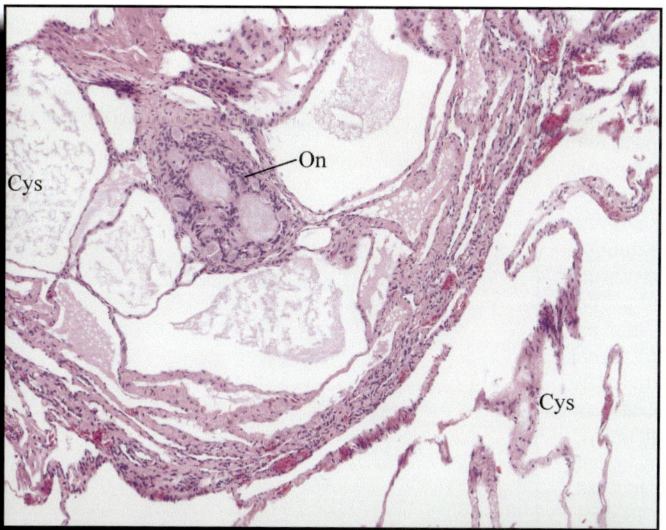

Figure 8.104 Gopher Tortoise, *Gopherus Polyphemus*. Testudinidae. Wild female. Polycystic dysplastic folliculogenesis. Ovarian tissue is confirmed by the presence of an oogonial nest (On). Normal progression of folliculogenesis is not present and follicles are replaced by cysts (Cys) lined by cells of epithelial origin.

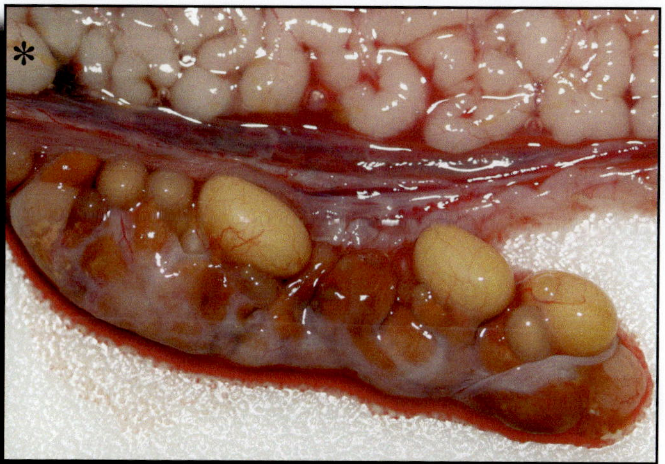

Figure 8.105 Rhinoceros Ratsnake, *Gonyosoma boulengeri*. Colubridae. Captive female. Simultaneous acute regression of vitellogenic follicles resulting in oophoritis (the dark brown follicles), and regional coelomitis (pale regions around the follicles). Anterior (*). (Photo by Tim Walsh. Courtesy of The Smithsonian Institution.)

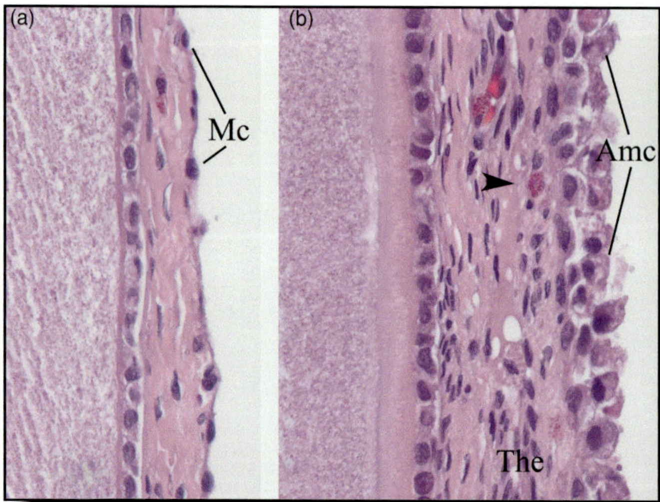

Figure 8.106 Eastern Box Turtle, *Terrapene carolina carolina*. Emydidae. Captive female. Coelomic surface of vitellogenic follicles, with and without intracoelomic yolk phagocytosis. (a) Mesothelial cells (Mc) on the outer surface of normal follicles without yolk phagocytosis. (b) The outer mesothelial surface of the follicle during yolk phagocytosis coated by activated mesothelial cells (Amc) containing phagocytized yolk. Low numbers of infiltrating heterophils (arrowhead) are present in the theca externa (The) of this follicle.

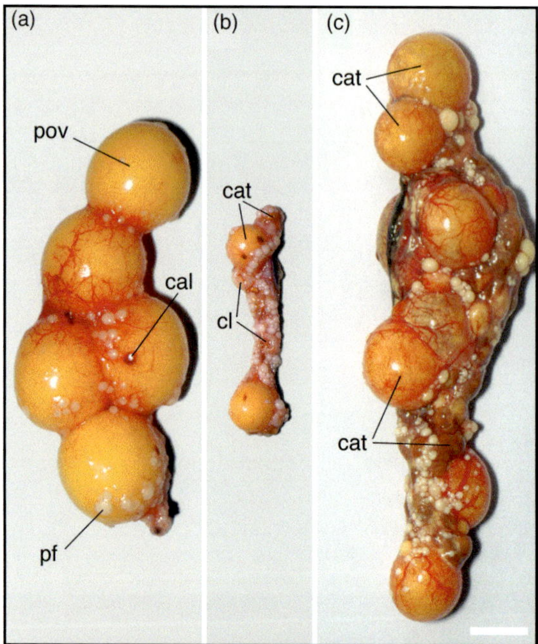

Figure 8.107 Tuatara. *Sphenodon punctatus*. Sphenodontidae. Wild and captive females. (a) Right ovary from wild female in preovulatory condition with five enlarged (18–19 mm in diameter) preovulatory follicles. Immature white or creamy pre-vitellogenic follicles up to 4 mm diameter and small flat brown scars, presumptive corpora albicans) are also visible. (b) Left ovary from wild gravid female with at least seven corpora lutea up to 5 mm with visible ovulation apertures. Note several yolk follicles from 5–12 mm diameter (presumed corpora atretica), numerous pre-vitellogenic follicles, and several presumed corpora albicantia. For histology of the corpora lutea of this female, see female M52 of Guillette and Cree (1997). Uterine eggs of this female were radio-opaque but incompletely calcified. (c) Abnormal ovary of overweight captive female with exposure to a male for several years. This ovary contains enlarged follicles up to 17 mm diameter with signs of atresia (uneven or brown coloration, sometimes with internal fluid), as well as numerous pre-vitellogenic follicles. Abnormal fatty deposits were present on the liver and peritoneum. *Abbreviations:* cal, corpus albicans; cat, corpus atreticum; cl, corpus luteum; pf, pre-vitellogenic follicle; pov, preovulatory follicle. Scale bar, 1 cm in all. (Courtesy of Alison Cree (a, b), Marcus Simons (c), from Cree, 2015. *Reproductive Biology and Phylogeny of Lizards and Tuatara*. With permission.)

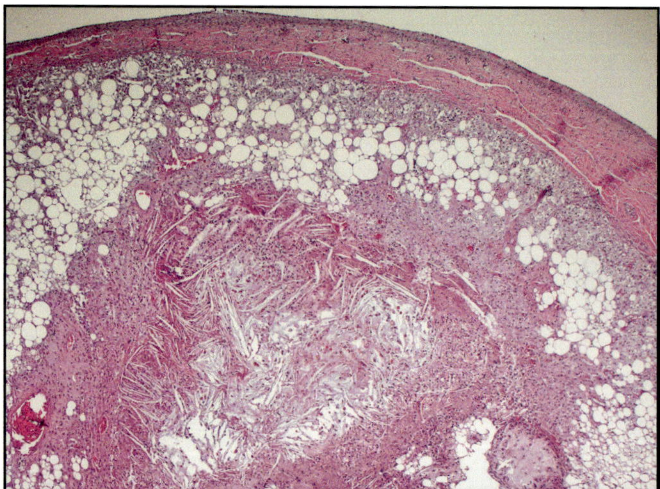

Figure 8.108 Viper Boa, *Candoia aspera*. Boidae. Captive female, geriatric. Ovary. Degenerated vitellogenic follicle with large central area of cellular debris containing cholesterol clefts surrounded by granulomatous inflammation. H&E stain.

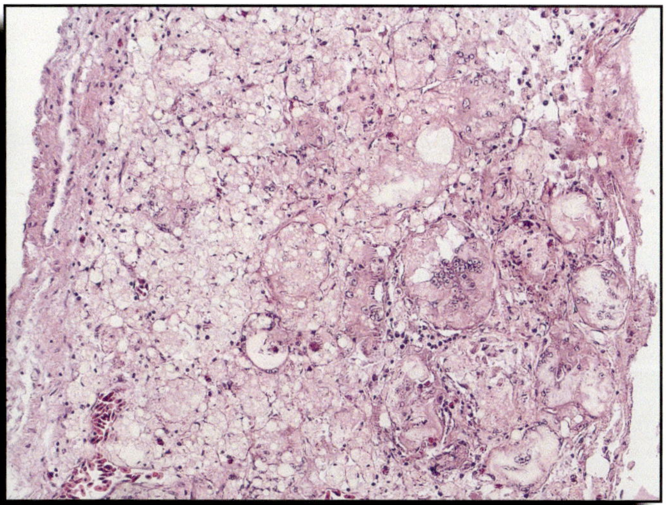

Figure 8.109 Aggasiz's Desert Tortoise, *Gopherus agassizii*. Testudinidae. Captive female. Ovary with inflamed atretic follicle (late Wang Stage IV) in a geriatric moribund individual (Wang et al., 2013). Xanthogranulomatous inflammation accompanies the degenerative change. H&E stain.

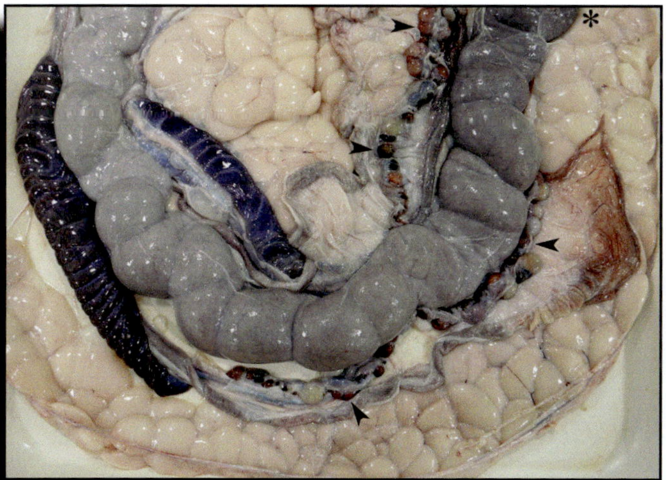

Figure 8.110 Reticulated Python, *Malayopython reticulatus*. Pythonidae. Captive obese female. Follicular atresia (arrowheads) in right and left ovaries. Anterior (*).

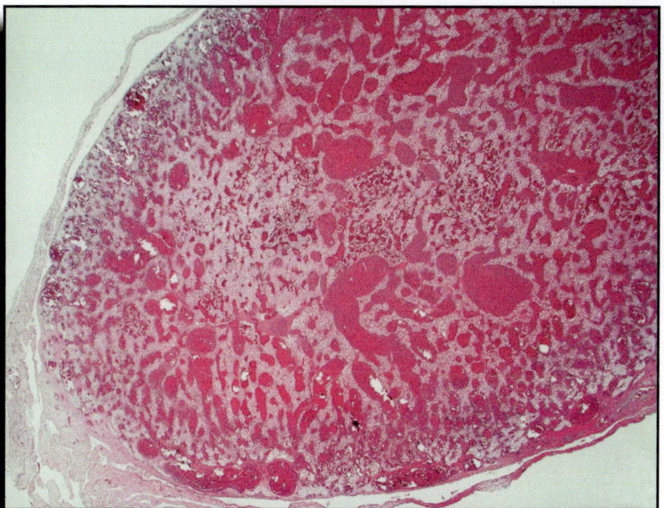

Figure 8.111 Reticulated Python, *Malayopython reticulatus*. Pythonidae. Captive Female, Figure 8.110. Ovary. Vitellogenic follicle. Early stage of diffuse follicular atresia occurring in an obese female with concurrent pneumonia.

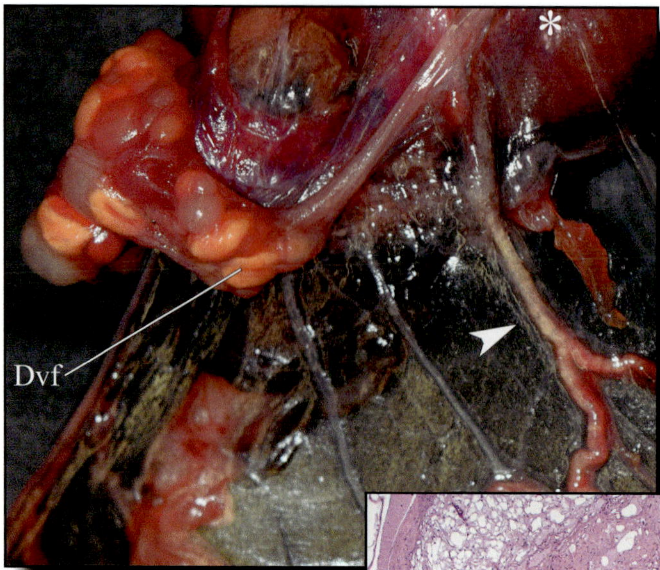

Figure 8.112 Savannah Monitor, *Varanus exanthematicus*. Varanidae. Captive female. Systemic vascular mineralization (arrowhead) due to calcium:phosphorus imbalance. Ovary with chronic delayed atresia and degeneration of vitellogenic follicles (Dvf). Anterior (*).

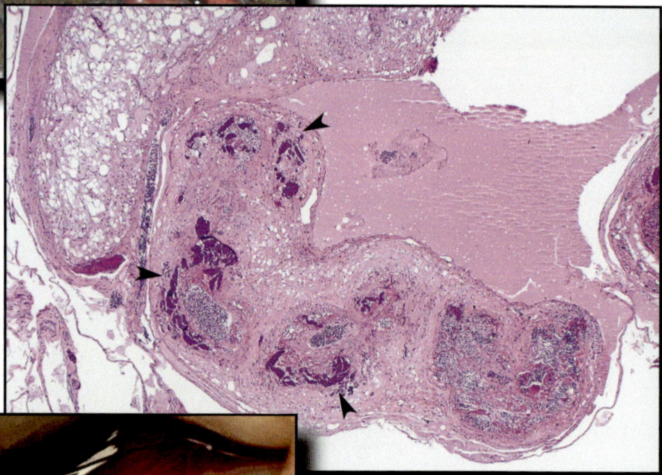

Figure 8.113 Savannah Monitor, *Varanus exanthematicus*. Varanidae. Captive female. This monitor had metastatic mineralization of vessels resulting in decreased perfusion of ovaries. Ovaries also have foci of mineralization (arrowheads). H&E stain.

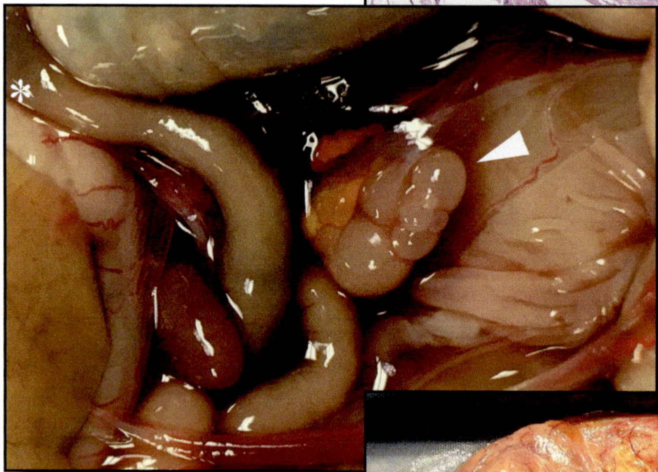

Figure 8.114 African Fat-Tailed Gecko, *Hemitheconyx caudicinctus*. Eublepharidae. Captive female. Ovary with cystic follicles (arrowhead). Anterior (*). (Courtesy of Brad Minson.)

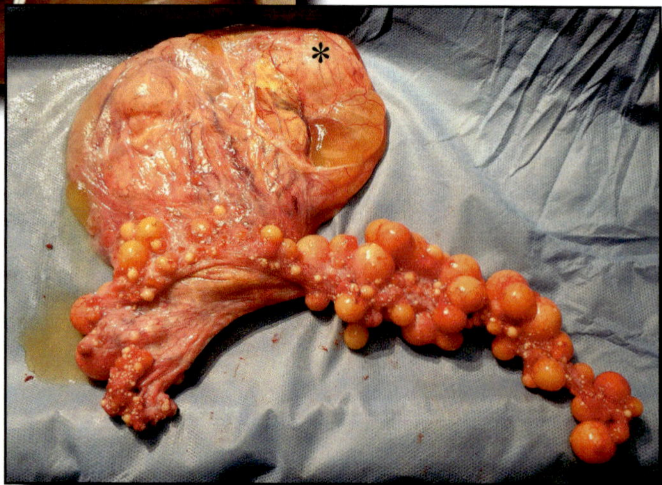

Figure 8.115 Agassiz's Desert Tortoise, *Gopherus agassizii*. Testudinidae. Captive female. Large Ovarian cyst (*), (Courtesy of Thomas Boyer.)

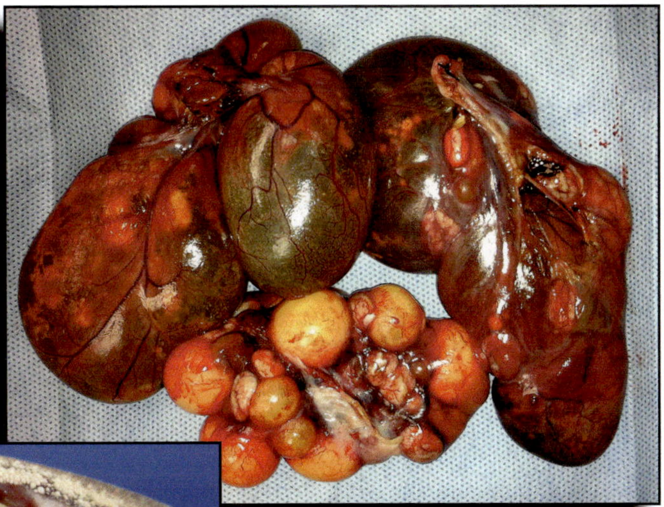

Figure 8.116 Gila Monster, *Heloderma suspectum*. Helodermatidae. Captive female. Oophoritis with severe follicular cystic change. (Courtesy of Darin Collins.)

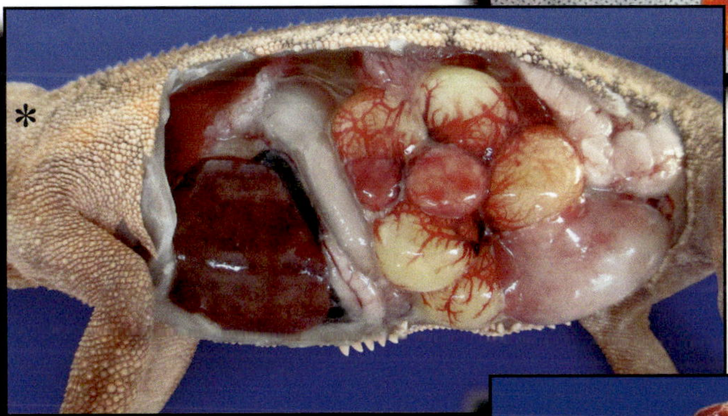

Figure 8.117 Panther Chameleon, *Furcifer pardalis*. Captive female. Chamaeleonidae. Oophoritis. Ovarian follicles are atretic and multiple large vitellogenic follicles are pale and have surface vascular congestion. The coelomic cavity contains fibrin and free fluid (coelomitis). Anterior (*). (Courtesy of Karina Mathes.)

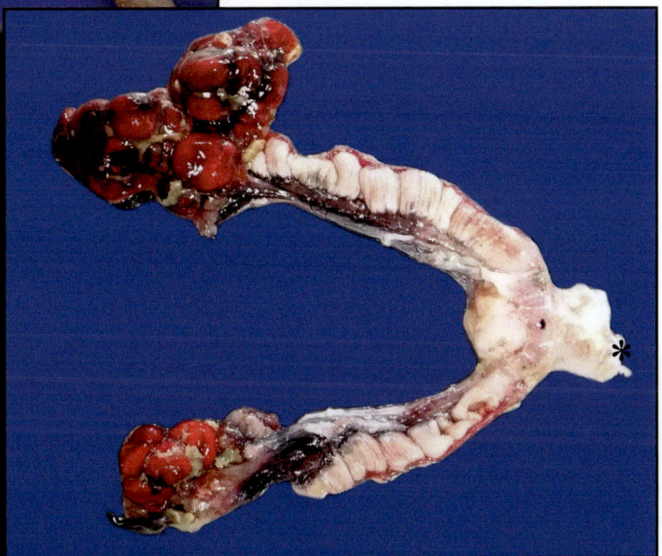

Figure 8.118 Central Bearded Dragon, *Pogona vitticeps*. Agamidae. Captive female. Bilateral oophoritis likely secondary to preovulatory stasis. (Courtesy of Michael Garner and Christie Buie.)

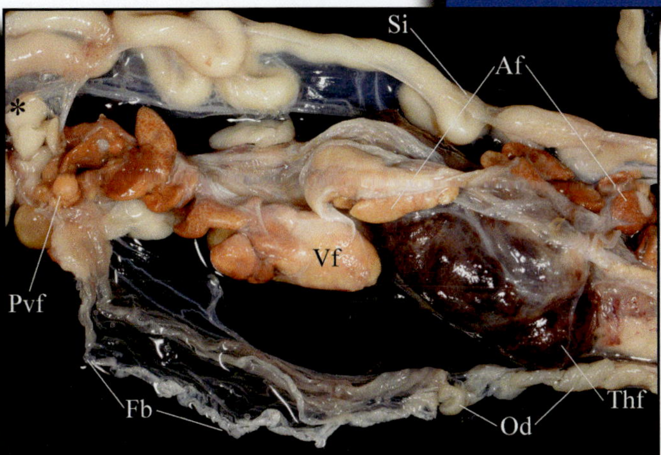

Figure 8.119 Central American Ratsnake, *Senticolis triaspis mutabilis*. Colubridae. Captive female. Ovarian Inflammation secondary to ohronic hepatic fibrosis and vena caval thrombosis. Adipose foci (Af) are atrophied. The ovary contains previtellogenic follicles (Pvf), a vitellogenic follicle (Vf) and an thrombosed hemorrhagic follicle (Thf) with pooled blood. The oviduct (Od) and fimbria (Fb) are devoid of follicles. The small intestine (Si) is empty. Anterior (*). Formalin fixed specimen. (Courtesy of Elliott Jacobson.)

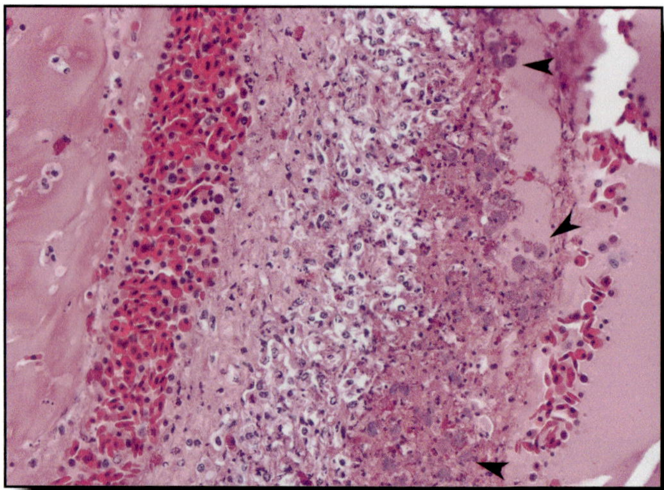

Figure 8.120 Central American Ratsnake, *Senticolis triaspis mutabilis*. Captive female. Colubridae. Ovarian hemorrhage and congestion with bacterial overgrowth and bacterial phagocytosis (arrowheads) in a vitelline follicle secondary to degenerative hepatic disease and vena caval thrombosis. H&E stain.

Figure 8.121 Green Iguana, *Iguana iguana*. Iguanidae. Captive female. Ovaries with preovulatory stasis at surgery and postsurgical evaluation of right and left ovary. A section of coelomic mesenteric adipose tissue (arrowhead) with focal necrosis and steatitis was also removed. (Courtesy of Charles Innis.)

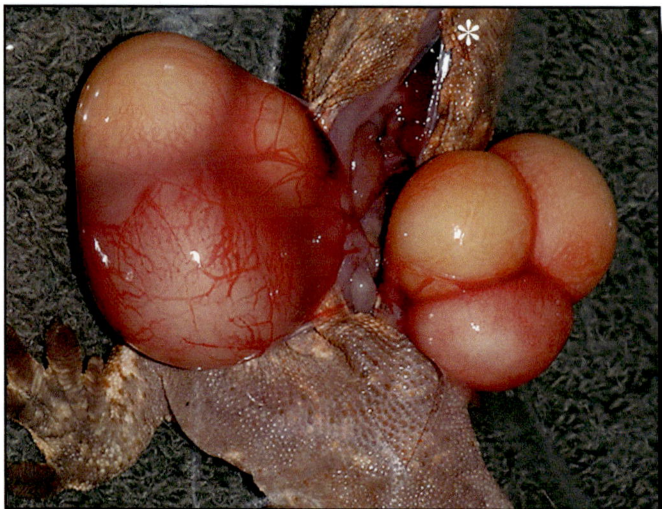

Figure 8.122 Crested Gecko, *Correlophus ciliatus*. Diplodactylidae. Captive female. This gecko presented with preovulatory follicular stasis and was treated by bilateral ovariectomy with uneventful recovery. Anterior (*). (Courtesy of Stephen Barten.)

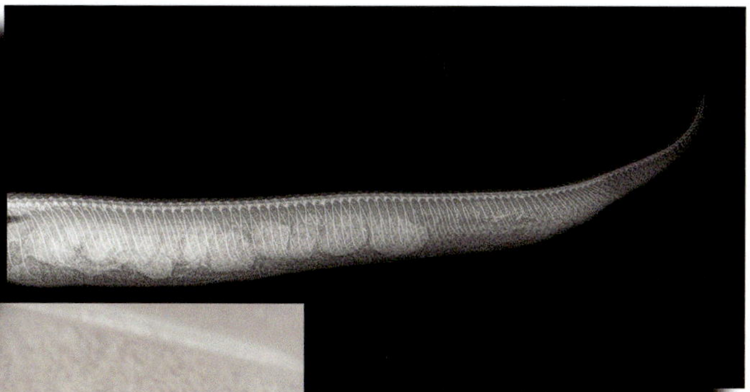

Figure 8.123 Ridge-Nosed Rattlesnake, *Crotalus willardi*. Viperidae. Captive female. Lateral caudal radiograph. Chronic preovulatory egg retention. Retained follicles at various stages from multiple cycles are inspissated and radiopaque. (Courtesy of Stephen Barten.)

Figure 8.124 Ridge-Nosed Rattlesnake, *Crotalus willardi*. Viperidae. Captive female. Chronic preovulatory egg retention. Retained follicles from multiple cycles at various stages of inspissation are present (see Figure 8.123). Bilateral ovariectomy was curative. Anterior (*). (Courtesy of Stephen Barten.)

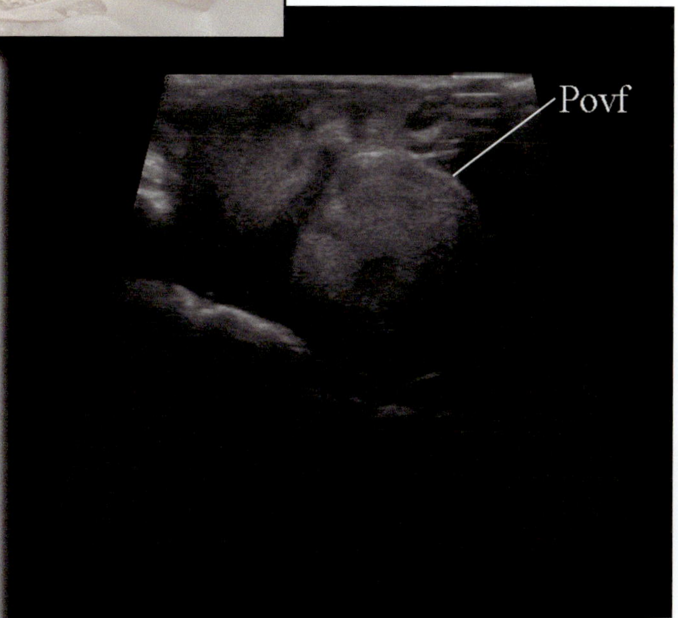

Figure 8.125 Central Bearded Dragon, *Pogona vitticeps*. Agamidae. Captive female. Preovulatory follicular stasis. Clustered preovulatory vitelline follicles (Povf) have variable hyperechogenicity. Ultrasound. (Courtesy of Karina Mathes.)

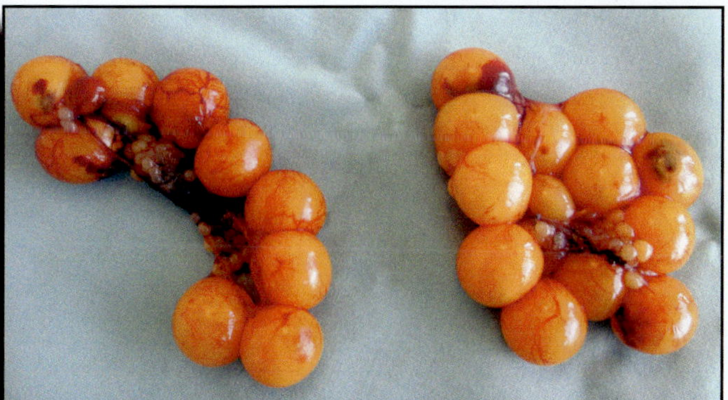

Figure 8.126 Central Bearded Dragon, *Pogona vitticeps*, Agamidae. Captive female. Preovulatory follicular stasis (see Figure 8.125). The left and right ovaries consist of multiple follicles in different sizes and stages of development including multiple previtellogenic and vitellogenic follicles. (Courtesy of Karina Mathes.)

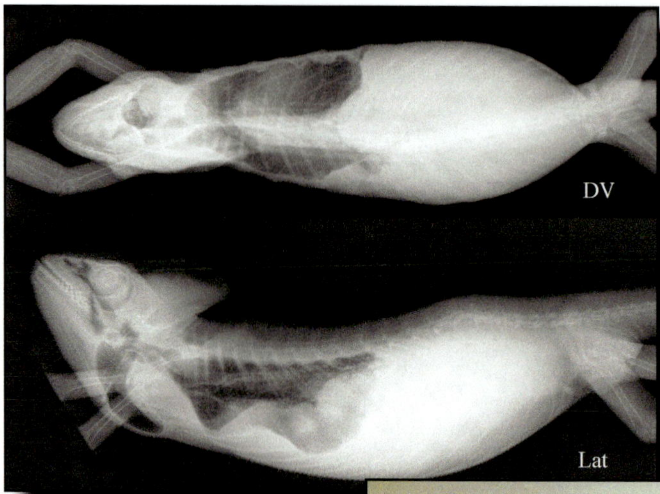

Figure 8.127 Veiled Chameleon, *Chamaeleo calyptratus*. Chamaeleonidae. Captive female. Preovulatory follicular stasis (POFS). The ceolmoic cavity is diffusely radiopaque without definition of viscera. Calcification, characteristic of postovulatory eggs, is not observed. Dorsal-ventral (DV) and lateral (Lat) radiographs. (Courtesy of Karina Mathes.)

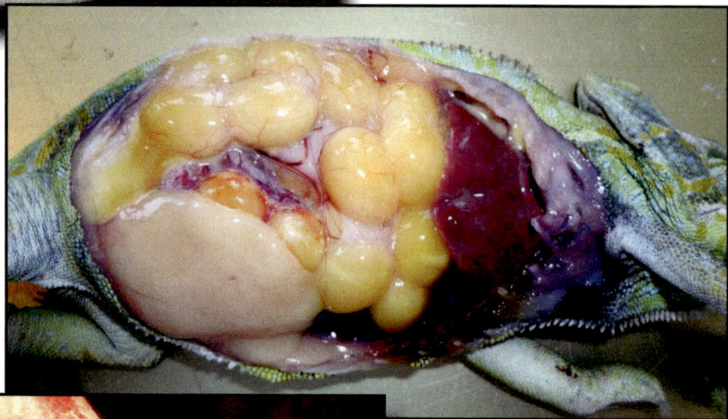

Figure 8.128 Veiled Chameleon, *Chamaeleo calyptratus*. Chamaeleonidae. Captive female. Preovulatory follicular stasis (POFS) diagnosed by radiograph (Figure 8.127) prior to euthanasia and necropsy. Necropsy reveals fibrinous adhesions and multiple large oval preovulatory follicles. (Courtesy of Karina Mathes.)

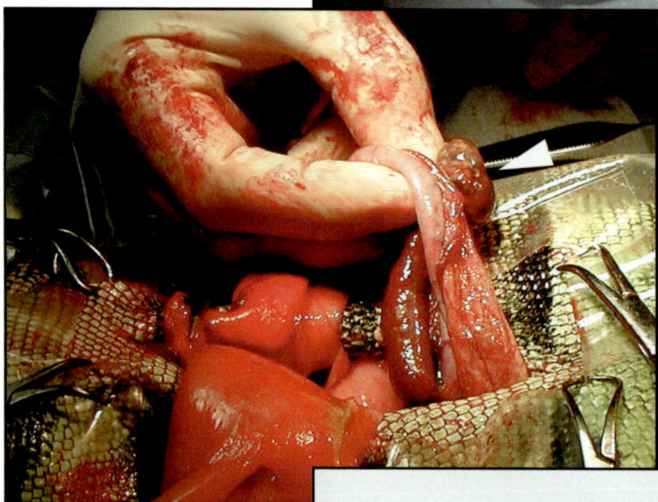

Figure 8.129 Green Iguana, *Iguana iguana*. Iguanidae. Captive female. Celiotomy and surgical removal of ectopic vitellogenic follicle (arrowhead) adhered to the mesentery. (Courtesy of Charles Innis.)

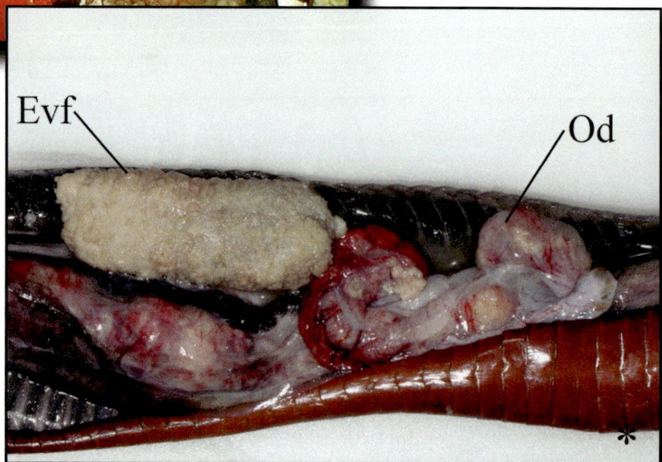

Figure 8.130 Long-Nosed Viper, *Vipera ammodytes*. Viperidae. Captive female. Ectopic vitellogenic follicles (Evf) with dystrophic mineralization. The oviduct (Od) contains degenerated follicles. Anterior (*). (Courtesy of Elliott Jacobson.)

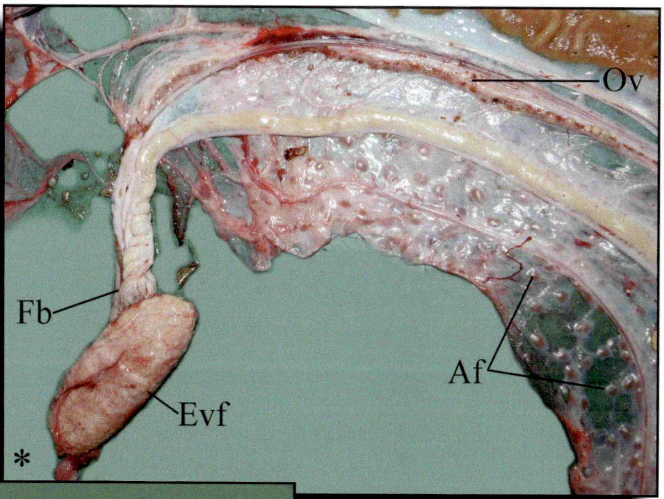

Figure 8.131 Burmese Python, *Python bivittatus*. Pythonidae. Captive female. Ectopic vitellogenic follicle (Evf) adhered to the fimbria (Fb). Note the static ovary (Ov) and atrophied coelomic adipose foci (Af). This python was necropsied at peak of breeding season in its native environment and should have been in full vitellogenesis. Anterior (*).

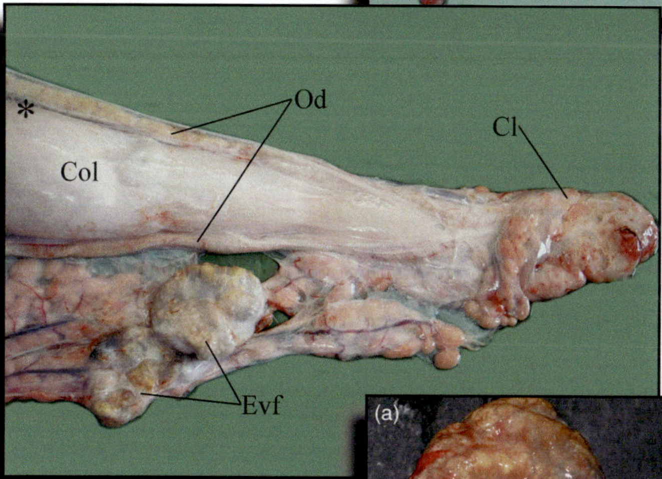

Figure 8.132 Burmese Python, *Python bivittatus*. Pythonidae. Captive female. Ectopic vitellogenic follicles (Evf) with dystrophic mineralization adhered to the mesentery near the cloaca (Cl). Note paired non-gravid oviducts (Od) adjacent to the colon (Col). Anterior (*).

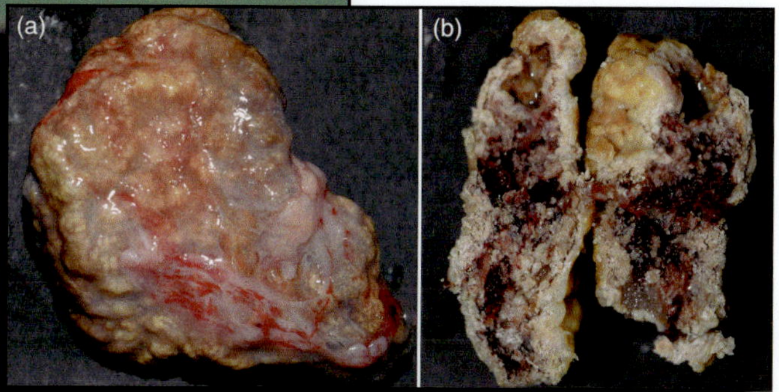

Figure 8.133 Burmese Python, *Python bivittatus*. Pythonidae. Captive female. Ectopic degenerated vitellogenic follicles with dystrophic mineralization, whole (a) and sectioned (b).

Figure 8.134 Northern Pine Snake, *Pituophis melanoleucus melanoleucus*. Colubridae. Captive female during oviposition. One infertile slug and three normal eggs. (Courtesy of Lauren Augustine.)

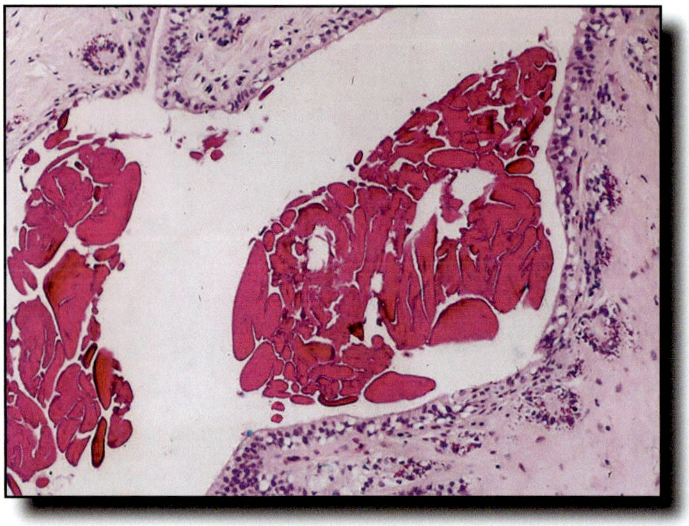

Figure 8.135 Green Iguana, *Iguana iguana*. Iguanidae. Captive female. Lumen of the oviduct with inspissated proteinaceous debris. Oviduct glands are atrophied. H&E stain. (Courtesy of Peernel Zwart.)

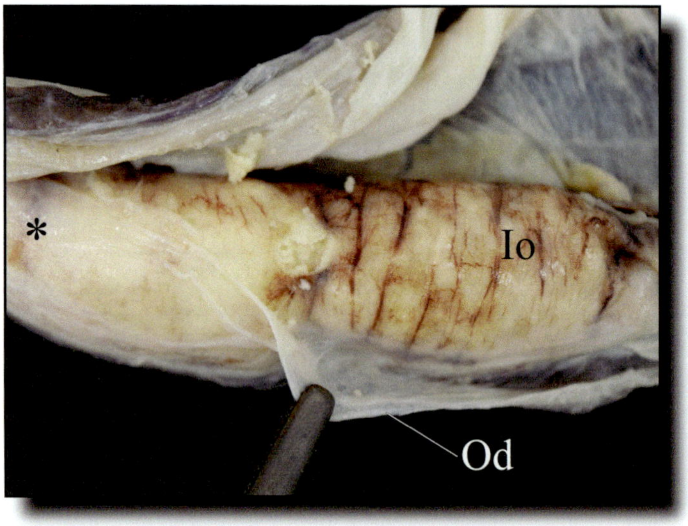

Figure 8.136 Long-Nosed Viper, *Vipera ammodytes*. Viperidae. Captive female. The wall of the oviduct (Od) is thin. The oviduct in this snake contains infertile ova (Io) (slugs). This snake also had an ectopic vitelline follicle (Figure 8.131). Formalin fixed specimen. Anterior (*). (Courtesy of Elliott Jacobson.)

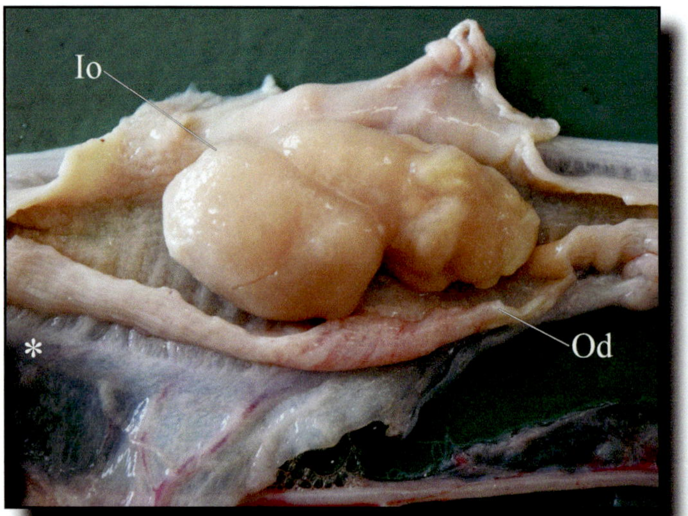

Figure 8.137 Reticulated Python, *Malayopython reticulatus*. Pythonidae. Captive female. Chronic salpingitis. Partially resorbed infertile ova (Io) with chronic inflammation and thickening of oviduct (Od) wall. Anterior (*).

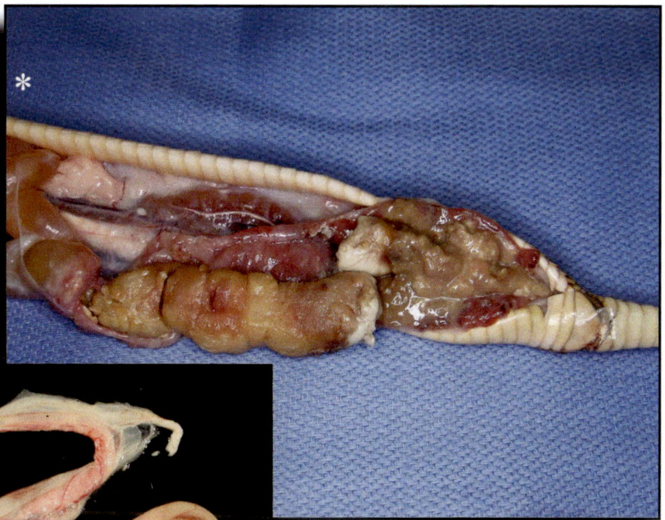

Figure 8.138 Common Garter Snake, *Thamnophis sirtalis*. Colubridae. Captive female. This viviparous snake presented moribund with a swollen caudal coelom. There were multiple retained infertile necrotic ova in the caudal part of one oviduct. (Courtesy of Stephen Barten.)

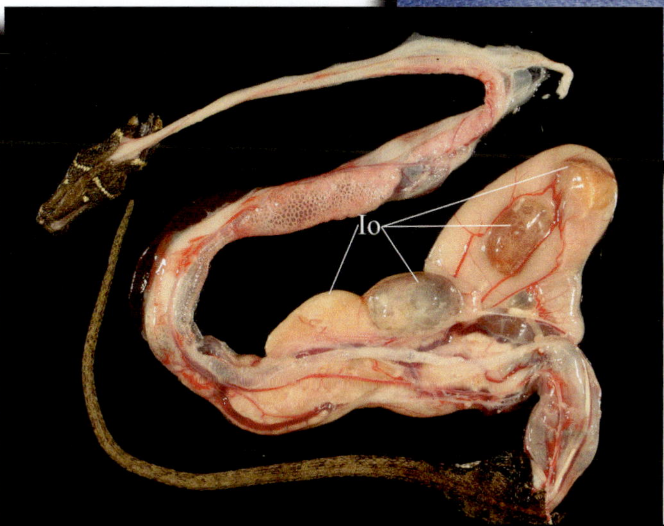

Figure 8.139 Tentacled Snake, *Erpeton tentaculatum*. Homalopsidae. Captive female. Salpingitis. The oviduct wall is thin and transparent with four infertile ova (Io) in the lumen. (Courtesy of The Smithsonian Institution.)

Figure 8.140 Veiled Chameleon, *Chamaeleo calyptratus*. Chamaeleonidae. Captive female. Postovulatory egg stasis (POES). This female has coelomic swelling and stress color pattern typical of POES. (Courtesy of Karina Mathes.)

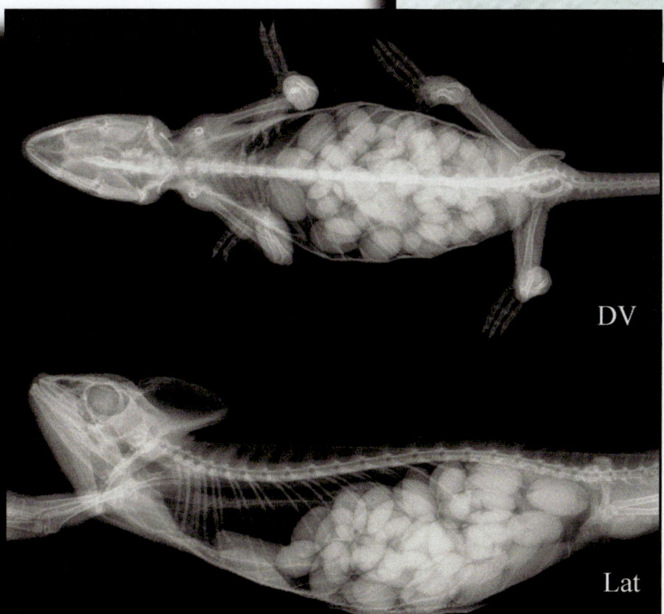

Figure 8.141 Veiled Chameleon, *Chamaeleo calyptratus*. Chamaeleonidae. Captive female. Postovulatory egg stasis (POES) (see Figure 8.141). Dorsal-ventral (DV) and lateral (Lat) radiographs. Note multiple radiodense oval eggs. (Courtesy of Karina Mathes.)

Figure 8.142 Parson's Chameleon, *Calumma parsonii.* Chamaeleonidae. Captive female. Postovulatory egg stasis (POES) treated by ovariosalpingectomy. Excised oviduct and eggs. The chameleon has temporary yellow pigmentation cause by anesthesia. Recovery was uneventful. (Courtesy of Stephen Barten.)

Figure 8.143 Agassiz's Desert Tortoise, *Gopherus agassizii.* Testudinidae. Wild moribund female. Note probable egg stasis. The left oviduct (Lod) has congested vasculature visible from the serosa surface (arrow). Anterior (*).

Figure 8.144 Green Iguana, *Iguana iguana.* Iguanidae. Captive female with yolk coelomitis. Coelomic surface of oviduct with phagocytosis of yolk by activated mesothelial cells (Amc). H&E stain. (Courtesy of Peernel Zwart.)

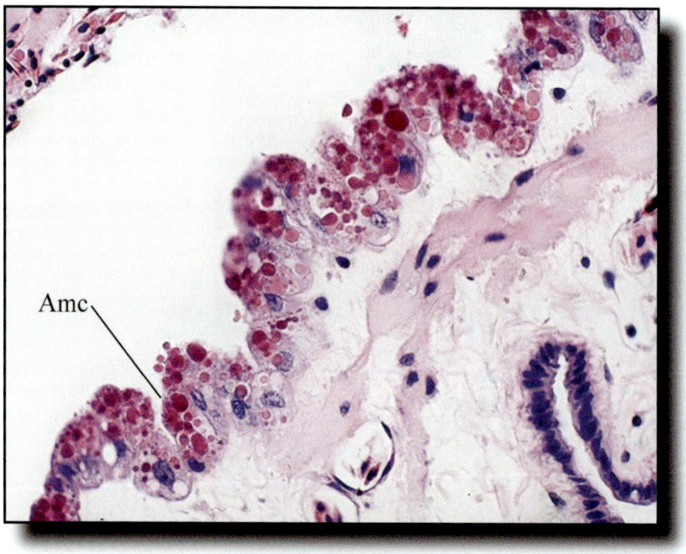

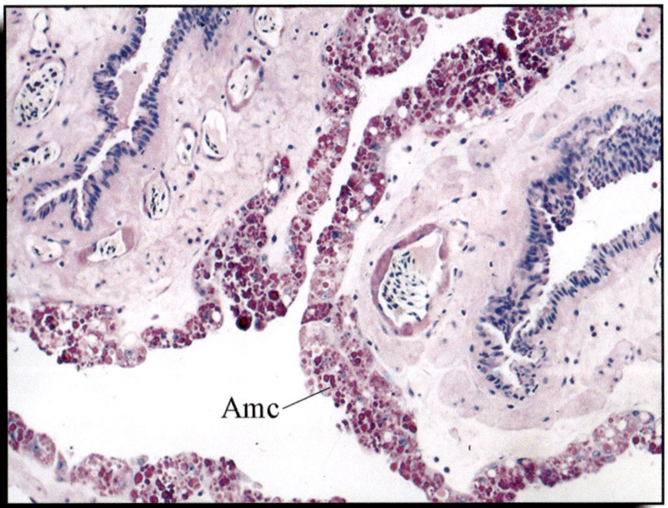

Figure 8.145 Green Iguana, *Iguana iguana*. Iguanidae. Captive female with yolk coelomitis. Vitelline from phagocytized yolk in activated mesothelial cells (Amc) stains pink with application of periodic acid-Schiff (PAS) stain. PAS stain. (Courtesy of Peernel Zwart.)

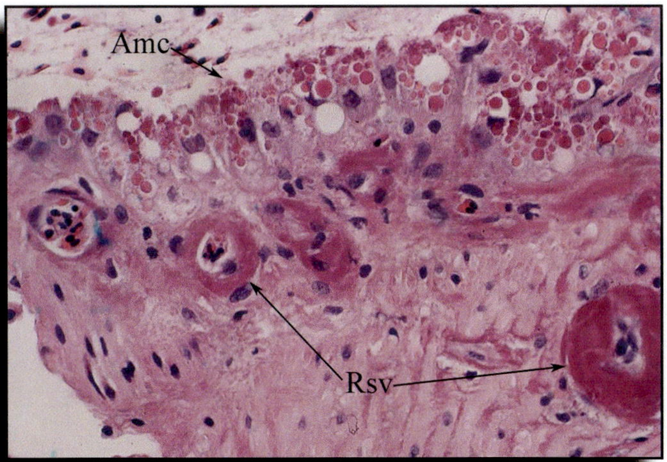

Figure 8.146 Green Iguana, *Iguana iguana*. Iguanidae. Captive female with yolk coelomitis. Coelomic surface of oviduct. Phagocytosis of yolk by activated mesothelial cells (Amc) and fibrinoid necrosis of subserosal vessels (Rsv). H&E stain. (Courtesy of Peernel Zwart.)

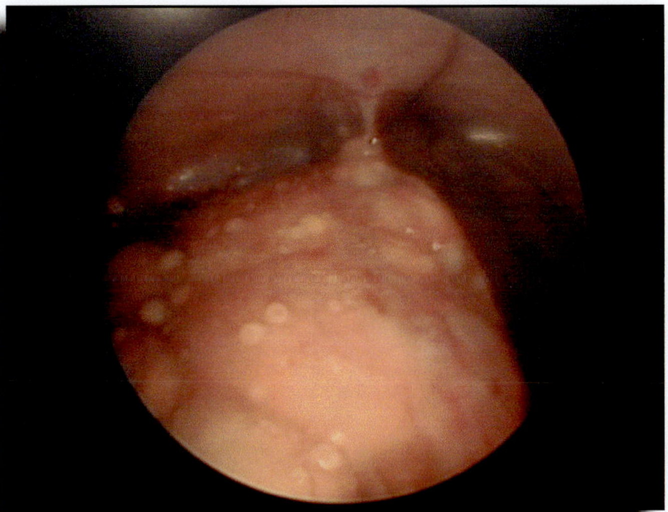

Figure 8.147 Impressed Tortoise, *Manouria impressa*. Testudinidae. Captive live female. Chronic yolk coelomitis confirmed endoscopically. Multifocal deposition of yolk on the mesothelial surface of visceral organs likely occurred following recurrent rupture of late or atretic vitelline follicles. This may occur during postovulatory follicular stasis (POFS). (Courtesy of Charles Innis.)

Figure 8.148 Fiji Banded Iguana, *Brachylophus fasciatus*. Iguanidae. Captive female. Coelomic cavity coated with globular yellow to white friable material adhered to visceral organs including the ovaries, oviducts, kidney and parietal body wall. Chronic yolk leakage in this species likely occurs during atresia of large vitellogenic follicles. Anterior (*). (Courtesy of Brian Stacy.)

Figure 8.149 Red-Eared Slider, *Trachemys scripta elegans*. Emydidae. Captive female. Severe exudative yolk coelomitis likely complicated by secondary bacterial infection. Anterior (*). (Courtesy of Charles Innis.)

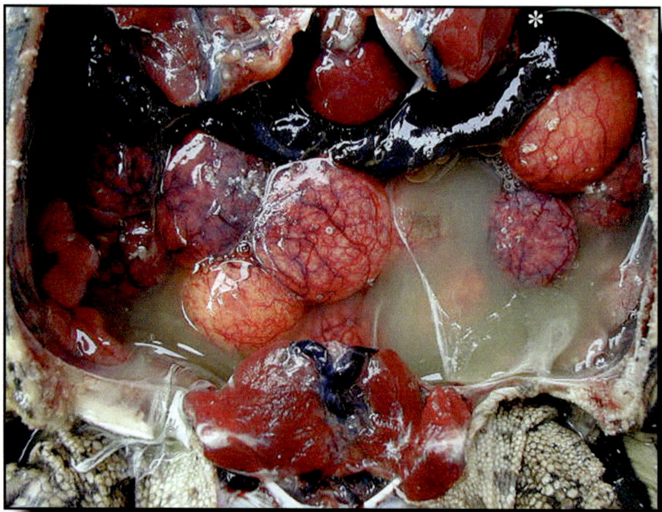

Figure 8.150 Savannah Monitor, *Varanus exanthematicus*. Varanidae. Captive female. Follicular atresia and rupture of vitelline follicles with associated severe yolk coelomitis and secondary bacterial infection. Anterior (*).

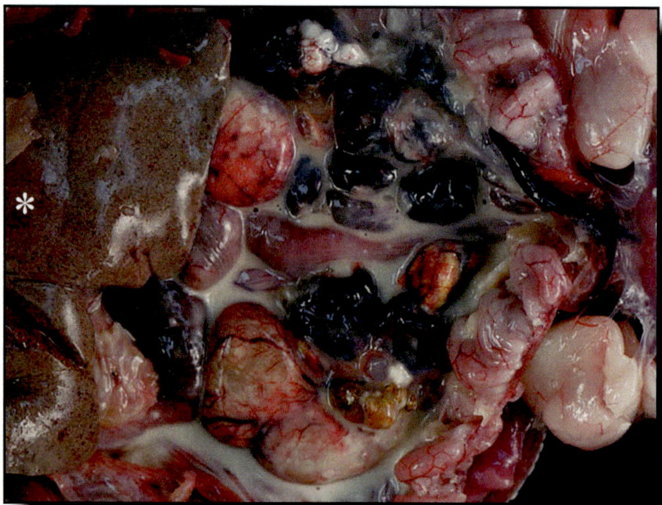

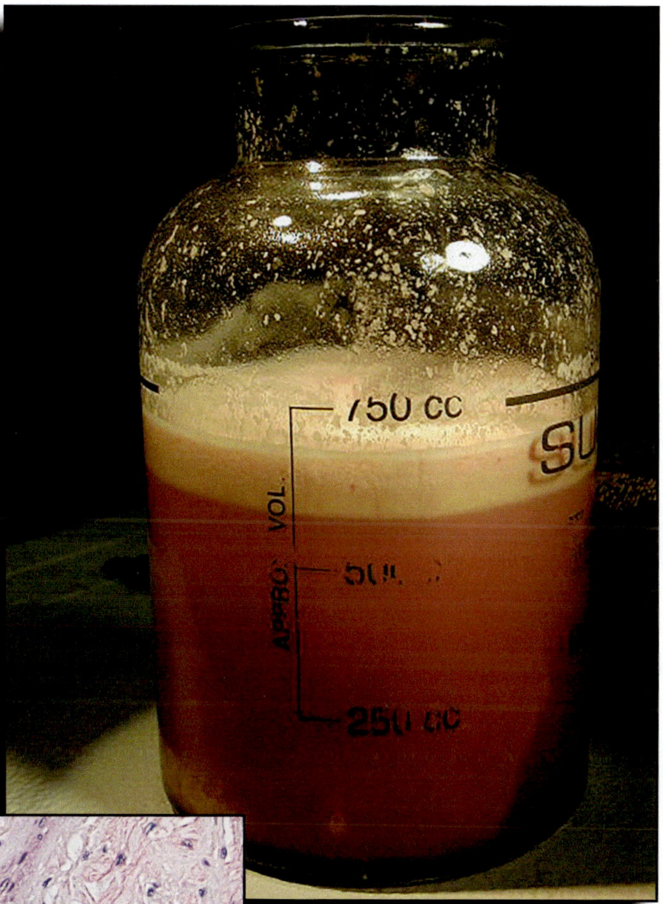

Figure 8.151 Rhinoceros Iguana, *Cyclura cornuta*. Iguanidae. Fluid collected from the coelomic cavity of a captive female with severe chronic yolk coelomitis. (Courtesy of Charles Innis.)

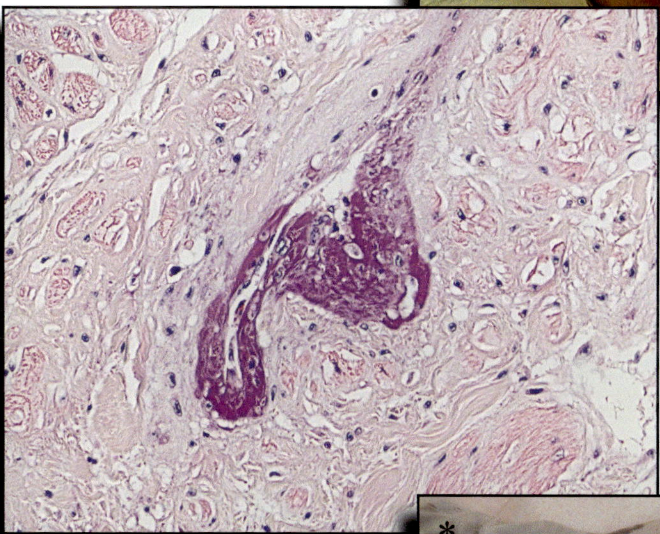

Figure 8.152 Komodo Dragon, *Varanus komodoensis*. Varanidae. Captive female. Yolk embolism in vasculature of cardiac muscle. H&E stain. (Courtesy of Fabiano Oliveira.)

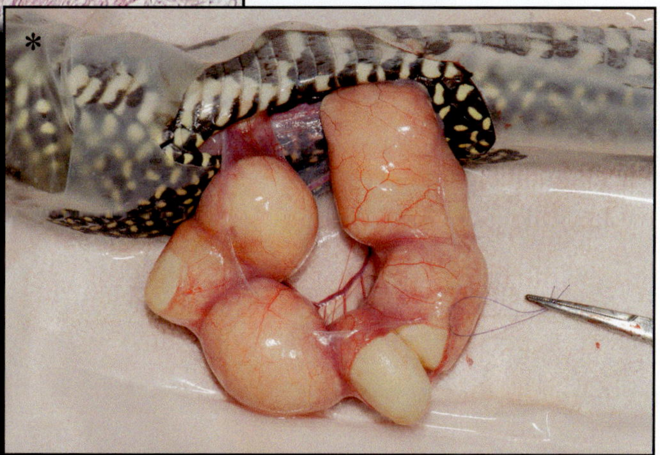

Figure 8.153 Speckled Kingsnake, *Lampropeltis getula holbrooki*. Colubridae. Captive female. Dystocia. Multiple variably sized eggs adhered to the mucosa of the oviduct. The eggs were removed by salpingotomy. Anterior (*). (Courtesy of Stephen Barten.)

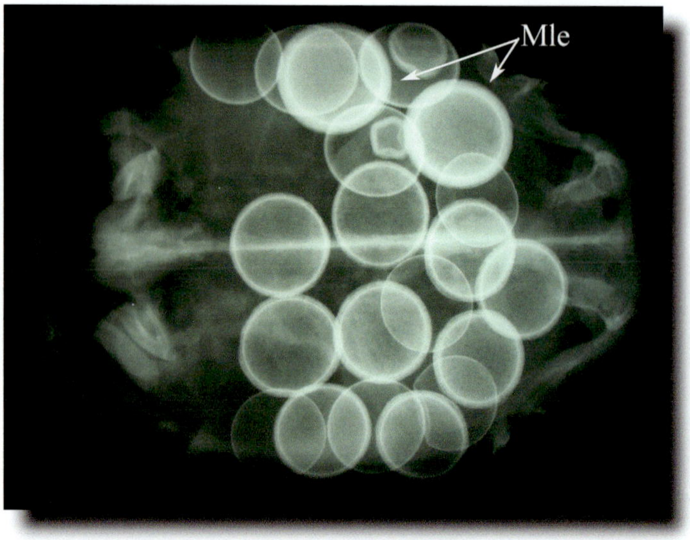

Figure 8.154 Yellow-Footed Tortoise, *Chelonoidis denticulata*. Testudinidae. Captive female. D-V radiograph. Obstructive dystocia caused by large multilaminar eggs (Mle) (Figure 8.156). (Courtesy of Charles Innis.)

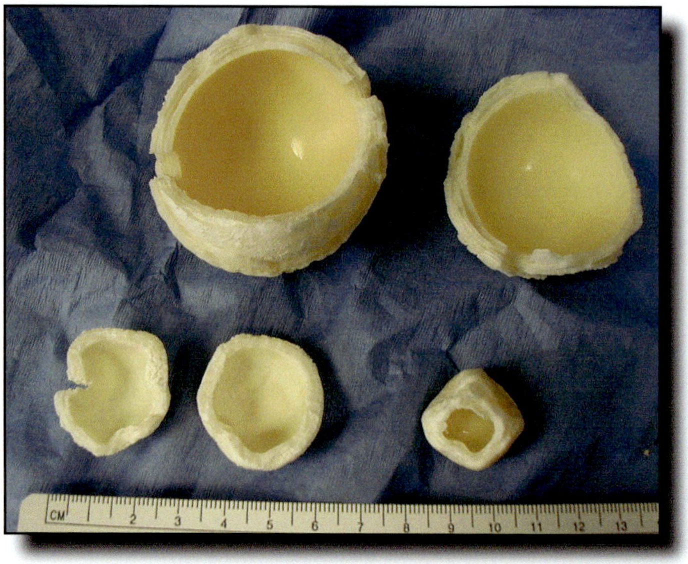

Figure 8.155 Yellow-Footed Tortoise, *Chelonoidis denticulata*, Testudinidae. Captive female. Variable shaped multilaminar eggs after surgical removal (Figure 8.154). The tortoise had prolonged egg retention, which predisposes to shell thickening. (Courtesy of Charles Innis.)

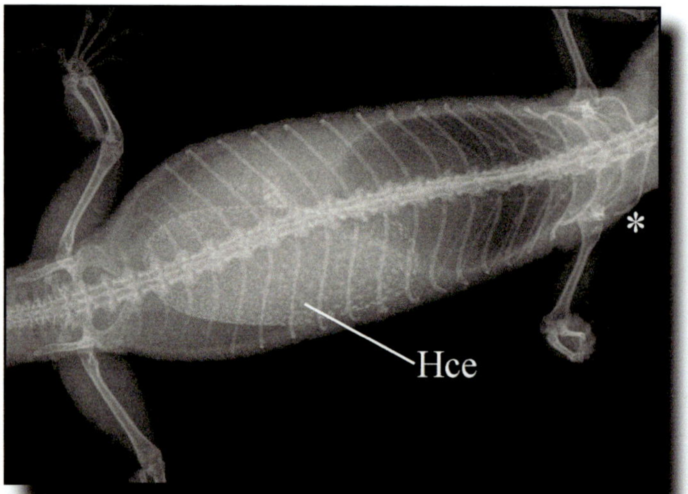

Figure 8.156 Leopard Gecko, *Eublepharis macularius*. Eublepharidae. Captive female. D-V radiograph of a three-year-old gecko with a large single hypercalcified egg (Hce). Anterior (*). (Courtesy of Karina Mathes.)

Figure 8.157 Leopard Gecko, *Eublepharis macularius*. Eublepharidae. Captive female. Large retained egg that was surgically removed from the Leopard Gecko in Figure 8.156. The shell has multiple raised yellow to brown coalescing foci of hypercalcification. (Courtesy of Karina Mathes.)

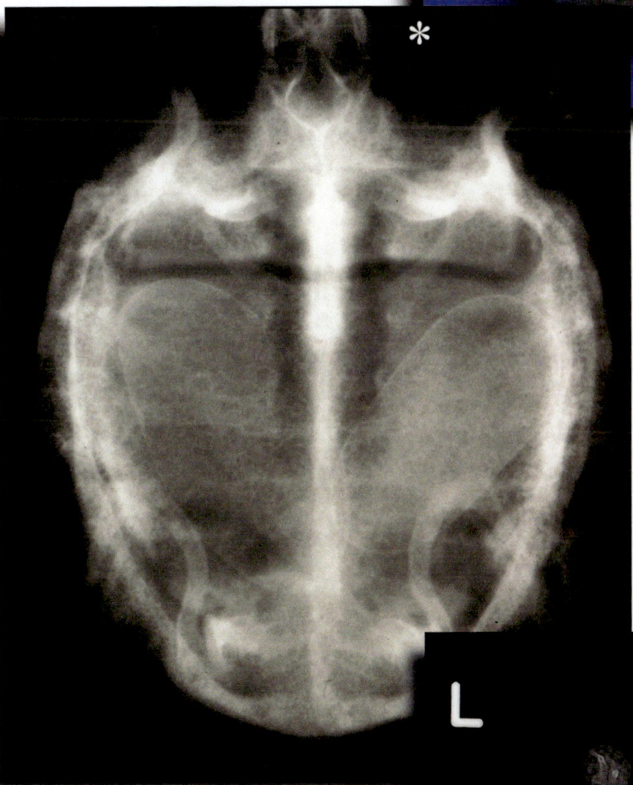

Figure 8.158 Eastern Box Turtle, *Terrapene carolina carolina*. Emydidae. Captive female. D-V radiograph. Note abnormally large eggs lodged near the pelvic canal. Anterior (*). (Courtesy of Charles Innis.)

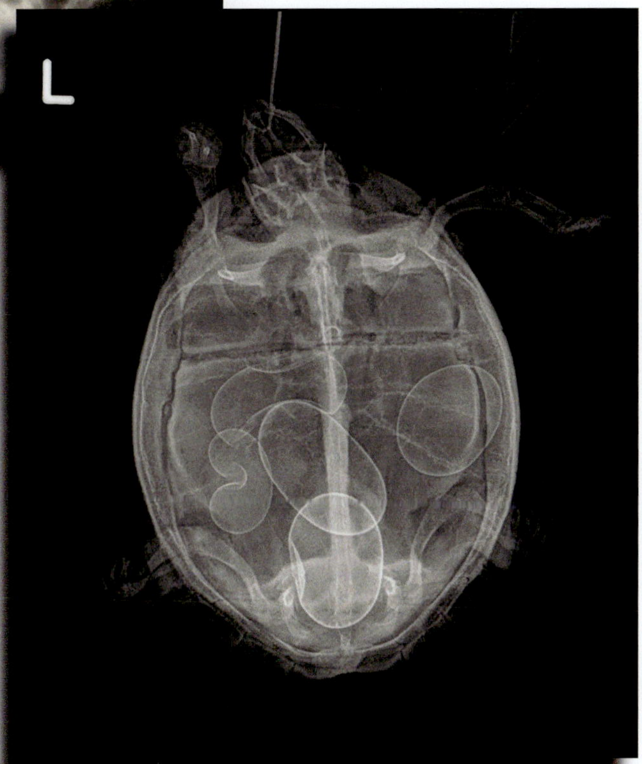

Figure 8.159 Eastern Box Turtle, *Terrapene carolina carolina*. Emydidae. Captive female. D-V radiograph revealing misshapen eggs and a large egg trapped in the pelvic canal. The egg in the pelvic canal was collapsed using a syringe and large bore needle advanced through the cloaca. Subsequent injections of calcium and oxytocin resulted in oviposition of the remaining eggs. (Courtesy of Stephen Barten.)

Figure 8.160 Speckled Rattlesnake, *Crotalus mitchelli*. Viperidae. Captive female. Dystocia in a snake that previously delivered three normal neonates. The coelom cranial to the cloaca was distended and suggested a retained embryo, which was confirmed in this lateral radiograph (arrowhead). Anterior (*). (Courtesy of Stephen Barten.)

Figure 8.161 Speckled Rattlesnake, *Crotalus mitchelli*. Viperidae. Captive female. Retained embryo in Figure 8.160 that was removed by salpingotomy. Anterior (*). (Courtesy of Stephen Barten.)

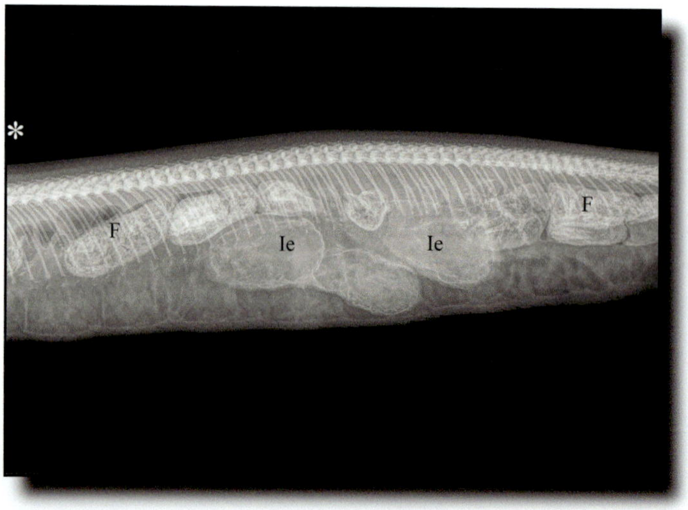

Figure 8.162 Olive Python, *Liasis olivaceus*. Pythonidae. Captive female. Chronic dystocia. Lateral radiograph revealing three inspissated eggs (Ie) in the oviduct, with feces (F) in the colon. The eggs were removed by salpingotomy and recovery was uneventful. Anterior (*). (Courtesy of Stephen Barten.)

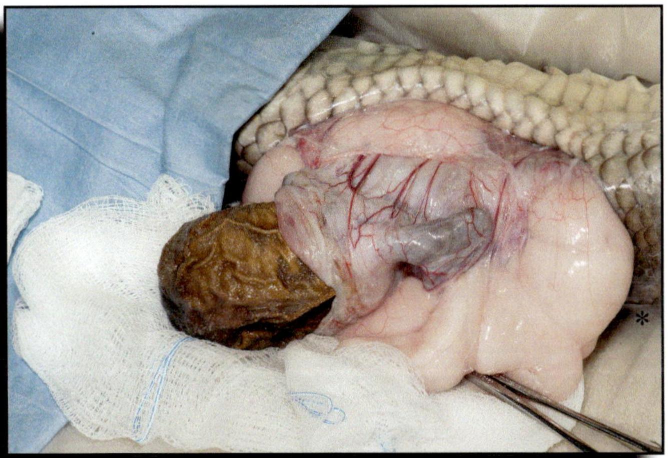

Figure 8.163 Olive Python, *Liasis olivaceus*. Pythonidae. Captive female. Chronic dystocia. Retained inspissated eggs in python in Figure 8.162 removed by salpingotomy. Anterior (*). (Courtesy of Stephen Barten.)

Figure 8.164 Eastern Indigo Snake, *Drymarchon couperi*. Colubridae. Captive female. Physiological dystocia. The body wall anterior to the cloaca is distended with three eggs (arrowheads) that failed to be oviposited; a fourth egg is present in the proximal oviduct (arrowhead).

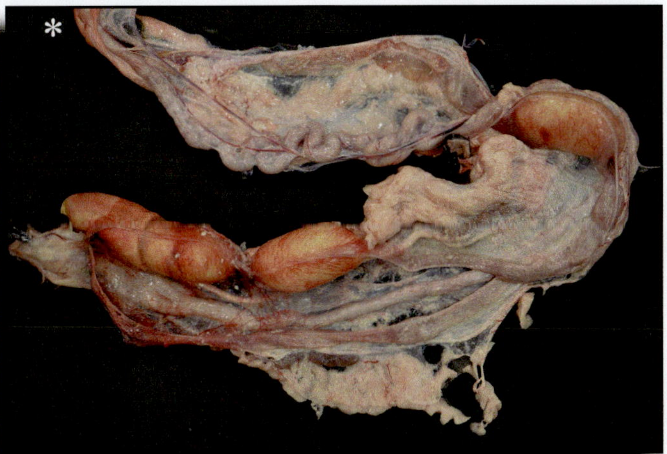

Figure 8.165 Eastern Indigo Snake, *Drymarchon couperi*. Colubridae. Captive female. Physiological dystocia of Eastern Indigo Snake in Figure 8.164. Three eggs that failed to pass were clustered near the cloaca and a fourth egg was in the proximal oviduct. Anterior (*).

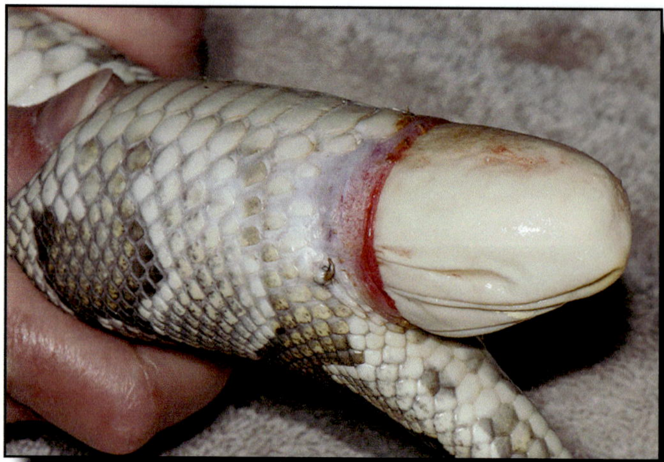

Figure 8.166 Ball Python, *Python regius*. Pythonidae. Captive female. Two retained eggs in distal oviduct adjacent to the cloaca were manually expressed while the snake was anesthetized. (Courtesy of Stephen Barten.)

Figure 8.167 Ball Python, *Python regius*. Pythonidae. Captive female. The two nonviable retained eggs from the Ball Python in Figure 8.166 contained inspissated yolk. (Courtesy of Stephen Barten.)

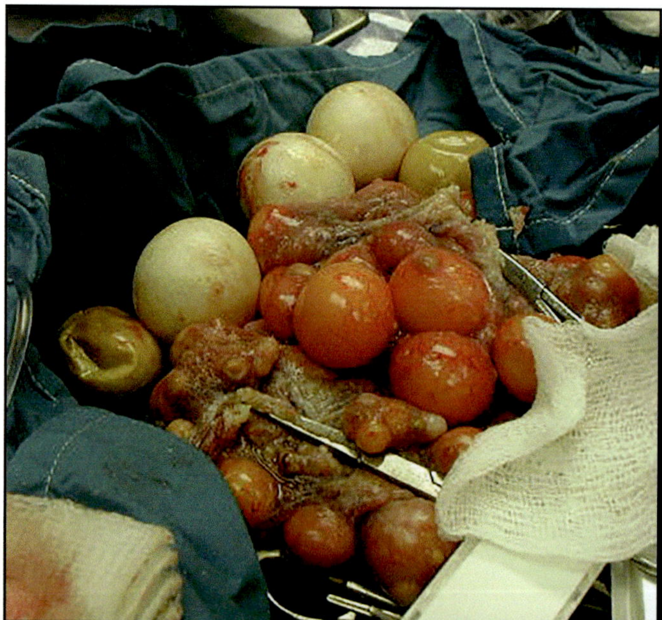

Figure 8.168 Yellow-Footed Tortoise, *Chelonoidis denticulata*. Testudinidae. Captive female. Ectopic eggs in the coelomic cavity after reverse peristalsis of the oviduct associated with obstructive dystocia. (Courtesy of Charles Innis.)

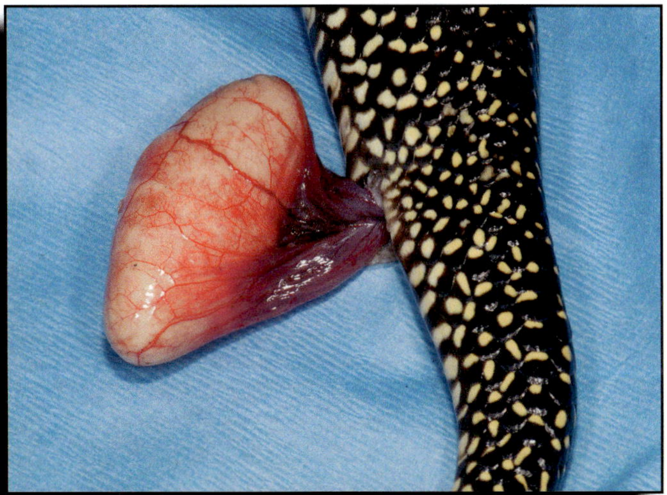

Figure 8.169 Speckled Kingsnake, *Lampropeltis getula holbrooki*, Colubridae. Captive female. Dystocia with oviductal prolapse. The prolapsed oviduct contained a single egg. An incision was made in the prolapsed tissue to remove the egg and the prolapse was reduced. A celiotomy incision was made and the remaining eggs were removed by salpingotomy (see Figure 8.153). (Courtesy of Stephen Barten.)

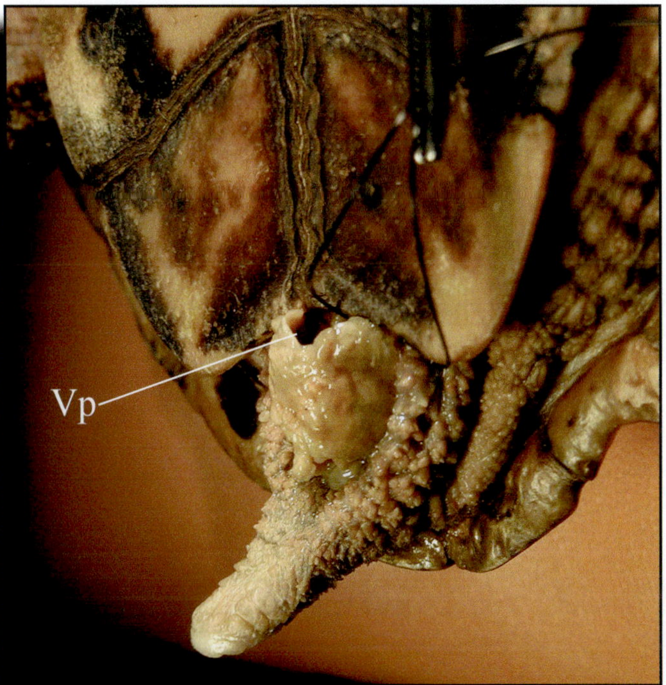

Figure 8.170 Four-Eyed Turtle, *Sacalia quadriocellata*. Geoemydidae. Vaginal prolapse (Vp) of the oviduct. (Courtesy of Charles Innis.)

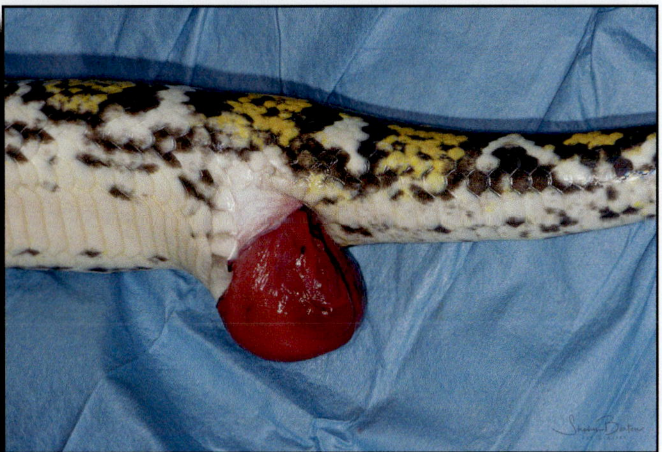

Figure 8.171 Reticulated Python, *Malayopython reticulatus*. Pythonidae. Captive female. Cloacal prolapse after ovipositing a clutch of eggs. The prolapse was reduced and held in place with two cloacal sutures that narrowed the cloacal opening; it did not reoccur after suture removal 3 weeks later. (Courtesy of Stephen Barten.)

Figure 8.172 Chuckwalla, *Sauromalus* sp. Iguanidae. Captive female. Lymphangiectasis of the oviduct associated with lymphatic obstruction, as would occur with a mesenteric mass, adhesion, or a prolapsed oviduct. (Courtesy of Michael Garner.)

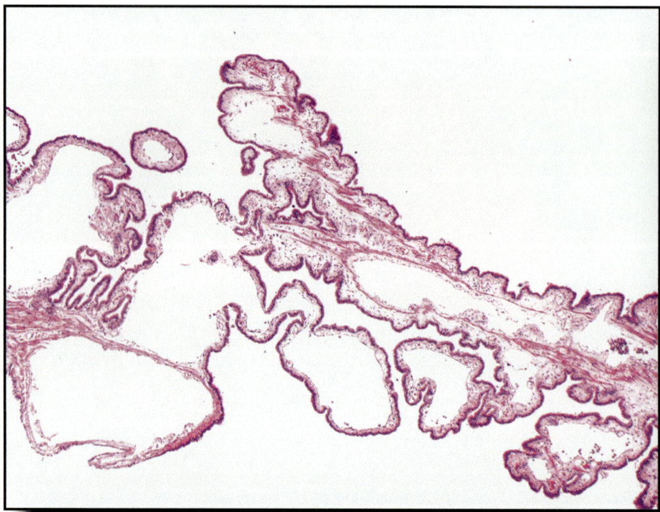

Figure 8.173 Cuban Crocodile, *Crocodylus rhombifer.* Crocodylidae. Captive female. Faintly visible sperm on the previtelline membrane. Fluorescent microscopy. (Courtesy of Lauren Augustine.)

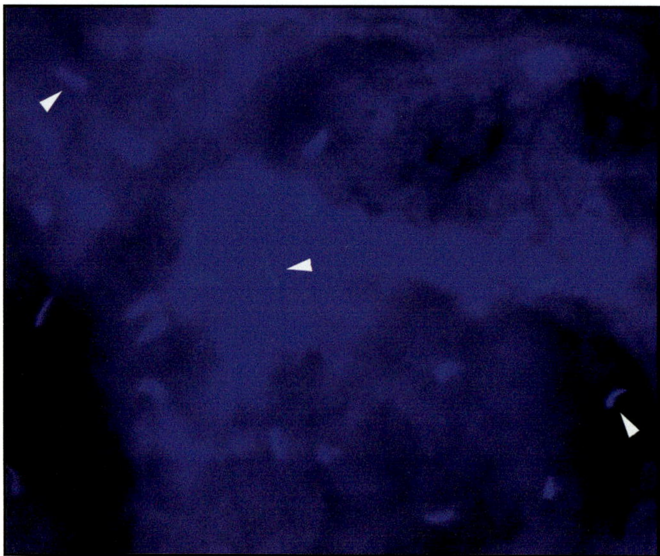

Figure 8.174 Ball Python, *Python regius*. Pythonidae. Captive female. Fertile egg with developed placental vessels (left) compared to a nonfertile egg (right). (Courtesy David Barker and Tracy Barker, from Barker DG and Barker TM, *Ball Pythons*, Copyright 2006, VPI library. With permission.)

Figure 8.175 Cuban Crocodile, *Crocodylus rhombifer.* Crocodylidae. Captive female. Candled fertile egg with banding indicative of fertility. (Courtesy of Lauren Augustine.)

Figure 8.176 Cuban Crocodile, *Crocodylus rhombifer.* Crocodylidae. Captive female. Candled infertile egg without banding. (Courtesy of Lauren Augustine.)

Figure 8.177 Cuban Crocodile, *Crocodylus rhombifer.* Crocodylidae. Captive female. Infertile crocodile egg (Figure 8.176). (Courtesy of Lauren Augustine.)

Figure 8.178 Ball Python, *Python regius*. Pythonidae. Captive female. Small infertile eggs "Pearls". (Courtesy David Barker and Tracy Barker, from Barker DG and Barker TM, Ball Pythons, Copyright 2006, VPI library. With permission.)

Figure 8.179 Common Keelback Snake *Tropidonophis mairii*. Colubridae. Captive female. Fungal overgrowth in nonviable eggs. (Courtesy of Gregory Brown.)

Figure 8.180 Eastern Indigo Snake, *Drymarchon couperi*. Colubridae. Captive female. Misshapen egg with overgrowth of *Fusarium* sp. (arrowheads) on surface of eggshell.

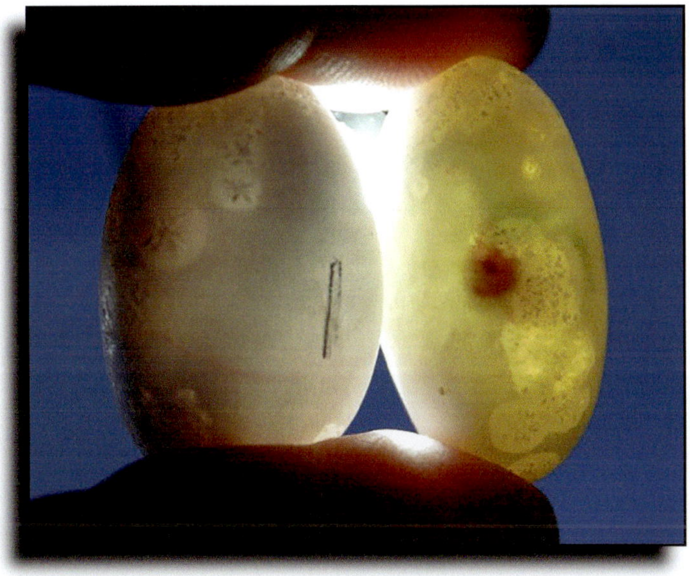

Figure 8.181 Common Keelback Snake, *Tropidonophis mairii*. Colubridae. Captive female. Fungal overgrowth in candled nonviable eggs. (Courtesy of Gregory Brown.)

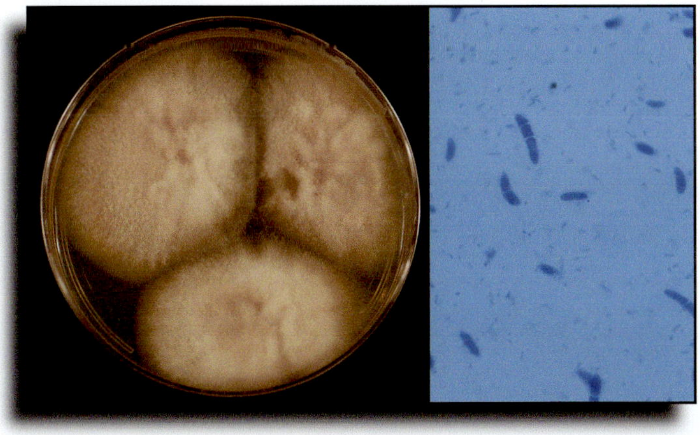

Figure 8.182 Eastern Indigo Snake, *Drymarchon couperi*. Colubridae. Captive female. Fungal plate and cytologic appearance of *Fusarium* sp. isolated from late-term infertile eggs.

Figure 8.183 Red-Footed Tortoise, *Chelonoidis carbonarius*. Testudinidae. Captive female. Infertile egg with bacterial colonization. (Courtesy of The Smithsonian Institution.)

Figure 8.184 Ball Python, *Python regius*. Pythonidae. Captive female. Radiograph of late stage eggs near hatching. The infertile egg (Ie) does not contain vertebra. Radiograph. (Courtesy of Brad Minson.)

Figure 8.185 Eastern Indigo Snake, *Drymarchon couperi*. Colubridae. Captive female. Late-term embryo death. Overgrowth of *Fusarium* sp.

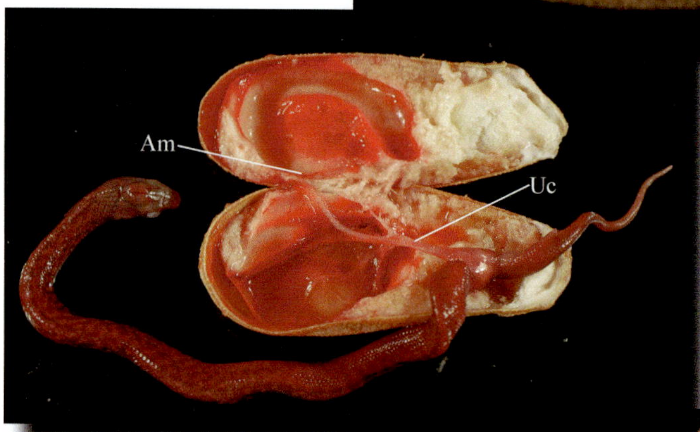

Figure 8.186 Eastern Indigo Snake, *Drymarchon couperi*. Colubridae. Captive female. Late fetal death. Yolk is overgrown with *Fusarium* sp. Note the umbilical cord (Uc) and attachment to the amnion (Am).

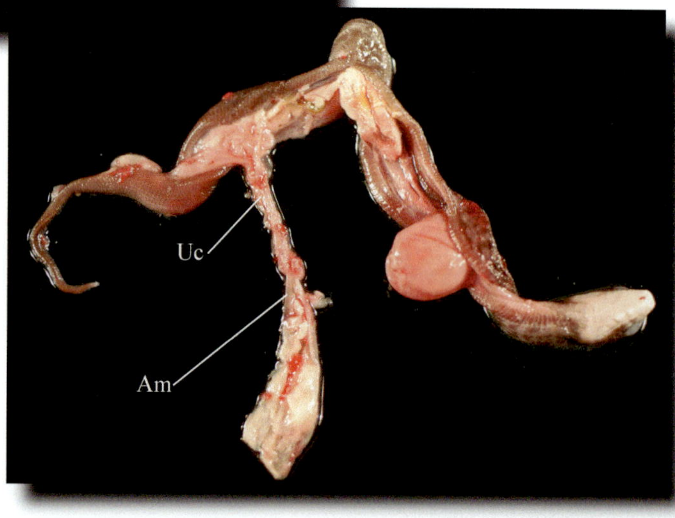

Figure 8.187 Eastern Indigo Snake, *Drymarchon couperi*. Colubridae. Unsexed neonate from captive female. Schistosomus reflexus-like birth defect. There is a failure of the body cavity to close, and viscera and umbilicus are externalized. The umbilical cord (Uc) is attached to a remnant of the amnion (Am).

Figure 8.188 Cuban Crocodile, *Crocodylus rhombifer*. Crocodylidae. Unsexed neonate from captive female. The yolk sac is not resorbed. The misshapened jaws and head, and the retained yolk represent developmental defects that may be due to the position of the embryo within the egg, improper incubation conditions, or innate genetic defects. (Courtesy of The Smithsonian Institution.)

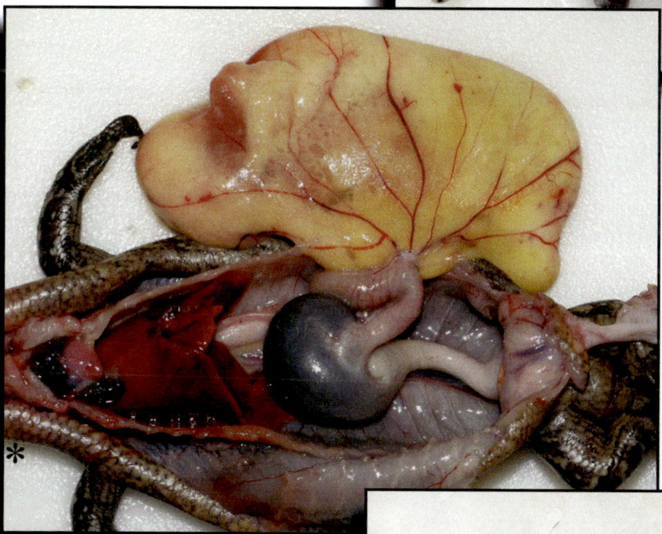

Figure 8.189 Solomon Islands Skink, *Corucia zebrata*. Scincidae. Unsexed neonate from captive female. Unabsorbed yolk sac. Anterior (*). (Courtesy of The Smithsonian Institution.)

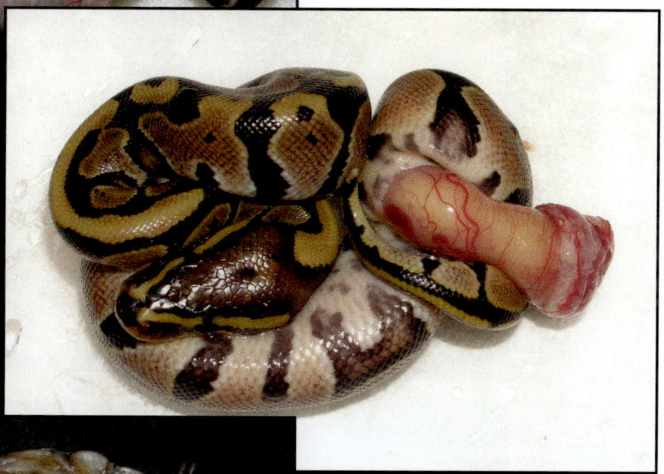

Figure 8.190 Ball Python, *Python regius*. Pythonidae. Neonate hatched from captive female. Externalized yolk sac. (Courtesy David Barker and Tracy Barker, from Barker DG and Barker TM, *Ball Pythons*, Copyright 2006, VPI library, with permission.)

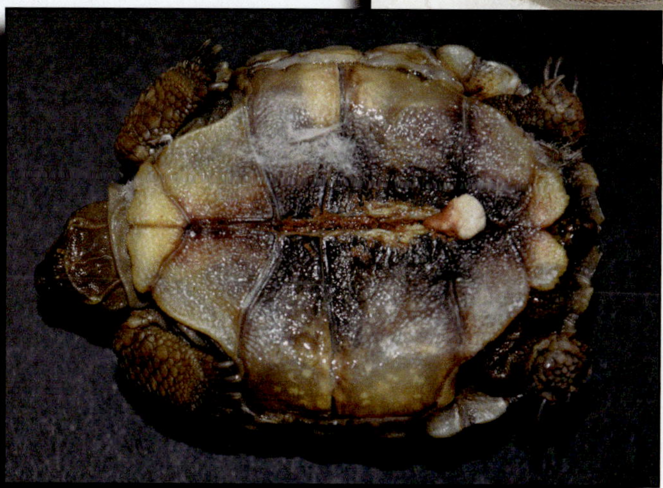

Figure 8.191 Gopher Tortoise, *Gopherus polyphemus*. Testudinidae. Unsexed neonate from wild female. Open umbilicus. The neonate also has postmortem mycotic overgrowth on the plastron.

Figure 8.192 Arizona Green Rat Snake, *Senticolis triaspis intermedia*. Colubridae. Unsexed neonate from captive female. Congenital defect in closure of umbilicus and omphalitis. The scales in area of the umbilicus are not normal in size or shape. (Courtesy of Elliott Jacobson.)

Figure 8.193 Ball Python, *Python regius*. Pythonidae. Unsexed neonate from captive female. Distension of the caudal ceolomic cavity by a retained yolk sack. An image of the infected yolk sack following surgical removal is in Volume 1, Chapter 11. (Courtesy of Elliott Jacobson.)

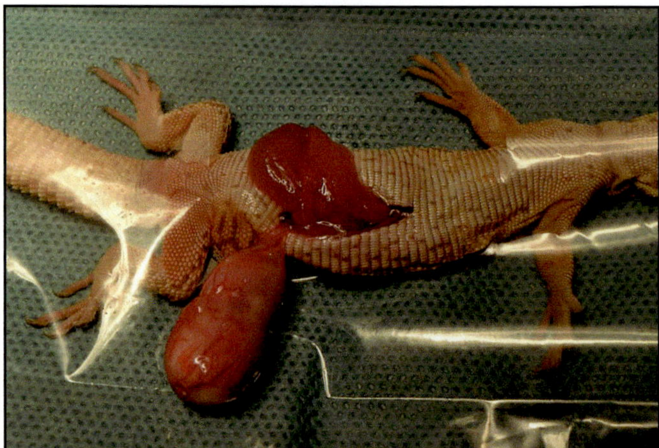

Figure 8.194 Spiny-Tailed Monitor, *Varanus acanthurus*. Varanidae. Unsexed neonate from captive female. Surgical removal of retained yolk sac. Anterior (*). (Courtesy of Shane Simpson.)

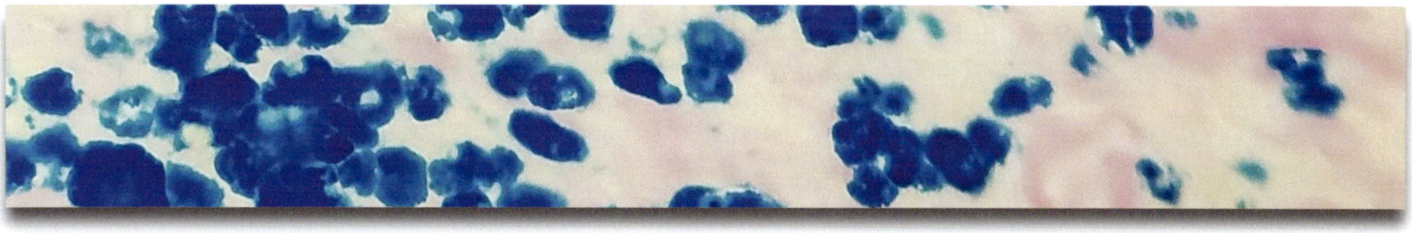

9

REPTILE CARDIOVASCULAR SYSTEM

JAMES E. BOGAN JR. AND JOAQUIN ORTEGA

Contents

9.1 Introduction

Studies on reptile hearts date back to 350 BC with Aristotle, who described the presence of the heart and blood vessels in multiple animal species, including turtles and snakes (Aristotle, 350 BC). In the early seventeenth century, Harvey described blood flow and the beating heart of various reptiles, noting the loss of crimson color to the ventricle with each heartbeat (Harvey, 1628). A little over 200 years later, Brücke laid the foundation of understanding the ventricular anatomy and coined the current terminology for the three ventricular compartments (Brücke, 1852). Detailed study from multiple researchers over the last century have been built on these foundations, greatly enhancing our current knowledge of reptilian cardiovascular anatomy and physiology (Farrell et al., 1998; Hicks 1998; Jensen et al., 2014).

Cardiac output (CO) is defined as the product of the stroke volume (SV) and heart rate (HR). CO is much lower in reptiles than in birds and mammals, consistent with their 10-fold lower metabolism (Burggren et al., 1997; Hulbert and Else, 2004). The decreased CO is from the lower HR, as it has been shown that the SV in reptiles approximates that of birds and mammals (Burggren et al., 1997; Hicks and Wang, 1998; Lillywhite et al., 1999; Seymour and Baylock, 2000; Overgaard et al., 2002; Skovgaard et al., 2005a, 2007, 2009; Galli et al., 2005a, 2007; Taylor et al., 2009).

Cardiac work is the product of the stroke volume, mean arterial pressure, and heart rate (SV × MAP × HR). In reptiles, cardiac work is much lower than that of birds and mammals. The lower cardiac work is reflected in a smaller relative heart size (Jensen et al., 2014). A typical avian or mammalian ventricle has a mass 0.4% of body mass and a heart mass 0.8% of body mass (Seymour, 1987; Poupa, 1994; Seymour and Baylock, 2000). In contrast, a typical reptile ventricle mass is roughly 0.2% body mass and heart mass 0.3% body mass (Hess, 1921; Poupa and Lindström, 1983; Farrell et al., 1998; Jensen et al., 2014). However, more athletic reptiles, like the

Tree Snake (*Dispholidus typus*), may have a larger heart mass, closer to 0.45% of body mass (Poupa and Lindström, 1983; Farrell et al., 1998).

9.2 Anatomy and Physiology

The heart of non-crocodilian reptiles has four chambers: sinus venosus, right atrium, left atrium, and ventricle (**Figures 9.1–9.5**). The ventricle can be further subdivided into three major sub-chambers: cavum venosum, cavum arteriosum, and cavum pulmonale. The heart of crocodilians, however, has five chambers: sinus venosus, right atrium with auricle, right ventricle, left atrium with auricle, and left ventricle (**Figure 9.6**). The chambers are lined by endocardium which is continuous with the endothelial lining of the blood vessels (Farrell et al., 1998).

The reptilian heart is surrounded by a pericardial sac. The serous membrane lining the pericardial sac is a reflection of epicardium (Farrell et al., 1998). Between the epicardium and endocardium are muscle and connective tissue, collectively called myocardium. In chelonians, crocodilians, and lizards, the apex of the heart is attached to the pericardial sac by the gubernaculum cordis (**Figure 9.7**) (Farrell et al., 1998).

9.2.1 Anatomy of Chelonia and Squamata

Although Chelonia are phylogenetically more closely related to Crocodylia (Crawford et al., 2012), the chelonian cardiovascular anatomy and physiology are more closely related to Squamata. To simplify the discussion, the cardiovascular anatomy and physiology for Chelonia will be combined with Squamata (**Figures 9.8–9.12**).

In most chelonians, the heart is located just caudal to the thoracic girdle, ventral to the lungs. In most lizards, the heart is located within the thoracic girdle. Varanids, however, have their heart located more caudally. In snakes, the heart location is variable depending on snake's environmental status. In terrestrial and arboreal species, it is located 15%–25% body length and in aquatic species, 25%–45% body length (Seymour, 1987; Lilywhite, 1987). A cranially located heart reduces hydrostatic pressure above the heart, stabilizing cerebral blood pressure (Seymour, 1987).

The sinus venosus is the chamber receiving cardiac inflow from the systemic circulation and is comprised of the left precaval vein and the confluent right precaval and postcaval veins. The sinus venosus has the thinnest walls of all the chambers and is attached to dorsal wall of the ventricle by a fibrous strand, the dorsal ligament (Farrell et al., 1998; Jensen et al., 2010). The walls of sinus venosus contain cardiac muscle and act as a pacemaker (Buchanan, 1956).

The right atrium is larger than the left in snakes and chelonians, and the interatrial septum is very thin in all reptiles (Farrell et al., 1998; Jensen et al., 2010). The left atrium is not only smaller than the right but also less muscular (Farrell et al., 1998). The larger right atrium may allow accommodation of increased venous return associated with right-to-left shunting during apnea (Farrell et al., 1998).

Blood flows into the right atrium from the sinus venosus through an oblique opening with a pair of membranous flap-like leaflets, the sinoatrial valve (Farrell et al., 1998; Jensen et al., 2010). This valve prevents blood from flowing backward out of the right atrium. The lumen of the right atrium is filled with a trabecular network of muscle fibers and pillars that is best described as sponge-like (Farrell et al., 1998). The left atrium receives blood from the pulmonary vein in the dorsal wall near the interatrial septum, not far from sinoatrial opening (Farrell et al., 1998), and is not protected by valves (Jensen et al., 2010). Both atria extend a short distance into the ventricle, forming an inverted truncated cone of muscle called the atrioventricular funnel (Farrell et al., 1998).

The atrioventricular funnel is the location of the fibromyxomatous atrioventricular valves. Both AV valves are highly efficient and have two leaflets: a septal cusp and marginal cusp (Johansen, 1959; Farrell et al., 1998). The septal cusps are the main functional portions of AV valves in reptiles (Farrell et al., 1998) and are fused medially with the interatrial septum and each other (Jensen et al., 2010). On dorsal and ventral sides, the AV valves are fused with the ventricular musculature (Jensen et al., 2010). The marginal cusps are very small and easily overlooked (Farrell et al., 1998).

The ventricle has a compact, cortical myocardium surrounding a more spongy, medullary myocardium (Farrell et al., 1998). The cortical layer is a thin outer layer running longitudinally from apex to base. The medullary layer is comprised of two spirally oriented sheets of muscle fibers (MacKinnon and Heatwole, 1981). In two-dimensional histologic section, this layer appears sinusoidal or spongiform (**Figure 9.4**). The ventricle is divided by three prominent septa: the vertical septum, the bulbuslamelle, and the muscular ridge (Jensen et al., 2010). The most prominent of these septa is the muscular ridge, which has also been called the horizontal septum or Muskelleiste (Farrell et al., 1998; Hicks, 1998; Jensen et al., 2010). The muscular ridge separates the cavum pulmonale from the cavum venosum and is a band of muscle arising from the base of the aortic arches and spirals toward the apical part of the dorsal ventricle where it joins the dorsal portion of the vertical septum (Farrell et al., 1998). The vertical septum lies across the caudal edge of the interatrial septum and forms a ridge between cavum venosum and cavum arteriosum (Farrell et al., 1998). The space between the free margin of the vertical septum and the atrioventricular valves is termed the interventricular canal. This canal connects the cavum venosum and cavum arteriosum (Farrell et al., 1998).

Three arterial trunks (the common pulmonary artery and the left and right aortae) exit the ventricle, twisting dextrally around one another at 135°–180° (Farrell et al., 1998; Jensen et al., 2010c). Each artery is guarded by a pair of semilunar valves (Jensen et al., 2010c). The pulmonary artery arises from the cavum pulmonale, while the aortae arise from the cavum venosum (Farrell et al., 1998). Similar to some mammalian ungulates, some squamates may have cartilage within the heart base referred to as the cartilago cordis (Young et al., 1994). The cartilago cordis is believed to be a vestigial remnant of the conus arteriosus (Young et al., 1994; Farrell et al., 1998).

9.2.2 Anatomy of Sphenodontia

There is limited study of the cardiovascular system of Sphenodontia. Most of the anatomy is similar to that of lizards, but there are some attributes to their anatomy that suggest it is a more primitive system. Like lizards, Tuatara hearts have four chambers: sinus venosus, right and left atria, and the ventricle. Unlike lizards, the ventricle is only subdivided into two sections, the cavum dorsale and the cavum ventrale (Simons, 1965). The cavum ventrale is analogous to the cavum pulmonale in lizards. The two cava are separated by the horizontal septum, and the cavum dorsale is not further subdivided into cavum arteriosum and cavum venosum (Grigg and Simons, 1972). Although there is less division in the ventricle, the timing of the contractions seems to be sufficient in minimizing the mixing of pulmonary and systemic blood (Grigg and Simons, 1972). A gubernaculum cordis may or may not be present. When present, it does contain a vessel (Simons, 1965).

9.2.3 Anatomy of Crocodylia

The crocodilian cardiovascular system is more sophisticated than those of other reptiles and may be an indication of endothermic ancestors (Seymour et al., 2004; Summers, 2005). There are few anatomical differences among the crocodilian species (Webb, 1979) and the heart is located on midline in the midbody and has five chambers: sinus venosus, right atrium, right ventricle, left atrium, and left ventricle (**Figures 9.13–9.15**). The chambers are lined by endocardium which is continuous with the endothelial lining of the blood vessels (Farrell et al., 1998).

The sinus venosus has the thinnest walls of all the heart chambers and is shaped as an elongated pyramid with apex pointed caudally (Webb, 1979). Just as in other reptiles, the sinus venosus can be defined as a confluence of the precaval vein, the hepatic vein, and the postcaval vein (Grigg, 1989). Both atria are equal in size and the atrial funnel extends well into ventricle (Farrell et al., 1998). The atrioventricular valves are bicuspid, fibromyxomatous, and have a septal and a marginal cusp. Both cusps are membranous but the marginal cusp of right atrioventricular valve in crocodilians has a thick, lip-like structure (Farrell et al., 1998).

Unlike other reptiles, the crocodilian heart contains two ventricles which are demarcated externally by the interventricular sulcus. The ventricles are completely separated by the interventricular septum. The interventricular septum is membranous at the heart base and continuous with the interatrial septum, but the apical third is more muscular and heavily trabeculated. These trabeculae are tiny holes (~1.5 mm) that allow communication between the two ventricles but are of unknown significance (Farrell et al., 1998).

There are three major vessels exiting the ventricles, each with a bicuspid valve and bound by a heavy connective tissue sheath. This bundle of vessels is referred to as the truncus arteriosus, and a conus arteriosus is detectable as a small bend of muscle around the pulmonary trunk (Farrell et al., 1998). The right aorta exits the left ventricle dividing cranially into the carotid and subclavian arteries and caudally into the dorsal aorta. The coronary artery also arises from the right aorta (Greenfield and Morrow, 1961).

The left aorta and the pulmonary trunk exit the right ventricle. Near the heart base a muscular ridge (the aorticopulmonary septum) divides the right ventricle into ventral and lateral compartments, separating the pulmonary trunk from the left aorta. This septum transitions into a membranous septum as it courses cranially and is also continuous with the interatrial septum (Farrell et al., 1998). In addition, the walls of the right ventricle near the pulmonary trunk contain connective tissue blocks that fit together like teeth in a cog, forming an additional valve apical to the membranous pulmonary bicuspid valve (Axelsson et al., 1996). The pulmonary trunk divides into the left and right pulmonary arteries and supplies the right and left lungs (Farrell et al., 1998). The left aorta supplies the celiac artery and then connects with the right aorta through a small anastomosis (Farmer et al., 2008).

Right and left aortae are connected in two places: proximally at the foramen of Panizza and distally by a small anastomosis. The foramen of Panizza is a hole in the interaortic septum just above (distal to) the aortic valves at the heart base (Panizza, 1833). In both aortic valves, the cusps associated with the interaortic septum (medial cusps) are longer than the lateral cusps and cover the foramen of Panizza during ventricular contraction (Farrell et al., 1998; Farmer et al., 2008). The size of the foramen of Panizza is controlled by smooth muscle under hormonal regulation (Axelsson and Franklin, 2001). When the diameter of the foramen increases in size, the septal cusps of the aortic valves cannot cover the hole, allowing for a right-to-left shunt (Axelsson and Franklin, 2001). Distally the aortae are connected by a small anastomosis caudal to the celiac artery (Axelsson and Franklin, 2011). Since the left aorta arises from the right ventricle, its blood is rich in carbon dioxide and lower in pH than the blood within the right aorta. The left aorta also supplies blood to the digestive tract through the celiac artery where this hypercapneic blood assists in lowering the stomach pH, aiding in digestion (Farmer et al., 2008).

9.2.4 Cardiac Blood Supply

Unlike mammalian hearts where the myocardium receives its entire oxygen supply from the coronary vessels, the reptilian myocardium receives oxygen from both coronary circulation and venous blood held within the chambers (Grant and Regnier, 1926; Farrell et al., 1998). The inner spongy myocardium, which makes up 60%–80% of ventricular mass, gets its oxygen from diffusion through its trabecular nature (Burggren et al., 1997; Hagensen et al., 2008). The outer compact myocardium receives its oxygen from epimural and intramural coronary vessels (MacKinnon and Heatwole, 1981, p. 168; Farrell et al., 1998; Jensen et al., 2010c).

The coronary artery arises from the right aortic trunk and bifurcates into ventral and dorsal branches (Farrell et al., 1998; Jensen et al., 2010c). The ventral branch is smaller and supplies the ventral right side of the ventricle. The dorsal branch follows the coronary sulcus to the dorsal side of the heart where the post-truncal artery branches off the main stem coronary artery. The post-truncal artery supplies the atria, aortic arches, and left ventral ventricle. The main stem coronary artery continues to feed the dorsal surface of the ventricle, the atria, and the sinus venosus (Farrell et al., 1998; Jensen et al., 2010a,c).

The coronary vessels are poorly developed in Chelonia, and the turtle heart is known for hypoxia tolerance (Wasser, 1995; Overgaard et al., 2005). In contrast, squamate hearts are very sensitive to lack of oxygen and may be the reason for well-developed coronary vessels in these species (Zaar et al., 2006). The necessity of the coronary artery in all squamates, however, is questionable. One study in the South American Rattlesnake (*Crotalus durissus*) found no change in heart rate, blood pressure, or electrocardiogram tracings after ligation of the coronary artery (Hagensen et al., 2008). Myocardial necrosis was also absent at necropsy of these snakes (Hagensen et al., 2008). This was not a long-term study, however, and coronary circulation may be necessary for cardiac blood supply during periods of hypoxia and/or digestion, as seen in the Rainbow Trout (*Oncorhynchus mykiss*) (Steffensen and Farrell et al., 1998; Hagensen et al., 2008). Descriptions of coronary veins is limited but it is noted that they drain into the sinus venosus (Farrell et al., 1998).

9.2.5 Neuroanatomy

Dual autonomic innervation of the reptile heart has been well established (Furness and Moore, 1970; Yamauchi and Chiba, 1973; Hedberg and Nilsson, 1976a; Wang et al., 2001a,b). The sympathetic innervation arises from the glossopharyngeal and spinal nerves, while the parasympathetic innervation arises from the vagus nerve (Adams, 1958; Donald and Lillywhite, 1988; Donald et al., 1990a; Davies and Donald, 1992; van Vliet and West, 1994; Filogonio et al., 2014). The sympathetic and parasympathetic nerves join each other near the head and run to the cardiac pacemaker as a mixed "vagosympathetic" nerve (Hedberg and Nilsson, 1976a).

Initially, the reptilian pacemaker was thought to be located in the walls of the sinus venosus (Buchanan, 1956). Recent work by Jensen et al. (2017) has shown the pacemaker to be located within a ring of tissue adjacent to the sinoatrial junction. From here, specialized conduction fibers travel into the atria and extend into the atrioventricular funnel, but there are no Purkinje fibers extending into the ventricle as in mammals (Robb, 1953; Buchanan, 1956; Cranefield, 1965; Burggren, 1976; Leeson, 1981; Koplra, 1987; Farrell et al., 1998). In non-crocodilian reptiles, it appears that the topographical arrangement of the muscle fibers at junction sites between fibrous tissue, rather than specialized fibers, account for the pauses occurring between the contraction of one chamber and the next (Davies and Francis, 1941; Farrell et al., 1998; Stecyk et al., 2013). This arrangement of muscle fibers for conduction is similar to what occurs in embryonic mammals (Jensen et al., 2012). In crocodilians, it has been shown that specialized conduction fibers are present in the ventricles as well as in the atria (Kvasilova et al., 2015).

Only a few reports deal with the intramyocardial spread of electrical activation in reptilian hearts (Kaplan and Schwartz, 1963; Mullen, 1967; Valentinuzzi et al., 1969a,c; Shmakov and Roshchevskii, 1982) and it appears that in reptiles the conduction is from base to apex (Lewis, 1916; Harris, 1941; Mullen, 1967; Valentinuzzi et al., 1969b; McDonald and Heath, 1971; Zhao-Xian et al., 1991; Heaton-Jones and King, 1994; Christian and Grigg, 1999; Jensen et al., 2012), and from endocardium to epicardium (Kharin and Shmakov, 2009). Some studies have shown a left-to-right conduction in turtles with normally broad cardiac anatomy (Gray, 1950; Burggren, 1978).

The sympathetic innervation to the heart causes positive inotropic and chronotropic effects through beta-adrenoceptors (Hedberg and Nilsson, 1976a; Nilsson, 1983; Chiu and Sham, 1985; Donald and Lillywhite, 1988; Donald et al., 1990a), and epinephrine has been shown to be a positive inotrope in reptiles (Yamauchi and Chiba, 1973; Hedberg and Nilsson, 1976b; Ball and Hicks, 1996; Nielsen and Gesser, 2001). In turtles, the dorsal portion of the ventricle responds more strongly to epinephrine than the ventral portion (Ball and Hicks, 1996). Unlike chelonians, both sides of the ophidian heart respond with similar strength to epinephrine (Zaar et al., 2006). The atria, however, respond more vigorously to epinephrine than the ventricle (Hedberg and Nilsson, 1976b; Zaar et al., 2006).

The parasympathetic innervation to the heart causes negative inotropic and chronotropic responses (Nilsson, 1983; Ball and Hicks, 1996). In *Crotalus*, it has been shown that the heart operates under a degree of inhibitory vagal tone which overrides a smaller sympathetic tone (Ball and Hicks, 1996; Wang et al., 2001a,b). Acetylcholine has been shown to be the major parasympathetic neurotransmitter in the reptilian heart and pulmonary vasculature, and it is antagonized by atropine sulfate (Burnstock, 1969; Hedberg and Nilsson, 1976a; Berger and Burnstock, 1979; Donald et al., 1990a,b; Morris and Nilsson, 1994; Hicks, 1998). It has also been shown that the atria respond more vigorously than the ventricle to acetylcholine, suggesting that the ventricle is not under vagal control (Hedberg and Nilsson, 1976b; Zaar et al., 2006).

9.2.6 Blood Flow and Shunting

In mammals, blood flows from the right side of the heart to the lungs and then to the left side of the heart before continuing to the body. With the pulmonary and systemic pumps arranged in series, any change in blood flow in one circuit must be matched in the other circuit. In animals with intermittent breathing, as in most reptiles, the use of cardiac energy to perfuse the pulmonary system is not physiologically economical. When taking this into consideration, the advantage of having the pulmonary and systemic pumps in a parallel arrangement is more apparent (White, 1976; Nilsson, 1994; Hicks, 1998).

9.2.6.1 Non-Crocodilians During breathing, oxygen-depleted blood flows from the coronary, hepatic, precaval, and postcaval veins into the sinus venosus. From there the blood passes through the sinoatrial ostium into the right atrium. As the atria contract, the blood then flows through the right atrioventricular valve and into the cavum venosum of the ventricle. With the aid of the ventricular muscular ridges, this blood then flows from the cavum venosum into the cavum pulmonale. During ventricular contraction, this blood will be ejected through the pulmonary artery to the lungs where it becomes oxygenated. The oxygen-rich blood returns from the lungs through the pulmonary veins into the left atrium. When the atria contract, this blood is ejected through the left atrioventricular valve into the cavum arteriosum of the ventricle. During ventricular contraction, the oxygen-rich blood is ejected from the cavum arteriosum, through the cavum venosum, and into both aortae to circulate back into the body (Jensen et al., 2010a). So, the ventricle operates as a single pressure pump supplying both systemic and pulmonary circulations (**Figure 9.16**) (Hicks, 1998). To summarize:

$$\text{body} \rightarrow \text{SV} \rightarrow \text{RA} \rightarrow \text{CV} \rightarrow \text{CP} \rightarrow \text{PA} \rightarrow \text{lungs} \rightarrow$$
$$\text{LA} \rightarrow \text{CA} \rightarrow \text{CV} \rightarrow \text{RAo/LAo} \rightarrow \text{body}$$

This blood flow pattern facilitates carbon dioxide elimination in the lung, minimizes ventilation/perfusion mismatching, and improves systemic oxygen transport and myocardial oxygenation (Wang et al., 1997; Hicks, 2002). Timing of contractions prevents left-to-right shunting, as the muscular ridge and the bulbuslamelle abut during contraction. This abutment closes the opening between the cavum venosum and cavum pulmonale, preventing oxygen-rich blood from entering the pulmonary circulation (Webb et al., 1971; Wang et al., 2002, 2003a; Jensen and Wang, 2009; Jensen et al., 2010a).

Very little is known about the normal patterns of cardiac shunting in reptiles, and although some species possess prominent ventricular pressure separation (Wang et al., 2003a), others can exhibit sizeable cardiac shunts (Lillywhite and Donald, 1989). All squamate hearts have essentially the same design, but varanid lizards and pythons have evolved in parallel a pronounced ventricular separation (Jensen et al., 2014). Echocardiography has revealed that the atrioventricular valves in pythons descend deep into the ventricle during ventricular filling and thereby greatly reduce the communication between the systemic (cavum arteriosum) and pulmonary (cavum pulmonale) ventricular chambers during diastole. Angioscopy and echocardiography have shown how the muscular ridges and the bulbuslamelle contract against each other in systole and provide a functional division of the ventricle. Although washout shunts are inevitable in the subdivided reptile ventricle, it has been shown that the site of shunting, the cavum venosum, is very small throughout the cardiac cycle. It is concluded that the python ventricle is incapable of the pronounced and variable shunts of other snakes, because of its architecture and valvular mechanics (Jensen et al., 2010b).

In contrast, Chelonia and most Squamata do not possess a pressure separation between the systemic and pulmonary arteries (Galli et al., 2005a,b; Joyce et al., 2016), and large right-to-left cardiac shunts at low temperatures have been inferred from arterial oxygen levels (Wang et al., 2000). During a right-to-left shunt, the blood flowing from the right atrium fills the cavum venosum of the ventricle. This blood then ejects directly through the aortae instead of the pulmonary artery, bypassing the lung (Hicks, 2002). Independent flow between circuits is dependent upon relative vascular resistance in the two circuits and a partial pulmonary bypass is possible. In some circumstances, a right-to-left shunt can approach 100% (Lillywhite and Donald, 1989; Overgaard et al., 2002). During ventilation, however, rapid pulmonary perfusion occurs to optimize gas exchange (White, 1976). This bypassing of the lung has an advantage during apnea by saving cardiac energy, "metering" lung oxygen stores, reducing carbon dioxide flux into the lung, reducing plasma filtration into the lung, facilitating warming, triggering hypometabolism, and facilitating stomach acid secretion and digestion (Hicks, 2002).

During apnea, there is an increase in parasympathetic tone resulting in bradycardia and an increase in pulmonary resistance (Taylor et al., 2009). This causes a decrease in pulmonary blood flow and an increase in right-to-left shunting of blood into the aortae (**Figure 9.17**) (Hicks, 2002). During ventilation, there is a decrease in parasympathetic tone resulting in tachycardia and a decrease in pulmonary resistance. This lessens the right-to-left shunt and allows an increase in pulmonary blood flow (Farmer, 1999; Hicks, 2002; Galli et al., 2005a; Axelsson et al., 2007).

9.2.6.2 Crocodilians While at rest and breathing air, the crocodilian cardiovascular blood flow pattern is similar to that of birds and mammals (Nilsson, 1994). Deoxygenated blood from the body enters the sinus venosus which is pumped into the right atrium and then pumped into the right ventricle. During ventricular contraction, the deoxygenated blood is pumped through the pulmonary trunk and to the lungs where it is oxygenated. The oxygenated blood returns from the lungs to the left atrium where it is pumped into the left ventricle. During ventricular contraction, the oxygenated blood is pumped through the right aorta to the vital organs. The septal leaflet of the right aortic valve covers the foramen of Panizza during systole, preventing left ventricular blood from flowing into the left aorta. During diastole, however, some of this blood may flow through the foramen due to elastic recoil (**Figure 9.18**) (Farrell et al., 1998). Therefore, the blood in the left aorta originates from the right aorta via the foramen of Panizza and the distal anastomosis while the crocodilian is at rest, breathing air (Nilsson, 1994; Farmer et al., 2008).

The semilunar bicuspid valves of the great vessels are leaflets. When the ventricular pressure exceeds the arterial pressure, the valve opens and blood is ejected into the artery. In the right ventricle, there are two exits for blood ejection—the left aorta and the pulmonary trunk. When resting and breathing air, the pulmonary vascular pressure is much lower than the aortic pressure and consequently the pressure in the right ventricle does not exceed systemic pressure, and the valve to the left aorta remains closed and all of the right ventricular output goes to the lungs.

During breath-holding, the pressure in the pulmonary system increases. This increase in the pulmonary system allows the right ventricular pressure to also increase, allowing up to 25% of the stroke volume to eject into the left aorta (Farrell et al., 1998). This triggers a right-to-left shunting of blood away from the lungs, metering the lung oxygen stores (Hicks, 2002). In addition,, the "cog valve" portion of the right ventricular outflow tract may close under certain circumstances, allowing a more dramatic right-to-left shunt (**Figure 9.19**). Epinephrine and stress have been shown to cause the cog valve to open, decreasing the right-to-left shunt (Franklin and Axelsson, 2000). Factors shown to close the cog valve and increasing a right-to-left shunt include acetylcholine, gastrin-releasing peptide, sotalol, substance P, and histamine (Farmer et al., 2008).

9.2.7 Blood Pressure

Typical atrial pressures range from 3–5 mm Hg and the typical ventricle systolic pressures range from 40–70 mm Hg (Johansen, 1959). The typical systemic pressures in the aortae range from 50–60 mm Hg (systolic) and 40–50 mm Hg (diastolic) which is slightly higher than amphibians but roughly 50% that of birds and mammals (Johansen, 1959; Seymour and Baylock, 2000; Hagensen et al., 2008). The pulmonary arterial systolic pressure is usually equal to the aortic systolic pressure, but the pulmonary diastolic pressure is often lower than the aortic systemic pressure by 20–30 mm Hg (Johansen, 1959; Heisler and Glass, 1985; Lillywhite, 1987; Lillywhite and Donald, 1989; Stinner and Ely, 1993; Hicks, 1998; Overgaard et al., 2002; Filogonio et al., 2014). In Varanidae and Pythonidae this is different due to a well-developed muscular ridge dividing the two sides of the ventricle, and systemic pressure can greatly exceed pulmonary pressure (Burggren and Johansen, 1982; Wang et al., 2002, 2003a; Zaar et al., 2006; Jensen et al., 2010a; Enok et al., 2014; Joyce et al., 2016). It seems that the high systemic pressures correlate with the high activity and metabolic demands of varanids (Burggren and Johansen, 1982). Pythonidae are ambush predators and are not particularly active but have been shown to sustain high metabolic rates and oxygen consumption during digestion (Wang et al., 2003b). The functional division in the heart of the *Python molurus* may also be due to the high metabolic need during shivering thermogenesis while brooding (Wang et al., 2002, 2003a).

In the absence of external disturbance, the resting blood pressure in reptiles is relatively stable (Lillywhite and Donald, 1994), yet the arterial blood pressure can differ sevenfold in various species (Johansen, 1959; Lillywhite and Seymour, 1978; Lillywhite and Pough, 1983; Seymour, 1987; Stinner and Ely, 1993; Wang et al., 2001a, 2003a; Seymour and Arndt, 2004). Multiple factors that can affect blood pressure of reptiles include habitat, activity, hunting, digestion, and temperature.

As in mammals, regulation of arterial pressure is accomplished in reptiles by moment-to-moment neural activation of baroreflexes involving cardiovascular mechanoreceptors, in addition to chronic adjustments of fluid balance to preserve the blood volume (Lillywhite and Smith, 1981, 1984; Smits and Lillywhite, 1985). There are no carotid sinus baroreceptors in reptiles as in mammals (Adams, 1958; Hohnke, 1975; Seymour and Barker, 1983; Berger, 1987). In reptiles, the baroreceptors are present in the truncus arteriosus and central arteries (**Figures 9.20–9.21**) (Kozubowski, 1985; Backhouse et al., 1989; Lillywhite and Donald, 1994). These baroreceptors are essential in neuroendocrine regulation of blood pressure and shunting.

9.2.8 Neuroendocrine Function

As in mammals, there are two main pressor systems in the reptiles: the alpha-adrenergic system and the renin-angiotensin system (Chiu et al., 1986). The alpha-adrenergic system is the major influencing pressor system and is composed primarily of catecholamines and acetylcholine. Catecholamines stimulate the heart through beta-adrenoceptors causing positive inotropic and chronotropic effects (Hedberg and Nilsson, 1976a,b; Nilsson, 1983; Donald et al., 1990a; da Silva Braga et al., 2016), while acetylcholine acts as a negative inotropic and chronotropic agent on the heart (Nilsson, 1983).

As in mammals, catecholamines from the autonomic nervous system stimulate reptilian vessel alpha adrenoceptors and cause vasoconstriction (Lillywhite and Seymour, 1978; Lillywhite and Smith, 1981; Lillywhite and Gallagher, 1985; Lillywhite, 1987).

In addition to this alpha-adrenergic system, the renin-angiotensin system also plays an important role in reptilian blood pressure regulation (Yung and Chiu, 1985; Chiu et al., 1986; Breno and Picarelli, 1992; Borgheresi et al., 1996; Breno et al., 2001). Just as in mammals, renin converts angiotensinogen into angiotensin I, which is converted into angiotensin II by the angiotensin-converting enzyme (ACE) (Nakayama et al., 1977; Lazari et al., 1994). Angiotensin II will increase the blood pressure by stimulating catecholamine release, promoting sodium chloride reabsorption in the kidneys, promoting potassium excretion by the kidneys, directly stimulating arterial constriction, stimulating arginine vasopressin secretion by the pituitary gland, and stimulating aldosterone secretion by the adrenal gland (Lazari et al., 1994). Arginine vasopressin increases blood pressure by causing constriction of capillaries and arterioles. Little work has been done on aldosterone in reptiles, but it has been shown to behave in the freshwater snake, *Natrix cyclopion*, as in mammals by stimulating sodium resorption and subsequent water resorption in the renal tubules (Larsen et al., 2014). Whether this action is mirrored in other reptiles needs further research. Use of an ACE-inhibitor, such as captopril, will prevent the conversion of angiotensin I to angiotensin II and in turn lower the blood pressure in squamates (Chiu et al., 1986; Breno and Picarelli, 1992).

In addition, other physiological chemicals, such as kinins and histamines, play an important role in cardiovascular regulation (Abdalla et al., 1989; Wang et al., 2000a). Bradykinin has positive inotropic and chronotropic effects on beta-adrenergic receptors in the heart and on alpha-adrenergic receptors in peripheral vessels (Wang et al., 2003a). Substance P (a tachykinin), causes a decrease in blood pressure and a subsequent increase in the cardiac output (Wang et al., 2000a,b). Endothelins function in vascular homeostasis by causing vascular contraction, increasing either pulmonary or systemic resistance and increasing blood pressure (Agapitov and Haynes, 2002; Mesquita et al., 2008; Skovgaard et al., 2008). Endothelins have been shown to be present in Chelonia, Crocodylia, and Squamata, but appear to be absent in less evolved snakes (colubridae) (Poder et al., 1991; Platzack et al., 2002; Skovgaard et al., 2005b; Skovgaard and Wang, 2006; Mesquita et al., 2008). Atrial natriuretic peptide, a potent vasodilating hormone, has also been demonstrated in ophidian cardiocytes (Mifune et al., 1996). In addition, histamine has been shown to be a positive chronotrope in pythons after a meal (Enok et al., 2012). The source of histamine has not been determined, but it does not appear to originate from mast cells, and it is not associated with gastrin or cholecystokinin (Enok et al., 2012).

9.2.8.1 Giant Constrictors Interestingly, giant constrictors like the Burmese Python (*Python molurus*) and Yellow Anaconda (*Eunectes notaeus*) can develop dramatic cardiac hypertrophy during digestion of a large meal (Wang et al., 2001a; Andersen et al., 2005; Zerbe et al., 2011; Enok et al., 2013). This hypertrophic growth only occurs after a meal that is high in protein and not after a meal that is high in lipids (Henriksen et al., 2014; McCue et al., 2015). This may change the magnitude of cardiac shunts (Starck, 2009), prioritizing visceral perfusion with increased superior mesenteric and hepatic portal blood flow (Secor and White, 2010). Once the digestive process has been completed, the cardiac hypertrophy resolves and the heart returns to its previous size (Jensen et al., 2011). These findings, however, are often inconsistent and not always reproducible (Jensen et al., 2011), and the hypertrophy may instead reflect stressed conditions (Slay et al., 2014).

9.3 Cardiovascular Diseases

Cardiovascular diseases are underdiagnosed in reptiles and can be attributed to multiple causes. A method to organize these causes is by using the DAMNIT mnemonic (Gibbons, 2009): Degenerative, Deposition, Developmental Anomalies, Metabolic, Neoplastic, Nutritional, Idiopathic, Infectious, and Traumatic.

9.3.1 Degenerative Diseases

Degenerative diseases are progressive disorders where there is a deterioration in the quality or function of a tissue or organ. The most common degenerative cardiovascular disease in reptiles is arteriosclerosis. Arteriosclerosis literally means *hardening of the arteries*, and is more fully defined as chronic arterial change consisting of hardening, loss of elasticity, and luminal narrowing resulting usually from proliferative and degenerative, rather than inflammatory, changes of the media and intima (**Figures 9.22–9.25**) (Robinson et al., 2016). Reptilian arteriosclerosis commonly is identified in captive animals and has been associated with inappropriate diet (Schuchman and Taylor, 1970; Frye, 1991b), although aging, chronic renal disease, and previous cardiovascular insults may be important in the pathogenesis of this condition. Arteriosclerosis can predispose to thrombosis and possibly also contributes to aortic aneurysms, especially in squamates (**Figure 9.25**).

There are few documented cases of acute or non-fibrosing degenerative cardiovascular diseases in reptiles. In Florida, 58% of captive Leatherback Turtle (*Dermochelys coriacea*) hatchlings and 41% of wild leatherbacks had cardiac degeneration associated with systemic disease (Miller et al., 2009). Heart failure from undefined case of cardiomyopathy was also reported in a Deckert's Rat Snake (*Elaphe obsoleta deckertii*) where the myocardial fibers were degenerated and necrotic (Jacobson et al., 1979).

Myocardial fibrosis is common in reptiles, generally occurring circumferentially in the ventricle and primarily at the interface of the spongiform and compact muscular tunics, and cardiac fibrosis is somewhat common in relatively young giant tortoises (**Figure 9.26**). Endocardiosis frequently accompanies myocardial fibrosis, especially involving the atrioventricular valve (**Figures 9.27–9.29**). Epicardial and pericardial fibrosis, with fibrous epicardial adhesions to the pericardium, suggest chronic pericardial effusion due to cardiac insufficiency or trauma, and the adhesions can be strikingly obvious at necropsy, especially in the giant tortoises (**Figures 9.30–9.31**). Evidence of chronic hemorrhage (xanthogranulomatous inflammation or hemosiderophages) in these lesions suggests the possibility that previous cardiac blood draws or other trauma could possibly contribute to lesions of this nature as well.

9.3.2 Deposition Diseases

Atherosclerosis is the accumulation of lipids, foamy macrophages, lipids, or cholesterol crystals and sometimes concurrent mineral or bone in vessel walls (Ellulu et al., 2016; Robinson et al., 2016). Atherosclerosis is a common disease in birds, but is poorly documented in reptiles (Schilliger et al., 2010b, 2019). In a study of 200 reptiles with arterial disease, primary lipid-containing intimal lesions were present in less than 5% of cases. It is suggested that the low incidence of atherosclerosis in reptiles may be related to their comparatively low systemic blood pressure with resultant minimal stress injury to the arterial wall (Finlayson and Woods, 1977). However, there is a relationship between obesity and cardiovascular disease in snakes. Obese snakes were found to have elevated cholesterol, triglyceride, and beta-lipoprotein levels when compared to non-obese snakes and they died at younger age with an aortic aneurysm, ventricular aneurysm, and/or medial mineralization of arteries (Bauer and Jacobson, 1989; Paré and Lentini, 2010). Chronic hypercholesterolemia was shown to cause changes in the ECG of Chelonia that included increases in amplitude, irregularities in wave forms, and increases in conduction time (Ferguson, 1942).

Atherosclerosis with and without pericardial effusion have been reported in Central Bearded Dragons (*Pogona vitticeps*). Segmental to circumferential plaque-like thickening of the intima was observed close to the base of the major arteries and they were composed of mature, well-vascularized fibrous connective tissue separated by aggregates of lipid-laden macrophages, erythrocytes, and/or hemosiderin-laden macrophages (Schilliger et al., 2010b; 2019) (see also Chapter 3). Atheromatous plaques were also observed in the aorta of a 20-year-old Two-Banded Monitor (*Varanus salvator*) (Vastesaeger et al., 1965).

Mineralization of the great vessels is a common finding in captive reptiles at necropsy. Tissue mineralization has been associated with metastatic mineralization resulting from renal failure or hypervitaminosis D (Frye, 1991b; Kik and Mitchell, 2005; Schmidt and Reavill, 2010). In lizards it was often associated with nutritional bone disease (Finlayson and Woods, 1977). Interestingly, a recent ultrasound study showed that more than 75% of apparently healthy green iguanas had mineralization of the great vessels (**Figures 9.32–9.33**) (Gustavsen et al., 2014). Other metabolic disturbances that cause derangements in calcium metabolism, such as paraneoplastic syndrome or ovarian disease, may also contribute to metastatic mineralization. Dystrophic mineralization, or localized mineralization of diseased tissue in the absence of hypercalcemia, involving the myocardium and arterial wall is common in reptiles and likely due to many different factors, including aging, trauma, previous episodes of vasculitis or myocarditis, or hypoxia. Dystrophic mineralization has been reported in several species of reptiles, with a considerably higher prevalence in Squamata than in Chelonia and Crocodylia. Its occurrence in snakes was considered a sequel of a parasitic arteritis (Finlayson and Woods, 1977).

Gout is the deposition of uric acid precipitates in renal, visceral, articular, and other tissues. Uric acid is derived from the breakdown of purine bases from both endogenous and exogenous sources of nucleic acids. Gout can be caused by several factors, including dehydration, nutritional imbalance, toxicosis, and infectious disease, these processes initially affecting the kidney, which in turn leads to hyperuricemia. Gout has been reported in tortoises, snakes, and lizards. A common site for visceral urate deposition is the pericardial sac (**Figures 9.34–9.35**), although urates can deposit on the endocardial surface, myocardial sinusoidal surfaces and the valves, and on the surfaces and within the walls of the great vessels. Urate deposition in the smaller vessels of various tissues presents as segmental to circumferential, full thickness fibrinoid change, sometimes associated with fibrin or urate thrombi (**Figure 9.36**).

Hemochromatosis, also known as iron overload, indicates accumulation of iron in the body from any cause. Hemochromatosis has been widely described in mammals and birds, but there are also some reports in reptiles (Stolk, 1961; Vivaldo et al., 2009). One of the authors has seen a case of hemochromatosis in a gecko, with large accumulations of iron in the liver and spleen, but also in the lung and heart. Areas of degeneration and necrosis of cardiomyocytes were observed only in regions of iron accumulation. It is unclear if that gecko had a hereditary mutation that resulted in the accumulation of iron or if the dietary intake was excessive.

9.3.3 Developmental Anomalies

Anomalies, or congenital lesions, are common in reptiles, but those affecting the cardiovascular system are uncommon.

In Chelonia, a persistent patent ductus arteriosus has been described in the Greek Tortoise (*Testudo graeca*) (Burggren, 1976). In Crocodylia, a congenital bilateral aortic stenosis and an interventricular septal defect have been described in American Alligators (*Alligator mississippiensis*) (Brockman and Kennedy, 1962; McIntosh et al., 1967).

In Squamata, there have been more cardiovascular anomalies recorded than in the other reptile orders. Some dicephalic snakes have been documented with two hearts (**Figures 9.37–9.38**) (de Albuquerque et al., 2010, 2013). A bifid ventricle has been described in a Ball Python (*Python regius*) (**Figure 9.39**) (Jensen and Wang, 2009) and an ectopic cordis in a Rainbow Boa (*Epicrates cenchria*) (Mitchell, 2009). An atrial septal defect in a Komodo Dragon (*Varanus komodoensis*) was the cause of heart failure and congestion of several organs (Pizzi et al., 2009). Malformed or incompletely formed valves have also been described in multiple squamate species (Frye, 1991a,b; Clippinger, 1993; Kik and Mitchell, 2005; Mitchell, 2009; Bogan, 2017).

9.3.4 Metabolic Diseases

Metabolic derangements have also been noted to cause cardiac disease in reptiles. Renal disease can produce metabolic acidosis and hyperkalemia. This disturbance exerts negative inotropic effects because the excess protons from the acidosis compete with calcium ion binding (Orchard and Kentish, 1990) and cause life-threatening arrhythmias. In several species, however, severe acidosis does not seem to affect contractility so long as the reptile is not undergoing anaerobic metabolism (Wojtaszek, 1979; Gesser and Poupa, 1983; Wasser et al., 1990; Overgaard et al., 2005; Zaar et al., 2006). As renal disease progresses to renal failure, circulating uric acid levels will begin to rise, eventually precipitating into tissues, including the great vessels, pericardium, epicardium, and myocardium (Jacobson, 1976; Frye, 1991b; Schmidt and Reavill, 2010) (see Section 9.3.2).

Hyperviscosity-like syndrome is an uncommon condition seen in Squamata (Gyimesi et al., 2008; LaDouceur and Garner, 2017). Hyperviscosity in domestic animals is typically associated with erythrocytosis and hypergammaglobulinemia secondary to hematopoietic neoplasia (Newkirk et al., 2012). The most common clinical sign in lizards and snakes is sudden death. Common gross lesions include coelomic effusion, pulmonary edema, marked congestion, and sometimes hemorrhage (**Figures 9.40–9.41**). Histologically, animals consistently had deeply amphophilic fluid in the vasculature and interstitium of various organs, most commonly in the lungs, heart, liver, and kidney. Females are over-represented, and concurrent disease was identified in several cases, including sepsis, neoplasia, and yolk coelomitis. Possible predisposing factors include hypothermia, hyperthermia, sepsis, and endocrine abnormalities (LaDouceur and Garner, 2017).

9.3.5 Neoplastic Diseases

Reptilian neoplastic diseases are described in Chapter 1. Although primary and metastatic neoplasms have been described in multiple species of Chelonia and Squamata, cardiac neoplasia has not been described in either Sphenodontia or Crocodylia. Fibropapillomatosis is commonly seen in Green Turtles (*Chelonia mydas*) of the Caribbean and the southern United States, and typical lesions have been found in the hearts of stranded Green Turtles (Aguirre et al., 2002; Work et al., 2004). Fibropapillomas in Green Turtles are associated with herpesvirus infection (see Volume 1, Chapter 10). A Loggerhead Turtle (*Caretta caretta*) had multicentric lymphoma that involved cardiac tissue (Oros et al., 2001). Primary cardiac neoplasms reported in Squamata include hemangiosarcoma, fibrosarcoma, and rhabdomyosarcoma (Elkan and Cooper, 1976; Hruban et al., 1992; Gumber et al., 2010; and Stumpel et al., 2012). Other neoplasms of Squamata with metastasis to the heart include spindle cell sarcoma (**Figures 9.42–9.43**), leukemia, dermal mastocytoma, oviduct adenocarcinoma, disseminated chondosarcoma, splenopancreatic ductal adenocarcinoma, and lymphoma (Zwart and Harshbarger, 1972; Goldberg and Holshuh, 1991; Ramsay et al., 1996; Schumacher et al., 1998; Pereira and Viner, 2008, Schilliger et al., 2011; Reavill and Schmidt, 2012; Mathews et al., 2014).

9.3.6 Nutritional Diseases

Nutritional secondary hyperparathyroidism is perhaps the most important nutritional disorder in captive reptiles, particularly in herbivorous lizards and chelonians. It can be caused by a deficiency in dietary calcium and/or vitamin D3 and/or inadequate exposure to UVB radiation, leading to hypocalcemia and bone resorption (Mans and Braun, 2014). Affected animals routinely present with generalized muscle fasciculation and tremors. Although it affects mostly the skeletal muscles, the direct effect that hypocalcemia has on the cardiac muscle should also be considered as a cause of myofiber hypercontraction or cardiac arrhythmia (Kik and Mitchell, 2005).

There are few reports of white muscle disease in reptiles (vitamin E/selenium deficiency), and typically the final diagnosis is tentative or corroborated with a combination of clinical signs, histologic examination of heart, skeletal muscle and fat, and if appropriate, response to treatment. The consistent histologic finding in these cases is generalized degenerative myopathy, which may also involve the heart (Farnsworth et al., 1986; Les Gabor, 2005; Cole et al., 2008; Mans and Braun, 2014). As in mammals and birds, the presence of fat necrosis or necrotizing steatitis in conjunction with the myopathy aids in distinguishing this disease from the more common stress-related or exertional myopathies.

9.3.7 Idiopathic Cardiovascular Diseases

Aneurysms are well known in humans, and several cases have been described in mammals and birds (mostly turkeys), but they are poorly documented in reptiles (Robinson et al., 2016). Aneurysm is an abnormal and localized dilation to more than 50% of the normal diameter of any vessels (usually arteries are affected, especially large elastic arteries, but the lesion can also occur in veins) or the heart (Goldfinger et al., 2014; Miller and Gal, 2016). In reptiles, aneurysms have been described in Bearded Dragons, a Sail Tail Lizard (*Hydrosaurus amboinensis*), a Burmese Python (*Python molurus bivittatus*), an Amethyst Python (*Liasis amethistinus*), a Western Diamondback Rattlesnake (*Crotalus atrox*), and in a Spanish Pond Turtle (*Mauremys leprosa*) (Ardlie and Schwartz, 1965; Finlayson and Woods, 1977; Griner, 1983; Bauer and Jacobson, 1989; Rush et al., 2001; Barten et al., 2006; Sweet et al., 2009). The anatomical differences in the heart of Varanidae and Pythonidae (a well-developed muscular ridge dividing the two sides of the ventricle) cause a higher systemic pressure and may be associated to the caudal aneurysms observed in these species. The authors have observed aneurisms and/or aortic dissection (without focal dilation of the vessel) They were located mostly in the coelomic cavity (near the heart base), but they were also found behind the eye or in the femoral arteries. Dissecting aneurysm was observed in nine animals (75%) and was characterized by focally extensive hemorrhage that dissected the tunica media, separating bundles of smooth myofibers and creating a cavity within the arterial wall. All arteries presented some degree of wall degeneration, indicating that arteriosclerosis may play a role in the development of aneurysms in bearded dragons (Ortega et al., unpublished data). Genetic predisposition, hypertension, or trauma to the arteries during feeding have been also proposed as possible causes of aneurysm (**Figures 9.44–9.50**) (Barten et al., 2006; Raiti, 2012).

9.3.8 Infectious Diseases

The primary infectious diseases of reptiles are described in detail in Volume 1. Cardiac lesions of sepsis or other infectious disease processes reflect typical thromboemboli, inflammatory infiltrates, and granulomas likely to be seen in other tissues. Vegetative endocarditis is common in reptiles, and a thorough gross examination of the heart at necropsy is recommended.

9.3.8.1 Chelonia There are few reported primary infectious diseases of the cardiovascular system in Chelonia. A chlamydiosis epizootic outbreak resulted in hundreds of deaths of juvenile (4- to 5-year-old) Green Sea Turtles (*Chelonia mydas*). The most severe and consistent lesion was necrotizing myocarditis, but pericarditis, endocarditis, and mesenteric arteritis were also reported (Homer et al., 1994).

Fatal septicemia caused by *Helicobacter* infection was described in a Pancake Tortoise which presented with regional cellulitis and edema of the head and neck, and pericarditis (Stacy and Wellehan, 2010). Pericardial effusion in a California Desert Tortoise (*Gopherus agassizii*) with osteomyelitis, and vegetative endocarditis in a Leatherback Turtle due to *Vibrio damsela* (Obendorf et al., 1987; Penninck et al., 1991) have also been reported (**Figures 9.51–9.52**). In a study of 93 sea turtles from Spain, only 4 (4.30%) presented with lesions in the cardiovascular system, identified as fibrinous exudative pericarditis. Half of these were caused by *Staphylococcus* sp. and the other half had no bacteria isolates (Oros et al., 2005). It is unclear if any of these represent primary bacterial pathogens, or if these are opportunists targeting the heart as a manifestation of sepsis.

Blood flukes belonging to the family Spirorchiidae (Digenea) represent a major threat to sea turtle health and are considered the most important parasitic cause of turtle strandings and mortality worldwide (Wolke et al., 1982; Rand and Wiles, 1985; Glazebrook et al., 1989; Cordero-Tapia et al., 2004; Stacy et al., 2010; Santoro et al., 2017) (see also Volume 1, Chapter 13). The pathologic changes associated with the spirorchiids and their eggs include aneurysms, arteritis of great vessels with dark nodular thickenings, endocarditis, hemorrhagic lesions, thrombi, vasculitis, and granulomatous reactions (**Figures 9.53–9.56**) (Santoro et al., 2007). Resolution of thrombi was observed to occur via a combination of granuloma formation around indigestible components (spirorchid fluke eggshells) and exteriorization through the vessel wall, which resulted in granulomatous nodules on the adventitial surface (Gordon et al., 1998). Spirorchiid eggs have also been described in the myocardium of Freshwater Turtles (*Trachemys scripta elegans* and *Chrysemys picta*) (Johnson et al., 1998). Lesions in the vasculature of the brain and meninges can be especially severe, and possibly contribute to hydrocephalus.

9.3.8.2 Squamata Inclusion body disease (IBD) is a disease affecting mainly boid snakes and has been associated with coinfection by various reptarenavirus species (family Arenaviridae) (Keller et al., 2017). Recent studies showed that boas are more susceptible than pythons to IBD. More than 80% of clinically healthy boa constrictors tested positive for IBD, which could reflect the possibility that boas are natural hosts of these viruses in the wild (Chang et al., 2016; Stenglein et al., 2017). Although eosinophilic intracytoplasmic inclusion bodies have been found in cardiac myocytes and endothelial cells of affected snakes, the role of reptarenavirus in heart disease is undetermined and may be minimal (Turchetti et al., 2013) (see also Volume 1, Chapter 10).

Bacterial infections often lead to septicemia, causing microemboli which can seed bacteria in the heart (**Figures 9.57–9.60**). Multiple opportunistic gram-negative bacteria have been shown to cause infections within the ophidian heart (Jacobson et al., 1991; Woodhouse et al., 2014). These infections can infect either the cardiac muscle or colonize the epicardial or endocardial lining and heart valves and form vegetative lesions (Jacobson et al., 1991; Schroff et al., 2010; Wernick et al., 2015). As in mammals, vegetative endocarditis can cause valvular insufficiency and subsequent heart failure (Jacobson et al., 1991; Rishniw and Carmel, 1999; Schroff et al., 2010; Shilliger et al., 2010a; Wernick et al., 2015). In addition, granulomatous pericarditis and myocarditis due to chlamydial infection has been reported in Puff Adders (*Bitis arietans*) and Emerald Tree Boas (*Corallus caninus*) (Jacobson et al., 1989, 2002). Severe myocardial hematoma and myocarditis was reported as the cause of death in an anaconda (**Figure 9.61**) (A. Wuenschmann, personal communication).

Snake fungal disease (SFD) is an emerging and common skin infection of wild snakes in North America and Europe caused by *Ophidiomyces ophidiicola* (Lorch et al., 2015; Franklinos et al., 2017). Disseminated visceral infections are uncommon, perhaps because the lesion remains localized to skin in most animals, or others succumb to secondary disease processes such as bacterial sepsis or dehydration. However, lymphoplasmacytic to lymphohistiocytic or granulomatous inflammation affecting many internal organs, including the heart, has been reported (Lorch et al., 2015, 2016).

Microsporidia are a group of spore-forming unicellular organisms, once considered to be protozoa but are now known to be fungi (Han and Weiss, 2017). In Bearded Dragons (*Pagona vitticeps*) systemic microsporidial infections have been described (**Figures 9.62–9.63**) (Jacobson et al., 1998; Richter et al., 2013). These obligate intracellular pathogens are gram positive and acid fast and have a small polar granule that stains using a periodic acid-Schiff reaction (**Figure 9.64**). On hematoxylin and eosin-stained tissue, Microsporidia present as clusters of light basophilic intracytoplasmic organisms.

Parasitic diseases can extend into the cardiovascular system of squamates (**Figure 9.65**). Migration of parasites through the arteries can cause injury to the endothelium, resulting in aortic plaques, saccular aneurysms, and thrombosis (Ardlie and Schwartz, 1965; Frye, 1991b). Various genera of filarid worms can be found in reptiles (see Volume 1, Chapter 13). Microfilaremia is usually an incidental finding but heavy infections can cause thrombosis resulting in edema and necrosis (Strik et al., 2007). Protozoal myocarditis caused by *Sarcocystis* spp. has also been observed in snakes (Frye, 1991b).

9.3.8.3 Sphenodontia There is little information describing cardiovascular infections in Sphenodontia.

The parasite *Hepatozoon tuatarae* (Apicomplexa) was identified infecting leukocytes and erythrocytes of Tuataras, but both forms of infection were rarely detected in the same individual. Although infection prevalence and intensity were low in this study, its pathogenic role remains unknown (Godfrey et al., 2011).

9.3.8.4 Crocodylia Recent outbreaks of infection with West Nile virus in human populations and in farmed alligators in the United States have placed the research emphasis on crocodiles and alligators as susceptible species and reservoirs for this zoonotic disease (Klenk et al., 2004; Ariel, 2011). The main lesions observed in the cardiovascular system are pericardial effusion, myocarditis, myocardial degeneration, and necrosis (**Figure 9.66**) (Miller et al., 2003; Jacobson et al., 2005).

Systemic bacterial infection can cause severe lesions in the heart, as observed in a Nile Crocodile (*Crocodylus niloticus*) with septicemia caused by *Citrobacter braakii* that presented with severe fibrinous mural endocarditis and myocarditis in the right ventricle (**Figure 9.67**) (Rosell et al., 2019). Also, in Nile Crocodiles, severe pentastomid infection caused by *Leiperia cincinnalis* was observed in heart, pulmonary artery, and aorta, causing endothelial ulceration and inflammation (Junker et al., 1999, 2016). Mycoplasmosis has been associated with severe outbreaks and mortality in alligators. *Mycoplasma alligatoris* was identified as the causative agent, and the most consistent lesion was pericardial effusion, but also fibrinous pericarditis, myocarditis, and endocarditis have been observed (**Figure 9.68**; Volume 1, **Figures 11.71–11.72**) (Clippinger et al., 2000; Brown et al., 2001).

9.3.9 Trauma

Trauma in wild reptiles was the most common cause for presentations at a wildlife center (Brown and Sleeman, 2002). Most traumatic injuries involving the heart are usually instantly fatal; however, necrotic lesions were reported in the heart of a Boa Constrictor (*Boa constrictor constrictor*) with cardiomegaly. Trauma was suspected as the inciting cause, as there was a circular skin lesion consistent with a puncture on the ventral scutes at the level of the heart (Gray, 2004).

Iatrogenic trauma to the heart during cardiac blood draws may also lead to cardiac disease in snakes. Although multiple blood collections via cardiocentesis have been reported to be safe in the snake (Isaza et al., 2004; Brown, 2010; Divers, 2013), the authors have identified lesions to the contrary (see Section 9.3.1). In addition, snakes with underlying cardiac disease may develop cardiac tamponade and death following cardiocentesis (**Figures 9.69–9.70**) (Selleri and Girolamo, 2012).

A recent study in Loggerhead Sea Turtles (*Caretta caretta*) showed that underwater entrapment in fishing gear followed by rapid decompression may cause gas bubble formation within the bloodstream (embolism) and tissues leading to organ injury, impairment, and even mortality in some bycaught individuals. Gas was found within the atria (especially the right atrium) and the sinus venosus, as well as within abdominal, mesenteric, gastric, pancreatic, hepatic, and renal veins (García-Párraga et al., 2014). The likelihood of fatal decompression increases with increasing depth of gear deployment, establishing a direct relationship between depth, risk, and severity of embolism, which has not been previously demonstrated in any breath-hold diving species (**Figures 9.71–9.72**). Gas emboli have also occurred in captive reptiles due to supersaturation of the aquatic environment associated with pump failures (Lung and Garner, 2000). Propeller trauma can penetrate the lung and cause coelomic emphysema and gas embolization. (Fahlman et al., 2017)

Yolk embolism (see Chapter 8) has been reported in several species of reptiles, and can be due to trauma, bacterial oophoritis, or folliculostasis (Garner et al., 2003). Yolk embolism seems to be particularly common in Komodo Dragons. Evidence of yolk embolization was found in 10 sea turtles with globular eosinophilic material within the cardiovascular system, most frequently within the heart or vessels of the brain (Stacy et al., 2013). Mechanical injury resulting in vascular damage and traumatic introduction of yolk into the bloodstream is the most obvious potential mechanism of yolk embolization (**Figure 9.73**). Occasionally, air or lipid emboli may be encountered in the glomeruli. Distinguishing air from lipid can be difficult; displaced nuclei can be endothelial or adipocyte. Oil Red O staining is most helpful in differentiating these emboli. Glycerin emboli were fatal following administration of an enema in a Bearded Dragon (**Figures 9.74–9.76**).

References

Abdalla FMF, Hiraichi E, Picarelli ZP, and Prezoto BC. 1989. Kallikrein-kinin system in the plasma of the snake *Botrops jararaca*. *Br J Pharmacol* 98:252–258.

Adams WE. 1958. *Comparative Morphology of the Carotid Body and Carotid Sinus*. Thomas, Springfield, IL.

Agapitov AV and Haynes WG. 2002. Role of endothelin in cardiovascular disease. *J Renin-Angiotensin-Aldosterone System* 3:1–15.

Aguirre AA, Balazs GH, Spraker TR, Murakawa SK, Zimmerman B. 2002. Pathology of oropharyngeal fibropapillomatosis in green turtles *Chelonia mydas*. *J Aquatic Animal Health* 14:298–304.

Andersen JB, Rourke BC, Caiozzo VJ, Bennett AF, and Hicks JW. 2005. Postprandial cardiac hypertrophy in pythons. *Nature* 434:37–38.

Ardlie NG and Schwartz CJ. 1965. Arterial pathology in the Australian reptile. *J Pathol Bacteriol* 90:487–494.

Ariel E. 2011. Viruses in reptiles. *Vet Res* 42:100.

Aristotle. 350BC. On the Parts of Animals. Translated by Ogle, Accessed W. May 13, 2017. http://classics.mit.edu/Aristotle/parts_animals.html.

Axelsson M, Dang Q, Pitsillides K, Munns S, Hicks J, Kassab GS. 2007. A novel, fully implantable, multichannel biotelemetry system for measurement of blood flow, pressure, ECG, and temperature. *J App Physiol* 102:1220–1228.

Axelsson M and Franklin CE. 2001. The calibre of the foramen of Panizza in *Crocodylus porosus* is variable and under adrenergic control. *J Comp Physiol B* 171:341–346.

Axelsson M and Franklin CE. 2011. Elucidating the responses and role of the cardiovascular system in crocodilians during diving: Fifty years on from the work of CG Wilber. *Comp Biochem Physiol A* 160:1–8.

Axelsson M, Franklin CE, Löfman CO, Nilsson S, and Grigg GC. 1996. Dynamic anatomical study of cardiac shunting in crocodiles using high-resolution angioscopy. *J Exp Biol* 199:359–365.

Backhouse SS, Harper AA, BN van Vliet, West NH. 1989. Localization and conduction velocities of arterial baroreceptors in the anesthetized snake. *J Physiol* 418:135P.

Ball DC and Hicks JW. 1996. Adrenergic and cholinergic response of ventricular muscle from the turtle, *Trachemys (Pseudemys) scripta*. *Comp Biochem Physiol*. 113A:135–141.

Barten S, Wyneken J, Mader D, and M Garner. 2006. Aneurysm in the dorsolateral neck of two bearded dragons (*Pogona vitticeps*), in *Proc Assoc Rept Amphib Vet*, pp. 43–44.

Bauer JE and Jacobson ER. 1989. Hyperlipidemia and cardiovascular disease in obese rattlesnakes, in *Proceedings of the Third International Colloquium on the Pathology of Reptiles and Amphibians*, 1989 Jan 13–15, Orlando, Florida, p. 11.

Berger PJ. 1987. The reptilian baroreceptor and its role in cardiovascular control. *Am Zool* 27:111–120.

Berger PJ and G Burnstock. 1979. Autonomic nervous system, in *Biology of the Reptilia. Neurology*, vol. 10B. Gans C, Northcutt RG, and Ulinski P (Eds.), Academic Press, New York, NY, pp. 1–57.

Bogan JE. 2017. Ophidian cardiology – A review. *J Herp Med Surg* 27:62–77.

Borgheresi RAMB, Lucca JD, Carmona E, and Picarelli ZP. 1996. Isolation and identification of angiotensin-like peptides from the plasma of *Bothrops jararaca*. *Comp Biochem Physiol, Part B Biochem Mol Biol* 113:467–473.

Breno MC and Picarelli ZP. 1992. The vasopressor action of angiotensin in the snake *Bothrops jacaraca*. *Comp Biochem Physiol* 101A:819–825.

Breno MC, Porto CS, and Picarelli ZP. 2001. Angiotensin receptor in the heart of *Bothrops jaracara* snake. *Eur J Pharmacol* 417:27–35.

Brockman HL and Kennedy JP. 1962. Interventricular septal defect in *Alligator mississipiensis* Daudin. *Texas Reports Biol Med* 20:719.

Brown C. 2010. Cardiac blood sample collection from snakes. *Lab Anim* 39:208–209.

Brown DR, Nogueira MF, Schoeb TR, Vliet KA, Bennett RA, Pye GW, and Jacobson ER. 2001. Pathology of experimental mycoplasmosis in American alligators. *J Wildl Dis* 37 (4):671–679.

Brown JD and Sleeman JM. 2002. Morbidity and mortality of reptiles admitted to the Wildlife Center of Virginia, 1991 to 2000. *J Wildl Dis* 38:699–705.

Brücke E. 1852. *Beiträge zur vergleichenden Anatomie und Physiologie des Gefäss-Systems*. Kaiserlich-Königliche Hof-und Staatsdruckerei, Wein.

Buchanan JG. 1956. The gross and minute anatomy of the heart of the lizard, *Leiolopisma grande*. *Trans Roy Soc New Zealand* 84:103–120.

Burggren WW. 1976. The persistence of a patent ductus arteriosus in an adult specimen of the tortoise *Testudo graeca*. *Copeia*, 1976:405–407.

Burggren WW. 1978. Influence of intermittent breathing on ventricular depolarization patterns in chelonian reptiles. *J Physiol* 278:349–364.

Burggren WW, Farrell AP, Lillywhite HB. 1997. Vertebrate cardiovascular systems, in *Handbook of Comparative Physiology*, Dantzler W (Ed.), Oxford University Press, New York, NY. pp. 215–308.

Burggren WW and Johansen K. 1982. Ventricular haemodynamics in the monitor lizard *Varanus exanthematicus*: Pulmonary and systemic pressure separation. *J Exp Biol* 96:343–354.

Burnstock G. 1969. Evolution of the autonomic innervation of the visceral and cardiovascular systems in vertebrates. *Pharmacol Rev* 21:247–324.

Chang L, Fu D, Stenglein MD, Hernandez JA, DeRisi JL, and Jacobson ER. 2016. Detection and prevalence of boid inclusion body disease in collections of boas and pythons using immunological assays. *Vet J* 218:13–18.

Chiu KW and Sham JS. 1985. Adrenergic receptors of isolated snake atria. *Comp Biochem Physiol C* 81:445–450.

Chiu KW, Wong VCK, Chan MY, and Pang PKT. 1986. Blood pressure homeostasis in the snake, *Ptyas korros*. *Gen Comp Endocrinol* 64:300–304.

Christian E and Grigg GC. 1999. Electrical activation of the ventricular myocardium of the crocodile *Crocodylus johnstoni*: A combined microscopic and electrophysical study. *Comp Biochem Physiol A*: 123:17–23.

Clippinger TL. 1993. Aortic stenosis and atrio/ventricular dilatation in a green iguana (*Iguana iguana*), in *Proc Am Assoc Zoo Vet*, pp. 390–393.

Clippinger TL, Bennett AR, Johnson CM et al. 2000. Morbidity and mortality associated with a new mycoplasma species from captive American alligators (*Alligator mississippiensis*). *J Zoo Wildl Med* 31:303–314.

Cole G, Rao DB, Steinberg H et al. 2008 Suspected vitamin E and selenium deficiency in a veiled chameleon (*Chamaeleo calyptratus*). *J Herpetol Med Surg* 18(3/4):113–116.

Cordero-Tapia AA, Gardner SC, Arellano-Blanco J, and Inohuye-Rivera RB. 2004. Learedius learedi infection in black turtles (*Chelonia mydas agassizii*), Baja California Sur, Mexico. *J Parasitol* 90:645–647.

Cranefield PF. 1965. The atrioventricular node and the ventricular conducting system in the non-mammalian vertebrate heart. *Ann New York Acad Sci* 127:145–150.

Crawford NG, Faircloth BC, JE McCormack, Brumfield RT, Winker K, and Glenn TC. 2012. More than 1000 ultraconserved elements provide evidence that turtles are the sister group of archosaurs. *Biol Lett* 8:783–786.

da Silva Braga VH, Armelin VA, Teixeira MT, Abe AS, Rantin FT, and Florindo LH. 2016. The effects of feeding on cardiac control of broad-nosed caiman (*Caiman latirostris*): The 3 role of autonomic nervous system and NANC factors 4. *J Exp Zool A* 325:524–531.

Davies F and Francis ETB. 1941. The heart of a salamander (*Salamandra salamandra*) with special reference to the conducting (connecting) system and its bearing on the phylogeny of the conducting systems of mammalian and avian hearts. *Phil Trans Roy Soc, Ser B* 231:99–130.

Davies PJ and Donald JA. 1992. Distribution and colocalisation of neuropeptides in the large systemic arteries and veins of the ratsnake, *Elaphe obsoleta*. *Cell Tissue Res* 269:495–504.

de Albuquerque NR, Arruda WS, Costa AS, Galharte RCV, Vargas LGH, and Moreno IH. 2010. A dicephalic yellow anaconda snake, *Eunectes notaeus* (Serpentes:Boidae), from Southern Pantanal, Brazil. *J Nat Hist* 44:1989–1994.

de Albuquerque NR, Piatti L, and Wallach V. 2013. Dicephalism in the green racer snake, *Philodryas patagoniensis* (Serpentes, Colubridae), from southeastern Brazil. *Herp Notes* 6:85–87.

Divers SJ. 2013. Reptile clinical pathology. *Brit Sm Anim Vet Assoc Cong* 2013.

Donald JA and Lillywhite HB. 1988. Adrenergic innervation of the large arteries and veins of the semiarboreal snake *Elaphe obsoleta*. *Cell Tissue Res* 269:495–504.

Donald JA, O'Shea JE, and Lillywhite HB. 1990a. Somatostatin and the innervation of the heart of the ratsnake, *Elaphe obsoleta*. *J Comp Physiol* 159:677–685.

Donald JA, O'Shea JE, and Lillywhite HB. 1990b. Neural regulation of the pulmonary vasculature in a semi-arboreal snake, *Elaphe obsoleta*. *J Comp Physiol* 159B:677–685.

Elkan E and Cooper JE. 1976. Tumours and pseudotumours in some reptiles. *J Comp Pathol* 86:337–348.

Ellulu MS, Patimah I, Khaza'ai H, Rahmat A, Abed Y, and Ali F. 2016. Athersclerotic cardiovascular disease: A review of initiators and protective factors. *Inflammopharmacology* 24:1–10.

Enok S, Simonsen LS, Pedersen S, and Skovgaard N. 2012. Humeral regulation of heart rate during digestion in pythons (*Python molurus* and *Python regius*). *AJP Reg Integ Comp Physiol* 302:R1176–R1183.

Enok S, Simmons LS, and Wang T. 2013. The contribution of gastric digestion and ingestion of amino acids on the postprandial rise in oxygen consumption, heart rate and growth of viscera organs in pythons. *Comp Biochem Physiol A:Physiol* 165:46–53.

Enok S, Slay C, Abe AS, Hicks JW, and Wang T. 2014. Intraspecific scaling of arterial blood pressure in Burmese python. *J Exp Biol* 217:2232–2234.

Fahlman A, Crespo-Picazo JL, Sterba-Boatwright B, Stacy BA, and Garcia-Parraga D. 2017. Defining risk variables causing gas embolism in loggerhead sea turtles (*Caretta caretta*) caught in trawls and gillnets. *Sci Rep* 7(1):2739.

Farmer CG. 1999. Evolution of the vertebrate cardio-pulmonary system. *Annu Rev Physiol* 61:573–592.

Farmer CG, Uriona TJ, Olsen DB, Steenblik M, and Sanders K. 2008. The right-to-left shunt of crocodilians serves digestion. *Physiol Biochem Zool* 81:125–137.

Farnsworth RJ, Brannian RE, Fletcher KC, Klassen SASA: 1986, Vitamin E–selenium responsive condition in the Green Iguana. *J Zoo Anim Med* 17:42–45.

Farrell AP, Gamperl AK, and Francis ETB. 1998. Comparative aspects of heart morphology, in *Biology of Reptilia, Vol 19 (Morphology G)*. Gans C and Gaunt AS (Eds.), Society for the Study of Amphibians and Reptiles, Ithaca, NY, pp. 375–424.

Ferguson J. 1942. Cholesterol and the heart: Its effect on the activity of the heart of the frog and the turtle. *CMAJ* 47:60.

Filogonio R, Taylor EW, Carreira LBT, Leite GSPC, Abe AS, and CAC Leite. 2014. Systemic blood flow relations in conscious South American rattlesnakes. *S Am J Herp* 9:171–176.

Finlayson R and Woods SJ. 1977. Arterial disease of reptiles. *J Zool* 183:397–410.

Franklin CE and Axelsson M. 2000. An actively controlled heart valve. *Nature* 406:847–848.

Franklinos LHV, Lorch JM, Bohuski E et al. 2017. Emerging fungal pathogen Ophidiomyces ophiodiicola in wild European snakes. *Sci Rep* 7 (1):3844.

Frye F. 1991a. Characteristics of cardiomyopathy in two pythons: Aortic valvular stenosis and secondary cardiomyopathy in a Children's python, *Liasis childreni*, and ventricular wall hypoplasia, first-degree heart block, and plasmacytic pericardititis in a juvenile Burmese python, *Python bivittatus. Proc IV Int Coll Path Reptiles Amphib.* Bad Nauheim, Germany.

Frye FL. 1991b. Common pathologic lesions & disease processes, in *Biomedical and Surgical Aspects of Captive Reptile Husbandry*, vol. 2. Frye FL (Ed.), Krieger Publishing Co., Malabar, FL, pp. 529–619.

Furness AP and Moore J. 1970. The adrenergic innervation of the cardio-vascular system of the lizard Trachysaurus rugosus. *Z Zellforsch.* 108:150–176.

Galli GL, Skovgaard N, Abe AS, Taylor EW, and Wang T. 2005a. The role of nitric oxide in the regulation of the systemic and pulmonary vasculature of the South American rattlesnake *Crotalus durissus terrificus. J Comp Physiol* 175:201–208.

Galli GL, Skovgaard N, Abe AS, Taylor EW, Conlon MJ, and Wang T. 2005b. Cardiovascular actions of the rattlesnake bradykinin ([Val[1], Thr[6]] bradykinin) in the anesthetized South American rattlesnake *Crotalus durissus terrificus. Am J Physiol – Reg Intg Comp Physiol* 288:R456–R465.

Galli GL, Skovgaard N, Abe AS, Taylor EW, and Wang T. 2007. The adrenergic regulation of the cardiovascular system in the South American rattlesnake *Crotalus durissus. J Comp Physiol* 148:510–520.

García-Párraga, D, Crespo-Picazo JL, de Quirós YB, Cervera V, Martí-Bonmati L, Díaz-Delgado J, Arbelo M, Moore MJ, Jepson PD, and A Fernández. 2014. Decompression sickness ('the bends') in sea turtles. *Dis Aquat Organ* 111(3):191–205.

Garner MM, Gyimesi ZS, Rasmussen JM et al. 2003. An amyloid-like deposition disorder in Komodo dragons, in *Proc Assoc Reptile Amphibian Vet*, Minneapolis, MN, pp. 50–51.

Gesser H and Poupa O. 1983. Acidosis and cardiac muscle contractility: Comparative aspects. *Comp Biochem Physiol.* 76A:559–566.

Gibbons PM. 2009. Problem-oriented exotic companion animal practice. *J Exotic Pet Med* 18:181–186.

Glazebrook JS, Campbell RSF, and D Blair. 1989. Studies on cardiovascular fluke (Digenea: Spirorchiidae) infections in sea turtles from the Great Barrier Reef, Queensland, Australia. *J Comp Pathol* 101:231–250.

Godfrey SG, Nelson NJ, and Bull CM. 2011. Ecology and dynamics of the blood parasite, Hepatozoon tuatarae (Apicomplexa), in tuatara (*Sphenodon punctatus*) on Stephens Island, New Zealand. *J Wildl Dis* 47:126–139.

Goldberg SR and Holshuh HJ. 1991. A case of leukemia in the desert spiny lizard (Sceloporus magister). *J Wildl Dis* 27:521–525.

Goldfinger JZ, Halperin JL, Marin ML, Stewart AS, Eagle KA, and V Fuster. 2014. Thoracic aortic aneurysm and dissection. *J Am Coll Cardiol* 64 (16):1725–1739.

Gordon AN, Kelly WR, and Cribb TH. 1998. Lesions caused by cardiovascular flukes (Digenea: Spirorchidae) in stranded green turtles (*Chelonia mydas*). *Vet Pathol* 35:21–30.

Grant RT and M Regnier. 1926. The comparative anatomy of the cardiac coronary vessels. *Heart* 13:285–317.

Gray SW. 1950. Pattern of contraction and relaxation in tortoise ventricle. *Am J Physiol* 162:249–258.

Gray T. 2004. Cranial body swelling in a boa constrictor, *Boa constrictor constrictor. J Herp Med Surg* 14:22–24.

Greenfield LJ and Morrow AG. 1961. The cardiovascular hemodynamics of Crocodylia. *J Surg Res* 1:97–103.

Grigg GC. 1989. The heart and patterns of cardiac outflow in Crocodilia. *Proc Austal Physiol Pharmacol Soc* 20:43–57.

Grigg GC and Simons JR. 1972. Preferential distribution of left and right auricular blood into the arterial arches of the tuatara, Spenodon punctatus. *J Zool Lond* 167:481–486.

Griner L. 1983. Systemic diseases. Sub-order Sauria, in *Pathology of Zoo Animals.* Griner L (Ed.), Zoological Society of San Diego, San Diego, CA, pp. 32–35.

Gumber S, Nevarez JG, and Cho DY. 2010. Endocardial fibrosarcoma in a reticulated python (*Python reticularis*). *J Vet Diagn Invest* 22:1013–1016.

Gustavsen KA, Saunders AB, Young BD, Winter RL, and Hoppes SM. 2014. Echocardiographic and radiographic findings in a cohort of healthy adult green iguanas *(Iguana iguana)*. *J Vet Cardiol* 16:185–196.

Gyimesi ZS, Garner MM, Burns RB. 2008. Goiter and thyroid disease in captive Kirtland's snakes, *Clonophis kirtlandii*, *J Herp Med Surg*, 18(3/4):75–80.

Hagensen MK, Abe AS, Falk E, and Wang T. 2008. Physiological importance of the coronary arterial blood supply to the rattlesnake heart. *J Exper Biol* 211:3588–3593.

Han B and Weiss LM. 2017. Microsporidia: Obligate intracellular pathogens within the fungal kingdom. *Microbiology Spectrum* 5:1–26.

Harris AS. 1941. The spread of excitation in turtle, dog, cat and monkey ventricles. *J Am Physiol* 134:319–332.

Harvey W. 1628. On the motion of the heart and blood in animals. Available at http://www.fordham.edu/halsall/mod/1628 harvey-blood.asp Accessed 13.5.2017.

Heaton-Jones TG and King RR. 1994. Characterization of the electrocardiogram of the American alligator (*Alligator mississippiensis*). *J Zoo Wildl Med* 25:40–47.

Hedberg A and Nilsson S. 1976a. Vago-sympathetic innervation of the heart of puff adder, *Bitis arietans*. *Comp Biochem Physiol* 53C:3–8.

Hedberg A and Nilsson S. 1976b. Inotropic responses to drugs of isolated heart strips from a snake. *Acta Physiol Scand* 96:35A.

Heisler N and Glass ML. 1985. Mechanisms and regulation of central vascular shunts in reptiles, in *Cardiovascular Shunts: Phylogenetic, Ontogenetic and Clinical Aspects*. Johansen K and Burggren WW (Ed.), Munksgaard, Copenhagen, pp. 334–354.

Henriksen PS, Enok A, Overgaard J, and Wang T. 2014. Food composition influences metabolism, heart rate and organ growth during digestion in *Python regius*. *Comp Biochem Physiol A* 183:36–44.

Hess R. 1921. Das herzgewicht der wirbeltiere. *Zool Jb.* 38:243–364.

Hicks JW. 1998. Cardiac shunting in reptiles: Mechanisms, regulation, and physiological functions, in *Biology of Reptilia, Vol 19 (Morphology G)*. Gans C, and Gaunt AS (Eds.), Society for the Study of Amphibians and Reptiles, Ithaca, NY, pp. 425–483.

Hicks JW. 2002. The physiological and evolutionary significance of cardiovascular shunting patterns in reptiles. *News Physiol Sci* 17:241–245.

Hicks JW and Wang T. 1998. Cardiovascular regulation during anoxia in the turtle: An *in vivo* study. *Physiol Zool* 71:1–14.

Hohnke LA. 1975. Regulation of arterial blood pressure in the common green iguana. *Am J Physiol – Legacy Content* 228:386–391.

Homer BL, Jacobson ER, Schumacher J, and Scherba G. 1994. Chlamydiosis in mariculture-reared green sea turtles (*Chelonia mydas*). *Vet Path* 31:1–7.

Hruban Z, Vardiman J, Meehan T, Frye F, and Carter WE. 1992. Haematopoietic malignancies in zoo animals. *J Comp Path* 106:15–24.

Hulbert AJ and Else PL. 2004. Basal metabolic rate: History, composition, regulation, and usefulness. *Physiol Biochem Zool* 77:869–876.

Isaza R, Andrews G, Coke R, and Hunter R. 2004. Assessment of multiple cardiocentesis in ball pythons (*Python regius*). *Contemp Top Lab Anim Sci* 43:35–38.

Jacobson ER. 1976. Gentamicin-related visceral gout in two boid snakes [Uric acid]. *Vet Med Sm An Clin* 71:361–363.

Jacobson ER, Gaskin JM, and Mansell J. 1989. Chlamydial infection in puff adders (*Bitis arietans*). *J Zoo Wildl Med* 20:364–369.

Jacobson ER, Ginn PE, Troutman JM et al. 2005. West Nile virus infection in farmed American alligators (*Alligator mississippiensis*) in Florida. *J Wildl Dis* 41:96–106.

Jacobson ER, Green DE, Undeen AH, Cranfield M, and Vaughn KL. 1998. Systemic microsporidiosis in inland bearded dragons (*Pogona vitticeps*). *J Zoo Wildl Med* 29:315–323.

Jacobson ER, Homer B, and Adams W. 1991. Endocarditis and congestive heart failure in a Burmese python. *J Zoo Wildl Med* 22:245–248.

Jacobson E, Origgi F, Heard D, and Detrisac C. 2002. Immunohistochemical staining of chlamydial antigen in emerald tree boas (*Corallus caninus*). *J Vet Diagn Invest* 14:487–494.

Jacobson ER, Seely JC, Novilla MN, and Davidson JP. 1979. Heart failure associated with unusual hepatic inclusions in a Deckert's rat snake. *J Wildl Dis* 15:75–81.

Jensen B, Abe AS, Andrade DV, and Wang T. 2010a. The heart of the South American rattlesnake, *Crotalus durissus*. *J Morphol* 271:1066–1077.

Jensen B, Nielsen JM, Axelsson M, Pedersen M, Löfman C, and Wang T. 2010b. How the python heart separates pulmonary and systemic blood pressures and blood flows. *J Exp Biol* 213:1611–1617.

Jensen B, Nyengaard JR, Pedersen M, and Wang T. 2010c. Anatomy of the python heart. *Anat Sci Int* 85:194–203.

Jensen B, Boukens BJD, Postma AV et al. 2012. Identifying the evolutionary building blocks of the cardiac conduction system. *PLOS ONE*.7(9):e44231.

Jensen B, Larsen CK, Nielsen JM, and Wang T. 2011. Change in cardiac function, but not form, in postprandial pythons. *Comp Biochem Physiol A* 160:35–42.

Jensen B, Moorman AF, and Wang T. 2014. Structure and function of the hearts of lizards and snakes. *Biol Rev* 89:302–336.

Jensen B, Vesterskov S, Boukens BJ, Nielsen JM, Moormann AFM, Christffels VM, and Wang T. 2017. Morpho-functional characterization of the systemic venous pole of the reptile heart. *Sci Rep* 7:1–12.

Jensen B and Wang T. 2009. Hemodynamic consequences of cardiac malformations in two juvenile ball pythons (*Python regius*). *J Zoo Wildl Med* 40:752–756.

Johansen, K. 1959. Circulation in the three-chambered snake heart. *Circ Res* 7:828–832.

Johnson CA, Griffith JW, Tenorio P, Hytrek S, and Lang CM. 1998. Fatal trematodiasis in research turtles. *Comp Med* 48:340–343.

Joyce W, Axelsson M, Altimiras J, and Wang T. 2016. In situ cardiac perfusion reveals interspecific variation of intraventricular flow separation in reptiles. *J Exp Biol* 219:2220–2227.

Junker K, Boomker J, and Bolton LA. 1999. Pentastomid infections in Nile crocodiles (*Crocodylus niloticus*) in the Kruger National Park, South Africa, with a description of the males of Alofia simpsoni. *Onderstepoort J Veterinary Res* 66:65.

Junker K, Calitz F, Govender D, Krasnov BR, and Boomker J. 2016. Pentastome assemblages of the Nile crocodile, *Crocodylus niloticus* Laurenti (Reptilia: Crocodylidae), in the Kruger National Park, South Africa. *Folia Parasitol* 63:040.

Kaplan HM and Schwartz C. 1963. Electrocardiography in turtles. *Life Sci* 9:637–645.

Keller S, Hetzel U, Sironen T, Korzyukov Y, Vapalahti O, Kipar A, and Hepojoki J. 2017. Co-infecting reptarenaviruses can be vertically transmitted in boa constrictor. *PLoS Pathog* 13(1):e1006179.

Kharin S and Shmakov D. 2009. A comparative study of contractility of the heart ventricle in some ectothermic vertebrates. *Acta Herpetol* 4:57–71.

Kik, KJL and Mitchell MA. 2005. Reptile cardiology: A review of anatomy and physiology, diagnostic approaches, and clinical disease. *Sem Av Ex Pet Med* 14:52–60.

Klenk K, Snow J, Morgan K et al. . 2004. Alligators as West Nile virus amplifiers. *Emerg Infect Dis* 10:2150–2155.

Koplra EC. 1987. The ultrastructure of alligator conductive tissue: An electron microscopic study of the sino-atrial node. *Acta Physiol Hung* 69:27–84.

Kozubowski MM. 1985. Cardiovascular mechanoreceptors in the rat snake (*Elaphe obsoleta*) and the rattlesnake (*Crotalus viridis belleri*): Location and characteristics. *PhD diss* University of Kansas, Lawrence.

Kvasilova A, Jensen B, Christoffels V, and Sedmera D. 2015. Development of the ventricular conduction system of the crocodilian heart. *FASEB J* 29 supplement number 557.6. https://www.fasebj.org/doi/10.1096/fasebj.29.1_supplement.557.6.

LaDouceur EEB and Garner MM. 2017. Hyperviscosity-like syndrome in reptiles, in *Proc 49th AAZV Ann Conf*, pp. 79–80.

Larsen EH, Deaton LE, Onken H, M O'Donnell, Grosell M, Dantzler WH, and Weihrauch D. 2014. Osmoregulation and excretion. *Compr Physiol* 4:405–573.

Lázari MFM, Breno MC, Abreu LC, and Picarelli ZP. 1994. Some functional aspects of the renin-angiotensin system in the snake *Bothrops jararaca*. *Comp Biochem Phyiol* 108A:145–152.

Leeson TS. 1981. The fine structure of snake myocardium. *Acta Anat* 109:252–269.

Les Gabor J. 2005. Nutritional degenerative myopathy in a population of captive bred *Uroplatus phantasticus* (satanic leaf-tailed geckoes). *J Vet Diagn Invest* 17(1):71–73.

Lewis T. 1916. The spread of the excitatory process in the vertebrate heart. Parts 1–5. *Phil Trans Roy Soc Lond* 207:221–310.

Lillywhite HB. 1987. Circulatory adaptations of snakes to gravity. *Am Zool* 27:81–95.

Lillywhite HB and Donald JA. 1989. Pulmonary blood flow regulation in an aquatic snake. *Science* 245:293–295.

Lillywhite HB and Donald JA. 1994. Neural regulation of arterial blood pressure in snakes. *Physiol Zool* 67:1260–1283.

Lillywhite HB and Gallagher KP. 1985. Hemodynamic adjustments to head-up posture in the partly arboreal snake, *Elaphe obsoleta*. *J Exp Zool* 235:325–334.

Lillywhite HB and Pough FH. 1983. Control of arterial pressure in aquatic sea snakes. *Am J Physiol* 244: R66–R73.

Lillywhite HB and Seymour RS. 1978. Regulation of arterial blood pressure in Australian tiger snakes. *J Exp Biol* 75:65–79.

Lillywhite HB and Smith LH. 1981. Haemodynamic responses to haemorrhage in the snake *Elaphe obsoleta obsoleta*. *J Exp Biol* 94:275–283.

Lillywhite HB and Smits AW. 1984. Lability of blood volume in snakes and its relation to activity and hypertension. *J Exp Biol* 110:267–274.

Lillywhite HB, Zippel KC, and Farrell AP. 1999. Resting and maximal heart rates in ectothermic vertebrates. *Comp Biochem Physiol A: Physiol* 124:369–382.

Lorch JM, Knowles S, Lankton JS et al. . 2016. Snake fungal disease: An emerging threat to wild snakes. *Philos Trans R Soc Lond B Biol Sci* 371(1709).

Lorch JM, Lankton J, Werner K, Falendysz EA, K McCurley, and Blehert DS. 2015. Experimental infection of snakes with ophidiomyces ophiodiicola causes pathological changes that typify snake fungal disease. *MBio* 6(6): e01534–15.

Lung NP and Garner MM. 2000. Gas bubble disease in Malaysian painted terrapins, *Callagur borneoensi*, in *Proceedings of the Assoiciation of Reptilian and Amphibian Veterinarians, 7th Annual Conference*, Reno, NV.

MacKinnon MR and Heatwole H. 1981. Comparative cardiac anatomy of the reptilia. iv. The coronary arterial ciculation. *J Morph* 170:1–27.

Mans C and Braun J. 2014. Update on common nutritional disorders of captive reptiles. *Vet Clin North Am Exot Anim Pract* 17(3):369–395.

Mathews ME, Eshar D, Delk KW, Pohlman L, and Andrews G. 2014. Splenopancreatic ductal adenocarcinoma with multiorgan metastasis in a northern water snake (*Nerodia sipedon*). *J Zoo Wildl Med* 45:437–440.

McCue MD, Guzman RM, and Passement CA. 2015. Digesting pythons quickly oxidize the proteins in their meals and save the lipids for later. *J Exp Biol* 218:2089–2096.

McDonald HS and Heath JE. 1971. Electrocardiographic observations on the tuatara, *Sphenodon punctatus*. *Comp Biochem Physiol A* 40:881–892.

McIntosh HD, Morris JJ Jr, Whalen RE, Hernandez RR, and Hackel DB. 1967. Bilateral functional subaortic stenosis in the alligator. *Trans Am Clin Climatol Assoc* 78:119.

Mesquita LSM, Frias FT, Carmona E, and Borgheresi RAMB. 2008. Differences in endothelial receptor types in the vasculature of *Bothrops vararaca* (Viperidae) and Oxyrhopus guibei (Colubridae) snakes. *Comp Biochem Physiol Part C Tox & Pharmacol* 148:61–67.

Mifune H, Suzuki S, Nokihara K, and Noda Y. 1996. Distribution of immunoreactive atrial and brain natriuretic peptides in the heart of the chicken, quail, snake and frog. *Exp Anim* 45:125–133.

Miller DL, Mauel MJ, Baldwin C, Burtle G, Ingram D, Hines ME, and Frazier KS. 2003. West Nile virus in farmed alligators. *Emerg Infect Dis* 9 (7):794–799

Miller DL, Wyneken J, Rajeev S et al. 2009. Pathologic findings in hatchling and posthatchling leatherback sea turtles (*Dermochelys coriacea*) from Florida. *J Wildl Dis* 45:962–971.

Miller LM and Gal A. 2016. Cardiovascular system and lymphatic vessels, in *Pathologic Basis of Veterinary Disease*. Zachary JF (Ed.), pp. 561–616.

Mitchell MA. 2009. Reptile cardiology. *Vet Clin North Am Exot Anim Pract* 12:65–79.

Morris JL and Nilsson S. 1994. The circulatory system, in *Comparative Physiology and Evolution of the Autonomic Nervous System*. Nilsson S and Holmgren S (Eds.), Harwood Academic Publishers, Newark, NJ, pp. 193–246.

Mullen RK. 1967. Comparative electrocardiography of the Squamata. *Physiol Zool* 40:14–126.

Nakayama T, Nakajima T, and Sokabe H. 1977. Comparative studies on angiotensins. IV. structure of snake (*Elaphe climocophora*) angiotensin. *Chem Pharm Bull* 25:3255–3260.

Newkirk K, Brannick E, and Kusewirr D. 2012. Neoplasia and tumor biology, in *Pathologic Basis of Veterinary Disease*, 6th ed. Zachary J (Ed.), Elsevier, St. Louis (MO), pp. 286–323.

Nielsen JS and Gesser H. 2001. Effects of high extracellular [K^+] and adrenaline on force development, relaxation and membrane potential in cardiac muscle from freshwater turtle and raibbow trout. *J Exp Biol* 2004:261–268.

Nilsson, S. 1983. Autonomic nerve function in vertebrates, in *Zoophysiology*, vol. 13. Farner DS et al. (Eds.), Springer-Verlag, Berlin.

Nilsson, S. 1994. The crocodilian heart and central hemodynamics. *Cardioscience* 5:163–166.

Obendorf DL, Carson J, and McManus TJ. 1987. Vibrio damsela infection in a stranded leatherback turtle (*Dermochelys coriacea*). *J Wildl Dis* 23:666–668.

Orchard CH and Kentish JC. 1990. Effects of changes of pH on the contractile function of cardiac muscle. *Am J Physiol* 258:C967–C981.

Orós J, Torrent A, AE de los Monteros et al. 2001. Multicentric lymphoblastic lymphoma in a loggerhead sea turtle (*Caretta caretta*). *Vet Path* 38:464–467.

Orós J, Torrent A, Calabuig P, and Déniz S. 2005. Diseases and causes of mortality among sea turtles stranded in the Canary Islands, Spain (1998–2001). *Dis Aq Org* 63:13–24.

Overgaard J, Stecyk JAW, Farrell AP, and Wang T. 2002. Adrenergic control of the cardiovascular system in the turtle Trachemys scripta. *J Exp Biol* 205:3335–3345.

Overgaard J, Wang T, Nielsen OB, and Gesser H. 2005. Extracellular determinants of contractile force in painted turtle (*Chrysemys picta*): Role of anoxia, temperature and acclimation. *Physiol Biochem Zool* 78:976–995.

Panizza B. 1833. Sulla structura del coure e sulla circolazione del sonque del Crocodilus lucius. *Biblioth Ital* 70:87–91.

Paré JA and Lentini AM. 2010. Reptile geriatrics. *Vet Clin North Am Exot An Prac* 13:15–25.

Penninck DG, Stewart JS, Paul-Murphy J, and Pion P. 1991. Ultrasonography of the California desert tortoise (*Xerobates agassizi*): Anatomy and application. *Vet Rad* 32:112–116.

Pereira ME and Viner TC. 2008. Oviduct adenocarcinoma in some species of captive snakes. *Vet Pathol* 45:693–697.

Pizzi R, Pereira YM, Rambaud YF et al. 2009. Secundum atrial septal defect in a Komodo dragon (*Varanus komodoensis*). *Vet Rec* 164:472–473.

Platzack B, Wang Y, Crossley D, Lance V, Hicks JW, and Conlon JM. 2002. Characterization and cardiovascular actions of endothelin-1 and endothelin-3 from the American alligator. *Am J Physiol – Reg Int Comp Physiol* 282:R594–R602.

Poder TC, Silberberg SD, and Rampe D. 1991. Contraction of reptile, amphibian, and fishblood vessels by endothelin-1. *Can J Physiol Pharm* 69:215–217.

Poupa O. 1994. Cardiac muscle and its blood supply: Paleophysiological notes. *Cell Mol Biol Res* 40:153–165.

Poupa O and Lindström L. 1983. Comparative scaling aspects of heart and body weights with reference to blood supply of cardiac fibers. *Comp Biochem Physiol* 76:413–421.

Raiti P. 2012. Husbandry, diseases, and veterinary care of the bearded dragon (*Pogona vitticeps*). *J Herp Med Surg* 22:117–131.

Ramsay EC, Munson L, Lowenstine L, and Fowler ME. 1996. A retrospective study of neoplasia in a collection of captive snakes. *J Zoo Wildl Med* 27:28–34.

Rand TG and Wiles M. 1985. Histopathology of Infections by Learedius learedi Price, 1934 and Neospirorchis schistosomatoides Price, 1934 (Digenea: Spirorchiidae) in Wild Green Turtles, *Chelonia mydas* L, from Bermuda. *J Wildl Dis* 21:461–463.

Reavill DR and Schmidt RE. 2012. Metastatic chondrosarcoma in a corn snake (*Pantherophis gutattus*). *J Herp Med Surg* 22:67–69.

Richter B, Csokai J, Graner I, Eisenberg T, Pantchev N, Eskens HU, and Nedorost N. 2013. Encephalitozoonosis in two inland bearded dragons (*Pogona vitticeps*). *J Comp Pathol* 148:278–282.

Rishniw M and Carmel BP. 1999. Atrioventricular valvular insufficiency and congestive heart failure in a carpet python. *Australian Vet J* 77:580–583.

Robb JS. 1953. Specialized (conducting) tissue in the turtle heart. *Am J Physiol* 172:7–13.

Robinson, Wayne F, and Nicholas Robinson A. 2016. Cardiovascular system, in *Jubb, Kennedy, and Palmer's Pathology of Domestic Animals*. Grant Maxie M (Ed.), pp. 1–101.

Rosell J, Barragán A, Carbonell MD et al. 2019. Pathology in practice. *J Am Vet Med Assoc* 255:1349–1352.

Rush EM, Donnelly TM, and J Walberg. 2001. What's your diagnosis? Cardiopulmonary arrest in a Burmese python. Aortic aneurysm. *Lab Anim (NY)* 30:24–27.

Santoro M, Di Nocera F, Iaccarino D et al. 2017. Pathology and molecular analysis of Hapalotrema mistroides (Digenea: Spirorchiidae) infecting a Mediterranean loggerhead turtle Caretta caretta. *Dis Aquat Organ* 124(2):101–108.

Santoro M, Morales JA, and Rodríguez-Ortíz B. 2007. Spirorchiidiosis (Digenea: Spirorchiidae) and lesions associated with parasites in Caribbean green turtles (*Chelonia mydas*). *Vet Rec* 161(14):482–486.

Schilliger L, Lemberger K, Chai N, Bourgeois A, and Charpentier M. 2010a. Atherosclerosis associated with pericardial effusion in a central bearded dragon (*Pogona vitticeps*). *J Vet Diagn Invest* 22:789–792.

Schilliger L, Paillusseau C, Gandar F et al. . 2019. Hypertensive heart disease and encephalopathy in a central bearded dragon (*Pogona vitticeps*) with severe atherosclerosis and first-degree atrioventricular block. *J Zoo Wildl Med* 50(2):482–486.

Schilliger L, Selleri P, and Frye FL. 2011. Lymphoblastic lymphoma and leukemic blood profile in a red-tail boa (*Boa constrictor constrictor*) with concurrent inclusion body disease. *J Vet Diagn Invest* 23:159–162.

Schilliger L, Tréhiou –Sechi E, Petit A, Misbach C, and Chetboul V. 2010b. Double valvular insufficiency in a Burmese python (*Python molurus bivittatus*, Linnaeus, 1758) suffering from concomitant bacterial pneumonia. *J Zoo Wildl Med* 41:742–744.

Schmidt R and Reavill D. 2010. Cardiopulmonary disease in reptiles, in *2010 Proceed ARAV*, pp. 90–98.

Schroff S, Schmidt V, Kiefer I, Krautwald-Junghanns ME, and Pees M. 2010. Ultrasonographic diagnosis of an endocarditis valvularis in a Burmese python (*Python molurus bivittatus*) with pneumonia. *J Zoo Wildl Med* 41(4):721–724.

Schuchman SM and Taylor DO. 1970. Arteriosclerosis in an iguana (*Iguana iguana*). *J Am Vet Med Assoc* 157:614.

Schumacher JR, Bennet A, Fox LE, Deem SL, Neuwirth L, and Fox JH. 1998. Mast cell tumor in an eastern kingsnake (*Lampropeltis getulus getulus*). *J Vet Diag Invest* 10:101–104.

Secor SM and White SE. 2010. Prioritizing blood flow: Cardiovascular performance in response to the competing demands of locomotion and digestion of the Burmese python, *Python molurus*. *J Exp Biol* 213:78–88.

Selleri P and di Girolamo N. 2012. Cardiac tamponade following cardiocentesis in a cardiopathic boa constrictor imperator (*Boa constrictor imperator*). *J Sm An Pract* 53:487.

Seymour RS. 1987. Scaling of cardiovascular physiology in snakes. *Amer Zool* 27:97–107.

Seymour RS and Arndt JO. 2004. Independent effects of heart-head distance and caudal blood pooling on blood pressure in aquatic and terrestrial snakes. *J Exp Biol* 207:1305–1311.

Seymour RS and Barker SJ. 1983. Evolution of blood pressure regulation in snakes. *Proc Int Union Physiol Sci*, 15:312.

Seymour RS and Baylock AJ. 2000. The principle of Laplace and scaling of ventricular wall stress and blood pressure in mammals and birds. *Physiol Biochem Zool* 73:389–405.

Seymour RS, Bennett-Stamper CL, Johnston SD, Carrier DR, and Grigg GC. 2004. Evidence for endothermic ancestors of crocodiles at the stem of archosaur evolution. *Physiol Biochem Zool* 77:1051–1067.

Shmakov KN and Roshchevskii MP. 1982. Intramural chronotopography of cardiac ventricular depolarization in reptiles. *Zh Evol Biokhim Fizol* 18:164–169.

Simons JR. 1965. The heart of the tuatara (*Sphenodon punctatus*). *J Zool Lond* 146:451–466.

Skovgaard N, Abe AS, Andrade DV, and Wang T. 2005a. Hypoxic pulmonary vasoconstriction in reptiles: A comparative study of four species with different lung structures and pulmonary blood pressures. *Am J Physiol - Reg Intg Comp Physiol* 289:R1280–R1288.

Skovgaard N, Warren DE, Jackson DC, and Wang T. 2005b. Endothelin-1 causes systemic vasodilation in anesthetized turtles (*Trachemys scripta*) through activation of ETB-receptors. *J Exper Biol* 208:3739–3746.

Skovgaard N, Conlon JM, and Wang T. 2007. Evidence that neurotensin mediates postprandial intestinal hyperemia in the python, *Python regius*. *Am J Physiol – Reg Intg Comp Physiol* 293:R1393–R1399.

Skovgaard N, Møller K, Gesser H, and Wang T. 2009. Histamine induces postprandial tachycardia through a direct effect on cardiac H_2 receptors in pythons. *Am J Physiol – Reg Intg Comp Physiol* 296:R774–R785.

Skovgaard N and Wang T. 2006. Local control of pulmonary blood flow and lung structure in reptiles: Implications for ventilation perfusion matching. *Resp Physiol Neurobiol* 154:107–117.

Skovgaard N, Zibrandtsen H, Laursen BE, Simonsen U, and Wang T. 2008. Hypoxia-induced vasoconstriction in alligator (*Alligator mississippiensis*) intrapulmonary arteries: A role for endothelin-1? *J Exp Biol* 211:1565–1570.

Slay CE, Enok S, Hicks JW, and Wang T. 2014. Reduction of blood oxygen levels enhances postprandial cardiac hypertrophy in Burmese pythons (*Python bivittatus*). *J Exp Biol* 217:1784–1789.

Smits AW and Lillywhite HB. 1985. Maintenance of blood volume in snakes: Transcapillary shifts of extravascular fluids during acute hemorrhage. *J Comp Physiol* 155B:305–310.

Stacy BA, Foley AM, Greiner E, Herbst LH, Bolten A, Klein P, Manire CA, and Jacobson ER. 2010. Spirorchiidiasis in stranded loggerhead *Caretta caretta* and green turtles *Chelonia mydas* in Florida (USA): Host pathology and significance. *Dis Aquat Organ* 89(3):237–259..

Stacy BA, Foley A, Garner MM, and Mettee N. 2013. Yolk embolism associated with trauma in vitellogenic sea turtles in Florida (USA): A review of 11 cases. *J Zoo Wildl Med* 44(4):1043–1048.

Stacy BA and Wellehan JF. 2010. Fatal septicemia caused by Helicobacter infection in a pancake tortoise (*Malacochersus tornieri*). *J Vet Diagn Invest* 22(4):660–662.

Starck JM. 2009. Functional morphology and patterns of blood flow in the heart of *Python regius*. *J Morphol* 270:673–687.

Stecyk JA, Couturier CS, and Nilsson GE. 2013. Expression of "pacemaker" channels in the turtle heart: Effects of anoxia and cold-acclimation, in *Proceedings of The Physiological Society*, The Physiological Society.

Steffensen JF and Farrell AP. 1998. Swimming performance, venous oxygen tension and cardiac performance of coronary-ligated rainbow trout, *Oncorhynchus mykiss*, exposed to progressive hypoxia. *Comp Biochem Physiol* 142A:205–214.

Stenglein MD, SanchezMigallon Guzman D, Garcia VE, Layton ML, Hoon-Hanks LL, Boback SM, Keel MK, Drazenovich T, Hawkins MG, and DeRisi JL. 2017. Differential disease susceptibility in experimentally reptarenavirus infected boa constrictors and ball pythons. *J Virol* 91:e00451-17.

Stinner JN and Ely DL. 1993. Blood pressure during routine activity, stress, and feeding in black racer snakes (*Coluber constrictor*). *Am J Physiol* 264:R79–R84.

Stolk, A. 1961. Haemochromatosis in the lizard Lacerta viridis. *Proc K Ned Akad Wet C.* 64:464–477.

Strik NI, Alleman AR, and Harr KE. 2007. Circulating inflammatory cells, in *Infectious Diseases and Pathology of Reptiles: Color Atlas and Text.* Jacobson ER (Ed.), CRC Press, Boca Raton, Florida, pp. 167–218.

Stumpel JBG, del-Pozo J, French A, and Eatwell K. 2012. Cardiac hemangioma in a corn snake (*Pantherophis guttatus*). *J Zoo Wildl Med* 43:360–366.

Summers AP. 2005. Evolution: Warm-hearted crocs. *Nature* 434:833–834.

Sweet C, Linnetz E, Golden E, and Mayer J. 2009. What is your diagnosis? *J Am Vet Med Assoc* 234:1259–1260.

Taylor EW, Andrade DV, Abe AS, Leite CAC, and Wang T. 2009. The unequal influences of the left and the right vagi on the control of the heart and pulmonary artery in the rattlesnake, Crotalus dirussis. *J Exper Biol* 212:145–151.

Turchetti AP, Tinoco HP, Malta M, da Costa M, Pessanha AT, Soave SA, Paixão TA, Santos RL. 2013. Inclusion body disease in a Corallus hortulanus. *Braz J Vet Pathol*;6:15–18.

Valentinuzzi ME, Hoff HE, and Geddes LA. 1969a. Observations of the electrical activity of the snake heart. *J Electrocard* 2:39–50.

Valentinuzzi ME, Hoff HE, and Geddes LA. 1969b. Electrocardiogram of the snake: Effect of the location of the electrodes and cardiac vectors. *J Electrocard* 2:245–252.

Valentinuzzi ME, Hoff HE, and Geddes LA. 1969c. Electrocardiogram of the snake: Intervals and durations. *J Electrocard* 2:343–352.

van Vliet BN and West NH. 1994. Physiologic trends in the barareceptor control of arterial blood pressure. *Physiol Zool* 67:1284–1304.

Vastesaeger MM, Delcourt R, and Gillot PH. 1965. Spontaneous atherosclerosis in fishes and reptiles, in *Comparative Atherosclerosis*, Roberts JC, Straus R (Eds.) Hoeber Harper and Row, New York, NY. pp. 129–149.

Vivaldo SG, Marquez LJG, Sarabia DO, Garcia JLV, and Casas FC. 2009. Pathology in the olive ridley turtles (*Lepidochelys olivacea*) that arrived to the shores of Cuyutlan, Colima, Mexico. *Vet Mex* 40:69–78.

Wang T, Altimiras J, and Axelsson M. 2002. Intracardiac flow separation in an *in situ* perfused heart of Burmese python *Python molurus*. *J Exp Biol* 205:2715–2723.

Wang T, Altimiras J, Klein W, and Axelsson M. 2003a. Ventricular haemodynamics in *Python molurus*: Separation of pulmonary and systemic pressures. *J Exp Biol* 206:4241–4245.

Wang T, Zaar M, Arvedsen S, Vedel-Smith C, and Overgaard J. 2003b. Effects of temperature on the metabolic response to feeding in *Python molurus*. *Comp Biochem Physiol* 133A:519–527.

Wang T, Axelsson M, Jensen J, and Conlon JM. 2000. Cardiovascular actions of python bradykinin and substance P in the anesthetized python, *Python regius*. *Am J Physiol Reg Integ Comp Physiol* 279:R531–R538.

Wang T, Krosniunas EH, and Hicks JW. 1997. The role of cardiac shunts in the regulation of arterial blood gases. *Amer Zool* 37:12–22.

Wang T, Taylor EW, Andrade D, and Abe AS. 2001a. Autonomic control of heart rate during forced activity and digestion in the snake Boa constrictor. *J Exp Biol* 204:3553–3560.

Wang T, Warburton SJ, Abe AS, and Taylor EW. 2001b. Vagal control of heart rate and cardiac shunts in reptiles: Relation to metabolic state. *Exp Physiol* 86:777–786.

Wasser JS. 1995. Anoxia and ischemia tolerance in turtle heart. *Braz J Med Biol Res* 28:1233–1240.

Wasser JS, Freund EV, Gonzalez LA, and Jackson DC. 1990. Force and acid-base state of turtle cardiac tissue exposed to combined anoxia and acidosis. *Am J Physiol* 259:R15–R20.

Webb GJW. 1979. Comparative cardiac anatomy of the Reptilia III. The heart of crocodilians and an hypothesis on the completion of the interventricular septum of crocodilians and birds. *J Morph* 161:221–240.

Webb G, Heatwole H, and De Bavay J. 1971. Comparative cardiac anatomy of the Reptilia. I. The chambers and septa of the varanid ventricle. *J Morphol* 134:335–350.

Wernick MB, Novo-Matos J, Ebling A, Kühn K, Ruetten M, Hilbe M, Howard J, Chang R, Prohaska S, and Hatt JM. 2015. Valvulopathy consistent with endocarditis in an Argentine boa (*Boa constrictor occidentalis*). *J Zoo Wildl Med* 46(1):124–129.

White FN. 1976. Circulation, in *Biology of the Reptilia, Vol 5(Physiology A)*. Gans C (Ed.), Academic Press, London, pp. 275–334.

Wojtaszek, J. 1979. Motor reactivity of isolated heart of grass snake (Natrix natrix L.) effects of temperature and pH. *Acta Physiol Pol* 30:399–409.

Wolke RE, Brooks DR, and George A. 1982. Spirorchidiasis in loggerhead sea turtles (*Caretta caretta*): Pathology 1. *J Wildl Dis* 18:175–185.

Woodhouse SJ, Fitzgerald SD, Lim A, and Bolin SR. 2014. Disseminated *Mycobacterium haemobilium* infection in an Assam trinket snake (*Elaphe frenata*). *J Zoo Wildl Med* 45:966–969.

Work TM, Balazs GH, Rameyer RA, and Morris RA. 2004. Retrospective pathology survey of green turtles *Chelonia mydas* with fibropapillomatosis in the Hawaiian Islands, 1993–2003. *Dis Aquat Org* 62:163–176.

Yamauchi T and Chiba T. 1973. Adrenergic and cholinergic innervation of the turtle heart ventricle. *Z Zellforsch* 143:485–493.

Young BA, Street SL, and Wassersug RJ. 1994. Anatomical and gravitational influences on cardiac displacement in snakes (Lepidosauria, Serpentes). *Zoomorph* 114:169–175.

Yung WH and Chiu KW. 1985. Contractile responses of the isolated dorsal aorta of the snake to angiotensin II and norepinephrine. *Gen Comp Endocrinol* 60:259–265.

Zaar M, Overgaard J, Gesser H, and Wang T. 2006. Contractile properties of the functionally divided python heart: Two sides of the same matter. *Comp Biochem Physiol A* 146:163–173.

Zerbe P, Glaus T, Clauss M, and Steinmetz HW. 2011. Ultrasonographic evaluation of postprandial heart variation in juvenile Paraguay anacondas (*Eunectes notaeus*). *Am J Vet Res* 72:1253–1258.

Zhao-Xian W, Sun NZ, Mao WP, Chen JP, and Huang GQ. 1991. The breathing pattern and heart rates of *Alligator sinensis*. *Comp Biochem Physiol A* 98:77–87.

Zwart P and Harshbarger JC. 1972. Hematopoietic neoplasms in lizards: Report of a typical case in *Hydrosaurus amboinensis* and of a probable case in *Varanus salvator*. *Intern J Cancer* 9:548–553.

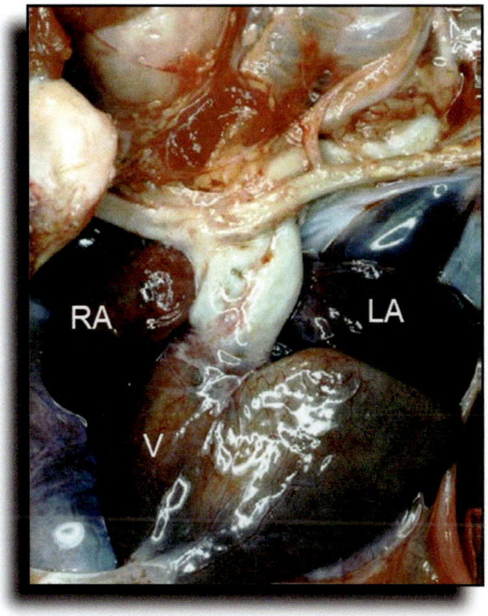

Figure 9.1 Green Turtle, *Chelonia mydras*. Cheloniidae. Normal heart. Ventral view. RA = right atrium, LA = left atrium, V = ventricle. (Courtesy of Jeanette Wyneken.)

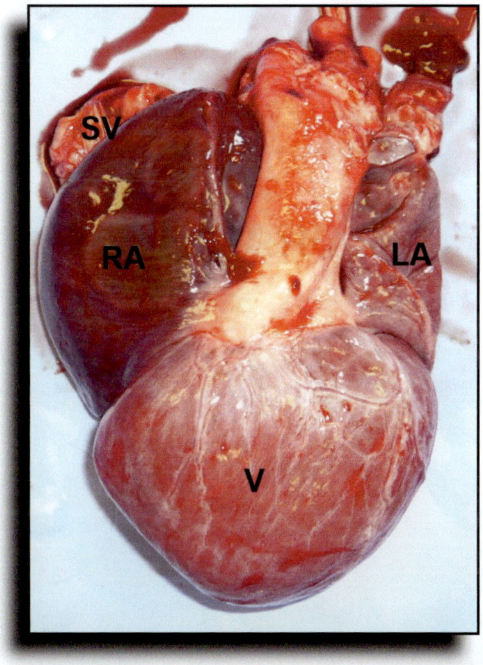

Figure 9.2 Komodo Dragon, *Varanus komodoensis*. Varanidae. Normal heart. Ventral view. SV = sinus venosus, RA = right atrium, LA = left atrium, V = ventricle. (Courtesy of Kelly Helmick.)

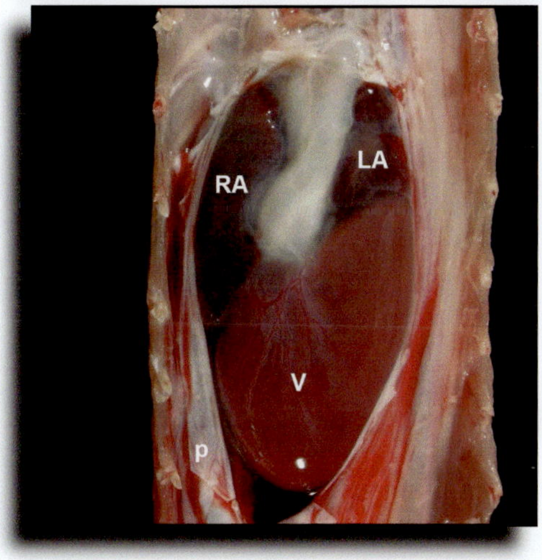

Figure 9.3 Indigo Snake, *Drymarchon couperi*. Colubridae. Normal heart. Ventral view. RA = right atrium, LA = left atrium, V = ventricle, p = pericardium (reflected). (Courtesy of Arno Wuenschmann.)

Figure 9.4 Louisiana Pine Snake, *Pituophis ruthveni*. Colubridae. Histology of a normal heart in a 1-day old. v = ventricle (ventricular lumen), vs = ventricular sinus, mt = trabecular myocardium, mc = compact myocardium, a = atrium, avv = atrioventricular valve, sv = sinus venosum, gv = great vessels at cardiac outflow tract, av = aortic valve. (Courtesy of Michael Garner.)

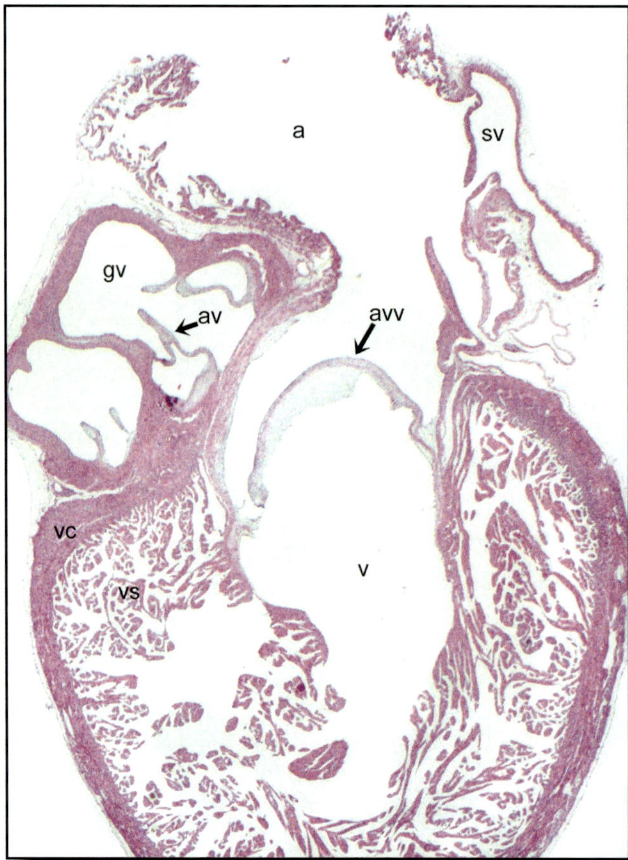

Figure 9.5 Boa Constrictor, *Boa constrictor*. Boidae. Normal heart. Ventral view. RA = right atrium, LA = left atrium, V = ventricle. (Courtesy of Anatomic Pathology Service, School of Veterinary Medicine, UC Davis.)

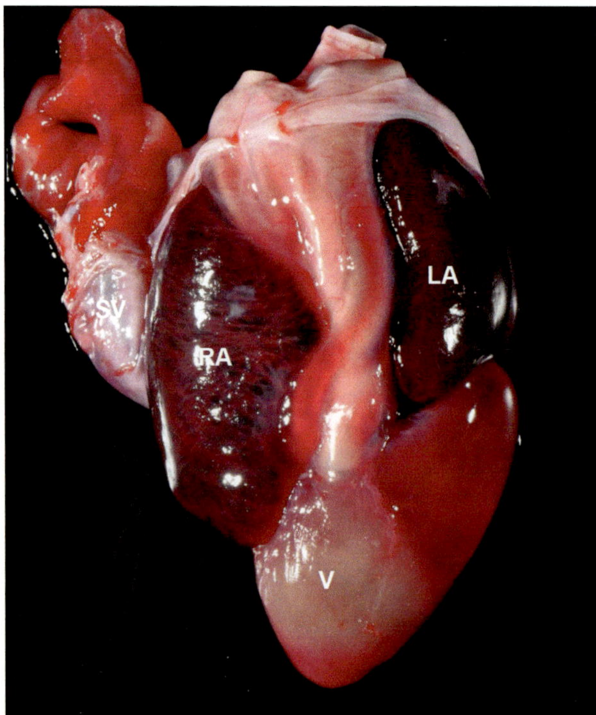

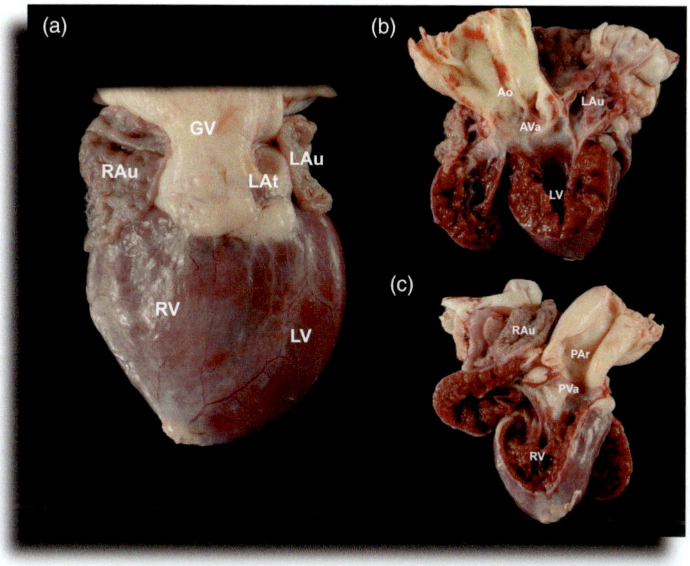

Figure 9.6 American Alligator, *Alligator mississippiensis*, Alligatoridae. Normal heart. Ventral view. (a) RV = right ventricle, LV = left ventricle, LAt = left atrium, LAu = left auricle, RAu = right auricle, GV = great vessels at cardiac outflow tract. Adipose tissue around the great vessels masks the venus venosum in this image. (b) LV = left ventricle, LAu = left auricle, AVa = left atrioventricular valve, Ao = aorta. (c) RV = right ventricle, RAu = right auricle, PVa = pulmonic valve, PAr = pulmonic artery. (Images courtesy of Albert Leon and Robert Ossiboff; Plate courtesy of Michael Garner and Christie Buie.)

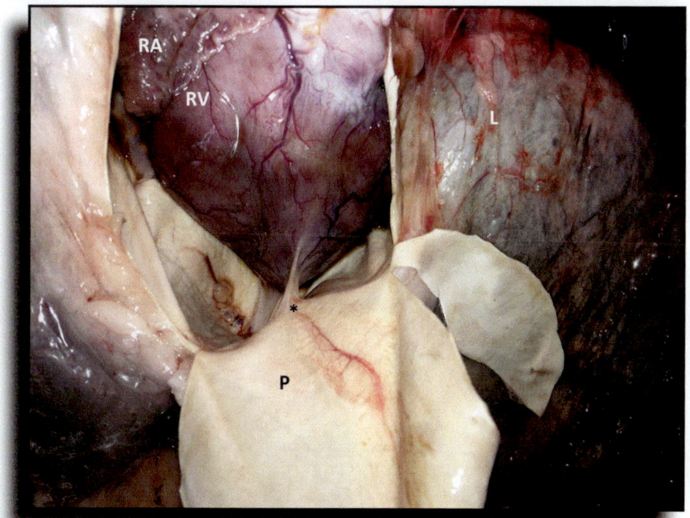

Figure 9.7 Black Caiman, *Melanosuchus niger.* Alligatoridae. Gubernaculum cordis. Pericardium has been reflected caudally over the liver. L = liver, P = pericardium, RA = right atrium, RV = right ventricle, * = gubernaculum cordis.

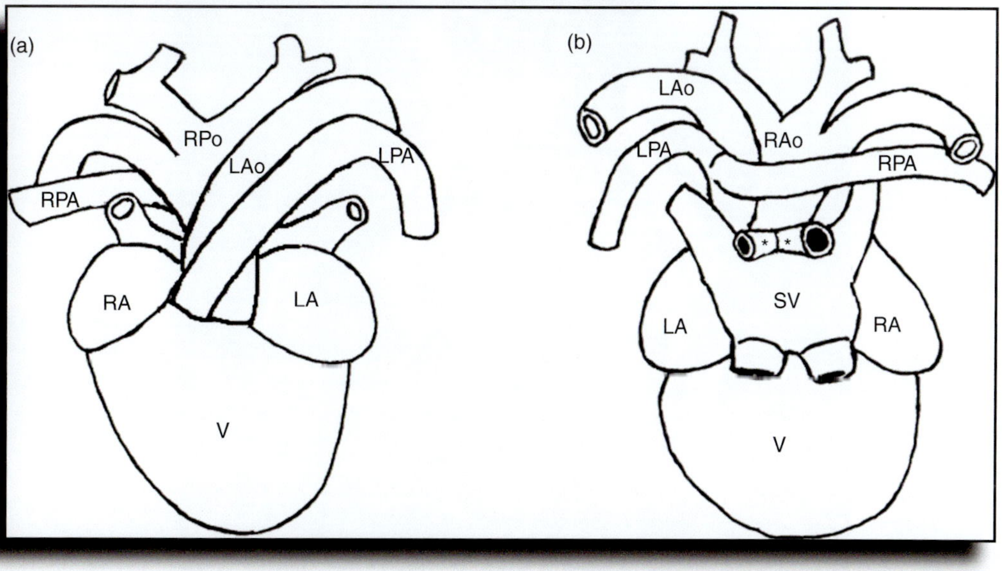

Figure 9.8 Diagram of external anatomy of a reptilian (non-crocodilian) heart. Ventral (a) and dorsal (b) aspects. LA = left atrium, RA = right atrium, V = ventricle, SV = sinus venosus, PT = pulmonary trunk, RPA = right pulmonary artery, LPA = left pulmonary artery, RAo = right aorta, LAo = left aorta, * = pulmonary veins.

Figure 9.9 Diagram of internal anatomy of a reptilian (non-crocodilian) heart. LA = left atrium, RA = right atrium, V = ventricle, PT = pulmonary trunk, RPA = right pulmonary artery, LPA = left pulmonary artery, RAo = right aorta, LAo = left aorta, PV = pulmonary veins, CV = cavum venosum, CP = cavum pulmonale, CA = cavum arteriosum, MR = muscular ridge.

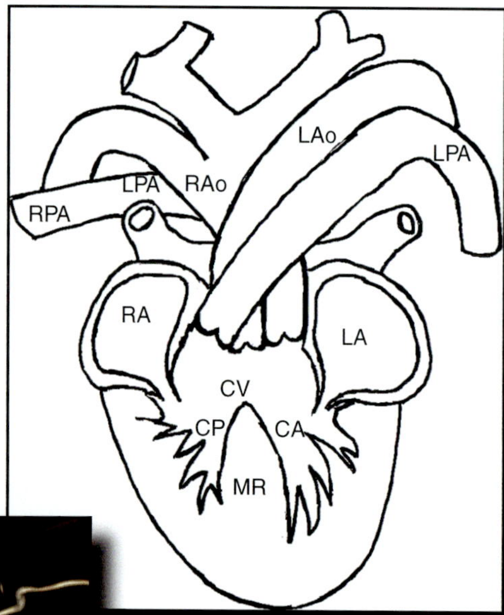

Figure 9.10 African Spurred Tortoise, *Centrochelys sulcata*. Testudinidae. Normal Heart: ventral (a) and dorsal (b) views. Cardiovascular system perfused with BriteVu® contrast agent (Murray, UT, USA). Scanned at 200 μm on a Epica Vimago clinical CT. (Courtesy of Dr. Scott Echols, The Grey Parrot Anatomy Project and Scarlet Imaging, LLC. www.scarletimaging.com.)

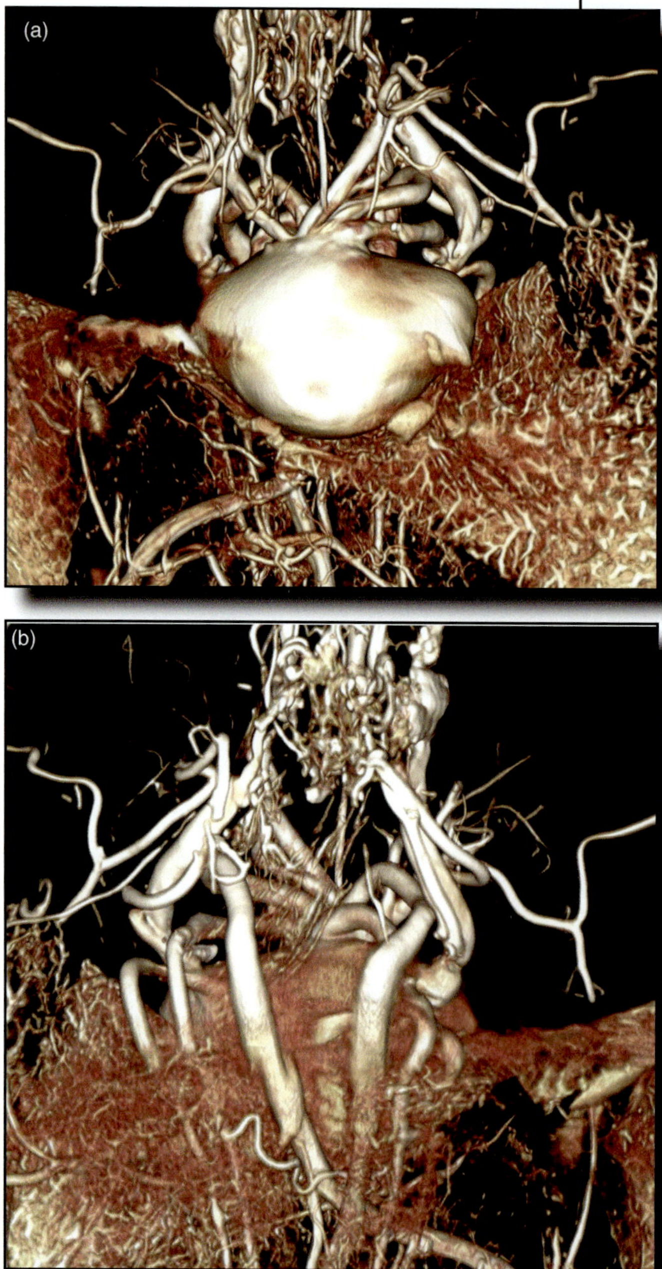

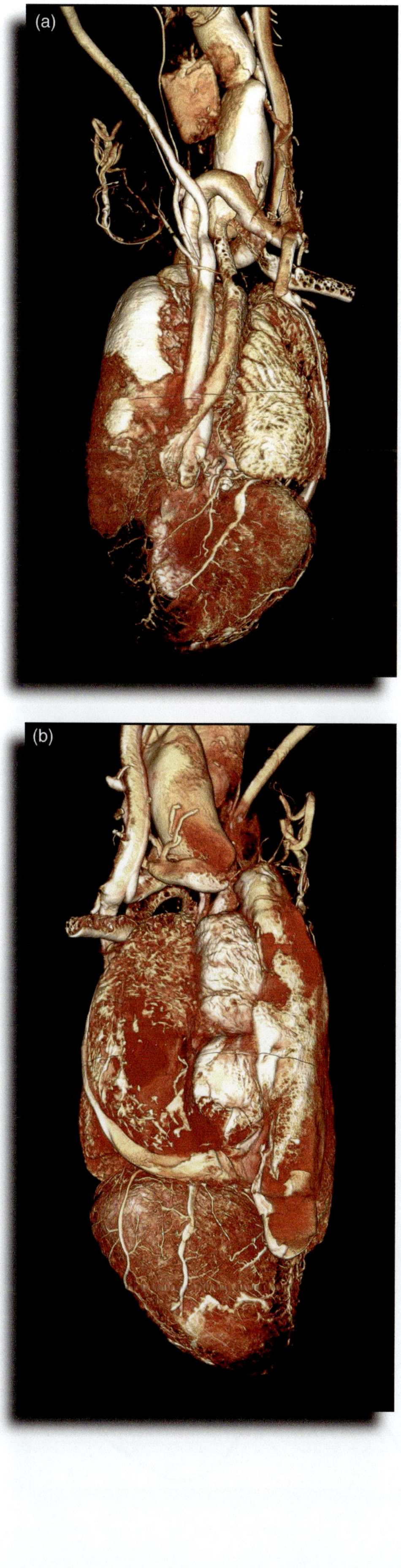

Figure 9.11 Ball Python, *Python regius.* Pythonidae. Normal heart: ventral (a) and dorsal (b) views. Cardiovascular system perfused with BriteVu contrast agent (Murray, UT, USA). Scanned at 100 μm on a Siemens Inveon micro CT. (Courtesy of Dr. Scott Echols, The Grey Parrot Anatomy Project and Scarlet Imaging, LLC. www.scarletimaging.com.)

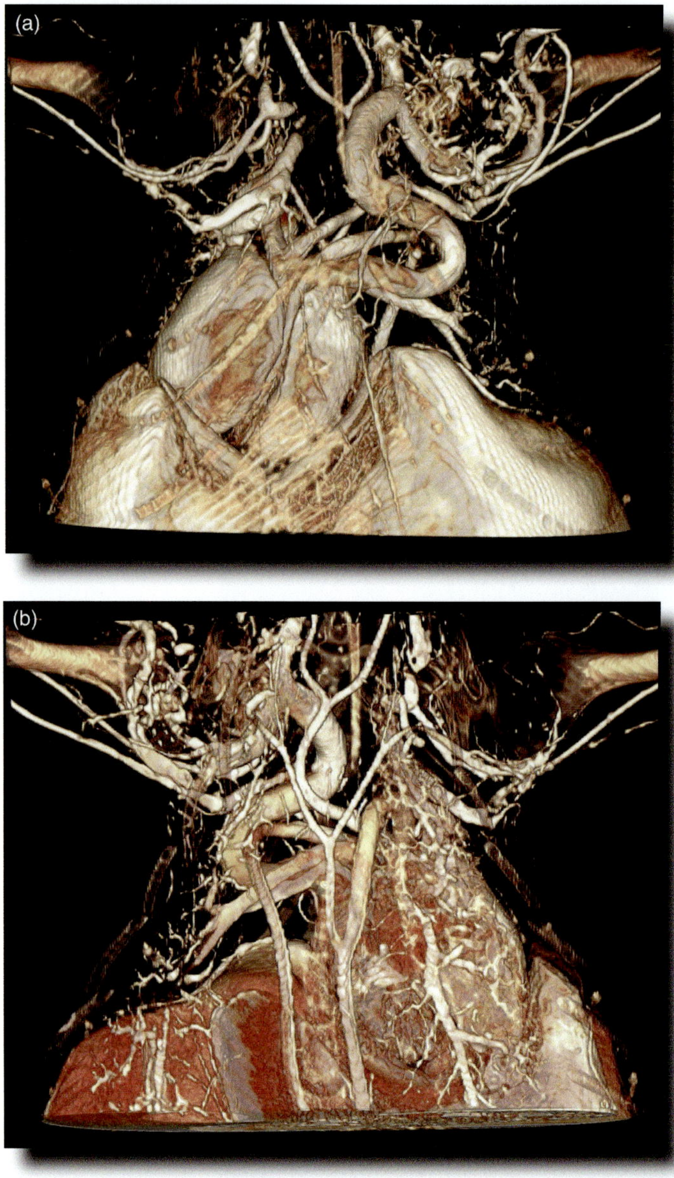

Figure 9.12 White-Throated Monitor, *Varanus albigularis.* Varanidae. Normal heart: ventral (a) and dorsal (b) views. Cardiovascular system perfused with BriteVu contrast agent (Murray, UT, USA). Scanned at 0.6 mm on a Siemens Definition Flash clinical CT. (Courtesy of Dr. Scott Echols, The Grey Parrot Anatomy Project and Scarlet Imaging, LLC. www.scarletimaging.com.)

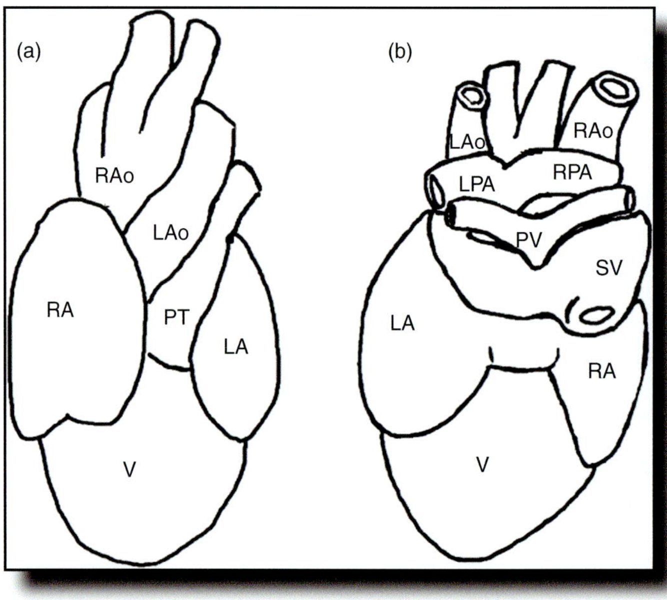

Figure 9.13 Diagram of external anatomy of a crocodilian heart. Ventral (a) and dorsal (b) aspects. LA = left atrium, RA = right atrium, V = ventricle, SV = sinus venosus, PT = pulmonary trunk, PA = pulmonary arteries, RAo = right aorta, LAo = left aorta, PV = pulmonary veins, LPA = left pulmonary artery, RPA = right pulmonary artery.

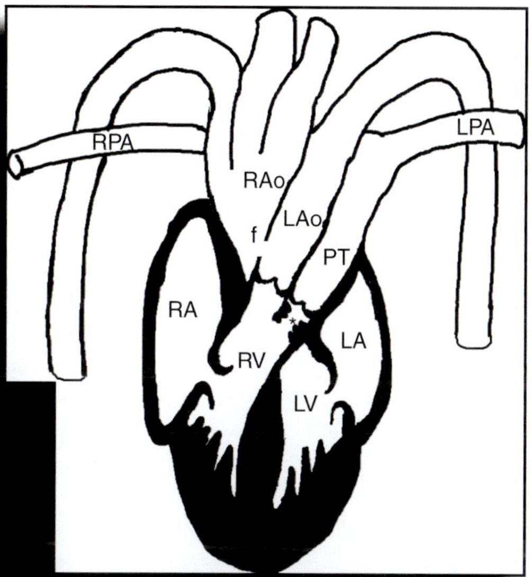

Figure 9.14 Diagram of internal anatomy of a crocodilian heart. LA = left atrium, RA = right atrium, RV = right ventricle, LV = left ventricle, PT = pulmonary trunk, RPA = right pulmonary artery, LPA = left pulmonary artery, RAo = right aorta, LAo = left aorta, f = foramen of Panizza, * = cog valve.

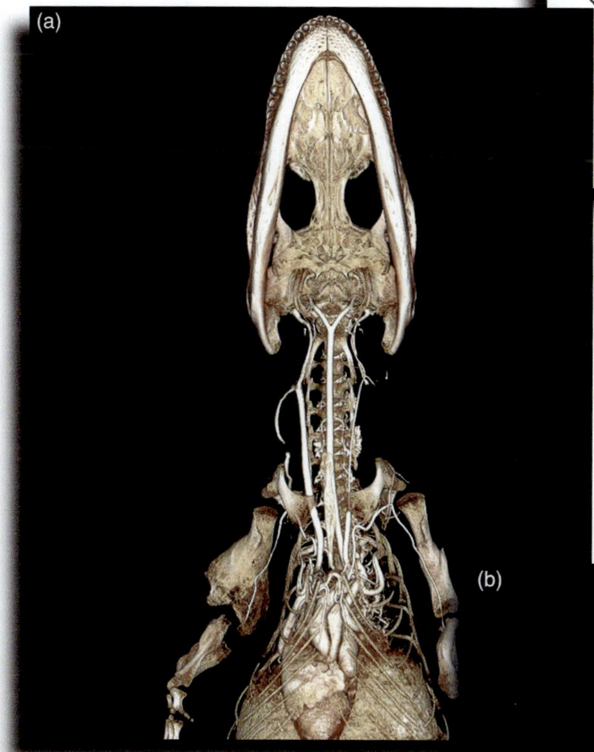

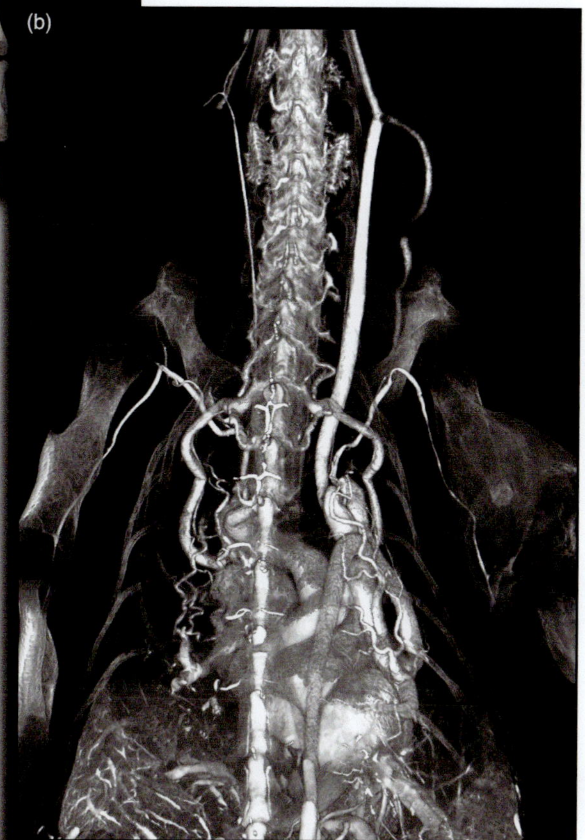

Figure 9.15 American Alligator, *Alligator mississippiensis*. Alligatoridae. Normal heart: ventral (a) and dorsal (b) views. Cardiovascular system perfused with BriteVu contrast agent (Murray, UT, USA). Scanned at 100 μm on a Siemens Inveon micro CT. (Courtesy of Colleen Farmer, University of Utah and Scott Echols, The Grey Parrot Anatomy Project and Scarlet Imaging, LLC. www.scarletimaging.com.)

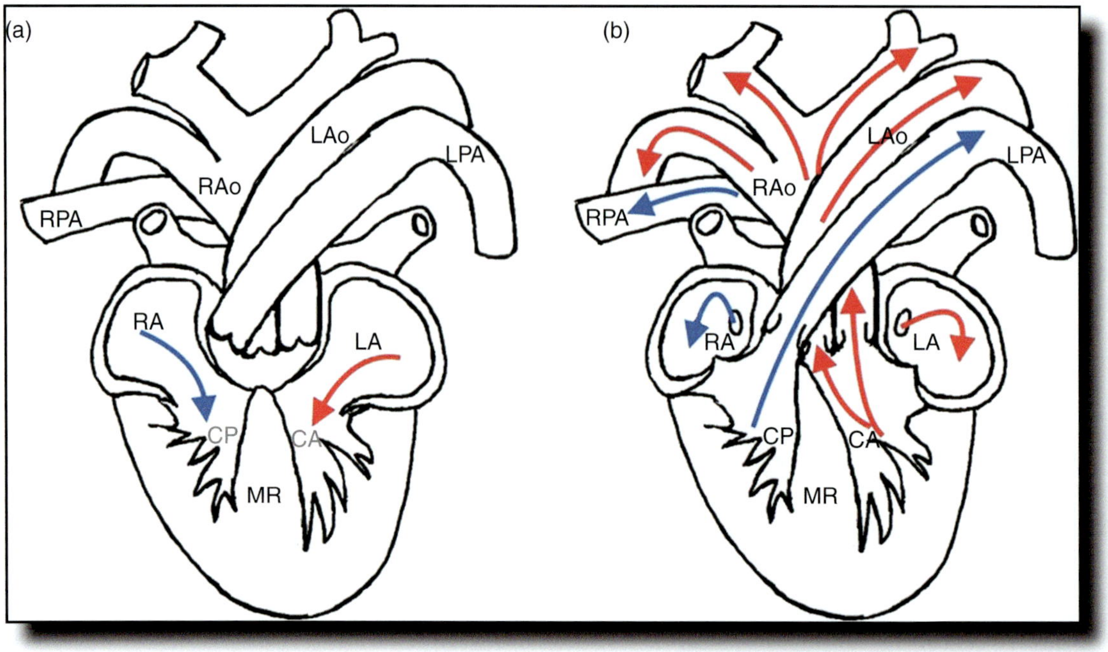

Figure 9.16 Diagrammatic representation of blood flow in a non-crocodilian heart. (a) Ventricular diastole. Deoxygenated blood flows from the right atrium, through the cavum venosum, and into the cavum pulmonale. Oxygenated blood flows from the left atrium into the cavum arteriosum. (b) Ventricular systole. Deoxygenated blood flows from the cavum pulmonale into pulmonary trunk. Oxygenated blood is pumped from the cavum arteriosum, through the cavum venosum, and out through the aortae. Deoxygenated blood enters the right atrium from the sinus venosus and oxygenated blood enters the left atrium from the pulmonary veins. Blue arrows indicate deoxygenated blood, red arrows indicate oxygenated blood. RAo = right aorta, LAo = left aorta, RPA = right pulmonary artery, LPA = left pulmonary artery, RA = right atrium, LA = left atrium, CV = cavum venosum, CP = cavum pulmonale, CA = cavum arteriosum, MR = muscular ridge.

Figure 9.17 Diagrammatic representation of a right-to-left shunt in a non-crocodilian heart. During apnea, there is an increase in pulmonary resistance, causing a decrease in pulmonary blood flow (blue dashed line) and an increase in right-to-left shunting of blood into the aortae. Blue arrows indicate deoxygenated blood, red arrows indicate oxygenated blood, purple arrows indicate a mixture of oxygenated and deoxygenated blood.

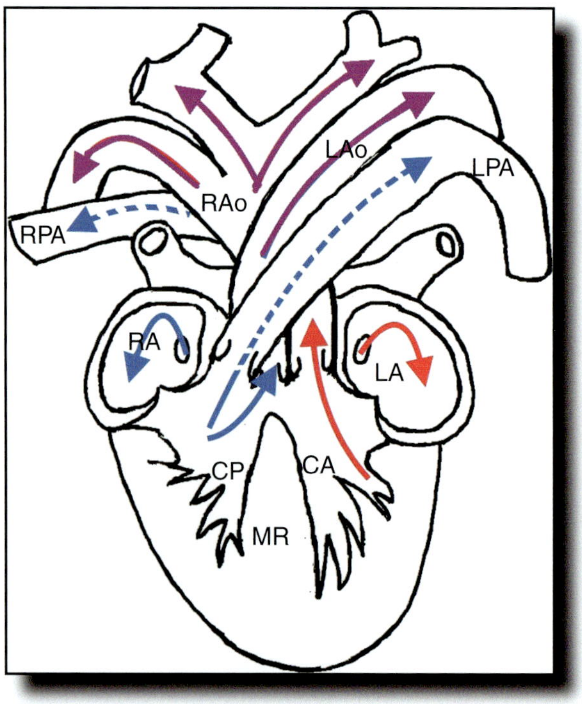

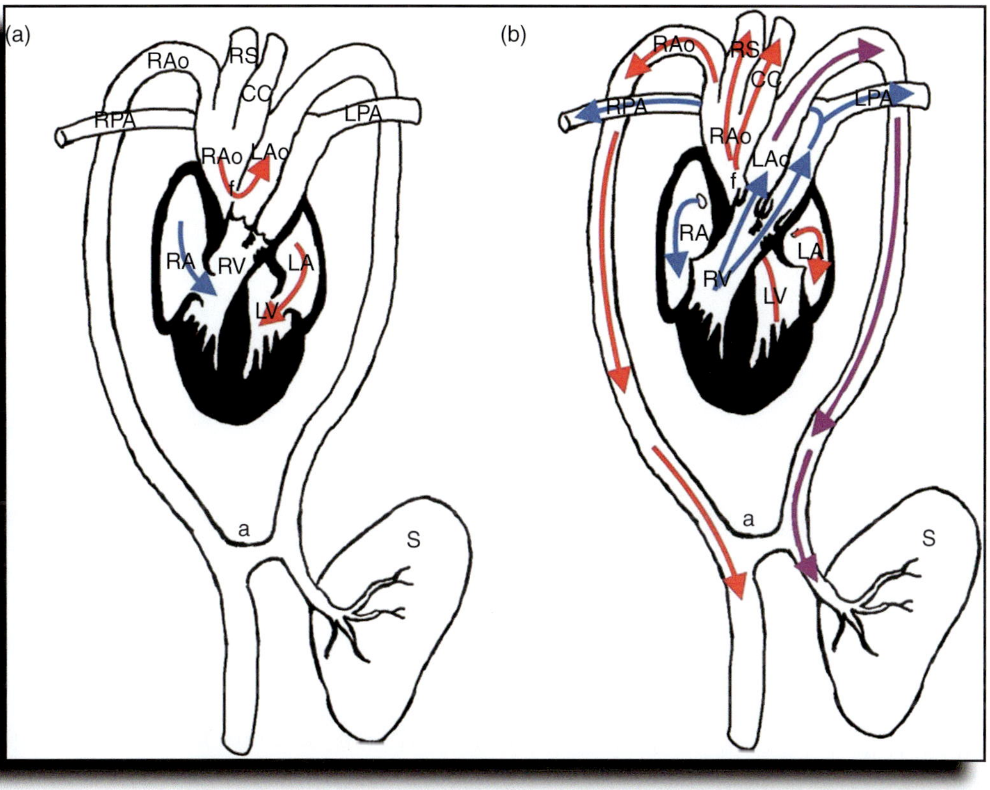

Figure 9.18 Diagrammatic representation of blood flow in a crocodilian heart. (a) Ventricular diastole. Deoxygenated blood flows from the right atrium into the right ventricle. Oxygenated blood flows from the left atrium into the left ventricle. Due to elastic recoil during diastole, some of the blood in the right aorta may flow through the foramen of Panizza and into the left aorta. (b) Ventricular systole. Deoxygenated blood enters the right atrium from the sinus venosus and oxygenated blood enters the left atrium from the pulmonary veins. Oxygenated blood is pumped from the left ventricle through the right aorta. The right subclavian and common carotid arteries branch off the right aorta. Deoxygenated blood flows from the right ventricle into the pulmonary trunk and left aorta. In the left aorta, deoxygenated blood may mix with oxygenated blood that passed through the foramen of Panizza during diastole. Note how the aortic valve leaflets obstruct the foramen of Panizza, which prevents shunting of blood during systole. Blue arrows indicate deoxygenated blood, red arrows indicate oxygenated blood, purple arrows indicate a mixture of oxygenated and deoxygenated blood. RAo = right aorta, RS = right subclavian artery, CC = common carotid artery, LAo = left aorta, RPA = right pulmonary artery, LPA = left pulmonary artery, RA = right atrium, LA = left atrium, RV = right ventricle, LV = left ventricle, f = foramen of Panizza, a = anastomosis, S = stomach.

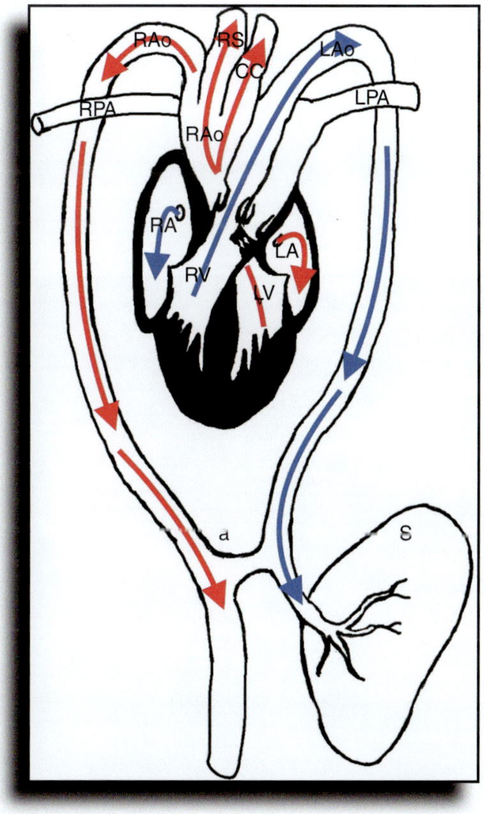

Figure 9.19 Diagrammatic representation of a right-to-left shunt in the crocodilian heart. During apnea, there is an increase in pulmonary resistance and a closing of the cog valve, ceasing pulmonary blood flow and causing a right-to-left shunting of blood into the left aorta. This diverts deoxygenated (hypercarbic) blood to the digestive system, aiding in creating a low pH environment within the stomach. Oxygenated blood is preferentially diverted to the vital organs. Blue arrows indicate deoxygenated blood, red arrows indicate oxygenated blood, purple arrows indicate a mixture of oxygenated and deoxygenated blood. RAo = right aorta, RS = right subclavian artery, CC = common carotid artery, LAo = left aorta, RPA = right pulmonary artery, LPA = left pulmonary artery, RA = right atrium, LA = left atrium, RV = right ventricle, LV = left ventricle, f = foramen of Panizza, a = anastomosis, S = stomach.

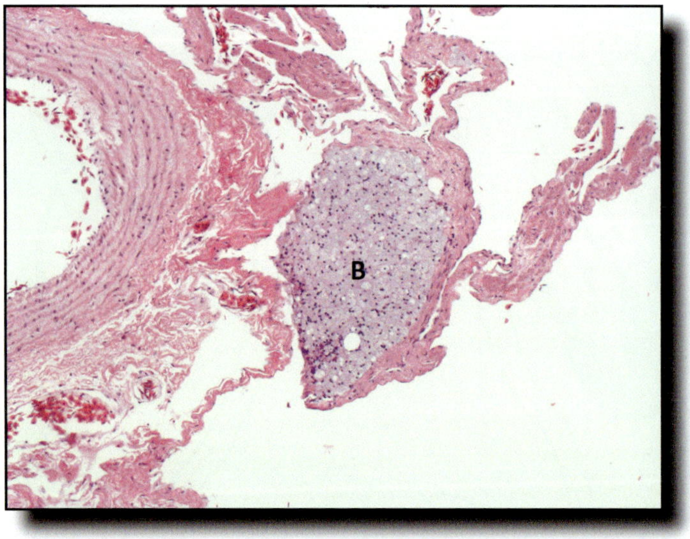

Figure 9.20 Common Musk Turtle, *Sternotherus odoratus.* Note heart-based baroreceptor (B). (Courtesy of Michael Garner.)

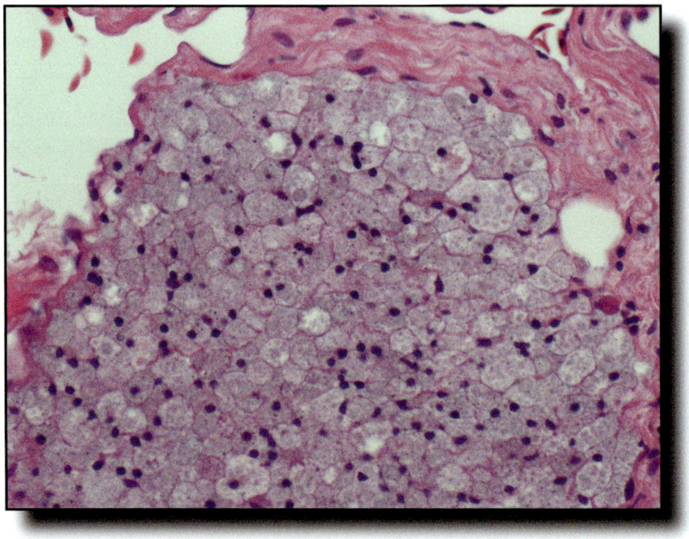

Figure 9.21 Common Musk Turtle, *Sternotherus odoratus.* Higher magnification of Figure 9.20. Note the round, granulated neuroendocrine cells that comprise the baroreceptor. These cells resemble the chromaffin cells of the adrenal gland. (Courtesy of Michael Garner.)

Figure 9.22 Bearded Dragon, *Pagona vitticeps.* Agamidae. Arteriosclerosis of the aorta. The aorta has been incised revealing numerous plaques and thickening.

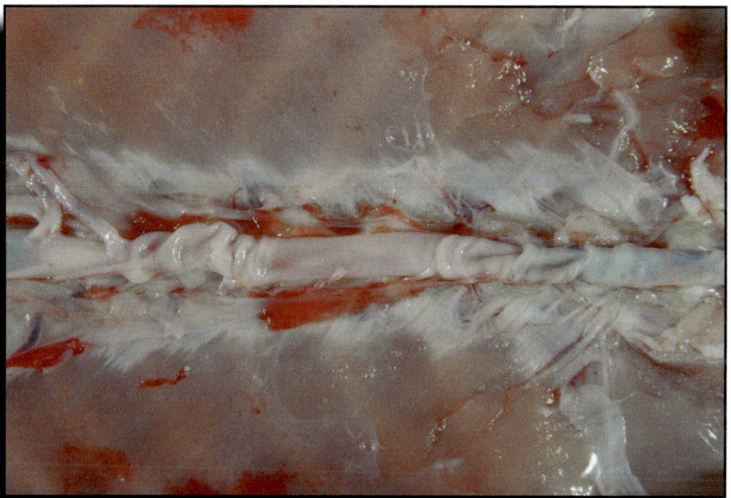

Figure 9.23 Gila Monster, *Heloderma suspectum*. Helodermatidae. Arteriosclerosis of the aorta. (Courtesy of Anatomic Pathology Service, School of Veterinary Medicine, UC Davis.)

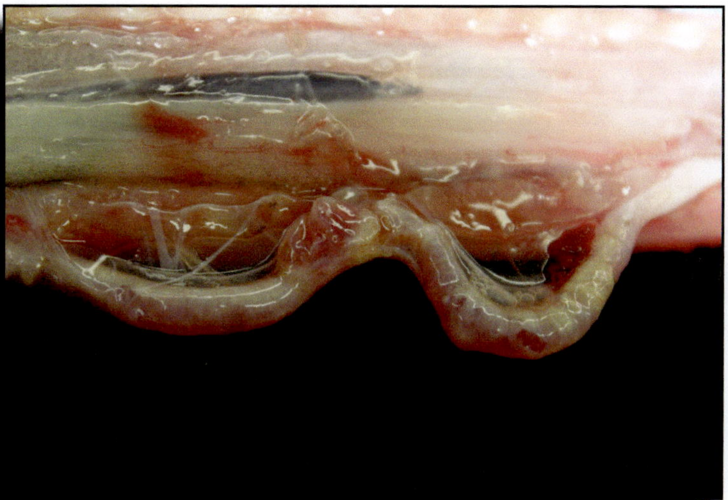

Figure 9.24 East African Green Mamba, *Dendroaspis angusticeps*. Elapidae. Arteriosclerosis of aorta. Note the associated aneurysm.

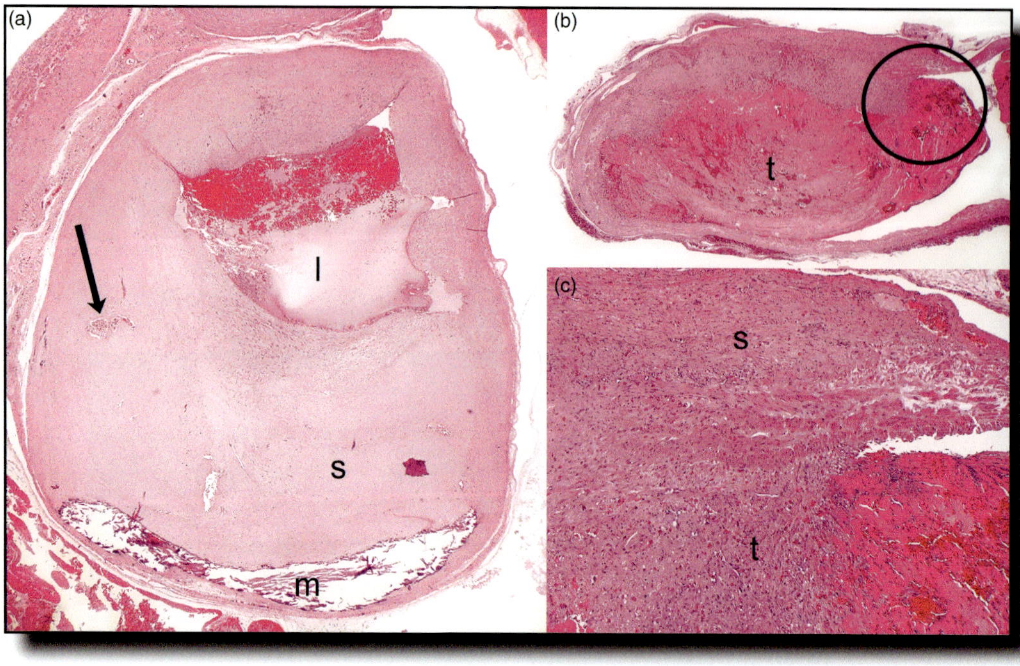

Figure 9.25 Gray Banded Kingsnake, *Lampropeltis alterna*. Colubridae. (a) Arteriosclerosis in the aorta at the outflow tract. Note luminal attenuation (l) due to severe irregular full thickness sclerosis (s), and a small focus of recanalization (arrow). (b) full thickness thrombus (t) is in the sclerotic aorta at mid-body. (c) higher magnification of the circled region in (b) illustrating the sclerotic media (s) and the interface of the partially organized thrombus (t) and new fibrin deposition. (Courtesy of Michael Garner and Christie Buie.)

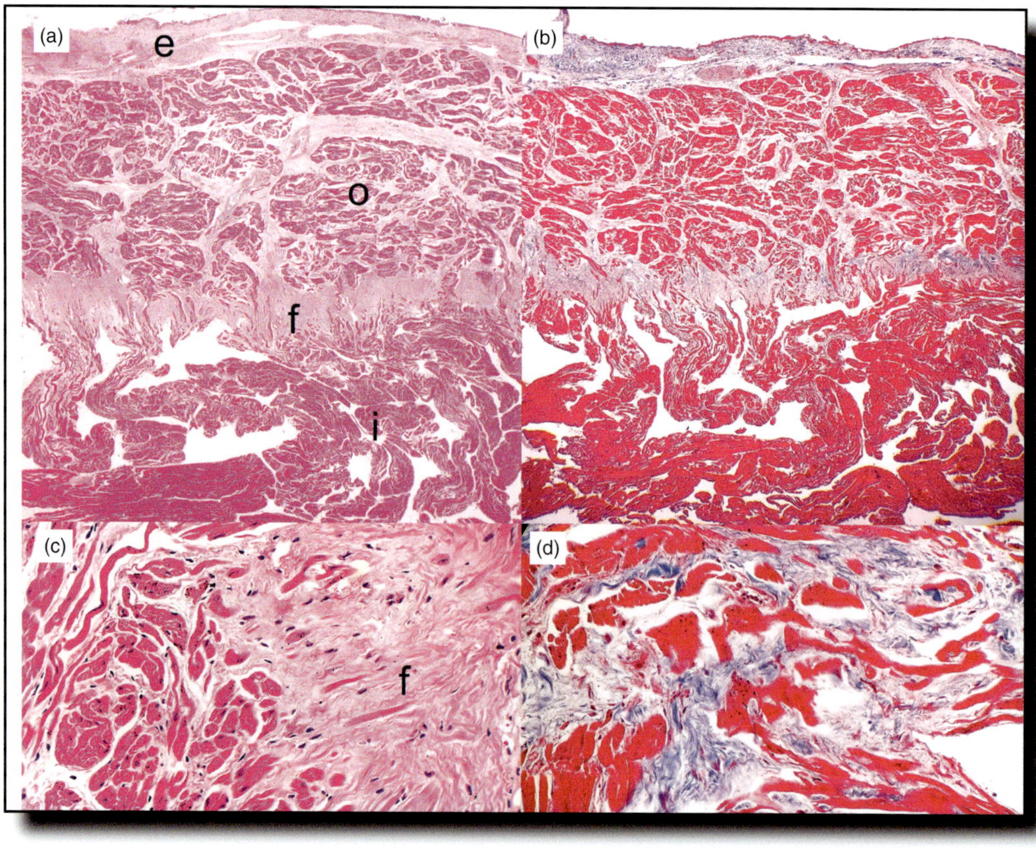

Figure 9.26 Twenty-five-year-old Aldabra Tortoise, *Aldabrachelys gigantea*. Testudines. Premature degenerative cardiac disease. (a) Full thickness image of ventricle showing epicardium (e), outer muscular tunic of myocardium (0), zone of fibrosis (f), and inner muscular tunic of myocardium (i), H&E stain. (b) Same image as (a), stained with Masson's trichrome to accentuate the fibrosis, which stains blue. (c) Higher image of (a) showing fibrosis (f) in myocardium, H&E stain. (d) Higher image of (b), showing the blue staining fibrosis, Masson's trichrome stain. (Courtesy of Michael Garner and Christie Buie.)

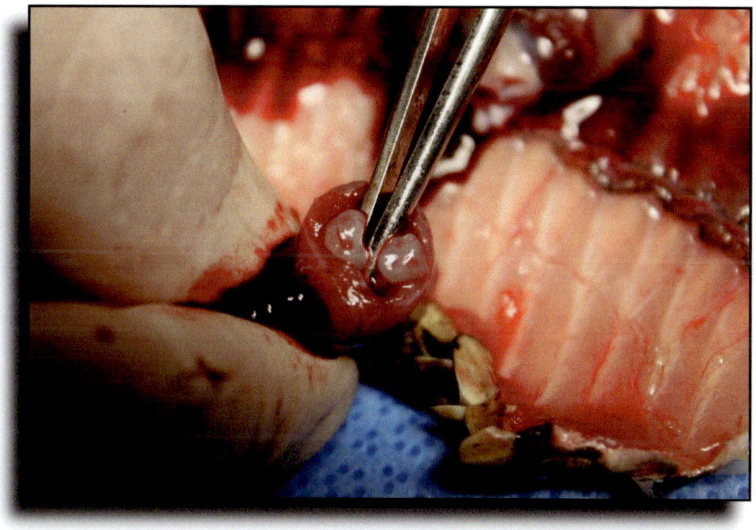

Figure 9.27 Fer-de-Lance, *Bothrops asper*. Viperidae. Endocardiosis of atrioventricular valves.

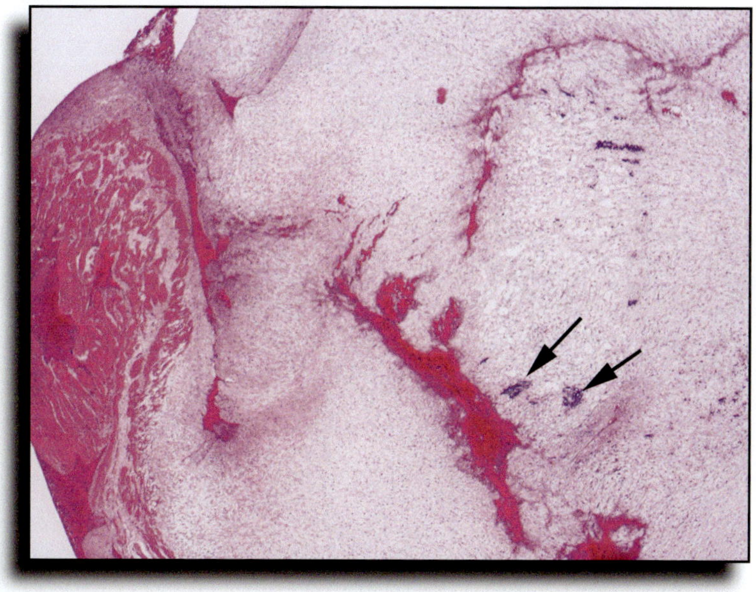

Figure 9.28 Eastern Indigo Snake, *Drymarchon couperi*. Colubridae. Atrioventricular valvular endocardiosis. The valve has marked thickening and disorganization of the fibromxyatous interstitium, with congested irregular vessels and foci of mineralization (arrows). (Courtesy of Michael Garner.)

Figure 9.29 Eastern Indigo Snake, *Drymarchon couperi*. Colubridae. Higher magnification of Figure 9.28. (Courtesy of Michael Garner.)

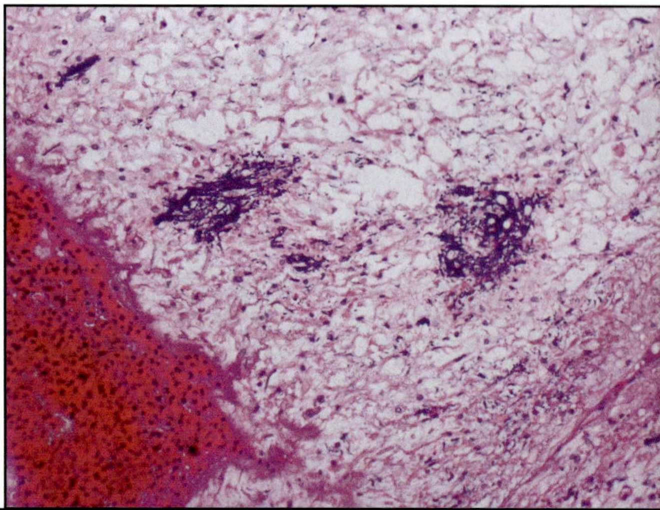

Figure 9.30 Galapagos Tortoise, *Chelonoidis nigra*. Testudinidae. Note white striations through the ventricle, corresponding to severe fibrosis of the myocardium. Some endocardiosis is also in the atrioventricular valve. (Courtesy of Kim Cook.)

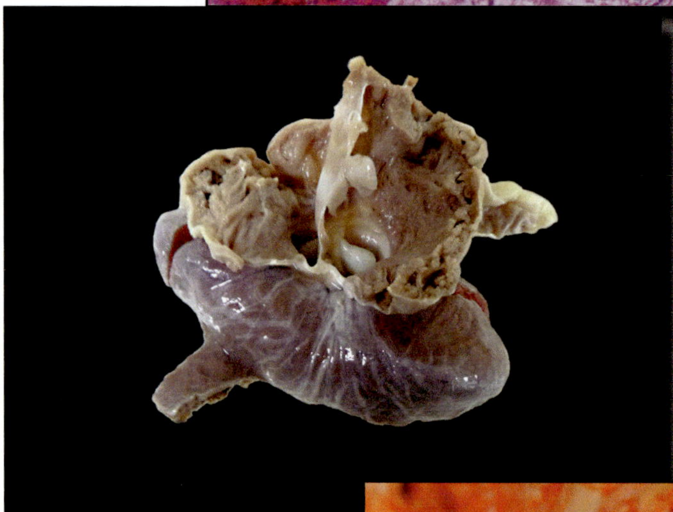

Figure 9.31 Galapagos Tortoise, *Chelonoidis nigra*. Testudinidae. Note epicardial fibrous adhesions associated with degenerative cardiac disease. (Courtesy of Angelique Leone.)

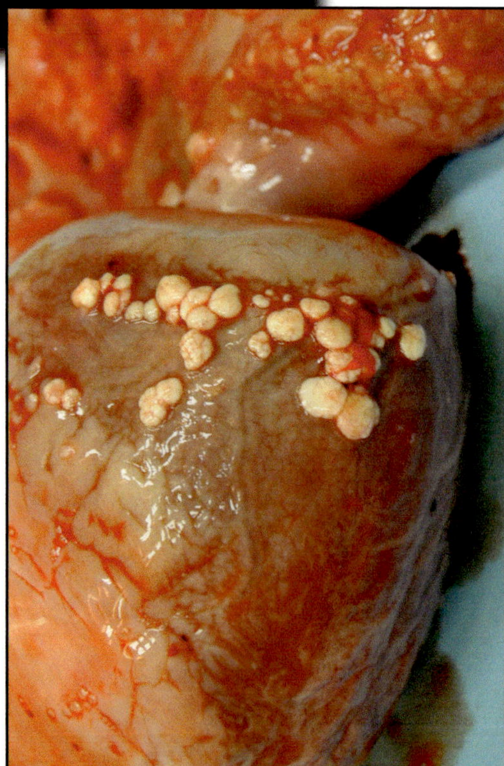

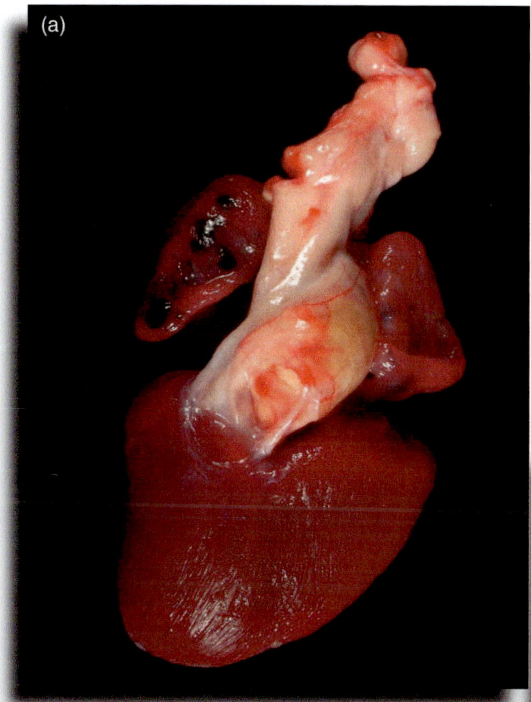

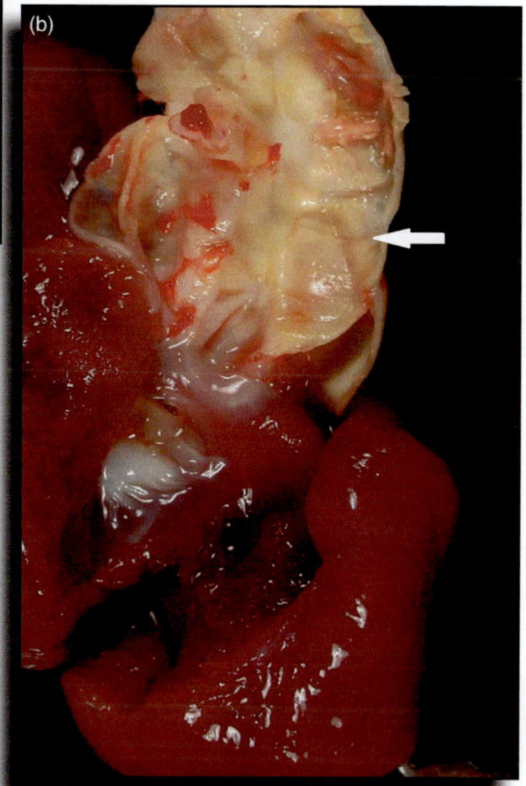

Figure 9.32 Green Iguana, *Iguana iguana*. Iguanidae. Great vessel mineralization: (a) ventral view revealing an enlarged pulmonary trunk. (b) internal view of pulmonary trunk which has multiple thickened areas of mineralization (arrow). (Courtesy of Arno Wuenschmann.)

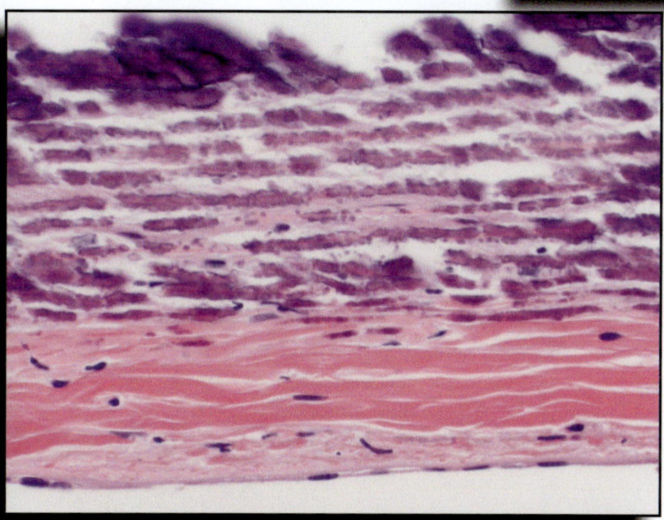

Figure 9.33 Green Iguana, *Iguana iguana*. Iguanidae. Aortic mineralization. H&E stain. (Courtesy of Arno Wuenschmann.)

Figure 9.34 Boa Constrictor, *Boa constrictor.* Boidae. Urate deposition on the epicardial surface of the heart in a boa with visceral gout. (Courtesy of Francisco Uzal.)

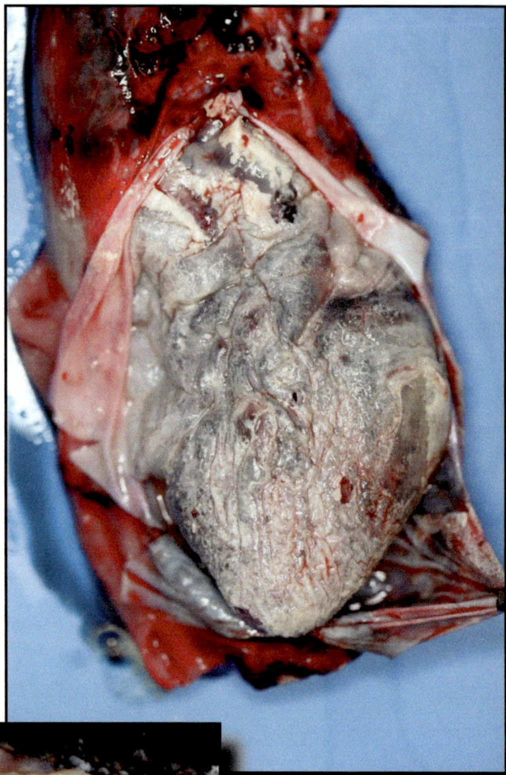

Figure 9.35 Oustalets's Chameleon, *Furcifer oustaleti.* Chamaeleonidae. Urate deposition on the epicardial surface of the heart in a chameleon with visceral gout. (Courtesy of Arno Wuenschmann.)

Figure 9.36 Rock Iguana, *Cyclura nubile.* Iguanidae. (a) Uric acid deposition (visceral gout) in a great vessel. Note irregular circumferential disruption of the intima and media, and luminal attenuation. (b) Higher magnification, showing clefts delineated by inflammatory cells, typical of urate deposition. The crystals have mostly dissolved during formalin fixation, leaving the silhouette of the crystal. (Courtesy of Michael Garner.)

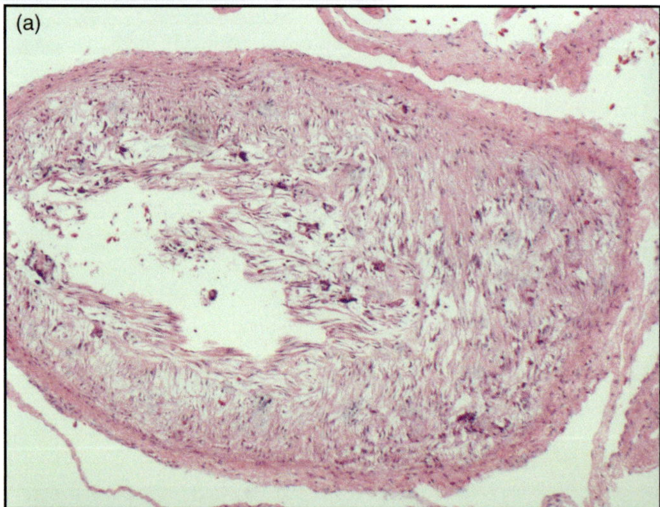

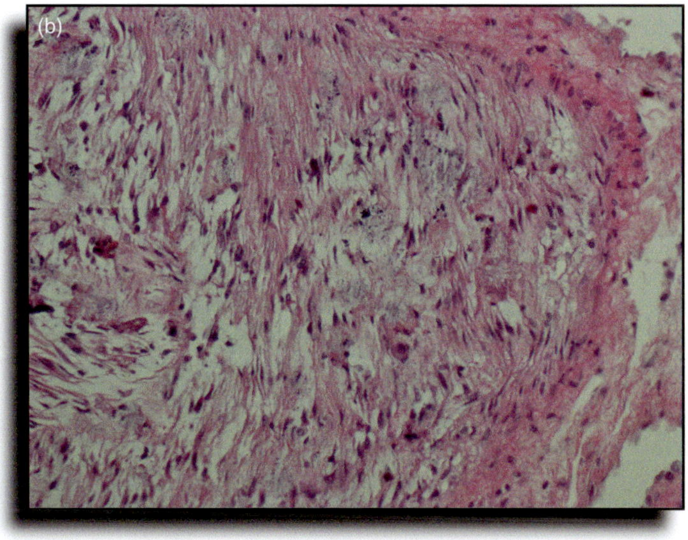

Figure 9.37 Yellow Anaconda, *Eunectes notaeus.* Boidae. (a) Dicephalism. (b) Gross image depicting two hearts in this dicephalic snake. (Courtesy of Nelson R. de Albuquerque.)

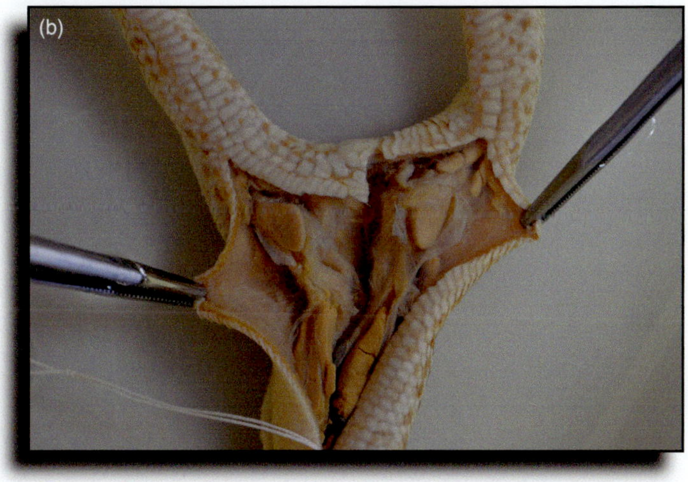

Figure 9.38 Green Racer Snake, *Philodryas patagoniensis*. Colubridae. Two hearts present in a dicephalic snake. Formalin fixed tissues. (Courtesy of Nelson R. de Albuquerque.)

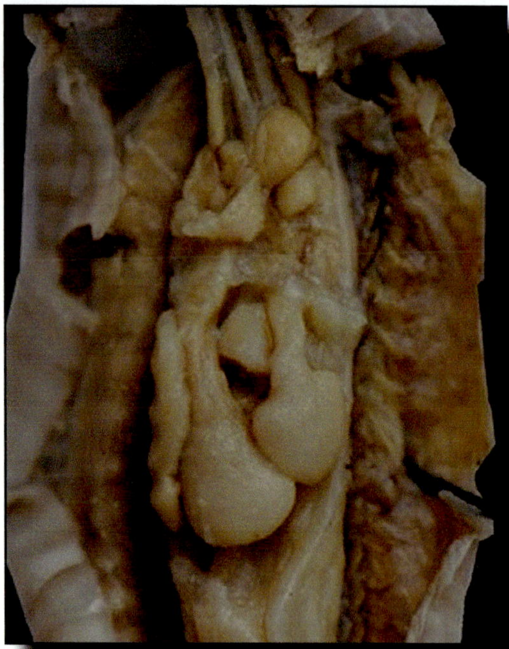

Figure 9.39 Ball Python, *Python regius*. Boidae. Bifid ventricle. (a) Ventral view of in situ beating heart, demonstrating the septal defect. The atria are enlarged, and the ventricle is abnormally broad with bulging chambers. CA = cavum arteriosum, CP = cavum pulmonale, LA = left atrium, LAo = left aorta, PA = pulmonary artery, RA = right atrium. (b) 2 mm transverse section of the ventricular base. The muscular ridge is blunt (arrow) and less developed than normal, leading to a large communication between the cavum pulmonale (which gives rise to the pulmonary artery) and the cavum venosum (which gives rise to the aortae). MR = muscular ridge, CA = cavum arteriosum, avv = atrioventricular valve, CV = cavum venosum, CP = cavum pulmonale. (Courtesy of Bjarke Jensen.)

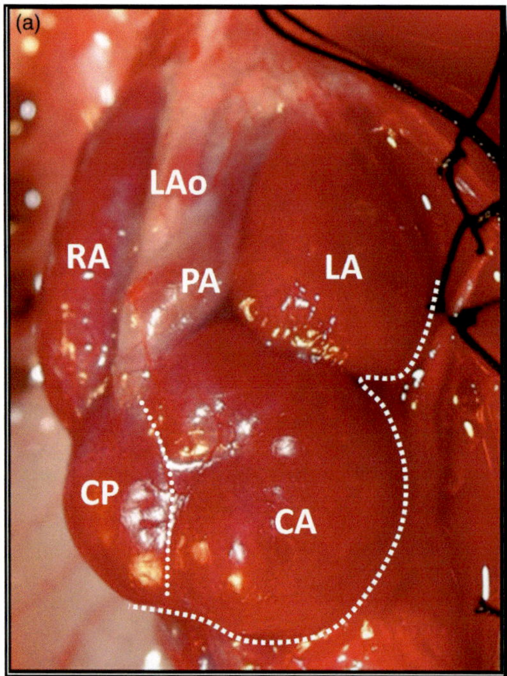

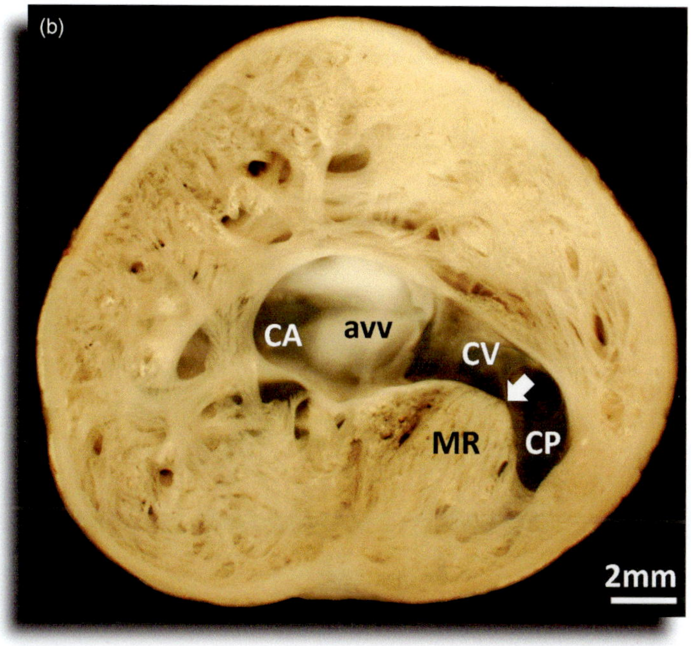

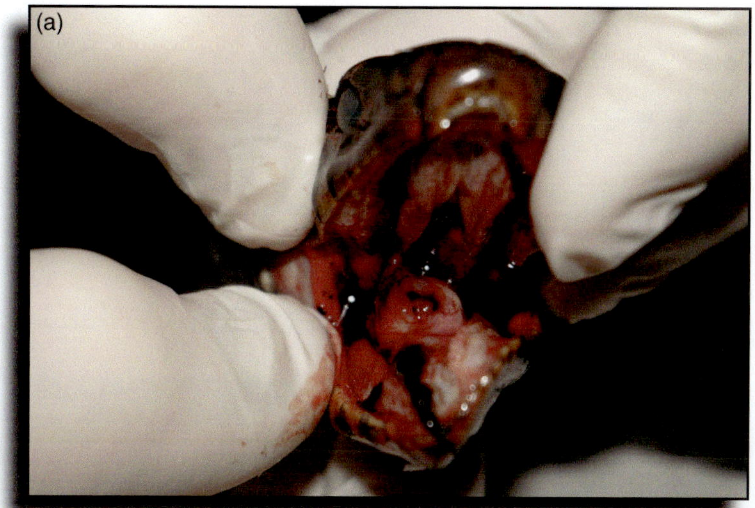

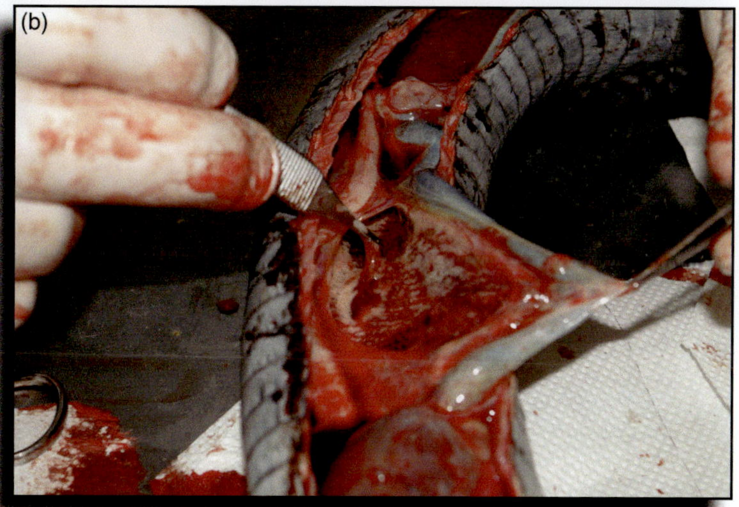

Figure 9.40 Eastern Indigo Snake, *Drymarchon couperi*. Colubridae. Hyperviscosity-like syndrome. Note hemorrhagic fluid in the oral cavity (a) and faveolar lung (b). (Courtesy of Kurt Volle.)

Figure 9.41 Eastern Indigo Snake, *Drymarchon couperi*. Colubridae. Hyperviscosity-like syndrome. Note intensely proteinaceous fluid in the lumen and surrounding tissues of the trachea (a), faveolae and capillaries of the lung (b), and glomeruli and peritubular capillaries of the kidney (c).

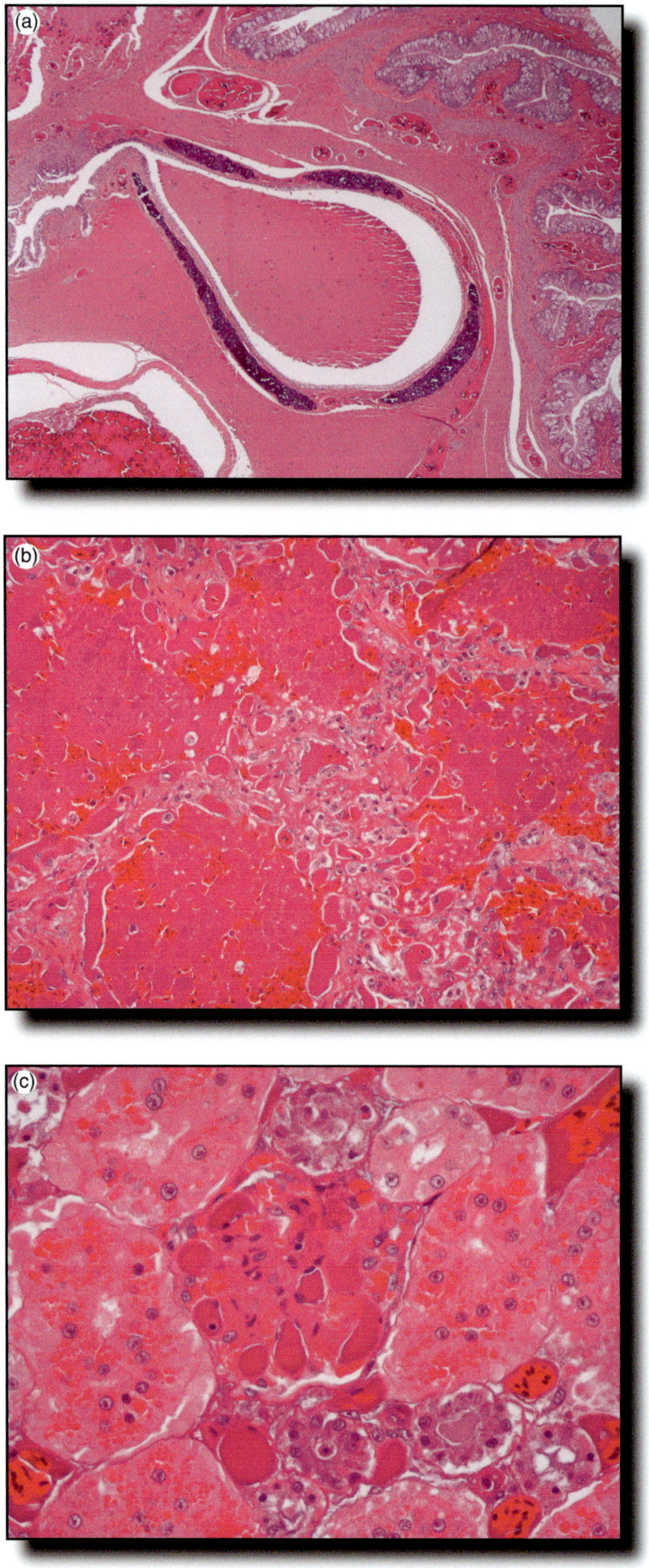

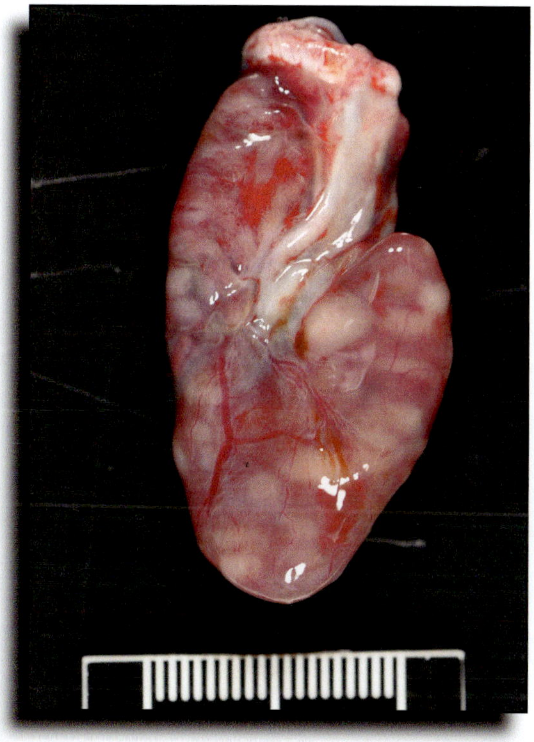

Figure 9.42 Yellow Ratsnake, *Pantherophis alleghaniensis*. Colubridae. Cardiac spindle cell sarcoma. (Courtesy of Anatomic Pathology Service, School of Veterinary Medicine, UC Davis.)

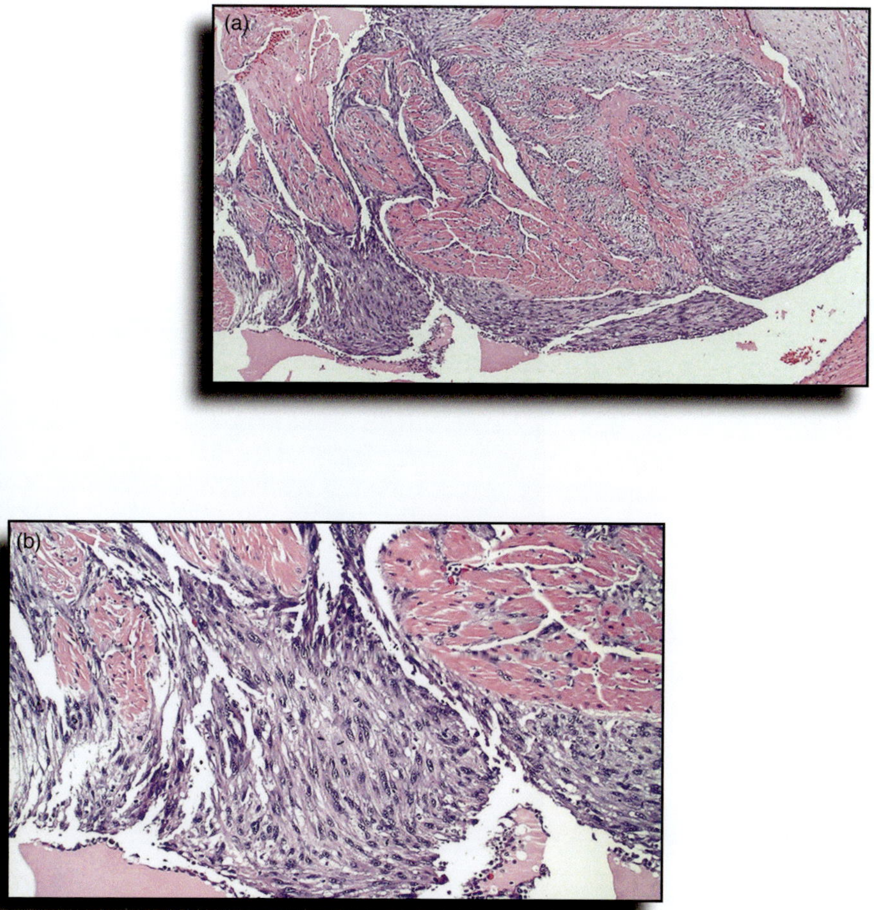

Figure 9.43 Western Diamondback Rattlesnake, *Crotalus atrox*. Viperidae. Cardiac spindle cell sarcoma. (a) Multiple bundles of spindle cells are infiltrating and replacing the myocardium. (b) Higher magnification. H&E stain. (Courtesy of David Gasper.)

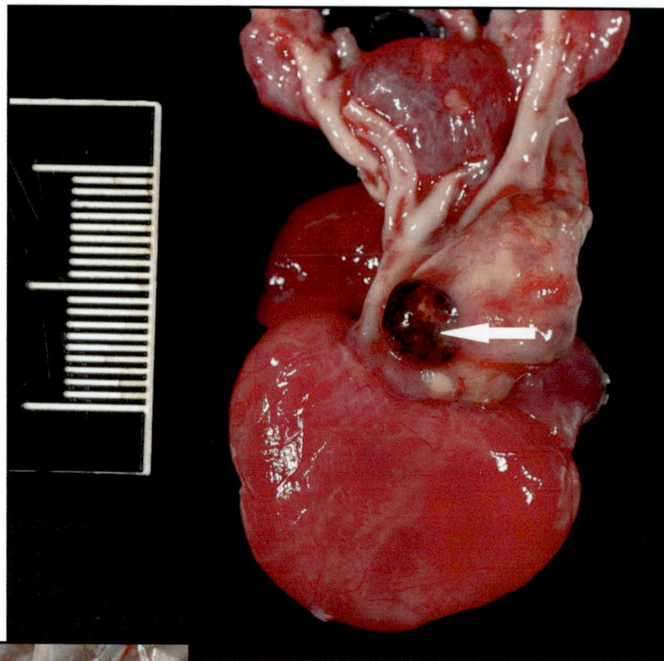

Figure 9.44 Elongated Tortoise, *Indotestudo elongate.* Testudinidae. Focal dilation of the aorta (aneurysm) (Arrow). (Courtesy of Anatomic Pathology Service, School of Veterinary Medicine, UC Davis.)

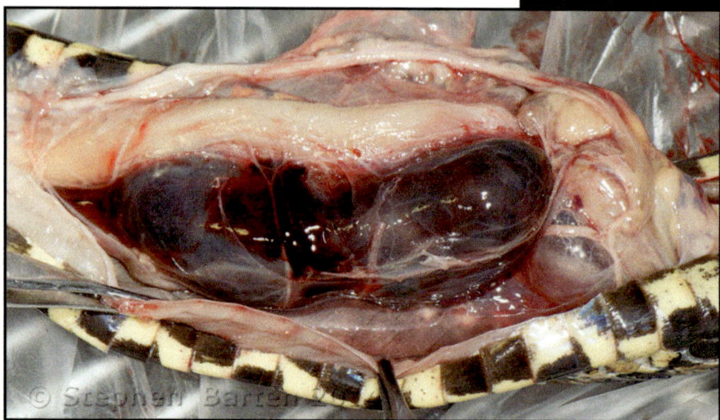

Figure 9.45 Fox Snake, *Pantherophis gloydi.* Colubridae. Aortic aneurysm. (Courtesy of Steve Barten.)

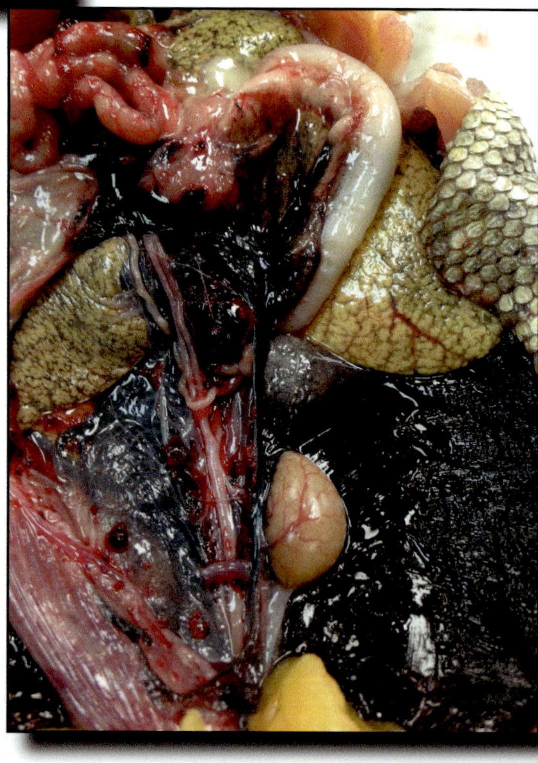

Figure 9.46 Bearded Dragon, *Pagona vitticeps.* Agamidae. Aortic aneurysm in descending aorta. Note intracoelomic hemorrhage associated with rupture of an aneurysm in the descending aorta. The rupture site is not recognizable in this image. (Courtesy of Sue Horton.)

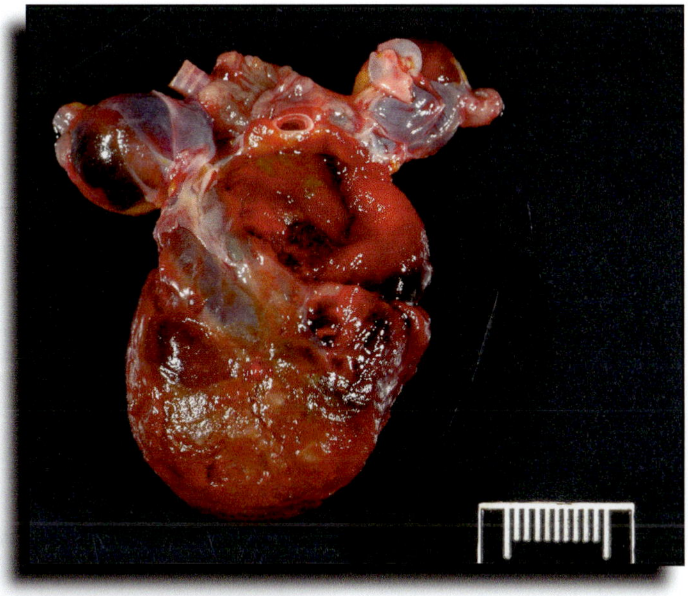

Figure 9.47 Bearded Dragon, *Pagona vitticeps.* Agamidae. Heart base aortic aneurysm. (Courtesy of Anatomic Pathology Service, School of Veterinary Medicine, UC Davis.)

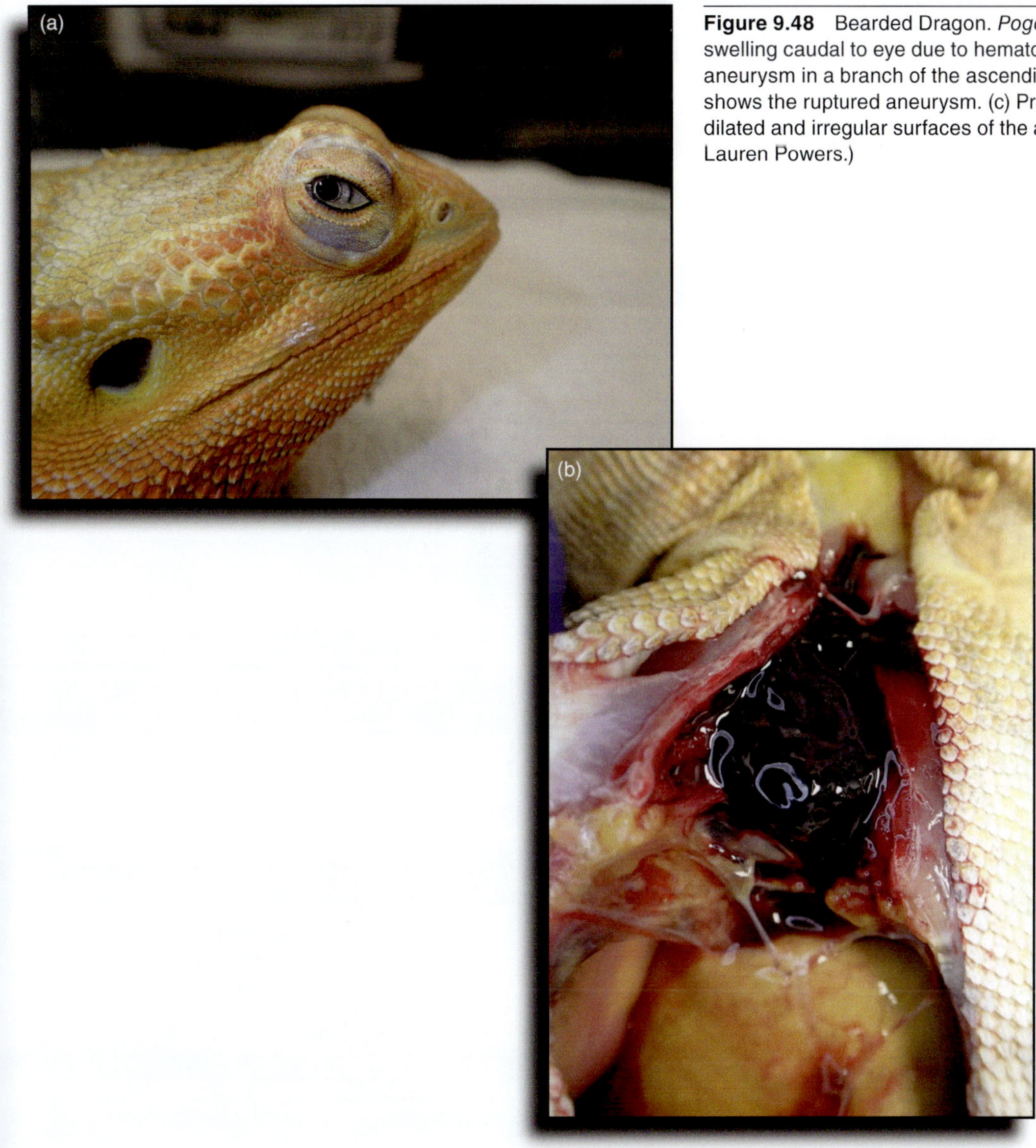

Figure 9.48 Bearded Dragon. *Pogona vitticeps*. Agamidae. (a) Unilateral swelling caudal to eye due to hematoma formation associated with an aneurysm in a branch of the ascending aorta. (b) Prosected aortic branches shows the ruptured aneurysm. (c) Prosection of the aneurysm shows the dilated and irregular surfaces of the affected aortic bifurcation. (Courtesy of Lauren Powers.)

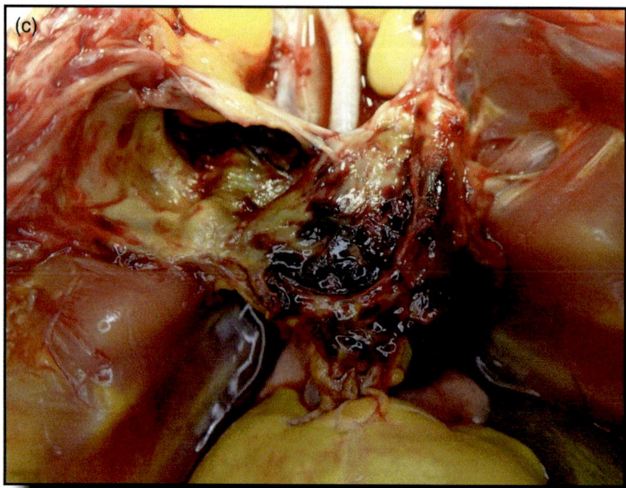

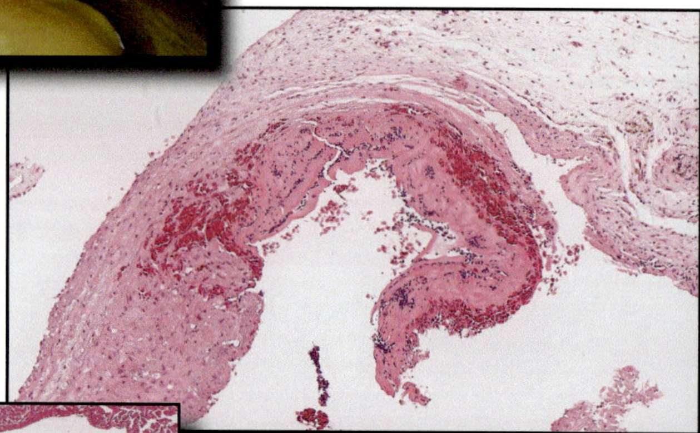

Figure 9.49 Bearded Dragon, *Pagona vitticeps.* Agamidae. Dissecting aortic aneurysm. H&E stain. (Courtesy of Michael Garner.)

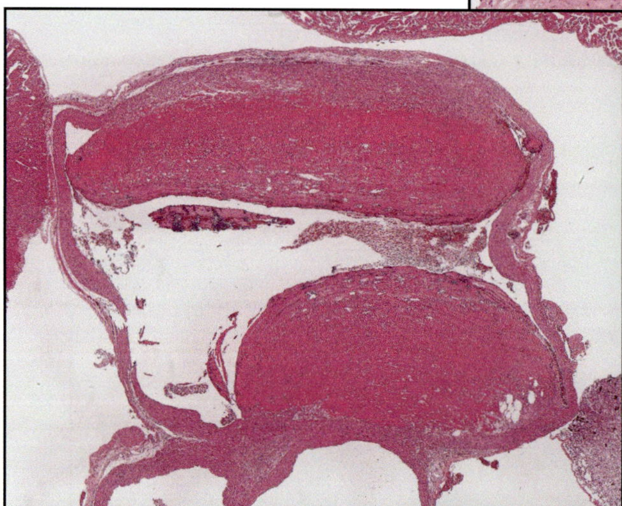

Figure 9.50 Bearded Dragon, *Pagona vitticeps.* Agamidae. Dissecting aortic aneurysm with thrombosis. H&E stain. (Courtesy of Michael Garner.)

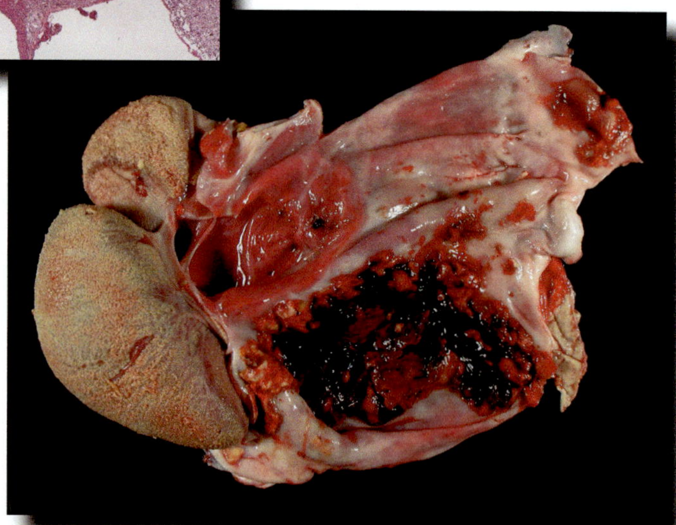

Figure 9.51 Galapago Tortoise, *Chelonoidis nigra.* Testudinidae. Fibrinous pericarditis. (Courtesy of Arno Wuenschmann.)

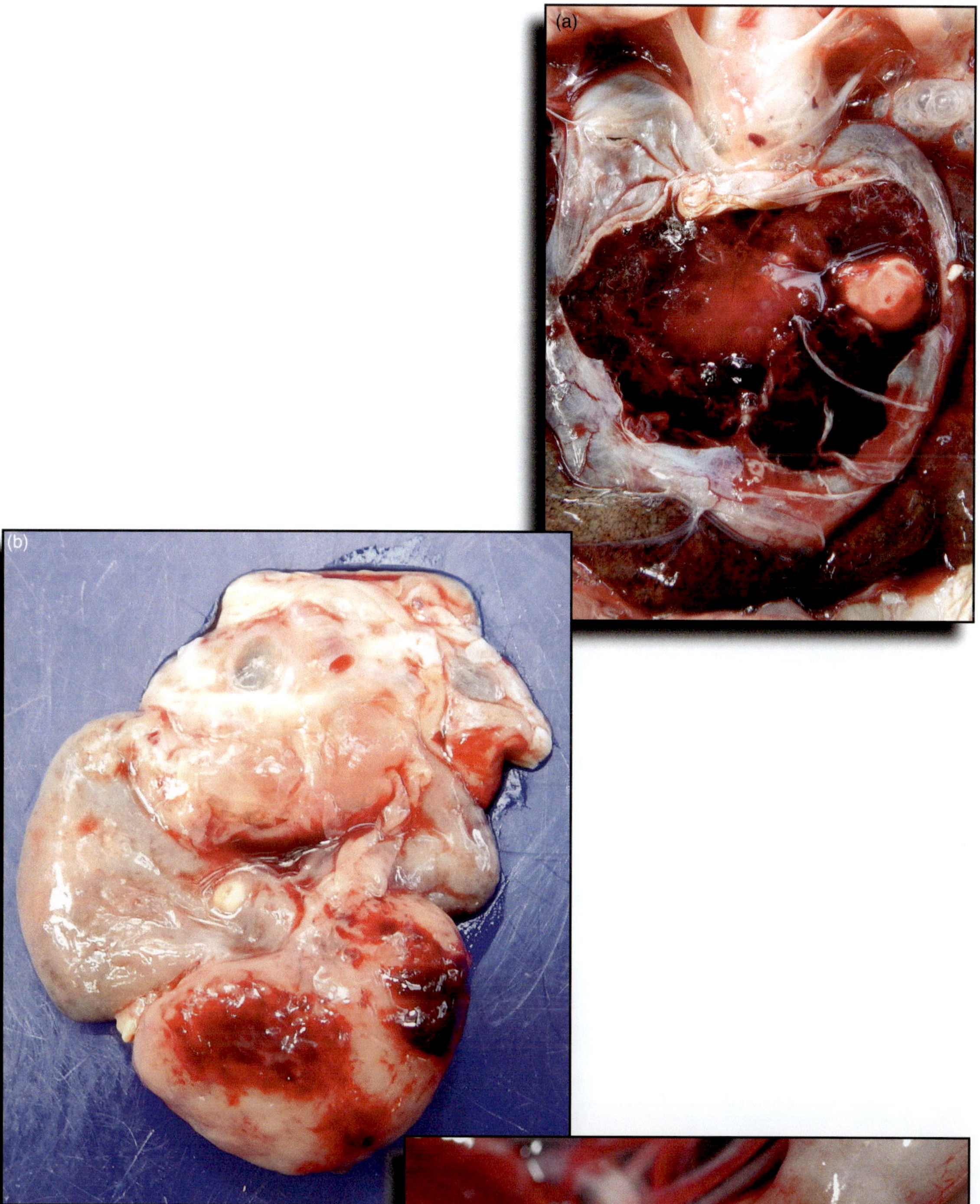

Figure 9.52 Sulawesi Tortoise, *Indotestudo forsteni*. Testudinidae. Pericardial hemorrhagic effusion (a) and epicarditis (b). (Courtesy of Paul Gibbons.)

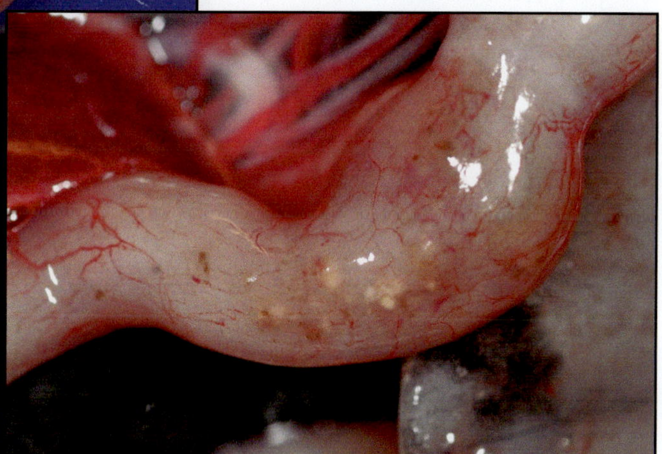

Figure 9.53 Loggerhead Turtle, *Caretta caretta*. Cheloniidae. Spirorchid egg granulomas in the intestinal serosa. Spirorchids are a group of vascular trematodes of fresh water and marine turtles that can spread hematogenously to almost every organ. (Courtesy of José Luis Crespo.)

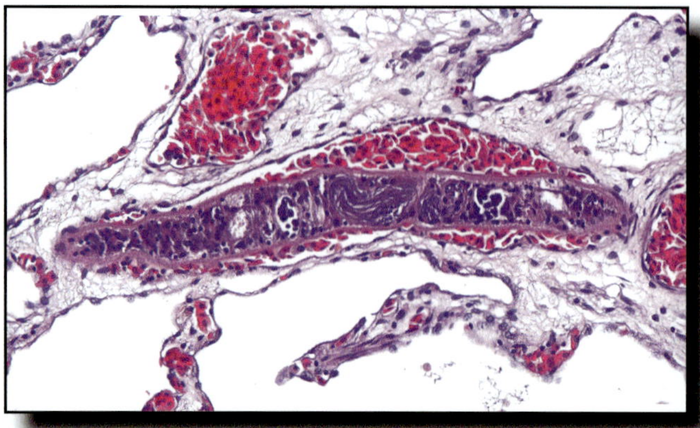

Figure 9.54 Loggerhead Turtle, *Caretta caretta.* Cheloniidae. Spirorchiid adult in a pulmonary vessel. H&E stain. (Courtesy of Juan Manuel Corpa.)

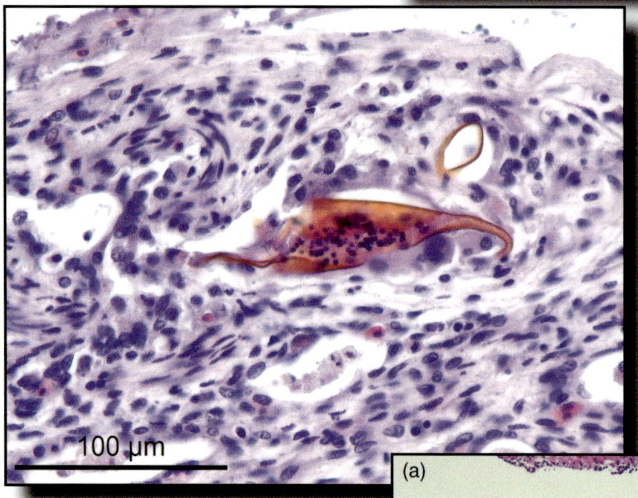

Figure 9.55 Loggerhead Turtle, *Caretta caretta*. Cheloniidae. Spirorchiid egg granulomas in the intestine. H&E stain.

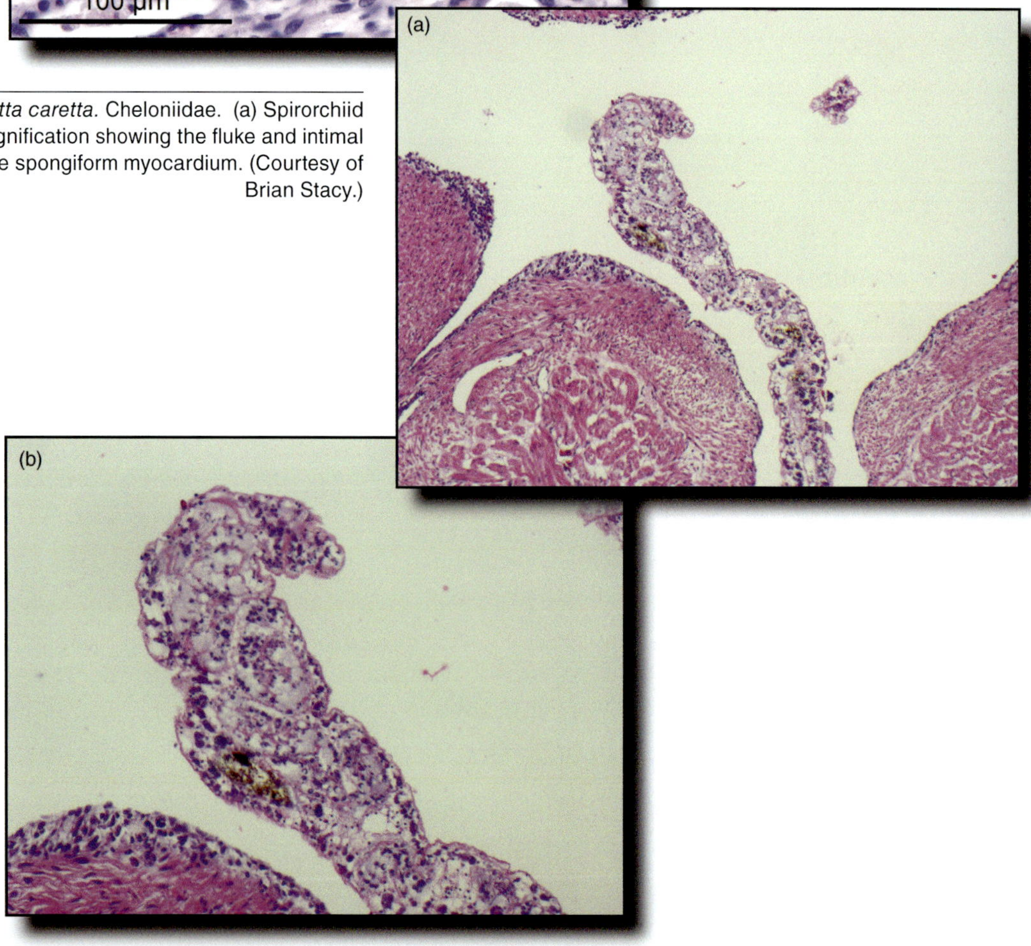

Figure 9.56 Loggerhead Turtle, *Caretta caretta.* Cheloniidae. (a) Spirorchiid fluke larva in the heart. (b) Higher magnification showing the fluke and intimal hyperplasia lining the spongiform myocardium. (Courtesy of Brian Stacy.)

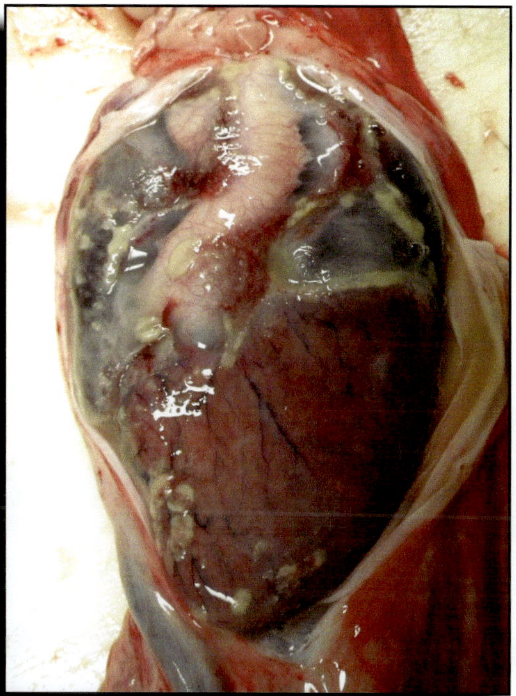

Figure 9.57 Anaconda, *Eunectes* sp. Boidae. Pericarditis and epicarditis. Note thickened pericardial sac and fibrinous exudate on epicardium. (Courtesy of Nancy Stedman.)

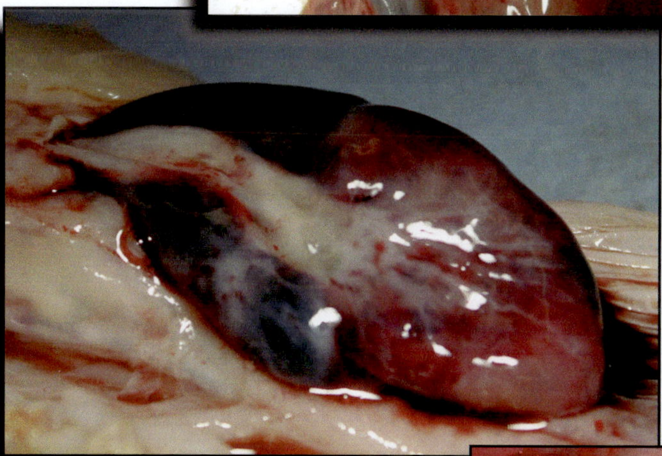

Figure 9.58 Cottonmouth, *Agkistrodon piscivorus.* Viperidae. Bacterial granulomata in the heart.

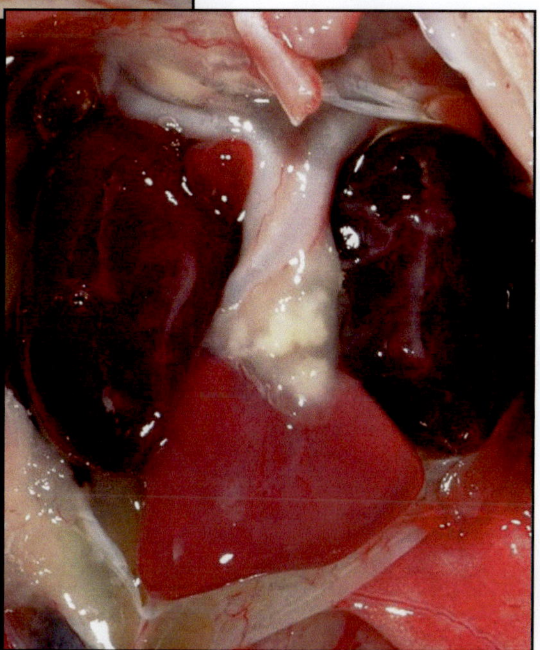

Figure 9.59 Green Iguana, *Iguana iguana.* Iguanidae. Bacterial granuloma at the heart base involving the great vessels. (Courtesy of Anatomic Pathology Service, School of Veterinary Medicine, UC Davis.)

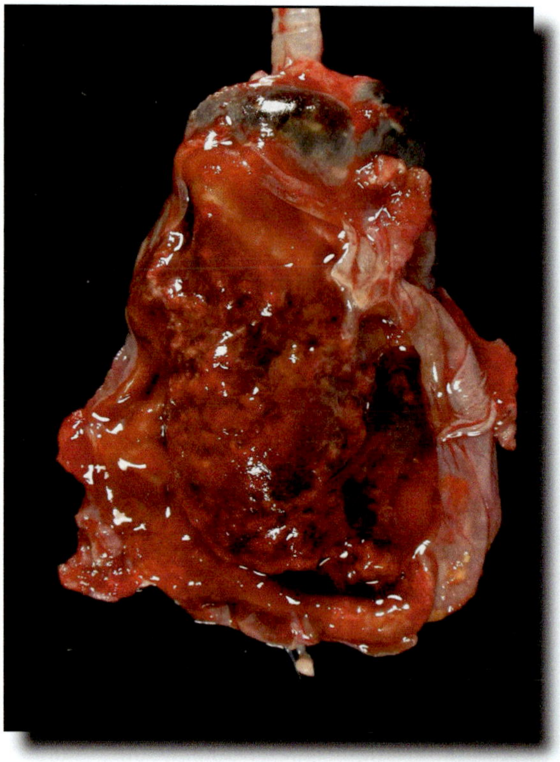

Figure 9.60 Bearded Dragon, *Pagona vitticeps.* Agamidae. Bacterial pericarditis and epicarditis. Note thickened pericardium and fibrinohemorrhagic exudate on the surface of the epicardium. (Courtesy of Arno Wuenschmann.)

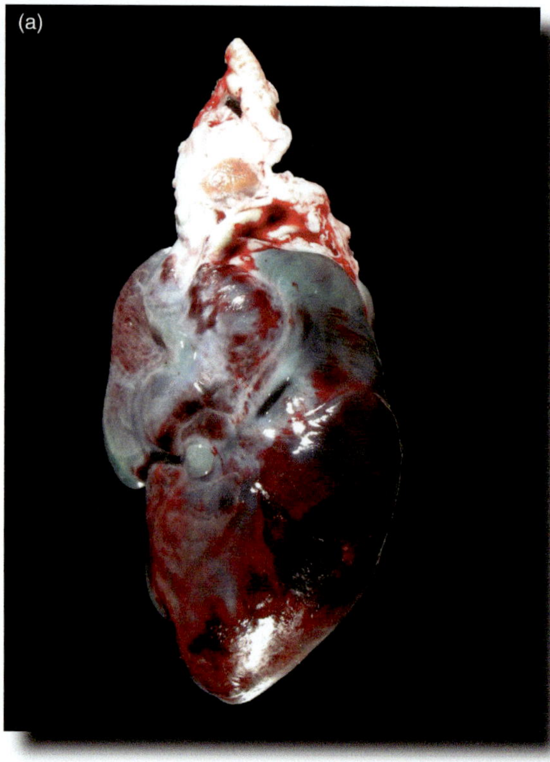

Figure 9.61 Green Anaconda, *Eunectes murinus.* Boidae. Ventricular myocardial hematoma, ventral external (a) and ventral internal (b) views. (Courtesy of Arno Wuenschmann.)

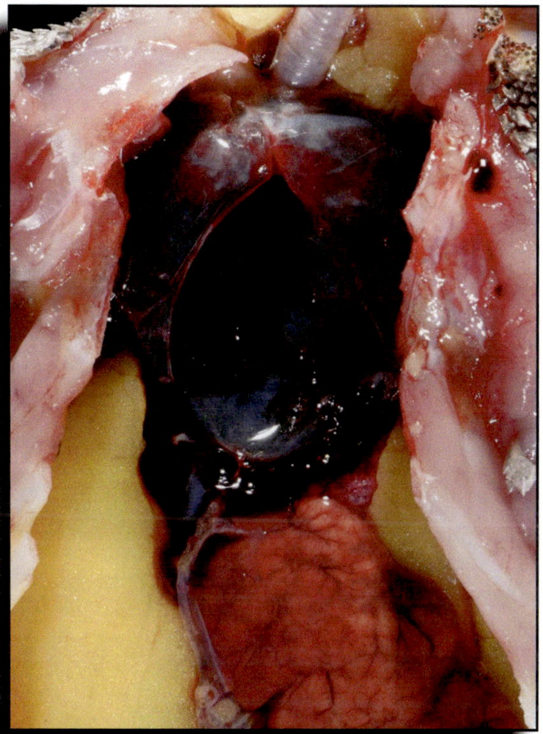

Figure 9.62 Bearded Dragon, *Pagona vitticeps*. Agamidae. Microsporidiosis and hemopericardium. (Courtesy of Arno Wuenschmann.)

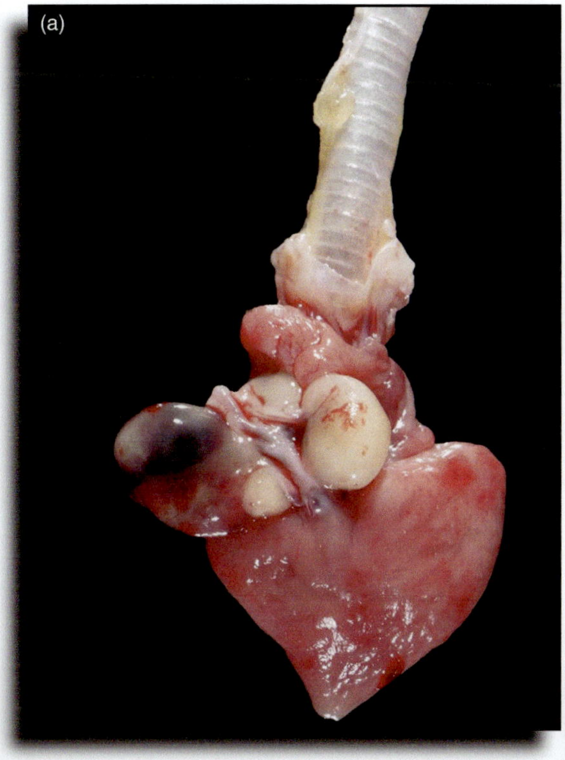

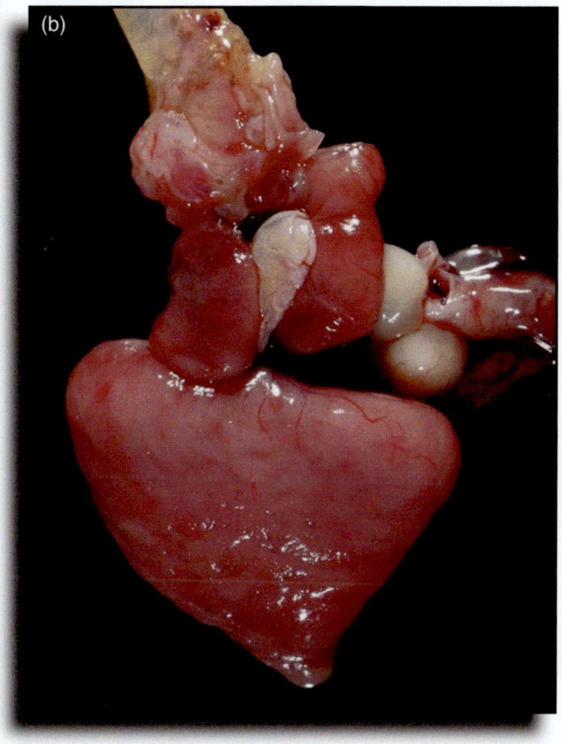

Figure 9.63 Bearded Dragon, *Pagona vitticeps.* Agamidae. Microsporidial aortic vasculitis: (a) dorsal view and (b) ventral view. (Courtesy of Arno Wuenschmann.)

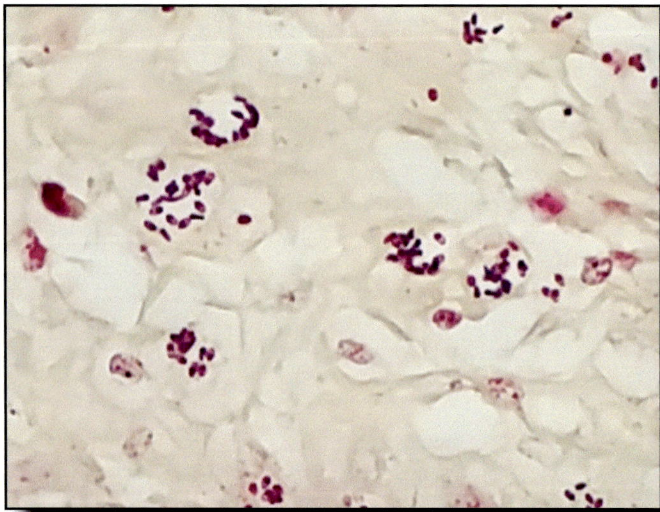

Figure 9.64 Bearded Dragon, *Pagona vitticeps*. Agamidae. Aorta. Note clusters of gram-positive microsporidial spores within the cytoplasm of macrophages. Brown and Brenn Stain. (Courtesy of Michael Garner.)

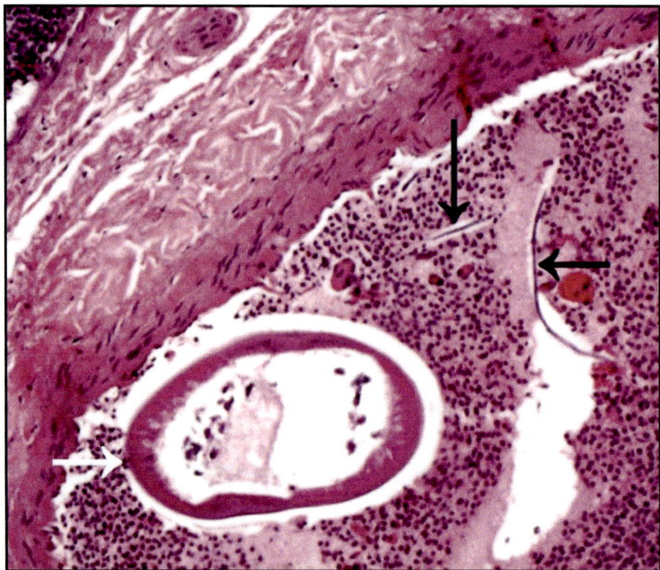

Figure 9.65 Mexican Beaded Lizard, *Heloderma horridum*. Helodermatidae. Adult filaria and microfilaria (black arrows) in a pulmonary vein. (From Orós J. 2011 *Colour Atlas of Diseases of Reptiles*, Editorial Inter-Medica S.A. With permission.)

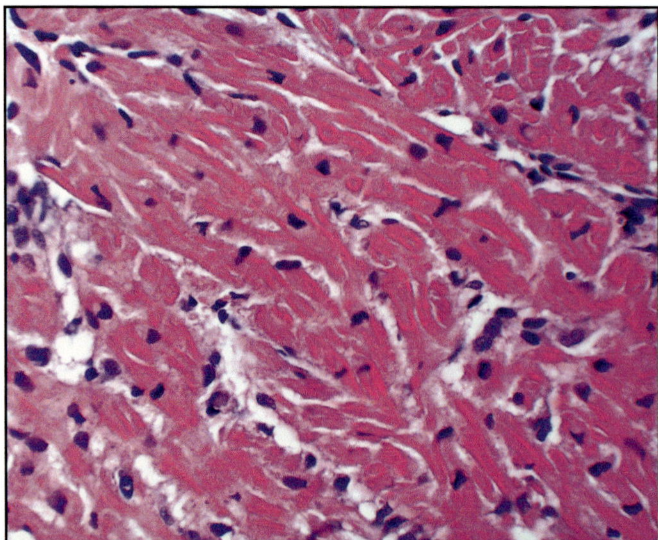

Figure 9.66 American Alligator, *Alligator mississippiensis*. Alligatoridae. West Nile virus infection. Note mild myocardial necrosis. (Courtesy of Elliott Jacobson.)

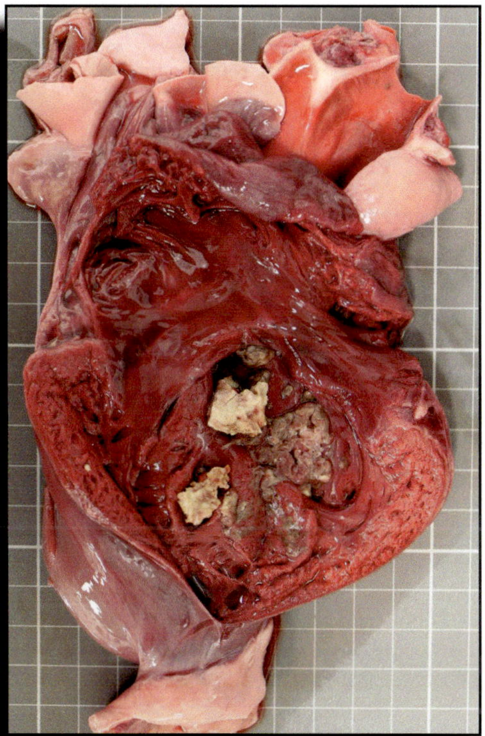

Figure 9.67 Nile Crocodile, *Crocodylus niloticus.* Crocodylidae. Bacterial endocarditis. Note exudate adhered to the endocardium of the right ventricle. (Courtesy of Juan Manuel Corpa.)

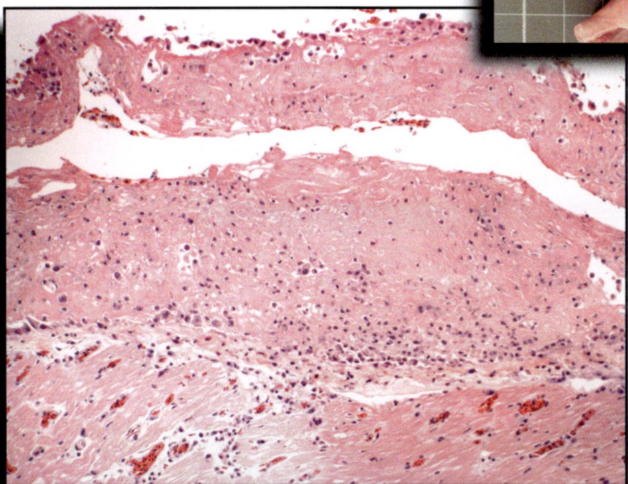

Figure 9.68 American Alligator, *Alligator mississippiensis*. Alligatoridae. Histologic images of mycoplasma-associated myocarditis and mycoplasma-associated epicarditis. Note mat of fibrin and inflammatory cells on epicardial surface. (Courtesy of Elliott Jacobson.)

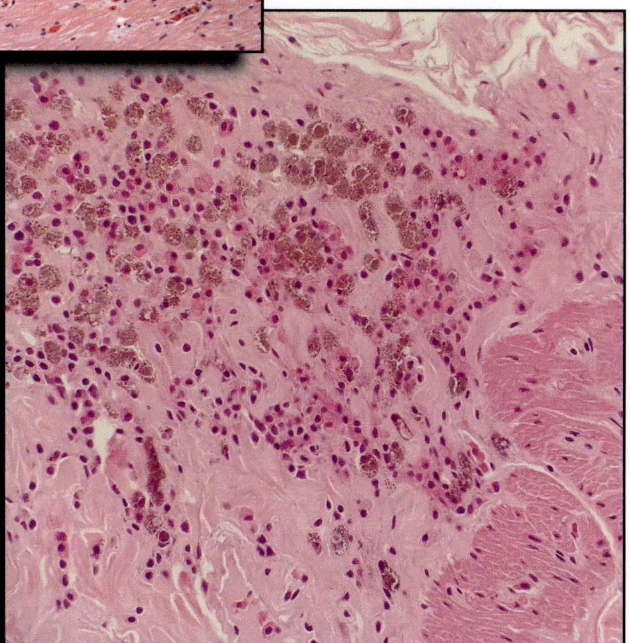

Figure 9.69 Gopher Snake, *Pituophis catenifer.* Colubridae. Epicardial hemosiderosis. Note infiltrate of histocytes containing intracytoplasmic brown pigment, and associated fibrosis attributed to chronic hemorrhage and trauma associated with cardiac blood draw. H&E. (Courtesy of Michael Garner.)

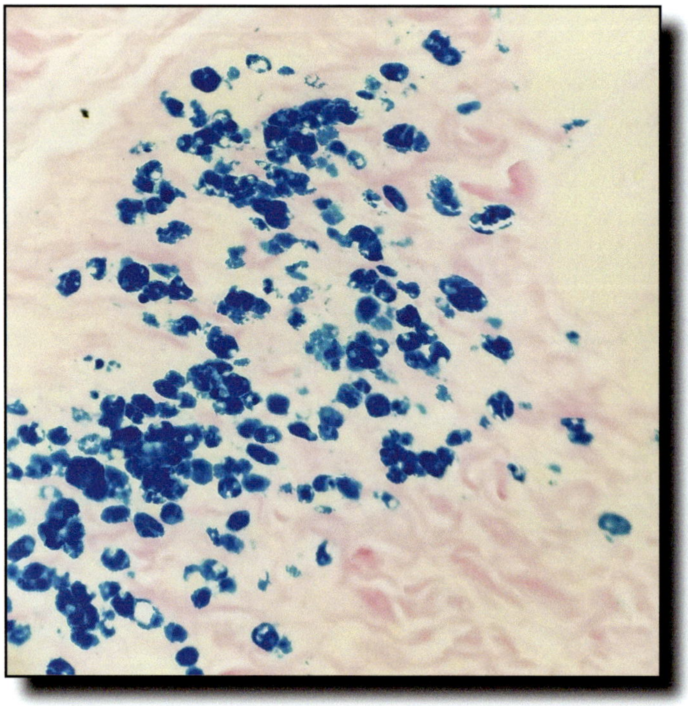

Figure 9.70 Gopher Snake, *Pituophis catenifer*. Colubridae. Epicardial hemosiderosis. Pigmented material in the histiocytes stains intensely positive for iron, consistent with hemosiderosis of chronic hemorrhage. Prussian blue. (Courtesy of Michael Garner.)

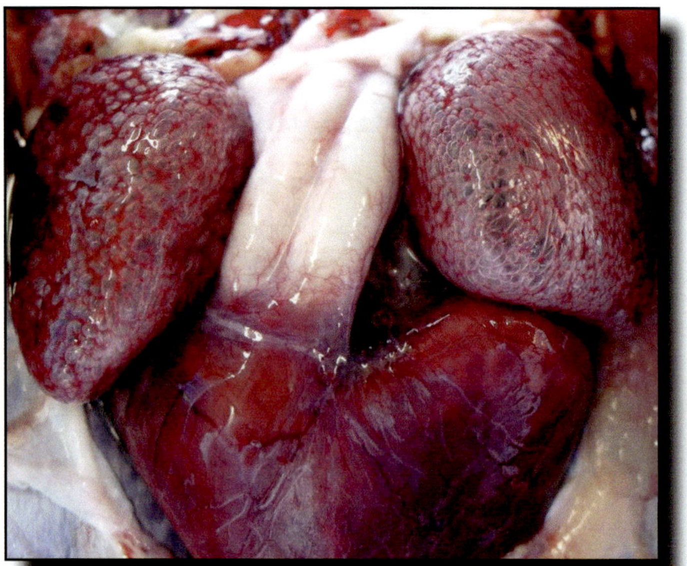

Figure 9.71 Loggerhead Turtle, *Caretta caretta*. Cheloniidae. Gas bubble accumulation within the atriae. (Courtesy of Oceanografic Valencia.)

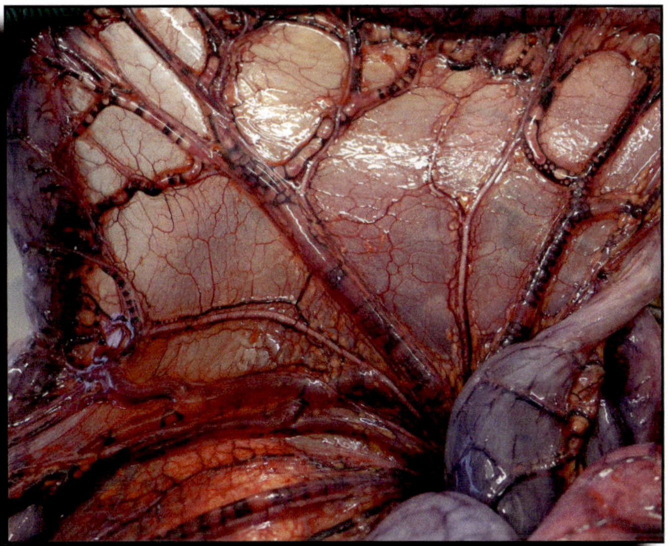

Figure 9.72 Loggerhead Turtle, *Caretta caretta.* Cheloniidae. Gas bubble accumulation within the mesenteric veins. (Courtesy of Oceanografic Valencia.)

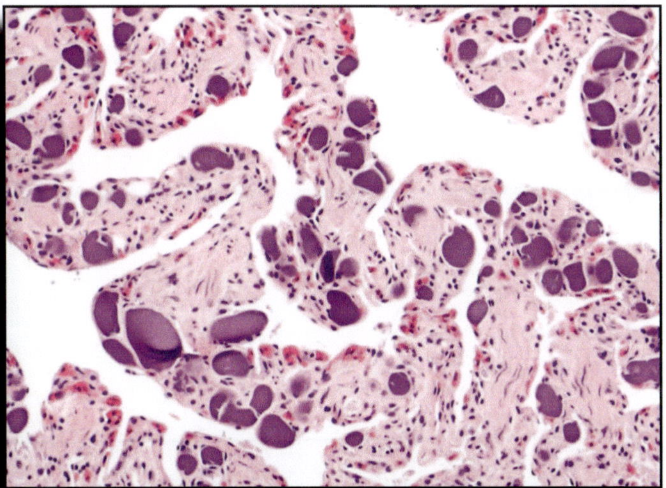

Figure 9.73 Corn Snake, *Pantherophis guttatus.* Colubridae. Yolk material within the pulmonary vessels. These patients have a peracute disease course and may die before the onset of histologic evidence of a tissue reaction. Note that pulmonary tissue is otherwise histologically within normal limits. (Courtesy of Anatomic Pathology Service, School of Veterinary Medicine, UC Davis.)

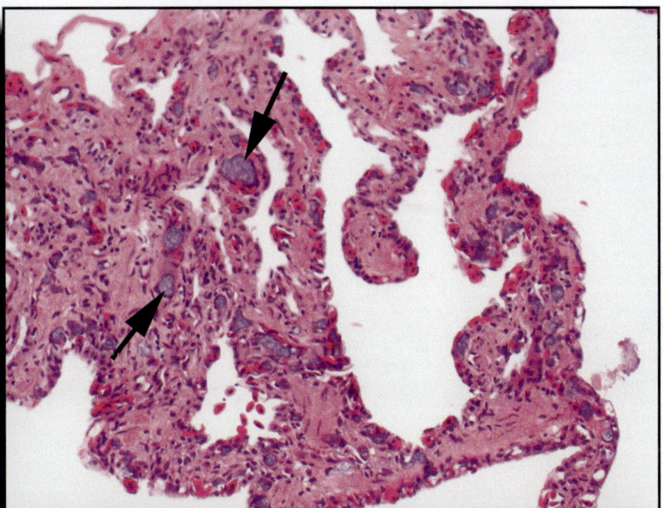

Figure 9.74 Bearded Dragon, *Pagona vitticeps.* Agamidae. Glycerin emboli in lung. Note extracellular amorphous basophilic material occluding the pulmonary capillaries (arrows), and lack of associated tissue response. (Courtesy of Michael Garner.)

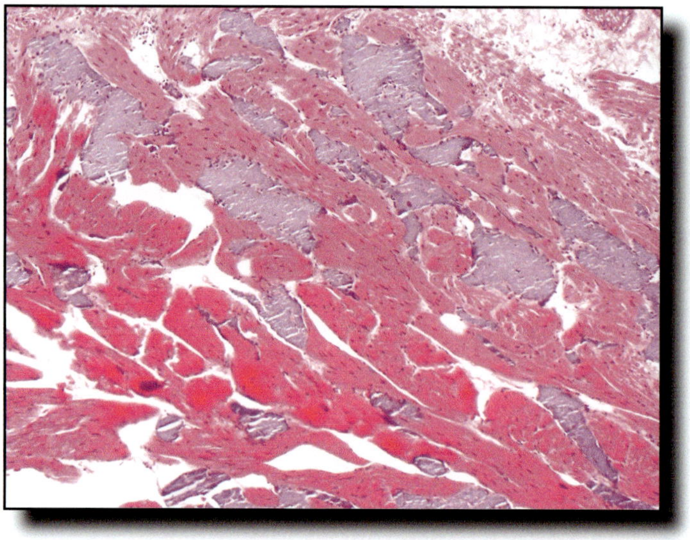

Figure 9.75 Bearded Dragon, *Pagona vitticeps*. Agamidae. Glycerin emboli in heart. Note accumulations of amorphous basophilic material in the ventricular sinuses. (Courtesy of Michael Garner.)

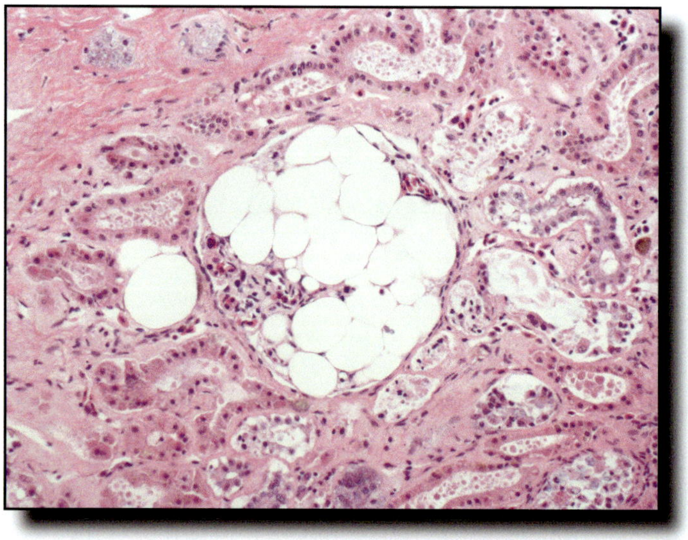

Figure 9.76 Anole, *Anolis* sp. Dactyloidae. Glomerular lipid or gas emboli. Note marked expansion of the glomerular capillaries by clear luminal spherical structures. A lipid stain would be needed to differentiate lipid from gas in this lesion.

10

BONE ALTERATION BY DISEASE

Its Appearance, Phylogeny, and Penetrance through Geologic Time

BRUCE ROTHSCHILD

Contents

10.1 Introduction and General Background

There are four traditional approaches to pathology: clinical, radiologic, histologic, and laboratory testing, each providing its own window into the nature of disease. To those approaches is added macroscopic examination of bones as isolated events, skeletal phenomenon, and as epidemiologic patterns (Rothschild and Martin, 2006). Disease presents as a spectrum, not only as to severity, but also as to the extent of skeletal involvement and penetrance of various signs. Identification of lesions in isolated bones has limits, as more than one disease can have the same appearance. Examination of skeletons allows more confidence in diagnosis, but it is the spectrum as represented in the population sample that is highly diagnostic (Rothschild and Martin, 2006).

Bone has a limited response repertoire, but osseous manifestations of disease are a trans-phylogenetic phenomenon, at least when de-fleshed bones are examined macroscopically. Names may differ, and there are semantic differences in their application across interdisciplinary fields. Collections of infectious detritus in mammals are referred to as abscesses; those in reptiles have been referred to as fibrisceses (Huchzermeyer and Cooper, 2000), although this term is not currently in common usage. Mammalian abscesses contain necrotic material with pus (neutrophilic leukocytes) and are referred to as pyogenic, while those in reptiles are described as masses of necrotic heterophilic leukocytes or fibrin (Huchzermeyer and Cooper, 2000); however, the appearance of the bone alterations is not distinguishable macroscopically (Rothschild and Martin, 2006), although that distinction has semantic overtones. Montali (1988) reported fibrin deposits surrounded by the same cells found in mammals and referred to it as a heterophilic granuloma. Sometimes referred to as pseudotumors (Elkin and Cooper, 1976) with caseous debris (Elkin, 1981), such obsolete descriptions are more characteristic of tubercular and other granulomatous diseases than pyogenic infections in mammals. For further information see Volume 1, Chapter 5.

10.2 Descriptives

Macroscopic recognition of bone lesions is predicated upon the character of alterations and patterns in individual bones in the whole individual and as part of a spectrum in the afflicted population. The approach is to identify what doesn't seem to belong, and what is missing. Bumps in bone are generally manifestations of stress fractures and infection (e.g., abscesses, by any other name); rarely, neoplasia. Stress fractures are located in highly reproducible locations, facilitating their recognition. Phalanges are especially prone to this affliction in dinosaurs, with specific location (i.e., manus or pes) determined by behavior. Fanciful illustration of sauropods (giant dinosaurs) rearing on their hind legs is documented as such by total absence of forefoot and vertebral stress fractures, while hindfoot involvement is common. Easily recognized radiologically on the basis of a linear cleft, macroscopic examination of the bone surface may reveal a cortical defect if there has been insufficient remodeling. Plastic deformation is the term used when remodeling is prominent. The differential consideration when a cortical defect (e.g., cleft) is recognized is osteomalacia (aka rickets), related to vitamin D deficiency. However, bumps are not found with vitamin D deficiency unless there has actually been a full fracture with callus formation. Bumps caused by stress fractures (**Figure 10.1**) tend to be focal and aligned across the direction of responsible forces, while infection-related bumps tend to be somewhat symmetrical, as do those related to benign neoplasia. Malignant neoplasia tends to produce focal bone enlargement, rather than an actual bump. Warped bones (revealing torque) may be a sign of vitamin D deficiency, but also may reflect other metabolic influences/inborn errors of metabolism (e.g., mucopolysaccharidosis). Heavy bones may reflect fluorosis, while light bones may indicate osteoporosis or be taphonomic (fossilization) in derivation. Alterations of joint surfaces include exogenous bone formation and defects. Osteophytes are localized lateral to the articular surface. Enthesial (connective tissue between tendon/ligament and bone) ossification/calcification generally parallels the long axis of the bone and follows the direction of stress forces. Calcification on articular surfaces (**Figure 10.2**) identifies crystalline arthritis, in the form of calcium pyrophosphate deposition and hydroxyapatite deposition disease. Defects in articular surfaces may reflect failure of endochondral bone formation from original cartilage, and can be a residuum of previous articular fracture or the result of compromised vascular supply to the subchondral bone, allowing its collapse.

10.3 Congenital Anomalies

Congenital abnormalities are relatively rare, in the absence of environmental stress (e.g., temperature), toxicosis or a genetic bottleneck. Brachycephaly, supernumerary limbs, and limb and tail loss are sporadically noted in crocodilians; cleft palate, in lizards (Jeweled Lizard [*Timon lepidus*, formerly *Lacerta lepida*] and European Common Lizard [*Zootoca vivipara*]), Savannah Monitor (*Varanus exanthematicus*), and snakes (European Viper [*Viper berus*]), duplications in modern turtles (**Figure 10.3**), and head and neck duplication (derodymus dicephaly) in Lower Cretaceous champosaur, vertebral and rib duplication in Jurassic *Aelodon* (*Crocodilus*) *priscus* and *Rhacheosaurus gracilis* and modern crocodilians (American Alligator [*Alligator mississippiensis*] and American Crocodile [*Crocodylus acutus*]) and Beaded Lizard (*Heloderma horridum*) and Gila Monster (*Heloderma suspectum*) (Buffetaut et al., 2006; Rothschild et al., 2012a,b,c). Tusk duplication has been recognized in mammal-like reptilian dicynodonts *Eodicynodon oosthuizeni*, *Emudops oweni*, and *Kannemeyeria siocephalus* (Fröbisch and Reisz, 2008; Vega and Maish, 2014) and congenital deficiency (unilateral tusks), in *Odontocyclops whaitsi* and *Tropidostoa dubium*.

Hemivertebrae have been noted as isolated phenomenon in the Gharial (*Gavialis gangeticus*), African Rock (*Python sebae*) and Burmese Pythons (*Python molurus*), European Grass Snakes (*Natrix* [formerly *Tropidonotus*] *natrix*), Yellow-Bellied Sea Snake (*Pelamis platura*), Racer (*Coluber* sp.), Ring-Necked Snake (*Diadophis punctatus*), Rat Snake (*Elaphe* sp.), King Snake (*Lampropeltis* sp.), Pine Wood Snake (*Rhadinaea* sp.), and Garter Snake (*Thamnophis* sp.) (Baur, 1891; Rothschild et al., 2012a,b,c). Among dinosaurs, a wedge-shaped vertebrae was found in an embryonic Cretaceous *Hypacrosaurus* (Duckbill Dinosaur) and hemivertebrae, in a Late Jurassic iguanodon *Dysalotosaurus*, and in the synchronic plesiosaur *Cimoliasaurus plicatus* (Lydekker, 1889; Janensch, 1934; Witzmann et al., 2008; Tanke and Rothschild, 2014), and congenital caudal fusion in the carnosaur *Allosaurus* (Madsen, 1976; Rothschild, 1997a). Spina bifida was observed in an Early Triassic mammal-like reptile, the therapsid *Myosaurus gracilis* (Fröbisch et al., 2016); a unilateral neural arch, in the Jurassic sauropod *Camarasaurus* (McIntosh et al., 1996); and hip dysplasia, in a beaked Triassic reptile, the rhynchosaur *Hyperodapedon sajuanensis* (Trotteyn and Martínez, 2013).

While rarely afflicted, turtles are subject to a variety of congenital defects ranging from duplication to absent or supernumerary bones or limbs, kyphosis, kinked tails, and especially to carapace and plastron deformities. A variety of duplication patterns have been observed in modern turtles: plastron to plastron, side to side, or posterior to posterior, as well as deformities in Miocene Sea Turtles *Caretta*, *Procolpochelys grandaeva*, and *Syllomus* (Zangerl and Turnbull, 1955; Weems, 1974). Bone deficiencies (frontal and septum orbitotemporale) were noted in the Eocene Side-Necked Turtle *Ummulisani rutgersensis* (Gaffney and Tong, 2008; Rothschild et al., 2012a,b,c). Analogously, fused scutes were present in an Upper Triassic eagle lizard, the crocodile-like aetiosaur *Paratypothorax andressorum* (Lucas, 2000). For additional information on congenital malformations, see Chapter 4

10.4 Osteochondrosis versus Osteochondritis

The terms *osteochondrosis* and *osteochondritis* are often used interchangeably in the human and veterinary medical literature (Rothschild and Martin, 2006; Rothschild et al., 2012a,b,c). Teleologically, they actually define very different processes. Osteochondrosis is a focal failure of cartilage to be replaced by (endochondral) bone during normal growth, leaving a "divot" (Resnick, 2002; Rothschild and Martin, 2006). Examination of affected bone reveals an articular surface defect with smooth, non-elevated, but sharply defined borders (Wegener and Heje, 1992; Thorp et al., 1995; Brogdon and Crotty, 1999, Resnick, 2002; Rothschild and Martin, 2006). As cartilage generally is not preserved in fossils (Briggs, 1991), it appears as a "divot" (**Figure 10.4**). Osteochondritis is the result of focal loss of bone vitality (aseptic avascular necrosis, to distinguish it from infection-related avascular necrosis) because of compromise of its blood supply (Resnick, 2002; Rothschild and Martin, 2006). This process results in irregular margins and bases (Rejno and Stromberg, 1978; Resnick, 2002; Rothschild and Martin, 2006). The issue has been even more complex in human knees. Most of what has been referred to as osteochondritis in human knees actually appears to be subchondral fractures (Vellet et al., 1991; Yu et al., 1998; Bohndorf, 1999; Brogdon and Crotty, 1999; Resnick, 2002; Rothschild and Martin, 2006).

Osteochondrosis has been recognized in the Orinoco Crocodile (*Crocodylus intermedius*) and American Crocodile (*C. acutus*) (Blanco, 1997) and artificially produced in modern reptiles by injection of calcitonin (a calcium regulatory hormone) (Belanger et al., 1973). It was present in the phalanges of 23% of *Edmontosaurus* hadrosaurs (Duckbilled Dinosaurs) (**Figure 10.4**) and has been observed in *Iguanodon* and sauropods *Janenschia robusta*, *Lufengosaurus*, *Brachiosaurus brunai*, and camarsaurids (Rothschild and Tanke, 2006; Xing et al., 2015; Tschopp et al., 2016).

10.5 Osteoarthritis or Osteoarthrosis

Osteoarthritis is a disease of diarthrodial joints. The semantics of its designation reflects tradition versus pathophysiology. Traditionally referred to as osteoarthritis, identifying a specific variety of arthritis, the "itis" in that designation is in conflict with perspectives of its pathophysiology. Recognized as a non-inflammatory form of arthritis (in distinction to the very inflammatory rheumatoid arthritis), the "itis" suffix is teleologically suspect. Thus, some suggest the term osteoarthrosis to emphasize its non-inflammatory nature.

Osteoarthritis is recognized macroscopically in bone on the basis of development of osteophytes, outgrowths perpendicular to their articular surface (Resnick, 2002; Rothschild and Martin, 2006), and therein has been a major source of confusion with respect to mammalian pathogenesis. Mammals have intervertebral disks separating vertebrae, and those disk junctions are technically not diarthrodial joints. This is in contrast to zygapophyseal and costovertebral joints which are diarthrodial, and at least the former is occasionally subject to arthritis (Resnick, 2002; Rothschild and Martin, 2006). As vertebral centra osteophytes do not involve diarthrodial joints, they do not actually identify arthritis, let alone osteoarthritis. The term spondylosis deformans more actually describes the phenomenon of vertebral centra osteophytes. Spondylosis deformans appears to be a phenomenon of aging (Rothschild, 2015a). It does not produce symptoms or disability unless the bone overgrowth actually impinges on the spinal cord or spinal nerves. The issue is more complex in reptiles, many of which actually have synovial-lined intervertebral "diarthrodial" joints rather than disks separating vertebral centra (Rothschild et al., in press). However, the direction of overgrowth still is diagnostic. Overgrowths of bone that are parallel to vertebral endplates identify osteophytes and therefore osteoarthritis (**Figure 10.5**). Many of us (including the author of this chapter) long thought that dinosaurs and the contemporary marine reptiles (at least mosasaurs) had disk spaces and therefore had spondylosis deformans (Rothschild, 1990; Brochu, 2003). It is now known that they had synovial-lined joints and therefore that the osteophytes were actually indicative of osteoarthritis of the spine—although still presuming them to be asymptomatic (Rothschild, 2015a). Thus, past study of osteoarthritis in dinosaurs was limited to the peripheral skeleton and it was identified in only two of the large number of individuals, genera, and families examined (Rothschild, 1990). The only discovered peripheral joint osteoarthritis was in the ankles of two individuals from a herd of 39 *Iguanodon* discovered in a coal mine a mile beneath the surface of the city of Brussels. This low prevalence is analogous to that of mammals. Only 1% of free-ranging mammals manifest osteoarthritis of peripheral joints, an affliction that is quite common in captive conspecifics (Rothschild and Woods, 1993; Rothschild et al., 1999; Rothschild, 2003). Primary osteoarthritis is rare in free-ranging animals, in the absence of injury or other form of joint damage or malposition. The lone example in a sauropod is that of the *Apatosaurus* (formerly, and now perhaps again, called *Brontosaurus*) "Bertha" (Rothschild and Feldkeller, 2016). The other alleged osteoarthritis in the fossil record is that suggested in a plesiosaur (Wells, 1964), but that attribution is uncertain as it was neither described nor illustrated.

10.6 Diffuse Idiopathic Skeletal Hyperostosis

Diffuse idiopathic skeletal hyperostosis (DISH) has been recognized in sauropods (i.e., 50% of *Diplodocus* and *Apatosaurus*, formally called *Brontosaurus*, and 25% of *Camarasaurus*) (**Figure 10.6**) and in a lambosaurine hadrosaur (duckbill dinosaur) (Rothschild, 1987a; Rothschild and Berman, 1991; Tanke and Rothschild, 2014). This tendency to ossification of ligaments, tendon attachments, and capsules most prominently presents as ossification of the anterior longitudinal ligament of the vertebral column. It is of unknown cause, although it can be mimicked by hypervitaminosis A. DISH is associated with Haversian bone; while hypervitaminosis A, with wormian bone. The latter characteristic of is intrasutural bones or sutural bones, which are extra bone pieces that can occur within a suture (joint) in the skull.

10.7 Stress Fractures

Stress fractures (**Figure 10.1**) are common in ceratopsians, predominantly affecting foot phalanges, but also noted in their jugals (Rothschild, 1988a; Rothschild and Martin, 2006; Rothschild and Tanke, 2006; Tanke and Rothschild, 2010). Manual theropod stress fractures are occasionally noted in tyranosaurids *Albertosaurus*, *Tarbosaurus*, and *Tyrannosaurus* (Rothschild et al., 2001a; Rothschild and Molnar, 2008); pedal, in sauropods *Apatosaurus*, *Barosaurus*, *Brachiosaurus*, *Camarasaurus*, *Diplodocus*, and *Nurosaurus* hindfoot phalanges (Rothschild and Molnar, 2005), and in both manus and pes of *Allosaurus*, dromaeosaurids, *Saurornitholestes*, and *Chirostenotes* (Rothschild et al., 2001a; Rothschild and Tanke, 2005). They have also been noted in the ribs of the mosasaur *Prognathodon* (Schulp et al., 2004)

10.8 Exostoses, Osteochondroma, and Avulsions

Exostosis and avulsions are present in the humerus of the *Tyrannosaurus* "Sue" (Rothschild and Tanke, 2005) and in *Albertosaurus* (Molnar, 2001). Osteochondroma were noted in tyrannosaurids, including *Gorgosaurus* (Rothschild and Tanke, 2005; Rothschild and Molnar, 2008).

10.9 Infection

Infections affecting bone usually produce quite distinctive changes in afflicted areas. These include cortical defects, disorganization of trabeculae in underlying bone, and new bone formation on the bone surface. These are not necessarily pathognomonic, in contrast to filigree surface reaction patterns which appear to be specific to the diagnosis of infection. Some infections produce very characteristic bone alterations, sometimes even pathognomonic for a specific infection (e.g., brucellosis) or a specific variety of infection (e.g., fungal disease) (Resnick, 2002; Rothschild and Martin, 2006). Destruction of articular surfaces may produce erosion, flail joints, or ankylosis.

Snakes and lizards occasionally develop calcific/osseous masses adjacent to and adherent to ribs and vertebrae (**Figures 10.7–10.8**). Evaluation of a similar lesions in the Blue-Tail Monitor (*Varanus doreanus*) identified this phenomenon as a complication of osteomyelitis of the adjacent rib(s) (Rothschild, 2014). It is likely that the phenomenon in snakes has a similar origin, suggesting the result of penetrating trauma or hematogenous spread of a septic process.

The cause of osteomyelitis in turtle carapace and plastron (**Figure 10.9**) has been difficult to assess and problematic for distinguishing infection from simple colonization. Pyogenic bacteria (e.g., *Pseudomonas*, *Citrobacter*, *Klebsiella*), mycobacteria, fungi (e.g., *Mucor*), and algae have been isolated (Jacobson, 1994; Arvy and Fertard, 2002; Barnett, 2003). Damage, commonly referred to as shell rot, has thus been attributed to *Mucorales*, *Rusarium*, *Geotrichum*, *Trichosporon*, *Caniothyrium*, algae, and *Beneckea chitinovora*, and most recently *Emydomyces testavorans* (Garner et al., 1997; Woodburn et al., 2019). For specifics as to the breadth and depth of shell disease, see Rothschild et al. (2012a,b,c). For additional information on infectious causes, see Volume 1, Chapters 11 and 12.

Skeletal and joint infections have been similarly attributed to specific organisms, their isolation precluding surface contamination, allowing more confidence in attributing causation to pyogenic bacteria, mycobacteria, and fungi (Rhodin et al., 1990; Jacobson, 1994). Spinal infection in snakes attributed to *Salmonella arizonae* infection is especially common (**Figures 10.7–10.8**) (Isaza et al., 2000; Grupka et al., 2006). Peripheral abscesses and joint erosions have been noted in a Brown Tortoise (*Manouria emys*), Leatherback Turtle (*Dermochelys coriacea*), Radiated Tortoise (*Astrochelys radiata*), Agassiz's Desert Tortoise (*Gopherus agassizii*), Atlantic Ridley Turtle (*Lepidochelys kempii*), Indian Star Tortoise (*Geochelone elegans*), and Big-Headed Pantanal Swamp Turtle (*Acanthochelys macrocephala*) (Ogden et al., 1981; Jacobson, 1994; Rothschild et al., 2012a,b,c). Infections in Atlantic Ridley Turtles were attributed to *Mycobacterium chelonae* and *Nocardia* (Harms et al., 2002; Greer et al., 2003).

Infections are so common in modern reptiles as to diminish their being reported. Their apparent infrequency in the Mesozoic has perhaps made for more facile publication of what might otherwise be considered case reports (Tanke and Rothschild, 2002, 2014; Burnham et al., 2012; Redelstorff et al., 2014).

The oldest examples of osteomyelitis were in a Permian Captorhinid *Labidosaurus hamatus* and a Triassic Dicynodont (Synapsida, Therapsida) *Stahleckeria potens* and

Rhynchocephalian Tuatara (*Sphenodon punctatus*) ancestor *Clevosaurus brasiliensis* (Reisz et al., 2011; Vega and Maisch, 2014; Romo-de-Vivar-Martínez et al., 2016). Infections were localized to the mandible in these and in more recent reptiles, an Upper Jurassic pliosaur *Pliosaurus* and a Paleozoic reptile *Lapidosaurus hamatus* (Reisz et al., 2011; Sassoon et al., 2012). A tuberculosis-mimicking infection was noted in Middle Triassic reptiles *Proneusticosaurus silesiacus* and *Pistosaurus longaevus* (Surmik et al., 2017; submitted-a).

Dinosaur lesions often elicit publication, perhaps explaining more numerous citations. Infections were commonly noted in theropods (e.g., *Allosaurus*, *Deinonychus antirrhopus*, *Herrerasaurus*, and *Tyrannosaurus*) feet, often associated with fractures (Laws, 1995; Marshall et al., 1998; Molnar, 2001; Hanna, 2000; Tanke and Rothschild, 2002; Vittore and Henderson, 2013). The theropod *Dilophosaurus* was also subject to peripheral joint infection, and an osteomyelitis-related cauliflower-like overgrowth in an *Allosaurus* humerus (Rothschild and Tanke, 2005) was originally mistaken for cancer (Taylor, 1992). Osteomyelitis has been reported in Late Cretaceous tyrannosaurids *Daspletosaurus* and *Tyrannosaurus*, hadrosaurs, hypsilophodontids, sauropods *Apatosaurus* and *Baurutitan brito*, the ceratopsian *Pachyrhinosaurus* vertebrae, *Camptosaurus* ilium (penetrating injury suspected), and in stegosaur spikes (Moodie, 1923; Gross et al., 1993; Rothschild, 1997a; Carpenter, 1998; Williamson and Carr, 1999; McWhinney et al., 2001; Tanke and Rothschild, 2002, 2010, 2014; Rothschild and Molnar, 2008; Anné et al., 2016; García et al., 2016).

Facial damage from conspecific interactions is common in the tyrannosaurids, but infection has yet to be documented. Rega and Brochu (2001) and Wolff et al. (2009) erroneously claimed fungal infection as responsible for mandibular holes in the *Tyrannosaurus rex* Sue, but such lesions are common in that species. In contrast to the expansile nature of fungal disease, the lesions in *Tyrannosaurus rex* are simply holes with infilling by new bone formation, indistinguishable from healing trephinations in humans (Rothschild, 2010; Verano, 2016) and from vascular channels. Infection, however, has been noted in a parietal bone of a Troodon skull (parietal) (Rothschild, 1997a). Jaw infections in mosasaurs *Mosasaurus* may occur in isolation (Schulp et al., 2006) or associated with fracture (Lingham-Soliar, 2004) and infection is responsible for some instances of vertebral fusion in mosasaurs (Rothschild and Everhart, 2015).

10.10 Avascular Necrosis

Necrosis of bone from compromised circulation produces radiolucent zones on x-ray and subsidence of overlying articular surfaces (**Figure 10.10**), both of which identify diving behavior in Mesozoic mosasaurs, ichthyosaurs, and plesiosaurs and their prey, turtles (Rothschild and Martin, 1987; Rothschild, 1988b; Martin and Rothschild, 1989; Rothschild and Storrs, 2003; Rothschild, 2012).

It was first recognized in mosasaurs and proved to be genus rather than locality dependent (Rothschild, 1988b, 2008b). Prevalence in Kansas, Alabama (US), and Belgium mosasaurs was identical. If a genus was afflicted, all adults of that genus were affected. *Platecarpus* had the highest proportion of affected vertebrae, approximately five times that of the other affected genera, including *Tylosaurus*, *Mosasaurus*, *Plioplatecarpus*, *Prognathodon*, and *Hainosaurus*. *Clidastes*, *Ectenosaurus*, *Halisaurus*, *Kolposaurus*, and *Globidens* were spared this disorder (Rothschild, 1988b, 2008b). Basal sauropterygians were mostly spared (Surmik et al., 2017), while the more advanced (and later) plesiosaurs (Cryptoclididae excepted) were commonly affected (Rothschild and Storrs, 2003).

Attribution of avascular necrosis to decompression syndrome from diving was confirmed by its absence in terrestrial turtles and its presence only in marine turtles, including *Trionyx* from the Mediterranean (Rothschild, 1988b). Present in one of the oldest turtles, *Odontochelys* (Rothschild and Naples, 2015), it was common in the Cretaceous (Rothschild, 1991). Prevalence ranged from 41% in the Cretaceous (affecting Desmatochelyidae, Toxochelyidae, Protostegidae, and Pleurosternidae), to 9% in the Eocene and 5% in the Oligocene, becoming negligible in the Holocene (0.3%) (Rothschild, 1987b; 1988a; 1991). Currently, classic evidence of avascular necrosis has been found as an isolated phenomenon limited to Ridley's Turtle (*Lepidochelys olivacea*), Common and Alligator Snapping Turtles (*Chelydra serpintina* and *Macroclemys temmincki*, respectively), and Mud Turtles (*Kinosternon* spp.) (Eckert et al., 1986, 1989; Rothschild, 1987b).

10.11 Metabolic Disease

Reports of metabolic disease have predominantly been related to vitamin D deficiency as might occur with osteomalacia, parathyroid disease, crystalline forms of arthritis (predominantly gout and calcium pyrophosphate deposition disease) and Paget's disease.

10.11.1 Vitamin D–Related Disease

Osteomalacia and rickets (the term used prior to epiphyseal closure) are predominantly disorders of captive animals, related to vitamin D deficiency, although non-vitamin-related dietary exposure (e.g., phosphorus excess or calcium deficiency) and renal disease may have a causal role (Frye, 1994; Rothschild, 2008a; Rothschild et al., 2012a,b,c). Bone becomes softened and less resistant to biomechanical factors, resulting in deformity, bone resorption, and fibrous metaplasia (**Figure 10.13**) (Wallach, 1971; Glazebrook, 1980). Epiphyseal prominence is also noted. Such has been reported in Painted Turtles (*Chrysemys picta*), Ornate Box Turtles (*Terrapene ornata*), Common Snapping Turtle, Red-Eared Sliders (*Trachemys scripta elegans*), South

American Red-Legged Tortoises (*Chelonoidis carbonarius*, Leopard Tortoises (Stigmochelys [formerly *Geochelone*] *pardalis*), African Spurred Tortoises (*Centrochelys* [formerly *Geochelone*] *sulcate*), Mediterranean Spur-Thighed Tortoises (*Testudo graeca*), Hermann's Tortoises (*Testudo hermanni*), and Aldabra Tortoises (*Aldabrachelys* [formerly *Dipsochelys*]) (Hunt, 1957; Reichenbach-Klinke and Elkan, 1965; Reichenbach-Klinke, 1977). For additional information on nutritional bone disease, see Chapter 2.

10.11.2 Parathyroid-Related Disease

Tumors or hyperplasia of the parathyroid gland have been suggested as causes for carapace and plastron deformity in South American Red-Legged, Mediterranean Spur-Thighed, and Agassiz's Desert (*Gopherus agassizii*) Tortoises (Frye and Carney, 1975; Frye, 1981).

10.11.3 Paget's Disease

Paget's disease was apparently present in a Jurassic dinosaur (**Figure 10.11a,b**) (Witzmann et al., 2011); however, the mosaic pattern of bone development attributed to Paget's disease in humans is a normal pattern of bone development in reptiles and should not be overinterpreted as an abnormality in the histogenesis of bone (**Figure 10.12**).

10.11.4 Gout

Gout is a systemic disease that takes two forms that are usually mutually exclusive: visceral (kidney and other visceral organ urate crystal deposition) or as a disease of joints. It is recognized on the basis of classic erosions with sclerotic margins, appearing punched out, but with an unusual rim, different in color and texture from the surrounding bone (**Figure 10.13**). Crocodilians (American Alligator [*Alligator mississippiensis*], Spectacled Caiman [*Caiman crocodilus*, formerly *Caiman sclerops*], American Crocodile [*Crocodilus acutus*], Morelet's Crocodile [*C. moreletii*], New Guinea Crocodile [*C. novaeguineae*], Freshwater Crocodile [*C. johnsonii*], Nile Crocodile [C. *niloticus*] and Indo-Pacific Crocodile [*C. porosus*], False Gharial [*Tomistoma schlegelii*], and true Gharial [*Gavialis gangeticus*]) are especially prone to articular gout. Articular gout has also been found in Tegu Lizards (*Tupinambis* spp.), Savannah Monitor (*Varanus exanthematicus*), Green Iguana (*Iguana iguana*), African Fat-Tailed Gecko (*Hemitheconyx caudicinctus*) (**Figure 10.16**), and turtles. The latter include Mediterranean Pond Turtle (*Mauremys leprosa*), Red-Eared and Yellow-Bellied Slider (*Trachemys scripta elegans* and *T. s. scripta*, respectively) turtles, Radiated Tortoise (*Geochelone radiata*), African Spurred Tortoise, California Desert Tortoise, Rain Forest (*Chelonoidis* [formerly *Testudo*] *denticulata*), Turkish (Spur-Thighed), and Hermann's Tortoises (Rothschild, 2008a; Rothschild et al., 2012a,b,c). Gout has been recognized in tyrannosaurs, including *Tyrannosaurus rex* (Rothschild et al., 1997). For additional information on gout, see Chapters 2 and 3.

10.11.5 Calcium Pyrophosphate Deposition Disease

Calcium pyrophosphate deposition disease (CPPD), also referred to as articular pseudogout or calcinosis circumscripta, is a disorder involving deposition of calcium crystals (**Figure 10.2**). In humans, it presents in a dichotomous manner, with erosive disease and with deposition of new calcific material on joint surfaces (Rothschild et al., 1992). The erosions have a crumbling appearance, facilitating distinguishing them from those of rheumatoid arthritis and spondyloarthropathy.

CPPD can be a primary disorder, familial, or can occur secondary to parathyroid and thyroid disease, hypomagnesemia, and certain inborne errors of metabolism (e.g., homocystinuria). CPPD in the Komodo Dragon (*Varanus komodiensis*) (Rothschild et al., 2012a,c) appears to be a phenomenon occurring solely as a complication of another form of arthritis, spondyloarthropathy (see next section). Involvement in other reptiles has not been so related. These have included an Egyptian Spiny-Tailed Lizard (*Uromastyx* sp.), Red-Eared Slider, Cooter (*Pseudemys* spp.), Agassiz's Desert Tortoise, and Mediterranean Spur-Thighed Tortoise (Frye and Dutra, 1976; Brogard, 1987; Frye, 1994; López del Castillo, 1998; Rothschild et al., 1992, 2012a,b,c,d; Rothschild and Bruno, 2020).

The macroscopic findings on bone examination are indistinguishable from those of hydroxyapatite deposition, identifiable on crystalline examination and found in Red-Belly Short-Necked Turtles (*Emydura subglobos* [formerly *Emydura albertisii*]) (Wenker et al., 1999). For additional information on pseudogout, see Chapter 3.

10.12 Spondyloarthropathy

Spondyloarthropathy is the term for a specific variety of arthritis, one that past publications often failed to distinguish from rheumatoid arthritis, and spondyloarthropathy was not listed in their differential considerations as an alternative to the rheumatoid (mis)-diagnosis. "Rheumatoid arthritis" has often been used as a generic term for any inflammatory arthritis (Carr and Michels, 1997; Rothschild and Martin, 2006). Thus, literature requires careful examination, not for diagnosis, but rather for the manifestations. Close review reveals no actual evidence of rheumatoid arthritis, exclusive of the human condition (Rothschild and Martin, 2006). The damage reported by Carr and Michels (1997) is quite different from what is observed in human rheumatoid arthritis (Rothschild and Martin, 2006). The noted reactive new bone formation and subchondral erosions are not found among humans with rheumatoid arthritis, but actually are attributable to another disease, spondyloarthropathy (Rothschild and Woods, 1991; Rothschild, 1997b).

The challenge is also exacerbated by terminology. Spondyloarthropathy also has been referred to as "spondyloarthritis," and even as "spondarthritis" (Cisneros et al., 2010), although the latter term has been applied to spinal osteophytes and even infection (Rothschild, 2008a; Rothschild et al., 2012a,b,c). The axial skeletal involvement of spondyloarthropathy is quite unique both in distribution and character. Targeting of zygapophyseal and costovertebral joints with erosion and fusion is only otherwise noted as an isolated phenomenon with infection. When present as a diffuse phenomenon (affecting multiple joints), such involvement is classic and diagnostic for spondyloarthropathy (Rothschild and Woods, 1991). The classic vertebral centra finding is ossification of the anulus fibrosus, resulting in fusion/ankylosis. Peripheral joint involvement is characterized by erosions of the subchondral articular surface (**Figure 10.14a,b**).

Spondyloarthropathy is rare in modern reptiles, with the exception of Crocodylidae and Varanidae, especially the Crocodile Monitor (*Varanus salvadorii*), Gould's Goanna (*Varanus gouldii*), and the Komodo Dragon (Rothschild, 2008a; Rothschild et al., 2012a,b,c). In these animals, a condition resembling hypertrophic osteopathy may occur, with or without visceral abscesses or neoplasia (**Figure 10.15a–c**). Shoulder involvement reported as osteoarthritis by Raidal et al. (2006) in a Loggerhead Turtle (*Caretta caretta*) is likely spondyloarthropathy, on the basis of the subchondral erosions illustrated (Rothschild et al., 2012a,b,c; Rothschild, 2015b). An interesting form of ascending idiopathic proliferative osteoarthropathy occurs in female Green Iguanas. The lesion is first seen in the mid to distal caudal vertebral level and over time ascends to involve the pelvic and lumbar vertebrae. The disease is associated with marked fusion of articular surfaces, ankylosis, and formation of circumferential zones of reactive subperiosteal bone and cartilage. The cause for the disease has not been determined, but its occurrence only in females suggests hormonal derangements may contribute to the pathogenesis (**Figure 10.16a–e**) (Innis and Garner, 2006).

The representation of spondyloarthropathy in the fossil record is extensive, including crocodiles *Tomistoma dowsoni* in the Miocene and *Leidyosuchus formidabilis* in the Paleocene and related phytosaur *Angistorhinus ruetimeyeri* in the Triassic (Ruffer, 1917; Sawyer and Erickson, 1998; Rothschild, 2008a; Witzmann et al., 2014). It has been recognized in *Dimetrodon*, *Diadectes*, and *Ctenorhachis* from the Permian (Rothschild, 2008b).

Vertebral fusion is present in sauropods *Apatosaurus* and *Uberabatitan ribeiroi* (Lovelace, 2014; Martinelli et al., 2014; Redelstorff et al., 2015); zygapophyseal erosions, in *Camarasaurus* (Rothschild et al., 2002, 2013); and syndesmophytes in hadrosaurs and ceratopsia (Tanke and Rothschild, 2010, 2014) identifying Mezosoic spondyloarthropathy in dinosaurs, as was recognized in Cretaceous marine lizards, the mosasaurs (Rothschild and Everhart, 2015), and a large, basal archosaurian Lower Triassic reptile (Cisneros et al., 2010). Fusion of caudal vertebrae was noted in the theropod *Neovenator* and syndesmophytes were noted in the *Tyrannosaurus rex*, affectionately named Sampson (Tanke and Rothschild, 2002). It was quite prevalent in the short-lived (relative to geologic time) Paleocene (60 million ybp) corythodon (mammal) *Barylambda faberi* (Rothschild et al., 2002). It was relatively rare in Oligocene mammals, with the exception of rhinoceros (5%) and horse (1%) ancestors in the Oligocene, increasing geometrically through the Miocene to the Pleistocene, and today afflicts 35% of rhinoceros (independent of species) and 8% of horses (Rothschild et al., 2001b). Vertebral fusion/ankylosis is occasionally noted in modern day squamates, especially snakes, likely related to infection (**Figure 10.5**).

10.13 Benign Neoplasia

Benign tumors affecting bone in extant reptiles are uncommonly reported, usually as isolated observations. The record in modern reptiles includes cartilaginous tumors in the Bengal Monitor (*Varanus bengalensis* [formerly *Varanus dracaena*]) and Mangrove Monitor (*Varanus indicus*), enchondromas in the Rhinoceros Iguana (*Cyclura cornuta*), European Green Lizard (*Lacerta viridis*), and African Rock Python, chondromas in the European Green Lizards, chondro-oseofibroma in Bengal Monitors, Nile Monitor (*V. niloticus*), Green Iguana, Caiman Islands Ground Iguana (*Cyclura nubila*), Hardwick's Spiny-Tailed Lizard (*Uromastyx hardwickii*), Black-Neck Spitting Cobra (*Naja nigricollis*), osteochondroma in Hermann's Tortoise and the extinct Middle Triassic dicynodont, *Stahleckeria potens* (Stolk, 1958; Rothschild et al., 2012a,b,c; Vega and Maisch, 2014) tracheal chondromas in Ball Pythons (*Python regius*) and osteomas in the Water Monitor (*Varanus salvator*), European Green Lizards, and Hardwick's Spiny-Tailed Lizard (Drew et al., 1999; Rothschild et al., 2012a,b,c).

Older reports in the fossil record are more complex to assess. Terminology and more recent analyses of alleged disease in modern animals has provided clarification requiring reconsideration of some past impressions. Moodie's (1918) assertion that an osteoma was present on a mosasaur vertebra is now recognized as actually a hamartoma (Rothschild, 2008b).

Histological examination of a tumorous bony growth that was originally interpreted by Weems (1974) in the Miocene Sea Turtle (*Syllomus aegyptiacus*) actually had a characteristic fine trabecular pattern characteristic of an osteoblastoma (Resnick, 2002; Rothschild et al., 2012a,b,c).

Radiologic surveys, however, have been most informative. Systematic evaluation of North American dinosaur vertebrae revealed lesions limited to one family, hadrosaurs, in which certain diseases were surprisingly common. Isolated instances of osteoma, osteoblastoma, and desmoplastic fibroma were found, but hemangiomas were present in 5% of hadrosaurs (i.e., *Edmontosaurus*) surveyed (Rothschild et al., 2003). Subsequently, both osteoma and hemangioma were identified in a sauropod (Barbosa et al., 2016) and an ameloblastoma, in a *Tyrannosaurus* (Dumbrava et al., 2016).

10.14 Malignant Neoplasia

Malignant tumors are commonly reported in modern squamates but are uncommon in chelonians and rare in crocodilians; however, malignant tumors of bone and cartilage are uncommon in all groups. Examples include a chondrosarcoma in a Red Ratsnake (*Pantherophis guttatus* [formerly *Elaphe guttata*]), osteosarcoma in California Mountain Kingsnake (*Lampropeltis zonata*), Burmese Python, Rufous Beak Snake (*Rhamphiophis rostratus*), European Green Lizard, and Green Iguana, osteochondrosarcoma in Black-Neck Spitting Cobra, melanoma in a Pine Snake (*Pituophis melanoleucus*), a neurilemmal sarcoma in a Hermann's Tortoise, and lymphoblastic lymphoma in the Loggerhead Turtle (Cooper et al., 1983; Rothschild et al., 2012a,b,c). Metastatic carcinoma was found in a Common Golden Tegu (*Tupinambis tequixn*) and Hermann's Tortoise (Cooper et al., 1983; Orós et al., 2001; Mauldin and Done, 2006; Rothschild et al., 2012a,c). Metastatic cancer in a Green Iguana was accompanied by diffuse periosteal reaction involving the distal diaphysis of the long bones, a known cancer-related disorder referred to as hypertrophic osteoarthropathy (Rothschild and Martin, 2006).

The fossil record is more succinct. Radiologic skeletal surveys revealed no difference in prevalence from current rates (Natarajan et al., 2007). There is one instance of metastatic carcinoma in a hadrosaur *Edmontosaurus* (Rothschild et al., 2003). There is also a lapidary example, wherein a slice of bone from an unknown dinosaur had damage attributable to metastatic cancer (Rothschild and Tanke, 1999). The *Allosaurus* humeral cauliflower-like overgrowth attributed by Taylor (1992) to cancer was on closer examination identified as osteomyelitis (Rothschild and Tanke, 2005). For additional information on benign and malignant neoplasia of bone and cartilage, see Chapter 1.

10.15 Summary

Bone disease is transphylogenetic in character. While there is variation in some of its pathophysiology, the end results are indistinguishable between upper and lower vertebrates and have not been substantially altered through geologic time. Trauma, infection, metabolic disease, and neoplasia have species specificity as to rate of affliction (prevalence), but their manifestations are unchanged through time (geologic) and space (phylogeny). While many of the diseases are recognizable on macroscopic visual examination of skeletons, radiologic examination often provides etiologic confirmation and identifies additional occurrence. Noting that macroscopic evidence of metastatic disease is found in approximately one-fourth of individuals with cancer, it is pertinent to note that radiologic examination identifies a significant number of additional occurrences (Rothschild and Martin, 2006). Systematic skeletal radiologic examination is recommended to establish true population disease penetrance.

References

Anné J, Hedrick BP, and Schein JP. 2016. First diagnosis of septic arthritis in a dinosaur. *R Soc Open Sci* 3:160222. Doi: 10.1098/rsos.160222

Arvy C, and Fertard B. 2002. Pathologie des tortues: étude synthétique. *Bulletin de la Société Herpétologique de France* 100:1–152.

Barbosa FH, Pereira PV, Bergqvist LP, and Rothschild BM. 2016. Multiple neoplasms in a single sauropod dinosaur from the Upper Cretaceous of Brazil. *Cretac Res* 62:13–17.

Barnett S. 2003. Shell infections: When there are chinks in the armor. *Newsletter Mid-Atlantic Turtle and Tortoise Society: Terrapin Tales (MATTS)* October:1–8.

Baur G. 1891. On intercalation of vertebrae. *J Morphol* 4:329–336.

Belanger LF, Dimond MT, and Copp DH. 1973. Histological observations on bone and cartilage in growing turtles treated with calcitonin. *Gen Comp Endocrinol* 20:297–304.

Blanco PA. 1997. Enfermedades degererativas oseas y articulares en caiman del Orinoco (*Crocodylus intermedius*) and caiman de la costa (*Crocodylus acutus*) presentacion des tres (3) casos, in *Proceedings of the 4th Regional Meeting of the Crocodile Specialist Group of South America and the Caribean Region*, Centro de Innovación Agroindustrial, SC, Villahermosa, Tabasco, pp. 26–37.

Bohndorf K. 1999. Imaging of acute injuries of the articular surfaces (chondral, osteochondral and subchondral fractures). *Skeletal Radiol* 28:545–560.

Briggs DE. 1991. Extraordinary fossils. *Am Sci* 79:130–141.

Brochu CA. 2003. Osteology of Tyrannosaurus rex: Insights from a nearly complete skeleton and high-resolution CT analysis of the skull. *J Vertebr Paleontol* 22(Suppl):1–138.

Brogard J. 1987. *Les maladies des reptiles*. Point Vétérinaire, Maisons-Alfort, 334 pp.

Brogdon BG, and Crotty JM. 1999. The hidden divot: A new type of incomplete fracture? *Am J Roentgenol* 172:789–791.

Buffetaut E, Li J, Tong H, and Zhang H. 2006. A two-headed reptile from the Cretaceous of China. *Biol Lett* 3:80–81.

Burnham D, Rothschild BM, and Martin LD. 2012. Plesiosaurs had a taste for birds. *J Vertebr Paleontol* 2012: S71.

Carpenter K. 1998. Evidence for predatory behavior by carnivorous dinosaurs. *Gaia* 15:135–144.

Carr AP, and Michels G. 1997. Identifying noninfectious erosive arthritis in dogs and cats. *Veterin Med* 92:804–810.

Cisneros JC, Cabral UG, de Beer F, Damiani R, and Fortier DC. 2010. Spondarthritis in the Triassic. *PLOS ONE* 5(10):e13425. Doi: 10.1371/journal.pone.0013425

Cooper JE, Jackson OF, and Harshbarger JC. 1983. A neurilemmal sarcoma in a tortoise (*Testudo hermanni*). *J Comp Pathol* 93:541.

Drew ML, Phalen DN, Berridge BR, Johnson TL, Bouley D, Weeks BR, Miller HA, and Walker MA. 1999. Partial tracheal obstruction due to chondromas in ball pythons (*Python regius*). *J Zoo Wildl Med* 30:151–157.

Dumbrava MD, Rothschild BM, Weishampel DB et al. 2016. A dinosaurian facial deformity and the first occurrence of ameloblastoma in the fossil record. *Sci Rep* 6:29271. doi:10.1038/srep29271

Eckert SA, Eckert KK, Ponganis P, and Kooyman GL. 1989. Diving and foraging behavior of leatherback sea turtles (*Dermatochelys coriacea*). *Can J Zool* 67:2834–2840.

Eckert SA, Nellis DW, Eckert KK, and Kooyman GL. 1986. Diving patterns of two leatherback sea turtles (*Dermatochelys coriacea*) during internesting intervals at Sandy Point, St. Croix, U.S. Virgin Islands. *Herpetologica* 42:381–388.

Elkin E. 1981. Pathology and histopathological techniques, in *Diseases of the Reptilia*, Vol. 1. Cooper JE, and Jackson OF (Eds.), Academic Press, London, pp. 75–91.

Elkin E, and Cooper JE. 1976. Tumors and pseudotumors in some reptiles. *J Comp Pathol* 86:337–348.

Fröbisch J, Jansen M, Aschenbach T, Rothschild BM, and Witzman F 2016. Spina bifida in the Early Triassic dicynodont therapsid *Myosaurus gracilis*. *J Vertebr Paleontol* 36:106.

Fröbisch J, and Reisz RR 2008. New species of *Emydops* (Synapsida, Anomodontia) and a discussion of dental variability and pathology in dicynodonts. *J Vertebr Paleontol* 28:770–787.

Frye FL. 1981. *Biomedical and Surgical Aspects of Captive Reptile Husbandry*. Vol 1. Krieger Publishing Company, Malabar, Florida, 325 pp.

Frye FL. 1994. Diagnosis and Surgical Treatment of Reptilian Neoplasma with a compilation of Cases 1966–1993. *In Vivo* 8:885–892.

Frye FL, and Carney JD. 1975. Parathyroid adenoma in a tortoise. *Vet Med Small Anim Clin* 70:582–584.

Frye FL, and Dutra FR. 1976. Articular pseudogout in a turtle (*Chrysemys* [*Trachemys*] *p. elegans*). *Vet Med Small Anim Clin* 71:655–659.

García RA, Cerda IA, Heller M, Rothschild BM, and Zurriaguz V. 2016. The first evidence of osteomyelitis in a sauropod dinosaur. *Lethaia* 50:227–236. Doi: 10.1111/let.12189.

Garner MM, Herrington R, Howerth EW et al. 1997. Shell disease in river cooters (*Pseudemys concinna*) and yellow-bellied turtles (*Trachemys scripta*) in a Georgia (USA) lake. *J Wildl Dis* 33:78–86.

Gaffney ES, and Tong H. 2008. Redescription of the skull of Ummulisani rutgersensis Gaffney, Tong, and Meylan, 2006, a bothremydid side-necked turtle from the Eocene of Morocco. *Am Mus Novit* 3615:1–20.

Glazebrook JS. 1980. Diseases in farmed sea turtles, in *Management of Turtle Resources*. Glazebrook (Ed.), James Cook University, North Queensland, pp. 42–55.

Greer LL, Strandberg JD, and Whitaker BR. 2003. Mycobacterium chelonae osteoarthritis in a Kemp's Ridley sea turtle (*Lepidochelys kempii*). *J Wildl Dis* 39:736–741.

Gross JD, Rich TH, and Vickers-Rich P. 1993. Bone infection, chronic osteomyelitis in a hypsilophodontid dinosaur In Early Cretaceous Polar Australia. *Natl Geogr Res Explor* 9:286–293.

Grupka LM, Ramsay EC, and Bemis DA. 2006. *Salmonella* surveillance in a collection of rattlesnakes (*Crotalus* spp.). *J Zoo Wildl Med* 37:306–312.

Hanna RR. 2000. Dinosaurs got hurt too, in *The Scientific American Book of Dinosaurs*. Paul GS (Ed.), St. Martin's Press, New York, pp. 119–126.

Harms CA, Lewbar GA, and Beasley J. 2002. Medical management of mixed nocardial and unidentified fungal osteomyelitis in a Kemp's Ridley sea turtle, *Lepidochelys kempii*. *J Herp Med Surg* 12(3):21–26.

Huchzermeyer FW, and Cooper JE. 2000. Fibriscess, not abscess, resulting from a localized inflammatory response to infection in reptiles and birds. *Vet Rec* 147:515–517.

Hunt TJ. 1957. Notes on diseases and mortality in testudines. *Herpetologica* 13:19–23.

Innis CJ, and Garner MM. 2006. Spinal osteoarthropathy in green iguanas (*Iguana iguana*), in *Proceedings, Association of Reptile and Amphibian Veterinarians*, Baltimore, MD.

Isaza R, Garner M, and Jacobson E. 2000. Proliferative osteoarthritis and osteoarthrosis in 15 snakes. *J Zoo Wildl Med* 31:20–27.

Jacobson ER. 1994. Causes of mortality and diseases in tortoises—A review. *J Zoo Wildl Med* 25:2–17.

Janensch W. 1934. Eine halbseitige überzählige Wirbelbildung bei einem Dinosaurier. *Sitzungsberichte Gesellschaft Naturforschender Freunde Berlin* 1934:458–462.

Laws RR. 1995. Description and analysis of the multiple pathological bones of a sub-adult *Allosaurus fragilis* (MOR 693), in *Geological Society of America Program, Rocky Mountain Section, 47th Annual Meeting*, May 18–19, 1995, Bozeman, Montana, Vol 27(4): p. 43.

Lingham-Soliar T. 2004. Palaeopathology and injury in the extinct mosasaurs (Lepidosauromorpha, Squamata) and implications for modern reptiles. *Lethaia* 37:255–262.

López del Castillo C. 1998. Metabolic diseases: Gout and articular pseudogout. *Reptilia (GB)* 3:13–16.

Lovelace DM. 2014. Developmental failure of segmentation in a caudal vertebra of *Apatosaurus* (Sauropoda). *Anat Rec* 297:1262–1269.

Lucas SG. 2000. Pathological aetosaur armor from the Upper Triassic of Germany. *Stuttgarter Beiträge zur Naturkunde. Serie B (Geologie und Paläontologie)* 281; 1–6.

Lydekker R. 1889. *Catalogue of the Fossil Reptilia and Amphibia in the British Museum. Part 2. Orders Ichthyopterygia and Sauropterygia*. British Museum of Natural History, London, United Kingdom.

Madsen Jr JH. 1976. *Allosaurus fragilis: A revised osteology*. Utah Geological and Mineral Survey Bulletin 109. Geological and Mineral Survey, Utah Department of Natural Resources, Salt Lake Utah.

Marshall CD, Brinkman RL, and Bowman K. 1998. Fracture and osteomyelitis in PII of the second pedal digit of Deinoychus antirrhopus (Ostrom) an Early Cretaceous "raptor" dinosaur, in *Palaontological Association Annual Meeting Abstracts, non-paginated*.

Martin LD, and Rothschild BM. 1989. Paleopathology and diving mosasaurs. *Am Sci* 77:460–467.

Martinelli AG, Teixeira VP, Marinho TS et al. 2014. Fused mid-caudal vertebra in the titanosaur *Uberabatitan ribeiroi* from the Late cretaceous of Brazil and other bone lesions. *Lethaia* 48:456–462. doi: 10.111/let.12117.

Mauldin GN, and Done LB. 2006. Oncology, in *Reptile Medicine and Surgery*. Mader DR (Ed.), Saunders, Philadelphia, pp. 299–322.

McIntosh JS, Miles CA, Cloward KC, and Parker JR. 1996. A nearly complete skeleton of *Camarasaurus*. *Bull Gunma Museum Nat Hist* 1:1–87.

McWhinney LA, Rothschild BM, and Carpenter K. 2001. Post-traumatic osteomyelitis in Stegosaurus dermal spikes, in *The Armored Dinosaurs*. Carpenter K (Ed.), Indiana University Press, Bloomington, pp. 141–156.

Molnar RE. 2001. Theropod paleopathology: A literature survey, in *Mesozoic Vertebral Life*. Tanke DH, and Carpenter K (Eds.), Indiana University Press, Bloomington, pp. 337–363.

Montali RJ. 1988. Comparative pathology of inflammation in the higher vertebrates (reptiles, birds and mammals). *J Comp Pathol* 99:1–26.

Moodie RL. 1918. Paleontological evidences of the antiquity of disease. *Scientific Monthly* 1918:264–281.

Moodie RL. 1923. *Paleopathology*. University of Chicago Press, Chicago.

Natarajan LC, Melott AL, Rothschild BM, and Martin LD. 2007. Bone cancer rates in dinosaurs compared to modern vertebrates. *Trans Kans Acad Sci* 110:155–158.

Ogden JA, Rhodin AG, Conlogue BJ, and Light TR. 1981. Pathobiology of septic arthritis contiguous osteomyelitis in a leatherback turtle (*Dermochelys coriacea*). *J Wildl Dis* 17(2):277–287.

Orós J, Torrent A, and de los Monteros AE. 2001. Multicentric lymphoblastic lymphoma in a loggerhead sea turtle (*Caretta caretta*). *Vet Pathol* 38:464–467.

Raidal SR, Shearer PL, and Prince R. 2006. Chronic shoulder osteoarthritis in a loggerhead turtle (*Caretta caretta*). *Aust Vet J* 84:231–234.

Redelstorff R, Hayashi S, Rothschild B, and Chinsamy-Turan A. 2014. Multifocal bone infection in stegosaurs from Como Bluff, Wyoming. *Lethaia* 48:47–55. doi: 10.1111/let.12086.

Redelstorff R, Hayashi S, Rothschild B, and Chinsamy-Turan A. 2015. Vertebral fusion in two Early Jurassic sauropodomorph dinosaurs from the Lufeng Formation of Yunnan, China. *Acta Palaeontol Pol* 60:643–649.

Rega EA, and Brochu C. 2001. Paleopathology of a mature Tyrannosaurus rex. *J Vertebr Paleontol* 21(Suppl):92A.

Reichenbach-Klinke H, and Elkan E. 1965. *The Principal Diseases of Lower Vertebrates*. XII + 600pp.; Academic Press, New York.

Reichenbach-Klinke H-H. 1977. *Krankheiten der Reptilien*. 2. Aufl. Gustav Fischer Verlag, Stuttgart, 228pp.

Reisz RR, Scott DM, Pynn BR, and Modesto SP. 2011. Osteomyelitis in Paleozoic reptile: Ancient evidence for bacterial infection and its evolutionary significances. *Naturwissenschaften* 98:551–555.

Rejno S, and Stromberg B. 1978. Osteochondrosis in the Horse. II. Pathology. *Acta Radiol* 358(Suppl. 12):153–178.

Resnick D. 2002. *Diagnosis of Bone and Joint Disorders*. Saunders, Philadelphia.

Rhodin AG, Anders GJ, Russell A, and Mittermeier RA. 1990. Acanthochelys macrocephala. *Cat Am Amphib Reptil* 481:1–2.

Romo-de-Vivar-Martínez PR, Martinelli AG, Neto VE, and Soares MB. 2016. Evidence of osteomyelitis in the dentary of the late Triassic rhynchocephalian *Clevosaurus brasiliensis* (Lepidosauria: Rhynchocephalia) from southern Brazil and behavior implications. *Historical Biol* 29:320–327. doi: 10.1080/089/08912963.2016.115828

Rothschild B, Hao-Wen T, Tao D, and Xiaoting Z. 2013. Arthritis through Geologic Time and its Environmental Implications. *Journal of Ancient Diseases & Preventive Remedies* 1:1–3.

Rothschild B, Helbling M, and Miles C. 2002. Spondyloarthropathy in the Jurassic. *Lancet* 360:1454.

Rothschild BM. 1987a. Diffuse idiopathic skeletal hyperostosis as reflected in the paleontologic record: Dinosaurs and early mammals. *Semin Arthritis Rheum* 17:119–125.

Rothschild BM. 1987b. Decompression syndrome in fossil marine turtles. *Annls Carnegie Mus* 56:253–258.

Rothschild BM. 1988a. Stress fracture in a ceratopsian phalanx. *J Paleontol* 62:302–303.

Rothschild BM. 1988b. Avascular necrosis in diving vertebrates: The turtle test. *J Vertebr Paleontol* 8(suppl. 3):25A.

Rothschild BM. 1990. Radiologic assessment of osteoarthritis in dinosaurs. *Annls Carnegie Mus* 59:295–301.

Rothschild BM. 1991. Stratophenetic analysis of avascular necrosis in turtles: Affirmation of the decompression syndrome hypothesis. *Comp Biochem Physiol* 100A:529–535.

Rothschild BM. 1997a. Dinosaurian paleopathology, in *The Complete Dinosaur*. Farlow JO, and Brett-Surman MK (Eds.), Indiana University Press, Bloomington, pp. 426–448.

Rothschild BM. 1997b. Two faces of "rheumatoid arthritis": Type A versus type B disease. *J Clin Rheumatol* 3:334–338.

Rothschild BM. 2003. Osteoarthritis as a complication of artificial environment: The *Cavia* (guinea pig) story. *Ann Rheum Dis* 62:1022–1023.

Rothschild BM. 2008a. Scientifically rigorous reptile and amphibian osseous pathology: Lessons for forensic herpetology from comparative and paleo-pathology. *Applied Herpetology* 10:39–116.

Rothschild BM. 2008b. Vertebral pathology in mosasaurs, in *Proceedings of the Second Mosasaur Meeting, Fort Hays Studies Special Issue 3*. Everhart M (Ed.), Fort Hays State University, Hays, Kansas, pp. 90–94.

Rothschild BM. 2010. Actualistic approach to Ceratopsian cranial ontogeny and scars: The flip the other side of the jugal. *J Vertebr Paleontol* 30:173A.

Rothschild BM. 2012. Perspectives on decompression syndrome and the methodology of science. *Naturwissenschaften* 99:673–674.

Rothschild BM. 2014. Paravertebral masses in blue-tailed monitor, *Varanus dorianus*, indicative of soft-tissue infection with associated osteomyelitis. *J Zoo Wildl Med* 45:47–52.

Rothschild BM. 2015a. Lumbar Spondylosis. eMedicine from WebMD. Available at: http://emedicine.medscape.com/article/249036

Rothschild BM. 2015b. Gout. eMedicine Orthopedics. http://ww.emebdicine.com/orthoped/topic124.htm

Rothschild BM, and Berman D. 1991. Fusion of caudal vertebrae in Late Jurassic sauropods. *J Vertebr Paleontol* 11:29–36.

Rothschild BM, and Bruno MA. 2020. Calcium pyrophosphate deposition disease. *eMed Radiol* http://www.emedicine.com/radio-TOPIC125.HTM

Rothschild BM, DePalma RA, Burnham DA, and Martin LD. Anatomy of a dinosaur—Clarification of vertebrae in vertebrate anatomy. *Anat Histol Embryol* (in press).

Rothschild BM, and Everhart MJ. 2015. Co-ossification of vertebrae in mosasaurs (Squamata, Mosasauridae): Evidence of habitat interactions and susceptibility to bone disease. *Trans Kans Acad Sci* 118:265–275.

Rothschild B, and Feldkeller E. 2016. Spondyloarthritis in antiquity and history, in *Oxford Textbook of Axial Spondyloarthritis.* Oxford University Press, Oxford, pp. 1–12.

Rothschild BM, Hong N, and Turnquist JE. 1999. Skeletal survey of Cayo Santiago rhesus macaques: Osteoarthritis and apical plate excrescences. *Semin Arthritis RHeum* 29:100–111.

Rothschild BM, and Martin LD. 1987. Avascular necrosis: Occurrence in diving Cretaceous mosasaurs. *Science* 236:75–77.

Rothschild BM, and Martin LD. 2006. *Skeletal Impact of Disease.* New Mexico Museum of Natural History Press, Albuquerque.

Rothschild BM, and Molnar RE. 2005. Sauropod stress fractures as clues to activity, in *Thunder Lizards.* Tidwell V, and Carpenter K (Eds.), Indiana University Press, Bloomington, pp. 281–392.

Rothschild BM, and Molnar RE. 2008. Tyrannosaurid pathologies as clues to nature and nurture in the Cretaceous, in *Tyrannosaurus Rex: The Tyrant King.* Larson P, and Carpenter K (Eds.), Indiana University Press, Bloomington, pp. 287–304.

Rothschild BM, and Naples V. 2015. Decompression syndrome and diving behavior in *Odontochelys,* the first turtle. *Acta Palaeontol Pol* 601:163–167. Doi: 10.4202/app.2012.0113

Rothschild BM, and Storrs GW. 2003. Decompression syndrome in plesiosaurs (Sauropterygia: Reptilia). *J Vertebr Paleontol* 23:324–328.

Rothschild BM, Schultze H-P, and Pelligrini R. 2012a. *Herpetological Osteopathology: Annotated Bibliography of Amphibians and Reptiles.* Springer-Verlag, Heidelberg, Germany.

Rothschild BM, Schultze H-P, and Pellegrini R. 2012b. Osseous and other hard tissue pathologies in turtles, in *Morphology and Evolution of Turtles.* Brinkman D, Gardner J, and Holroyd P (Eds.), Springer, New York, pp. 501–534.

Rothschild M, Xiaoting Z, and Martin LD. 2012c. Adaptations for marine habitat and the effect of Triassic and Jurassic predator pressure on development of decompression syndrome in ichthyosaurs. *Naturwissenschaften* 99:443–448.

Rothschild BM, and Tanke D. 1999. Mesozoic neoplasia, from Moodie to the lapidary experience. *J Vert Paleontol* 18:73A.

Rothschild BM, and Tanke DH. 2005. Theropod paleopathology; State of the art review, in *The Carnivourous Dinosaurs.* Carenter K (Ed.), Indiana University Press, Bloomington, pp. 351 365.

Rothschild BM, and Tanke DH. 2006. Osteochondrosis in Late Cretaceous Hadrosauria: A manifestation of ontologic failure, in *Horns and Beaks: Ceratopsian and Ornithopod Dinosaurs.* Carpenter K (Ed.), Indiana University Press, Bloomington, pp. 171–183.

Rothschild BM, Tanke D, and Carpenter K. 1997. Gout in *Tyrannosaurus rex* and other tyrannosaurids. *Nature* 387:357.

Rothschild BM, Tanke DH, and Ford TL. 2001a. Theropod stress fractures and tendon avulsions as a clue to activity, in *Mesozoic Vertebrate Life.* Tanke DH, and Carpenter K (Eds.), Indiana University Press, Bloomington, pp. 331–336.

Rothschild BM, Prothero DR, and Rothschild C. 2001b. Origins of spondyloarthropathy in Perissodactyla. *Clin Exp Rheum* 19:628–632.

Rothschild BM, Tanke DH, Helbling M, and Martin LD. 2003. Epidemiologic study of tumors in dinosaurs. *Naturwissenschaften* 90:495–500.

Rothschild BM, and Woods RJ. 1991. Spondyloarthropathy: Erosive arthritis in representative defleshed bones. *Amer J Phys Anthropol* 85:125–134.

Rothschild BM, and Woods RJ. 1993. Arthritis in New World monkeys: Osteoarthritis, calcium pyrophosphate deposition disease and spondyloarthropathy. *Int J Primatol* 14:61–78.

Rothschild BM, Woods RJ, and Rothschild C. 1992. Calcium pyrophosphate deposition disease: Description in defleshed skeletons. *Clin Exp Rheumatol* 10:557–564.

Ruffer MA. 1917. A pathological specimen dating from the lower Miocene period. (Extrait de 'Contributions à l'étude des Vertébrés miocène de l'Egypte.) *Cairo Survey Department* 1917:1–7, Le Caire.

Sassoon J, Noè LF, and Benton MJ. 2012. Cranial anatomy, taxonomic implications and palaeopathology of an Upper Jurassic pliosaur (Reptilia: Sauropterygia) from Westbury, Wiltshire, UK. *Palaeontology* 55:43–773.

Sawyer GT, and Erickson BR. 1998. *Palepathology of the Paleocene crocodile Leidyosuchus (=Borealosuchus) formidabilis).* Science Museum of Minnesota, St. Paul Monograph 4:1–38.

Schulp AS, Walenkamp GH, Hofman PA, Rothschild BM, and Jagt JW. 2004. Rib fracture in *Prognathodon saturator* (Mosasauridae, Late Cretaceous). *Netherlands Journal of Geological Science* 83:251–254.

Schulp AS, Walenkamp GH, Hofman PA, Stuip Y, and Rothschild BM. 2006. Chronic bone infection in the jaw of *Mosasaurus hoffmanni* (Squamata). *Oryctos* 6:41–52.

Stolk A. 1958. Tumors of Reptiles. 4. Multiple osteomas in the lizard, Lacerta viridis. *Beaufortia* 7:1–9.

Surmik D, Rothschild BM, Dulski M, and Janiszeswska K. 2017. Two types of bone necrosis in the Middle Triassic *Pistosaurus longaevus* bones – The result of integrated studies. *R Soc Open Sci* 4(7):170204, doi: 10.1098/rsos.170204.

Surmik D, Szczygielski T, Janiszewska K, and Rothschild BM. 2018. Tuberculosis-like respiratory infection in 245-million-year-old marine reptile suggested by bone pathologies. *Royal Society Open Science* 5(6):180225. Doi: 10.1098/rsos.180225.

Tanke DH, and Rothschild BM. 2002. Dinosores: An Annotated Bibliography of Dinosaur Paleopathology and Related Topics – 1938–2001. *New Mexico Museum of Natural History and Science Bulletin* 20:1–96.

Tanke DH, and Rothschild BM. 2010. Paleopathologies in Albertan ceratopsids and their behavioral significance. in *New Perspectives on Horned Dinosaurs; The Royal Tyrrell Museum Ceratopsian Symposium.* Ryan MJ, Chinnery-Allgeir BJ, and Eberth DA (Eds.), Indiana University Press, Bloomington, pp. 355–384.

Tanke DH, and Rothschild BM. 2014. Paleopathologies in Late Cretaceous hadrosauridae from Alberta, Canada, in *Hadrosaurs*. Eberth DA, and Evans DC (Eds.), Indiana University Press, Bloomington, pp. 540–571.

Taylor P. 1992. Doctors try to diagnose dinosaur cancer. *Toronto Globe and Mail*, January 5: a1–a2.

Thorp BH, Ekman S, Jakowlew SB, and Goddard C. 1995. Porcine osteochondrosis: Deficiencies in Transforming Growth Factor-Beta and Insulin-Like Growth Factor I. *Calcific Tissue International* 56:376–381.

Trotteyn MJ, and Martínez RN. 2013. Primer registro de dysplasia coxofemoral en un rincosaurio del triásico superior. *Aeghiniana* 50:217–226.

Tschopp E, Wings O, Frauenfelder T, and Rothschild BM. 2016. Pathological phalanges in a camarasaurid sauropod dinosaur and implications on behavior. *Acta Palaeontol Pol* 61:125–134.

Vega CS, and Maisch MW. 2014. Pathological features in Upper Permian and idle Triassic dicynodonts (Synapsida, Therapsida), Vertebrate Paleobiology and Paleoanthropology, in *Early Evolutionary History of the Synapsida*. Kammerer CF, Angielczyk KD, and Fröbisch J (Eds.), Springer, New York, pp. 151–161.

Vellet AD, Marks PH, Fowler PJ, and Munro TG. 1991. Occult post-traumatic osteochondral lesions of the knee: Prevalence, classification, and short-term sequelae evaluated with MR imaging. *Radiology* 178:271–276.

Verano JW. 2016. Holes in the Head: The Art and Archaeology of Trepanation in Ancient Peru. *Dumbarton Oaks Pre-Columbian Art and Archaeology Studies Series* 38:1–352.

Vittore CP, and Henderson MD. 2013. Brodie abscess involving a tyrannosaur phalanx: Imaging and implications, in *Tyrannosaurid Paleobiology*. Parrish JM, Molnar RE, Currie PJ, and Koppelhus EB (Eds.), University of Indiana Press, Bloomington, pp. 223–236.

Wallach JD. 1971. Environmental and nutritional diseases of captive reptiles. *J Am Vet Med Assoc* 159:1632–1643.

Weems RE. 1974. Middle Miocene sea turtles (*Syllomus*, *Procolpochelys*, *Psephorus*) from the Calvert Formation. *J Paleontol* 48: 278–303.

Wegener KM, and Heje NI. 1992. Dyschondroplasia (osteochondrosis) in articular-epiphyseal cartilage complexes of three calves from 24 to 103 days of age. *Vet Pathol* 29:562–563.

Wells C. 1964. Pathological epipodials and tarsus in *Stretosaurus macromerus* from the Kimmeridge Clay, Stretha, Cambridgeshire. *QJ Geol Soc Lond* 120:299–304.

Wenker CJ, Bart M, Guscetti F, Hatt J-M, and Isenbugel E. 1999. Periarticular hydroxyapatite deposition disease in two red-bellied short-necked turtles (*Emydura albertisi*). *Annual Proceedings of the American Association of Zoo Veterinarians* 1999:23–26.

Williamson TE, and Carr TD. 1999. A new tyrannosaurid (Dinosauria: Theropoda) partial skeleton from the Upper Cretaceous Kirtland Formation, San Juan Basin, NM. *New Mexico Geology* 21(2):42–43.

Witzmann F, Asbach P, Remes K, Hampe O, Hilger A, and Paulke A. 2008. Vertebral pathology in an ornithopod dinosaur: A hemivertebra in *Dysalotosaurus lettowvorbecki* from the Jurassic of Tanzania. *Anat Rec* 291:1149–1155.

Witzmann F, Claeson CM, and Hampe O et al. 2011. Paget disease of bone in a Jurassic dinosaur. *Curr Biol* 21:R647–R648.

Witzmann F, Schwarz-Wings D, Hampe O, Fritsch G, and Asbach P. 2014. Evidence of spondyloarthropathy in the spine of a phytosaur (Reptilia: Archosauriformes) from the Late Triassic of Halberstadt, Germany. *PLOS ONE* 9:e85511

Wolff ED, Salisbury SW, Horner JR, and Varricchio DJ. 2009. common avian infection plagued the tyrant dinosaurs. *PLOS ONE* 4(9):e7288. doi: 10.1371/journal.pone.0007288

Woodburn DB, Miller AN, Allender MC, Maddox CW, and Terio KA. 2019. *Emydomyces testavorans*, a new genus and species of onygenalean fungus isolated from shell lesions of freshwater aquatic turtles. *J Clin Microbiol* 57:e00628–18. doi: 10.1128/JCM.00628-18

Xing L, Rothschild BM, Ran H et al. 2015. Vertebral fusion in two early Jurassic sauropodomorph dinosaurs from the Lufeng Formation of Yunnan, China. *Acta Palaeontol Pol* 60:643–649.

Yu JS, Greenway G, and Resnick D 1998. Osteochondral defect of the glenoid fossa: Cross-sectional imaging features. *Radiology* 206:35–40.

Zangerl R, and Turnbull W. 1955. *Procolpochelys grandaeva* (Leidy), an early carettine sea turtle. *Fieldiana-Zool* 37:345–382.

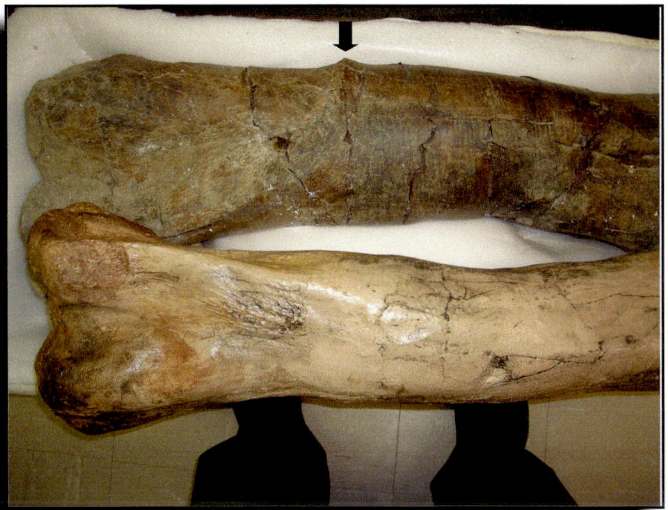

Figure 10.1 Tyrannosauridae. *Tyrannosaurus rex*. Anterior view of distal femurs. Bump (arrow) containing crack in upper femur contrasts with normal femur below.

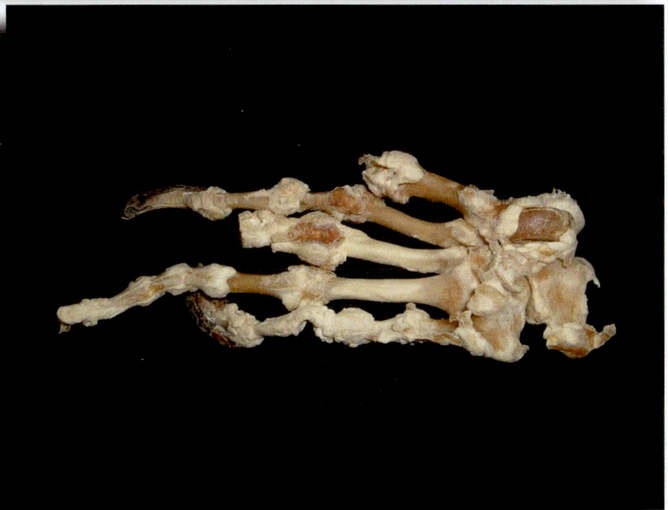

Figure 10.2 Komodo Dragon, *Varanus komodoensis*. Varanidae. Posterior view of pes (distal portion of hindlimb). Calcium pyrophosphate deposits are prominent over joints.

Figure 10.3 Pancake Tortoise, *Malacochersus tornieri.* Testudinidae. Dorsal view of conjoined individuals. (Courtesy of R. Funk.)

Figure 10.4 Hadrosaur, *Edmantosaurus*. Hadrosauridae. Superior view of proximal phalanges. Surface defects (divots) produced by osteochondrosis.

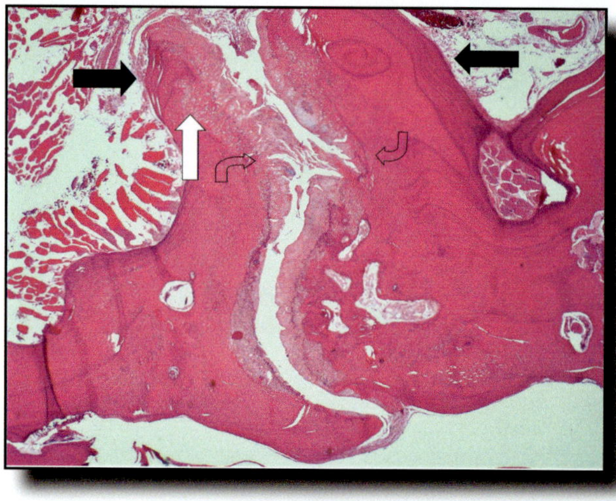

Figure 10.5 Gopher Snake, *Pituophis catenifer catenifer.* Colubridae. Degenerative arthritis involving vertebral articulations, including condyle (clear arrows) and dorsal facet. Note the cartilaginous erosions (white arrow), fibrillation, and fibrocartilagenous exostoses (black arrow) typical of age-related degenerative arthritis in snakes. H&E stain. (Courtesy of Michael Garner.)

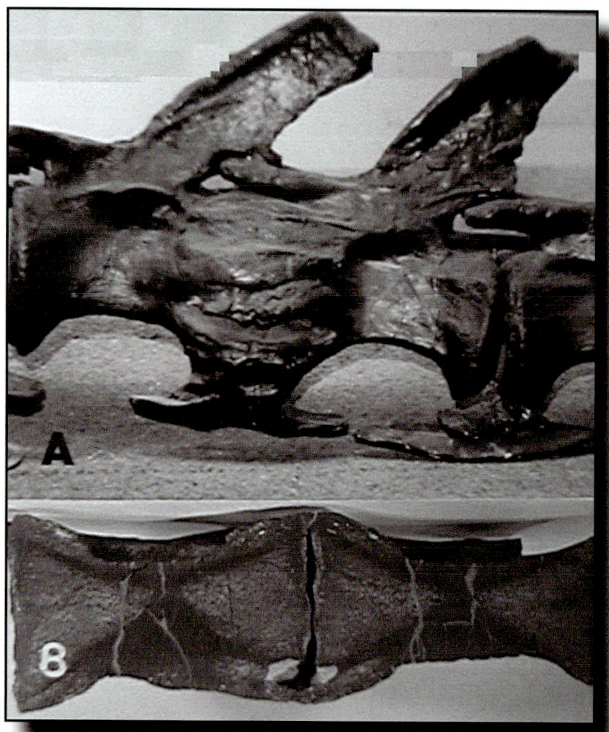

Figure 10.6 Sauropod dinosaur, *Diplodicus*. Diplodocidae. Lateral view of fused caudal vertebrae. Cross-section reveals ligamentous ossification from DISH.

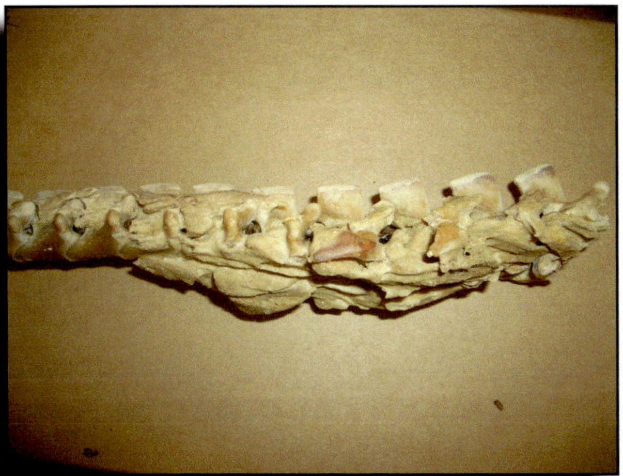

Figure 10.7 Monitor Lizard, *Varanus doriensis*. Varanidae. Lateral view of vertebral column. Infectious bone proliferation on inferior and lateral surfaces.

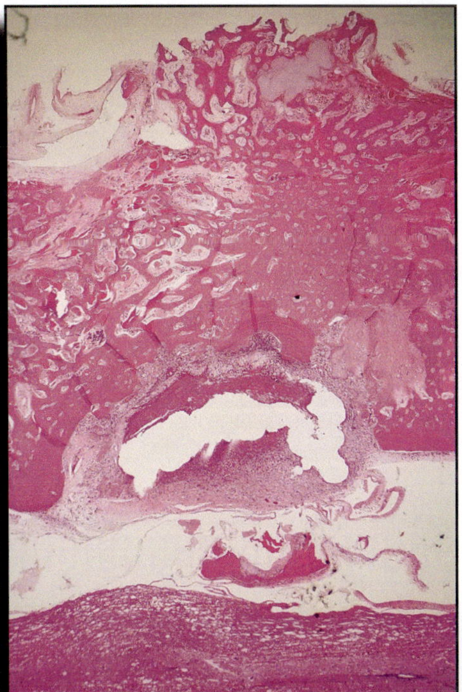

Figure 10.8 Mexican Moccasin (Cantil), *Agkistodon bilineatus*. Viperidae. Focus of osteomyelitis in a vertebral body associated with marked subperiosteal reactive "sunburst" bone formation and ankylosis. Inflammation extends into the spinal canal and is associated with cord displacement and myelinic edema. H&E stain. (Courtesy of Michael Garner.)

Figure 10.9 Loggerhead Turtle, *Caretta caretta*. Cheloniidae. Superior view of carapace. Multiple surface defects indicating shell disease.

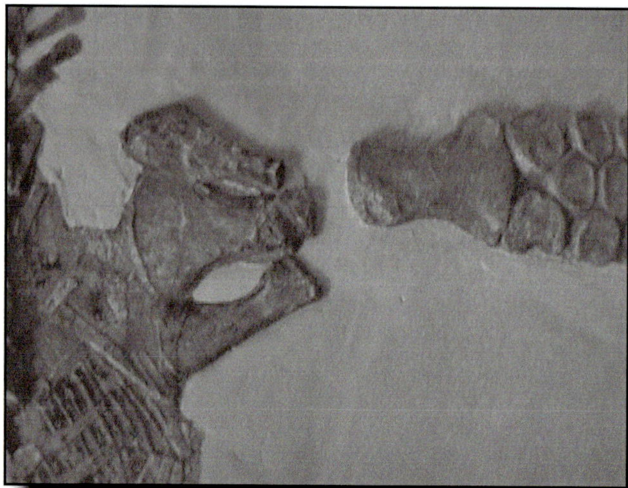

Figure 10.10 Extinct marine reptile, *Ichthyosaurus*. Ichthyosauridae. Superior view of hip joint. Defect in proximal humerus caused by avascular necrosis.

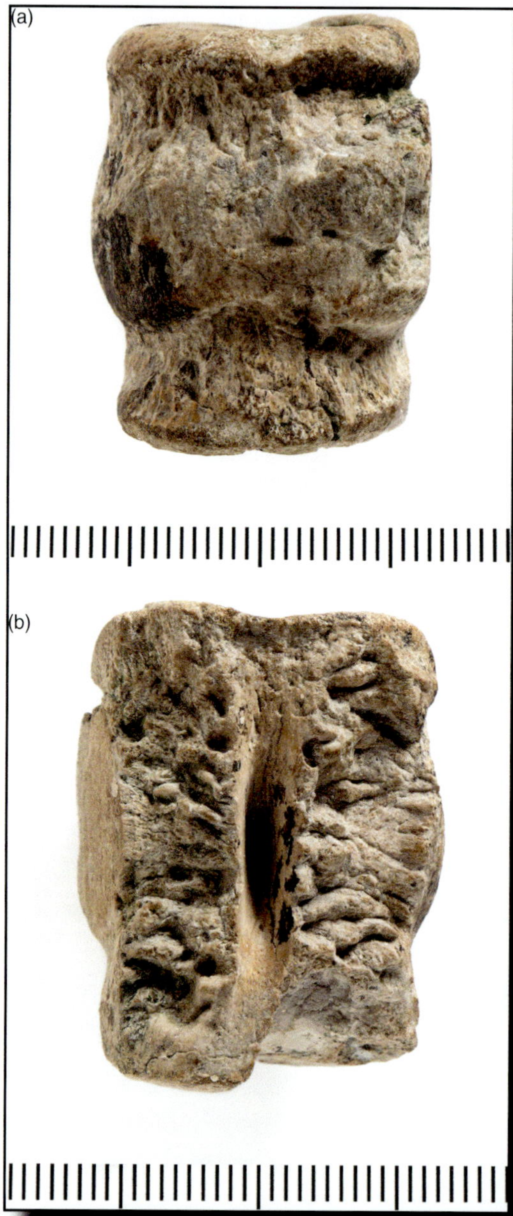

Figure 10.11 Dryosaurid Iguanodontian Dinosaur, *Dysalotosaurus lettowvorbecki.* Iguanodontidae. Macroscopic (a) and longitudinal cross-sectional view (b) of Braila Museum, Braila (MBR) 1336 with Paget's disease. (Courtesy of Carola Radke, photographer from the Museum für Naturkunde Berlin [MfN].)

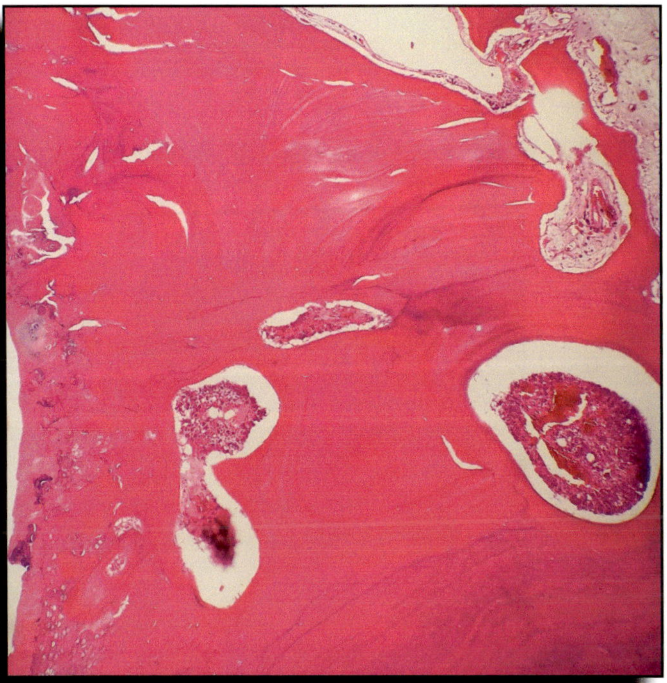

Figure 10.12 Children's Python, *Antaresia childreni*. Pythonidae. Longitudinal section through a vertebral body, showing the mosaic pattern of mature bone, a normal feature in the histogenesis of bone in snakes. H&E. (Courtesy of Michael Garner.)

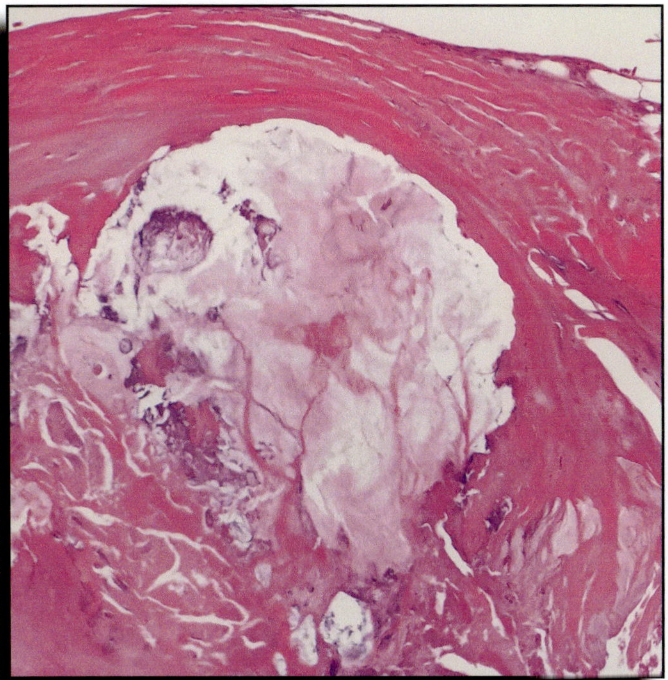

Figure 10.13 African Fat-Tailed Gecko, *Hemitheconyx caudicinctus*. Eublepharidae. Articular gout. Intervertebral fibrocartilaginous joint capsule is expanded and disrupted by urate deposits. (Courtesy of Michael Garner.)

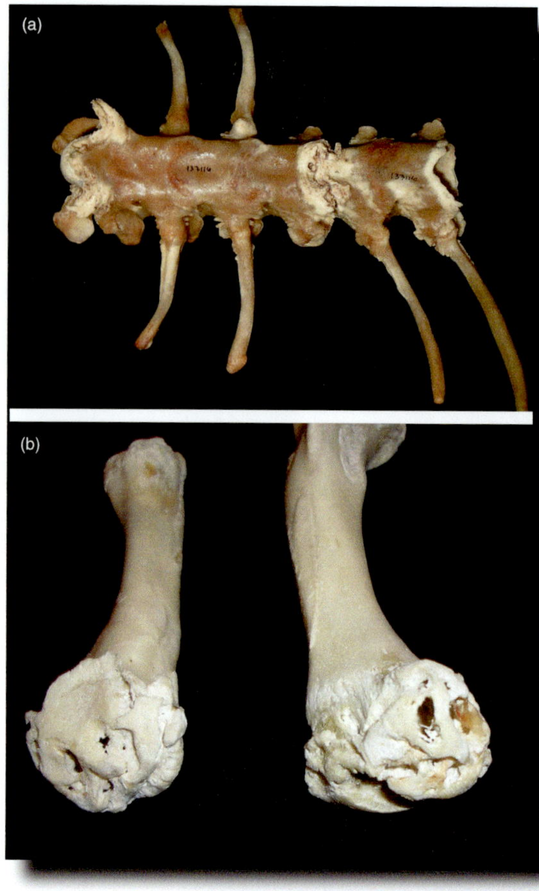

Figure 10.14 Komodo Dragon, *Varanus komodoensis*. Varanidae. (a) Anterior view of thoracic vertebrae. Fusion through anulus fibrosus caused by spondyloarthropathy. (b) Anterior oblique views of metacarpals. Subchondral erosions caused by spondyloarthropathy.

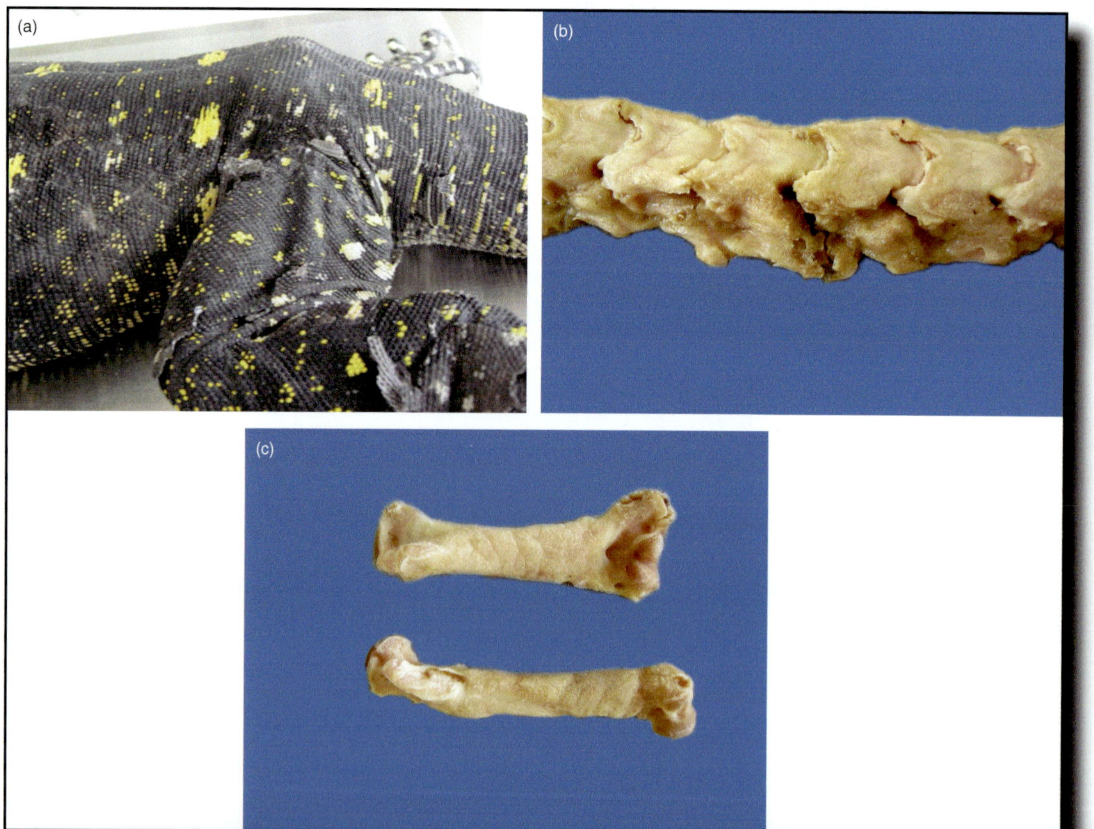

Figure 10.15 Crocodile Monitor, *Varanus salvadorii*. Varanidae. Hypertrophic osteopathy. (a) Note irregular sacral malformation. (b) Macroscopic anterior-oblique view of fused vertebrae associated with proliferative bone, and fracture at a site of ankylosis. (c) Proliferative bone along the entire length of the femoral diaphysis. (Courtesy of Chriss Miller.)

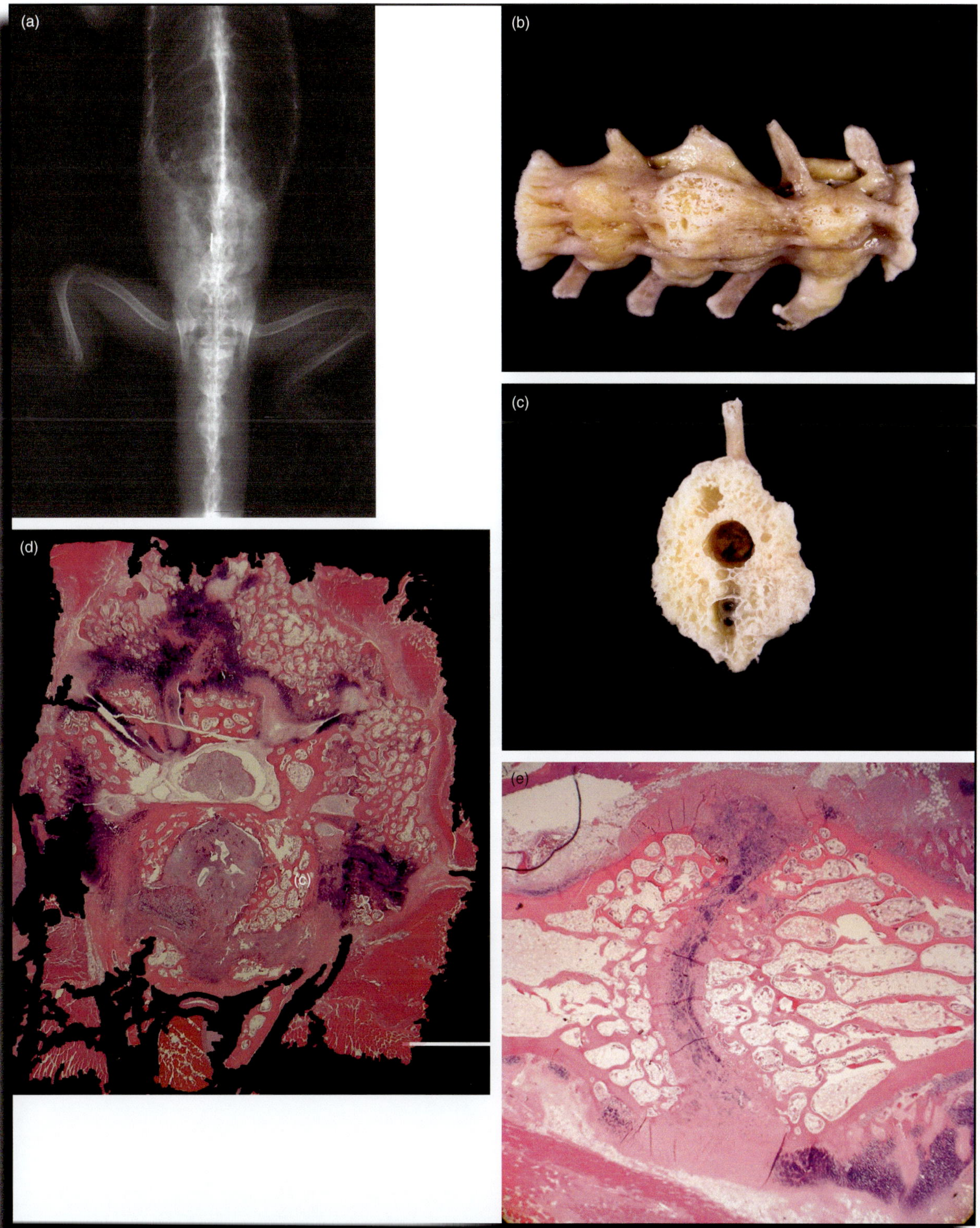

Figure 10.16 Green Iguana, *Iguana iguana*. Iguanidae. Idiopathic proliferative vertebral osteopathy. (a) Dorsal-ventral radiographic image showing fusion of vertebrae associated with proliferative bone, at this stage having ascended from the tail into the lumbosacral and thoracic spine. (Courtesy of Charles Innes.) (b) Dorsal ventral image of fused tail vertebrae. (c) A cross-section through a caudal vertebral body. A narrow incomplete zone of cortical bone is disrupted by marked circumferential subperiosteal new bone. (d) Sub-gross photomicrograph, showing the changes described in (c). (e) Longitudinal section through vertebral articulation showing fused articular surface, and ankylosis with formation of marked metaphyseal subperiosteal new bone and cartilage. H&E stain. (Images (b–e) courtesy of Michael Garner and Christie Buie.)

INDEX

M

N

O

R

For Product Safety Concerns and Information please contact our EU representative GPSR@taylorandfrancis.com Taylor & Francis Verlag GmbH, Kaufingerstraße 24, 80331 München, Germany

Batch number: 10203134

Printed by Printforce, the Netherlands